# PREPARATIVE CARBOHYDRATE CHEMISTRY

# PREPARATIVE CARBOHYDRATE CHEMISTRY

edited by
**STEPHEN HANESSIAN**
*University of Montreal*
*Montreal, Quebec, Canada*

Marcel Dekker, Inc.   New York•Basel•Hong Kong

**Library of Congress Cataloging-in-Publication Data**

Hanessian, Stephen.
   Preparative carbohydrate chemistry / Stephen Hanessian.
     p.  cm.
   Includes index.
   ISBN 0-8247-9802-3 (alk. paper)
   1. Carbohydrates—Derivatives.  I. Title.
QD321.H288  1996
547'.780459—DC21

                                                                                       96-39338
                                                                                                CIP

The publisher offers discounts on this book when ordered in bulk quantities. For more information, write to Special Sales/Professional Marketing at the address below.

This book is printed on acid-free paper.

**Copyright © 1997 by MARCEL DEKKER, INC. All Rights Reserved.**

Neither this book nor any part may be reproduced or transmitted in any form or by any means, electronic or mechanical, including photocopying, microfilming, and recording, or by any information storage and retrieval system, without permission in writing from the publisher.

MARCEL DEKKER, INC.
270 Madison Avenue, New York, New York 10016

Current printing (last digit):
10 9 8 7 6 5 4 3 2 1

**PRINTED IN THE UNITED STATES OF AMERICA**

# Preface

Carbohydrate chemistry has been an important and vital subdiscipline of organic chemistry ever since the pioneering discoveries of Emil Fischer. Stereochemical features, conformational aspects, and stereoelectronic principles dealt with in organic chemistry in general are deeply rooted in molecules we generally refer to as sugars. Over the years, carbohydrate chemistry has served as an important link between organic chemistry, medicinal chemistry, and biology.

In recent years, the general areas of carbohydrate chemistry and biochemistry have enjoyed unprecedented popularity. A clear indication of this resurgence of interest is the large contingent of "noncarbohydrate" chemists by training, who have flocked to the area with new ideas and exciting applications. This, coupled with the increasingly important role that sugar molecules are playing in glycobiology; in anti-infective therapy as components of antibiotics, antitumor, and antiviral agents; and in related biomedical areas, makes this old subdiscipline of organic chemistry a vibrant and rejuvenated area in which to work. What has been sorely missing, however, is an authoritative monograph that describes the preparation of some of the more important carbohydrate derivatives and related molecules in an up-to-date and concise manner.

This volume, written by authorities on the subject, is a compendium of classic procedures for the synthesis and utilization of carbohydrate-related molecules. Representative, state-of-the-art procedures provide even newcomers to the field with ready access to commonly used carbohydrate derivatives for a variety of applications.

A total of 28 chapters have been grouped under 7 themes. Each chapter consists of introduction, discussion, and experimental sections that cover the particular "method" in a thorough manner. The reader will thus be introduced to the subject matter pertaining to a general method, a specific reaction, or type of derivative, as well as to the experimental procedures performed in the author's laboratory and described in the literature, whenever pertinent.

The first four chapters, which come under the general theme of sugar derivatives, represent methods for the transformation of sugar molecules into synthetically useful derivatives, such as acetals, dithioacetals, ethers, and related compounds. The following six chapters explore selected reactions of sugar derivatives, in which some of the most important bond-forming reactions in the modifications of sugars are discussed.

The third theme is concerned with the chemical synthesis of *O*- and *N*-glycosyl compounds, and of oligosaccharides, and is the subject of ten chapters. The most widely used methods for glycoside synthesis are discussed, together with the inclusion of conceptually new approaches to anomeric activation and glycoside synthesis.

The use of enzymes in carbohydrate chemistry is covered in two chapters on enzymatic synthesis of sialic acid, KDO, and related deoxyulosonic acids, and oligosaccharides.

The theme of *C*-glycosyl compounds is covered by two chapters on free radical– and Lewis acid–mediated transformations that address the stereocontrolled synthesis of carbon substituents at the anomeric position.

The sixth theme, carbocycles from carbohydrates, explores how functionalized cyclopentanes and cyclohexanes can be prepared from carbohydrate precursors, thereby extending the usefulness of sugars for the synthesis of mainstream organic compounds.

The last theme, total synthesis of sugars from nonsugars, groups two chapters that are concerned with how amino sugars, deoxy sugars, and sugars in general are synthesized from amino acids and related compounds, as well as by de novo methods.

In short introductory commentaries on each theme, I have attempted to provide some insight into the topic, reflect on its evolution over the years, and put recent developments in perspective.

The foregoing themes and the specific chapters represent some of the most preparatively useful methods in modern carbohydrate chemistry. Although coverage is restricted to a selection of topics, the most important aspects of preparative carbohydrate chemistry have been dealt with in an expert manner. As a consequence, it is my hope that this volume will have lasting value.

I am greatly indebted to all the authors, who responded with great enthusiasm to my initial proposal by providing chapters. I would also like to acknowledge the expert assistance of Carol St-Vincent Major and Michelle Piché in the preparation of my own chapters, and Gurijala V. Reddy, Olivier Rogel, and Benoit Larouche for producing artwork for them.

Finally, I hope the preparative methods described in this monograph will be of service to present and future generations of investigators in the pursuit of their individual research objectives.

*Stephen Hanessian*

# Contents

| | |
|---|---|
| *Preface* | *iii* |
| *Contributors* | *xi* |

**PART I   SUGAR DERIVATIVES**                                          1
   *Commentary by Stephen Hanessian*

1. **Synthesis of Isopropylidene, Benzylidene, and Related Acetals**      3
    *Pierre Calinaud and Jacques Gelas*

    I.   Introduction                                                     3
    II.  Methods                                                          6
    II.  Experimental Procedures                                         15
         References                                                      28

2. **Dialkyl Dithioacetals of Sugars**                                   35
    *Derek Horton and Peter Norris*

    I.   Introduction                                                    36
    II.  Methods                                                         39
    III. Experimental Procedures                                         43
         References                                                      50

3. **Regioselective Cleavage of *O*-Benzylidene Acetals to Benzyl Ethers**  53
    *Per J. Garegg*

    I.   Introduction                                                    54
    II.  Regioselective Reductive Cleavage of *O*-Benzylidene Acetals to
         Benzyl Ethers                                                   57
    III. Mechanistic Considerations                                      61
    IV.  Experimental Procedures                                         62
         References                                                      65

4. **Selective *O*-Substitution and Oxidation Using Stannylene Acetals and
    Stannyl Ethers**                                                     69
    *Serge David*

    I.   Introduction                                                    69
    II.  Methods                                                         70

## PART II  SELECTED REACTIONS IN CARBOHYDRATE CHEMISTRY     85
*Commentary by Stephen Hanessian*

**5. $S_N2$-Type Halogenation and Azidation Reactions with Carbohydrate Triflates**     87
*Edith R. Binkley and Roger W. Binkley*

    I. Triflate Synthesis and Reactivity     88
    II. Introducing Azido and Halogeno Groups by Triflate Displacement     90
    III. Reactions of Carbohydrate Triflates     93
    IV. Carbohydrate Imidazolylsulfonates     96
    V. Experimental Procedures     97
        References     102

**6. Direct Halogenation of Carbohydrate Derivatives**     105
*Walter A. Szarek and Xianqi Kong*

    I. Introduction     106
    II. General Methods for the Direct Halogenation of Alcohols     107
    III. Experimental Procedures     116
        References     123

**7. Nucleophilic Displacement Reactions of Imidazole-1-Sulfonate Esters**     127
*Jean-Michel Vatèle and Stephen Hanessian*

    I. Introduction     127
    II. Methods     130
    III. Experimental Procedures     136
        References     145

**8. Free Radical Deoxygenation of Thiocarbonyl Derivatives of Alcohols**     151
*D. H. R. Barton, J. A. Ferreira, and J. C. Jaszberenyi*

    I. Introduction     151
    II. Methods     153
    III. Experimental Procedures     157
        Notes and References     168

**9. Thiazole-Based One-Carbon Extension of Carbohydrate Derivatives**     173
*Alessandro Dondoni and Alberto Marra*

    I. Introduction     174
    II. Methods     174
    III. Experimental Procedures     188
        References     196

III. Experimental Procedures     75
    References     82

| | | |
|---|---|---|
| **10.** | **Selected Methods for Synthesis of Branched-Chain Sugars**<br>*Yves Chapleur and Françoise Chrétien* | **207** |
| | I. Introduction | 207 |
| | II. Methods | 211 |
| | III. Experimental Procedures | 240 |
| | References | 252 |

| | | |
|---|---|---|
| **PART III** | **CHEMICAL SYNTHESIS OF *O*- AND *N*-GLYCOSYL COMPOUNDS, AND OF OLIGOSACCHARIDES**<br>*Commentary by Stephen Hanessian* | **263** |
| **11.** | ***O*- and *N*-Glycopeptides: Synthesis of Selectively Deprotected Building Blocks**<br>*Horst Kunz* | **265** |
| | I. Introduction | 265 |
| | II. Methods | 268 |
| | III. Experimental Procedures | 273 |
| | References | 279 |
| **12.** | **Oligosaccharide Synthesis with Trichloroacetimidates**<br>*Richard R. Schmidt and Karl-Heinz Jung* | **283** |
| | I. Introduction | 283 |
| | II. The Trichloroacetimidate Method | 289 |
| | III. Experimental Procedures | 296 |
| | References and Notes | 308 |
| **13.** | **Oligosaccharide Synthesis from Glycosyl Fluorides and Sulfides**<br>*K. C. Nicolaou and Hiroaki Ueno* | **313** |
| | I. Introduction | 314 |
| | II. Methods | 314 |
| | III. Experimental Procedures | 329 |
| | References | 336 |
| **14.** | **Oligosaccharide Synthesis by *n*-Pentenyl Glycosides**<br>*Bert Fraser-Reid and Robert Madsen* | **339** |
| | I. Introduction | 339 |
| | II. Methods | 341 |
| | III. Experimental Procedures | 348 |
| | References and Notes | 354 |

## 15. Chemical Synthesis of Sialyl Glycosides    357
*Akira Hasegawa and Makoto Kiso*

    I. Introduction    358
    II. Regio- and α-Stereoselective Sialyl Glycoside Syntheses Using Thioglycosides of Sialic Acids in Acetonitrile    359
    III. Applications to Systematic Synthesis of Gangliosides and Sialyloligosaccharides    364
    IV. Experimental Procedures    370
    References    375

## 16. Glycoside Synthesis Based on the Remote Activation Concept: An Overview    381
*Stephen Hanessian*

    I. Introduction    381
    II. The Challenges of the Glycosidic Bond    382
    III. The Remote Activation Concept    383
    IV. New Generations of Glycosyl Donors    386
    References    387

## 17. Glycoside and Oligosaccharide Synthesis with Unprotected Glycosyl Donors Based on the Remote Activation Concept    389
*Boliang Lou, Gurijala V. Reddy, Heng Wang, and Stephen Hanessian*

    I. Introduction    390
    II. Glycoside and Oligosaccharide Synthesis Using 3-Methoxy-2-pyridyloxy (MOP) *O*-Unprotected Glycosyl Donors    391
    III. Experimental Procedures    398
    References    410

## 18. Oligosaccharide Synthesis by Remote Activation: *O*-Protected 3-Methoxy-2-pyridyloxy (MOP) Glycosyl Donors    413
*Boliang Lou, Hoan Khai Huynh, and Stephen Hanessian*

    I. Introduction    414
    II. *O*-Protected 3-Methoxy-2-pyridyloxy Glycosyl Donors    415
    III. Applications to the Synthesis of T Antigen and Sialyl Le[x]    419
    IV. Experimental Procedures    422
    References    427

## 19. Oligosaccharide Synthesis by Remote Activation: *O*-Protected Glycosyl 2-thiopyridylcarbonate Donors    431
*Boliang Lou, Hoan Khai Huynh, and Stephen Hanessian*

    I. Introduction    432
    II. Methods: Glycosyl 2-thiopyridylcarbonates (TOPCAT) as Glycosyl Donors    434
    III. Experimental Procedures    439
    References    447

Contents ix

20. **Oligosaccharide Synthesis by Selective Anomeric Activation with MOP- and TOPCAT-Leaving Groups**    449
    *Boliang Lou, Elisabeth Eckhardt, and Stephen Hanessian*

    | | | |
    |---|---|---|
    | I. | Introduction | 450 |
    | II. | Experimental Procedures | 457 |
    | | References | 464 |

## PART IV  ENZYMATIC SYNTHESIS OF SIALIC ACID, KDO, AND RELATED DEOXYULOSONIC ACIDS, AND OF OLIGOSACCHARIDES    467
*Commentary by Stephen Hanessian*

21. **Enzymatic Synthesis of Carbohydrates**    469
    *Claudine Augé and Christine Gautheron-Le Narvor*

    | | | |
    |---|---|---|
    | I. | Introduction | 469 |
    | II. | Methods | 471 |
    | III. | Experimental Procedures | 478 |
    | | References | 482 |

22. **Oligosaccharide Synthesis by Enzymatic Glycosidation**    485
    *Wolfgang Fitz and Chi-Huey Wong*

    | | | |
    |---|---|---|
    | I. | Introduction | 486 |
    | II. | Enzymatic Glycosidation | 486 |
    | III. | Experimental Procedures | 494 |
    | | References | 502 |

## PART V  SYNTHESIS OF C-GLYCOSYL COMPOUNDS    505
*Commentary by Stephen Hanessian*

23. ***C*-Glycosyl Compounds from Free Radical Reactions**    507
    *Bernd Giese and Heinz-Georg Zeitz*

    | | | |
    |---|---|---|
    | I. | Introduction | 507 |
    | II. | Intermolecular Methods | 510 |
    | III. | Intramolecular Methods | 516 |
    | IV. | Experimental Procedures | 517 |
    | | References | 524 |

24. **Synthesis of Glycosylarenes**    527
    *Keisuke Suzuki and Takashi Matsumoto*

    | | | |
    |---|---|---|
    | I. | Introduction | 528 |
    | II. | Methods | 530 |
    | III. | Experimental Procedures | 535 |
    | | References | 541 |

## PART VI  CARBOCYCLES FROM CARBOHYDRATES — 543
*Commentary by Stephen Hanessian*

### 25. Functionalized Carbocylic Derivatives from Carbohydrates: Free Radical and Organometallic Methods — 545
*T. V. RajanBabu*

- I. Introduction — 546
- II. Cyclopentanes — 546
- III. Cyclohexanes — 554
- IV. Functionalized Carbocylic Compounds via Organometallic Methods — 555
- V. Experimental Procedures — 558
- References and Notes — 565

### 26. The Conversion of Carbohydrates to Cyclohexane Derivatives — 569
*Robert J. Ferrier*

- I. Introduction — 570
- II. Methods — 571
- III. Experimental Procedures — 585
- References — 590

## PART VII  TOTAL SYNTHESIS OF SUGARS FROM NONSUGARS — 593
*Commentary by Stephen Hanessian*

### 27. Total Synthesis of Amino Sugars — 595
*Janusz Jurczak*

- I. Introduction — 595
- II. Methods — 596
- III. Experimental Procedures — 601
- References — 611

### 28. Total Synthesis of Sugars — 615
*Aleksander Zamojski*

- I. Introduction — 615
- II. Methods — 617
- III. Experimental Procedures — 622
- References and Notes — 634

Index — 637

# Contributors

**Claudine Augé**  Institut de Chimie Moléculaire d'Orsay, Université Paris-Sud, Orsay, France

**D. H. R. Barton**  Department of Chemistry, Texas A&M University, College Station, Texas

**Edith R. Binkley**  Center for Carbohydrate Study, Oberlin, Ohio

**Roger W. Binkley**  Center for Carbohydrate Study, Oberlin, and Department of Chemistry, Cleveland State University, Cleveland, Ohio

**Pierre Calinaud**  École Nationale Supérieure de Chimie de Clermont-Ferrand, Aubière, France

**Yves Chapleur**  Institut Nancéien de Chimie Moléculaire, URA CNRS 486, Université Henri Poincaré—Nancy I, Vandoeuvre, France

**Françoise Chrétien**  Institut Nancéien de Chimie Moléculaire, URA CNRS 486, Université Henri Poincaré—Nancy I, Vandoeuvre, France

**Serge David**  I.C.M.O., Laboratoire de Chimie Organique Multifunctionelle, Université Paris-Sud, Orsay, France

**Alessandro Dondoni**  Dipartimento di Chimica, Università di Ferrara, Ferrara, Italy

**Elisabeth Eckhardt**  Boehringer Mannheim GmbH, Penzberg, Germany

**J. A. Ferreira**  Department of Chemistry, Texas A&M University, College Station, Texas

**Robert J. Ferrier**  Department of Chemistry, Victoria University of Wellington, Wellington, New Zealand

**Wolfgang Fitz**  Department of Chemistry, The Scripps Research Institute, La Jolla, California

**Bert Fraser-Reid**  Department of Chemistry, Duke University, Durham, North Carolina

**Per J. Garegg**  Department of Organic Chemistry, Arrhenius Laboratory, Stockholm University, Stockholm, Sweden

**Christine Gautheron-Le Narvor**  Institut de Chimie Moléculaire d'Orsay, Université Paris-Sud, Orsay, France

**Jacques Gelas**  École Nationale Supérieure de Chimie de Clermont-Ferrand, Aubière, France

**Bernd Giese**  Department of Chemistry, University of Basel, Basel, Switzerland

**Stephen Hanessian**  Department of Chemistry, University of Montreal, Montreal, Quebec, Canada

**Akira Hasegawa**  Department of Applied Bioorganic Chemistry, Gifu University, Gifu, Japan

**Derek Horton**  Department of Chemistry, The American University, Washington, D.C.

**Hoan Khai Huynh**  Department of Chemistry, University of Montreal, Montreal, Quebec, Canada

**J. C. Jaszberenyi**  Department of Organic Chemical Technology, Technical University of Budapest, Budapest, Hungary

**Karl-Heinz Jung**  Fakultät für Chemie, Universität Konstanz, Konstanz, Germany

**Janusz Jurczak**  Department of Chemistry, Warsaw University and Institute of Organic Chemistry, Polish Academy of Sciences, Warsaw, Poland

**Makoto Kiso**  Department of Applied Bioorganic Chemistry, Gifu University, Gifu, Japan

**Xianqi Kong**  Department of Chemistry, Queen's University, Kingston, Ontario, Canada

**Horst Kunz**  Institut für Organische Chemie, Johannes Gutenberg-Universität Mainz, Mainz, Germany

**Boliang Lou**  Department of Chemistry, Cytel Corporation, San Diego, California

**Robert Madsen**  Department of Chemistry, Duke University, Durham, North Carolina

**Alberto Marra**  Dipartimento di Chimica, Università di Ferrara, Ferrara, Italy

**Takashi Matsumoto**  Department of Chemistry, Tokyo Institute of Technology, Tokyo, Japan

**K. C. Nicolaou**  The Scripps Research Institute and University of California at San Diego, La Jolla, California

# Contributors

**Peter Norris**  Department of Chemistry, Youngstown State University, Youngstown, Ohio

**T. V. RajanBabu**  Department of Chemistry, The Ohio State University, Columbus, Ohio

**Gurijala V. Reddy**  Department of Chemistry, University of Montreal, Montreal, Quebec, Canada

**Richard R. Schmidt**  Fakultät für Chemie, Universität Konstanz, Konstanz, Germany

**Keisuke Suzuki**  Department of Chemistry, Tokyo Institute of Technology, Tokyo, Japan

**Walter A. Szarek**  Department of Chemistry, Queen's University, Kingston, Ontario, Canada

**Hiroaki Ueno**  The Scripps Research Institute and University of California at San Diego, La Jolla, California

**Jean-Michel Vatèle**  Department of Chemistry, Université Claude Bernard, Villeurbanne, France

**Heng Wang**  Department of Chemistry, University of Montreal, Montreal, Quebec, Canada

**Chi-Huey Wong**  Department of Chemistry, The Scripps Research Institute, La Jolla, California

**Aleksander Zamojski**  Institute of Organic Chemistry, Polish Academy of Sciences, Warsaw, Poland

**Heinz-Georg Zeitz**  Department of Chemistry, University of Basel, Basel, Switzerland

# 1

# Sugar Derivatives

**THEMES**: *Acetals, Dithioacetals, Ethers, Site-Selective Oxidations*

Because of their abundance and their endowment with unique stereochemical and functional features, carbohydrates can be considered as one of nature's better gifts to the synthetic organic chemist. Although this personal opinion was appreciated and shared by a few sugar chemistry aficionados 30 years ago, there was reluctance on the part of the general community of synthetic organic chemists to venture into the field as explorers or exploiters. Today this is no longer true.

One of the main deterrents to this practice was a combination of "sugarophobia" and conservatism on the part of the noncarbohydrate chemists, who preferred to tread along the charted waters of terpenes and related traditional natural products. Indeed, what better carbon frameworks to study mechanistic organic chemistry and to advance the state of the art of synthesis? Major insights, rules, and theories were being advanced throughout the 1960s and 1970s. Meanwhile the carbohydrate chemist, born and bred on a sugar-rich diet, had become a bit of an isolationist, content to work in and around the periphery of a sugar molecule, and with good reason. In fact, a large number of natural products with antibiotic, antitumor, and antiviral properties contained unusual sugar moieties that had to be isolated, their structures elucidated and synthesized. Who would be better suited for such a task but the carbohydrate chemist?

The notion that sugar molecules need not be just sugar molecules was popularized in the mid-1960s and early 1970s. In recent years, this school of thought has had a large following from the community of synthetic organic chemists in general. I do not think that chemists label themselves as sugar, terpene, or alkaloid people any longer. Much of this has had to do with changing attitudes in the classroom with revised curricula, in the laboratory with the need to synthesize enantiomerically pure molecules, and in people's psyches in general.

A major operational problem in working with polyhydroxylated molecules, is their derivatization in ways that render them synthetically useful as starting materials or as intermediates in synthesis. Thus, the availability of sugars, with a plethora of stereochemical and functional features, can be an amenity as well as a problem.

Fortunately, the rules of chemical reactivity and conformational analysis, coupled with the laws of thermodynamics, join forces to allow us to functionalize polyhydroxy aldehydes and ketones (aldoses and alduloses) in a selective and predictable fashion.

Water-soluble sugars disguised as hemiacetals, become organic–solvent-soluble as

*O*-protected cyclic or acyclic carbon frameworks. The choice of acetals or ethers as derivatives allows a systematic manipulation of diols and polyols. Kinetic control and a lesser affinity for protonation on sulfur compared with oxygen allows the transformation of cyclic hemiacetals into acyclic dialkyl dithioacetals. Acetal, ether, and dithioacetal derivatives are some of the pivotal intermediates needed to explore various applications of carbohydrates in synthesis.

Selectivity can be an overriding commodity in cases where reactivity is dictated by logic and accepted concepts. Such is the case with stannylene acetals of diols and trialkylstannyl ethers of alcohols. Enhanced nucleophilicity of oxygen attached to tin and well-documented stereoelectronic effects associated with methine carbon atoms of trialkyltin ethers lead to remarkably selective reactions of *O*-substitution and oxidation in polyhydroxy compounds.

The following four chapters offer insight and experimental details in the selective derivatization of sugar molecules.

*Stephen Hanessian*

# 1
# Synthesis of Isopropylidene, Benzylidene, and Related Acetals

**Pierre Calinaud and Jacques Gelas**
École Nationale Supérieure de Chimie de Clermont-Ferrand, Aubière, France

| | | |
|---|---|---|
| I. | Introduction | 3 |
| II. | Methods for Preparation of Acetals in Carbohydrate Chemistry | 6 |
| | A. General methods | 6 |
| | B. Mechanistic and structural aspects | 11 |
| III. | Experimental Procedures | 15 |
| | A. Acyclic sugars | 15 |
| | B. Pentoses | 16 |
| | C. Hexoses | 18 |
| | D. Aminosugars | 23 |
| | E. Deoxysugars | 24 |
| | F. Oligosaccharides | 25 |
| | G. Acetalation of *trans*-vicinal diols | 27 |
| | References | 28 |

## I. INTRODUCTION

The condensation of aldehydes and ketones with alcohols and polyols is one of the first reactions of the organic chemistry. Following the pioneering work by Wurtz [1] (acetaldehyde and ethylene glycol), and by Meunier [2] (catalysis with acids), Emil Fischer [3] described as early as 1895 the formation of acetals* of glycoses (first from D-fructose and acetone). Since then, this protecting group has been extensively used in organic chemistry, in general, and in carbohydrate chemistry, in particular. These developments concern not

---

*Following the recommendation of IUPAC (rule C-331.1) the term *acetal* should be given to the compounds obtained through the reaction of a carbonyl group of an *aldehyde* as well as from a *ketone*.

only acyclic and cyclic acetals, but also analogues in which the oxygen atoms have been replaced by other heteroatoms, the sulfur atom being of particular importance (thio- and dithio-acetals). This chapter will consider only the most popular and useful acetals, with some comments concerning related acetals and extension to oligosaccharides. The case where the acetal involves the anomeric center (glycosides) falls outside the scope of this chapter. A later chapter deals with acyclic dithioacetals, and these can be found elsewhere in this monograph.

Several reviews have already been published on the subject, for example, the acetalation of alditols [4], of aldoses and aldosides [5,6], and of ketoses [7]. Some aspects of the stereochemistry of cyclic acetals have been discussed in a review dealing with cyclic derivatives of carbohydrates [8], also in a general article [9] and, more recently, in a chapter of a monograph devoted to the stereochemistry and the conformational analysis of sugars [10]. Aspects on predicting reactions patterns of alditol–aldehyde reactions are reviewed within a general series of books on carbohydrates [11]. The formation and migration of cyclic acetals of carbohydrates have also been reviewed [12,13].

The evident success of the transformation of polyols into cyclic acetals as a method for temporary protection, is mainly due to the following features: (1) accessibility and cheapness of the reagents; (2) ease of the procedure leading quickly and in high yield to the protected derivatives; (3) inertness of the protecting group to a large variety of reagents used in the structural modifications of the substrate; (4) ease and high-yielding step for deprotection. Usually, reagents for acetalation are quite common chemicals that are essentially nontoxic; their uses are well established and straightforward. Some representative procedures of the various methods will be presented here, especially for the most important derivatives; namely, *O*-isopropylidene and *O*-benzylidene sugars. For example, 1,2:5,6-di-*O*-isopropylidene-D-glucofuranose **1**, 1,2:3,4-di-*O*-isopropylidene-D-galactopyranose **2**, and methyl 4,6-*O*-benzylidene-α-D-glucopyranoside **3**, continue to be used extensively by sugar chemists.

Only short comments will be given for other acetal derivatives that are less popular. Chart 1 presents a list of formulae of cyclic acetals, mainly, those with five- and six-membered rings (1,3-dioxolanes and 1,3-dioxanes). Seven-membered ring acetals are omitted because they are scarcely represented in carbohydrate chemistry. The special case of spiroacetals and cyclohexane-1,2-diacetal-protecting groups, which have been reported recently, will be presented in Part II.

The essential justifications for the choice of one type of acetal among the various possibilities are probably (1) the structure of the acetal obtained (i.e., dioxolane or dioxane type; with or without involvement of the anomeric hydroxyl group; obtention of a furanoid or a pyranoid protected form of the sugar, especially when one starts from a free one); (2) the respective reactivity of these acetals as far as the deprotecting step is concerned. A brief discussion of the point (1) will be given in the next paragraph. Relative to the deprotection of cyclic acetals, generally their cleavage, regenerating a diol, is obtained using very similar acidic aqueous conditions [4–7]. However, a selective removal of one acetal in the presence of the same (or different) functions, at distinct positions in the same molecule,

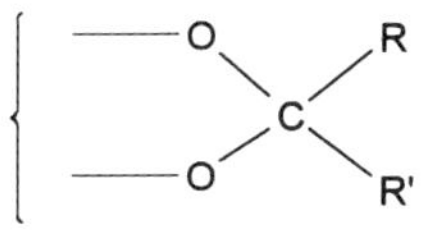

# Isopropylidene, Benzylidene, and Related Acetals

| | R | R' |
|---|---|---|
| O-methylene | H | H |
| O-ethylidene | Me | H |
| O-cycloalkylidenes<br>n=4 cyclopentylidene<br>n=5 cyclohexylidene | (CH$_2$)$_n$ | |
| O-isopropylidene | Me | Me |
| O-benzylidene | Ph | H |
| O-benzylidene substituted<br>Y= o-NO$_2$, p-OMe, p-NMe$_2$ | Y-C$_6$H$_4$ | H |

**Chart 1** Most common cyclic acetals used in carbohydrate chemistry

is possible and has been quite often observed. As examples, one can recall that generally a 1,2-O-isopropylidene group is more resistant to acid hydrolysis than the same group at any other position. trans-Fused 4,6-O-benzylidene acetals of hexopyranosides are hydrolyzed faster than the corresponding cis-fused acetals and a para-anisylidene group can be removed without loss of a benzylidene group in the same molecule by graded acid hydrolysis. A list of representative examples of this kind of selective removal within a multifunctional carbohydrate derivative can be found in a review partly devoted to acetals [14].

Finally, it should be emphasized, even if it is paradoxical, that this excellent protecting group can, under special conditions, behave as a real functional group with its own reactivity. During these last 20 years, reactions have opened the way for the development of strategies for structural modifications, thereby amplifying the interest for acetals. Among these reactions one can briefly recall: (1) oxidation (ozonolysis, action of potassium permanganate); (2) photolysis; (3) halogenation (N-bromosuccinimide, triphenylmethylfluoroborate, and halide ions; hydrogen bromide in acetic acid; dibromomethylmethylether; miscellaneous reagents); (4) hydrogenolysis (mixed hydride reagents); (5) action of strong bases (ring opening with butyllithium, other strong bases); (6) formation of esters induced by peroxides and (7) cleavage with Grignard reagents. This reactivity has been the subject of a review [15] that demonstrated the versatility of acetals

Chart 2 shows some protective groups closely related to cyclic acetals, and it may be useful to comment briefly about them as they will not be discussed further here. The first example corresponds to the O-cyanoalkylidene group, especially the O-cyanoethylidene group, which actually has been introduced in carbohydrate chemistry as a method for activation of the anomeric center in oligosaccharide synthesis [16]. Other examples are less closely related to acetals and result from the substitution of the acetal carbon atom by an heteroatom (Si, Sn, or B) or correspond to the presence of three heteroatoms (O or N) on this center. Thus, use of 1,3-dichloro-1,1,3,3-tetraisopropyldisiloxane in basic medium has been introduced for the simultaneous protection of the 3'- and the 5'- OH groups in nucleosides [17]; this strategy has been extended to the monosaccharides and the migration

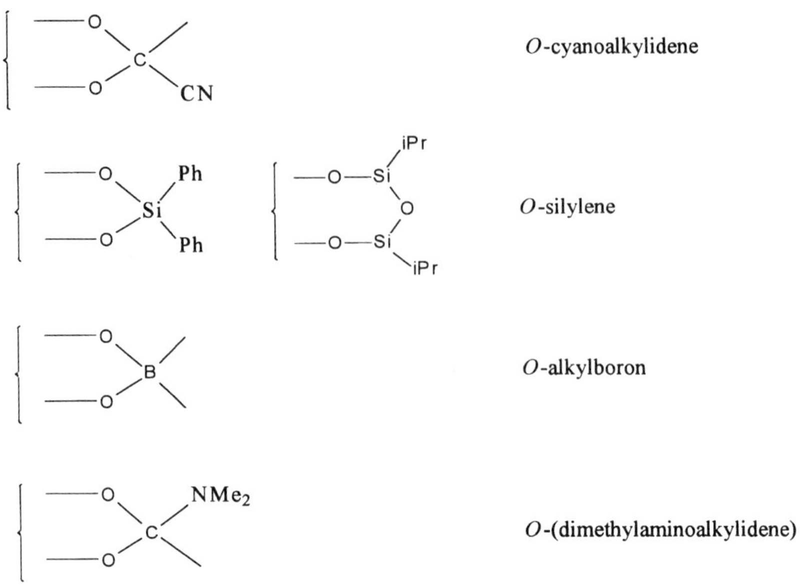

**Chart 2** Derivatives related to acetals used in carbohydrate chemistry

O-cyanoalkylidene

O-silylene

O-alkylboron

O-(dimethylaminoalkylidene)

of the silyl-protecting group has been studied [18]. A slightly different silyl group has also been suggested for the selective protection of sucrose, even if the interosidic acetals (1′,2-silylene and 1′,2:6,6′-disilylene derivatives), resulting from the action of dimethoxydiphenylsilane in the presence of acid, are obtained in low yield [19]. More interesting from the preparative point of view is the introduction in carbohydrate chemistry of the reaction of dibutyltin oxide giving dibutylstannylene derivatives (or stannoxane) [20]. Their reactivity with electrophiles gives predominantly monosubstituted products, usually with a high regioselectivity [21], as exemplified by a monoalkylation [22]. Another example is offered by cyclic boronates, which have been used to a limited extent owing to their high sensitivity to hydrolytic conditions [23]. However, the O-ethylboron derivatives have been especially developed to give special assistance in various controlled reactions of monosaccharides [24]. The last example is concerned with protecting groups closer to *ortho*-esters than to acetals. The selective formation of *ortho*-esters at nonanomeric positions has been recently described [25]. Amide acetals have been used particularly in carbohydrate chemistry in the α-(dimethylamino)-ethylidene and -benzylidene acetal series [26]. Their general properties have been considered, especially the acid hydrolysis to monoesters, which is of value in the ribofuranoside series for oligonucleotide synthesis.

## II. METHODS FOR PREPARATION OF ACETALS IN CARBOHYDRATE CHEMISTRY

### A. General Methods

Fundamentally, we can classify the different methods into two categories, depending on the experimental conditions: (1) acid or neutral medium; (2) basic conditions. Less common procedures will be presented in a third section.

## Isopropylidene, Benzylidene, and Related Acetals

*Acetalation in Acidic or Neutral Conditions*

**Direct Condensation of a Carbonyl Derivative.** Historically, this is the first procedure and generally the sugar and an aldehyde (or a ketone) are simply mixed either directly (the reagent, for instance propanone, used in a large excess, also being the solvent) or in solution in a solvent (*N,N*-dimethylformamide is the most frequently used, dimethylsulfoxide being encountered far less) and eventually in the presence of a catalyst. The latter can be either a soluble acid (practically all kinds of organic and inorganic acids have been tested, and the most frequently used are sulfuric acid, *p*-toluenesulfonic acid, camphorsulfonic acid, or hydrogen chloride) or an insoluble one (Amberlyst resins, Montmorillonite K10). An idealized representation of the mechanism of the reaction is given in Scheme 1, but it does not necessarily give the exact nature of all possible intermediates (see Sec. II.B).

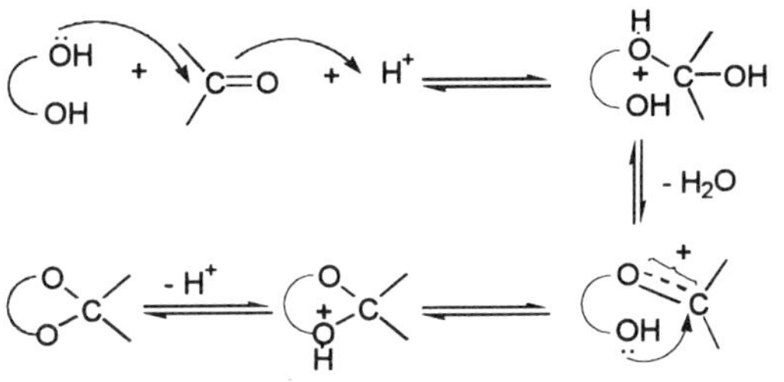

**Scheme 1**

The use of a Lewis acid (e.g., triethylfluoroborate, zinc chloride, stannous chloride, titanium chloride, iron(III)chloride) and other reagents (e.g., iodine, trimethylsilane, trifluoromethane-sulfonylsilane) have also been recommended. Exhaustive lists of catalysts and conditions can be found in reviews devoted to carbohydrates [5–7], or to general organic chemistry [27,28]. However, one can add the new catalyst, which has been introduced for the smooth formation of *p*-methoxybenzylidene acetals and *p*-methoxyphenylmethyl methyl ether [29], namely 2,3-dichloro-5,6-dicyano-*p*-benzoquinone (DDQ), and has been applied very recently [30] to the synthesis of isopropylidene mixed acetals.

Obviously, the condensation of a carbonyl group with a diol produces 1 mol of water and because of the reversibility of the reaction (hydrolysis of the acetal), yields are lowered if this by-product is not removed. For such a purpose, there are essentially two possibilities: (1) the continuous removal of water by an azeotropic distillation with a solvent mainly chosen for its boiling point (petroleum ether, benzene, toluene, xylene, for instance); (2) the presence of a desiccant (the most commonly taken is copper(II)sulfate, but sodium sulfate or molecular sieves have been also used); molecules known to be water scavengers, such as *ortho*-esters or dialkylsulfites, have also been suggested, even if they are seldom used in carbohydrate chemistry.

Important in this quite general strategy is that, for practically all instances, the reaction is under thermodynamic control, and the control of the stoichiometry is extremely difficult. It follows that only the more stable acetals are produced (see Sec. II.B) and usually multiacetals are obtained if several hydroxyl groups are available within the same molecule. This has been a major concern in acetalation reactions in neutral conditions. For instance, use of copper(II)sulfate either in acetone alone or in *N,N*-dimethylformamide without any additional catalyst, leads to acetals with structures that differ from those resulting from reactions in the presence of an acid. The reaction depends on the temperature [31]; however, the strict neutrality of a medium in which copper(II)sulfate and polyols are interacting can be questioned.

**Transacetalation.** This strategy, based on an acetal exchange in acid conditions, has been introduced more recently in carbohydrate chemistry [32–34]. It offers several advantages over the direct condensation of the corresponding free carbonyl group: (1) anhydrous conditions can be strictly followed, as the only by-product is 2 mol of the alcohol (e.g., MeOH) used to prepare the reagent (Scheme 2); this alcohol can even be removed by

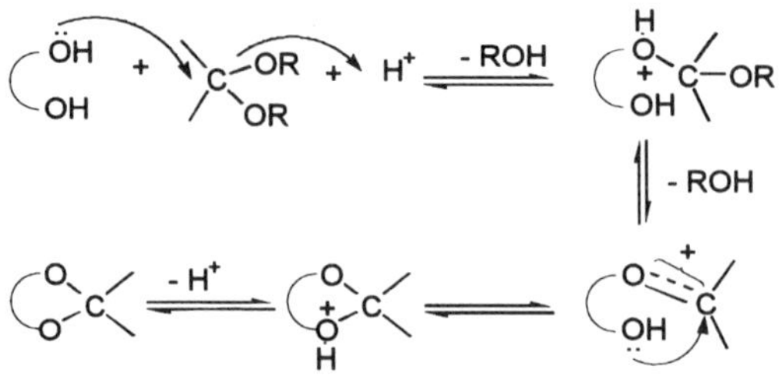

**Scheme 2**

diminished pressure to displace the equilibrium if necessary; (2) the stoichiometric of the reaction can be controlled; (3) in some instances, it is possible to obtain acetal(s) under kinetically controlled conditions, even if many sugars (especially free sugars) still react to give the more stable structures (see Sec. II.B); (4) also there is a possibility of obtaining strained acetals, such as those resulting from the acetalation of 2,3-*trans*-diols of pyranosides, although yields are generally low.

Thus, the formation of *O*-isopropylidene derivatives using 2,2-dimethoxy-propane–*N*,*N*-dimethylformamide–*p*-toluenesulfonic acid has become one of the most popular ways to protect diols. This strategy has been applied to many sugars and is compatible with aminosugars [35] and oligosaccharides such as sucrose [36], maltose, laminaribiose, cellobiose, and gentobiose [37]. It has been extended to *O*-benzylidene derivatives for which the use of α,α-dimethoxytoluene can advantageously replace benzaldehyde [38,39]. Its application to oligosaccharides is also possible and has been described, for instance, for maltooligosaccharides [40]. A slight modification of the classic procedure (the transacetalation is followed by a partial hydrolysis of the crude mixture to remove unstable acyclic acetals) offers a convenient route to an interosidic, eight-membered, cyclic benzylidene acetals [41].

Once again, efforts have been made to find neutral conditions that can modulate the course of the reaction. For instance, use of 2,2-dimethoxypropane in solution in 1,2,-dimethoxyethane (which probably plays a role through its interaction with polyols) has been suggested as a reagent for acetalation in neutral conditions (no catalyst) of D-mannitol [42] and D-glucitol [43].

For transacetalation reactions, it is worth noting here the recent strategy introduced for selective protection of vicinal diols (and especially with a *trans* configuration) in carbohydrates: a double exchange involving the acetal functions of 1,1,2,2-dimethoxy-cyclohexane gave a dispiroacetal, structurally related to a 1,4-dioxane, stabilized by the axial position of the methoxyl groups [44]. This method completes the preceding one using enol ethers, leading to an analogue of 1,4-dioxane [45] (see following section).

**Acetalation with Enol Ethers Under Kinetically Controlled Conditions.** The first mention of the use of an enol ether to protect the hydroxyl group of an alcohol was developed by Paul [46], who introduced the reaction with dihydropyran to give tetrahydropyranyl ethers, which is still used 60 years later. In spite of some noticeable developments, such as the preparation of 2′,3′-*O*-alkylidene derivatives of nucleosides [33]; the synthesis of 4,6-*O*-ethylidene-α-D-glucopyranoside with use of methylvinylether [47]; the intra-

# Isopropylidene, Benzylidene, and Related Acetals

molecular addition of an hydroxyl group on monovinylethers of 1,2-diols [48]; and its application to the synthesis of 1,2-$O$-isopropylidene-α-D-galactose [49], this strategy was underestimated until it was shown that the use of 2-alkoxypropene in $N,N$-dimethylformamide was a simple and efficient method of *acetonation* under exclusive kinetically controlled conditions. In many instances, the products differed from those prepared under thermodynamic control [50]. The reaction (Scheme 3) is characterized mainly by the

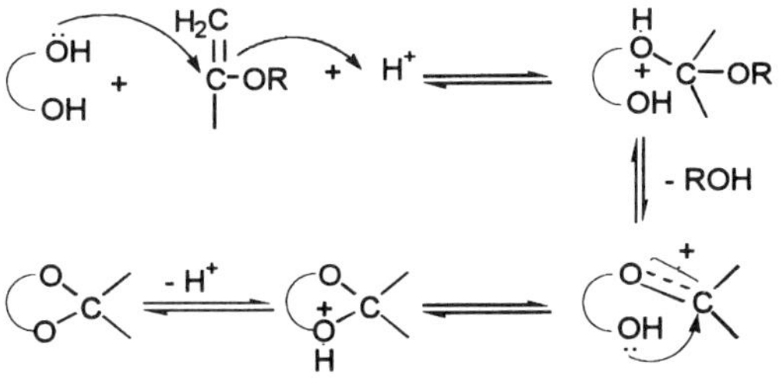

**Scheme 3**

following considerations. (1) The favored site for initial attack by the reagent is at a primary hydroxyl group, even with free sugars in solution; thus, D-glucose [51], D-mannose [52], D-allose, and D-talose [53] give the 4,6-$O$-isopropylidene derivative exclusively, and D-galactose gives essentially the same type of acetal [53]. (2) If the favored tautomeric form of the sugar in solution does not have a primary hydroxyl group, the attack of the most reactive secondary group leads to 1,3-dioxolanes without tautomerization: for example, one observes exclusive formation of 3,4-$O$-isopropylidene-D-arabinopyranose, and formation of 3,4-$O$-isopropylidene-D-ribopyranose as major products [54]; on the other hand, a tautomerization leads to the formation of the more stable 2,3-$O$-isopropylidene-D-lyxofuranose [55] (also compare the acetonation products of D-fucose and D-rhamnose [56]). (3) In the initial process, the anomeric hydroxyl group does not take part in the reaction. (4) Access to either mono- or diacetals is permitted by careful stoichiometric control, as in the preparation of mono- and di-$O$-isopropylidene-D-mannopyranoses [52]. (5) The method can be applied to the acetonation of oligosaccharides with the same characteristics and without cleavage of the glycosidic bond. For example, it is possible to effect a selective monoacetonation of α,α-trehalose [57] and to obtain acetonides of sucrose [58], lactose [59], and maltose [60]. (6) The method permits an efficient access to a strained ring, as in the acetonation of *trans* vicinal diols [61], the formation of medium-sized acetals (interosidic acetals of oligosaccharides [58–60], and to obtain 1,5-$O$-isopropylidene-D-ribofuranose [54]. The reaction has been used in a variety of contexts covering a large assortment of sugars. For instance, it has been applied to the selective acetonation of alditols, such as D-mannitol [62] and 1,4-anhydropolyols [63], ketoses [64], diethyldithioacetals of monosaccharides [65], as well as thiosugars (5-thio-D-xylopyranoside [66]). It can also be used to obtain *aldehydo* derivatives of monosaccharides, exemplified by *aldehydo*-2,3:4,5-di-$O$-isopropylidene-D-xylose [55,67].

Extension of this strategy to other vinyl ethers using essentially 2-methoxypropene has been described. Minor structural variations in the reagent used are possible. Thus, entirely comparable results giving cyclohexylidene acetals are observed using 1-alkoxycyclohexene [68]. 2-Trimethylsilyloxypropene has been used for the acetalation of 1,2-cyclohexanediol [69]. On the other hand, acyclic acetals (6-substituted mixed acetals) are obtained when methyl 2,3-$O$-benzyl-α-D-glucopyranoside is allowed to react with 2-benzyloxypropene or 2-benzyloxy-3-fluoropropene [70]. More important structural variations are concerned with (1) the smooth preparation of ethylenic acetals (monomers for polymerization) with ethylenic enol ethers [71]; (2) the selective reaction of the vinyl ether unit in ketonic enol ethers, leading to an acetal substituted with a ketonic chain [72]; (3) the introduction of a bis-dihydropyran (namely 3,3′,4,4′-tetrahydro-6,6′-bi-2$H$-pyran) as a reagent useful for the selective protection of *trans* (diequatorial) vicinal diols in monosaccharides for which there is an evident competition between the acetalation of 1,2-*cis*, 1,2-*trans*, and 1,3-diols; the formation of an unique dispiroacetal (a tetraoxadispiro-[5.0.5.4]hexadecane) is explained by the anomeric effect, which stabilized the structure of a 1,4-dioxane substituted by two axial C−O bonds originating from this bis-dihydropyran [45].

Finally, all these reactions are catalyzed by *p*-toluenesulfonic acid, or camphorsulfonic acid, or pyridinium salts. Use of pyridinium *p*-toluenesulfonate is now well established as a mild catalyst. We have already noted the recent use of DDQ which has recently proved to be effective with 2-methoxypropene [30].

*Acetalation in Basic Conditions*
The search for kinetically controlled conditions has stimulated the study of basic media for acetalation. Essentially, methylene and benzylidene acetals have been prepared according to reactions corresponding to Scheme **4**:

**Scheme 4**

Thus, dichloro- or dibromomethane in the presence of sodium hydride in solution in $N,N$-dimethylformamide gives $O$-methylene derivatives [73,74]. Other conditions are also possible, for instance; use of potassium hydroxide and dimethylsulfoxyde [75], but an interesting development is the application of the phase-transfer catalysis technique, by which dibromomethane and sodium hydroxide in water, in the presence of an appropriate ammonium salt, leads to a *cis*-2,3-$O$-methylenation of methyl-4,6-$O$-benzylidene-α-D-mannopyranoside, [76] and similar conditions afford the *trans*-2,3-$O$-methylenation of methyl-4,6-$O$-benzylidene-D-hexopyranosides [77]. Other examples have been published [78].

Concerning the $O$-benzylidene derivatives, the main objective is to obtain stereoisomeric control of the chirality introduced at the acetal carbon atom. Classic methods of benzylidenation (using benzaldehyde or α,α-dimethoxytoluene) are based on acidic catalysis and give the more stable compound, with a thermodynamically controlled acetalic configuration. For instance, the known [79] free-energy for the *equatorial* preference of a phenyl substituent at position 2 on a 1,3-dioxane (structurally comparable with a 4,6-$O$-benzylidene derivative) is such that no diastereoisomer corresponding to the *axial* position can be isolated in acidic medium. In contrast both diastereoisomers of 4,6-$O$-benzylidene acetals are actually obtained [80] using α,α-dimethoxytoluene and potassium *tert*-butoxide. Even if the reaction has not yet found practical applications owing to its rather low yield, benzaldehyde itself can react in the presence of potassium *tert*-butoxide with methyl-2,3-di-$O$-(toly-*p*-sulfonyl)-α-D-glucopyranoside to give 4,6-$O$-benzylidene acetals [81]. In fact, it has been demonstrated that a strong basic medium is unnecessary. Thus, an alternative method of benzylidenation is the reaction of α, α-dihalotoluenes simply in pyridine at reflux. When 1,3-dioxolanes are obtained, the formation of both *exo-* and *endo-*phenyl-substituted derivatives is observed, but for a 1,3-dioxane, only the most stable isomer (equatorial phenyl) is obtained [82]. However, in noncarbohydrates, pyridinium chloride can catalyze acetalations [83]. It has recently been shown that 4,6-*O*-isopropylidene-sucrose can be conveniently obtained using 2-methoxypropene in solution in pyridine and, in the presence of pyridinium, *p*-toluenesulfonate [84]; thus, it is not surprising to observe the formation of the more stable compound in the preceding examples of benzylidenation.

*Miscellaneous Methods*

Several less general methods of acetalation are known [28], and a few of them have found some application in carbohydrate chemistry. Because they can represent exceptional alternatives to classic procedures, they will be briefly presented here.

**Hydrogenolysis of *ortho*-Esters.** A two-stage procedure for converting methoxyethylidene derivatives by using boron trifluoride, followed by reduction with lithium aluminium hydride, has been used to prepare *exo-* and *endo*-diastereoisomers of methyl 3,4-*O*-ethylidene-β-L-arabinopyranoside [85]. A similar approach has been described using diborane [86]. Other reducing agents, such as "mixed hydrides", prepared by mixing lithium aluminium hydride and aluminium chloride (see ref. [15] and references cited therein), have been useful to directly reduce *ortho*-esters (methoxyethylidene derivatives) to methylene, ethylidene, and benzylidene acetals [74].

*Ortho*-esters at position 1,2- of sugars are more easily prepared than the corresponding acetals; as an exchange of both functional groups is possible, 1,2-*O*-alkylidene derivatives can be prepared by the reaction of these *ortho*-esters with the appropriate carbonyl reagent in strictly anhydrous conditons and in the presence of an acid [87].

In fact, these reactions likely involve 1,3-dioxocarbenium ions, which can also be prepared from pyranosyl chlorides and reduced to acetals; thus *endo-* and *exo-*1,2-*O*-ethylidene-α-D-allopyranoses have been prepared from penta-*O*-acetyl-β-D-allopyranosyl chloride (reaction with sodium borohydride) [88]. These types of 1,2-acetoxonium ions are also known to react with dialkylcadmium to give 1,3-diacetals [89].

**Action of *N*-Bromosuccinimide in Dimethylsulfoxide.** An alternative to the classic methylenation of diols has been offered by the simple procedure using *N*-bromosuccinimide in dimethylsulfoxide and has been applied, for instance, to the acetalation of D-mannitol and D-ribofuranosides [90].

## B. Mechanistic and Structural Aspects

The formation and the hydrolysis of acyclic and cyclic acetals have been studied in rather great detail [91]. Several reviews on this topic are available [92] and some comments have been made [13] concerning the carbohydrate series. We have shown in Schemes **1**, **2**, and **3** that a common feature of this reaction seems to be the intermediacy of an oxocarbenium ion. However, the cyclization of such an intermediate has been questioned more recently [93] in the light of the Baldwin's rules for ring closure [94]. At least for the five-membered ring, an $S_N2$-type displacement mechanism for the protonated form (**B**) of the hemiacetal (**A**) (favorable 5-exo-tet cyclization) has been proposed; rather than the unfavorable 5-endo-trig cyclization of the oxocarbenium ion (**C**); (Scheme **5**). Except when the formation of the enol ether (**D**) is structurally impossible, the intermediacy of such a compound remains feasible.

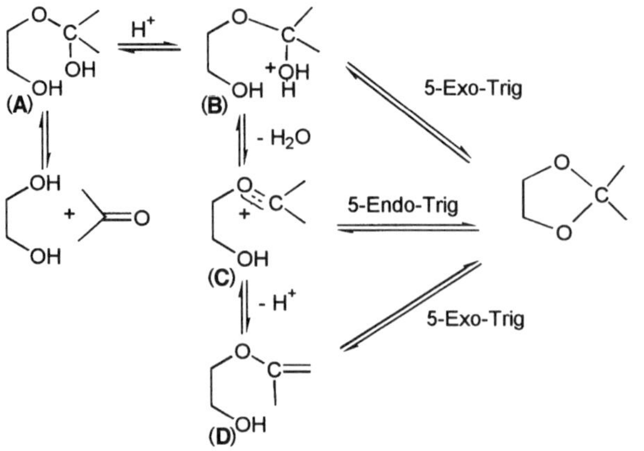

Scheme 5

*Kinetic Control Versus Thermodynamic Control*

The major mechanistic and structural aspect of the acetalation process is its orientation toward derivatives obtained either under thermodynamically controlled conditions or under kinetically controlled conditions. We will not discuss here all structural factors concerning the relative stabilities of acyclic and cyclic acetals of polyols and monosaccharides, because such a discussion has been extensively reviewed and adequately commented on [8,10,12 –14]. However, it is important to focus here on the main consequences of these relative stabilities in relation to the various experimental conditions to orientate the choice of specific conditions, particularly for the most important monosaccharides (D-glucose, D-mannose, and D-galactose).

Concerning the most popular derivatives, one can say (Scheme 6) that a better kinetic control is obtained using successively acetone, or 2,2-dialkoxypropanes, or 2-alkoxypropenes.

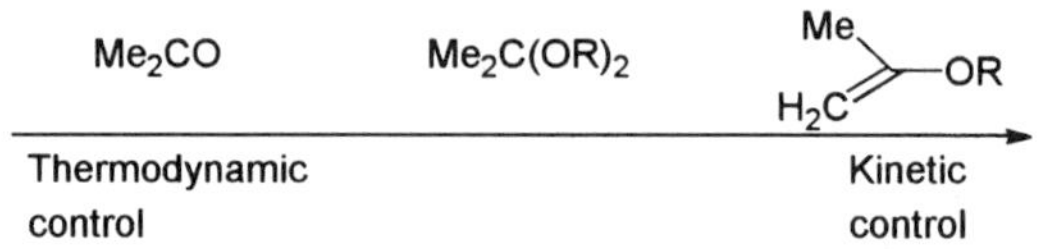

Scheme 6

Practically all examples in the literature show that the use of acetone leads to the more stable acetals and, conversely, that 2-methoxypropene generally allows an access to structurally quite different products (kinetic compounds). An intermediate behavior is observed for the transacetalation process involving 2,2-methoxypropane, which gives results either similar to those obtained with acetone or similar to those obtained with enol ethers. A good choice for the elucidation of whether a reaction is under thermodynamic or kinetic control is the study of the acetonation of free monosaccharides, which are subject to the tautomerization phenomena. Two examples of the "mixed behavior" of 2,2-dimethoxypropane are given in Schemes 7 and 8. The reaction of D-glucose with acetone and acid [95] gives the classic diacetone glucose **1** (see Scheme 7). On the other hand, use of 2,2-dimethoxypropane [96], or 2-methoxypropene [51], gives a high yield of the less stable pyranoid monoacetal **4** (kinetic and stoichiometric controls). It is possible to confirm the

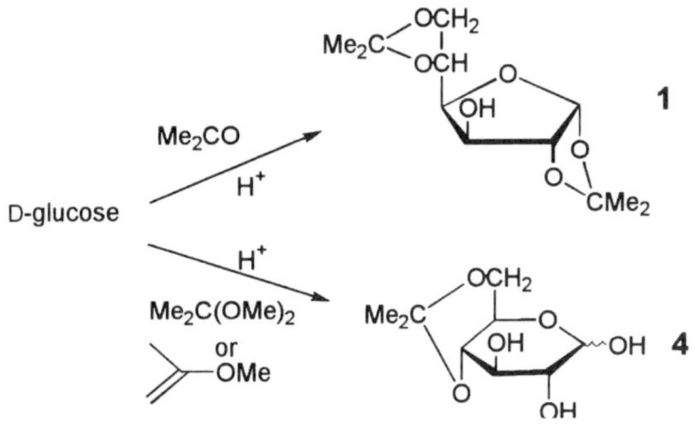

# Isopropylidene, Benzylidene, and Related Acetals

**Scheme 7**

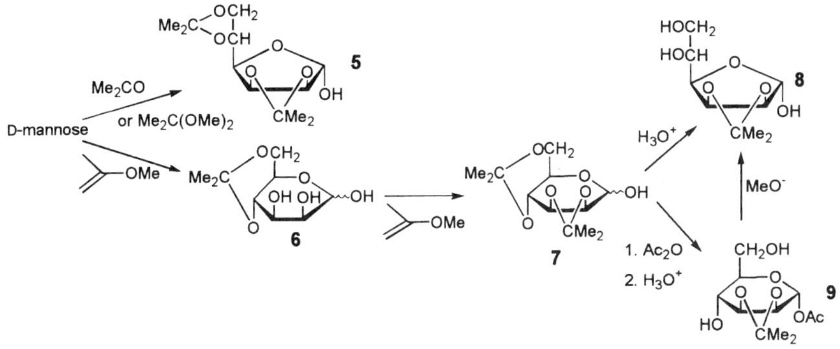

**Scheme 8**

relative stabilities of acetals **1** and **4** by the easy transformation of the latter into **1** by treatment in acidic acetone [51].

The second example concerns the study of acetonation of D-mannose (see Scheme **8**) and allows a clear distinction between the use of 2,2-dimethoxypropane and 2-methoxypropene. Thus, whereas D-mannose gives 2,3:5,6-di-$O$-isopropylidene-D-mannofuranose **5** by reaction of the free sugar with acetone [5,6] as well as with 2,2-dimethoxypropane [96], the major compound (more than 85%) obtained with 2-methoxypropene is 4,6-$O$-isopropylidene-D-mannopyranose **6** [52]. Once again, a confirmation of the better stability of furanoid acetals in this series is given by the selective hydrolysis of the 2,3:4,6-di-$O$-isopropylidene-D-mannopyranose **7** (by-product of the preceding reaction or quantitatively obtained by action of 2-methoxypropene on acetal **6**), which gives the furanoid monoacetal **8**. Actually, the pyranoid monoacetal **9** can be easily prepared as soon as the anomeric hydroxyl group is protected by acetylation [52].

The main characteristics of the use of 2-methoxypropene for the acetonation of sugars have already been summarized (see Sec. II.A), and one of them is the initial attack on the primary hydroxyl group, if any, in the preferred tautomeric form. Although this is confirmed by easily obtaining 4,6-isopropylidene-D-galactopyranose **10** (Scheme **9**) in 67% yield, the 3,4-$cis$-diol, as 3,4-$O$-isopropylidene-D-galactopyranose **11** is also isolated (14% yield) along with traces of 5,6-$O$-isopropylidene-D-galactofuranose [53]. This result is still in clear contrast with the classic acetonation of D-galactose, which gives [95] the well known 1,2:3,4-di-$O$-isopropylidene-D-galactopyranose **12** exclusively.

The difficulty in obtaining pyranoid 4,6-$O$-isopropylidene derivatives is due to the

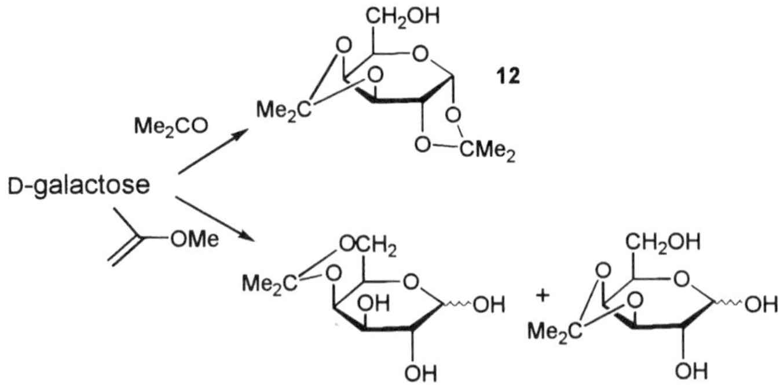

Scheme 9

presence of an axial methyl group at position 2 of the 1,3-dioxane system. A strong *syn*-axial interaction [98] results, which has been confirmed by the evaluation of the free energy of such an axial methyl group and is estimated to more than 3 kcal/mol [99]. Obviously, using aldehydes (benzaldehyde most frequently) instead of acetone, suppresses this interaction, and pyranoid derivatives are thus easily obtained.

The acetonation under kinetically controlled conditions is also useful for the protection of vicinal *trans*-diols, which are quite reluctant to cyclization into five-membered rings. Although use of 2-methoxypropene has been successful in this objective [61,66], one should recommend the recently discovered uses of reagents that minimized the ring strain by obtaining six-membered rings from vicinal *trans*-diols, which are protected (Scheme **10**) as 1,4-dioxanes (dispiroacetals, *trans*-decalinic system) stabilized by an anomeric effect.

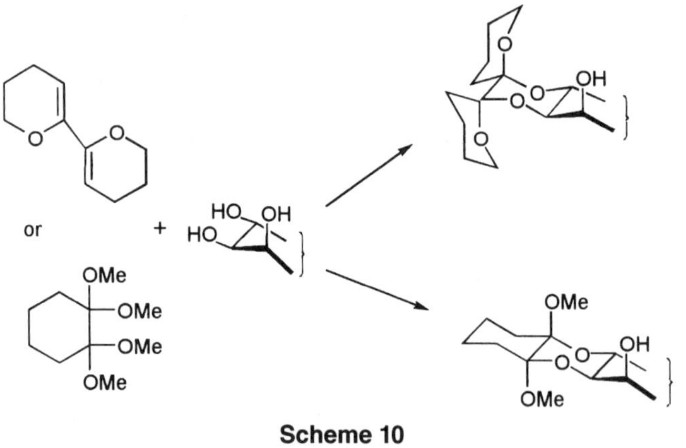

Scheme 10

*Selective Hydrolysis of Diacetals*

One of the interesting properties of sugars protected by more than one cyclic acetal group is the possibility for them to experience a selective hydrolysis, which may be of great potential for practical applications in synthesis. The regioselectivity of hydrolysis of multiacetals (mainly diacetals) is governed particularly by the eventual implication of the anomeric center and the structure of the bicyclic ring system that essentially can be either (1) a 1,3-dioxolane or a 1,3-dioxane, or (2) fused to a furanose or a pyranose, or (3) covalently independent from the sugar ring. Most of these factors and their consequences have already been reviewed and discussed [see Refs. 5,6,14]. Thus, one can briefly summarize some well-established observations:

1. Usually 2,2-disubstituted 1,3-dioxanes (for instance, acetonides) are hydrolyzed more easily than corresponding 1,3-dioxolanes (essentially owing to the strong *syn*-axial interaction operative in the six-membered ring [79,98].
2. Most of 1,2:5,6-di-*O*-isopropylidene acetals of the aldohexoses may be selectively (or partially) hydrolyzed to 1,2-*O*-isopropylidene derivatives.
3. For sugars in which the acetal function does not involve the anomeric center, a 1,3-dioxolane *cis*-fused to a furanose or a pyranose is more stable than the 1,3-dioxolane which involves a side chain.
4. *trans*-Fused 4,6-*O*-benzylidene acetals of hexopyranosides (*trans*-decalinic system) are generally hydrolyzed faster than the corresponding *cis*-fused acetals (*cis*-decalinic system).

Two specific examples of selected hydrolysis of diacetals are given in the experimental section in the D-*gluco*-furanose and the D-mannopyranose series.

## III. EXPERIMENTAL PROCEDURES

The following procedures have been arbitrarily chosen as representative of classic acetals extensively used as versatile starting materials for synthesis [102]. It covers aspects of the chemistry of acyclic and cyclic monosaccharides and some disaccharides. Procedures from other laboratories have been reproduced from the original publication and their authors are acknowledged.

### A. Acyclic Sugars

*Aldehydo-2,3:4,5-di-O-isopropylidene-D-xylose [67]*

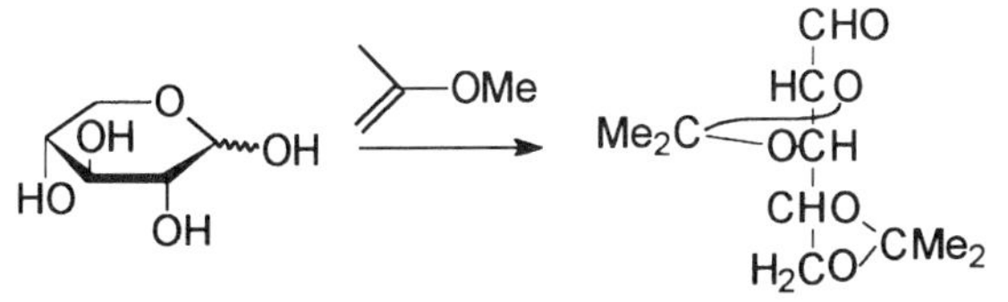

*p*-Toluenesulfonic acid hydrate (40 mg) was added with stirring to a solution of D-xylose (10 g, 0.067 mol) and 2-methoxypropene (14.4 g, 0.2 mol) in DMF (130 mL) at 0°C. After 8 h at 0–5°C, the xylose has reacted (thin-layer chromatography [TLC], 3:2 ethyl acetate/hexane), and three spots were evident by TLC ($R_f$ 0.61, 0.30, and 0.28). The acid was neutralized by stirring with dried Amberlite IRA-400 resin ($OH^-$ form). The resin was removed, washed with $CH_3OH$, and the extracts and reaction mixture were evaporated under vacuum (1 mm, < 40°C) to give a syrup (14.4 g) that was thoroughly extracted with dry hexane. The insoluble residue (5.9 g, TLC) 3:2 ethyl acetate/hexane, $R_f$ 0.28, major; 0.30, minor; 0.61, trace) has inappreciable amounts of 3,5-*O*-isopropylidene-D-xylofuranose or 1,2-*O*-isopropylidene-D-xylopyranose and may been been made up of acyclic monoisopropylidenation products. Vacuum evaporation of the hexane-soluble fraction gave 8.5 g (51% yield, $R_f$ 0.61) of the free aldehyde 13.

## Acetonation of Diethyl Dithioacetals of Monosaccharides

As the well-known transformation of free monosaccharides to diethyl dithioacetals is probably the best access to open-chain sugar derivatives, the preparation of acetals [65] of such compounds has been studied using either conventional methods [for instance see Ref. 101 and references cited therein, for the cupric sulfate catalyzed isopropylidenation with acetone] or kinetically controlled conditions. Thus the synthesis of cyclohexylidene acetals (using 1-ethoxycyclohexene) or isopropylidene acetals (using 2-methoxypropene) of diethyl dithioacetals of D-arabinose, D-xylose, D-glucose, and D-galactose has been described [65]. As a specific example we reproduce here only the reaction involving D-glucose diethyl dithioacetal.

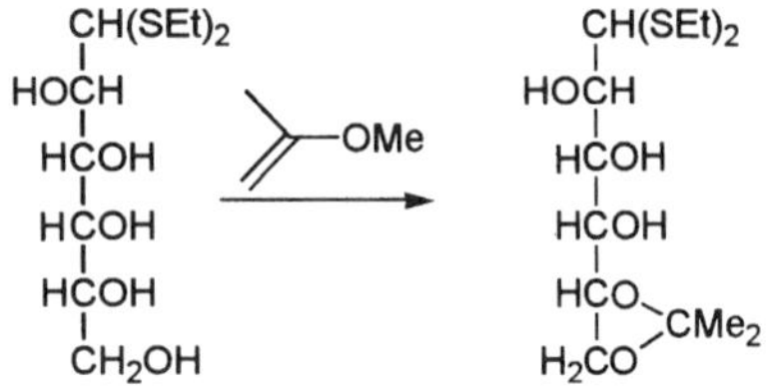

A solution of the dry D-glucose diethyl dithioacetal (10.725 g; 37.5 mmol) 2-methoxypropene (3.245 g; 45 mmol) in anhydrous DMF (130 mL) and $p$-toluenesulfonic acid (375 mg) was kept for 68 h at 0°C. The homogeneous mixture was kept with exclusion moisture until TLC indicated that all the starting material had reacted, and it was then poured into a solution of sodium hydrogenocarbonate (2% w/v, 60 mL). This mixture was extracted with ether (4 × 30 mL). The combined ether extracts were washed with water (2 × 30 mL), dried (magnesium sulfate), and evaporated, giving yellowish crystals (8.275 g, 68%) that were recrystallized twice from dichloromethane petroleum ether to give colorless crystals of **14**: yield 5.735 g (47%), mp 73.5–74.5°C, $[\alpha]_D -11°$ ($c$ 2.027, methanol).

## B. Pentoses

*Furanoses*

*Methyl 2,3-O-isopropylidene-β-D-ribofuranoside [100]*

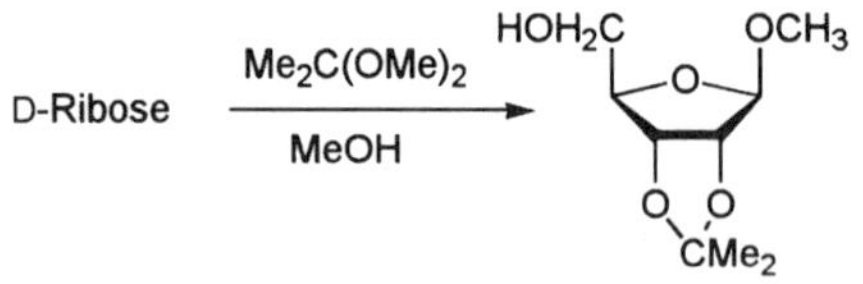

A solution of 50 g (330 mmol) of dry D-ribose in 1.0 L of acetone, 100 mL of 2,2-dimethoxypropane, and 200 mL of methanol containing 20 mL of methanol saturated with hydrogen chloride at 0°C was stirred at 25°C overnight. The resulting orange solution was neutralized with pyridine and evaporated to a yellow oil. This oil was partitioned between 500 mL of water and 200 mL of ether. The water layer was extracted twice with 200-mL portions of ether, and the combined ether extracts were dried. Evaporation yielded a pale yellow oil, which was distilled at 0.3 mm and 75°C to give 47 g (70%) of the colorless, protected glycoside: $n_D$ 1.4507, $[\alpha]_D -82.2°$ ($c$ 2, chloroform).

## 2,3-O-Isopropylidene-α-D-lyxofuranose [55]

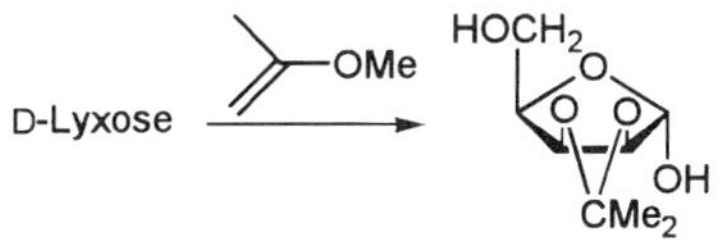

D-Lyxose →(2-methoxypropene/OMe)→ **16** (HOCH$_2$, furanose with 2,3-O-CMe$_2$, OH)

To a solution of D-lyxose (1.5 g; 10 mmol) in anhydrous DMF (30 mL) at 0°C was added 2 Eq of 2-methoxypropene and a catalytic amount of *p*-toluenesulfonic acid. After 3 h at 0°C, the mixture was made neutral. The filtrate was evaporated under diminished pressure at 40°C to afford 2.0 g of crude product (yield 80–85%), which was purified on a column of silica gel (EtOAc) to afford 1.3 g (68%) of **16** mp 80–82°C, [α]$_D$ +23 → +18° (final, H$_2$O).

## Pyranoses

### 3,4-O-Isopropylidene-β-D-ribopyranose [54]

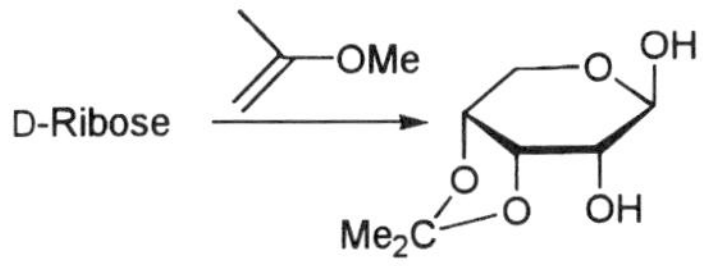

D-Ribose →(2-methoxypropene/OMe)→ **17**

To a solution of D-ribose (7.5 g, 50 mmol) in dry DMF (30 mL) containing 1 g of Drierite and maintained below 5°C with an ice bath, 2-methoxypropene (100 mmol) and *p*-toluenesulfonic acid (20 mg) were added. The mixture was stirred magnetically at 0–5°C until monitoring by TLC indicated that all starting material had disappeared (3–4 h), whereupon anhydrous sodium carbonate (5 g) was added and the cooling mixture was stirred vigorously for 1 h more. In subsequent experiments, 3,4-O-isopropylidene-D-ribopyranose **17** was obtained directly by evaporating the neutralized reaction mixture to remove DMF, extracting the residue with ethyl acetate, adding ether to the extract and nucleating; yields were in the range 40–50%.

3,4-O-Isopropylidene-D-ribopyranose **17** obtained by this procedure had an mp of 115–117°C (from ethyl acetate), [α]$_D$ −85° initial → −82° (final 24 h; *c* 1.1, water).

### 3,4-O-isopropylidene-β-D-arabinopyranose [54]

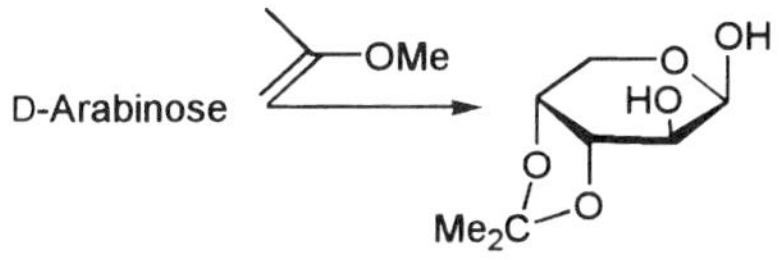

D-Arabinose →(2-methoxypropene/OMe)→ **18**

To a solution of D-arabinose (7.5 g, 50 mmol) in dry DMF (150 mL; the slightly turbid mixture became clear after 1 min of reaction) containing 1 g of Drierite, and maintained below 5°C with an ice-bath, 2-methoxypropene (100 mmol) and *p*-toluenesulfonic acid (20 mg) were added. The mixture was stirred magnetically at 0–5°C until monitoring by TLC indicated that all starting material had disappeared (3–4 h), whereupon anhydrous sodium carbonate (5 g) was added, and the cooling mixture was stirred vigorously for 1 h more. The mixture was filtered, poured into ice water (50 mL), and extracted with dichloromethane (3 × 30 mL), and the combined organic extracts were washed with water (3 × 20 mL). The aqueous phase and the combined aqueous extracts were freeze-dried. Column chromatography (silica gel 150 g; 4:1 ethyl acetate/methanol) gave pure **2**; yield 4.8 g (63%). In a direct procedure, the original neutralized reaction-mixture was evaporated directly in vacuo and the resultant syrup dissolved in ethyl acetate. Addition of ether and a crystal nucleus affored solid **2** in 60–70% yield.

3,4-*O*-Isopropylidene-α-D-arabinopyranose **18** thus obtained had a m.p. of 75–76°C. Slow evaporation of a solution in 1:1 ethyl acetate–methanol gave white crystals, mp 82–84°C, $[\alpha]_D$ −156° (initial, extrapolated) → −128° (10–12 min) → −111° (final, 24 h; *c* 1.1, water).

## C. Hexoses

*Furanoses*

*1,2:5,6-Di-O-isopropylidene-D-glucofuranose [95]*

D-Glucose $\xrightarrow{\text{Me}_2\text{CO}}_{\text{H}_2\text{SO}_4}$ [structure of 1,2:5,6-di-O-isopropylidene-D-glucofuranose]

Anhydrous α-D-glucose (200 g), powdered in a Waring blender, is stirred vigorously with 4 L of acetone in an ice bath. Sulfuric acid (96%, 160 mL) is added in 20-mL portions at 10–15-min intervals, while maintaining the temperature at 5–10°C. After the addition of the sulfuric acid, the vigorous stirring is continued for 5 h, allowing the temperature to rise gradually to 20–25°C. The solution is cooled again (ice bath), and 50% sodium hydroxide solution (245 g of NaOH in 300 mL of water) is added with stirring to near neutrality. The addition is made slowly to avoid heating. A small amount of sodium hydrogen carbonate is added to maintain the solution near neutrality. After standing overnight, the salts are removed by filtration, and the acetone solution is concentrated under reduced pressure to a thick syrup that solidifies on standing. The mixture is dissolved in chloroform on a water bath, and the solution is extracted with water. The chloroform solution is then washed with chloroform or dichloromethane. The respective water and chloroform solutions are combined. The chloroform solution contains the di-*O*-isopropylidene derivative and the water the mono-*O*-isopropylidene derivative. The solutions are concentrated under reduced pressure to syrups. The mono-*O*-isopropylidene derivative is crystallized from ethyl acetate; yield 37 g, mp 160°C. The di-*O*-isopropylidene derivative is recrystallized from cyclohexane; yield 121 g, mp 110°C.

*1,2-O-Isopropylidene-D-glucofuranose [95]*

The procedure described in the foregoing is followed until evaporation of the acetone solution to a syrup. Water (2.5 L) is added, and the mixture is distilled under reduced pressure at 60–70°C to 1600 mL to remove acetone and acetone condensation products. The final alkaline aqueous misture is adjusted to pH 2 with concentrated hydrochloric acid and heated 4 h at 40°C with constant stirring. The hydrolysate is neutralized to pH 8 with sodium hydroxide and filtered from 1.7 g of insoluble material. The filtrate is concentrated under reduced pressure to incipient crystallization of 1,2-O-isopropylidene-α-D-glucofuranose. The product is removed by filtration, washed with cold ethanol, and air-dried; yield 81.2 g, mp 161°C, [α]$_D$ −12° (c 8.3, water). Concentration of the mother liquor gives a second crop; yield 33.6 g (total yield 55%). Evaporation of the final mother liquor to near dryness gives 83 g of crude product, mp 148–152°C, which is difficult to purify and, therefore, is added to the hydrolysate of a succeeding run.

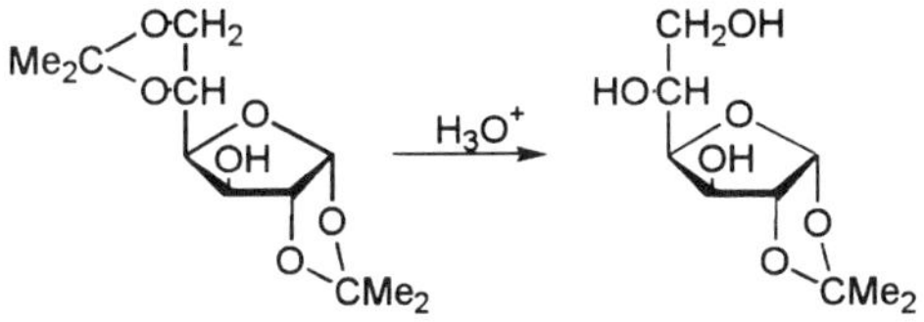

*Pyranoses*

*Methyl 4,6-O-benzylidene-α- and β-D-glucopyranoside [38]*

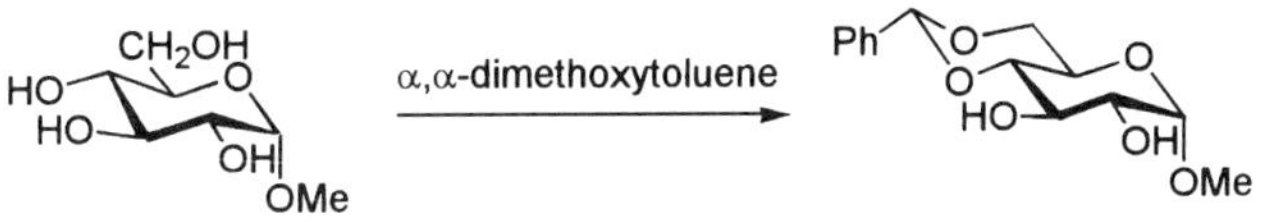

Methyl-α-D-glucopyranoside (9.7 g), α,α-dimethoxytoluene (7.6 g), DMF (40 mL), and *p*-toluenesulfonic acid (0.025 g) were placed in a 250-mL, round-bottomed flask; this was then attached to a Büchi evaporator, rotated, evacuated, and lowered into a water bath at 60 ± 5°C, so that DMF refluxed in the vapor duct. After 1 h, a short-path evaporation adaptor (description of which is given in the reference) was fitted between the flask and the vapor duct, and the DMF was evaporated, the temperature of the water bath being raised to 100°C. When no more DMF distilled over, the flask was cooled and removed from the evaporator. A solution of sodium hydrogen carbonate (1 g) in water (50 mL) was added to the residue, and the mixture was heated at 100°C until the product was finely dispersed. The mixture was cooled to 20°C, and the product was filtered off, washed thoroughly with water, and dried for 4 h at 30°C and then overnight in vacuo over phosphorus pentaoxide and paraffin wax to give **19** (11.6 g, 82.4%); mp 166–167°C. Recrystallization from propyl alcohol (28 mL) gave **19** (8.95 g, 63.5%); mp 167.5–168.5°C, [α]$_D$ +105° (c 1.1, chloroform).

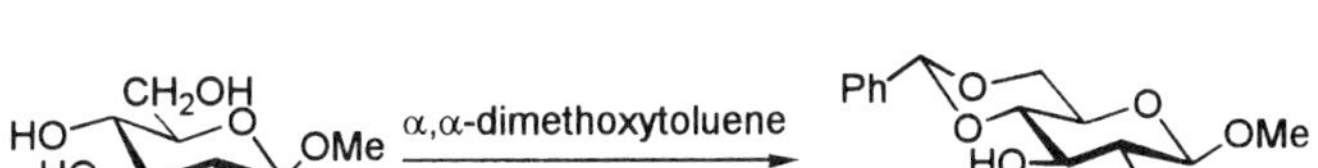

Methyl-β-D-glucopyranoside (9.7 g) was benzylidenated as described in the foregoing. After removal of the solvents, the cake of product was broken up with a spatula, and dissolved in a solution of sodium hydrogen carbonate (1 g) in water (150 mL) and ethanol (150 mL) by heating on a boiling-water bath. The solution was cooled to 4°C, and compound **20** was filtered off, washed well with water, and dried for 30 h at 30°. The product (8.2 g, 58%) had an m.p. of 207–208.5°C (unchanged by recrystallization from ethyl alcohol) and [α]$_D$ −76° (c 1.0, methanol).

*Methyl 2,3-O-acetyl 4,6-O-benzylidene-α-D-glucopyranoside [82]*

Methyl α-D-glucopyranoside (7.6 g), in dry pyridine (100 mL) and benzyl chloride (7.6 g), was refluxed with a calcium chloride tube fitted to the top of the condenser for 9 h. After cooling to room temperature, acetic anhydride (20 mL) was added, and the solution allowed to stand overnight. Excess water was added, and the mixture was extracted with benzene. The benzene layer was washed with, in turn, ice cold 1 M sulfuric acid, saturated aqueous sodium bicarbonate, and finally water. The benzene solution was dried over magnesium sulfate, filtered and concentrated. The dark-colored residue was recrystallized to yield methyl 2,3-di-O-acetyl-4,5-O-benzylidene-α-D-glucopyranoside **21**: mp 101–104°C, $[\alpha]_D$ +75° (c 1.0, chloroform). Part of the material was deacetylated with 1.67% ammonia in methanol to yield methyl 4,6-O-benzylidene-α-D-glucopyranoside **19**: mp 161–163°C undepressed on admixture with an authentic sample.

*Methyl 2,6-di-O-acetyl-3,4-O-benzylidene-β-D-galactopyranoside [82]*

A solution of 1.77 g of methyl 6-O-acetyl-β-D-galactopyranoside **22** (obtained from β-D-galactopyranoside by a sequence of (1) selective 6-O tritylation; (2) benzylation at O-3, 4, and 5; (3) detritylation; (4) O-acetylation; and (5) catalytic hydrogenolysis) and 1.61 g of α,α-dichlorotoluene in 25 mL of pyridine is heated at the reflux temperature for 5 h. More α,α-dichlorotoluene (1.61 g) is added and the solution is heated to reflux for an additional 3 h. Acetic anhydride (5 mL) is added to the still warm solution, which is then allowed to stand at 20–25°C overnight. The solution is diluted with toluene, and the toluene solution is shaked with water, aqueous sodium hydrogenocarbonate, and then water. After drying with sodium sulfate and filtration, the solution is concentrated to dryness under reduced pressure. Pyridine is removed by repeated codistillation with toluene under reduced pressure. The crude crystalline product is washed with light petroleum (60–80°C) to remove excess α,α-dichlorotoluene. The remaining product (2.74 g) is purified by silica gel column chromatography (column length 50 cm, diameter 5 cm) using 9:1 v/v chloroform/ethyl ether as eluent to give **23**, yield 1.60 g (58%), mp 113–117°C.

*4,6-O-Isopropylidene-D-glucopyranose [51]*

To a solution of D-glucose (5.4 g, 30 mmol) in DMF (100 mL, dried over Drierite) kept below 5°C in an ice bath was added 2-methoxypropene (5.2 g, 60 mmol) and p-toluenesulfonic acid (~10 mg). The mixture was stirred magnetically at 0–5°C until TLC monitoring indicated that all starting material had disappeared (about 5–6 h), whereupon sodium

## Isopropylidene, Benzylidene, and Related Acetals

D-Glucose → [structure] **24**

carbonate (~5 g) was added, with energic stirring of the cold mixture for 1 h. The mixture was refrigerated overnight and then filtered, and the filtrate was poured into ice water (50 mL). The resultant solution was extracted with methylene chloride (3 × 50 mL), and the combined organic extracts were washed with water (4 × 20 mL). The aqueous phase and the combined aqueous extracts were freeze-dried (48 h) to yield **24** as a white solid (6.25 g, 95%) that was homogeneous by TLC and, after trimethylsilylation, showed no appreciable peaks for components other than the α- and β-anomers of **24**. Recrystallization could be effected from ethanol–hexane to give small, white granules; mp 169.5–170.5°C $[\alpha]_D$ +24° (initial, extrapolated) → +8.5° ($c$ 0.5 9 min) → −7.3° (final, 48 h; $c$ 2.1, water).

### 4,6-O-Isopropylidene-D-galactopyranose [53]

D-Galactose → [structure] **25** major + [structure] **26** minor

To a slightly turbid mixture of D-galactose (which became clear after ~5 min of reaction); (9.0 g, 50 mmol) containing 1 g of Sikkon (Fluka dehydrating agent) maintained at 0–5°C (ice bath) are added 2-methoxypropene (7.2 g, 100 mmol) and p-toluenesulfonic acid (30–50 mg). The mixture is stirred magnetically at 0–5°C until monitoring by TLC indicates that practically all of the starting material has disappeared (~4 h), whereupon anhydrous sodium carbonate (~5 g) is added, and the cold mixture is stirred vigorously for 1 h more. The mixture is filtered, and the filtrate poured into ice water (50 mL). The resultant solution is extracted with dichloromethane (3 × 30 mL), and the extracts are combined, extracted with water (3 × 30 mL), and dried (sodium sulfate). The aqueous phase is combined with the water extracts, and the entire solution is freeze-dried. The freeze-dried aqueous extract gave 9.8 g of an amorphous solid, TLC of which showed a major component ($R_f$ 0.30; 3:1 benzene/ethanol), a minor one ($R_f$ 0.37), and traces of a third component ($R_f$ 0.45). These products were separated by column chromatography (440 g of silica gel, 3:1 benzene/ethanol) to give successively 5,6-O-isopropylidene-D-galactofuranose (0.10 g, yield 1–2%) and acetal **26** (1.5 g, yield 14%) and **25** (7.5 g, yield 67%). Directly on evaporation of the eluates, these crystalline products were obtained pure. 4,6-O-Isopropylidene-D-galactopyranose **25** (7.4 g, yield 67%) had an mp of 141–142°C $[\alpha]_D$ +92° (3 min) → +118° (48 h, $c$ 0.1, water) and 3,4-O-isopropylidene-D-galactopyranose **26**: [mp 99–103°C, $[\alpha]_D$ +105° (3 min) → +44° (48 h; $c$ 0.1, water).

### 1,2:3,4-Di-O-isopropylidene-D-galactopyranose [95]

In a 4- to 6-L, wide-necked bottle, equipped with a ground-glass stopper, are placed 90 g (0.5 mol) of finely powdered anhydrous D-galactose (200 g, 1.25 mol) of powdered

D-Galactose $\xrightarrow[\text{CuSO}_4,\text{H}_2\text{SO}_4]{\text{Me}_2\text{CO}}$

anhydrous cupric sulfate, 10 mL of concentrated sulfuric acid, and 2 L (27.4 mol) of anhydrous acetone. The mixture is shaken 24 h on a mechanical shaker. The cupric sulfate is removed by filtration and washed with anhydrous acetone; the washings are combined with the original filtrate. The combined washings and filtrate are neutralized by shaking with 94 g (1.27 mol) of powdered calcium hydroxide until the solution is neutral to Congo red. The unreacted calcium hydroxide and calcium sulfate are filtered and washed with dry acetone, and the filtrate is concentrated by distillation of the acetone at atmospheric pressure. After a thin syrup has been obtained, the major portion of the remaining acetone is removed by distillation at 50°C and 15 mm (water aspirator). The last traces of acetone are finally removed by distillation at 100°C and 6 mm. The residual light yellow oil is crude 1,2:3,4-di-O-α-D-galactopyranose: yield 100–120 g, (76–92%), $[\alpha]_D$ −55° (c 3.5, chloroform).

### 4,6-O-Isopropylidene-D-mannopyranose [52]

A solution of D-mannose (5.4 g, 30 mmol) in dry DMF (20 mL) containing Drierite (1 g) was maintained below 5°C (ice bath) and 2-methoxypropene (4.3 g, 60 mmol) and p-toluenesulfonic acid (~20 mg) were added. The mixture was stirred magnetically at 0–5°C until monitoring by TLC indicated that all starting material had disappeared (~5 h), whereafter anhydrous sodium carbonate was added, and the cold mixture was stirred vigorously for 1 more hour. The mixture was filtered, and the filtrate poured into ice water (50 mL). The product was extracted with dichloromethane (4 × 20 mL), and the extracts were combined, and washed with water (4 × 20 mL). The aqueous phase and the combined, aqueous extracts were freeze-dried, to yield an amorphous solid (mono-O-isopropylidene derivatives; fraction A, 6.0 g). Evaporation of the dried (sodium sulfate) dichloromethane extract gave a syrup (di-O-isopropylidene derivatives, fraction B; 0.7 g). Fraction A was essentially a mixture of the anomers of the 4,6-O-isopropylidene-D-mannopyranose 27 with the α-anomer strongly preponderant. The amorphous solid (yield 91%) was recrystallized twice from ethyl acetate to give the α-anomer as a microcrystalline, white powder; yield 5.4 g (82%), mp 156–157°C $[\alpha]_D$ −1° (3 min) → −16° (5 min) → −24° (final, 48 h; c 1.2, water).

### 1-O-Acetyl-2,3:4,6-di-O-isopropylidene-D-mannopyranose and Its Selective Hydrolysis to 1-O-acetyl-2,3-O-isopropylidene-D-mannopyranose [52]

A solution of D-mannose (5.4 g, 30 mmol) in dry DMF (20 mL), containing Drierite (1 g) was maintained at ~ −10°C, and 2-methoxypropene (4.3 g, 60 mmol) and p-toluenesulfonic acid (~20 mg) were added. The mixture was stirred magnetically for ~3 h, and then a further

## Isopropylidene, Benzylidene, and Related Acetals

D-Mannose $\xrightarrow{\substack{(1) \text{ CH}_2=\text{C(Me)OMe} \\ (2) \text{ Ac}_2\text{O}}}$ **28** $\xrightarrow{\text{AcOH-H}_2\text{O}}$ **29**

amount of ether (4.3 g, 60 mmol) was added dropwise over ~2 h, the temperature being kept at ~ −10°C. The TLC (ethyl acetate) indicated that slow-migrating components ($R_f < 0.5$) were absent. The mixture was then treated exactly as described for the foregoing procedure. In the present experiment, the aqueous phase contained only traces of the monoacetals. Evaporation of the dichloromethane extract gave an amorphous solid the properties of which in TLC (1:2, ethyl acetate/petroleum ether) were very similar to those of the syrup (fraction B) described for structure 27 for the preparation of the monoacetal, except for the presence of very minor, fast-migrating contaminants. The mixture was acetylated conventionally with acetic anhydride and pyridine. Evaporation of the solvents, and nucleation (nucleation was not needed in subsequent preparations) gave a solid. One recrystallization from methanol–water gave reasonably pure compound; a second recystallization afforded analytically pure 1-*O*-acetyl-2,3:4,6-di-*O*-isopropylidene-α-D-mannopyranose **28**: yield 5.9 g (65% from D-mannose); mp 145–147°C (methanol–water), $[\alpha]_D$ +3° (*c* 1.0, chloroform). If necessary the acetate could be deacetylated conventionally to the free sugar [52]. A suspension of diacetal **28** (0.5 g) in 1:3 acetic acid/water (20 mL) was stirred at room temperature until dissolution was complete (~1 h). The solution was then refrigerated (~0°) overnight. Use of TLC then indicated the presence of a major component ($R_f$ 0.43, ethyl acetate). The solution was freeze-dried to give 1-*O*-acetyl-2,3-*O*-isopropylidene-α-D-mannopyranose **29** as a microcrystalline powder that could be effectively purified by recrystallization from ethyl acetate; yield 0.32 g (74%), mp 130–131°C, $[\alpha]_D$ −24.5° (*c* 0.9, chloroform).

Note that if the anomeric hydroxyl group was not acetylated as it is in the acetal **28**, a transformation of the monoacetal to its isomer 2,3-*O*-isopropylidene-D-mannofuranose could be observed.

### D. Aminosugars

*Acetonation of 2-acetamido-2-deoxy-D-glucose [35]*

2-acetamido-2-deoxy-D-glucose $\xrightarrow{\text{Me}_2\text{C(OMe)}_2}$ **30**

It had been demonstrated that the result of the reaction was dependent on the temperature at which it was conducted.

1. At 80–85°C during 15 min, a stirred solution of 2-acetamido-2-deoxy-D-glucose (9.5 g, 43 mmol) and *p*-toluenesulfonic acid monohydrate (100 mg) in dry DMF (130 mL) was heated to 80–85°C, and then 2,2-dimethoxypropane (20 mL) was added; stirring was continued for 15 min at 80–85° (the starting material was then no longer detectable by TLC). The mixture was cooled and treated with

Amberlite IRA-410 (OH⁻) ion-exchange resin to remove the acid. After filtration, the filtrate was evaporated at 60° (bath). Crystallization was spontaneous, and when evaporation was complete, the mass was cooled, stirred with chloroform, and the product removed by filtration. The crystalline product (6 g, 54%) was identified as 2-acetamido-2-deoxy-4,6-*O*-isopropylidene-D-glucopyranose **30**: mp 189–190°C [α]$_D$ +57.5° (*c* 0.99, methanol).

2. At room temperature during 2 h, the reaction was conducted with 9.5 g of 2-acetamido-2-deoxy-D-glucose by the same method and procedure [reported in Ref. 96]. After recrystallization of the product (10g, 90%), the final filtrate was chromatographed on a column of silicic acid (30 g) with 30:1 chloroform/methanol.

### E. Deoxysugars

*Acetonation of L-fucose and L-rhamnose [56]*

3,4-*O*-Isopropylidene-L-fucopyranose:

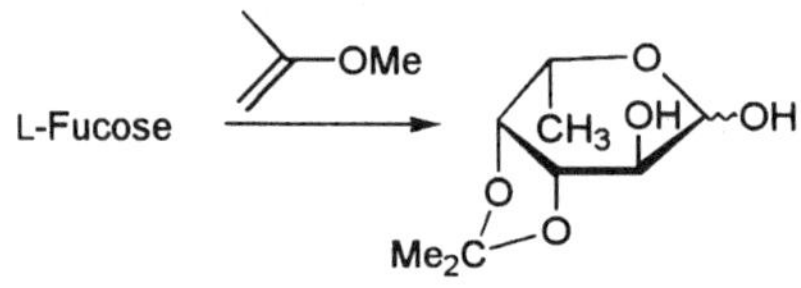

A solution of L-fucose (4.92 g, 30 mmol) in anhydrous DMF (50 mL) was stirred with a desiccant (Drierite or Sikkon, 1 g) at 0°C (ice bath), and 2-methoxypropene (2.16 g, 30 mmol) was added followed by *p*-toluenesulfonic acid (~20 mg). After 1 h at 0°C, an additional stoichiometric amount of reagent (2.16 g) was added, and stirring was continued for 2 h at 0°C. Sodium carbonate (~5 g) was added, and the mixture was stirred further 1 h at room temperature. The solids were filtered off, and the filtrate was evaporated under diminished pressure at 40°C to a syrup that contained (TLC, ethyl acetate) one major product plus minor, fast-migrating components. Rapid chromatography of the product on a column of silica gel gave pure, crystalline 3,4-*O*-isopropylidene-L-fucopyranose **31**; yield 3.7 g (~60%); mp 110–111°C, [α]$_D$ −90° → −70° (24 h, equil.; *c* 0.2, water).

2,3-*O*-Isopropylidene-L-rhamnopyranose

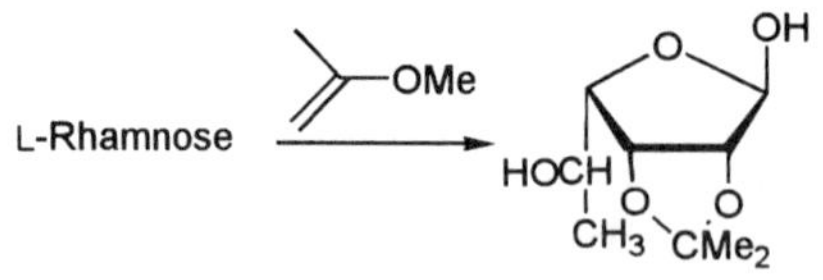

The procedure used for L-fucose was applied with L-rhamnose (4.22 g, 30 mmol) except that twice the stoichiometric amount of 2-methoxypropene (4.32 g, 60 mmol) was added directly at the beginning of the reaction. The TLC of the crude amorphous residue (that remained after removal of the solvent) showed essentially only one spot (purity > 95% by NMR) corresponding to the attempting acetal (yield 5.2 g, 85%). Purification by rapid column chromatography (1:1 ethyl acetate/petroleum ether) gave pure, syrupy 2,3-*O*-isopropylidene-L-rhamnopyranose **32** (4.9 g, 80%) that eventually crystallized by slow evaporating from the chromatography solvent; mp 90–91°C, $[\alpha]_D$ +10° (equil., 24 h, c 0.1 water).

## F. Oligosaccharides

### Benzylidenation of Sucrose [97]

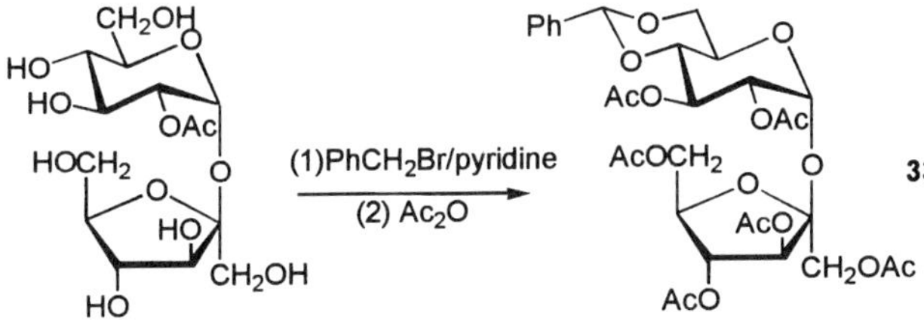

A solution of sucrose (2.5 g) in dry pyridine (50 mL) was treated with benzylidene bromide (2.8 mL) at 85°C for 1.5 h. After a further addition of benzylidene bromide (1 mL), the reaction mixture was heated at 95°C for 0.5 h, treated with acetic anhydride (5 mL) at 0°C, and then stored at room temperature for 5 h. The solution was poured into ice water and extracted with dichloromethane, and the organic layer was washed with water and dried ($Na_2SO_4$). The TLC (4:1 ether/light petroleum) showed a mixture of four products. The $R_f$ of the slow-moving spot was identical with that of sucrose octa-acetate, and the second, fast-moving spot was the major product. The solution was concentrated and fractionated on a column of silica gel (200 g), using 1:1 ether/light petroleum. 1′,2,3,3′,4′,6′-Hexa-O-acetyl-4,6-O-benzylidene-sucrose 33 (1.7 g, 35%), which crystallized in the tubes of the fraction collector, had an m.p. of 155–157°C, $[\alpha]_D$ +44.3° (c 0.82, chloroform).

### Acetonation of Sucrose [58]

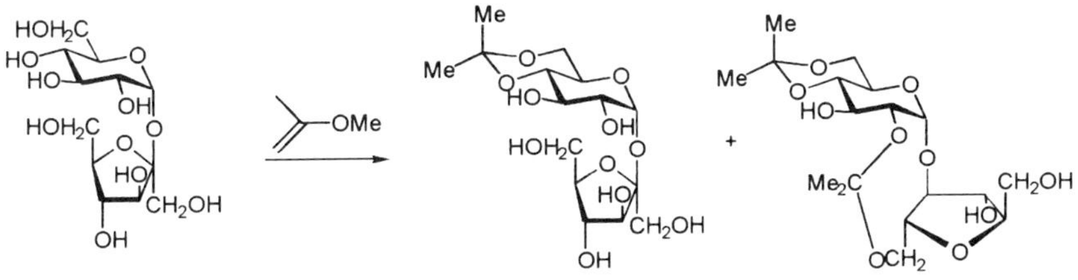

A solution of sucrose (34.2 g, 0.1 mol) in dry DMF (400 mL) containing molecular sieve pellets (1/16 in., type 3 Å) was stirred with 2-methoxypropene (12.1 mL, 0.13 mol) in the presence of dry p-toluenesulfonic acid (25 mg) for 40 min at 70°C, cooled to room temperature, and made neutral with anhydrous sodium carbonate. The inorganic residue was filtered off and the filtrate evaporated to a syrup. Elution of the syrup from a column of silica gel with 1:1 ethyl acetate/acetone afforded the diacetal 2,1′:4,6-di-O-isopropylidene-sucrose 35 as a syrup: 3 g (7%); $[\alpha]_D$ +25.5° (c 1, methanol). Further elution gave the major product 4,6-O-isopropylidene-sucrose 34: yield 23 g (60%); white powder; $[\alpha]_D$ +45.4° (c 1.0, methanol).

## Acetonation of α,α-Trehalose [57]

α,α-trehalose → **36**

Reagents: (1) =C(OMe)Me; (2) Ac$_2$O

To a stirred mixture of anhydrous α,α-trehalose (3.42 g, 10 mmol), dry DMF (40 mL) below 5°C, and Sikkon (Fluka dehydrating agent, 1 g) were added 2-methoxypropene (1.29 g, 15 mmol) and p-toluenesulfonic acid (~20 mg). The mixture was stirred for 4 h at 0–5°C, sodium carbonate (~2 g) was added, and the mixture was stirred vigorously for 1 h, filtered, and concentrated at 40°C/< 1 torr. To a solution of the residue (4.9 g) in pyridine (5 mL) was added a solution of acetic anhydride (15 mL) in pyridine (15 mL) with stirring at 0°C. The mixture was stirred overnight at room temperature and then poured onto ice. The product was extracted with dichloromethane, and the extract was concentrated to give an amorphous solid (5.6 g). Use of TLC (1:1 ethyl acetate/light petroleum) revealed three products ($R_f$ 0.69, 0.56, and 0.42). Elution (ethyl acetate/light petroleum, 1:1) from silica gel (400 g) gave, first the 4,6:4′6′-diacetal tetraacetate (0.8 g): mp 187–189°C, $[α]_D$ +142° (c 1, chloroform). Eluted second was the monoacetal hexaacetate **36** (2.4 g, 38%): mp 79–80°C; $[α]_D$ +150.5° (c 1.1, chloroform). Eluted third was octa-O-acetyl-α-α-trehalose (1.0 g): mp 97–98°C, $[α]_D$ +157° (c 1.1, chloroform).

## Acetonation of Maltose [37]

The investigation of various conditions for the reaction of 2,2-dimethoxypropane has been conducted on several disaccharides [37]. The nature of the acetals thus obtained (mono- or multiacetals and pyranosylpyranose or pyranosylacyclic hexose forms of the protected disaccharide) depends largely on the temperature (room temperature, 40°C, 80°C), the solvent (DMF or 1,4-dioxane), and the amount of the acid catalyst that are chosen for the procedure. As a specific example, the synthesis of a tetraacetal of maltose is given here.

maltose → **37** (via Me$_2$C(OMe)$_2$)

To a stirred solution of maltose (300 mg, 0.88 mmol) in 1,4-dioxane (3 mL) was added p-toluenesulfonic acid (3 mg) and then 2,2-dimethoxypropane (0.9 mL, 8.3 mol/mol of maltose). The mixture was stirred for 15 h at 80°C, and then treated with Amberlite IRA-410 ($^-$OH) resin, to remove the acid. The resin was filtered off and washed with methanol. The filtrate and washings were combined, and evaporated, and the syrupy residue was chromatographed on a column of silicic acid (10 g) with chloroform and then 100:1 chloroform/methanol. The latter eluate yielded a syrup of the tetraacetal **37** (280 mg, 63%) which could be acetylated to give the syrupy diacetate: $[α]_D$ +68°, (c 0.114, chloroform).

## G. Acetalation of *trans*-Vicinal Diols

*Isopropylidene Acetals*

*Methyl 2,3:4,6-di-O-isopropylidene-D-glucopyranoside [61]*

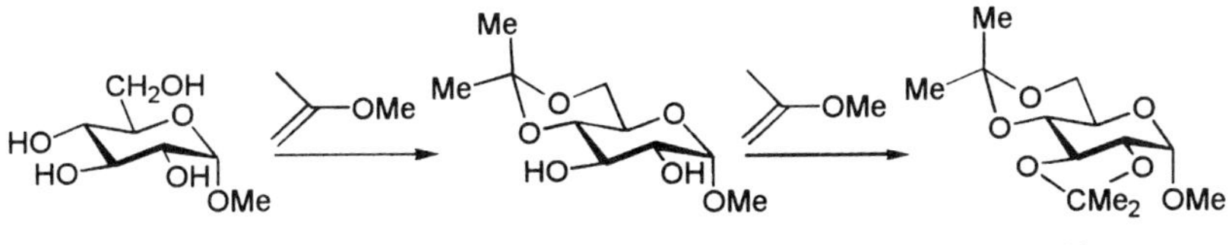

To a solution of 4,6-*O*-isopropylidene-D-glucopyranoside **38** (2.3 g, 10 mmol) in dry DMF (30 mL) at 0°C was added desiccant (Sikkon, 5 g) followed by methoxypropene (1.8 g, 25 mmol) in dry DMF (10 mL) and a catalytic amount (5 mg) of *p*-toluenesulfonic acid, and the mixture was agitated vigorously by magnetic stirring for 4 h at 0°C, with exclusion of moisture. Sodium carbonate (5 g) was added, and the mixture was further stirred for 1 h at room temperature. After filtration and evaporation of the filtrate, the crude product was obtained amorphous in almost quantitative yield. It could be used directly for further transformations. Crystallization of the product from hexane gave methyl 2,3:4,6-di-*O*-isopropylidene-D-glucopyranoside **39** as plates (1.9 g, 73%): mp 84–85°C, $[\alpha]_D$ +95° (*c* 0.1, chloroform). The same product could be obtained by direct acetonation of methyl α-D-glucopyranoside (1.9 g, 10 mmol) with 2-methoxypropene (2.9 g, 40 mmol) under essentially the same conditions, but the yield was lower (~65%) than that obtained by the two-stage procedure.

*Dispiroacetals*

Two new methods have been discovered for the selective protection of *trans*-diequatorial vicinal diols [44,45]. One of them uses 3,3′,4,4′-tetrahydro-6,6′-bi-2*H*-pyran (bis-DHP) **40** to transform diols into dispiroacetals and the second uses 1,1,2,2-tetrahydromethoxycyclohexane **41** to obtain cyclohexane-1,2-diacetal-protected sugars. These two methods have been compared [44,45] and we give here, as representative examples, one procedure for each strategy.

*Dispiroacetal of Methyl α-L-fucopyranoside [45]*

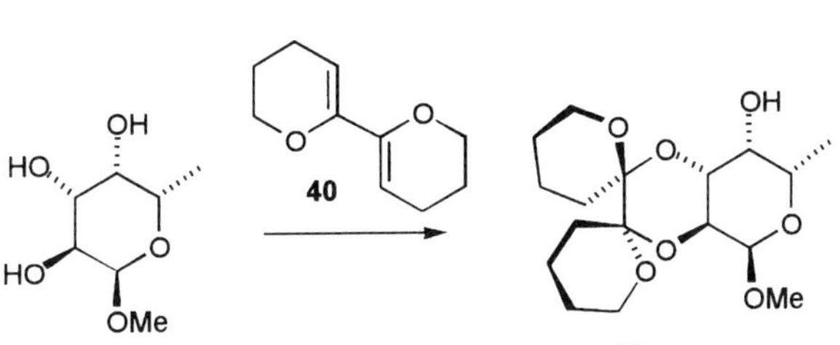

*dl*-Camphorsulfonic acid (15 mg) was added to a stirred solution of methyl α-L-fucopyranoside (1.42 mmol) and bis-dihydropyran (**40**) (3.12 mmol), in dry chloroform (25 mL), and the mixture was heated under reflux for between 2 and 8 h. Anhydrous ethylene glycol (9 mmol) was added and heating continued for a further 0.5 h. The resultant solution was diluted with dichloromethane (40 mL) and added to saturated aqueous sodium hydrogenocarbonate (20 mL), extracted with dichloromethane (3 × 10 mL), dried (anhydrous magnesium sulfate), filtered, and concentrated in vacuo. Purification by column chromatography on silica gel (25:50 ethyl acetate/petroleum ether) gave methyl 2,3-O-(6,6'-octahydro-6,6'-bi-2H-pyran-2,2'-diyl)-α-L-fucopyranoside **42**: yield 76%, [α]$_D$ +2.7° (c 1.0, chloroform).

*Dispiroacetals of Methyl α-D-mannopyranoside [44]*

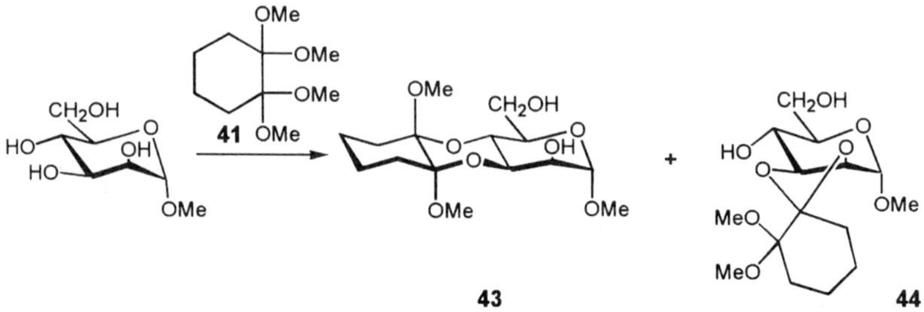

dl-Camphorsulfonic acid (308 mg, 1.33 mmol) was added to a stirred solution of methyl α-D-mannopyranoside (3.46 g, 17.8 mmol), 1,1,2,2-tetramethoxycyclohexane (4.60 g, 24.4 mmol) and trimethyl *ortho*-formate (2.0 mL) in dry methanol (25 mL), and the mixture was heated under reflux for 16 h. After neutralization with sodium hydrogenocarbonate (~0.5 g), the solvent was removed in vacuo, and the crude material was purified by column chromatography on silica gel to furnish **44** (666 mg, 11%), as an off-white foam, and slightly impure **43**, which was further purified by slow crystallization from diethyl ether to give clean **43** (2.83 g, 48%): mp 168°C, [α]$_D$ +191° (c 0.94, chloroform).

## REFERENCES

1. A. Wurtz, Sur une combinaison d'aldéhyde et d'oxyde d'éthylène, *Compt. Rend.* 53:378 (1861); *Annales de Chimie et de Physique (Ann.)* 120:328 (1861).
2. J. Meunier, Sur les acétals benzoïques de la mannite et de ses homologues: action décomposante de l'aldéhyde benzoïque, *Compt. Rend* 107:910 (1888).
3. E. Fisher, Ueber die verbindungen der zucker mit den alkoholen und ketonen, *Berichte der deutschen chemischen Gesellschaft (Ber.)* 28:1145 (1895).
4. S. A. Barker and E. J. Bourne, Acetals and ketals of the tetritols, pentitols and hexitols, *Adv. Carbohydr. Chem.* 7:137 (1952).
5. A. N. de Belder, Cyclic acetals of the aldoses and aldosides, *Adv. Carbohydr. Chem.* 20:219 (1965).
6. A. N. de Belder, Cyclic acetals of the aldoses and aldosides, *Adv. Carbohydr. Chem. Biochem.* 34:179 (1977).
7. R. F. Brady, Jr., Cyclic acetals of ketoses, *Adv. Carbohydr. Chem. Biochem.* 26:197 (1971).
8. J. A. Mills, The stereochemistry of cyclic derivatives of carbohydrates, *Adv. Carbohydr. Chem.* 10:1 (1955).
9. R. J. Ferrier and W. G. Overend, Novel aspects of the stereochemistry of carbohydrates, *Q. Rev. Chem. Soc.* 8:265 (1959).
10. J. F. Stoddart, *Stereochemistry of Carbohydrates*, Wiley Interscience, New York, 1971, p. 186.

11. A. B. Foster, Cyclic acetals derivatives of sugars and alditols. *The Carbohydrates: Chemistry, Biochemistry*, Vol. 1A, (W. Pigman and D. Horton, eds.), Academic Press, New York, 1972,p. 391.
12. R. U. Lemieux, Rearrrangements and isomerization in carbohydrate chemistry. *Molecular Rearrangements*, Part II (P. de Mayo, ed.). Wiley-Interscience, New York, 1963, p. 723.
13. D. M. Clode, Carbohydrate cyclic acetal formation and migration, *Chem. Rev. 79*:491 (1979).
14. A. H. Haines, The selective removal of protecting groups in carbohydrate chemistry, *Adv. Carbohydr. Chem. Biochem. 39*:13 (1981).
15. J. Gelas, The reactivity of cyclic acetals of aldoses and aldosides, *Adv. Carbohydr. Chem. Biochem. 39*:71 (1981).
16. A. F. Bochkov and N. K. Kochetkov, A new approach to the synthesis of oligosaccharides. Synthesis of 1,2-*O*-(1-cyanoaklylidene) sugar derivatives, *Carbohydr. Res 39*:355 (1975); V.I. Betaneli, M.V. Ovchinnikov, L. Backinowski, and N.K. Kochetkov, Synthesis of 1,2-*O*-(1-cyanoalkylidene) sugar derivatives *Carbohydr. Res. 68*:C11 (1979).
17. W. T. Markiewicz, Tetrahydroisopropyldisiloxan a group for simultaneous protections, *J. Chem. Res. (S)*, 24 (1979); W. T. Markiewicz and M. Wieirowski, *Nucl. Acids Res.* 185 (1978).
18. C. H. M. Verdegaal, P. L. Jansse, J. F. M. de Roolj, and J. H. van Boom, Acid-catalysed isomerization of the tetraisopropyl-disiloxane-1,3-diyl group. Simultaneous protection of the secondary alcoholic functions, *Tetrahedron Lett.*, 1571 (1980).
19. M. R. Jenner and R. Khan, Use of dimethoxydiphenylsilane, *N,N*-dimethylformamide, and toluene-*p*-toluene sulphonic acid as a novel acetalating reagent, *J. Chem. Soc. Chem. Commun.* 50 (1980).
20. I. D. Jenkins, J. P. H. Verheyden, and J. G. Moffatt, Synthesis of the nucleoside antibiotic nucleocidin, *J. Am. Chem Soc. 93*:4323 (1971); S. David, Conversion sélective de diols-α cyclaniques en cétols (acyloïne) par traitement au brome de leur dérivé organostannique, *C. R. Acad. Sci. Ser. C 278*:1051 (1974); D. Wagner, J. P. H. Verheyden, and J. G. Moffatt, Preparation and synthetic utility of some organotin derivatives of nucleosides, *J. Org. Chem. 39*:24 (1974); C. Augé, S. David, and A. Veyrières, Complete regiospecificity in the benzylation of a *cis*-diol by the stannylidene procedure, *J. Chem. Soc. Chem. Commun.* 375 (1976); M. Nashed and L. Anderson, Selective substitution at the 3-position in a 3,4-*O*-dibutylstannylene-D-galactose derivatives. An improved synthesis of 2,4,6-tri-*O*-benzyl-D-galactose, *Carbohydr. Res. 56*:419 (1977); T. Ogawa and M. Matsui, A new approach to regioselective acylation of polyhydroxy compounds, *Carbohydr. Res. 56*:C1 (1977); T. Ogawa and M. Matsui, Regioselective stannylation. Acylation of carbohydrates: coordination control, *Tetrahedron 37*:2363 (1981).
21. S. David and S. Hanessian, Regioselective manipulation of hydroxyl groups via organotin derivatives, *Tetrahedron 41*:643 (1985).
22. F. Dasgupta and P. J. Garegg, Monoalkykation of tributyltin activated methyl 4,6-*O*-benzylidene-α-D-gluco- and -galactopyranosides, *Synthesis* 1121 (1994).
23. R. J. Ferrier, Carbohydrate boronates, *Adv. Carbohydr. Chem. Biochem. 35*:31 (1978).
24. W. V. Dahloff and R. Köster, Some new *O*-ethylboron-assisted carbohydrate transformation, *Heterocycles 18*:421 (1982).
25. M. Bouchra, P. Calinaud, and J. Gelas, A new method of orthoesterification at non-anomeric position under kinetic control. Application to D-glucose and D-mannose series and selective hydrolysis of the corresponding orthoesters, *Carbohydr. Res. 267*:227 (1995); M. Bouchra, P. Calinaud, and J. Gelas, Application of the orthoesterification under kinetically controlled conditions to the selective acylation in D-gluco-, D-ribo-, and D-xylo-furanose series, *Synthesis* 561 (1995).
26. S. Hanessian and E. Moralioglu, Chemistry of 1-(dimethylamino)ethylidene and α-(dimethylamino)benzylidene acetals. Utility as blocking groups for diols and for selective esterification, *Tetrahedron Lett.* 813 (1971); S. Hanessian and E. Moralioglu, Chemistry and synthetic utility of α-(dimethylamino)benzylidene and 1-(dimethylamino)ethylidene acetals, *Can. J. Chem.*

50:233 (1972); R. F. Abdulla and R. S. Brinkmeyer, The chemistry of formamide acetals, *Tetrahedron* 1675 (1979).
27. C. B. Reese, Protection of alcoholic hydroxyl groups and glycol systems. *Protective Groups in Organic Chemistry* (J. F. W. McOmie, ed.) Plenum Press, London, 1973, p. 95.
28. T. W. Greene and P. J. M. Wuts, *Protective Groups in Organic Synthesis*, John Wiley & Sons, New York, 1991.
29. Y. Oikawa, T. Yoshioka, and O. Yonemitsu, Protection of hydroxy groups by intramolecular oxidative formation of methoxybenzylidene acetals with DDQ, *Tetrahedron Lett.* 889 (1982); Y. Oikawa, T. Nishi, and O. Yonemitsu, Kinetic acetalation for 1,2- and 1,3-diol protection by the reaction of *p*-methoxyphenylmethyl methyl ether with DDQ, *Tetrahedron Lett.* 4037 (1983).
30. O. Kjølberg and K. Neumann, Synthesis of acyclic carbohydrate isopropylidene mixed acetals using 2,3-dichloro-5,6-dicyano-*p*-benzoquinone as a catalyst, *Acta Chem. Scand.* 48:80 (1994)
31. S. Morgenlie, Isopropylidene derivatives of α-D-galatofuranose, *Acta Chem. Scand.* 27:3609 (1973); S. Morgenlie, Oxidation of carbohydrate derivatives with silver carbonate on celite. X. Identification of three mono-*O*-isopropylidene derivatives of D-galactose, *Acta Chem. Scand. B 29*:367 (1975); S. Morgenlie, Reaction of some aldoses with anhydrous cupric sulphate–acetone, *Carbohydr. Res.* 41:77 (1975).
32. A. Hampton, A new procedure for the conversion of ribonucleosides to 2',3'-*O*-isopropylidene derivatives, *J. Am. Chem. Soc.* 83:3640 (1961).
33. S. Chládek and J. Smrt, Oligonucleotidic compounds. V. 2',3'-*O*-alkylidene derivatives of ribonucleosides, *Coll. Czech. Chem. Commun.* 28:1301 (1963).
34. M. E. Evans and F. W. Parrish, Methyl 2,3:4,6-di-*O*-isopropylidene-α-D-glucopyranoside, *Tetrahedron Lett.* 3805 (1966); Evans and F. W. Parrish, and L. Long, Jr, Acetal exchange reactions, *Carbohydr. Res.* 3:453 (1967).
35. A. Hasegawa and M. Kiso, Acetonation of 2-(acylamino)-2-deoxy-D-glucose, *Carbohydr. Res.* 63:91 (1978).
36. R. Khan and S. Mufti, Synthesis of 1',2:4,6-di-*O*-isopropylidene-sucrose, *Carbohydr. Res.* 43:247 (1975); R. Khan, S. Mufti, and M. R. Jenner, Synthesis and reactions of 4,6-acetals of sucrose, *Carbohydr. Res.* 65:109 (1978).
37. Y. Ueno, K. Hori, R. Yamauchi, M. Kiso, A. Hasegawa, and K. Kato, Reaction of maltose with 2,2-dimethoxypropane, *Carbohydr. Res.* 89:271 (1981); Y. Ueno, K. Hori, R. Yamauchi, M. Koso, A. Hasegawa, and K. Kato, Reaction of some D-glucobioses with 2,2-dimethoxypropane, *Carbohydr. Res.* 96:65 (1981).
38. M. E. Evans, Methyl 4,6-*O*-benzylidene-α- and β-D-glucosides, *Carbohydr. Res.* 21:473 (1972).
39. D. Horton and W. Weckerle, A preparative synthesis of 3-amino-2,3,6-trideoxy-L-*lyxo*-hexoses (daunosamine) hydrochloride from D-mannose, *Carbohydr. Res.* 44:227 (1975); D. Horton and W. Weckerle, *Am. Chem. Soc. Symp. Ser.* 39:22 (1977).
40. N. Sakairi, M. Hayashida, and H. Kuzuhara, 1,6-Anhydro-β-maltotriose: Preparation from pullulan, and regioselective partial protection reactions, *Carbohydr. Res.* 185:91 (1989).
41. N. Sakairi and H. Kuzuhara, Chemical behavior of benzylidene acetal groups bridging the contiguous glucose residues in maltooligosaccharide derivatives, *Carbohydr. Res.* 246:61 (1993).
42. G. J. F. Chittenden, Isopropylidenation of D-mannitol under neutral conditions, *Carbohydr. Res.* 87:219 (1980).
43. G. J. F. Chittenden, Some aspects of the isopropylidenation of D-glucitol under neutral conditions, *Carbohydr. Res.* 108:81 (1982).
44. S. V. Ley, H. W. M. Priepke, and S. L. Warriner, Cyclohexane-1,2-diacetals (CDA): A new protective group for vicinal diols in carbohydrates, *Angew. Chem. Int. Ed. Engl.* 33:2290 (1994).
45. S. V. Ley, R. Leslie, P. D. Tiffin, and M. Woods, Dispiroketals in synthesis (Part 2): A new

group for the selective protection of diequatorial vicinal diols in carbohydrates, *Tetrahedron Lett.* 4767 (1992); S. V. Ley, G.-J. Boons, R. Leslie, M. Woods, and D. M. Hollinshead, Dispiroketals in synthesis (part 3): Selective protection of diequatorial vicinal diols in carbohydrates, *Synthesis* 689 (1993).

46. R. Paul, Recherches sur le noyau hydropyranique. I. Aldéhyde δ-oxyvalérianique et ses dérivés, *Bull. Soc. Chim. Fr. 1*:971 (1934).
47. M. L. Wolfrom, A. Beattie, and S. Bhattacharjee, Reaction of alkylvinyl ethers with methyl α-D-glucopyranoside, *J. Org. Chem. 33*:1067 (1968).
48. H. S. Hill and L. M. Pidgeon, Mechanism of the acetal reaction. The explosive rearrangement of hydroxyethylvinyl ether into ethylidene glycol, *J. Am. Chem. Soc. 50*:2718 (1928).
49. R. Gigg and C. D. Warren, The allyl ether as a protecting group in carbohydrate chemistry. Part II. *J. Chem. Soc.* 1903 (1968).
50. J. Gelas and D. Horton, Acetonation of carbohydrates under kinetic control by use of 2-alkoxypropenes. *Heterocycles 16*:1587 (1981).
51. M. L. Wolfrom, A. B. Diwadkar, J. Gelas, and D. Horton, A new method of acetonation. Synthesis of 4,6-*O*-isopropylidene-α-D-glucopyranose, *Carbohydr. Res. 35*:87 (1974).
52. J. Gelas and D. Horton, Kinetic acetonation of D-mannose: preparation of 4,6-mono and 2,3:4,6-di-*O*-isopropylidene-D-mannose, *Carbohydr. Res. 67*:371 (1978).
53. J. Gelas and D. Horton, Kinetic acetonation of D-galactose, D-allose and D-talose with alkyl propenyl ethers as a preparative route to the 4,6-*O*-isopropylidene aldohexopyranoses, *Carbohydr. Res. 71*:103 (1979).
54. J. Gelas and D. Horton, Acetonation of D-ribose and D-arabinose with alkyl isopropenyl ethers, *Carbohydr. Res. 45*:181 (1975).
55. J. Barbat, J. Gelas, and D. Horton. Reactions of D-lyxose and D-xylose with 2-methoxypropene under kinetic conditions, *Carbohydr. Res. 219*:115 (1991).
56. J. Barbat, J. Gelas, and D. Horton, Acetonation of L-fucose, L-rhamnose and 2-deoxy-D-*erythro*-pentose under kinetically controlled conditions, *Carbohydr. Res. 116*:312 (1983).
57. J. Defaye, H. Driguez, B. Henrissat, J. Gelas, and E. Bar-Guilloux, Asymmetric acetalation of α-α-trehalose: synthesis of α-D-galactopyranosyl and 6-deoxy-6-fluoro-α-D-glucopyranosyl-α-D-glucopyranoside, *Carbohydr. Res. 63*:41 (1978).
58. E. Fanton, J. Gelas, D. Horton, K. Karl, R. Khan, C. Kuan Lee, and G. Patel, Kinetic acetonation of sucrose: preparative access to a chirally substituted 1,3,6-trioxacyclooctane system, *J. Org. Chem. 46*:4057 (1981).
59. M. Alonso-Lopez, J. Barbat, E. Fanton, A. Fernandez-Mayoralas, J. Gelas, D. Horton, M. Martin-Lomas, and S. Penades, The acetonation of lactose and benzyl β-lactoside with 2-methoxypropene, *Tetrahedron 43*:1169 (1987).
60. M. C. Cruzado and M. Martin-Lomas, The heterogeneous, catalytic, transfer hydrogenolysis of tri-*O*-benzyl derivatives of 1,6-anhydro-β-D-hexopyranoses, *Carbohydr. Res. 170*:249 (1987).
61. J. L. Debost, J. Gelas, D. Horton, and O. Mols, Preparative acetonation of pyranoid, vicinal *trans*-glycols under kinetic control: methyl 2,3:4,6-di-*O*-isopropylidene-α- and β-D-glucopyranoside, *Carbohydr. Res. 125*:329 (1984).
62. J. L. Debost, J. Gelas, and D. Horton, Selective preparation of mono- and diacetals of D-mannitol, *J. Org. Chem. 48*:1381 (1983).
63. A. Duclos, C. Fayet, and J. Gelas, A simple conversion of polyols into anhydroalditols, *Synthesis* 1087 (1994).
64. E. Fanton, J. Gelas, and D. Horton, Novel modes for selective protection of ketose sugars and oligosaccharides of biological and industrial importance, *J. Chem. Soc. Chem. Commun.* 21 (1980).
65. T. B. Grindley, C. J. P. Cote, and C. Wickramage, Kinetic cyclohexylidenation and isopropylidenation of aldose diethyl dithioacetals, *Carbohydr. Res. 140*:215 (1985).
66. D. Horton, Y. Li, V. Barberousse, F. Bellamy, P. Renaut, and S. Samreth, Synthesis of the 2- and the 4-monomethyl ethers and the 4-deoxy-4-fluoro derivative of cyanophenyl 1,5-dithio-β-D-xylopyranoside as potential antithrombic agents, *Carbohydr. Res. 249*:39 (1993).

67. M. Koos and H. S. Mosker, A convenient synthesis of apiose, *Carbohydr. Res. 146*:335 (1986).
68. P. J. Garegg, T. Iversen, and T. Norberg, A practical synthesis of *p*-nitrophenyl-β-D-mannopyranoside, *Carbohydr. Res. 73*:313 (1979); T. Iversen and D. R. Bundle, Antigenic determinants of *Salmonella* serogroups A and D1. Synthesis of trisaccharide glycosides for use as artificial antigens. *Carbohydr. Res. 103*:29 (1982).
69. G. L. Larson and A. Hernandez, The reaction of trimethylsilyl enol ethers with diols, *J. Org. Chem. 38*:3935 (1973).
70. T. Mukaiyama, M. Ohshima, and M. Murakami, 2-Benzyloxy-propene: A novel protective reagent of hydroxyl groups, *Chem. Lett.* 265 (1984); T. Mukaiyama, M. Ohshima, H. Nagaoka, and M. Murakami, A novel acid-resistant acetal-type protective group for alcohols, *Chem. Lett.* 615 (1984).
71. E. Fanton, C. Fayet, J. Gelas, D. Jhurry, A. Deffieux, and M. Fontanille, Ethylenic acetals of sucrose and their copolymerization with vinyl monomers, *Carbohydr. Res. 226*:337 (1992).
72. C. Fayet and J. Gelas, Reaction of mono- and oligo-saccharides with keto or ethylenic enol ethers as a route to functionalized acetals and monomers for polymerization, *Carbohydr. Res. 239*:177 (1993).
73. J. S. Brimacombe, A. B. Foster, B. D. Jones, and J. J. Willard, Reaction of the cyclohexane-1,2-diols and methyl 4,6-*O*-benzylidene-α-D-gluco- and galactopyranoside with methylene halides, *J. Chem. Soc. C.* 2404 (1967).
74. S. S. Battacharjee and P. A. J. Gorin, Hydrogenolysis of cyclic and acyclic orthoesters of carbohydrates with lithium aluminium hydride–aluminium trichloride, *Carbohydr. Res. 12*:57 (1970).
75. A. Lipták, V. A. Oláh, and J. Kerékgyártó, A convenient synthesis of carbohydrate methylene acetals, *Synthesis* 421 (1982).
76. P. Di Cesare and B. Gross, Préparation d'éthers et d'acétals en série glucidique au moyen de la catalyse par transfert de phase, *Carbohydr. Res. 48*:271 (1976).
77. K. S. Kim and W. A. Szarek, Methylenation of carbohydrates using phase-transfer catalysis. *Synthesis* 48 (1978).
78. D. G. Norman, C. B. Reese, and H. T. Serafinowska, 2′,3′-*O*-Methylene derivatives of ribonucleosides, *Synthesis* 751 (1985).
79. E. L. Eliel, Insights gained from conformational analysis in heterocyclic systems, *Pure Appl. Chem. 25*:509 (1971).
80. N. Baggett, J. M. Duxbury, A. B. Foster, and J. M. Webber, Diastereoisomers of methyl-4,6-*O*-benzylidene-2,3-di-*O*-methyl-α-D-gluco- and galactopyranoside, *Chem. Ind.* (Lond.) 1832 (1964); N. Baggett, J. M. Duxbury, A. B. Foster, and J. M. Webber, Diastereoisomeric forms of methyl 4,6-*O*-benzylidene-2,3-di-*O*-methyl-α-D-gluco- and β-D-galactopyranoside, *Carbohydr. Res. 1*:22 (1965).
81. N. Baggett, M. D. Mosihuzzaman, and J. M. Webber, Benzylidenation in basic medium: The reaction of methyl 2,3-di-*O*-methyl-6-toluene-*p*-sulfonyl-α-D-glucopyranoside and 3-chloro-1-propanol with benzaldehyde, *Carbohydr. Res. 11*:263 (1969).
82. P. J. Garegg, L. Maron, and C. G. Swahn, An alternative preparation of *O*-benzylidene acetals, *Acta Chem. Scand. 26*:518 (1972); P. J. Garegg, and C. G. Swahn, An alternative preparation of *O*-benzylidene acetals, part II, *Acta Chem. Scand. 26*:3895 (1972); K. Eklind, P. J. Garegg, and B. Gotthammar, An alternative synthesis of 3,6-dideoxy-D-*xylo*-hexose (abequose), *Acta Chem. Scand. B 29*:633 (1975); P. J. Garegg and C. G. Swahn, Benzylidenation of diols with α-α-dihalotoluenes in pyridine, *Methods Carbohydr. Chem. 8*:317 (1980).
83. J. Gelas, Recherches dans la série des acétals cycliques. VIII. Synthèse du méthyl-1 dioxa-2,8 thiabicyclo[3.2.1]octane, *Tetrahedron Lett.* 509 (1971); J. Egyed, P. Demerseman, and R. Royer. Réactions induites par le chlorhydrate de pyridine. X. Synthèse d'acétals cycliques, *Bull. Soc. Chim. Fr.* 2287 (1972).
84. C. Fayet and J. Gelas, unpublished results.
85. J. G. Buchanan and A. R. Edgar, A new synthesis of methyl 3,4-*O*-ethylidene-β-L-arabinopyranoside by reduction of an acetoxonium ion salt, *Chem. Commun.* 29 (1967).

86. J. G. Buchanan and A. R. Edgar, Acetoxonium ions from acetoxyoxiranes and orthoesters: Their conversion into ethylidene acetals, rearrangement, and solvolysis, *Carbohydr. Res. 49*:289 (1975).
87. R. U. Lemieux and D. H. Detert, 1,2-Alkylidene and 1,2-ortholactone derivatives of 3,4,6-tri-*O*-acetyl-D-glucose, *Can. J. Chem. 46*:1039 (1968).
88. W. E. Dick, D. Weisleder, and J. E. Hodge, *O*-Ethylidene-D-allopyranoses: 1,2-*O*-, 1,2:4,6- and 2,3:4,6-di-*O*-ethylidene derivatives, *Carbohydr. Res. 42*:65 (1975).
89. R. G. Rees, A. R. Tatchell, and R. D. Wells, 1,2,-*O*- Alkylidene-α-D-glycopyranoses. Part I. The diastereoisomeric 1,2-*O*-(1-methyl-propylidene)-α-D-glucopyranose, *J. Chem. Soc. C* 1768 (1967).
90. S. Hanessian, P. Lavallée, and A. G. Pernet, A new synthesis of methylene acetals, *Carbohydr. Res. 26*:258 (1973).
91. H. D. Adkins and A. E. Broderick, Hemiacetal formation and refractive indices and densities of mixture of certain alcohols and aldehydes, *J. Am. Chem. Soc. 50*:499 (1928); T. H. Fife and L. K. Jao, Substituent effects in acetal hydrolysis, *J. Org. Chem. 30*: 1492 (1965); P. M. Collins, The kinetic of the acid-catalysed hydrolysis of some isopropylidene furanoses, *Tetrahedron 21*:1809 (1965); R. H. De Wolfe, K. M. Ivanetich, and N. F. Perry, General acidic catalysis in benzophenone ketal hydrolysis, *J. Org. Chem. 34*:848 (1969).
92. E. Schmitz and I. Eichorn, *The Chemistry of the Ether Linkage* (S. Pataï, ed.) Interscience, New York, 1967, p. 309; E. H. Cordes, *Prog. Phys. Org. Chem.* Vol. 4, (A. Streiweiser, Jr. and R. W. Taft, eds.) Interscience, New York, 1967, p.1; E. H. Cordes and H. G. Bull, Mechanism and catalysis for hydrolysis of acetals, *Chem. Rev. 74*:581 (1974); T. H. Lowry and K. S. Richardson, *Mechanism and Theory in Organic Chemistry*, Harper & Row, New York, 1976, p. 424.
93. C. P. Reddy and R. Balaji Rao, Mechanism of cyclic acetal formation, *Tetrahedron 38*:1825 (1982).
94. J. E. Baldwin, Endo-Trigonal reactions: A disfavoured ring closure, *J. Chem. Soc. Chem. Commun.* 736 (1976).
95. O. T. Schmidt, Isopropylidene derivatives, *Methods Carbohydr. Chem. 2*:318 (1963).
96. A. Hasegawa and H. G. Fletcher, Jr., The behavior of some aldoses with 2,2-dimethoxypropane-*N,N*-dimethylformamide-*p*-toluene-sulfonic acid, *Carbohydr. Res. 29*:209 (1973).
97. R. Khan, Sucrochemistry. Part XIII. Synthesis of 4,6-*O*-benzylidene-sucrose, *Carbohydr. Res. 32*:375 (1974).
98. H. C. Brown, J. H. Brewster, and H. Shechter, An interpretation of the chemical behavior of five- and six-membered ring compounds, *J. Am. Chem. Soc. 76*:467 (1954).
99. E. L. Eliel and E. Juaristi, Anomeric effect origin and consequences, *Am. Chem. Soc. Ser. 87*:95 (1979).
100. N. J. Leonard and K. L. Carraway, 5-Amino-5-deoxyribose derivatives. Synthesis and use in the preparation of "reversed" nucleosides, *J. Heterocycl. Chem. 3*:485 (1966).
101. H. Zinner, G. Rembarz, and H. Klocking, Isopropyliden-verbindungen der D-arabinose-mercaptale, *Chem. Ber. 90*:2688 (1957); P. A. J. Gorin, Acetonation of aldose diethyl dithioacetals, *Can. J. Chem. 43*:2078 (1965); D. G. Lance and J. K. N. Jones, Acetonation of D-xylose diethyl dithioacetal, *Can. J. Chem. 45*:1533 (1967).
102. S. Hanessian, Total synthesis of natural products: the "Chiron" approach, *Organic Chemistry Series*, Vol. 3, Pergamon Press, New York, 1983.

# 2
# Dialkyl Dithioacetals of Sugars

**Derek Horton**
The American University, Washington, D.C.

**Peter Norris**
Youngstown State University, Youngstown, Ohio

| | | |
|---|---|---|
| I. | Introduction | 36 |
| | A. Other acyclic derivatives: hydrazones, osazones, and oximes | 36 |
| | B. Cyanohydrin and nitromethane adducts | 36 |
| | C. Dithioacetals and their synthetic applications | 37 |
| II. | Methods | 39 |
| | A. Introduction | 39 |
| | B. Scope and limitations | 39 |
| III. | Experimental Procedures | 43 |
| | A. D-Arabinose diethyl dithioacetal | 43 |
| | B. D-Xylose diethyl dithioacetal | 43 |
| | C. D-Ribose diphenyl dithioacetal | 44 |
| | D. D-Lyxose ethylene dithioacetal | 44 |
| | E. D-Glucose diethyl dithioacetal | 45 |
| | F. D-Galactose diethyl dithioacetal | 45 |
| | G. D-Fructose diethyl dithioacetal | 46 |
| | H. L-Sorbose dimethyl dithioacetal | 46 |
| | I. 2-Deoxy-D-arabinohexose diethyl dithioacetal | 47 |
| | J. 6-Deoxy-L-mannose (L-rhamnose) diphenyl dithioacetal | 47 |
| | K. 2-Acetamido-2-deoxy-D-glucose diethyl dithioacetal | 48 |
| | L. 2-Amino-2-deoxy-D-galactose diethyl dithioacetal | 48 |
| | M. 2-Amino-2-deoxy-D-glucose propan-1,3-diyl dithioacetal hydrochloride | 49 |
| | N. Sodium D-glucuronate diethyl dithioacetal | 49 |
| | References | 50 |

## I. INTRODUCTION

Acyclic derivative of sugars have played a significant role in the area of synthetic carbohydrate chemistry, permitting numerous useful transformations that are not possible with the parent sugars, which exist almost exclusively in the hemiacetal form. Trapping of aldoses in the acyclic form as their dialkyl dithioacetals, by treatment with thiols in the presence of acid, has been a synthetically important method ever since Emil Fischer's first report some 100 years ago [1], and remains an important tool in modern synthetic carbohydrate chemistry.

### A. Other Acyclic Derivatives: Hydrazones, Osazones, and Oximes

Diethyl dithioacetals have proved useful for characterizing sugars because many of them are readily obtained in crystalline form [2]. They are stable products, do not exhibit tautomerism, and can be readily reconverted into the parent sugar. Although acyclic phenylhydrazones may be formed when reducing sugars are treated with phenylhydrazine [3], the treatment of aldoses and 2-ketoses with an excess of phenylhydrazine generally affords 1,2-bis(phenylhydrazones) (osazones) [4]. Osazones and phenylhydrazones may still exhibit tautomerism and are less useful for characterization purposes than the derived phenylosotriazoles obtained through oxidation of osazones with copper(II) sulfate [5]. Reaction of reducing sugars with hydroxylamine affords oximes, and these likewise readily interconvert between acyclic and cyclic forms, as judged from the products of acetylation [6]. Acetylation of the oximes derived from aldoses in the presence of NaOAc leads by an acyclic hexaacetate to the corresponding acyclic peracetylated aldononitrile; this sequence is part of the Wohl degradation [7] (Scheme 1).

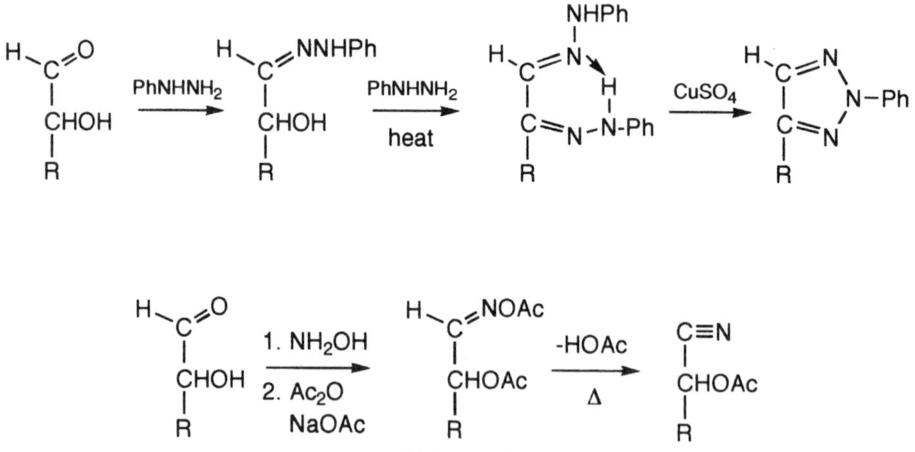

**Scheme 1**

### B. Cyanohydrin and Nitromethane Adducts

Reactions of significance in the extension of monosaccharide carbon chains include treatment of reducing sugars with strong carbon nucleophiles, such as cyanide and anions derived from nitromethane. Such reactions are generally irreversible and convert the parent sugar completely into an acyclic product through the small equilibrium concentration of its acyclic form. The former reaction, which constitutes the key step of the Fischer–Kiliani chain ascent process [8], results in acyclic cyanohydrin formation that yields two products because a new asymmetric center is formed. These nitriles are generally hydrolyzed to the epimeric aldonic acids [8] or their lactones, or are reduced in situ, under controlled conditions (sodium amalgam [9] or $H_2$/Pd/$BaSO_4$ [10], to the corresponding aldoses. The aldol reaction of aldoses with nitromethane in basic medium results in epimeric pairs of 1-deoxy-1-nitroalditols, which can be hydrolyzed by means of a Nef reaction to the corresponding one-carbon homologated aldoses [11] (Scheme 2).

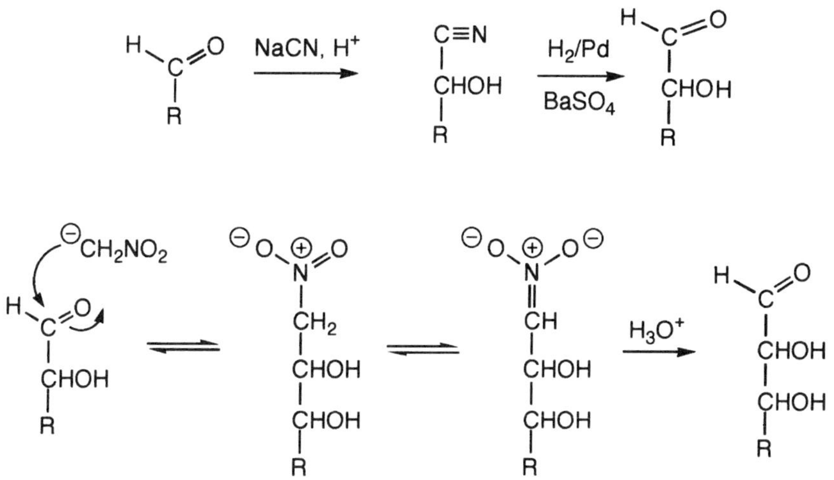

**Scheme 2**

## C. Dithioacetals and Their Synthetic Applications

Since the first description of sugar diethyl dithioacetals by Fischer [1], such compounds have become the most frequently reported acyclic sugar derivatives, and they have arguably become the most useful and synthetically versatile of all acyclic carbohydrate compounds. The facile conversion of a wide variety of aldoses, ketoses, and deoxy and aminodeoxy sugars into acyclic forms as their dialkyl dithioacetals, permits a range of preparatively useful conversions, not only by means of hydroxyl group chemistry, but also through interconversions possible with the dithioacetal moiety, such as demercaptalation to *aldehydo* sugars, deprotonation–alkylation ("umpolung") chemistry, and kinetically controlled demercaptalation to afford thiofuranosides and other furanose derivatives.

Besides acting as a convenient protecting group for sugar carbonyls, the dithioacetal functionality is capable of a variety of important transformations relevant to the carbohydrate field. Dithioacetals constitute the most useful precursors for *aldehydo* sugars through protection of the chain hydroxyl groups and subsequent removal of the dithioacetal function under neutral conditions, usually with Hg(II) salts [12] (Scheme 3). Similar deprotection in alcohol solvent is a facile route to dialkyl acetal derivatives of sugars [13]. These acyclic *aldehydo* sugar derivatives have great synthetic versatility in all aspects of general carbonyl chemistry; for example, as intermediates in the synthesis of chain-extended enoate derivatives by means of the Wittig olefination method. The resultant alkenes have been used in extensive studies of chirality transfer in Diels–Alder reactions with various dienes [14], and such studies have led to useful procedures for the synthesis of enantiomerically pure carbocycles from simple carbohydrate precursors [15].

Dialkyl dithioacetals of free aldoses are precursors for a useful chain-descent se-

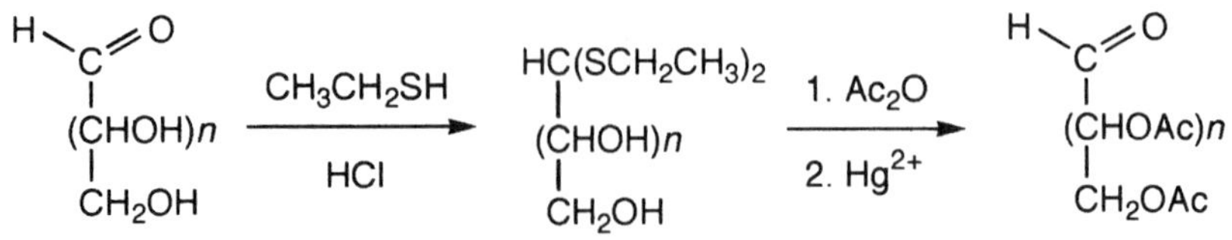

$$\underset{\substack{|\\(\text{CHOH})n\\|\\ \text{CH}_2\text{OH}}}{\text{H}\diagdown\text{C}\diagup\text{O}} \xrightarrow[\text{HCl}]{\text{CH}_3\text{CH}_2\text{SH}} \underset{\substack{|\\(\text{CHOH})n\\|\\ \text{CH}_2\text{OH}}}{\text{HC}(\text{SCH}_2\text{CH}_3)_2} \xrightarrow[\text{2. Hg}^{2+}]{\text{1. Ac}_2\text{O}} \underset{\substack{|\\(\text{CHOAc})n\\|\\ \text{CH}_2\text{OAc}}}{\text{H}\diagdown\text{C}\diagup\text{O}}$$

**Scheme 3**

quence (McDonald–Fischer degradation). Oxidation with peroxides affords disulfones, which readily undergo elimination to alkenes through loss of the α-hydroxyl function [16]. Base-induced degradation of the disulfones occurs by a retroaldol sequence and results in a one-carbon descent for aldoses [17].

A useful method for chain extension involves removal of the activated proton α to the sulfur substituents, followed by the addition of electrophiles. Alkylation with simple alkyl halides, followed by hydrolytic removal of the dithioacetal group, results in 1-deoxyglycos-2-uloses [18], whereas acylation with N,N-dimethylformamide yields glycosulose derivatives [19]. Sugar dithioacetals bearing an oxygen functionality at C-2 undergo elimination in these presence of strong bases, to afford ketene dithioacetal derivatives [20]. These ketene derivatives can be reduced (LiAlH$_4$) to the corresponding 2-deoxy-dialkyl dithioacetals, which undergo Hg(II)-promoted hydrolysis to afford 2-deoxyaldoses [21]. The deprotonation–alkylation method is a useful route to 1,3-dideoxy-2-ketoses from dithioacetals of 2-deoxyaldoses [22]. Reduction of dithioacetal derivatives of carbohydrates with Raney nickel provides a simple route to 1-deoxyalditols (Wolfrom–Karabinos reaction; Scheme **4**) [23].

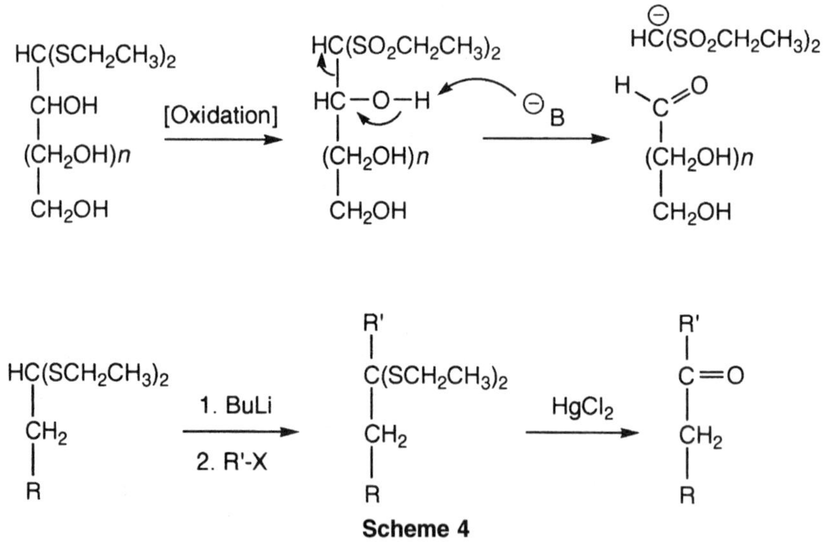

**Scheme 4**

A further useful reaction of dialkyl dithioacetals involves treatment with bromine in an aprotic solvent, which effects monobromination at C-1. These brominated intermediates react with such nucleophiles as alcohols to form mixed acetal products [24]. This method has proved especially useful for the synthesis of nucleoside analogues having acyclic sugar chains attached to the heterocycle [25] (Scheme **5**).

More recently, dithioacetals of sugars have found use as important intermediates in the synthesis of such compounds as carba sugars, in which the ring oxygen of the parent

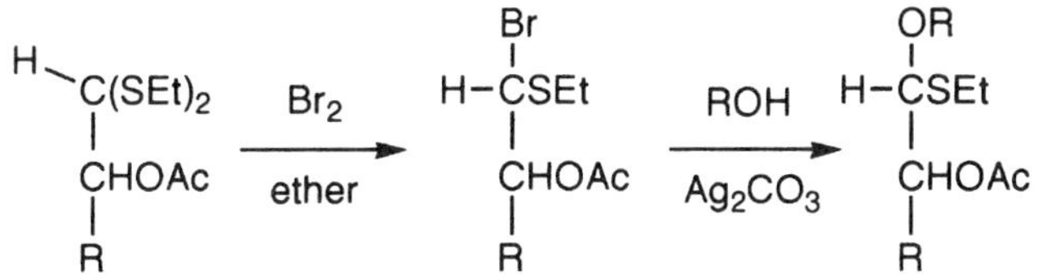

$$\underset{\text{R}}{\underset{|}{\text{CHOAc}}}\!\!\!\!-\!\!\!\!\underset{}{\overset{\text{H}}{\text{C(SEt)}_2}} \xrightarrow[\text{ether}]{\text{Br}_2} \underset{\text{R}}{\underset{|}{\text{CHOAc}}}\!\!\!\!-\!\!\!\!\underset{}{\overset{\text{Br}}{\underset{|}{\text{H-CSEt}}}} \xrightarrow[\text{Ag}_2\text{CO}_3]{\text{ROH}} \underset{\text{R}}{\underset{|}{\text{CHOAc}}}\!\!\!\!-\!\!\!\!\underset{}{\overset{\text{OR}}{\underset{|}{\text{H-CSEt}}}}$$

**Scheme 5**

sugar is replaced by carbon [26], as well as applications in natural product synthesis. As will be detailed in the following sections, dithioacetal derivatives of sugars are readily available from simple starting materials. The range of thiols used in the preparation of dithioacetals has been extended since the initial studies made by Emil Fischer [1] to include not only the simple alkanethiols (MeSH, EtSH, and so on), but also thiophenols, benzylthiols, and dithiols (e.g., $HSCH_2CH_2SH$). Ethanethiol, as a low-boiling (albeit exceptionally malodorous) liquid, retains procedural convenience, although use of the gaseous methanethiol affords products for which nuclear magnetic resonance (NMR) spectra are greatly simplified.

The acyclic dithioacetal derivatives formed from reactions with such thiols offer a variety of reaction modes for the synthesis of modified sugars, as well as for the total synthesis of enantiomerically pure natural products and, as such, are attractive and versatile intermediates in synthetic carbohydrate chemistry.

## II. METHODS

### A. Introduction

Fischer's seminal investigation into acyclic sugar-derived dialkyl dithioacetals [1] stemmed from his earlier observation that aldoses react with alcohols to give glycosides, rather than acyclic acetals. While turning his attention to the reactions of aldoses with ethanethiol in concentrated hydrochloric acid, Fischer supposed that thioglycosides would form, but he found that the reaction products, many of which crystallized directly from the reaction mixtures, were in fact the diethyl dithioacetal derivatives. Although Fischer's studies used the simple alkanethiols, the range of thiols used in these mercaptalation reactions has expanded to include benzenethiols, α-toluenethiols, and alkanedithiols. With the continuing popularity of sugars as starting materials in total syntheses of chiral natural products, the use of dithioacetals remains a principal method for protecting the carbonyl groups of sugar-derived synthetic intermediates.

### B. Scope and Limitations

Emil Fischer's initial studies [1] focused on dithioacetal formation from a variety of simple aldopentoses, aldohexoses, and aldoheptoses. The products were obtained by simply treating the sugar with an excess of ethanethiol in concentrated hydrochloric acid at low temperature (~0°C). Many of these dithioacetals have low water solubility, crystallize spontaneously from the reaction mixtures, and can be isolated by simple filtration and washing with cold water. Fischer's key paper [1] also outlined most of the major reaction modes of the diethyl dithioacetal derivatives, later developed in detail, and thus paved the way for subsequent studies into this important class of sugar derivative.

One of the major drawbacks of the original Fischer method is that isolation of the products relies on their low solubility in aqueous solution. Direct isolation of water-soluble dithioacetals thus was not possible. This problem has since been addressed in detail, and a variety of methods have been developed to enable isolation of such products. In the case of such water-soluble products as D-xylose diethyl dithioacetal, addition of solid lead carbonate to the reaction mixture, and subsequent filtration and evaporation, leads to the product as a syrup that can be obtained crystalline through an acetylation–deacetylation sequence [27]. Extraction of the excess thiol from the reaction mixture with an organic solvent, such as ether [28], also aids in inducing crystallization of products. The formation and isolation of diethyl dithioacetal derivatives of a wide variety of aldoses, ranging from trioses to octoses, proceeds smoothly [29], although reported yields are not uniformly high.

The mechanism of formation of the alkyl dithioacetals of simple sugars has been considered [30] to involve (1) initial, acid-catalyzed, addition of one equivalent of thiol to the sugar carbonyl group, (2) displacement of water from the thus-formed monothiohemiacetal, and (3) deprotonation to afford the dithioacetal product. The reaction temperature and certain functional groups present within the sugar can affect the nature of the product formed; however, for the simple aldoses, the product isolated after reaction for short times at low temperature (~0°C), and in the presence of strong mineral acid, is generally the dithioacetal (Scheme 6).

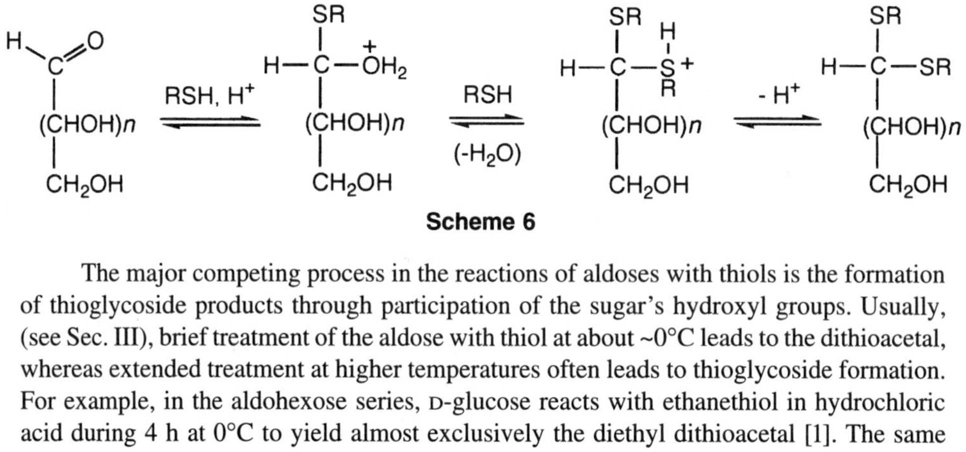

**Scheme 6**

The major competing process in the reactions of aldoses with thiols is the formation of thioglycoside products through participation of the sugar's hydroxyl groups. Usually, (see Sec. III), brief treatment of the aldose with thiol at about ~0°C leads to the dithioacetal, whereas extended treatment at higher temperatures often leads to thioglycoside formation. For example, in the aldohexose series, D-glucose reacts with ethanethiol in hydrochloric acid during 4 h at 0°C to yield almost exclusively the diethyl dithioacetal [1]. The same reaction run at room temperature affords principally ethyl-1-thio-α-D-glucopyranoside [31] (Scheme 7). A detailed chromatographic investigation of this reaction [32] revealed that the

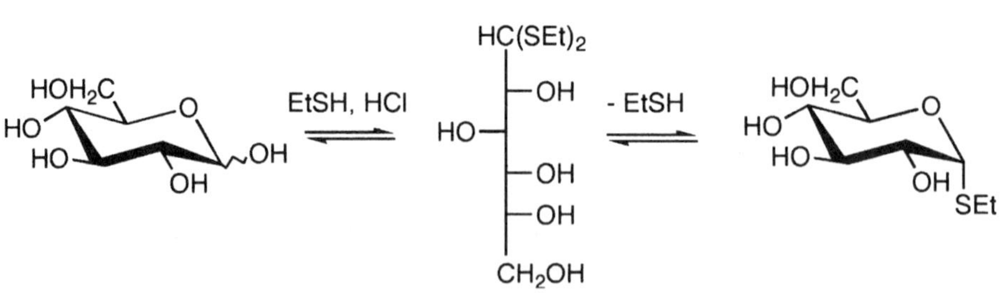

**Scheme 7**

dithioacetal is formed rapidly at room temperature, but undergoes conversion to the thiopyranoside(s) over several hours. Similarly, D-mannose yields the dithioacetal (63%) after 5 min at 0°C [33], and the ethyl thiopyranosides (31%) after 16 h at room temperature

[34]. L-Arabinose diethyl dithioacetal appears to be more thermodynamically stable than the thioglycosides, for it is the only product isolated from mercaptalation of L-arabinose, even after extended exposure to the reagents at room temperature [34]. In general, however, the best yields of dithioacetal products are obtained after short reaction times at room temperature or lower, with good technique to remove aqueous acid rapidly from the product during the isolation stage.

The reaction of reducing sugars with thiols has been expanded to include most of the biologically important classes of aldoses and ketoses. Amino sugars, either in N-protected form, or as their hydrochloride salts, undergo mercaptalation with a variety of thiols to afford acyclic dialkyl dithioacetals; forcing conditions are required for the amine-salt forms to generate a doubly protonated reactant. For example, D-glucosamine hydrochloride reacts with 1,2-ethanedithiol, initially at 0°C and then at room temperature, in *fuming* hydrochloric acid, to give an essentially quantitative yield of the ethylene dithioacetal hydrochloride [35]. In contrast, N-acetyl-D-glucosamine reacts with ethanethiol in ordinary, concentrated hydrochloric acid (37%) to afford the crystalline dithioacetal in 81% yield [36] (Scheme **8**). Treatment of a variety of 3-amino-3-deoxy sugars with ethanethiol leads not only to diethyl dithioacetals, but also to varying amounts of thioglycoside products [37]. In the case of 3-amino-3-deoxy-D-mannose, a detailed analysis of the reaction indicated that the dithioacetal is formed under kinetic control, followed by equilibration to afford a mixture of products, the relative amounts of which are dependent on their thermodynamic stabilities [32].

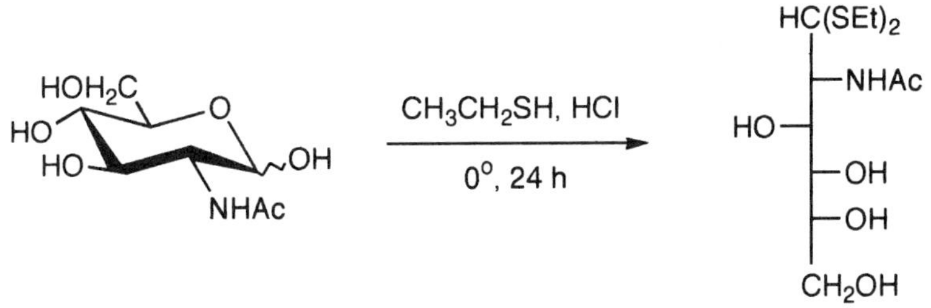

**Scheme 8**

Dialkyl dithioacetal derivatives can be obtained from deoxy sugars under conventional mercaptalation conditions, although the acid lability of 2-deoxyaldoses may diminish yields if proper precautions are not taken. For example, 2-deoxy-D-*arabino*-hexose affords its crystalline diethyl dithioacetal on treatment with ethanethiol in hydrochloric acid, but the 40% yield reported [38] has been improved to 69% using essentially the same procedure [18]. The biologically important 6-deoxy-L-mannose (L-rhamnose) undergoes facile mercaptalation in the presence of benzenethiol, to afford the corresponding diphenyl dithioace-

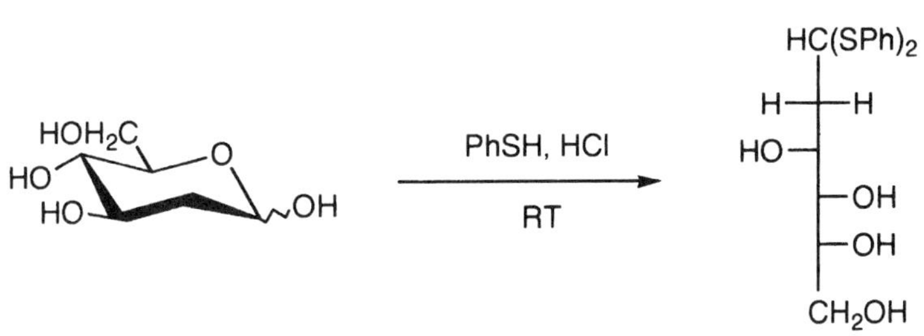

**Scheme 9**

tal in satisfactory (55%) yield [39]. In general, forcing conditions (fuming HCl) are needed to form the *diphenyl* dithioacetals in satisfactory yield [20] (Scheme 9).

Dialkyl dithioacetal derivatives of ketoses, such as D-fructose and L-sorbose, are inaccessible directly from the parent sugars, the ketose undergoing extensive decomposition under the conditions employed for mercaptalation of aldoses. Such derivatives can, however, be prepared by indirect methods. Acetylation of D-fructose [40] and L-sorbose with acetic anhydride and zinc chloride [41] leads to good yields of acyclic pentaacetates in which the ketose carbonyl is not involved in a cyclic acetal. Subsequent treatment of these acetylated derivatives with thiols affords the acetylated dialkyl dithioacetals in satisfactory yields, and conventional deacetylation affords the unprotected dialkyl dithioacetals [40,41] (Scheme 10).

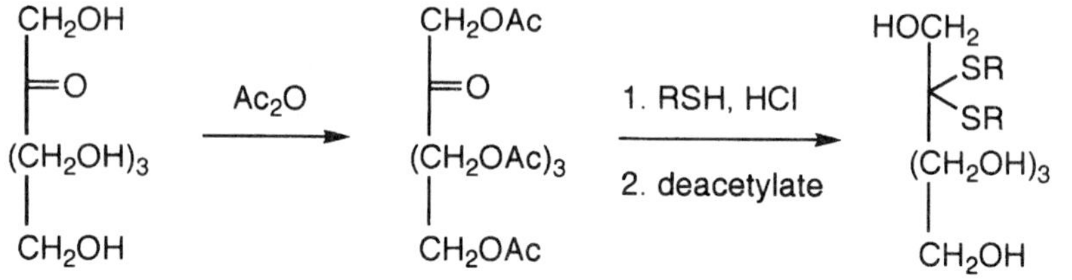

**Scheme 10**

Aldos-2-uloses, which have the ketonic function at C-2, undergo mercaptalation under the usual conditions, to afford only 1,1-(dialkyl dithioacetal) derivatives. The ketone function in these compounds is distinctly unreactive toward thiols in acidic medium, probably as a consequence of electronic and steric effects [42] (Scheme 11). Examples of 1,2-bis(dialkyl dithioacetal) derivatives have been prepared by indirect methods [43]. Dialdoses, on the other hand, in which the two carbonyl groups are separated by the sugar backbone, readily undergo mercaptalation at both carbonyl centers to afford bis(dialkyl dithioacetal) derivatives [44].

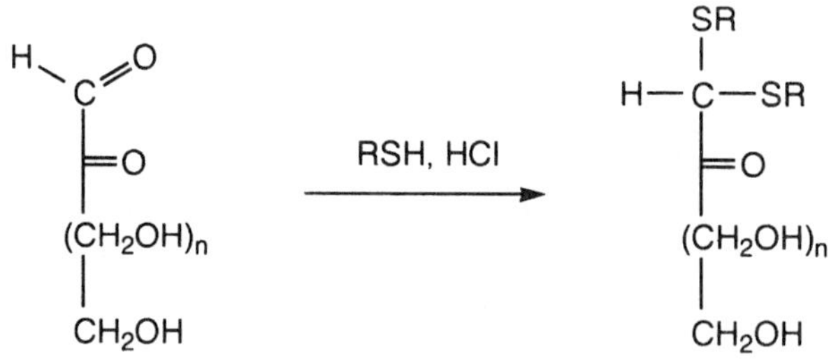

**Scheme 11**

Glycuronic acids also undergo mercaptalation in the presence of thiols and acid. Thus, D-glucurono-3,6-lactone reacts readily with ethanethiol in concentrated hydrochloric acid to afford, after neutralization with methanolic sodium hydroxide, sodium D-glucuronate diethyl dithioacetal [45] (Scheme 12). In a similar reaction, mercaptalation and subsequent acetylation of D-glucurono-3,6-lactone yields 2,4,5-tri-*O*-acetyl-D-glucurono-3,6-lactone diethyl dithioacetal, which serves as a precursor to D-glucuronamide diethyl dithioacetal [45]. The biologically significant *N*-acetylneuraminic acid (sialic acid) also affords a crystalline diethyl dithioacetal on treatment with ethanethiol in concentrated hydrochloric acid [46].

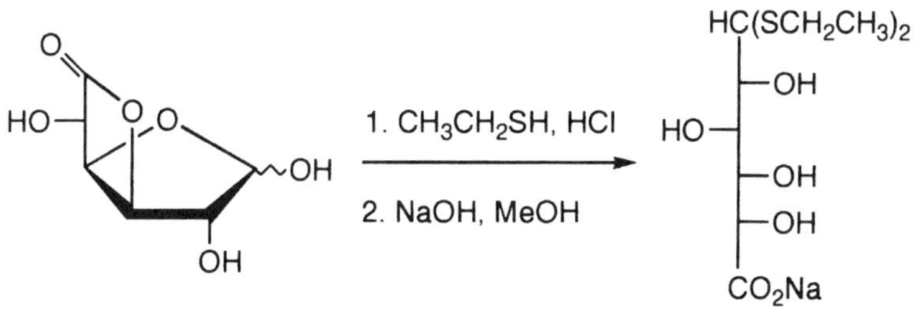

## Dialkyl Dithioacetals of Sugars

**Scheme 12**

A practical consideration in working with ethanethiol is the pervasive stench of this and other volatile thiols, especially as such thiols are used in minute concentration as odor markers for natural gas. It is not easy to perform the standard preparative procedures, during which transfer and filtration operations are performed, in a closed system, and vapors carried through a venting system are detectable at considerable distances. In small-scale operations, it may be possible to employ a sodium hypochlorite trap to convert the thiols into nonvolatile, oxidized products.

The chemical transformations of dialkyl dithioacetals have been reviewed in detail [47] and offer routes to a variety of useful carbohydrate derivatives. Dialkyl dithioacetal derivatives of sugars continue to play an important role in modern synthetic carbohydrate chemistry through reactions of the dithioacetal function and manipulation of the sugar hydroxyl groups. Dithioacetals also provide a convenient method for temporary protection of sugar carbonyl groups in the synthesis of noncarbohydrate natural products.

## III. EXPERIMENTAL PROCEDURES

### A. D-Arabinose Diethyl Dithioacetal [1,48]

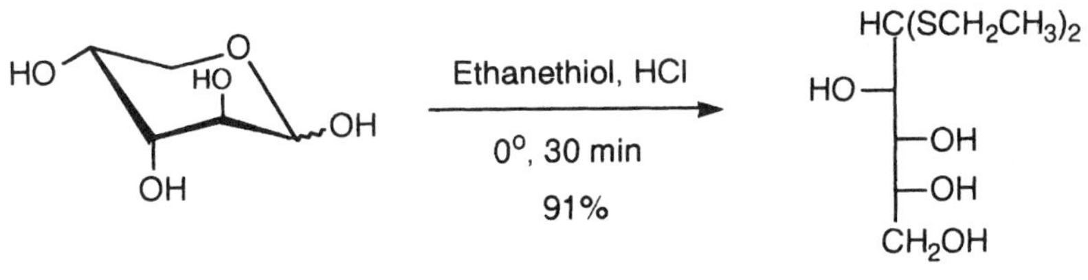

(1)

This is a slight modification of the method of Fischer [1]. D-Arabinose (250 g, 1.66 mol) is dissolved at room temperature in 37% hydrochloric acid (250 mL) in a 1-L Erlenmeyer flask and cooled at once to 0°C. Technical ethanethiol (250 mL) is added, the flask is stoppered, and the two layers are shaken vigorously. Copious crystallization occurs after 15 min and, after 30 min, the crude product is collected by filtration and washed with ice cold water. Recrystallization from water affords the pure dithioacetal (380 g, 91%): mp 124–125°C, $[\alpha]_D$ 0° (c 3.0, pyridine).

### B. D-Xylose Diethyl Dithioacetal [27,48]

D-Xylose (50 g, 0.3 mol) is dissolved in 37% aqueous hydrochloric acid (50 mL) in an Erlenmeyer flask, and the solution is cooled to 0°C. Ethanethiol (50 mL) is added, the flask stoppered, and the biphasic mixture shaken at 0°C for 1 h. The resultant purple solution is then added, in portions, to a stirred suspension of lead carbonate (90 g) in methanol (100 mL). After the addition is complete the mixture is stirred for an additional 16 h and then decolorizing charcoal is added. The mixture is filtered, and the colorless filtrate is evaporated to a syrup. Addition of ether (150 mL) and methanol (10 mL) causes crystallization, and the product is collected by filtration (30.4 g, 42%): mp 61–63°C, [α]$_D$ −30.8° (c 0.4, water).

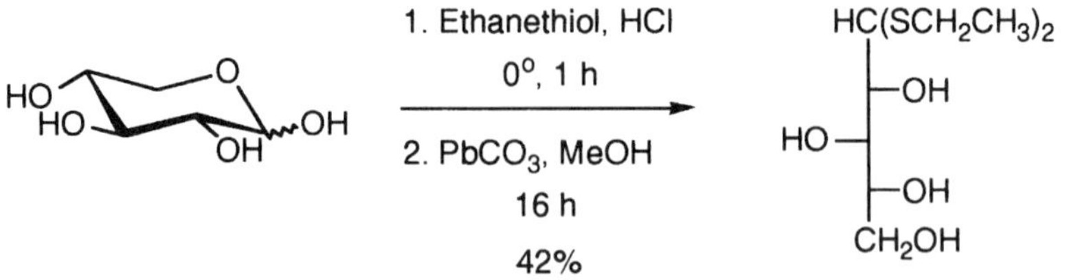

## C. D-Ribose Diphenyl Dithioacetal [20]

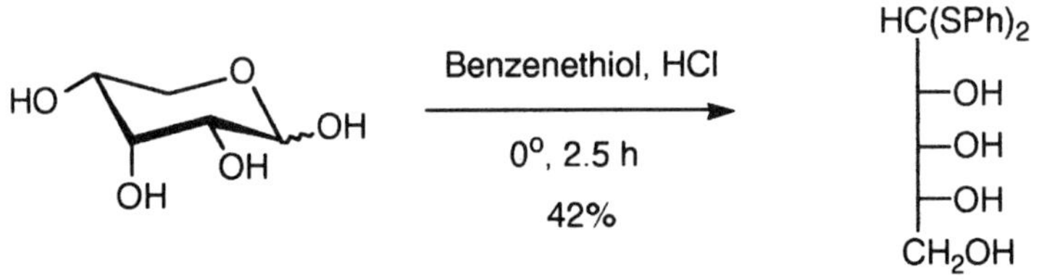

Concentrated hydrochloric acid (9.0 mL) is cooled to 0°C and saturated with hydrogen chloride gas. D-Ribose (3.0 g, 20 mmol) and benzenethiol (5.6 mL) are added, and the mixture is shaken for 2 h at 0°C and then for 20 min at room temperature. The solution is then poured into cold water (100 mL), and the resultant syrup is decanted from the water layer. This syrup is then dissolved in ethyl acetate (50 mL), the solution dried over sodium carbonate, and evaporated. After dissolving the syrup in benzene (100 mL), the product crystallizes slowly (~2 days) at 0°C and is collected by filtration (3.0 g, 40%, collected in two crops). Recrystallization is effected from ethanol–ether and then ethanol–water to afford the pure diphenyl dithioacetal: mp 101.5–102.0°C, [α]$_D$ +42.3° (c 1, pyridine).

## D. D-Lyxose Ethylene Dithioacetal [51]

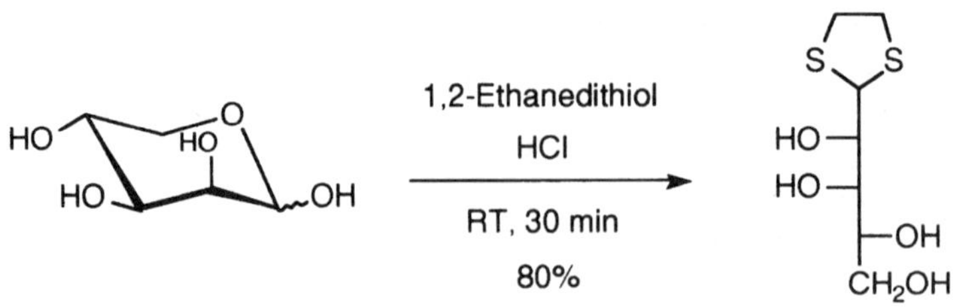

D-Lyxose (3.0 g, 20.0 mmol) and 1,2-ethanedithiol (1.08 mL, 10.0 mmol) are placed in concentrated hydrochloric acid (3.0 mL) and the mixture is shaken in a stoppered bottle for 10 min at room temperature. The mixture is poured into ice water (20 mL), and the homogeneous solution is passed through a column (1.8 × 20 cm) of anion-exchange resin (e.g., Dowex 1, Amberlite IRA-400, Wofatit L, OH form). The first 20 mL of effluent is discarded; the subsequent 200 mL contains the product and is evaporated under diminished pressure at 40–50°C to a syrup that crystallizes on standing and scratching with a glass rod. Recrystallization from isopropyl alcohol affords the dithioacetal as colorless leaflets (3.6 g, 80%): mp 142°C $[\alpha]_D$ +17.5° (c 1.82, methanol).

### E. D-Glucose Diethyl Dithioacetal [1, 49]

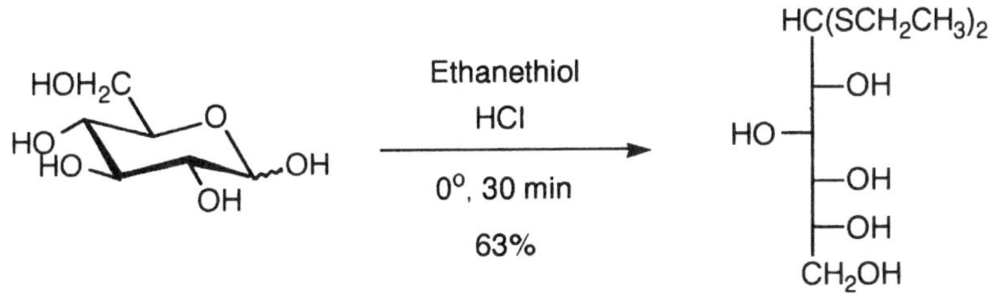

Anhydrous D-glucose (100 g, 0.56 mol) is dissolved in 37% hydrochloric acid (85 mL) at room temperature in a glass-stoppered bottle. Technical ethanethiol (100 mL) is added, and the mixture is shaken vigorously at room temperature, opening the stopper occasionally to allow release of pressure. The temperature is maintained at ~20°C by periodically adding small amounts of ice and by immersing the vessel in an ice bath. The shaking is continued until copious crystallization occurs, and the bottle is then cooled for 30 min in an ice–salt bath. After suction filtration, the solid mass is washed with ice water, and immediately recrystallized from hot water, containing a small amount of sodium hydrogencarbonate. After allowing the flask to cool, the solid is filtered off with suction and washed with a small amount of ice water, followed by a small amount of ethanol and ether (yield ~100 g, 63%): mp 127°C, $[\alpha]_D$ −30° (water).

### F. D-Galactose Diethyl Dithioacetal [50]

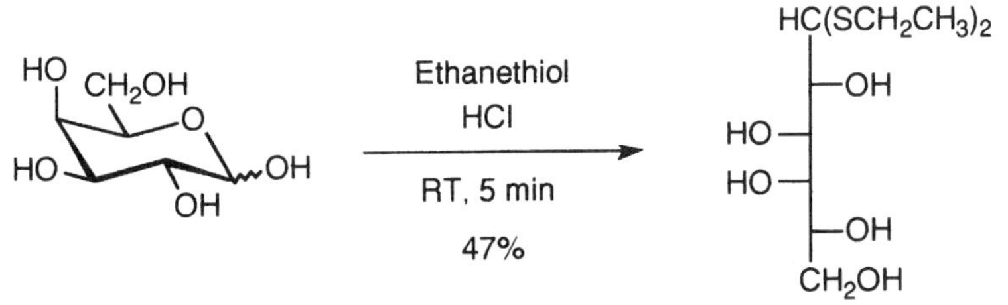

D-Galactose (50 g, 0.278 mol) is dissolved at room temperature in 37% hydrochloric acid (75 mL) in a 500-mL, glass-stoppered, wide-mouthed bottle. Technical ethanethiol (50 mL) is added, and the mixture is shaken vigorously at room temperature, the pressure being released occasionally. A temperature increase becomes evident after a few minutes, and a small amount of ice and ice water is added, which causes the reaction mixture to solidify almost immediately. More ice water is added and the crystalline mass filtered immediately and washed with a small amount of ice water. Recrystallization from absolute ethanol and then again from water affords pure diethyl dithioacetal (37 g, 47%): mp 140–142°C, $[\alpha]_D$ −3.5° (pyridine).

## G. D-Fructose Diethyl Dithioacetal [40]

$$\begin{array}{c} CH_2OH \\ {=\!\!=}O \\ HO{-\!\!\!-} \\ {-\!\!\!-}OH \\ {-\!\!\!-}OH \\ CH_2OH \end{array} \xrightarrow{Ac_2O} \begin{array}{c} CH_2OAc \\ {=\!\!=}O \\ AcO{-\!\!\!-} \\ {-\!\!\!-}OAc \\ {-\!\!\!-}OAc \\ CH_2OAc \end{array} \xrightarrow[\substack{0°,\,4\,h \\ \text{2. NH}_3,\,\text{MeOH}}]{\substack{\text{1. Ethanethiol} \\ \text{ZnCl}_2}} \begin{array}{c} HOH_2C{\diagdown}SCH_2CH_3 \\ {\diagup}SCH_2CH_3 \\ HO{-\!\!\!-} \\ {-\!\!\!-}OH \\ {-\!\!\!-}OH \\ CH_2OH \end{array} \quad (7)$$

Freshly fused zinc chloride (8.0 g) is dissolved in ethanethiol (60 mL), which has previously been dried, in a stoppered flask. The solution is cooled in an ice–salt bath and anhydrous sodium sulfate (15 g) and 1,3,4,5,6-penta-O-acetyl-keto-D-fructose (20 g, 51.3 mmol) is added with shaking. The flask is stoppered and then stored in the freezing mixture for 4 h. The mixture is then poured carefully into saturated sodium hydrogencarbonate solution (~150 mL). The resultant precipitate is filtered off and extracted into warm chloroform; the original filtrate is also extracted with chloroform. The combined extracts are dried and evaporated to a syrup, from which petroleum ether is evaporated repeatedly until crystallization occurs. The solid can be recrystallized from ether–petroleum ether (10.9 g, 43%), mp 76–78°C (83°C after further recrystallization), $[\alpha]_D$ +20° (c 3.7, chloroform).

The dithioacetal pentaacetate (12.0 g, 24.0 mmol) is dissolved in absolute methanol (100 mL) and the flask is cooled in an ice–salt bath while a rapid stream of ammonia gas is bubbled through for about 45 min. The solution is kept in a freezer overnight and then evaporated to a syrup at room temperature. This material is dissolved in methanol, the solution evaporated to a small volume, and ether (2 volumes) added to cause crystallization. The solid is filtered off and washed with ether (5.4 g, 79%): mp 55–60°C. After repeated recrystallizations from methanol–ether the product has mp 65–67°C, $[\alpha]_D$ +35.8° (c 4.0, methanol).

## H. L-Sorbose Dimethyl Dithioacetal [41]

$$\begin{array}{c} CH_2OH \\ {=\!\!=}O \\ HO{-\!\!\!-} \\ {-\!\!\!-}OH \\ HO{-\!\!\!-} \\ CH_2OH \end{array} \xrightarrow[\text{ZnCl}_2]{Ac_2O} \begin{array}{c} CH_2OAc \\ {=\!\!=}O \\ AcO{-\!\!\!-} \\ {-\!\!\!-}OAc \\ AcO{-\!\!\!-} \\ CH_2OAc \end{array} \xrightarrow[\substack{\text{2. Ba(OMe)}_2, \\ \text{MeOH}}]{\substack{\text{1. MeSH, ZnCl}_2 \\ -15°}} \begin{array}{c} HOH_2C{\diagdown}SCH_3 \\ {\diagup}SCH_3 \\ HO{-\!\!\!-} \\ {-\!\!\!-}OH \\ HO{-\!\!\!-} \\ CH_2OH \end{array} \quad (8)$$

L-Sorbose (3.60 g, 20.0 mmol) is added to a stirring solution of anhydrous zinc chloride (0.36 g) and acetic anhydride (70 mL) at 0°C. After allowing the reaction to warm to room temperature, the mixture is stirred for 3 h, allowed to stand for a further 30 h, and then poured into water (300 mL) with stirring. The mixture is stirred for 3 h to decompose excess acetic anhydride, and then the solid mass is filtered and recrystallized from ethanol to yield the pentaacetate (5.4 g, 70%): mp 96.5–97.5°C, $[\alpha]_D$ +2.8° (c 2.14, chloroform).

Anhydrous zinc chloride (1.60 g) is dissolved in methanethiol (12 mL) held at −15°C.

Penta-*O*-acetyl-L-sorbose (3.90 g, 10.0 mmol) and anhydrous sodium sulfate (3.00 g) are added, and the mixture is kept at 0°C, with occasional stirring, for 3 h. After allowing it to stand at room temperature for 5 days, the solution is mixed with saturated sodium hydrogencarbonate (40 mL), which causes the product, along with inorganic salts, to precipitate. After filtration the solid is extracted with warm chloroform, and the filtrate is extracted with three portions of chloroform. The combined extracts are dried with sodium sulfate and evaporated to yield a syrup, which solidifies on standing. The product is recrystallized from absolute ethanol (2.75 g, 59%): mp 116°C, $[\alpha]_D$ $-12.3°$ (*c* 2.32, chloroform).

To a solution of penta-*O*-acetyl-L-sorbose dimethyl dithioacetal (4.68 g, 10.0 mmol) in methanol (100 mL) is added barium methoxide (60 mL, 10 mM in methanol) and the mixture is allowed to stand for 16 h at room temperature. The mixture is then neutralized with sulfuric acid (0.2 M), shaken with barium carbonate and activated charcoal, then filtered, and evaporated to a syrup, which is dried under high vacuum (2.51 g, 97%): $[\alpha]_D$ $-37°$ (*c* 1.36, chloroform).

### I. 2-Deoxy-D-arabinohexose Diethyl Dithioacetal [18,38]

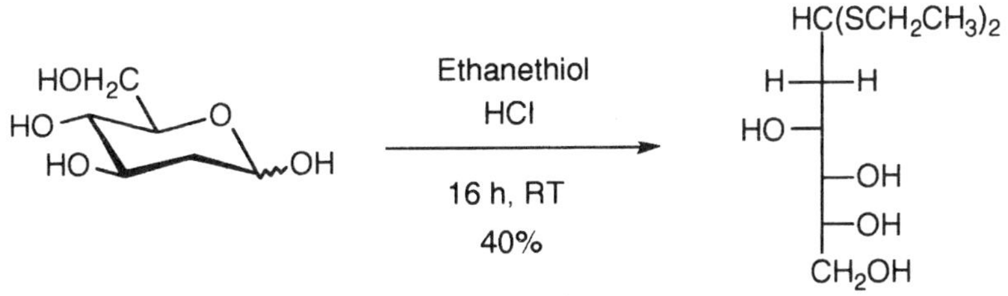

$$\text{2-deoxy-D-arabino-hexopyranose} \xrightarrow[\text{16 h, RT}]{\substack{\text{Ethanethiol} \\ \text{HCl} \\ 40\%}} \text{HC(SCH}_2\text{CH}_3)_2\text{-CH(H)-CH(OH)-CH(OH)-CH(OH)-CH}_2\text{OH} \qquad (9)$$

2-Deoxy-D-*arabino*-hexose (0.8 g, 4.88 mmol) is dissolved in concentrated hydrochloric acid (5 mL) and ethanethiol (2 mL), and the mixture is shaken overnight at room temperature. Ethanol (50 mL) is added, and the solution is neutralized with lead carbonate. Insoluble salts are filtered and washed with ethanol (20 mL). Evaporation affords a syrup (0.5 g, 40%), which crystallizes slowly and can be recrystallized from acetone–ether: mp 132°C, $[\alpha]_D$ $-8.8°$ (*c* 2.5, methanol). A later procedure, employing essentially the same conditions, reports a yield of 69% [18].

### J. 6-Deoxy-L-mannose (L-Rhamnose) Diphenyl Dithioacetal [39]

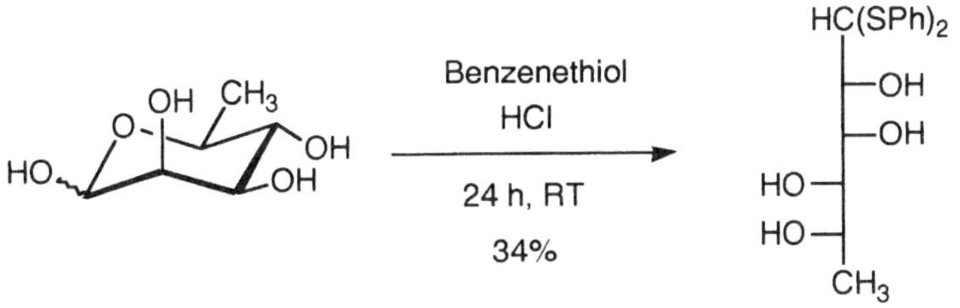

$$\text{L-rhamnopyranose} \xrightarrow[\text{24 h, RT}]{\substack{\text{Benzenethiol} \\ \text{HCl} \\ 34\%}} \text{HC(SPh)}_2\text{-CH(OH)-CH(OH)-CH(HO)-CH(HO)-CH}_3 \qquad (10)$$

Concentrated hydrochloric acid (40 mL) is saturated with hydrogen chloride at 0°C, and then L-rhamnose monohydrate (25 g, 0.152 mol) and benzenethiol (45 mL) are added, and the mixture is shaken for 24 h at room temperature. The red mixture is poured into ice water

(500 mL), and the precipitate is filtered and washed with water (300 mL) and ether (100 mL) to give the product (26.5 g, 55%). Recrystallization from boiling water (~3 L) gives the title compound as colorless plates (15.5 g, 34%): mp 124–125.5°C, $[\alpha]_D$ +74.8° (c 1.1, ethanol).

### K. 2-Acetamido-2-deoxy-D-glucose Diethyl Dithioacetal [36]

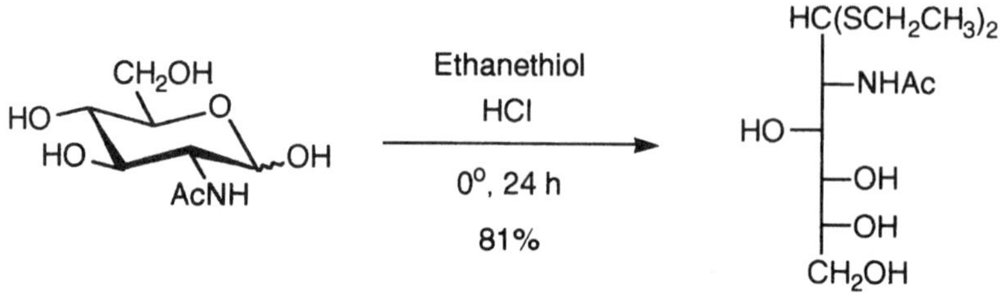

$$\text{N-Acetyl-D-glucosamine} \xrightarrow[\text{0°, 24 h}]{\text{Ethanethiol, HCl}} \text{product} \quad (11)$$

(81%)

N-Acetyl-D-glucosamine (30.g, 0.135 mol) is dissolved in concentrated hydrochloric acid (120 mL), and technical grade ethanethiol (120 mL) is added in a stoppered flask at 0°C. The reaction is stirred at that temperature for 24 h and then neutralized with excess basic lead carbonate (700 g). The mixture is diluted with water (500 mL), and the solids are filtered and washed with water (1 L). After evaporation of the combined filtrate and washings, the resultant residue is crystallized from methanol–ether (36 g, 81%). Further recrystallization from methanol–chloroform–ether gives pure product: mp 129.5–130.5°C, $[\alpha]_D$ −35° (c 4, water).

### L. 2-Amino-2-deoxy-D-galactose Diethyl Dithioacetal [45,53]

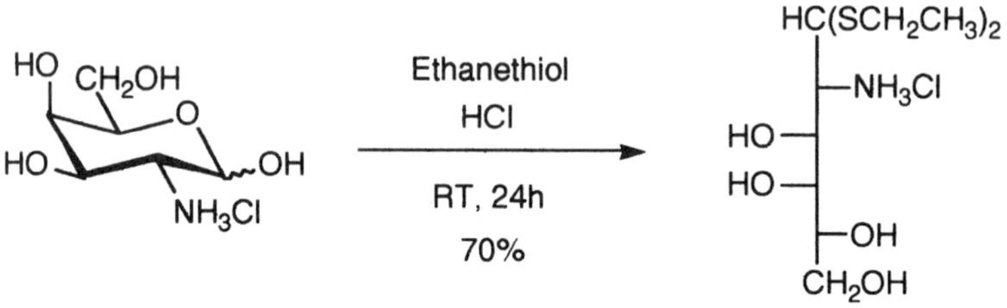

$$\text{2-Amino-2-deoxy-D-galactose·HCl} \xrightarrow[\text{RT, 24h}]{\text{Ethanethiol, HCl}} \text{product} \quad (12)$$

(70%)

This is a modification of the original method described by Wolfrom and Onodera [45]. Concentrated hydrochloric acid (15 mL) is cooled to 0°C and then saturated with hydrogen chloride gas. 2-Amino-2-deoxy-D-galactose hydrochloride (5.0 g, 23.2 mmol) and ethanethiol (16 mL) are added, the flask is sealed, and the mixture stirred vigorously for 24 h at room temperature. After this time, a trap containing 5% sodium hypochlorite (1 L) is attached to the vessel and nitrogen gas is bubbled through the reaction mixture to remove excess ethanethiol. The mixture is then placed onto a column (20 × 5 cm) of Dowex 1-X8 (OH⁻, 200–400 mesh) anion-exchange resin and eluted with 50% aqueous ethanol. The effluent is concentrated, dissolved in 95% ethanol, and refrigerated. Suction filtration affords the product as white needles, and additional product is obtained from the filtrate by repeated crystallization with 95% ethanol (total product 3.8 g).

Further amounts of product are afforded from the mother liquors through separation on a 90 × 2.5-cm column of Dowex 50W-4X (H⁺, 200–400 mesh) cation-exchange resin.

The column is eluted with a linear 400-mL gradient from 0.3 to 1.67 M hydrochloric acid to afford, first, two mobile components (after ~1900 and ~2050 mL of eluent, respectively), followed by the title compound (~3000 mL of eluent) as colorless crystals after evaporation and crystallization (1.2 g). The total amount of product collected from the crystallization and column separation is 5 g (70%): mp 157°C, $[\alpha]_D$ +29° (c 0.86, water).

## M. 2-Amino-2-deoxy-D-glucose Propan-1,3-diyl Dithioacetal Hydrochloride [35]

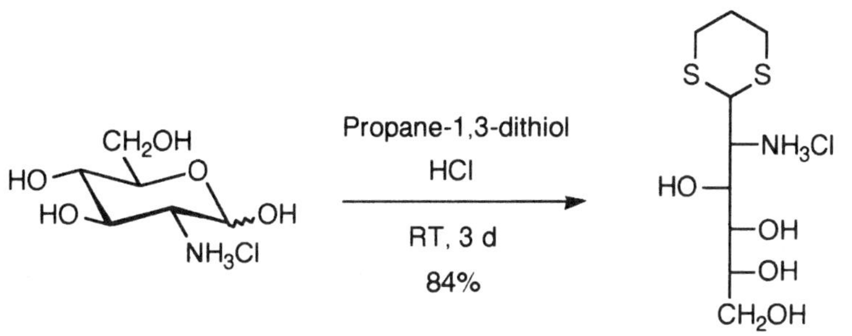

2-Amino-2-deoxy-D-glucose hydrochloride (15 g, 69.6 mmol) is dissolved in fuming hydrochloric acid (concentrated hydrochloric acid saturated with hydrogen chloride at 0°C, 120 mL), propane-1,3-dithiol (18 g, 167 mmol) is added with vigorous stirring at 0°C, and the resultant mixture stirred at room temperature for 3 days. Ethanol is added, the solution neutralized with lead carbonate, and the salts filtered. Evaporation affords a solid, which is recrystallized from ethanol–water–ether (18 g, 84%): mp 204–206°C, $[\alpha]_D$ −8.9° (c 1.45, water).

## N. Sodium D-Glucuronate Diethyl Dithioacetal [45]

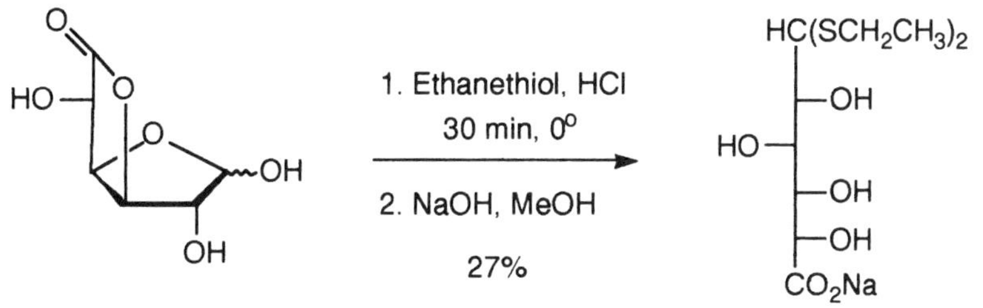

D-Glucurono-3,6-lactone (3.0 g, 17.0 mmol) is suspended in concentrated hydrochloric acid (3 mL) and shaken with ethanethiol (3 mL) for 30 min at 0°C. The reaction mixture is poured into ice water (6 mL), and the solution is then extracted with ethyl acetate (2 × 30 mL). Evaporation of the extracts yields a thick syrup, which is dissolved in methanol, and again evaporated to a syrup, which is then dried in a vacuum dessicator. The syrup is again dissolved in methanol (20 mL), and the solution is neutralized by the slow addition of a concentrated solution of sodium hydroxide in methanol. Scratching the sides of the flask promotes crystallization and the product is collected by filtration and washed with methanol

(1.58 g, 27%): mp 115–118°C, $[\alpha]_D$ $-37°$ ($c$ 4.1, water). This compound is a monohydrate; the anhydrous form is obtained by drying at 80°C at 1–2 mm, mp 115–118°C.

## REFERENCES

1. E. Fischer, Über die Verbindungen der Zuckerarten mit den Mercaptanen, *Ber.* 27:673 (1894).
2. M. L. Wolfrom and J. V. Karabinos, The identification of aldose sugars by their mercaptal acetates, *J. Am. Chem. Soc.* 67:500 (1945).
3. H.-H. Stroh and K. Milde, Über die Reaktion von 2-Desoxy-zuckern mit Arylhydrazinen, *Ber.* 98:941 (1965).
4. H. El Khadem, Chemistry of osazones, *Adv. Carbohydr. Chem.* 20:139 (1965).
5. R. M. Hann and C. S. Hudson, The action of copper sulfate on phenylosazones of the sugars. Phenyl-D-glucosotriazole, *J. Am. Chem. Soc.* 66:735 (1944).
6. H. Jacobi, Über die Oxime einiger Zuckerarten, *Ber.* 24:696 (1891).
7. A. Wohl, Abbau des Traubenzuckers, *Ber.* 26:730 (1893).
8. H. Kiliani, Oxydation der Galactosecarbonsäure, *Ber.* 22:521 (1889) and references therein.
9. E. Fischer, Reduktion von Säuren der Zuckergruppe I, *Ber.* 22:2204 (1889).
10. R. Kuhn and H. Grassner, Mechanismus der Katalytischen Hydrierung von Hydroxynitriles, *Ann.* 612:55 (1958); R. Kuhn and P. Klesse, Darstellung von L-Glucose und L-Mannose, *Ber.* 91:1989 (1958).
11. J. C. Sowden, The nitromethane and 2-nitroethanol syntheses, *Adv. Carbohydr. Chem.* 6:291 (1951).
12. M. L. Wolfrom, The acetate of the free aldehyde form of glucose, *J. Am. Chem. Soc.* 51:2188 (1929); B.-T. Gröbel and D. Seebach, Umpolung of the reactivity of carbonyl compounds through sulfur-containing reagents, *Synthesis* 357 (1977).
13. M. L. Wolfrom and S. W. Waisbrot, The dimethyl acetal of *d*-glucose, *J. Am. Chem. Soc.* 60:854 (1938).
14. D. Horton and T. Usui, Transformations of unsaturated acyclic sugars into enantiomerically pure norbornene derivatives, *Carbohydr. Res.* 216:51 (1991).
    D. Horton and D. Koh, Stereocontrol in Diels–Alder cycloaddition to unsaturated sugars: Reactivities of acyclic seven-carbon *trans* dienophiles derived from aldopentoses, *Carbohydr. Res.* 250:249 (1993).
15. D. Horton, D. Koh, Y. Takagi, and T. Usui, Diels–Alder cycloadditions to unsaturated sugars, *Cycloadditions in Carbohydrate Chemistry, ACS Symp. Ser.* 494:66 (1992).
16. R. Barker and D. L. McDonald, The disulfones derived from the dithioacetals of certain hexoses, *J. Am. Chem. Soc.* 82:2297 (1960); D. L. McDonald and H. O. L. Fischer, The degradation of ketoses by the disulfone method, *J. Am. Chem Soc.* 77:4348 (1955).
17. D. L. McDonald and H. O. L. Fischer, The degradation of sugars by means of their disulfones, *J. Am. Chem. Soc.* 74:2087 (1952); D. L. McDonald and H. O. L. Fischer, Degradation of aldoses by means of their sulfones, *Biochim. Biophys. Acta* 12:203 (1953).
18. D. Horton and R. A. Markovs, Carbon-chain extension through C-1 of 2-deoxyaldose dithioacetal derivatives: A route to 1,3-dideoxy-2-ketoses, *Carbohydr. Res.* 78:295 (1980).
19. P. Di Cesare and D. Horton, Synthetic routes to higher-carbon sugars. 1-C-formylation of a 2-deoxyaldose via the anion of its diethyl dithioacetal, *Carbohydr. Res.* 107:147 (1982).
20. D. Horton and J. D. Wander, Diphenyl dithioacetals of D-ribose, D-xylose, and D- and L-arabinose. Conformational studies and formation of a ketene diphenyl dithioacetal, *Carbohydr. Res.* 13:33 (1969).
21. M. Y. H. Wong and G. R. Gray, 2-Deoxypentoses. Stereoselective reduction of ketene dithioacetals, *J. Am. Chem. Soc.* 100:3548 (1978).
22. B. Berrang, D. Horton, and J. D. Wander, Formation and reactions of ketene diphenyl dithioacetals derived from aldoses, *J. Org. Chem.* 38:187 (1973).

23. M. L. Wolfrom and J. V. Karabinos, Carbonyl reduction by thioacetal hydrogenolysis, *J. Am. Chem. Soc. 66*:909 (1944).
24. F. Weygand, H. Ziemann, and H. J. Bestmann, Eine neue Methode zur Darstellung von Acetobromzuckern, *Ber. 91*:2534 (1958).
25. D. Horton, Thio sugars: Stereochemical questions and synthesis of antimetabolites, *Pure Appl. Chem. 42*:301 (1975).
26. K. Tadano, H. Maeda, M. Hoshino, Y. Iimura, and T. Suami, A novel transformation of four aldoses to some optically pure pseudohexopyranoses and a pseudopentofuranose, carbocyclic analogs of hexopyranoses and pentofuranoses, *J. Org. Chem. 52*:1946 (1987).
27. M. L. Wolfrom, M. R. Newlin, and E. E. Stahly, *Aldehydo-d*-xylose tetra-acetate and the mercaptals of xylose and maltose, *J. Am Chem. Soc. 53*:4379 (1931).
28. W. G. Overend, M. Stacey, and J. Staněk, Deoxy-sugars. Part VII. A study of the reactions of some derivatives of 2-deoxy-D-glucose, *J. Chem. Soc.* 2841 (1949).
29. H. W. Arnold and W. L. Evans, The mechanism of carbohydrate oxidation. Part XXII. The preparation and reactions of glyceraldehyde diethyl mercaptal, *J. Am. Chem. Soc. 58*:1950 (1936); R. M. Hann, A. T. Merrill, and C. S. Hudson, Proof of the configurations of D-gluco-L-gala, D-gluco-L-talo, and D-gala-L-gluco-octoses, *J. Am. Chem. Soc. 66*:1912 (1944)
30. E. Campaigne and J. R. Leal, α,β-Unsaturated sulfides. The reaction of thiophenol with certain ketones, *J. Am. Chem. Soc. 76*:1272 (1954).
31. E. Pacsu and E. J. Wilson, The formation of α-ethylthioglucopyranoside from glucose ethylmercaptal, *J. Am. Chem. Soc. 61*:1930 (1939).
32. M. L. Wolfrom, D. Horton, and H. G. Garg, Thioglycosides of 3-amino-3-deoxy-D-mannose, *J. Org. Chem. 28*:1569 (1963).
33. P. A. Levene and G. M. Meyer, Pentamethyl-*d*-mannose and pentamethyl-*d*-galactose and their dimethyl acetals, *J. Biol. Chem. 74*:695 (1927).
34. J. Fried and D. E. Waltz, Ethyl thioglycosides of D-mannose and D-galactose and a new synthesis of styracitol, *J. Am. Chem. Soc. 71*:140 (1949).
35. P. Angibeaud, C. Bosso, J. Defaye, D. Horton, C. Cohen-Addad, and M. Thomas, Vicinal participation in the nitrous acid deamination of chiral 1-thio-2-aminoalkanepolyols: the deamination of 2-amino-2-deoxy-D-glucose ethylene and propan-1,3-diyl dithioacetals, *J. Chem. Soc. Perkins Trans. 1*:1583 (1979).
36. M. L. Wolfrom and A. Thompson, 2-Amino-2-deoxy-α-D-xylose hydrochloride, *Methods Carbohydr. Chem. 1*:209 (1962); M. L. Wolfrom and K. Anno, Acetylated thioacetals of D-glucosamine, *J. Am. Chem. Soc. 74*:6150 (1952).
37. J. Jary, Z. Kefurtová, and J. Kovár, Amino sugars. XX. Synthesis of 3-amino- and 6-aminodeoxyhexoses of the *gluco* and *allo* configuration, *Collect. Czech. Chem. Commun. 34*:1452 (1969); M. von Saltza, J. D. Dutcher, J. Reid, and O. Wintersteiner, Nystatin. IV. The stereochemistry of mycosamine, *J. Org. Chem. 28*:999 (1963).
38. I. W. Hughes, W. G. Overend, and M. Stacey, Deoxy-sugars. Part VIII. The constitution of αβ-methyl-2-deoxy-D-glucofuranoside, *J. Chem. Soc.* 2846 (1949).
39. D. Horton and J. D. Wander, Conformation of acyclic derivatives of sugars, *Carbohydr. Res. 15*:271 (1970).
40. M. L. Wolfrom and A. Thompson, Keto-fructose pentaacetate, *J. Am. Chem. Soc. 56*:880 (1934).
41. H. Zinner and U. Schneider, Mercaptale and Tritylverbindungen der L-sorbose, *Ber. 96*:2159 (1963).
42. A. Schönberg and O. Schütz, Organic sulfur compounds. VI. Relation between constitution and heat stability of organic compounds. Thermal decomposition of mercaptols, *Ann. 454*:47 (1927) and references cited therein.
43. F. Weygand, E. Klieger, and H. J. Bestmann, D-Glucosone-*aldehydo*-mercaptale, *Ber. 90*:645 (1957).
44. S. Bayne, Osone dithioacetals, *Proc. Chem. Soc.* 170 (1958).
45. M. L. Wolfrom and K. Onodera, Dithioacetals of D-glucuronic acid and 2-amino-2-deoxy-D-galactose, *J. Am. Chem. Soc. 79*:4737 (1957).

46. R. Kuhn and R. Brossmer, Die Konfiguration der Lactaminsäure, *Angew. Chem. 69*:534 (1957).
47. D. Horton and J. D. Wander, Dithioacetals of sugars, *Adv. Carbohydr. Chem. Biochem. 32*:15 (1976).
48. S. J. Eitelman, Ph.D. Dissertation, The Ohio State University (1975).
49. M. L. Wolfrom and A. Thompson, Acyclic monosaccharide derivatives, *Methods Carbohydr. Chem. 2*:427 (1963).
50. M. L. Wolfrom, The fifth pentaacetate of galactose, its alcoholate and aldehydrol, *J. Am. Chem. Soc. 52*:2464 (1930).
51. H. Zinner, H. Brandner, and G. Rembarz, Derivate der Zucker-mercaptale, II: Acyl- und Tritylverbindungen der D-Lyxose- und der D-Arabinose-mercaptale, *Ber. 89*:800 (1956).
52. H. Zinner, Notiz über die Isolierung der Wasserlöslichen Aldosemercaptale mit hilfe von Ionenaustauschern, *Ber. 86*:495 (1953).
53. W. Dills, Jr. and T. R. Covey, A convenient synthesis of 1-deoxy-D-tagatose, *Carbohydr. Res. 89*:338 (1981).

# 3

# Regioselective Cleavage of *O*-Benzylidene Acetals to Benzyl Ethers

Per J. Garegg
*Stockholm University, Stockholm, Sweden*

|  |  |  |
|---|---|---|
| I. | Introduction | 54 |
|  | A. Protecting groups: general considerations | 54 |
|  | B. Regioselective openings of cyclic derivatives | 55 |
|  | C. Regioselective openings of benzylidene acetals | 56 |
| II. | Regioselective Reductive Cleavage of *O*-Benzylidene Acetals to Benzyl Ethers | 57 |
| III. | Mechanistic Considerations | 61 |
| IV. | Experimental Procedures | 62 |
|  | A. Benzyl 2,3,4-tri-*O*-benzyl-β-D-glucopyranoside | 62 |
|  | B. Methyl 3-*O*-benzyl-4,6-*O*-β-D-mannopyranoside | 63 |
|  | C. Methyl 2-*O*-benzyl-4,6-*O*-benzylidene-β-D-mannopyranoside | 63 |
|  | D. Methyl 2,3,6-tri-*O*-benzyl-β-D-glucopyranoside | 63 |
|  | E. Methyl 2,3-di-*O*-benzoyl-4-*O*-benzyl-α-D-glucopyranoside | 64 |
|  | F. Methyl 2,3-di-*O*-benzoyl-6-*O*-benzyl-α-D-glucopyranoside | 64 |
|  | G. Methyl *O*-(2,3,4,6-tetra-*O*-benzoyl-β-D-glucopyranosyl-(1→3)-2-*O*-benzoyl-1-thio-β-D-glucopyranoside | 64 |
|  | H. Methyl 2,3-di-*O*-benzyl-4-*O*-(methoxybenzyl)-α-D-glucopyranoside | 65 |
|  | I. Methyl 2,3-di-*O*-benzyl-6-*O*-(4-methoxybenzyl)-α-D-glucopyranoside | 65 |
|  | References | 65 |

## I. INTRODUCTION

### A. Protecting Groups: General Considerations

From the viewpoint of a synthesis chemist, carbohydrates would appear to be severely overfunctionalized. Thus, in a hexopyranose, one has to contend with five hydroxyl groups distributed over six carbon atoms. Furthermore, four of the hydroxyl groups are chiral. Obviously, to carry out synthetic manipulations on such molecules one has to learn to protect hydroxyl groups (or amino groups in the case of aminodeoxy sugars) to leave free only those destined for reactions. Therefore, a rich repertoire of protecting group manipulations for this purpose has evolved [1,2]. Table 1 shows some of the more common ones in current use.

Because these various groups are attached and removed under different conditions, many avenues to partially protected carbohydrate intermediates are possible. The subject has been extensively reviewed [3–6]. Extensive schemes have been worked out using allyl ethers as O-protecting groups (Table 2) [7]. The variety in stability to acid, base, tert-butoxide anion, and Wilkinson's catalyst opens up a variety of protecting groups strategies. Common N-protective groups for amino sugars include acyl, arylidene [8], carbobenzoxyl, phthaloyl [9]; also the use of temporarily masking of an amino function as an azido group.

The following outlines a few unequivocal general routes to hexopyranoses and -pyranosides, O-protected at all positions, but one: A free OH at C-2 can be obtained by converting an acetobromohexose into the corresponding 1,2-ortho-ester, then converting this into the 3,4,6-tri-O-benzyl derivative, which is then converted into a 2-O-acetyl-3,4,6-tri-O-benzylglycosyl halide, and then made into a suitable glycoside which, in turn, is deacetylated to liberate the 2-OH [10,11]. In the galactose and glucose series 1,2:5,6-di-O-isopropylidene acetals give immediate access to a free 3-OH group. A free 4-OH is traditionally obtained from a pyranosidic 4,6-O-benzylidene acetal by first blocking the 2,3-positions, then removing the 4,6-acetal, substituting the 6-position (primary OH) with a sterically demanding substituent, such as a trityl group [12] or a diphenyl-tert-butylsilyl group [13]. A free 6-OH is generally obtained from a pyranoside by first substituting at the 6-position with one of these two sterically demanding substitutents, then blocking the 2,3,4-positions and, finally, removing the trityl or silyl group to obtain a free 6-OH.

**Table 1**  Some Commonly Used O-Protecting Groups

| Group | Attachment | Removal |
|---|---|---|
| OAc | $Ac_2O$ or AcCl/pyr | $NaOCH_3/CH_3OH$ |
| OBz | BzCl/pyr | $NaOCH_3/CH_3OH$ |
| OBn | BnBr/DMF/NaH | $H_2$/Pd |
| $OCH_2CH=CH_2$ | $ClCH_2CH=CH_2$/DMF/NaH | $(Ph_3P)RhCl$, then mild acid |
| $OCPh_3$ | $ClCPh_3$/pyridine | Mild acid |
| $OSiPh_2Bu^t$ | $ClSiPh_2Bu^t$ | $Bu_4N^+F^-$ |
| -O\\C(CH_3)_2/-O | $(CH_3)_2(OCH_3)_2$/TsOH | Mild acid |
| -O\\CH/-O/Ph | PhCHO/TsOH | Mild acid or $H_2$/Pd |

**Table 2** Allyl Ethers as O-Protecting Groups

| Structure | Stability in acid and base | Product with KOtertBu-DMSO | Product with (Ph$_3$P)$_3$RhCl |
|---|---|---|---|
| ROCH$_2$CH=CH$_2$ | Acid- and base-stable | ROCH=CHCH$_3$ (fast) | ROCH=CHCH$_3$ (fast) |
| ROCH=CHCH$_3$ | Base-stable only | No reaction | No reaction |
| ROCH$_2$CH=CHCH$_3$ | Acid- and base-stable | ROH (very fast) | ROCH=CHCH$_2$CH$_3$ |
| ROCH$_2$C(CH$_3$)$_2$=CH$_2$ | Acid- and base-stable | ROCH=C(CH$_3$)$_2$ (slow) | ROCH=C(CH$_3$)$_2$ (fast) |
| ROCH$_2$CH=C(CH$_3$)$_2$ | Acid- and base-stable | ROH (fast) | No reaction |

Because nuclear magnetic resonance (NMR) is particularly advantageous for carbohydrates, protecting group routes may often be shortened by the use of partial substitution reactions [4–6], followed by chromatographic separations and characterization by means of NMR.

## B. Regioselective Openings of Cyclic Derivatives

Treatment of a cyclic 2,3- or 3,4-*ortho*-ester in a pyranoside under mild acidic conditions will cause regioselective cleavage so that a free *equatorial* OH is produced adjacent to an *axial O*-acyl group [14] (Scheme 1). The reasons for this have engendered considerable theoretical treatment based on considerations of stereoelectronic effects [15].

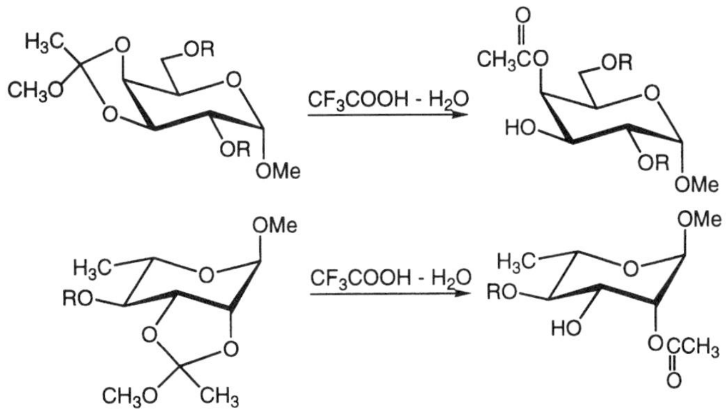

**Scheme 1**

By contradistinction, the corresponding stannylidene acetals, after acylation or alkylation will yield products with a free *axial* OH adjacent to an acylated or alkylated *equatorial O*-substituent (Scheme 2) [16,17]. These two routes are obviously of considerable use in protective group schemes.

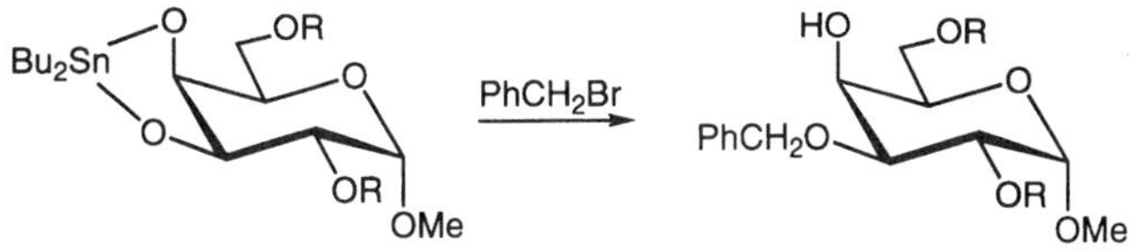

**Scheme 2**

## C. Regioselective Openings of Benzylidene Acetals

A most useful route leading regioselectively to 6-bromo-6-deoxy derivatives involves the treatment of pyranosidic 4,6-*O*-benzylidene acetals with *N*-bromosuccinimide [18]. An acyl group is produced in the 4-position (Scheme 3) [19].

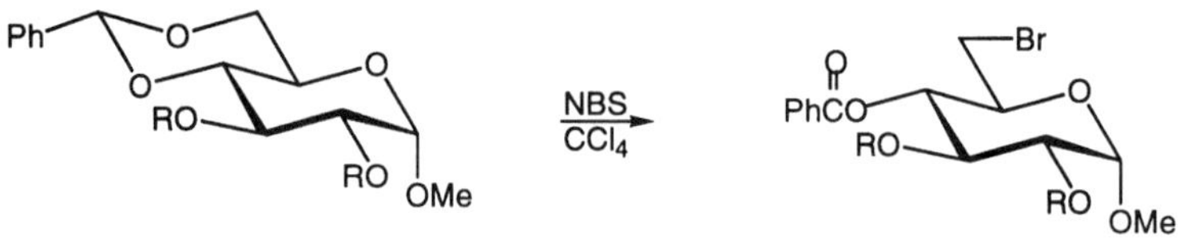

R = Alkyl, benzyl or H

**Scheme 3**

Benzylidene *dioxolane* derivatives, under the same conditions except for the presence of water, give rise to monohydroxy benzoates with an *equatorial* hydroxyl group adjacent to an *axial O*-benzoyl group (Scheme 4) [20].

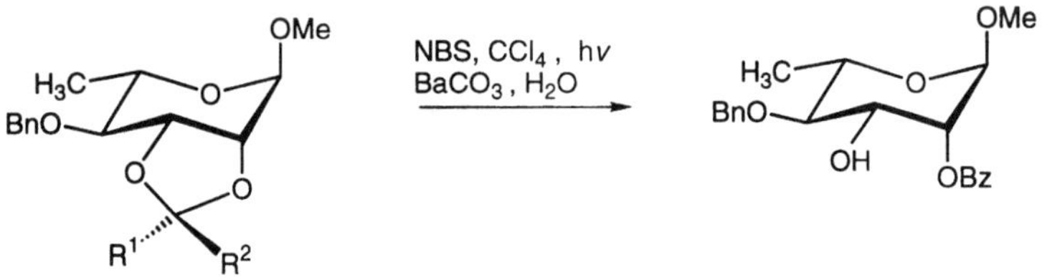

$R^1 = C_6H_5, R^2 = H$
$R^1 = H, R^2 = C_6H_5$

**Scheme 4**

Treatment with butyllithium, with otherwise suitably protected 2,3-*O*-benzylidene acetals of hexopyranosides, gives deoxyulosides (Scheme 5) [21].

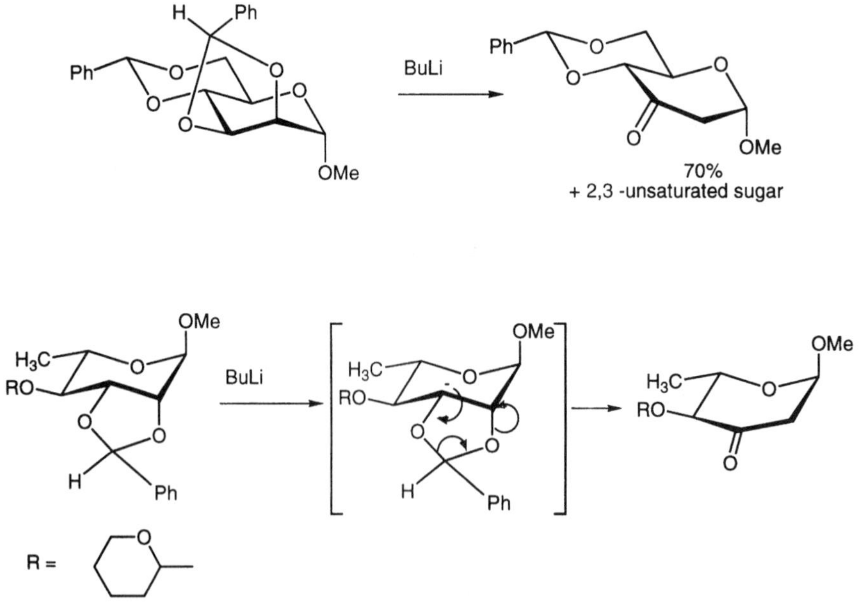

70%
+ 2,3-unsaturated sugar

**Scheme 5**

The presence in the 4- and 6-positions of groups such as hydroxyl and O-benzyl groups that are ionized by butyllithium must be avoided because they give rise to other products [22,23].

## II. REGIOSELECTIVE REDUCTIVE CLEAVAGE OF O-BENZYLIDENE ACETALS TO BENZYL ETHERS

Reductive opening of carbohydrate benzylidene acetals with LiAlH$_4$/AlCl$_3$ was first described by Bhattacharjee and Gorin [24]. It was subsequently extensively investigated by Lipták and co-workers [18, 25]. For *dioxane* rings (i.e., pyranosidic 4,6-acetals with an equatorial phenyl group), the predominant regioselectivity was directed to a free 6-OH, with an O-benzyl group at C-4, especially when a bulky group was present at C-3. For *dioxolane* rings the regioselectivity depended on the stereochemistry at the benzylidene acetalic carbon atom (Scheme **6**).

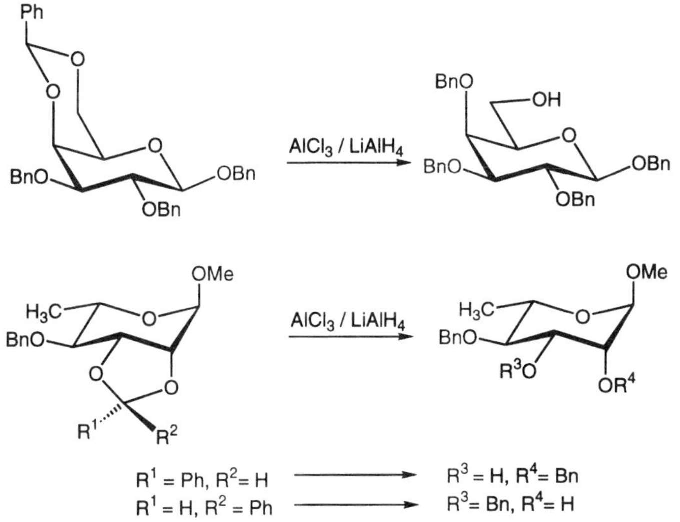

**Scheme 6**

A report by Horne and Jordan [26], describing treatment with sodium borohydride–HCl in methanol of the phenylethylidene acetal of ethylene glycol (Scheme **7**), led us to examine this reaction in the carbohydrate series [27]. Surprisingly, for pyranosidic 4,6-

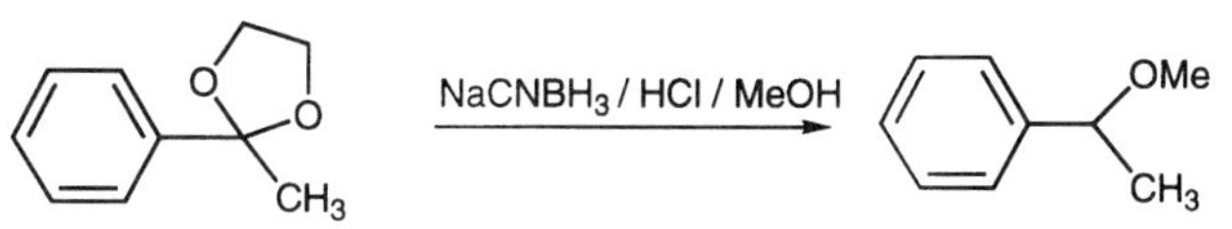

**Scheme 7**

benzylidene acetals, a regioselectivity complementary to that in the Lipták–Nánási experiments was observed, in that the main product invariably was the 6-O-benzyl ether with the 4-OH free. Furthermore, by contradistinction to the LiAlH$_4$–AlCl$_3$ reaction, the NaCNBH$_3$/HCl treatment allowed the presence of acyl and acetamido groups at the other positions in the pyranoside ring. Several examples are given in Scheme **8** [27].

One advantageous result of these findings is that, after the reductive opening, one can

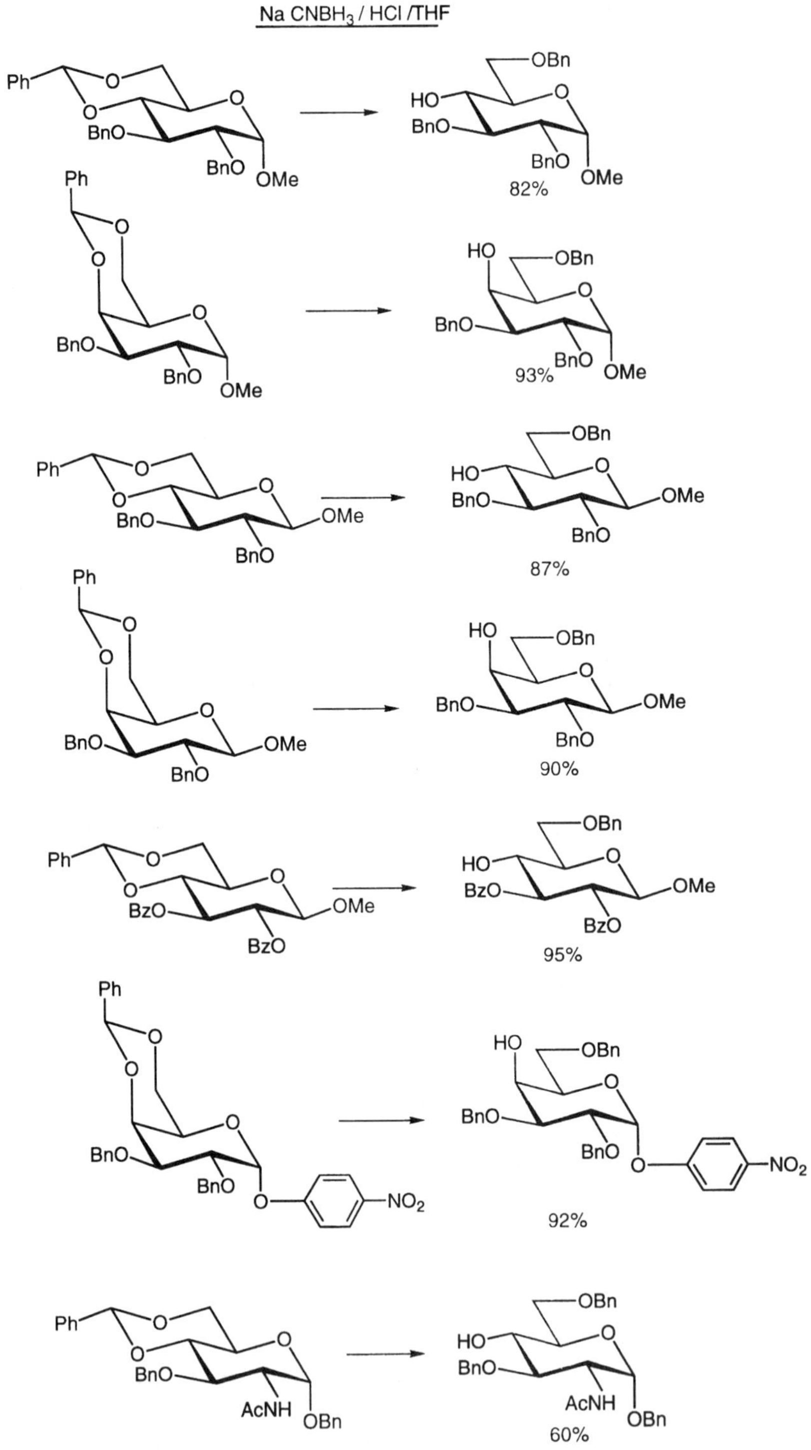
Na CNBH3 / HCl /THF
Ph
BnO
BnO
OMe
HO
OBn
82%
93%
87%
90%
BzO
95%
NO2
92%
AcNH
60%

**Scheme 8**

# Cleavage of O-Benzylidene Acetals

have a free 4-OH, a benzyl group in the 6-position, and acyl groups in the 2,3-positions. This opens up new possibilities for further protecting group manipulation. More simply, compared with the foregoing strategy for obtaining a single free 4-OH in a pyranosidic ring, the present method represents a substantial advance.

That the reductive cleavage is compatible with the presence of O-acyl groups raises the question of whether the regioselectivity may be changed back to that of Lipták–Nánási. Indeed, when the solvent is changed from tetrahydrofuran to toluene and the reduction system to $Me_3NBH_3$–$AlCl_3$, a free 6-OH and O-benzyl at C-6 is produced (Scheme **9**) [27].

The regioselectivity of the reductive openings of five-membered (oxolane-type) benzylidene acetals has been examined. The same regioselectivity as that observed for the $LiAlH_4$–$AlCl_3$ openings was observed (Scheme **10**) [27].

The reductive openings of benzylidene acetals, as outlined in the foregoing, have found extensive use in oligosaccharide synthesis.

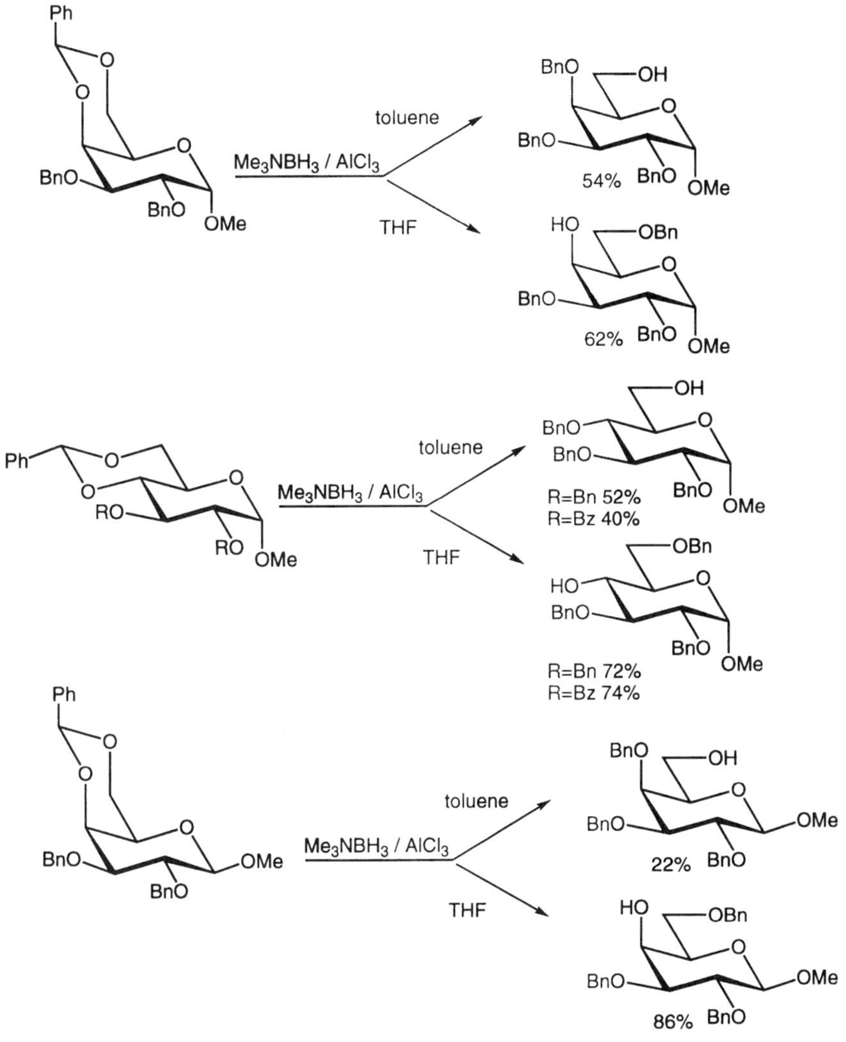

**Scheme 9**

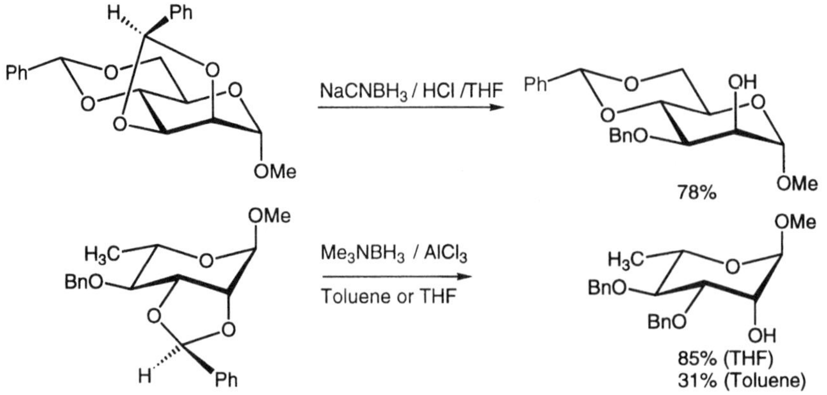

**Scheme 10**

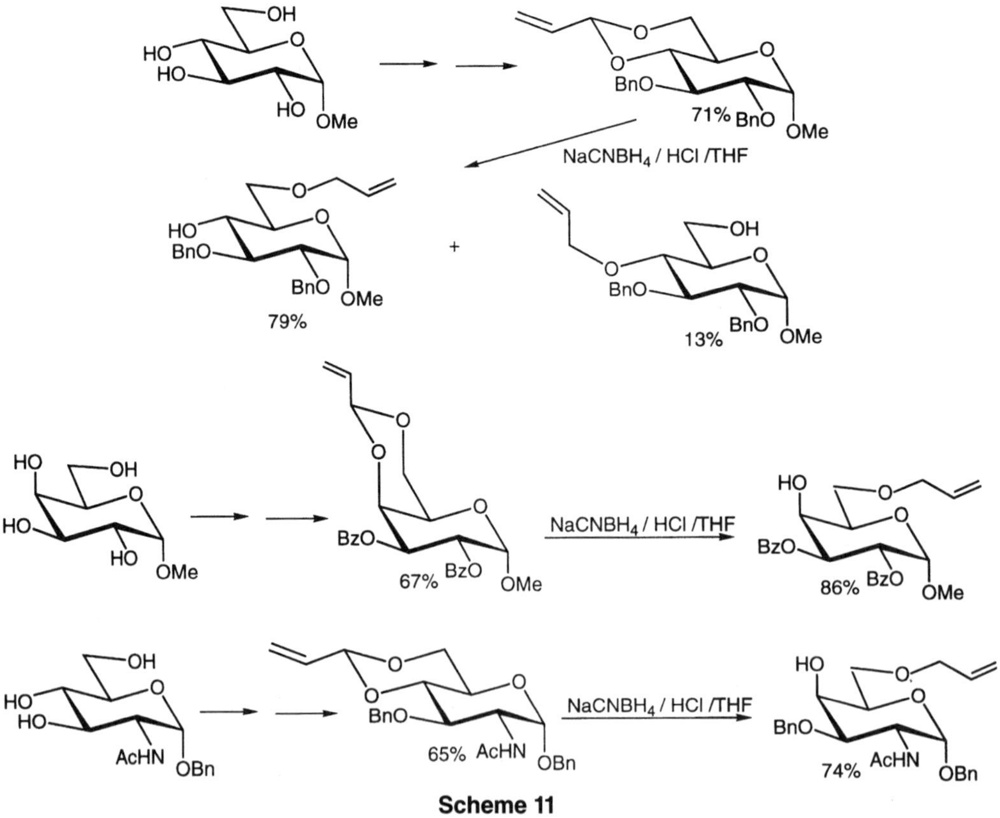

**Scheme 11**

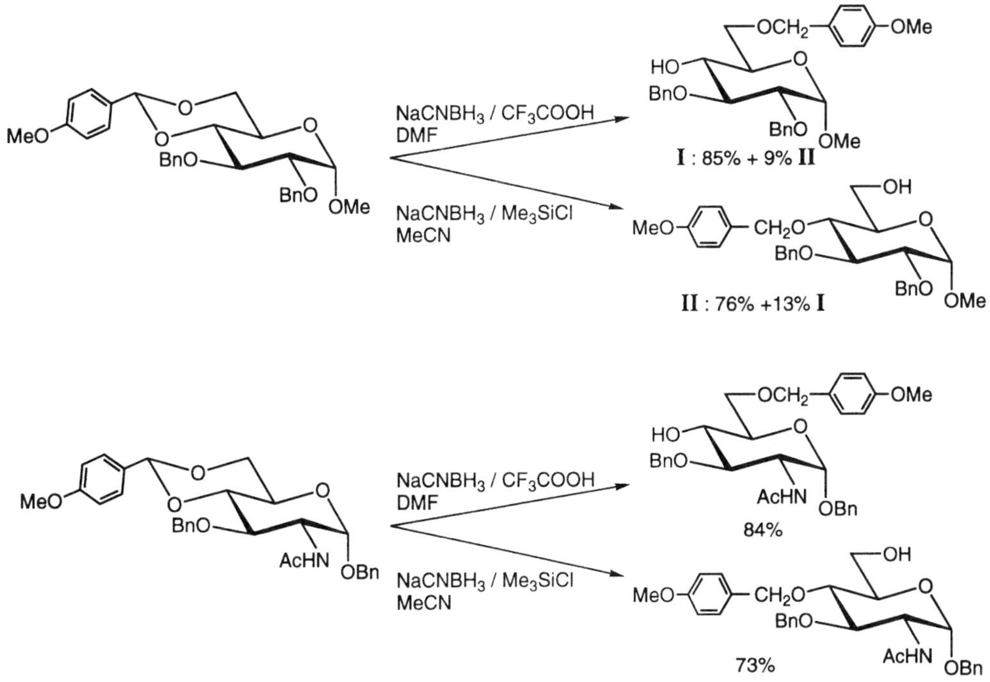

**Scheme 12**

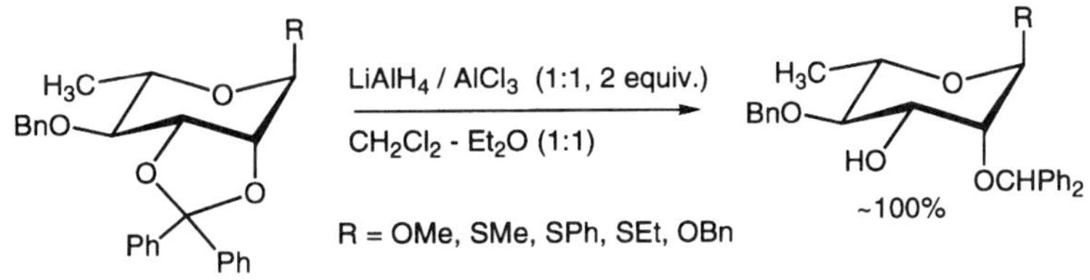

**Scheme 13**

Reductive cleavage, using the NaCNBH$_3$–HCl–THF system to produce 2′-propenylidene acetals, produced the same regioselectivity as that found in benzylidene acetals [27] (Scheme 11), and also for the *p*-methoxybenzylidene acetals (Scheme 12) [27]. These observations open the way to further strategies for protecting group manipulations.

In extensions of the original work, Lipták and coworkers [28] have shown that reductive opening of dioxolane diphenylmethylene acetals by LiAlH$_4$–AlCl$_3$ regioselectively produces the *axial* diphenylmethyl ethers with an adjacent *equatorial* hydroxyl group (Scheme 13).

## III. MECHANISTIC CONSIDERATIONS

Because detailed physicochemical examinations, such as, for example, rate studies, isotope effects, effects of substituents in the phenyl ring, and solvent effects, have not been carried out, and the only available information is product analysis, mechanistic interpretation must necessarily be highly speculative. However, there would seem to be two major mechanistic possibilities. These are shown in Scheme 14.

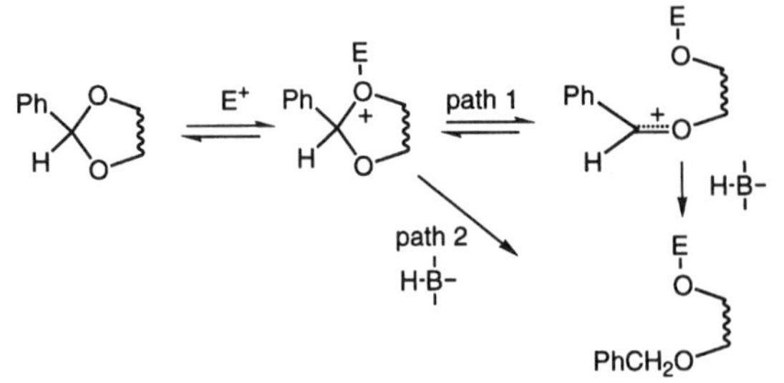

**Scheme 14**

In comparisons between the LiAlH$_4$–AlCl$_3$ system and the NaCNBH$_3$–HCl one, there is an obvious difference in the steric bulk of AlCl$_3$ and that of a proton. This would direct the electrophilic attack of AlCl$_3$ to O-6 (primary position), with ensuing reductive cleavage producing the 4-benzyl ether. This explanation presupposes that the first step is rate-determining.

Further difficulties arise in explaining the solvent dependence in opening the rings using the Me$_3$NBH$_3$–AlCl$_3$ system. The expected stronger solvation of AlCl$_3$ in tetrahydrofuran, versus toluene or dichloromethane, would indicate preferential electrophilic attack at O-6 in tetrahydrofuran, and the production of the 4-benzyl ether, contrary to what is observed.

If the reduction step is the rate-determining one, path **1** (see Scheme **14**), which appears to be the plausible one, fails to explain the different selectivity for the *exo*- and *endo*-oxolane acetals, because these would then proceed through the same oxo-carbenium ion intermediate.

Clearly, only with thorough mechanistic studies can one clarify these points.

## IV.  EXPERIMENTAL PROCEDURES

### A.  Benzyl 2,3,4-Tri-*O*-benzyl-β-D-glucopyranoside [29]

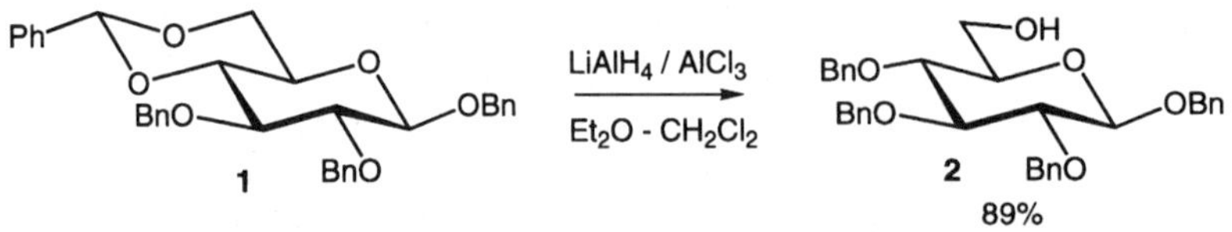

To a solution of **1** [30] (3.00 g, 5.57 mmol) in 1:1 dichloromethane/ether (60 mL), LiAlH$_4$ (1 g, 26.3 mmol) was added in three portions with stirring. The mixture was slowly heated to reflux temperature. A solution of AlCl$_3$ (3 g, 22.5 mmol) in ether (30 mL) was added to the hot solution during 30 min. After boiling under reflux for 1.5–2 h, thin-layer chromatography (TLC) showed the absence of starting material. The mixture was cooled, excess LiAlH$_4$ was decomposed with ethyl acetate (10 mL), and Al(OH)$_3$ was precipitated by adding water (15 mL). After addition of and extraction with ether (50 mL), the organic layer was separated, and the residue was washed with ether. The combined organic phase was washed with water (3 × 20 mL), dried, and concentrated to give **2** (2.68 g, 89%), mp 104–

105°C, $[\alpha]_D$ −11.5° (c 0.5, $CHCl_3$); reported values [31] mp 105–106°C, $[\alpha]_D$ −9.2° ($CHCl_3$).

### B. Methyl 3-O-benzyl-4,6-O-benzylidene-β-D-mannopyranoside [25]

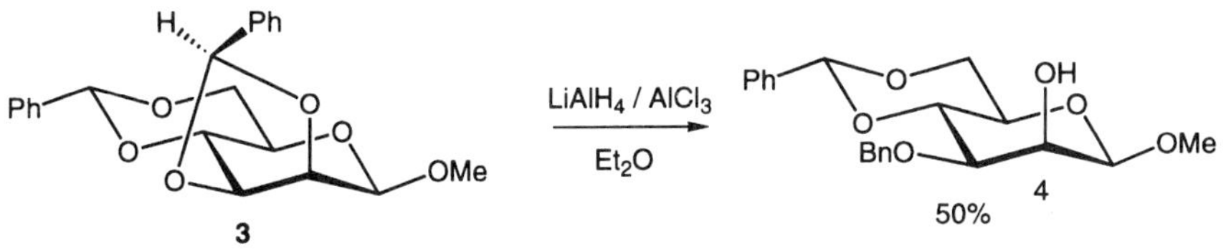

$$\text{3} \xrightarrow{\text{LiAlH}_4 \text{ / AlCl}_3, \text{Et}_2\text{O}} \text{4} \quad (50\%) \quad (2)$$

To a solution of **3** [25] (500 mg, 1.35 mmol) in 1:1 dichloromethane/ether (40 mL), $LiAlH_4$ (52 mg, 1.37 mmol) and a solution of $AlCl_3$ (200 mg, 1.50 mmol) in ether (40 mL) were added. The mixture was stirred at room temperature for 90 min. The product was workedup as described for Eq. (1) 2 to give a crystalline mixture (490 mg, 97%) of **4** and a minor component in a ratio of 95:5 (TLC, silica gel 19:1 dichloromethane/acetone). Three recrystallizations from ethanol (5 mL) yielded **4** (250 mg, 50%), mp 119–120°C, $[\alpha]_D$ −32° (c 0.93, $CHCl_3$); reported values [32] mp 119.5–120°C, $[\alpha]_D$ −32° ($CHCl_3$).

### C. Methyl 2-O-benzyl-4,6-O-benzylidene-β-D-mannopyranoside [25]

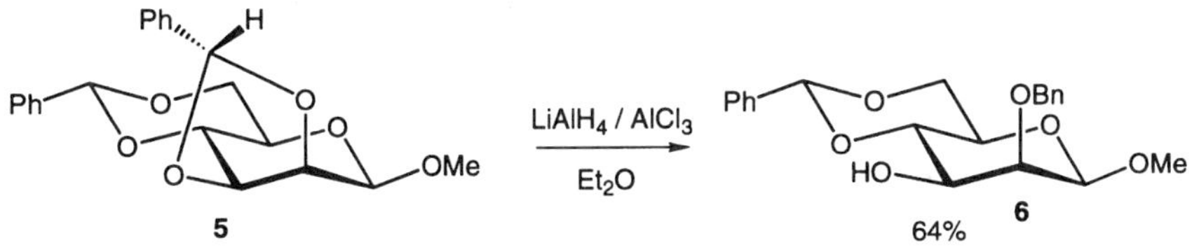

$$\text{5} \xrightarrow{\text{LiAlH}_4 \text{ / AlCl}_3, \text{Et}_2\text{O}} \text{6} \quad (64\%) \quad (3)$$

Compound **5** [25] (500 mg, 1.35 mmol) was hydrogenolyzed, as described for **3**. After 6 min no starting material remained (TLC). A major and two minor products had formed. Crystallization from ethanol (10 mL) gave pure **6** (320 mg, 64%), mp 150–152°C, $[\alpha]_D$ −128° (c 0.93, $CHCl_3$); reported values [32] mp 153.5–154°C, $[\alpha]_D$ −131° ($CHCl_3$).

### D. Methyl 2,3,6-tri-O-benzyl-β-D-glucopyranoside [33]

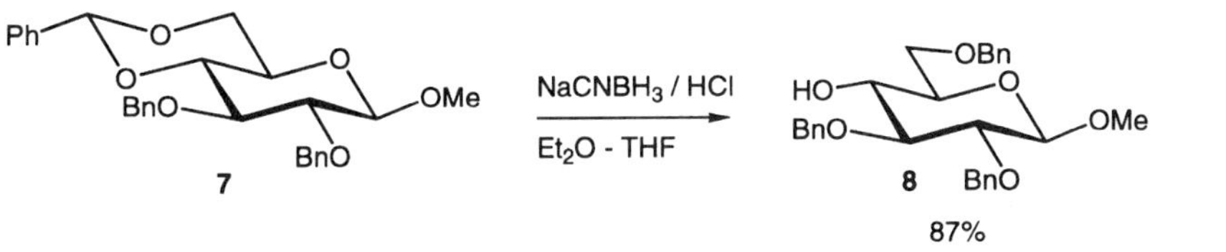

$$\text{7} \xrightarrow{\text{NaCNBH}_3 \text{ / HCl}, \text{Et}_2\text{O - THF}} \text{8} \quad (87\%) \quad (4)$$

Hydrogen chloride in diethyl ether at room temperature was added to a mixture of **7** [34] (1.00 g, 2.16 mmol) and sodium cyanoborohydride (1.70 g, 27.0 mmol) in tetrahydrofuran (30 mL, distilled from $LiAlH_4$) containing 3-A molecular sieves, until the evolution of gas ceased. Use of TLC [silica gel, light petroleum (bp 40–60°)/ethyl acetate 5:1] after 5 min, indicated complete reaction. The mixture was diluted with $CH_2Cl_2$ (50 mL) and water, filtered, and the solution was extracted with water and then with saturated aqueous

NaHCO$_3$. The organic layer was dried (Na$_2$SO$_4$) and concentrated. The resulting syrup was applied to a column of silica gel that was eluted with light petroleum (mp 40–60°)/ethyl acetate 2:1 to yield **8**, which was crystallized from ethyl ether/light petroleum (0.87 g, 87%), mp 64–65°C, [α]$_D$ −17° (c 1.0, CHCl$_3$).

### E. Methyl 2,3-di-O-benzoyl-4-O-benzyl-α-D-glucopyranoside [35]

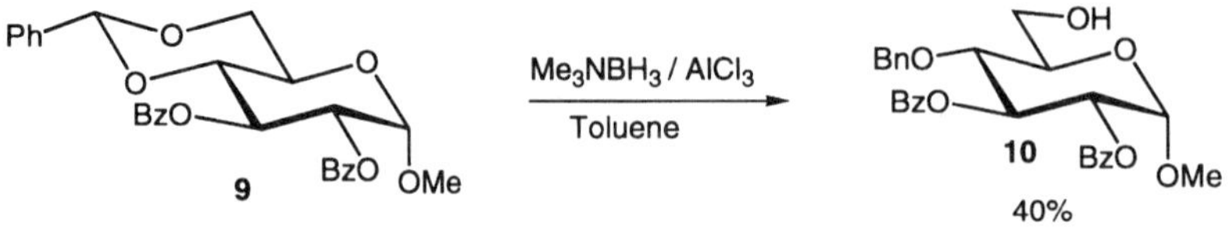

Aluminum chloride (AlCl$_3$; 817 mg, 6.12 mmol) was added to a mixture of **9** [36] (500 mg, 1.02 mmol), Me$_3$NBH$_3$ (446 mg, 6.12 mmol) in toluene (20 ml) containing 4-A molecular sieves, which had been stirred at room temperature for 30 min. When TLC (silica gel, toluene/ethyl acetate 2:1) indicated complete reaction (several hours), the mixture was filtered, and concentrated. The residue was coconcentrated three times with methanol and **10** was crystallized from ethyl ether/light petroleum (200 mg, 40%), mp 117–118°C, [α]$_D$ +132° (c 1.49, CHCl$_3$) after silica gel chromatography (toluene/ethyl acetate, 4:1).

### F. Methyl 2,3-di-O-benzoyl-6-O-benzyl-α-D-glucopyranoside [35]

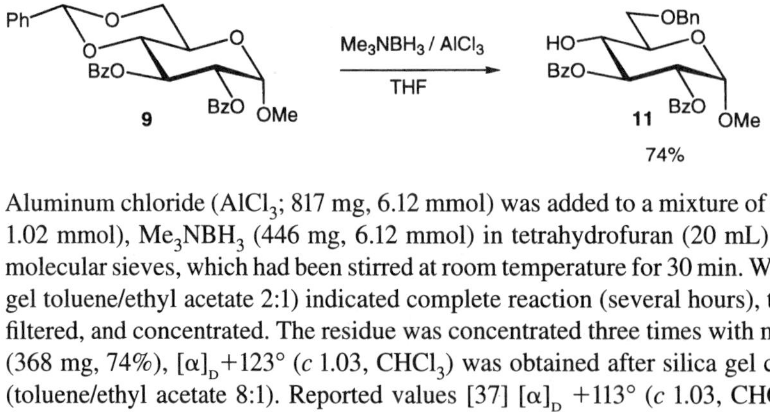

Aluminum chloride (AlCl$_3$; 817 mg, 6.12 mmol) was added to a mixture of **9** [36] (500 mg, 1.02 mmol), Me$_3$NBH$_3$ (446 mg, 6.12 mmol) in tetrahydrofuran (20 mL) containing 4 A molecular sieves, which had been stirred at room temperature for 30 min. When TLC (silica gel toluene/ethyl acetate 2:1) indicated complete reaction (several hours), the mixture was filtered, and concentrated. The residue was concentrated three times with methanol, and **11** (368 mg, 74%), [α]$_D$ +123° (c 1.03, CHCl$_3$) was obtained after silica gel chromatography (toluene/ethyl acetate 8:1). Reported values [37] [α]$_D$ +113° (c 1.03, CHCl$_3$).

### G. Methyl O-(2,3,4,6-tetra-O-benzoyl-β-D-glucopyranosyl-(1→3)-2-O-benzoyl-1-thio-β-D-glucopyranoside [38]

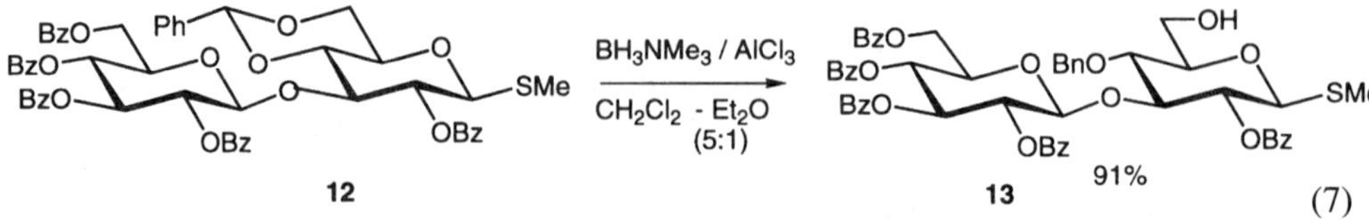

A solution of AlCl$_3$ (2.67 g, 20.0 mmol) in ether (30 mL) was added to a stirred mixture of **12** [38] (4.90 g, 5.0 mmol), Me$_3$NBH$_3$ (14.59 g, 200 mmol), and 4-A molecular sieves in

$CH_2Cl_2$ (100 mL) and ether (20 mL) during 15 min at 0°C. After 30 min, the mixture was filtered through Celite and the solids were washed with $CH_2Cl_2$ (100 mL). The combined filtrate and washings was stirred with M $H_2SO_4$ (250 mL) for 30 min. The organic layer was washed with aqueous $NaHCO_3$, then water, dried ($MgSO_4$), filtered, and concentrated. Purification by silica gel chromatography (85:15 toluene/ethyl acetate) gave **13** (4.47 g, 91%), $[\alpha]_D$ +5° (*c* 1.2, $CHCl_3$).

### H. Methyl 2,3-di-*O*-benzyl-4-*O*-(4-methoxybenzyl)-α-D-glucopyranoside [39]

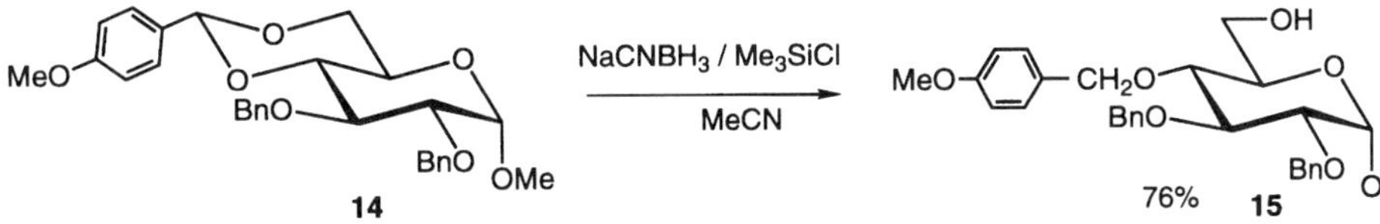

To a stirred mixture of **14** [39] (493 mg, 1.0 mmol), $NaCNBH_3$ (383 mg, 6.0 mmol) and 3-A molecular sieves in MeCN (20 mL) at room temperature was added as solution, kept at 0°C, of $Me_3SiCl$ (652 mg, 6.0 mmol) in MeCN (6 mL). The reaction mixture was stirred for 5 h at room temperature, filtered through Celite, and poured into ice-cold saturated aqueous $NaHCO_3$. The aqueous phase was repeatedly extracted with $CH_2Cl_2$. The combined extracts were washed with saturated aqueous $NaHCO_3$, dried ($MgSO_4$), filtered, and concentrated. The residue was subjected to silica gel column chromatography (toluene/ethyl acetate 2:1) to yield **15** (375 mg, 76%), $[\alpha]_D$ +19.3° (*c* 1.0, $CHCl_3$). Regioisomer **16** (13%) was also obtained.

### I. Methyl 2,3-di-*O*-benzyl-6-*O*-(4-methoxybenzyl)-α-D-glucopyranoside [39]

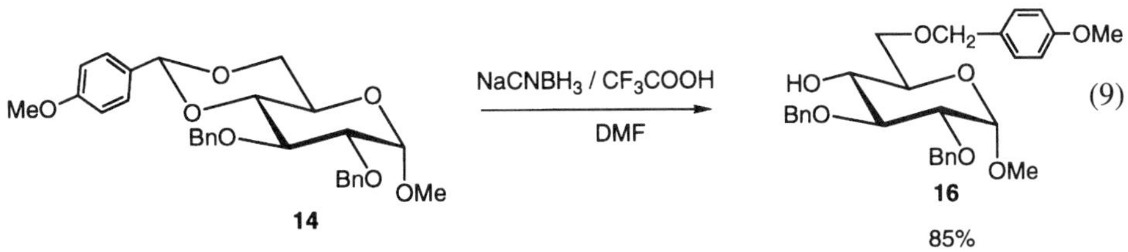

To a stirred mixture of **14** [39] (493 mg, 1.0 mmol), $NaCNBH_3$ (314 mg, 5.0 mmol) and 3-A molecular sieves in $HCONMe_2$ (8 mL) at room temperature was added as solution, kept at 0°C, of $CF_3COOH$ (1.14 g, 10 mmol) in $HCONMe_2$ (6 mL). After 7 h at room temperature, the product was workedup and subjected to chromatography, as described for regioisomer **15**, to yield **16** (419 mg, 85%) $[\alpha]_D$ + 7.7° (*c* 1.0, $CHCl_3$). Regioisomer **15** (9%) was also obtained.

## REFERENCES

1. T. W. Greene and P. G. M. Wuts, *Protective Groups in Organic Synthesis*, 2nd ed., John Wiley & Sons, New York, 1991.

2. P. J. Kociénski, *Protecting Groups*, Georg Thieme Verlag, Stuttgart, 1994.
3. A. F. Bochkov and G. E. Zaikov, *Chemistry of the O-glycosidic Bond: Formation and Cleavage*, Pergamon Press, Oxford, 1979.
4. A. H. Haines, Relative reactivities of hydroxyl groups in carbohydrates, *Advances in Carbohydrate Chemistry and Biochemistry, 33* (R. S. Tipson and D. Horton, eds.), Academic Press, New York, 1976, p. 11.
5. A. H. Haines, The selective removal of protecting groups in carbohydrate chemistry, *Advances in Carbohydrate Chemistry and Biochemistry, 39* (R. S. Tipson and D. Horton, eds.), Academic Press, New York, 1981, p. 13.
6. J. Staněk, Jr., Preparation of selectively alkylated saccharides as synthetic intermediates, *Top. Curr. Chem. 154*:209 (1990).
7. R. Gigg, The allyl ether as protecting group in carbohydrate chemistry. Part II. The 3-methylbut-2-enyl ("Prenyl") group, *J. Chem. Soc. Perkin Trans. 1*:738 (1980) and references therein.
8. S. Umezawa, Structure and synthesis of aminoglycoside antibiotics, *Adv. Carbohydr. Chem. Biochem. 30*:111 (1974).
9. R. U. Lemieux, T. Takeda, and B. Y. Chung, Synthetic methods for carbohydrates. *Am. Chem. Soc. Symp. Ser. 39*:90 (1976).
10. P. A. J. Gorin and A. S. Perlin, Configuration of glycosidic linkages in oligosaccharides. IX Synthesis of α- and β-D-mannopyranosyl disaccharides, *Can. J. Chem. 39*:2474 (1961).
11. H. B. Borén, G. Ekborg, K. Eklind, P. J. Garegg, Å. Pilotti, and C. G. Swahn, Benzylated orthoesters in glycoside synthesis, *Acta Chem. Scand. 27*:2639 (1973).
12. G. R. Barker, Triphenylmethyl ethers, *Methods Carbohydr. Chem. 2*:168 (1963).
13. S. Hanessian and P. Lavallée, The preparation and synthetic utility of *tert*-butyldiphenylsilyl ethers, *Can. J. Chem. 53*:2975 (1975).
14. R. U. Lemieux and H. Driguez, The chemical synthesis of 2-*O*-(α-L-fucoyranosyl)-3-*O*-(α-D-galactopyranosyl)-D-galactose. The terminal structure of the blood-group B antigenic determinant, *J. Am. Chem. Soc. 97*:4069 (1975).
15. P. Deslongchamps, Stereoelectronic effects in organic chemistry, *Organic Chemistry Series, 1* (J. E. Baldwin, ed.), Pergamon Press, Oxford, 1983.
16. C. Augé, S. David, and A. Veyrières, Complete regiospecificity in the benzylation of a *cis*-diol by the stannylidene procedure. *J. Chem. Soc., Chem. Commun.* 375 (1976).
17. M. A. Nashed and L. Anderson, Organotin derivatives and the selective acylation of the equatorial hydroxy group in a vicinal, equatorial–axial pair. *Tetrahedron Lett.* 3503 (1976).
18. J. Gelas, The reactivity of cyclic acetals of aldoses and aldosides, *Adv. Carbohyd. Chem. Biochem. 39*:71 (1981).
19. S. Hanessian, Approaches to the synthesis of halodeoxy sugars, *Adv. Chem. Ser. 74*:159 (1968).
20. R. W. Binkley, G. S. Goewey, and J. Johnston, Regioselective ring opening of benzylidene acetals, A photochemically initiated reaction for partial deprotection of carbohydrates, *J. Org. Chem. 49*:992 (1984).
21. A. Klemer and G. Rodemeyer, Eine einfache Synthese von Methyl-4,6-*O*-benzyliden-2-desoxy-α-D-*erythro*-hexopyranosid-3-ulose, *Chem. Ber. 107*:2612 (1974).
22. A. Klemer and D. Balkan, Synthese der methyl-2,6-dideoxy-4-*O*-(tetrahydro-2-*H*-pyran-2-yl)-α-L-*erythro*-hexopyranosid-3-ulose und der 5-Desoxy-2,3-*O*-isopropyliden-1-*O*-(tetrahydro-2*H*-pyran-2-yl)-β-D-*threo*-hexulo-5-enopyranose, *J. Chem. Res.* 3823 (1978).
23. D. M. Clode, D. Horton, and W. Weckerle, Reaction of derivatives of methyl 2,3-*O*-benzylidene-6-deoxy-α-L-mannopyranoside with butyllithium: Synthesis hof methyl 2,6-dideoxy-α-L-*erythro*-hexopyranosid-3-ulose, *Carbohydr. Res. 49*:305 (1976).
24. S. S. Bhattacharjee and P. A. J. Gorin, Hydrogenolysis of carbohydrate acetals, ketals and cyclic orthoesters with lithium aluminium hydride–aluminium trichloride, *Can. J. Chem. 47*:1195 (1969).
25. A. Lipták, J. Imre, J. Harangi, P. Nánási, and A. Neszmélyi, Chemo-, stereo- and regioselective hydrogenolysis of carbohydrate benzylidene acetals. Synthesis of benzyl ethers of benzyl α-D-,

methyl β-D-mannopyranosides and benzyl α-D-rhamnopyranosides by ring cleavage of benzylidene derivatives with the $LiAlH_4$–$AlCl_3$ reagent, *Tetrahedron 38*:3721 (1982) and references therein.

26. D. A. Horne and A. Jordan, An efficient reduction of acetals and ketals to methyl ethers, *Tetrahedron Lett.* 1357 (1978).
27. P.J. Garegg, Some aspects of regio-, stereo- and chemoselective reactions in carbohydrate chemistry, *Pure Appl. Chem. 56*:845 (1984) and references therein.
28. A. Borbás, J. Hajkò, M. Kajtár-Peredy, and A. Lipták, Hydrogenolysis of dioxolane-type diphenylmethylene acetals by $AlClH_2$ to axial diphenylmethyl ethers, *J. Carbohydr. Chem. 12*: 191 (1993) and references therein.
29. A. Lipták, I. Jodál, and P. Nánási, Stereoselective ring-cleavage of 3-*O*-benzyl- and 2,3-di-*O*-benzyl-4,6-*O*-benzylidenehexopyranoside derivatives with the $LiAlH_4$–$AlCl_3$ reagent, *Carbohydr. Res. 44*:1 (1975).
30. A. Klemer, Syntheses eines Trisaccharides mit verzweigter Struktur (4-α,6-Bis-D-glucosido-D-glucose), *Chem. Ber. 92*:218 (1959).
31. E. Zissis and H. G. Fletcher, Jr., Benzyl 2,3,4-tri-*O*-benzyl-β-D-glucopyranosiduronic acid and some related compounds, *Carbohydr. Res. 12*:361 (1970).
32. P. J. Garegg, Migration of an acetyl group between the C-2 and C-3 hydroxyl groups in methyl mannopyranosides, *Arkiv Kemi 23*:255 (1964).
33. P. J. Garegg, J. Hultberg, and S. Wallin, A novel reductive ring-opening of carbohydrate benzylidene acetals, *Carbohydr. Res. 108*:97 (1982).
34. J. C. Dennison and O. I. McGilary, 4:6-Dimethyl β-methylglucoside and some derivatives of β-methylglucoside, *J. Chem. Soc.* 1616 (1951).
35. M. Ek, P. J. Garegg, H. Hultberg, and S. Oscarson, Reductive ring openings of carbohydrate benzylidene acetals using borane-trimethylamine and aluminium chloride. Regioselectivity and solvent dependance, *J. Carbohydr. Chem. 2*:305 (1983).
36. E. G. Ansell and J. Honeyman, Sugar nitrates. Part I. Removal of nitrate and toluene-*p*-sulphonyl groups from esters of substituted methyl-D-glucosides, *J. Chem. Soc.* 2778 (1952).
37. P. J. Garegg and H. Hultberg, A novel, reductive ring-opening of carbohydrate benzylidene acetals, with unusual regioselectivity, *Carbohydr. Res. 93*:C10 (1981).
38. P. Fügedi, W. Birberg, P. J. Garegg, and Å. Pilotti, Synthesis of a branched heptasaccharide having phytoalexin-elicitor activity, *Carbohydr. Res. 164*:297 (1987).
39. R. Johansson and B. Samuelsson, Regioselective reductive ring-opening of 4-methoxybenzylidene acetals of hexopyranosides. Access to a novel protecting-group strategy. Part I., *J. Chem. Soc., Perkin Trans. 1*:2371 (1984).

# 4

# Selective O-Substitution and Oxidation Using Stannylene Acetals and Stannyl Ethers

## Serge David
*Université Paris-Sud, Orsay, France*

| | | |
|---|---|---|
| I. | Introduction | 69 |
| II. | Methods | 70 |
| | A. Formation and constitution of stannylenes | 70 |
| | B. Practical aspects of stannylene chemistry | 72 |
| | C. Tributylstannyl ethers | 73 |
| | D. Commentary on the experiments | 74 |
| III. | Experimental Procedures | 75 |
| | A. Benzoylation in benzene | 76 |
| | B. Toluenesulfonylation in methanol | 76 |
| | C. Selective benzylation in benzene, catalyzed by tetrabutylammonium iodide | 76 |
| | D. Allylation of an unprotected disaccharide glycoside | 77 |
| | E. Regioselective sulfation of a partially protected pentasaccharide | 77 |
| | F. Regioselective silylation through a minor stannylene isomer | 78 |
| | G. Preparation of hydroxy ketones | 79 |
| | H. Electrophilic substitution of a tributyltin ether | 82 |
| | References | 82 |

## I. INTRODUCTION

Acylation of sugars is generally carried out with acyl chlorides or anhydrides, in pyridine solution, at room temperature. Allylation and benzylation is most efficiently achieved with the corresponding halides in anhydrous $N,N$-dimethylformamide, in the presence of a base, to generate the alcoholate. Sodium hydride is the most convenient one, but its use may result in *N*-alkylation of acetamido sugars. Then benzylation may be achieved in the presence of a mixture of barium oxide and barium hydroxide octahydrate. In this way, benzyl 2-acetamido-3-*O*-benzyl-4,6-*O*-benzylidene-2-deoxy-α-D-glucopyranoside **1** was prepared in 87% yield in 15 min at room temperature [1].

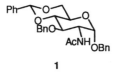

1

Benzylation and allylation can also be achieved in the presence of an acidic catalyst, with benzyl and allyl trichloroacetimidates, and trifluoromethylsulfonic acid (50 µL/g of starting hydroxy component) [2]. Allyl ethers, RO-CH$_2$-CH=CH$_2$, may be obtained from the easily prepared mixed carbonates, R-O-CO-CH$_2$-CH=CH$_2$ by exposition to 2 mol% of Pd (PPh$_3$)$_4$ in oxolane solution at 60°–70°C [3]. Selective substitution does not appear to be generally possible with these methods but, in some special cases, the outcome of the reaction may be satisfactory [4]. The more useful reagent in this context is probably triphenylchloromethane (trityl chloride). Its reactions are selective for the primary position, provided there is no large excess of reagent and an unduly long contact time. As protected derivatives, trityl ethers are not ideal. The added weight is 222 and, although their lability in mildly acidic conditions allows easy deprotection, it may be a serious inconvenience in extended reaction schemes.

Selective substitution may sometimes be achieved by indirect methods. Thus, treatment of 4,6-*O*-benzylidene acetals with cyanoborohydride and ethereal hydrogen chloride converts them into the 6-*O*-benzyl ether, with a free hydroxyl group at the 4-position (see Chap. 3). Isopropylidene and benzylidene ketals are opened by trimethylaluminium to give good yields of α-hydroxy *tert*-butyl ethers, and α-hydroxy 1-phenylethyl ethers, respectively [5]. The scope of the latter method has not yet been explored.

The method of substitution that will be presently described [6] involves, first, the conversion of dihydroxy derivatives into the so-called stannylenes. This is a convenient name, but it is incorrect from the point of view of the nomenclature of organotin derivatives. For instance, the compounds with the five-membered ring containing the tin atom should be viewed as derivatives of "2,2-dibutyl-1,3,2-dioxastannolane." The name stannylene was coined by the first investigators, probably by analogy with isopropylidene. However, despite some superficial similarity in structure, it was soon realized that "dibutylstannylenes" share no properties with isopropylidene derivatives. Some use of the tributylstannyl ethers of sugars will be also considered in this chapter.

The synthetic applications involving tin–oxygen bonds in organic chemistry (including carbohydrates) have been reviewed in a book on the use of tin in organic synthesis [7].

## II. METHODS

### A. Formation and Constitution of Stannylenes

The very inexpensive material sold under the name "dibutyltin oxide" is an amorphous, insoluble, polymeric powder with the composition Bu$_2$SnO. A suspension of this powder in an appropriate organic solvent in the presence of a 1,2- or 1,3-diol becomes clear more or less quickly because of the formation of a stannylene. These derivatives are soluble in most solvents, polar or nonpolar. Equation (1) indicates the stoichiometry of the process:

## O-Substitution and Oxidation by Stannylenes

$$\begin{array}{c} R' \\ CHOH \\ (CH_2)_n \\ CHOH \\ R \end{array} + Bu_2SnO \longrightarrow \begin{array}{c} R' \\ (CH_2)_n \\ R \end{array}\!\!\!\!\!\!\!\!\begin{array}{c} O \\ SnBu_2 \\ O \end{array} + H_2O \qquad (1)$$

$$n = 0, 1$$

However, there are as yet no proofs that the five- or six-membered ring on the right-hand side of Eq. (1) is capable of independent existence. In the solid state, in the known cases, these rings are associated in dimers such as **2**. In these dimers, the tin atom is five-coordinate, and one of the oxygen atoms tricoordinate. These units may exist as such, or be associated through an hexacoordinate tin atom and a tricoordinate oxygen atom into oligomers or infinite ribbons. The $^{119}$Sn nuclear magnetic resonance (NMR) measurements indicate that tin is five- or six-coordinate in solutions of stannylenes. In the gas phase, peaks corresponding to only dimers were observed in the field desorption mass spectra of the stannylenes of 1,2-cyclohexanediols. The driving forces to dimerization are the two electronegative substituents at the tin atom and the presence of a ring. The consequence of the length of the tin covalent radius is that, in any conformation of this ring, "normal" angles at carbon or oxygen atoms involve a very small angle at the tin component, which can be achieved only with $d$ orbital participation. In nonpolar solvents, polymerization is the only possibility available to the monomeric unit to relieve strain, but another path seems possible in polar solvents; that is, coordination of the tin atom to a molecule of solvent to give a trigonal bipyramid, such as **3**. In structures such as **2** or **3**, the oxygen atoms of the diol are in equatorial and apical positions, and the "ideal" angle at tin, 90° although greater than the actual one, nevertheless, is nearer to it than the tetrahedral angle.

**2**      **3**

The treatment of stannylenes by acid chlorides, methyl, allyl, or benzyl halides under proper conditions generally gives products of monosubstitution [Eq. (2)]:

$$\begin{array}{c} A \\ CHOH \\ (CH_2)_n \\ CHOH \\ B \end{array} \xrightarrow{Bu_2SnO} \left[ \begin{array}{c} A \\ (CH_2)_n \\ B \end{array}\!\!\!\!\!\!\!\!\begin{array}{c} O \\ SnBu_2 \\ O \end{array} \right]_p \xrightarrow[\text{"H}^+\text{"}]{RX} \begin{array}{c} A \\ CHOH \\ (CH_2)_n \\ CHOR \\ B \end{array} \qquad (2)$$

This could be explained by the constitution of stannylenes, as depicted in formula **2** or **3**. The two oxygen atoms of the parent diol are differentiated in the stannylene, one at the equatorial, and the other one at the apical, position in a trigonal bipyramid centered on the tin atom. It seems logical that the apical oxygen is the reactive one in selective substitution, in view of the known tendency of the more electronegative ligand to adopt this position in a trigonal bipyramid. Furthermore, the reactivity of the equatorial oxygen atom in dimers is abolished by tricoordination.

However, this monosubstitution is also regiospecific: only one derivative is usually isolated. In principle, any pair of hydroxyl groups in a sugar molecule can give two different stannylenes, such as **2** or **3**, depending on which hydroxyl group of the pair adopts the apical position. Again, it appears, from reactivity of the crude product, that usually only one is formed. Prediction of the favored one does not yet appear possible.

When there are more than two hydroxyl groups on the same molecule, the tin atom may span several different pairs of oxygen atoms. Apparently, for either kinetic or thermodynamic reasons, one is preferred. This is clear in competitive experiments [8]. For instance, refluxing an equimolecular mixture of glycosides **4**, **6**, and $Bu_2SnO$ appears to give overwhelmingly the stannylene of **4**, for subsequent benzoylation gave the benzoate **5** in 98% yield. Finally, there is evidence that, at least in some cases, a minor, imperceptible stannylene in a polyhydroxylated molecule is in rapid equilibrium with the major one.

**4**  R = H
**5**  R = Bz

**6**

## B. Practical Aspects of Stannylene Chemistry

Stannylenes may be prepared by refluxing equimolecular mixtures of sugar and $Bu_2SnO$ in benzene or toluene, with a Dean-Stark separator, until the system is clear. Unfortunately, the more toxic benzene appears more convenient. The toxicity of organotin compounds has been reviewed [7]. In the handling of stannylene, elaborate precautions to avoid moisture are by no means necessary. Their sensitivity to protic solvents vary widely. Actually, some may be crystallized from methanol or ethanol. In any event, they are much more easily prepared than the tributylstannyl ethers $Bu_3Sn$-O-R, which are exceedingly sensitive to moisture. Dibutylstannylenes are hydrolyzed on silica gel thin-layer chromatography (TLC) plates, and migrate as the parent diols. After the end of the reaction, the solution is concentrated to a small volume, or evaporated to dryness if it is necessary to change the solvent, which is rarely the case. Isolation of pure stannylenes is never necessary.

Stannylenes have been prepared efficiently from some fairly dilute (20–40 mM) solutions of nucleotides in methanol, by refluxing for 30 min in the presence of an equimolar amount of $Bu_2SnO$, without removal of water [9]. In these conditions, the soluble methyl ether $CH_3$-O-$SnBu_2$-O-$SnBu_2$-$OCH_3$ may be the reactive species. This is a useful technique for compounds insoluble in benzene or toluene, provided the stannylene is stable in methanol, which is not always so. Problems may arise with *trans*-diequatorial 1,2-diols.

Usually, treatment of the crude stannylene with benzoyl chloride for a few minutes at room temperature gives one of the two possible benzoates, in very high yield, in a regioselective manner. Allyl and benzyl halides do not react under these conditions, and they react only sluggishly in refluxing nonpolar solvents. Substitution proceeds in a satisfactory manner in the presence of 5–10 mol% of a quaternary ammonium bromide or iodide. Alternatively, the stannylene may be transferred into *N,N*-dimethylformamide, in which substitution occurs at 80°C. Again, allylation and benzylation of polyhydroxylated compounds is regioselective by these methods.

## O-Substitution and Oxidation by Stannylenes

An intramolecular reaction (i.e., the displacement of a mesylate in the presence of cesium fluoride), a key step in the total synthesis of octosyl acid A [Eq. (3)], has been reported in preliminary form [10]. Equation (4) describes ring-closures to oxolane rings, the

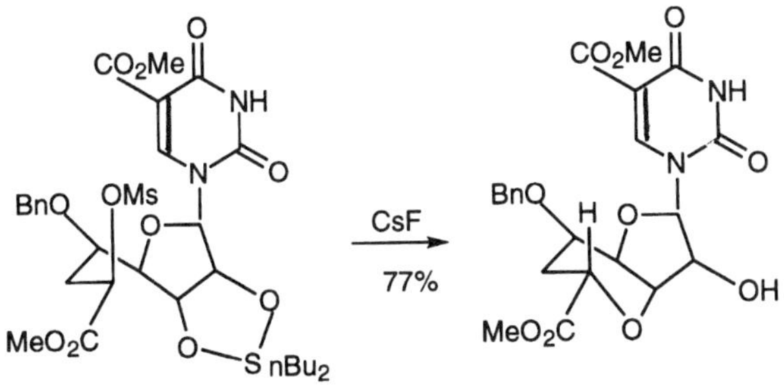

$$\text{[Scheme, Eq. (3): BnO/OMs nucleoside with } \text{SnBu}_2 \text{ acetal} \xrightarrow{\text{CsF, 77\%}} \text{cyclized product]} \quad (3)$$

$$\text{[Scheme, Eq. (4): D-galactal } \xrightarrow{\text{1) Bu}_2\text{SnO; 2) I}_2 \text{ (CH}_2\text{Cl}_2\text{)}} \text{1,6-anhydro (57\%) + 1,4-anhydro (9\%)]} \quad (4)$$

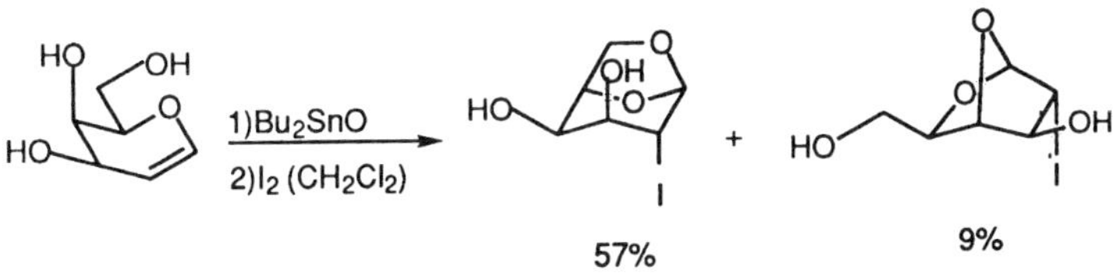

1,6- and 1,4-anhydro derivatives of 2-deoxy-2-iodo-β-D-galactopyranose [11]. Presumably, attack of D-galactal at C-2 by iodine generates a carbenium ion at C-1, which quickly adds to one of the tin-substituted oxygens.

Stannylenes are oxidized to keto-alcohols by dropwise addition of bromine in dichloromethane. The reaction proceeds at room temperature at the speed of a titration. The attention of chemists who want to check their reaction by infrared (IR) examination in situ is drawn to the fact that the ketone may be chelated to tin in these conditions ($\nu_{CO}$ 1685 cm$^{-1}$). The reaction is regiospecific, giving only one of the two possible keto-alcohols. It has been used in total synthesis; for instance, in the synthesis of (+)-spectinomycin (described under Sec. III.G) [12,13]. The replacement of bromine by N-bromosuccinimide as oxidant has been reported [14].

A problem of stannylene procedures, which has not yet found a uniformly satisfactory solution, is the disposal of the tin by-products, which are insoluble in water. In a few very favorable cases, treatment with methanol converted them to the crystalline compound Br-Bu$_2$Sn-O-SnBu$_2$-OH, which could be removed by filtration. In the other cases, silica gel column chromatography is always successful. The highly lipophile, dibutyltin compounds move to the front in nonpolar eluents. However, chromatography is a great inconvenience in large-scale work.

### C. Tributylstannyl Ethers

Bis(tributyltin) oxide, Bu$_3$SnOSnBu$_3$, is a high-boiling oil, bp 180°C/2 mm, with a penetrating, unpleasant, sweetish smell. It is readily available. Mixing with an alcohol R-OH gives a tributylstannyl ether according to Eq. (5), a reversible reaction that may be drawn to completion by azeotropic removal of water. These ethers are moisture-sensitive. They are undoubtedly covalent derivatives, yet their reactions with electrophiles are, broadly speaking, those of metal alcoholates. For instance, treatment with benzyl bromide gives a benzyl ether, according to Eq. (6).

$$R\text{-}OH + (Bu_3Sn)_2O \rightleftharpoons {}_2R\text{-}O\text{-}SnBu_3 + H_2O \qquad (5)$$

$$R\text{-}O\text{-}SnBu_3 + BnBr \rightarrow R\text{-}O\text{-}Bn + Bu_3SnBr \qquad (6)$$

However, with polyhydroxylated substrates, their reactions are not that simple. A regiospecificity of substitution is observed, sometimes very good. This may be explained by a rapid equilibrium between all possible monoethers. The reaction may occur at the most stable position, or substitution at a preferred site could occur concurrently with a rapid displacement of the equilibrium. Alternatively, the ether function may be stabilized at some particular position by coordination to the tin atom an oxygen atom of the substrate in a favorable position.

The substitution of allyl and benzyl halides is also catalyzed by quaternary ammonium halides. In fact, the method was discovered with tributylstannyl ethers [15]. The conversion of tributylstannyl ethers to ketones with $N$-bromosuccinimide is an alternative to the bromine oxidation of stannylenes [12,13]. Again, the disposal of the tin by-product $Bu_3SnX$ may be a problem in large-scale work. It may be fairly efficiently extracted with hexane from a solution of the reaction mixture in acetonitrile.

## D. Commentary on the Experiments

Experiment A [see Eq. (7)] describes the preparation of a stannylene in a nonpolar solvent, followed by benzoylation. Experiment B [see Eq. (8)] is typical of the use of methanol as solvent. Despite the great dilution, the reactions are fairly rapid. The dibutylstannylene of adenosine prepared in this way, could be isolated in the crystalline state, but this is not necessary, and tosylation was achieved in good yield by direct treatment of the methanol solution. The inertness of the primary alcoholic function of adenosine in these conditions is remarkable. In experiment C [see Eq. (9)], again we come back to benzene for selective benzylation, but with quaternary ammonium salts as catalysts. There is no useful reaction in their absence. On the other hand, benzylation of the stannylene proceeds normally without catalyst if the solvent is exchanged for $N,N$-dimethylformamide. The regioselectivity is generally less in this solvent, but has occasionally been better. The β-D-galactopyranoside residue in methyl β-lactoside reacts similarly to benzyl β-D-galactopyranoside [see experiment D; Eq. (10)]. The hydroxyl groups of the D-*gluco* residue are inert, so that only one hydroxyl group is allylated out of seven. In this synthesis, the tin by-product could be readily eliminated as a methanol-insoluble, crystalline compound. After its removal, disaccharide **14** [see Eq. (10)] could be obtained directly from the reaction medium by crystallization, without chromatography. Thus, there is no upper limit to the scale-up of the preparation of this very important starting material for glycoconjugate synthesis. The preparation described involves forcing conditions to obtain the stannylene. Methyl β-$N$-acetyllactosaminide reacts in a similar manner. Experiment E [see Eq. (11)] shows that the sulfating reagent $SO_3NMe_3$ behaves toward stannylenes in a manner similar to the other electrophiles considered so far. Actually, the preparation in 83% yield of the sulfate of the alcoholic function at position 3 of the galactose residue of an unprotected $N$-acetyllactosaminide has been reported, in preliminary form, among other examples [16]. Independently, Lubineau and Lemoine [17], in the course of a program [18] on the preparation of some sulfated higher oligosaccharides, reported the preparation, by the stannylene, of the 3'-$O$-sulfated Lewis[a] trisaccharide. There is currently tremendous interest in the sulfates of the higher oligosaccharides, for compounds such as **16** are among the most powerful ligands for selectins, the animal lectins involved in inflammatory conditions. The synthesis of **16** [see Eq. (12)] was abridged, by four steps, by the stannylene procedure. Interestingly, direct sulfation of the same pentasaccharide **15** [see Eq. (11)] (R=H) with the complex $SO_3NMe_3$ gave the primary sulfate in 74% yield. An extensive study of the scope of sulfation by stannylenes has been reported [19].

At first sight, the outcome of experiment F [see Eq. (13)] is paradoxical. Regioselective allylation and benzylation at the 3-position in unprotected β-D-galactopyranosides residue can be interpreted only by the formation of a stannylene spanning the 2–3- or the 3–4-position, whereas we observe substitution at the primary position with a bulky silylating reagent. One possible explanation is that there is rapid equilibrium between two stannylenes, the minor one activating the primary alcoholic function. This minor component is the only one that can react with the bulky chlorosilane reagent, so that all the reaction proceeds through this channel. Also, we cannot exclude the possible involvement of the solvent oxolane.

The generation of hydroxyketones by the brominolysis of stannylenes has been used several times in total synthesis. Experiment G [see Eq. (15)] describes a key step in the total synthesis of the antibiotic (+)-spectinomycin [12,13]. It is remarkable that the two oxygen atoms bound to the tin atom originate from hydroxyl groups, which are part of different functions, a hemiketal and a secondary alcohol. The oxidation is selective for one alcoholic function out of three. The same product was obtained by N-bromosuccinimide oxidation of the tributylstannyl ether.

Experiment G [see Eq. (16)] shows that a stannylene may be prepared from a molecule protected by acetylation.

Experiment H [see Eq. (18)] gives an example of the use of a tributylstannyl ether in regioselective substitution [20]. This is an especially interesting case, for the dibutylstannylene of diol **28** gives only a 70% yield of the ether **29**, the rest being the starting material. This stopping at 70% completion could be explained by the existence of a trimeric stannylene, the 2,2-dibutyl-1,3,2-dioxastannoxane, corresponding to the unreactive sugar molecule, being buried between the two others, to which it could be associated by an hexacoordinate tin atom and two oxygen atoms deactivated by threefold coordination.

## III. EXPERIMENTAL PROCEDURES

### A. Benzoylation in Benzene [21]

(7)

*Benzyl 3-O-benzoyl-4,6-O-benzylidene-β-D-galactopyranoside*

A mixture of benzyl 4,6-O-benzylidene-β-D-galactopyranoside (7.16 g, 20 mmol) and $Bu_2SnO$ (5.4 g) in benzene (0.7 L) was refluxed for 16 h, with azeotropic removal of water. The clear (or almost clear) solution was then concentrated to 100 mL and cooled to room temperature. Benzoyl chloride (2.4 mL, 22 mmol) was added, and the mixture was stirred for 1 min, and then evaporated to dryness. Chromatography of the residue on a 3.3 × 46-cm silica gel column (chloroform/methanol, 99:1) gave the 3-*O*-benzoate (8.8 g, 96%), mp 179–180°C (2-propanol); reported [22] mp 178–179°C.

### B. Toluenesulfonylation in Methanol [9]

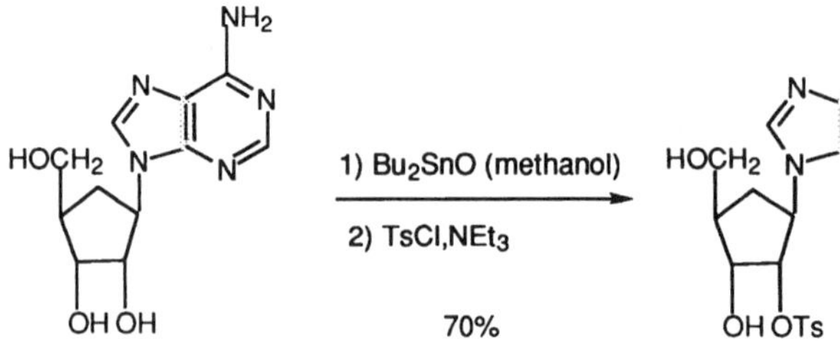

$$\text{9} \xrightarrow[\text{2) TsCl, NEt}_3]{\text{1) Bu}_2\text{SnO (methanol)}} \text{10} \quad 70\% \tag{8}$$

*2′,3′-O-(Dibutylstannylene)adenosine*

A mixture of adenosine (267 mg, 1 mmol), dibutyltin oxide (250 mg, 1 mmol), and methanol (25 mL) was heated under reflux for 30 min and then evaporated to dryness. Two crystallizations from ethanol–acetone gave 350 mg (70%): mp 154–156°C; $\lambda_{max}$ (MeOH) 259 nm (*e* 14,700).

*2′-O-p-Toluenesulfonyladenosine* **(10)**

Triethylamine (10.5 mL, 75 mmol) and *p*-toluenesulfonyl chloride (14.25 g, 75 mmol) were added to a solution of the foregoing stannylene (5 mmol) prepared in situ in methanol (100 mL). After 5 min at room temperature the solvent was evaporated, and the residue was partitioned between water and ether. The aqueous phase was concentrated and stored at 4°C, giving 1.47 g (70%) of **10**: mp 229–230°C; $\lambda_{max}$ (pH 2) 229 nm (*e* 12,900), 257 nm (12,400); $\lambda_{max}$ (pH 11) 228 nm (*e* 12,300), 261 nm (12,700) [9].

### C. Selective Benzylation in Benzene, Catalyzed by Tetrabutylammonium Iodide [23]

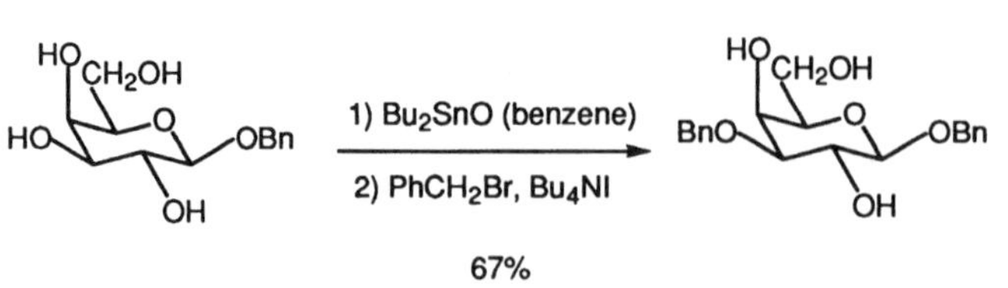

$$\text{11} \xrightarrow[\text{2) PhCH}_2\text{Br, Bu}_4\text{NI}]{\text{1) Bu}_2\text{SnO (benzene)}} \text{12} \quad 67\% \tag{9}$$

*Benzyl 3-O-benzyl-β-D-galactopyranoside* **(12)**

A mixture of benzyl-β-D-galactopyranoside **11** (1 mmol) and dibutyltin oxide (1 mmol) in benzene was refluxed for 16 h, with azeotropic removal of water. The solution was evaporated to approximately 25 mL, tetrabutylammonium iodide (1 mmol) and benzyl bromide (0.25 mL) were added, and the mixture was refluxed for 2 h. Evaporation to dryness gave a residue that was processed by column chromatography on silica gel (dichloromethane/ethyl acetate, 3:2), to give benzyl 3-O-benzyl-β-D-galactopyranoside **12** (0.67 mmol) as crystals; mp 105°C (ether), $[\alpha]_D$ −20° (c 2.0 in $CH_2Cl_2$).

## D. Allylation of an Unprotected Disaccharide Glycoside [24,25]

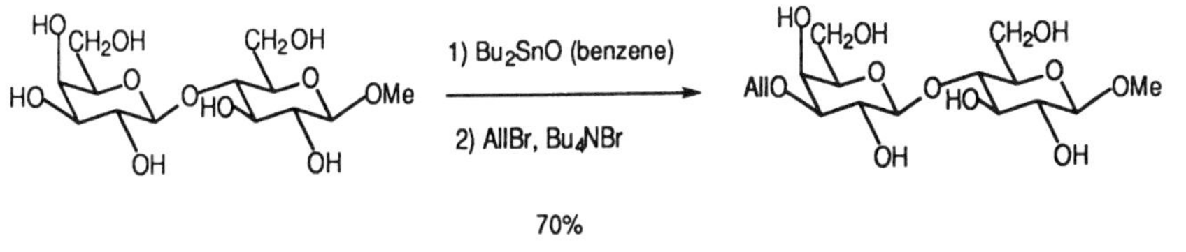

$$\text{13} \xrightarrow[\text{2) AllBr, Bu}_4\text{NBr}]{\text{1) Bu}_2\text{SnO (benzene)}} \text{14} \quad (10)$$

70%

*Methyl 4-O-(3-O-allyl-β-D-galactopyranosyl)-β-D-glucopyranoside* **(14)**

A suspension of methyl β-lactoside **13** (5.0 g, 14 mmol) in benzene (150 mL) was refluxed for 17 h in the presence of $Bu_2SnO$ (4.18 g, 16.6 mmol) in a flask equipped with a Dean-Stark separator. Then allyl bromide (20 mL) and tetrabutylammonium bromide (2 g, 6.2 mmol) were added, the solution was refluxed for 3 h, and the volatiles were evaporated. The residue was dissolved in water and washed twice with ethyl acetate. The aqueous phase was evaporated to dryness, benzene (150 mL) and $Bu_2SnO$ (3.83 g) were added to the residue and the mixture was refluxed as before for 17 h. Then allyl bromide (15 mL) and tetra-butylammonium bromide (1.5 g) were added, and the solution was again refluxed for 50 min. The volatiles were evaporated, and the residue was taken over in methanol. The crystalline precipitate of $BrBu_2SnOSnBu_2(OH)$, mp 80–81°C, was removed by filtration and the mother liquor was evaporated. Addition of ethyl acetate to the residue gave the allyl ether **14** as a crystalline compound (3.88 g, 70%): mp 107–112°C; $[\alpha]_D$ + 20° (c 1.0 in water).

## E. Regioselective Sulfation of a Partially Protected Pentasaccharide [18]

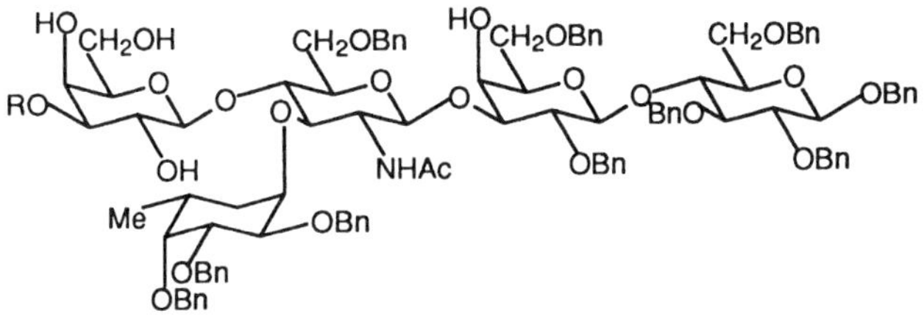

**15**

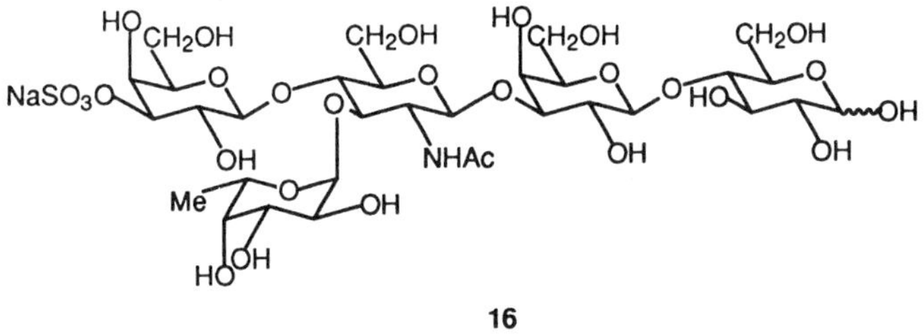

**16**

**15** (R = H) $\xrightarrow[\text{2) SO}_3\cdot\text{Me}_3\text{N}]{\text{1) Bu}_2\text{SnO (toluene)}}$ **15** (R = NaSO$_3$)  (11)

73–78%

**15** (R = NaSO$_3$) $\xrightarrow{\text{H}_2\text{, Pd/C (methanol-water)}}$ **16**  (12)

94%

*Benzyl O-(3-O-sulfo-β-D-galactopyranosyl)-(1→4)-O-[(2,3,4-tri-O-benzyl-α-L-fucopyranosyl)-(1→3)]-O-(2-acetamido-6-O-benzyl-2-deoxy-β-D-glucopyranosyl)-(1→3)-O-(2,6-di-O-benzyl-β-D-galactopyranosyl)-(1→4)2,3,6-tri-O-benzyl-β-D-glucopyranoside, sodium salt (15, R = NaSO$_3$)*

A mixture of the partially protected pentasaccharide (**15**, R = H) (50 mg, 28.5 μmol) and Bu$_2$SnO (8 mg, 31 μmol) in toluene (5 mL) was refluxed for 15 h, with azeotropic removal of water. Toluene was evaporated, and 0.2 mL of a 0.48-M solution of the complex sulfur trioxide–trimethylamine in $N,N$-dimethylformamide was added. The mixture was stirred for 7 h at room temperature, then diluted with ethyl acetate, and washed with a saturated aqueous solution of sodium hydrogencarbonate. The aqueous phase was extracted with ethyl acetate, and the combined organic layers were washed with water, dried (MgSO$_4$), and evaporated. The residue was purified by flash chromatography (silica 60A.C.C., Chromagel, 6–35 μm, S.D.S., France) and eluted with (19:1) ethyl acetate/methanol, and afterward passed through the cation-exchange resin AG50 WX8 (Na$^+$), with methanol as eluent. Evaporation of the methanol solution gave the sulfated pentasaccharide (38 mg, 73%) as a foam: $[\alpha]_D$ −23° ($c$ 2.4 in CH$_2$Cl$_2$).

*O-(3-O-sulfo-β-D-galactopyranosyl)-(1→4)-O-[(α-L-fucopyranosyl)-(1→3)]-O-β-D-galactopyranosyl-(1→4)-D-gluco-pyranose, sodium salt* **(16)**

A 20-mM solution of pentasaccharide **15** (R = NaSO$_3$), in 2.5:1 methanol/water, was shaken for 60 h at room temperature in an hydrogen atmosphere (1 bar) in the presence of 10% palladium on charcoal (150 mg). The mixture was filtered over a Celite bed, the solution was evaporated, and the residue was dissolved in water. This aqueous solution was passed through a column of cation-exchange resin AG50 WX8 (Na$^+$). Pentasaccharide **16** was eluted with water and recovered by freeze-drying: $[\alpha]_D$ −7° ($c$ 1.54 in water, after a 24-h standing); $^1$H NMR (D$_2$O): δ 4.324 (1H, dd, $J_{2e,3e}$ 9.87, $J_{3e,4e}$ 3.25 Hz, H-3e).

### F. Regioselective Silylation Through a Minor Stannylene Isomer [26]

*Ethyl 4-O-(6-O-t-butyldimethylsilyl-β-D-galactopyranosyl)-1-thio-β-D-glucopyranoside* **(18)**

A solution of ethyl 1-thio-β-D-galactopyranoside (5 g, 12.9 mmol) and dibutyltin oxide (3.2 g, 13 mmol) in MeOH (500 mL) was boiled under reflux for 4 h. Removal of the solvent gave a white powder that was dissolved in dry THF (500 mL). $t$-Butyldimethylsilylchloride (1.95 g, 13 mmol) was added to the solution and left to stir for 24 h at room temperature, after which time TLC (9:1, CH$_2$Cl$_2$/MeOH) showed that no starting material remained.

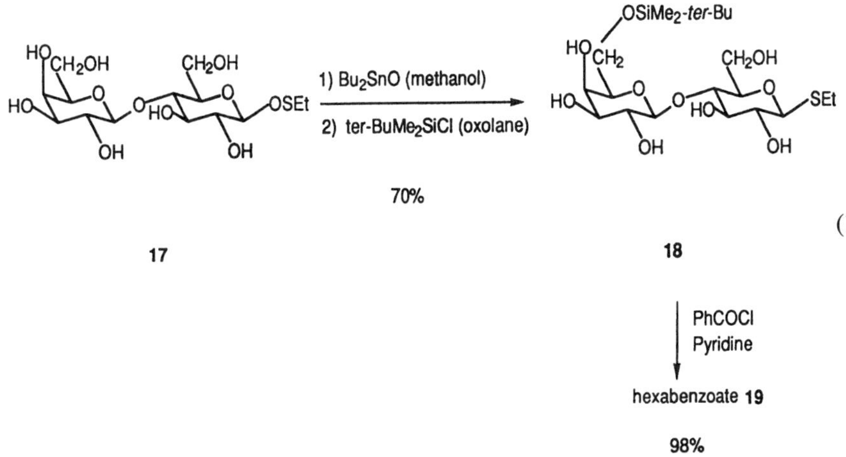

## O-Substitution and Oxidation by Stannylenes

(13)

17 → 18, 70%

1) Bu₂SnO (methanol)
2) ter-BuMe₂SiCl (oxolane)

↓ PhCOCl, Pyridine

hexabenzoate 19
98%

Removal of the solvent under reduced pressure gave a syrup, which was subjected to column chromatography (5% MeOH–CH$_2$Cl$_2$) to give **18** as a white crystalline solid: (6.25 g, 96%; mp 126–128°C; [α]$_D$ −33° (c 1.56, MeOH) which, owing to problems of instability associated with cleavage of the silyl ether, was immediately benzoylated.

### Ethyl 2,3,6-tri-O-benzoyl-4-O-(2,3,4-tri-O-benzoyl-6-O-t-butyldimethylsilyl-β-D-galactopyranosyl)-1-thio-β-D-glucopyranoside (19)

Benzoyl chloride (5 mL, 43 mmol) was added to a stirred solution of **18** (2 g, 4 mmol) in pyridine (20 mL). Stirring was continued for 48 h to give a pink solution that was washed with aqueous Na$_2$CO$_3$ and dried over MgSO$_4$. Evaporation of solvent gave a syrup that was subjected to column chromatography (9:1, hexane/EtOAc) to give **19**: 4.39 g, 3.9 mmol, 98%; mp 79–81°C); [α]$_D$ + 76° (c 1.3, CH$_2$Cl$_2$).

### G. Preparation of Hydroxy Ketones

*Benzyl 2,3-di-O-benzyl-α-D-xylo-hexopyranos-4-uloside* **(21)** [27]

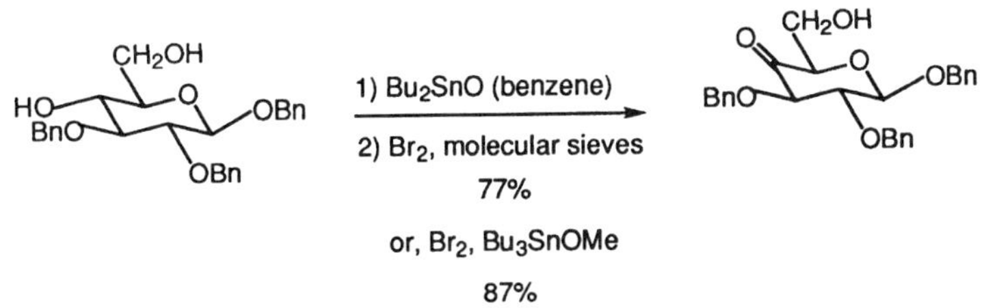

20 → 21, (14)

1) Bu₂SnO (benzene)
2) Br₂, molecular sieves
77%

or, Br₂, Bu₃SnOMe
87%

### General procedure for brominolyses

A mixture of the diol (2 mmol) and dibutyltin oxide (545 mg; 2.2 mmol) in benzene was refluxed overnight with azeotropic removal of water by a Dean-Stark condenser. Evaporation of the solvent at 100°C then gave the crude stannylene derivative, which was used without further purification. Brominolyses were carried out either on solutions of the stannylene derivatives in benzene (5 mL) in the presence of 4-Å molecular sieves (2 g) (method A) or on solutions in dichloromethane (8 mL), in the presence of tributyltin methoxide (320 mg; 1 mmol) (method B). In both methods, bromine (320 mg, 2 mmol) in dichloromethane (4 mL) was added dropwise at room temperature to the well-stirred solution of the stannylene derivative as long as the discoloration was rapid (about 0.5 h). Thin-layer chromatography then indicated the presence of only two components in the reaction mixture, possibly as organotin derivatives: the hydroxyketone and the starting diol. After filtration (if necessary), the solution was evaporated to dryness, and the hydroxyketone was isolated from the residue by column chromatography on a silica gel column (35 × 1.5 cm) with the eluant mixtures given in the following.

*Benzyl 2,3-di-O-benzyl-α-D-xylo-hexopyranose-4-uloside* **(21)**

This was prepared from the diol **20** (eluant chloroform), yields 77% (A) and 87% (B) as a syrup; $\nu_{max}$ (film) 1740 (CO) and 3500 cm$^{-1}$ (OH). When either molecular sieves or tributyltin methoxide were omitted, the yield dropped to 57% [27].

*The Penultimate Step in the Total Synthesis of (+)-Spectinomycin [12,13]*

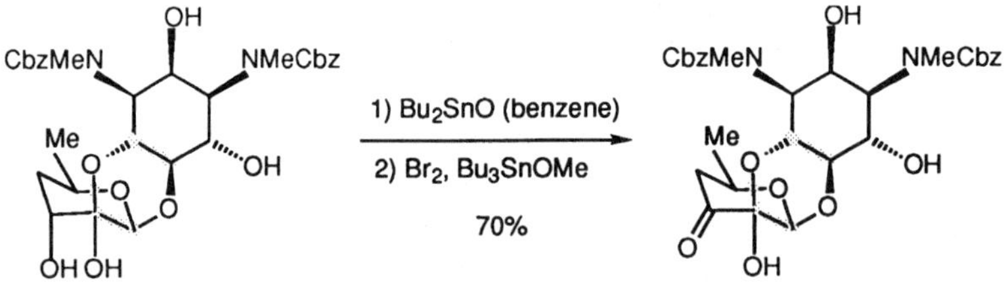

(15)

N,N'-[(benzyloxy)carbonyl] spectinomycin **(23)**

A mixture of **22** (200 mg, 0.33 mmol) and Bu$_2$SnO (90 mg, 1.1 Eq) was refluxed in methanol (10 mL) for 1 h, evaporated, benzene added, and the solution evaporated again to give a syrup. The stannylene derivative was dissolved in 5 mL of dichloromethane, containing 0.12 mL (1.1 Eq) of tri-*n*-butyltin methoxide (**24**), and the solution was cooled to 0°C. This was treated with a solution of bromine (1.1 Eq), 2% in CH$_2$Cl$_2$ dropwise, over a period of 1 h. Residual bromine was destroyed with cyclohexene and the crude product was isolated by precipitation with hexanes to give an amorphous solid (184 mg, 92%). Chromatography on silica gel (hexanes/EtOAc, 2:3) gave pure product, identical with a sample prepared from the antibiotic: $[\alpha]_D$ − 4.6° (*c* 0.28, CHCl$_3$; IR $\nu_{max}$ (KBr): 3460 (OH), 1685 (Cbz) cm$^{-1}$.

*Oxidation of a Molecule Partially Protected by Esterification [28]*

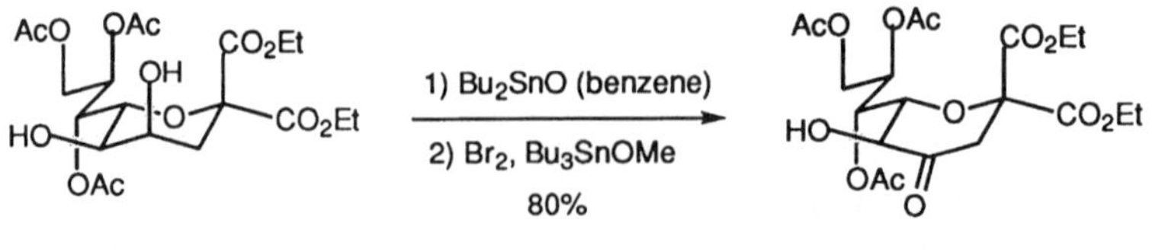

(16)

### Ethyl 7,8,9-tri-O-acetyl-2,6-anhydro-2,3-dideoxy-2-C-ethoxycarbonyl-D-*glycero*-D-*talo*-non-4-ulosonate (25)

Diol **24** (872 mg, 1.82 mmol) and dibutylstannyloxide (498 mg, 2 mmol) were refluxed in toluene (50 mL) during 4 h with azeotropic removal of water. The solvent was then partially removed by distillation (until a ~ 2-mL volume was reached), and the mixture was cooled to room temperature. Anhydrous dichloromethane (2 mL) was added, and tributylstannyl methoxide (610 mg, 1.9 mmol) was added in one portion. A solution of bromine in dichloromethane (0.25 M, 8 mL, 2 mmol) was then added dropwise as long as decoloration occurred. Evaporation to dryness and flash chromatography of the residue (1:1 AcOEt/hexane) allowed isolation of the pure hydroxyketone **25** (369 mg, 80%) as a pale yellow oil: $[\alpha]_D$ $-17°$ (*c* 2.8, $Cl_2CH_2$).

### *Preparation of Hydroxyketones by Oxidation with N-Bromosuccinimide [14]*

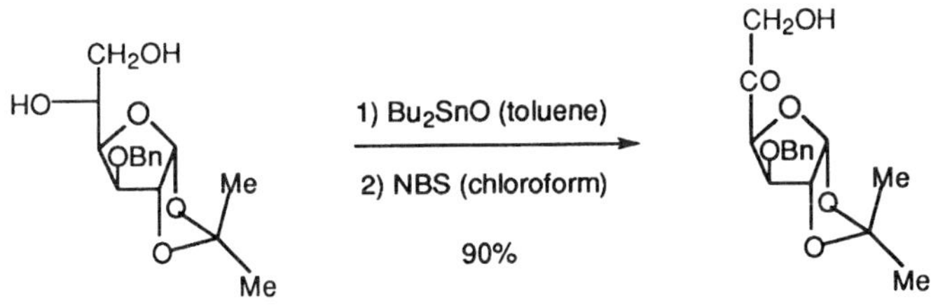

(17)

### General methods for oxidation of dibutylstannylene acetals with N-bromosuccinimide

The diol (1 mmol) was refluxed with dibutyltin oxide (1 Eq) for 12 h in toluene (20 mL) in an apparatus for the continuous removal of water. The toluene was removed on a vacuum line at 20°C, and the residue was dried for 30 min under reduced pressure (0.1 torr). The residue was taken up in dry chloroform (10 mL), and *N*-bromosuccinimide (NBS; 1 Eq) was added. The stirred reaction mixture was monitored by TLC. The reaction was complete at times ranging from 2 to 30 min. The mixture was poured directly onto a column for separation using the eluant listed for each compound.

### *3-O-Benzyl-1,2-O-isopropylidene-*α*-D-xylo-hexofuranos-5-ulose* (27)

Oxidation of the dibutylstannylene acetal derived from 3-*O*-benzyl-1,2-*O*-isopropylidene-α-D-glucofuranose (**26**) with NBS was finished within 5 min. The eluant for column separation was the mixture of hexane and ethyl acetate (from 4:1 to 2:1). Evaporation of solvent from the corresponding fractions afforded **27** (280 mg, 90%). Colorless crystals were obtained from ethyl acetate–hexane: mp 117–118°C; $[\alpha]_D$ $-114.5°$ (*c* 1.00, chloroform); IR (Nujol) 3491 $cm^{-1}$ (OH), 1725 $cm^{-1}$ (C=O), $^{13}$C-NMR ($CDCl_3$, 62.90 MHz) δ 208.2 (C-5).

## H. Electrophilic Substitution of a Tributyltin Ether [20]

$$\underset{\mathbf{28}}{\text{[structure]}} \xrightarrow[\text{2) BnBr, Bu}_4\text{NBr}]{\text{1) (Bu}_3\text{Sn})_2\text{O (toluene)}} \underset{\mathbf{29}}{\text{[structure]}} \quad 90\% \quad (18)$$

*Benzyl 3,6-di-O-benzyl-2-deoxy-2-phthalimido-β-D-glucopyranoside* (27)

A solution of **28** (4.2 g, 8.6 mmol) in anhydrous toluene (100 mL) was heated at reflux for 2 h in the presence of bis(tributyltin) oxide (4.4 mL, 8.6 mmol), with azeotropic removal of water. The solution was cooled to room temperature, tetrabutylammonium iodide (3.2 g, 8.6 mmol) and benzyl bromide (4.9 mL, 41 mmol) were added, and the mixture was heated at reflux for 5 h. After evaporation of volatiles, silica gel column chromatography (ethyl acetate/hexane, 1:1 v/v) gave compound **29** as an oil: 4.5 g, 89%. $[\alpha]_D$ −9.4° ($c$ 1.0 in chloroform).

## REFERENCES

1. J. C. Jacquinet, J. M. Petit, and P. Sinay, Synthèses du benzyl 2-acetamido-3,6-di-O-benzyl-2-désoxy-α-D-glucopyranoside, *Carbohydr. Res.* 38:305 (1974).
2. H. P. Wessel, T. Iversen, and D. R. Bundle, Acid-catalysed benzylation and allylation by alkyl trichloroacetimidates, *J. Chem. Soc. Perkin Trans.* 1:2247 (1985).
3. F. Guibé and Y. Saint M'Leux, The allyloxycarbonyl group for alcohol protection: Quantitative removal or transformation into allyl protecting group via π-allyl complexes of palladium, *Tetrahedron Lett.* 22:3591 (1981).
4. A. H. Haines, Relative reactivities of hydroxyl groups in carbohydrates, *Adv. Carbohydr. Chem. Biochem.* 33:11 (1976).
5. S. Takano, T. Ohkawa, and K. Ogasawara, Regioselective formation of 3-*tert*-alkoxy-1,2-glycols from 2,3-O-alkylidenetriols with trimethylaluminium, *Tetrahedron Lett.* 29:1823 (1988).
6. S. David and S. Hanessian, Regioselective manipulation of hydroxyl groups via organotin derivatives, *Tetrahedron* 41:643 (1985).
7. M. Pereyre, J. P. Quintard, and A. Rahm, *Tin in Organic Synthesis*, Butterworth, London, 1987.
8. S. David and A. Malleron, unreported experiments.
9. D. Wagner, J. P. H. Verheyden, and J. G. Moffatt, Preparation and synthetic utility of some organotin derivatives of nucleosides, *J. Org. Chem.* 39:24 (1974).
10. S. Danishefsky and R. Hungate, Total synthesis of octosyl acid A: A new departure in organostannylene chemistry, *J. Am. Chem. Soc.* 108:2486 (1986).
11. C. Leteux and A. Veyrières, Synthesis of α-C-glycopyranosides of D-galactosamine and D-glucosamine via iodocylisation of corresponding glycals and silver tetrafluoroboramide-promoted alkynylation at the anomeric centre, *J. Chem. Soc. Perkin Trans.* 1:2647 (1994).
12. S. Hanessian and R. Roy, Synthesis of (+)-spectinomycin, *J. Am. Chem. Soc.* 101:5839 (1979).

13. S. Hanessian and R. Roy, Chemistry of (+)-spectinomycin: Its total synthesis, stereocontrolled rearrangement and analogs, *Can J. Chem.* 63:163 (1985).
14. X. Kong and T. Bruce Grindley, An improved method for the regioselective oxidation of stannylene acetals and dimerisation of the α-hydroxyketone products, *J. Carbohydr. Chem.* 12:557 (1993).
15. J. Alais and A. Veyrières, Synthesis of O-β-D-galactopyranosyl-(1-4)-O-2-acetamido-2-deoxy-β-D-glucopyranosyl-(1-3)-D-mannose, a postulated trisaccharide of human erythrocyte membrane sialoglycoprotein, *J. Chem. Soc. Perkin Trans.* 1:377 (1981).
16. B. Guilbert, N. J. Davis, and S. Flitsch, Regioselective sulfation of disaccharides using dibutylstannylene acetals, *Tetrahedron Lett.* 35:6563 (1994).
17. A. Lubineau and R. Lemoine, Regioselective sulfation of galactose derivatives through the stannylene procedure. New synthesis of the 3'-O-sulfated Lewis[a] trisaccharide. *Tetrahedron Lett.* 35:8795 (1994).
18. R. Lemoine, Synthèse d'oligosaccharides impliqués dans les processus de reconnaissance cellulaire, Thèse, Université de Paris-Sud, Orsay, 1995.
19. S. Langston, B. Bernet, and A. Vasella, Temporary protection and activation in the regioselective synthesis of saccharide sulfates, *Helv. Chim. Acta* 77:2341 (1994).
20. D. Lafont, P. Boullanger, and B. Fenet, Réactivité comparée de divers accepteurs de la D-glucosamine lors de la synthèse de précurseurs du chitobiose, *J. Carbohydr. Chem.* 13:565 (1994).
21. S. David and A. Malleron, unpublished data.
22. G. J. F. Chittenden, An improved preparation of benzyl 3-O-benzoyl-4,6-O-benzylidene-β-D-galactopyranoside, *Carbohydr. Res.* 16:495 (1971).
23. S. David, A. Thieffry, and A. Veyrières, A mild procedure for the regiospecific benzylation and allylation of poly-hydroxy-compounds via their stannylene derivatives in non-polar solvents, *J. Chem. Soc. Perkin Trans.* 1:1796 (1981).
24. J. Alais, A. Maranduba, and A. Veyrières, Regioselective mono-O-alkylation of disaccharide glycosides through their dibutylstannylene complexes, *Tetrahedron Lett.* 24:2383 (1983).
25. A. Maranduba, Synthèse d'oligosaccharides dérivés du lactose: Une contribution à l'étude de la biosynthèse des antigènes Ii, Université Paris-Sud, Orsay (1985).
26. A. Glen, D. A. Leigh, R. P. Martin, J. P. Smart, and A. M. Truscello, The regioselective *tert*-butyldimethylsilylation of the 6'-hydroxyl group of lactose derivatives via their dibutylstannylene acetals, *Carbohydr. Res.* 248:365 (1993).
27. S. David and A. Thieffry, The brominolysis reaction of stannylene derivatives: A regiospecific synthesis of carbohydrate-derived hydroxy-ketones, *J. Chem. Soc. Perkin Trans.* 1:1568 (1979).
28. A. Lubineau, H. Arcostanzo, and Y. Queneau, Stereochemical study of cycloadditions using erythrose and threose based dienes as sources of 2-nonulosonic acid analogs, *J. Carbohydr. Chem.* 14:1307 (1995).

# II
# Selected Reactions in Carbohydrate Chemistry

**THEMES**: *Displacements of Sulfonates; Halogenation of Alcohols; Deoxygenation; C-Branching; Chain Extension*

Sugar molecules can be considered as a playground for exploratory organic synthesis, particularly in the area of functional group manipulation. The combination of five- and six-membered conformationally predisposed frameworks, with hydroxy and hemiacetal groups (with an option for a C-2 amino group), including stereochemical variations and unique spatial dispositions, allow the exploration of many synthetic transformations. Thus, sugars are ideal substrates to test out $S_N2$ nucleophilic displacement reactions with a variety of heteroatom and halogen-based nucleophiles. These can be tried on primary and secondary hydroxy groups at different sites and with different steric or stereoelectronic dispositions. By the same token, sugar derivatives are excellent candidates for the development and testing of novel leaving groups (nucleofuges). Anyone with a super nucleophile or nucleofuge combination of reagents should test it out on an $S_N2$ displacement reaction at C-3 of 1,2:5,6-di-*O*-isopropylidene-α-D-glucofuranose, or at C-2 of an alkyl α-D-glucopyranoside. Even with excellent leaving groups (e.g., triflates), the threat of β-elimination accompanying substitution will loom over anybody's head in the aforecited examples.

Peripheral manipulation of hydroxyl groups by $S_N2$ displacement reactions of appropriately activated derivatives (alkylsulfonates, chlorosulfates, phosphonium salts, and related methods) is an indispensable tool for the introduction of substituents directly attached to the sugar framework. These reactions invariably take place with inversion of configuration, but vary in efficiency, depending on the site of substitution, the nature of the nucleophile, the leaving group, and the polarity of the solvent. In other words, the ABCs of bimolecular nucleophilic displacement reactions are operative, except that with sugar molecules, stereoelectronic nonbonded interactions can raise their ugly head. Frequently, the introduction of oxygen, nitrogen, sulfur, and halogen nucleophiles can be effected with excellent control, the notable examples cited in the foregoing being the benchmarks. It is also relatively easy to replace a hydroxy group by a hydrogen, which gives a deoxy sugar. This operation has become all the more practical since the advent of free radical-mediated reactions of thiocarbonyl esters (Barton reaction), and of halides that must be prepared from the alcohol by derivatization or an $S_N2$ displacement reaction.

By far, the most challenging task in bond-forming reactions with sugar derivatives is to effect C—C branching. The $S_N2$ nucleophilic displacement reactions with carbon nucleophiles of appropriate leaving groups are, to a large extent, confined to primary positions. It would be unthinkable, for example, to displace a C-3 sulfonate (or halide) of 1,2:5,6-di-*O*-isopropylidene-α-D-glucofuranose with an organometallic methyl source. Elimination would be the predominant, if not exclusive, pathway. Even the best combination of high nucleophilicity and low basicity (e.g., cuprates, rather than alkyl lithium reagents) would fail miserably. This is another example for which carbohydrates can be used as a testing ground for new reagents and reactions.

The C-branched derivatives on secondary carbon sites are easily accessible by other modes of activation, such as the opening of epoxides, attack on carbonyl sugar derivatives, free radical allylations (Keck reaction), and related methods. Such modified sugar derivatives, starting with naturally occurring compounds, such as D-glucose, are versatile and indispensable building blocks (chirons) for the synthesis of carbohydrate or noncarbohydrate target molecules.

The Kiliani cyanohydrin synthesis has served carbohydrate chemists admirably since its discovery in 1885. This important C−C bond-forming reaction allows chain extension, with the creation of a new stereogenic center. There have been several "newer" methods and variations of the original reaction over the years. The advent of anion equivalents (Seebach's umpolung concept) has produced yet another one-carbon synthon for stereocontrolled chain elongation, with the option to iterate the process, thus building "taller" sugars, with their "heads" always pointing upward.

Natural sugar frameworks offer unique platforms for chemical manipulation and conversion into sugar or nonsugar target molecules. The following six chapters offer some of the most useful methods for functionalizing sugar molecules.

*Stephen Hanessian*

# 5

# $S_N2$-Type Halogenation and Azidation Reactions with Carbohydrate Triflates

**Edith R. Binkley**
*Center for Carbohydrate Study, Oberlin, Ohio*

**Roger W. Binkley**
*Center for Carbohydrate Study, Oberlin, and Cleveland State University, Cleveland, Ohio*

|      |                                                                                      |     |
|------|--------------------------------------------------------------------------------------|-----|
| I.   | Triflate Synthesis and Reactivity                                                    | 88  |
|      | A. Reactivity of sulfonic acid esters in substitution reactions                      | 88  |
|      | B. Synthesis of triflates                                                            | 89  |
| II.  | Introducing Azido and Halogeno Groups by Triflate Displacement                       | 90  |
|      | A. Deoxyhalogeno sugars                                                              | 90  |
|      | B. Azidodeoxy sugars                                                                 | 93  |
| III. | Reactions of Carbohydrate Triflates                                                  | 93  |
|      | A. Substitution reactions                                                            | 93  |
|      | B. Elimination reactions                                                             | 94  |
|      | C. Rearrangement reactions                                                           | 96  |
|      | D. Effect of anion nucleophilicity on reactivity                                     | 96  |
| IV.  | Carbohydrate Imidazolylsulfonates                                                    | 96  |
| V.   | Experimental Procedures                                                              | 97  |
|      | A. Synthesis of triflate esters with triflic anhydride                               | 97  |
|      | B. Synthesis of a triflate ester with triflyl chloride                               | 97  |
|      | C. Synthesis of a reactive triflate ester in the presence of a nonnucleophilic base, 2,6-di-*tert*-butyl-4-methylpyridine | 98  |
|      | D. Triflate Displacement by bromide, chloride, and iodide ions                       | 98  |
|      | E. Triflate displacement by bromide, chloride, and iodide ions from a nucleoside     | 99  |
|      | F. Triflate displacement by fluoride ion                                             | 100 |
|      | G. Triflate displacement by azide ion                                                | 101 |
|      | References                                                                           | 102 |

## I. TRIFLATE SYNTHESIS AND REACTIVITY

### A. Reactivity of Sulfonic Acid Esters in Substitution Reactions

Substitution reactions that involve organic compounds use a variety of leaving groups. Those leaving groups frequently used, however, represent only a small subset of the larger collection. Table 1 contains a list of some common leaving groups with their reactivities expressed as the relative rates of solvolysis of the corresponding 1-phenylethyl derivatives [1]. Among the groups listed in Table 1, esters of sulfonic acids are the most reactive, and consequently, are particularly attractive choices where displacement from carbohydrates is concerned. ($S_N2$ substitution reactions of carbohydrates are often difficult to achieve and, even when possible, can require vigorous reaction conditions.) Another reason for selecting sulfonic acid esters as leaving groups for substitution reactions of carbohydrates is that these esters are easily prepared from compounds containing hydroxyl groups.

Given the assumption that the relative reactivities shown in Table 1 should be similar when carbohydrates are the substrates, triflates (trifluoromethanesulfonates) should be orders of magnitude more reactive than tosylates (*p*-toluenesulfonates) and mesylates (methanesulfonates). Experience has shown that this assumption is justified [2–4]. The greater reactivity of triflates reduces reaction times and allows selection of milder reaction conditions; for example, the transformations shown in Eqs. (1) and (2) occur more rapidly

R= Tf   (1 h, 70 °C, 78%)
R= Ts   (3 days, 100 °C )

Tf = $CF_3SO_2$
Ts = *p*- $CH_3C_6H_4SO_2$

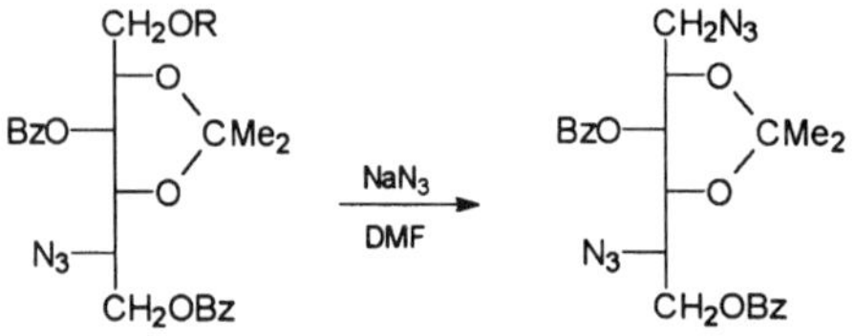

(1)

R = Tf   (30 min, 25 °C, 91%)
R = Ts   (10 h, 120 °C)

(2)

and at lower temperature when triflates replace tosylates as reactants [5]. Greater triflate reactivity becomes critical when other leaving groups cannot be displaced [Eq. (3); 6].

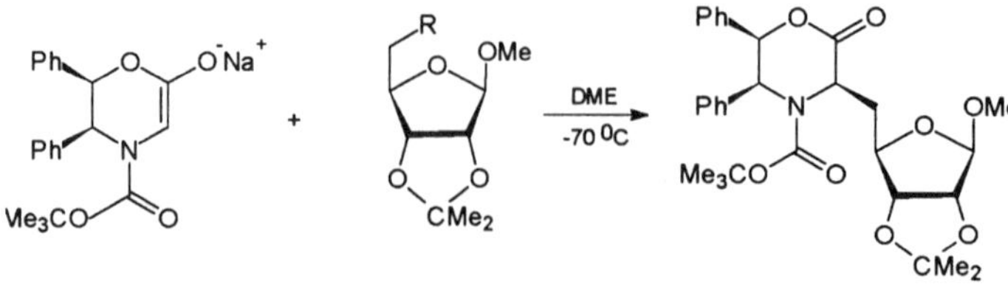

R = OTf   40 % yield
R = OMs   no reaction
R = I     no reaction

DME = $CH_3OCH_2CH_2OCH_3$
Ms = $CH_3SO_2$—

(3)

**Table 1** Relative Rates of Solvolysis of 1-Phenylethyl Esters and Halides

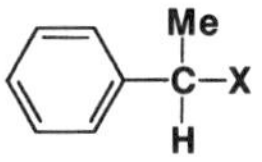

Ph-CH(Me)-X

| X | $k_{rel}$ | X | $k_{rel}$ |
|---|---|---|---|
| Cl | 1 | 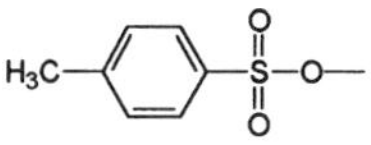 4-MeC$_6$H$_4$-SO$_2$-O— | 37,000 |
| 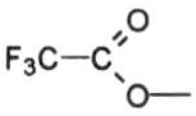 F$_3$C-C(=O)-O— | 2 | 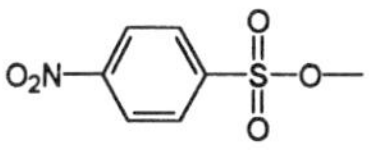 4-O$_2$N-C$_6$H$_4$-SO$_2$-O— | 440,000 |
| Br | 14 | Imidazolyl-SO$_2$-O— | Data not available |
| I | 91 | F$_3$C-SO$_2$-O— | 14 million |
| H$_3$C-SO$_2$-O— | 30,000 | | |

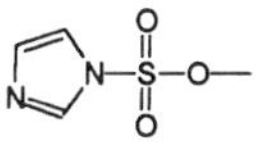

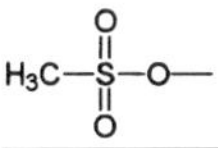

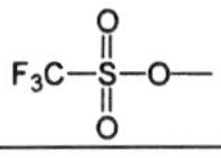

## B. Synthesis of Triflates

Section V contains specific examples of procedures for triflate synthesis; some general comment about these procedures is worthwhile. Triflic anhydride is the reagent of choice for preparing triflates derived from sugars, whereas triflyl chloride is used more often when nucleosides are involved. Triflate formation with triflic anhydride requires addition of a base (usually pyridine) to the reaction mixture to neutralize the triflic acid produced [Eq. (4); 7]. Some triflates are reactive enough that pyridine can function as a nucleophile in the substitution process. In these instances, replacement of pyridine with a non-nucleophilic base, such as 2,6-di-*t*-butyl-4-methylpyridine, avoids this undesired reaction (Scheme 1) [7].

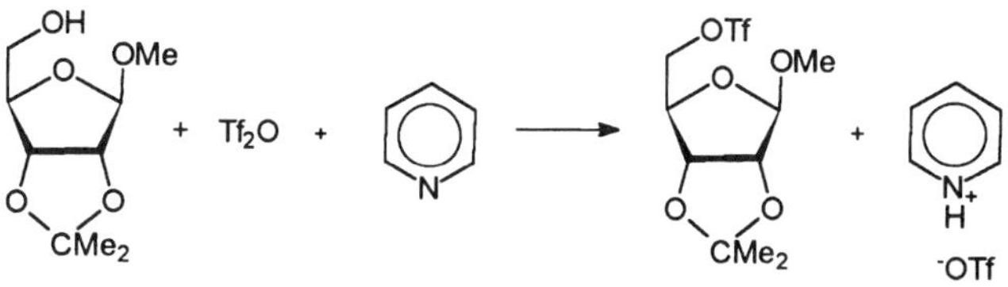

Sugar-OH + Tf$_2$O + pyridine ⟶ Sugar-OTf + pyridinium⁺ ⁻OTf    (4)

Triflyl chloride is a less reactive triflating agent than triflic anhydride. When triflyl chloride is used, deprotonating the participating hydroxyl group is a common procedure for increasing its reactivity before triflyl chloride addition (Scheme 2) [8]. Formation of an alkoxy anion has the advantage that addition of an organic base to the reaction mixture is no longer necessary for acid neutralization; however, anion formation requires decidedly more basic reaction conditions than those created by the presence of pyridine. An alternative to prior alkoxide formation is to add a base stronger than pyridine (e.g., *N,N*-dimethylamino-pyridine [DMAP]) to the reaction mixture to increase the concentration of alkoxide ions [Eq. (5); 9]. Practical factors that should be considered with the use of triflyl chloride include appropriate procedures for handling a volatile liquid (bp 29–32°C) and recognition that the chloride ion generated during triflate formation may become an undesired nucleophile.

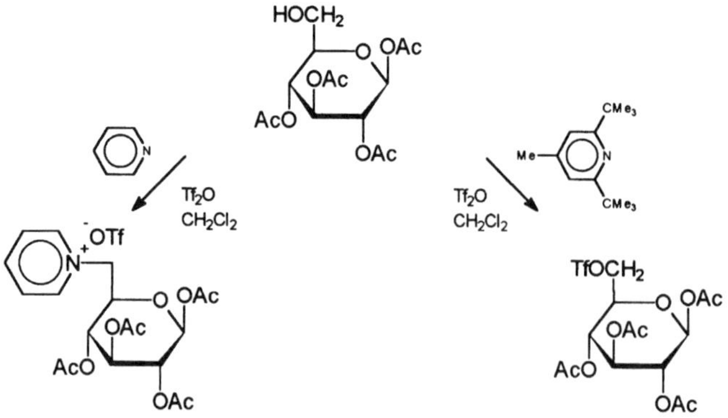

**Scheme 1** Triflate reaction with pyridine.

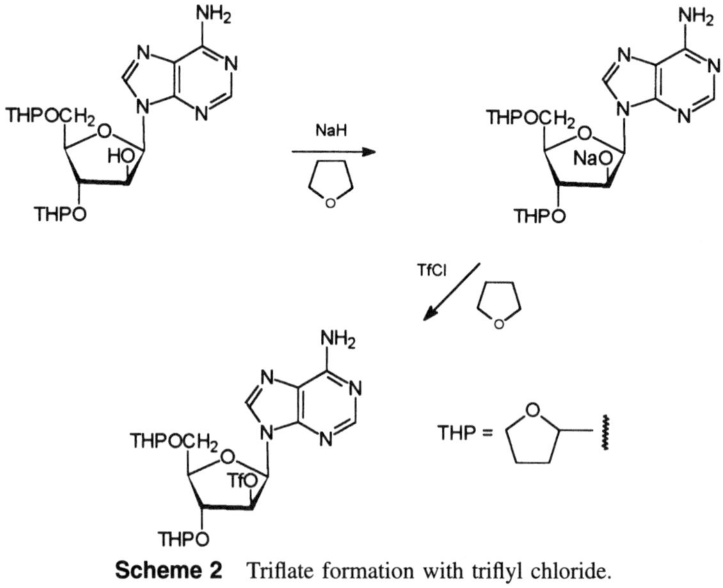

**Scheme 2** Triflate formation with triflyl chloride.

## II. INTRODUCING AZIDO AND HALOGENO GROUPS BY TRIFLATE DISPLACEMENT

### A. Deoxyhalogeno Sugars

During the past two decades, triflate displacement has become a common method for introducing heteroatoms into carbohydrates. One group of compounds easily prepared from triflates is the deoxyhalogeno sugars [Eq. (6); 10]. Synthesizing halogenated carbohydrates by $S_N2$ substitution reactions has important ramifications owing to the biological activity and synthetic usefulness of these compounds. For example, [$^{18}$F]2-deoxy-2-fluoro-D-glucose is a radiopharmaceutical used in positron emission tomography (PET) [11]. One important method for synthesis of this deoxyfluoro sugar involves the substitution reaction shown in Eq. (7) [12]. In a different type of example, triflate formation and displacement by iodide ion provides a means for regioselectively replacing primary hydroxyl groups with iodine under mild conditions. Regioselective iodination has become part of a procedure for

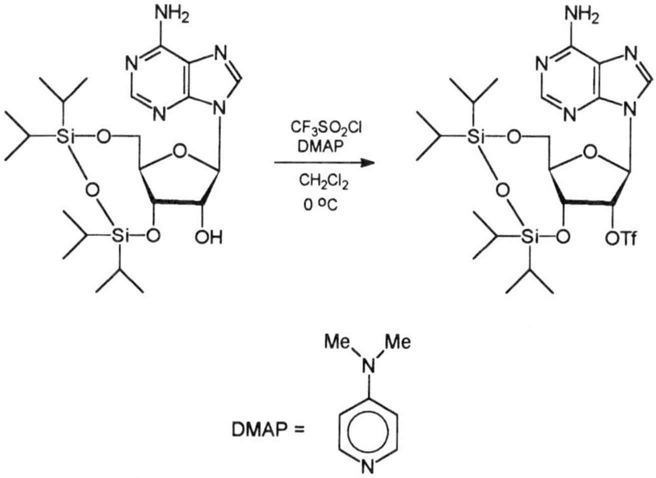

# $S_N2$-Type Halogenation and Azidation Reactions

$$(5)$$

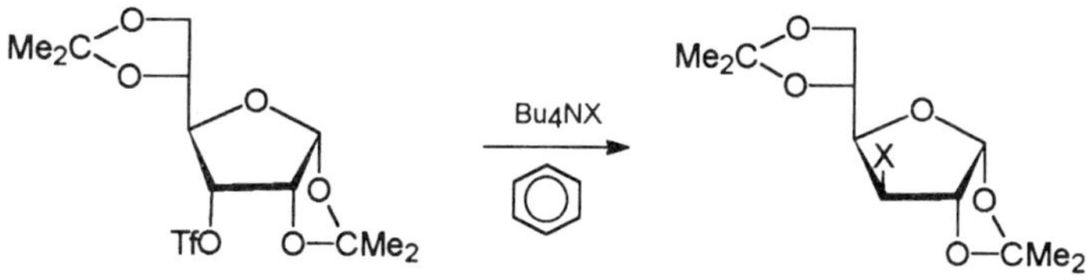

DMAP =

$$(6)$$

X = I    87%
X = Br   96%
X = Cl   85%

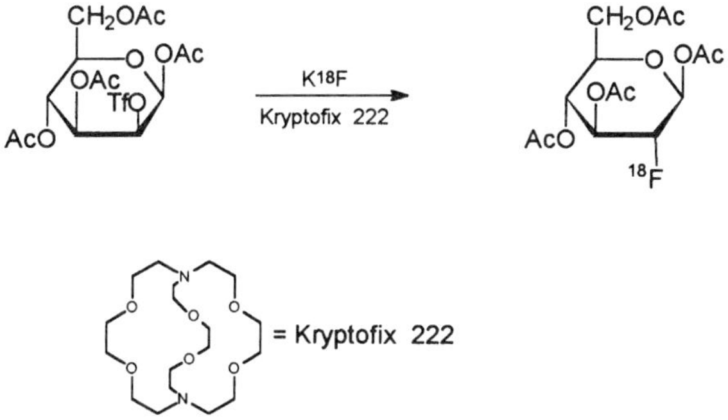

$$(7)$$

= Kryptofix 222

the selective, nonhydrolytic cleavage of glycosidic linkages [Eq. (8); 13]. This process is useful in determining structures of complex carbohydrates [14].

Although many deoxyhalogeno sugars have been synthesized by tosylate and mesylate displacement (and other reactions) [15–17], halogenated carbohydrates have become more readily available because they can be synthesized from triflates. Triflate displacement permits the synthesis of a number of compounds that are difficult or impossible to prepare in other ways. Because deoxyhalogeno sugars serve as intermediates in the synthesis of other sugar derivatives, improved methods for deoxyhalogeno sugar formation translate into better procedures for preparation of other carbohydrates. For example, triflate displacement by iodide or bromide ions, when followed by replacement of a halogen atom with a hydrogen atom, provides a pathway for formation of deoxy sugars (Scheme 3) [10,18].

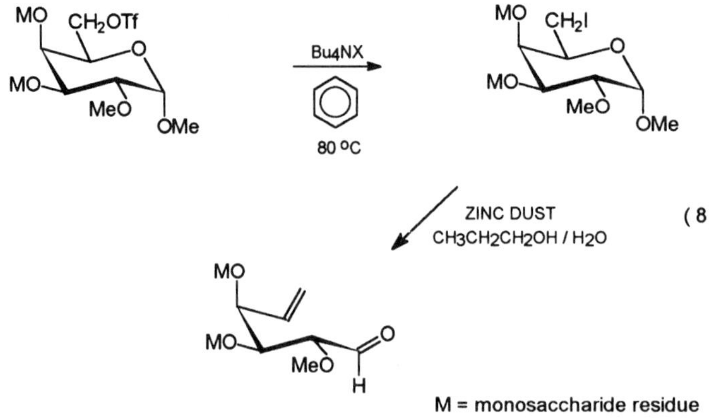

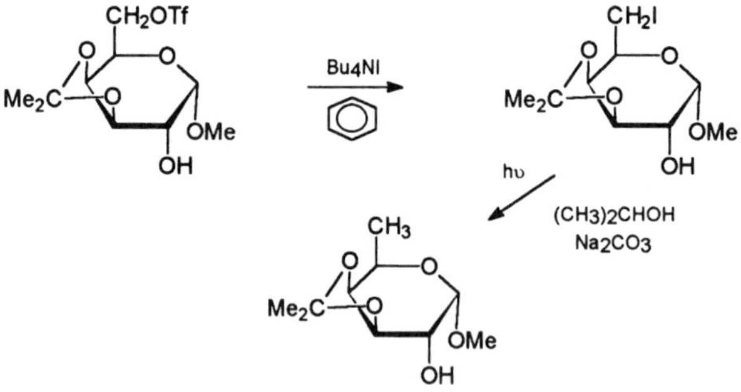

**Scheme 3** Synthesis of deoxy sugars.

(Although deoxy sugars can also be formed directly from triflates by reaction with hydride reagents [Eq. (9); 19], molecular rearrangement [Eq. (10); 20] sometimes accompanies this pathway.)

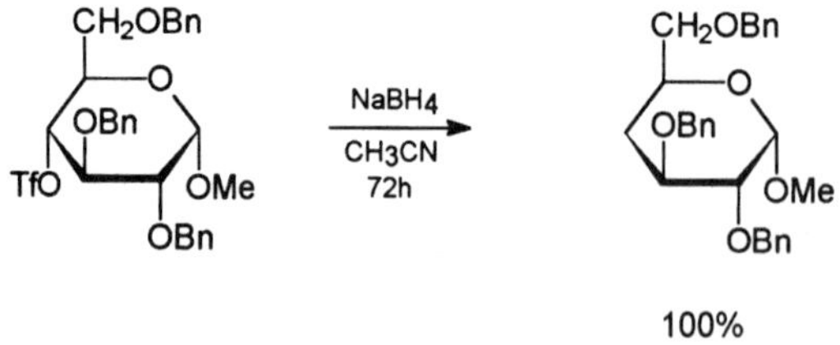

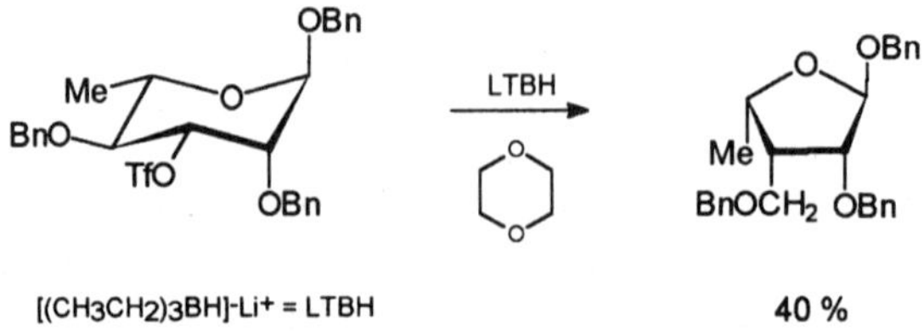

## B. Azidodeoxy Sugars

Sulfonate displacement by nitrogen nucleophiles is a powerful technique for introducing nitrogen atoms into carbohydrates [21]. Because the ultimate goal of such reactions is usually the synthesis of aminodeoxy sugars, azide ion is often a reactant. Azide ion is a good choice because it is a potent nucleophile and because azides are readily converted into the corresponding aminodeoxy sugars (Scheme 4) [21,22]. Azides also can be intermediates

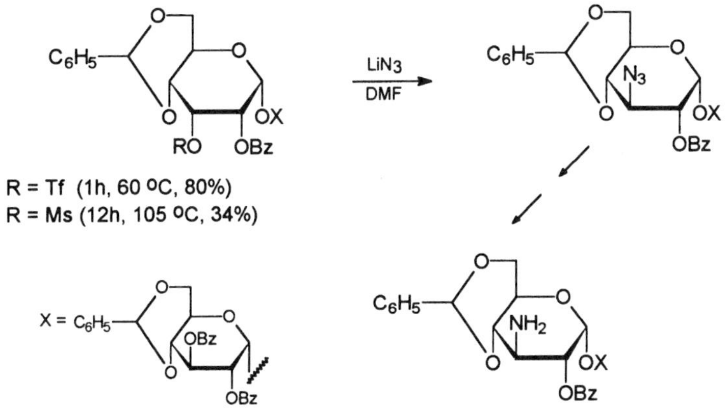

R = Tf (1h, 60 °C, 80%)
R = Ms (12h, 105 °C, 34%)

X = C$_6$H$_5$

**Scheme 4** Synthesis of aminodeoxy sugars.

in the synthesis of carbonyl compounds. Primary azides produce aldehydes in average to good yields on photolysis [Eq. (11); 23], but formation of ketones from secondary azides is generally a low-yield process [24].

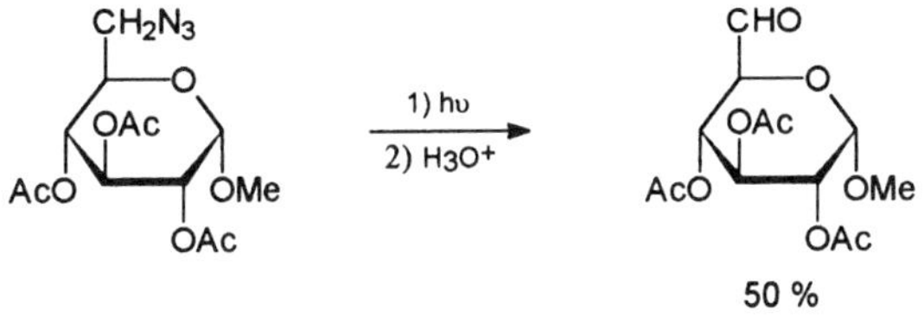

$$\qquad \qquad (11)$$

50 %

## III. REACTIONS OF CARBOHYDRATE TRIFLATES

### A. Substitution Reactions

For sulfonic acid esters of most organic compounds, substitution reactions occur under relatively mild conditions, even when less reactive sulfonates are the leaving groups. Esters as reactive as triflates typically are necessary only in those situations for which substitution is difficult; for example, when a leaving group is attached to an sp$^2$-hybridized carbon atom, or when the leaving group must depart from a carbon atom α to a carbonyl group. For highly functionalized molecules, such as carbohydrates, the situation is different. For carbohydrates, displacement reactions with less reactive sulfonic acid esters will generally take place from primary carbon atoms—although vigorous reaction conditions may be required—but substitution at secondary carbon atoms is sometimes impossible. In contrast, when carbohydrate triflates are the substrates, S$_N$2 substitution reactions at primary and at secondary carbon atoms are usually possible under mild conditions. [Examples appear in Eqs. (1), (2), (6), (7), (8), and in Schemes 3 and 4.] The broad range of compounds prepared from carbohydrate triflates stands as evidence that triflate displacement is a reaction particularly well-suited for carbohydrates [2].

## B. Elimination Reactions

Even though triflates provide the needed level of reactivity for substitution processes in carbohydrates, other reactions, such as elimination, can compete with substitution. Factors that promote competing reactions can usually be identified; for example, strong bases and weak nucleophiles favor $E_2$ elimination. Because among halide ions the strongest base is the fluoride ion, its reaction with triflates generates elimination products most often [4]. Although triflate displacement by fluoride ion without competing reactions does occur [Eq. (12); 25], substitution by this anion often is accompanied by elimination (Table 2) [26] and,

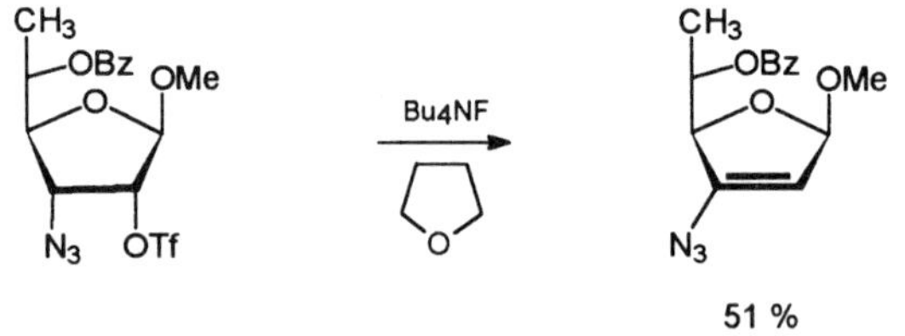

sometimes, elimination products are the only ones formed [Eq. (13); 27]. Elimination reactions, when other halide ions are the nucleophiles, are much less common. Comparative studies, in which halide ions react with a triflate that structurally favors elimination, show that fluoride ion causes the most elimination reactions to occur (Table 3) [28]. Azide ion also is sufficiently basic to promote elimination reactions [Eq. (14); 29]; however,

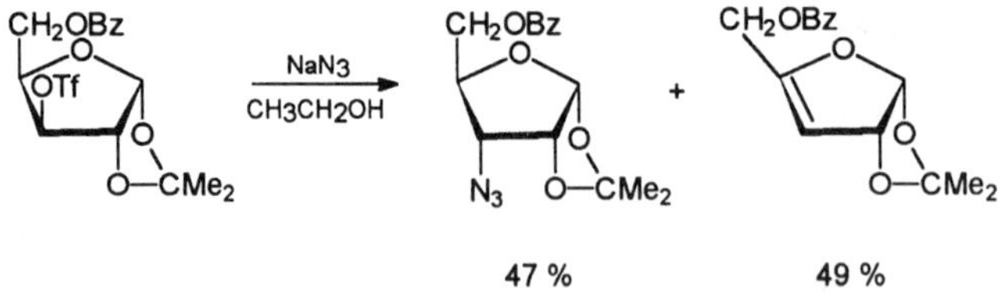

because azide ion is an effective nucleophile, substitution is usually the dominant or exclusive reaction [Eq. (15); 30].

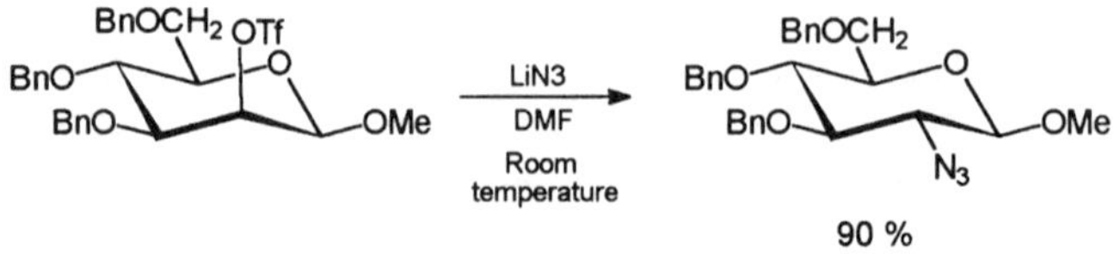

**Table 2** Substitution and Elimination Reactions by Fluoride Ion

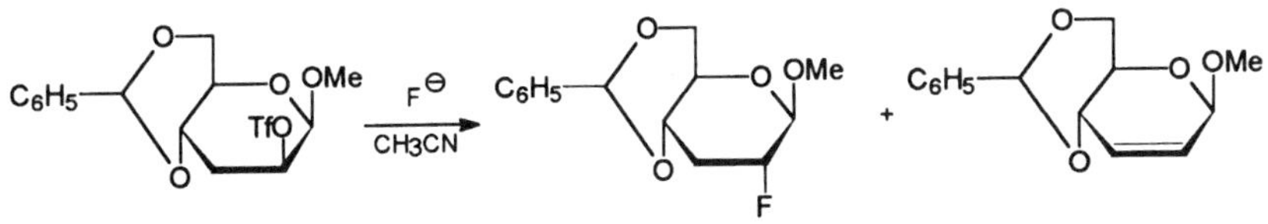

| Compound | Substitution product yield (%) | Elimination product yield (%) |
|---|---|---|
| Bu$_4$NF | 40 | 27 |
| KF | 0 | 0 |
| KF–crown ether | 48 | 37 |
| CsF | 11 | 21 |

**Table 3** Competing Substitution and Elimination Reactions

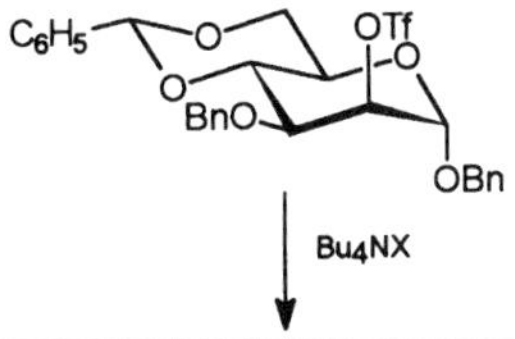

| | I | Br | Cl | F | N$_3$ |
|---|---|---|---|---|---|
| 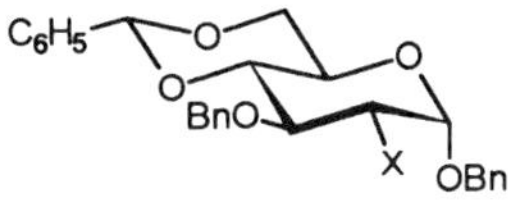 (substitution product) | 79% | 26% | 10% | — | 38% |
| 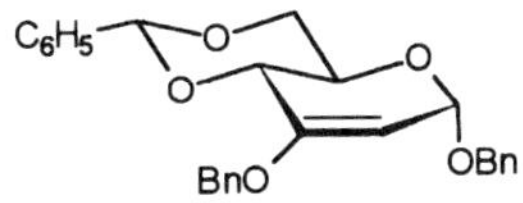 (elimination product 1) | — | 24% | 11% | 88% | 49% |
| (ketone product 1) | — | 21% | — | — | — |
| (enone product) | — | 24% | 54% | — | — |

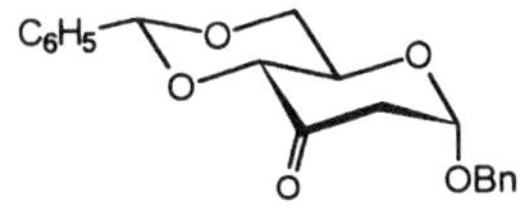

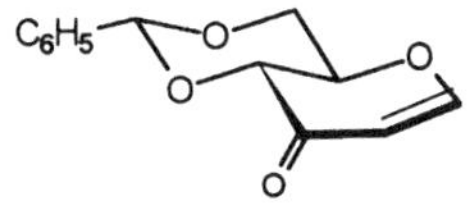

## C. Rearrangement Reactions

Although molecular rearrangements of carbohydrate triflates are rare, they are not unknown [2,21,31]. Many of these rearrangements are ring-contraction reactions associated with treatment of triflates with lithium triethylborohydride [see Eq. (10); 20,21]. Reported rearrangement reactions occurring when halide or azide ions are the nucleophiles are quite rare and are restricted to epimerization at centers next to a carbonyl group. Such reactions have been reported only for the more basic fluoride [32] and azide ions (Scheme 5) [33].

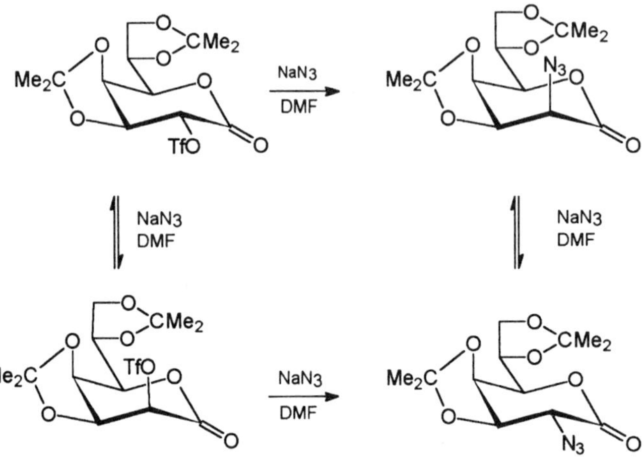

**Scheme 5** Epimerization by azide ion.

## D. Effect of Anion Nucleophilicity on Reactivity

Increasing the effective nucleophilicity of an ion allows $S_N2$ substitution reactions to occur under milder conditions. An anion will become a better nucleophile when it is less effectively solvated and when it is further separated from its counterion. Methods that can achieve these changes include selection of a tetraalklammonium counterion [see Eqs. (6) and (8)], addition of a crown ether or a cryptand [see Eq. (7)], and use of a solvent that effectively solvates cations [see Eqs. (1) and (2)].

The nucleophile for which increased reactivity is most critical is the fluoride ion [3,4]. Water molecules bind tightly to this ion, and their presence dramatically reduces its effective nucleophilicity. A variety of fluoride ion sources have been used in an effort to improve product yields in deoxyfluoro sugar synthesis [26,34]. The yields of substitution and elimination products generated from reactions with fluoride ion from several sources are listed in Table 2 [26]. Currently, the most attractive source of fluoride ion is tris(dimethylamino)-sulfur (trimethylsilyl)difluoride (TASF), which is soluble in a variety of organic solvents and produces an anhydrous fluoride ion [35].

## IV. CARBOHYDRATE IMIDAZOLYLSULFONATES

Carbohydrate imidazolylsulfonates (imidazylates) [36] are a type of sulfonic acid ester that deserves special comment because these compounds react readily with nucleophiles in $S_N2$ substitution processes. Although imidazylates are not used as frequently as triflates in substitution reactions, they appear to be comparable in reactivity [Eq. (16); 37,38].

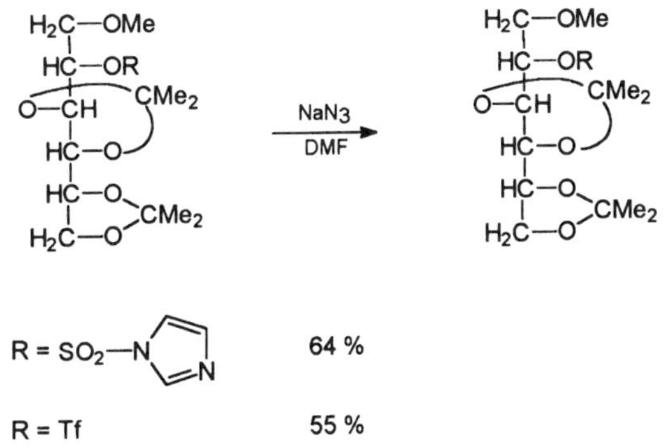

## V. EXPERIMENTAL PROCEDURES*

### A. Synthesis of Triflate Esters with Triflic Anhydride (General Procedure) [Eq. (17); 10]

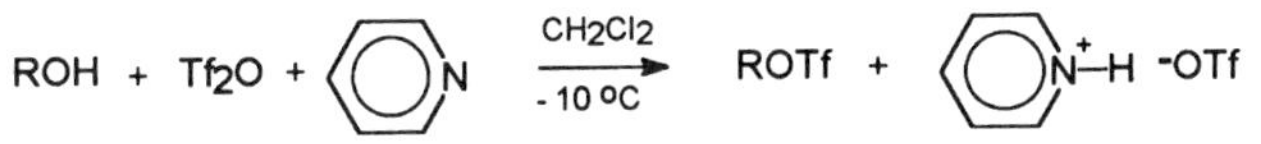

A 100-mL, two-neck, round-bottom flask, equipped with two addition funnels was charged with pyridine (0.43 mL, 5.5 mmol) and 20 mL of methylene chloride. A solution of triflic anhydride (0.86 mL, 5.11 mmol) dissolved in 10 mL of methylene chloride was placed in one addition funnel. The sugar (2.55 mmol) dissolved in 10 mL of methylene chloride was placed in the other addition funnel. The flask was cooled to $-10°C$ in an ice–acetone bath, and the triflic anhydride solution added dropwise. A thick white precipitate began to form during the addition. After addition was complete, the suspension was allowed to stir for an additional 10 min. The sugar solution was added dropwise and stirring continued for an additional 1.5 h. The reaction mixture was poured into 50 mL of ice water, the layers were separated, and the aqueous layer was extracted with two 50-mL portions of methylene chloride. The combined extracts were dried over sodium sulfate, and the solvent was removed in vacuo. Hexane extraction, followed by in vacuo solvent removal, gave the triflate ester [10].

### B. Synthesis of a Triflate Ester with Triflyl Chloride [9]

*2'-O-(Trifluoromethanesulfonyl)-3',5'-O-(1,1,3,3-tetraisopropyl-1,3-disiloxanyl)adenosine [Eq. (18) 2]*

Trifluoromethanesulfonyl chloride (202 mg, 1.20 mmol) was added to a cold (0°C), stirred solution of **1** [39] (509 mg, 1.00 mmol), and DMAP (366 mg, 3.00 mmol), in anhydrous $CH_2Cl_2$ (5 mL). The yellow solution was stirred for 10 min and partitioned between ice-cold $AcOH/H_2O$ (1:99, 150 mL) and $CH_2Cl_2$ (2 × 75 mL). The combined organic phase was washed with ice-cold, saturated $NaHCO_3/H_2O$ (150 mL), brine (150 mL), and dried $Na_2SO_4$), filtered, and evaporated. The residue (619 mg) was crystallized ($CHCl_3$/hexanes, 1:2) to give **2** (422 mg). An additional 54 mg was obtained by preparative thin-layer chromatography (TLC; $CHCl_3$/MeOH, 19:1) of the concentrated mother liquor to give **2** (476 mg, 74%) as a colorless solid: mp 152–154°C dec.

---

*Reprinted with permission from *J. Org. Chem.* 45:4387 (1980). Copyright © 1980 American Chemical Society.

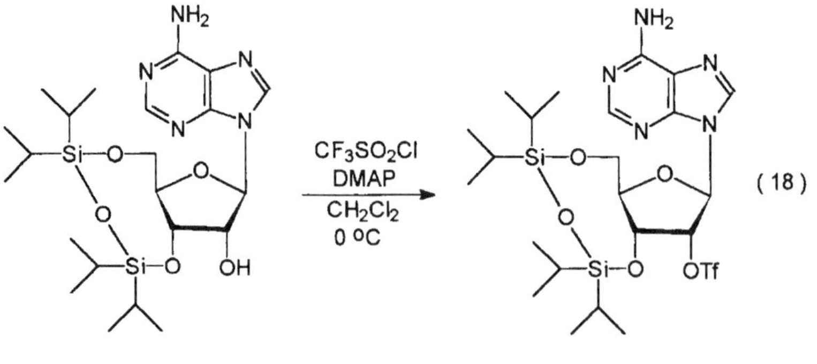

[Structures of compounds 1 and 2 with reagents CF₃SO₂Cl, DMAP, CH₂Cl₂, 0 °C]

## C. Synthesis of a Reactive Triflate Ester in the Presence of a Nonnucleophilic Base, 2,6-Di-*tert*-butyl-4-methylpyridine [7]*

*Preparation of 1,2,3,4-Tetra-O-[(trifluoromethyl)sulfonyl]-β-D-glucopyranose [Eq. (19); 4]*

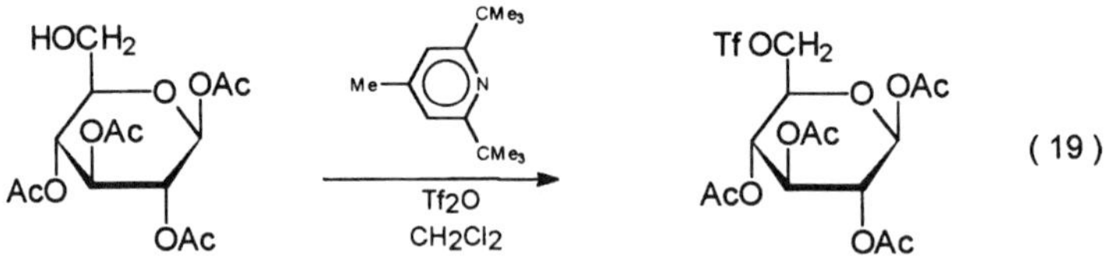

[Structures of compounds 3 and 4 with reagents Tf₂O, CH₂Cl₂, and 2,6-di-*tert*-butyl-4-methylpyridine] (19)

Trifluoromethanesulfonic (triflic) anhydride (0.32 mL, 1.90 mmol) and 2,6-di-*t*-butyl-4-methylpyridine (0.396 g, 1.93 mmol) were dissolved in 20 mL of anhydrous CH$_2$Cl$_2$ at 25°C, and to this solution was added 0.475 g (1.36 mmol) of 1,2,3,4-tetra-O-acetyl-β-D-*gluco*-pyranose (**3**) [40] in 10 mL of CH$_2$Cl$_2$. After stirring at 25°C for 1 h, the reaction mixture was poured into 200 mL of ice water containing 2 g of NaHCO$_3$ and shaken vigorously. The layers were separated, and the aqueous layer was extracted with 20 mL of CH$_2$Cl$_2$. The combined organic extracts were dried over Na$_2$SO$_4$, and the solvent was removed by distillation under reduced pressure. Chromatography produced **4** in 84% yield, mp 85°C dec.

## D. Triflate Displacement by Bromide, Chloride, and Iodide Ions [41]

*Syntheses of 1,3,4,6-Tetra-O-acetyl-2-deoxy-2-halogeno-β-D-glucopyranoses [Eq. (20); 6, 7, and 8]*

The triflate **5** (0.593 g, 1.70 mmol) was combined with the appropriate tetrabutylammonium halide (3.34 mmol) in 30 mL of anhydrous benzene and heated at reflux for 2 h. After cooling to room temperature, the benzene was removed under reduced pressure. The residue was dissolved in the minimum amount of dichloromethane, chromatographed *on a 2.5 × 10 cm column of 230–400 mesh silica gel with a 3:2 ratio of ethyl ether/hexane*

---

*Reprinted with permission from *J. Org. Chem* 48:676–677 (1983). Copyright © 1983 American Chemical Society.

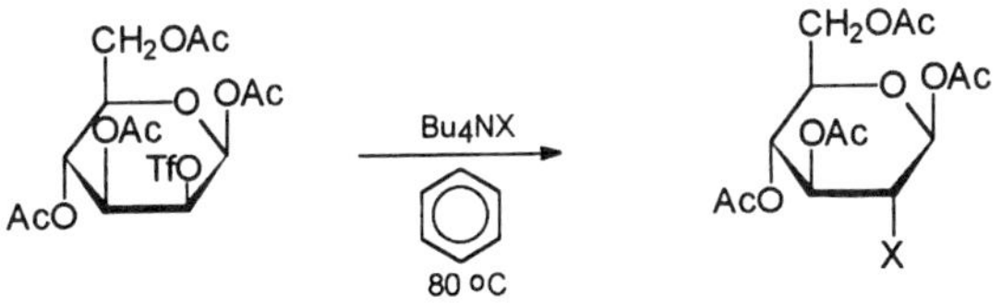

# $S_N2$-Type Halogenation and Azidation Reactions

(general procedure [41]) and recrystallized from ethanol. The reaction products [see Eq. (20) **6**, **7**, and **8**] were identified by analysis of their NMR spectra, and melting point comparison with values reported in the literature: compound **6** mp 95–96°C; compound **7** mp 110–111°C; compound **8**, mp 113.5–114.5°C.

## E. Triflate Displacement by Bromide, Chloride, and Iodide Ions from a Nucleoside [42]

*5-Benzoyl-4-(dibenzoylamino)-7-[3,5-O-1,1,3,3-tetraisopropyldisiloxanyl)-2-deoxy-2-bromo-β-D-arabinofuranosyl]pyrrolo[3,2-d]-pyrimidine [Eq. (21); **10**]*

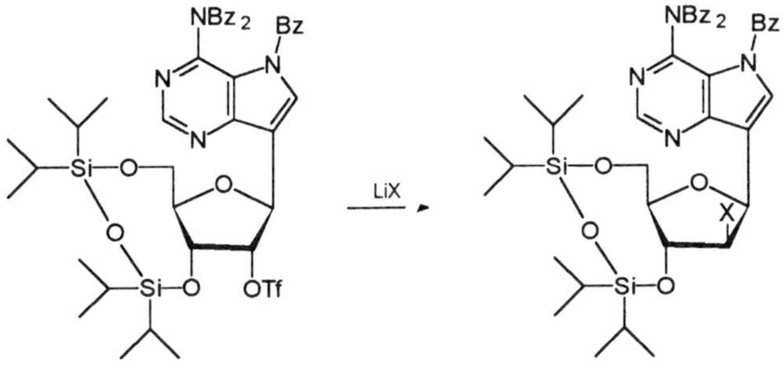

A mixture of **9** (1.50 g, 1.57 mmol) and lithium bromide (0.284 g, 3.27 mmol) in dry hexamethylphosphoramide (HMPA; 16 mL) was stirred for 4 h in a sealed vessel under an argon atmosphere. The mixture was poured over ice water (~100 g). The precipitate formed, was separated by filtration, washed thoroughly with $H_2O$, and dried in vacuo at 60°C over $P_2O_5$ for 4 h. The crude product (2.04 g) was purified by flash column chromatography using $CH_2Cl_2$/MeOH (200:1) as the eluent to give **10** as a white solid (1.16 g, 57%; mp 135–136°C).

*5-Benzoyl-4-(dibenzoylamino)-7-[3,5-O-1,1,3,3-tetraisopropyldisiloxanyl)-2-deoxy-2-chloro-β-D-arabinofuranosyl]pyrrolo[3,2-d]-pyrimidine [see Eq. (21); **11**]*

A mixture of **9** (1.31 g, 1.3 mmol) and lithium chloride (1.50 mmol) in dry HMPA (10 mL) was stirred under argon in a sealed vessel at 25°C for 1 h. The mixture was poured over ice water. The light-yellow precipitate formed was separated by filtration, washed thoroughly with $H_2O$, and dried over $P_2O_5$ in vacuo to give a crude product, which was purified by preparative TLC (five plates) using $CH_2Cl_2$/MeOH (100:1) as the developing agent. The plates were developed twice. The desired product was extracted with $CH_2Cl_2$:MeOH (100:2), filtered, and the filtrate concentrated in vacuo to give **11** as a white solid (0.525 g, 45%; mp 124–126°C).

*5-Benzoyl-4-(dibenzoylamino)-7-[3,5-O-1,1,3,3-tetraisopropyldisiloxanyl)-2-deoxy-2-iodo-β-D-arabinofuranosyl]pyrrolo[3,2-d]-pyrimidine [see Eq. (21); 12]*

A mixture of **9** (2.85 g, 2.99 mmol) and potassium iodide (0.496 g, 2.99 mmol) in dry HMPA (30 mL) under argon in a sealed vessel, was sonicated for 10 min and then magnetically stirred at 25°C for 24 h. The mixture was then poured over crushed ice and the resulting precipitate filtered, washed thoroughly with $H_2O$, and dissolved in $CH_2Cl_2$. The $CH_2Cl_2$ solution was washed with $H_2O$, dried ($Na_2SO_4$), and evaporated in vacuo. The resulting residue was subjected to flash column chromatography using $CH_2Cl_2$/MeOH (300:1) as the eluent. Fractions containing the pure product were pooled and concentrated in vacuo to give **12** as a white solid (1.30 g, 46%; mp 124–126°C).

### F. Triflate Displacement by Fluoride Ion [7]*

**13** → **14** (22)

(Reagents: $Bu_4NF$, $CH_2Cl_2$)

*Synthesis of 1,1,3,4-Tetra-O-acetyl-β-D-glucopyranose Derivatives*

Triflic anhydride (0.34 mL, 1.9 mmol) and compound **13** [Eq. (22)] (0.41 g, 2.0 mmol) were dissolved in 20 mL of anhydrous $CH_2Cl_2$. To this solution was added 0.49 g (1.4 mmol) of tetraacetate in 10 mL of $CH_2Cl_2$ and the reaction mixture stirred for 60 min. A solution of 4.0 mmol of tetrabutylammonium fluoride in 20 mL of $CH_2Cl_2$ then was added and the mixture stirred for 3 h. [When $Bu_4NF\cdot3H_2O$ was used as the nucleophile, the triflate **13** was isolated. This reaction required the $CH_2Cl_2$ to be distilled at this point, benzene (50 mL) added, and the reaction mixture heated to reflux for 5 min.] The reaction mixture next was poured into 200 mL of cold 1% $NaHCO_3$ solution and shaken vigorously in a separatory funnel. The organic phase was removed and the aqueous phase extracted with 20 mL of $CH_2Cl_2$. The combined organic layers were washed with 50 mL of 5% $NaHSO_3$ and 50 mL of saturate $NaHCO_3$ and dried over anhydrous $Na_2SO_4$. After distillation of the solvent, under reduced pressure, the residue was chromatographed on a 2.5 × 10 cm column of 230- to 400-mesh silica gel with a 3:2 ratio of ether/pentane. [General procedure from Ref. 7] Compound **14**, (27%; mp 126.5–127.5°C) was isolated.

---

*Reprinted with permission from *J. Org. Chem.* 48:676–677 (1983). Copyright © 1983 American Chemical Society.

## G. Triflate Displacement by Azide Ion [9,43]

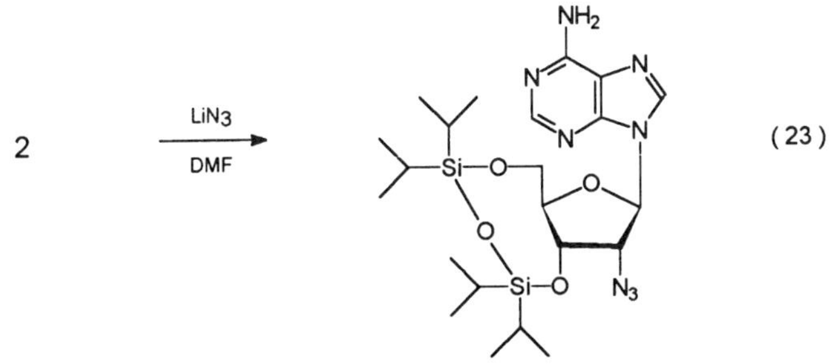

*2'-Azido-2'-deoxy-3',5'-O-(1,1,3,3-tetraisopropyl-1,3-disiloxanyl)adenosine [Eq. (23); 15]*

A solution of crude **2** (688 mg) and LiN$_3$ (245 mg, 5.0 mmol) in anhydrous DMF (10 mL) was stirred at ambient temperature for 2 h. Water (50 mL) was added and the mixture extracted (EtOAc, 2 × 100 mL). The combined organic phase was washed with brine (100 mL), dried (Na$_2$SO$_4$), and evaporated. Crystallization (98% EtOH) of the resulting solid foam (607 mg) gave **15** (350 mg, two crops); 66%, as colorless rods: mp 175–177°C.

*4-O-Allyl-1,6-anhydro-2-azido-3-O-benzyl-2-deoxy-β-D-glucopyranose [Eq. (24); 17]*

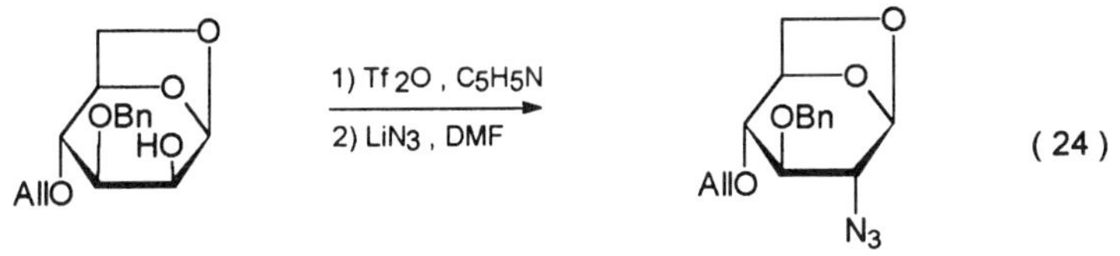

Trifluoromethanesulfonic anhydride (0.54 mL, 3.22 mmol) in dry 1,2-dichloroethane (2 mL) was added under an atmosphere of oxygen-free nitrogen at −10°C to a stirred solution of pyridine (0.30 mL, 3.75 mmol) and dry 1,2-dichloroethane (10 mL). After 10 min, a solution of compound **16** (0.47 g, 1.61 mmol) in 1,2-dichloroethane (3 mL) was added to the mixture and stirring was continued for 1 h at 0°C. Then TLC analysis indicated complete conversion of starting compound **16** ($R_f$ 0.19) into the C-2 triflyl ester ($R_f$ 0.68). Aqueous sodium bicarbonate (10 mL, 10%, w/v) and dichloromethane (50 mL) were added, the organic layer was separated, and the aqueous layer extracted with dichloromethane. The combined extracts were washed with water, dried (MgSO$_4$), evaporated, and coevaporated with toluene (10 mL). Crude triflate was dissolved in dry N,N-dimethylformamide (20 mL), lithium azide (0.91 g, 16 mmol) was added, and the mixture stirred at 20°C for 4 min, after which time TLC analysis showed complete conversion of the triflyl ester ($R_f$ 0.68) into the azido sugar **17** ($R_f$ 0.65). The solution was evaporated to dryness, dissolved in dichloromethane (50 mL), washed with water (20 mL), dried (MgSO$_4$), and concentrated in vacuo. The residual oil was applied to a column of silica gel (10 g, eluant dichloromethane): yield 0.43 g (85%); $R_f$ 0.28 (dichloromethane): [α]$_D$ +74° (c 1.0, chloroform).

## REFERENCES

1. P. J. Stang, M. Hanack, and L. R. Subramanian, Perfluoroalkanesulfonic esters: methods of preparation and applications in organic chemistry, *Synthesis* 85 (1982).
2. R. W. Binkley and M. G. Ambrose, Synthesis and reactions of carbohydrate trifluoromethanesulfonates (carbohydrate triflates), *J. Carbohydr. Chem. 3*:1 (1984).
3. P. J. Card, Synthesis of fluorinated carbohydrates, *J. Carbohydr. Chem. 4*:451 (1985).
4. T. Tsuchiya, Chemistry and developments of fluorinated carbohydrates, *Adv. Carbohydr. Chem. Biochem. 48*:91 (1990).
5. T. Usui, Y. Takagi, T. Tsuchiya, and S. Umezawa, Synthesis of 3-*O*-acetyl-2,6-diazido-4-*O*-benzyl-2,6-dideoxy-L-idopyranosyl chloride, a glycosyl halide for the synthesis of neomycin B, *Carbohydr. Res. 130*:165 (1984).
6. L. W. Dudycz, Stereoselective synthesis of 6-amino-5,6-dideoxy-D-*ribo*-heptafuranouronates with the aid of glycine chiral templates, *Nucleosides Nucleotides 10*:329 (1991).
7. M. G. Ambrose and R. W. Binkley, Synthesis of deoxyhalogeno sugars. Reaction of halide ions with 1,2,3,4-tetra-*O*-acetyl-6-*O*-[(trifluoromethyl)sulfonyl-β-D-glucopyranose, *J. Org. Chem. 48*:674 (1983).
8. M. Ikehara and H. Miki, Studies of nucleosides and nucleotides. LXXXII. Cyclonucleosides. (39). Synthesis and Properties of 2′-halogeno-2′-deoxyadenosines, *Chem. Pharm. Bull. 26*: 2449 (1978).
9. M. J. Robins, S. D. Hawrelak, A. E. Hernandez, and S. F. Wnuk, Nucleic acid related compounds. 71. Efficient general synthesis of purine (amino, azido, and triflate)-sugar nucleosides, *Nucleosides Nucleotides 11*:821 (1992).
10. R. W. Binkley, M. G. Ambrose, and D. G. Hehemann, Synthesis of deoxyhalogeno sugars. Displacement of the (trifluoromethanesulfonyl)oxy (triflyl) group by halide ion, *J. Org. Chem. 45*:4387 (1980).
11. T. J. Tewson, Procedures, pitfalls and solutions in the production of [$^{18}$F]2-deoxy-2-fluoro-D-glucose: A paradigm in the routine synthesis of fluorine-18 radiopharmaceuticals, *Nucl. Med. Biol. 16*:533 (1989).
12. K. Hamacher, H. H. Coenen, and G. Stocklin, Efficient stereospecific synthesis of no-carrier-added 2-[$^{18}$F]fluoro-2-deoxy-D-glucose using aminopolyether supported nucleophilic substitution, *J. Nucl. Med. 27*:235 (1986).
13. G. O. Aspinall, D. Chatterjee, and L. Khondo, The hex-5-enose degradation: zinc dust cleavage of 6-deoxy-6-iodo-α-D-galactopyranosidic linkages in methylated di- and trisaccharides, *Can. J. Chem. 62*:2728 (1984).
14. G. O. Aspinall, Chemical modification and selective fragmentation of polysaccharides, *Acc. Chem. Res. 20*:114 (1987).
15. R. S. Tipson, Sulfonic esters of carbohydrates, *Adv. Carbohydr. Chem. 8*:107 (1953).
16. D. H. Ball and F. W. Parrish, Sulfonic esters of carbohydrates: Part I, *Adv. Carbohydr. Chem. 23*:233 (1968).
17. D. H. Ball and F. W. Parrish, Sulfonic esters of carbohydrates: Part II, *Adv. Carbohydr. Chem. 24*:139 (1969).
18. R. W. Binkley and D. G. Hehemann, Photolysis of deoxyiodo sugars, *Carbohydr. Res. 74*:337 (1979).
19. E. P. Barrette and L. Goodman, Convenient and stereospecific synthesis of deoxy sugars. Reductive displacement of trifluoromethanesulfonates, *J. Org. Chem. 49*:176 (1984).
20. V. Pozsgay and A. Neszmelyi, Ring contraction by carbon participation in a hexopyranoside ring: Formation of benzyl 2-*O*-benzyl-3,5-dideoxy-3-*C*-benzyloxymethyl-α-L-*lyxo*-pentofuranoside, *Tetrahedron Lett. 21*:211 (1980).
21. H. H. Baer, Recent synthetic studies in nitrogen-containing and deoxygenated sugars, *Pure Appl. Chem. 61*:1217 (1989).
22. H. H. Baer and A. J. Bell, The synthesis of 3-amino-3-deoxy-α-D-glucopyranosyl α-D-

glucopyranoside (3-amino-3-deoxy-α, α-trehalose), *Carbohydr. Res. 75*:175 (1979).
23. D. Horton, A. E. Luetzow, and J. C. Wease, Generation of an aldehyde group in a protected sugar by photolysis of a primary azide, *Carbohydr. Res. 8*:366 (1968).
24. D. M. Clode and D. Horton, Photolysis of secondary azides of sugars, *Carbohydr. Res. 14*:405 (1970).
25. L. A. Mulard, P. Kovac, and C. P. J. Gilaudemans, Synthesis of specifically monofluorinated ligands related to the *O*-polysaccharide of *Shigella dysenteriae* type 1, *Carbohydr. Res. 259*:21 (1994).
26. T. Tsuchiya, Y. Takahaski, M. Endo, S. Umezawa, and H. Umezawa, Synthesis of 2'3'-dideoxy-2'-fluorokanamycin A, *J. Carbohydr. Chem. 4*:587 (1985).
27. A. Hasegawa, M. Goto, and M. Kiso, An unusual behavior of methyl or benzyl 3-azido-5-*O*-benzoyl-3,6-dideoxy-α-L-talofuranoside with (dimethylamino) sulfur trifluoride; migration of the alkoxyl group from the C-1 to the C-2 position, *J. Carbohydr. Chem. 4*:627 (1985).
28. W. Karpiesiuk, A. Banaszek, and A. Zamojski, Nuceophilic displacement of the triflate group in benzyl 3-*O*-benzyl-4,6-*O*-benzylidene-2-trifluoromethanesulfonyl-α-D-mannopyranoside, *Carbohydr. Res. 186*:156 (1989).
29. M. Yamashita, Y. Kawai, I. Uchida, T. Komori, M. Kohsaka, H. Imanaka, K. Sakane, H. Setoi, and T. Teraji, Structure and total synthesis of chrysandin, a new antifungal antibiotic, *Tetrahedron Lett. 25*:4689 (1984).
30. J. N. Vox, J. H. van Boom, C. A. A. van Boeckel, and T. Beetz, Azide substitution at mannose derivatives. A potential new route to glucosamine derivatives, *J. Carbohydr. Chem. 3*:117 (1984).
31. B. Nawrot, K. W. Pankiewicz, R. A. Zepf, and K. A. Watanabe, Synthesis and reactivity of benzyl 2-*O*-trifluoromethylsulfonyl- and benzyl 3-*O*-trifluoromethylsulfonyl-β-D-ribofuranoside—the first evidence of trifluoromethylsulfonyl (triflyl) migration in carbohydrates, *J. Carbohydr. Chem. 7*:95 (1988).
32. R. Albert, K. Dax, S. Seidl, H. Sterk, and A. E. Stutz, 5-Deoxy-5-fluoro-D-glucofuranose and L-iodofuranose synthesis and NMR studies, *J. Carbohydr. Chem. 4*:513 (1985).
33. I. Bruce, G. W. J. Fleet, A. Girdhar, M. Haraldsson, J. M. Peach, and D. J. Watkin, Retention and apparent inversion during azide displacement of α-triflates of 1,5-lactones, *Tetrahedron 46*:19 (1990).
34. T. Haradahira, M. Maeda, Y. Kai, H. Omae, and M. Kojima, Improved synthesis of 2-deoxy-2-fluoro-D-glucose using fluoride ion, *Chem. Pharm. Bull. 33*:165 (1985).
35. B. Doboszewski, G. W. Hay, and W. A. Szarek, The rapid synthesis of deoxyfluoro sugars using tris(dimethylamino)sulfonium difluorotrimethylsilicate (TASF), *Can. J. Chem. 65*:412 (1987).
36. S. Hanessian and J.-M. Vatele, Design and reactivity of organic functional groups: imidazolylsulfonate (imidazylate)—an efficient and versatile leaving group, *Tetrahedron Lett. 22*:3579 (1981).
37. C. Auge, S. David, and A. Malleron, An inexpensive route to 2-azido-2-deoxy-D-mannose and its conversion into an azido analog of *N*-acetylneuraminic acid, *Carbohydr. Res. 188*:201 (1989).
38. M. Kiso, A. Yasui, and A. Hasegawa, Synthesis of 2-azido-2-deoxy-3,4:5,6-di-*O*-isopropylidene-*aldehydo*-D-mannose dimethyl acetal and 4-azido-4-deoxy-2,3:5,6-di-*O*-isopropylidene-*aldehydo*-D-galactose dimethyl acetal, *Carbohydr. Res. 127*:137 (1984).
39. M. J. Robins, J. S. Wilson, and F. Hansske, Nucleic acid related compounds. 42. A general procedure for efficient deoxygenation of secondary alcohols. Regiospecific and stereoselective conversion of ribonucleosides to 2'deoxynucleosides, *J. Am. Chem. Soc. 105*:4059 (1983).
40. D. D. Reynolds and W. L. Evans, β-D-Gluco-1,2,3,4-tetraacetate. in *Organic Synthesis* Vol. 3 (E. C. Horning, ed.), John Wiley & Sons, New York, 1955, p. 432.
41. R. W. Binkley, M. G. Ambrose, and D. G. Hehemann, Reactions of per-*O*-acetylated carbohydrate triflates with halide ions, *J. Carbohydr. Chem. 6*:203 (1987).
42. K. V. B. Rao, W.-Y. Ren, J. H. Burchenal, and R. S. Klein, Nucleosides 137. Synthetic

modifications at the 2′-position of pyrrolo[3,2-*d*]pyrimidine and thieno[3,2-*d*]pyrimidine C-nucleosides. Synthesis of "2′-deoxy-9-deazaadenosine" and of "9-deaza *ara*-A," *Nucleosides Nucleotides 5*:539 (1986).

43. M. Kloosterman, M. P. de Nijs, and J. H. van Boom, Synthesis of 1,6-anhydro-2-*O*-trifluoromethanesulphonyl-β-D-mannopyranose derivatives and their conversion into the corresponding 1,6-anhydro-2-azido-deoxy-β-D-glucopyranoses: A convenient and efficient approach, *J. Carbohydr. Chem. 5*:215 (1986).

# 6

# Direct Halogenation of Carbohydrate Derivatives

### Walter A. Szarek and Xianqi Kong
*Queen's University, Kingston, Ontario, Canada*

| | | |
|---|---|---|
| I. | Introduction | 106 |
| II. | General Methods for the Direct Halogenation of Alcohols | 107 |
| | A. Synthetic usefulness of deoxyhalogeno sugars | 109 |
| | B. Reaction of sulfuryl chloride with carbohydrate derivatives | 111 |
| III. | Experimental Procedures | 116 |
| | A. Methyl α-D-glucopyranoside 2,3,4,6-tetra(chlorosulfate) | 116 |
| | B. Methyl 6-chloro-6-deoxy-α-D-glucopyranoside 2,3,4-tri(chlorosulfate) | 117 |
| | C. Methyl 4,6-dichloro-4,6-dideoxy-α-D-galactopyranoside | 117 |
| | D. Reaction of methyl β-D-glucopyranoside with sulfuryl chloride | 118 |
| | E. Reaction of methyl pentofuranosides with sulfuryl chloride: synthesis of 5-chloro-5-deoxypentoses | 118 |
| | F. Reductive dechlorination of methyl 4,6-dichloro-4,6-dideoxy-α-D-galactopyranoside | 119 |
| | G. Reductive dechlorination of methyl 3,6-dichloro-3,6-dideoxy-β-D-allopyranoside | 120 |
| | H. Synthesis of 3,6-dideoxy-D-*ribo*-hexose (paratose) | 121 |
| | I. Methyl 4,6-diazido-4,6-dideoxy-α-D-glucopyranoside | 121 |
| | J. Methyl 2,3-dideoxy-β-D-*glycero*-hex-2-enopyranosid-4-ulose | 122 |
| | References | 123 |

## I. INTRODUCTION

Deoxyhalogeno sugars are carbohydrate derivatives in which hydroxyl groups at positions other than the anomeric center have been replaced by halogen atoms. The search for new methods of synthesis of this long-known class of carbohydrate derivatives has been an active area of investigation for many years. The compounds are of use as synthetic intermediates, for example, in the synthesis of other rare sugars, such as deoxy and aminodeoxy sugars, and many of them are biologically important. Thus, for example, the expanding application of deoxyfluoro sugars for the study of carbohydrate metabolism and transport in both normal and pathological states has stimulated interest in their chemical and biological properties [1–4]. Furthermore, it has actuated intensive efforts to develop improved methods of synthesis, especially, procedures suitable for the preparation of $^{18}$F-labeled carbohydrates for use in medical imaging [3]. Included in the approaches taken to this end are addition reactions of such reagents as molecular fluorine, xenon difluoride, and acetyl hypofluorite, reaction of free hydroxyl groups with diethylaminosulfur trifluoride (DAST), nucleophilic ring-openings with potassium hydrogen fluoride, and nucleophilic displacement of good leaving groups by a range of fluoride salts. The first naturally occurring derivative of a fluoro sugar is the nucleoside antibiotic nucleocidin (4'-fluoro-5'-O-sulfamoyladenosine; **1**), for which synthesis has been achieved [5] (Fig. 1).

A major part of this chapter is concerned with a discussion of a method for the synthesis of chlorodeoxy sugars by direct replacement of hydroxyl groups by chlorine

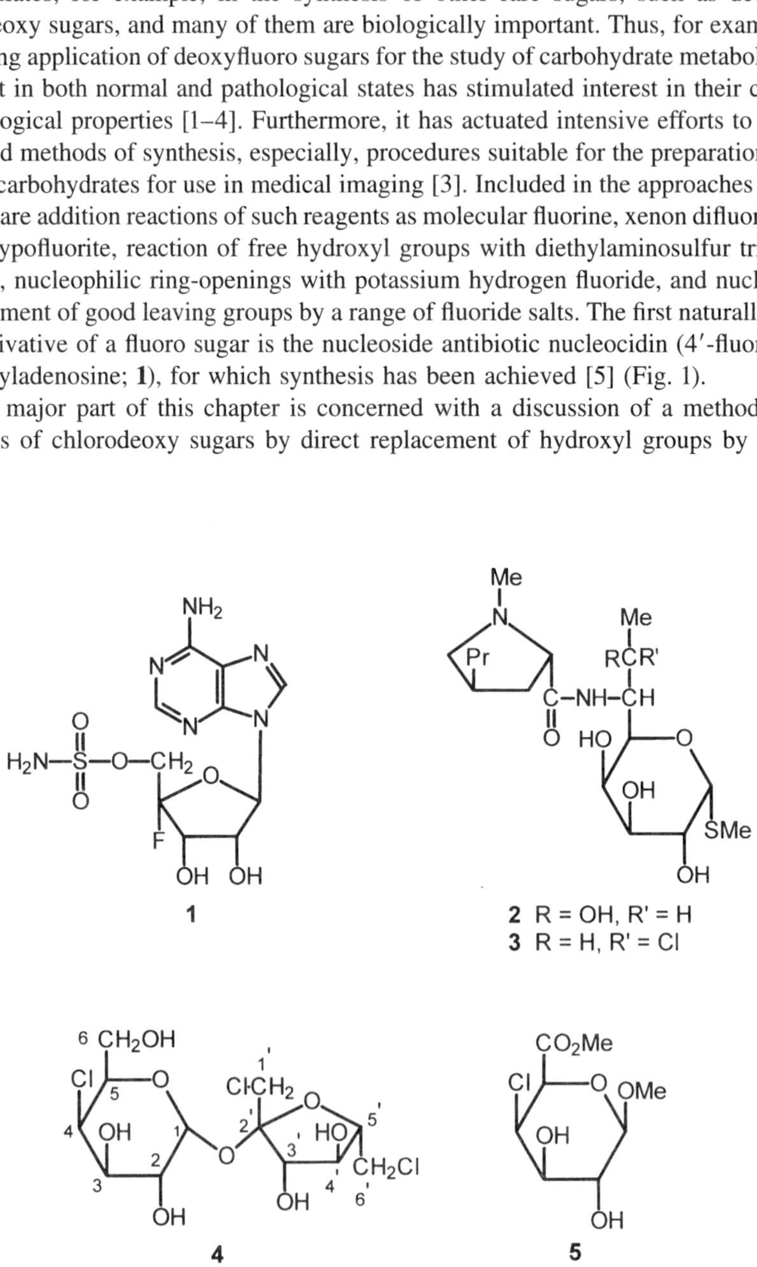

**Figure 1** Structure of some biologically important derivatives of deoxyhalogeno sugars.

atoms. A dramatic demonstration of the importance of such an approach was provided by the discovery that treatment of lincomycin **2**, an antibiotic containing an aminodideoxy-octose, with thionyl chloride in carbon tetrachloride, or more satisfactorily with triphenylphosphine dichloride or triphenylphosphine–carbon tetrachloride, resulted in replacement of the 7-hydroxyl group of the carbohydrate moiety by chlorine, to give a significantly more active antibiotic, namely, clindamycin **3** [6]. Also, a chlorinated kanamycin derivative possesses strong inhibitory activity against several kinds of bacteria [7]. Strong enhancement of sweetness of sucrose has been achieved by replacing specific hydroxyl groups, namely those at C-4, C-1', C-4', and C-6' in the glucosyl and fructosyl units of sucrose, by highly lipophilic chlorine atoms [8]; of particular interest is 4,1',6'-trichloro-4,1',6'-trideoxy-*galacto*-sucrose (sucralose; **4**), which is 650 times as sweet as sucrose. Recently, methyl (methyl 4-chloro-4-deoxy-β-D-galactopyranosid)uronate (**5**) was found to exhibit a concentration-dependent inhibition of both hepatocyte cellular glycosaminoglycan and protein synthesis [9].

## II. GENERAL METHODS FOR THE DIRECT HALOGENATION OF ALCOHOLS

A variety of methods are available for the introduction of halogen into a carbohydrate. These include displacement reactions, direct replacement of hydroxyl groups, additions to unsaturated carbohydrates, and several miscellaneous methods, such as reaction of *O*-benzylidene sugars with *N*-bromosuccinimide, cleavage of carbohydrate oxiranes by halogen-containing reagents, and cleavage of epithio and epimino sugars. Much information on the synthesis and chemistry of deoxyhalogeno sugars is contained in reviews and monographs [1–4,10–14].

The synthesis of deoxyhalogeno sugars by the direct replacement of hydroxyl groups by halogen atoms is an attractive general approach and continues to be of interest. Some of the methods using this approach are discussed briefly in this section, and a detailed description of a method for the synthesis of chlorodeoxy sugars is given later in this chapter. This last method involves the reaction of sulfuryl chloride ($SO_2Cl_2$) with carbohydrate derivatives containing free hydroxyl groups. The reaction gives fully substituted derivatives containing both chlorodeoxy and chlorosulfate groups. The chlorodeoxy groups are formed by bimolecular displacement with liberated chloride ions of certain of the chlorosulfonyloxy groups. The chlorosulfate groups can be easily removed using sodium iodide to give the corresponding hydroxyl groups with retention of configuration. The method has led to facile syntheses of difficultly accessible chlorodeoxy, deoxy, and aminodeoxy sugars.

A general and important method for activating alcohols toward nucleophilic substitution is by converting them into alkoxyphosphonium ions:

$$R'_3P + E-Y \longrightarrow R'_3P\begin{smallmatrix}E\\Y\end{smallmatrix} \rightleftharpoons R'_3\overset{\oplus}{P}-E + \overset{\ominus}{Y}$$

$$R'_3\overset{\oplus}{P}-E + ROH \longrightarrow R'_3\overset{\oplus}{P}-OR + HE$$

$$R'_3\overset{\oplus}{P}-OR + \overset{\ominus}{Nu} \longrightarrow R'_3P=O + R-Nu$$

This chemistry has been employed for replacing hydroxyl groups in carbohydrates by halogen atoms. For example, there has been considerable interest in the method of conversion of alcohols, by their reaction with carbon tetrachloride and tertiary phosphines (usually triphenylphosphine) into the corresponding chlorides. The reaction proceeds by way of in situ generation of a chlorophosphonium ion; this ion then reacts with the alcohol to give an alkoxyphosphonium ion, which is converted into the chloride. The process has been applied [15] to the difficult case of the 1-position in a 2-hexulose. Both 2,3:4,5-di-$O$-isopropylidene-D-fructose and 2,3:4,6-di-$O$-isopropylidene-L-sorbose react readily with triphenylphospine–carbon tetrachloride to yield the corresponding 1-chloro-1-deoxy derivatives. The replacement of a secondary hydroxyl group in a monosaccharide by chlorine with this method has also been reported. Some limitations in the application of this chlorination procedure have been noted [15]. For example, with 1,2:5,6-di-$O$-isopropylidene-α-D-glucofuranose, a rearrangement of the 5,6-acetal linkage accompanies chlorination at C-6, to give 6-chloro-6-deoxy-1,2:3,5-di-$O$-isopropylidene-α-D-glucofuranose.

The use of triphenylphosphine–carbon tetrachloride to convert lincomycin **2** into clindamycin **3** has already been mentioned; the 7-bromo and 7-iodo analogues of **3** were also prepared by treatment of lincomycin hydrochloride with triphenylphosphine and carbon tetrabromide or carbon tetraiodide, with acetonitrile as the solvent [6].

There are several methods for the conversion of alcohols into iodides using phosphorus-containing reagents. Two important reagents (Rydon reagents) are methyltriphenoxyphosphonium iodide **6** and iodotriphenoxyphosphonium iodide **7**, respectively formed

$$(PhO)_3\overset{\oplus}{P}\text{—Me } I^{\ominus} \qquad\qquad (PhO)_3\overset{\oplus}{P}\text{—I } I^{\ominus}$$
$$\quad\quad\text{6} \qquad\qquad\qquad\qquad\quad \text{7}$$

by the reaction of triphenyl phosphite with methyl iodide and with iodine. The reaction of **6** with alcohols is considered to involve nucleophilic attack on phosphorus, with expulsion of phenol and formation of the alkoxyphosphonium salt **8**, which then affords the alkyl

$$\text{6 + ROH} \longrightarrow (PhO)_2\underset{\underset{\text{Me}}{|}}{\overset{\oplus}{P}}\text{—OR } I^{\ominus} \longrightarrow RI + (PhO)_2\underset{\underset{O}{\parallel}}{P}\text{—Me}$$
$$\qquad\qquad\qquad\qquad\text{8}$$

iodide and diphenyl methylphosphonate. According to this mechanism, the conversion of an alcohol into the corresponding iodide should occur with inversion of configuration; however, because of the possibility of nucleophilic attack by iodide ion on the initially formed iodides, the inversion can be accompanied by some racemization.

There are many examples of the reaction of carbohydrates with Rydon reagents [16]; the reaction is controlled by steric factors. Thus, no reaction occurred between 1,2-$O$-isopropylidene-5,6-di-$O$-methyl-α-D-glucofuranose and either **6** or bromotriphenoxyphosphonium bromide, presumably because of the steric hindrance caused by the trioxabicyclo [3.3.0]octane ring-system, whereas methyl 2,5,6-tri-$O$-methyl-β-D-glucofuranoside reacted with **6** to give a 3-deoxy-3-iodo derivative in 31% yield.

A variety of other phosphorus-containing reagents [11,12,17] have been employed for replacing hydroxyl groups in carbohydrates by halogen atoms. Two other methods that have been applied widely are the reaction with (halogenomethylene)dimethyliminium halides [11,12] and the introduction of fluorine by diethylaminosulfur trifluoride (DAST reagent) [3,18]. $N,N$-Dimethylformamide reacts with halides of inorganic acids (phosgene, phosphoryl chloride, phosphorus trichloride, and thionyl chloride) to form an active intermediate, namely (chloromethylene)dimethyliminium chloride **9**; **9** and its bromine

$$\text{ROH} + (\text{Me}_2\overset{\oplus}{\text{N}}=\text{CHX})\overset{\ominus}{\text{X}} \longrightarrow [\text{R}-\text{O}-\text{CH}=\overset{\oplus}{\text{N}}\text{Me}_2 \ \overset{\ominus}{\text{X}}] \longrightarrow \text{RX} + \text{Me}_2\text{NCHO}$$

**9** X = Cl  
**10** X = Br  
**11**

analogue **10** have been found to be highly effective for replacing hydroxyl groups by halogen. The method involves the thermal decomposition in solution of the adduct, such as **11**, formed from the alcohol and the strongly electrophilic (halogenomethylene)dimethyliminium halide. Hanessian and Plessas [19] were the first workers to apply the method to the synthesis of deoxyhalogeno sugars. The DAST reagent reacts with a hydroxyl group to give an unstable intermediate, $C-OSF_2NEt_2$, with liberation of HF; attack by a fluoride ion then forms a $C-F$ bond, with inversion of configuration. The $^{18}F$-labeled DAST has been prepared [20]. Many examples of the applications of the DAST reagent in carbohydrates have been discussed by Tsuchiya [3]. Other methods for the direct replacement of hydroxyl groups in carbohydrates by halogen atoms have been reviewed also [12].

## A. Synthetic Usefulness of Deoxyhalogeno Sugars

Deoxyhalogeno sugars are particularly useful in the synthesis of other rare sugars, such as deoxy and aminodeoxy sugars. Deoxyhalogeno sugars are susceptible to nucleophilic attack, leading to displacement, elimination, or anhydro-ring formation. The ease of displacement decreases in the order I > Br > Cl > F; the iodo, bromo, and chloro derivatives have been widely used, and have undergone a variety of nucleophilic substitution reactions. The reactivity of halo groups in pyranoid derivatives can be predicted in an approximate and qualitative way by a consideration of the steric and polar factors affecting the formation of the transition state. A summary of the stereoelectronic factors affecting displacement reactions has been published [21,22]; some of these aspects are considered and illustrated in Section II.B.

Deoxyhalogeno sugars are useful intermediates in the synthesis of deoxy sugars [23–25], an important class of carbohydrates that occur quite widely in nature. The iodo and bromo derivatives can be reduced to form the deoxy sugars by a variety of reducing agents, including zinc in acetic acid, sodium amalgam in aqueous ether or ethanol, lithium aluminum hydride, and hydrogen in the presence of palladium-on-carbon or Raney nickel. Hanessian [23] has described many examples of the reductive dehalogenation of deoxyhalogeno sugars. Before 1969, the reduction of chlorodeoxy sugars was rare [7,26]; however, in that year, it was reported [27] that these derivatives can be reduced by a particularly active form of Raney nickel catalyst. Thus, hydrogenation over Raney nickel of methyl 4,6-dichloro-4,6-dideoxy-3-O-methyl-D-galactopyranoside **13** (prepared by reaction of methyl 3-O-methyl-D-glucopyranoside **12** with sulfuryl chloride, followed by dechlorosulfation of the product by use of sodium iodide) gave a product that, on acid-catalyzed hydrolysis, afforded the antibiotic sugar D-chalcose (**14**; Scheme 1). A facile synthesis of 4,6-dideoxy-D-xylo-hexose has been reported [28] by an analogous route in an overall yield of 65% from the commercially available methyl α-D-glucopyranoside.

In the last-cited work [28], methyl 4,6-dichloro-4,6-dideoxy-α-D-galactopyranoside

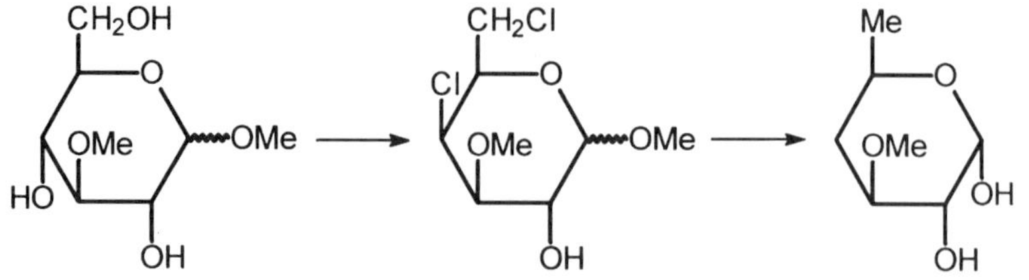

**Scheme 1**

was converted into methyl 4,6-dideoxy-α-D-xylo-hexopyranoside by hydrogenation over Raney nickel in the presence of potassium hydroxide. However, when triethylamine was substituted for potassium hydroxide, a selective, reductive dechlorination occurred at C-4 to give methyl 6-chloro-4,6-dideoxy-α-D-xylo-hexopyranoside [29]. Similarly, hydrogenation of methyl 3,4,6-trichloro-3,4,6-trideoxy-α-D-allopyranoside **15** over Raney nickel in the presence of potassium hydroxide affords methyl 3,4,6-trideoxy-α-D-*erythro*-hexopyranoside **16**, whereas hydrogenation in the presence of triethylamine leads to methyl 6-chloro-3,4,6-trideoxy-α-D-*erythro*-hexopyranoside [30] (**17**; Scheme 2). One example of the usefulness of this procedure was provided by the ready synthesis of a 4-deoxyhexose [29].

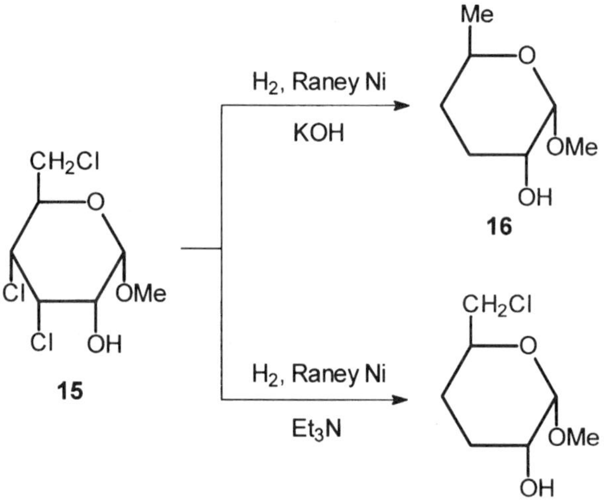

**Scheme 2**

The reduction of chlorodeoxy sugars has been achieved [31,32] in high yield by means of tributyltin hydride in the presence of 2,2′-azobis(2-methylpropionitrile). The reaction with methyl 2,3-di-*O*-acetyl-4,6-dichloro-4,6-dideoxy-α-D-galactopyranoside at 60°C gave [31] methyl 2,3-di-*O*-acetyl-6-chloro-4,6-dideoxy-α-D-xylo-hexopyranoside as the main product. A free-radical mechanism has been proposed [33] for the reduction of alkyl halides by organotin hydrides. In accordance with this proposal, the presence of the radical initiator 2,2′-azobis(2-methylpropionitrile) was essential for the reduction of chlorodeoxy sugars; moreover, the relative reactivities of the two chlorine atoms in methyl 2,3-di-*O*-acetyl-4,6-dichloro-4,6-dideoxy-α-D-galactopyranoside follow a free-radical order.

Hanessian and Plessas [19] have reported another example of a selective dehalogenation in the carbohydrate field. Thus, catalytic hydrogenation of methyl 4-*O*-benzoyl-3-

# Halogenation of Carbohydrate Derivatives

bromo-2,6-dichloro-2,3,6-trideoxy-α-D-mannopyranoside over palladium-on-carbon in the presence of barium carbonate gave methyl 4-*O*-benzoyl-6-chloro-2,3,6-trideoxy-α-D-*erythro*-hexopyranoside. The presence of a bromine atom at C-3 apparently leads to the selective reduction of the halogen atoms on C-2 and C-3, relative to the chlorine atom on C-6, because in methyl 3,4-*O*-benzylidene-2,6-dichloro-2,6-dideoxy-α-D-altropyranoside the two chlorine atoms were inert to catalytic hydrogenation.

Chlorodeoxy sugars have been reduced [15] also with lithium aluminum hydride. In one experiment, 3-deuterio-1,2:5,6-di-*O*-isopropylidene-α-D-allofuranose was converted into 3-chloro-3-deoxy-3-deuterio-1,2:5,6-di-*O*-isopropylidene-α-D-glucofuranose by treatment with triphenylphosphine–carbon tetrachloride; reduction with lithium aluminum hydride gave 3-deoxy-3-deuterio-1,2:5,6-di-*O*-isopropylidene-α-D-*ribo*-hexofuranose with retention of configuration at C-3.

In Section II.B a discussion of the reaction of sulfuryl chloride with carbohydrate derivatives is presented, and in Section III are given some experimental procedures representative of this method for the synthesis of chlorodeoxy sugars and procedures of simple processes for the conversion of the chlorodeoxy sugars into other rare sugars.

## B. Reaction of Sulfuryl Chloride with Carbohydrate Derivatives

The reaction of sulfuryl chloride ($SO_2Cl_2$) with carbohydrates containing free hydroxyl groups has become a well-established method for the preparation of chlorodeoxy sugars [11,12]. As indicated in Section II, the process involves the initial formation of chlorosulfate groups, followed by bimolecular displacement of certain of these by chloride ion liberated during the chlorosulfation. The displacement occurs only at those centers where the steric and polar factors are favorable for an $S_N2$ reaction.

The reaction of sulfuryl chloride with carbohydrate derivatives was investigated first by Helferich and co-workers [34–36], and was extensively studied by Jones and his colleagues [37–45]. The latter studies elucidated the stereochemical principles involved in the various transformations, and they have made available a convenient and effective procedure for the preparation of chlorodeoxy sugars.

In 1921 Helferich [34] found that treatment of methyl α-D-glucopyranoside **18** with sulfuryl chloride in a mixture of pyridine and chloroform at 5°C afforded a compound, the structure of which was established several years later [37,38] to be that of methyl 4,6-dichloro-4,6-dideoxy-α-D-galactopyranoside 2,3-cyclic sulfate (**19**; Scheme 3). The cyclic sulfates are readily cleaved by dilute alkalis, or by methanolic ammonia, to give the salt of a monosulfate; the hemi-ester is then desulfated by acid to yield the chlorodeoxy sugar [35,36]. Both of these reactions occur with retention of configuration [37].

When only a minimum proportion of pyridine is employed, the reaction of sulfuryl

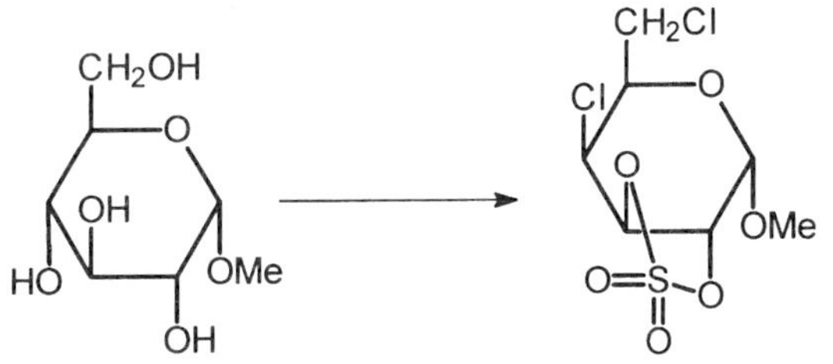

Scheme 3

chloride with carbohydrates produces chlorosulfuric esters, instead of cyclic sulfates; thus, methyl 4,6-dichloro-4,6-dideoxy-α-D-galactopyranoside 2,3-di(chlorosulfate) **20** was obtained by application of the reaction to methyl α-D-glucopyranoside **18**, followed by isolation of the product at room temperature [40]. Compound **20** reacts with pyridine to give the 2,3-cyclic sulfate **19**. Thus, in the presence of an excess of pyridine, appropriately oriented chlorosulfates can be converted into cyclic sulfates. Presumably, one of the chlorosulfate groups is first hydrolyzed with S—O bond scission and, hence, with retention of configuration; the free hydroxyl group so formed then attacks the remaining chlorosulfate group with S—Cl bond scission, to give the cyclic sulfate. As the two chlorosulfate groups in **20** are equatorially oriented, the spatial requirement for the latter process is very favorable. The presence of axially attached chlorosulfate groups can lead to the formation either of anhydro sugars or of complex products resulting from elimination reactions. Also, whereas methyl 4,6-O-benzylidene-α-D-glucopyranoside 2,3-di(chlorosulfate) is readily converted by pyridine at 0°C into the 2,3-cyclic sulfate, with a strong base, such as sodium methoxide, it forms methyl 2,3-anhydro-4,6-O-benzylidene-α-D-allopyranoside [40].

Chlorosulfate groups can be readily removed to give the corresponding hydroxyl groups, with retention of configuration, by treatment of a solution of the carbohydrate chlorosulfate in methanol with sodium iodide in aqueous methanol [40]; immediate liberation of iodine and evolution of sulfur dioxide occurs. A possible mechanism for the dechlorosulfation reaction involves displacement by iodide at the chlorine atom; the initially formed iodine monochloride would react with iodide ion to give iodine and chloride ion. Alternatively, an unstable iodosulfate ion could be formed as an intermediate [46].

$$I^{\ominus} + Cl-S(=O)_2-OR \longrightarrow RO^{\ominus} + SO_2 + ICl$$

The manner of formation of chlorodeoxy sugars by the reaction of sulfuryl chloride with carbohydrates containing free hydroxyl groups was discerned by the series of reactions outlined in Scheme **4** [41]. Treatment of methyl α-D-glucopyranoside **18** with sulfuryl chloride and pyridine in chloroform solution, and isolation of the product at room temperature, afforded methyl 4,6-dichloro-4,6-dideoxy-α-D-galactopyranoside 2,3-di(chlorosulfate) **20**. Isolation of the product at 0°C gave methyl 6-chloro-6-deoxy-α-D-glucopyranoside 2,3,4-tri(chlorosulfate) **21**, and isolation at −70°C yielded methyl α-D-glucopyranoside 2,3,4,6-tetra(chlorosulfate) **22**. Furthermore, compound **22** was converted into the 6-chloro-6-deoxy derivative **21** on treatment with 1 mole of pyridinium chloride per mole, and compound **21** was converted into the 4,6-dichloro-4,6-dideoxy derivative **20** on treatment with an excess of pyridinium chloride. Thus, the chlorodeoxy groups are formed by bimolecular displacement of certain of the chlorosulfonyloxy groups by chloride ion liberated during the chlorosulfation.

It is often possible to predict the reactivity of a chlorosulfonyloxy group by a consideration of the steric and polar factors affecting the formation of the transition state [21,43,44]. The presence of a vicinal, axial substituent or of a β-*trans*-axial substituent on a pyranoid ring inhibits replacement of a chlorosulfonyloxy group; also, a chlorosulfate group at C-2 is deactivated to nucleophilic substitution by chloride ion.

An example of the inhibitory effect of a vicinal, axial substituent was provided by the observation [41] that, whereas methyl 4,6- dichloro-4,6-dideoxy-β-D-glucopyranoside 2,3-

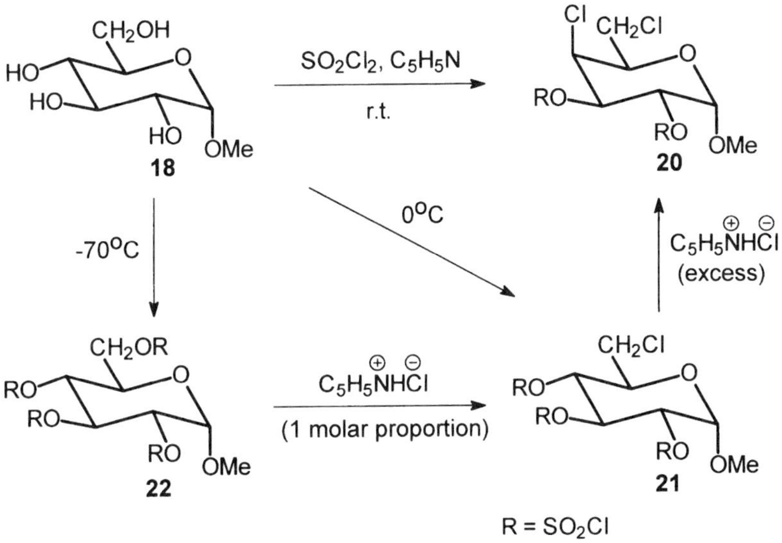

## Halogenation of Carbohydrate Derivatives

**Scheme 4**

di(chlorosulfate) readily undergoes displacement of the chlorosulfonyloxy group at C-3 by chloride ion, the corresponding galactopyranoside, having the chloro group at C-4 in an axial orientation, is resistant to replacement.

As indicated in the foregoing, the reaction of methyl α-D-glucopyranoside **18** with sulfuryl chloride and pyridine in chloroform solution, and isolation of the product at room temperature, gave methyl 4,6-dichloro-4,6-dideoxy-α-D-galactopyranoside 2,3-di(chlorosulfate) **20**. In this example, the lack of substitution by chloride ion of the chlorosulfonyloxy group at C-3 is attributed to the presence of the β-*trans*-axial methoxyl group at C-1; also, a chlorosulfonyloxy group at C-2 is deactivated to nucleophilic substitution by chloride ion. Consistent with this rationalization is the observation [47] that the reaction with methyl β-D-glucopyranoside **23**, having the methoxyl group at C-1 in an equatorial orientation, followed by dechlorosulfation of the product using sodium iodide, afforded both methyl 4,6-dichloro-4,6-dideoxy-β-D-galactopyranoside **24** and methyl 3,6-dichloro-3,6-dideoxy-β-D-allopyranoside (**25**; Scheme 5).

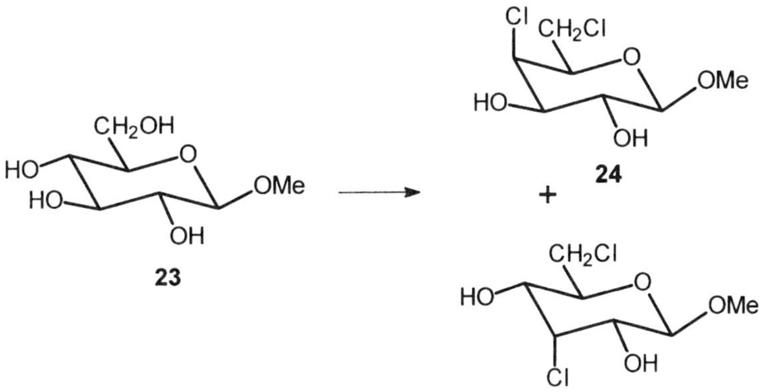

**Scheme 5**

Another example of the β-*trans*-axial substituent effect is provided by the reaction of methyl α-D-mannopyranoside **26** with sulfuryl chloride, which gave methyl 6-chloro-6-deoxy-α-D-mannopyranoside 2,3,4-tri(chlorosulfate) **26a**; even treatment of the product with pyridinium chloride for 12 h at 50°C did not effect further substitution [41] (Scheme **6**).

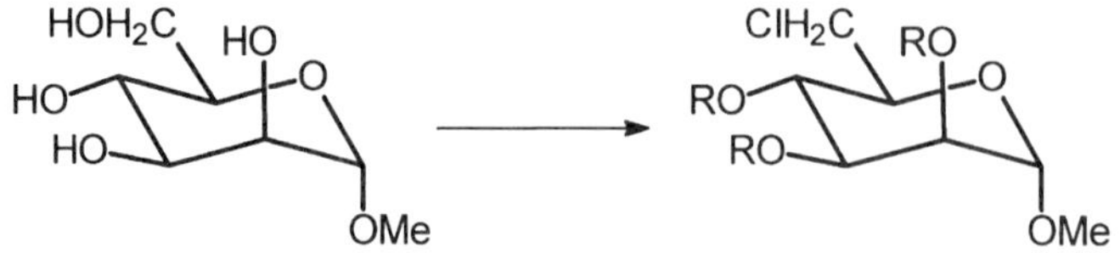

**Scheme 6**

In this case, the lack of displacement of the chlorosulfonyloxy group at C-4 is attributed to the presence of an axial group at C-2, and the lack of displacement at C-3 is attributed to the presence of the axial methoxyl group at C-1.

The reaction of sulfuryl chloride with reducing sugars yields glycosyl chlorides containing both chlorosulfate and chlorodeoxy groups [39]. Thus, the reactions with D-glucose and D-xylose, with isolation of the products at room temperature, afford 4,6-dichloro-4,6-dideoxy-D-galactopyranosyl chloride 2,3-di(chlorosulfate) and 4-chloro-4-deoxy-L-arabinopyranosyl chloride 2,3-di(chlorosulfate), respectively. If the product of reaction with D-xylose is isolated at low temperature, substitution of the chlorosulfonyloxy group at C-4 by chloride ion is inhibited, and D-xylopyranosyl chloride 2,3,4-tri-(chlorosulfate) is obtained in good yield; the actual experimental conditions involved treatment of D-xylose with sulfuryl chloride and pyridine in chloroform solution at −70°C for 2 h, followed by allowing the temperature to rise to −10°C, and then maintaining it between −10 and 0°C for 30 min [48]. Under these conditions [49], crystalline α-D-xylopyranose yields crystalline β-D-xylopyranosyl chloride 2,3,4-tri(chlorosulfate) **27**, and crystalline β-D-lyxopyranose is converted into α-D-lyxopyranosyl chloride 2,3,4-tri(chlorosulfate) **28**.

These results suggest that the replacement of the hydroxyl group at C-1 by chlorine occurs by way of an intermediate chlorosulfate, resulting in an overall inversion of configuration; moreover, they show that anomerization does not occur to any significant extent under the reaction conditions.

The chlorosulfated glycosyl chlorides **27** and **28** were converted into 2-chloro-2-deoxypentoses on treatment with aluminum chloride [50]. In the case of **27**, the use of 1.5 mol-equivalents gave exclusively the 2-chloro-2-deoxy-α-D-lyxopyranosyl chloride 3,4-di(chlorosulfate) **29**; treatment of **29** with sodium iodide yielded crystalline 2-chloro-2-deoxy-D-lyxose. Similarly, **28** afforded 2-chloro-2-deoxy-α-D-xylopyranosyl chloride 3,4-di(chlorosulfate) (**30**; Scheme **7**).

The formation of **29** from **27** has been interpreted [50] as occurring by an intramolecular displacement of the chlorosulfate at C-2, by participation of the chlorine atom at

## Halogenation of Carbohydrate Derivatives

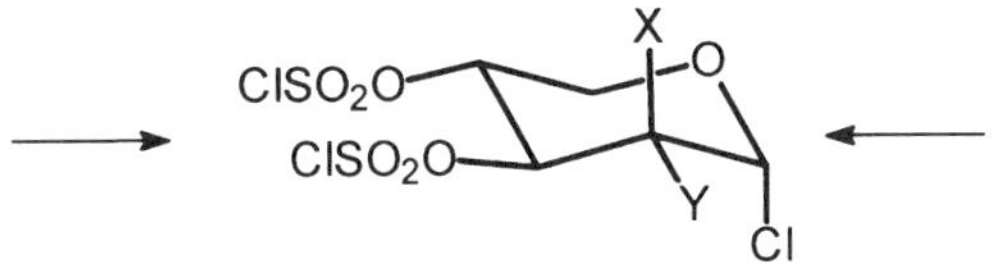

**29** X = Cl, Y = H
**30** X = H, Y = Cl

**Scheme 7**

C-1, to give initially a chloronium ion, as shown. Favored attack by chloride ion at the more highly reactive center (C-1) would result in a net inversion of configuration at C-2 of **27**, to yield **29**. In the reaction of α-D-lyxopyranosyl chloride 2,3,4-tri(chlorosulfate) **28** with aluminum chloride, a 1,2-*cis* product **30** was preponderant, whereas the proposed mechanism should yield a 1,2-*trans* product; this result is, presumably, due to the anomerization of the thermodynamically less-stable β anomer of **30** to the (more stable) 1,2-*cis* product **30**.

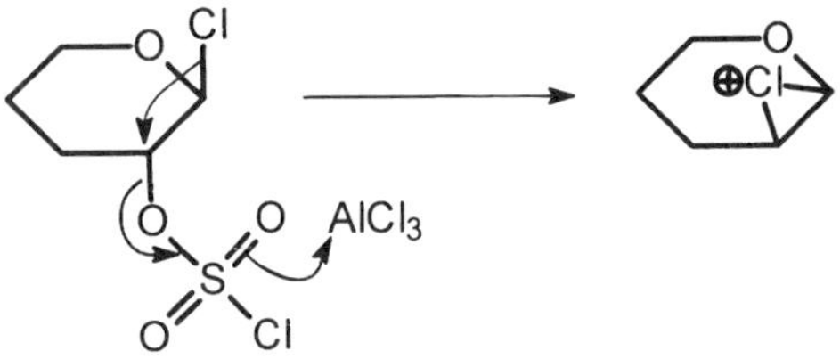

Pyranoid derivatives having a chlorosulfonyloxy group in a *trans*-diaxial relation with a ring proton may undergo an elimination reaction to yield unsaturated compounds [40,42]. Thus, on treatment with sulfuryl chloride and pyridine, followed by heating with a solution of pyridinium chloride in chloroform, methyl α-L-arabinopyranoside gave crystalline methyl 3,4-dichloro-3,4-dideoxy-β-D-ribopyranoside 2-chlorosulfate **31**, which, on standing in pyridine, underwent the loss of chlorosulfuric acid to give crystalline methyl 3,4-dichloro-2,3,4-trideoxy-β-D-*glycero*-pent-2-enopyranoside [42] (**32**; Scheme **8**).

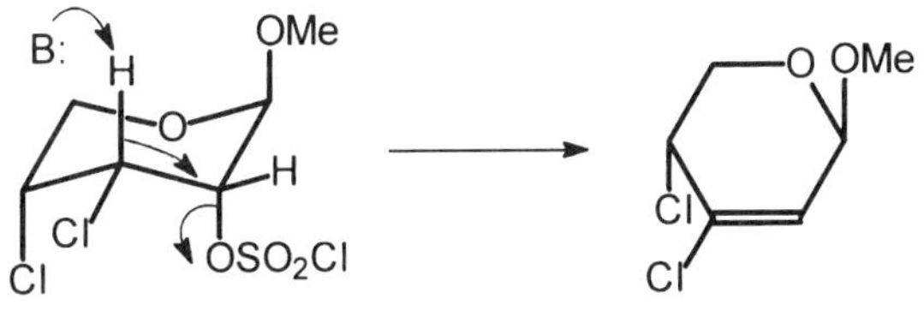

**Scheme 8**

The reaction of sulfuryl chloride with furanoid derivatives has been studied [51,52], and the reaction has been applied to disaccharides, for example, methyl β-maltoside [53] and sucrose [54,55], to yield chlorodeoxy derivatives.

In Section II.A of this chapter the usefulness of chlorodeoxy sugars in the synthesis of deoxy sugars was indicated. A synthesis [56] of the biologically important sugar paratose (3,6-dideoxy-D-*ribo*-hexose), by a route involving reductive dechlorination by hydrogenation over Raney nickel catalyst, provides a noteworthy demonstration of the versatility of the method of preparation of chlorodeoxy sugars by use of sulfuryl chloride. Thus, methyl 4,6-*O*-benzylidene-3-chloro-3-deoxy-β-D-allopyranoside **33** was initially prepared by reaction of methyl 4,6-*O*-benzylidene-β-D-glucopyranoside with sulfuryl chloride, followed by dechlorosulfation of the product by use of sodium iodide; acid-catalyzed *O*-debenzylidenation of compound **33** then gave methyl 3-chloro-3-deoxy-β-D-allopyranoside **34**. Treatment of **34** with sulfuryl chloride, followed by dechlorosulfation of the product, afforded methyl 3,6-dichloro-3,6-dideoxy-β-D-allopyranoside **25**. The nonsubstitution by chloride ion of the intermediate chlorosulfonyloxy group at C-4 is attributed to the presence of the vicinal, axial substituent at C-3; also, a chlorosulfonyloxy group at C-2 is deactivated to nucleophilic substitution by chloride ion. Hydrogenation of **25**, in the presence of potassium hydroxide, over Raney nickel gave methyl 3,6-dideoxy- β-D-*ribo*-hexopyranoside **35** which, on acid-catalyzed hydrolysis, afforded paratose (**36**; Scheme 9). Paratose has been isolated from lipopolysaccharides elaborated by gram-negative bacteria. Experimental procedures for the synthesis of **36** are given in Section III.

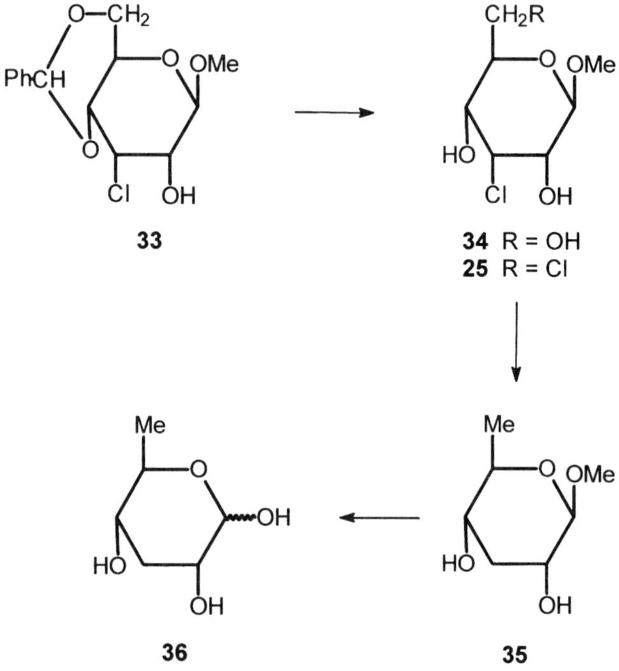

**Scheme 9**

Section III also gives details of an example of a simple synthesis of azidodeoxy sugars by displacement with azide ion of chloro groups. The displacement of chloro groups in sugars and reduction of the resultant azido derivatives constitutes a convenient, high-yielding procedure for synthesis of amino sugars [29].

## III. EXPERIMENTAL PROCEDURES*

### A. Methyl α-D-Glucopyranoside 2,3,4,6-Tetra(chlorosulfate) [41]

Methyl α-D-glucopyranoside (**18**, 10 g), previously dried over phosphoric oxide, was partially dissolved in pyridine (40 mL). Chloroform (100 mL), dried over anhydrous

---

*Optical rotations were measured at 22°–25°C.

### Halogenation of Carbohydrate Derivatives

**18** (HO-CH₂OH-HO-HO-OMe pyranose) → SO₂Cl₂, C₅H₅N, CHCl₃, −70°C → **22** R = SO₂Cl (RO-CH₂OR-RO-RO-OMe)

sodium sulfate, was added to the pyridine solution. The heterogeneous reaction mixture was cooled in a dry ice–acetone bath (about −70°C), and an excess of redistilled sulfuryl chloride (26 mL) was added dropwise over of period of 0.5 h with vigorous stirring [39]. Cooling was continued for a further 2 h, and the cold reaction mixture, while still at the temperature of the dry ice–acetone bath, was quickly poured into a vigorously stirred solution of ice-cold 10% sulfuric acid (1000 mL). The product (8.7 g, 29%) was recrystallized from chloroform–petroleum ether (bp 50°–60°C) and gave **22** as colorless needles (7 g, 23%): mp 118°C, $[\alpha]_D$ +92° ($c$ 0.8, CHCl₃).

### B. Methyl 6-Chloro-6-deoxy-α-D-glucopyranoside 2,3,4-Tri(chlorosulfate) [41]

**18** → SO₂Cl₂, C₅H₅N, CHCl₃, −70 to 0°C → **21** R = SO₂Cl

The reaction in A was performed using methyl α-D-glucopyranoside (**18**, 10 g), except that the reaction mixture was allowed to warm up slowly to 0°C and then was poured, at this temperature, into a vigorously stirred solution of ice-cold 10% sulfuric acid (1000 mL). The crystalline chloroform-soluble product was recrystallized twice from chloroform–petroleum ether (bp 35°–60°C) and gave **21** as large colorless prisms (15.5 g, 59%): mp 90°–91°C, $[\alpha]_D$ +115° ($c$ 0.9, CHCl₃).

### C. Methyl 4,6-Dichloro-4,6-dideoxy-α-D-galactopyranoside [28,39–41]

**18** → 1) SO₂Cl₂, C₅H₅N, CHCl₃, −70°C to r.t.; 2) NaI, MeOH → **37**

Methyl α-D-glucopyranoside (**18**, 40 g) was converted into methyl 4,6-dichloro-4,6-dideoxy-α-D-galactopyranoside 2,3-di(chlorosulfate) (**20**, 78 g), by the method of Jennings and Jones [39,40]. The syrupy material was dissolved in methanol (1200 mL), and a solution of sodium iodide [63 g in 160 mL of methanol–water (1:1, v/v)] was added. The solution was neutralized with sodium hydrogen carbonate. The insoluble material was removed by filtration, and the filtrate was concentrated to dryness. The residue was extracted with hot chloroform (3 × 200 mL), and the dried (MgSO$_4$) extracts were concentrated to give a crystalline product that was recrystallized from chloroform–petroleum ether (bp 40°–60°C); yield 36 g (76% based on **18**). The physical constants were in agreement with those reported previously [37,40] for compound **37**; reported [40] mp 158°C, [α]$_D$ +179° (c 2.0, H$_2$O).

### D. Reaction of Methyl β-D-Glucopyranoside with Sulfuryl Chloride [47]

To a mixture, cooled in a dry ice–acetone bath, of methyl β-D-glucopyranoside (**23**, 5.0 g, 26 mmol) in dry pyridine (20 mL) and chloroform, was added, dropwise with stirring, sulfuryl chloride (13 mL, 170 mmol) over a period of 30 min. The reaction mixture was stirred for a further 2 h at the low temperature, and then for 2 h at room temperature. The mixture was poured into ice and water, and the chloroform layer was separated; the aqueous solution was extracted several times with chloroform. The combined chloroform solutions were washed with sodium hydrogen carbonate solution and then with water. Concentration of the dried (Na$_2$SO$_4$) chloroform solution gave a syrup (13.5 g). The syrup was dissolved in methanol. To the stirred methanol solution was added sodium hydrogen carbonate (75 g) and then, dropwise, sodium iodide (0.55 g) in methanol; the progress of the dechlorosulfation was monitored by thin-layer chromatography (TLC) [Silica Gel G; 3:1 (v/v) pentane–ethyl acetate]. At the end of the reaction, the mixture was filtered through Celite, and the filtrate was concentrated. The residue was shown by TLC [Silica Gel G; 3:1 (v/v) ethyl acetate-chloroform] to contain two new, major components having $R_f$ 0.40 and $R_f$ 0.53. These two components were isolated by column chromatography [Silica Gel 60, 70–230 mesh, E. Merck; 3:1 (v/v) ethyl acetate–chloroform], and then recrystallized from chloroform–petroleum ether (bp 60°–80°C), to give methyl 4,6-dichloro-4,6-dideoxy-β-D-galactopyranoside (**24**; 1.22 g, 21%): mp 154°–155°C, [α]$_D$ +8° (c 0.8, H$_2$O); reported [39] mp 154°C, [α]$_D$ −8° (c 0.8, H$_2$O); and methyl 3,6-dichloro-3,6-dideoxy-β-D-allopyranoside (**25**; 2.86 g, 48%): mp 164°–165°C, [α]$_D$ −54° (c 0.6, CHCl$_3$); reported [57] mp 163°–164°C, [α]$_D$ −54° (CHCl$_3$).

### E. Reaction of Methyl Pentofuranosides with Sulfuryl Chloride: Synthesis of 5-Chloro-5-deoxypentoses [51]

To a cooled (dry ice–acetone) solution of the methyl pentofuranoside (1.7 g, 10 mmol) in dry pyridine (8 mL) and chloroform (20 mL) was added sulfuryl chloride (4 mL, 30 mmol) dropwise, with stirring. Cooling was continued for an additional 2 h, and then the reaction mixture was stirred for a further 1 h at 0°C. The mixture was diluted with chloroform (150 mL), and the solution was washed successively with water, 3% hydrochloric acid, 5% sodium hydrogen carbonate solution, and water, dried (MgSO$_4$), and evaporated. The mixture of chlorosulfation products was resolved by chromatography on silica gel. The products were dechlorosulfated in the usual manner [28,40] by treatment of a solution of the carbohydrate chlorosulfate in methanol with sodium iodide in aqueous methanol.

This general method has been employed for the preparation of methyl 5-chloro-5-deoxy-α-D-ribofuranoside, methyl 5-chloro-5-deoxy-β-D-ribofuranoside, methyl 5-chloro-5-deoxy-α-D-xylofuranoside, and methyl 5-chloro-5-deoxy-β-D-xylofuranoside.

## F. Reductive Dechlorination of Methyl 4,6-Dichloro-4,6-dideoxy-α-D-galactopyranoside

*In the Presence of Potassium Hydroxide [28]*

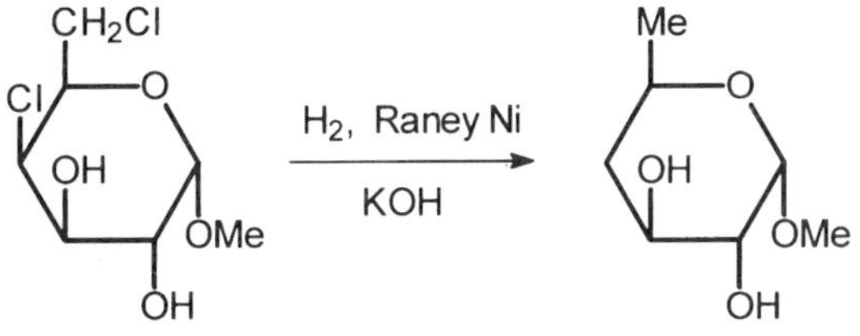

A solution of methyl 4,6-dichloro-4,6-dideoxy-α-D-galactopyranoside [28] (**37**, 4.62 g) in absolute ethanol (50 mL) containing potassium hydroxide (4.4 g) and W-4 Raney nickel catalyst [27,58] (10 g) was subjected to a hydrogen pressure of 45 psig for 4 h. TLC [Silica Gel G, E. Merck; 9:1 (v/v) chloroform–methanol] showed that the starting material had all reacted. The solution was filtered free of catalyst, and the filtrate was neutralized with 1 *M* hydrochloric acid and concentrated to dryness. The residue was extracted with hot chloroform (4 × 50 mL), and the dried (MgSO$_4$) extracts were concentrated to give a syrup (3 g, 91%), which was distilled at 120°C (bath)/0.1 torr; the distillate crystallized on standing. Recrystallization from ethyl acetate–petroleum ether (bp 60°–80°C) gave methyl 4,6-dideoxy-α-D-xylo-hexopyranoside (**38**) as long needles; mp 70°–75°C, [α]$_D$ +172° (*c* 1.0, MeOH); recrystallization from ethyl acetate gave the compound as prisms: mp 107°–109°C, [α]$_D$ +171° (*c* 1.0, MeOH); reported [59] mp 104°–106°C, [α]$_D$ +182° (MeOH); reported [32] mp 105°–108°C, [α]$_D$ +171° (*c* 1.0, MeOH).

*In the Presence of Triethylamine [29]*

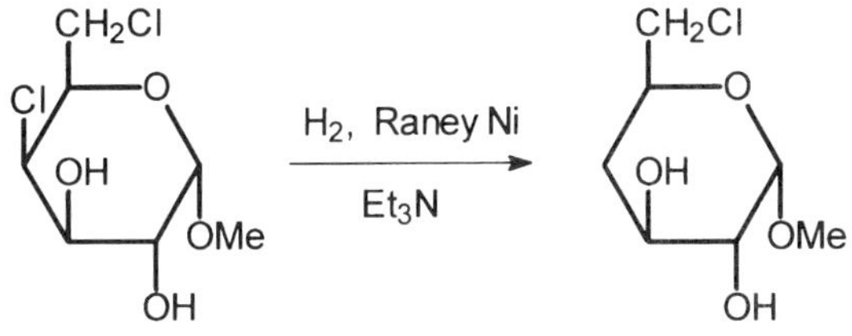

A solution of methyl 4,6-dichloro-4,6-dideoxy-α-D-galactopyranoside [28] (**37**, 1 g) in absolute ethanol (30 mL) containing triethylamine (1.3 mL) and W-4 Raney nickel catalyst

[58] (2 g) was subjected to a hydrogen pressure of 45 psig for 30 h. The filtered solution was neutralized with 2 $M$ hydrochloric acid and concentrated to a residue, which was partitioned between chloroform and water. The dried ($MgSO_4$) chloroform solution was concentrated to a residue that crystallized from ethyl acetate. Fractionation on silica gel, with 5% methanol in ethyl acetate as eluent, and then recrystallization from ethyl acetate–petroleum ether (bp 60°–80°C) gave methyl 6-chloro-4,6-dideoxy-α-D-xylo-hexopyranoside **39** as colorless needles (about 90%): mp 110°–111°C, $[\alpha]_D$ +165° ($c$ 1.02, MeOH).

## G. Reductive Dechlorination of Methyl 3,6-Dichloro-3,6-dideoxy-β-D-allopyranoside [60]

*In the Presence of Potassium Hydroxide*

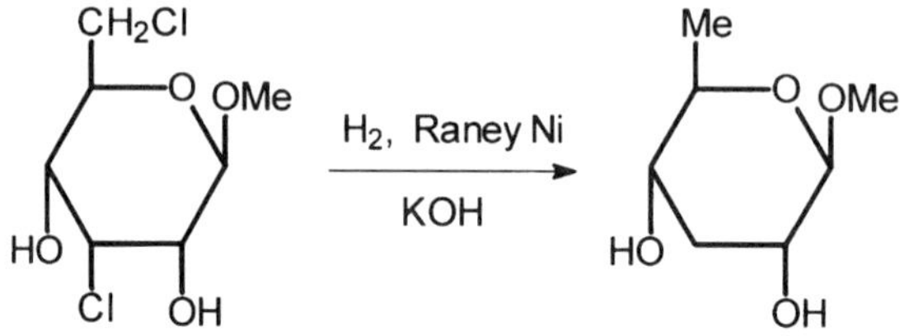

A solution of methyl 3,6-dichloro-3,6-dideoxy-β-D-allopyranoside [47] (**25**, 1.5 g) in ethanol (150 mL) containing W-4 Raney nickel catalyst [27,58] (30 mL ethanol slurry) and potassium hydroxide (1.5 g) was subjected to a hydrogen pressure of 45 psig for 4 h. TLC [Silica Gel G, Brinkmann; 2:1 (v/v) ethyl acetate–petroleum ether (bp 60°–80°C)] showed that the starting material ($R_f$ 0.42) had all reacted and revealed the presence of a new component having an $R_f$ of 0.14. The catalyst was removed by filtration, and the filtrate was neutralized with 1 $N$ sulfuric acid. The mixture was filtered through Celite, and the dried ($MgSO_4$) filtrate was concentrated to yield a yellow oil (1.2 g), which was distilled at 90°–100°C/1 torr; the colorless distillate (0.81 g, 77%) crystallized on standing. Recrystallization from ethyl acetate–petroleum ether (bp 60°–80°C) afforded methyl 3,6-dideoxy-β-D-*ribo*-hexopyranoside **35** as colorless crystals (0.65 g, 62%): mp 62°–64°C, $[\alpha]_D$ −65.5° ($c$ 1.2, $H_2O$).

*In the Presence of Triethylamine*

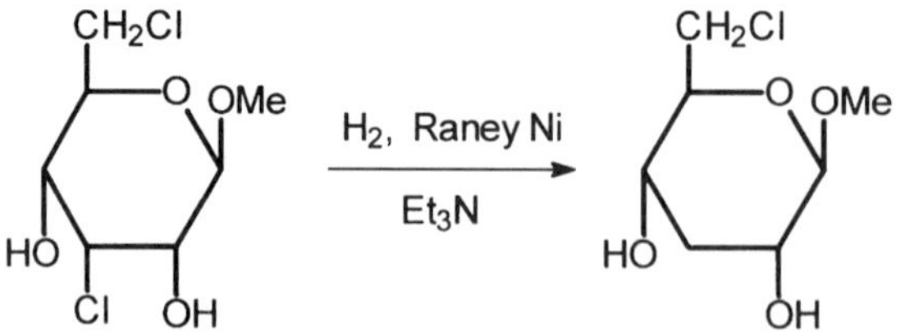

A solution of methyl 3,6-dichloro-3,6-dideoxy-β-D-allopyranoside [47] (**25**, 1.0 g) in ethanol (30 mL) containing W-4 Raney nickel catalyst [27,58] (20 mL ethanol slurry) and triethylamine (1.3 mL) was subjected to a hydrogen pressure of 45 psig for 36 h. TLC [Silica Gel G, Brinkmann; 2:1 (v/v) ethyl acetate–petroleum ether (bp 60°–80°C)] showed that the starting material ($R_f$ 0.42) had all reacted and revealed the presence of a new component having an $R_f$ of 0.23. The filtered solution was concentrated to a syrup, which

## Halogenation of Carbohydrate Derivatives

was chromatographed on silica gel, with ethyl acetate as eluent. Methyl 6-chloro-3,6-dideoxy-β-D-*ribo*-hexopyranoside **40** was obtained as a colorless syrup that crystallized on standing; yield 0.68 g (80%), mp 41°–43°C, $[\alpha]_D$ −57.3° (c 1.2, H$_2$O).

### H. Synthesis of 3,6-Dideoxy-D-*ribo*-hexose (Paratose) [56]

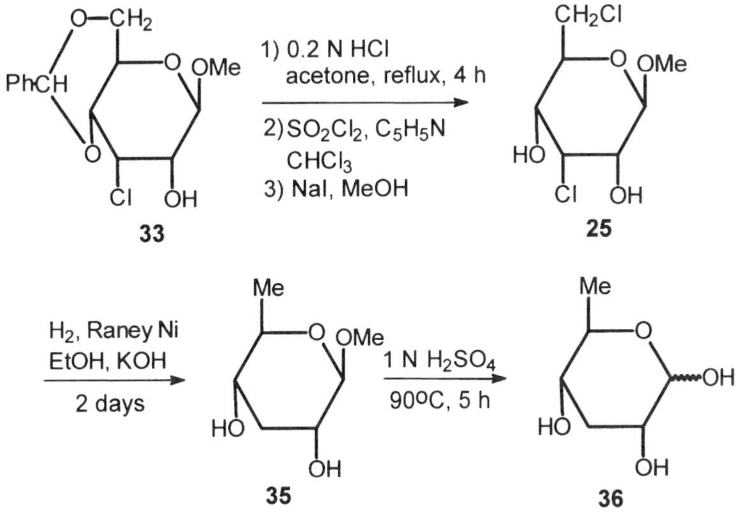

A solution of methyl 4,6-*O*-benzylidene-3-chloro-3-deoxy-β-D-allopyranoside [29] (**33**, 2.45 g) in acetone (40 mL) and 0.2 N hydrochloric acid (13 mL) was heated at reflux temperature for 4 h. The acetone was removed by distillation, and the aqueous solution was neutralized with Duolite A-4(OH⁻) ion-exchange resin and concentrated to yield methyl 3-chloro-3-deoxy-β-D-allopyranoside as a chromatographically homogeneous (TLC) syrup [41] (1.56 g), $R_f$ 0.5 [6:2:1 (v/v/v) 1-butanol–ethanol–water], $[\alpha]_D$ −49° (c 1.0, H$_2$O). This syrupy product (1.39 g) was converted into methyl 3,6-dichloro-3,6-dideoxy-β-D-allopyranoside **25** as described by Cottrell et al. [44]. The crystalline product was recrystallized from chloroform–petroleum ether to give the final product (0.5 g, 33%): mp 162°–163°C, $[\alpha]_D$ −43° (c 0.6, CHCl$_3$); reported [44] mp 154°–156°C, $[\alpha]_D$ −45° (c 1.0, CHCl$_3$).

A solution of methyl 3,6-dichloro-3,6-dideoxy-β-D-allopyranoside (**25**, 148 mg) in ethanol (50 mL) containing potassium hydroxide (70 mg) and W-4 Raney nickel [27,58] (40 mg) was shaken in an atmosphere of hydrogen for 2 days. A solid product was isolated in the usual manner; recrystallization from ethyl acetate–petroleum ether gave methyl 3,6-dideoxy-β-D-*ribo*-hexopyranoside (**35**, 89 mg, 86%): mp 63°–65°C, $[\alpha]_D$ −60° (c 1.5, MeOH). A solution of methyl 3,6-dideoxy-β-D-*ribo*-hexopyranoside (**35**, 86 mg) in 1 N sulfuric acid (10 mL) was heated at 90°C for 5 h. The cooled solution was neutralized with Duolite A-4(OH⁻) ion-exchange resin and concentrated to yield paratose **36** as a homogeneous (TLC) syrup (61 mg, 78%): $R_f$ 0.21 [Silica Gel G; 6:1 (v/v) chloroform–methanol], $[\alpha]_D$ +7° (c 1.1, H$_2$O), in agreement with values reported previously [61,62] for paratose.

### I. Methyl 4,6-Diazido-4,6-dideoxy-α-D-glucopyranoside [29]

A stirred mixture of methyl 2,3-di-*O*-acetyl-4,6-dichloro-4,6-dideoxy-α-D-galactopyranoside (**41**, 1 mmol) [prepared by acetylation (acetic anhydride–pyridine) of methyl 4,6-dichloro-4,6-dideoxy-α-D-galactopyranoside **37**], sodium azide (4 mmol), and dry *N*,*N*-

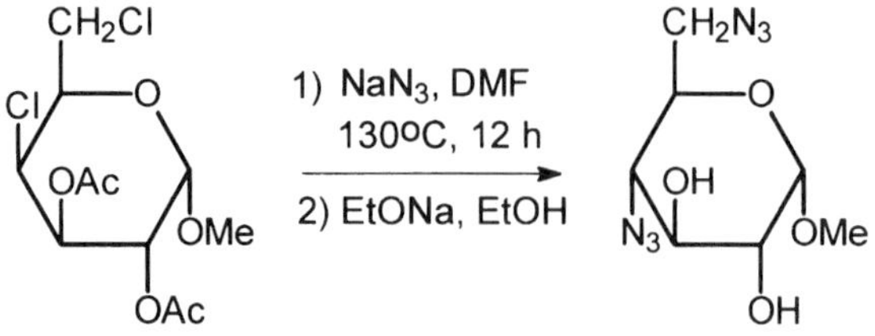

dimethylformamide was maintained at 130°C for 12 h. The solvent was removed under reduced pressure, and the residue was partitioned between chloroform and water. The chloroform solution was dried over anhydrous sodium sulfate and concentrated to leave a syrupy product which was de-esterified by treatment with ethanolic sodium ethoxide [63] to give methyl 4,6-diazido-4,6-dideoxy-α-D-glucopyranoside **42** (90%): mp 85°–87°C, $[\alpha]_D$ +108° (*c* 1.05, CHCl$_3$); reported [63] mp 86°–88°C, $[\alpha]_D$ +110° (*c* 1.18, CHCl$_3$).

## J. Methyl 2,3-Dideoxy-β-D-*glycero*-hex-2-enopyranosid-4-ulose [64]

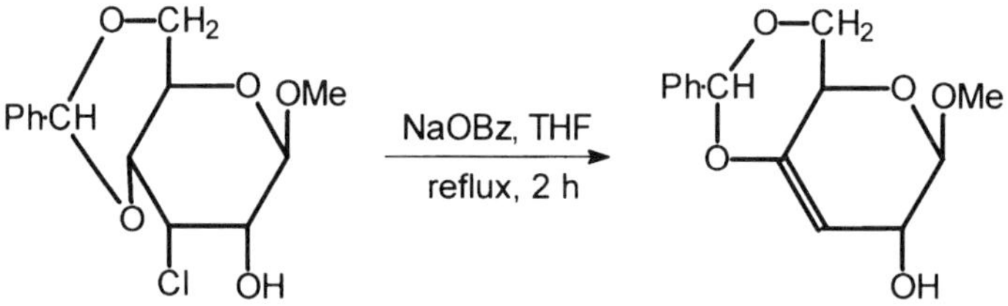

A solution of methyl 4,6-*O*-benzylidene-3-chloro-3-deoxy-β-D-allopyranoside [29] (**33**, 12.6 g) in dry tetrahydrofuran (500 mL) containing sodium benzoate (12 g) was boiled for 2 h under reflux. The cooled suspension was filtered, and the filtrate was evaporated to dryness. The residue was extracted with chloroform, and the extract was washed with water, dried (MgSO$_4$), and evaporated to dryness. Crystallization of the product from chloroform–petroleum ether gave methyl 4,6-*O*-benzylidene-β-D-*erythro*-hex-3-enopyranoside **43**, yield 8.8 g (85%), mp 134°–135°C, $[\alpha]_D$ −43° (*c* 1.58, CHCl$_3$).

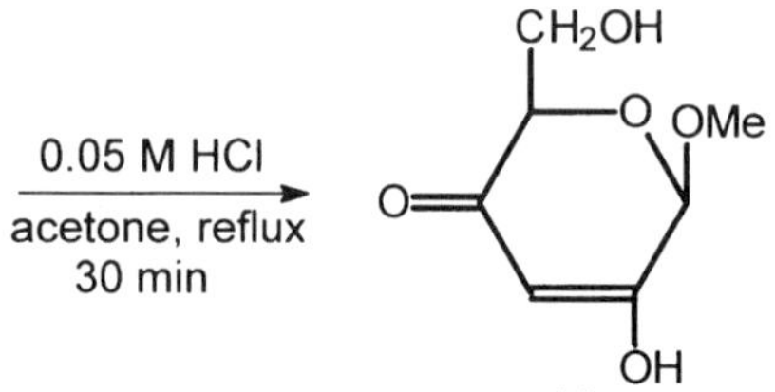

A solution of methyl 4,6-*O*-benzylidene-β-D-*erythro*-hex-3-enopyranoside (**43**, 4 g) in acetone (60 mL) and 0.05 *M* hydrochloric acid (40 mL) was boiled for 30 min under reflux. The solution was cooled, made neutral with barium carbonate, and the mixture was filtered. The acetone was evaporated, and the resulting aqueous solution was washed with petroleum ether, and then extracted with chloroform; the extracts were washed with a small amount of water (to remove the material that did not migrate in TLC), dried (MgSO$_4$), and evaporated, to give methyl 2,3-dideoxy-β-D-*glycero*-hex-2-enopyranosid-4-ulose **44** as a chromatographically homogeneous syrup; yield 1.7 g (71%), $R_f$ 0.67 [Silica Gel G; 1.9 (v/v) methanol–ethyl acetate], $[\alpha]_D$ −14.3° (c 0.8, Me$_2$CO).

## REFERENCES

1. A. A. E. Penglis, Fluorinated carbohydrates, *Adv. Carbohydr. Chem. Biochem. 38*:195 (1981).
2. N. F. Taylor, *Fluorinated Carbohydrates. Chemical and Biochemical Aspects*, ACS Symp. Ser. *374* (1988).
3. T. Tsuchiya, Chemistry and developments of fluorinated carbohydrates, *Adv. Carbohydr. Chem. Biochem. 48*:91 (1990).
4. Special issue on fluoro sugars, *Carbohydr. Res. 249*(1) (1993).
5. I. D. Jenkins, J. P. H. Verheyden, and J. G. Moffatt, Synthesis of the nucleoside antibiotic nucleocidin, *J. Am. Chem. Soc. 93*:4323 (1971).
6. R. D. Birkenmeyer and F. Kagan, Lincomycin. XI. Synthesis and structure of clindamycin, a potent antibacterial agent, *J. Med. Chem. 13*:616 (1970).
7. T. Tsuchiya and S. Umezawa, Studies of antibiotics and related substances. XXI. The synthesis of deoxy and chlorodeoxy derivatives of kanamycin, *Bull. Chem. Soc. Jpn. 38*:1181 (1965).
8. T. Suami, L. Hough, M. Tsuboi, T. Machinami, and N. Watanabe, Molecular mechanisms of sweet taste. V. Sucralose and its derivatives, *J. Carbohydr. Chem. 13*:1079 (1994).
9. S. S. Thomas, J. Plenkiewicz, E. R. Ison, M. Bols, W. Zou, W. A. Szarek, and R. Kisilevsky, Influence of monosaccharide derivatives on liver cell glycosaminoglycan synthesis: 3-deoxy-D-xylo-hexose (3-deoxy-D-galactose) and methyl (methyl 4-chloro-4-deoxy-β-D-galactopyranosid)uronate, *Biochim. Biophys. Acta 1272*:37 (1995).
10. J. E. G. Barnett, Halogenated carbohydrates, *Adv. Carbohydr. Chem. 22*:177 (1967).
11. S. Hanessian, Some approaches to the synthesis of halodeoxy sugars, *Adv. Chem. Ser. 74*:159 (1968).
12. W. A. Szarek, Deoxyhalogeno sugars, *Adv. Carbohydr. Chem. Biochem. 28*:225 (1973).
13. W. A. Szarek, General carbohydrate synthesis, in *MTP International Review of Science, Organic Chemistry Series One*, Vol. 7, *Carbohydrates*, G. O. Aspinall, ed., Butterworths, London, 1973, p. 71.
14. W. A. Szarek and D. M. Vyas, General carbohydrate synthesis, in *MTP International Review of Science, Organic Chemistry Series Two*, Vol. 7, *Carbohydrates*, G. O. Aspinall, ed., Butterworths, London, 1976, p. 89.
15. C. R. Haylock, L. D. Melton, K. N. Slessor, and A. S. Tracey, Chlorodeoxy and deoxy sugars, *Carbohydr. Res. 16*:375 (1971).
16. N. K. Kochetkov and A. I. Usov, The reaction of carbohydrates with triphenyl phosphite methiodide and related compounds. A new synthesis of deoxy sugars, *Tetrahedron 19*:973 (1963).
17. P. J. Garegg, Some aspects of regio-, stereo-, and chemoselective reactions in carbohydrate chemistry, *Pure Appl. Chem. 56*:845 (1984).
18. M. Hudlický, Fluorination with diethylaminosulfur trifluoride and related aminofluorosulfuranes, *Org. React. 35*:513 (1988).
19. S. Hanessian and N. R. Plessas, Reactions of carbohydrates with (halomethylene)dimethyliminium halides and related reagents. Synthesis of some chlorodeoxy sugars, *J. Org. Chem. 34*:2163 (1969).
20. M. G. Straatmann and M. J. Welch, Fluorine-18-labeled diethylaminosulfur trifluoride (DAST): A fluorine-for-hydroxyl fluorinating agent, *J. Nucl. Med. 18*:151 (1977).
21. A. C. Richardson, Nucleophilic replacement reactions of sulphonates. Part VI. A summary of steric and polar factors, *Carbohydr. Res. 10*:395 (1969).
22. A. C. Richardson, Amino and nitro sugars, in *MTP International Review of Science, Organic*

*Chemistry Series One*, Vol. 7. *Carbohydrates*, G. O. Aspinall, ed., Butterworths, London, 1973, p. 105.
23. S. Hanessian, Deoxy sugars, *Adv. Carbohydr. Chem. 21*:143 (1966).
24. S. Hanessian, ed., *Deoxy Sugars, Adv. Chem. Ser. 74* (1968).
25. R. F. Butterworth and S. Hanessian, Tables of the properties of deoxy sugars and their simple derivatives, *Adv. Carbohydr. Chem. Biochem. 26*:279 (1971).
26. L. Vargha and J. Kuszmann, 2-Chloro-2-deoxy-D-arabinose and -D-ribose. A new synthesis of 2-deoxy-D-*erythro*-pentose, *Chem. Ber. 96*:411 (1963).
27. B. T. Lawton, D. J. Ward, W. A. Szarek, and J. K. N. Jones, Synthesis of D-chalcose, *Can. J. Chem. 47*:2899 (1969).
28. B. T. Lawton, W. A. Szarek, and J. K. N. Jones, A facile synthesis of 4,6-dideoxy-D-xylo-hexose, *Carbohydr. Res. 14*:255 (1970).
29. B. T. Lawton, W. A. Szarek, and J. K. N. Jones, Synthesis of deoxy and aminodeoxy sugars by way of chlorodeoxy sugars, *Carbohydr. Res. 15*:397 (1970).
30. B. T. Lawton, W. A. Szarek, and J. K. N. Jones, unpublished results.
31. H. Arita, N. Ueda, and Y. Matsushima, The reduction of chlorodeoxy sugars by tri-*n*-butyltin hydride, *Bull. Chem. Soc. Jpn. 45*:567 (1972).
32. H. Paulsen, B. Sumfleth, and H. Redlich, Selektive Synthese der D-Aldgarose aus Aldgamycin E und F, *Chem. Ber. 109*:1362 (1976).
33. L. W. Menapace and H. G. Kuivila, Mechanism of reduction of alkyl halides by organotin hydrides, *J. Am. Chem. Soc. 86*:3047 (1964).
34. B. Helferich, Two new derivatives of α- and β-methyl glucoside, *Berichte der deutschen chemischen Gesellschaft 54*:1082 (1921).
35. B. Helferich, A. Löwa, W. Nippe, and H. Riedel, Two new derivatives of trehalose and mannitol and an α-methyl glucoside dichlorohydrin, *Berichte der deutschen chemischen Gesellschaft 56*:1083 (1923).
36. B. Helferich, G. Sprock, and E. Besler, *d*-Glucose 5,6-dichlorohydrin, *Berichte der deutschen chemischen Gesellschaft 58*:886 (1925).
37. P. D. Bragg, J. K. N. Jones, and J. C. Turner, The reaction of sulphuryl chloride with glycosides and sugar alcohols. Part I, *Can. J. Chem. 37*:1412 (1959).
38. J. K. N. Jones, M. B. Perry, and J. C. Turner, The reaction of sulphuryl chloride with glycosides and sugar alcohols. Part II, *Can. J. Chem. 38*:1122 (1960).
39. H. J. Jennings and J. K. N. Jones, The reaction of sulphuryl chloride with reducing sugars. Part I, *Can. J. Chem. 40*:1408 (1962).
40. H. J. Jennings and J. K. N. Jones, The reaction of chlorosulphate esters of sugars with pyridine, *Can. J. Chem. 41*:1151 (1963).
41. H. J. Jennings and J. K. N. Jones, Reaction of sugar chlorosulfates. Part V. The synthesis of chlorodeoxy sugars, *Can. J. Chem. 43*:2372 (1965).
42. H. J. Jennings and J. K. N. Jones, Reaction of sugar chlorosulfates. Part VI. The structure of unsaturated chlorodeoxy sugars, *Can. J. Chem. 43*:3018 (1965).
43. A. G. Cottrell, E. Buncel, and J. K. N. Jones, Chlorosulphate as a leaving group: The synthesis of a methyl tetrachloro-tetradeoxy-hexoside, *Chem. Ind. (Lond.)*: p. 522 (1966).
44. A. G. Cottrell, E. Buncel, and J. K. N. Jones, Reactions of sugar chlorosulfates. VII. Some conformational aspects, *Can. J. Chem. 44*:1483 (1966).
45. S. S. Ali, T. J. Mepham, I. M. E. Thiel, E. Buncel, and J. K. N. Jones, Reactions of sugar chlorosulfates. Part VIII. D-Ribose and its derivatives, *Carbohydr. Res. 5*:118 (1967).
46. E. Buncel, Chlorosulfates, *Chem. Rev. 70*:323 (1970).
47. D. M. Dean, W. A. Szarek, and J. K. N. Jones, A reinvestigation of the reaction of methyl β-D-glucopyranoside with sulfuryl chloride, *Carbohydr. Res. 33*:383 (1974).
48. H. J. Jennings, Synthesis of 6-*O*-α- and β-D-xylopyranosyl-D-mannopyranose. (Glycosidation of α- and β-D-xylopyranosyl chloride, 2,3,4-tri(chlorosulfate)), *Can. J. Chem. 46*:2799 (1968).
49. H. J. Jennings, Conformations and configurations of some chlorosulfated α- and β-D-pentopyranosyl chlorides, *Can. J. Chem. 47*:1157 (1969).

50. H. J. Jennings, Stereospecific synthesis of two epimeric 2-chlorodeoxy pentoses involving anomeric chlorine participation, *Can. J. Chem. 48*:1834 (1970).
51. B. Achmatowicz, W. A. Szarek, J. K. N. Jones, and E. H. Williams, Reaction of methyl pentofuranosides with sulfuryl chloride, *Carbohydr. Res. 36*:C14 (1974).
52. H. Parolis, The synthesis of chlorodeoxyhexofuranoid derivatives, *Carbohydr. Res. 114*:21 (1983).
53. P. L. Durette, L. Hough, and A. C. Richardson, The chemistry of maltose. Part I. The reaction of methyl β-maltoside with sulphuryl chloride, *Carbohydr. Res. 31*:114 (1973).
54. J. M. Ballard, L. Hough, A. C. Richardson, and P. H. Fairclough, Sucrochemistry. Part XII. Reaction of sucrose with sulphuryl chloride, *J. Chem. Soc. Perkin I*:1524 (1973).
55. L. Hough, S. P. Phadnis, and E. Tarelli, The preparation of 4,6-dichloro-4,6-dideoxy-α-D-galactopyranosyl 6-chloro-6-deoxy-β-D-fructofuranoside and the conversion of chlorinated derivatives into anhydrides, *Carbohydr. Res. 44*:37 (1975).
56. E. H. Williams, W. A. Szarek, and J. K. N. Jones, Synthesis of paratose (3,6-dideoxy-D-*ribo*-hexose) and tyvelose (3,6-dideoxy-D-*arabino*-hexose), *Can. J. Chem. 49*:796 (1971).
57. R. G. Edwards, L. Hough, A. C. Richardson, and E. Tarelli, A reappraisal of the selectivity of the mesyl chloride–*N,N*-dimethylformamide reagent. Chlorination at secondary positions, *Tetrahedron Lett.*: p. 2369 (1973).
58. A. A. Pavlic and H. Adkins, Preparation of a Raney nickel catalyst, *J. Am. Chem. Soc. 68*:1471 (1946).
59. G. Siewert and O. Westphal, Substitution of secondary *p*-tolylsulfonyloxy groups by iodine. Synthesis of 4-deoxy- and 4,6-dideoxy-D-xylo-hexose, *Liebigs Annalen der Chemie 720*:161 (1969).
60. W. A. Szarek, A. Zamojski, A. R. Gibson, D. M. Vyas, and J. K. N. Jones, Selective, reductive dechlorination of chlorodeoxy sugars. Structural determination of chlorodeoxy and deoxy sugars by $^{13}C$ nuclear magnetic resonance spectroscopy, *Can. J. Chem. 54*:3783 (1976).
61. O. Westphal and O. Lüderitz, 3,6-Dideoxyhexoses—chemistry and biology, *Angew. Chem. 72*:881 (1960).
62. C. Fouquey, J. Polonsky, and E. Lederer, Synthesis of three 3,6-dideoxyhexoses. Determination of the structure of the natural sugars tyvelose, ascarylose, and paratose, *Bull. Soc. Chim. Fr.* p. 803 (1959).
63. J. Hill, L. Hough, and A. C. Richardson, Nucleophilic replacement reactions of sulphonates. Part I. The preparation of derivatives of 4,6-diamino-4,6-dideoxy-D-glucose and -D-galactose, *Carbohydr. Res. 8*:7 (1968).
64. E. H. Williams, W. A. Szarek, and J. K. N. Jones, Preparation of unsaturated carbohydrates from methyl 4,6-*O*-benzylidene-3-chloro-3-deoxy-β-D-allopyranoside, and their utility in the synthesis of sugars of biological importance, *Carbohydr. Res. 20*:49 (1971).

# 7

# Nucleophilic Displacement Reactions of Imidazole-1-Sulfonate Esters

**Jean-Michel Vatèle**
Université Claude Bernard, Villeurbanne, France

**Stephen Hanessian**
University of Montreal, Montreal, Quebec, Canada

|     |                                                                       |     |
|-----|-----------------------------------------------------------------------|-----|
| I.  | Introduction                                                          | 127 |
| II. | Methods                                                               | 130 |
|     | A. Preparation and nucleophilic substitution reactions of imidazole-1-sulfonates | 130 |
|     | B. Formation of imidazole-1-sulfonates from alcohols                  | 130 |
|     | C. Reactions of carbohydrate imidazole-1-sulfonates                   | 131 |
|     | D. Conclusion                                                         | 135 |
| III.| Experimental Procedures                                               | 136 |
|     | A. Preparation of imidazole-1-sulfonates                              | 136 |
|     | B. Substitution reactions                                             | 139 |
|     | References                                                            | 145 |

## I. INTRODUCTION

Since the pioneering work of Karl Freudenberg on displacements of carbohydrate *p*-toluenesulfonates [1–5], bimolecular nucleophilic substitutions became one of the most employed and useful reactions in carbohydrate chemistry. Indeed $S_N2$-type reactions have allowed the introduction of a variety of heteroatoms (halogens, N-, O-, S-) into carbohydrates, and the resulting compounds have been used in many synthetic and biological contexts [6].

Deoxy halogeno sugars are of particular interest because halogenated carbohydrates and nucleosides have biologically interesting properties [7,8–14]. They are also excellent precursors to deoxy sugars, an important class of biologically occurring carbohydrates [9,15,16]. Introduction of nitrogen substituents by nucleophilic substitution is the most effective way to prepare certain amino sugars [17,18], which are also of widespread occurrence in nature [19].

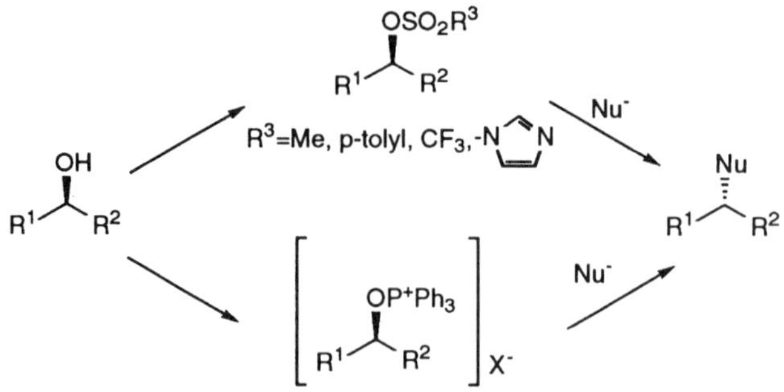

Scheme 1

Substitution of primary and secondary alcohols by $S_N2$ reactions is generally effected according to two protocols, as shown in Scheme 1. The most common approach involves the formation of a sulfonate ester which is an excellent leaving group. In the second method, a direct conversion from the alcohol can be achieved by a transient species, usually an alkoxyphosphonium salt.

Sulfonic esters of sugar derivatives provide a versatile and simple method for activating hydroxy groups for a bimolecular displacement reactions with nucleophiles. Until the late 1970s, the most common sulfonate leaving groups consisted of either p-toluenesulfonates or methanesulfonates [20–22]. The use of a p-bromobenzenesulfonate (brosylate), a leaving group ten times more reactive than a p-toluenesulfonate group [23], has been occasionally reported [24], and has allowed, for example, substitutions under mild conditions at the C-4 position of D-gluco- and D-galactopyranoside derivatives with a variety of nucleophiles [25,26]. The introduction of the trifluoromethanesulfonate group (triflate) [27,28] in the field of carbohydrate chemistry, greatly improved the efficiency of displacement reactions [29,30], which were sluggish or impossible when other sulfonic esters were used [22,31,32]. Nucleophilic displacement reactions of secondary sulfonates have been discussed in terms of polar and electronic effects [32]. In general, $S_N2$ substitutions at C-2 in alkyl α-D-glucopyranosides are difficult to achieve. It is thus not easy to prepare 2-substituted alkyl α-D-mannopyranosides by displacement of 2-arylsulfonates, for example. Even with the more reactive triflates, a temperature of 80°C (5 h) was needed to effect a displacement with benzoate ion [33]. The corresponding reaction with a mesylate leaving group was not possible even at 153°C (120 h) [34] (Scheme 2).

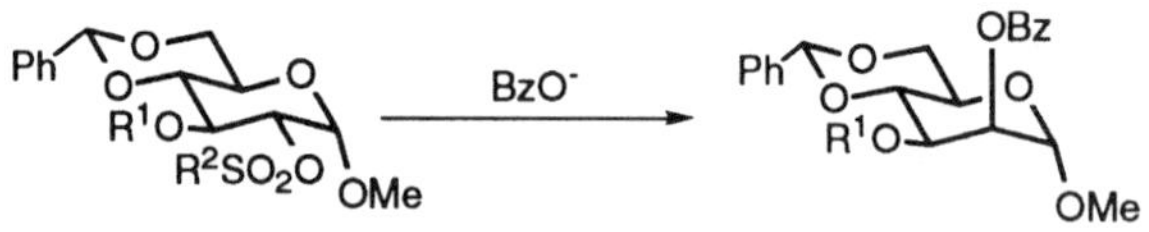

$R^2=CH_3$    120h, 153°C    3.5% yield [34]
$R^2=CF_3$    5h, 80°C       82% yield [33]

Scheme 2

A variety of phosphorus-containing reagents have been employed in sugar chemistry in the direct $S_N2$ replacement of hydroxy groups mainly by halide ions [35]. Most of these reagents used a combination of triphenylphosphine with an electrophilic halogen source such as N-halosuccinimide [36,37], carbon tetrahalides [38–41], and more recently, trihaloimidazole or iodine and imidazole [42–45] or related reagents [46]. The Mitsunobu reaction, which relies on the diethyl azodicarboxylate (DEAD)–triphenylphosphine redox system, is the most versatile and synthetically useful of phosphorus (III)-based reagents, and it has been extensively used in the synthesis of ester-, amino-, halo-, and sulfur-substituted carbohydrates [47,48]. An example of the efficiency of the Mitsunobu reaction compared with p-toluenesulfonate esters to promote nucleophilic displacements is depicted in Scheme 3 [49,50].

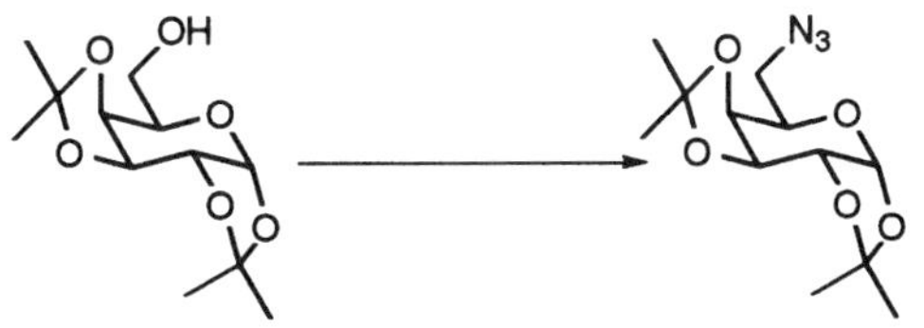

| p-TsCl, then NaN$_3$, DMF | 120°C, 19h | 84% yield [49] |
|---|---|---|
| DIAD, PPh$_3$, Zn(N$_3$)$_2$ | r.t., 0.5-2h | 82% yield [50] |

**Scheme 3**

The original Rydon reagent [51], methyl triphenoxyphosphonium iodide [(PhO)$_3$P$^+$-MeI$^-$], which was used extensively for the preparation of deoxyiodo sugars [52–54] and nucleosides [41,55–59] is still a reliable and efficient protocol.

Other reagents, such as sulfuryl chloride [60], the Vilsmeier-Haack reagent [(Me$_2$N$^+$=CHX)X$^-$] [61] and, more recently, diethylaminosulfur trifluoride (DAST) [11–13] are useful for the direct replacement of hydroxy groups by halogen atoms.

Many reagents known for the direct conversion of alcohols into halides have been used to achieve site-selective nucleophilic substitution in polyhydroxy compounds [44–46,62]. In some cases, it is possible to effect selective displacements of primary alcohols in the presence of secondary ones, and to achieve differential substitutions of secondary alcohols in unprotected sugars [44,45,64,65] (Scheme 4).

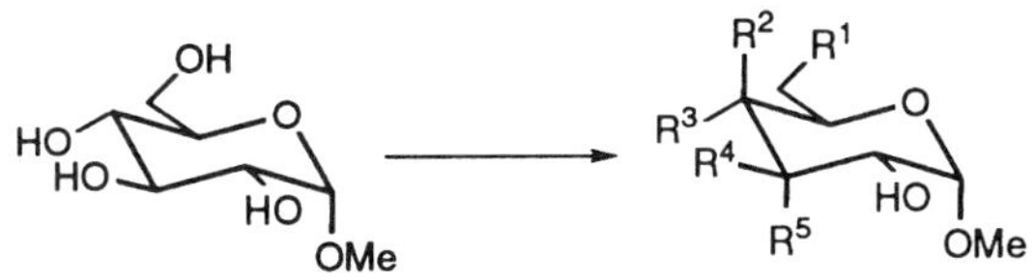

| PPh$_3$, I$_2$, imidazole | R$^1$=R$^2$=I, R$^4$=OH, R$^3$=R$^5$=H | 70% yield [45] |
|---|---|---|
| PPh$_3$, tribromoimidazole | R$^1$=R$^5$=Br, R$^3$=OH, R$^2$=R$^4$=H | 75% yield [44, 45] |

**Scheme 4**

Even though phosphorus-based reagents and sulfonate esters have found extensive applications as leaving groups in carbohydrate chemistry, they have some drawbacks. Thus, triflate esters are usually prepared from triflic anhydride, a relatively expensive reagent, which could preclude its use on a large scale. Because of their high reactivity, triflate esters have a limited shelf-life. Their preparation occasionally requires the use of specific sophisticated bases such as 2,6-di-t-butyl-4-methylpyridine [63]. The major difficulty experienced with the triphenylphosphine-based reagents and Mitsunobu reactions is the removal of redox by-products such as triphenylphosphine oxide and diethoxycarbonylhydrazine, respectively. Triphenylphosphine-based reagents are known to undergo rearrangements in certain cases [35,40,66,67].

## II. METHODS

### A. Preparation and Nucleophilic Substitution Reactions of Imidazole-1-sulfonates

The imidazole-1-sulfonate (imidazylate, Imz) group was a "designed" leaving group for $S_N2$ displacement reactions by virtue of its inherent structure [68]. It represents a different notion of reactivity because, in addition to the inherent sulfonate-type character, its reactivity can be enhanced by remote activation involving a nitrogen atom (Scheme 5).

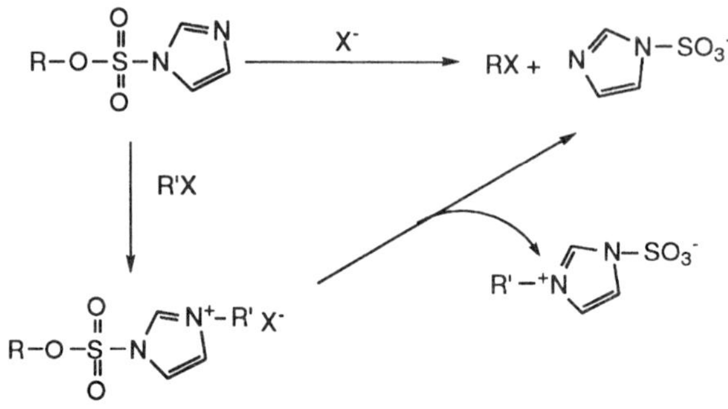

**Scheme 5**

### B. Formation of Imidazole-1-sulfonates from Alcohols

Two methods have been developed for preparing carbohydrate imidazole-1-sulfonates [68] (Scheme 6). The first, (method A) consists in allowing a partially protected sugar to react with 1.5 eq. of sulfuryl chloride in dimethylformamide (DMF) at −40°C in the presence

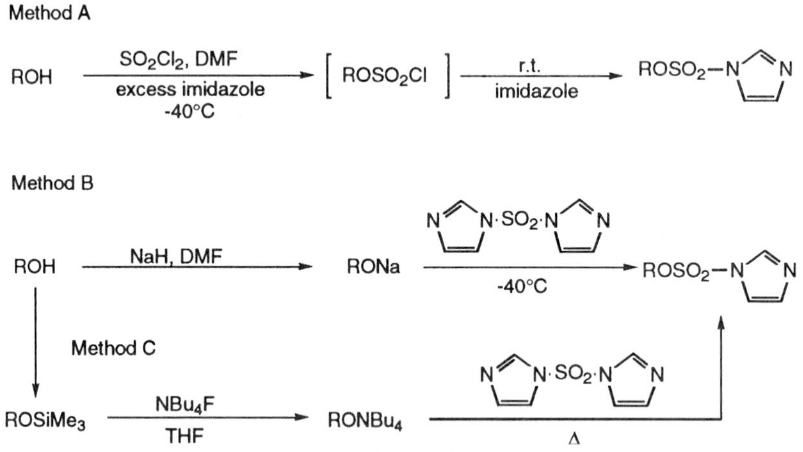

**Scheme 6**

of 6 eq. of imidazole, to form initially a chlorosulfate ester [69,70], in which the chlorine atom is substituted by imidazole at room temperature. With few exceptions, ester formation was completed within 1 h. In the second method (method B), the alkoxide, generated by treatment with NaH in DMF reacts readily at −40°C with an excess of $N,N'$-sulfuryldiimidazole, a stable and highly crystalline reagent [71]. An alternative method, in which the use of NaH is avoided, consists in generating the alkoxide from the corresponding trimethylsilyl ether with tetrabutylammonium fluoride [68] followed by addition of $N,N'$-sulfuryldiimidazole.

Most carbohydrate imidazylates have good crystallization properties, and they are stable at room temperature. Common functionalities encountered in carbohydrates are compatible under the conditions of formation of imidazylates. For imidazolylsulfonylation of vicinal secondary diols, the obtained product depends on the method used. Thus, in the sulfuryl chloride method (method A), a sugar disulfonate was obtained [72], whereas the use of NaH/$N,N'$-sulfuryldiimidazole led to a cyclic sulfate [73].

*Scope and Limitations*

With primary or nonsterically hindered secondary alcohols, method A led exclusively to the formation of chlorinated sugars [68]. The mildness and efficiency of this chlorination method was used to advantage in the synthesis of a dichloro aminoglycoside, a precursor of seldomycin factor 2 [68] (Scheme **7**).

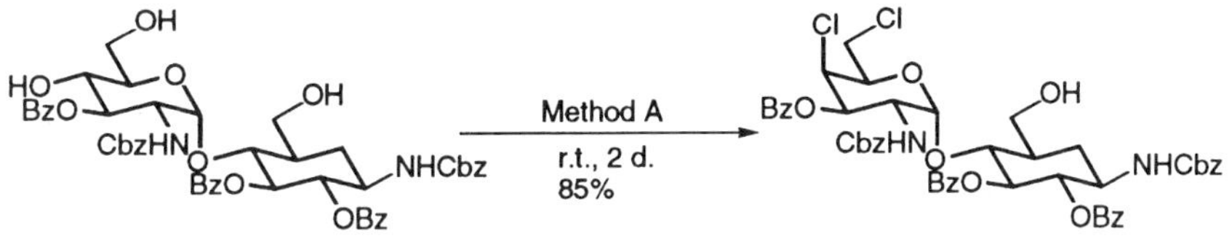

**Scheme 7**

Imidazolesulfonylation of the C-4 axial oxygen of galactopyranoside derivatives, involving $N,N'$-sulfuryldiimidazole **1** and NaH, must be carried out a low temperature ($-30°C$, 7 h). At room temperature, a product of β-elimination was obtained [74] by a favored *anti* elimination process [75], presumably induced by the sodium salt of imidazole present in the medium.

## C. Reactions of Carbohydrate Imidazole-1-sulfonates

Until the introduction of triflate or imidazylate esters in carbohydrates, $S_N2$-displacements of carbohydrate sulfonates with charged nucleophiles in certain positions of hexopyranose or furanose derivatives were not possible, or gave low yields of substituted products owing to the predominance of elimination or rearrangement reactions [6,22,32,33].

*Substitution Reactions*

**Displacements of 6-Mesylate or Tosylate Esters of D-Galactopyranose Derivatives.** These reactions are known to be particularly sluggish compared with analogous reactions in the D-glucopyranose series [20,77,76], presumably because of the polar, repulsive forces in the transition state involving lone pairs of electrons on the axial O-4 and on the ring-oxygen atom [32]. In contrast, reaction of the 6-imidazylate ester of 1,2:3,4-di-*O*-isopropylidene-α-D-galactopyranose at room temperature with a variety of nucleophiles (azide, halides) proceeded in good yields (75–81%) at room temperature [68]. Given the inherent functional design of imidazylate group, treatment with methyl iodide gave the corresponding iodide in good yield [68,78]. The difference of reactivity between imidazylate and tosylate groups is illustrated in Scheme **8**.

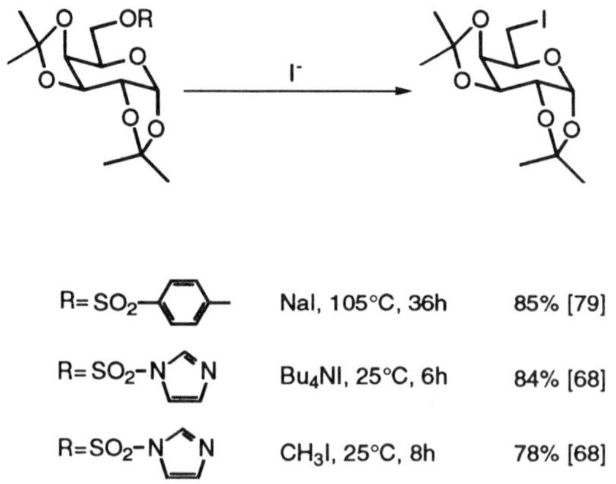

## Scheme 8

R=SO$_2$-C$_6$H$_4$-    NaI, 105°C, 36h    85% [79]

R=SO$_2$-Im    Bu$_4$NI, 25°C, 6h    84% [68]

R=SO$_2$-Im    CH$_3$I, 25°C, 8h    78% [68]

**Imidazylate Displacements at the C-2 Position of Sugar Derivatives.** Attempted displacement of 2-tosylate or mesylate esters situated at C-2 of a pyranoside ring with charged nucleophiles is normally unsuccessful [22,32,33]. In contrast, the 2-imidazylate ester of α- or β-D-glucopyranoside derivatives could be readily displaced under mild conditions with halide, carboxylate, and N-nucleophiles in excellent yields [68,80]. Introduction of benzoate and azide groups by nucleophilic displacement reactions of 2-imidazylate esters in β-D-glucopyranoside derivatives has been used in a new synthetic strategy for the synthesis of disaccharides having β-D-mannopyranosyl and 2-acetamido-2-deoxy-β-D-mannopyranosyl units [81,82], frequently found in bacterial polysaccharides and glycoproteins [83] (Scheme **9**).

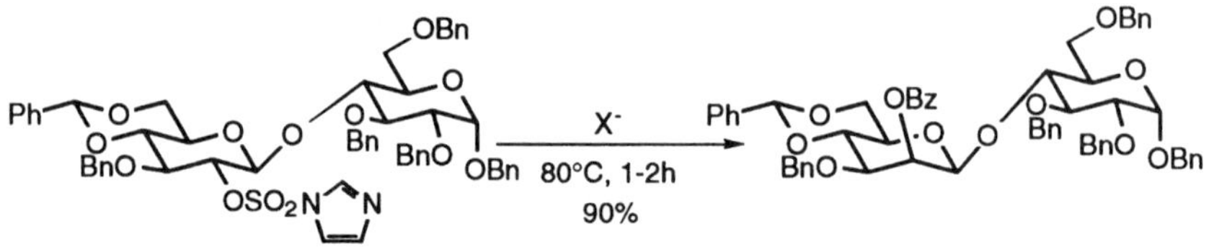

## Scheme 9

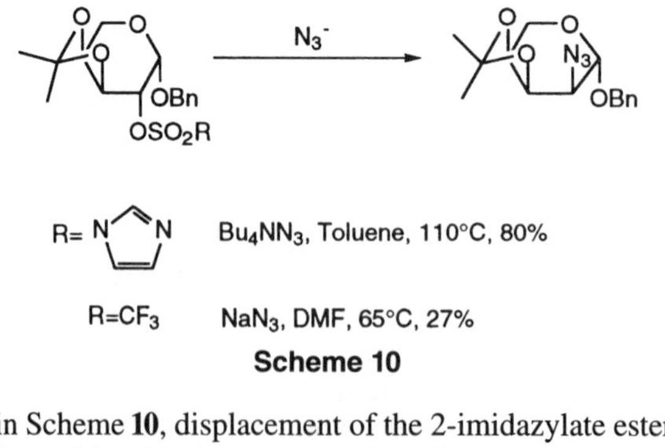

## Scheme 10

R= Im    Bu$_4$NN$_3$, Toluene, 110°C, 80%

R=CF$_3$    NaN$_3$, DMF, 65°C, 27%

As depicted in Scheme **10**, displacement of the 2-imidazylate ester of a β-L-arabinopyranoside derivative with azide ions occurred readily in refluxing toluene to afford a 2-azido-2-deoxy-L-ribopyranoside [84]. Treatment of the corresponding triflate with sodium azide gave the same azido compound in only 27% yield [84].

2-O-Sulfonates of α-D-mannopyranoside derivatives are usually resistant to displacement with charged nucleophiles [85]. Treatment of the 2-imidazylate of methyl 3-O-

## Displacement of Imidazole-1-Sulfonate Esters

benzyl-4,6-O-benzylidene-α-D-mannopyranoside with tetrabutylammonium azide in refluxing toluene gave the expected 2-azido-2-deoxy-gluco derivative in only 23% yield [86]. The major product was that resulting from β-elimination. As expected, the *trans* relation between H-3 and the leaving group favors elimination over substitution. Nevertheless, in spite of the modest yield of the 2-azido product, the 2-imidazylate ester emerges as a useful nucleofugal group here.

As in the pyranoside series, the substitution at C-2 of furanoside derivatives is also difficult [87]. Treatment of the 2-imidazylate ester of a benzyl 5-deoxy-α-D-hexofuranoside derivative with tetrabutylammonium azide or benzoate in refluxing toluene led to the formation of substitution products in 82 and 53% yields, respectively [88] (Scheme 11).

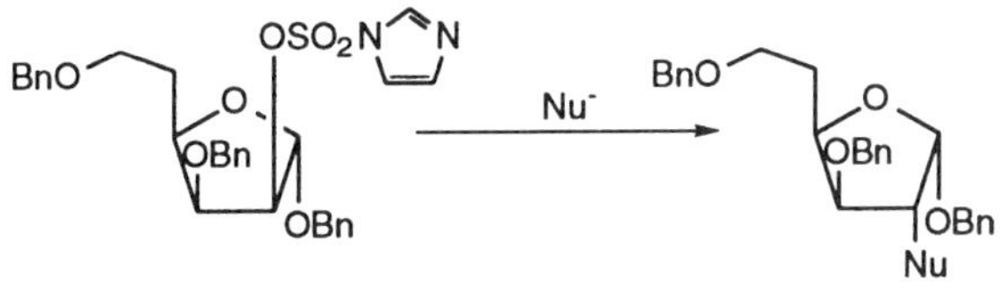

Bu$_4$NN$_3$, 110°C, 16h    82% yield
Bu$_4$NOBz, 110°C, 24h   53% yield

**Scheme 11**

Another interesting use of the imidazylate group in the furanoside series has been described by Tann et al. [89]. Thus, treatment of the C-2 imidazylate ester of the α-D-ribofuranoside tribenzoate derivative with KHF$_2$ and HF, at 160°C, led to the 2-deoxy-2-fluoro sugar in 63% yield, presumably through the intermediacy of the fluorosulfate ester (HPLC). The 2-fluoro compound is a key intermediate in the preparation of 2'-fluoro-2'-deoxy D-arabinofuranosyl pyrimidine nucleosides, which exhibit powerful antiherpetic activity [89] (Scheme 12).

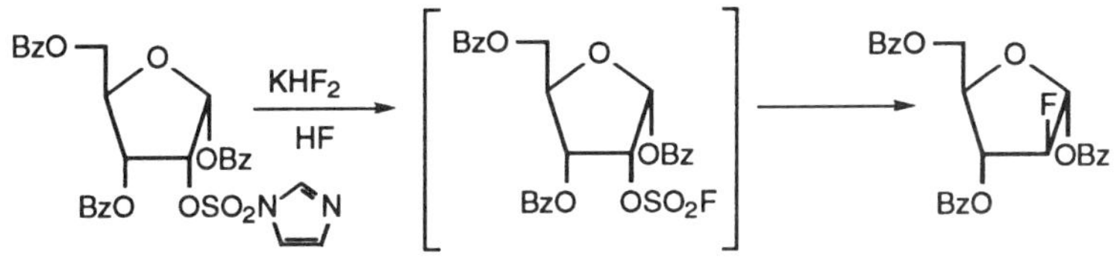

**Scheme 12**

**Imidazole-1-sulfonate Displacements at the C-3 and C-4 Positions of Sugar Derivatives.**   Displacement reactions of tosylates at the C-3 position of α-D-glucopyranoside derivatives proceed with difficulty because of a nonbonded 1,3-diaxial interaction between the aglycone and the approaching nucleophile [32], as well as complications arising from neighboring group participation [90].

Treatment of methyl 4,6-O-benzylidene-2-benzyloxycarbonylamino-2-deoxy-3-O-(imidazole-1-sulfonyl)-α-D-glucopyranoside with tetrabutylammonium iodide in refluxing toluene (3 h) gave the corresponding iodo derivative in 90% yield [68] (Scheme 13). When

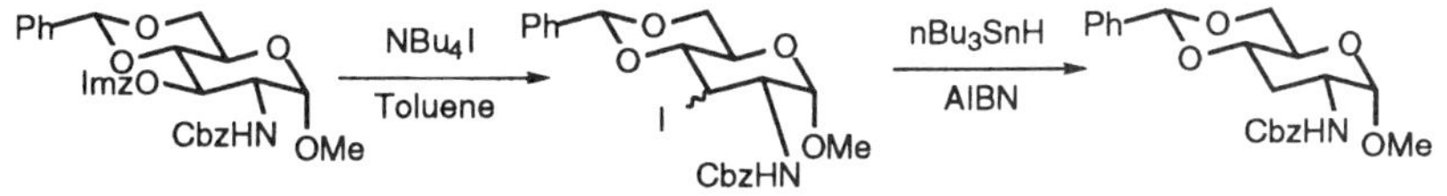

**Scheme 13**

the tosylate ester was used as leaving group, much more drastic conditions were necessary (144 h, reflux in DMF-butanone), resulting in a lower yield (40%) [91].

The facile formation of the iodo compound, and the subsequent reduction with tributyltin hydride opens a new access to 2-amino-2,3-dideoxy-D-glucose (D-lividosamine) present in various aminoglycoside antibiotics [92]. The same strategy been used in the synthesis of C-10'-C-1 fragment of boromycin [93]. Displacement of the C-3 imidazylate ester of a 2-azido-2-deoxy-α-D-altropyranose derivative with benzoate occurred readily [94], compared with the corresponding tosylate [95].

The C-3 tosylate ester of 1,2:5,6-di-O-isopropylidene-α-D-glucofuranose is particularly resistant to displacement with charged nucleophiles [96]. The β-elimination product is usually observed because of the *trans* relation between the leaving group and H-4. The C-3 imidazylate ester is displaced by several nucleophiles under mild conditions and in good yields [68] (Scheme 14). The facile displacement with benzoate ion [97] is note-

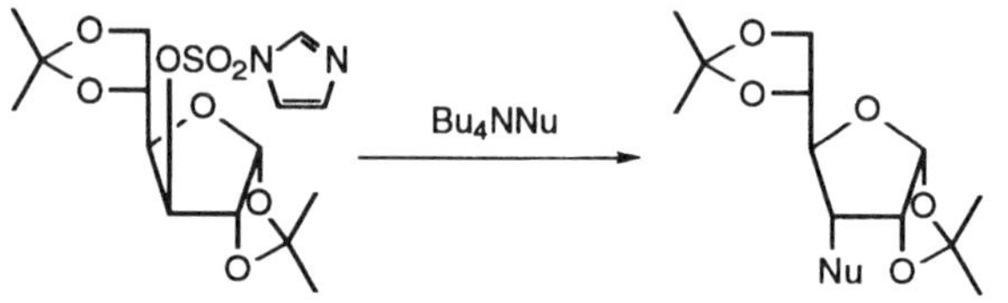

| | | |
|---|---|---|
| 80°C,5h | Nu=N₃ | 62% [68] |
| 80°C,72h | Nu=I | 72% [68] |
| 100°C,2h | Nu=OBz | 97% [98] |

**Scheme 14**

worthy, and it provides an alternative synthesis of D-allofuranose derivatives, which is normally done by an oxidation–reduction sequence [98].

Inversion of configuration at C-4 of a sedoheptulosan derivative, involving imidazylate as leaving group, is a key step in the synthesis of validamine, a known inhibitor of α-D-glucosidase [99]. Thus, despite the presence of the C-2 axial benzoate, the imidazylate group at C-4 undergoes a facile displacement with benzoate anion (3 h, 100°C) to afford the corresponding β-D-ido derivative in 75% yield (Scheme 15).

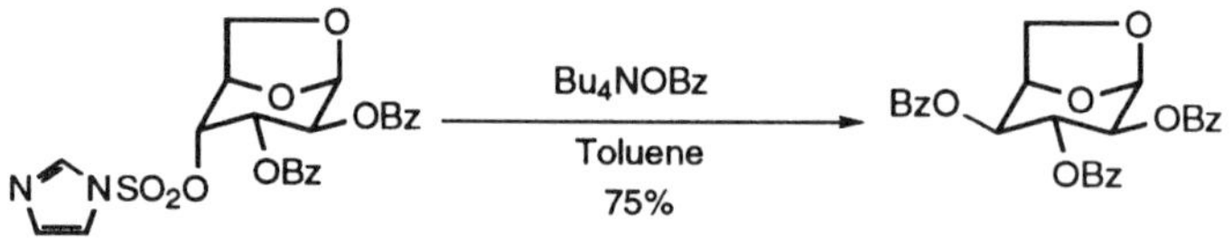

**Scheme 15**

Substitution of a 4'-O-imidazylate derivative of a β-lactoside derivative with benzoate in 95% yield has been described [74].

*Other Reactions of Imidazole-1-sulfonates*

As with other sulfonate esters, vicinal diimidazylates can give rise to the corresponding unsaturated sugars by the procedure of Tipson and Cohen (Zn, NaI, DMF) [72].

6-O-Imidazole-1-sulfonyl esters of α-D-galactopyranoside derivatives are easily transformed to their corresponding 3,6-anhydro derivatives [78] at room temperature in dry DMF, by intramolecular participation of 3-O-methyl substituent (Scheme 16). Formation of

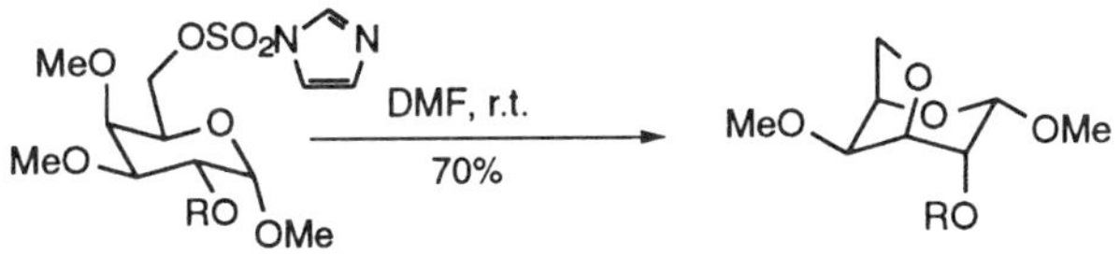

### Displacement of Imidazole-1-Sulfonate Esters

**Scheme 16**

R=(2,3,4-tri-O-methyl-β-D-xylopyranosyl)

3,6-anhydro-D-galactopyranosides during solvolysis of 6-sulfonates with alkoxy group participation is precedented [101,102].

Attempted displacement of the 2-O-imidazole-1-sulfonyl ester of methyl 3,4,6-tri-O-methyl β-D-galactopyranoside with azide ion led to the formation of the corresponding 2,5-anhydro sugar [86], as illustrated in Scheme 17. The predominance of a ring-contraction

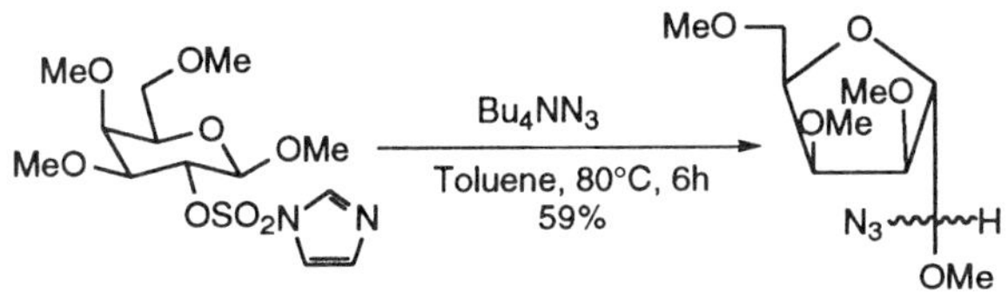

**Scheme 17**

reaction over an $S_N2$-substitution reaction can be explained by a steric interaction between the incoming nucleophile and the axial C-4 substituent [32], and by the favored antiparallel disposition of the C-1–O-5 bond and the equatorial leaving groups at C-2.

The excellent nucleofugal properties of the imidazylate group has been used in the intramolecular cyclization of derivatives of N-substituted L-serine derivatives, leading to β-lactams [105,106] (Scheme **18**).

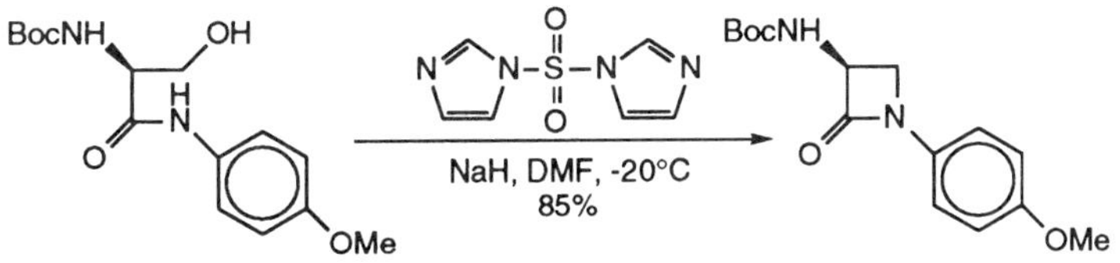

**Scheme 18**

### D. Conclusion

The imidazole-1-sulfonate (imidazylate) group is a versatile leaving group that allows $S_N2$ substitutions in carbohydrates, with a variety of nucleophiles, at positions where other alkyl and aryl sulfonate esters are known to be ineffective. The imidazylate group complements the triflate leaving group for "difficult" substitution reactions of carbohydrate derivatives. Imidazylates have the advantage over triflates of having a longer shelf life, of crystallinity, and of being compatible with chromatographic purification. They are hydrolytically much more stable than triflates, and they can be prepared from readily available and inexpensive reagents.

## III. EXPERIMENTAL PROCEDURES*

### A. Preparation of Imidazole-1-sulfonates

*Method A ($SO_2Cl_2$, Imidazole)*

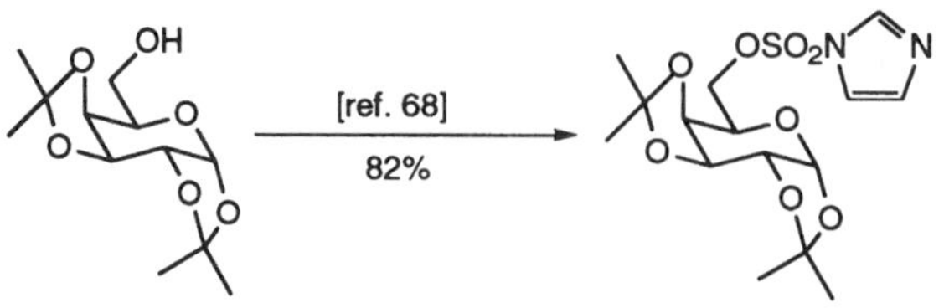

**General Procedure [68].** To a solution of a partially protected sugar (1.0 mmol) and imidazole (0.41 g, 6.0 mmol) in 5 mL of DMF, cooled to −40°C, was added sulfuryl chloride (0.12 mL, 1.5 mmol, 15 eq.) under nitrogen. The solution was stirred for 1 h at −40°C at 1 h at room temperature, then water and ether (or $CH_2Cl_2$) were added, depending on the solubility of sulfonate. The aqueous layer was extracted with ether (or $CH_2Cl_2$) and the combined organic extracts were washed twice with water, dried ($Na_2SO_4$), and evaporated. Flash chromatography of the residue afforded the desired imidazylate. For compound **2** (ether–petroleum ether, 2:1), oil; (0.97 g, 83%); $[\alpha]_D$ −55° (c 1.16, $CH_2Cl_2$) starting from 0.8 g of **1**.

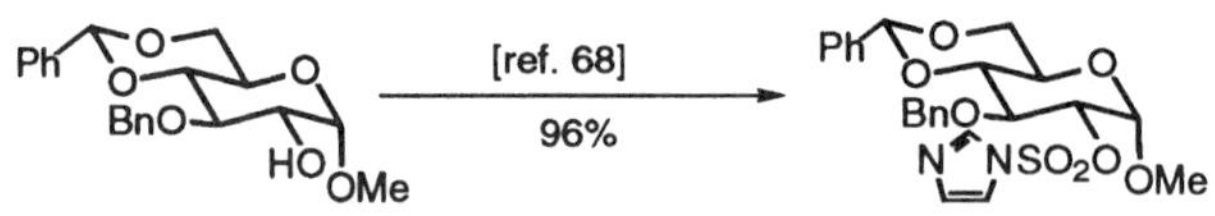

Flash chromatography of the residue obtained from 1 g of **3** (ether–petroleum ether, 3:1) gave the sulfonate **4** (1.3 g, 96%) as an oil; $[\alpha]_D$ +29.2° (c 0.59, $CH_2Cl_2$).

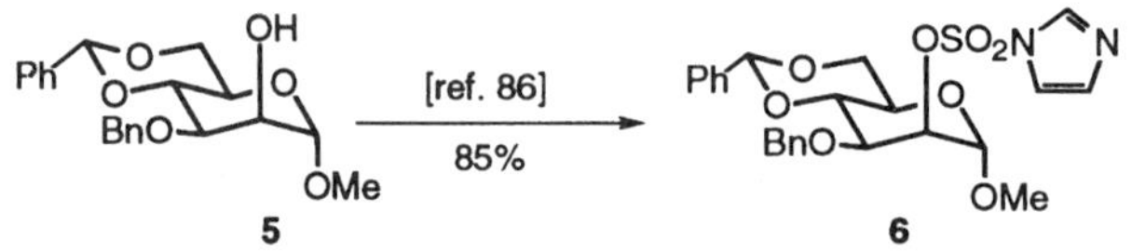

Chromatography on silica gel of the residue obtained from 3 g (8 mmol) of **5** (ether–petroleum ether, 3:1) furnished compound **6** (3.42 g, 85%) as a foam: $[\alpha]_D$ −1.5° (c 1.3, EtOH).

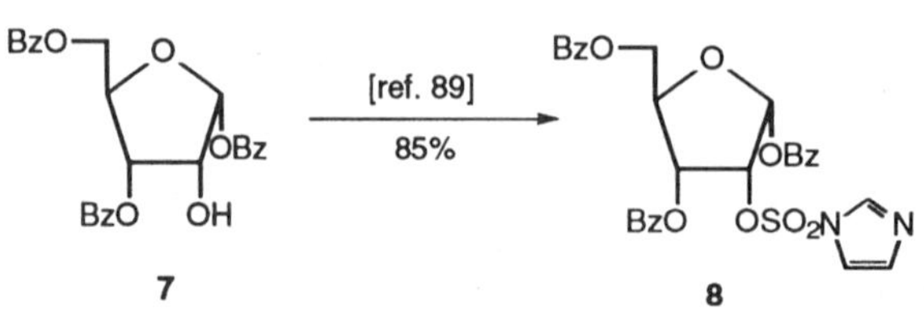

---

*Optical rotations were measured at 20–22°C.

## Displacement of Imidazole-1-Sulfonate Esters

For **7**, the reaction was effected on 85 g (0.184 mol) of **7** in 700 mL of $CH_2Cl_2$. After washing the reaction mixture with water, addition of Skellysolve B caused the precipitation of compound **8**. Recrystallization from 2-propanol gave the pure imidazylate **8** (93 g, 85%): mp 129°–130.5°C.

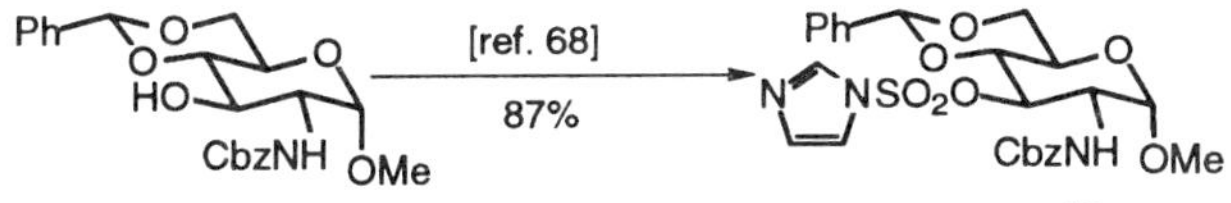

**9** → [ref. 68], 87% → **10**

The residue obtained from 1 g of compound **9** was chromatographed on silica gel (chloroform–EtOAc, 4:1) to give the crystalline imidazylate **10** (1.14 g, 87%): mp 128°–129°C (EtOAc), $[\alpha]_D$ +18.2° (c 0.33, $CH_2Cl_2$).

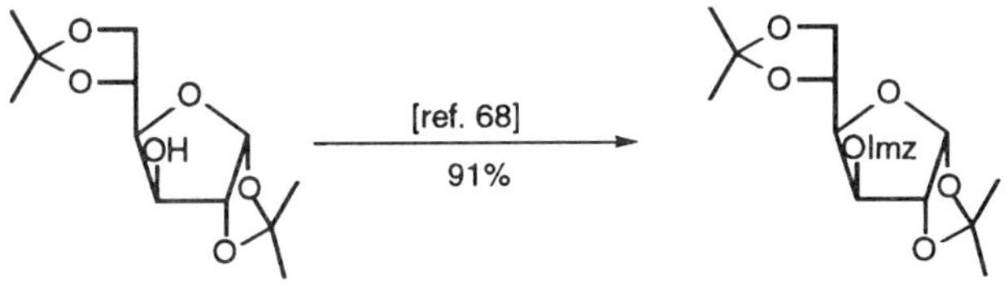

**11** → [ref. 68], 91% → **12**

Flash chromatography of the residue obtained from 1 g (3.8 mmol) of **11** (ether–petroleum ether, 2:1) afforded the product **12** (1.36 g, 91%): mp 98°–99°C (ether–hexane), $[\alpha]_D$ −76.3° (c 0.8, $CH_2Cl_2$).

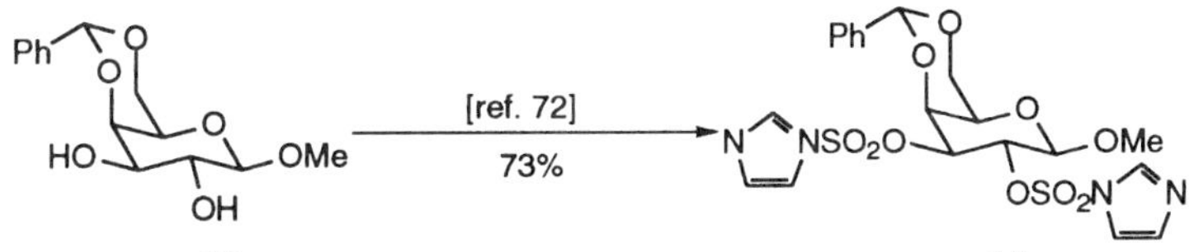

**13** → [ref. 72], 73% → **14**

To the diol **13** (2 g, 7.1 mmol) in 25 mL of dry DMF, cooled to −40°C, was added sulfuryl chloride (2.3 mL, 28 mmol, 4 eq.). The temperature was allowed to reach −30°C within 30 min and then to 25°C in 15 min. Imidazole (11.6 g, 142 mmol) was added with cooling over a period of 3 min. After stirring 3 h at room temperature, the reaction mixture was worked up as usual to give the diimidazylate **14** (2.8 g, 73%) as a semicrystalline compound.

*Method B (NaH, DMF/N,N'-sulfuryldiimidazole 15) [68]*

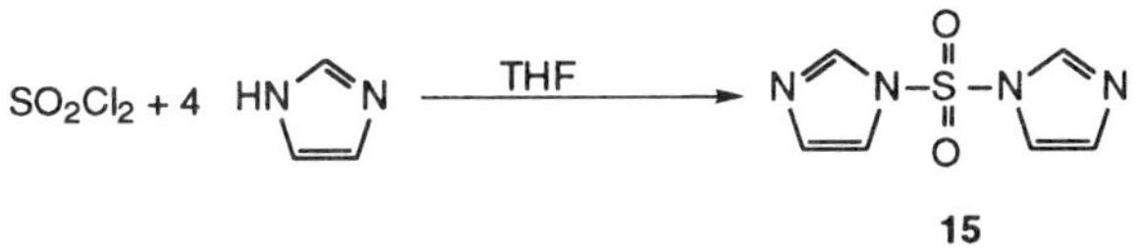

$SO_2Cl_2$ + 4 HN⌒N → THF → **15**

**N,N'-Sulfuryldiimidazole [71].** To a well-stirred solution of imidazole (19.9 g, 292 mmol) in 360 mL of THF, cooled in an ice bath, was added $SO_2Cl_2$ (9.83 g, 73 mmol) in 5 mL of toluene. After stirring 1 h at room temperature, the precipitate of imidazolium chloride was filtered and the filtrate concentrated. Recrystallization from ethanol gave **15** (9.91 g, 68%): mp 141°–141.5°C.

**General Procedure [68].** To an ice-chilled solution of partially protected sugar (1 mmol) in 5 mL of DMF, was added, under nitrogen, NaH (60% dispersion in mineral oil, 0.06 g, 1.5 eq.). The suspension was stirred at room temperature for 30 min and cooled to −40°C. A solution of $N,N'$-sulfuryldiimidazole **15** (0.30 g, 1.5 Eq) in 3 mL of DMF was added, and the mixture stirred for 30 min at −40°C. After addition of MeOH (0.2 mL), and stirring for 30 min at −40°C, the reaction mixture was poured into cold water and extracted twice with ether. The combined ethereal extracts were washed with water until pH 7, dried ($Na_2SO_4$), and evaporated to dryness.

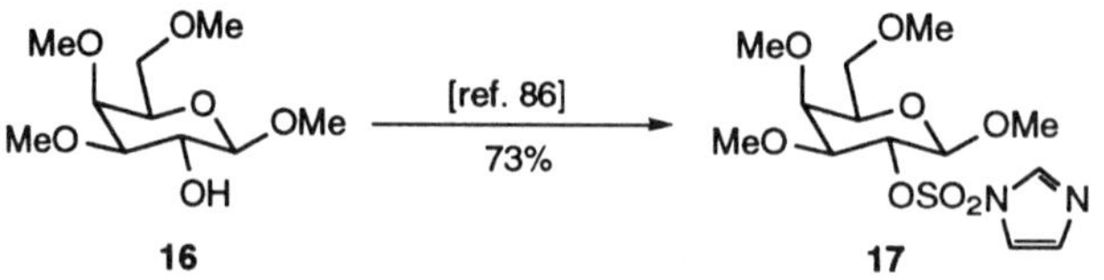

Flash chromatography of the residue obtained from 3.82 g (16.2 mmol) of **16** (ether-MeOH, 97:3), gave the pure sulfonate **17** (4.33 g, 73%): mp 125°C (ether), $[\alpha]_D$ +24° ($c$ 1.6, $CH_2Cl_2$).

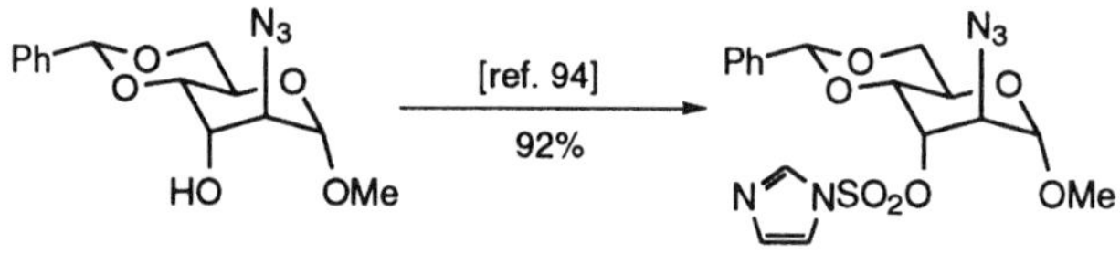

The residue, obtained from 4.9 g (0.02 mol) of **18**, gave a solid, which was crystallized from ethyl acetate–hexane to give **19** (6.7 g, 92%): mp 111°–112°C $[\alpha]_D$ +50.2° ($c$ 1.01, $CHCl_3$).

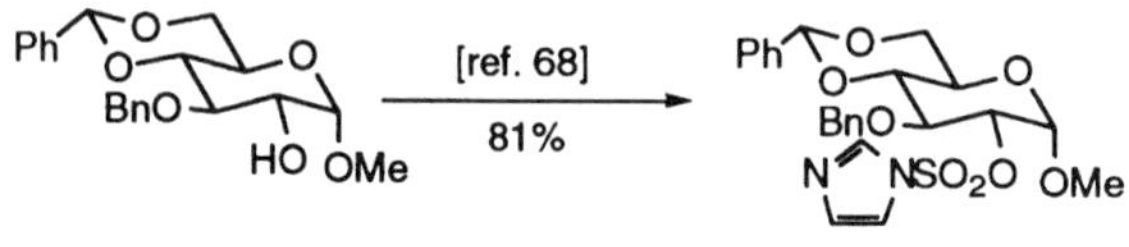

Flash chromatography of the product obtained from **3** (ether–petroleum ether, 3:1) gave the sulfonate **4** (81%): $[\alpha]_D$ +28.7° ($c$ 0.42, $CH_2Cl_2$).

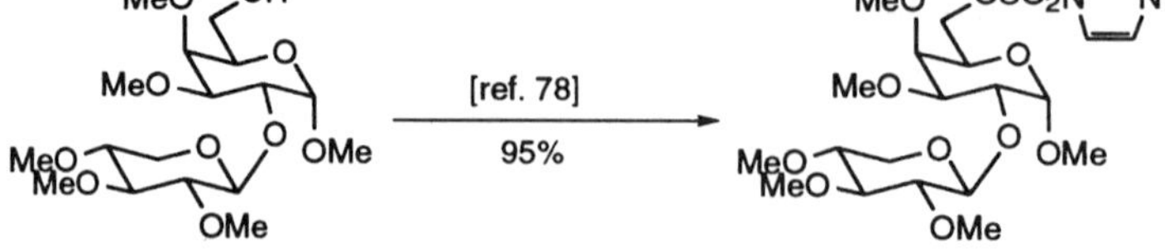

The dried organic extract obtained from (0.40 g, 1 mmol) of **20** was concentrated to a syrup, which crystallized on storage at 0°C. Recrystallization from ether gave the disaccharide imidazylate **21** (0.5 g, 95%): mp 95°–99°C.

The residue obtained from **22** (0.497 g, 0.64 mmol), was purified by chromatography

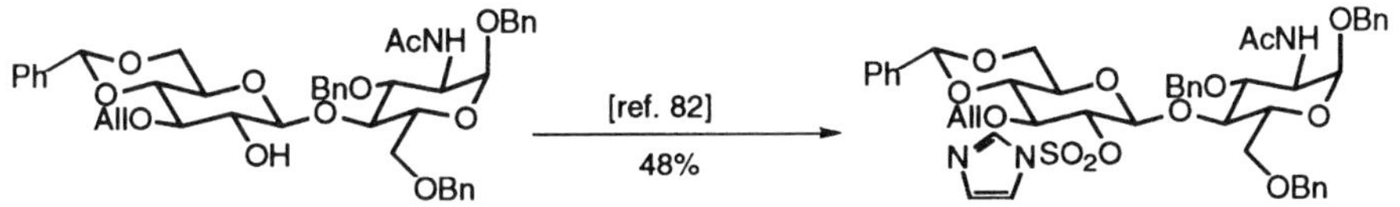

# Displacement of Imidazole-1-Sulfonate Esters

on silica gel (toluene-EtOAc, 2:1) to afford **23** (0.283 g, 48%) as a syrup: $[\alpha]_D$ +54° (c 1.0, $CHCl_3$).

*From Trimethylsilyl Ethers [68]*

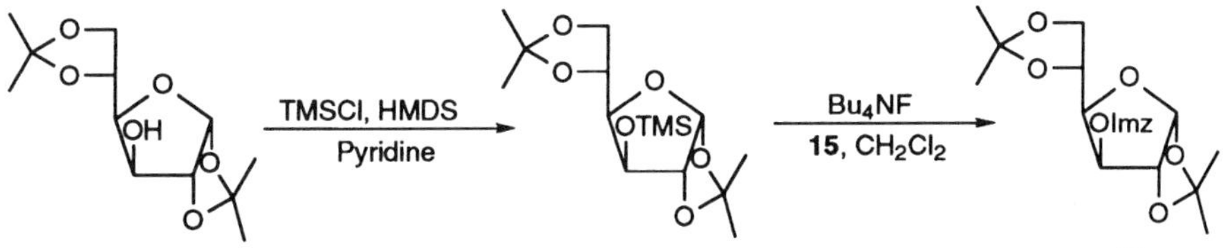

To a solution of diacetone glucose **11** (0.4 g, 1.5 mmol) in 5 mL of pyridine were successively added 1 mL of hexamethyldisilazane and a 0.5 mL of chlorotrimethylsilane. The solution was stirred for 30 min at room temperature and evaporated to dryness. To a solution of the crude silylated ether in $CH_2Cl_2$ (10 mL) were added 2.5 mL of $Bu_4NF$ (1 $M$ in THF) and $N,N'$-sulfuryldiimidazole (0.46 g, 1.5 eq.). The solution was refluxed for 4 h, diluted with $CH_2Cl_2$, washed twice with water, dried ($Na_2SO_4$), and evaporated. Flash chromatography of the residue (ether–petroleum ether, 2:1) gave the imidazylate **12** (0.55 g, 91%): mp 98.5°–99.5°C (ether–hexane).

## B. Substitution Reactions

*Halide Nucleophiles [68]*

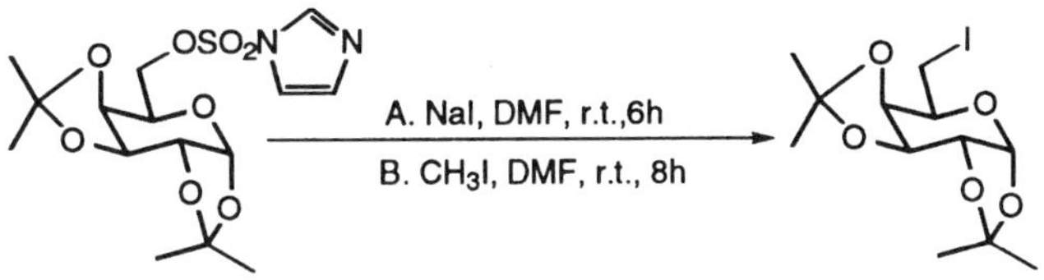

1. A mixture of imidazylate **2** (0.25 g, 0.6 mmol) and NaI (0.25 g, 2.5 Eq) in solution in 2.5 mL of DMF was stirred at room temperature for 6 h. The solution was diluted with ether and washed with a saturated solution of $Na_2S_2O_3$ then water. The organic layer was dried ($Na_2SO_4$) and evaporated. The oily residue was purified on a pad of silica gel (ether–petroleum ether, 1:4) to give the iodo derivative **24** (0.19 g, 81%) as an oil, which crystallized slowly on standing: mp 70°C; $[\alpha]_D$ −47.3° (c 0.82, $CH_2Cl_2$); reported [107], mp 70°C; $[\alpha]_D$ −50°.

2. To a solution of **2** (0.25 g, 0.6 mmol) in 2.5 mL of DMF were added imidazole (0.04 g, 0.6 mmol, 1 eq.) and iodomethane (0.5 mL, 12 eq.). The solution was stirred for 8 h and worked up as for 1 to give **24** (0.184 g, 78%) which was identical in all respects to the product prepared in method 1.

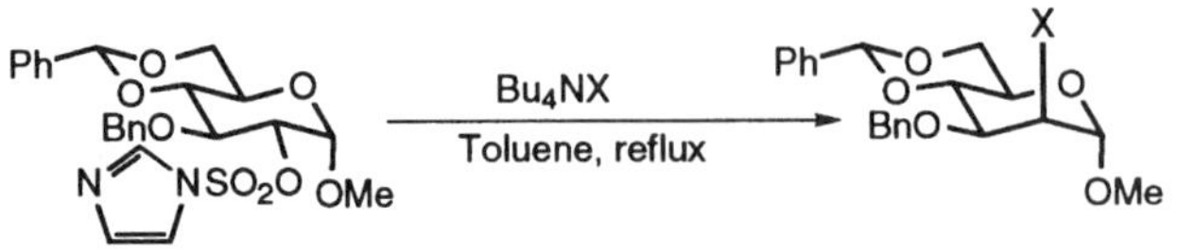

**General Procedure [68].** To a solution of imidazlyate **4** (0.5 g, 1 mmol) in 10 mL of toluene was added tetrabutylammonium halide (X = Cl, 0.5 g; X = I, 1.1 g) (3 mmol). The solution was refluxed 3 h (X = Cl$^-$) or 18 h (X = I$^-$), diluted with toluene, washed with water, dried (Na$_2$SO$_4$), and evaporated. The residue was purified by chromatography on a silica gel (ether–petroleum ether, 1:4).

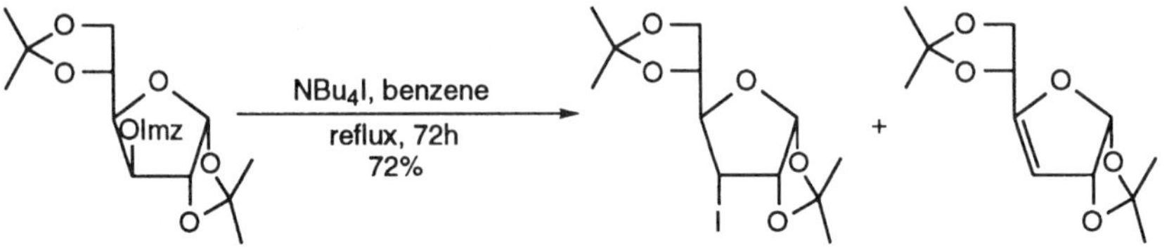

For the chloro sugar **25** (0.18 g, 78%), oil: $[\alpha]_D$ +3.1° (c 0.29, CH$_2$Cl$_2$); for the iodo sugar **26** (0.39 g, 81%): $[\alpha]_D$ −34.2° (c 0.5, CH$_2$Cl$_2$).

To a solution of compound **12** (0.2 g, 0.5 mmol) in 5 mL of benzene was added tetrabutylammonium iodide (0.55 g, 1.5 mmol, 3 eq.). After refluxing the solution for 72 h, usual workup gave a crude product. Flash chromatography of the residue (ether–petroleum ether, 1:3) gave first the unsaturated sugar **28** (0.018 g, 12%): mp 51°C. The second fraction was the desired iodo derivative **27** (0.133 g, 72%): $[\alpha]_D$ +67.2° (c 0.58, CH$_2$Cl$_2$); reported $[\alpha]_D$ +66.3° (c 2.2, CHCl$_3$).

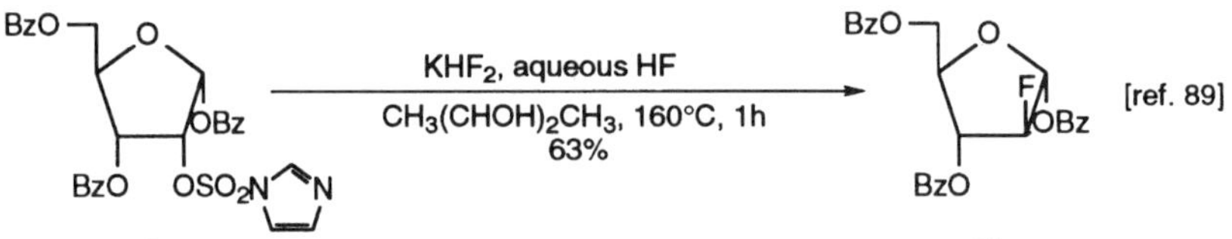

To a mixture of **8** (100.8 g, 0.17 mol), KHF$_2$ (53.1 g, 0.68 mol) in 250 mL of 2,3-butanediol, was added HF (50% in H$_2$O, 23.5 mL, 0.68 mol) under mechanical stirring at 160°C. After 1 h of heating, the reaction was quenched (150 mL of ice and 150 mL of brine) and extracted with CH$_2$Cl$_2$. The extract was washed with brine, H$_2$O, saturated NaHCO$_3$, dried (NaSO$_4$). After filtration through a pad of silica gel, the solvent was removed to give an oil, which was crystallized from 250 mL warm 95% EtOH giving **29** (48.6 g, 63%): mp 82°C.

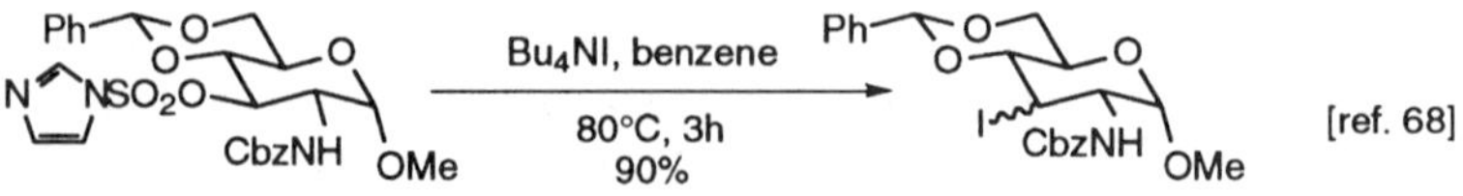

## Displacement of Imidazole-1-Sulfonate Esters

A solution of imidazylate **10** (0.4 g, 0.74 mmol) and tetrabutylammonium iodide (0.84 g, 2.3 mmol, 3 eq.) in 8 mL of benzene was refluxed during 3 h. After a usual workup, flash chromatography of the residue (EtOAc–CHCl$_3$, 2:98) afforded the iodo derivative **30** (0.345 g, 90%) as a foam, which crystallized slowly from ethanol: mp 110°–111°C.

### N-*Nucleophiles*

**General Procedure [68].** A solution of imidazylate (1 mmol) and tetrabutylammonium azide [109] (0.85 g, 3 mmol, 3 eq.) in 8 mL of toluene was heated during 1–6 h, depending on the substrate. The reaction mixture was washed with water, dried (Na$_2$SO$_4$), and evaporated.

Alteratively, tetrabutylammonium azide, which is not available commercially, can be replaced by tetrabutylammonium chloride and sodium azide according to the following protocol:

A mixture of imidazylate (1 mmol), tetrabutylammonium chloride (0.84 g, 3 mmol, 3 eq.) and sodium azide (0.23 g, 3.5 mmol, 3.5 eq.) in 8 mL of toluene was stirred at room temperature for 30 min and then heated, as described in the foregoing, to give the azido derivative.

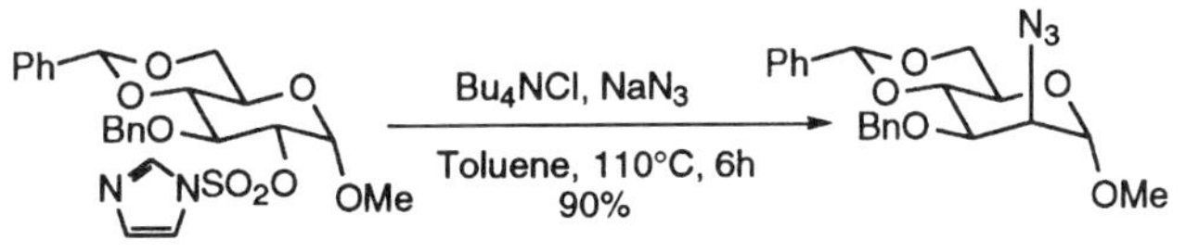

Flash chromatography of the residue, obtained from 0.55 g (1.1 mmol) of **4** (ether–petroleum, 1:3) gave the azido derivative **31** (0.39 g, 90%) as an oil: [α]$_D$ +30.1° (*c* 5.4, CH$_2$Cl$_2$).

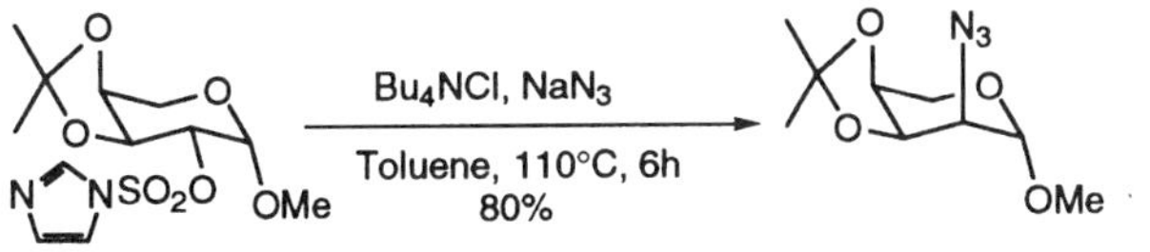

The reaction mixture was heated at 110°C for 6 h. After a usual workup, the residue, obtained from 0.41 g (1.0 mmol) of crude **32**, was purified by flash chromatography to afford **33** (0.245 g, 80%) as a syrup: [α]$_D$ +66.8° (*c* 1.1, CHCl$_3$).

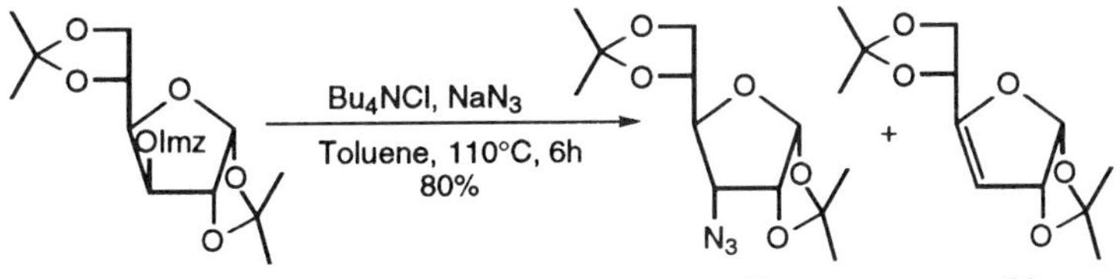

The reaction mixture was heated at 80°C for 5 h. Flash chromatography (ether–petroleum ether, 1:3) of the residue, obtained from 0.3 g (0.8 mmol) of **12**, gave the first unsaturated sugar **28** (0.055 g, 29%): mp 51°C; reported [96], mp 51°C, followed by the

3-azido derivative **34** (0.136 g, 62%) obtained as an oil: $[\alpha]_D$ +74.6° (c 1.90, $CH_2Cl_2$); reported [96], $[\alpha]_D$ +72.0° (c 1.0, $CHCl_3$).

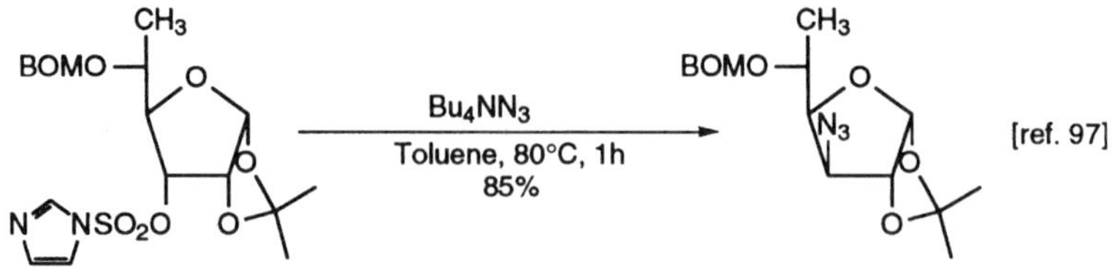

The reaction mixture was heated at 80°C for 1 h. Flash chromatography (hexane–ether, 19:1) of the residue, obtained from 5.95 g (21 mmol) of crude **35**, afforded **36** (2.08 g, 85%): $[\alpha]_D$ +2° (c 0.3, $CHCl_3$).

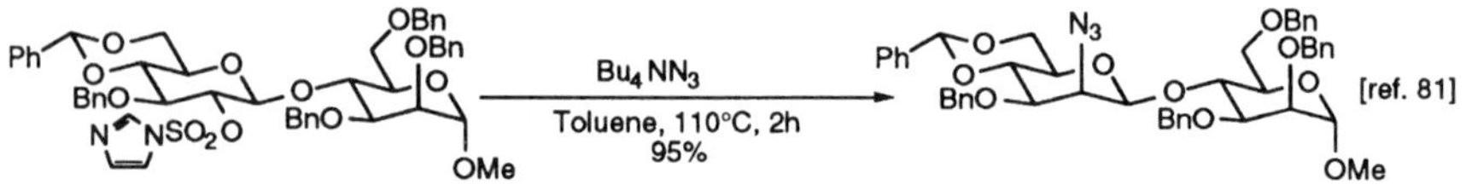

The reaction mixture was heated at 110°C for 3 h. The crude product crystallized from ethanol to give **38** (95%): mp 121°C; $[\alpha]_D$ +15.5° (c 1.1, $CHCl_3$).

*O-Nucleophiles*

Benzoate anion was used as an oxygen nucleophile for imidazylates **4, 12, 41, 43**; acetate ion was used for imidazylate **19** (acetate), and nitrite ion for imidazylate **22**.

**General Procedure with Tetrabutylammonium Benzoate [68].** A solution of imidazylate (1 mmol) and tetrabutylammonium benzoate (1.1 g, 3 mmol, 3 eq.) in 10 mL of toluene was heated for a period. The reaction mixture was cooled, washed twice with water, dried ($Na_2SO_4$), and evaporated to dryness.

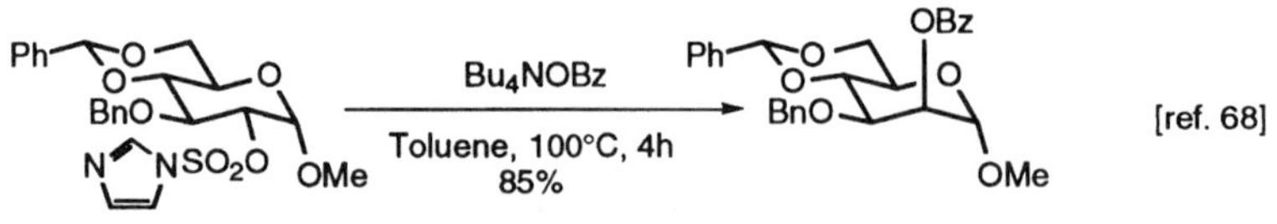

The reaction mixture was heated for 4 h at 100°C. Flash chromatography (ether–petroleum ether, 1:4) of the residue, obtained from 0.25 g (0.5 mmol) of **4**, gave **39** (0.2 g, 85%) as a foam: $[\alpha]_D$ −56.7° (c 0.37, $CH_2Cl_2$); reported [110]; $[\alpha]_D$ −52.9° (c 1, $CHCl_3$).

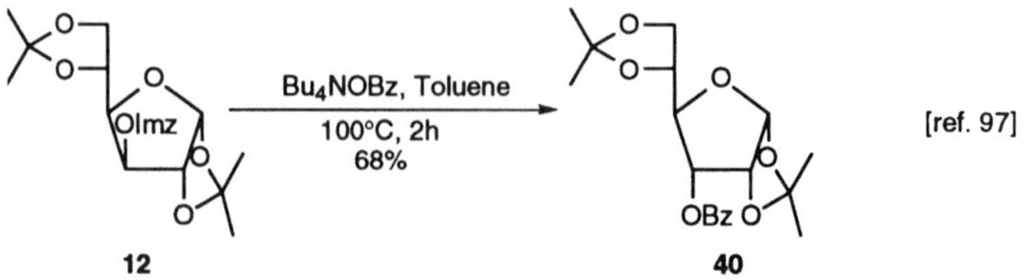

The reaction mixture was heated for 2 h at 100°C. Flash chromatography (hexane–ether, 5:1) of the residue, obtained from 7.8 g (20 mmol) of crude **12**, afforded the benzoate **40** (4.95 g, 68%): mp 72°–73°C (CH$_2$Cl$_2$–hexane); reported [111]; mp 75°C.

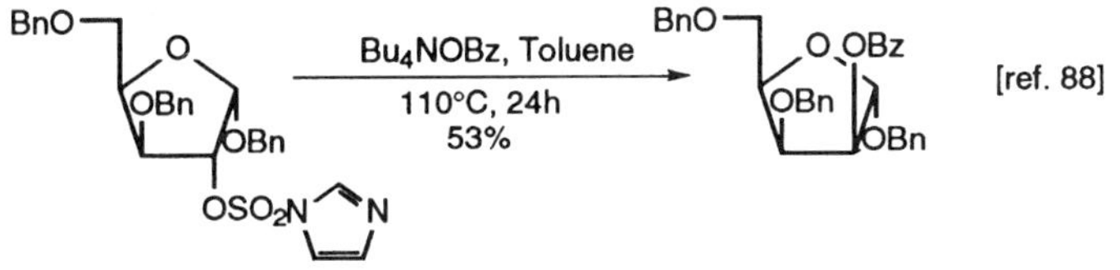

The reaction mixture was heated for 24 h at 110°C. Chromatography of the residue (hexane–EtOAc, 95:5), obtained from 1.05 g (1.18 mmol) of crude **41**, afforded the amorphous benzoate **42** (0.38 g, 53%): [α]$_D$ +61° (c 1.3, CH$_2$Cl$_2$).

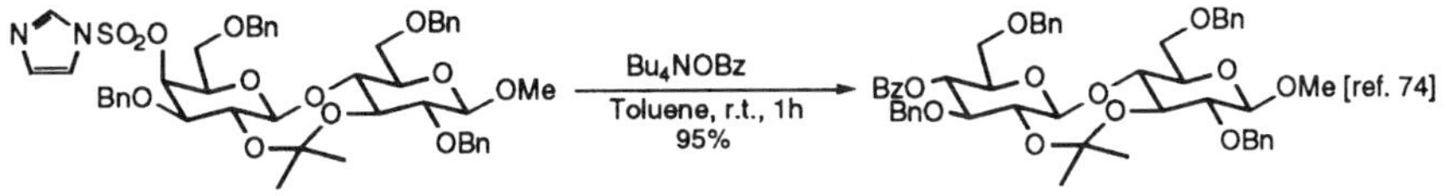

A mixture of **43** (0.15 g, 0.15 mmol) and Bu$_4$NOBz (0.21 g, 0.58 mmol) in 3 mL of toluene was stirred at room temperature for 1 h, diluted with toluene, washed with water, dried (Na$_2$SO$_4$), and evaporated to give **44** (0.12 g, 95%): mp 168°–170°C (ethanol–acetone: [α]$_D$ −43° (c 1.0, CH$_2$Cl$_2$).

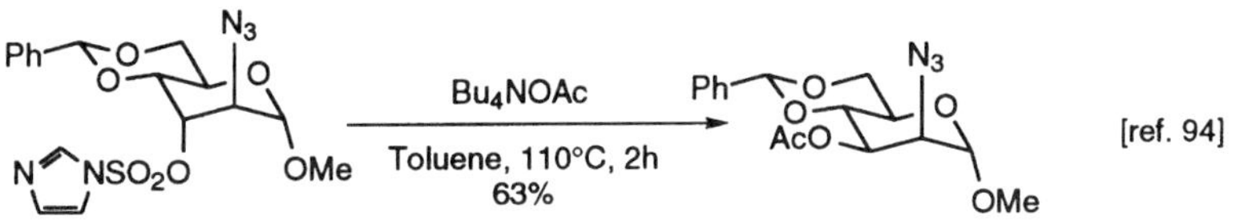

A solution of **19** (3.95 g, 9.0 mmol) and tetrabutylammonium acetate (15.3 g, 50 mmol, 5.5 eq.) in 350 mL of toluene was boiled under reflux for 2 h. The mixture was partitioned between EtOAc and water, the organic phase was washed with water, dried (Na$_2$SO$_4$), and evaporated. The residue was purified on a Lobar column (hexane–EtOAc, 3:1) to give **45** (1.93 g 63%) as a syrup: [α]$_D$ +39.1° (c 1.0, CHCl$_3$).

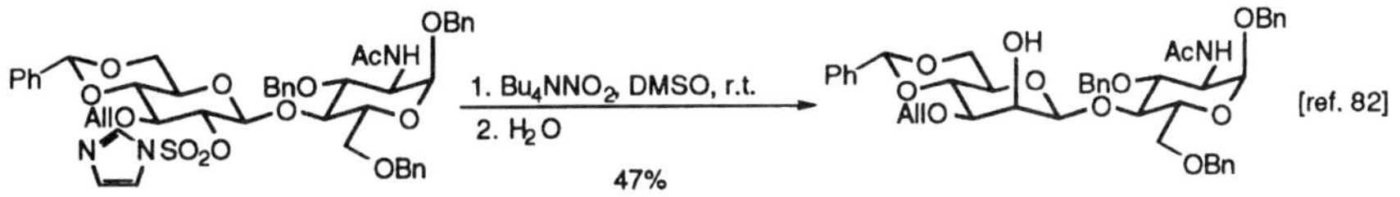

A mixture of imidazylate **22** (0.055 g, 0.06 mmol) and tetrabutylammonium nitrite (0.483 g, 1.68 mmol) in 2 mL of DMSO was stirred at room temperature. The reaction mixture was diluted with water and extracted with ether. Flash chromatography on silica gel of the residue (petroleum ether–EtOAc, 1:4) afforded **46** (0.022 g, 47%): [α]$_D$ +74.0° (c 0.25, CHCl$_3$).

*Other Reactions*

**Ring Contraction Reactions: Synthesis of a D-talitol Derivative.** A mixture of disaccharide **47** (4.74 g, 6.4 mmol) and tetrabutylammonium benzoate (7 g, 19 mmol, 3 eq.) in 50 mL of toluene was heated 6 h at 80°C and then was evaporated to dryness. Flash chromatography of the residue (ether–hexane, 3:4) afforded a mixture of epimers **48a, b** (6 g, 83%). To a solution of this mixture (3.5 g, 4.9 mmol) in 40 mL of tetrahydrofuran (THF) was added, at room temperature, LiAlH$_4$ (0.56 g, 15 mmol). After stirring 10 min, excess hydride was destroyed by addition of EtOAc (3 drops). Ether was added (200 mL), followed by water (1 mL). After filtration and evaporation of solvents, the residue was purified by column chromatography on silica gel (ether–hexane, 3:4) to give **49** (0.88 g, 88%): $[\alpha]_D$ +4.35 (c 1.8, CH$_2$Cl$_2$). The second fraction was identified as the glucose derivative **50**, (1.10 g, 78%).

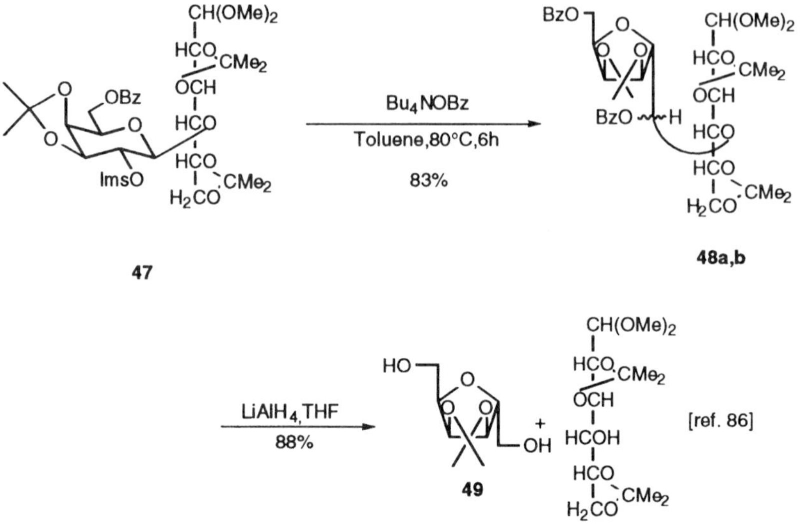

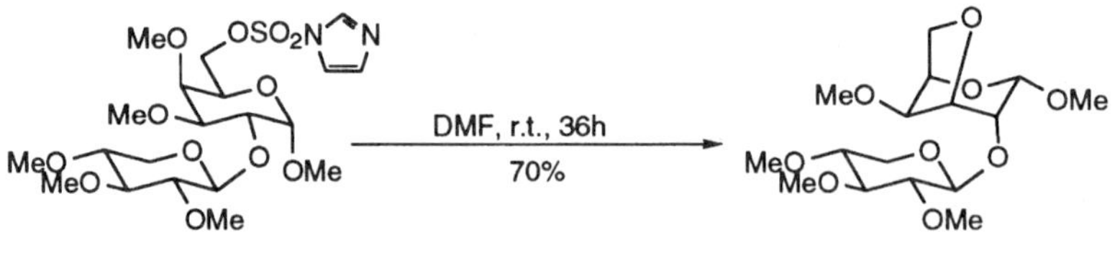

**Formation of 3,6-Anhydro-D-galactopyranoside Derivatives.** Imidazylate **51** (0.036 g, 0.06 mmol) was kept in dry DMF (2 mL) for 36 h. The acidic solution was neutralized with NaHCO$_3$, diluted with water, and extracted with CHCl$_3$. The extract was processed as usual, and chromatography on silica gel (CHCl$_3$–acetone, 4:1) of the residue gave **52** (0.016 g, 70%) as a syrup.

**Synthesis of a β-Lactam.** A solution of compound **53** (0.5 g, 1.6 mmol) in 7 mL of DMF was added at 0°C to a suspension of sodium hydride (0.097 g, 1.5 eq.; 60% suspension in mineral oil) in 2 mL of DMF. The mixture was cooled at −20°C and N,N'-sulfuryl-diimidazole (0.48 g, 1 Eq) in 4 mL of DMF was added. After stirring 1 h to −20°C,

## Displacement of Imidazole-1-Sulfonate Esters

[ref.105]

methanol and chloroform were added and the mixture was washed with brine, then with water. After evaporation of volatiles, the residue was crystallized from EtOAc to give **54** (0.41 g, 85%): mp 180°–180.5°C; $[\alpha]_D$ +48.5° (*c* 1.30, CHCl$_3$).


## REFERENCES

1. K. Freudenberg and F. Brauns, Zur kenntnis der Aceton-zucker, I: Umwandlungen der Diaceton-Glucose, *Chem. Ber.* 55:3233 (1922).
2. K. Freudenberg and R. M. Hixon, Zur kenntnis der Aceton-zucker, IV: Versuche mit Galaktose and Mannose, *Chem. Ber.* 56:2119 (1923).
3. K. Freudenberg and A. Doser, Zur kenntnis der Aceton-zucker, V: Die Synthese von Amino Hexosen aus Galaktose, *Chem. Ber.* 58:294 (1925).
4. K. Freudenberg, O. Burkhart, and E. Braun, Zur kenntnis der Aceton-zucker, VIII: Eine neue Amino-Glucose, *Chem. Ber.* 59:714 (1926).
5. K. Freudenberg and K. Raschig, Zur kenntnis der Aceton-Zucker, XII: Umwandlung der *d*-Galaktose in *d*-Fucose, *Chem. Ber.* 60:1633 (1927).
6. L. Hough and A. C. Richardson, The monosaccharides: Pentoses, hexoses, heptoses and higher sugars, in *Rodd's Chemistry of Carbon Compounds*, Vol. 1F, G. Coffey, ed., Elsevier, Amsterdam, 1967, pp. 448–566.
7. J. E. Barnett, Halogenated sugars, *Adv. Carbohydr. Chem.* 22:177 (1967).
8. W. A. Szarek, Deoxyhalogeno sugars, *Adv. Carbohydr. Chem. Biochem.* 28:255 (1973); see Chap. 6.
9. S. Hanessian, Some approaches to the synthesis of halodeoxy sugars, *Adv. Chem. Ser.* 74:159 (1968).
10. A. A. E. Penglis, Fluorinated carbohydrates, *Adv. Carbohydr. Chem. Biochem.* 38:195 (1981).
11. P. J. Card, Synthesis of fluorinated carbohydrates, *J. Carbohydr. Chem.* 4:451 (1985); see Ref. 65.
12. P. Herdewijn, A. Van Aerschot, and L. Kerremans, Synthesis of nucleosides fluorinated in the sugar moiety, *Nucleosides Nucleotides* 8:65 (1989).
13. T. Tsuchiya, Chemistry and developments of fluorinated carbohydrates, *Adv. Carbohydr. Chem. Biochem.* 48:91 (1990): see also M. Hudlicky, Fluorination with diethylaminosulfur trifluoride and related aminofluorosulfuranes, *Org. React.* 35:513 (1988).
14. Fluorinated carbohydrates, chemical and biochemical aspects, *ACS Symp. Ser.* 374, N. F. Taylor, ed., American Chemical Society, Washington DC, 1988.
15. S. Hanessian, Deoxy sugars, *Adv. Carbohydr. Chem.* 21:143 (1966).
16. J. Thiem and W. Klaffke, Synthesis of deoxy oligosaccharide, *Top. Curr. Chem.* 154:285 (1990).
17. L. Hough and A. C. Richardson, The monosaccharides: Pentoses, hexoses, heptoses and higher sugars, in *Rodd's Chemistry of Carbon Compounds*, Vol. 1F, G. Coffey, ed., Elsevier, Amsterdam, 1967, pp. 453–456.
18. R. W. Jeanloz, ed., *The Amino Sugars*, Vol. 1A, Academic Press, New York, 1969.

19. E. A. Balaz and R. W. Jeanloz, eds., *The Amino Sugars*, Vol. 2A, Academic Press, New York, 1965.
20. R. S. Tipson, Sulfonic esters of carbohydrates, *Adv. Carbohydr. Chem. 8*:107 (1953).
21. D. H. Ball and F. W. Parrish, Sulfonic esters of carbohydrates: Part I, *Adv. Carbohydr. Chem. Biochem. 23*:233 (1968).
22. D. H. Ball and F. W. Parrish, Sulfonic esters of carbohydrates: Part II, *Adv. Carbohydr. Chem. Biochem. 24*:139 (1969).
23. R. E. Robertson, Solvolysis in water, *Prog. Phys. Org. Chem. 4*:213 (1967).
24. M. C. Wu, L. Anderson, C. W. Slife, and L. J. Jensen, Effect of solvent temperature, and nature of the sulfonate group on the azide displacement reaction of sugar sulfonates, *J. Org. Chem. 39*:3014 (1974).
25. A. F. Cook and W. G. Overend, Nucleosides derived from 3- and 4-deoxy-D-*xylo*-hexose, *J. Chem. Soc.* (C) 1549 (1966).
26. A. Maradufu and S. Perlin, Synthesis of analogs of methyl β-D-galactopyranoside modified at C-4, *Carbohydr. Res. 32*:261 (1974).
27. A. Streitwieser, Jr., C. L. Wilkins, and E. Kiehlman, Kinetic and isotope effects in solvolyses of ethyl trifluoromethanesulfonate, *J. Am. Chem. Soc. 90*:1598 (1968).
28. T. M. Su, W. F. Slilwinski, and P. V. R. Schleyer, The solvolysis of highly unreactive substrates using the trifluoromethanesulfonate (triflate) leaving group, *J. Am. Chem. Soc. 91*:5386 (1969).
29. P. J. Stang, M. Hanack, and L. R. Subramanian, Perfluoroalkanesulfonic esters: Methods of preparation and applications in organic chemistry, *Synthesis* p. 85 (1982).
30. R. W. Binkley and M. G. Ambrose, Synthesis and reactions of trifluoromethanesulfonates (carbohydrate triflates), *J. Carbohydr. Res. 3*:1 (1984); see Chap. 5.
31. L. Hough and A. C. Richardson, The monosaccharides: Pentoses, hexoses, heptoses and higher sugars, in *Rodd's Chemistry of Carbon Compounds*, Vol. 1F, G. Coffey, ed., Elsevier, Amsterdam, 1967, p. 403.
32. A. C. Richardson, Nucleophilic replacement reactions of sulphonates. Part VI. A summary of steric and polar factors, *Carbohydr. Res. 10*:395 (1969).
33. Y. Ishido and N. Sakairi, Nucleophilic substitution reactions at C-2 of methyl-3-*O*-benzoyl-4,6-*O*-benzylidene-2-*O*-(trifluoromethylsulfonyl)-α-D-glucopyranoside, *Carbohydr. Res. 97*:151 (1981).
34. M. Miljkovic, M. Glicorijvic, and D. Glisin, Steric and electrostatic interactions in reactions of carbohydrates. III. Direct displacement of the C-2 sulfonate of methyl 4,6-*O*-benzylidene-3-*O*-methyl-2-*O*-methylsulfonyl-β-D-gluco- and mannopyranosides, *J. Org. Chem. 39*:3223 (1974).
35. B. R. Castro, Replacement of alcoholic hydroxy groups by halogens and other nucleophiles via oxyphosphonium intermediates, *Org. React. 29*:1 (1983).
36. S. Hanessian, M. M. Pompipom, and P. Lavallée, Procedures for the direct replacement of primary hydroxyl groups in carbohydrates by halogen, *Carbohydr. Res. 24*:45 (1972).
37. S. Hanessian and P. Lavallée, Selective substitution reactions of α,α-trehalose: Preparation of 6-monofunctional derivatives, *Carbohydr. Res. 38*:303 (1973).
38. P. C. Crofts and I. M. Downie, A novel oxidation of triethylphosphite, *J. Chem. Soc.* 2559 (1963).
39. J. B. Lee and T. J. Nolan, Sugar esters. IV. The preparation of chloro esters under essentially neutral conditions, *Can. J Chem. 44*:1331 (1966).
40. C. R. Haylock, L. D. Melton, K. N. Slessor, and A. S. Tracey, Chlorodeoxy and deoxy sugars, *Carbohydr. Res. 16*:375 (1971).
41. J. P. Verheyden and J. G. Moffatt, Halo sugar nucleosides. III. Reactions for the chlorination and bromination of nucleoside hydroxy groups, *J. Org. Chem. 37*:2289 (1972).
42. P. J. Garegg and B. Samuelsson, Novel reagent system for converting a hydroxy-group into an iodo-group in carbohydrates with inversion of configuration, *J. Chem. Soc. Chem. Commun.* p. 978 (1979).

43. P. J. Garegg and B. Samuelsson, Novel reagent system for converting a hydroxy-group into an iodo-group in carbohydrates with inversion of configuration, Part 2, *J. Chem. Soc. Perkin Trans. 1*:2866 (1980).
44. B. Classon, P. J. Garegg, and B. Samuelsson, Conversion of hydroxy groups into bromo groups in carbohydrates with inversion of configuration, *Can. J. Chem. 59*:339 (1981).
45. P. J. Garegg, R. Johansson, C. Orthega, and B. Samuelsson, Novel reagent system for converting a hydroxy-group into an iodo-group in carbohydrates with inversion of configuration, *J. Chem. Soc. Perkin Trans. 1*:681 (1982).
46. B. Classon, Z. Liu, and B. Samuelsson, New halogenation reagent systems useful for the mild one-step conversion of alcohols into iodides or bromides, *J. Org. Chem. 53*:6126 (1988).
47. O. Mitsunobu, The use of diethyl azodicarboxylate and triphenylphosphine in synthesis and transformation of natural products, *Synthesis* p. 1 (1981).
48. D. L. Hughes, The Mitsunobu reaction, *Org. React. 42*:335 (1992).
49. W. A. Szarek and J. K. N. Jones, Carbohydrates containing nitrogen in a five-membered ring and an attempted synthesis of a carbohydrate with nitrogen in a seven-membered ring, *Can. J. Chem. 43*:2345 (1965).
50. M. C. Viaud and P. Rollin, Zinc azide mediated Mitsunobu substitution. An expedient method for the one-pot azidation of alcohols, *Synthesis* p. 130 (1990).
51. H. N. Rydon and B. L. Tonge, The organic chemistry of phosphorus. Part III. The nature of the compounds of triaryl phosphites and the halogens, *J. Chem. Soc. 3043* (1956) and references cited therein.
52. J. B. Lee and M. M. El Sawi, Synthesis of deoxy sugars, *Chem. Ind.* p. 839 (1960).
53. N. K. Kochetkov and A. I. Usov, Reaction of carbohydrates with triphenylphosphite methiodide and related compounds. Synthesis of deoxy sugars, *Tetrahedron 19*:973 (1963).
54. K. Kefurt, J. Jary, and Z. Samek, Reaction of methyl 2,3-*O*-isopropylidene-α-L-rhamnopyranoside with triphenylphosphite methiodide, *J. Chem. Soc. Chem. Commun.* p. 213 (1969).
55. J. P. H. Verheyden and J. G. Moffatt, Direct iodination of the sugar moiety in nucleosides, *J. Am Chem. Soc. 86*:2093 (1964).
56. G. A. R. Johnston, Approaches to the preparation of 3′-deoxynucleosides, *Aust. J. Chem. 21*:513 (1968).
57. J. P. H. Verheyden and J. G. Moffatt, Halo sugar nucleosides. I. Iodination of the primary group of nucleosides with methyltriphenoxyphosphonium iodide, *J. Org. Chem. 35*:2319 (1970).
58. J. P. H. Verheyden and J. G. Moffatt, Halo sugar nucleosides. II. Iodination of the secondary hydroxy groups of nucleosides with methyltriphenoxyphosphonium iodide, *J. Org. Chem. 35*:2868 (1970).
59. J. P. H. Verheyden and J. G. Moffatt, Halo sugar nucleosides. IV. Synthesis of some 4′,5′-unsaturated pyrimidine nucleosides, *J. Org. Chem. 39*:3573 (1974).
60. W. A. Szarek, Deoxyhalogeno sugars, *Adv. Carbohydr. Chem. Biochem. 28*:230 (1973).
61. W. A. Szarek, Deoxyhalogeno sugars, *Adv. Carbohydr. Chem. Biochem. 28*:250 (1973).
62. A. H. Haines, Relative reactivities of hydroxy groups in carbohydrates, *Adv. Carbohydr. Chem. Biochem. 33*:11 (1976).
63. M. G. Ambrose and R. W. Binkley, Synthesis of deoxyhalogeno sugars. Reactions of halide ions with 1,2,3,4-tetra-*O*-acetyl-6-*O*-[(trifluoromethyl)sulfonyl]-β-D-glucopyranose, *J. Org. Chem. 48*:674 (1983).
64. K. Weinges, S. Haremsa, and W. Maurer, The Mitsunobu reaction on methyl glycosides as alcohol component, *Carbohydr. Res. 164*:453 (1987).
65. P. J. Card and G. S. Reddy, Fluorinated carbohydrates. 2. Selective fluorination of gluco- and mannopyranosides. Use of 2-D NMR for structural assignments, *J. Org. Chem. 48*:4734 (1983).
66. N. K. Kochetkov, A. I. Usov, and K. S. Adamyants, Synthesis and nucleophilic substitution reactions of some iodo-deoxy sugars, *Tetrahedron 27*:549 (1971).
67. H. Kunz and P. Schmidt, Synthese und Reaktionen der 3-*O*-Phosphoniogluco-und Allofuranosen, *Liebigs Ann. Chem.* p. 1245 (1982).

68. S. Hanessian and J. M. Vatèle, Design and reactivity of organic functional groups: Imidazolyl-sulfonate (imidazylate). An efficient and versatile leaving group, *Tetrahedron Lett. 22*:3579 (1981).
69. B. Helferich, Über zwei neue Derivate von-α-und β-Methyl Glucosid, *Chem. Ber. 56*:1082 (1921).
70. H. J. Jennings and J. K. N. Jones, Reactions of sugar chlorosulfates. (V). Synthesis of chlorodeoxy sugars, *Can. J. Chem. 43*:3018 (1965).
71. H. A. Staab and K. Wendel, 1,1″-Thionyl-di-imidazol and 1,1′-Sulfuryl-di-imidazol, *Ann. Chem. 694*:86 (1966).
72. K. Bock and M. Meldal, Synthesis of tetrasaccharides related to the O-specific determinants of *Salmonella* serogroups A, B, and $D_1$, *Acta Chem. Scand. B. 38*:255 (1984).
73. T. J. Tewson and M. Soderlind, 1-Propenyl 4,6-*O*-benzylidene-β-D-mannopyranoside-2,3-cyclic sulfate: A substrate for the synthesis of [F-18] 2-deoxy-2-fluoro-D-glucose, *J. Carbohydr. 4*:529 (1985).
74. M. Bernabe, A. Fernandez-Mayorabas, J. Jimenez-Barbero, M. Martin-Lomaas, and A. Rivera, The conformation of eight-membered 3,2′-*O*-isopropylidene acetals of some common disaccharides, *J. Chem. Soc. Perkin Trans. 2*:1865 (1989).
75. P. Deslongchamps, Reactions on $sp^2$ type unsaturated systems, in *Stereo-Electronic Effects in Organic Chemistry*, Pergamon Press, Oxford, 1983, p. 252.
76. S. Nadkarni and N. R. Williams, Displacement reactions of galactose 6-sulphonate derivatives, *J. Chem. Soc.* p. 3496 (1965).
77. J. M. Sugihara and W. J. Teerlink, Stereochemical effects in the nucleophilic displacement reactions of primary carbohydrate benzene sulfonate esters with sodium iodide, *J. Org. Chem. 29*:550 (1964).
78. G. O. Aspinall, D. Chatterjee, and L. Khondo, The hex-5-enose degradation: Zinc dust cleavage of 6-deoxy-6-iodo-α-D-galactopyranosidic linkages in methylated di- and trisaccharides, *Can. J. Chem. 62*:2728 (1984).
79. A. L. Raymond and E. F. Schroeder, Synthesis of some iodo-sugar derivatives, *J. Am. Chem. Soc. 70*:2785 (1948).
80. C. Augé, S. David, C. Gautheron, A. Malleron, and B. Cavayé, Preparation of six naturally occurring sialic acids with immobilized acylneuraminate pyruvate lyase, *New J. Chem. 12*:733 (1988).
81. S. David, A. Malleron, and C. Dini, Preparation of oligosaccharides with β-D-mannopyranosyl and 2-azido-2-deoxy-β-D-mannopyranosyl residues by inversion at C-2 after coupling, *Carbohydr. Res. 188*:193 (1989).
82. H. Paulsen, R. Wilkens, F. Reck, and I. Brockhausen, Synthese von verzweigten Tetrasaccharid- und Pentasaccharid-Strukturene von *N*-Glycoprotein, Methyliert and 4′-OH des Vezweigungsgliedes, *Liebigs Ann. Chem.* 1303 (1992).
83. J. Montreuil, Primary structure of glycoproteins glycans: Bases for the molecular biology of glycoproteins, *Adv. Carbohydr. Chem. Biochem. 37*:153 (1980).
84. H. Hashimoto, K. Araki, Y. Saito, M. Kawa, and Y. Yoshimura, Preparation of 2-azido-2-deoxypentose derivatives, *Bull Chem. Soc. Jpn. 59*:3131 (1986).
85. A. Olesker, A. Dessinges, T. T. Thang, and G. Lukacs, Synthèse stéréospécifique de désoxy-2-fluoro-2-hexopyranosides en vue de leur utilisation en médecine nucléaire, *C. R. Hebd. Séances Acad. Sci. II 295*:575 (1982).
86. F. M. El Sayed Ahmed, S. David, and J. M. Vatèle, Réactivité des *N*-imidazolylsulfonates en C-2 *d*′-α-D-manno- et galactopyranosides, *Carbohydr. Res. 155*:19 (1986).
87. R. Ranganathan, Modification of the 2′-position of purine nucleosides: Syntheses of 2′-α-substituted-2′-deoxyadenosine analogs, *Tetrahedron Lett.* p. 1291 (1977).
88. S. David, A. Malleron, and B. Cavayé, Aldolases in organic synthesis: Acylneuraminate–pyruvate lyase accepts furanoses as substrates, *New J. Chem. 16*:751 (1972).
89. C. H. Tann, P. R. Brodfuehrer, S. P. Brundidge, C. Sanino, Jr., and H. G. Howell, Fluorocarbohydrates in synthesis. An efficient synthesis of 1-(2-deoxy-2-fluoro-β-D-arabinofuranosyl)-thymine (β-FMAU), *J. Org. Chem. 50*:3644 (1985).

90. P. G. Gross, K. Brendel, and H. K. Zimmerman, Über eine neue oxazolidon. Synthese und die darstellung von 2,6-Diamino-2,6-dideoxy-D-gulose, *Angew. Chem. 76*:377 (1964).
91. R. Khan and L. Hough, Nucleophilic replacement reactions of sulfonates. Part VII. Selective tosylation of methyl 2-benzamido-2-deoxy-α-D-glucopyranoside and a synthesis of methyl 2-benzamido-2,3,6-trideoxy-α-D-ribo-hexopyranoside, *Carbohydr. Res. 24*:141 (1972).
92. H. Umezawa, S. Umezawa, T. Tsuchiya, and Y. Okazaki, 3′,4′-Dideoxy kanamycin B active against kanamycin-resistant *Escherichia coli* and *Pseudomonas aeruginosa, J. Antibiot. 24*:485 (1971).
93. S. Hanessian, P. C. Tyler, G. Demailly, and Y. Chapleur, Total synthesis and stereochemical identity of the $C_{18}H_{32}O_5$ degradation product of boromycin, *J. Am. Chem. Soc. 103*:6243 (1981).
94. T. Sugawara and K. Igarashi, Synthesis of a trisaccharide component of the capsular polysaccharide of *Streptococcus pneumoniae* Type 19F, *Carbohydr. Res. 172*:195 (1988).
95. R. D. Guthrie and D. Murphy, Nitrogen-containing carbohydrate derivatives. Part IX. Synthesis of 2,3-diamino-2,3-dideoxy derivatives, *J. Chem. Soc.* p. 6956 (1965).
96. U. G. Nayak and R. L. Whistler, Nucleophilic displacement in 1,2:5,6-di-*O*-isopropylidene-3-*O*-(*p*-tolylsulfonyl)-α-D-glucofuranose, *J. Org. Chem. 34*:3819 (1969).
97. J. Alais and S. David. A precursor to the β-pyranosides of 3-amino-3,6-dideoxy-D-mannose-(mycosamine), *Carbohydr. Res. 230*:79 (1992).
98. D. C. Baker, D. Horton, and C. G. Tindall, Jr., Preparation of mono- and disaccharides, in *Methods in Carbohydrate Chemistry*, Vol 7, R. L. Whistler and J. N. BeMiller, eds., Academic Press, Orlando, 1976, pp. 3–6.
99. H. Satoshi and H. Fukase, Inosose derivatives and production thereof, *Eur. Patent* 240175A1, 1987 (*CA 109*:55166b).
100. J. S. Brimacombe and O. A. Ching, Nucleophilic displacement reactions in carbohydrates. Part IX. The solvolysis of methyl 6-*O*-Methanesulphonyl-2,3-di-*O*-methyl-β-D-galactopyranoside: A methoxy group participation. *Carbohydr. Res. 9*:287 (1969).
101. L. V. Volkova, M. G. Luchinskaya, N. G. Morozova, N. B. Rozanova, and R. P. Evstigneeva, Synthesis of 6-iodo-6-deoxy substituted carbohydrates, *J. Gen. Chem. (USSR) 42*:2101 (1972).
102. J. Defaye, 2,5-Anhydrides of sugars and related compounds, *Adv. Carbohydr. Chem. Biochem. 25*:181 (1970).
103. J. Defaye, Désamination nitreuse de la D-galactosamine. Préparation des 2,5-anhydro-D-talose et 2,5-anhydro-D-talitol, *Bull. Soc. Chim. Fr.* p. 999 (1964).
104. E. Venkata Rao, J. G. Buchanan, and J. Baddiley, Type-specific substance from pneumococcus type 10A—structure of the dephosphorylated repeating unit, *Biochem. J. 100*:801 (1966).
105. S. Hanessian, C. Couture, and H. Wyss, Design and reactivity of organic functional groups. Utility of imidazolylsulfonates in the synthesis of monobactams and 3-amino nocardicinic acid, *Can. J. Chem. 63*:3613 (1985).
106. S. Hanessian, S. P. Sahoo, C. Couture, and H. Wyss, Novel synthetic approaches to monocyclic β-lactam antibiotics, *Bull Soc. Chim. Belg. 93*:571 (1984).
107. O.T. Schmidt, 6-Deoxy-α-D-galactose (α-D-fucose), in *Methods of Carbohydrate Chemistry*, Vol. 1, R. L. Whistler and M. L. Wolfrom, eds., Academic Press, New York, 1962, p. 191.
108. H. Kunz and P. Schmidt, Eine neue Synthese für 3-Desoxy-3-iod-zucker über Alkoxyphosphoniumsalze, *Tetrahedron Lett.* p. 2123 (1979).
109. A. Brändström, B. Lamm, and I. Palmertz, The use of tetrabutylammonium azide in the Curtius rearrangement, *Acta Chem. Scand. Ser. B 28*:699 (1974).
110. M. A. Nashed, An improved method for selective substitution on O-3 of D-mannose. Applications to the synthesis of methyl 3-*O*-methyl- and 2-*O*-methyl-α-D-mannopyranosides, *Carbohydr. Res. 60*:200 (1978).
111. M. J. Bessman, J. R. Lehman, J. Adler, S. B. Zimmerman, E. S. Simms, and A. Kornberg, Enzymic synthesis of deoxyribonucleic acid. III. Incorporation of pyrimidine and purine analogs into deoxyribonucleic acid, *Proc. Natl. Acad. Sci. USA 44*:633 (1958).

# 8
# Free Radical Deoxygenation of Thiocarbonyl Derivatives of Alcohols

**D. H. R. Barton and J. A. Ferreira**
*Texas A&M University, College Station, Texas*

**J. C. Jaszberenyi**
*Technical University of Budapest, Budapest, Hungary*

| | | |
|---|---|---|
| I. | Introduction | 151 |
| II. | Methods | 153 |
| | A. Radical chain deoxygenations | 153 |
| | B. Mechanistic considerations | 154 |
| | C. Alternative hydrogen atom sources and chain carriers | 155 |
| | D. Variations in the functionality of the thiocarbonyl group | 156 |
| | E. Summary | 156 |
| III. | Experimental Procedures | 157 |
| | A. Deoxygenation with tributyltin hydride | 157 |
| | B. Deoxygenation with dimethyl phosphite and related reagents | 161 |
| | C. Deoxygenation with organosilanes | 163 |
| | D. Deoxygenation with carbocyclization | 167 |
| | Notes and References | 168 |

## I. INTRODUCTION

Deoxygenation of various natural products plays an important role in the synthetic transformation of these compounds, including various carbohydrates and antibiotics. Deoxy sugars, as well as their deoxyamino sugar counterparts, are useful groups of compounds [1]. Deoxy sugars are important in the chemistry of various antitumor compounds and other bioactive molecules. Methods for selective removal or replacement of one or more hydroxyl (or amino) group(s) in these complex carbohydrates, aminoglycoside antibiotics, and the like, are important for the synthesis of novel semisynthetic derivatives. These compounds show a different, sometimes improved, biological profile and often enhanced bioactivity. These derivatives remain similar to their parent molecules; consequently, they can still be recognized by the specific target enzymes involved in their bioactivity. However, with certain functional groups removed, the enzymes involved in the deactivation of the parent bioactive molecules are often unable to deactivate the new deoxy or deamino compounds. Various polydeoxy monoamino sugar components are found in important anticancer drugs (such as anthracycline antibiotics) and their semisynthetic derivatives [2]. These deoxygenated carbon sites are also less vulnerable to undesired enzymatic biotransformations than their hydroxylated counterparts. The various methods leading to polydeoxy and polydeoxy-monoamino sugars are covered in an excellent book by Pelyvás and coauthors [3]. Various aspects of the chemistry of anthracycline antibiotics, aureolic acids, cardiac glycosides, and other antibiotics, and the chemistry of orthosomycine antibiotics was discussed by Thiem and Klaffke [4]. Selected syntheses leading to deoxy sugars are discussed by Collins and Ferrier in their recent book [5]. These methods include opening of an epoxide ring, reduction of suitable functional groups, additions to unsaturated compounds, and other methods, including synthesis from noncarbohydrate compounds and various degradation and chain-elongation procedures.

The classic ionic methods for the removal of an unwanted hydroxyl group are summarized in detail in the book by Larock [6]. One method involves—for primary and unhindered secondary alcohols—the synthesis of the corresponding mesylates or tosylates. These compounds are prepared readily and then transformed into the corresponding deoxy compounds by reduction [7]. Alternatively, introduction of a thiolate or halogen by a nucleophilic reaction can also be used. These compounds can then be readily desulfurized or dehalogenated (Scheme 1). Tertiary alcohols present no problem either, because a

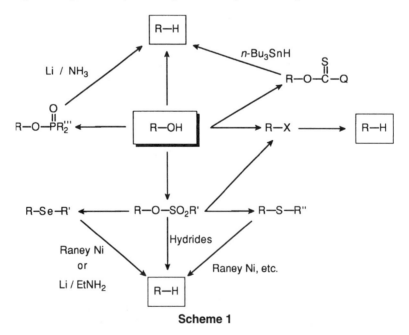

Scheme 1

dehydration–reduction sequence leads to the required deoxy compound in high yield [7]. However, deoxygenation of secondary alcohols with hindered hydroxyl groups may be difficult. Often the $S_N2$ reaction is disfavored, and other methods of deoxygenation are needed. Various polyhydroxy compounds, such as carbohydrates and other carbohydrate-containing natural products, can be found in this group.

## II. METHODS

### A. Radical Chain Deoxygenations

Radical chain chemistry is often employed for the transformation of an alcohol to the corresponding deoxy derivative. The secondary alcohol **1** is first converted into a suitable thiocarbonyl derivative. The first derivatives investigated were thioxobenzoates **2**, xanthates **3**, and thiocarbonylimidazolides **4** (Scheme 2). On reduction with tributyltin hydride, these derivatives afforded a good yield of the appropriate deoxy compounds [8–10].

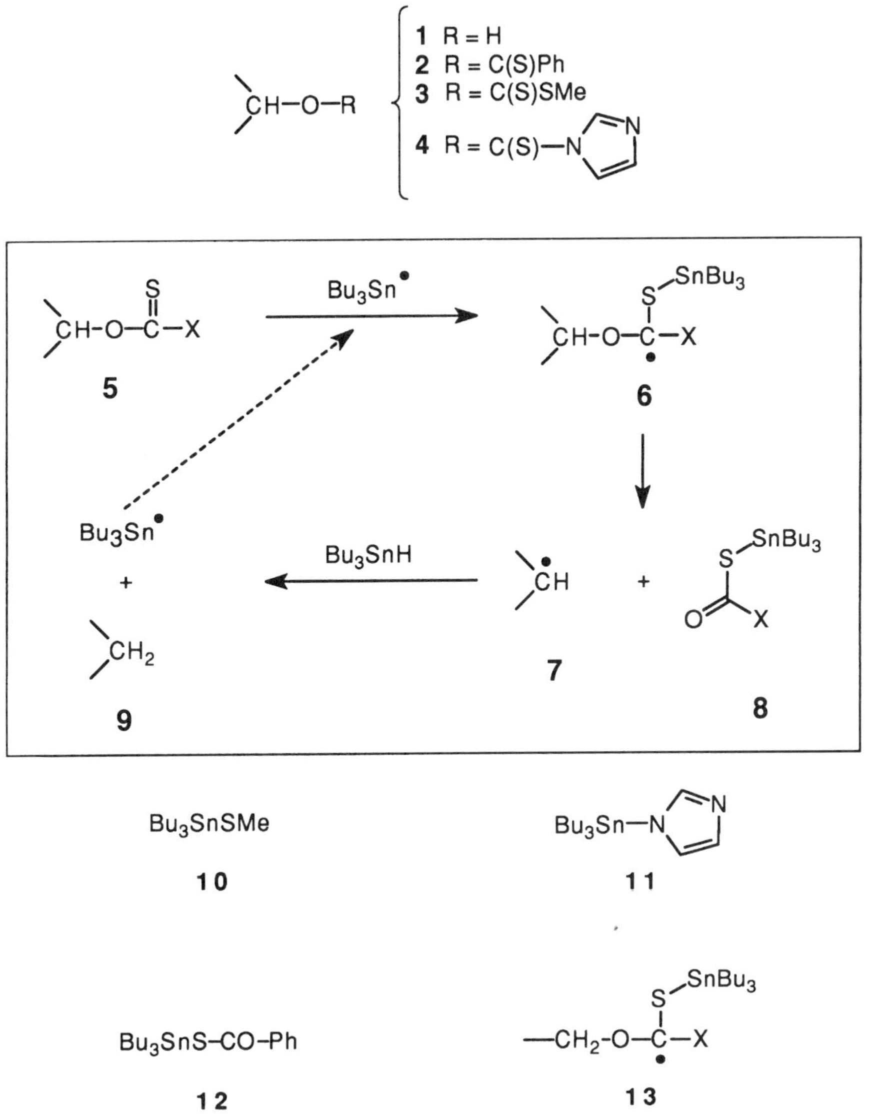

**Scheme 2**

The mechanism of the reaction is summarized in Scheme **2**. The thiocarbonyl derivative **5**, on attack by the tributyltin radical, affords, with formation of a tin–sulfur bond, the intermediate radical **6**. This then fragments into the desired carbon radical **7** and the thiocarbonyltin derivative **8**. Finally, radical **7** is reduced by hydrogen atom transfer to the desired product **9** with reformation of the tributyltin radical. The derivatives of type **8** are not stable. At room temperature they lose COS and afford the tin compounds **10** and **11**. This does not apply to the thioxobenzoates where the analogue of **8** is **12**, which is stable.

Primary alcohols can also be deoxygenated by radical chemistry [11a,11b]. However, it requires a higher temperature to break the carbon–oxygen bond in radical **13**. Secondary alcohols can be conveniently deoxygenated in benzene under reflux (80°C), but primary alcohols require toluene or xylene under reflux.

The xanthates of tertiary alcohols are unstable compounds. However, they can be prepared [11c,11d] and even characterized by microanalysis. Their deoxygenation is relatively simple because of the weak tertiary carbon–oxygen bond.

The work so far described has been dependent on the Sn–H bond in tributyltin hydride. Triphenyltin hydride can also be employed [11e,11f]. The unusual efficiency of triphenyltin hydride for the desulfurization of thiocarbonyl compounds has only recently been reported [12]. There are specialized texts on organotin compounds [13] in which the properties of the Sn–H bond are discussed at length.

The Barton–McCombie deoxygenation reaction was invented for use in the manipulation of aminoglycoside antibiotics. It has become a popular method because of the mild conditions employed. Radical reactions have advantages over ionic reactions for carbohydrate chemistry. In this context, there is little neighboring group interference in cationic reactions and little elimination compared with normal nucleophilic displacement reactions.

The first sugar derivative to be examined [10] was 1,2:4,6-di-*O*-isopropylideneglucofuranose **14**. This was converted, in excellent yield, to the corresponding methyl xanthate **15**. Reduction with tributyltin hydride in toluene under reflux afforded the desired 3-deoxy-derivative **16** in 80–90% isolated yield (Scheme **3**). Hitherto, this had been a difficult and

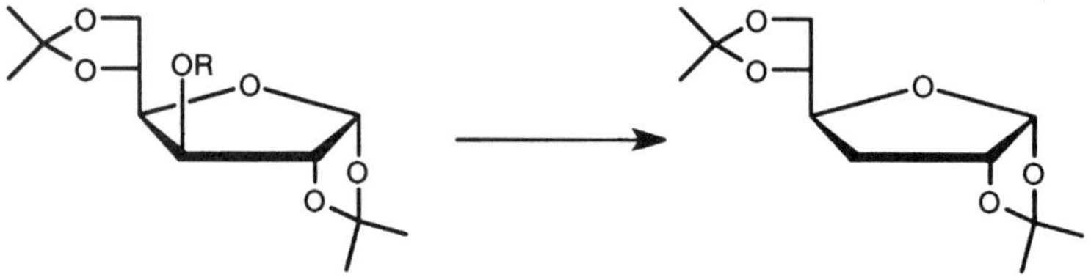

**14** R = H
**15** R = C(S)-SMe

**16**

**Scheme 3**

cumbersome transformation, with low yields. All subsequent syntheses of 3-deoxyglucose derivatives have used this, or an equivalent, radical procedure [14].

### B. Mechanistic Considerations

The original conception of Barton and McCombie [10], which is summarized in Scheme **2**, was later questioned [15]. The alternative mechanism summarized in Scheme **4** was suggested. However, a low-temperature study using the nuclear magnetic resonance (NMR) of $^{119}$Sn confirmed the original proposal [10]. The evidence for the correctness of Scheme **2** is summarized in several communications [16–20] so that further discussion is not needed.

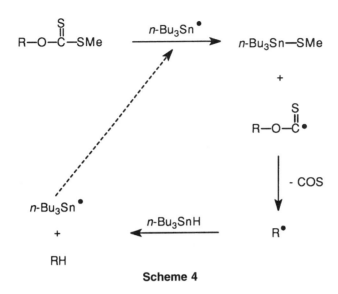

**Scheme 4**

## C. Alternative Hydrogen Atom Sources and Chain Carriers

Although tributyltin and triphenyltin hydrides are excellent reagents for radical chemistry, they have the disadvantage that they are costly and of relatively high molecular weight. For small operations, in the laboratory, this is not serious. However, there is a major additional problem, particularly for industrial use, because of the toxicity of these tin compounds and because of the difficulty of removing "tin dimers" that are formed as by-products [13]. Similar comments apply to germanium hydrides, which are even more expensive [21].

Of the various elements with weak M–H bonds that combine the ease of homolytic rupture [22] with little or no toxicity and reasonable price, silicon has attracted most attention. The organic chemistry of silicon [23–25] has provided many potential reagents. However, tris(trimethylsilyl)silane **17** was the first silane shown to be a good replacement for tin hydride [26]. The mechanism of this type of reaction is exactly the same as that which applies for tributyltin or triphenyltin hydride (see Scheme 2). There have been many applications of tris(trimethylsilyl)silane **17** recently [27–30], and authoritative summaries have been published [31,32]. However, this reagent has the disadvantage of being of high molecular weight and its expense per hydrogen atom delivered is prohibitive for normal organic synthesis. It is, in fact, a valuable tool, not a reagent.

Diphenylsilane is a convenient alternative [33]. Triphenylsilane and phenylsilane are also quite suitable [34,35]. Even triethylsilane can be used if it serves as solvent [36], but it is not as efficient as other silanes because the Si–H bond strength is too great.

Another ingenious way in which to use triethylsilane is by polarity reversal, using a thiol. Other trialkylsilanes of higher boiling point can also be used to advantage [37].

(Me$_3$Si)$_3$—Si—R        Me$_3$Si—Si(Me)(SH)—SiMe$_3$

**17** R = H
**18** R = SH                **19**

Reductions with tris(trimethylsilyl)silanethiol **18** [38] and with heptamethyltrisilane-2-thiol **19** [39] are also effective, but again the cost of these reagents would be prohibitive per hydrogen delivered for larger-scale synthesis.

The P–H bond is also weak enough to serve in thione-based radical chemistry. The first success was achieved with dialkyl phosphites [40a]. However, hypophosphorous acid and its salts proved to be even better [40b]. In a final paper [40c], the various phosphorus-based reagents were compared. The crystalline salt $N$-ethylpiperidine hypophosphite was very convenient and has been commercialized [40c]. The use of hypophosphorous acid has the advantages of nontoxicity, cheapness, and ease of removal from the organic reduction products. It already has several industrial applications.

## D. Variations in the Functionality of the Thiocarbonyl Group

The first studies on the deoxygenation of secondary alcohols were carried out with thioxobenzoates, methyl xanthates, and imidazoline thiocarbonyl derivatives. Thioxobenzoates are prepared using Vilsmeier chemistry; however, this is a two-stage process. Likewise, methyl xanthates require the reaction of the anion of an alcohol with $CS_2$ followed by methylation with an excess of methyl iodide. In general, this is a simple one-pot procedure. Thiocarbonyl-bis-imidazoline is a convenient reagent that affords the thiocarbonyl derivative directly in one step. Another one-step reagent was introduced by Robins. This is phenoxythiocarbonyl chloride, PhO-CS-Cl. It has the advantage of reacting with alcohols under mild basic conditions at room temperature [41]. Pentafluorophenoxythiocarbonyl chloride is another useful reagent that is preferred for the deoxygenation of hindered secondary hydroxyl groups [42,43]. The analogous 4-fluorophenoxythiocarbonyl chloride [33a] is also a useful and less expensive reagent. All three of these reagents are commercially available.

The relative rates of acylation and of deoxygenation have been determined with these various reagents [44]. As expected, the pentafluoro reagent reacts the fastest with an alcohol under standard conditions, followed by the 4-fluoro reagent, and the phenyl derivative is the slowest. However, for the deoxygenation reaction the fastest group is the methyl xanthate. The slowest is the pentafluorophenyl derivative. This is not important because all of the thiocarbonyl derivatives mentioned give very fast radical reactions [44].

A useful variation for the deoxygenation reaction is to react the alcohol with thiophosgene and then to treat the resultant thiocarbonyl group with the appropriate phenol [43].

## E. Summary

This chapter shows how radical chemistry based on thiocarbonyl derivatives of secondary alcohols can be useful in the manipulation of natural products and especially in the deoxygenation of carbohydrates. From the original conception in 1975, the variety of thiocarbonyl derivatives used has increased, but the methyl xanthate function still remains the simplest and cheapest, when other functionality in the molecule does not interfere. Otherwise, selective acylation with aryloxythiocarbonyl reagents is important. Many of the functional groups present in carbohydrates and other natural products do not interfere with radical reactions.

Whereas the nature of the thiocarbonyl function has hardly changed over the years, the type of reducing reagent has improved greatly. Although small-scale work with tin hydrides may continue, synthesis on a multigram or kilogram scale will be done with Si–H or P–H reagents. The use of hypophosphorous acid salts would seem to be a major advantage given their nontoxicity and cheapness. The authors have been informed of a deoxygenation, repeatedly carried out in this way on a 300-g scale, in an industrial laboratory.

There remains an area where progress is still to be made. This is in the invention of new, inexpensive, and temperature-variable initiators. The present reliance on azoisobutyronitrile (AIBN) and dibenzoyl peroxide limits the temperature range, and there is always some danger with peroxides. The use of triethylborane and oxygen, originally introduced by Brown some decades ago, has recently been appreciated better because it permits radical initiation at low temperatures.

## III. EXPERIMENTAL PROCEDURES*

### A. Deoxygenation with Tributyltin Hydride

*Typical Procedure for Deoxygenation with Tributyltin Hydride Without a Radical Initiator [10]*

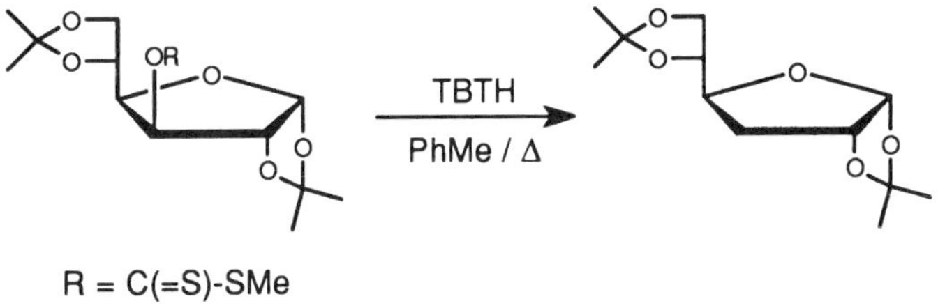

$$20 \xrightarrow{\text{TBTH, PhMe}/\Delta} 21 \quad (1)$$

R = C(=S)-SMe

**20**　　　　**21**

The $S$-methyl dithiocarbonate [Eq. (1)] **20** (1.75 g) in toluene (40 mL) was added over 1 h to tributylstannane (TBTH, 2.1 g) in toluene (30 mL) under argon and at reflux. Refluxing was continued overnight, and the solvent was then removed at 50°C and 15 mmHg. The product was chromatographed over silica gel [elution with light petroleum (bp 40°–60°C) containing an increasing proportion of ether (5% increments)]. After elution of tin compounds, followed by a minor by-product (a carbohydrate derivative also containing a tributyltin residue), the desired deoxy compound **21** was obtained as an oil (1.04 g, 85%); $[\alpha]_D$ −7.5° ($c$ 10, EtOH).

*Typical Procedure for Deoxygenation with a Polymer-Supported Tin Hydride [13a]*

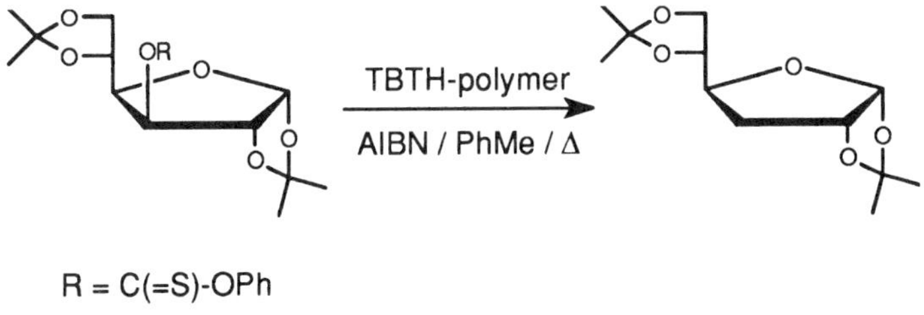

$$22 \xrightarrow{\text{TBTH-polymer, AIBN / PhMe}/\Delta} 23 \quad (2)$$

R = C(=S)-OPh

**22**　　　　**23**

---

*Optical rotations were measured at 22°–25°C.

To a solution of 0.79 g (2 mmol) 1,2:5,6-diisopropylidene-3-phenoxythiocarbonato-α-D-glucose [Eq. (2)] **22**, in 50 mL of dry toluene under an argon atmosphere, 3.3 g (4 mmol SnH) of the polymer and 5 mg (30 μmol) of AIBN were added. The mixture was stirred slowly (magnetic stirrer, about 60 r/min) and heated to 80°C for 9 h. After 4 h an additional 5 mg portion of AIBN was added. After cooling, the polymer was filtered off and washed twice with 20 mL of toluene. The combined filtrates were evaporated to dryness, and the foamy residue was chromatographed (silica Si-60; $CH_2Cl_2$–$Et_2O$, 9:1). The isolated solid was recrystallized from ether–petroleum ether to give 0.41 g (85%) of 1,2:5,6-diisopropylidene-3-deoxy-α-D-glucose **23**.

*Typical Procedure for Deoxygenation with the N-Ethylpiperidine Salt of Hypophosphorous Acid [40b,40c]*

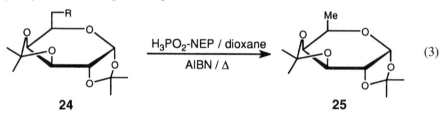

R = O(C=S)$OC_6H_4$-4F

The solution of 1,2:3,4-di-*O*-isopropylidene-D-galactopyranose-6-*O*-(4-fluorophenyl)-thionocarbonate [Eq. (3)] **24** (0.166 g, 0.4 mmol) and the *N*-ethylpiperidine salt of hypophosphorous acid ($H_3PO_2$–NEP, 0.72 g, 4.0 mmol) in dioxane (3 mL) under argon was treated with 150 μL of AIBN solution (0.2176 g of AIBN in 3 mL of dioxane) seven times (every 30 min) under reflux. The solution was washed with water and dried over anhydrous $MgSO_4$. After evaporation of the solvent, the residue was analyzed by nuclear magnetic resonance (NMR) to give 91% of the corresponding deoxy product 1,2:3,4-di-*O*-isopropylidene-6-deoxy-D-galactopyranose **25**.

*Deoxygenation of 1,2:3,4-di-O-isopropylidene-D-galactopyranose-5-O-(4-fluorophenyl)thionocarbonate [40a,40c]*

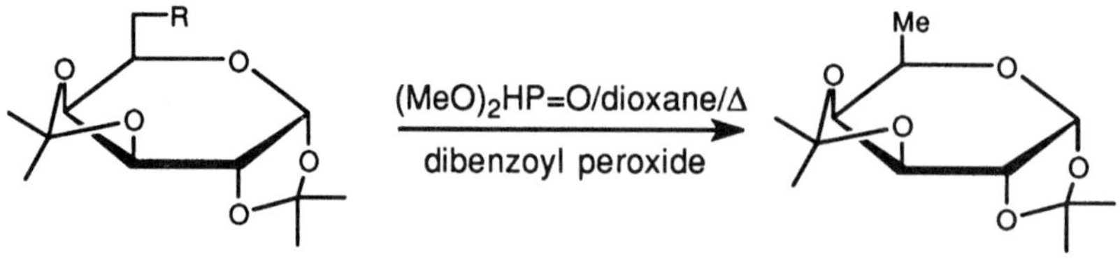

R = O(C=S)$OC_6H_4$-4F

**24**                                    **25**

The solution of 1,2:3,4-di-*O*-isopropylidene-D-galactopyranose-6-*O*-(4-fluorophenyl)-thionocarbonate as in A [Eq. (4)] **24** (0.197 g, 0.475 mmol) and dimethyl phosphite (0.22 mL, 2.38 mmol) in dioxane (3 mL) under argon was treated with 150 μL of dibenzoyl peroxide solution (0.387 g of benzoyl peroxide in 3 mL of dioxane) three times at 30 min intervals under reflux. After evaporation of the solvent, the residue was analyzed by NMR to give 90% of the deoxy product **25**.

### 1,6-Anhydro-2-deoxy-3,4-O-isopropylidene-D-galactose 27 from the Methylxanthate 26 with TBTH [10]

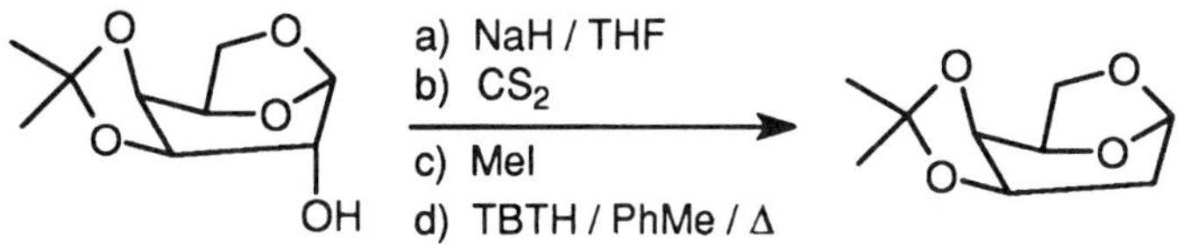

a) NaH / THF
b) $CS_2$
c) MeI
d) TBTH / PhMe / Δ

(5)

**26**  94% overall  **27**

A mixture of 1,6-anhydro-3,4-O-isopropylidene-β-D-galactose, [Eq. (5)] **26** (900 mg), sodium hydride dispersion (80%; 270 mg), imidazole (5 mg), and dry tetrahydrofuran (12 mL) was stirred for 0.5 h at room temperature. Carbon disulfide (2 mL) was added and stirring was continued for 1 h. Methylation [MeI (0.5 mL)] and the usual workup gave a yellow oil, which was heated under reflux in toluene (40 mL) during addition, over 1 h, of a solution of tributylstannane (TBTH, 1.6 g) in toluene (30 mL) under an argon atmosphere. Refluxing was continued for 16 h, the solvent was evaporated, and the residue was chromatographed over silica gel. Evaporation of the pure fractions gave the deoxy compound **27** (780 mg, 94%). Distillation gave an analytical sample of bp 75°C at 2 mmHg: $[\alpha]_D$ −141°.

### Preparation of 1,2:5,6-Di-O-isopropylidene-3-O-(methylthio)thiocarbonyl-α-D-glucofuranose 29 [45]

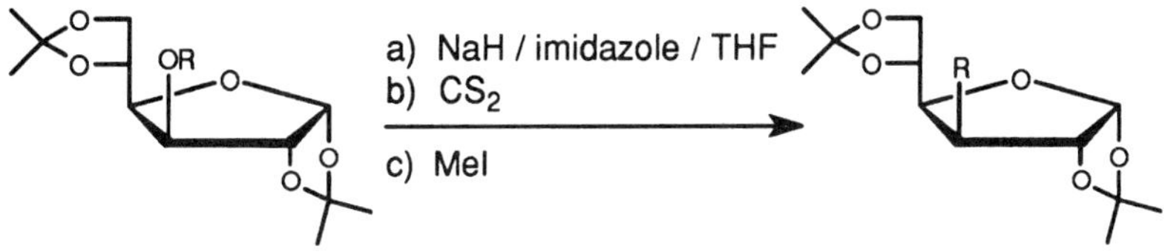

a) NaH / imidazole / THF
b) $CS_2$
c) MeI

(6)

R = H    R = O(C=S)SMe

**28**    **29**

The solution of NaH (1.53 g, 60% dispersion, 38.4 mmol), diacetone-α-D-glucose, [Eq. (6)] **28** (5 g, 19.2 mmol), and imidazole (65 mg, 0.96 mmol) in THF (40 mL) was stirred for 2 h under argon. Carbon disulfide (5.77 mL, 96 mmol) was added and the mixture was stirred for 12 h. MeI (6 mL, 96 mmol) was added, and the solution was stirred for another 2 h. The organic layer was washed with 1 M HCl, saturated $NaHCO_3$, brine, and dried over anhydrous $MgSO_4$. After evaporation of the solvent, the residue was recrystallized (EtOH/$H_2O$) to afford 5.51 g (82%) of the xanthate **29**: mp 58°–59°C.

### Deoxygenation of 1,2:5,6-Di-O-isopropylidene-α-D-glucofuranose xanthate 29 with Diethyl Phosphite [40a,40c]

The solution of 1,2:5,6-di-O-isopropylidene-α-D-glucofuranose xanthate, [Eq. (7)] **29** (0.28 g, 0.8 mmol) and diethyl phosphite (0.52 mL, 4 mmol) in dioxane (3 mL) under argon

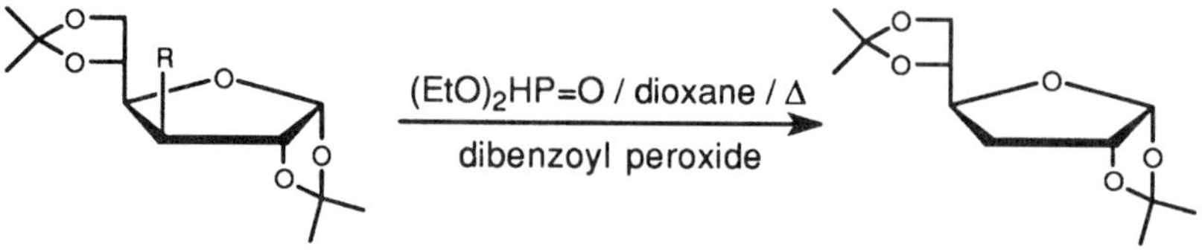

R = O(C=S)SMe

**29** → **30**  (Eq. 7, (EtO)$_2$HP=O / dioxane / Δ, dibenzoyl peroxide)

was treated with 38.7 mg of dibenzoyl peroxide twice at 30-min intervals under reflux. After evaporation of the solvent, the residue was analyzed by NMR to give 90% of the deoxy product **30**.

## 1,2:5,6-Di-O-isopropylidene-α-D-glucofuranose-3-O-(pentafluorophenyl)thionocarbonate 31 [42]

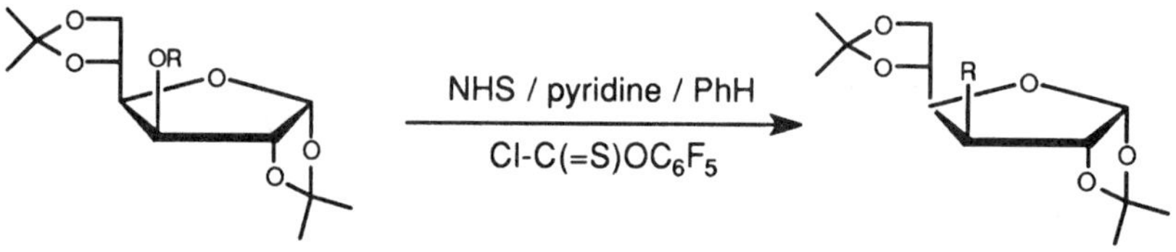

**28** (R = H) → **31** (R = O(C=S)OC$_6$F$_5$)  (Eq. 8, NHS / pyridine / PhH, Cl-C(=S)OC$_6$F$_5$)

To a solution of diacetone-α-D-glucofuranose [Eq. (8)] **28** (1 g, 3.84 mmol), N-hydroxysuccinimide (NHS, 0.044 g, 0.384 mmol), and dry pyridine (0.93 mL, 11.52 mmol) in C$_6$H$_6$ (20 mL) was added pentafluorophenyl chlorothionoformate (1.54 mL, 9.6 mmol) dropwise at room temperature under argon. The solution was stirred for an additional 2 h. The organic layer was washed with 1 M HCl, saturated NaHCO$_3$, brine, and dried over anhydrous MgSO$_4$. After evaporation of the solvent under reduced pressure, the thiocarbonate by-product was precipitated with hexanes. After filtration and evaporation, the crude product was purified by column chromatography over silica gel (eluting with n-hexane–EtOAc 8:2) to afford 1.45 g (77%) of the thionocarbonate **31**: mp 65°–66°C.

## Preparation of 1,2:5,6-Di-O-isopropylidene-3-O-(4-fluorophenoxy)-thiocarbonyl-α-D-glucofuranose 32 [33a]

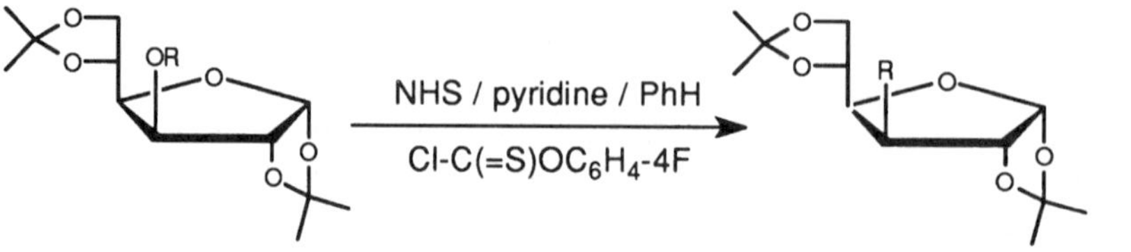

**28** (R = H) → **32** (R = O(C=S)OC$_6$H$_4$-4F)  (Eq. 9, NHS / pyridine / PhH, Cl-C(=S)OC$_6$H$_4$-4F)

To a solution of diacetone-α-D-glucose [Eq. (9)] **28** (2.60 g, 10 mmol), *N*-hydroxysuccinimide (NHS, 0.115 g, 1 mmol), and dry pyridine (2.43 mL, 30 mmol) in THF (50 mL) was added 4-fluorophenyl chlorothionoformate (3.81 g, 20 mmol) dropwise at room temperature under argon. The solution was stirred for an additional 2 h. The organic layer was washed with 1 *M* HCl, saturated NaHCO$_3$, brine, and dried over anhydrous MgSO$_4$. After evaporation of the solvent under reduced pressure, the thionocarbonate by-product was precipitated with hexanes. After filtration and evaporation, the crude product was purified by column chromatography over silica gel (eluting with *n*-hexane–CH$_2$Cl$_2$, 7:3) to afford 3.13 g (76%) of the thionocarbonate **32**: mp 82°-83°C (EtOH–H$_2$O); [α]$_D$ −33° (*c* 1, CHCl$_3$).

## B. Deoxygenation with Dimethyl Phosphate and Related Agents

*Deoxygenation of 1,2:5,6-Di-O-isopropylidene-α-D-glucofuranose-3-O-(pentafluorophenyl)thionocarbonate 31 with Dimethyl Phosphite [40a, 40c]*

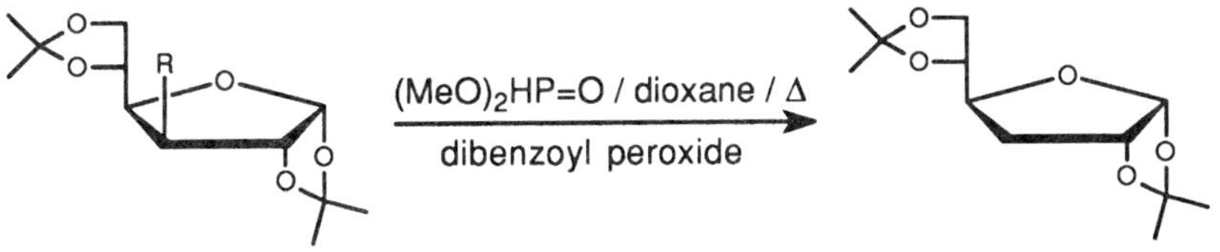

$$\text{31} \xrightarrow[\text{dibenzoyl peroxide}]{(MeO)_2HP{=}O \,/\, \text{dioxane} \,/\, \Delta} \text{30} \qquad (10)$$

R = O(C=S)OC$_6$F$_5$

**31**  **30**

The solution of 1,2:5,6-di-*O*-isopropylidene-α-D-glucofuranose-3-*O*-(pentafluorophenyl)-thionocarbonate) [Eq. (10)] **31** (0.195 g, 0.4 mmol) and dimethyl phosphite (0.18 mL, 2 mmol) in dioxane (3 mL) under argon was treated with 150 μL of benzoyl peroxide solution (0.387 g of benzoyl peroxide in 3 mL of dioxane) four times at 30-min intervals and under reflux. After evaporation of solvent, the residue was analyzed by NMR to give 100% of the deoxy product **30**.

*Deoxygenation of 1,2:5,6-Di-O-isopropylidene-α-D-glucofuranose-3-O-(4-fluorophenyl)thionocarbonate 32 [40a, 40c]*

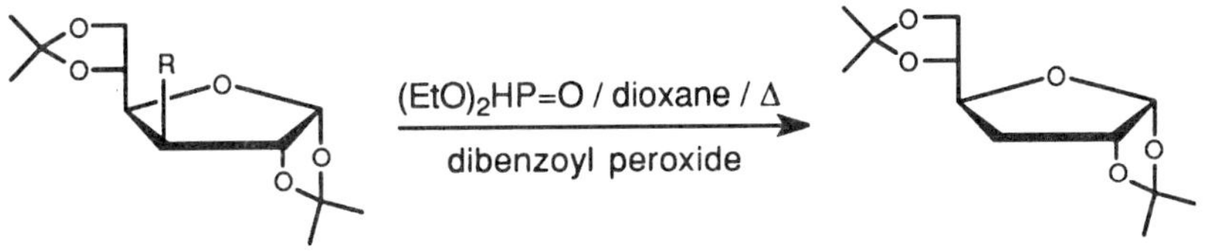

$$\text{32} \xrightarrow[\text{dibenzoyl peroxide}]{(EtO)_2HP{=}O \,/\, \text{dioxane} \,/\, \Delta} \text{30} \qquad (11)$$

R = O(C=S)OC$_6$H$_4$-4F

**32**  **30**

The solution of 1,2:5,6-di-*O*-isopropylidene-α-D-glucofuranose-3-*O*-(4-fluorophenyl)-thionocarbonate [Eq. (11)] **32** (83 mg, 0.2 mmol) and diethyl phosphite (138 μL, 1 mmol) in dioxane (1.5 mL) under argon was treated with 10 mg of dibenzoyl peroxide twice at 30-min intervals and under reflux. After evaporation of the solvent, the residue was analyzed by NMR to give 91% of the deoxy product **30**.

*Deoxygenation of 1,2:5,6-Di-O-isopropylidene-α-D-glucofuranose-3-O-(4-fluorophenyl)thionocarbonate 32 with Hypophosphorous Acid/Triethylamine/AIBN [40b, 40c]*

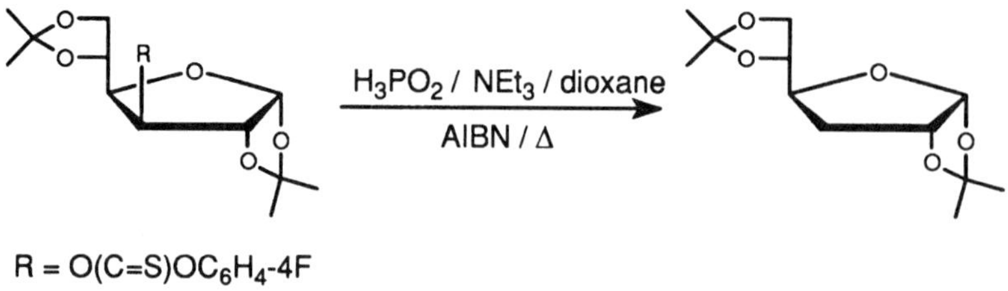

$$\text{32} \xrightarrow[\text{AIBN} / \Delta]{\text{H}_3\text{PO}_2 / \text{NEt}_3 / \text{dioxane}} \text{30} \quad (12)$$

R = O(C=S)OC$_6$H$_4$-4F

The solution of 1,2:5,6-di-*O*-isopropylidene-α-D-glucofuranose-3-*O*-(4-fluorophenyl)-thionocarbonate [Eq. (12)] **32** (83 mg, 0.2 mmol), hypophosphorous acid (0.1 mL, 1 mmol), and triethylamine (0.154 mL, 1.1 mmol) in dioxane (1.5 mL) under argon was treated with AIBN (6.5 mg) twice at 20 min intervals and under reflux. After completion, the reaction was washed with water and dried over anhydrous MgSO$_4$. After evaporation of the solvent, the residue was analyzed by $^1$H NMR to give 100% of the deoxy product **30**.

*Deoxygenation of 2-Deoxy-6,7-isopropylidene-3,5-Di-O-Methyl-L-Manno-Heptose-Trimethylene Dithioacetal 33 [46]*

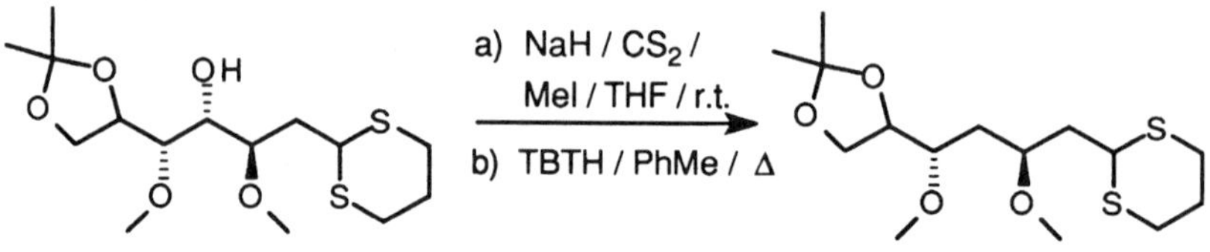

$$\text{33} \xrightarrow[\text{b) TBTH / PhMe / } \Delta]{\text{a) NaH / CS}_2 / \text{MeI / THF / r.t.}} \text{34} \quad (13)$$

A mixture of NaH (64 mg, 2.66 mmol) and the acetonide [Eq. (13)] **33** (500 mg, 1.42 mmol) in dry THF (15 mL) was stirred for 30 min and then treated with CS$_2$ (0.65 mL, 2.14 mmol). After 1 h stirring at 20°C, MeI (0.29 mL, 4.46 mmol) was added. The suspension was stirred for 30 min (TLC-control), ice water (1 mL) was added, and the solvent was removed under reduced pressure. The residue was dissolved in water and extracted with diethyl ether (3 × 20 mL). The combined ethereal extracts were dried over Na$_2$SO$_4$, filtered, and evaporated to dryness under reduced pressure. The crude xanthogenate was dissolved in toluene (5 mL) and this solution was added dropwise under argon to a boiling solution of tributyltin hydride (TBTH, 0.65 mL, 2.41 mmol) in toluene (25 mL). After completion of the addition, the mixture was refluxed for 14 h (TLC-control) and evaporated to dryness under reduced pressure. The product was chromatographed over silica gel [25 g, elution with light petroleum (bp 30°–70°C), containing an increasing proportion of ether (5% increments)]. After elution of organic tin compounds, the 2,4-dideoxy-6,7-isopropylidene-3,5-di-*O*-methyl-L-*xylo*-heptose trimethylene dithioacetal **34** was obtained as an oil (310 mg, 81% yield): [α]$_D$ −23.7° (*c* 1.13, CHCl$_3$).

## C. Deoxygenation with Organosilanes

*2,4-Dideoxygenation of 1,6-Anhydro-D-Glucose by the 2,4-bis-Phenylthionocarbonate 35 with tris(Trimethylsilyl)silane and AIBN [47, 48]*

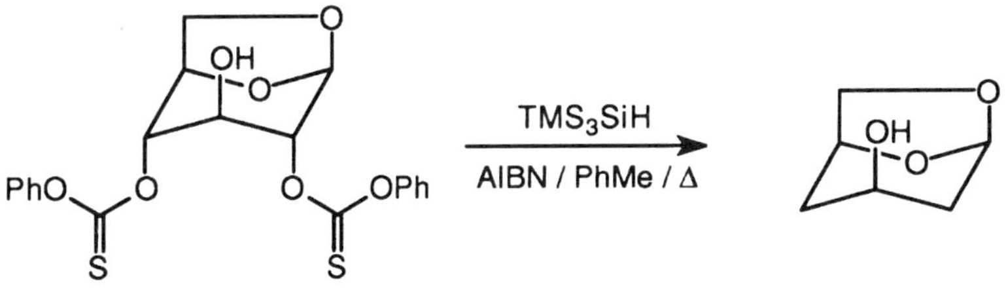

$$\mathbf{35} \xrightarrow[\text{AIBN / PhMe / }\Delta]{\text{TMS}_3\text{SiH}} \mathbf{36} \tag{14}$$

A solution of the starting 2,4-bis-thionocarbonate [Eq. (14)] **35** (87 mg, 0.20 mmol), tris(trimethylsilyl)silane (136 μL, 0.44 mmol), and AIBN (1.6 mg, 5 mmol %) in toluene (1.5 mL) was refluxed under argon for 1 h. After evaporation of the solvent under reduced pressure, the residue was analyzed by $^1$H NMR to give 87% of the 2,4-dideoxy product **36**.

*2,4-Dideoxygenation of 1,6-Anhydro-D-glucose by the 3-O-Trimethylsilyl-2,4-bis-phenylthionocarbonate 37 Derivative with Diphenylsilane and AIBN [47]*

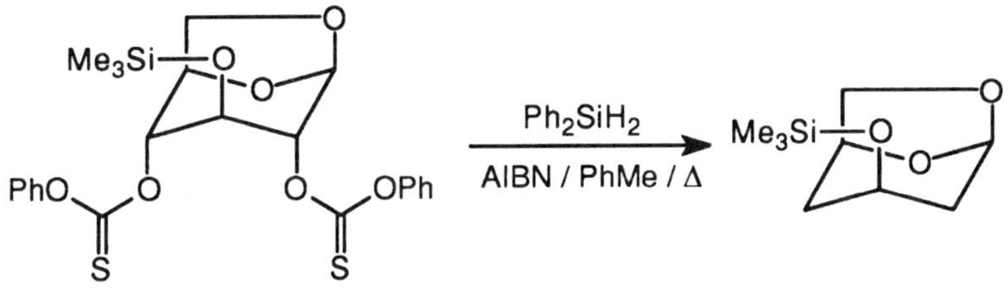

$$\mathbf{37} \xrightarrow[\text{AIBN / PhMe / }\Delta]{\text{Ph}_2\text{SiH}_2} \mathbf{38} \tag{15}$$

A solution of the starting 2,4-bis-thionocarbonate [Eq. (15)] **37** (174 mg, 0.40 mmol), chlorotrimethylsilane (0.31 mL, 2.4 mmol), and triethylamine (0.5 mL, 3.6 mmol) in benzene (3 mL) was stirred for 1 h at room temperature. After filtration, the solvent was evaporated. The residue was dissolved in toluene (1 mL) and diphenylsilane (294 μL, 1.6 mmol) was added. The reaction mixture was heated to reflux and treated under argon with 150 mL portions of an AIBN solution (262 mg of AIBN in 3 mL of dioxane) five times (at 20 min intervals) under reflux. After evaporation of the solvent under reduced pressure, the residue was analyzed by $^1$H NMR to give 85% of the 2,4-dideoxy-3-trimethylsilyl-1,6-anhydro-D-glucose product **38**.

*Monodeoxygenation of Methyl-2,3-di-O-methyl-α-D-glucopyranoside-4,5-dithiocarbonate to Methyl-4-deoxy-2,3-di-O-methyl-α-D-glucopyranoside 39 with TBTH and AIBN [49]*

Methyl 2,3-di-*O*-methyl-α-D-glucopyranoside 4,6-dithiocarbonate [Eq. (16)] **39** (0.264 g), tributyltin hydride (0.582 g), and AIBN (0.015 g) in dry toluene (20 mL) were successively added dropwise to refluxing toluene (20 mL) under argon over a period of 45 min. Subsequent additions of the tin hydride (2 × 0.292 g), together with AIBN (2 × 0.01 g) after 2 and 4 h, were necessary. The reaction was complete after 6 h. The solution was hydrolyzed with aqueous sodium hydroxide (40°C, 12 h). Workup (separation, extraction, and evaporation) gave the 4-deoxygenated product **40** (0.125 g, 61%) as an oil: $[\alpha]_D$ +70° (c 1.0, MeOH), $v_{max}$ (Nujol) 3605 cm$^{-1}$ (OH).

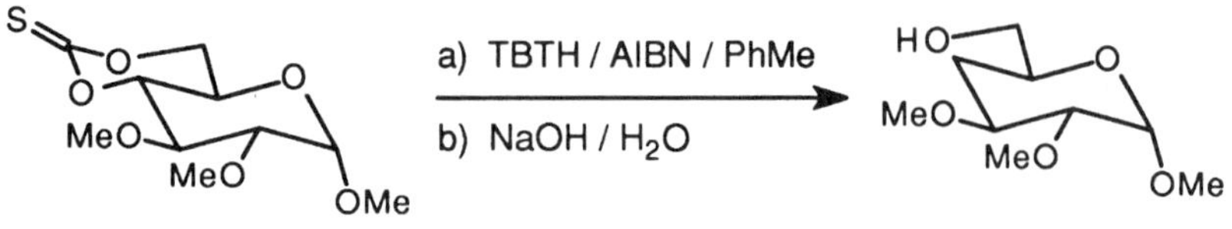

*Transformation of Methyl-{methyl 5-acetamido-3,5,9-trideoxy-9-iodo-4,7-bis-O-[(4-methylphenoxy)thiocarbonyl]-8-O-[(methylthio)carbonyl]-β-D-glycero-D-galacto-2-nonulopyranosid}onate 41 to Methyl-{methyl-5-acetamido-3,4,5,7,9-pentadeoxy-8-O-[(methylthio)carbonyl]-β-D-lyxo-2-nonulopyranosid}onate 42 with TBTH and AIBN [50]*

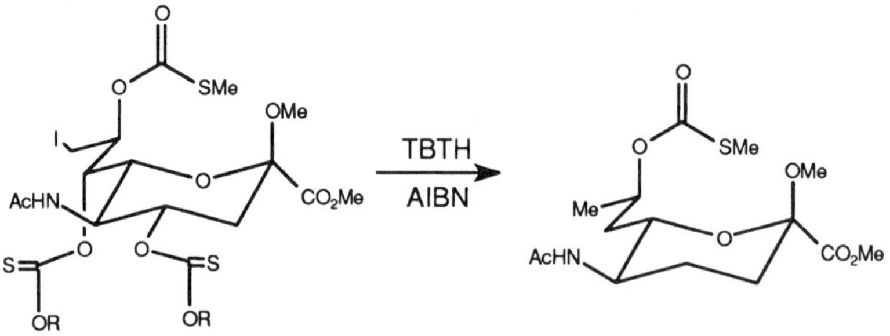

To a solution of **41** (448 mg) [Eq. (17)] in toluene (10 mL) was added TBTH (3 mmol) and a trace amount of AIBN. The solution was then heated under argon for 2.5 h at 110°C. After evaporation of the solvent, the residual foam was dissolved in acetonitrile (30 mL), and the solution was extracted three times with hexanes to remove several tin compounds. The acetonitrile was evaporated, and the residue was purified by chromatography over silica gel (40 g) to yield 165 mg (83%) of **42**.

*Transformation of Methyl-{methyl-5-acetamido-4-O-benzoyl-3,5,9-trideoxy-9-iodo-7-O-[(4-methylphenoxy)thiocarbonyl]-8-O-[(methylthio)carbonyl]-β-D-glycero-D-galacto-2-nonulopyranosid}onate 43 to Methyl-{methyl-5-acetamido-4-O-benzoyl-3,5,7,9-tetradeoxy-8-O-[(methylthio)carbonyl]-β-D-galacto-2-nonulopyranosid}onate 44 with TBTH and AIBN [50]*

By application of the method described for the synthesis of **42**, [see Eq. (17)], 465 mg of **43** [Eq. (18)] yielded 250 mg (86%) of **44**.

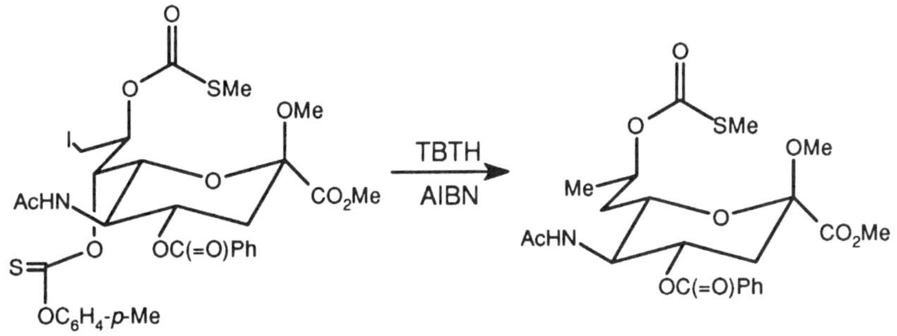

Free Radical Deoxygenation of Thiocarbonyls    165

(18)

**43** → **44**

*Synthesis and Deoxygenation of 3,4,6-Tri-O-benzyl-2-O-[(pentafluorophenoxy)thiocarbonyl]-β-glycoside 46 [43a]*

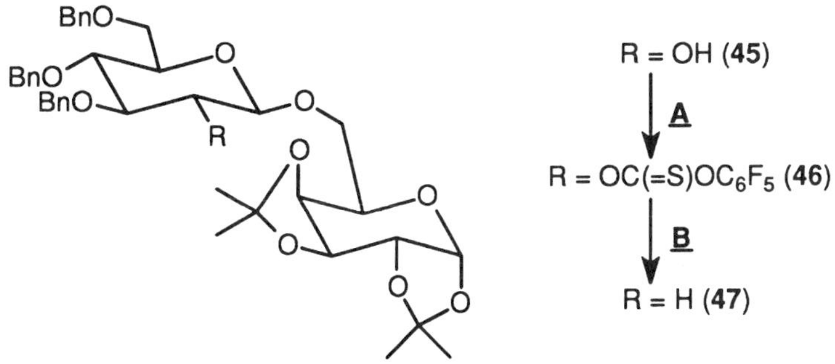

R = OH (**45**)

↓ **A**

R = OC(=S)OC$_6$F$_5$ (**46**)    (19)

↓ **B**

R = H (**47**)

1. *Preparation of the Pentafluorophenylthionocarbonate* **46**: The disaccharide **45**: [Eq. (19)] (0.1 mmol) was diluted with 1 mL of dry toluene, and N-hydroxysuccinimide (0.1 mmol) was added. Pentafluorophenyl chlorothionoformate (0.12 mmol) was then added dropwise, and finally anhydrous pyridine (0.5 mmol) was added. The yellow reaction mixture was heated to 80°C until TLC indicated that the reaction was complete. The product was purified by placing the entire reaction mixture on a 2 × 16-cm bed of silica gel and eluting with 4:1 petroleum ether–diethyl ether to give O-(3,4,6-tri-O-benzyl-2-O-[(pentafluorophenoxy)thiocarbonyl]-β-D-glucosyl)-(1→6)-1,2:3,4-di-O-isopropylidene-α-D-galactopyranose **46** in 77% yield and as a colorless oil: $[\alpha]_D$ −53.8° (c 0.40, CHCl$_3$).

2. *Preparation of the 2-deoxy-β-glycoside* **47** *with triphenyltin hydride*: The thionocarbonate **46** (0.1 mmol) [see Eq. (19)] was diluted with anhydrous toluene and AIBN was added (0.025 mmol as a standard solution in toluene). The reaction mixture was purged with argon and then heated to 110°C for 5 min, after which triphenyltin hydride (0.2 mmol) was added dropwise as a solution in toluene. The reaction mixture was heated for 1 h, cooled, and purified by placing the entire reaction mixture on a 2 × 16-cm column of silica gel. Elution with 85:15 hexanes–ethyl acetate afforded O-(3,4,6-tri-O-benzyl-2-deoxy-β-D-glucopyranosyl)-(1→6)-1,2:3,4-di-O-isopropylidene-α-D-galactopyranose **47** [see Eq. (19)] in 90% yield and as a clear liquid: $[\alpha]_D$ −45.1° (c 0.57, CHCl$_3$).

3. *Deoxygenation of a phenyl glycoside by the same method*: Application of procedure 1 to phenyl glycoside **48** [Eq. (20)] gave phenyl 3,4,6-tri-O-benzyl-2-O-[(pentafluorophenoxy)-thiocarbonyl]-β-D-*gluco*-pyranoside **49** in 93% yield and as a slightly yellow oil: $[\alpha]_D$ −6.52° (c 0.23, CHCl$_3$); application of procedure 2 to **49** gave phenyl 3,4,6-tri-

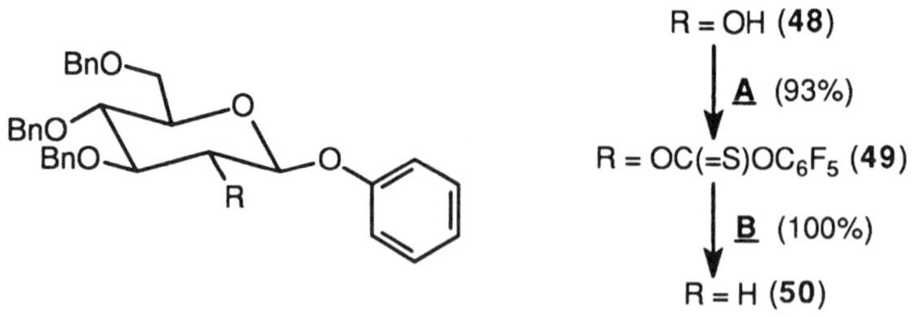

[Structure of tri-O-benzyl glucopyranoside derivative shown]

R = OH (**48**)

↓ **A** (93%)

R = OC(=S)OC$_6$F$_5$ (**49**)      (20)

↓ **B** (100%)

R = H (**50**)

$O$-benzyl-2-deoxy-β-D-glucopyranoside **50** in quantitative yield and as a clear liquid: $[\alpha]_D$ −6.32° ($c$ 1.29, CHCl$_3$).

### 3'-Deoxy-per-N-ethoxycarbonylseldomycin Factor 5 52 from the 3'-Thiocarbonylimidazolide 51 with TBTH [51]

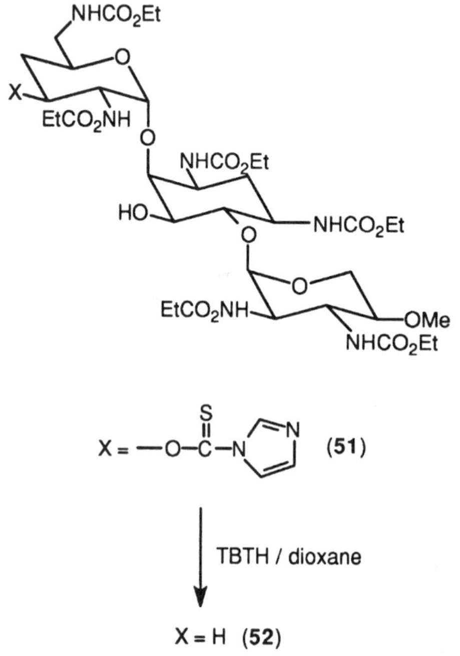

[Structure of per-N-ethoxycarbonylseldomycin factor 5 trisaccharide shown with substituents NHCO$_2$Et, EtCO$_2$NH, OMe, HO, and X]      (21)

X = —O—C(=S)—N(imidazolyl) (**51**)

↓ TBTH / dioxane

X = H (**52**)

A solution of per-$N$-ethoxycarbonylseldomycin factor 5-3'-thiocarbonylimidazolide **51** [Eq. (21)] (12.5 g, 12.58 mmol) in anhydrous dioxane (750 mL) was added dropwise to a refluxing suspension of tri-$n$-butylstannane (TBTH, 14.0 g, 12.7 mL, 48 mmol) in anhydrous dioxane (1200 mL) under a nitrogen atmosphere. After 2.5 h, the solvent was removed under reduced pressure, and the residue was chromatographed over silica gel to yield 9.8 g (90%) of 3'-deoxy-per-$N$-ethoxycarbonylseldomycin factor 5 **52**.

### Radical Chain Deoxygenation of Adenosine 53 by 2'-O-phenoxythiocarbonyl-3',5'-O-(1,1,3,3-tetraisopropyldisilox-1,3-diyl)adenosine with TBTH and AIBN [41a]

1. To 267 mg (1 mmol) of dried adenosine **53** [Eq. (22)], suspended in 10 mL of dry pyridine, was added 320 μL (316 mg, 1 mmol) of 1,3-dichloro-1,1,3,3-tetraisopropyldisiloxane (TPDSCl$_2$), and the mixture was stirred at room temperature for 3 h. Pyridine

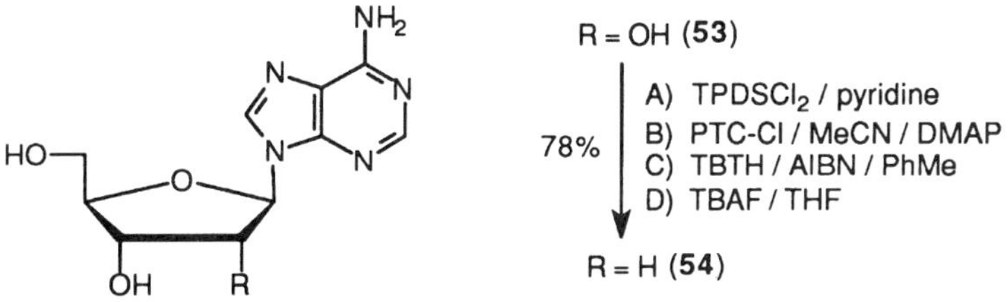

## Free Radical Deoxygenation of Thiocarbonyls

(22)

was evaporated, and the residue was partitioned between EtOAc and H₂O. The organic phase was washed with 2 × 20 mL of cold 1 M HCl, H₂O, saturated NaHCO₃, dried over Na₂SO₄, filtered, and evaporated. The resulting amorphous solid was of sufficient purity for direct use in the next step.

2. To the vacuum-dried residue obtained in step 1 was added 15 mL of anhydrous MeCN, 250 mg (2.05 mmol) of DMAP, and 200 µL (1.1 mmol) of phenyl chlorothionocarbonate (PTC-Cl). The solution was stirred at room temperature for 16 h. The solvent was evaporated and the residue was worked up as before (see Sec. 5, procedure 1 for converting **45** to **46**). The resulting product was sufficiently pure to be used directly in the reduction step 3. (The thionocarbonate can be isolated in this step on a silica column in 91% yield).

3. For the deoxygenation the crude thionocarbonate was dissolved in 20 mL of distilled toluene, and 32 mg (0.2 mmol) of AIBN and 400 µL (1.5 mmol) of TBTH were added. The solution was degassed with oxygen-free N₂ for 20 min, and then heated at 75°C for 3 h. The solvent was evaporated and the residue was purified over silica gel to give the deoxy compound (370 mg, 75%): mp 113°–114.5°C.

4. The crude deoxy compound (before chromatography) can be used for deprotection with TBAF and THF.

Treatment of 1.068 g (4 mmol) of adenosine **53** [Eq. (22)] by the four-step sequence described in the foregoing gave 780 mg (78%) of 2′-deoxyadenosine **54** after recrystallization.

### D. Deoxygenation and Carbocyclization

*Deoxygenation and Cyclization of the Carbohydrate-Derived Thiocarbonylimidazolide 55 with TBTH and AIBN [52]*

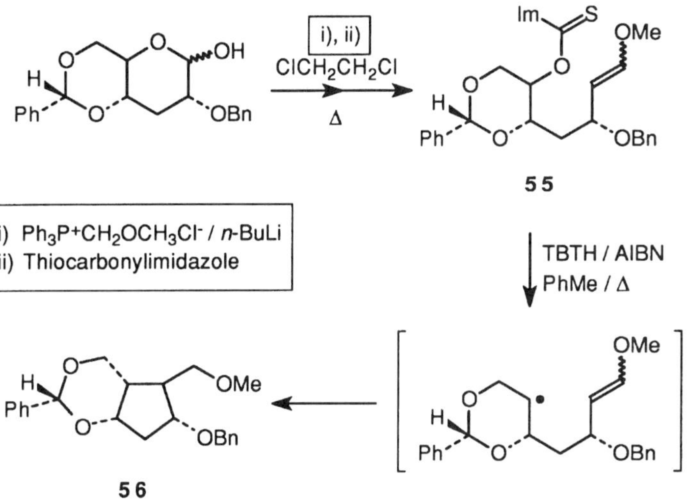

(23)

A solution of 6.13 g (12.8 mmol) of the thiocarbonyl-imidazolide **55** [Eq. (23)], 5.15 mL (19.1 mmol) of tributyltin hydride (TBTH), and 0.12 g of AIBN in 120 mL of dry toluene was refluxed for 1 h. Additional TBTH (0.5 Eq) and AIBN (60 mg) were added and the refluxing was continued for 1 h. The reaction mixture was added to 400 mL of ether, and washed with 80 mL each of saturated KF, 1 M HCl, and saturated $NaHCO_3$. The organic layer was washed with three more 50-mL portions of saturated KF solution and dried over anhydrous $MgSO_4$. Concentration and chromatography of the crude mixture yielded 3.59 g (58% for the two steps) of the cyclopentane derivative **56**: $[\alpha]_D$ $-23.8 \pm 0.8°$.

*Deoxygenation of Diacetone-α-D-glucose 28 via 1,2:5,6-Di-O-isopropylidene-3-O-(imidazolyl)thiocarbonyl-α-D-glucofuranose 57 [53]*

$$R = OH \ (\mathbf{28})$$

A) TCDI / THF / Δ

$$R = -O-\overset{S}{\underset{\|}{C}}-N\underset{}{\overset{}{\diagup}}\!\!\diagdown N \quad (\mathbf{57}) \tag{24}$$

B) TBTH / PhMe / Δ

$$R = H \ (\mathbf{30})$$

1. Solid *N,N*′-thiocarbonyldiimidazole (TCDI, 6 mmol) was added to a solution a diacetone-α-D-glucose **28** (3 mmol) [Eq. (24)] in THF (15 mL). The reaction mixture was gently refluxed under nitrogen atmosphere until TLC indicated complete consumption of the starting material. The solution was cooled, concentrated under vacuum, and flash-chromatographed (eluent ethyl acetate–hexane 1:1 v/v) to give 1,2:5,6-di-*O*-isopropylidene-3-*O*-(imidazolyl)thiocarbonyl-α-D-glucofuranose **57** as an oil (2.79 mmol, 93 % yield): $[\alpha]_D$ $-49.6°$ ($c$ 0.9, $CHCl_3$).

2. A mixture of the thiocarbonylimidazolide **57** (3 mmol) [Eq. (24)] in dry toluene (50 mL) was added dropwise over 30 min to a stirred, refluxing solution of toluene (200 mL) and tributyltin hydride (TBTH, 4.6 mmol) under nitrogen atmosphere. Refluxing was continued until TLC indicated complete reduction of the starting material. The solution was cooled and concentrated under reduced pressure. The residue was extracted with hot acetonitrile (3 × 50 mL) and the combined extracts were washed with hexanes (4 × 50 mL) to remove tin-containing compounds. The acetonitrile layer was concentrated under reduced pressure and flash-chromatographed to give pure 1,2:5,6-diisopropylidene-3-deoxy-α-D-glucose **30** as an oil (2.22 mmol, 74% yield): $[\alpha]_D$ $-6.9°$ ($c$ 2.6, $CHCl_3$).

## NOTES AND REFERENCES

1. (a) D. A. Cox, K. Richardson, and B. C. Ross, The aminoglycosides, in *Topics in Antibiotic Chemistry*, Vol. 1, P. G. Sammes and E. Horwood, eds., Halsted Press, New York, 1977; (b) T. Hayashi, T. Iwaoka, N. Takeda, and E. Ohki, Deoxysugar synthesis. IV. Deoxygenation of

aminoglycoside antibiotics through reduction of their dithiocarbonates, *Chem. Pharm. Bull.*, p. 1786 (1978); (c) D. H. R. Barton, G. Bringmann, G. Lamotte, W. B. Motherwell, R. S. Hay-Motherwell, and A. E. A. Porter, Reactions of relevance to the chemistry of aminoglycoside antibiotics. Part 14. A useful radical-deamination reaction, *J. Chem. Soc. Perkin Trans. 1*:2657 (1980); (d) D. H. R. Barton, G. Bringmann, and W. B. Motherwell, Reactions of relevance to the chemistry of aminoglycoside antibiotics. Part 15. The selective modification of neamine by radical-induced deamination, *J. Chem. Soc. Perkin Trans. 1*:2665 (1980).

2. Many examples can be found in W. Priebe, ed., *Anthracycline Antibiotics. New Analogues, Methods of Delivery, and Mechanisms of Action*, ACS Symp. Ser. 574, American Chemical Society, Washington, DC, 1995.

3. I. F. Pelyvás, C. Monneret, P. Herczegh, *Synthetic Aspects of Aminodeoxy Sugars of Antibiotics*, Springer-Verlag, Berlin, 1988.

4. J. Thiem and W. Klaffke, Synthesis of deoxy oligosaccharides, in *Topics in Current Chemistry*, 154, Carbohydrate Chemistry, J. Thiem, ed., Springer-Verlag, Berlin, 1990, pp. 285–332.

5. P. M. Collins and R. Ferrier, *Monosaccharides: Their Chemistry and Their Roles in Natural Products*, John Wiley & Sons, Chichester, 1995, pp. 206–216 and references there cited.

6. R. A. Larock, *Comprehensive Organic Transformations. A Guide to Functional Group Preparations*, VCH Publishers, New York, 1989.

7. S. W. McCombie, Reduction of saturated alcohols and amines to alkanes, in *Comprehensive Organic Synthesis*, Vol. 8, B. M. Trost and I. Fleming, eds., Pergamon Press, Oxford, 1991, pp. 811–833.

8. (a) B. Giese, *Radicals in Organic Synthesis: Formation of Carbon–Carbon Bonds*, Pergamon Press, Oxford, 1986; (b) W. B. Motherwell and D. Crich, *Free Radical Chain Reactions in Organic Synthesis*, Academic Press, London, 1992.

9. (a) D. P. Curran, The design and application of free radical chain reactions in organic synthesis: Part 1, *Synthesis*, p. 417 (1988); (b) D. P. Curran, The design and application of free radical chain reactions in organic synthesis: Part 2, *Synthesis*, p. 489 (1988); (c) C. P. Jasperse, D. P. Curran, and T. L. Fevig, Radical reactions in natural product synthesis, *Chem. Rev.* 91:1237 (1991).

10. D. H. R. Barton and S. W. McCombie, A new method for the deoxygenation of secondary alcohols, *J. Chem. Soc. Perkin Trans. 1*:1574 (1975).

11. Primary alcohols: (a) D. H. R. Barton, W. B. Motherwell, and A. Stange, Radical-induced deoxygenation of primary alcohols, *Synthesis*, p. 743 (1981); (b) D. H. R. Barton, P. Blundell, J. Dorchak, D. O. Jang, and J. C. Jaszberenyi, The invention of radical reactions. Part XXI. Simple methods for the radical deoxygenation of primary alcohols, *Tetrahedron* 47:8969 (1991).
    Tertiary alcohols: (c) D. H. R. Barton, W. Hartwig, R. S. Hay-Motherwell, W. B. Motherwell, and A. Stange, Radical deoxygenation of tertiary alcohols, *Tetrahedron Lett.* 23:2019 (1982); (d) D. H. R. Barton, S. I. Parekh, and C. L. Tse, On the stability and radical deoxygenation of tertiary xanthates, *Tetrahedron Lett.* 34:2733 (1993); (e) W. Hartwig, Modern methods for the radical deoxygenation of alcohols, *Tetrahedron* 39:2609 (1983); (f) M. Ramaiah, Radical Reactions in Organic Synthesis, *Tetrahedron* 43:3541 (1987).

12. K. C. Nicolaou, M. Sato, E. A. Theodorakis and N. D. Miller, Conversion of thionoesters and thionolactones to ethers; a general and efficient radical desulfurization, *J. Chem. Soc. Chem. Commun.*, p. 1583 (1995).

13. (a) W. P. Neumann, *The Organic Chemistry of Tin*, John Wiley & Sons, New York, 1970; (b) R. C. Poller, *The Chemistry of Organotin Compounds*, Logos Press, London, 1970; (c) M. Pereyre, J.-P. Quintard, and A. Rahm, *Tin in Organic Synthesis*, Butterworths, London, 1987; (d) A. G. Davies and P. J. Smith, Tin, in *Comprehensive Organometallic Chemistry*, Vol. 2, G. Wilkinson, F. G. A. Stone, and E. W. Abel, eds., Pergamon Press, Oxford, 1982, pp. 519–627; (e) W. P. Neumann, Tri-*n*-butyltin hydride as reagent in organic synthesis, *Synthesis*, 665 (1987).

*Other approaches involving the use of organotin hydrides:*

J. T. Groves, S. Kittisopikul, Dehalogenation reactions catalyzed by tri-*n*-butyltin chloride. Competition for carbon radicals by borohydride and tin hydride, *Tetrahedron Lett. 49*:4291 (1977).

N. M. Weinshenker, G. A. Crosby, and J. Y. Wong, Polymeric Reagents. IV. Synthesis and utilization of an insoluble polymeric organotin dihydride reagent, *J. Org. Chem. 40*:1966 (1975).

U. Gerigk, M. Gerlach, W. P. Neumann, R. Vieler, and V. Weintritt, Polymer-supported organotin hydrides as immobilized reagents for free-radical synthesis, *Synthesis*, p. 448 (1990).

M. Gerlach, F. Jördens, H. Kuhn, W. P. Neumann, and M. Peterseim, A polymer-supported organotin hydride and its multipurpose application in radical organic synthesis, *J. Org. Chem. 56*:5971 (1991).

W. P. Neumann and M. Peterseim, Elegant improvement of the deoxygenation of alcohols using a polystyrene-supported organotin hydride, *Synlett*, p. 801 (1992).

14. (a) C. Copeland and R. V. Stick, A synthesis of abequose (3,6-dideoxy-D-*xylo*-hexose), *Aust. J. Chem. 30*:1269 (1977); (b) J. J. Patroni and R. V. Stick, A synthesis of paratose (3,6-dideoxy-D-*ribo*-hexose), *Aust. J. Chem. 31*:445 (1978); (c) J. J. Patroni and R. V. Stick, Stereoselective reduction of 3-*O*-hexofuranosyl *S*-methyl dithiocarbonates with tributyltin deuteride, *J. Chem. Soc. Chem. Commun.*, p. 449 (1978); (d) J. J. Patroni and R. V. Stick, Stereoselective reduction of 3-*O*-hexofuranosyl *S*-methyl dithiocarbonates with tributyltin deuteride, *Aust. J. Chem. 32*:411 (1979).

15. P. J. Barker and A. L. J. Beckwith, E.S.R. Identification of alkoxythiocaronyl radicals as possible intermediates in Barton deoxygenation of alcohols, *J. Chem. Soc. Chem. Commun.* p. 683 (1984).

16. M. D. Bachi and E. Bosch, On the mechanism of reductive degradation of dithiocarbonates by tributylstannane, *J. Chem. Soc. Perkin Trans. 1*: 1517 (1988).

17. D. Crich, On the use of *S*-(4-alkenyl)-dithiocarbonates as mechanistic probes in the Barton–McCombie radical deoxygenation reaction, *Tetrahedron Lett. 29*:5805 (1988).

18. (a) K. Nozaki, K. Oshima, and K. Utimoto, Facile reduction of dithiocarbonates with *n*-Bu$_3$SnH-Et$_3$B. Easy access to hydrocarbons from secondary alcohols, *Tetrahedron Lett. 29*:6125 (1988); (b) K. Nozaki, K. Oshima, and K. Utimoto, Synthesis of lactones by intramolecular addition of alkoxythiocarbonyl free radicals to acetylenes, *Tetrahedron Lett. 29*:6127 (1988).

19. J. E. Forbes and S. Z. Zard, A novel radical chain reaction of xanthic anhydrides. Further observations on the intermediacy of alkoxythiocarbonyl radicals in the Barton-McCombie reaction, *Tetrahedron Lett. 30*:4367 (1989).

20. D. H. R. Barton, D. O. Jang, and J. C. Jaszberenyi, On the mechanism of deoxygenation of secondary alcohols by tin hydride reduction of methyl xanthates and other thiocarbonyl derivates, *Tetrahedron Lett. 31*:3991 (1990).

21. P. Rivière, M. Rivière-Baudet, and J. Satgé, Germanium, in *Comprehensive Organometallic Chemistry*, Vol. 2, G. Wilkinson, F. G. A. Stone, and E. W. Abel, eds., Pergamon Press, Oxford, 1982, pp. 399-518.

22. (a) R. Walsh, Bond dissociation energy values in silicon-containing compounds and some of their implications, *Acc. Chem. Res. 14*:246 (1981); (b) J. M. Kanabus-Kaminska, J. A. Hawari, D. Griller, and C. Chatgilialoglu, Reduction of silicon-hydrogen bond strengths, *J. Am. Chem. Soc. 109*:5267 (1987); (c) C. Chatgilialoglu, D. Griller, and M. Lesage, Rate constants for the reactions of tris(trimethylsilyl) radicals with organic halides, *J. Org. Chem. 54*:2492 (1989); (d) M. Ballestri, C. Chatgilialoglu, K. B. Clark, D. Griller, B. Giese, and B. Kopping, Tris(trimethylsilyl)silane as a radical-based reducing agent in synthesis, *J. Org. Chem. 56*:678 (1991); (e) C. Chatgilialoglu, A. Guerrini, and M. Lucarini, The Me$_3$Si substituent effect on the reactivity of silanes. Structural correlations between silyl radicals and their parent silanes, *J. Org. Chem. 57*:3405 (1992).

23. D. A. Armitage, Organosilanes, in *Comprehensive Organometallic Chemistry*, Vol. 2, G. Wilkinson, F. G. A. Stone, and E. W. Abel, eds., Pergamon Press, Oxford, 1982, pp. 1–203.

24. P. D. Magnus, T. Sarkar, and S. Djuric, Organosilicon compounds in organic synthesis, in *Comprehensive Organometallic Chemistry*, Vol. 7, G. Wilkinson, F. G. A. Stone, and E. W. Abel, eds., Pergamon Press, Oxford, 1982, pp. 614–626.
25. I. Fleming, Organic silicon chemistry, in *Comprehensive Organic Chemistry*, Vol. 3, D. N. Jones, ed., Pergamon Press, Oxford, 1979, pp. 561–576.
26. C. Chatgilialoglu, D. Griller, and M. Lesage, Tris(trimethylsilyl)silane. A new reducing agent, *J. Org. Chem. 53*:3641 (1988).
27. M. Lesage, C. Chatgilialoglu, and D. Griller, Tris(trimethylsilyl)silane: A catalyst for radical mediated reduction reactions, *Tetrahedron Lett. 30*:2733 (1989).
28. B. Giese, B. Kopping, and C. Chatgilialoglu, Tris(trimethylsilyl)silane as mediator in organic synthesis via radicals, *Tetrahedron Lett. 30*:681 (1989).
29. K. J. Kulicke and B. Giese, Hydrosilylation and cyclization reactions of alkenes and ketones with tris(trimethylsilyl)silane, *Synlett.* p. 91 (1990).
30. D. Schummer and G. Höfle, Tris(trimethylsilyl)silane as a reagent for the radical deoxygenation of alcohols, *Synlett.* p. 705 (1990).
31. C. Chatgilialoglu, Organosilanes as radical-based reducing agents in synthesis, *Acc. Chem. Res. 25*:188 (1992).
32. C. Chatgilialoglu, Silanes as new reducing agents in organic synthesis, in *Free Radicals in Synthesis and Biology*, F. Minisci, ed., Kluwer Academic Publishers, Dordrecht, 1989, pp. 115–123.
33. (a) D. H. R. Barton, D. O. Jang, and J. C. Jaszberenyi, An improved radical chain procedure for the deoxygenation of secondary and primary alcohols using diphenysilane as hydrogen atom donor and triethylborane–air as initiator, *Tetrahedron Lett. 31*:4681 (1990); (b) D. H. R. Barton, D. O. Jang, and J. C. Jaszberenyi, The invention of radical reactions. Part XXXI. Diphenylsilane: A reagent for deoxygenation of alcohols via their thiocarbonyl derivatives, deamination via isonitriles and dehalogenation of bromo- and iodo-compounds by radical chain chemistry, *Tetrahedron 49*:7193 (1993).
34. D. H. R. Barton, D. O. Jang, and J. C. Jaszberenyi, Radical deoxygenation of secondary and primary alcohols with phenylsilane, *Synlett*, p. 435 (1991).
35. D. H. R. Barton, D. O. Jang, and J. C. Jaszberenyi, The invention of radical reactions. Part XXIX. Radical mono- and dideoxygenations with silanes, *Tetrahedron 49*:2793 (1993).
36. (a) D. H. R. Barton, D. O. Jang, and J. C. Jaszberenyi, Radical mono- and dideoxygenations with the triethylsilane + benzoyl peroxide system, *Tetrahedron Lett. 32*:7187 (1991); (b) C. Chatgilialoglu, C. Ferreri, and M. Lucarini, A comment on the use of triethylsilane as a radical-based reducing agent, *J. Org. Chem. 58*:249 (1993).
37. (a) R. P. Allen, B. P. Roberts, and C. R. Willis, Thiols as polarity reversal catalysts for hydrogen-atom transfer from organosilanes to alkyl radicals: Reduction of alkyl halides by triethylsilane, *J. Chem. Soc. Chem. Commun.* p. 1387 (1989); (b) J. N. Kirwan, B. P. Roberts, and C. R. Willis, Deoxygenation of alcohols by the reactions of their xanthate esters with triethylsilane: An alternative to tributyltin hydride in the Barton–McCombie reaction, *Tetrahedron Lett. 31*:5093 (1990); (c) S. J. Cole, J. N. Kirwan, B. P. Roberts, C. R. Willis, Radical chain reduction of alkyl halides, dialkyl sulphides and *O*-alkyl-*S* methyl dithiocarbonates to alkanes by trialkylsilanes, *J. Chem. Soc. Perkin Trans. 1*:103 (1991).
38. M. Ballestri, C. Chatgilialoglu, and G. Seconi, $(Me_3Si)_3SiSH$: A new radical-based reducing agent, *J. Organomet. Chem, 408*:C1–C4 (1991).
39. J. Daroszewski, J. Lusztyk, M. Degueil, C. Navarro, and B. Maillard, Heptamethyltrisilane-2-thiol-mediated free-radical-chain reduction of organic halides, *J. Chem. Soc. Chem. Commun.* p. 586 (1991).
40. (a) D. H. R. Barton, D. O. Jang, and J. Jaszberenyi, Radical deoxygenations and dehalogenations with dialkyl phosphites as hydrogen atom source, *Tetrahedron Lett. 33*:2311 (1992); (b) D. H. R. Barton, D. O. Jang, and J. Jaszberenyi, Hypophosphorous acids and its salts: New reagents for

radical chain deoxygenation, dehalogenation and deamination, *Tetrahedron Lett. 33*:5709 (1992); (c) D. H. R. Barton, D. O. Jang, and J. C. Jaszberenyi, The invention of radical reactions. 32. Radical deoxygenations, dehalogenations, and deaminations with dialkyl phosphites and hypophosphorous acid as hydrogen sources, *J. Org. Chem. 58*:6838 (1993); (The novel reagent *N*-ethylpiperidine hypophosphite is available from the Aldrich Chemical Co., Inc.: Catalogue No. 43,617-8)

41. (a) M. J. Robins and J. S. Wilson, Smooth and efficient deoxygenation of secondary alcohols. A general procedure for the conversion of ribonucleosides to 2'-deoxynucleosides, *J. Am. Chem. Soc. 103*:932 (1981); (b) M. J. Robins, J. S. Wilson, and F. Hansske, Nucleic acid related compounds. 42. A general procedure for the efficient deoxygenation of secondary alcohols. Regiospecific and stereoselective conversion of ribonucleosides to 2'-deoxynucleosides, *J. Am. Chem. Soc. 105*:4059 (1983).

42. D. H. R. Barton and J. C. Jaszberenyi, Improved methods for the radical deoxygenation of secondary alcohols, *Tetrahedron Lett. 30*:2619 (1989).

43. (a) J. Gervay and S. Danishefsky, A stereospecific route to 2-deoxy-β-glycosides, *J. Org. Chem. 56*:5448 (1991); (b) J. T. Link, M. Gallant, S. Danishefsky, and S. Huber, The first synthesis of a fully functionalized core structure of staurosprine: Sequential indolyl glycosidation by endo and exo glycals, *J. Am. Chem. Soc. 115*:3782 (1993).

44. D. H. R. Barton, J. Dorchak, and J. C. Jaszberenyi, The invention of radical reactions. Part XXIV. Relative rates of acylation and radical deoxygenation of secondary alcohols, *Tetrahedron, 48*:7435 (1992).

45. A. K. Sanyal and C. B. Purves, Some xanthate methyl esters of glucose, *Can. J. Chem. 34*:426 (1956).

46. K. Krohn, S. Grignard, and G. Börner, From sugars to carbocycles. 4. Exclusive seven-membered ring formation from D-glucose, *Tetrahedron Asymmetry 5*:2485 (1994).

47. D. H. R. Barton, D. O. Jang, and J. C. Jaszberenyi, Tris(trimethylsilyl)silane and diphenylsilane in the radical chain dideoxygenation of 1,6-anhydro-D-glucose: A comparative study, *Tetrahedron Lett. 33*:6629 (1992).

48. P. Boquel, C. Loustau Cazalet, Y. Chapleur, S. Samreth, and F. Bellamy, An expeditious enantiospecific synthesis of a precursor of the lactonic portion of mevinic acids, *Tetrahedron Lett. 33*:1997 (1992).

49. D. H. R. Barton and R. Subramanian, Reactions of relevance to the chemistry of aminoglycoside antibiotics. Part 7. Conversion of thiocarbonates into deoxy-sugars, *J. Chem. Soc. Perkin Trans. 1*:1718 (1977).

50. E. Schreiner, R. Christian, and E. Zribal, Synthesis of 9-deoxy-, 7,9-dideoxy-, and 4,7,9-trideoxy-*N*-acetylneuraminic acid and their behaviour towards CMP-sialate synthase, *Liebigs Ann. Chem.* p. 93 (1990).

51. R. E. Carney, J. B. McAlpine, M. Jackson, R. S. Stanaszek, W. H. Washburn, M. Cirovic, and S. L. Mueller, Modification of seldomycin factor 5 at C-3', *J. Antibiot. 31*:441 (1978).

52. T. V. RajanBabu, From carbohydrates to carbocycles. 2. A free radical route to Corey lactone and other prostanoid intermediates, *J. Org. Chem. 53*:4522 (1988).

53. J. R. Rasmussen, C. J. Slinger, R. J. Kordish, and D. D. Newman-Evans, Synthesis of deoxy sugars. Deoxygenation by treatment with *N,N'*-thiocarbonyldiimidazole/tri-*n*-butylstannane, *J. Org. Chem. 46*:4843 (1981).

# 9

# Thiazole-Based One-Carbon Extension of Carbohydrate Derivatives

**Alessandro Dondoni and Alberto Marra**
*Università di Ferrara, Ferrara, Italy*

|  |  |  |
|---|---|---|
| I. | Introduction | 174 |
| II. | Methods | 174 |
|  | A. One-carbon homologation: The state of the art | 174 |
|  | B. One-carbon homologation: The thiazole-based approach | 178 |
|  | C. Conclusions | 187 |
| III. | Experimental Procedures | 188 |
|  | A. Addition of 2-(trimethylsilyl)thiazole to aldehydes | 188 |
|  | B. Inversion of configuration of the hydroxyl group | 188 |
|  | C. Preparation of polyalkyoxyaldehydes | 189 |
|  | D. One-carbon homologation of dialdoses | 189 |
|  | E. Synthesis of *N*-benzyl nitrones | 190 |
|  | F. Addition of 2-lithiothiazole to nitrones | 190 |
|  | G. Addition of 2-lithiothiazole to Lewis acid nitrone complexes | 191 |
|  | H. Reduction of *N*-benzyl hydroxylamines | 191 |
|  | I. Preparation of α-aminoaldehydes | 192 |
|  | J. Aminohomologation of dialdoses | 192 |
|  | K. Addition of 2-lithiothiazole to sugar lactones | 193 |
|  | L. Deoxygenation of thiazolyl ketol acetates | 194 |
|  | M. Preparation of *C*-formyl glycosides | 194 |
|  | N. *N*-Glycosidation of thiazolyl ketol acetates | 195 |
|  | O. Preparation of glycosyl α-azidoesters | 195 |
|  | P. Preparation of glycosyl α-aminoesters | 196 |
|  | Notes and References | 196 |

## I. INTRODUCTION

The emergence of information on the role that carbohydrates play in a variety of molecular recognition processes in biological systems [1] and the awareness of the convenient exploitation of these compounds as precursors to chiral building blocks for organic synthesis [2], have stimulated increased demand for practical synthetic methods in carbohydrate chemistry. In response, new chemical and biological synthetic strategies have been developed in recent years for the stereocontrolled synthesis of natural oligosaccharides and glycoconjugates and their unnatural analogues [3]. Among monosaccharides, considerable attention has been given to the synthesis of modified sugars and to compounds containing seven or more carbon atoms, the so-called higher-carbon sugars [4]. Syntheses have been carried out by various approaches: (1) from nonsugar precursors [5]; (2) direct coupling of two monosaccharide subunits (C-glycosidation), or construction of a second unit on a preexisting one [6]; (3) chain-elongation of either sugar or nonsugar starting materials by installation of a carbon chain bearing an apparent or masked functional group [7]. Each of these synthetic strategies has its own validity, provided it involves chemically and stereochemically efficient steps. However, prominent approaches are those based on the homologation of readily available natural sugars because, in principle, they should permit accessing various types of compounds and higher-carbon sugars of any required length. Quite interesting, although multicarbon homologation appears attractive and spectacular because of the rapid growth of the chain, sometimes iterative one-carbon homologation is more convenient, for it permits a full range of stereochemical variations and the introduction of differentially protected functional groups. While moving from one-carbon homologation based on the cyanohydrin and nitromethane syntheses (i.e., the roots of a great deal of carbohydrate chemistry), we will first shortly review recent methods reported from other laboratories. For a critical and comparative evaluation, only those methods will be described wherein the whole sequence, starting from an aldose or a ketose, and ending with the corresponding one-carbon higher homologue has been completed. However, formal homologations (i.e., methods dealing with intermediates that have been previously transformed in aldoses or ketoses) will be also considered. We will then describe in some more detail our own method that is based on the application of the thiazole–aldehyde synthesis [8].

## II. METHODS

### A. One-Carbon Homologation: The State of the Art

Among the classic methods for the extension of the aldose chain by one carbon atom from the reducing end [9], the Kiliani–Fischer cyanohydrin synthesis [10] is a milestone in carbohydrate chemistry. However after 110 years from discovery and numerous applications [11], including the preparation of carbon and hydrogen isotopically labeled compounds for mechanistic and structural studies [12], there are still several drawbacks that make the method impractical. These are the low and variable degree of selectivity and the harsh reaction conditions that are required to reveal the aldose from either the aldonic acid or directly from the cyanohydrin. Synthetic applications that have appeared in recent times confirmed these limitations. For instance, a quite low selectivity was registered [13] in the addition of the cyanide ion to the D-*galacto*-hexodialdo-1,5-pyranose derivative **1**

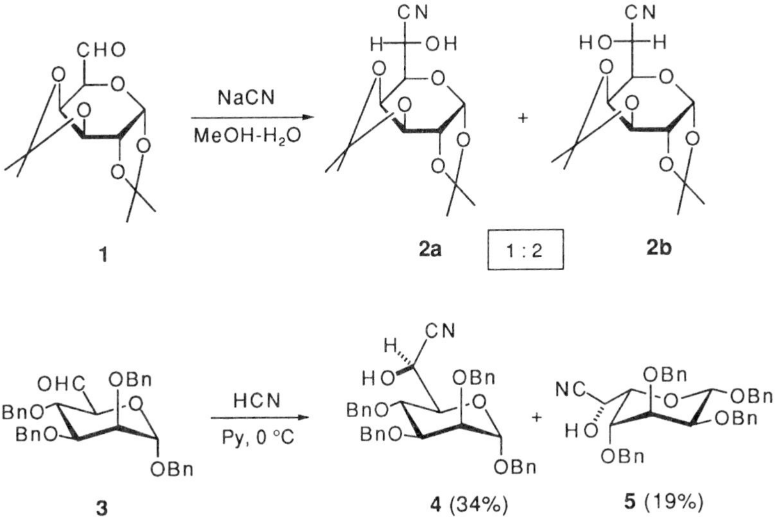

### Thiazole-Based One-Carbon Extension

**Scheme 1**

(Scheme 1) whereas the reaction with the D-*manno*-hexodialdo-1,5-pyranose derivative **3** produced, in low yield, the cyanohydrin **4** having an L-*glycero*-D-*manno* configuration and the stereoisomeric epimer at C-5, rather than at C-6 [14]. With an aim toward the synthesis of D- and L-*glycero*-D-*manno*-heptoses, the reaction of **3** with other reagents (2-methylfuran, alkyl magnesium chlorides) was also explored, without any substantial improvement of either stereoselectivity or yield.

Although the reaction of sodium cyanide with 2,6-anhydro-D-*glycero*-L-*manno*-heptose (**6**) (Scheme 2), a *C*-formyl galactopyranoside, seemed to occur with high level of

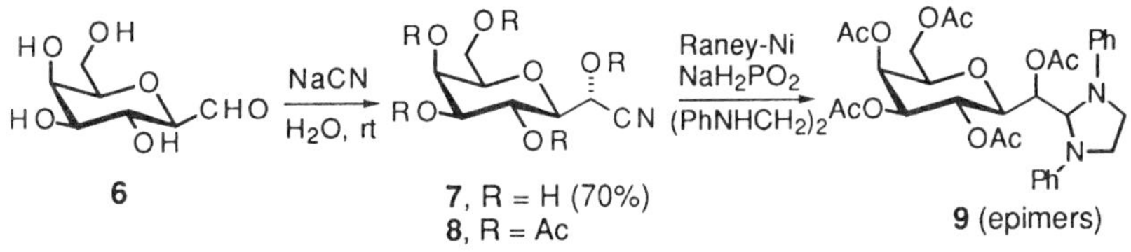

**Scheme 2**

diastereoselectivity on the basis of the isolation of 3,7-anhydro-D-*threo*-L-*talo*-octonitrile **7** as a single product, this was actually the result of the preferential crystallization of this compound, rather than a real stereochemical control [15]. The reaction with hydrogen cyanide in pyridine led to a 1:1 mixture of cyanohydrin epimers. Moreover, the reduction of the peracetylated pure cyanohydrin **8** by the method of Moffatt caused epimerization at C-2 leading to **9** as a mixture of stereoisomers in a 2.4:1 ratio. Although recent chemical [16] and enzymatic [17] methods have been described for the asymmetric addition of hydrogen cyanide and trimethylsilyl cyanide to carbonyl compounds, these approaches have not yet been employed for carbohydrate homologation.

An interesting variation of the cyanohydrin synthesis is based on the application of the Strecker reaction [18], or an appropriate modification of it, to protected dialdo sugars to give glycosyl α-aminonitriles, with good levels of diastereoselectivity (ds > 90%) [19]. This approach was employed for a formal synthesis of the amino octose lincosamine from the dialdose **1** [20] although D-*glycero*-D-*galacto* aminonitrile (**10a**), the target stereoisomer required for the synthesis, was obtained as the minor product (Scheme 3).

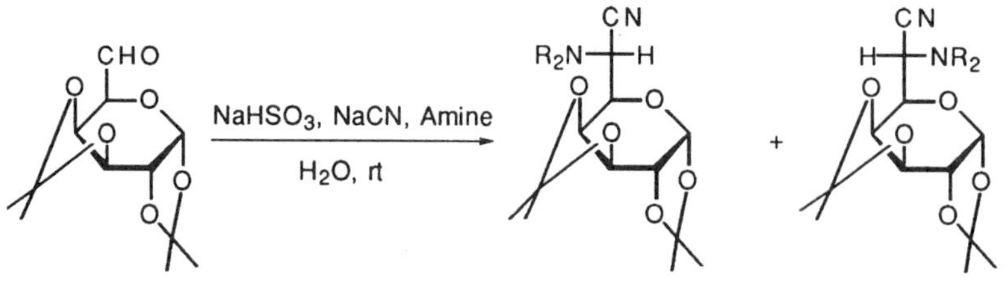

**Scheme 3**

The Sowden homologation [21], based on the nitroaldol condensation (Henry reaction) [22] between the *aldehydo* sugar and nitromethane in basic medium, followed by the Nef decomposition [23] of the resultant nitronate in strongly acidic conditions, has been employed in a more limited number of cases than the cyanohydrin synthesis. A recent example in this area is shown by the stepwise homologation of N-acetyl-D-mannosamine (**11**) into N-acetylneuraminic acid (**12**) [24] (Scheme 4). Also, this procedure has found

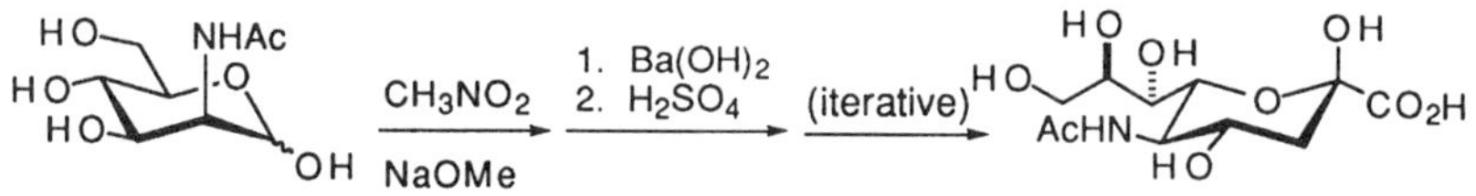

**Scheme 4**

application for the synthesis of isotopically labeled compounds [24]. Unfortunately, the efficiency of this synthesis was somewhat diminished by the numerous protection–deprotection steps of the hydroxyl groups. These manipulations were required when it was found that nitromethane did not always undergo base-catalyzed addition to unprotected hemiacetalic glycoses, whereas the reaction appears to be of general application with acyclic *aldehydo* sugars.

Examples of sugar homologation based on the use of 2-lithio-1,3-dithiane (2-LDT, **13**) are surprisingly rare relative to the popularity of this reagent [25]. Stereoselective chain extension of a partially protected hexose, 2,3:5,6-di-*O*-isopropylidene-α-D-mannofuranose (**14**), into a D-*glycero*-D-*galacto*-heptose derivative (**16**) was reported by Paulsen and his co-workers several years ago [26] (Scheme 5).

In an earlier report, the same author described the synthesis of branched carbohy-

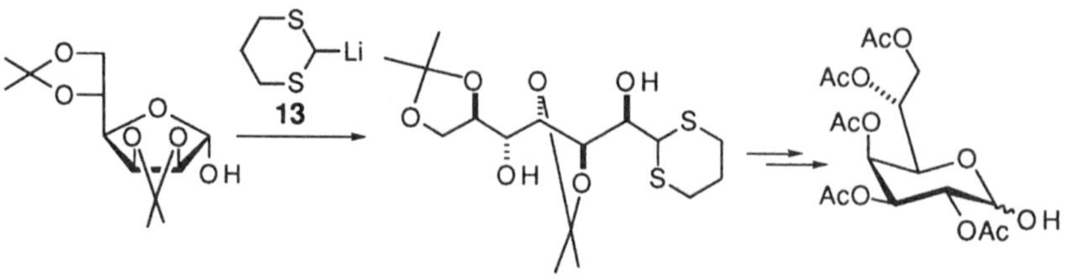

**Scheme 5**

drates bearing aldehyde side chains by addition of 2-LDT (**13**) to blocked cyclic uloses followed by Hg(II)-mediated hydrolysis of the 1,3-dithiane ring [27]. Stereoselectivity based on isolated products ranged between 62 and 89%. In the same instance it was pointed out that the addition of **13** to an acyclic blocked *aldehydo* sugars occurred with scarce diastereoselectivity as proved by the reaction with 2,3:4,5-di-*O*-isopropylidene-D-arabino-pyranose, which afforded a mixture of adducts with D-*gluco* and D-*manno* configuration. By contrast, good levels of *anti* selectivity were registered by Chikashita and co-workers [28] in an iterative homologation sequence (Scheme **6**) starting from 2,3-*O*-cyclohexyl-idene-D-glyceraldehyde (**17**). On the other hand *syn* selectivity remained substantially low.

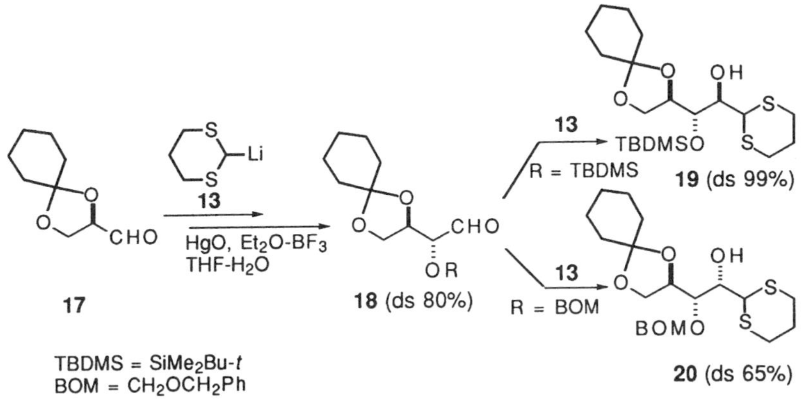

**Scheme 6**

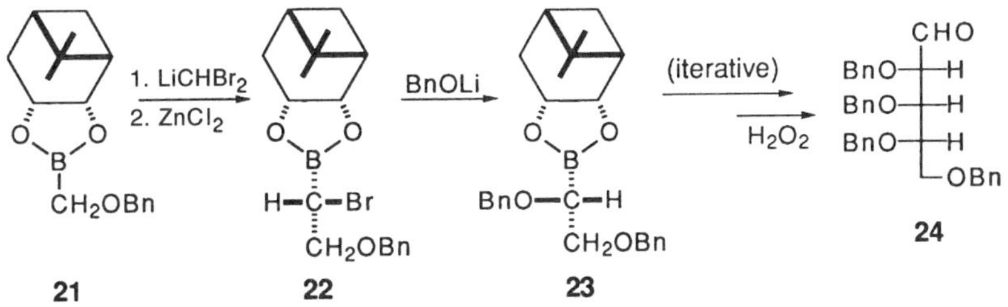

**Scheme 7**

An iterative chain-elongation process employing chiral boronic esters as templates has been described by Matteson [29] (Scheme **7**). The homologative cycle involves highly stereoselective reaction of (*S*)-pinanediol [(benzyloxy)-methyl]boronate (**21**) with (di-bromo-ethyl)lithium to give the [2-(benzyloxy)-1-bromoethyl]boronate (**22**) and replacement of the α-bromine by benzyl oxide. The sequential installation of three benzyloxymethylene groups appeared to have been the limit of the repetitive homologation sequence since L-ribose was the higher homolog accessible by this methodology. Quite correctly, it has been emphasized that the synthetic utility of this strategy is limited to the preparation of specifically labeled sugars [30].

Other methods have been reported which, however, lacked generality. One of these described the approach to L-*glycero*-D-*manno*-heptopyranosides by stereoselective chain extension of suitably protected hexodialdo-1,5-pyranosides with silylmethylmagnesium chlorides [31]. Another method, quite reminiscent of the Masamune–Sharpless approach [7g], reported the synthesis of four stereoisomeric tetrose derivatives by homologation of vinylsilanes to *E*-butenes, using phenyl (trimethylsilyl)methyl sulfide, followed by the Sharpless kinetic resolution and epoxidation [32].

## B. One-Carbon Homologation: The Thiazole-Based Approach

For a simple, yet seemingly trivial, operation, such as the stepwise construction of a polyhydroxylated aldehyde by adding one-carbon unit at a time, one needs to use a good reagent that serves as the formyl anion synthon for the iterative addition to the carbonyl of the aldehyde. If one wishes to proceed through an efficient synthetic route, the formyl anion equivalent must feature some essential characteristics, including (1) high stability, ensuring ease of preparation and storage; (2) high and stereocontrolled reactivity, giving rise to short reaction times and allowing the creation of new hydroxymethylene groups with the required configuration; (3) compatibility with various hydroxyl protective groups; (4) tolerance to synthetic manipulations of the polyhydroxylated carbon chain partially or completely constructed; (5) facile releasing of the formyl group under mild and neutral reaction conditions. Very few reagents of the old [33] and new repertoire [34] of formyl anion equivalents appear to qualify for such a role. Also newcomers show serious drawbacks and limited applications [35,36]. Hence, because we have been working for several years with 2-metalated thiazoles (**25**; Fig. 1), it is now convenient to evaluate their synthetic usefulness as formyl anion equivalents in the specific area of carbohydrate homologation.

*Acyclic Homologation*

Previous work from our laboratory [37] showed the use of 2-(trimethylsilyl)thiazole (2-TST) (**25a**, R = SiMe$_3$) as a stable and easily storable thiazolyl carbanion equivalent in reactions with various carbon electrophiles. All reactions occurred without the need for any added catalyst. Although a mechanistic rationale for this unique organosilicon reagent was later provided [38], we were initially stimulated to examine the reaction with 2,3-*O*-isopropylidene-D-glyceraldehyde (**26**) as an approach to a new one-carbon homologation method of sugars [39] (Scheme **8**). The system proved to work quite well both in our and other hands [40]. The addition of **25a** to **26** occurred with a high level of *anti* selectivity to give the alcohol **27** in almost quantitative yield. This product was protected as *O*-benzyl ether and then converted by one-pot cleavage of the thiazole ring into the D-erythrose **30**. Alternatively, the intermediates **28** and **29** of the unmasking sequence (*N*-methylation, reduction, metal-assisted hydrolysis) were isolated and characterized [41]. Later on we reported [42] an improvement of the original unmasking protocol wherein methyl triflate replaced methyl iodide in the *N*-methylation step and copper(II) chloride was used instead

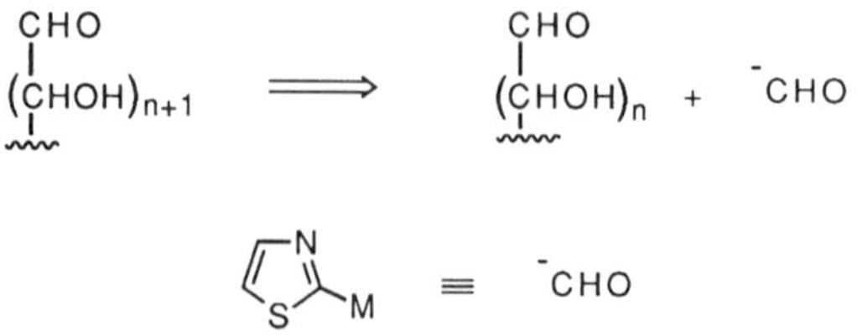

$$\begin{array}{c} \text{CHO} \\ | \\ (\text{CHOH})_{n+1} \end{array} \Longrightarrow \begin{array}{c} \text{CHO} \\ | \\ (\text{CHOH})_n \end{array} + \ ^-\text{CHO}$$

[thiazole-M structure] ≡ $^-$CHO

**25**

Figure 1

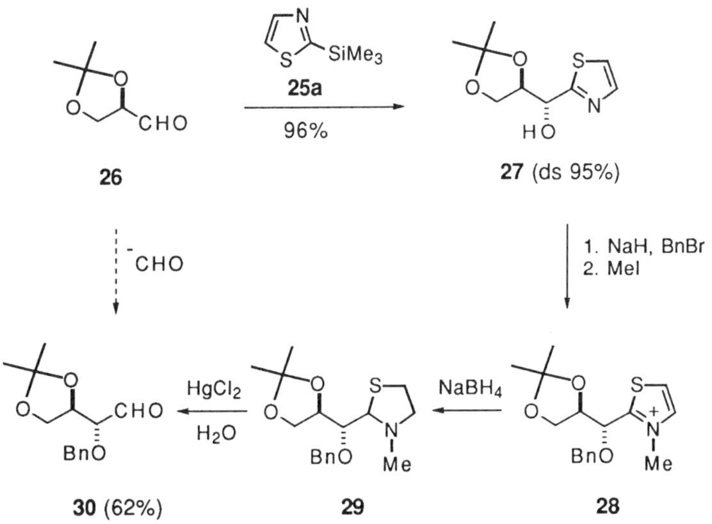

# Thiazole-Based One-Carbon Extension

**Scheme 8**

of mercury(II) chloride to assist the hydrolysis of the thiazolidine in the final step. It is worthwhile at this stage to emphasize that both the original and the modified thiazolyl-to-formyl unmasking protocols involve reactions that proceed under almost neutral conditions, thereby ensuring the configurational integrity of stereocenters and the stability of acid- and base-sensitive protective groups.

Although the synthetic equivalence of 2-TST (**25a**) to the formyl anion synthon was evident from the foregoing model study, the challenge lay in demonstrating the scope and limit of this strategy as a homologation technique. Hence, the iterative addition and unmasking protocols were repeated over several consecutive cycles so that the chain elongation of the triose **26** was brought up to the nonose **31** through the series of lower homologues having an all-*anti* configuration in the 1,2-polyol units (Scheme 9). This

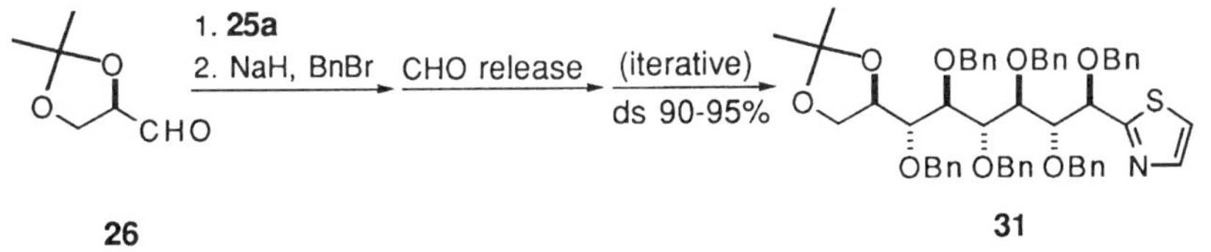

**Scheme 9**

sequential assembly of benzyloxymethylene groups does not appear to have been pushed to the limit, for high chemical yields (60–80%) and levels of selectivity (90–95%) were maintained over the whole iterative sequence.

Extension of the scope of the method by full control of the stereoselectivity failed because the use of Lewis acids as chelating agents to induce the *syn* addition of **25a** to aldehydes produced instead a substantial desilylation of the reagent. A remedy to this limitation was provided by the conversion of the *anti* adduct into the *syn* isomer by an oxidation–reduction sequence [43] (Scheme 10).

Therefore, a more complex homologative cycle was set up involving the addition of **25a** to alkoxy aldehydes and the inversion of the hydroxyl group. By this reaction sequence, the aldehyde **26** was converted into the two tetroses **30** and **34** and four pentoses **35**, **37**,

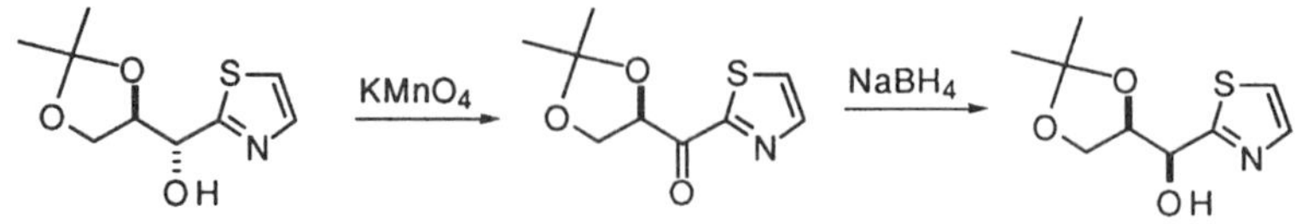

**Scheme 10**

**41**, and **42** [44], with high stereoselectivity in each step (Scheme **11**), with the exception of one reaction showing an unexpected stereochemical outcome (Scheme **12**). A quite similar approach has been recently reported by Chikashita and co-workers using 2-LDT (**13**) as a formyl anion equivalent (Scheme **6**) [28].

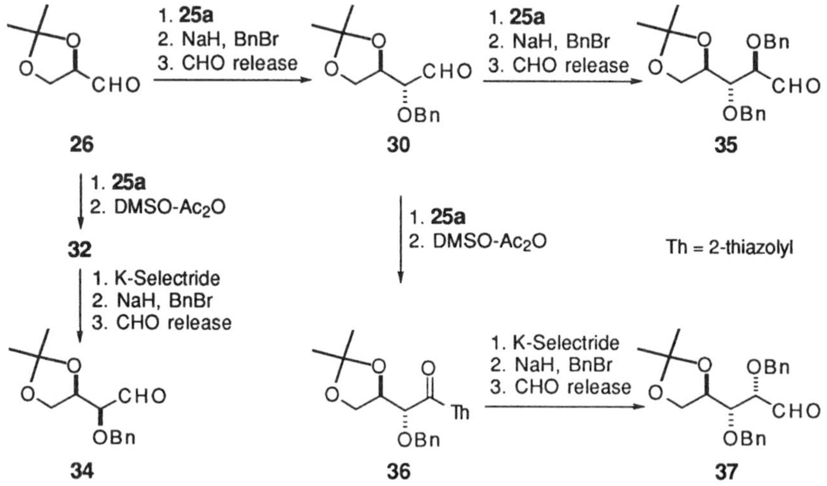

Th = 2-thiazolyl

**Scheme 11**

Th = 2-thiazolyl

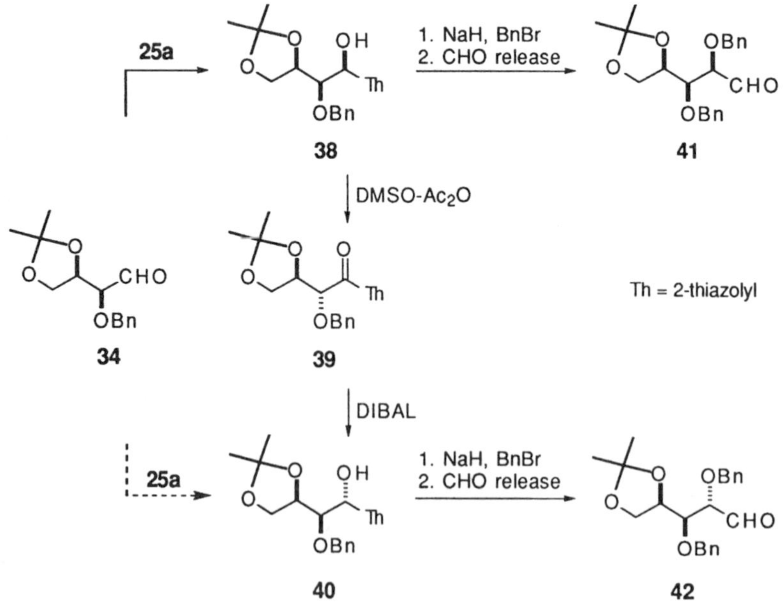

**Scheme 12**

## Thiazole-Based One-Carbon Extension

The direct homologation technique was then extended to the synthesis of various uncommon carbohydrate structures. Thus, higher sugars of the L-series were obtained starting from 2,3-*O*-isopropylidene-4-*O*-benzyl-L-threose (**43**) [39c] (Scheme 13), and the amino tetrose **47** and pentose **48** were prepared from the α-amino aldehyde **46** derived from L-serine [46a] (Scheme 14). These amino sugars were used as chiral building blocks for the

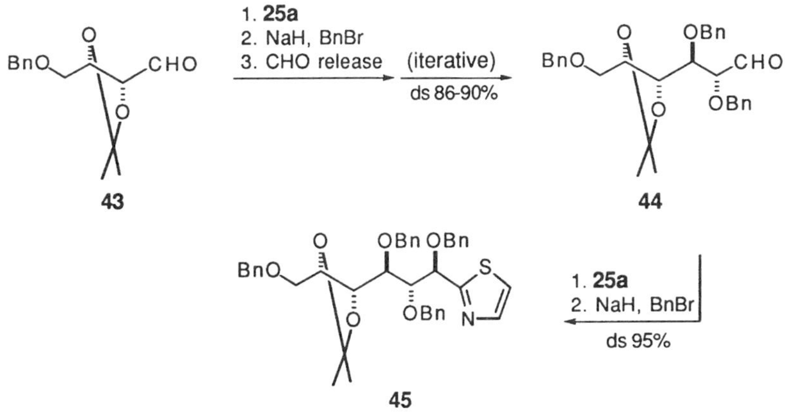

**Scheme 13**

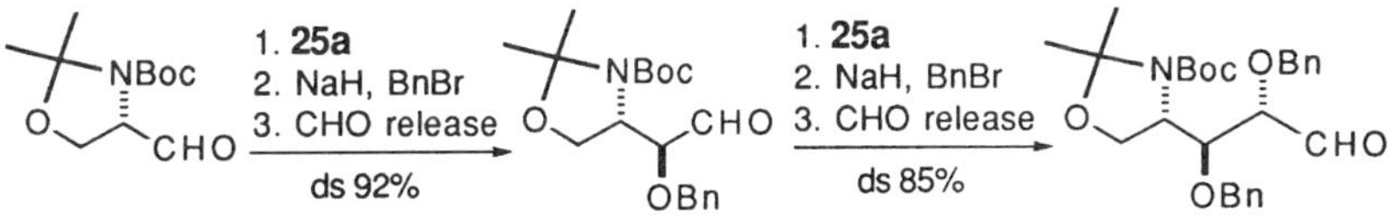

**Scheme 14**

synthesis of sphingosines. The *anti* selectivity observed for the addition of 2-TST (**25a**) to the *N,N*-diprotected amino aldehyde **46** can be reversed to *syn* selectivity by using a *N*-monoprotected derivative [46b]. Tunable stereoselectivity by the *N*-protecting group control has been achieved in other cases, thus extending considerably the scope of the α-amino aldehyde homologation with the use of **25a**.

The chain-extension of dialdoses provided a faster access to higher-carbon sugars. The building up of a stereochemically defined polyhydroxylated carbon chain at C-5 of a pyranose ring, is a quite important operation for the construction of the sugar moieties that are present in various antibiotics, such as hikizimycin [47] and tunicamycin [48]. Indeed, the application of the foregoing iterative addition–unmasking sequence to pentodialdo-1,4-furanose **49** and hexodialdo-1,5-pyranose **1** (Scheme 15) led to the corresponding series of one-carbon higher homologues in good yield and stereoselectivity in each cycle [39c,49].

The results of the one-carbon homologation of **49** to **50** and **1** to **52** were confirmed by recent work of Momenteau [50] and Aspinall [51] and their co-workers. However, with the aim of extending this homologation technique to other dialdoses, methyl 2,3,4-tri-*O*-benzyl-α-D-*manno*-hexodialdo-1,5-pyranoside (**54**; Scheme 16) reacted with 2-TST (**25a**) with poor diastereoselectivity to give the adducts **55a** and **55b** in 55 and 37% yield, respectively [51]. Poor selectivity was also observed with the *gluco* analogue of **54**. Conclusions on the reasons of the discrepancy between our and Aspinall's observations should be drawn after comparison of compounds having the same hydroxyl protective groups.

Results of this section illustrate quite well the efficiency and some limitations of the

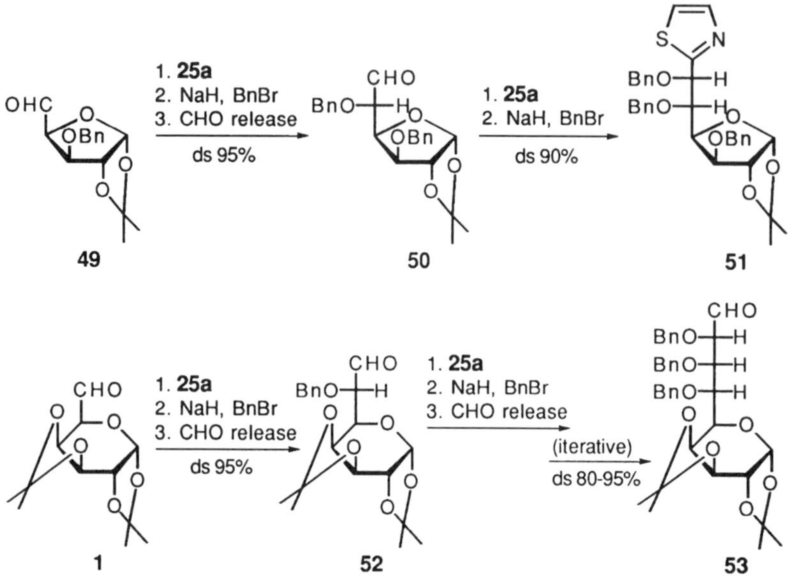

Scheme 15

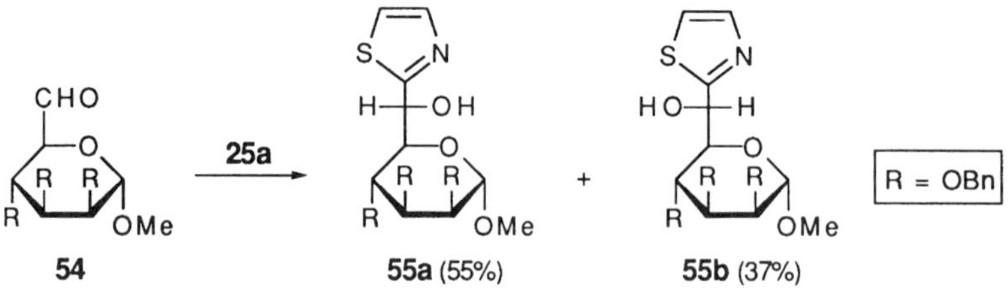

Scheme 16

homologation technique of *aldehydo* sugars employing 2-TST (**25a**) as a formyl anion equivalent. Several advantages over the existing methods arise from the intrinsic properties of this reagent which, in fact, combines high stability for easy storage and manipulation with high reactivity under different conditions. Chemical yields and selectivities, in general, were satisfactory, if not excellent, in most of the studies. A limitation stems from the failure to obtain total stereochemical control of the addition to the carbonyl by the intervention of external agents. Finally, the thiazole-to-formyl conversion constitutes a simple and highly reliable operation that can be carried out without affecting protective groups and stereogenic centers. The potential and synthetic value of this new homologation method has been graciously recognized [52].

*Acyclic Aminohomologation*

A remarkable extension of the foregoing method came from the idea that the reaction of a 2-metalated thiazole **25** with a nitrone derived from an α-alkoxy aldehyde, would give an hydroxylamine adduct that could be converted into an α-amino β-alkoxy aldehyde homologue (Fig. 2). The chain elongation of *aldehydo* sugars and installation of an amino group in the same homologation cycle (aminohomologation) appeared relevant to the synthesis of amino sugars, particularly higher-carbon members. Also in this approach, the aim was to developing a strategy that could overcome the limitations of the cyano amination [19,20] and possibly show higher efficiency than other syntheses of amino sugars [53].

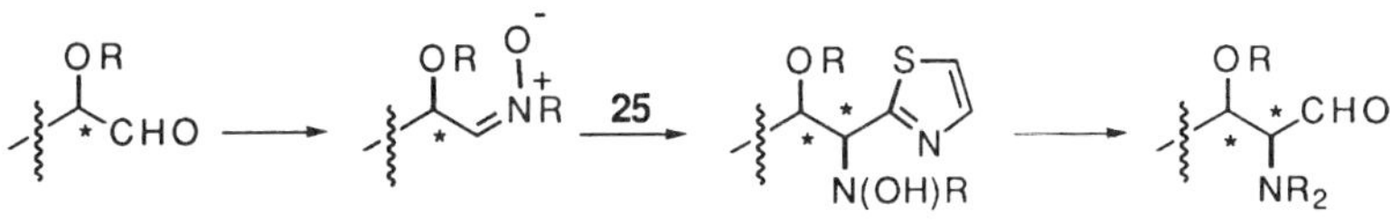

**Figure 2**

The choice of nitrones as iminium cation equivalents stemmed from the ease of synthesis and purification of compounds derived from a large set of polyalkoxy aldehydes, and their stability to handling and storage [54]. Moreover, it was expected that the high polarity of the nitrone group and the interaction with external additives [55] could be effectively exploited to affect both reactivity and stereoselectivity. The model aminohomologation of 2,3-O-isopropylidene-D-glyceraldehyde (**26**) illustrates the details of this method [56] (Scheme 17). Having observed that the silyl derivative **25a** was inert toward

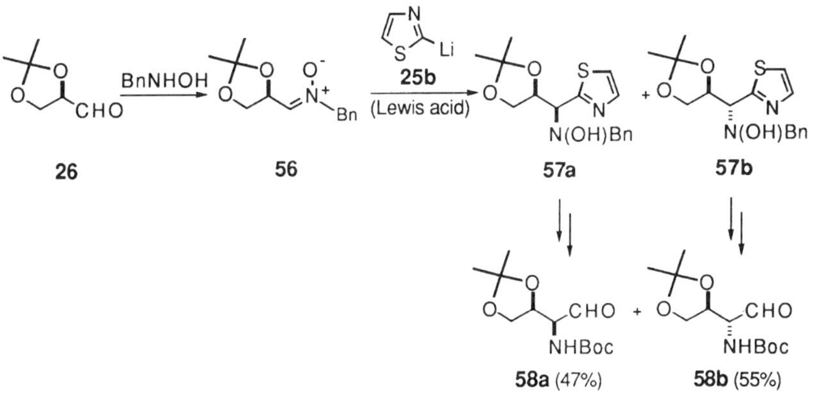

**Scheme 17**

the N-benzyl nitrone **56** derived from **26**, the reaction was successfully carried out with 2-lithiothiazole (2-LTT, **25b**) [57]. The addition of this organometal to **56** was highly stereoselective (ds 92%) giving rise to the *syn*-N-benzylhydroxylamine **57a** as major product. Also the reaction with **56** precomplexed with Lewis acids (Et$_2$AlCl, TiCl$_4$) was highly stereoselective (ds 95%), but in this case the *anti*-epimer **57b** was the main product [58]. The chemical yields of isolated hydroxylamines **57a** and **57b** were 82 and 84%, respectively. The conversion of these compounds to the corresponding α-amino aldehydes **58a** (2-deoxy-2-amino-D-threose) and **58b** (2-deoxy-2-amino-D-erythrose) was carried out by a high-yield reaction sequence involving the reductive dehydroxylation and debenzylation (TiCl$_3$ in MeOH–H$_2$O) of the hydroxylamino to amino group, the protection of the latter as N-Boc derivative, and finally the usual thiazolyl-to-formyl deblocking.

Having set up a protocol for the aminohomologation of various *aldehydo* sugars, the value of the method was tested by the synthesis of simple natural products. The first example involved [57a] the conversion of 2,3:4,5-di-O-isopropylidene-D-arabinose **59** (Scheme 18) through the nitrone **60** into the N-acetyl-D-mannosamine diacetonide **61** and the deprotected compound **11**, both well-known key intermediates for the synthesis of N-acetylneuraminic acid (Neu5Ac) [59,60]. Unfortunately, in this case the addition of 2-LTT (**25b**) to the nitrone **60** occurred with modest selectivity (ds 75% to the best); therefore, the overall yield of **61** was quite low (29%).

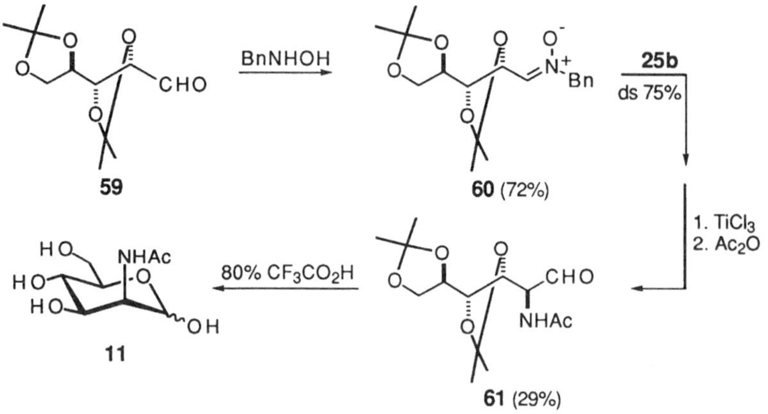

**Scheme 18**

The stereoselective aminohomologation of the dialdose **1** [61] (see Scheme **12**) was employed for the formal synthesis of destomic acid **62** and lincosamine **63**, the sugar constituents of the antibiotic natural products destomycin [62] and lincomycin [63], respectively. This approach was suggested by the observation that both compounds **62** and **63** (Fig. 3) feature a polyhydroxylated carbon chain with the *galacto* configuration that bears an aminomethylene group with *S* and *R* configuration, respectively.

Stereocontrolled addition of 2-LTT (**25b**) to the nitrone **64** (Scheme **19**), derived from

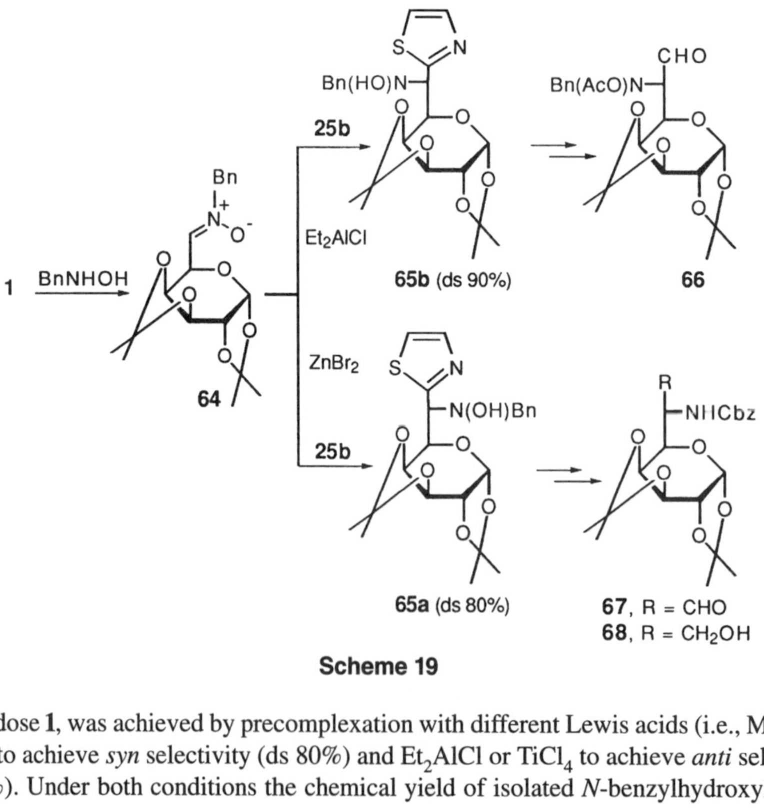

**Scheme 19**

the dialdose **1**, was achieved by precomplexation with different Lewis acids (i.e., MgBr$_2$ or ZnBr$_2$) to achieve *syn* selectivity (ds 80%) and Et$_2$AlCl or TiCl$_4$ to achieve *anti* selectivity (ds 90%). Under both conditions the chemical yield of isolated *N*-benzylhydroxylamines

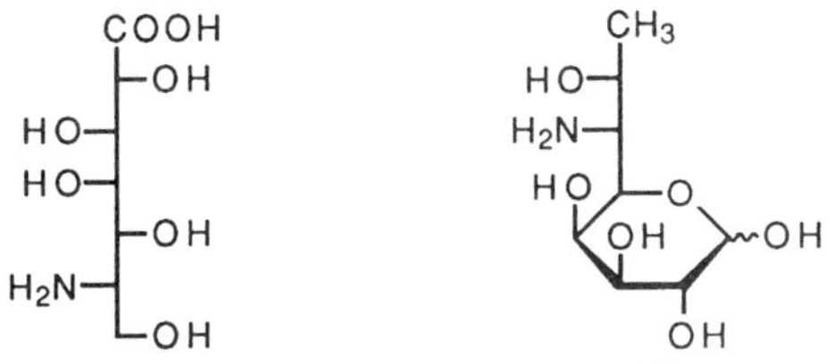

## Thiazole-Based One-Carbon Extension

**Figure 3**

65a and 65b was about 90%. The elaboration of these adducts afforded the diastereo-isomeric galactosyl α-amino aldehydes **66** and **67**. The aldehyde **66** and the amino alcohol **68** obtained by reduction of **67** had been previously prepared by other routes and converted to lincosamine **63** [64] and destomic acid **62** [65], respectively.

The synthesis of the aza sugar D-nojirimycin (**71** Scheme 20) [56], a well-known

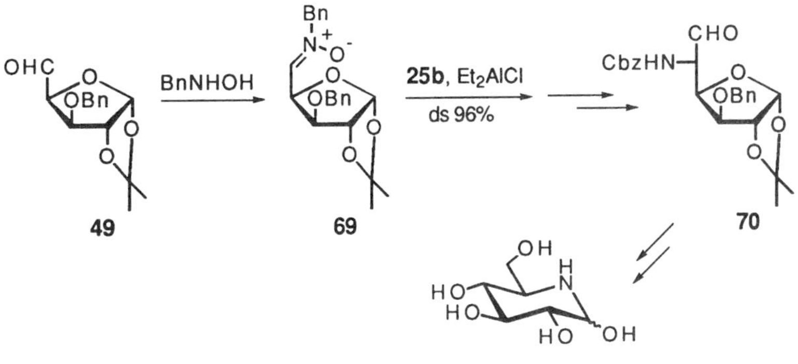

**Scheme 20**

member of a class of sugar analogues that are attracting increasing interest as glycosidase inhibitors and potential antiviral agents [66], completes the illustration of the conspicuous synthetic potential of the aminohomologation strategy.

The glycosyl α-amino aldehyde **70** obtained from the dialdose **49** through the nitrone **69** appeared nicely tailored for a rapid conversion to **71**. This simply involved the reduction of the formyl group and removal of the hydroxyl and amino protective groups. This approach should be extendible to the synthesis of various aza sugars containing suitable structural modifications.

### Homologation of Furanoses and Pyranoses

Because considerable attention has been focused in very recent years on C-formyl glycosides (FGL) (i.e., stereochemically well-defined 2,5- or 2,6-anhydro sugars) as key intermediates to C-glycosyl amino acids and a variety of modified glycoconjugates [67], the one-carbon homologation of sugars in their cyclic form has become an issue of increasing importance. Retrosynthetic analysis (Fig. 4) indicates that the construction of a C-formyl glycoside corresponds essentially to the displacement of the hydroxyl group at the anomeric carbon by a formyl anion synthon. The execution of this seemingly simple operation

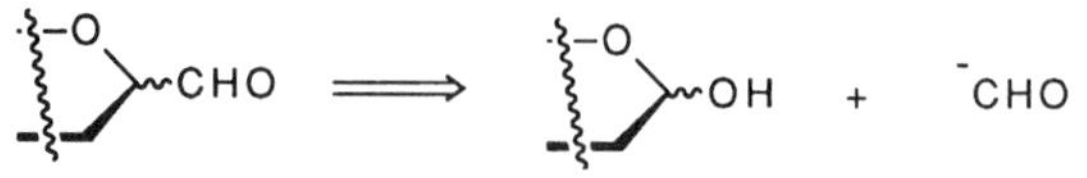

**Figure 4**

through the thiazole–aldehyde synthesis was quite challenging, both to confirm the synthetic value of the method [68] and to overcome the limitations of the very few existing syntheses of *C*-formyl glycosides [69]. Although synthetic methods of various *C*-glycoides have become available in recent years [70], the installation at the anomeric carbon of a simple, yet very important, functionality such as the formyl group was still an open problem.

After unsuccessful attempts of C1(sugar)–C2(thiazole) bond formation by electrophilic glycosylation of thiazole, using a pyranosyl acetate or trichloroacetimidate in the presence of Lewis acids, and by nucleophilic displacement of anomeric nitrate or tosylate by 2-LTT (**25b**), the problem was solved through an efficient, although slightly longer, route involving sugar lactones as activated substrates [71]. The example shown (Scheme 21) illustrates this synthetic method. The addition of **25b** to 2,3,4,6-tetra-*O*-benzyl-D-

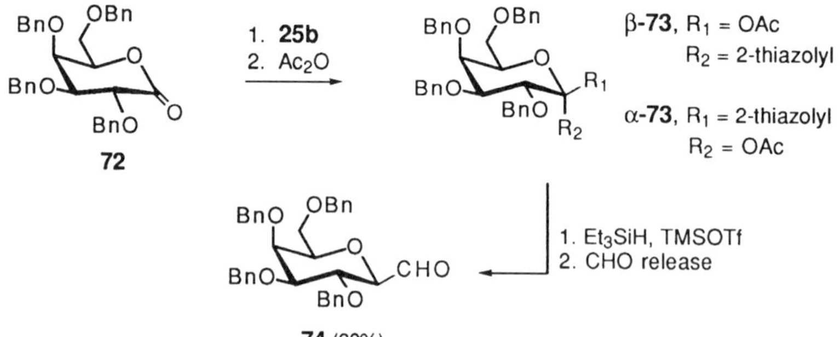

**Scheme 21**

galactopyranolactone (**72**) at $-80°C$, followed by treatment with acetic anhydride at the same temperature, afforded the ketol acetate β-**73** (75% yield). On the other hand, the acetylation at room temperature of the ketol, obtained after an aqueous quenching of the reaction mixture, gave as a major product the ketol acetate α-**73** (78% yield). The different distribution of the α- and β-epimer ketol acetates **73** appeared to reflect the kinetic and thermodynamic control on the configuration of the anomeric carbon. In other words, the stereoselective addition of **25b** to the less-hindered face of the carbonyl of **72** at low temperature should lead to a β-ketol that equilibrates at higher temperature to the more stable α-isomer. The reductive removal of the acetoxy group [72] from either α- or β-**73** or a mixture of them in the presence of Et$_3$SiH and TMSOTf [73] afforded the same β-linked thiazolyl *C*-glycoside which, subjected to the usual thiazolyl-to-formyl deblocking sequence, produced the corresponding *C*-formyl glycoside **74** in 60% isolated yield from the lactone **72**.

This homologation method was successfully applied to other pyrano-(*gluco*, *manno*, 2-deoxy-2-azido-*galacto*) and furano- (*manno*, *ribo*) lactones bearing various hydroxyl protective groups. The addition of 2-LTT (**25b**) always proceeded smoothly to give either the α- or β-epimer ketol acetate, depending on the quenching procedure of the reaction mixture. Similarly, the reduction of the ketol acetates produced the thiazolyl *C*-glycosides, whose configuration was in agreement with the hydride addition to the less hindered side of the pyran or furan oxonium ion. In all cases, the level of diastereoselectivity and chemical yields were good, if not excellent. Hence, a general and efficient reaction method that allows access to C-formyl glycosides bearing various glycosidic units appears to have been established. Subsequent to our first report [71a], a quite similar approach employing 2-lithio-1,3-dithiane (**13**) as formyl anion equivalent [74] has been applied to **72** to give the C-glycosyl aldehyde **74**. A second example dealing with the C-formylation of a mannofuranose derivative was also reported. In both cases the yields were much lower than those obtained by the thiazole–aldehyde synthesis. More promising, particularly for the synthesis of α-linked C-formyl glycosides, is the approach through alkynyl C-glycosides [74,75]. A reaction sequence involving the reduction of the alkynyl group to the alkenyl group and cleavage of the latter by ozonolysis was employed to reveal the formyl group. This procedure may be a serious limitation of the method since common hydroxyl protective groups such as the benzyl group do not tolerate strong oxidizing conditions [76].

The remarkable stability of the thiazole ring allowed synthetic manipulations of the thiazolyl ketol acetates, which extended considerably the scope of the above C-formylation method of furanoses and pyranoses. Instead of the reductive removal of the acetoxy group, the N-glycosidation of either α- or β-anomer **73** with TMSN$_3$ afforded stereoselectively the azido galactopyranoside **75** in 88% isolated yield (Scheme 22) [77]. The cleavage of the

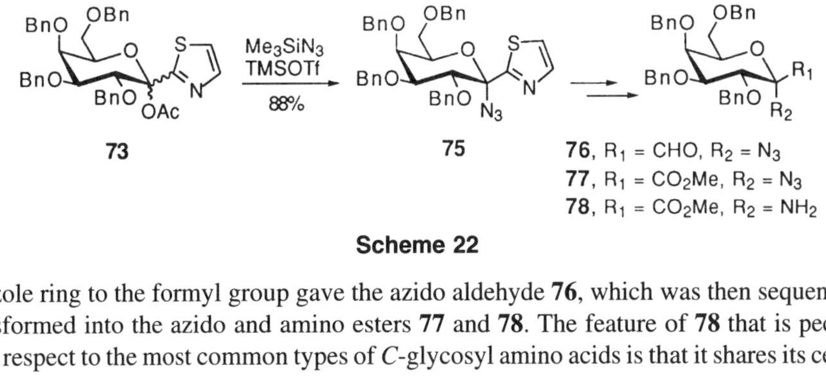

Scheme 22

thiazole ring to the formyl group gave the azido aldehyde **76**, which was then sequentially transformed into the azido and amino esters **77** and **78**. The feature of **78** that is peculiar with respect to the most common types of C-glycosyl amino acids is that it shares its central carbon atom with the sugar moiety.

The same type of reaction sequence was employed for the conversion of the thiazolyl ketol acetate **79** into the azido aldehyde **80** (Scheme 23) [77], a key precursor of the natural product hydantocidin **81** [78].

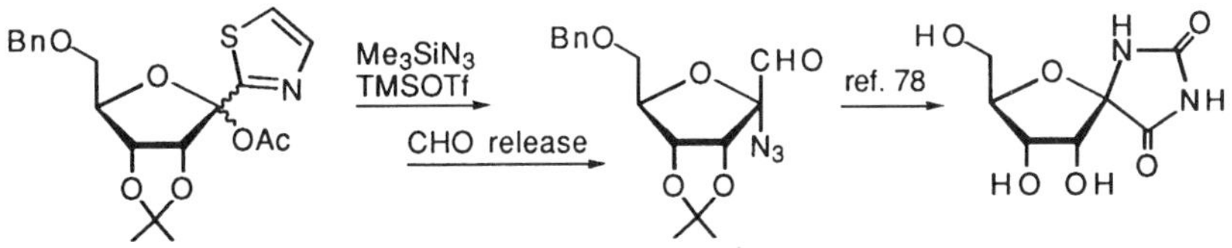

Scheme 23

## C. Conclusions

Owing to the combination of a high chemical stability and easy cleavage, the thiazole ring has emerged as a valuable synthetic auxiliary in different types of one-carbon homologation methods of carbohydrates. 2-Metalated thiazoles serve as efficient formyl anion equivalents. Although only 2-(trimethylsilyl)thiazole (**25a**) and 2-lithiothiazole (**25b**) have been extensively employed, one can foresee the preparation of other members of this class of reagents and their application in stereocontrolled homologation methodologies.

## III. EXPERIMENTAL PROCEDURES

### A. Addition of 2-(Trimethylsilyl)thiazole to Aldehydes [44]

$$\mathbf{30} \xrightarrow[\text{THF, -30 °C to r. t.}]{\mathbf{25a}} \mathbf{82} \quad 72\% \tag{1}$$

To a cooled ($-30°C$) and stirred solution of aldehyde [Eq. (1)] **30** [40] (1.25 g, 5.0 mmol) in dry THF (25 mL) was added dropwise a solution of 2-(trimethylsilyl)thiazole (1.18 g, 7.5 mmol) in dry THF (15 mL). Stirring was continued at room temperature overnight, then the solvent was removed under vacuum. The residue was dissolved in THF (40 mL) and treated at room temperature with a 1-$M$ solution of Bu$_4$NF in THF (7.5 mL, 7.5 mmol). After 2 h the solution was concentrated, the residue was dissolved in CH$_2$Cl$_2$ (40 mL), and washed with H$_2$O (2 × 20 mL). The organic phase was dried (Na$_2$SO$_4$) and concentrated. The residue was purified by column chromatography on silica gel (3:2 diethyl ether-hexanes) to give **82** (1.21 g, 72%) as a white solid: mp 105°–107°C (from hexane–diethyl ether), [α]$_D$ +10.7° ($c$ 0.2, CHCl$_3$).

### B. Inversion of Configuration of the Hydroxyl Group [44]

$$\mathbf{82} \xrightarrow[88\%]{\text{Ac}_2\text{O-DMSO}} \xrightarrow[\text{THF, -78 °C} \atop 90\%]{\text{K-Selectride}} \mathbf{83} \tag{2}$$

A solution of **82** [Eq. (2)] (0.67 g, 2.0 mmol) in DMSO (9.0 mL) and acetic anhydride (4.0 mL) was kept at room temperature overnight, then was diluted with saturated aqueous NaHCO$_3$ (100 mL). The mixture was stirred for 30 min, then extracted with Et$_2$O (50 mL). The organic phase was washed with H$_2$O (2 × 20 mL), dried (Na$_2$SO$_4$), and concentrated. The residue was purified by column chromatography on silica gel (3:2 hexanes–diethyl ether) to give the ketone (0.59 g, 88%) as an oil: [α]$_D$ +13.9° ($c$ 0.3, CHCl$_3$).

To a cooled ($-78°C$) and stirred solution of the ketone (0.59 g, 1.8 mmol) in dry THF (90 mL) was added dropwise potassium tri-$sec$-butylborohydride (K-Selectride, 3.6 mL, 3.6 mmol, of a 1-$M$ solution in THF). The mixture was stirred at $-78°C$ for an additional hour, then was treated with H$_2$O (7 mL), warmed to room temperature and concentrated. The residue was suspended in CH$_2$Cl$_2$ (100 mL), washed with brine, dried (Na$_2$SO$_4$), and concentrated. The crude product was eluted from a column of silica gel with 3:2 diethyl ether–hexanes to afford **83** (0.53 g, 90%) as an oil: [α]$_D$ +19.5° ($c$ 1.8, CHCl$_3$).

## C. Preparation of Polyalkyoxyaldehydes [44]

$$\text{82} \xrightarrow[\substack{\text{THF, r. t.} \\ 90\%}]{\text{NaH, BnBr}} \xrightarrow[\substack{80\%}]{\substack{1.\ \text{MeI} \\ 2.\ \text{NaBH}_4 \\ 3.\ \text{HgCl}_2,\ \text{H}_2\text{O}}} \text{35} \quad (3)$$

A mixture of alcohol **82** (1.34 g, 4.0 mmol) [Eq. (3)], sodium hydride (0.18 g, 4.4 mol, of a 60% dispersion in mineral oil), and dry THF (50 mL) was refluxed for 20 min, then cooled to room temperature and treated with Bu$_4$NI (0.15 g, 0.4 mmol) and benzyl bromide (520 μL, 4.4 mmol). The mixture was stirred at room temperature overnight, then concentrated, diluted with saturated aqueous NaHCO$_3$ (40 mL), and extracted with CH$_2$Cl$_2$ (2 × 50 mL). The combined extracts were dried (Na$_2$SO$_4$) and concentrated. The crude product was eluted from a column of silica gel with 3:2 hexanes–diethyl ether to afford the benzyl ether (1.53 g, 90%) as an oil: $[\alpha]_D$ +75.0° (c 1.4, CHCl$_3$).

A solution of the benzyl ether (1.53 g, 3.6 mmol) and iodomethane (2.2 mL, 36 mmol) in acetonitrile (30 mL) was refluxed overnight, then cooled to room temperature, and concentrated. The residue was treated with Et$_2$O to precipitate the *N*-methylthiazolium salt, which was collected by filtration. To a stirred solution of the salt in MeOH (30 mL) was added portionwise NaBH$_4$ (0.27 g, 7.2 mmol). The mixture was stirred at room temperature for an additional 30 min, then concentrated, diluted with brine (30 mL), and extracted with CH$_2$Cl$_2$ (2 × 40 mL). The combined extracts were dried (Na$_2$SO$_4$) and concentrated. To a solution of the thiazolidine in 5:1 CH$_3$CN–H$_2$O (50 mL) was added HgCl$_2$ (0.98 g, 3.6 mmol). The mixture was stirred at room temperature for 20 min and then filtered through Celite. Acetonitrile was evaporated, the residue was suspended in CH$_2$Cl$_2$ (100 mL) and washed with 2% aqueous KI (2 × 20 mL). The organic layer was dried (Na$_2$SO$_4$) and concentrated to give the crude aldehyde, which was eluted from a short column of silica gel with 3:2 hexanes–diethyl ether to afford **35** (1.07 g, 80%) as an oil: $[\alpha]_D$ +45.1° (c 1.2, CHCl$_3$); reported [79] $[\alpha]_D$ +46.3° (c 0.7, MeOH).

## D. One-Carbon Homologation of Dialdoses [49]

$$\text{49} \xrightarrow[\substack{\text{CH}_2\text{Cl}_2,\ \text{r. t.} \\ 80\%}]{\substack{\text{25a} \\ \text{SiMe}_3}} \xrightarrow[\substack{\text{THF, r. t.}}]{\text{NaH, BnBr}} \xrightarrow[\substack{74\%}]{\substack{1.\ \text{MeI} \\ 2.\ \text{NaBH}_4 \\ 3.\ \text{HgCl}_2,\ \text{H}_2\text{O}}} \text{50} \quad (4)$$

To a stirred solution of aldehyde **49** [Eq. (4)] [80] (0.70 g, 2.5 mmol) in dry CH$_2$Cl$_2$ (20 mL) was added dropwise a solution of 2-(trimethylsilyl)thiazole (0.59 g, 3.7 mmol) in dry CH$_2$Cl$_2$ (10 mL). The solution was kept at room temperature overnight, then the solvent was removed under vacuum. The residue was dissolved in THF (30 mL) and treated at room temperature with a 1-*M* solution of Bu$_4$NF in THF (2.5 mL, 2.5 mmol). After 2 h the solution was concentrated, the residue was dissolved in CH$_2$Cl$_2$ (40 mL), and washed with H$_2$O (2 × 20 mL). The organic phase was dried (Na$_2$SO$_4$) and concentrated. The residue was purified by column chromatography on silica gel (1:1 petroleum ether–ethyl acetate) to give the thiazolyl adduct (0.73 g, 80%) as a syrup.

A mixture of the alcohol (0.73 g, 2.0 mmol), sodium hydride (88 mg, 2.2 mmol, of a 60% dispersion in mineral oil), and dry THF (25 mL) was refluxed for 20 min, then cooled to room temperature and treated with $Bu_4NI$ (74 mg, 0.2 mmol) and benzyl bromide (260 μL, 2.2 mmol). The mixture was stirred at room temperature overnight, then concentrated, diluted with saturated aqueous $NaHCO_3$ (20 mL), and extracted with $CH_2Cl_2$ (2 × 30 mL). The combined extracts were dried ($Na_2SO_4$) and concentrated. The crude product was eluted from a column of silica gel with 7:3 petroleum ether–ethyl acetate to afford the benzyl ether as a syrup.

A solution of the benzyl ether and iodomethane (1.2 mL, 20 mmol) in acetonitrile (20 mL) was refluxed overnight, then cooled to room temperature and concentrated. To a cooled (−10°) and stirred solution of the N-methylthiazolium salt in MeOH (20 mL) was added portionwise $NaBH_4$ (0.15 g, 4.0 mmol). The mixture was stirred at room temperature for an additional 30 min, then diluted with acetone, and concentrated. The residue was diluted with brine (20 mL) and extracted with AcOEt (2 × 30 mL). The combined extracts were dried ($Na_2SO_4$) and concentrated. To a solution of the thiazolidine in 5:1 $CH_3CN–H_2O$ (20 mL) was added $HgCl_2$ (0.54 g, 2.0 mmol). The mixture was stirred at room temperature for 15 min and then filtered through Celite. Acetonitrile was evaporated, the residue was suspended in $CH_2Cl_2$ (50 mL), and washed with 20% aqueous KI (2 × 10 mL). The organic layer was dried ($Na_2SO_4$) and concentrated to give almost pure **50** (0.46 g, 74%) as an oil.

### E. Synthesis of N-Benzyl Nitrones [54]

$$26 \xrightarrow[\text{CH}_2\text{Cl}_2, \text{r.t.} \quad 86\%]{\text{BnNHOH}} 56 \tag{5}$$

A mixture of aldehyde **26** [Eq. (5)] [81] (2.60 g, 20.0 mmol), N-benzylhydroxylamine [82] (2.46 g, 20.0 mmol), anhydrous $MgSO_4$ (2.41 g, 20.0 mmol) and $CH_2Cl_2$ (150 mL) was stirred at room temperature for 4 h, then filtered through Celite and concentrated. The residue was eluted from a column of silica gel with 8:1 ethyl acetate–hexanes to give **56** (4.04 g, 86%) as a solid: mp 88°C, $[\alpha]_D$ +96.7° (c 0.5, $CHCl_3$).

### F. Addition of 2-Lithiothiazole to Nitrones [56]

$$56 + 25b \xrightarrow[\text{Et}_2\text{O-THF, -80 °C} \quad 74\%]{} 57a \tag{6}$$

To a cooled (−80°C) and stirred solution of nBuLi (10.0 mL, 16.0 mmol, of a 1.6-M solution in hexanes) in dry $Et_2O$ (30 mL) was added dropwise a solution of freshly distilled 2-bromothiazole (2.46 g, 15.0 mmol) in dry $Et_2O$ (15 mL), over a 30-min period (the temperature of the solution was not allowed to rise above −70°C). The yellow solution was stirred at −80°C for 20 min, then cooled to −90°C. A solution of nitrone **56** [Eq.(6)] (1.17 g, 5.0 mmol) in dry THF (60 mL) was added slowly to keep the temperature of the reaction mixture below −80°C. After an addition 15 min at −80°C, the mixture was quenched with saturated aqueous NH$_4$Cl (15 mL), stirred at room temperature for 10 min, and diluted with Et$_2$O (25 mL). The layers were separated and the aqueous layer extracted with Et$_2$O (3 × 50 mL). The combined organic extracts were washed with brine, dried (MgSO$_4$), and concentrated. The residue was eluted from a column of silica gel with 7:3 hexanes–diethyl ether to give **57a** (1.18 g, 74%) as an oil: [α]$_D$ −7.8° (c 0.7, CHCl$_3$).

## G. Addition of 2-Lithiothiazole to Lewis Acid Nitrone Complexes [56]

To a cooled (−80°C) and stirred solution of nBuLi (10.0 mL, 16.0 mmol, of a 1.6-M solution in hexanes) in dry Et$_2$O (30 mL) was added dropwise a solution of freshly distilled 2-bromothiazole (2.46 g, 15.0 mmol) in dry Et$_2$O (15 mL), over a 30-min period (the temperature of the solution was not allowed to rise above −70°C). The yellow solution was stirred at −80°C for 20 min, then cooled to −90°C. To a stirred solution of nitrone **56** [Eq. (7)] (1.17 g, 5.0 mmol) in dry Et$_2$O (100 mL) was added Et$_2$AlCl (5.0 mL, 5.0 mmol, of a 1-M solution in hexanes) in one portion at room temperature and stirring was continued for 15 min. The mixture was transferred under argon atmosphere into a dropping funnel and added slowly to the solution of 2-lithiothiazole to keep the temperature of the reaction mixture below −80°C. After an additional 30 min at −80°C, the mixture was quenched with 1-N aqueous NaOH (100 mL), stirred at room temperature for 10 min, and extracted with Et$_2$O (3 × 50 mL). The combined organic extracts were washed with brine, dried (MgSO$_4$), and concentrated. The residue was eluted from a column of silica gel with 7:3 hexanes–diethyl ether to give **57b** (1.30 g, 81%) as a white solid: mp 157–159°C, [α]$_D$ −9.0° (c 0.4, CHCl$_3$).

## H. Reduction of N-Benzyl Hydroxylamines [56]

A solution of hydroxylamine **57a** [Eq. (8)] (1.28 g, 4.0 mmol) in MeOH (50 mL) was treated with 20% w/w aqueous solution of TiCl$_3$ (1.55 g, 10.0 mmol, in 6.2 mL of H$_2$O) at room temperature for 15 min, then 5-M of aqueous NaOH was added, and stirring was continued for an additional 5 min. The mixture was extracted with ethyl acetate (4 × 25 mL), the combined extracts were washed with brine, dried (MgSO$_4$), and concentrated to afford the crude amine, which was used in the next step without purification. A solution of the crude amine and di-*tert*-butyl dicarbonate (1.53 g, 7.0 mmol) in 1,4-dioxane (30 mL) was kept at room temperature for 12 h, then saturated aqueous $NaHCO_3$ (80 mL) and $CH_2Cl_2$ (50 mL) were added. The organic layer was separated and the aqueous layer was extracted with $CH_2Cl_2$ (2 × 25 mL). The combined extracts were dried ($MgSO_4$) and concentrated. The residue was purified by column chromatography on silica gel (6:4 hexanes–diethyl ether) to give **84** (0.93 g, 74%) as a white solid: mp 75–76°C, $[\alpha]_D$ −18.6° (*c* 0.9, $CHCl_3$).

## I. Preparation of α-Aminoaldehydes [56]

$$\mathbf{84} \xrightarrow[64\%]{\text{1. MeOTf} \atop \text{2. NaBH}_4 \atop \text{3. CuCl}_2, \text{H}_2\text{O}} \mathbf{58a} \tag{9}$$

A mixture of **84** [Eq. (9)] (0.94 g, 3.0 mmol), activated 4-Å powdered molecular sieves (6.0 g), and dry $CH_3CN$ (50 mL) was stirred at room temperature for 10 min, and then methyl triflate (373 μL, 3.3 mmol) was added. The suspension was stirred for 15 min and then concentrated to dryness. The crude *N*-methylthiazolium salt was suspended in MeOH (50 mL), cooled to 0°C, and treated with $NaBH_4$ (252 mg, 6.6 mmol). The mixture was stirred at room temperature for an additional 10 min, diluted with acetone, filtered through Celite, and concentrated. To the solution of the crude thiazolidine in 10:1 $CH_3CN$-$H_2O$ (50 mL) was added CuO (0.72 g, 9.0 mmol) and $CuCl_2 \cdot 2H_2O$ (0.56 g, 3.3 mmol). The mixture was stirred at room temperature for 10 min and then filtered through Celite. Acetonitrile was evaporated (bath temperature not exceeding 30°C), the residue was diluted with $Et_2O$ (80 mL), and washed with brine (80 mL). The aqueous layer was extracted with $Et_2O$ (2 × 50 mL), the combined extracts were washed with saturated aqueous EDTA (disodium salt) and brine, dried ($MgSO_4$), and filtered through a short pad of Florisil (100–200 mesh) to afford **58a** (0.50 g, 64%) as an oil: $[\alpha]_D$ +8.1° (*c* 0.5, $CHCl_3$).

## J. Aminohomologation of Dialdoses [56]

$$\mathbf{49} \xrightarrow[\substack{CH_2Cl_2, \text{r.t.} \\ 86\%}]{BnNHOH} \mathbf{69} \xrightarrow[\substack{Et_2AlCl \\ Et_2O, -80\,°C \\ 86\%}]{\mathbf{25b}, \text{1. TiCl}_3 \atop \text{2. BnOCOCl} \atop 81\%} \xrightarrow[\substack{\text{1. MeOTf} \\ \text{2. NaBH}_4 \\ \text{3. CuCl}_2, H_2O \\ 76\%}]{} \mathbf{70} \tag{10}$$

A mixture of aldehyde **49** [Eq. (10)] [80] (1.39 g, 5.0 mmol), *N*-benzylhydroxylamine [82] (0.61 g, 5.0 mmol), anhydrous $MgSO_4$ (0.60 g, 5.0 mmol), and $CH_2Cl_2$ (40 mL) was stirred at room temperature for 4 h, then filtered through Celite, and concentrated. The residue was eluted from a column of silica gel with 4:1 diethyl ether–hexanes to give the *N*-benzyl nitrone **69** (1.65 g, 86%) as a solid: mp 96°C, $[\alpha]_D$ −135.7° (*c* 0.4, $CHCl_3$).

To a cooled ($-80°C$) and stirred solution of nBuLi (8.7 mL, 13.8 mmol, of a 1.6-$M$ solution in hexanes) in dry $Et_2O$ (30 mL) was added dropwise a solution of freshly distilled 2-bromothiazole (2.11 g, 12.9 mmol) in dry $Et_2O$ (15 mL) over a 30-min period (the temperature of the solution was not allowed to rise above $-70°C$). The yellow solution was stirred at $-80°C$ for 20 min, then cooled to $-90°C$. To a stirred solution of nitrone **69** (1.65 g, 4.3 mmol) in dry $Et_2O$ (100 mL) was added $Et_2AlCl$ (4.3 mL, 4.3 mmol, of a 1-$M$ solution in hexanes) in one portion at room temperature and stirring was continued for 15 min. The mixture was transferred under argon atmosphere into a dropping funnel and slowly added to the solution of 2-lithiothiazole to keep the temperature of the reaction mixture below $-80°C$. After an additional 30 min at $-80°C$, the mixture was quenched with 1-$N$ aqueous NaOH (100 mL), stirred at room temperature for 10 min, and extracted with $Et_2O$ (3 × 50 mL). The combined organic extracts were washed with brine, dried ($MgSO_4$), and concentrated. The residue was eluted from a column of silica gel with 6:4 hexanes–diethyl ether to give the hydroxylamine (1.73 g, 86%) as an oil: $[\alpha]_D$ $-26.5°$ ($c$ 0.5, $CHCl_3$).

A solution of the hydroxylamine (1.73 g, 3.7 mmol) in MeOH (50 mL) was treated with a 20% w/w aqueous solution of $TiCl_3$ (1.43 g, 9.2 mmol, in 5.7 mL of $H_2O$) at room temperature for 15 min, then 5-M aqueous NaOH was added, and stirring was continued for an additional 5 min. The mixture was extracted with ethyl acetate (4 × 25 mL), the combined extracts were washed with brine, dried ($MgSO_4$), and concentrated to afford the crude amine, which was used in the next step without purification. To a cooled (0°C) and stirred mixture of the crude amine and 7% aqueous $NaHCO_3$ (20 mL) in 1,4-dioxane (50 mL) was added benzyl chloroformate (0.57 mL, 4.0 mmol). Stirring was continued at 0°C for 20 min, then $H_2O$ (80 mL) and $CH_2Cl_2$ (50 mL) were added. The organic layer was separated, and the aqueous layer was extracted with $CH_2Cl_2$ (2 × 25 mL). The combined extracts were dried ($MgSO_4$) and concentrated. The residue was purified by column chromatography on silica gel (6:4 hexanes–diethyl ether) to give the $N$-benzyloxycarbonyl derivative (1.49 g, 81%) as a syrup: $[\alpha]_D$ $-4.9°$ ($c$ 0.8, $CHCl_3$).

A mixture of the $N$-benzyloxycarbonyl derivative (1.49 g, 3.0 mmol), activated 4-Å powdered molecular sieves (6.0 g), and dry $CH_3CN$ (50 mL) was stirred at room temperature for 10 min, and then methyl triflate (373 μL, 3.3 mmol) was added. The suspension was stirred for 15 min and then concentrated to dryness. The crude $N$-methylthiazolium salt was suspended in MeOH (50 mL), cooled to 0°C, and treated with $NaBH_4$ (252 mg, 6.6 mmol). The mixture was stirred at room temperature for an additional 10 min, diluted with acetone, filtered through Celite, and concentrated. To the solution of the crude thiazolidine in 10:1 $CH_3CN-H_2O$ (50 mL) was added CuO (0.72 g, 9.0 mmol) and $CuCl_2·2H_2O$ (0.56 g, 3.3 mmol). The mixture was stirred at room temperature for 10 min and then filtered through Celite. Acetonitrile was evaporated (bath temperature not exceeding 30°C) and the residue was diluted with $Et_2O$ (80 mL) and washed with brine (80 mL). The aqueous layer was extracted with $Et_2O$ (2 × 50 mL), the combined extracts were washed with saturated aqueous EDTA (disodium salt) and brine, dried ($MgSO_4$), and filtered through a short pad of Florisil (100–200 mesh) to afford **70** (1.01 g, 76%) as an oil: $[\alpha]_D$ $-26.1°$ ($c$ 3, $CHCl_3$).

### K. Addition of 2-Lithiothiazole to Sugar Lactones [71]

To a cooled ($-80°C$) and stirred solution of nBuLi (7.6 mL, 12.1 mmol, of a 1.6-$M$ solution in hexanes) in dry $Et_2O$ (17 mL) a solution of freshly distilled 2-bromothiazole (1.80 g, 11.1 mmol) in dry $Et_2O$ (4.3 mL) was added dropwise over a 30-min period (the temperature

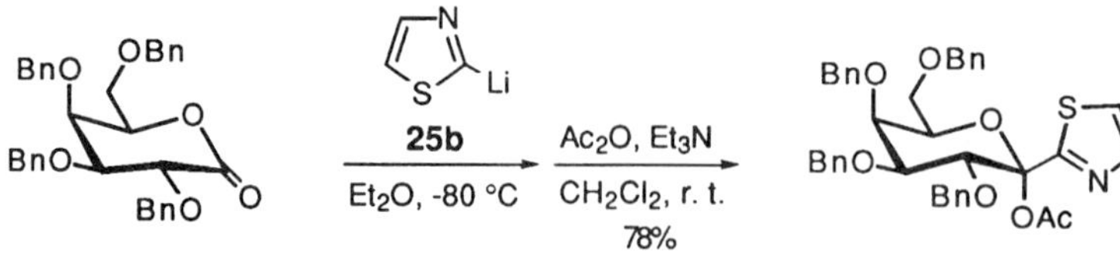

[Scheme, Eq. (11): 72 + 25b (thiazolyl-Li), Et₂O, −80 °C; then Ac₂O, Et₃N, CH₂Cl₂, r.t., 78% → α-73]

of the solution was not allowed to rise above −70°C). After the yellow solution had been stirred at −80°C for 20 min, a solution of 2,3,4,6-tetra-*O*-benzyl-D-galactonolactone **72** [Eq. (11)] [71] (5.00 g, 9.29 mmol) in dry THF (17 mL) was added slowly (25 min). After an additional 20 min, the mixture was allowed to warm to −65°C in 30 min and poured into 200 mL of a 1-*M* phosphate buffer (pH = 7). The layers were separated, and the aqueous layer was extracted with CH₂Cl₂ (2 × 100 mL). The combined organic layers were dried (Na₂SO₄) and concentrated. Flash chromatography (5:2 petroleum ether–ethyl acetate) of the residue gave 4.50 g (78%) of thiazolyl ketose as a syrup: $[\alpha]_D$ +6.4° (*c* 1, CHCl₃).

To a solution of thiazolyl ketose (4.50 g, 7.21 mmol) in CH₂Cl₂ (17 mL) were added at room temperature Et₃N (7 mL) and Ac₂O (7 mL). After standing overnight at room temperature the solution was concentrated to give 4.72 g (100%) of α-**73** as a syrup: $[\alpha]_D$ +34.7° (*c* 1, CHCl₃).

## L. Deoxygenation of Thiazolyl Ketol Acetates [71]

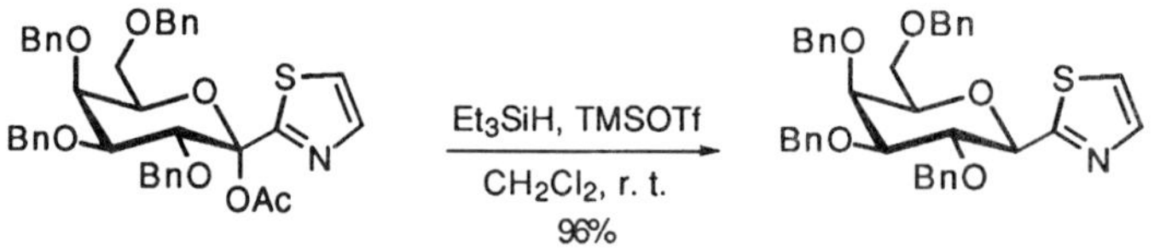

[Scheme, Eq. (12): α-73 → 85 with Et₃SiH, TMSOTf, CH₂Cl₂, r.t., 96%]

To a stirred mixture of α-**73** [Eq. (12)] (1.00 g, 1.5 mmol), activated 4-Å powdered molecular sieves (1.0 g), dry CH₂Cl₂ (12 mL), and triethylsilane (2.4 mL, 15.0 mmol) was added dropwise trimethylsilyl triflate (775 μL, 4.2 mmol). The mixture was stirred at room temperature for 30 min and then neutralized with Et₃N, diluted with CH₂Cl₂, filtered through Celite, and concentrated. The residue was eluted from a column of silica gel with 5:2 petroleum ether–ethyl acetate to give **85** (0.87 g, 96%) as a syrup: $[\alpha]_D$ 0°, $[\alpha]_{436}$ −3.9° (*c* 1, CHCl₃).

## M. Preparation of *C*-Formyl Glycosides [71]

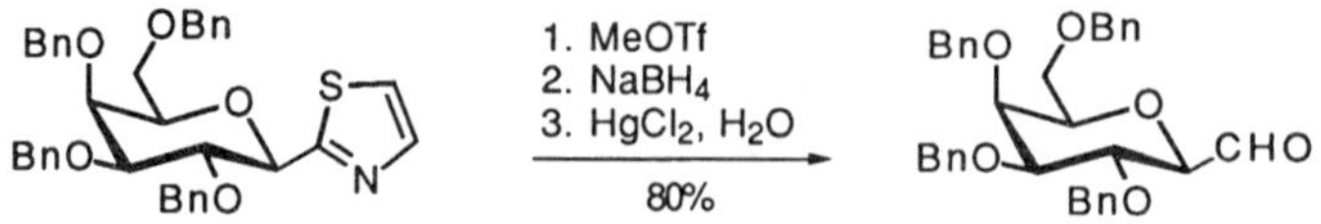

[Scheme, Eq. (13): 85 → 74 via 1. MeOTf; 2. NaBH₄; 3. HgCl₂, H₂O, 80%]

A mixture of **85** [Eq. (13)] (0.92 g, 1.5 mmol), activated 4-Å powdered molecular sieves (3.0 g), and dry CH₃CN (14 mL) was stirred at room temperature for 10 min, and then methyl triflate (223 μL, 2.0 mmol) was added. The suspension was stirred for 15 min and then concentrated to dryness. The crude *N*-methylthiazolium salt was suspended in MeOH

(14 mL), cooled to 0°C, and treated with NaBH$_4$ (126 mg, 3.3 mmol). The mixture was stirred at room temperature for an additional 10 min, diluted with acetone, filtered through Celite, and concentrated. To the solution of the crude thiazolidine in 10:1 CH$_3$CN–H$_2$O (14 mL) was added HgCl$_2$ (0.41 g, 1.5 mmol). The mixture was stirred for 15 min and then filtered through Celite. Acetonitrile was evaporated (bath temperature not exceeding 40°C). The residue was suspended in CH$_2$Cl$_2$ (100 mL) and washed with 20% aqueous KI (3 × 20 mL) and water (20 mL); the organic layer was dried (Na$_2$SO$_4$) and concentrated to give a syrup that was diluted with 40 mL of Et$_2$O and filtered through a short pad of Florisil (100–200 mesh) to afford a colorless solution. After a further washing of Florisil with AcOEt (15 mL), the organic phase was concentrated to yield 0.67 g (80%) of almost pure **74** (NMR analysis) as a syrup. The crude aldehyde decomposed partially when submitted to silica gel chromatography; nonetheless, this product was suitable for further elaborations.

### N. *N*-Glycosidation of Thiazolyl Ketol Acetates [77]

A mixture of α- or β-**73** [Eq. (14)] (1.00 g, 1.5 mmol), activated 4-Å powdered molecular sieves (0.6 g), trimethylsilyl azide (300 μL, 2.2 mmol), and dry CH$_2$Cl$_2$ (6 mL) was stirred at room temperature for 10 min, and then trimethylsilyl triflate (136 μL, 0.75 mmol) was added. The mixture was stirred at room temperature for 20 min, then neutralized with Et$_3$N, diluted with CH$_2$Cl$_2$, filtered through Celite, and concentrated. The residue was eluted from a column of silica gel with 6:1 cyclohexane–ethyl acetate to give **75** (0.86 g, 88%) as a syrup: [α]$_D$ +33.7° (*c* 0.7, CHCl$_3$).

### O. Preparation of Glycosyl α-Azidoesters [77]

A mixture of **75** [Eq. (15)] (0.70 g, 1.08 mmol), activated 4-Å powdered molecular sieves (1.0 g), and dry CH$_3$CN (3.6 mL) was stirred at room temperature for 10 min, then methyl triflate (159 μL, 1.40 mmol) was added. The suspension was stirred for 15 min and then concentrated to dryness (bath temperature not exceeding 40°C). To the crude *N*-methylthiazolium salt was added cold (0°C) MeOH (3 mL), and then NaBH$_4$ (90 mg, 2.37 mmol). The mixture was stirred at room temperature for an additional 10 min, diluted with acetone (8 mL), filtered through Celite, and concentrated. To a solution of the crude thiazolidine in 10:1 CH$_3$CN–H$_2$O (8 mL) was added HgCl$_2$ (176 mg, 0.65 mmol). The mixture was stirred for 15 min and then filtered through Celite. Acetonitrile was evaporated (bath temperature not exceeding 40°C) to afford a residue that was suspended in CH$_2$Cl$_2$ (20 mL) and washed with 20% aqueous KI (3 × 10 mL) and water (10 mL); the organic layer was dried ($Na_2SO_4$) and concentrated to give 0.47 g of crude azido aldehyde, which was used in the next step without further purification.

To a vigorously stirred mixture of silver nitrate (367 mg, 2.16 mmol), NaOH (173 mg, 4.32 mmol), and water (10 mL) was added a solution of the crude aldehyde (0.47 g) in freshly distilled THF (20 mL). Stirring was continued for 36 h at room temperature, then acetic acid was added up to pH 5, and the mixture was filtered through Celite. The solution was concentrated and the residue was dissolved in 1:1 MeOH–$Et_2O$ (20 mL), treated with an etheral solution of diazomethane at 0°C for 20 min, then concentrated. Flash chromatography (5:2 cyclohexane–ethyl acetate) of the residue afforded **77** (0.36 g, 54% from **75**) as a syrup: $[\alpha]_D$ +34.9° (c 1.6, $CHCl_3$).

### P. Preparation of Glycosyl α-Aminoesters [77]

$$\text{77} \xrightarrow[\text{t-BuOH-H}_2\text{O, r. t.}]{\text{H}_2, \text{Pd/C}} \text{78} \quad (16)$$

(77: tetra-O-benzyl glycosyl azide methyl ester; 78: tetra-O-benzyl glycosyl amine methyl ester; 62%)

A vigorously stirred mixture of **77** [Eq. (16)] (300 mg, 0.48 mmol) and 10% palladium on activated carbon (50 mg) in 9:1 $t$BuOH–$H_2O$ (10 mL) was degased under vacuum and saturated with hydrogen (by a $H_2$-filled balloon) three times. The suspension was stirred for an additional hour at room temperature under a slightly positive pressure of $H_2$ (balloon), then filtered through a plug of cotton, and concentrated. Flash chromatography (5:2 cyclohexane–ethyl acetate) of the residue gave first unreacted **77** (36 mg, 12%). Eluted second was **78** (178 mg, 62%) as a syrup: $[\alpha]_D$ +26.0° (c 0.6, $CHCl_3$).

## ACKNOWLEDGMENT

This review collects a part of the results of work on the thiazole–aldehyde synthesis that has been carried out over several years by various individuals whose names are reported in the references. One of us (A.D.) would like to express to all of them his gratitude and appreciation for their contribution. Financial support came from the Progetto Finalizzato Chimica Fine e Secondaria n. 2 (Rome), the Consiglio Nazionale delle Ricerche (CNR, Rome), the Ministero della Università e della Ricerca Scientifica (MURST, Rome). One of us (A. M.) is grateful to the University of Ferrara for having been appointed as Lecturer in Organic Chemistry at the Faculty of Engineering.

## NOTES AND REFERENCES

1. For reviews see:
   P. M. Collins and R. J. Ferrier, *Monosaccharides, Their Chemistry and Their Roles in Natural Products*, Wiley, Chichester, 1995.
   S. David, *Chimie Moléculaire et Supramoléculaire des Sucres*, InterEditions-CNRS Editions, Paris, 1995.
   A. Giannis, The sialyl Lewis[x] group and its analogues as ligands for selectins: Chemoenzymatic syntheses and biological functions, *Angew. Chem. Int. Ed. Engl. 33*:178 (1994).

C. Unverzagt, Got the flu? Try a designer agent derived from a sugar, *Angew. Chem Int. Ed. Engl. 32*:1691 (1993).

A. Kobata, Glycobiology: An expanding research area in carbohydrate chemistry, *Acc. Chem. Res. 26*:319 (1993).

T. Feizi and R. A. Childs, Carbohydrates as antigenic determinants of glycoproteins, *Biochem. J. 245*:1 (1987).

M. Höök, L. Kjellén, S. Johansson, and J. Robinson, Cell-surface glycosaminoglycans, *Annu. Rev. Biochem. 53*:847 (1984).

J. F. Kennedy and C. A. White, *Bioactive Carbohydrates in Chemistry, Biochemistry and Biology*, Ellis Horwood, Chichester, 1983.

R. Schauer, Chemistry, metabolism and biological functions of sialic acids, *Adv. Carbohydr. Chem. Biochem. 40*:131 (1982).

W. Reutter, E. Köttgen, C. Bauer, and W. Gerok, Biological significance of sialic acid, in *Sialic Acids. Chemistry, Metabolism, and Function*, R. Schauer, ed., Springer-Verlag, Vienna, 1982, p. 263.

S. Hakomori, Glycosphingolipids in cellular interaction, differentiation, and oncogenesis, *Annu. Rev. Biochem. 50*:733 (1981).

J. Montreuil, Primary structure of glycoprotein glycans: Basis for the molecular biology of glycoproteins, *Adv. Carbohydr. Chem. Biochem, 37*:157 (1980).

W. Pigman and D. Horton, eds., *The Carbohydrates. Chemistry and Biochemistry*, Academic Press, New York, 4 volumes, 1970–1980.

N. Sharon, *Complex Carbohydrates, Their Biochemistry, Biosynthesis, and Functions*, Addison Wesley, Reading, 1975.

See also:

M. L. Phillips, E. Nudelman, F. C. A. Gaeta, M. Perez, A. K. Singhal, S. Hakomori, and J. C. Paulson, ELAM-1 mediates cell adhesion by recognition of a carbohydrate ligand, sialyl-Le[x], *Science 250*:1130 (1990).

G. Waltz, A. Arufto, W. Kolanus, M. Bevilacqua, and B. Seed, Recognition by ELAM-1 of the sialyl-Le[x] determinant on myeloid and tumor cells, *Science 250*:1132 (1990).

J. B. Lowe, L. M. Stoolman, R. P. Nair, R. D. Larsen, T. L. Berhend, and R. M. Marks, ELAM-1-dependent cell adhesion to vascular endothelium determinant by a transfected human fucosyltransferase cDNA, *Cell 63*:475 (1990).

K. Drickamer, Two distinct classes of carbohydrate-recognition domains in animal lectins, *J. Biol. Chem. 263*:9557 (1988).

2. For reviews see:

F. W. Lichtenthaler, Enantiomerically pure building blocks from sugars: Efficient preparation and utilization in natural product synthesis, in *New Aspects of Organic Chemistry*, I, Z. Yoshida, T. Shiba, and Y. Ohshiro, eds., Kodansha, Tokyo, and VCH, Weinheim, 1989, p. 351.

S. Hanessian, *Total Synthesis of Natural Products: The "Chiron" Approach*, Pergamon Press, Oxford, 1983.

A. Vasella, Chiral building blocks in enantiomer synthesis -ex sugars, in *Modern Synthetic Methods*, R. Scheffold, ed., Otto Salle, Verlag, Frankfurt, 1980, p. 173.

See also: T. D. Inch, Formation of convenient chiral intermediates from carbohydrates and their use in synthesis, *Tetrahedron 40*:3161 (1984).

3. For reviews see:

H. J. M. Gijsen, L. Qiao, W. Fitz, and C.-H. Wong. Recent advances in the chemoenzymatic synthesis of carbohydrates and carbohydrate mimetics, *Chem. Rev. 96*:443 (1996).

W. Klaffke, Application of enzymes in the synthesis of saccharides and activated sugars, *Carbohydr. Eur. 10*:9 (1994).

C. H. Wong and G. M. Whitesides, Synthesis of glycoside bonds, in *Enzymes in Synthetic Organic Chemistry*, Pergamon Press, Oxford, 1994, p. 252.

K. Toshima and K. Tatsuta, Recent progress in O-glycosylation methods and its application to natural products synthesis, *Chem. Rev. 93*:1503 (1993).

J. E. Heidlas, K. W. Williams, and G. M. Whitesides, Nucleoside phosphate sugars: Syntheses on practical scales for use as reagents in the enzymatic preparation of oligosaccharides and glycoconjugates, *Acc. Chem. Res. 25*:307 (1992).

P. J. Garegg, Saccharides of biological importance: Challenges and opportunities for organic synthesis, *Acc. Chem Res. 25*:575 (1992).

P. Sinaÿ, Recent advances in glycosylation reactions, *Pure Appl. Chem. 63*:519 (1991).

R. R. Schmidt, New methods for the synthesis of glycosides and oligosaccharides. Are there alternatives to the Koenigs–Knorr method? *Angew. Chem. Int. Ed. Engl. 25*:212 (1986).

H. Paulsen, Advances in selective chemical syntheses of complex oligosaccharides, *Angew. Chem. Int. Ed. Engl. 21*:155 (1982).

See also:

T. Müller, G. Hummel, and R. R. Schmidt, Glycosyl phosphites as glycosyl donors. A comparative study, *Liebigs Ann. Chem.* p. 325 (1994).

H. Kondo, S. Aoki, Y. Ichikawa, R. L. Halcomb, H. Ritzen, and C. H. Wong, Glycosyl phosphites as glycosylation reagents: Scope and mechanism, *J. Org. Chem. 59*:864 (1994).

Y. Watanabe, C. Nakamoto, T. Yamamoto, and S. Ozaki, Glycosylation using glycosyl phosphite as a glycosyl donor, *Tetrahedron 50*:6523 (1994).

B. Fraser-Reid, U. E. Udodong, Z. Wu. H. Ottosson, J. R. Merritt, C. S. Rao, C. Roberts, and R. Madsen, *n*-Pentenyl glycosides in organic chemistry: A contemporary example of serendipity, *Synlett.* p. 927 (1992).

D. H. G. Crout, S. Singh, B. E. P. Swoboda, P. Critchley, and W. T. Gibson, Biotransformations in carbohydrate synthesis. *N*-Acetylgalactosaminyl transfer on to methyl *N*-acetyl-β-D-glucosaminide (methyl 2-acetamido-2-deoxy-β-D-glucopyranoside) and methyl *N*-acetyl-α-D-glucosaminide (methyl 2-acetamido-2-deoxy-α-D-glucopyranoside) catalysed by a α-*N*-acetylgalactosaminidase from *Aspergillus oryzae*, *J. Chem. Soc. Chem. Commun.* p. 704 (1992).

A. Marra, J. Esnault, A. Veyrières, and P. Sinaÿ, Isopropenyl glycosides and congeners as novel classes of glycosyl donors: Theme and variations, *J. Am. Chem. Soc. 114*:6354 (1992).

4. For reviews see:

G. Casiraghi, F. Zanardi, G. Rassu, and P. Spanu, Stereoselective approaches to bioactive carbohydrates and alkaloids—With a focus on recent syntheses drawing from the chiral pool, *Chem. Rev. 95*:1677 (1995).

C. Casiraghi and G. Rassu, Aspects of modern higher carbon sugar synthesis, in *Studies in Natural Product Chemistry*, Vol. 11, A.-ur Rahman, ed., Elsevier, Amsterdam, 1992, p. 429. J. S. Brimacombe, Higher carbon sugars, in *Studies in Natural Product Chemistry*, Vol. 4, part C, A.-ur Rahman, ed., Elsevier, Amsterdam, 1989, p. 157.

See also: J. S. Brimacombe, R. Hanna, A. K. M. S. Kabir, F. Bennett, and I. D. Taylor, Higher-carbon sugars. Part 1. The synthesis of some octose sugars via the osmylation of unsaturated precursors, *J. Chem. Soc. Perkin Trans. 1*, p. 815 (1986).

5. For reviews see:

T. Hudlicky, D. A. Entwistle, K. K. Pitzer, A. J. Thorpe, Modern methods of monosaccharides synthesis from non-carbohydrate sources, *Chem. Rev. 96*:1195 (1996).

S. J. Danishefsky and M. P. DeNinno, Totally synthetic routes to the higher monosaccharides, *Angew. Chem. Int. Ed. Engl. 26*:15 (1987).

R. R. Schmidt, Hetero-Diels–Alder reactions in highly functionalized natural product synthesis, *Acc. Chem. Res. 19*:250 (1986).

V. Jäger, I. Müller, R. Schohe, M. Frey, R. Ehrler, B. Häfele, and D. Schröter, Syntheses via isoxazolines. Old concepts and recent results on stereoselection at isoxazoline ring and adjacent centres, *Lect. Heterocyclic. Chem. 8*:79 (1985).

A. Zamojski, A. Banaszek, and G. Grynkiewicz, The synthesis of sugars from noncarbohydrate substrates, *Adv. Carbohydr. Chem. Biochem. 40*:1 (1982).

O. Achmatowicz, An approach to the synthesis of higher-carbon sugars, in *Organic Synthesis, Today and Tomorrow*, B. M. Trost and C. R. Hutchinson, eds., Pergamon Press, Oxford, 1981, p. 307.

See also:

R. M. Paton and A. A. Young, The nitrile oxide-isoxazoline route to higher-carbon dialdoses, *J. Chem. Soc. Chem. Commun.* p. 993 (1994).

S. Jeganathan and P. Vogel, Highly stereoselective total syntheses of octoses and derivatives, *J. Org. Chem.* 56:1133 (1991); P. Vogel, D. Fattori, F. Gasparini, and C. Le Drian, Optically pure 7-oxabicyclo[2.2.1]hept-5-en-2-yl derivatives ('naked sugars') as new chirons, *Synlett.* p. 173 (1990).

L. F. Tietze, Domino-reactions: The tandem-Knoevenagel-hetero-Diels–Alder reaction and its application in natural product synthesis, *J. Heterocyclic Chem.* 27:47 (1990).

6. H. Streicher, A. Geyer, and R. R. Schmidt, C-Disaccharides of ketoses, *Chem. Eur. J.* 2:502 (1996).

A. Dondoni, L. Kniezo, and M. Martinkova, A stereoselective hetero-Diels–Alder approach to carbon–carbon linked disaccharides, *J. Chem. Soc. Chem. Commun.* p. 1963 (1994).

D. Mazéas, T. Skrydstrup, O. Doumeix, and J.-M. Beau, Samarium iodide induced intramolecular C-glycoside formation: Efficient radical formation in the absence of an additive, *Angew. Chem. Int. Ed. Engl.* 33:1383 (1994).

A. Mallet, J.-M. Mallet, and P. Sinaÿ, The use of selenophenyl galactopyranosides for the synthesis of α- and β-(1-4)-C-disaccharides, *Tetrahedron Asymm.* 5:2593, (1994); Y. C. Xin, J.-M. Mallet, and P. Sinaÿ, An expeditious synthesis of a C-disaccharide using a temporary silaketal connection, *J. Chem. Soc. Chem. Commun.* p. 864 (1993); D. Rouzaud, and P. Sinaÿ, The first synthesis of a "C-disaccharide," *J. Chem. Soc. Chem. Commun.* p. 1353 (1983).

L. Lay, F. Nicotra, C. Pangrazio, L. Panza, and G. Russo, Synthesis of antimetabolites of sucrose, *J. Chem. Soc. Perkin Trans. 1*, p. 333 (1994); L. Lai, F. Nicotra, L. Panza, G. Russo, and E. Caneva, Synthesis of C-disaccharides through dimerization of *exo*-glycal, *J. Org. Chem.* 57:1304 (1992); A. Boschetti, F. Nicotra, L. Panza, G. Russo, and L. Zucchelli, Easy synthesis of a C-disaccharide, *J. Chem. Soc. Chem. Commun.* p. 1085 (1989).

D. P. Sutherlin and R. W. Armstrong, Stereoselective synthesis of dipyranyl C-disaccharides, *Tetrahedron Lett.* 34:4897 (1993).

O. R. Martin and W. Lai, Synthesis and Conformational Studies of β-(1-6)- and β,β-(1-1)-linked C-disaccharides, *J. Org. Chem.* 58:176 (1993).

Y. Wang, S. A. Babirad, and Y. Kishi, Preferred conformation of C-glycosides. 8. Synthesis of 1,4-linked carbon disaccharides, *J. Org. Chem.* 57:468 (1992).

A. Vasella, New reactions and intermediates involving the anomeric center, *Pure Appl. Chem.* 63:507 (1991).

R. R. Schmidt and R. Preuss, Synthesis of carbon bridged C-disaccharides, *Tetrahedron Lett.* 30:3409 (1989); B. Giese, M. Hoch, C. Lamberth, and R. R. Schmidt, Synthesis of methylene bridged C-disaccharides, *Tetrahedron Lett.* 29:1375 (1988).

S. Jarosz and B. Fraser-Reid, Synthesis of higher sugars via allyltltin derivatives of simple monosaccharides, *J. Org. Chem.* 54:4011 (1989).

B. Giese and T. Witzel, Synthesis of "C-disaccharides" by radical C–C bond formation, *Angew. Chem. Int. Ed. Engl.* 25:450 (1986).

Y. Fukuda, H. Sasai, and T. Suami, Synthetic approach toward antibiotic tunicamycins. 3. Methyl 3,4,7,8-tetra-*O*-acetyl-10-*O*-benzyl-2-benzyloxycarbonylamino-2,6-dideoxy-11,12,-*O*-isopropylidene-β-L-dodecodialdo-(12R)-furanose-(12,9)-pyranosides-(1,5), *Bull. Chem. Soc. Jpn.* 55:1574 (1982).

7. (a) While the one-carbon homologation will be reviewed in the next section, for a compilation of references concerning multicarbon homologation, *see*: G. Casiraghi, L. Colombo, G. Rassu, and P. Spanu, Ascent of the aldose series by four carbon atoms: Total synthesis of D-*glycero*-D-*talo*-L-*talo*-undecose pentaacetonide, *J. Org. Chem.* 56:2135 (1991).

For other recent work, see:

(b) J. A. Marshall and S. Beaudoin, Stereoselective synthesis of higher sugars by homologation of carbohydrate-derived enals with nonracemic γ-(silyloxy) allylic stannanes and substrate-directed hydroxylation, *J. Org. Chem.* 59:6614 (1994).

(c) C. H. Wong and G. M. Whitesides, C-C bond formation, in *Enzymes in Synthetic Organic Chemistry*, Pergamon Press, Oxford, 1994, p. 195.

(d) R. Fernández, C. Gasch, A. Gómez-Sánchez, J. E. Vílchez, A. L. Castro, M. J. Diánez, M. D. Estrada, and S. Pérez-Garrido, Lengthening of the carbon chain of sugars by the $CH(NO_2)$–$CH(OEt)_2$ fragment. A route to higher 2-amino-2-deoxyaldoses, *Carbohydr. Res. 247*:239 (1993).

(e) E. Kim, D. M. Gordon, W. Schmid, and G. M. Whitesides, Tin- and indium-mediated allylation in aqueous media: Application to unprotected carbohydrates, *J. Org. Chem. 58*:5500 (1993).

(f) T. Mukaiyama, The stereoselective synthesis of carbohydrates, in *Challenges in Synthetic Organic Chemistry*, Claredon Press, Oxford, 1990, p. 153.

(g) S. Masamune, W. Choy, J. S. Petersen, and L. R. Sita, Double asymmetric synthesis and a new strategy for stereochemical control in organic synthesis, *Angew. Chem. Int. Ed. Engl. 24*:1 (1985); S. Y. Ko, A. W. M. Lee, S. Masamune, L. A. Reed, III, K. B. Sharpless, and F. J. Walker, Total synthesis of the L-hexoses, *Tetrahedron 46*:245 (1990).

8. Thiazole-aldehyde synthesis: Preparation of aldehydes from C-2 substituted thiazoles by thiazolyl-to-formyl conversion, see: A. Dondoni, Carbohydrate synthesis via thiazoles, in *Modern Synthetic Methods*, R. Scheffold, ed., Verlag Helvetica Chimica Acta, Basel, 1992, p. 377; A. Dondoni, Acyclic diastereoselective synthesis using functionalized thiazoles. Routes to carbohydrates and related natural products, in *New Aspects of Organic Chemistry*, Z. Yoshida and Y. Ohshiro, eds., Kodansha, Tokyo, and VCH, Weinheim, 1992, p. 105.

9. L. Hough and A. C. Richardson, Synthesis of monosaccharides, in *The Carbohydrates. Chemistry and Biochemistry*, Vol. IA, W. Pigman and D. Horton, eds., Academic Press, New York, 1972, p. 113.

10. H. Kiliani, Über das Cyanhydrin der Lävulose, *Ber. Dtsch. Chem. Ges. 18*:3066 (1885); E. Fischer, Reduction von Säuren der Zuckergruppe, *Ber. Dtsch. Chem. Ges. 22*:2204 (1889).
    For recent historical review articles dealing with the work of Emil Fischer on the configuration of sugars, see:
    F. W. Lichtenthaler, 100 years 'Schlüssel-Schloss-Prinzip': What made Emil Fischer use this analogy? *Angew. Chem. Int. Ed. Engl. 33*:2364 (1994).
    F. W. Lichtenthaler, Emil Fischer's proof of the configuration of sugars: A centennial tribute, *Angew. Chem. Int. Ed. Engl. 31*:1541 (1992).
    M. Engel, A projection on Fischer, *Chem. Br.* 1106 (1992).

11. J. Stanek, M. Cerny, J. Kocourek, and J. Pacák, *The Monosaccharides*, Academic Press, New York, 1965, p. 144.

12. J. Wu, P. B. Bondo, T. Vuorinen, and A. S. Serianni, $^{13}C-^{13}C$ spin coupling constants in aldoses enriched with $^{13}C$ at the terminal hydroxymethyl carbon: Effect of coupling pathway structure on $J_{CC}$ in carbohydrates, *J. Am. Chem. Soc. 114*:3499 (1992); A. S. Serianni, H. A. Nunex, and R. Barker, Cyanohydrin synthesis: Studies with [$^{13}C$]cyanide, *J. Org. Chem. 45*:3329 (1980); A. S. Serianni and R. Barker, Isotopically-enriched carbohydrates: The preparation of [$^2H$]-enriched aldoses by catalytic hydrogenolysis of cyanohydrins with $^2H_2$, *Can. J. Chem. 57*:3160 (1979).

13. H. Saeki and E. Ohki, Synthesis of N-acetyllincosamine, *Chem. Pharm. Bull. 18*:789 (1970).

14. K. Dziewiszek and A. Zamojski, New syntheses of D- and L-*glycero*-D-*manno*-heptoses, *Carbohydr. Res. 150*:163 (1986).

15. H. Fritz, J. Lehmann, and P. Schlesselmann, Chain elongation by a seemingly stereospecific cyanohydrin synthesis: The preparation and configurational assignment of 3,7-anhydro-D-*threo*-L-*talo*- and L-*galacto*-octose, *Carbohydr. Res. 74*:309 (1979).

16. M. Hayashi, Y. Miyamoto, T. Inoue, and N. Oguni, Enantioselective trimethylsilylcyanation of some aldehydes catalyzed by chiral Schiff base–titanium alkoxide complexes, *J. Org. Chem. 58*:1515 (1993).
    H. Ohno, H. Nitta, K. Tanaka, A. Mori, and S. Inoue, A peptide-aluminum complex as a novel

chiral Lewis acid. Asymmetric addition of cyanotrimethylsilane to aldehydes, *J. Org. Chem. 57*:6778 (1992).

J. L. Garcia Ruano, A. M. Martin Castro, and J. H. Rodriguez, Reactions of chiral β-ketsulfoxides with Et₂AlCN: Asymmetric synthesis of cyanohydrin derivatives, *Tetrahedron Lett. 32*:3195 (1991).

R. Herranz, J. Castro-Pichel, and T. Garcia-López, Tributyltin cyanide, a novel reagent for the stereoselective preparation of 3-amino-2-hydroxy acids via cyanohydrin intermediates, *Synthesis* p. 703 (1989).

M. T. Reetz, M. W. Drewes, K. Harms, and W. Reif, Stereoselective cyanohydrin-forming reactions of chiral α-amino aldehydes, *Tetrahedron Lett. 29*:3295 (1988).

J. D. Elliott, V. M. F. Choi, and W. S. Johnson, Asymmetric synthesis via acetal templates. 5. Reactions with cyanotrimethylsilane. Enantioselective preparation of cyanohydrins and derivatives, *J. Org. Chem. 48*:2294 (1983).

17. F. Effenberger, Synthesis and reactions of optically active cyanohydrins, *Angew. Chem. Int. Ed. Engl. 33*:1555 (1994).

18. A. Hassner and C. Stumer, *Organic Syntheses Based on Name Reactions and Unnamed Reactions*, Elsevier, Oxford, 1994, p. 374.

19. S. Czernecki, A. Dieulesaint, and J.-M. Valery, Chain-extension of carbohydrates. I. Cyanoamination of 1,2:3,4-di-*O*-isopropylidene-α-D-*galacto*-hexodialdo-1,5-pyranose, *J. Carbohydr. Chem. 5*:469 (1986).

20. S. Czernecki and J.-M. Valery, Stereochemical studies of the synthesis of α-amino-nitriles related to lincosamine, *Carbohydr. Res. 184*:121 (1988).

21. J. C. Sowden and H. O. L. Fischer, Carbohydrate *C*-nitro alcohols: 1-Nitro-1-deoxy-D-mannitol, *J. Am. Chem. Soc. 66*:1312 (1944).

22. A. Hassner and C. Stumer, *Organic Syntheses Based on Name Reactions and Unnamed Reactions*, Elsevier, Oxford, 1994, p. 165.

23. A. Hassner and C. Stumer, *Organic Syntheses Based on Name Reactions and Unnamed Reactions*, Elsevier, Oxford, 1994, p. 273.

24. L. Benzing-Nguyen and M. B. Perry, Stepwise synthesis of *N*-acetylneuraminic acid and *N*-acetyl[1-¹³C]neuraminic acid, *J. Org. Chem. 43*:551 (1978).

25. For a recent review, see: P. C. Bulman Page, M. B. van Niel, and J. C. Prodger, Synthetic uses of the 1,3-dithiane grouping from 1977 to 1988, *Tetrahedron 45*:7643 (1989).
See also: D. Seebach, Nucleophile acyliierung mit 2-Lithium-1,3-dithianen bzw. -1,3,5-trithianen, *Synthesis* p. 17 (1969).

26. H. Paulsen, M. Schüller, M. A. Nashed, A. Heitmann, and H. Redlich, Synthese der L-*glycero*-D-*manno*-heptose, *Tetrahedron Lett. 26*:3869 (1985).

27. H. Paulsen, V. Sinnwell, and P. Stadler, Synthesis of branched carbohydrates with aldehyde sidechains. Simple synthesis of L-streptose and D-hamamelose, *Angew. Chem. Int. Ed. Engl. 11*:149 (1972).

28. H. Chikashita, T. Nikaya, and K. Itoh, Iterative and stereoselective one-carbon homologation of 1,2-*O*-cyclohexylidene-D-glyceraldehyde to aldose derivatives by employing 2-lithio-1,3-dithiane as a formyl anion equivalent, *Nat. Product Lett. 2*:183 (1993).

29. For a review see: D. S. Matteson, Asymmetric synthesis with boronic esters, *Acc. Chem. Res. 21*:294 (1988).
See also: D. S. Matteson and M. L. Peterson, Synthesis of L-(+)-ribose via (*s*)-pinanediol (α*S*)-α-bromo boronic esters, *J. Org. Chem. 52*:5116 (1987).

30. D. S. Matteson, A. A. Kandil, and R. Soundararajan, Synthesis of asymmetrically deuterated glycerol and dibenzylglyceraldehyde via boronic esters, *J. Am. Chem. Soc. 112*:3964 (1990).

31. G. J. P. H. Boons, G. A. van der Marel, and J. H. van Boom, A versatile and new highly stereoselective approach to the synthesis of L-*glycero*-D-*manno*-heptopyranosides, *Tetrahedron Lett. 30*:229 (1989).

32. B. Achmatowicz, P. Raubo, and J. Wicha, Synthesis of four stereoisomeric tetrose derivatives

from propargyl alcohol. One-carbon homologation of vinyl-silanes via α,β-epoxy silanes, *J. Org. Chem.* 57:6593 (1992).
33. D. J. Ager, Formyl and acyl anions, in *Umpoled Synthons*, T. A. Hase, ed., Wiley-Interscience, New York, 1987, p. 19.
34. A. Dondoni and L. Colombo, New formyl anion and cation equivalents, in *Advances in the Use of Synthons in Organic Chemistry*, Vol. 1, A. Dondoni, ed., JAI Press, Greenwich, 1993, p. 1.
35. A. Barco, S. Benetti, C. De Risi, G. P. Pollini, G. Spalluto, and V. Zanirato, 4-Isopropyl-2-oxazolin-5-one anion as a new convenient formyl anion equivalent for conjugate addition and aldol reactions, *Tetrahedron Lett.* 34:3907 (1993).
36. L. Colombo, M. Di Giacomo, G. Brusotti, and G. Delogu, Chiral 2-lithio-1,3-dioxolanes and 2-lithiooxazolidines: New formyl anion equivalents, *Tetrahedron Lett.* 35:2063 (1994).
J.-M. Lassaletta and R. Fernández, Michael addition of formaldehyde dimethyl-hydrazone to nitroolefins. A new formyl anion equivalent, *Tetrahedron Lett.* 33:3691 (1992).
37. A. Medici, G. Fantin, M. Fogagnolo, and A. Dondoni, Reactions of 2-(trimethylsilyl)thiazole with acyl chlorides and aldehydes. Synthesis of new thiazol-2-yl derivatives, *Tetrahedron Lett.* 24:2901 (1983); A. Dondoni, G. Fantin, M. Fogagnolo, A. Medici, and P. Pedrini, Synthesis of (trimethylsilyl)thiazoles and reactions with carbonyl compounds. Selectivity aspects and synthetic utility, *J. Org. Chem.* 53:1748 (1988).
38. A. Dondoni, A. W. Douglas, and I. Shinkai, Spirodioxolane intermediates in the reaction of 2-(trimethylsilyl)thiazole with aldehydes. Support for the 2-ylide mechanism, *J. Org. Chem.* 58:3196 (1993).
39. (a) A. Dondoni, M. Fogagnolo, A. Medici, and P. Pedrini, Diastereoselectivity in the 1,2-addition of silylazoles to chiral aldehydes. Stereocontrolled homologation of α-hydroxyaldehydes, *Tetrahedron Lett.* 26:5477 (1985); (b) A. Dondoni, G. Fantin, M. Fogagnolo, and A. Medici, Synthesis of long-chain sugars by iterative, diastereoselective homologation of 2,3-*O*-isopropylidene-D-glyceraldehyde with 2-(trimethylsilyl)thiazole, *Angew. Chem. Int. Ed. Engl.* 25:835 (1986); (c) A. Dondoni, G. Fantin, M. Fogagnolo, A. Medici, and P. Pedrini, Iterative, stereoselective homologation of chiral polyalkyoxyaldehydes employing 2-(trimethylsilyl)thiazole as a formyl anion equivalent. The thiazole route to the higher carbohydrates, *J. Org. Chem.* 54:693 (1989).
40. A. Dondoni and P. Merino, Diastereoselective homologation of D-(*R*)-glyceraldehyde acetonide using 2-(trimethylsilyl)thiazole: 2-*O*-benzyl-3,4,-*O*-isopropylidene-D-erythrose, *Org. Synth.* 72:21 (1993); The procedure has been checked by A. I. Meyers and G. P. Brengel, Colorado State University, USA.
41. Before our first report (see Ref 39a), there was only a single example of conversion of a 2-alkylthiazole into aldehyde. See: L. J. Altman and S. L. Richheimer, An aldehyde synthesis utilizing the thiazole ring system, *Tetrahedron Lett.* p. 4709 (1971).
42. A. Dondoni, A. Marra, and D. Perrone, Efficacious modification of the procedure for the aldehyde release from 2-substituted thiazoles, *J. Org. Chem.* 58:275 (1993).
43. A. Dondoni, G. Fantin, M. Fogagnolo, A. Medici, and P. Pedrini, Hydroxy group inversion in thiazole polyols by an oxidation–reduction sequence. An entry to *syn* 1,2-diol fragments in masked carbohydrates, *J. Org. Chem.* 54:702 (1989).
44. A. Dondoni, J. Orduna, and P. Merino, Construction of all *O*-alkoxy D-tetrose and D-pentose stereoisomers from 2,3-*O*-isopropylidene-D-glyceraldehyde using 2-(trimethylsilyl)thiazole as a formyl anion equivalent, *Synthesis* p. 201 (1992).
45. In contrast to the other cases, the addition of **25a** to the D-threose derivative **34** was *syn* selective (ds 81%) giving rise to a *xylo* instead of a *lyxo* configuration.
46. (a) A. Dondoni, G. Fantin, M. Fogagnolo, and P. Pedrini, Stereochemistry associated with the addition of 2-(trimethylsilyl)thiazole to differentially protected α-amino aldehydes. Applications toward the synthesis of amino sugars and sphingosines, *J. Org. Chem.* 55:1439 (1990).
(b) A. Dondoni, D. Perrone, and P. Merino, Chelation- and non-chelation-controlled addition of

2-(trimetilsilyl)thiazole to α-amino aldehydes: Stereoselective synthesis of the β-amino-α-hydroxy aldehyde intermediate for the preparation of the human immunodeficiency virus proteinase inhibitor Ro 31-8959, *J. Org. Chem.* 60:8074 (1995).

47. K. Uchida, T. Ichikawa, Y. Shimauchi, T. Ishikura, and A. Ozaki, Hikizimycin, a new antibiotic, *J. Antibiot.* 24:259 (1971).
48. A. Takatsuki, K. Arima, and G. Tamura, Tunicamycin, a new antibiotic. I. Isolation and characterization of tunicamycin, *J. Antibiot.* 24:215 (1971).
49. A. Dondoni, G. Fantin, M. Fogagnolo, and A. Medici, Stereospecific homologation of D-*xylo* and D-*galacto* dialdoses, *Tetrahedron* 43:3533 (1987).
50. P. Maillard, C. Huel, and M. Momenteau, Synthesis of new *meso*-tetrakis (glycosylated) porphyrins, *Tetrahedron Lett.* 33:8081 (1992).
51. N. K. Khare, R. K. Sood, and G. O. Aspinall, Diastereoselectivity in the synthesis of D-*glycero*-D-aldoheptoses by 2-(trimethylsilyl)thiazole homologation from hexodialdo-1,5-pyranose derivatives, *Can. J. Chem.* 72:237 (1994).
52. A. Hassner and C. Stumer, *Organic Syntheses Based on Name Reactions and Unnamed Reactions*, Elsevier, Oxford, 1994, p. 100. See also: M. Fieser, *Fieser and Fieser's Reagents for Organic Synthesis*, Vol. 16, Wiley, New York, 1992, p. 362.
53. J. Jurczak and A. Golebiowski, From α-amino acids to amino sugars, in *Studies in Natural Products Chemistry*, Vol. 4, A.-ur Rahman, ed., Elsevier, Amsterdam, 1989, p. 111.
    F. M. Hauser and S. R. Ellenberger, Syntheses of 2,3,6-trideoxy-3-amino- and 2,3,6-trideoxy-3-nitrohexoses, *Chem. Rev.* 86:35 (1986).
54. A. Dondoni, S. Franco, F. Junquera, F. L. Merchán, P. Merino, and T. Tejero, Synthesis of N-benzyl nitrones, *Synth. Commun.* 24:2537 (1994).
55. Z. Y. Chang and R. M. Coates, Diastereoselectivity of organometallic additions to nitrones bearing stereogenic N-substituents, *J. Org. Chem.* 55:3464 (1990); R. Huber and A. Vasella, *Tetrahedron* 46:33 (1990).
56. The results described in the following, part of which have been presented as a preliminary communication, are collected in a full paper. A. Dondoni, S. Franco, F. Junquera, F. L. Merchán, P. Merino, T. Tejero, and V. Bertolasi, Stereoselective homologation–amination of aldehydes by addition of their nitrones to C-2 metalated thiazoles. A general entry to α-amino aldehydes and amino sugars, *Chem. Eur. J.* 1:505 (1995).
57. (a) A. Dondoni, F. Junquera, F. L. Merchán, P. Merino, and T. Tejero, Stereoselective aminohomologation of chiral α-alkoxy aldehydes via thiazole addition to nitrones. Application to the synthesis of N-acetyl-D-mannosamine, *Tetrahedron Lett.* 33:4221 (1992). (b) A. Dondoni, S. Franco, F. L. Merchán, P. Merino, and T. Tejero, Stereocontrolled addition of 2-lithiothiazole to the nitrone derived from D-glyceraldehyde acetonide, *Tetrahedron Lett.* 34:5475 (1993).
58. The reversed configuration of these adducts that was mistakenly assigned in our first report (Ref 57a) was timely corrected in a second paper (Ref 57b). For a commentary to this reaction, see: A. Zamojski, Stereoselective aminohomologation of chiral α-alkoxy aldehydes via thiazole addition to nitrones. Application to the synthesis of N-acetyl-D-mannosamine, *Chemtracts Org. Chem.* 6:172; 368 (1993).
59. R. Csuk, M. Hugener, and A. Vasella, A new synthesis of N-acetylneuraminic acid, *Helv. Chim. Acta* 71:609 (1988).
60. U. Kragl, D. Gygax, O. Ghisalba, and C. Wandrey, Enzymatic two-step synthesis of N-acetylneuraminic acid in the enzyme membrane reactor, *Angew. Chem. Int. Ed. Engl.* 30:827 (1991).
    M.-J. Kim, W. J. Hennen, H. M. Sweers, and C.-H. Wong, Enzymes in carbohydrate synthesis: N-Acetylneuraminic acid aldolase catalyzed reactions and preparation of N-acetyl-2-deoxy-D-neuraminic acid derivatives, *J. Am. Chem. Soc.* 110:6481 (1988).
61. A. Dondoni, S. Franco, F. L. Merchán, P. Merino, and T. Tejero, Stereocontrolled addition of 2-thiazolyl organometallic reagents to C-galactopyranosylnitrone. A formal synthesis of destomic acid and lincosamine, *Synlett* p. 78 (1993).
62. M. Shimura, Y. Sekizawa, K. Iinuma, H. Naganawa, and S. Kondo, Structure of destomycin B,

*Agric. Biol. Chem. 40*:611 (1976); S. Kondo, K. Iinuma, H. Naganawa, M. Shimura, and Y. Sekizawa, Structural studies on destomycins A and B, *J. Antibiot. 28*:79 (1975). S. Kondo, K. Iinuma, H. Naganawa, M. Shimura, and Y. Sekizawa, Destomycin C, a new member of destomycin family antibiotics, *J. Antibiot. 28*:83 (1975).

63. R. E. Hornish, R. E. Gosline, and J. M. Nappier, Comparative metabolism of lincomycin in the swine, chicken, and rat, *Drug Metab. Rev. 18*:177 (1987).
64. T. Atsumi, T. Fukumaru, T. Ogawa, and M. Matsui, A stereoselective synthesis of *N*-acetyl-lincosamine derivatives, *Agric. Biol. Chem. 37*:2621 (1973).
65. A. Golebiowski, J. Kozak, and J. Jurczack, Syntheses of destomic acid and anhydrogalantinic acid from L-serinal, *J. Org. Chem. 56*:7344 (1991).
66. For reviews see:
    G. C. Look, C. H. Fotsch, and C.-H. Wong, Enzyme-catalyzed organic synthesis: Practical routes to aza sugars and their analogs for use as glycoprocessing inhibitors, *Acc. Chem. Res. 26*:182 (1993).
    M. L. Sinnott, Catalytic mechanisms of enzymic glycosyl transfer, *Chem. Rev. 90*:1171 (1990).
    G. W. J. Fleet, Homochiral compounds from sugars, *Chem. Br. 25*:287 (1989).
67. (a) H. Dietrich and R. R. Schmidt, Amino-substituted α-D-glucosylmethylbenzenes (benzyl α-*C*-glucosides) and an *N*-(*C*-α-D-glucosylmethyl)aniline (anilinomethyl α-*C*-glucoside); novel α-D-glucosidase inhibitors, *Carbohydr. Res. 250*:161 (1993).
    (b) C. R. Bertozzi, P. D. Hoeprich Jr., and M. D. Bednarski, Synthesis of carbon-linked glycopeptides as stable glycopeptide models, *J. Org. Chem. 57*:6092 (1992); C. R. Bertozzi, D. G. Cook, W. R. Kobertz, F. Gonzales-Scarano, and M. D. Bednarski, Carbon-linked galacto-sphingolipid analogs bind specifically to HIV-1 gp120, *J. Am. Chem. Soc. 114*:10639 (1992).
68. It is worth mentioning that the thiazole–aldehyde synthesis has also been employed to develop two- and three-carbon homologation methods by using well-designed thiazole-based reagents. See: Ref 8 and A. Dondoni, A. Marra, and P. Merino, Installation of the pyruvate unit in glycidic aldehydes via a Wittig olefination–Michael addition sequence utilizing a thiazole-armed carbonyl ylid, *J. Am. Chem. Soc. 116*:3324 (1994); A. Dondoni, A. Boscarato, and A. Marra, Synthesis of a sialic acid analog with the acetamido group at C-4, *Tetrahedron Asymm. 5*:2209 (1994); A. Dondoni, P. Merino, and D. Perrone, Totally chemical synthesis of azasugars via thiazole intermediates. Stereodivergent routes to (−)nojirimycin, (−)mannojirimycin and their 3-deoxy derivatives from serine, *Tetrahedron 49*:2939 (1993).
69. (a) Reductive hydrolysis of *C*-glycosyl nitrile: M. T. García López, F. G. De las Heras, and A. San Félix, Cyanosugars. IV. Synthesis of α-D-glucopyranosyl and α-D-galactopyranosyl cyanides and related 1,2-*cis*-*C*-glycosides, *J. Carbohydr. Chem. 6*:273 (1987); H. P. Albrecht, D. B. Repke, and J. G. Moffatt, *C*-glycosyl nucleosides. II. A facile synthesis of derivatives of 2,5-anhydro-D-allose, *J. Org. Chem. 38*:1836 (1973).
    (b) Swern oxidation of *C*-glycosyl carbinol; see Ref. 67a.
    (c) Ozonolysis of *C*-glycosyl allene: W. R. Kobertz, C. R. Bertozzi, and M. D. Bednarski, An efficient method for the synthesis of α- and β-*C*-glycosyl aldehydes, *Tetrahedron Lett. 33*:737 (1992); C. R. Bertozzi and M. D. Bednarski, The synthesis of 2-azido *C*-glycosyl sugars, *Tetrahedron Lett. 33*:3109 (1992).
70. M. H. D. Postema, *C-Glycoside Synthesis*, CRC Press, Boca Raton, 1995. P. C. Tang and D. E. Levy, *Chemistry of* C-*Glycosides*, Pergamon Press, Oxford, 1995.
    C. Jaramillo and S. Knapp, Synthesis of *C*-aryl glycosides, *Synthesis* p. 1 (1994). M. H. D. Postema, Recent developments in the synthesis of *C*-glycosides, *Tetrahedron 48*:8545 (1992).
71. (a) A. Dondoni and M.-C. Scherrmann, Thiazole-based synthesis of *C*-glycosyl aldehydes, *Tetrahedron Lett. 34*:7319 (1993). (b) A. Dondoni and M.-C. Scherrmann, Thiazole-based synthesis of formyl *C*-glycosides, *J. Org. Chem. 59*:6404 (1994).
72. Attempts to reductive removal of the hydroxyl group from the ketol by the use of silanes and Lewis acids (see Ref. 73) gave unaltered material. Also unsatisfactory were various attempts to dehydroxylate the ketol by radical procedures.

73. S. A. Babirad, Y. Wang, and Y. Kishi, Synthesis of *C*-disaccharides, *J. Org. Chem. 52*:1370 (1987).
74. M. L. E. Sánchez, V. Michelet, I. Besnier, and J. P. Genêt, Convenient methods for the synthesis of β-*C*-glycosyl aldehydes, *Synlett.* p. 705 (1994).
75. J. Désiré and A. Veyrières, Synthesis and reactions of an (α-D-glucopyranosyl)-phenylacetylene, *Carbohydr. Res. 268*:177 (1995).
76. A. Dondoni, A. Marra, and M.-C. Scherrmann, Furan-based synthesis of *C*-glycosyl carboxylates, *Tetrahedron Lett. 34*:7323 (1993).
77. A. Dondoni, M.-C. Scherrmann, A. Marra, and J.-L. Delépine, A general synthetic route to anomeric α-azido and α-amino acids and formal synthesis of (+)-hidantocidin, *J. Org. Chem. 59*:7517 (1994).
78. S. Mio, Y. Kumagawa, and S. Sugai, Synthetic studies on (+)-hydantocidin (3): A new synthetic method for construction of the spiro-hydantoin ring at the anomeric position of D-ribofuranose, *Tetrahedron 47*:2133 (1991).
79. H. Redlich, W. Bruns, W. Francke, V. Schurig, T. L. Payne, and J. P. Vité, Chiral building units from carbohydrates. XIII. Identification of the absolute configuration of *endo*-brevicomin from *Dendroctonus frontalis* and synthesis of both enantiomers from D-ribose, *Tetrahedron 43*:2029 (1987).
80. M. L. Wolfrom and S. Hanessian, The reaction of free carbonyl sugar derivatives with organometallic reagents. I. 6-Deoxy-L-idose and derivatives, *J. Org. Chem. 27*:1800 (1962).
81. C. R. Schmid and J. D. Bryant, D-(R)-Glyceraldehyde acetonide, *Org. Synth. 72*:6 (1993). C. R. Schmid, J. D. Bryan, M. Dowlatzedah, J. E. Phillips, D. E. Prather, R. D. Schantz, N. L. Sear, and C. S. Vianco, Synthesis of 2,3-*O*-isopropylidene-D-glyceraldehyde in high chemical and optical purity: Observations on the development of a practical bulk process, *J. Org. Chem. 56*:4056 (1991).
82. R. F. Borch, M. D. Bernstein, and H. D. Durst, The cyanohydridoborate anion as a selective reducing agent, *J. Am. Chem. Soc. 93*:2897 (1971).

# 10

## Selected Methods for Synthesis of Branched-Chain Sugars

### Yves Chapleur and Françoise Chrétien
*Université Henri Poincaré—Nancy I, Vandoeuvre, France*

|      |                                                         |     |
|------|---------------------------------------------------------|-----|
| I.   | Introduction                                            | 207 |
| II.  | Methods                                                 | 211 |
|      | A. Compounds with a C-C-X branch (type I)               | 211 |
|      | B. Compounds with a C-C-H (type II)                     | 215 |
|      | C. Compounds with a C=C branch (type III)               | 231 |
|      | D. Compounds with two branched chains                   | 235 |
|      | E. Conclusion                                           | 239 |
| III. | Experimental Procedures                                 | 240 |
|      | A. Compounds with a C-C-X branch (type I)               | 240 |
|      | B. Compounds with a C-C-H (type II)                     | 242 |
|      | C. Compounds with a C=C branch (type III)               | 249 |
|      | D. Compounds with two branched chains (types IV and V)  | 251 |
|      | References                                              | 252 |

## I. INTRODUCTION

Occurrence in nature of branched-chain carbohydrates has prompted interest in the syntheses of these complex structures and stimulated the preparation of analogues for biological evaluation. Consequently, new methods for the construction of these particular skeletons have been devised [1]. The use of carbohydrates as a cheap source of chiral starting materials [2–4] for the synthesis of complex, nonsugar molecules has prompted the emergence of new imaginative methods for formation of carbon–carbon bonds adapted to the particular reactivity of sugar moieties.

The purpose of this chapter is to provide details on some selected methods of C-branching of sugar, with a particular emphasis on the more recent methods of general use. This introduction will describe different strategies to branched-chain sugars, arranged by type of structure, and references to reviews and papers of general interest to the topic are given. The different methods are detailed with reference to recent pertinent work.

The methods have been classified according to the type of compounds they can form (Scheme 1); for example, sugars with an heteroatom (*type I*), with an hydrogen (*type II*), or

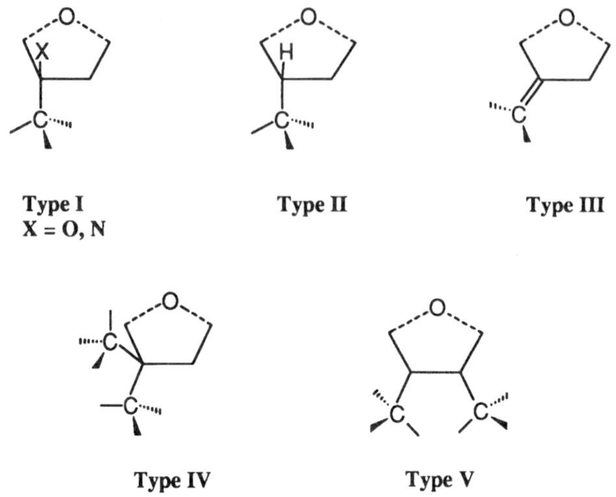

**Type I**
X = O, N

**Type II**

**Type III**

**Type IV**

**Type V**

**Scheme 1**

with a double bond at the branching point (*type III*). Some achievements in the formation of either geminal (*type IV*) or vicinal (*type V*) doubly branched-chain sugars have been reported and will be discussed in the last section.

As seen from examination of the literature, two main classes of starting carbohydrates are used, ketosugars and unsaturated sugars. Ketosugars have been and are still used in reactions with nucleophilic species, such as organometallics. Related reagents, such as Wittig-type reagents will also be discussed. On the other hand, unsaturated sugars are used for the direct formation of C–C bonds, which includes Michael addition, rearrangements, and more recently, the always growing field of free radical reactions. An excellent review on the topic appeared 10 years ago [1].

Naturally occurring branched-chain sugars are mainly sugars that have a heteroatom at the branching point (type I). Although the most commonly encountered are oxygen-substituted, some have a nitrogen substituent at the branching point. The biosynthetic pathway leading to these compounds has been reviewed [5]. Different biosyntheses from nucleotide-bound *glyco*-uloses are involved, depending on the substitution. Formyl and hydroxymethyl branched-chain sugars arise from ring contraction, with expulsion of one carbon, whereas methyl or two-carbon chain substitutions arise from transfer of the carbon chain from an appropriate donor.

From a chemical point of view, Type I branched-chain sugars can be obtained by simple condensation of the appropriate carbon nucleophile onto carbonyl derivatives. The main problem is the stereochemical control of the addition. The judicious choice of the protecting groups to induce a steric control, or the introduction of a conformational bias is used to solve this problem. On the other hand, Grignard derivatives are suitable reagents because of their ability to achieve high stereocontrol by chelated intermediates. The complexation of oxygen atoms with magnesium during Grignard condensation has been well illustrated in the chain elongation of 6-aldehydo sugars, in which the complexation

## Methods for Synthesis of Branched-Chain Sugars

between the carbonyl of the aldehyde and the ring oxygen is responsible for the high stereoselectivity of attack [6,7]. Coordination of magnesium to the carbonyl group and adjacent alkoxy groups may fix the conformation of the ketosugar and thus direct attack of the Grignard reagent. The chelation-controlled addition could give results complementary to those obtained with alkyllithium reagents. Some examples of such effects are described in Yoshimura's review [1].

Variations on this basic theme have been reported to reach more or less complex Type I branched-chain sugars (Scheme 2). A particular emphasis has been given on the use of

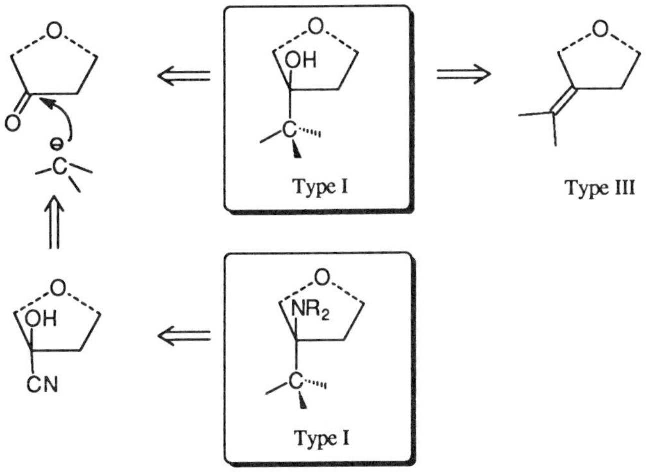

**Scheme 2**

unsaturated Grignard reagents or lithio derivatives. This allows subsequent chemical manipulations such as oxidative cleavage of the carbon–carbon double bond, giving the formyl derivatives, which can be reduced to hydroxymethyl derivatives. Hydroxylation of the double bond or Wacker oxidation or epoxidation are valuable transformations that permit further chain elongation. Functionalized alkyllithiums are less common, but some of them, as well as functionalized carbanions, such as dithiane, enolates, and such, have been successfully used. Some of these branched-chain sugars having an heteroatom at the branching point can be obtained by chemical manipulation of alkylidene sugars, which are readily available through Wittig and related reactions. Nitrogen-substituted type I branched-chain sugars have been prepared using the nucleophilic addition of cyanide ion on ketosugars, followed by aziridine formation. Other new routes to aziridine formation have been developed.

Deoxygenation of the tertiary alcohol of the preceding type I compounds (path **h** Scheme 3) should, in principle, lead to branched-chain sugars having a C-C-H branch (type II). This reaction is rather difficult, so direct formations of C–C bond on the sugar template have been developed. The main problem of this method is the control of the regio- and the stereoselectivity. This problem is well solved by the opening of epoxides with nucleophiles (path **a**). The reactivity of these species, easily obtained from diols or olefins, has been reviewed [8]. Carbon nucleophiles, such as organometallics and stabilized carbanions, have been used with success. Another fairly often used reaction is the 1,4-addition of carbon nucleophiles on activated double bonds. Organocopper derivatives are reagents of choice for this purpose, as well as dithiane and its derivatives. Usually, 1,4-addition on carbohydrate-derived enones occurs with high stereocontrol, axial approach of the reagent being favored (path **b**). By using enolones, it is possible to retain an oxygenated substituent, often a keto

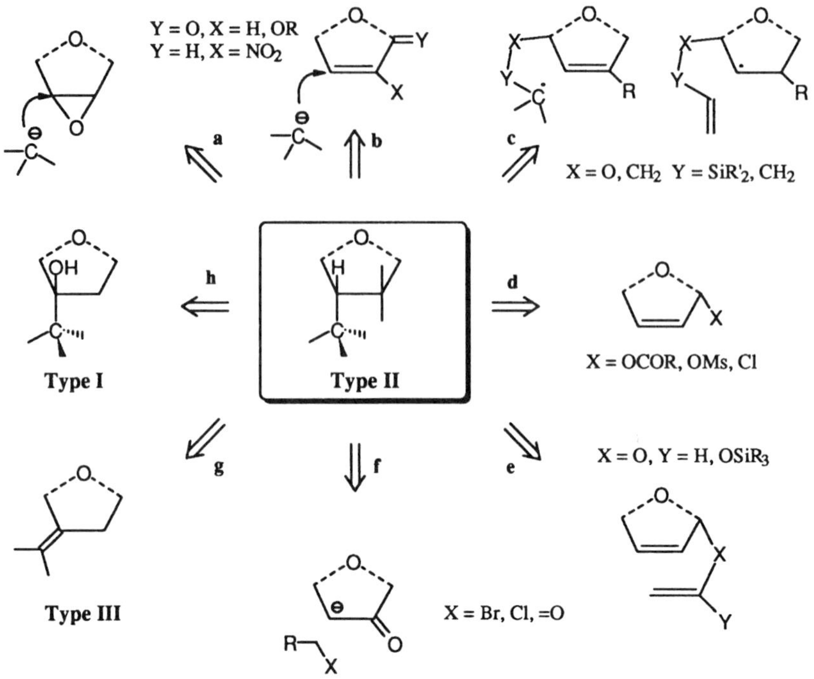

**Scheme 3**

group, at the α-position [9]. The synthesis of an aminated sugar α to the branching point has been accomplished by 1,4-addition of carbon nucleophiles on unsaturated nitro sugars.

Unsaturated sugars have been widely used for the synthesis of type II branched-chain sugars by direct introduction of carbon nucleophiles using palladium or copper catalysis (path **d**). Reduction of type III branched-chain sugars (path **g**) is also a practical way to form type II derivatives. 1,3-Dipolar cycloaddition of suitable dipoles, such as nitrones and nitrile oxides, on activated double bonds allows the simultaneous introduction of a carbon chain and an α-oxygenated group in a *cis* arrangement. But one of the interesting breakthroughs came recently from the use of free radical reactions (path **c**). These reactions have provided new highly stereoselective methods for the formation of carbon–carbon bonds, either by intramolecular cyclization or by intermolecular trapping of radicals [10]. Addition–fragmentation radical reactions have been also explored with success. Finally, the use of carbohydrates as nucleophiles has been explored recently. Sigmatropic rearrangement of unsaturated sugars has been used with success, in particular the Claisen rearrangement and its variants (path **e**). The chemistry of enolates derived from carbohydrates has been exploited in our group in alkylation [11] and aldolization reactions [12] (path **f**). This approach provides new branched-chain derivatives of type III (Scheme 4) that can be further used in new sets of reactions.

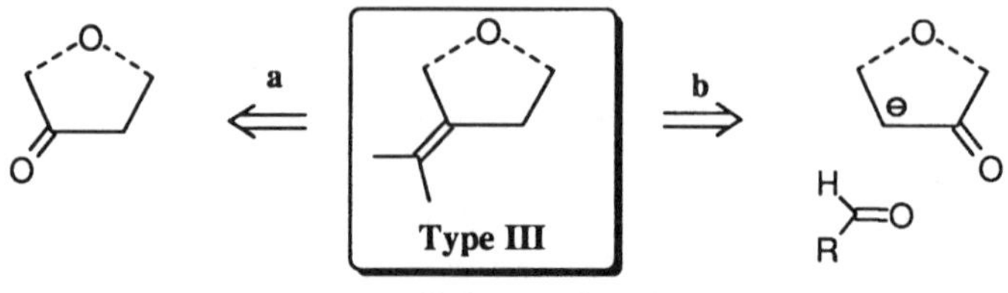

**Scheme 4**

# Methods for Synthesis of Branched-Chain Sugars

The synthesis of type III branched-chain sugars is based mainly on the use of ketosugars treated under Wittig-type conditions (see path **a**, Scheme **4**) [13]. Several other methods, such as aldolization–crotonization or direct alkylidenation at the α-position of the carbonyl group of a keto sugar have been developed (path **b**).

It is clear that all the preceding methods of formation of a C-branching in sugars can be applied several times on a substrate to give multibranched-chain sugars. Moreover, impressive developments of new specific methods for the synthesis of doubly and even triply branched-chain sugars of type IV and type V have been reported. Some examples of synthesis of such multibranched-chain sugars involving sigmatropic rearrangements or double alkylation reaction will be given at the end of this chapter.

The methods often used for the branching of carbon chain are also suitable for the chain extension of sugars at both ends (e.g., C-1 and C-5/C-6). Several developments of chain extension have recently culminated in the synthesis of long-chain carbohydrates of biological interest.

## II. Methods

### A. Compounds with a C-C-X Branch (Type I)

Among this group of branched-chain sugars, those having an oxygenated substituent are widely distributed, but other derivatives having a nitrogen at the branching point have been described. From a chemical point of view, the synthesis of these type I branched-chain carbohydrates consist of nucleophilic addition onto ketosugar derivatives. A few classes of carbon nucleophiles are well suited for this particular reaction. Accordingly, the use of Grignard reagents, alkyllithium, or functionalized carbanions is described in the following. These nucleophilic additions are often very stereoselective. Moreover, different selectivities may be observed on going from alkyllithium to Grignard reagents, mainly because of the possible complexation of magnesium with another oxygenated group of the molecule that directs the nucleophilic attack. This point has been already discussed [1,14].

*Use of Grignard Reagents*

The availability of these reagents is at the origin of numerous reports on their use in the carbohydrate field. For methyl branched-chain sugars obviously methyl Grignard or methyllithium are reagents of choice. An interesting procedure for the Swern oxidation of alcohols and subsequent use of the resulting highly reactive carbonyl compounds has been reported by Ireland and Norbeck [15]. Application of this oxidation process to sugar alcohols and subsequent reaction with methyl magnesium bromide gave excellent results (see Sec. III). One of the most fruitful methods is the condensation of unsaturated organomagnesium derivatives, because they can be further manipulated essentially by oxidative procedures. Ethynylation of ketosugars has been extensively explored. In this case, partial reduction of the triple bond is of interest to obtain vinylic derivatives [16]. A typical example is provided by the reaction of 1,2:5,6-di-*O*-isopropylidene-α-D-*ribo*-hex-3-ulo furanose **1** (Scheme **5**) with the Grignard derived from acetylene, recently reported by Kakinuma et al. [17]. For steric reasons, nucleophilic addition proceeds exclusively from the β-face of the furanose ring.

Alternatively, vinylation can be achieved in useful yields using readily available vinyl magnesium bromide. The allylic tertiary alcohol resulting from this condensation on a ketosugar can be further elaborated by suitable modification of the vinylic group.

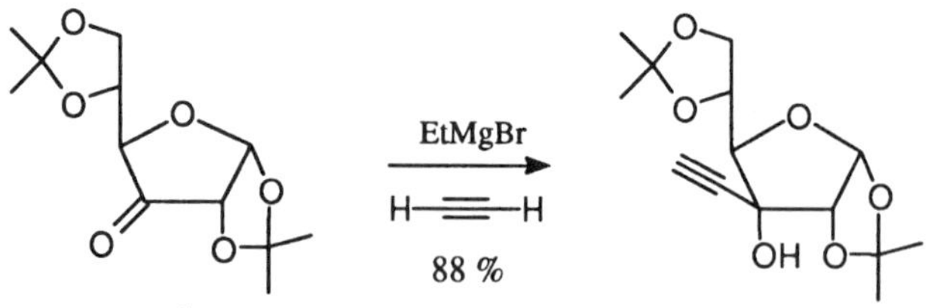

**Scheme 5**

Ozonolysis and subsequent reduction open the way to formyl and hydroxymethyl branched-chain derivatives [18]. High stereoselectivities are generally observed on Grignard addition on conformationally biased ketosugars such as **3** (Scheme 6), equatorial attack being strongly favored, giving the axial alcohol **4** as the major product.

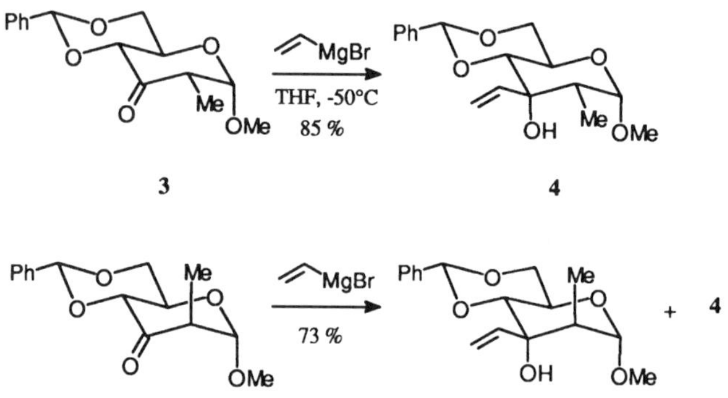

**Scheme 6**

This particular reaction has been reported recently by Lukacs in the context of complex molecules synthesis by Diels–Alder cycloaddition [19]. A typical example is given in the experimental section. Note that the reaction of **5** with vinyl magnesium bromide gave a mixture of **4** and **6** because of epimerization of the axial methyl group at C-2 owing to the basicity of the reagents [20].

*Use of Alkyllithium*

The use of alkyllithium is less documented, essentially because of the limited availability of these reagents, their often strong basicity, and the restricted number of functionalities compatible with the use of these reagents. Methoxyvinyllithium has been successfully introduced by Brimacombe's group for the introduction of acetyl and 2-hydroxyacetyl groups [21–23]. A typical procedure of 1-methoxylithium condensation on ketone **7** (Scheme 7) is given in the experimental section.

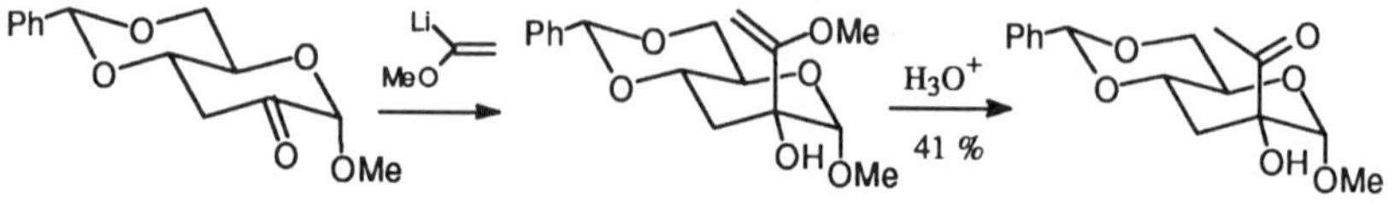

**Scheme 7**

## Methods for Synthesis of Branched-Chain Sugars

Another interesting approach is the use of 1,1-dimethoxy-2-lithio-2-propene, as demonstrated by Depezay's group, for the synthesis of C-methylene branched-chain sugars [24,25], intermediates in the synthesis of hamamelose G, a naturally occurring branched-chain pentose [26,27]. In this case, the condensation of this lithio derivative was effected on the aldehydo group of D-glyceraldehyde **10** (Scheme 8) to give a 3:7 mixture of **11a**

**Scheme 8**

and **11b**. The methylene group introduced in this process is stereoselectively epoxidized and the epoxide opened by hydroxide ion to give the hydroxymethyl derivatives **12a** and **12b**. This linear sugar was, in turn, transformed into C-2 branched-chain furanosides **13a** and **13b**.

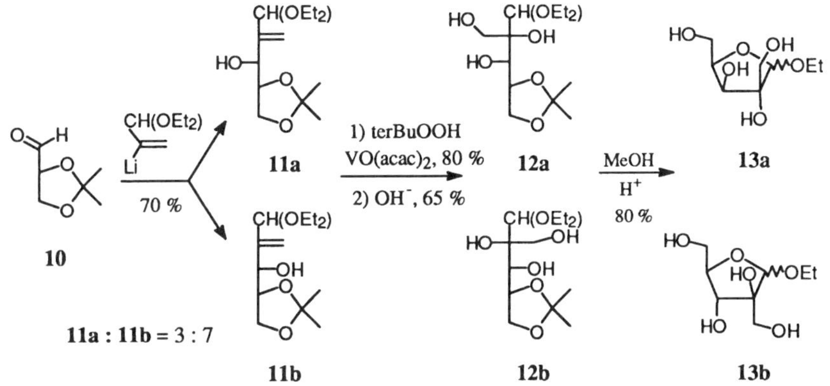

Spiro epoxides are also valuable intermediates for the synthesis of type I branched-chain sugars. These spiro epoxides are formed from ketosugars using diazomethane addition or sulfonium chemistry. Further ring opening of the epoxide allows the introduction of various nucleophiles [1]. Chloro spiroepoxide **15** (Scheme 9) has been prepared recently from ketosugar **1** dichloromethyllithium [28].

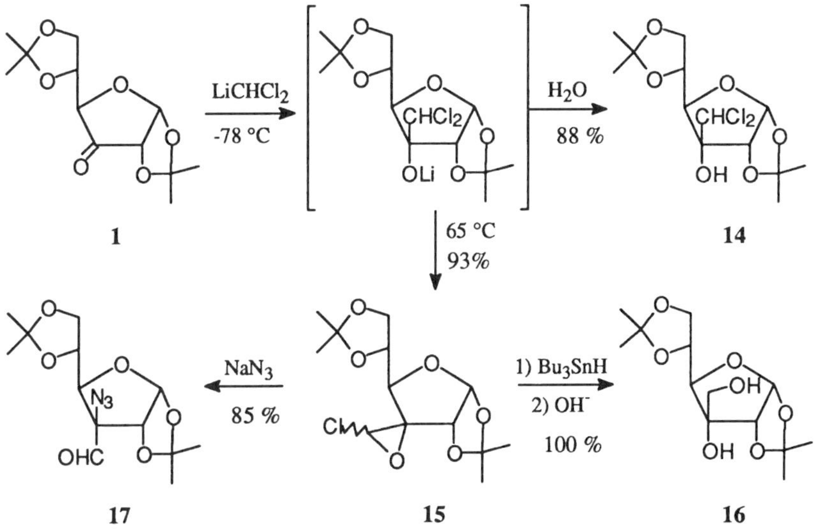

**Scheme 9**

Epoxide **15** can be reduced to the hydroxymethyl derivative **16** or elaborated to the azido aldehyde **17**. Some other applications to the synthesis of other types of branched-chain sugars have been described in the same paper. Azido aldehydes have been prepared and used in a formal synthesis of myriocin by Lucaks et al. [29].

*Use of Functionalized Carbanions*

**Dithiane.** Condensation of dithiane or substituted analogues [30] on ketosugars has been described by several groups [1]. Further deprotection of dithiane gave formyl derivatives. Acetyl or 2-hydroxyacetyl groups can also be obtained using suitably substituted dithiane derivatives. Extensive work by Paulsen et al. has been reported and typical procedures have been published [31].

**Enolates Type Anions.** The well-known Reformatsky reaction allows the introduction of a functionalized two-carbon chain. Modified experimental procedures have now been proposed, such as the reaction of ethylbromoacetate with ketosugar **18** (Scheme **10**) in the presence of zinc/silver graphite prepared from $C_8K$, giving **19** in excellent yield [32].

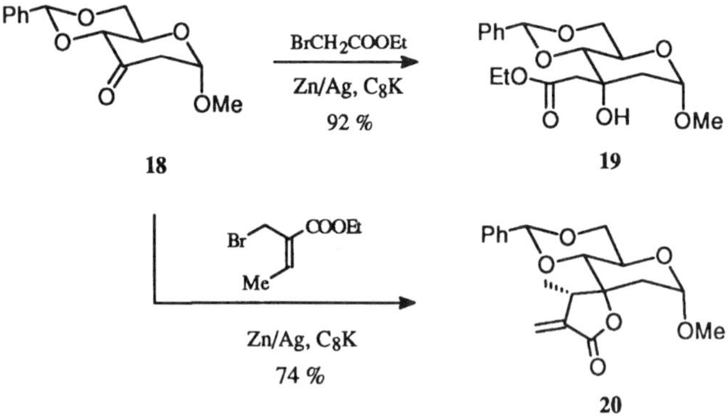

**Scheme 10**

A three-carbon unit can be introduced on ketosugars under Reformatsky conditions, as recently demonstrated by several groups [33,34]. The analogous Dreiding–Schmidt procedure has also been applied in this case with successful double stereodifferentiation [35]. This is exemplified on ketone **18** which yields lactone **20** as a single isomer (see Scheme **10**). The condensation on ketosugars of trimethylsilylacetate [36] or acrylate [33], in the presence of fluoride ion, has also been used with success for the synthesis of β-hydroxy acids or α-methylene-γ-lactones, respectively.

The Darzens condensation of chloromethyl *p*-tolysulfone has also been successfully employed for the synthesis of α,β-epoxysulfone **21** (Scheme **11**), which undergo nucleo-

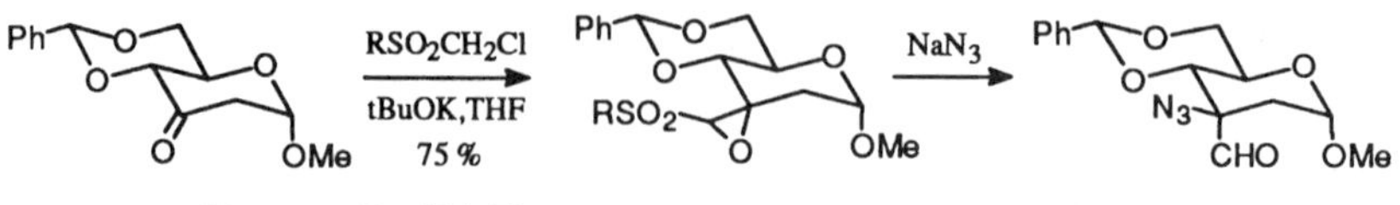

**Scheme 11**

philic ring opening by azide anion to furnish α-azido aldehyde **22** [37]. The stereoselectivity is excellent onto the conformationally biased systems **7** and **18**. This approach has been used in the formal synthesis of myriocin [29].

## Methods for Synthesis of Branched-Chain Sugars

*Miscellaneous Methods*

Alternative methods are available for the synthesis of type I branched-chain sugars. An interesting two-step procedure, using sequential methylenation-hydroxylation has been described [38,39]. It allows the formation of a tertiary alcohol having the configuration opposite that normally obtained using direct nucleophilic addition of methylmagnesium bromide on the corresponding ketone. This is illustrated in Scheme 12 by a sequence of

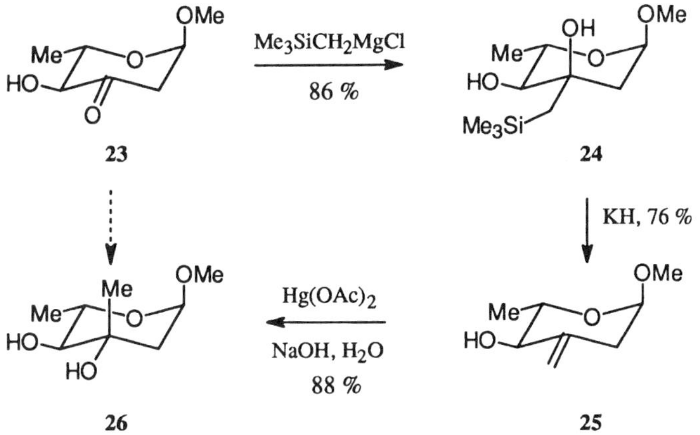

**Scheme 12**

olefination–hydroxylation of ketone 23 derived from L-rhamnose. Note that, in principle, methyl Grignard addition onto 23 would give the same stereochemistry as in 24 opposite that of 26.

Several routes to type I branched-chain sugars having a nitrogen at the branching point have been described. One strategy involves azidomercuration of exomethylene groups [40]. Spiro aziridines represent a method of choice to obtain branched-chain amino sugars [41,42]. They are generally obtained by mesylation of cyanohydrin derivatives obtained by condensation of cyanide ion on ketones. An alternative procedure for the formation of spiroaziridine has been described [43] according to Scheme 13. Reaction of

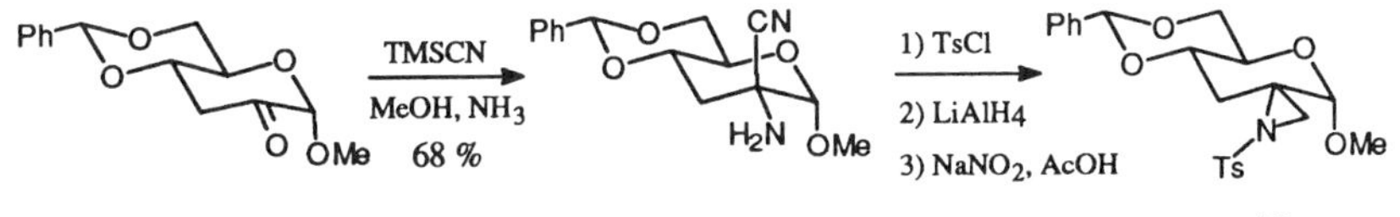

**Scheme 13**

trimethylsilyl cyanide with ketone 7 in the presence of ammonia gave amino nitrile 27, which was elaborated to the spiroaziridine 28 in three steps.

## B. Compounds with a C-C-H (Type II)

The synthesis of type II branched-chain sugars should seem a more difficult task because it needs activation of a carbon atom on the sugar template and a control of the stereochemical course of the carbon–carbon bond formation. Probably because of this apparent difficulty several methods have been devised in the last decade, in particular, in the field of organometallic and free radical reactions.

*Opening of Epoxides*

Major achievements have been reported in the ring opening of 2,3 or 3,4 epoxides of carbohydrates by a variety of carbon nucleophiles. An excellent review gives the state of the art [8]. The stereochemical course of these reactions is often clearcut, in particular for conformationally biased pyranosidic ring systems. The major product always results from an axial approach of the reagent, giving the well-known "*trans*-diaxial" arrangement of the branched chain and the vicinal alcohol according to Fürst-Plattner rules. Opening of 2,3-anhydro furanose is more dependent of anomeric configuration and of polar and steric effects [1,44,45].

**With Organometallic Reagents.** Grignard reagents have been used for the ring opening of epoxides [8]. However, they often gave by-products resulting from the opening of the epoxide by halogen and subsequent reductive elimination [46–48]. Anomalies in the stereochemistry of epoxide ring opening by Grignard reagents have been reported [49]. Some improvements have resulted from the introduction of copper salt catalysis in these reactions [50]. The development of new organocopper reagents gave new solutions to this problem [44,51,52]. One side reaction with $R_2CuLi$ [2RLi + CuI) reagents is the opening by iodide ion arising from the formation of lithium iodide [53]. Glycals may also be formed by reductive elimination of iodine at C-2 and of the aglycon. For the synthesis of biologically active β-lactones, such as tetrahydrolipstatin, we have recently investigated the formation of C-2 branched-chain sugars by opening one of the most widely used epoxides of *allo*-configuration with different organometallics [54]. As seen from Table 1 most of the combination of Grignard reagents with copper salts and Gilman reagents gave either glycal **31** or alcohol **32** (Scheme 14). However, the high-order cyanocuprate [55] derived from

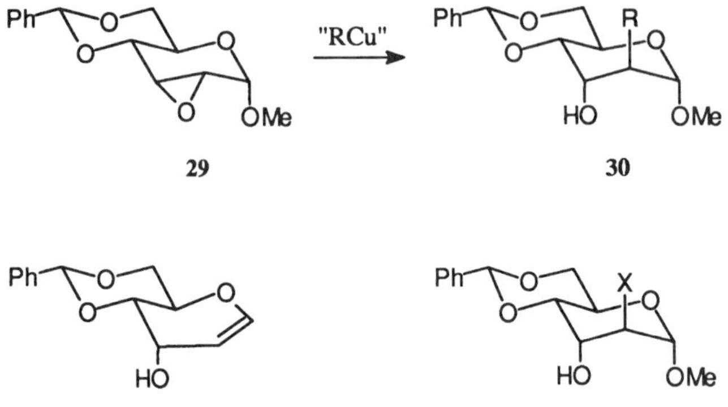

**Scheme 14**

hexyllithium and copper cyanide gave reproducible yields of the expected alcohol **30**. A typical example is given in the experimental section.

**With Functionalized Carbanions.** Other nucleophilic carbanions, such as lithiodithiane or sodiomalonate or acetylene anion react easily with epoxides.

**With Dithiane.** Dithiane is one of the most popular functionalized carbanion used for the ring opening of epoxides. This reaction has been studied by different groups [31,56]. The anion of bis(phenylthio)methane has been successfully used for the opening of epoxides derived from furanose [56]. This method is an excellent way to prepare formyl derivatives that are suitable for chain extension by olefination. Alternatively formyl derivatives can be reduced to hydroxymethyl derivatives. A recent application of epoxide **33**

**Table 1** Opening of Epoxide **29** with Organometallics

| Reagent (Eq.) | **30** (yield %) | **31** (yield %) | **32** (yield %) |
| --- | --- | --- | --- |
| HexMgCl, CuI (10%) | 0 | 90 | 0 |
| HexMgBr (2), CuCN (1) | 0 | 5 | 60 X = Br |
| HexLi, CuI (2), ether | 0 | 0 | 98 X = I |
| HexLi, CuI (2), THF | 0 | 98 | 0 |
| HexLi (2), CuCN (1), THF | 75 | 20 | 0 |

opening with dithiane itself is described by Scheme 15 and in the experimental section as a typical procedure [19].

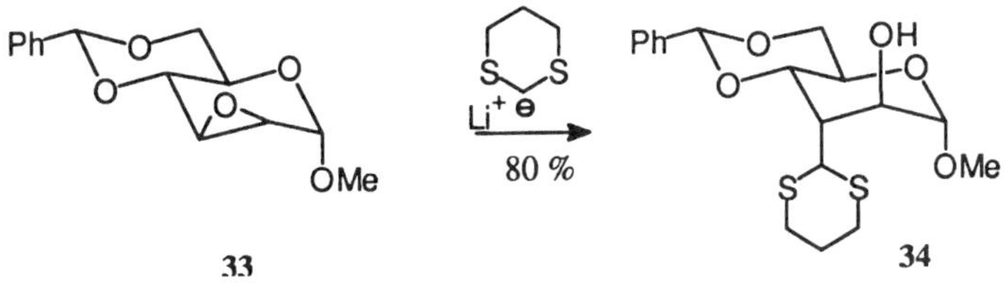

**Scheme 15**

*With Malonate.* Seminal examples of epoxide opening with malonates have been reported by Hanessian [57,58]. Once again high regio- and stereoselectivities were observed (Scheme 16).

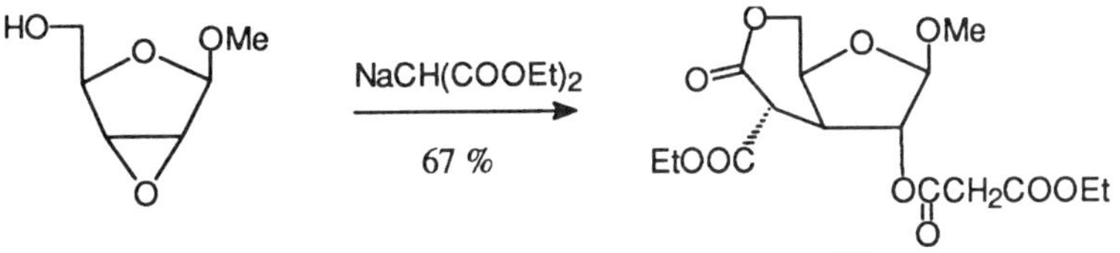

**Scheme 16**

*Conjugate Addition to Enones and Related Olefins*

With copper catalysis, organomagnesium derivatives can add in a 1,4 fashion [59]. The development of efficient organocopper derivatives has stimulated their use as reagents in the branching of sugars from unsaturated carbohydrates. For enone, the resulting branched-chain sugar is deoxygenated at the α-position [60]. The addition of these organometallic species is always regiospecific and highly stereoselective. Many examples are available from the literature [1]. One seminal example of addition of cuprate onto enone **37** is given by Scheme 17 [61]. Addition proceeds mainly from the axial direction giving only compound **38**.

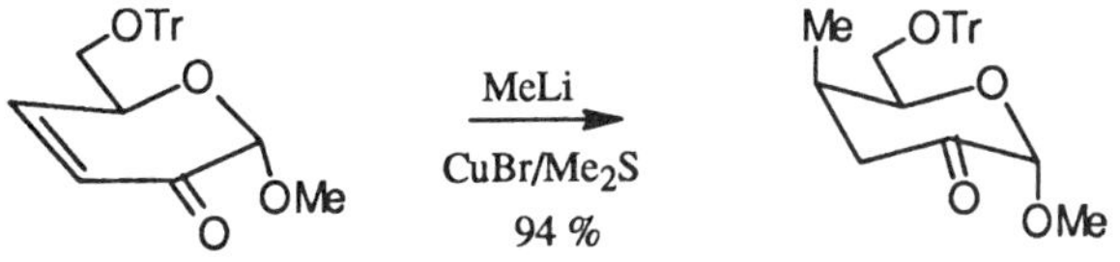

**Scheme 17**

An interesting reaction has been developed by Hanessian to reach the same type of branched-chain derivatives, without the need to prepare enones [62]. The enol acetate **38** (Scheme **18**), easily prepared by Ferrier rearrangement of 2-acetoxyglucal, underwent

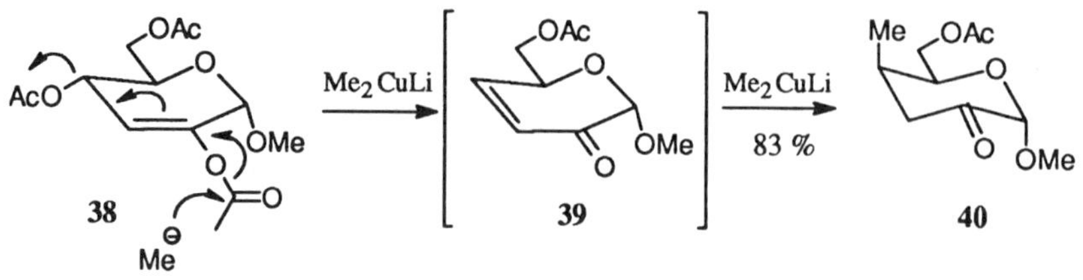

**Scheme 18**

successively: cleavage of the enol ester bond, formation of the corresponding ketone, and subsequent β-elimination of ester group at C-4, yielding the enone **39**. Subsequent 1,4-addition of cuprate led to the ketone **40** in excellent yield.

Along the same lines, enolones also undergo an interesting addition with organocuprates, as reported by Lichtenthaler [9]. For example, enolone **41** reacts with lithium dimethyl copper to give primarily 1,4-addition (e.g., compound **42**), which undergoes benzoyl migration to yield **43** (Scheme **19**).

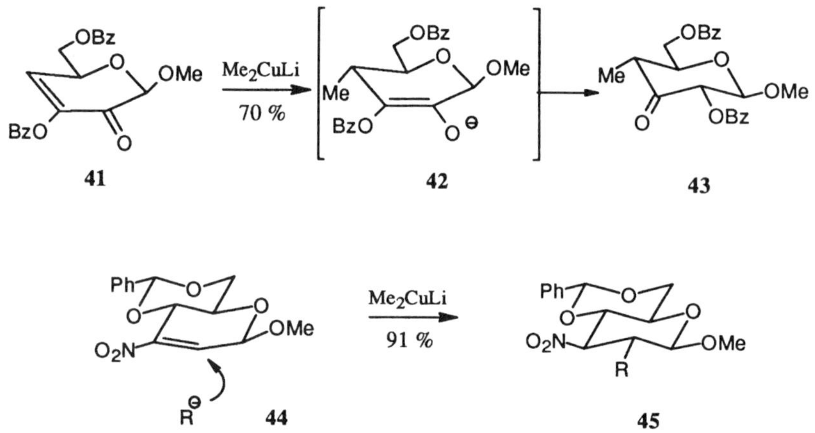

**Scheme 19**

Nitroolefins are good Michael acceptors and have been used as such in the carbohydrate series. One advantage of this approach would be the subsequent transformation into branched-chain amino sugars after reduction of the nitro group. Baer has reported the

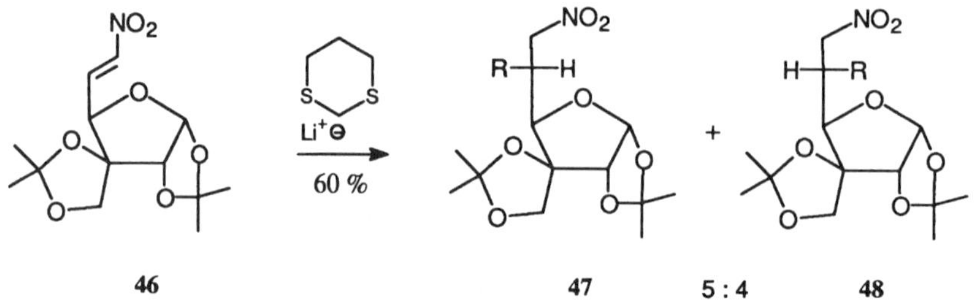

**Scheme 20**

## Methods for Synthesis of Branched-Chain Sugars

synthesis of C-2 branched-chain nitrosugar **45** by organocopper addition onto **44** which occurred *anti* to the aglycon [63] (see Scheme **19**).

As with epoxides, carbanions can add in a 1,4 fashion to enones or nitrosugars. Nitromethane anion has been used [64]. Dithiane anion has been successfully used in the addition to nitroolefins [65] and to enones [66]. Accordingly, C-5 branched-chain glucose derivatives **47** and **48** have been prepared from nitroolefin **46** (Scheme **20**) [67,68].

Sugar-derived enones have been also used as acceptors in free radical reactions to trap alkyl radicals as well as anomeric radicals (see Schemes **29** and **30**).

*Radical Reactions*

The recognition of the usefulness of radical reactions as a synthetic tool has prompted the exploration of this method for branched-chain sugars synthesis. Radical addition to an olefin is one of the most popular reactions yet investigated [10]. Two approaches have been devised: an *intramolecular* version, which is mostly a 5-*exo* cyclization and an *intermolecular* version in which the radical is trapped by an activated olefin.

**Intramolecular Reactions.** The intramolecular version is based on tethering, by carbon or heteroatom linkage, of an olefin to a carbon chain suitable for radical generation. The acetal linkage has been widely used as a "detachable radical" since its introduction by Ueno [69] and Stork [70]. In the carbohydrate field, two main pathways are investigated. In the first one, the radical is generated on the tether and added to a double bond located on the sugar template [Eq. (1), Scheme **21**]. The reverse principle is applied in the second version, for which the radical is generated on the sugar template and adds to an extracyclic olefin anchored to the sugar at the anomeric or at an allylic position by an oxygen atom [see Eq. (2), Scheme **21**].

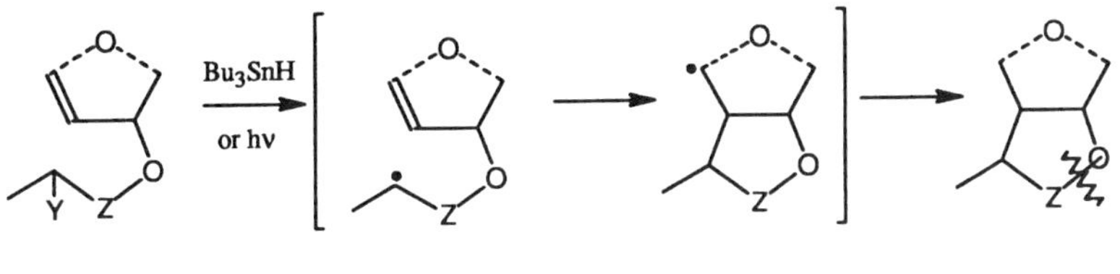

$Z = CHOR, SiR_2, CH_2, CH_2CH_2$

$Y = Br, I, SePh, OC(S)SR$ etc

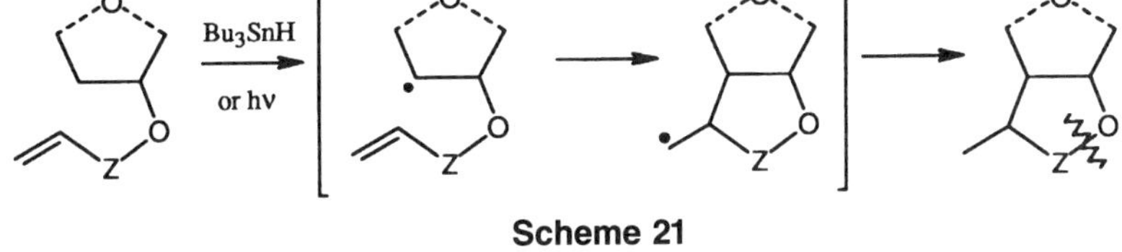

**Scheme 21**

The radical addition on unsaturated carbohydrates has been investigated independently by several groups [71–76]. We, and others, have studied several types of linkage. A glycosidic linkage is obviously used at the anomeric position. At other allylic positions we have used also acetal linkage [Z = CHOR; Eq. (1); 77], although a 1:1 mixture of epimers results from acetal formation [78]. The radical is generated from the corresponding bromide or iodide. The tin method has been used with success in the slow-addition mode. The configuration of the starting alcohol has no influence on the cyclization, which proceeds well in the 5-*exo* mode, with complete stereocontrol that always gives a *cis* ring fusion. The synthesis of C-2 and C-3 branched-chain structures **50** and **52** from **49** and **51**, respectively, is illustrated by Scheme **22**. Attempts to form six-membered rings are always sluggish and give extensive reduction of the radical precursor.

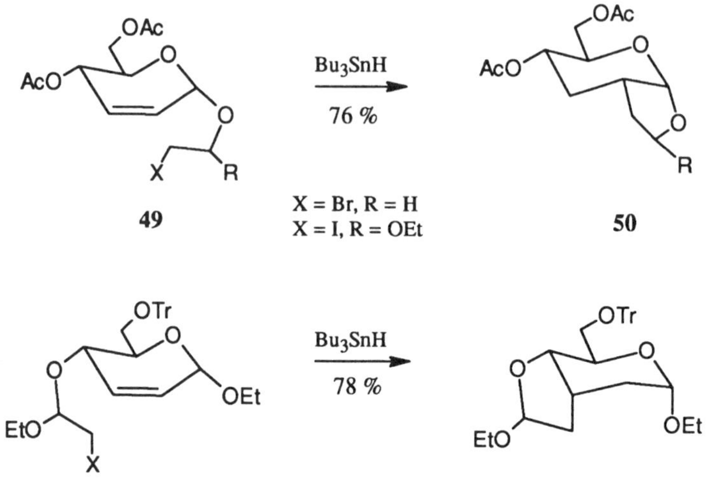

**Scheme 22**

This technique has been successfully used with 2,3- 3,4-, and 4,5- unsaturated sugars to form C-2, C-3, and C-4 branched-chain carbohydrates. Oxygen-substituted double bonds are also good radical acceptors [74,78]. Radicals also add to a glycal double bond to form C-2 branched-chain sugars [79].

Vinylic radicals, generated by tributyltin hydride addition on triple bonds, also add on 2,3 double bonds [80,81]. One advantage of this method is the presence of a tributylstannyl-

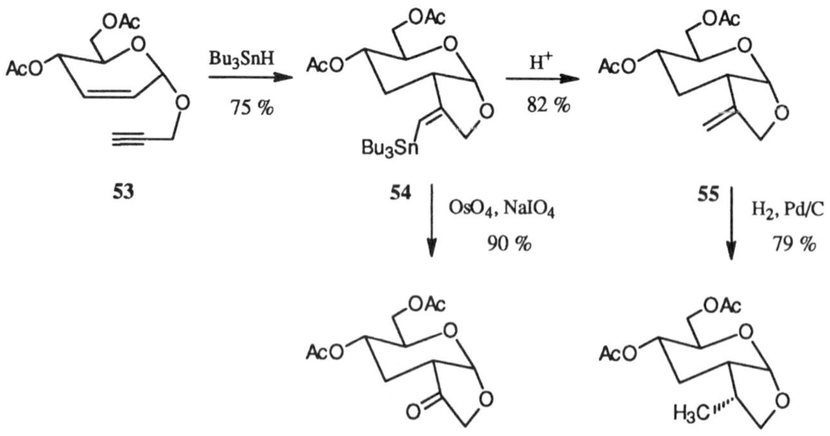

**Scheme 23**

## Methods for Synthesis of Branched-Chain Sugars

substituted vinylic bond in **54**, which can be destannylated to give **55** or oxidatively cleaved to the corresponding ketone **56** (Scheme 23). This is formally a unique acylation procedure at C-2 of a carbohydrate template. Reduction of the *exo*-methylene group in **55** gave only compound **57**.

The use of a silyl ether temporary linkage introduced by Nishiyama [82] and Stork [77,83] allows the facile cleavage of the five-membered ring formed in the cyclization by oxidation of the carbon–silicon bond. This procedure has been successfully used for the hydroxymethylation of sugars at position 3, 4, and 6 [84–86] (Scheme 24).

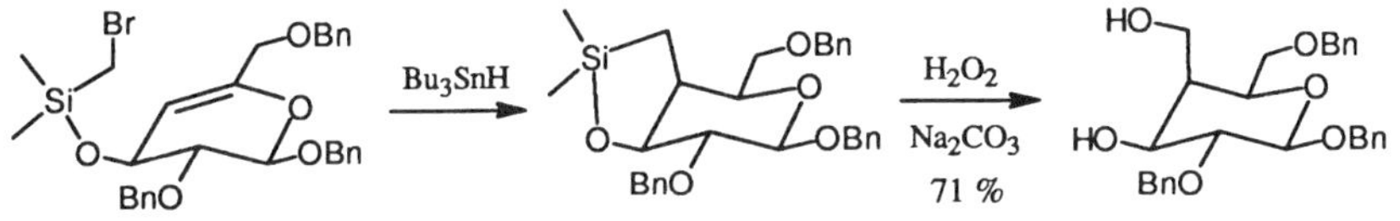

**Scheme 24**

In the second version of the intramolecular approach, the radical is generated on the sugar template and adds to an extracyclic olefin. As in the preceding approach, several variants have been proposed in the carbohydrate field and in the nucleoside field. In both cases, different linkages of the olefinic partner onto the sugars, such as acetal [87–89], ether [90,91], ester [92], carbon [93] or silyl linkages [94,95], have been used. This approach has been applied for the synthesis of C-2 branched-chain sugars, such as **62** and application to the connection of two sugar residues, such as **64** and **65**, has been reported by De Mesmaeker [96] (Scheme 25).

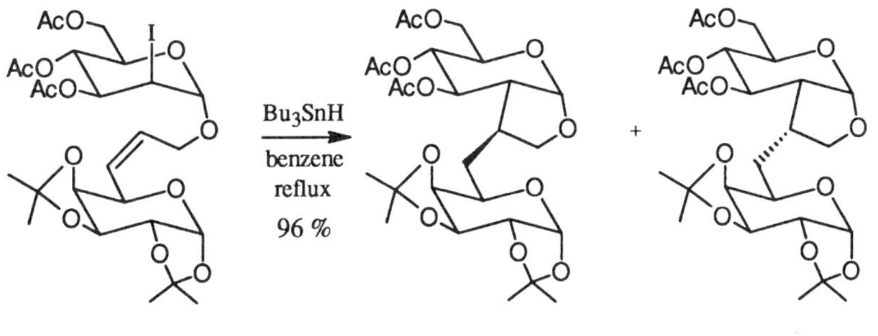

**Scheme 25**

The radical is generated on the sugar template using mostly the tin hydride method and halo derivatives [72,87,89], thionocarbonate [90–91], or selenium derivatives [91,94, 95]. Again, in these reactions, the kinetically controlled 5-*exo* cyclization is always preferred over the 6-*endo* cyclization. The question of the stereocontrol of this radical addition on the double bond is raised because a new stereogenic center may be formed in the process. The influence of the substituent of the olefin [96] as well as the configuration of the center bearing the tether have been discussed [91]. One relevant example in the field of nucleosides using phenylselenyl derivatives **67** as radical precursors is given in Scheme **26**.

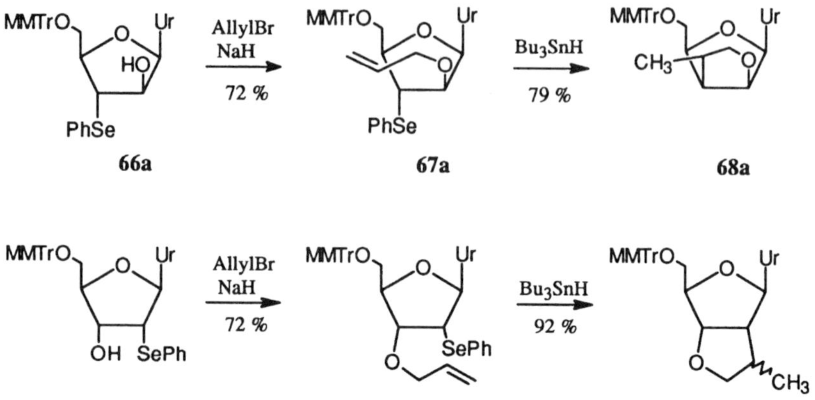

**Scheme 26**

The reaction has been applied to the synthesis of C-2′, C-3′ branched-chain uridine derivatives of either *lyxo* **67a** or *ribo* configuration **67b**. Interestingly, only one diastereoisomer **68a** is formed when the olefin chain is located on the β-face of the furanose ring (*lyxo* configuration), whereas a mixture of the two possible diastereoisomers **68b** is obtained when the olefin is located on the α-face (*ribo* configuration) [91] (see Scheme **26**).

The presence of a silicon atom in the tether strongly influences the cyclization mode. The 6-*endo* mode and even the 7-*endo* are favored over the 5-*exo* because of the greater Si–O bond length, compared with that of C–O [94,97].

The use of complex radical donors and radical acceptors, such as carbohydrates tethered by an acetal-containing linkage, has been reported by Sinaÿ [98]. Here, the radical addition of an anomeric radical to a 4-*exo*-methylene sugar derivative (compound **69**) proceeds in the 8-*endo* mode, with formation of a eight-membered acetal ring. The end product of this reaction, after removal of the tether and acetylation, is the C-disaccharide **70** (Scheme **27**).

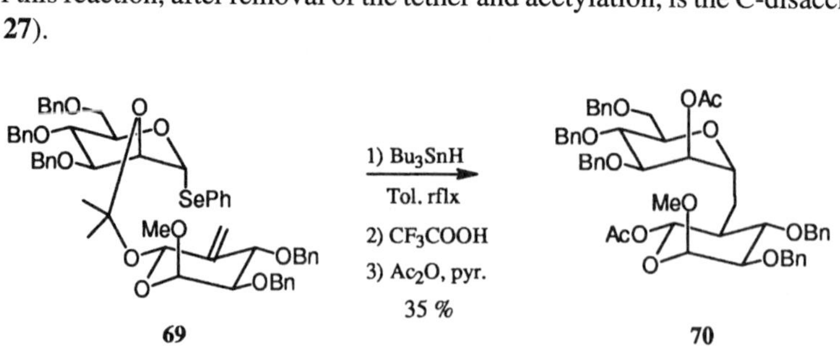

**Scheme 27**

Finally, this free radical approach is a good way to form carbocycles on the carbohydrate templates, provided the tether between the radical center and the olefin is a carbon chain [93,99–101], see also other contributions of this book describing the carbohydrate to carbocycle transformations.

## Methods for Synthesis of Branched-Chain Sugars

**Intermolecular Reactions.** The intermolecular version of free radical reactions of sugar-derived radicals consists mainly of addition onto suitably activated olefins, such as acrylonitrile, generally used in excess. This approach has been explored by Giese [102]. The stereochemical course of the reaction is dictated by steric effects of the vicinal substituents, as seen from the reaction of radical **71** where equatorial attack is favored over the axial with acrylonitrile (Scheme **28**). Only equatorial attack is observed using

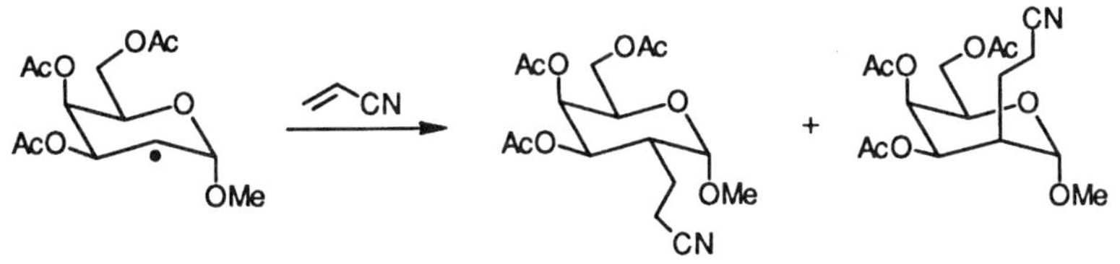

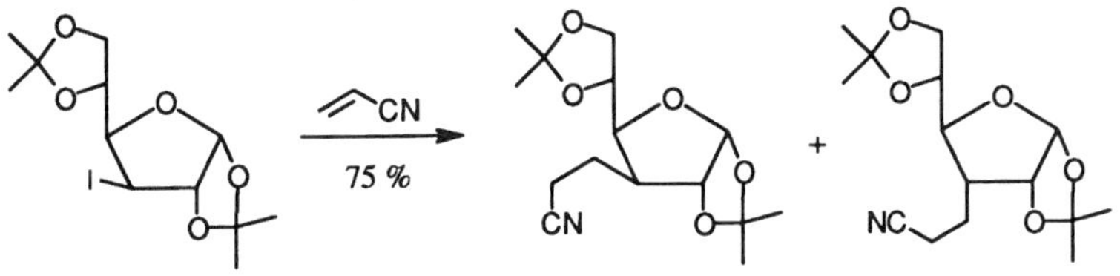

**Scheme 28**

fumaronitrile as the trapping olefin. More recently, the zinc–copper couple has been proposed to generate a radical from iodide **73**, which is trapped with good selectivity by acrylonitrile [103].

The photochemically induced radical addition of alcohols to enones has been described by Fraser-Reid [104–109]. Here again, the sense of addition depends on the steric effects of substituents, attack *anti* to the C-5 substituent being preferred [108,110]. Other uses of sugar-derived enones to trap radicals have been reported [111]. Enolone **77** gave interesting results in terms of selectivity [112]. In this instance, radical addition occurs with an equatorial selectivity, whereas cuprate addition occurs with an axial selectivity [9,62].

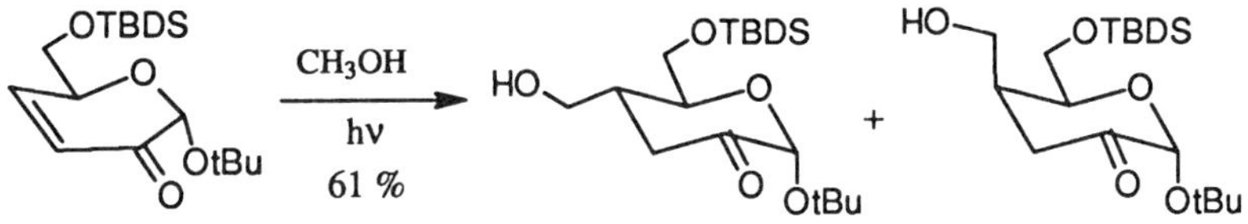

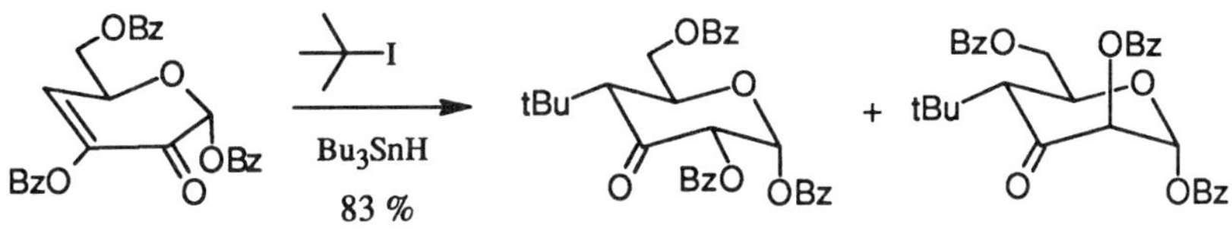

**Scheme 29**

Note also the radical migration of the benzoyl group from C-2 to C-3, as in the anionic version (Scheme **29**).

Anomeric radicals can be trapped by sugars having an *exo*-methylene group. This principle has been applied to C-2 *exo*-methylene lactones, opening the way to the interesting new class of carbon-linked disaccharides [113,114] (Scheme **30**).

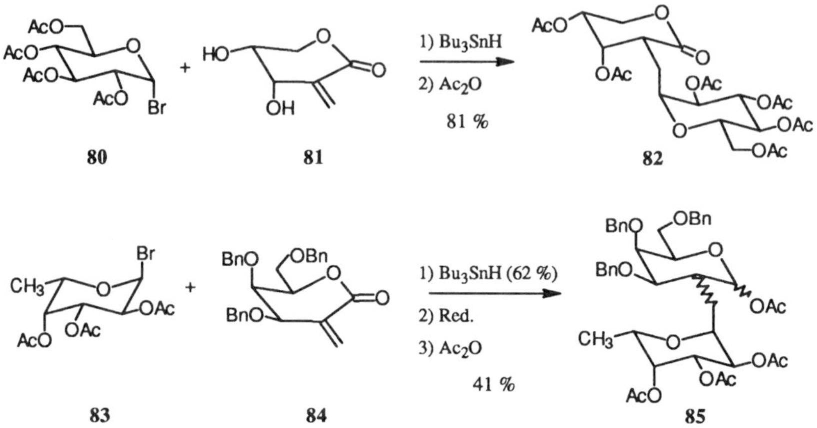

**Scheme 30**

In our group, we have explored the radical addition of alkyl radicals [97] and anomeric radicals [115] to sugar enones **134** having an *exo* double bond (see Schemes **46** and **47**). Comparison of radical and anionic 1,4-addition, shown here, yields strictly equivalent stereoselectivities. The selectivity also depends on the nature of the entering radical, and reversal of stereoselectivity is observed on going from primary alkyl radicals to the larger *t*-butyl radical [for analogous observations see also Ref. 116].

In all free radical cyclizations, a radical is formed on the sugar template (see Scheme **21** top line). Trapping of these intermediate radicals located at C-3 or C-2 [74] or at C-1 [79,85] with activated olefins has been reported and is a good way to obtain doubly vicinal branched-chain sugars in a single process.

Another interesting procedure is the free radical addition–fragmentation developed by Keck [117]. A radical is generated from a suitable carbohydrate precursor **86** and reacts with allyltributyltin to provide the allyl branched sugar **87** in good yield (Scheme **31**) [118].

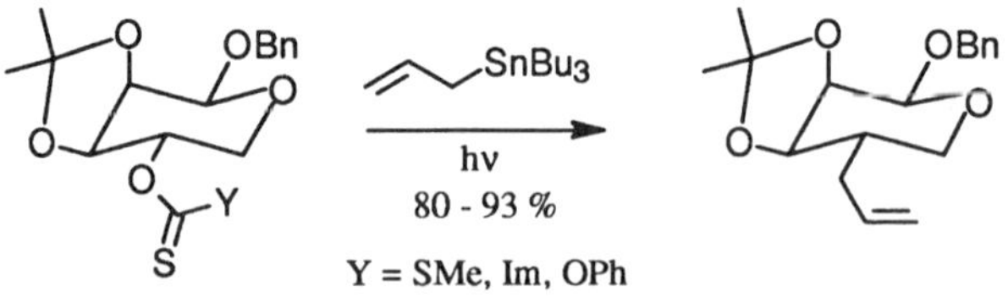

**Scheme 31**

This reaction has been also exploited for the branching at C-4 [99], and chain extension at C-6 of sugars [119].

Analogous addition of a sugar radical to an unsaturated sulfone has been described by the same authors in the course of pseudomonic acid synthesis. For this, UV irradiation of the iodosugar was used to generate a secondary radical [120].

*Direct Addition to Olefinic Compounds*

Allylic systems are interesting substrates for the investigation of new methods of C–C bond formation. Some of these methods have been adapted to the carbohydrate field. Enopyranosides are often crystalline compounds readily available, for example, using Ferrier rearrangement of tri-*O*-acetyl glycals with alcohols or its variants with carbon nucleophiles.

**Use of Organocopper Derivatives.** The reaction of allylic systems with organocopper, introduced by Crabbe [121], has been extensively studied by different groups using several combinations of organometallic species [55]. In our group some years ago, we investigated the application of this reaction to the carbohydrate field [122]. We found, for the first time, that carbohydrate allylic acetates react cleanly with cuprates obtained from alkyllithium and copper cyanide with complete regio- and stereocontrol. The use of cyanide as the ligand of copper [123] seems to be of tremendous importance. These now so-called low-order cuprates [55] gave excellent yields of C-2 branched-chain sugars in which the alkyl chain was introduced *anti* to the acetate leaving group. The reaction was a pure $S_N2'$ substitution.

The reaction is believed to proceed via a σ-allyl copper complex, in which the carbon–copper bond is formed at the γ-position, *anti* to the acetate leaving group. Reductive elimination of copper led to pure γ-substitution. With cyclic aliphatic allylic acetates, the selectivity is generally lower because the σ-allyl copper complex can isomerize to the π-allyl complex with loss of regioselectivity.

Our strategy has been applied to the well-known *erythro* derivative **88** (Scheme 32)

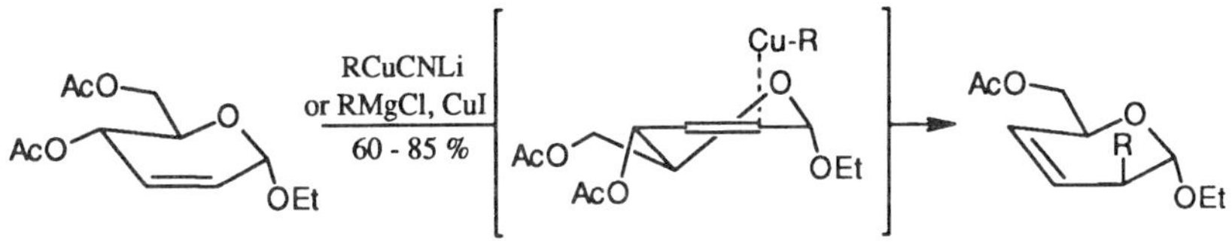

**Scheme 32**

using different alkyl groups, including phenyl and *t*-butyl, with the same complete regio- and stereoselectivity, whatever the alkyl group. More recently, in the context of tetrahydrolipostatin synthesis, we have investigated the reaction of copper iodide-catalyzed Grignard reagents. Although the C-6 acetate can be attacked by this reagent, it allows the introduction of long-chain aliphatic residues in good condition. This method is also efficient with C-glycosidic compounds and is one of the key reactions in Danishefsky's synthesis of avermectin $A_{1a}$ [124,125].

Other investigations along these lines have been reported using allylic benzothiazolyl (BTZ) ethers [126] and benzothiazolyl thioethers [127]. The reaction of copper-catalyzed Grignard reagents with the allylic systems such as **90** and **92** (Scheme 33) follow a different route. It proceeds with net γ-*syn*-substitution with alkyl group-derived organometallics. This is explained in terms of complexation of the copper reagent by the nitrogen atom of the benzothiazole ring system. This complexation forces attack of the copper reagent to occur *syn* to the leaving group. However, a phenyl group is introduced in the γ-*anti* fashion on compound **92**. This could be explained in terms of steric interactions that may result from the attack of the cuprate from the α-face of the molecule, *syn* to the aglycon. A case of direct γ-substitution has been reported by the same authors.

One may assume that all these reactions proceed by formation of a σ-allyl copper

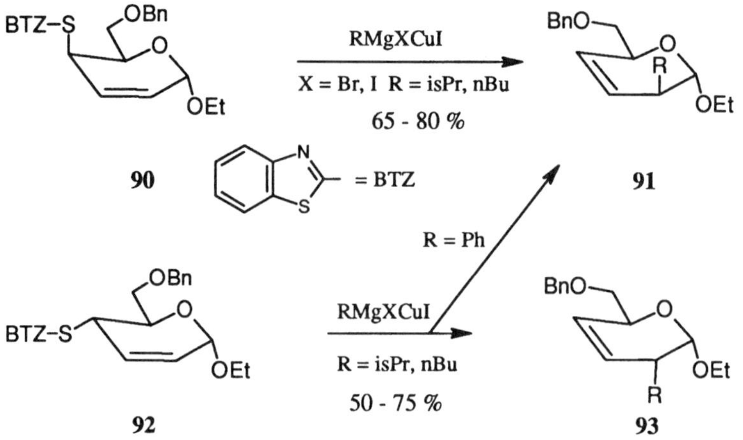

**Scheme 33**

complex and further reductive elimination. This σ-allyl bond is formed mainly *anti* to the leaving group [122,125], but when the leaving group is able to act as ligand of copper, such as BTZ, the σ-allyl complex is formed *syn* to this group. The initial formation of the σ-copper complex is supported by the fact that the epimer at C-4 of **88** did not react even with MeCuCNLi [128]. A strong steric interaction between the aglycon and the copper reagent precludes the formation of the copper complex at position C-2. In conclusion, this method, by a careful choice of the organometallic and of the activating group, allows the formation of carbon–carbon bond with a high degree of regio- and stereoselectivity.

**Using Palladium Chemistry.** Allylic halides and esters have been used in a palladium-catalyzed reaction with stabilized carbanions. The high stereospecificity of the reaction of dimethylmalonate that occurs with net retention of configuration (i.e., two inversions) is one of the main advantages of this approach [129]. Contrary to the use of organocopper reagent, the result is a direct replacement of the acetate group, with retention of configuration. This method has been used by Curran in a synthetic approach to pseudomonic acid [130,131]. The substitution of the acetate group by malonate anion is catalyzed by Pd(0)dppe and occurs with retention of the configuration and retention of the allylic system. This is exemplified on Scheme **34**.

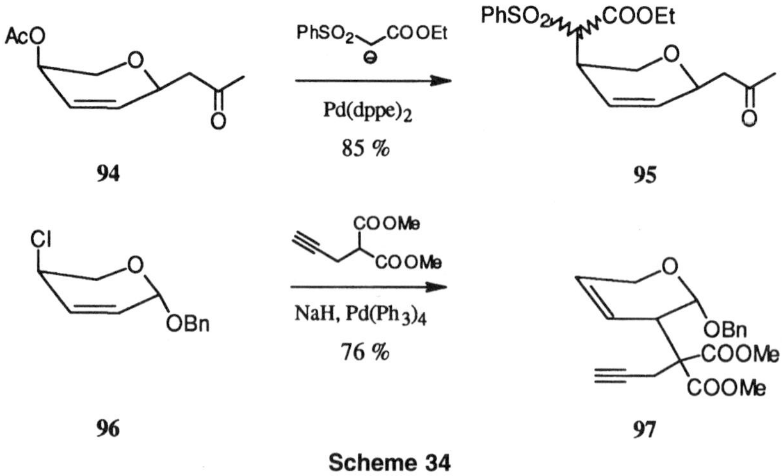

**Scheme 34**

Recently the use of allylic halide (see compound **96**, Scheme **34**) or sulfonate for the introduction of a malonyl residue at C-2 of pyranose derivatives using tetrakis(triphenylphosphine)palladium complex has been reported [132]. In this case, the malonyl chain is introduced *anti* to the leaving group at the γ-position.

**Nitrone–Nitrile Oxide Cycloaddition.** Unsaturated sugars have been used for the simultaneous formation of carbon–carbon and carbon–oxygen bonds in a *cis*-relation. One of the best ways to achieve this transformation is the cycloaddition of nitrone or nitrile oxides. The cycloaddition of nitrones with olefins has been reviewed [133]. The regioselectivity is almost complete when using activated double bonds, such as enone, enelactone (see compound **98**, Scheme **35**), or esters.

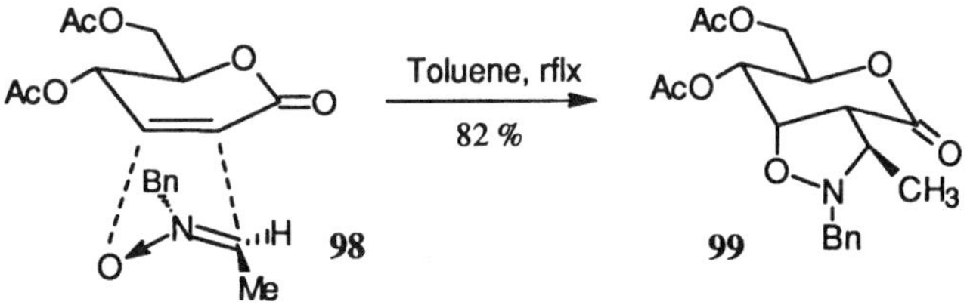

**Scheme 35**

This reaction has been discussed in terms of stereoselectivity, which is often very high on conformationally stable sugar templates [134]; a *cis*-fused isoxazolidine is always obtained. Subsequent cleavage of the N–O bond unravels the hydroxyl group on the sugar ring and the amino group on the branched chain [135]. With nitrones, a new stereogenic center is formed in the cycloaddition process (see compound **99**, Scheme **35**). The *endo* approach of the Z nitrone accounts for the observed configuration at this center [134,136]. One of the limitations of this approach is the often low reactivity of complex nitrones. Nitrile oxides have also been used [137,138]. An intramolecular version of this approach has been reported by Curran in his elegant synthesis of (−)-specionin from D-xylose [139,140].

*Rearrangements*

Molecular rearrangements, such as Wittig, Claisen, or Ireland–Claisen, have been applied to carbohydrate chemistry. The starting compound is generally an allylic alcohol. In a pioneering work, Ferrier and associates [141] describe the Claisen rearrangement of allyl vinyl ether derived from 2,3-unsaturated glycosides on heating at 180°C. A two-carbon chain, bearing an aldehyde function, is introduced at C-2 with a configuration dictated by that of the starting allylic alcohol, the rearrangement occurring in a *cis* fashion [141]. This strategy has been adapted to introduce the branch at C-3. Isomerization of allyl glycoside **100** gives the *O*-vinyl derivative **101** (R = H) which, on heating, rearranges to the C-3 branched glycal **102** (Scheme **36**) [142]. Other conditions have been proposed by Descotes, starting from 2,3-unsaturated glycosides of 1-hydroxypentanone **103** [143]. Photolysis of this glycoside yields the corresponding anomeric *O*-vinyl derivative **101**, which undergoes the thermic rearrangement. Glycoside **104** prepared from 2-(phenylselenyl) ethanol is also a valuable intermediate en route to *O*-vinyl glycosides [144]. The Still–Wittig rearrangement of stannyl ether has been used for the introduction of a hydroxymethyl group at C-4 of hex-2-enopyranosides [145].

The introduction of an *N,N*-dimethylacetamido function is possible from allylic

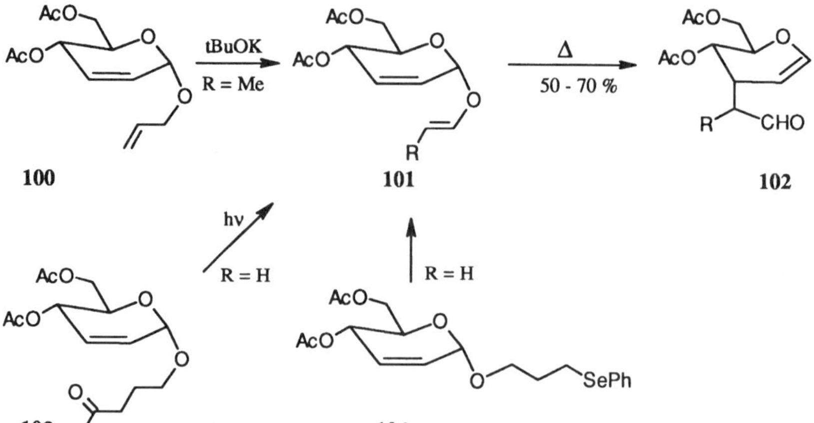


100 101 102 103 104

**Scheme 36**

alcohols, using the Eschenmoser variant of Claisen rearrangement [146] with amide acetals. This procedure has been applied to the synthesis of thromboxane $B_2$ [147]. The *ortho*-ester variant [5] of Claisen rearrangement has also been studied. For the same purpose, Pelyvas and Thiem recently investigated the Eschenmoser–Claisen and the *ortho*-ester rearrangements to introduce a two-carbon chain at C-4 [148]. Yields varying from 50 to 70% are obtained in the transformation of **105** into **106** using both methods (Scheme **37**). A typical procedure of this process is given in the experimental section.

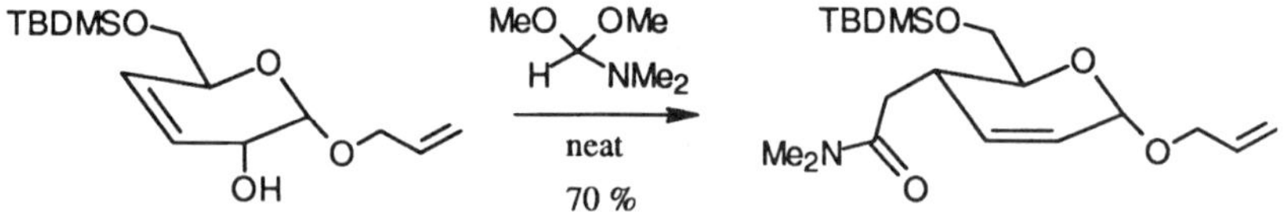


105 106

**Scheme 37**

Another interesting variant of Claisen rearrangement has been introduced by Ireland [149], and used by his group in carbohydrate chemistry. The starting compound is again an allylic alcohol that is esterified by a suitable carboxylic acid. This ester is enolized in basic medium, and quenching of the intermediate enolate at low temperature gives a ketene silyl

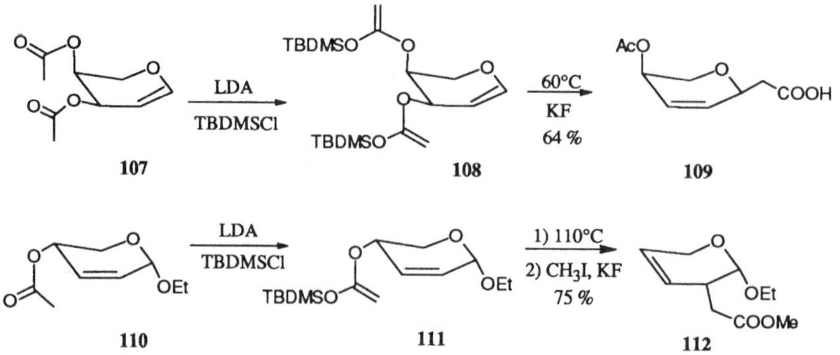


107 108 109 110 111 112

**Scheme 38**

# Methods for Synthesis of Branched-Chain Sugars

acetal, which rearranges on warming or heating. This tactic has been successfully used for the connection of two sugar units in the total synthesis of lasalocid A [150]. In the course of pseudomonic acid synthesis, Curran described the formation of C-glycoside **109** from glycal derivative **108** where two hydroxyl groups were protected as ketene silyl acetals [130]. The first rearrangement gave a second allylic system that did not rearrange under the same conditions [151]. The same strategy has been used in the synthesis of specionin [139,140]. The first step is the formation of the C-2 branched-chain sugar **112** from allylic acetate **110** [152] (Scheme **38**).

*Alkylation and Aldolization of Carbohydrate Enolates*

The chemistry of enolates has provided excellent routes to highly complex structures, in particular in the total synthesis of natural products. Because of the highly oxygenated structures of carbohydrates, enolate formation could easily result in β-elimination of a suitably located oxygenated group (ethers, esters, and such) to provide enone. For these reasons, the chemistry of carbohydrate enolates has been poorly documented.

One seminal example has been reported by Butterworth and associates, who described the methylation of a 2-deoxy 3-ketosugar with methyl iodide and barium oxide in dimethylformamide (DMF) [153].

Given the availability of compound **18** from the reaction of methyl di-*O*-benzylidene-D-mannopyranoside **113** with butyllithium [154,155], we reasoned that the intermediate enolate **114** should be sufficiently stable to be alkylated at low temperature (Scheme **39**). We

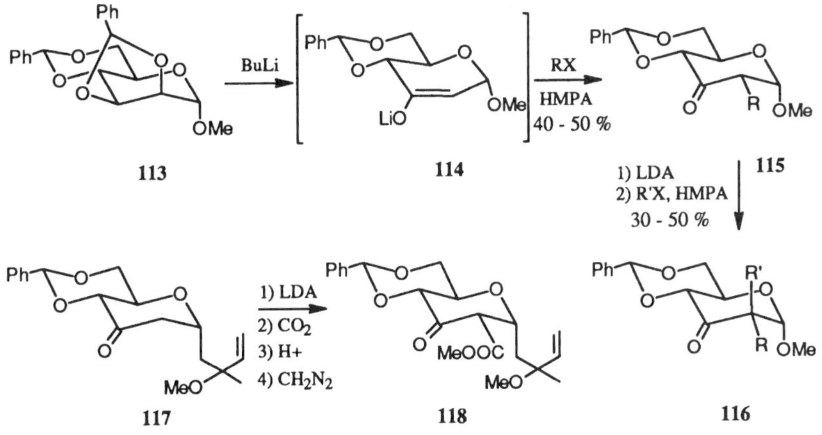

**Scheme 39**

have shown this was indeed true [11]. Alkylation occurs with complete stereoselectivity, giving the equatorially oriented derivatives branched at C-2 **115**. The process can be used a second time with these monoalkyl derivatives to provide dialkyl compound **116**, having a quaternary chiral center that can be obtained in a predictable configuration. Fraser-Reid's group has shown that the enolate alkylation reaction can be applied to C-glycosidic structures such as ketone **117** to give, for example, C-2 acyl derivatives **118** [156]. Further developments by this group in relation with the synthesis diquinanes structures [157], and by others, have been reported [158]. The unusual stability of the enolate **114** toward aglycon elimination has been explained in terms of stabilization of the negative charge at C-2.

The aldol reaction of enolate **114** has been investigated by us [12] and Fraser-Reid

[159–161]. Excellent β-facial selectivity on the enolate was observed, but there was a lower facial selectivity on the aldehyde partner. The cation was of tremendous importance, as seen from the reversal of selectivity when going from lithium to zinc or magnesium enolates [12] (Scheme **40**). This is explained by a Zimmerman–Traxler model in which a

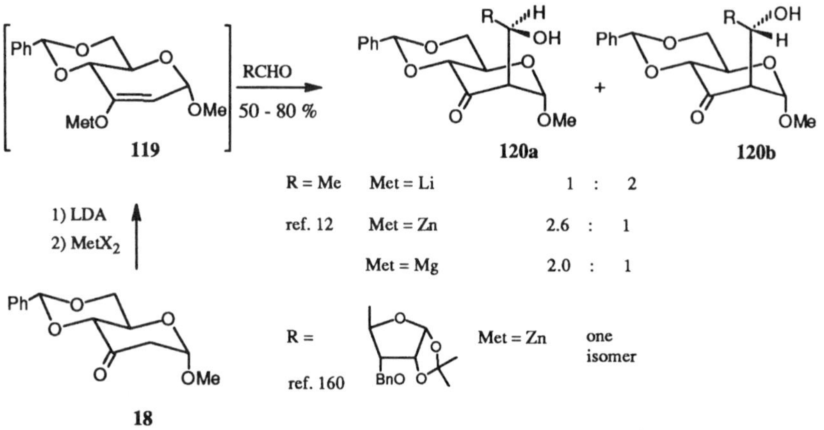

**Scheme 40**

chelation of the bivalent cation with the ring oxygen is assumed. Complex structures can be developed using this process, in particular a sugar aldehyde can be linked through this method with high diastereoselectivity [160,162].

This enolate approach is also efficient for methylation as well as hydroxymethylation of α-oxygenated enolates using methyl iodide or formaldehyde as the electrophile, as demonstrated in a series of papers by Klemer's group [163].

A related method involving the samarium iodide-mediated coupling of α-benzoyloxy lactones with ketones has been reported recently by Enholm [164,165].

One of the limitations of this approach is the necessary absence of a substituent at C-3 which could be eliminated. Probably, for that reason, the coupling of the assumed Sm(III) enolate derived from **121** (Scheme **41**) with simple ketones proceeds with modest facial

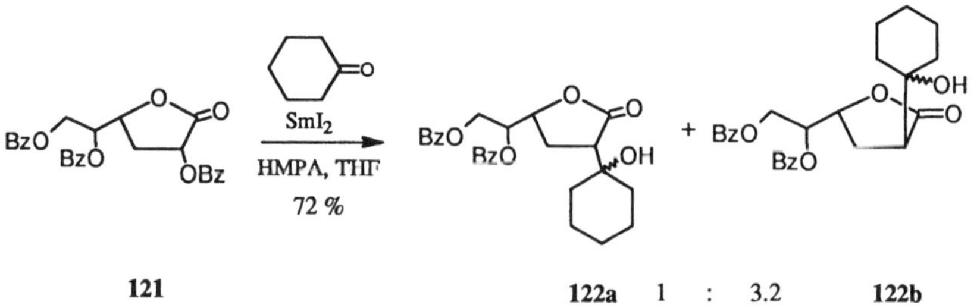

**Scheme 41**

selectivity on the enolate to give **122**, but with higher diastereoselectivity using complex ketones, such as (−)-methone (14:1) or dihydrocarvone (99:1).

*Miscellaneous Methods*

Other methods are available to prepare type II branched-chain sugars. In particular the classic reduction of the double bond of type III branched-chain sugars (see path **g**,

Scheme **3**) is a valuable process that can take advantage of steric effects to direct the reduction in the desired direction. Many examples are available in the literature [1].

## C. Compounds with a C=C Branch (Type III)

*From a Carbonyl Group*

As previously reported in the first part of this chapter, nucleophilic addition to keto sugars is the central point of the approaches to oxygen-substituted branched-chain sugars. Obviously, the availability of ketosugars has prompted their use as surrogates for the synthesis of branched-chain sugars having a double bond at the branching point (Scheme **42**). The

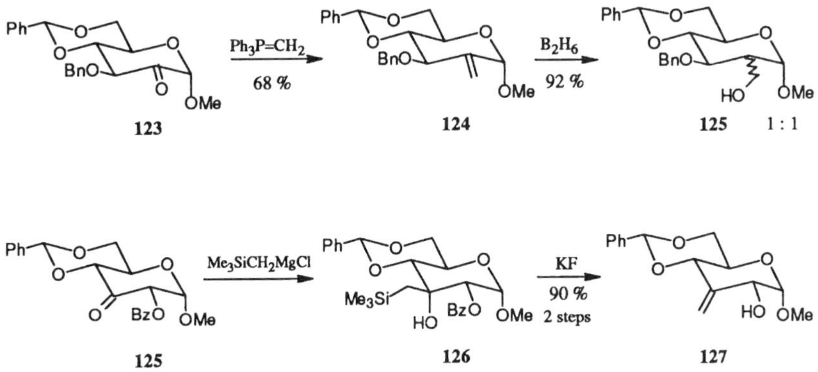


**Scheme 42**

main reaction used is the Wittig reaction and its variants (e.g., Wadworth–Emmons, Wittig–Horner, and others). Its use in carbohydrate chemistry has been reviewed [13]. Both stabilized and unstabilized Wittig-type reagents have been used. Methylenation has received much attention because it is an easy way to introduce a methyl group after hydrogenation or to obtain more oxygenated derivatives, for example, by dihydroxylation of the double bond. In this case, stereoselectivity opposite that resulting from nucleophilic attack on the carbonyl group may be observed [compare Ref. 166 and 167]. *exo*-Methylene derivatives are also good starting compounds for the synthesis of hydroxymethylated sugars by hydroboration of the methylene group [168,169]. The more recent Peterson olefination method has been investigated with success for the introduction of a methylene group [39,170]. An interesting use of the Tebbe's reagent for the synthesis of 4-C-formyl glucose derivative from the corresponding 4-keto derivative by the 4-*exo*-methylene derivative has been reported recently by Schmidt and Dietrich [169].

An interesting application to the direct methylenation of enol ester **128** (Scheme **43**)

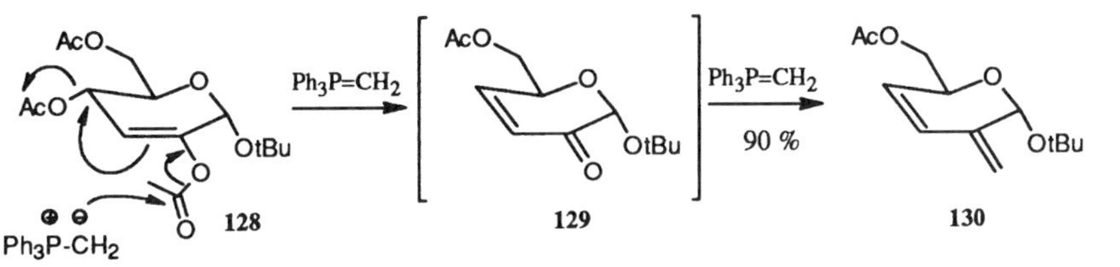


**Scheme 43**

has been described by Hanessian's group in the course of pheromone synthesis [171]. Diene **130** is formed in excellent yield through the intermediate enone **129**, which was not isolated.

Introduction of a two-carbon functionalized chain is easily achieved by the reaction of ketosugars with carboethoxymethylentriphenylphosphorane [172] or related reagents [173]. A classic example is given in Section III [172] (Scheme 44). Here, the problem is

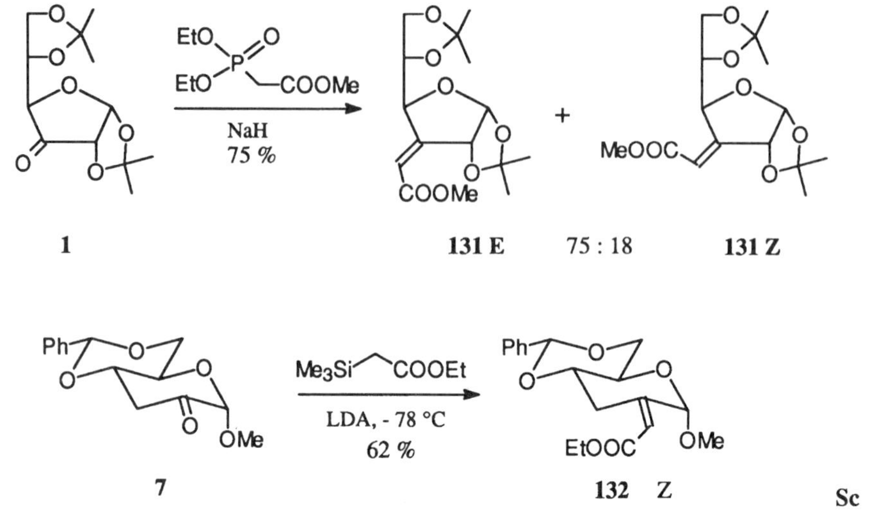

**Scheme 44**

obviously this *cis–trans* geometry control which can be achieved by careful choice of the solvent and reaction conditions. Again, in this case, reduction of the double bond or further manipulation of the double bond (e.g., 1,4-addition [174] or hydroxylation [175]) are possible with these branched-chain sugars to reach more complex branched-chain structures.

The Peterson olefination has been also investigated for the introduction of a two-carbon chain as in **132** [39,175,176]. This alternative procedure compares well with the reaction of stabilized Wittig-type reagents [compare both procedures in Ref. 175].

Another route to create a carbon–carbon double bond on a ketosugar is the Knoevenagel reaction and its variants. It has been applied by Szarek and Ali to prepare olefin **133** (Scheme 45) suitable for the formation of a doubly branched-chain sugar [177] [see Section II.D].

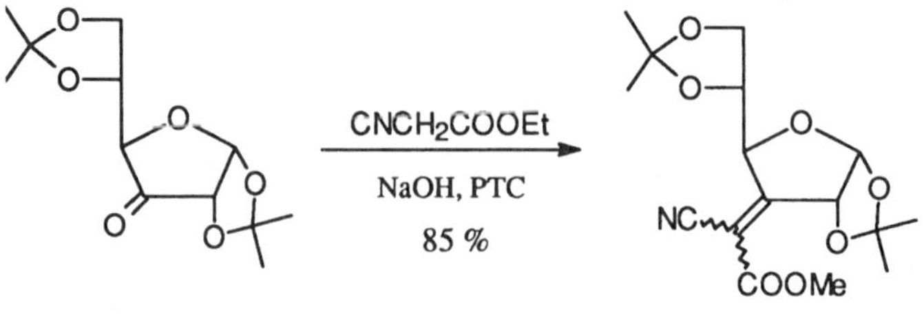

**Scheme 45**

*Aldolization–Crotonization*

Aldol reactions of sugar enolates have provided a good entry to C–C double bond branched-chain sugars [178]. As already mentioned, condensation of aldehydes, such as acetaldehyde or propionaldehyde, gives a mixture of aldol **120** of *R,S* configuration at the

## Methods for Synthesis of Branched-Chain Sugars

newly created chiral center (see Scheme 40). Facile epimerization of the axially oriented hydroxyalkyl chains readily occurs on column chromatography. Formation of an enone structure readily occurs on mesylation or acetylation of the aldol mixture by β-elimination. A single olefin **134** having the *E* configuration is obtained.

These activated double bonds can undergo various interesting reactions, providing new entries to more complex branched-chain structures. For example, the enone system **134** undergoes facile and clean hetero Diels–Alder reaction with electron-rich olefins under europium salt catalysis [178]. Only one isomer **135** is formed by an *endo*-transition state. This two-step procedure affords an easy way to incorporate new chiral centers in the branched-chain with often high regio- and stereoselectivities. Compounds **134** also undergo facile cuprate or radical 1,4-addition (Scheme 46). In both cases, the same type of regio-

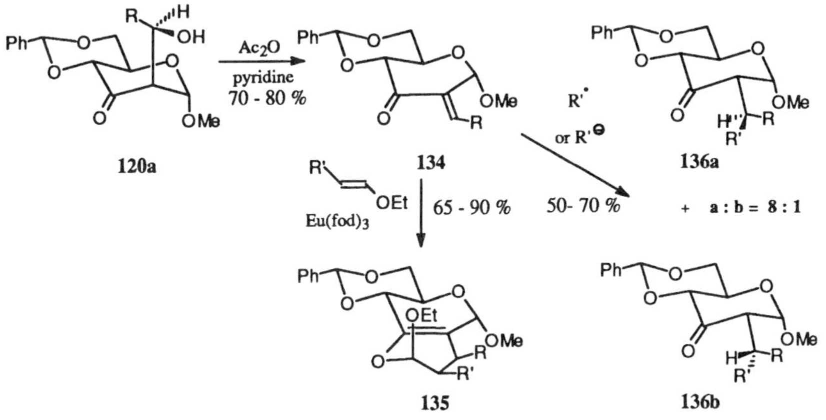

**Scheme 46**

selectivity is observed. The major compound **136a** results from attack of the nucleophile from the less-hindered β-face.

Intramolecular free radical reactions have been also exploited using silicon temporary attachment [97]. Reduction of enone **134**, gave the corresponding allylic alcohol, which is transformed into the bromomethyl-(dimethylsilyl) ether **137** (Scheme 47). Radical cy-

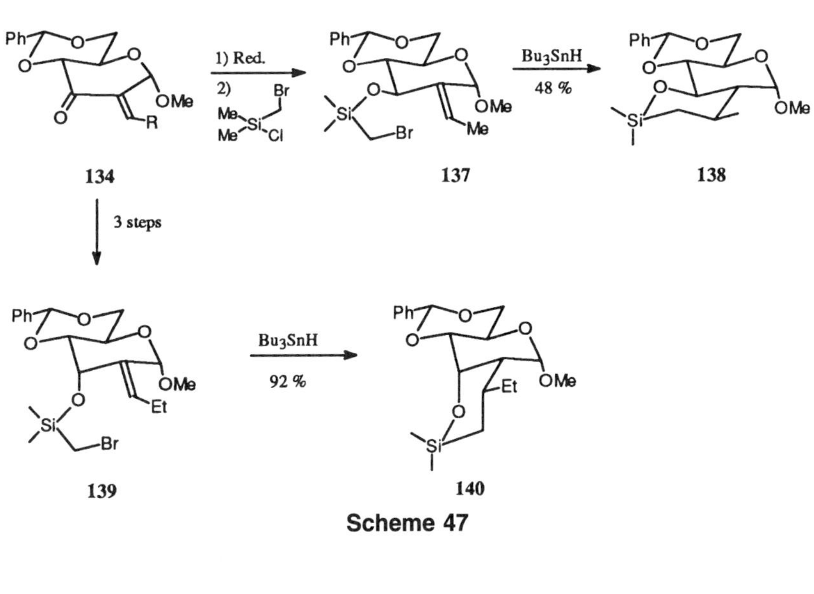

**Scheme 47**

clization gave the *trans*-fused six-membered ring compound **138**. However, the C-3 epimeric silyl ether **139** cyclizes readily to give the *cis*-fused six-membered ring **140**. The 6-*endo*-cyclization mode is the only observed pathway whatever the configuration at C-3.

Methiniminium salts have been used for the direct introduction of an alkylidene chain at the α-position of a carbonyl group [179]. 3-Chloro-3-phenylprop-2-enylidenedimethyliminium perchlorate reacts with 2- or 3-ketosugars **7** and **18** in the presence of DBU to form an *E–Z* mixture of compounds **141** and **142** (Scheme 48). Amide acetals also react with

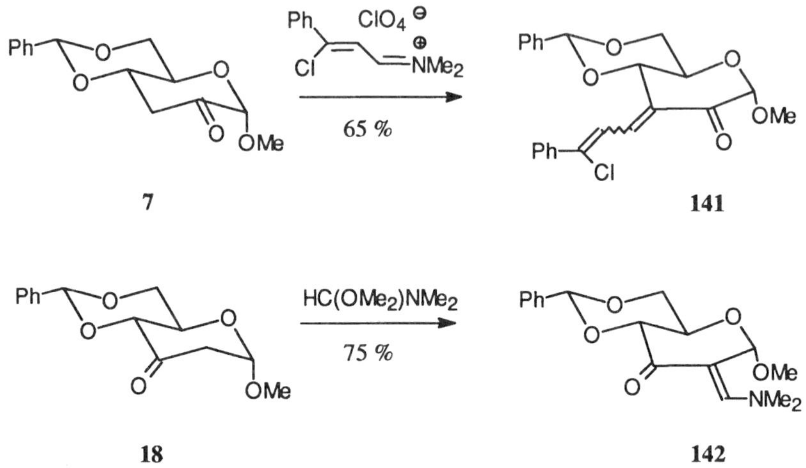

**Scheme 48**

ketone **18** to form the enamino ketone in excellent yields. These rather unstable compounds should be valuable intermediates for further chain extension on reaction with organometallics, for example [115].

Two new methods of formation of C–C bond at C-2 of glycals, with retention of the glycal double bond, have been published recently. Vilsmeier–Hack reaction is involved in the formylation of glycals [180]. A Friedel–Crafts reaction of acyl chlorides with glycals, in the presence of aluminium chloride, is required for the C-2 acylation of glycals (Scheme 49) [181].

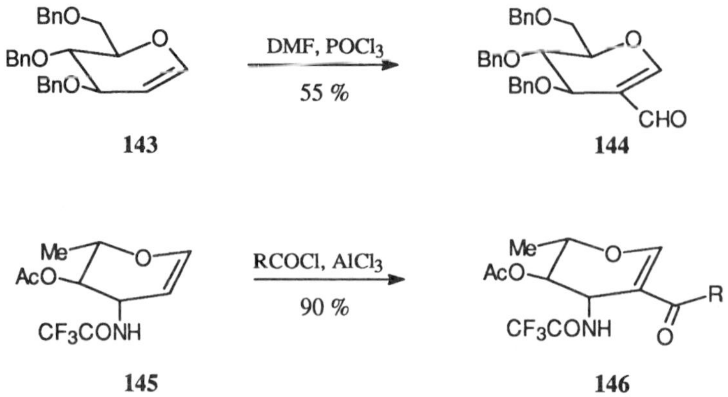

**Scheme 49**

## D. Compounds with Two Branched Chains

The synthesis of doubly branched-chain sugars should, in principle, follow the same principles through two successive formations of carbon–carbon bonds as described in the foregoing. However, additional methods especially designed to introduce two carbon chains on the same carbon (type IV) or on two vicinal carbons (type V) are now available.

*Geminal Doubly Branched-Chain Sugars (Type IV)*

The formation of quaternary chiral centers, which is of general interest in modern organic synthesis, has attracted interest from carbohydrate chemists. One of the earliest report came from Szarek and Ali, who reported nucleophilic addition of cyanide anion to olefins under phase-transfer catalytic conditions [177]. Nitromethylene derivatives also undergo nucleophilic addition of Grignard reagents to give the geminal branched-chain sugar **148** (Scheme **50**).

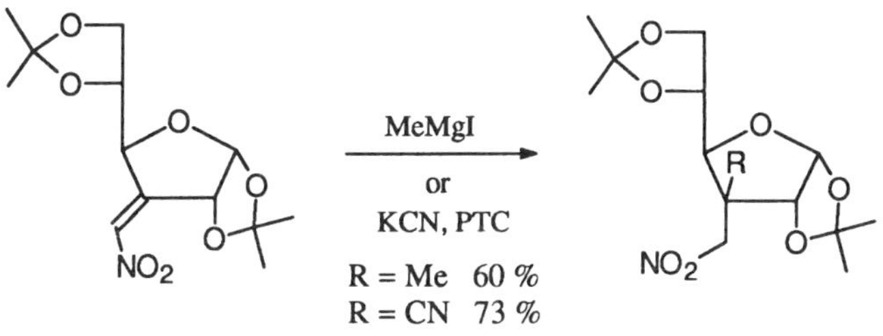

**Scheme 50**

As mentioned earlier, alkylation of sugar enolates is a method of choice to introduce two different carbon chains in a defined sequence to construct a quaternary chiral center with the desired configuration [11] (see Scheme **39**). Several other methods have been proposed, such as the ring opening of spirocyclopropanes [182] or the 1,4-addition of dimethylcuprate on a β-methyl-substituted enone [183].

A well-developed concept to achieve high stereocontrol in the formation of a quaternary chiral center has been introduced by Fraser-Reid, using the Claisen rearrangement along two lines. In the first approach a type III branched-chain sugar is prepared by Wittig

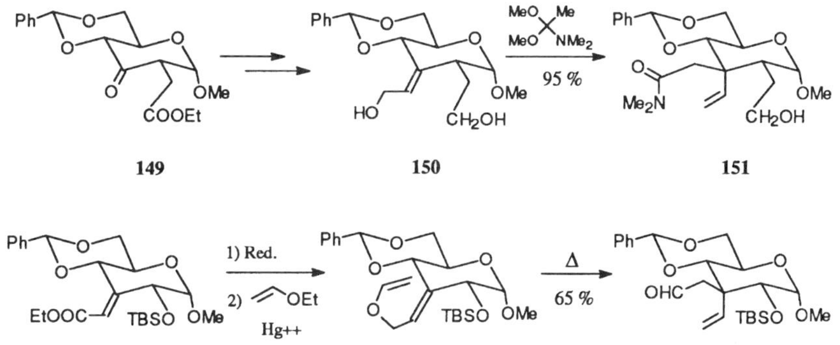

**Scheme 51**

reaction and subsequent reduction of the ester function to yield the allylic alcohol. Vinylation of this alcohol gave the required allyl–vinyl ether, which undergoes thermal rearrangement to provide the vinyl branched-chain type IV sugar (Scheme 51) [156,174,184].

The stereochemical course of this rearrangement depends on the position of the sugar ring at which the reaction is performed. Equatorial attack of the folded vinylic system is favored at position C-3, whereas axial attack is predominant at C-2 and C-4. This was explained in terms of stereoelectronic interactions of the oxygen ring lone pair and the spiro carbon center orbital [185]. Alternative procedures using the Eschenmoser–Claisen rearrangement [157] or *ortho*-ester–Claisen rearrangement [186] have been described by the same group. The strategies described in these papers combine the enolate alkylation method (see Sec. II.B) to obtain a type II branched-chain sugar, then Wittig reaction (see Sec. II.C), which gave a type III branched-chain sugar, and finally a rearrangement technology to reach complex structures with high stereoselectivity. The *ortho*-ester–Claisen rearrangement has also been investigated by Tadano's group on hexofuranoses [172,187,188]. For example, allyl alcohol **155** of *E* configuration is heated in ethyl orthoacetate at 135°C to provide **156** as a single isomer (Scheme 52). The corresponding *Z* allylic

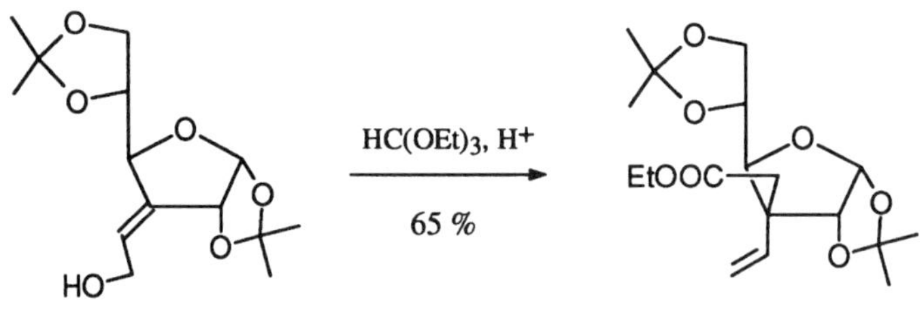

**Scheme 52**

alcohol gave a mixture of **156** and its epimer at C-3, albeit in low yield.

An interesting procedure has been developed by Vatèle to obtain C-5 branched-chain sugars [189]. It also takes advantage of the stereospecificity of the Claisen rearrangement of allyl (vinylsulfinyl) ether **158** derived from a 4,5-unsaturated sugar. An acrylic derivative **159** is formed in the process by elimination of sulfenic acid (Scheme 53).

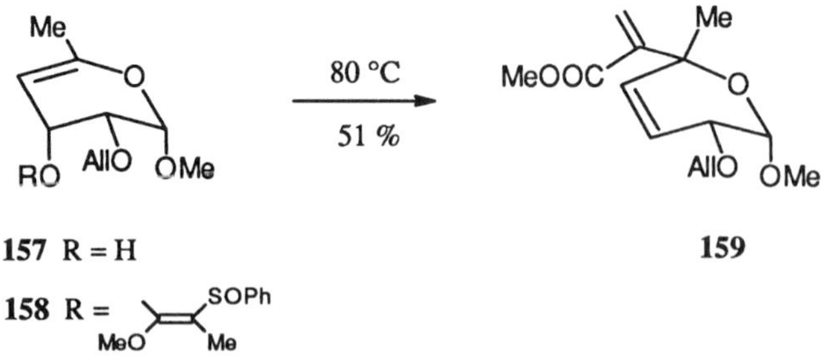

**Scheme 53**

*Vicinal Doubly Branched-Chain Sugars (Type V)*

The synthesis of vicinal doubly substituted carbohydrates can be achieved, as mentioned earlier, by successive formation of simple branched-chain carbohydrates. Other methods have emerged that are mostly based on the formation of 3-, 4-, 5-, or 6-membered ring fused to a pyranose–furanose ring. This process allows high stereocontrol on the two newly

# Methods for Synthesis of Branched-Chain Sugars

formed stereogenic centers. Some examples of these different approaches to cyclic structures will be examined in this section.

**Cyclopropane.** The formation of cyclopropane-containing sugars has been investigated by several groups. One interesting reaction is the direct cyclopropanation of epoxides using phosphonoacetate [190,191]. A synthesis of chrysanthemic acid, using cyclopropane **160**, has been devised using this strategy [192,193] (Scheme **54**).

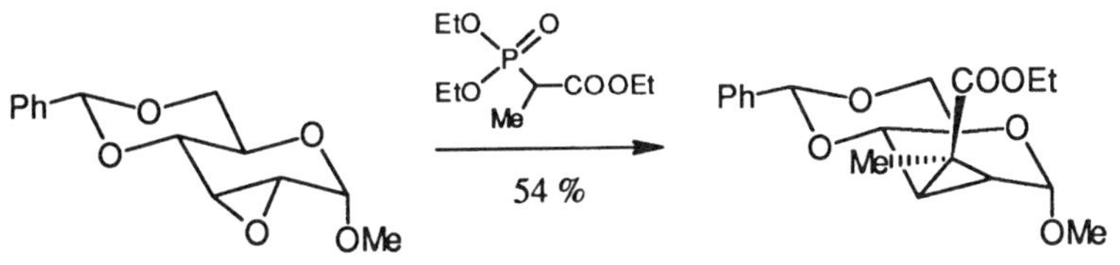

**Scheme 54**

Systematic studies of the dicyclopropanation of olefinic sugars have been published [194]. Dichlorocarbene, generated under phase-transfer catalysis reacts with 2,3-unsaturated pyranosides, such as **161** to yield a single isomer of the expected cyclopropane **162** (Scheme **55**). A typical procedure is given in Section III. The reaction is also possible with an enol

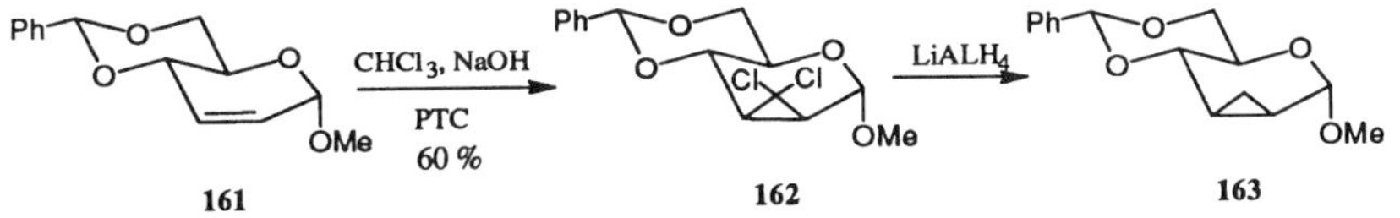

**Scheme 55**

ether double bond (3,4-eno-furanoside or 5,6-eno-pyranoside). The reduction of **162** to the cyclopropane derivative **163** has also been reported in the same studies [194].

**Cyclobutane.** A few examples of cyclobutane derivatives have been described in the carbohydrate series. Formation of this type of ring involves a 2+2 cycloaddition. Relevant examples of cycloaddition of dichloroketene on glucals, explored by Redlich [195] and Lallemand [196,197], and significant transformations of the four-membered ring, such as ketone **165a**, are given in Scheme **56**.

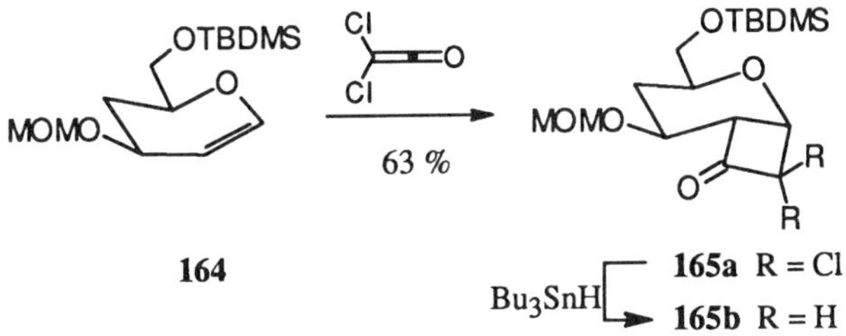

**Scheme 56**

**Cyclopentane.** Until recently there were only a limited number of reactions forming cyclopentane rings. The different strategies used to form cyclopentanes on sugar templates have been reviewed by Ferrier and Middleton [198]. Only recently published methods will be reported here. A strategy, based on the Pauson–Khand reaction of ene-yne [199], has been exploited recently by several groups. In this intramolecular strategy, a yne residue is anchored by an oxygenated tether to an unsaturated sugar. Thus, Lindsell et al.

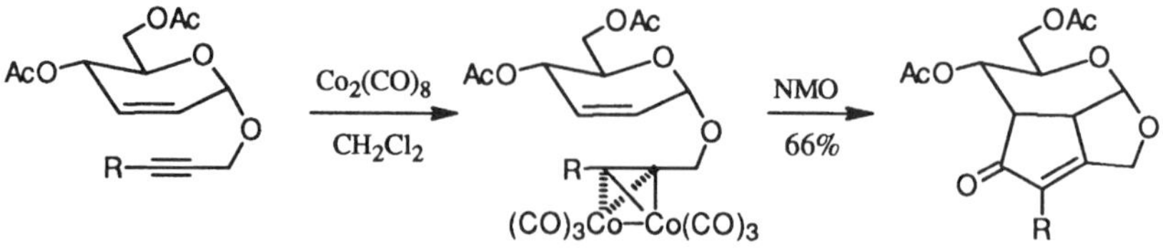

166 167 168

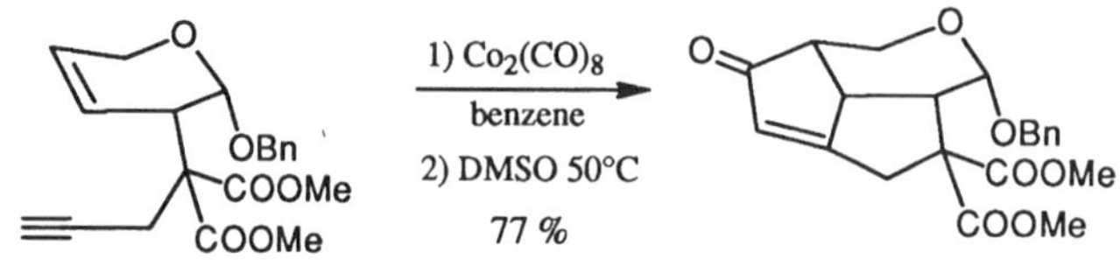

169 170

Scheme 57

[200] examined the reaction of the hex-2-enopyranosides such as **166** [80,81] (Scheme **57**) with octacarbonyldicobalt hexacarbonyl. However, if the hexacarbonyl dicobalt complexes with the sugar are formed, they do not undergo the expected Pauson–Khand reaction on $SiO_2$ or under CO atmosphere.

In a closely related work, Marco-Contelles succeeded in the transformation of **167** into cyclopentanone **168** using oxidative conditions. Different cyclopentanones anchored at C-1:C-2, C-2:C-3, or C-3:C-4 were obtained along this line [201].

More recently Voelter reported a related approach using a carbon tether to carry the yne derivatives [132]. The formation of these branched-chain sugars has been described in Section II.B. Formation of the hexacarbonyl dicobalt complex from ene-yne **169** in benzene, followed by heating in DMSO yields the biscyclopentano sugar derivative **170**.

**Cyclohexane.** Here again, a comprehensive review of the different methods used to form cyclohexanes in carbohydrate chemistry appeared recently [198]. Only the general strategies developed are given in the following. The formation of cyclohexane derivatives

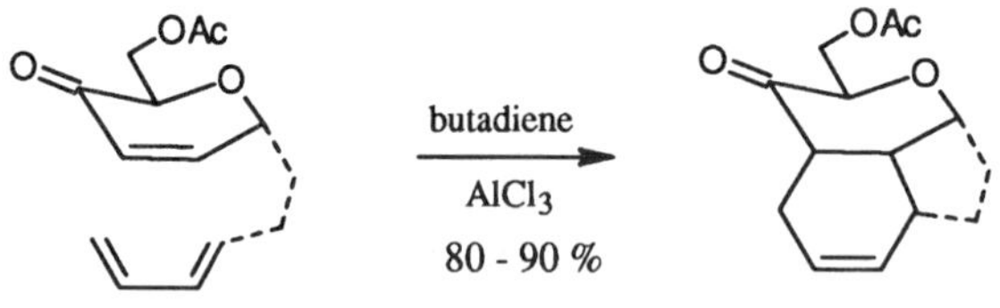

171 172

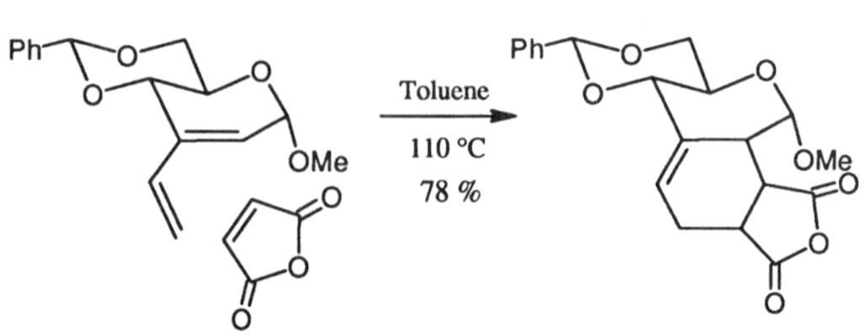

173 174

Scheme 58

# Methods for Synthesis of Branched-Chain Sugars

is by far the most studied, essentially using the Diels–Alder reaction, for which two main strategies have been adopted. In the first one, the unsaturated sugar (2,3-enoside) acts as the dienophile. More efficient results are obtained with activated double bond (enones) [202–205, see also Ref. 85 and references cited therein], or using an intramolecular Diels–Alder reaction (IMDA). In the latter approach, either an *O*-glycosidic tether [206] or a *C*-glycosidic one [85,207–209] has been successfully used to anchor the diene and the dienophile (see compound **171**, Scheme **58**). In the second strategy, the diene is first constructed on the sugar ring by appropriate C-branching of a vinyl appendage on an unsaturated sugar (compound **173**) and opposed an external dienophile, in general an activated double bond [210]. Either intermolecular [211–213] or intramolecular approaches can be used [214].

In the following we describe a typical example of the IMDA strategy taken from our recent investigations [215] toward the synthesis of the hexahydronaphthalene ring system of pravastatin, a potent HMGCoA reductase inhibitor [216]. The diene was anchored to the pyranoside system by means of a Ferrier rearrangement. Activation of the 2,3-double bond is necessary through oxidation at C-4. The cycloaddition of compound **175** proceeds well on heating in toluene in the presence of hydroquinone, giving the tricyclic derivative **176** as a single isomer (Scheme **59**).

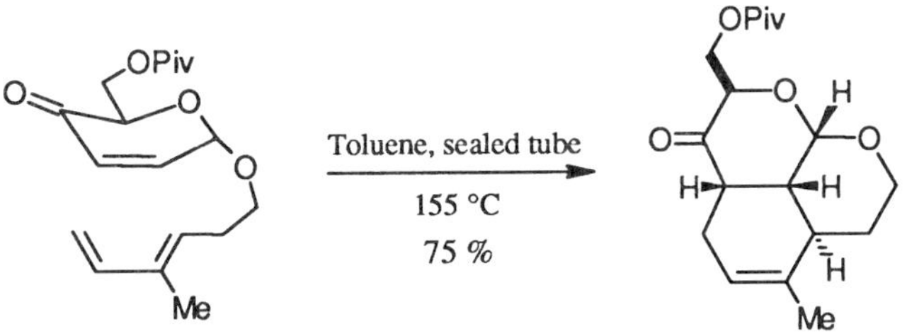

**Scheme 59**

## E. Conclusion

As a matter of fact, carbohydrate chemistry knows a renewed interest and, as seen from the foregoing examples, the synthesis of branched-chain sugars still continues to be a subject of intense investigations. Several reasons for this interest may be invoked. The large variety of sugar structures available, combined with the emergence of new synthetic methods promotes the application of many of these new processes to the carbohydrate field. A typical example can be found in the tremendous development of free radical reactions in the carbohydrate field, which started with the deoxygenation problem. New routes for carbon–carbon bond formation using free radicals have been discovered en route to branched-chain sugars and to more complex noncarbohydrate structures. The recognition that sugars can be nucleophiles in alkylation reaction is also of interest and has led to outstanding developments in relation to total synthesis. There is no doubt that the now growing field of glycobiology will stimulate the emergence of new methods. A particular need will certainly arise in the future in connection with the synthesis of new chimeric carbohydrates that may act as surrogates of biologically active carbohydrates. We will probably assist in the development of a "carbohydrate mimics" chemistry that will make large use of ancient methods of branching sugars, but will also stimulate the search for new selective processes for the formation of C–C bonds that are well adapted to carbohydrate substrates.

## III. EXPERIMENTAL PROCEDURES*

### A. Compounds with a C-C-X Branch (Type I)

*Tandem Swern Oxidation and Methyl Grignard Condensation [15]*

$$\mathbf{177} \xrightarrow[\substack{2)\ \text{MeMgBr, Et}_2\text{O} \\ -78,-50°C \\ 85\%}]{1)\ (\text{COCl})_2,\ \text{DMSO}} \mathbf{178} \quad (1)$$

To a solution of oxalyl chloride (67 μL, 0.77 mmol) in 2 mL of dry tetrahydrofuran (THF) at −78°C was added dimethylsulfoxide (DMSO; 57 μL, 0.81 mmol). This solution was allowed to warm to −35°C for 3 min and then recooled to −78°C. A solution of alcohol **177** (169 mg, 0.734 mmol) [Eq. (1)] in 1 mL of dry tetrahydrofuran was then added to the cool reaction mixture. The resulting solution was allowed to warm to −35°C and kept at this temperature for 15 min and then treated with triethylamine (0.51 mL, 3.7 mmol). The reaction mixture was allowed to warm briefly to room temperature and then recooled to −78°C. Methylmagnesium bromide (2.8-$M$ solution in ether, 1.31 mL, 3.67 mmol) was then added dropwise to the vigorously stirred reaction mixture. The temperature of the solution was allowed to warm to −50°C over 1 h, recooled to −78°C, and cautiously hydrolyzed with 0.5 mL of ethanol, and then 1 mL of a saturated aqueous solution of $NH_4Cl$ buffered to pH 8 with concentrated aqueous ammonia. The resulting mixture was warmed to room temperature and poured into 75 mL of the above buffer and extracted with ether (2 × 150 mL). The combined etheral layers were dried ($MgSO_4$). Concentration and chromatography (ether–petroleum ether, 1:1) afforded compound **178** (153 mg, 85%) as colorless oil: bp 100°C (0.005 mmHg), $[\alpha]_D$ +105° ($c$ 1.8, $CHCl_3$).

*Condensation of Ethylmagnesium Bromide [17]*

$$\mathbf{1} \xrightarrow[\substack{\text{H}\equiv\text{H} \\ \text{THF, r.t.} \\ 88\%}]{\text{EtMgBr}} \mathbf{2} \quad (2)$$

Into a 2-L three-necked flask equipped with a mechanical stirrer containing 650 mL of dry THF, purified, and dry acetylene gas was slowly introduced at room temperature for 1 h. While acetylene gas was continuously introduced, ethyl magnesium bromide (3-$M$ solution in ether, 100 mL, 300 mmol) was added dropwise over 5 h. The reaction mixture was stirred 1 h more after the addition. A solution of **1** (23.4 g, 90.7 mmol) [Eq. (2)] in 100 mL of THF was added dropwise to the reaction mixture over 40 min, and stirring was continued for 1 h. The reaction mixture was concentrated under vacuum to about 200 mL, and then saturated aqueous $NH_4Cl$ solution was added. Organic solvent was removed under reduced pressure, and the residual aqueous mixture was extracted with ether (4 × 200 mL). The combined

---

*Optical rotations were measured at 20–25°C.

etheral layers were washed with 1 N HCl, saturated aqueous NaHCO$_3$, and brine, and then dried (MgSO$_4$). Evaporation of the solvent provided a crude crystalline residue that was recrystallized from n-hexane–ether to give compound **2** (25 g, 88%): mp 105°C.

*Condensation of Vinylmagnesium Bromide [19]*

$$\text{3} \xrightarrow[\text{THF, -50°C}]{\text{CH}_2=\text{CHMgBr}} \text{4} \quad 92\% \tag{3}$$

Vinylmagnesium bromide (1-M solution in THF, 18 mL, 18 mmol) was added dropwise to a solution, at −78°C under argon, of ketosugar **3** or **5** (1 g, 3.6 mmol) [Eq. (3)]. After the addition was complete, the reaction mixture was kept at −50°C for 4 h and then quenched with a saturated solution de NH$_4$Cl (40 mL). This mixture was stirred for 1 h at room temperature and diluted with CH$_2$Cl$_2$ (150 mL). The layers were separated and the aqueous layer was extracted with CH$_2$Cl$_2$ (2 × 100 mL). The combined organic extracts were washed with water, dried (Na$_2$SO$_4$), and concentrated. The products were purified by column chromatography to give compound **4** (1.018 g, 92%) [Eq. (3)]; mp 163–167°C, $[\alpha]_D$ +114.2° (c 0.7, CHCl$_3$) or compound **6** (590 mg, 54%), mp 157–160°C, $[\alpha]_D$ +79.4° (c 0.6, CHCl$_3$).

*Condensation of Methoxyvinyllithium [21]*

$$\text{7} \xrightarrow[\text{THF, pentane, -60°C}]{\text{CH}_2=\text{C(OMe)Li}} \text{8} \xrightarrow[\text{dioxane}]{\text{HCl}} \text{9} \quad 41\% \tag{4}$$

To a stirred solution of 1-methoxyvinyl lithium [217] (22–30 mmol) in THF–pentane at −60°C under inert atmosphere was added dropwise a solution of compound **7** (2.1 g, 8 mmol) [Eq. (4)] in 40 mL of THF. The reaction mixture was stirred at this temperature for 30 min before it was allowed to warm slowly (0.5–1 h) to 0°C. It was then quenched with a saturated solution of NH$_4$Cl, and the resulting mixture was extracted with diethyl ether (3 × 100 mL). The combined organic layers were dried (MgSO$_4$) and concentrated to a syrup. The crude syrup, which contained compound **8**, was dissolved in 1,4-dioxane (100 mL) and treated with hydrochloric acid (0.04 M aqueous solution, 100 mL) at room temperature for 3 h. The aqueous solution was extracted with chloroform (3 × 150 mL), the combined organic extracts were washed with saturated solution of NaHCO$_3$ and dried (MgSO$_4$). Concentration under reduced pressure provided a crude crystalline residue that was recrystallized from diethyl ether–light petroleum to give compound **9** (1.05 g, 41%): mp 117–117.5°C, $[\alpha]_D$ +108° (c 1, CHCl$_3$); reported [218] mp 117°C, $[\alpha]_D$ +103° (c 1, CHCl$_3$).

*Reformatsky Reaction of Methylbromoacetate [32]*

To a suspension of C$_8$K (0.1 mol) freshly prepared in 15 mL of dry THF were added ZnCl$_2$ (14.95 g, 0.11 mol) and AgOAc (1.826 g, 0.011 mol) at room temperature. This suspension was heated at reflux for 30 min and then cooled to −78°C. To this cold suspension, a solution of ethylbromoacetate (16.7 g, 0.1 mol) and the ketone **18** (26.4 g, 0.1 mol) [Eq. (5)] in THF (400 mL) was added and the mixture was stirred at −78°C for 15 min. The mixture was then filtered, diluted with diethyl ether (400 mL), and washed with aqueous $Na_2HPO_4$ (100 mL). The organic layer was dried ($MgSO_4$) and evaporated under reduced pressure. The residue was purified by column chromatography to afford compound **19** (32.38 g, 92%): mp 90–91°C, $[\alpha]_D$ +72° (c 0.2, $CHCl_3$).

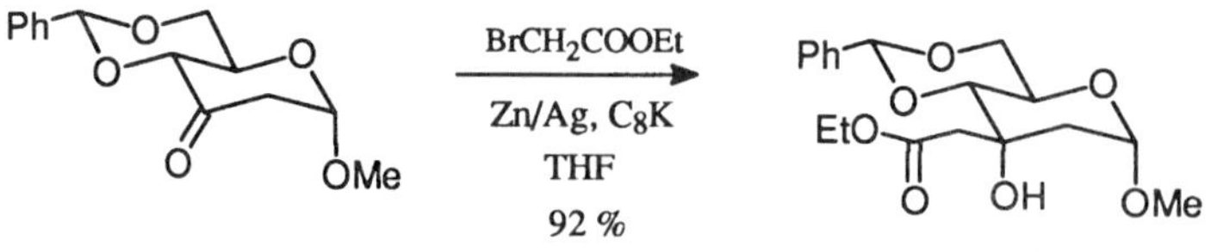

*Dreiding–Schmidt Condensation of 2-Bromomethylacrylate [35]*

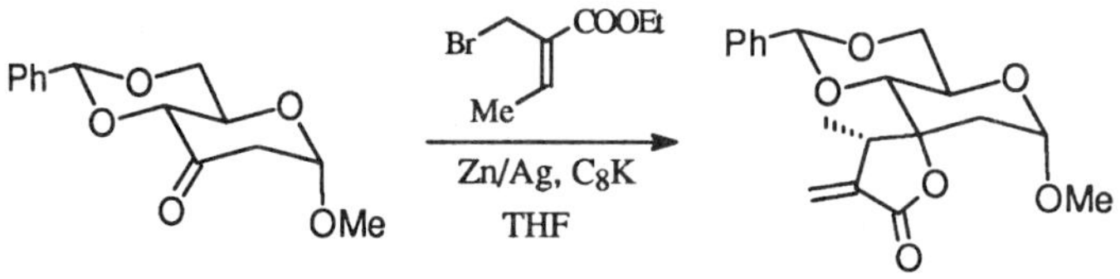

A mixture of degased graphite (1.56 g, 130 mmol) at 150°C for 30 min and clean, freshly cut potassium (0.66 g, 16.9 mmol) was stirred under argon at 150°C [32]. To the resulting bronze-colored $C_8K$ suspended in 100 mL of THF was added, in several portions, a mixture of anhydrous zinc chloride (1.1 g, 8.2 mmol) and silver acetate (0.12 g, 0.72 mmol) at room temperature with vigorous stirring. The addition of these salts caused the solvent to boil; heating with reflux was continued for an additional 25 min. The suspension was cooled to −30°C and the solution of compound **18** (1.90 g, 7.9 mmol) [Eq. (6)] and ethyl(2-bromomethyl)acrylate (1.70 g, 8.2 mmol) in 15 mL of THF was added slowly. After stirring for 7 h at −30°C, the mixture was filtered over a pad of Celite, diluted with ethyl acetate (150 mL), and washed with cold water (10 mL) and brine (10 mL). The organic layer was dried ($Na_2SO_4$) and concentrated under reduced pressure below 35°C. The residue was purified by column chromatography (AcOEt–hexane 10:1 ≥ 5:1) to afford compound **20** (1.84 g, 74%) as an oil: $[\alpha]_D$ +59° (c 1, $CHCl_3$).

## B. Compounds with a C-C-H (Type II)

*Opening of Epoxide with Dithiane [19]*

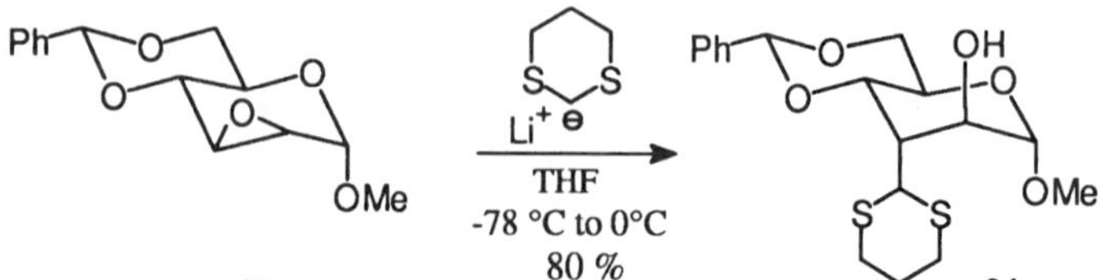

# Methods for Synthesis of Branched-Chain Sugars 243

To a solution of 1,3 dithiane (7 g, 60.8 mmol) in dry THF (300 mL) was added at −78°C butyl lithium (1.6-*M* solution in hexane, 35.6 mL, 57 mmol). After the addition, the reaction mixture was allowed to warm to −40°C and was stirred at this temperature for 1 h. The reaction mixture was recooled to −78°C and HMPA (15 mL, 183 mmol) was added dropwise to the solution and the mixture was warmed to −40°C for 30 min. The temperature was recooled to −78°C and crystalline epoxide **33** (5.02 g, 19 mmol) [Eq. (7)] was added. Stirring was maintained for 1 h at −78°C, and the reaction mixture was kept at 0°C for 60 h. The solution was washed with water (3 × 150 mL), the organic layer was dried ($Na_2SO_4$), and concentrated. The residue was purified by column chromatography (EtOAc–petroleum ether, 4:6) to give compound **34** (4.88 mg, 67%): mp 195°–197°C, $[\alpha]_D$ +112° (*c* 2.3, $CHCl_3$).

*Ring Opening of Epoxide with Organometallics [54]*

$$\text{29} \xrightarrow[\text{THF, 0°C}]{(C_6H_{13})_2CuCNLi_2} \text{30} \quad 76\% \tag{8}$$

To a suspension of CuCN (1.25 g, 14 mmol) in dry THF (150 mL) cooled at −78°C was added hexyl lithium (0.45-*M* solution in hexane, 60 mL, 27 mmol). The reaction mixture was allowed to warm to 0°C. Crystalline epoxide **29** (1.5 g, 5.62 mmol) [Eq. (8)] was added in small portions into the green, clear solution. The mixture was stirred at −10 to 0°C for 40 min to a dark, clear solution. Saturated aqueous $NH_4Cl$ solution (20 mL) was added, and the reaction mixture was diluted with $Et_2O$ (200 mL). The resulting suspension was filtered through a pad of Celite, the two phases were separated and the organic layer was washed with water (3 × 20 mL), dried ($MgSO_4$), and concentrated. The syrupy residue was purified by column chromatography (hexane–EtOAc, 6:1) to give compound **30** (1.5 g, 76%) as a gum: $[\alpha]_D$ +41.4° (*c* 1.5, $CHCl_3$).

*Conjugate Addition of Organocopper to Enone [61]*

$$\text{37} \xrightarrow[\text{Ether, -45 °C}]{MeLi, CuBr/Me_2S} \text{38} \quad 94\% \tag{9}$$

A suspension of dimethyl sulfide copper bromide (17 g, 82 mmol) in 600 mL of dry diethyl ether under argon was cooled to −78°C, and methyl lithium (1.6-*M* solution in diethyl ether, 100 mL, 160 mmol) was added. The reaction mixture was allowed to warm until clear and was then recooled to −78°C. A solution of compound **37** (16 g, 39 mmol) [Eq. (9)] in 150 mL of dry diethyl ether was added. After 0.5 h at −78°C the solution was allowed to warm to −40°C and maintained at this temperature for an additional 0.5 h. The reaction mixture was then washed with water (4 × 300 mL), dried ($MgSO_4$), and concentrated. Purification by column chromatography (AcOEt–petroleum ether, 10:90) gave compound **38** (15.69 g, 94%).

*Conjugate Addition of Dithiane to Nitroolefins [67]*

(10)

**179**

**180** $R_1$ = H, $R_2$ = dithianyl
**181** $R_1$ = dithianyl, $R_2$ = H

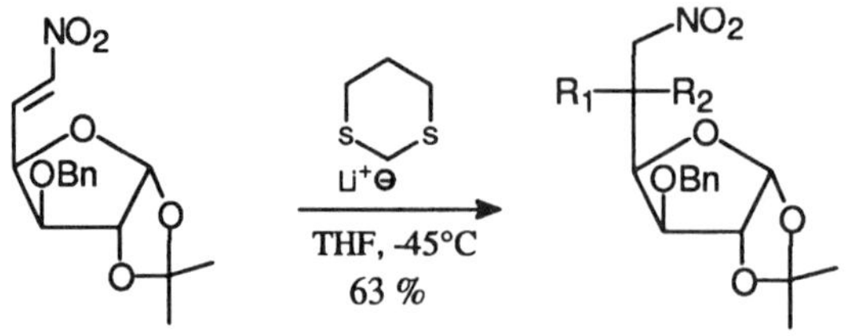

To a solution of **179** (3.2 g, 10 mmol) [Eq. (10)] in 15 mL of dry THF cooled at −45°C was added a solution of 2-lithio-1,3-dithian freshly prepared at −45°C from 1,3-dithian (1.68 g, 14 mmol) and butyl lithium (1.8-*M* solution in hexane, 7.7 mL, 14 mmol) in 20 mL of dry THF. The reaction mixture was kept at −45°C for 45 min and then acidified with acetic acid. After usual workup, the crude syrup (4.8 g) was washed three times with hot hexane, and compounds **180** and **181** were crystallized from methanol. Recrystallization from ethanol gave **180** as fine needles (1.5 g, 34%): mp 165°–166°C, $[\alpha]_D$ −34° (*c* 1.0, acetone). Chromatography on silica gel of the residue obtained after evaporation of the filtrate of **180** afforded **181** (1.3 g, 29%) as a syrup: $[\alpha]_D$ −45° (*c* 1.3, acetone)

*Radical Cyclization of Iodo Acetals [78]*

(11)

**51** **52**

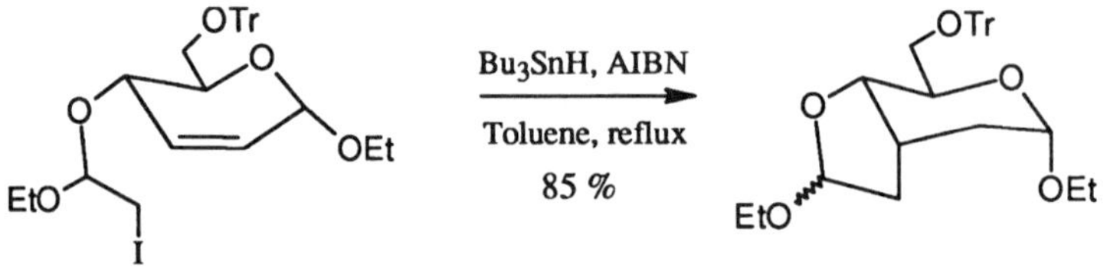

To a solution of iodo compound **51** (258 mg, 0.5 mmol) [Eq. (11)] and AIBN (8 mg, 0.05 mmol) in refluxing degased benzene (25 mL) under nitrogen was added $Bu_3SnH$ (0.15 mL, 0.55 mmol), and the mixture was stirred for 1 h. The solvent was removed and the resulting syrup was purified by column chromatography (hexane–EtOAc, 4:1) to afford compound **52a** (40 mg, 43%): $[\alpha]_D$ +83.3° (*c* 0.2, $CHCl_3$); and compound **52b** (39 mg, 42%): $[\alpha]_D$ +21.5° (*c* 0.3, $CHCl_3$).

*Cyclization of Vinylic Radicals on Sugar Olefins [81]*

(12)

**53** **54**

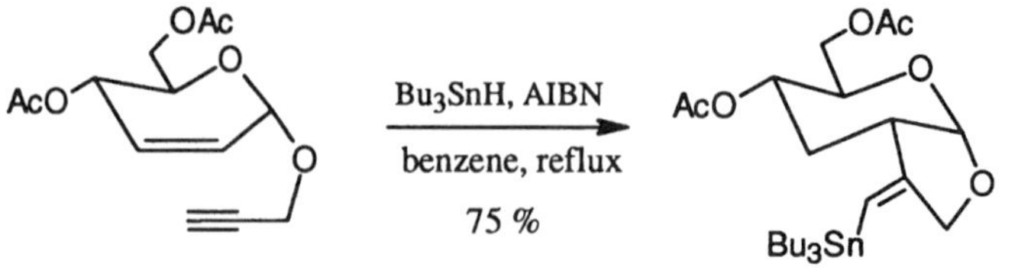

Compound **53** (1.15 g, 4.29 mmol) [Eq. (12)] was dissolved in degased benzene (200 mL) containing AIBN (52 mg, 0.43 mmol), and the solution was refluxed under argon. A solution of Bu$_3$SnH (1.3 mL, 4.72 mmol) in benzene (20 mL) was slowly added with a motor-driven syringe over 4 h. The mixture was stirred under these conditions until thin-layer chromatography (TLC) monitoring indicated complete reaction. After removal of the solvent, the crude residue was purified by column chromatography (hexane–EtOAc, 8:1) to give compound **54** (1.79 g, 75%) as a syrup: [α]$_D$ +78.4° (c 0.4, CHCl$_3$).

*Hydroxymethylation by Radical Cyclization of Silicon-Containing Tether [84]*

$$\begin{array}{c} \text{182} \end{array} \xrightarrow[\begin{array}{c} \text{2) Bu}_3\text{SnH, AIBN} \\ \text{benzene, reflux} \\ \text{3) H}_2\text{O}_2, \text{Na}_2\text{CO}_3 \end{array}]{\begin{array}{c} \text{1) BrCH}_2\text{Si(CH}_3)_2 \end{array}} \begin{array}{c} \text{60} \\ 71\% \end{array} \quad (13)$$

A solution of compound **182** (55.3 mg, 0.13 mmol) [Eq. (13)], (bromomethyl)chloro dimethyl silane (23 µL, 0.17 mmol), and triethylamine (35 µL, 0.26 mmol) in 1 mL of dry CH$_2$Cl$_2$ was stirred for 3 h at room temperature, and then the solvent was evaporated under reduced pressure. The residue was rapidly eluted from a column of basic silica gel (hexane–EtOAc, 3:1) to afford an unstable syrup. This silyl ether was dissolved under argon in 2 mL of dry degased benzene. This solution was heated under reflux, and a solution of tributyltin hydride (56 mg, 0.2 mmol) and AIBN (2%) in 1 mL of benzene was added dropwise over 1 h. The heating was continued for 3 h, and benzene was evaporated. Methanol (0.2 mL), THF (0.2 mL), 30% H$_2$O$_2$ (0.3 mL), and Na$_2$CO$_3$ (16 mg) were added to the residue. The mixture was refluxed for 4 h. Evaporation and column chromatography (hexanes–EtOAc, 1:1) gave the diol **60** (40 mg, 71%) as a gum: [α]$_D$ −12° (c 1, CHCl$_3$).

*Cyclization of Carbohydrate-Derived Radicals on Olefins [91]*

$$\begin{array}{c} \text{67a} \end{array} \xrightarrow[\text{benzene, reflux}]{\text{Bu}_3\text{SnH, AIBN}} \begin{array}{c} \text{68a} \\ 79\% \end{array} \quad (14)$$

The solution of compound **67a** (220 mg, 0.32 mmol) [Eq. (14)] in 15 mL of dry benzene was degased with a stream of argon for 10 min, and a mixture of Bu$_3$SnH (130 µL, 0.45 mmol) and AIBN (7.2 mg, 0.045 mmol) in dry benzene (7 mL) was added dropwise under reflux over 3 h. The reaction mixture was then refluxed for 2 h more, then allowed to cool to room temperature, and concentrated under reduced pressure. Purification of the crude residue by column chromatography afforded compound **68a** (137 mg, 79%).

*Intermolecular Radical 1,4-Addition on Enone [111]*

To a solution of compound **182** (945 mg, 7.5 mmol) [Eq. (15)] and iodomethane (1.4 mL, 22.5 mmol) in 1,2-dimethoxyethane (30 mL) under argon, refluxing over a 500-W heat lamp, Bu$_3$SnH (0.6 mL, 22.3 mmol) was added dropwise over 3 h. After the addition, the mixture was allowed to cool to room temperature, diluted with CH$_3$CN (150 mL), and extracted with light petroleum (5 × 30 mL). The acetonitrile layer was concentrated, and the residue was purified by flash chromatography [petroleum ether–EtOAc (1–3:1)], to give compound **183** (630 mg, 59%): $[\alpha]_D$ −290° (c 1, CHCl$_3$); reported [219] $[\alpha]_D$ −299.4 (Et$_2$O).

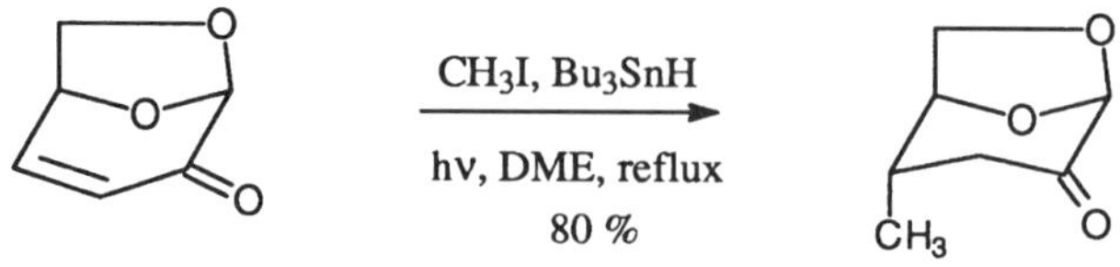

*Addition–Fragmentation of Allyltributyltin to Thionocarbonate [118]*

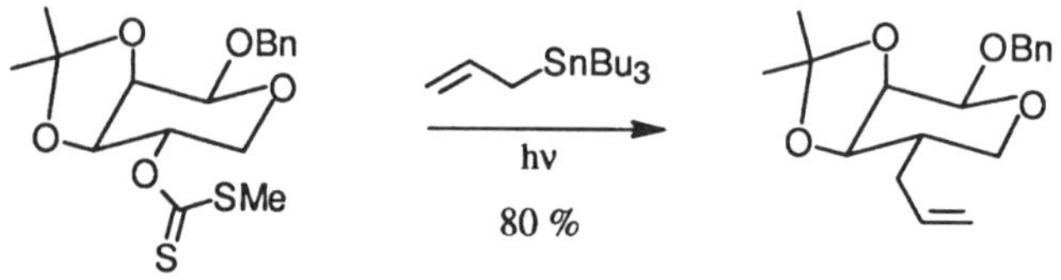

A solution of thionocarbonate **86** (11.16 mg, 26.8 mmol) [Eq. (16)] in 54 mL of toluene was placed in a Hanovia photolysis apparatus and allyltributylstannane (17.68 mg, 53.6 mmol) was added. After thoroughly degasing the solution with argon, it was irradiated during 65 h at room temperature with a 450-W Hanovia lamp with Pyrex filter. Solvent was removed under reduced pressure, and the crude product was purified by chromatography over silica gel (ether–hexane, 5:95) affording **87** (6.25 g, 80%) as a colorless oil: mp 43°–44°C; $[\alpha]_D$ −70.3° (c 0.08, CH$_2$Cl$_2$).

*Substitution of Allylic Esters by Organocopper Reagents [54]*

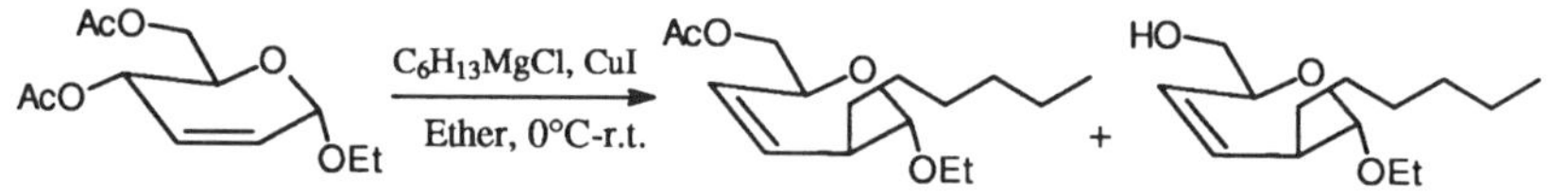

To a suspension of dry copper iodide (220 mg, 1.16 mmol) in anhydrous Et$_2$O (40 mL), cooled at 0°C under argon, was added hexyl magnesium chloride (1.22-M solution in THF, 6.67 mL, 11.6 mmol). After stirring at this temperature for 20 min, the solution was transferred, using a double-ended needle into a solution of compound **88** (1 g, 3.87 mmol) [Eq. (17)] in anhydrous Et$_2$O (40 mL). After stirring at room temperature for 1 h, the reaction was quenched with saturated aqueous NH$_4$Cl solution (20 mL), and the resulting mixture was extracted with CH$_2$Cl$_2$ (3 × 150 mL). The combined organic layers were washed with water, dried (MgSO$_4$), and concentrated. The residue was purified by column chromatography (hexane–EtOAc, 9:1) to afford compound **89a** (357 mg, 38%) as a gum: $[\alpha]_D$ +96.9° (c 1.6, CHCl$_3$); and compound **89b** (590 mg, 52%): $[\alpha]_D$ +142.1° (c 1.1, CHCl$_3$).

## Methods for Synthesis of Branched-Chain Sugars

*Palladium-Catalyzed Substitution of Allylic Halide [132]*

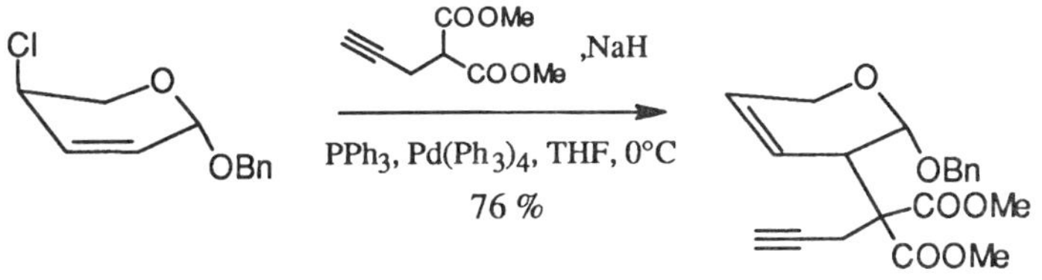

$$\text{96} \xrightarrow[\text{76 \%}]{\substack{\text{CH}\equiv\text{C-CH(COOMe)}_2, \text{NaH} \\ \text{PPh}_3, \text{Pd(Ph}_3)_4, \text{THF, 0°C}}} \text{97} \qquad (18)$$

A solution of compound **96** (1.0 g, 4.5 mmol) [Eq. (18)], triphenylphosphine (29.27 mg, 0.11 mmol) and tetrakis(triphenylphosphine)palladium (128.96 mg, 0.11 mmol) in dry THF (10 mL) was stirred at 0°C under argon for 20 min. In another flask, to a suspension of NaH (60% dispersion in mineral oil, 179 mg, 4.5 mmol) in THF (10 mL), dimethyl propargyl malonate (0.68 mL, 4.5 mmol) was added dropwise under argon at 0°C, and the reaction mixture was stirred for 20 min at this temperature. The first solution was added to the second one by a double-ended needle, and the reaction mixture was stirred at 0°C for 1 h. After completion of the reaction (TLC analysis), the solvent was evaporated. Purification by column chromatography ($CH_2Cl_2$–petroleum ether 0.5–4:10) afforded compound **97** (1.330 g, 76%).

*Nitrone Cycloaddition on Enelactone [136]*

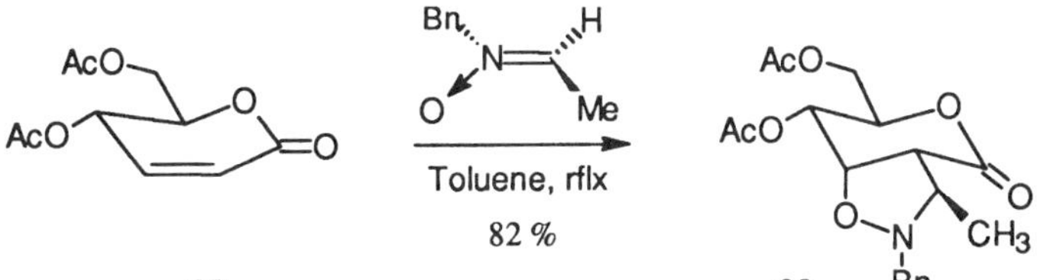

$$\text{98} \xrightarrow[\text{82 \%}]{\substack{\text{Bn-N(O)=CHMe} \\ \text{Toluene, rflx}}} \text{99} \qquad (19)$$

A mixture of (Z)-methyl-N-benzylnitrone (447 mg, 3 mmol) and compound **98** (684 mg, 3 mmol) [Eq. (19)] in methanol (5 mL) was refluxed for 1 h, and then the solvent was evaporated. The crude residue was purified by column chromatography (hexane–EtOAc, 1:1) to give compound **99** (923 mg, 82%) as a syrup: $[\alpha]_D$ −65.3° (c 1, $CH_2Cl_2$).

*Eschenmoser–Claisen Rearrangement [148]*

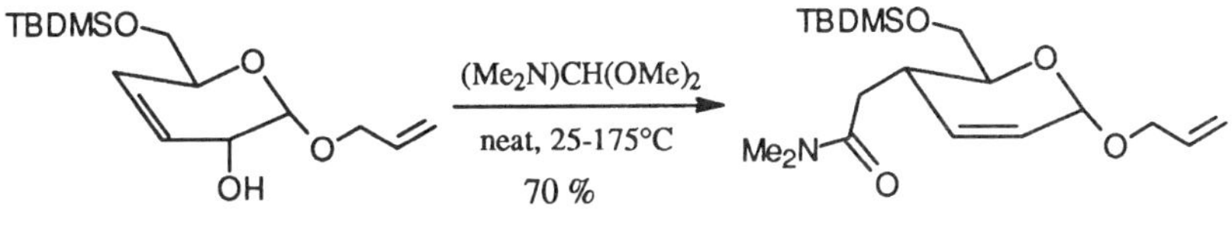

$$\text{105} \xrightarrow[\text{70 \%}]{\substack{(Me_2N)CH(OMe)_2 \\ \text{neat, 25-175°C}}} \text{106} \qquad (20)$$

A mixture of **105** (900 mg, 3 mmol) [Eq. (20)] and N,N-dimethylformamide dimethyl acetal (10.4 g, 78 mmol) was slowly heated, from 25° to 175°C (bath temperature) during 2 h, in the distillation flask of a microdistillation apparatus filled with a short Vigreux column. The distillate collected at 90°–110°C vapor temperature was recycled into the distillation flask, the column was replaced by a reflux condenser, and the mixture was heated at 175°–180°C for 3.5 h. The dark reaction mixture was evaporated and coevaporated with dry toluene (3 × 5 mL), the residue was dissolved in dry methanol (5 mL), and treated with a catalytic amount of sodium methoxide at room temperature for 16 h. The solvent was evaporated and the residue was dissolved in water (5 mL), the pH of the solution was adjusted to 7 by addition of HCl (2 N), and the neutral aqueous solution was then saturated with NaCl and extracted with $CH_2Cl_2$ (4 × 5 mL). The organic layers were combined, dried ($Na_2SO_4$), and concentrated. The crude product was purified by flash chromatography ($CH_2Cl_2$–MeOH, 99:1) to afford compound **106** (760 mg, 68%) as a syrup: $[\alpha]_D$ +77° (c 0.9, $CH_2Cl_2$).

*Alkylation of Carbohydrate Enolate [11]*

$$\text{113} \xrightarrow[\text{THF, -40°C}]{\text{BuLi}} \xrightarrow[\text{HMPA, -40°C}]{\text{BrCH}_2\text{COOMe}} \text{115} \quad (21)$$

To a solution of compound **113** (6.72 g, 20 mmol) [Eq. (21)] in dry THF (100 mL) was added butyl lithium (1.6-*M* solution in hexane, 31.25 mL, 50 mmol) at −40°C. The mixture was stirred at this temperature for 30 min, and a mixture of HMPA (9.78 g, 60 mL) and methyl bromoacetate (7.65 g, 50 mmol) was slowly added. The reaction mixture was stirred at −40°C until TLC ($Et_2O$–toluene, 1:1) indicated complete disappearing of the intermediate ketone. Saturated aqueous $NH_4Cl$ solution (20 mL) was added at −40°C. The reaction mixture was diluted with $Et_2O$ (100 mL), the two phases separated, and the organic layer was washed with water (2 × 15 mL), dried ($MgSO_4$), and concentrated. The crude syrup was purified by column chromatography to give compound **115** (4.7 g, 70%): mp 78°–80°C, $[\alpha]_D$ +82.6° (c 1.0, $CHCl_3$).

*Aldolization of Carbohydrate Enolate [161]*

$$\text{18} \xrightarrow[\text{THF, -78°C}]{\text{1) KN(SiMe}_3)_2} \xrightarrow{\text{2) ZnCl}_2, \text{RCHO}} \text{185} \quad 73\% \quad (22)$$

To a solution at 0°C of compound **18** (700 mg, 2.65 mmol) [Eq. (22)] in THF (50 mL), potassium bis(trimethylsilyl)amide (0.65-*M* solution in toluene; 4.9 mL, 3.18 mmol) was rapidly added. The pale yellow solution was stirred at 0°C for 30 min and then cooled to −78°C. To this cooled solution, $ZnCl_2$ (1-*M* solution in $Et_2O$, 3.2 mL, 3.2 mmol) was added, and the reaction mixture was stirred for 30 min. (*R*)-2,3-*O*-isopropylidene glyceraldehyde (450 mg, 3.46 mmol) was added in one portion and the resulting mixture was stirred at −78°C for 0.5 h, and then allowed to warm slowly to −40°C. After the reaction reached completion (1.5–2 h), saturated aqueous $NH_4Cl$ solution (15 mL) was added at −40°C. The reaction mixture was diluted with EtOAc (20 mL), the two phases separated, and the aqueous layer was extracted with EtOAc (3 × 15 mL). The combined organic layers were dried ($MgSO_4$) and concentrated. The crude product was purified by flash chromatography (EtOAc–petroleum ether, 1:1) to give compound **185** (760 mg, 73%): mp 146°–147°C, $[\alpha]_D$ +27.8° (*c* 1.2, $CHCl_3$).

## C. Compounds with a C=C Branch (Type III)

*Wittig Methylenation [168]*

$$\mathbf{123} \xrightarrow[\text{THF, r.t.}]{Ph_3P=CH_2} \mathbf{124} \quad 68\% \qquad (23)$$

Butyl lithium (1.6-*M* solution in hexane, 7.6 mL, 12.1 mmol) was added dropwise at −40°C to a solution of methyltriphenylphosphonium bromide (4.34 g, 12.1 mmol) in THF (60 mL), and the reaction mixture was stirred for 1 h at −40°C and then allowed to warm to 0°C. A solution of compound **123** (3 g, 8.1 mmol) [Eq. (23)] in THF (300 mL) was rapidly added. The mixture was left to warm up to room temperature during 1 h, and then washed with water (50 mL). The organic layer was dried ($Na_2SO_4$) and concentrated. The residue was purified by column chromatography ($CH_2Cl_2$) to give compound **124** (2.02 g, 68%): mp 113°–114°C, $[\alpha]_D$ +15° (*c* 2.2, $CHCl_3$).

*Peterson Methylenation [170]*

$$\mathbf{125} \xrightarrow[\text{Ether, toluene, r.t.}]{Me_3SiCH_2MgCl} \mathbf{126} \xrightarrow[\text{THF, reflux}]{KF} \mathbf{127} \quad 90\% \qquad (24)$$

Magnesium turnings (2.57 g, 106 mmol) were placed under dry argon in a 1-L three-necked, round-bottom flask equipped with a condenser, a dry ice condenser, and a dropping funnel, with a pressure equalization arm. A solution of (bromomethyl)trimethylsilane (0.841 g, 5 mmol) in anhydrous diethyl ether (75 mL) was then added. To this mixture a solution of chloromethyltrimethylsilane (14.2 g, 116 mmol) in 50 mL of diethyl ether was added at a rate sufficient to maintain reflux. The reaction mixture was stirred under reflux for an additional hour. The solution was then allowed to cool to room temperature, and a solution of the ketosugar **125** (6.33 g, 16.5 mmol) [Eq. (24)] in warm toluene (400 mL) was added dropwise. The solution was stirred for 3 h, and quenched with a saturated solution of $NH_4Cl$, extracted with diethyl ether (3 × 350 mL). The combined ethereal extracts were dried ($MgSO_4$) and evaporated under reduced pressure to give the crude compound **126** (8.85 g) as a syrup, which was used for the next step without further purification. A solution of the crude product **126** in anhydrous THF (250 mL) was added dropwise to a suspension of potassium hydride (8.5 g, 205 mmol) in THF (225 mL) under argon. The reaction mixture was then heated to reflux for 4 h. The resulting brown solution was then slowly poured into a mixture of saturated solution of $NH_4Cl$ (300 mL) and diethylether (500 mL). The layers were separated, and the aqueous layer was extracted with diethyl ether (2 × 200 mL). The combined extracts were dried ($MgSO_4$) and concentrated, under reduced pressure, to give the crude product **127**. Purification by recrystallization from dichloromethane–hexane afforded pure compound **127** (2.71 g, 58%): mp 194.5°–195°C, [α]$_D$ +145° (c 1, CHCl$_3$); reported [220] mp 192.5°–194°C (ethanol), [α]$_D$ +159° (c 1, CHCl$_3$).

*Condensation of Silylated Enolates on Ketone [175]*

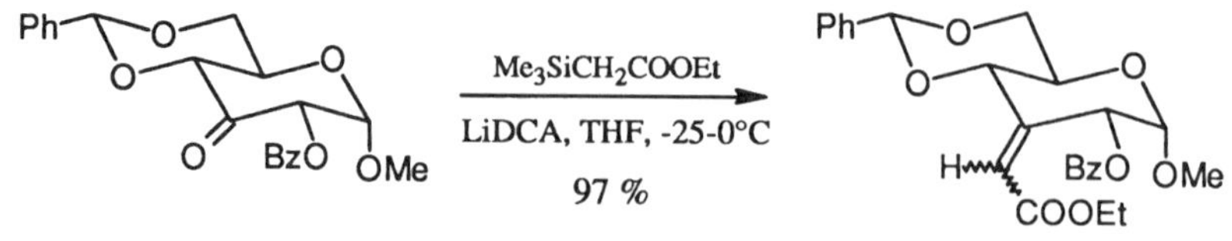

$$125 \xrightarrow[\text{LiDCA, THF, -25-0°C}]{\text{Me}_3\text{SiCH}_2\text{COOEt}} 186 \quad 97\% \tag{25}$$

To a solution of dicyclohexylamine (363 mg, 2 mmol) in THF (10 mL) under an atmosphere of argon at −78°C, butyl lithium (1.6-*M* solution in THF; 1.25 mL, 2 mmol) and triethylmethylsilylacetate (320 mg, 2 mmol) was added dropwise, and then a solution of compound **125** (384 mg, 1 mmol) [Eq. (25)] in THF (5 mL) was added over 10 min. The solution was kept at −78°C for 1 h, then at −25°C for 1 h, and finally at 0°C for 1 h. The reaction mixture was poured into a saturated solution of NaCl and then extracted with EtOAc (3 × 50 mL). The combined organic layers were washed with water, dried (MgSO$_4$), and evaporated to give a syrupy residue that was purified by flash chromatography (hexane–EtOAc, 3:1) to give compound **186** (440 mg, 97%) as a mixture of the Z and E isomers in 2.3 ratio.

*Wittig–Horner Olefination [172]*

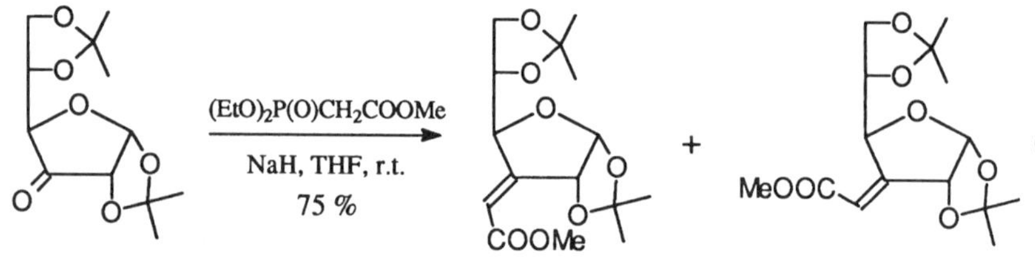

$$\mathbf{1} \xrightarrow[\text{NaH, THF, r.t.}]{(\text{EtO})_2\text{P(O)CH}_2\text{COOMe}} \mathbf{131\ E} + \mathbf{131\ Z} \quad 75\%\quad 75:18 \tag{26}$$

To a suspension of sodium hydride (60% in mineral oil, 340 mg, 8.5 mmol) in 20 mL of dry THF was added diethyl[(ethoxycarbonyl)methyl]phosphonate (3.308 g, 17 mmol). The reaction mixture was cooled to 0°C and stirred for 30 min, and a solution of the ketosugar **1** (2.19 g, 8.5 mmol) [Eq. (26)] in THF (40 mL) was added. The reaction mixture was allowed to warm to room temperature, stirred for 1 h, and then diluted with water (500 mL). The phases were separated, and the aqueous layer was extracted with CH$_2$Cl$_2$ (3 × 100 mL), the combined organic layers were dried (Na$_2$SO$_4$) and evaporated. The residue was chromatographed (hexane–EtOAc, 10:1 to 5:1) to afford compound **131Z** (504 mg, 18%): [α]$_D$ +220.7° (c 1, CHCl$_3$); and compound **131E** (2.104 g, 75%): [α]$_D$ +174.9° (c 1.4, CHCl$_3$).

*Aldolization–Crotonization of Ketone Enolate [178]*

To a solution of compound **120** (4.895 g, 15.2 mmol) [Eq. (27)] in a mixture of CH$_2$Cl$_2$ (60 mL) and pyridine (20 mL), cooled at 0°C, mesyl chloride (1.8 mL, 22.8 mmol) was added. The reaction mixture was stirred at room temperature for 24 h, and pyridine was

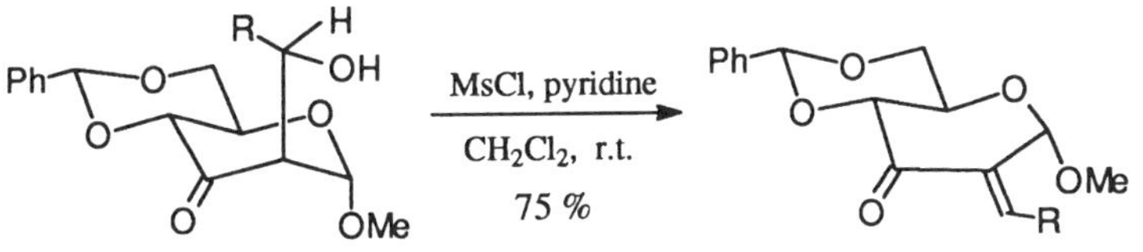

# Methods for Synthesis of Branched-Chain Sugars

$$\text{120} \xrightarrow[\text{CH}_2\text{Cl}_2, \text{r.t.}]{\text{MsCl, pyridine}} \text{134} \quad 75\% \quad (27)$$

removed under reduced pressure. The residue was diluted with CH$_2$Cl$_2$ (250 mL) and washed with water (2 × 20 mL). The organic layer was dried (MgSO$_4$) and concentrated to give a crude crystalline residue, which was recrystallized (CH$_2$Cl$_2$–hexane) to afford pure compound **134** (3.466 g, 75%): mp 178°–179°C, [α]$_D$ +84.6° (c 1.0, CHCl$_3$).

## D. Compounds with Two Branched Chains (Types IV and V)

*Geminal Doubly Branched-Chain Sugars by Eschenmoser–Claisen Rearrangement [157]*

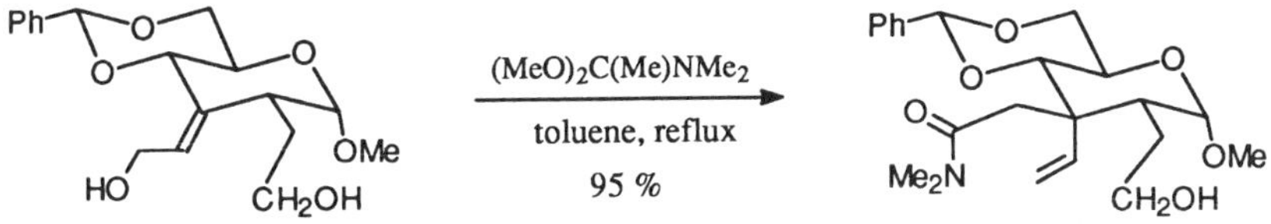

$$\text{150} \xrightarrow[\text{toluene, reflux}]{(\text{MeO})_2\text{C(Me)NMe}_2} \text{151} \quad 95\% \quad (28)$$

To a solution of the allylic alcohol **150** (16.7 g, 49.7 mmol) [Eq. (28)] in 500 mL of toluene, under argon, was added *N*,*N*-dimethylacetamide dimethyl acetal (33.117 g, 249 mmol). The reaction mixture was heated at reflux for 10 h, with continuous removal of methanol by molecular sieve (4 Å) trap. The reaction mixture was then cooled to room temperature and concentrated under reduced pressure. The residue was purified by flash chromatography on silica gel (EtOAc) to afford compound **151** (13.6 g, 68%) as a white foam. The corresponding acetate was obtained in 27% yield and quantitatively transformed into **151**: [α]$_D$ +2.7° (c 1.1, CHCl$_3$).

*Dichlorocyclopropanation Under Phase-Transfer Catalysis [194]*

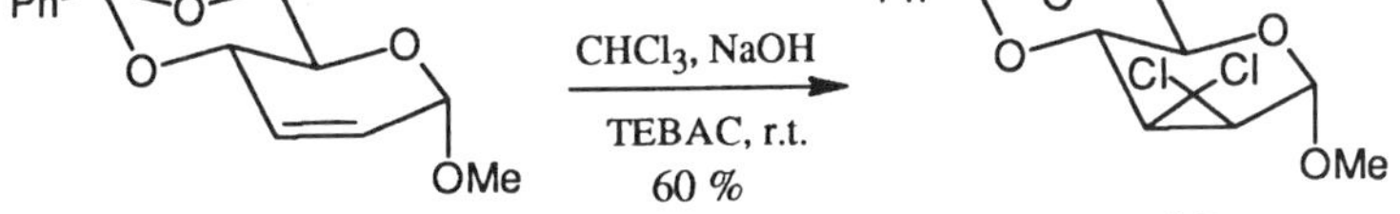

$$\text{161} \xrightarrow[\text{TEBAC, r.t.}]{\text{CHCl}_3, \text{NaOH}} \text{162} \quad 60\% \quad (29)$$

A mixture of a solution of **161** (744 mg, 3 mmol) [Eq. (29)] and benzyltriethylammonium chloride (6.8 mg, 0.03 mmol) in CHCl$_3$ (5 mL) and 50% aqueous NaOH solution (5 mL) was vigorously stirred at room temperature for 18 h. Water (10 mL) was then slowly added to the reaction mixture, and then extracted with diethyl ether (2 × 15 mL). The combined organic layer was washed with water (10 mL), dried (MgSO$_4$), filtered, and concentrated under vacuum. The residue was purified by column chromatography (hexane–Et$_2$O, 3:2) to afford compound **162** (596 mg, 60%): mp 134°–135°C, [α]$_D$ +30° (c 1.5, CHCl$_3$).

*Cyclopentanone Using Pauson–Khand Reaction [201]*

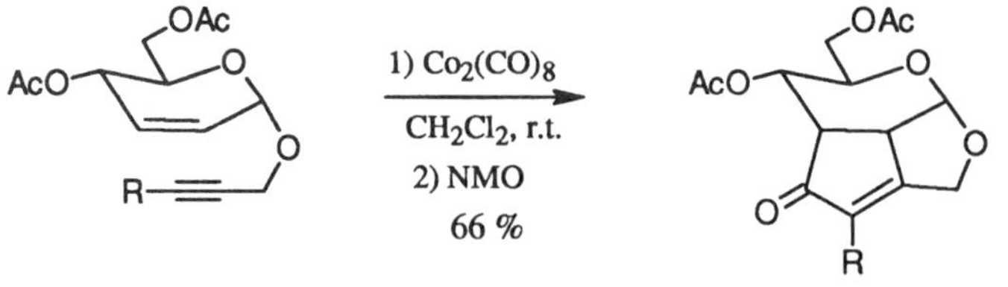

(30)

**166** **168**

To a solution of compound **166** (306 mg, 1 mmol) [Eq. (30)] in 12 mL of $CH_2Cl_2$ was added in one portion cobalt carbonyl (343 mg, 1.1 mmol) at room temperature. The mixture was stirred for 3 h, and then anhydrous 4-methylmorpholine $N$-oxide monohydrate (850 mg, 6.3 mmol) was slowly added and stirring was continued for 5 h more. Part of the solvent was evaporated, and the suspension was purified by flash chromatography (hexane–EtOAc) to provide compound **168** (194 mg, 66%): $[\alpha]_D$ −17° ($c$ 2.3, $CHCl_3$).

*Cyclohexane Formation Using Diels–Alder Cycloaddition [214]*

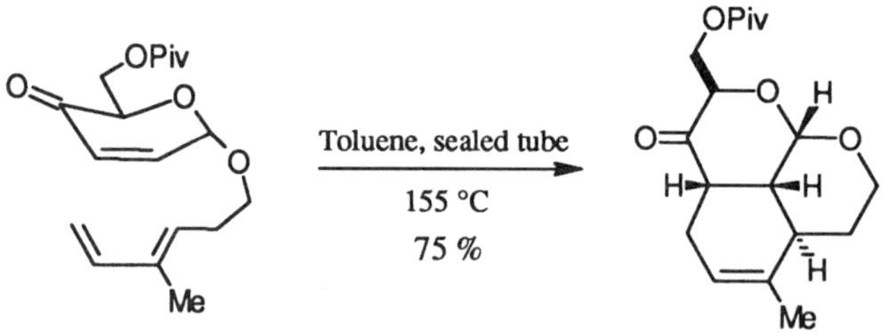

(31)

**175** **176**

A solution of the trienone **175** (400 mg, 1.24 mmol) [Eq. (31)] in 20 mL of dry, degased toluene containing hydroquinone (34 mg, 0.3 mmol) was heated in a sealed tube at 155°C over 14 h. Evaporation and chromatography on silica gel (hexane–EtOAc, 4:1) afforded the cycloadduct **176** (300 mg, 75%): $[\alpha]_D$ −343° ($c$ 0.3, $CHCl_3$).

## REFERENCES

1. J. Yoshimura, Synthesis of branched chain sugars, *Adv. Carbohydr. Chem. Biochem. 42*:69 (1984).
2. S. Hanessian, *Total Synthesis of Natural Products: The "Chiron" Approach*, J. E. Baldwin, ed., Pergamon Press, Oxford, 1983.
3. B. Fraser-Reid and R. C. Anderson, Carbohydrate derivatives in the asymmetric synthesis of natural products, *Fortschr. Chem. Org. Naturst. 39*:1 (1978).
4. T. D. Inch, Formation of convenient chiral intermediates from carbohydrates and their use in synthesis, *Tetrahedron 40*:3161 (1984).
5. H. Grisebach, Biosynthesis of sugar components of antibiotic substances, *Adv. Carbohydr. Chem. Biochem. 35*:81 (1978).
6. M. L. Wolfrom and S. Hanessian, Reaction of free carbonyl sugar derivatives with organometallics reagents. II 6-Deoxy-L-idose and a branched-chain sugar, *J. Org. Chem. 27*:2107 (1962).

7. M. Dasser, F. Chrétien, and Y. Chapleur, A facile and stereospecific synthesis of L-glycero-D-manno-heptose and some derivatives, *J. Chem. Soc. Perkin Trans. 1*:3091 (1990).
8. N. R. Williams, Oxirane derivatives of aldoses, *Adv. Carbohydr. Chem. Biochem.* 25:109 (1970).
9. F. W. Lichtenthaler, Sugar enolones: Synthesis, reactions of preparative interest and γ-pyrone formation, *Pure Appl. Chem.* 50:1343 (1978).
10. B. Giese, *Radical in Organic Synthesis: Formation of Carbon–Carbon Bonds*, Pergamon Press, Oxford, 1986.
11. Y. Chapleur, A short synthesis of 2-*C*-alkyl-2-deoxy sugars from D-mannose, *J. Chem. Soc. Chem. Commun.* p. 141 (1983).
12. Y. Chapleur, F. Longchambon, and H. Gillier, Aldolisation of a carbohydrate enolate: Stereochemical outcome and X-ray crystal structure determination of an aldol product, *J. Chem. Soc. Chem. Commun.* p. 564 (1988).
13. Y. A. Zhdanov, Y. E. Alexeev, and V. G. Alexeeva, The Wittig reaction in carbohydrate chemistry, *Adv. Carbohydr. Chem. Biochem.* 27:227 (1972).
14. A. Rosenthal and S. N. Mikhailov, Branched-chain sugars. Modifications in the reaction of 1,2:5,6-di-*O*-isopropylidene-α-D-ribo-hexofuranos-3-ulose with Grignard and organolithium reagents, *J. Carbohydr. Nucleosides Nucleotides* 6:237 (1979).
15. R. E. Ireland and D. W. Norbeck, Application of the Swern oxidation to the manipulation of highly reactive carbonyl compounds, *J. Org. Chem.* 50:2198 (1985).
16. D. Horton and E. K. Just, Stereospecific chain-branching by C-alkylation at the ketonic and enolic positions of 1,6-anhydro-2,3-*O*-isopropylidene-beta-D-*lyxo*-hexopyranos-4-ulose, *Carbohydr. Res.* 18:81 (1971). D. C. Baker, D. K. Brown, D. Horton, and R. G. Nickol, Synthesis of branched-chain sugar derivatives related to algarose, *Carbohydr. Res.* 32:299 (1974).
17. K. Kakinuma, Y. Iihama, I. Takagi, K. Ozawa, N. Yamauchi, N. Imamura, Y. Esumi, and M. Uramoto, Diacetone glucose architecture as a chirality template. II. Versatile synthon for the chiral deuterium labelling and synthesis of all diastereomers of chirally monodeuterated glycerol, *Tetrahedron* 48:3763 (1992).
18. J. Yoshimura, Stereoselective synthesis of branched-chain sugar derivatives, *Pure Appl. Chem.* 53:113 (1981).
19. J. C. Lopez, E. Lameignere, C. Burnouf, M. de Los Angeles Laborde, A. A. Ghini, A. Olesker, and G. Lukacs, Efficient routes to pyranosidic homologated conjugated enals and dienes from monosaccharides, *Tetrahedron* 49:7701 (1993).
20. S. Hanessian and G. Rancourt, Carbohydrates as chiral intermediates in organic synthesis. Two functionalized chemical precursors comprising eight of the ten chiral centers of erythronolide, *Can. J. Chem.* 55:1111 (1977).
21. J. S. Brimacombe, R. Hanna, A. M. Mather, and T. J. R. Weakley, Branched-chain sugars. Part 8. The synthesis of C-acetylpyranosides and a pillarose derivative using 1-methoxyvinyl-lithium, *J. Chem. Soc. Perkin Trans.* p. 273 (1980).
22. J. S. Brimacombe and A. M. Mather, Branched-chain sugars. Part 7. A route to sugars with two-carbon branches using 1-methoxyvinyl-lithium, *J. Chem. Soc. Perkin Trans. 1*:269 (1980).
23. J. S. Brimacombe, A. M. Mather, and R. Hanna, A synthesis of a derivative of pillarose, *Tetrahedron Lett.* 13:1171 (1978).
24. J.-C. Depezay and Y. Le Merrer, Sucres branchés: Synthèse par aldolisation dirigée des désoxy-2 méthylène-2C D-érythro et D-thréo pentoses, *Tetrahetron Lett.* 32:2865 (1978).
25. J.-C. Depezay and Y. Le Merrer, Synthèse par aldolisation dirigée des 2-désoxy-2-C-méthylène-D-érythro- et -D-thréo-pentoses, *Carbohydr. Res.* 83:51 (1980).
26. J.-C. Depezay and A. Duréault, Sucres branchées: Époxydation stéréospécifique des 2-désoxy-2-méthylène-2C-D-pentoses. Synthèse de méthyl glycosides d'epihamamélose et de l'hamamélose, *Tetrahedron Lett.* 32:2869 (1978).
27. J.-C. Depezay, A. Duréault, and M. Sanière, Synthèse stéréospécifique de 4-désoxytétroses et de D-xylo- et D-arabino-pentoses branchés sur le carbone 2, *Carbohydr. Res.* 83:273 (1980).

28. K. Sato, K. Suzuki, M. Uead, M. Katayama, and Y. Kajihara, A novel reagent for the synthesis of branched-chain functionalized sugars. dichloromethyllithium, *Chem. Lett.* p. 1469 (1991).
29. S. Deloisy, T. T. Thang, A. Olesker, and G. Lukacs, Synthesis of α-azido aldehydes. Stereoselective formal access to the immunosuppressant myriocin, *Tetrahedron Lett. 35*:4783 (1994).
30. P. C. Bulman Page, M. B. van Niel, and J. C. Prodger, Synthetic uses of 1,3-dithiane grouping from 1977 to 1988, *Tetrahedron 45*:7643 (1989).
31. H. Paulsen, V. Sinnwell, and J. Thiem, Applications of the 1,3-dithiane procedure for the synthesis of branched-chain carbohydrates, *Methods Carbohydr. Chem. 8*:185 (1980).
32. R. Csuk, A. Fürstner, and H. Weidmann, Efficient, low temperature Reformatsky reactions of extended scope, *J. Chem. Soc. Chem. Commun.* p. 775 (1986).
33. R. Csuk, A. Fürstner, H. Sterk, and H. Weidmann, Synthesis of carbohydrate derived α-methylene-γ-lactones by diastereoselective, low temperature Reformatsky-type reactions, *J. Carbohydr. Chem. 5*:459 (1986).
34. A. P. Rauter, J. A. Figueiredo, I. Ismael, M. S. Pais, A. G. Gonzales, J. Daiz, and J. B. Barrera, Synthesis of α-methylene-γ-lactone in furanosidic systems, *J. Carbohydr. Chem. 6*:259 (1987).
35. R. Csuk, B. I. Glänzer, Z. Hu, and R. Boese, Double stereodifferentiating Dreiding–Schmidt reactions, *Tetrahedron 50*:1111 (1994).
36. R. Csuk, A. Fürstner, and H. Weidmann, Branching of ketosugars by ethyl trimethylsilylacetate/tetra-*n*-butylammonium fluoride, *J. Carbohydr. Chem. 5*:77 (1986).
37. T. T. Thang, M. de Los Angeles Laborde, A. Olesker, and G. Lukacs, A new approach to the stereospecific synthesis of branched-chain sugars, *J. Chem. Soc. Chem. Commun.* p. 1581 (1988).
38. K. Kobayashi, K. Kakinuma, and H. G. Floss, Synthesis of chiral acetic acid by chirality transfer from D-glucose, *J. Org. Chem. 49*:1290 (1984).
39. R. M. Giuliano, Stereoselective synthesis of L-olivomycose, *Carbohydr. Res. 131*:341 (1984).
40. J. S. Brimacombe, J. A. Miller, and U. Zakir, An approach to the synthesis of branched chain amino sugars from *C*-methylene sugars, *Carbohydr. Res. 49*:233 (1976).
41. J.-M. Bourgeois, Synthèse de sucres aminés ramifiés II. Synthèse et réactions de spiroaziridines, *Helv. Chim. Acta 57*:2553 (1974).
42. J.-M. Bourgeois, Synthèse de sucres aminés ramifiés. IV. Synthèse de quelques dérivés nouveaux par l'intermédiaire d'une hexose-spiro-aziridine, *Helv. Chim. Acta 59*:2114 (1976).
43. R. J. Alves, S. Castillon, A. Dessinges, P. Herczegh, J. C. Lopez, G. Lukacs, A. Olesker, and T. T. Thang, A route to functionalized branched-chain amino sugars via nitrous acid promoted spiroaziridine formation, *J. Org. Chem. 53*:4616 (1988).
44. H. Yamamoto, H. Sasaki, and S. Inokawa, Reaction of lithium dimethyl cuprate with methyl 2,3-anhydro-5-deoxy-α-D-ribofuranoside. A new, convenient route for the preparation of 2,5-dideoxy-2-*C*-methyl-D-arabinofuranose derivatives, *Carbohydr. Res. 100*:C44 (1982).
45. J. A. Montgomery, M. C. Thorpe, S. D. Clayton, and H. J. Thomas, Some observations on the reaction of ammonia with methyl 2,3-anhydro-α-D-ribofuranoside, *Carbohydr. Res. 32*:404 (1974).
46. R. U. Lemieux, E. Fraga, and K. A. Watanabe, Preparation of unsaturated carbohydrates. A facile synthesis of methyl 4,6-*O*-benzylidene-D-hex-2-enopyranosides, *Can. J. Chem. 46*:61 (1968).
47. M. Sharma and R. K. Brown, The mechanism of the reaction of methyllithium with 2,3-anhydro-4,6-*O*-benzylidene-α-D-allopyranoside to form 4,6-*O*-benzylidene-1,2-didehydro-1,2-dideoxy-2-methyl-D-*ribo*-hexopyranose, *Can. J. Chem. 46*:757 (1968).
48. T. D. Inch and G. D. Lewis, The synthesis and degradation of benzyl 4,6-*O*-benzylidene-2,3-dideoxy-3-*C*-ethyl-2-*C*-hydroxymethyl-α-D-glucopyranoside and mannopyranoside, *Carbohydr. Res. 22*:91 (1972).
49. C. Brockway, P. Kocienski, and C. Pant, Unusual stereochemistry in the copper-catalysed ring

opening of a carbohydrate oxirane with vinylmagnesium bromide, *J. Chem. Soc. Perkin Trans.* p. 875 (1984).

50. P. J. Hodges and G. Procter, 1,6-Anhydroglucose in organic synthesis; preparation of fragments suitable for natural product synthesis, *Tetrahedron Lett. 26*:4111 (1985).

51. N. Tsuda, S. Yokota, T. Kudo, and O. Mitsunobu, A convenient method for the preparation of 4,6-*O*-benzylideneglycals from methyl 2,3-anhydro-4,6-*O*-benzylidene-α-D-hexopyranosides, *Chem. Lett.* p. 289 (1983).

52. J. Yoshimura, N. Kawauchi, T. Yasumori, K.-I. Sato, and H. Hashimoto, Synthesis of 2,3-anhydro- and 3,4-anhydro-hexopyranosides having a methylbranch on the oxirane ring, and their reactions with some lithium methylcuprate reagents, *Carbohydr. Res. 133*:255 (1984).

53. D. R. Hicks and B. Fraser-Reid, The 2- and 3-*C*-methyl derivatives of methyl 2,3-dideoxy-α-D-*erythro*-hex-2-enopyranoside-4-ulose, *Can. J. Chem. 53*:2017 (1975).

54. C. Loustau Cazalet, Nouvelles voies de synthèse de lactones bioactives, Thèse, Université de Nancy (France) (1994).

55. B. H. Lipshutz and S. Sengupta, Organocopper reagents: Substitution, conjugate addition, carbo/metallocupration and other reactions, *Org. React. 41*:135 (1992).

56. L. Castellanos, A. Gateau-Olesker, F. Panne-Jacolot, J. Cleophax, and S. D. Gero, Synthèse d'analogues de dérivés dioxaprostanoïques à partir du D- et du L-xylose, *Tetrahedron 37*:1691 (1981).

57. S. Hanessian and P. Dextraze, Preparative and exploratory carbohydrate chemistry. Carbanions in carbohydrate chemistry. Novel methods for chain extension and branching, *Can. J. Chem. 50*:226 (1972).

58. S. Hanessian, P. Dextraze, and R. Masse, Preparative and exploratory carbohydrate chemistry. Regiospecific and asymmetric introduction of functionalized branching in carbohydrates, *Carbohydr. Res. 26*:264 (1973).

59. P. Perlmutter, Conjugate addition reactions in organic synthesis, *Tetrahedron Organic Chemistry Series*, J. E. Baldwin and P. D. Magnus, eds., Pergamon Press, Oxford, 1992.

60. N. L. Holder, The chemistry of hexenuloses, *Chem. Rev. 82*:287 (1982).

61. D. E. Plaumann, B. J. Fitzsimmons, B. M. Richie, and B. Fraser-Reid, Synthetic route to 6,8-dioxabicyclo[3.2.1]octyl pheromones from D-glucose derivatives. 4. Synthesis of (−)-multistriatin, *J. Org. Chem. 47*:941 (1982).

62. S. Hanessian, P. C. Tyler, and Y. Chapleur, Reaction of lithium dimethylcuprate with conformationally biased acyloxy enol esters. Regio and stereocontrolled access to functionalized six-carbon chiral synthons, *Tetrahedron Lett. 22*:4583 (1981).

63. H. H. Baer and Z. S. Hanna, Synthesis of branched-chain sugars by use of organocopper reagents, *Carbohydr. Res. 85*:136 (1980).

64. A. C. Forsyth, R. M. Paton, and I. Watt, Highly selective base-catalyzed additions of nitromethane to levoglucosenone, *Tetrahedron Lett. 30*:993 (1989).

65. H. H. Baer and K. S. Ong, Reactions of nitro sugars. IX. The synthesis of branched-chain dinitro sugars by Michael addition, *Can. J. Chem. 46*:2511 (1968).

66. H. Paulsen and W. Koebernick, Synthese von verzweigten zuckern durch 1,4-addition an pyranosid-enone, *Carbohydr. Res. 56*:53 (1977).

67. M. Funabashi and J. Yoshimura, Branched chain sugars. Part 14. Synthesis of new branched-chain cyclitols having *myo*- or *scyllo*-, and *muco*-configuration from 3-*O*-benzyl-5,6-dideoxy-5-*C*-(1,3-dithian-2-yl)-6-nitro-L-idofuranose and D-glucofuranose, *J. Chem. Soc. Perkin Trans.* p. 1425 (1979).

68. M. Funabashi, H. Wakai, K. Sato, and J. Yoshimura, Branched-chain sugars. Part 15. Synthesis of 1-L-(1,2,3′,4,5/3,6)-3-hydroxymethyl-4,5-*O*-isopropylidene-3,3′-*O*-methylene-6-nitro-2,3,4,5-tetrahydrocyclohexanecarbaldehyde dimethyl acetal, a potential key compound for total synthesis of optically active tetrodotoxin, *J. Chem. Soc. Perkin Trans.* p. 14 (1980).

69. Y. Ueno, K. Chino, M. Watanabe, O. Moriya, and M. Okawara, Homolytic carbocyclisation by

69. use of heterogenous supported organotin catalyst. A new synthetic route to 2-alkoxytetrahydrofurans and γ-butyrolactones, *J. Am. Chem. Soc. 104*:5564 (1982).
70. G. Stork, R. Mook, Jr., S. A. Biller, and S. D. Rychnovsky, Free radical cyclization of bromoacetals. Use in the construction of bicyclic acetals and lactones, *J. Am. Chem. Soc. 105*:3741 (1983).
71. C. E. Mc Donald and R. W. Dugger, A formal total synthesis of (−)-isoavenaciolide, *Tetrahedron Lett. 29*:2413 (1988).
72. A. De Mesmaeker, P. Hoffmann, and B. Ernst, Stereoselective bond formation in carbohydrates by radical cyclization reactions, *Tetrahedron Lett. 29*:6585 (1988).
73. Y. Chapleur and N. Moufid, Radical cyclisation of some unsaturated carbohydrate derived acetals, *J. Chem. Soc. Chem. Commun.* p. 39 (1989).
74. R. J. Ferrier, P. M. Petersen, and M. A. Taylor, Radical cyclisation reactions leading to doubly branched carbohydrates and 6- and 8-oxygenated 2,9-dioxabicyclo[4.3.0]nonane derivatives, *J. Chem. Soc. Chem. Commun.* p. 1247 (1989).
75. R. Nouguier, C. Lesueur, E. De Riggi, M. P. Bertrand, and A. Virgili, Stereoselective free-radical cyclisation on a sugar template. The sulphonyl radical as a synthetic tool for functionalized glycosides, *Tetrahedron Lett. 31*:3541 (1990).
76. C. Lesueur, R. Nouguier, M. P. Bertrand, P. Hoffmann, and A. De Mesmaeker, Stereocontrol in radical cyclization on sugar templates, *Tetrahedron 50*:5369 (1994).
77. G. Stork and M. J. Sofia, Stereospecific reductive methylation via a radical cyclization-desilylation process, *J. Am. Chem. Soc. 108*:6826 (1986).
78. N. Moufid, Y. Chapleur, and P. Mayon, Radical cyclisation of some unsaturated carbohydrate-derived propargyl ethers and acetals, *J. Chem. Soc. Perkin Trans.* p. 999 (1992).
79. J. C. Lopez and B. Fraser-Reid, Serial radical reactions of enol ethers: Ready routes to highly functionalized C-glycosyl derivatives, *J. Am. Chem. Soc. 111*:3450 (1989).
80. N. Moufid and Y. Chapleur, Radical cyclization of allyl propargyl ethers and acetals: application to the stereocontrolled acylation of carbohydrates, *Tetrahedron Lett. 32*:1799 (1991).
81. N. Moufid, Y. Chapleur, and P. Mayon, Radical cyclisation of some unsaturated carbohydrate-derived acetals, *J. Chem. Soc. Perkin Trans.* p. 991 (1992).
82. H. Nishiyama, T. Kitajima, M. Matsumoto, and K. Itoh, Silylmethyl radical cyclization: New stereoselective method for 1,3-diol synthesis from allylic alcohols, *J. Org. Chem. 49*:2298 (1984).
83. G. Stork and M. Kahn, Control of ring junction stereochemistry via radical cyclisation, *J. Am. Chem. Soc. 107*:500 (1985).
84. V. Pedretti, J. M. Mallet, and P. Sinay, Silylmethylene radical cyclization—a stereoselective approach to branched sugars, *Carbohydr. Res. 244*:247 (1993).
85. J. C. Lopez, A. M. Gomez, and B. Fraser-Reid, Silicon-tethered radical cyclization and intramolecular Diels–Alder strategies are combined to provide a ready route to highly functionalized decalins, *J. Chem. Soc. Chem. Commun.* p. 762 (1993).
86. K. Augustyns, J. Rozenski, A. van Aerschot, R. Busson, P. Claes, and P. Herdewijn, Synthesis of a new branched chain hexopyranosyl nucleoside—1-[2′,3′-dideoxy-3′-C-(hydroxymethyl)-α-D-*erythro*-pentopyranosyl]- thymine, *Tetrahedron 50*:1189 (1994).
87. C. Audin, J. M. Lancelin, and J. M. Beau, Radical cyclization on carbohydrate pyranosides: A controlled formation of functionalized ring-fused bicyclic acetals, *Tetrahedron Lett. 29*:3691 (1988).
88. A. De Mesmaeker, P. Hoffmann, and B. Ernst, Stereoselective C–C bond formation in carbohydrates by radical cyclization reactions, *Tetrahedron Lett. 30*:57 (1989).
89. M. E. Jung and S. W. T. Choe, Stereospecific intramolecular formyl transfer via radical cyclization-fragmentation—preparation of alkyl 2-deoxy-2-α-formylglucopyranosides and similar compounds, *Tetrahedron Lett. 34*:6247 (1993).
90. G. V. M. Sharma and S. R. Vepachedu, A simple and stereoselective synthesis of avenaciolide from D-glucose, *Tetrahedron Lett. 31*:4931 (1990).

91. J. C. Wu, Z. Xi, C. Gioeli, and J. Chattopadhyaya, Intramolecular cyclisation-trapping of carbon radicals by olefins as means to functionnalised 2' and 3' carbon in β-D-nucleosides, *Tetrahedron 47*:2237 (1991).
92. S. Velazquez, S. Huss, and M.-J. Camarasa, Stereoselective synthesis of [3.3.0] fused lactones (γ-butyrolactones) of sugars and nucleosides by free radical intramolecular cyclization, *J. Chem. Soc. Chem. Commun.* p. 1263 (1991).
93. H. Hashimoto, K. Furuichi, and T. Miwa, Cyclopentane annelated pyranosides: A new approach to chiral iridoid synthesis, *J. Chem. Soc. Chem. Commun.* p. 1002 (1987).
94. X. Zhen, P. Agback, J. Plavec, A. Sandstrom, and J. Chattopadhyaya, New stereocontrolled synthesis of isomeric C-branched-β-D-nucleosides by intramolecular free-radical cyclization—opening reactions based on temporary silicon connection, *Tetrahedron 48*:349 (1992).
95. Z. Xi, J. Rong, and J. Chattopadhyaya, Diastereospecific synthesis of 2'- or 3'-C-branched nucleosides through intramolecular free-radical capture by silicon-tethered acetylene, *Tetrahedron 50*:5255 (1994).
96. A. De Mesmaeker, P. Hoffmann, T. Winkler, and A. Waldmer, Stereoselective C–C bond formation in carbohydrates by radical cyclization reactions, application for the synthesis of α-C(2)-branched sugars, *Synlett.* p. 201 (1990).
97. P. Mayon and Y. Chapleur, Exclusive 6-*endo* radical cyclizations of α-silyl radicals derived from carbohydrate allylic silylethers, *Tetrahedron Lett. 35*:3703 (1994).
98. Y. C. Xin, J. M. Mallet, and P. Sinay, An expeditious synthesis of a C-disaccharide using a temporary silaketal connection, *J. Chem. Soc. Chem. Commun.* p. 864 (1993).
99. K. S. Gröninger, K. F. Jäger, and B. Giese, Cyclization reactions with allyl-substituted glucose derivatives, *Liebigs Ann. Chem.* p. 731 (1987).
100. B. Fraser-Reid and R. Tsang, *Strategies and Tactics in Organic Synthesis* 2, T. Lindberg, ed., Academic Press, New York, 1989, pp. 123–162.
101. J. Marco-Contelles, P. Ruiz-Fernandez, and B. Sanchez, New annulated furanoses: A new free-radical isomerization of an *S*-methyl hex-5-enylxanthate to an *S*-(cyclopentylmethyl) *S*-methyl dithiocarbonate, *J. Org. Chem. 58*:2894 (1993).
102. B. Giese, J. A. Gonzalez-Gomez, and T. Witzel, The scope of radical CC-coupling by the "tin method," *Angew. Chem. Int. Ed. Engl. 23*:69 (1984).
103. P. Blanchard, A. D Da Silva, J.-L. Fourrey, A. S. Machado, and M. Robert-Gero, Zinc-copper couple promoted C-branching in the carbohydrate series, *Tetrahedron Lett. 33*:8089 (1992).
104. B. Fraser-Reid, N. L. Holder, and M. B. Yunker, Ground and excited state 1,4-addition reactions of some carbohydrate enones, *J. Chem. Soc. Chem. Commun.* p. 1286 (1972).
105. B. Fraser-Reid, N. L. Holder, D. R. Hicks, and D. L. Walker, Synthetic applications of the photochemically induced addition of oxycarbinyl species to α-enones. Part I. The addition of simple alcohols, *Can. J. Chem. 55*:3978 (1977).
106. B. Fraser-Reid, R. C. Anderson, D. R. Hicks, and D. L. Walker, Synthetic applications of the photochemically induced addition of oxycarbinyl species to α-enones. Part II. The addition of ketals, aldehydes, and polyfunctional species, *Can. J. Chem. 5*:3986 (1977).
107. B. Fraser-Reid, R. Underwood, M. Osterhout, J. A. Grossman, and D. Liotta, Some observations on the relative reactivities of α-enones of oxanes and cyclohexanes, *J. Org. Chem. 51*:2152 (1986).
108. U. E. Udodong and B. Fraser-Reid, Electrophilic amination as a route to deoxyamino sugars: Synthesis of the key intermediate for 1-β-methylcarbapenem, *J. Org. Chem. 53*:2132 (1988).
109. Z. Benkö, B. Fraser-Reid, P. S. Mariano, and A. L. Beckwith, Conjugate addition of methanol to α-enones: Photochemistry and stereochemical detail, *J. Org. Chem. 53*:2066 (1988).
110. B. Giese, The stereoselectivity of intramolecular free radical reactions, *Angew. Chem. Int. Ed. Engl. 28*:969 (1989).
111. R. Blattner and D. M. Page, Radical addition to levoglucosenone, synthesis of anhydrosugar herbicide analogues, *J. Carbohydr. Chem. 132*:27 (1994).

112. B. Giese and T. Witzel, Stereoselective radical reactions with enolones, *Tetrahedron Lett.* 28:2571 (1987).
113. B. Giese, M. Hoch, C. Lamberth, and R. R. Schmidt, Synthesis of methylene bridged C-disaccharides, *Tetrahedron Lett.* 29:1375 (1988).
114. R. M. Bimwala and P. Vogel, Synthesis of α-(1,2)-, α-(1,3)-, α-(1,4)-, and α(1,5)-C-linked disaccharides through 2,3,4,6-tetra-*O*-acetylglucopyranosyl radical additions to 3-methylene-7-oxabicyclo[2.2.1]-heptan-2-one derivatives, *J. Org. Chem.* 57:2076 (1992).
115. Unpublished results, P. Mayon and Y. Chapleur (1995).
116. B. Giese, W. Damm, T. Witzel, and H. G. Zeitz, The influence of substituents at prochiral centers on the stereoselectivity of enolate radicals, *Tetrahedron Lett.* 34:7053 (1993).
117. G. E. Keck and J. B. Yates, Carbon–carbon bond formation via the reaction of trialkylallylstannanes with organic halides, *J. Am. Chem. Soc. 104*:5829 (1982).
118. G. E. Keck, D. F. Kachensky, and E. J. Enholm, Pseudomonic acid C from L-lyxose, *J. Org. Chem.* 50:4317 (1985).
119. D. Mootoo, P. Wilson, and V. Jammalamadaka, Application of the Keck radical coupling reaction to the preparation of allylated C-5 furanosides and C-6 pyranosides, *J. Carbohydr. Chem. 13*:841 (1994).
120. G. E. Keck and A. M. Tafesh, Free-radical addition fragmentation reactions in synthesis: A second generation synthesis of (+)-pseudomonic acid C, *J. Org. Chem.* 54:5845 (1989).
121. R. Rona, L. Tokes, J. Tremble, and P. Crabbe, Synthesis of olefins and dienes: Steroids, *Chem. Commun.* p. 43 (1969).
122. Y. Chapleur and Y. Grapsas, Stereospecific formation of C–C bonds on ethyl 4,6-di-*O*-acetyl-2,3 dideoxy-D-*ribo*-hex-3-enopyranoside, *Carbohydr. Res. 141*:153 (1985).
123. J. Levisalles, H. Rudler-Chauvin, and H. Rudler, Réaction des organocuprates avec les acétates allyliques. Existence probable de complexes π-allyliques du cuivre, *J. Organomet. Chem. 136*:103 (1977).
124. S. J. Danishefsky, D. M. Armistead, F. E. Wincott, H. G. Selnick, and R. Hungate, The total synthesis of the aglycon of avermectin A[1a], *J. Am. Chem. Soc. 109*:8117 (1987).
125. S. J. Danishefsky, D. M. Armistead, F. E. Wincott, H. G. Selnick, and R. Hungate, The total synthesis of the aglycon of avermectin A[1a], *J. Am Chem. Soc. 111*:2967 (1989).
126. S. Valverde, M. Bernabe, S. Garcia-Ochoa, and A. M. Gomez, Regio- and stereochemistry of cross coupling of organocopper reagents with allyl ethers: Effect of the leaving group, *J. Org. Chem. 55*:2294 (1990).
127. S. Valverde, M. Bernabe, A. M. Gomez, and P. Puebla, Cross coupling reactions of 2-(allyloxy-(thio))benzothiazoles with organocopper reagents in dihydropyranoid systems—mechanistic implications of the substrate and the reagent—regiocontrolled and stereocontrolled access to branched-chain sugars, *J. Org. Chem.* 57:4546 (1992).
128. Unpublished results, Y. Chapleur and S. Kasri (1993).
129. H. H. Baer and Z. S. Hanna, The preparation of amino sugars and branched-chain sugars by palladium-catalysed allylic substitution of alkyl hex-2-enopyranosides, *Can. J. Chem.* 59:889 (1981).
130. D. P. Curran, An approach to the enantiocontrolled synthesis of pseudomonic acids via a novel mon-Claisen rearrangement, *Tetrahedron Lett.* 23:4309 (1982).
131. D. P. Curran and Y. Suh, Synthetic applications of a substituent controlled Claisen rearrangement. Preparation of advanced intermediates for the synthesis of pseudomonic acid, *Tetrahedron Lett.* 25:4179 (1984).
132. N. Naz, T. H. Al-Tel, Y. Al-Abed, and W. Voelter, Synthesis of polyfunctionalized bis-annulated pyranosides: Useful intermediates for triquinane synthesis, *Tetrahedron Lett.* 35:8581 (1994).
133. P. N. Confalone and E. M. Huie, The [3 + 2] nitrone-olefine cycloaddition reaction, *Org. React.* 36:1 (1988).
134. A. J. Blake, T. A. Cook, A. C. Forsyth, R. O. Gould, and R. M. Paton, 1,3 Dipolar Cycloaddition, *Tetrahedron 48*:8053 (1992).

135. I. Panfil, C. Belzecki, and M. Chmielewski, An entry to the optically pure β-lactam skeleton based on 1,3-dipolar cycloaddition of nitrones to 4,6-di-O-acetyl-2,3-dideoxy-D-threo-hex-2-enonolactone, *J. Carbohydr. Chem.* 6:463 (1987).
136. I. Panfil, C. Belzecki, Z. Urbanczyk-Lipkowska, and M. Chmielewski, 1,3-Dipolar cycloaddition of nitrones to sugars enelactones, *Tetrahedron* 47:10087 (1991).
137. H. Gnichtel and L. Autenreich-Ansorge, Stereo- und regio-chemie der 1,3-dipolaren cycloaddition von benzonitril an hex-2-enopyranoside und hex-2-enopyranosid-4-ulose, *Liebigs Ann. Chem.* p. 2217 (1985).
138. H. Gnichtel, L. Autenrieth-Ansorge, and J. Dachman, Cycloaddition mit ungesättigten zuckern, VI Die synthese diastereomerer 4H-pyrano[3,4-d]isoxazol-4-one, *J. Carbohydr. Chem.* 6:673 (1987).
139. D. P. Curran, P. B. Jacobs, R. L. Elliot, and B. H. Kim, Total synthesis of specionin, *J. Am. Chem. Soc.* 109:5280 (1987).
140. D. P. Curran and P. B. Jacobs, Combination Claisen–nitrile oxide annulation. A strategy for ring construction with rigid stereocontrol dictated by an allylic hydroxyl group, *Tetrahedron Lett.* 26:2031 (1985).
141. R. J. Ferrier and N. Vethaviyasar, Unsaturated carbohydrates. Part XVII. Synthesis of branched-chain sugar derivatives by application of the Claisen rearrangement, *J. Chem. Soc. Perkin Trans.* p. 1791 (1973).
142. K. Heyms and R. Hohlweg, [3.3]-Sigmatrope umlagerungen an glycalen und pseudoglycalen, *Chem. Ber.* 111:1632 (1978).
143. L. Cottier, G. Remy, and G. Descotes, Photochemical synthesis of O-vinyl glycosides and their transformation into C-branched sugars, *Synthesis* p. 711 (1979).
144. P. Rollin, V. Verez Bencomo, and P. Sinay, Use of selenium in carbohydrate chemistry: Formation of vinyl-glycosides, *Synthesis* p. 134 (1984).
145. A. G. Wee and L. Zhang, The Still–Wittig rearrangement in the preparation of a synthetically useful 4C-hydroxymethyl-hex-2-enopyranoside, *Synth. Commun.* 23:325 (1993).
146. A. E. Wick, D. Felix, K. Steen, and A. Eschenmoser, Claisen'sche umlagerungen bei allyl- und benzylalkoholen mit hilfe von acetalen des N,N-dimethylacetamids, *Helv. Chim. Acta* 47:2425 (1964).
147. E. J. Corey, M. Shibazaki, and J. Knolle, Simple stereocontrolled synthesis of thromboxane $B_2$, *Tetrahedron Lett.* p. 1625 (1977).
148. I. Pelyvas, T. Lindhorst, and J. Thiem, En route to thromboxane compounds from carbohydrates, I. Synthesis of the unsaturated sugar precursors, *Liebigs Ann. Chem.* p. 761 (1990).
149. R. E. Ireland and J. P. Vevert, A chiral total synthesis of (−) and (+) nonactic acids from carbohydrate precursors and the definition of the transition for the enolate Claisen rearrangement in heterocyclic systems, *J. Org. Chem.* 45:4259 (1980).
150. R. E. Ireland, R. C. Anderson, R. Badoud, B. J. Fitzsimmons, G. J. McGarvey, S. Thaisrivongs, and G. S. Wilcox, The total synthesis of ionophore antibiotics. A convergent synthesis of lasalocid A (X537A), *J. Am. Chem. Soc.* 105:1988 (1983).
151. D. P. Curran and Y. Suh, Substituent effects on the Claisen rearrangement. The accelerating effect of a 6-donor substituent, *J. Am. Chem. Soc.* 106:5002 (1984).
152. D. P. Curran and Y. Suh, Selective mono-Claisen rearrangement of carbohydrate glycals. A chemical consequence of the vinylogous anomeric effect, *Carbohydr. Res.* 171:161 (1987).
153. R. F. Butterworth, W. G. Overend, and N. R. Williams, A new route to branched-chain sugars by C-alkylation of methyl glycopyranosiduloses, *Tetrahedron Lett.* p. 3229 (1968).
154. A. Klemer and G. Rodemeyer, Eine einfache synthese von methyl-4,6-O-benzyliden-2-desoxy-α-D-erythro-hexopyranosid-3-ulose, *Chem. Ber.* 107:2612 (1974).
155. D. Horton and W. Weckerle, A preparative synthesis of 3-amino-2,3,6-trideoxy-L-lyxo-hexose hydrochloride from D-mannose, *Carbohydr. Res.* 44:227 (1975).
156. D. B. Tulshian and B. Fraser-Reid, Studies directed at the synthesis of verrucarol from D-glucose: The A-B moiety, *Tetrahedron* 40:2083 (1984).

157. J. K. Dickson, R. Tsang, J. M. Llera, and B. Fraser-Reid, Serial radical cyclisation of branched carbohydrates 1. Simple pyranosides diquinanes, *J. Org. Chem. 54*:5350 (1989).
158. D. Alker, D. N. Jones, G. M. Taylor, and W. W. Wood, Synthesis of homochiral trisubstituted γ-butyrolactones, *Tetrahedron Lett. 32*:1667 (1991).
159. K. L. Yu and B. Fraser-Reid, Facial-selective carbohydrate-based aldol additions, *J. Chem. Soc. Chem. Commun.* p. 1442 (1989).
160. S. Handa, R. Tsang, A. T. Mc Phail, and B. Fraser-Reid, The pyranoside ring as a nucleophile in aldol condensations, *J. Org. Chem. 52*:3489 (1987).
161. K. L. Yu, S. Handa, R. Tsang, and B. Fraser-Reid, Carbohydrate-derived partners display remarkably high stereoselectivity in aldol coupling reactions, *Tetrahedron 47*:189 (1991).
162. T. Haneda, P. G. Goekjian, S. H. Kim, and Y. Kishi, Preferred conformation of C-glycosides. 10. Synthesis and conformational analysis of carbon trisaccharides, *J. Org. Chem. 57*:490 (1992).
163. A. Klemer and H. Wilbers, Neue synthesen von derivaten der Antibiotikazucker 3-Amino-2,3,6-tridesoxy-3-C-methyl-L-*xylo*-hexopyranose, L-Vancosamin, D-Rubranitrose und von vorläufern der L-Decinitrose und D-Kijanose, *Liebigs Ann. Chem.* p. 815 (1987).
164. E. J. Enholm and S. Jiang, Highly stereoselective coupling of carbohydrate lactones with terpene ketones promoted by SmI2, *Tetrahedron Lett. 33*:6069 (1993).
165. E. J. Enholm, S. Jiang, and K. Abboud, Branched-chain carbohydrate lactones from a samarium(II) iodide promoted serial deoxygenation-carbonyl addition reaction, *J. Org. Chem. 58*:4061 (1993).
166. M. Matsuzawa and J. Yoshimura, Synthesis of methyl-4-deoxy-4-C-(hydroxymethyl)-5-O-methyl-2,3-O-methylene-L-idonate, *Carbohydr. Res. 81*:C5 (1980).
167. J. Yoshimura and M. Matsuzawa, Stereoselective synthesis of methyl 4-C-acetyl-6-deoxy-2,3-O-methylene-D-galactonate and -D-gluconate. Determination of the D-galacto configuration of methyl eurekanate by synthesis, *Carbohydr. Res. 85*:C1 (1980).
168. P. Sarda, A. Olesker, and G. Lukacs, Synthesis of methyl 2-C-acetamidomethyl-2-deoxy-α-D-glucopyranoside and its manno isomer, *Carbohydr. Res. 229*:161 (1992).
169. H. Dietrich and R. R. Schmidt, Synthesis of carbon-bridged C-lactose and derivatives, *Liebigs Ann. Chem.* p. 975 (1994).
170. F. A. Carey and W. C. Franck, Preparation of 3-C-methylene sugars by Peterson olefination, *J. Org. Chem. 47*:3548 (1982).
171. S. Hanessian, G. Demailly, Y. Chapleur, and S. Leger, Facile access to chiral 5-hydroxy 2-methyl hexanoic acid lactones (pheromones of Carpenter bee), *J. Chem. Soc. Chem. Commun.* p. 1125 (1981).
172. K. Tadano, Y. Idogaki, H. Yamada, and T. Suima, *ortho* Ester Claisen rearrangements of three 3-C(hydroxymethyl)methylene derivatives of hexofuranoses: Stereoselective introduction of a quaternary center on C-3 of D-*ribo*-, *lyxo*- and D-*arabino*-hexofuranoses, *J. Org. Chem. 52*:1201 (1987).
173. J. M. J. Tronchet and B. Gentile, 3-C(Acylméthylène)-3-désoxy-1,2:5,6-di-O-isopropylidène-α-D-*ribo*- et -*xylo*-hexofuranoses, *Carbohydr. Res. 44*:23 (1975).
174. B. Fraser-Reid, R. Tsang, D. B. Tulshian, and K. M. Sun, Highly stereoselective routes to functionalized geminal alkyl derivatives to carbohydrates, *J. Org. Chem. 46*:3767 (1981).
175. K. Hara, H. Fujimoto, K. Sato, H. Hashimoto, and J. Yoshimura, Synthesis of methyl 4-deoxy-3-C-[(S)-1,2-dihydroxyethyl]-α-D-*xylo*-hexopyranoside and methyl 2,2'-anhydro-3-C-[(S)-1,2-dihydroxyethyl]-α-D-glucopyranoside derivatives, *Carbohydr. Res. 159*:65 (1987).
176. R. Csuk, A. Fürstner, and H. Weidmann, Synthesis and rearrangement reactions of C-alkylidene carbohydrates, *J. Carbohydr. Chem. 5*:271 (1986).
177. Y. Ali and W. A. Szarek, Synthetic approaches to *gem*-di-C-alkyl derivatives of carbohydrates: Nucleophilic addition reactions of 3-C-methylene compounds derived from 1,2:5,6-di-O-isopropylidene-α-D-*ribo*-hexofuranos-3-ulose using phase transfer catalysis, *Carbohydr. Res. 67*:C17 (1978).

178. Y. Chapleur and M.-N. Euvrard, Hetero Diels–Alder reactions in carbohydrate chemistry: A new strategy for multichiral arrays synthesis, *J. Chem. Soc. Chem. Commun.* p. 884 (1987).
179. H. Feist, K. Peseke, and P. Köll, Synthesis of branched-chain sugars with methiniminium, *Carbohydr. Res. 247*:315 (1993).
180. N. G. Ramesh and K. K. Balasubramanian, Vilsmeier-Haack reaction of glycals. A short route to C-2 formyl glycals, *Tetrahedron Lett. 32*:3875 (1991).
181. W. Priebe, G. Grynkiewicz, and N. Neamati, One step C-acylation of glycals and 2-deoxyhexopyranoses at C-2, *Tetrahedron Lett. 33*:7681 (1992).
182. H. Hashimoto, N. Kawauchi, and J. Yoshimura, Two complementary methods for introduction of *gem*-dimethyl group in hexopyranoside ring, *Chem. Lett.* p. 965 (1985).
183. N. Kawauchi, K. Sato, J. Yoshimura, and H. Hashimoto, Synthesis of 2,2-di-*C*-methyl-2-deoxy- and 4,4-di-*C*-methyl-4-deoxypranosides via Michael addition of conjugated enopyranosiduloses, *Bull. Chem. Soc. Jpn 60*:1433 (1987).
184. D. B. Tulshian, R. Tsang, and B. Fraser-Reid, Out-of ring Claisen rearrangements are highly stereoselective in pyranoses: Routes to *gem*-dialkylated sugars, *J. Org. Chem. 49*:2347 (1984).
185. B. Fraser-Reid, D. B. Tulshian, and R. Tsang, Striking contrasts in the stereoselectivities of spiro-Claisen rearrangements of pyranosides versus carbocycles, *Tetrahedron Lett. 25*:4579 (1984).
186. H. Pak, J. K. Dickson, and B. Fraser-Reid, Serial radical cyclisation of branched carbohydrates 2, *J. Org. Chem. 54*:5357 (1989).
187. J. Ishihara, K. Tomita, K. Tadano, and S. Ogawa, Total syntheses of (−)-acetomycin and its 3-stereoisomers at C-4 and C-5, *J. Org. Chem. 57*:3789 (1992).
188. K. Tadano, Natural product synthesis starting with carbohydrates based on the Claisen rearrangement protocol, in *Studies in Natural Products Chemistry*, Atta-ur Rahman, ed., Elsevier Science, New York, 1992, pp. 405–455.
189. J.-M. Vatèle, Nouvelle voie d'accès stéréoselective aux esters acryliques et son application en chimie des sucres, *Carbohydr. Res. 136*:177 (1985).
190. W. Meyer zu Reckendorf and U. Kamprath-Scholtz, Synthesis of 2,3-(carboxymethylene)- and 2,3-(2-hydroxyethylidene)-2,3-dideoxy-D-mannose. Monosaccharides with a cyclopropane ring, *Chem. Ber. 105*:673 (1972).
191. B. Fraser-Reid and B. J. Carthy, Synthesis of some α-cyclopropylcarbonyl glycopyranosides, *Can. J. Chem. 50*:2928 (1972).
192. B. J. Fitzsimmons and B. Fraser-Reid, Annulated pyranosides as chiral synthons for carbocyclic systems. Enantiospecific routes to both (+) and (−) chrysanthemum carboxylic acids from a single progenitor, *J. Am. Chem. Soc. 101*:6123 (1979).
193. B. J. Fitzsimmons and B. Fraser-Reid, Annulated pyranosides V. An enantiospecific route to (+) and (−) chrysanthemum dicarboxylic acids, *Tetrahedron 40*:1279 (1984).
194. P. Duchaussoy, P. Di Cesare, and B. Gross, Synthesis of sugars containing the dihalocyclopropane moiety: A new access to the heptose series, *Synthesis* p. 198 (1979).
195. H. Redlich, J. B. Lenfers, and J. Kopf, Chiral cyclobutanones by [2 + 2] cycloaddition of dichloroketene to carbohydrate enol ethers, *Angew. Chem. Int. Ed. Engl. 28*:777 (1989).
196. I. Hanna, J. Pan, and J. Y. Lallemand, Optically active cyclobutanones from glycals: Preparation and regioselective cleavage, *Synlett* p. 510 (1991).
197. J. W. Pan, I. Hanna, and J. Y. Lallemand, Optically active cyclobutanones from glycals. 2. Synthesis of chiral cyclobutane derivatives by tetrahydropyran ring-opening, *Tetrahedron Lett. 32*:7543 (1991).
198. R. J. Ferrier and S. Middleton, The conversion of carbohydrate derivatives into functionalized cyclohexanes and cyclopentanes, *Chem. Rev. 93*:2779 (1993).
199. P. L. Pauson, Organometallics in organic synthesis, in *Aspects of Modern Interdisciplinary Field*, A. de Meijere and H. tom Dieck, eds., Springer, Berlin, 1987.
200. W. E. Lindsell, P. N. Preston, and A. B. Rettie, Synthesis and characterisation of hexacarbonyldicobalt complexes derived from 2-propynyl and 3-butynyl 4,6-di-*O*-acetyl-2,3-dideoxy-α-D-*erythro*-hex-2-enopyranosides, *Carbohydr. Res. 254*:311 (1994).

201. J. Marco-Contelles, The Paulson-Khand reaction on carbohydrate templates I. Synthesis of bis-heteroannulated-phyranosides, *Tetrahedron Lett. 35*:5059 (1994).
202. J. L. Primeau, R. C. Anderson, and B. Fraser-Reid, Annulated pyranosides: Some cyclohexano and cyclopentano-sugars, *J. Chem. Soc. Chem. Commun.* p. 6 (1980).
203. M. G. Essig, F. Shafizadeh, T. G. Cochran, and R. Stenkamp, The crystal structure of a septanose derived from levoglucosenone, *Carbohydr. Res. 129*:55 (1984).
204. H. Gnichtel, C. Gumprecht, and P. Luger, Die stereochemie der Diels Adler reaktion mit hex-2-enopyranosid und hex-2-enopyranoside-4-ulosen, *Liebigs Ann. Chem.* p. 1531 (1984).
205. K. M. Sun, R. Guiliano, and B. Fraser-Reid, Diacetone glucose derived dienes in Diels–Alder reactions. Products and transformations, *J. Org. Chem. 50*:4774 (1985).
206. B. Fraser-Reid, Z. Benkö, R. Guiliano, K. M. Sun, and N. Taylor, Complete stereospecificity in the intramolecular Diels–Alder reaction of an ester derived from diacetone glucose, *J. Chem. Soc. Chem. Commun.* p. 1029 (1984).
207. J. Herscovici, S. Delatre, and K. Antonakis, Enantiospecific naphthopyran synthesis by intramolecular Diels–Alder cyclisation of 4-keto 2,3-unsaturated *C*-glycosides, *Tetrahedron Lett. 32*:1183 (1991).
208. J. Herscovici, S. Delatre, L. Boumaiza, and K. Antonakis, Stereocontrolled routes to functionalized [1,8-*bc*]-naphthopyran—a study on the total synthesis of quassinoids and tetrahydronaphthalene antibiotics, *J. Org. Chem. 58*:3928 (1993).
209. R. Tsang and B. Fraser-Reid, Pyranose α-enones provide ready access to functionalized *trans*-decalins via bis annulated pyranosides obtained by intramolecular Diels–Alder reactions; a key intermediate for forskolin, *J. Org. Chem. 57*:1065 (1992).
210. J. C. Lopez and G. Lukacs, Pyranose-derived dienes and conjugated enals—preparation and Diels–Alder cycloaddition reactions. in *Cycloaddition Reactions in Carbohydrate Chemistry*, 1992, pp. 33–49.
211. R. M. Giuliano and J. H. Buzby, Synthesis and Diels–Alder reactions of dienopyranosides, *Carbohydr. Res. 158*:C1 (1986).
212. J. C. Lopez, E. Lameignere, and G. Lukacs, Stereospecificity in Diels–Alder reactions of dienes and dienophiles derived from methyl 4,6-*O*-benzylidene-α-D-glucopyranoside, *J. Chem. Soc. Chem. Commun.* p. 706 (1988).
213. B. H. Lipshutz, S. L. Nguyen, and T. R. Elworthy, Preparation and Diels–Alder reactions of a pyranoid vinyl glycal: Model studies for anthraquinone aglycone and carbohydrate syntheses, *Tetrahedron 44*:3355 (1988).
214. A. A. Ghini, C. Burnouf, J. C. Lopez, A. Olesker, and G. Lukacs, Intramolecular Diels–Alder reactions on pyranose trienes. Stereoselective access to bis-annulated pyranosides, *Tetrahedron Lett. 31*:2301 (1990).
215. C. Taillefumier and Y. Chapleur, An entry to enantiomerically pure *cis*-decalinic structures from carbohydrates, *J. Chem. Soc. Chem. Commun.* p. 937 (1995).
216. Y. Chapleur, The chemistry and total synthesis of mevinic acids, in *Recent Progress in the Chemistry of Antibiotics*, G. Lukacs and S. Ueno, eds., Springer, New York, 1993 pp. 829–937.
217. J. E. Baldwin, G. A. Höfle, and O. W. Lever, Jr., α-Methoxyvinyllithium and related metalated enol ethers. Practical reagents for nucleophilic acylation, *J. Am. Chem. Soc. 96*:7125 (1974).
218. H. Redlich, H. J. Neumann, and H. Paulsen, Radikalische desoxygenierung von verzweigten zuckern mit tri-*n*-butylzinnhydrid. Stereochemie der reaktion, *Chem. Ber. 110*:2911 (1977).
219. F. Shafizadeh and P. P. S. Chi, Preparation of 1,6-anhydro-3,4-dideoxy-β-D-glycero-hex-3-enopyranos-3-ulose and some derivatives thereof, *Carbohydr. Res. 58*:79 (1977).
220. D. E. Iley, S. K. Tam, and B. Fraser-Reid, 1,5-anhydro-4,6-*O*-benzylidene-1,2,3-trideoxy-3-*C*-methylene-D-*erythro*-hex-1-enitol: Six synthetic routes, *Carbohydr. Res. 55*:193 (1977).

# Chemical Synthesis of *O*- and *N*-Glycosyl Compounds, and of Oligosaccharides

**THEMES**: *O- and N-Glycopeptides; Oligosaccharide Synthesis from Glycosyl O-Trichloroacetimidates, Thioethers, Fluorides and 4-Pentenyl Glycosides; Sialyl Glycosides; Glycosides and Oligosaccharides by Remote Activation; Glycosides and Oligosaccharides from Unprotected Glycosyl Donors*

I believe it was Sir Derek Barton, who once remarked that half of carbohydrate chemistry was situated at the anomeric carbon atom. Indeed, the chemical synthesis of the glycosidic bond has been a continuous challenge ever since Emil Fischer's venerable acid-catalyzed glycosylation reaction of simple alcohols in 1893. Curiously, Fischer had been preceded by A. Michael, who in 1879, described the synthesis of a phenylglucoside under base-catalyzed conditions. In the ensuing century, the chemistry of the glycosidic linkage has enjoyed enormous research activity on many fronts, culminating with the emergence of a new subdiscipline, glycobiology, in which oligosaccharide chemistry plays a dominant role. Glycobiology can be considered as the meeting grounds of carbohydrate chemistry and biology. A panoply of important interactions at the molecular level, ranging from the mode of action of lifesaving antibiotics and antitumor agents, to cell surface phenomena associated with immunological responses that are vital for maintaining life processes, involve carbohydrate motifs. The importance of carbohydrate components of many antibiotics for their biological activity was recognized many years ago. Streptomycin, one of a handful of wonder drugs of the middle of the century, is an example. Several aminoglycoside and macrolide antibiotics soon followed, and were marketed as important therapeutic agents. Unusual naturally occurring nucleosides with antibiotic, antitumor, and antiviral activities emerged in the 1960s and 1970s. Nature continues to adorn its most complex chemical structures with glycosidic appendages, as in calicheamycin and avermectin $B_{1a}$. Some of the most interesting aminodeoxy, deoxy, and branched-chain sugars have been associated with natural products, originally found in soil samples or other exotic sources, then produced by fermentation on tonnage scale.

The animal and plant worlds have yielded some of the more fascinating carbohydrate-based polymers as homo- and heteropolysaccharides, as glycoproteins, as lipopolysaccharides, and as glycolipids, to mention a few. Blood group substances, antigenic determinants, and cell surface carbohydrates provide a fascinating array of oligosaccharide structures, which often include sialic acid as a terminal unit.

Many biologically important interactions at the molecular level involve carbohydrates and proteins, and there is no natural recognition mode that is more intimate and specific.

Although the entire topology of a sugar molecule is critical for its designated biological function, the α or β nature of the glycosidic linkage plays an equally, if not more important role. Marvel, for example, at the anomeric specificity of glycosidases and glycosyl transferases, on the one hand, and the requirement of a given anomeric configuration for maximal antibiotic or antigenic activity on the other.

The advent of glycobiology and the tremendous recent advances in understanding interactions at the molecular level, have instigated the need to devise stereocontrolled syntheses of oligosaccharides, of glycopeptides, and of related compounds. Significant progress has been made in the stereocontrolled assembly of oligosaccharides in single units, or by so-called block synthesis over the past 20 years. Still, there is no universal method for glycoside synthesis. Small variations in the nature of the donor or acceptor molecules may result in inefficient coupling or an altered α/β-ratio compared with a seemingly related case (e.g., D-gluco and D-galacto donors vis-a-vis the same acceptor). The necessity to use O-protected donors and selectively O-protected acceptors adds to the already tedious operations of synthesis and assembly.

Unlike solid-phase peptide synthesis, automated oligosaccharide assembly is still highly experimental, but it is being actively explored. Fortunately, in many instances, carbohydrate-based biological function manifests itself at the oligosaccharide level. Therefore, our present needs are for easy accessibility to stereochemically defined oligosaccharides, O- and N-glycopeptides, and related compounds of reasonable molecular weights. Such structures are currently accessible through existing methodology, even if it is sometimes at a premium.

The next ten chapters should provide the reader with state-of-the-art methods in glycoside and oligosaccharide synthesis from leading researchers in the field.

*Stephen Hanessian*

# 11

## O- and N-Glycopeptides: Synthesis of Selectively Deprotected Building Blocks

**Horst Kunz**
*Johannes Gutenberg-Universität Mainz, Mainz, Germany*

|      |                                                                                                    |     |
|------|----------------------------------------------------------------------------------------------------|-----|
| I.   | Introduction                                                                                       | 265 |
| II.  | Methods                                                                                            | 268 |
|      | A. The fluorenylmethoxycarbonyl group                                                              | 268 |
|      | B. The Fmoc/*tert*-butyl ester combination                                                         | 269 |
|      | C. Allyl ester protecting group                                                                    | 270 |
|      | D. The allyloxycarbonyl groups                                                                     | 271 |
|      | E. Establishing sufficient acid stability in the saccharide portion                                | 272 |
|      | F. Allylic anchors in the solid-phase synthesis of glycopeptides                                   | 272 |
| III. | Experimental Procedures                                                                            | 273 |
|      | A. Orthogonal deprotection of Fmoc O-glycopeptide benzyl esters                                    | 273 |
|      | B. Orthogonal deprotection of Fmoc O-glycopeptide *tert*-butyl esters                              | 274 |
|      | C. The Fmoc–allyl ester combination                                                                | 276 |
|      | D. Removal of the allyloxycarbonyl group                                                           | 277 |
|      | E. Exchange of ether-type for ester-type protection in the saccharide portion: Generation of stability against acids | 279 |
|      | References                                                                                         | 279 |

## I. INTRODUCTION

Results of cell biological, biochemical, and immunological research of the past decades have revealed that glycosylation is a very common posttranslational modification of proteins in eukaryotic cells [1]. The carbohydrate portions of glycoproteins obviously play important roles in the organized distribution of these macromolecules within the cells and within multicellular organisms. In particular, they are recognition signals involved in intercellular communication, for example, in cell adhesion, regulation of the cell growth, infectious processes, and immunological differentiations [2]. For structural elucidation of these biological selection processes, model glycopeptides of exactly specified carbohydrate and peptide structure (e.g., type **1** and **2** are of interest.

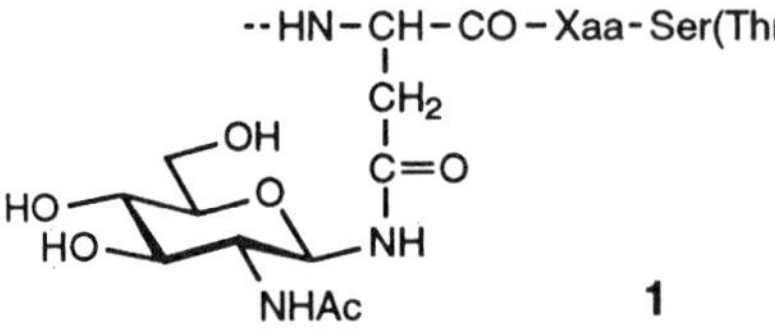

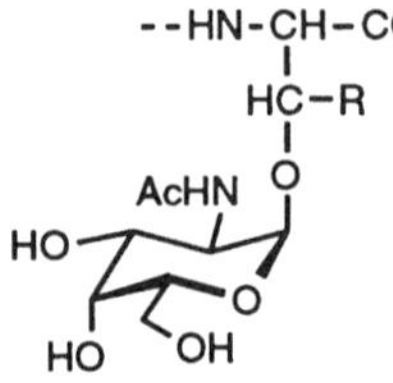

Until the late 1970s *O*- and *N*-glycosylated amino acids were constructed and used as standards for structural elucidations of the linkage regions between carbohydrate and peptide parts of glycoproteins [3,4]. However, the synthesis of glycopeptides demanded the development of versatile and selective protecting group techniques [5]. This holds true, in particular, for glycopeptides with complex oligosaccharide portions, which themselves must be formed in laborious multistep syntheses [6,7]. Much attention has to be paid to the glycosidic bond present in all natural glycoproteins. The acetal-type glycosidic and intersaccharidic bonds are potentially sensitive to strong acids and, for *O*-glycosyl serine and threonine derivatives (e.g., of **2**), also sensitive to bases. Therefore, controlled and selective deblocking of only one functional group of the polyfunctional glycosyl amino acids and glycopeptides is a critical problem in glycopeptide synthesis. Because the synthesis of oligosaccharides and glycosides has been described in a number of reviews [6,7] and is presented in succeeding chapters, this contribution is focused on protecting group techniques in glycopeptide chemistry.

A major progress in glycopeptide synthesis was achieved when it was demonstrated that the NH$_2$-terminal 9-fluorenylmethoxycarbonyl (Fmoc) group can be selectively removed from glycopeptides using the weak base morpholine (p$K_a$ 8.3). This holds true for *O*-glycosyl serine and threonine esters, which are sensitive to base-catalyzed β-elimination of the carbohydrate [10].

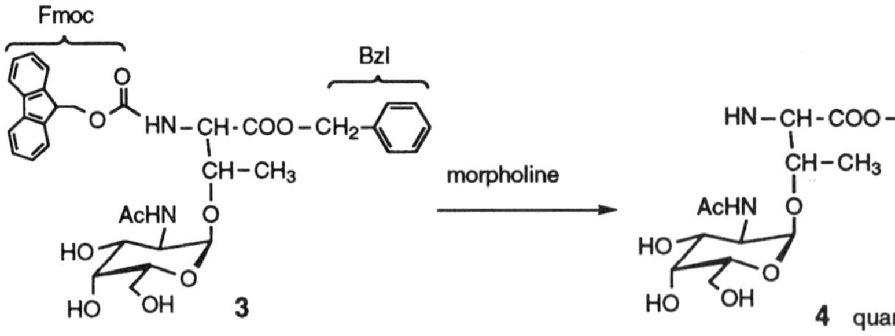

Because *O*-glycosylation can also be accomplished with active esters (e.g., pentafluorophenyl esters) [11] of Fmoc serine and threonine, the Fmoc technique provides a general method for the synthesis of glycopeptides. Thomsen–Friedenreich antigen glycopeptides and neoglycoproteins have been obtained by this method in preparative amounts [12]. In combination with acid-labile polymeric benzyl ester anchors, this Fmoc technique was applied to solid-phase syntheses of glycopeptides [11,13,14].

## Protecting Groups in Glycopeptide Synthesis

The Fmoc group is also useful in the synthesis of N-glycopeptides containing the $N^4$-glycosyl asparagine structure **1**. Because the N-glycosyl amide bond is relatively stable to acids and bases, the Fmoc group can be combined with a number of carboxyl-protecting groups in the synthesis of glycopeptides (e.g., the benzyl ester [15], the *tert*-butyl ester [16], as well as polymeric benzyl- [14] and allyl-type [17,18] ester anchors in solid-phase synthesis). However, problems arise even in N-glycopeptide synthesis, if sensitive glycosidic bonds (e.g., fucoside bonds) are present in the glycane portion [19].

Combinations of the benzyloxycarbonyl (Z) group with the *tert*-butyl ester [20] and of the *tert*-butyloxycarbonyl (Boc) group with the (polymeric) benzylester [21] have successfully been applied in glycopeptide synthesis. The sensitivity of the Boc group toward the acidic conditions of glycosylation reactions and the often sluggishly proceeding heterogeneous hydrogenation are unfavorable for a general application of these protecting group strategies. The allyl ester [22] and the allyloxycarbonyl group [23] were proved to be stable under conditions usually applied in peptide and glycopeptide synthesis. Nevertheless, these allylic protecting groups can be removed from O- [22,24] and N-glycopeptides [25] under practically neutral conditions by palladium(0)-catalyzed allyl transfer by using nucleophiles that irreversibly trap the allylic moiety.

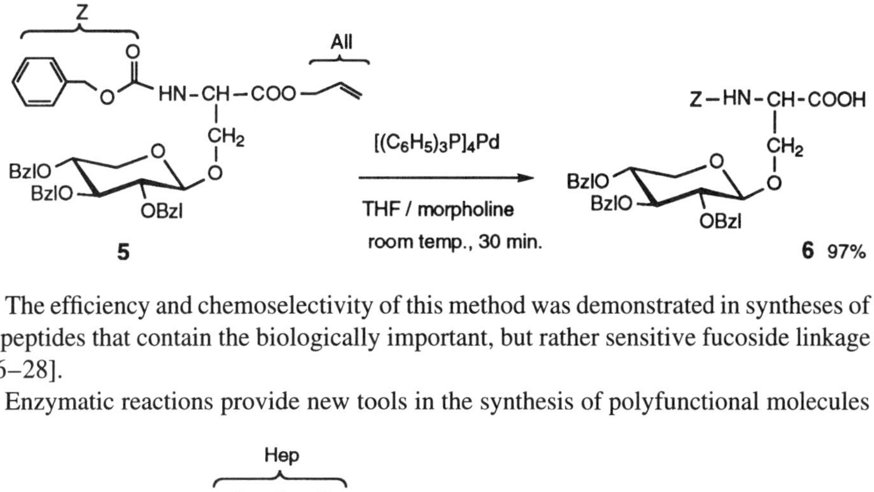

The efficiency and chemoselectivity of this method was demonstrated in syntheses of glycopeptides that contain the biologically important, but rather sensitive fucoside linkage [19,26–28].

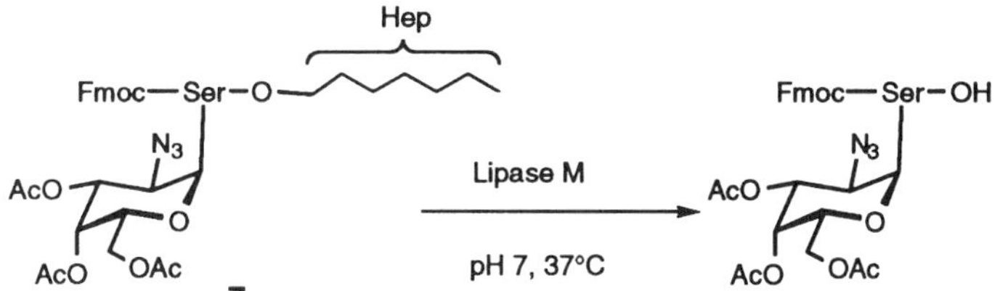

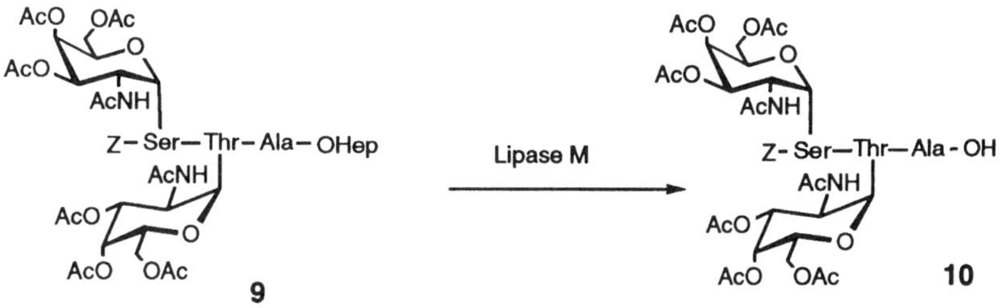

Enzymatic reactions provide new tools in the synthesis of polyfunctional molecules such as glycopeptides. For example, lipases exhibiting no protease activity selectively hydrolyze glycopeptide heptyl esters [29]. Under these conditions, the ester-protecting groups in the carbohydrate portion, the peptide bond, the amino protection, and the glycosidic bonds remain unaffected.

The choice of protecting groups also has an important effect on the stability of intersaccharidic and glycosidic bonds toward acids. If acidolytic cleavage of protecting groups is necessary during the synthesis of glycopeptides, the establishment of partial or complete acyl-type protection within the saccharide units is favorable [19,26]. For the construction of the O-glycosidic linkages to serine and threonine derivatives, glycosyl halides and promotion by mercury or silver salts, in particular by silver triflate [30], is generally efficient. Mucin-type glycopeptides are accessible using 2-azidogalactose donors [6]. Glycosyl trichloroacetimidates [7,22] and thioglycosides were also demonstrated to be useful donors in the synthesis of O-glycopeptides. Furthermore, electrophilic activation of glycals offers an interesting perspective in glycopeptide synthesis [31]. General access to N-glycopeptides is attained through glycosyl azides [4,25–28] that are reduced to glycosylamines, and the latter condensed with aspartic acid derivatives, these being selectively unblocked at the β-carboxyl group. Alternatively, N-acetylglucosamine derivatives react directly with ammonium hydrogencarbonate to form the corresponding glycosylamines [14].

## II. METHODS

Glycopeptides contain many functional groups of different reactivity as well as O- and N-glycosidic bonds. Therefore, the compatibility and chemoselectivity of the applied reactions is a fundamental prerequisite in glycopeptide synthesis. In this chapter, efficient and generally applicable methods and their combinations will be illustrated by examples [5,8,9].

### A. The Fluorenylmethoxycarbonyl Group

The Fmoc group, which is very useful in peptide synthesis [32], has proved an efficient tool in glycopeptide chemistry [10]. Because the Fmoc protection of the amino function is rather stable to acids, it can be combined with *tert*-butyl-type protecting groups and exposed to stereoselective glycosylations that require a more or less acidic milieu.

Glycosylation of the Fmoc serine benzyl ester **11** with 2,3,4-tri-O-benzoyl-α-D-xylopyranosyl bromide **12** promoted by silver triflate (30) afforded the β-xylosyl serine conjugate **13** in high yield [10]. The efficiency and compatibility of the Fmoc technique in glycopeptide synthesis was demonstrated on substrate **13** [10]. Treatment with morpholine (neat or diluted with dichloromethane 1:1 [18] or dimethylformamide [33]) resulted in quantitative and selective removal of the Fmoc group to give **14**, whereas hydrogenation selectively deblocked the carboxylic function resulting in **15**. The chemoselectivity of the amino deprotection was also shown for the glycopeptide **17**. Building blocks **14** and **15** can be used for alternative $NH_2$- or COOH-terminal chain extension to construct glycopeptides of a desired sequence. Removal of the O-acyl protection from the carbohydrate portion is achieved under weakly basic conditions, as is shown for the formation of **16** [10].

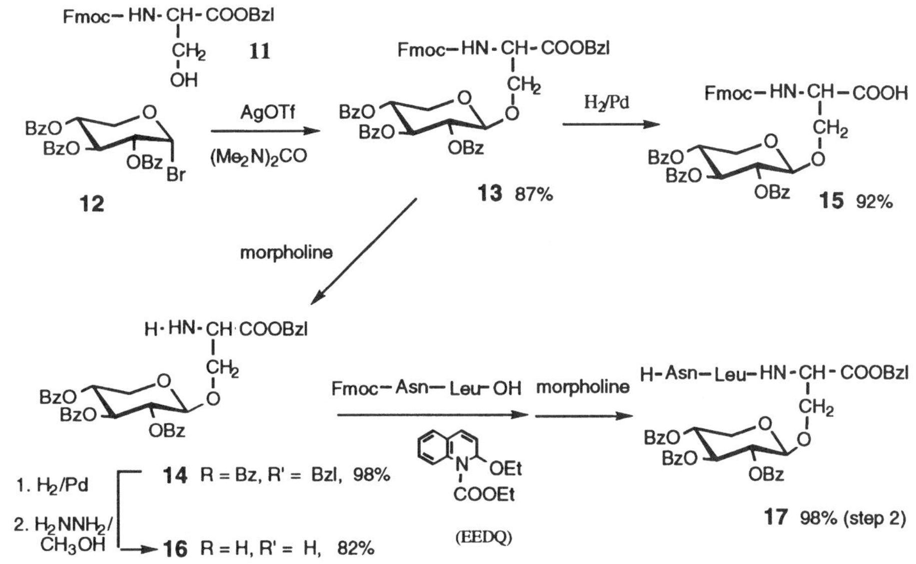

# Protecting Groups in Glycopeptide Synthesis

## B. The Fmoc/*tert*-Butyl Ester Combination

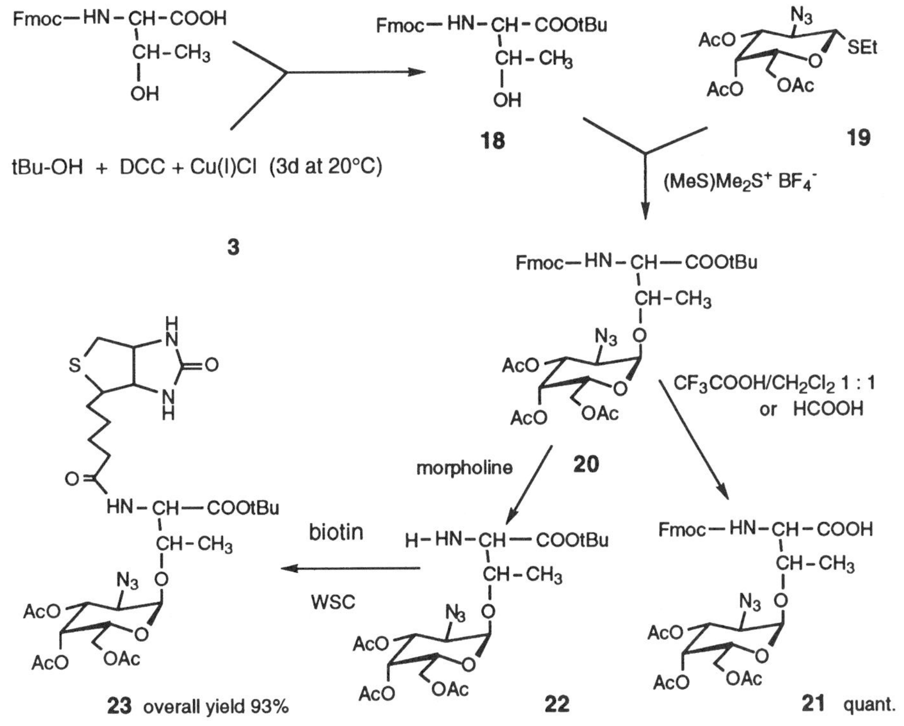

The Fmoc group can also be combined with the *tert*-butyl ester in *O*-glycopeptide synthesis [16,34]. The Fmoc serine and threonine *tert*-butyl esters with free β-hydroxyl groups are required as starting materials for this chemistry. These compounds (e.g., the threonine derivative **18**) become readily accessible by treatment of the Fmoc amino acid with a prereacted mixture of *tert*-butanol, carbodiimide, and copper(I) chloride [34]. Glycosylation of **18** using the thioglycoside donor **19** yielded mainly the desired α-anomer of conjugate **20** [34]. The pure anomers were obtained by flash chromatography. In the α- or, alternatively β-anomer of **20**, the *tert*-butyl ester group is selectively cleaved using formic acid or anhydrous trifluoroacetic acid/dichloromethane (1:1 v/v).

Alternatively, the Fmoc group is selectively eliminated with morpholine to give **22** [34]. Both reactions are the basis of chain-extending glycopeptide synthesis in solution as well as on solid phases [11,12,14]. Because the free amino group in compound **22** often gives rise to O → N acetyl transfer or β-elimination (at least in solution), subsequent peptide condensation should be carried out immediately. Condensation of **22** with biotin, using a water-soluble carbodiimide (WSC) yielded the biotinyl conjugate **23** containing a label (biotin) as well as a photoactivatable group (azide).

## C. Allyl Ester Protecting Group

The allyl ester [22] can be selectively removed in the presence of Z- [22], Boc- [25,28], or Fmoc amino protection [24]. For example, Fmoc threonine allyl ester **24** was glycosylated with 2-azido-3,4,6-tri-*O*-acetyl-2-deoxy-α-D-galactopyranosyl bromide **25** [6] to give the conjugate **26** in high yield [24]. Separation of the α-anomer and transformation of its

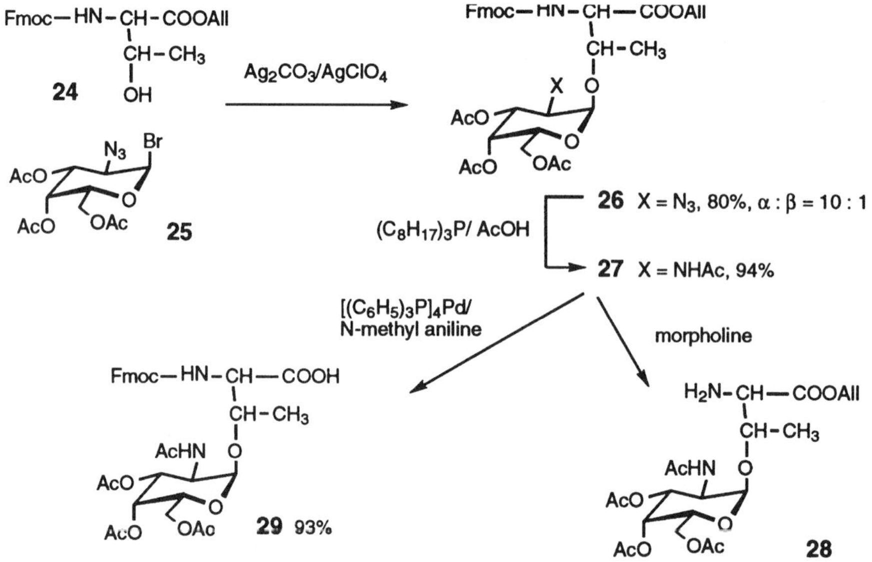

2-azido moiety to the acetamido group using either thioacetic acid or trioctylphosphine–acetic acid afforded the $T_n$ antigen structural element **27** [24]. Selective removal of the Fmoc group was achieved quantitatively by treatment with morpholine (see earlier discussion). Again, the amino-deblocked compound **28** is not very stable and should immediately be used for further condensation.

The alternative selective cleavage of the allyl ester was achieved by Pd(0)-catalyzed allyl transfer. Because the Fmoc group is sensitive to morpholine, a very weak base must be used as the allyl-trapping nucleophile. *N*-Methyl aniline was favorable, and its use as a scavenger nucleophile gave rise to the acid **29** in a high yield.

## D. The Allyloxycarbonyl Groups

In this chapter the efficiency of the allyloxycarbonyl (Aloc) amino-protecting group is illustrated in the synthesis of *N*-glycopeptides. In general, the *N*-glycosidic linkage of *N*-glycopeptides is formed by condensation of the corresponding 1-amino-*N*-acetylglucosamine derivative with a *N*-protected aspartic acid α-mono ester [4]. According to this principle, the chitobiosyl azide **30** selectively deprotected at O-6 [26] was coupled with *O*-benzylated α-fucosyl bromide **31** [35] under in situ anomerization conditions [36] to give the branched α-fucosyl chitobiosyl azide **32** [26].

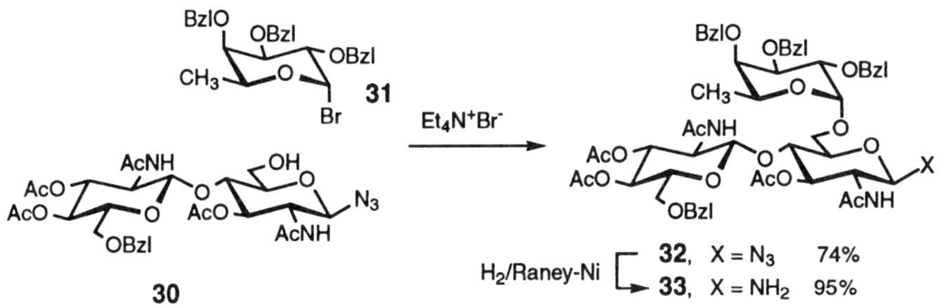

Subsequent hydrogenolysis of the azido group catalyzed by Raney nickel (washed to neutral reaction) furnished the glycosylamine **33**. Condensation of **33** with *N*-allyloxycarbonyl aspartic acid α-*tert*-butyl ester **34** was accomplished using ethyl 2-ethoxy-1,2-dihydroquinoline-1-carboxylate (EEDQ) [37] to give the fully protected *N*-glycosyl asparagine derivative **35** [26]. Similar to the allyl ester, the $NH_2$-terminal allyloxycarbonyl (Aloc) group can be removed under practically neutral conditions [23], as was demonstrated on fucosyl chitobiose asparagine conjugate **35** [19,26].

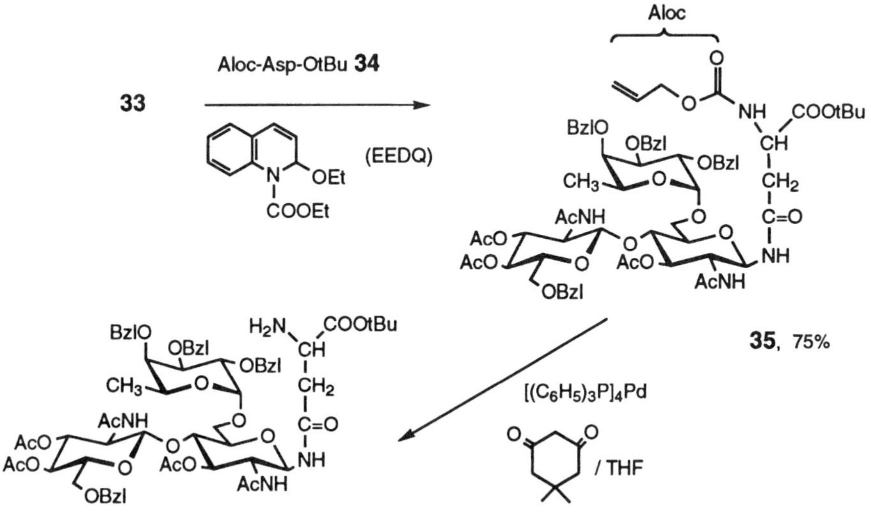

The Pd(0)-catalyzed transfer of the allyl moiety to dimedone [23], morpholine [22], or *N,N'*-dimethyl barbituric acid [27] resulted in the completely selective cleavage of the Aloc group, whereas the numerous other protecting groups and the glycosidic bonds, including the fucoside bond, remained unaffected [19,26]. It is obvious, that this selective and mild method is a promising tool for the synthesis of demanding glycopeptides (e.g., core N-glycopeptides containing the problematic β-mannoside bond) [38]. In contrast, the tert-butyl ester of **35** cannot be cleaved under acidic conditions without destroying the α-fucoside bond.

### E. Establishing Sufficient Acid Stability in the Saccharide Portion

If the acidolytic reactions often applied in peptide and solid-phase synthesis are to become adaptable to glycopeptide chemistry, a general stabilization of the saccharide portions of glycopeptides toward acids must be accomplished. Acetamido groups and O-acetyl-protecting groups in the carbohydrates exhibit an effective stabilizing influence on the glycosidic bonds. Therefore, the exchange of ether-type for acyl-type protection is of importance for a versatile glycopeptide synthesis, as is shown for fucosyl chitobiose derivative **37** [19,26].

The methoxybenzyl ether in **37** can be selectively cleaved, even in the presence of the anomeric azide, by oxidation with ceric ammonium nitrate. Subsequently, the N-phthaloyl group is removed with hydrazine, and the amino as well as the hydroxyl groups are acetylated to give **38**, which can now be used in glycopeptide synthesis. Asparagine conjugates carrying this type of saccharide side chain can be subjected to a selective acidolysis of tert-butyl esters and ethers without affecting the fucoside bond [19,26].

### F. Allylic Anchors in the Solid-Phase Synthesis of Glycopeptides

To illustrate the progress in glycopeptide synthesis and, in particular, the efficiency of the allylic-protecting method, the solid-phase synthesis of mucin-type glycopeptides on a polymeric support with an allylic anchor **39** is outlined [33]. Owing to the stability of the allylic ester to both acidic and basic conditions, the Fmoc as well as the Boc group can be used for temporary amino protection. One advantage of this versatility is the ability to switch from Fmoc to Boc protection on the level of the polymer-linked dipeptide to avoid diketopiperazine formation. Furthermore, the release of the glycopeptide carried out by palladium(0)-catalyzed allyltransfer to N-methylaniline proceeds almost quantitatively, and without affecting other protecting groups, to yield the glycononapeptide **40** in an overall yield (relative to **39**) of 95% and a purity of more than 95% [according to high-performance liquid chromatography (HPLC)]. Compound **40**, which is selectively de-blocked at the carboxyl group, can immediately be used in fragment condensations.

## Protecting Groups in Glycopeptide Synthesis

**39**

DMF/Morpholine 1:1; 3eq Boc-Pro-OH, DIC, HOBt, DMF; Ac₂O/Pyridine
CH₂Cl₂/TFA; Hünig Base; 4 eq Fmoc-Ala-OH, DIC, HOBt, DMF; Ac₂O
DMF/Morpholine 1:1; 3eq Fmoc-Pro-OH, DIC, HOBt, DMF; Ac₂O/Pyridine

[Ph₃P]₄Pd(0), Morpholine, DMF/DMSO; CH₂Cl₂

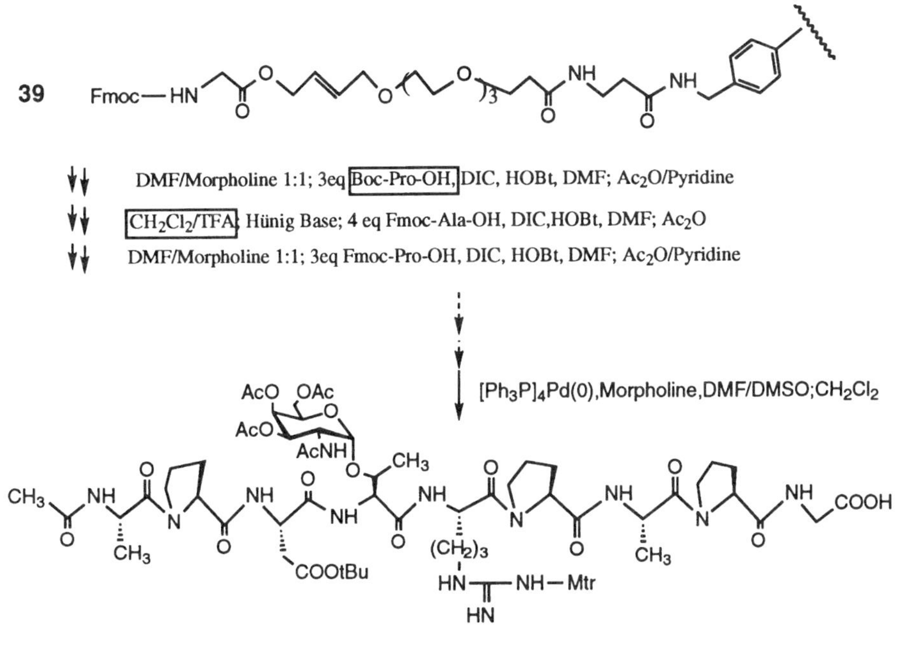

**40** overall yield (18 steps): 95%, purity > 95% (HPLC)

In conclusion, the selective-protecting group techniques [5,8,9] together with the versatile methods of oligosaccharide synthesis [6,7] make complex glycopeptides available, which are of interest in interdisciplinary investigations of biological selection processes. The choice of compatible methods determines the success of syntheses of these polyfunctional molecules. On the basis of the elaborated selective methods, the chemical synthesis of glycopeptides delivers preparative amounts of model compounds of exactly specified structure.

## III. EXPERIMENTAL PROCEDURES*

### A. Orthogonal Deprotection of Fmoc O-glycopeptide Benzyl Esters (10)

*N-(9-Fluorenylmethoxycarbonyl)-O-(2,3,4-tri-O-benzoyl-β-D-xylopyranosyl)-L-serine Benzyl Ester* **13**

To a solution of silver trifluoromethanesulfonate (7.71 g, 30 mmol) in dry dichloromethane (100 mL) at −40°C in the dark is added dropwise a solution of 9-fluorenylmethoxycarbonyl serine benzyl ester **11** (8.45 g, 20 mmol), 2,3,4-tri-O-benzoyl-α-D-xylopyranosyl bromide **12** (11.14 g, 21.2 mmol) and tetramethyl urea (3.65 g, 31.4 mmol) in dichloromethane (100 mL). After 18 h of stirring at room temperature, the precipitate is filtered off and washed with dichloromethane (200 mL). The organic solution is washed with water (200 mL), 1% KHCO₃ solution (twice 200 mL) and water, dried with Na₂SO₄, and concentrated in vacuo. The crude product is recrystallized from ethyl acetate–*n*-hexane. (If the reaction was not complete, chromatography on silica gel 60 in toluene/ethanol 9:1 is recommended). Yield: 15 g (87%); mp 136°C, [α]$_D$ −27.8° (*c* 1.3, CHCl₃), $R_f$ 0.64 (toluene/ethanol 26:1).

---

*Optical rotations were measured at 22°C.

*Removal of the Fmoc Group: General Procedure*

The protected *O*-glycosyl amino acid ester or *O*-glycosyl peptide ester (1 mmol) is stirred in morpholine (10 mL) or morpholine–dichloromethane (1:1) for 30 min. After addition of dichloromethane (100 mL), the solution is washed with diluted aqueous acid (citric acid or HCl pH 4, 50 mL) and with water (4 × 50 mL), dried with $Na_2SO_4$, and concentrated in vacuo. The crude product is dissolved in 2–5 mL of ethyl acetate. During chromatography on silica gel (50 g) with petrolum ether ethyl–acetate (2:1), *N*-(9-fluorenylmethyl)morpholine is eluted. Subsequently, the deblocked amino acid or peptide ester is eluted with methanol. *O*-(2,3,4-Tri-*O*-benzoyl-β-D- xylopyranosyl)-L-serine benzyl ester **14**: yield, 0.63 g (98%); mp 55°C; $[\alpha]_D$ −41.4 (*c* 0.5, $CH_3OH$). *N*-(L-Asparaginyl-L-leucyl-)-*O*-(2,3,4,-*O*-benzoyl-β-D-xylopyranosyl)-L-serine benzyl ester **17**: yield, 0.85 g (98%); amorphous; $[\alpha]_D$ −42.3° (*c* 0.5; $CH_3OH$).

*Removal of the Benzyl Ester*

**N-(9-Fluorenylmethoxycarbonyl)-*O*-(2,3,4-tri-*O*-benzoyl-β-D-xylopyranosyl)-L-serine 15.** The Fmoc *O*-xylosyl serine benzyl ester **13** (1.0 g, 1.2 mmol) is stirred in methanol (40 mL) at room temperature and subjected to hydrogenolysis for 18 h under atmospheric pressure using palladium–charcoal (0.2 g, 5%) as the catalyst. The educt **13** dissolves slowly. The catalyst is filtered off, and the solvent evaporated in vacuo. If the residue is not pure according to thin-layer chromatography (TLC), it is dissolved in 2 mL of ethyl acetate and purified by chromatography on a short column of silia gel 60. The by-products are eluted with petroleum ether–ethyl acetate; the product **15** with methanol: yield: 0.85 (92%); mp 109°C, $[\alpha]_D$ −12.6° (*c* 0.3, $CH_3OH$); $R_f$ 0.64 (toluene–ethanol, 1:2).

*Removal of the Carboxyl and O-Acyl Protection*

**O-(2,3,4-Tri-*O*-benzoyl-β-D-xylopyranosyl)-L-serine.** A solution of xylosyl serine benzyl ester **14** (0.5 g, 0.78 mmol) in methanol (20 mL) is hydrogenated for 18 h using palladium–charcoal (5%, 0.1 g) as the catalyst. The catalyst is filtered off, and the solvent evaporated to give the pure product: yield, 0.41 g (95%), mp 130°–133°C; $[\alpha]_D$ −29.2 (*c* 0.6, $CH_3OH$); $R_f$ 0.48 (ethyl acetate–methanol–water 13:5:2.4).

**O-(β-D-Xylopyranosyl)-L-serine 16.** To a solution of *O*-benzoyl protected xylosyl serine (0.23 g, 0.4 mmol) in methanol (20 mL) is added at room temperature hydrazine hydrate (100%, 20 mL). Monitoring by TLC (ethyl acetate–methanol–water 13:5:2.4) proves the reaction to be complete within 30 min. After 40 min, acetone (50 mL) is added to transform the hydrazine to the acetone azine (which is volatile). After stirring for 1 h, the solution is concentrated in vacuo, the residue is stirred with ethyl acetate to extract impurities and give a pure product: yield, 80 mg (86%), mp 215°–218°C (decomposition); reported [39] 230°–235°C, $[\alpha]_D$ −49.8° (*c* 0.3, $H_2O$), reported [39] $[\alpha]_D$ −12° (*c* 1.0, $H_2O$).

## B. Orthogonal Deprotection of Fmoc *O*-Glycopeptide *tert*-Butyl Esters

*N-(9-Fluorenylmethoxycarbonyl)*-L-*threonine* tert-*butyl ester* **18** [34]

A mixture of 1,3-dicyclohexyl carbodiimide (92.8 g, 0.45 mol) *tert*-butanol (43.4 g, 0.58 mol) and copper(I) chloride (1 g, 0.01 mol) is stirred in a tightly sealed flask at room temperature in the dark. After 5 days, dry dichloromethane (300 mL) is added. A solution of Fmoc threonine (47.16 g, 0.138 mol) in dry dichloromethane (300 mL) is added dropwise to the stirred mixture. After stirring for 4 h (TLC monitoring, toluene–ethanol 10:1), the mixture is filtered, and the filtrate concentrated in vacuo to about 150 mL. Newly precipitated dicyclohexyl urea is again removed by filtration. The filtrate is diluted with dichloromethane to a volume of 500 mL, washed with saturated NaHCO$_3$ solution (3 × 100 mL). In cases of unsatisfactory phase separation, the aqueous layer is re-extracted with dichloromethane. The organic layer is dried with MgSO$_4$ and the solvent evaporated in vacuo. The remainder is dissolved in a small volume of ethyl acetate, kept at $-28°C$ for some hours and filtered once more to remove any urea. The filtrate was concentrated in vacuo and the residue purified by flash chromatography on silica gel (500 g, 0.043–0.06 mm; E. Merck, Darmstadt, Germany) in petroleum ether–ethyl acetate (2:1), and subsequent recrystallization from ether–petroleum ether: yield, 42 g (77%); mp 74°C, reported [16] 83°C; $[\alpha]_D$ $-9.5$ ($c$ 1.05, CHCl$_3$); reported [16] $[\alpha]_D$ 9.0° ($c$ 1.15, CHCl$_3$).

Other $N$-protected threonine derivatives as well as $N$-protected serine derivatives can be converted to the corresponding *tert*-butyl esters using this method [34].

### N-(9-Fluorenylmethoxycarbonyl)-O-(3,4,6-tri-O-acetyl-2-azido-2-deoxy-α-D-galactopyranosyl)-L-threonine tert-Butyl Ester 20 [34]

Under careful exclusion of moisture and oxygen, a solution of ethyl 1-thio-3,4,6-tri-$O$-acetyl-2-deoxy-2-azido-β-D-galactopyranoside [40] (375 mg, 1 mmol) and Fmoc-Thr-OtBu **18** (600 mg, 1.5 mmol) in dry toluene (10 mL) is stirred with powdered 4-Å molecular sieves (500 mg) for 1 h at 20°C. After cooling to 5°C, a solution of dimethyl(methylthio)-sulfonium tetrafluoroborate [41] (520 mg, 2 mmol) in dry dichloromethane (10 mL) is added. After 1 h at 5°C and 16 h at 25°C, ethyl-diisopropylamine (130 mg, 1 mmol) is added. The mixture is stirred for 1 h, filtered, concentrated in vacuo, and toluene (10 mL) is distilled off in vacuo from the residue. Purification by flash-chromatography in toluene–acetone (9:2) gives the mixture of anomers: yield: 520 mg (86%), α/β = 3.1. Separation of the anomers is carried out by flash chromatography in dichloromethane–acetone 100:3 on silica gel (200 g): α-anomer **20**: yield, 380 mg (63%); $[\alpha]_D$ 84.6° ($c$ 1, CHCl$_3$; reported [16] $[\alpha]_D$ 69.3° ($c$ 1, CHCl$_3$).

β Anomer: $N$-(9-Fluorenylmethoxycarbonyl)-$O$-(3,4,6-tri-$O$-acetyl-2-azido-2-deoxy-β-D-galactopyranosyl)-L-threonine *tert*-Butyl Ester: yield, 115 mg (19%), $[\alpha]_D$ $-8.0°$ ($c$ 1, CHCl$_3$).

### N-(9-Fluorenylmethoxycarbonyl)-O-(3,4,6-tri-O-acetyl-2-azido-2-deoxy-β-D-galactopyranosyl)-L-threonine 21

To Fmoc $O$-glycosyl threonine *tert*-butyl ester **20** (290 mg, 0.4 mmol) dissolved in dry dichloromethane (5 mL) is added dry trifluoroacetic acid (3 mL) at 0°C. The mixture is allowed to warm to room temperature, and the reaction is monitored by TLC (dichloromethane–ethanol 10:1). After 3 h, the conversion is complete. The solvent is evaporated in vacuo. Toluene (10 mL) is codistilled in vacuo from the remainder. Small amounts of an unpolar impurity are separated by flash chromatography on silica gel (20 g) in dichloromethane–ethanol (15:1): yield 230 mg (90%); $[\alpha]_D$ 69.7° ($c$ 1, CHCl$_3$); $R_f$ 0.32 (CH$_2$Cl$_2$–ethanol 10:1).

*Removal of the Fmoc Group and Labeling with Biotin*

***O*-(3,4,6-Tri-*O*-acetyl-2-azido-2-deoxy-α-D-galactopyranosyl)-L-threonine *tert*-Butyl Ester 22.** A solution of Fmoc-Thr(αAc$_3$GalN$_3$)-OtBu **20** (215 mg, 0.3 mmol) in freshly distilled morpholine (2 mL) is stirred at room temperature for 20 min. After concentration in vacuo and codistillation with toluene (twice 3 mL) in vacuo, **22** is obtained quantitatively. It is immediately used in the following reaction.

***N*-(D-Biotinyl)-*O*-(3,4,6-tri-*O*-acetyl-2-azido-2-deoxy-α-D-galactopyranosyl)-L-threonine *tert*-Butyl Ester 23 [34].** A mixture of D-biotin (150 mg, 0.6 mmol), 1-ethyl-3-(3-dimethylaminopropyl)-carbodiimide (EDC: 580 mg, 3 mmol), and 1-hydroxybenzotriazole (HOBt: 540 mg, 4 mmol) in dimethylformamide (DMF: 2 mL) is stirred under exclusion of moisture at 22°C. After 45 min, the biotin is dissolved, and a solution of freshly prepared glycosyl threonine ester **22** (0.3 mmol, preceding procedure) in dichloromethane (2 mL) is added at 0°C. After stirring for 16 h at room temperature, the solvent is evaporated in vacuo, the remainder dissolved in dichloromethane (50 mL), extracted with ice-cold 0.2 *N* HCl (3 × 25 mL), water (25 mL), and saturated NaHCO$_3$ solution (2 × 25 mL), dried with MgSO$_4$, and concentrated in vacuo. Purification by flash chromatography on silica gel (20 g) in dichloromethane–ethanol (25:1) yields **23**; 200 mg (93%); [α]$_D$ 96.5° (*c* 1, CHCl$_3$); $R_f$ 0.29 (toluene–acetone 4:1).

## C. The Fmoc–Allyl Ester Combination [24]

*N-(Fluorenylmethoxycarbonyl)-L-threonine Allyl Ester* **24**

To a solution of L-threonine allyl ester hydrochloride [22,42] (6.0 g, 30.6 mmol, the corresponding hydrotrifluoroacetate or hydrotoluenesulfonate can also be used) in saturated NaHCO$_3$ solution (100 mL) and dioxane (100 mL) is added dropwise at 0°C a solution of 9-fluorenylmethyl chloroformate (10.8 g, 41.8 mmol) in dioxane (50 mL). After stirring for 24 h, the solvent is evaporated in vacuo, the remainder dissolved in ethyl acetate (200 mL), washed with 0.5 *N* HCl, saturated NaHCO$_3$ solution, and water (each 100 mL), dried with MgSO$_4$, and concentrated in vacuo. The crude product is subjected to chromatography on silica gel (300 g) in petroleum ether–ethyl acetate (4:1), and the obtained product is recrystallized from ethyl acetate–petroleum ether to give pure **24**: yield, 11.2 g (96%); mp 98°–100°C; [α]$_D$ −17.2° (*c* 1, dimethylformamide); $R_f$ 0.33 (petroleum ether–ethyl acetate 2:1).

*N-(9-Fluorenylmethoxycarbonyl)-O-(3,4,6-tri-O-acetyl-2-azido-2-deoxy-α-D-galactopyranosyl)-L-threonine Allyl Ester* **26**

Under argon atmosphere and exclusion of light and moisture, a mixture of Fmoc-Thr-OAll **24** (4.9 g, 12.9 mmol), Ag$_2$CO$_3$ (4.0 g, 14.5 mmol), and powdered 4-Å molecular sieves (2 g) in dry toluene (60 mL) and dichloromethane (90 mL) is stirred for 1 h. At room temperature, AgClO$_4$ (0.4 g, 1.9 mmol) is added. After 20 min, a solution of 3,4,6-tri-*O*-acetyl-2-azido-2-deoxy-α-D-galactopyranosyl bromide [43] **25** (4.2 g, 10.7 mmol) in toluene (90 mL)–dichloromethane (90 mL) is added dropwise within 1 h. After 24–40 h at room temperature (TLC monitoring; dichloromethane–acetone 45:1), dichloromethane (100 mL) is added. The mixture is filtered through Hyflow, and the filtrate is extracted with saturated NaHCO$_3$ solution (2 × 100 mL) and water. The organic layer is dried with MgSO$_4$ and the solvent evaporated in vacuo. Purification by flash chromatography on silica gel (100 g) in dichloromethane → dichloromethane–methanol (250:1) yields a mixture of the α and β anomers (α/β 20:1), 5.8 g (79%). Repeated flash chromatography gives the α anomer **26**: yield, 4.5 g (61%); mp 59°–61°C, $[\alpha]_D$ +66.7° (c 1, CHCl$_3$); $R_f$ 0.40 (CH$_2$Cl$_2$–acetone 45:1).

### N-(9-Fluorenylmethoxycarbonyl)-O-(2-acetamido-3,4,6-tri-O-acetyl-2-deoxy-α-D-galactopyranosyl)-L-threonine Allyl Ester 27

To a solution of Fmoc-Thr(Ac$_3$GalN$_3$)-OAll **26** (4.0 g, 5.8 mmol) in dry THF (50 mL) is added at 0°C acetic acid (1 mL) and, subsequently, a solution of tri-*n*-octylphosphine (3.4 mL, 7.4 mmol) in THF (10 mL). After 2 h, the mixture is concentrated in vacuo. To complete the *N*-acetylation, the remainder is stirred with acetic anhydride (2 mL)–pyridine (10 mL) for 6 h, concentrated in vacuo. Toluene (3 × 20 mL) is distilled in vacuo from the residue, which is subsequently purified by flash chromatography on silica gel (100 g) in petroleum ether–ethyl acetate → ethyl acetate. The product is recrystallized from ethyl acetate–petroleum ether: yield, 3.9 g (94%); mp 66–68° C; $[\alpha]_D$ +36.3° (c 1, CHCl$_3$); $R_f$ 0.56 (ethyl acetate).

The selective removal of the Fmoc group from **27** is carried out in strict analogy to Sections III.A and III.B for its removal. Similar to product **22**, the obtained **28** is not stable and must be subjected to further reaction immediately.

### N-(9-Fluorenylmethoxycarbonyl)-O-(2-acetamido-3,4,6-tri-O-acetyl-2-deoxy-α-D-galactopyranosyl)-L-threonine 29

To a solution of Fmoc-Thr(Ac$_3$GalNAc)-OAll **27** (2 g, 2.8 mmol) in dry THF (20 mL) under argon atmosphere is added *N*-methyl aniline (0.5 mL) and a catalytic amount (50 mg, 0.043 mmol; 1.5 mol%) tetrakis(triphenylphosphine)palladium(0). The mixture is stirred in the dark for 2 h and then concentrated in vacuo. Flash chromatography on silica gel (60 g) in ethyl acetate → methanol and recrystallization from methanol–petroleum ether and ethyl acetate–petroleum ether gives **29** as crystals: yield, 1.8 g (96%); mp 114°–116°C, reported [16] amorphous, obtained by a different procedure); $[\alpha]_D$ +75.8° (c 0.5, CH$_3$OH), reported [16] $[\alpha]_D$ +65.0° (c 1.45, CDCl$_3$); $R_f$ 0.46 (CH$_2$Cl$_2$–CH$_3$OH 1:1).

## D. Removal of the Allyloxycarbonyl Group [19,23,26]

### 2-Acetamido-4-O-(2-acetamido-3,4-di-O-acetyl-6-O-benzyl-2-deoxy-β-D-glucopyranosyl)-3-O-acetyl-6-O-(2,3,4-tri-O-benzyl-α-L-fucopyranosyl)-2-deoxy-β-D-glucopyranosyl Azide 32

Tetraethylammonium bromide (4 g, 19 mmol), 2-acetamido-4-*O*-(2-acetamido-3,4-di-*O*-acetyl-6-*O*-benzyl-2-deoxy-β-D-glucopyranosyl)-3-*O*-acetyl-2-deoxy-β-D-glucopyranosyl azide [26] **30** (1.25 g, 1.88 mol) and 4-Å molecular sieves (6 g) were stirred in 15 mL of dimethylformamide–dichloromethane (2:1) for 30 min. Using a syringe, 2,3,4-tri-*O*-benzyl-α-D-fucopyranosyl bromide [35] **31** (3 g, 6 mmol) dissolved in dichloromethane (5 mL) was added dropwise. After 4 days, the mixture was filtered through Celite, which subsequently was washed with dichloromethane (300 mL). The combined organic solutions were extracted with 1 *M* KHCO$_3$ solution (3 × 100 mL), dried with MgSO$_4$, concentrated in vacuo, and the remainder was dried in high vacuo. Chromatography (100 g silica gel) in petroleum ether–ethyl acetate 2:1 → dichloromethane–methanol 20:1, and by recrystallization from dichloromethane–diisopropyl ether gave **32**: 1.5 g (74%); mp 175°C; $[\alpha]_D$ −38.65° (c 0.5, chloroform; $R_f$ 0.4 (CHCl$_3$–MeOH 10:1).

*2-Acetamido-4-O-(2-acetamido-3,4-di-O-acetyl-6-O-benzyl-2-deoxy-β-D-glucopyranosyl)-3-O-acetyl-6-O-(2,3,4-tri-O-benzyl-α-L-fucopyranosyl)-2-deoxy-β-D-glucopyranosylamine* **33**

The trisaccharide azide **32** (1 g, 0.92 mmol) was dissolved in methanol (20 mL). After addition of Raney nickel (200 mg; E. Merck, Darmstadt, Germany), which was washed ten times with water, hydrogenation was performed for 3 h. After filtration and concentration in vacuo the glycosylamine **33** was obtained: yield, 927 mg (95%); $[\alpha]_D$ −24.8° (c 0.5, $CH_2Cl_2$); $R_f$ 0.2 ($CHCL_3$–MeOH 10:1).

*N-Allyloxycarbonyl-aspartic acid α-tert-Butyl Ester* **34**

To a solution of 1-*O*-*tert*-butyl aspartate [44] (3 g, 15.9 mmol) and $KHCO_3$ (3 g, 32 mmol) in water (50 mL) was added at 0°C allyl chloroformate (1.7 ml, 16 mmol; E. Merck, Darmstadt, Germany). After stirring for 1 h, the mixture was extracted with diethyl ether (100 mL). The aqueous layer was acidified to pH 2 using 1 *M* HCl and extracted with diethyl ether (4 × 50 mL). These ether solutions were combined, dried with $MgSO_4$. The solvent was evaporated in vacuo to give **34** as an oil; 3.87 g (89%), $[\alpha]_D$ +21° (c 1, $CH_2Cl_2$); $R_f$ 0.25 (petroleum ether–ethyl acetate 2:1).

*$N^2$-(Allyloxycarbonyl)-$N^4$-(2-acetamido-4-O-(2-acetamido-3,4-di-O-acetyl-6-O-benzyl-2-deoxy-β-D-glucopyranosyl)-3-O-acetyl-6-O-(2,3,4-tri-O-benzyl-α-L-fucopyranosyl)-2-deoxy-β-D-glucopyranosyl)-L-asparagine tert-Butyl Ester* **35** [26]

The trisaccharide amine **33** (0.5 g, 0.47 mmol), Aloc-Asp-OtBu **34** (0.2 g, 0.73 mmol), and EEDQ [37] (0.6 g, 2.4 mmol) were stirred in dimethylformamide (5 mL). After 3 days the solvent was distilled off in high vacuo. The remainder was purified by flash chromatography on silica gel (60 g) in petroleum ether–ethyl acetate 2:1 → dichloromethane–methanol 50:1 to give **35**: 466 mg (75%), $[\alpha]_D$ −28.4° (c 0.75, $CH_2Cl_2$); $R_f$ 0.38 ($CHCl_3$–MeOH 10:1).

The corresponding α-anomer was isolated by this chromatography as a by-product: *$N^2$-(Allyloxycarbonyl)-$N^4$-(2-acetamido-4-O-(2-acetamido-3,4-di-O-acetyl-6-O-benzyl-2-deoxy-β-D-glucopyranosyl)-3-O-acetyl-6-O-(2,3,4-O-benzyl-α-L-fucopyranosyl)-2-deoxy-α-D-glucopyranosyl)-L-asparagine tert-butyl ester*: yield, 62 mg (10%); $R_f$ 0.42 ($CHCl_3$–MeOH 10:1).

*$N^4$-[2-Acetamido-4-O-(2-acetamido-3,4-di-O-acetyl-6-O-benzyl-2-deoxy-β-D-glucopyranosyl)-3-O-acetyl-6-O-(2,3,4-tri-O-benzyl-α-L-fucopyranosyl)-2-deoxy-β-D-glucopyranosyl]-L-asparagine tert-Butyl Ester* **36**

To a solution of $N^2$-allyloxycarbonyl-$N^4$-[2-acetamido-4-*O*-(2-acetamido-3,4-di-*O*-acetyl-6-*O*-benzyl-2-deoxy-β-D-glucopyranosyl)-3-*O*-acetyl-6-*O*-(2,3,4-tri-*O*-benzyl-α-L-fucopyranosyl)-2-deoxy-β-D-glucopyranosyl]-L-asparagine *tert*-butyl ester [19,26] **35** [400 mg; 0.305 mmol; $[\alpha]_D$ −28.4° (c 0.75, $CH_2Cl_2$] and 5,5-dimethyl-cyclohexane-1,3-dion (dimedone, 1 g, 7.1 mmol) in oxygen-free tetrahydrofuran (15 mL) is added tetrakis(triphenylphosphine)-palladium(0) (50 mg, 0.043 mmol) under argon atmosphere. After stirring for 2 h in the dark, the solvent is evaporated in vacuo and the remainder dissolved in dichloromethane (200 mL). The organic layer is washed with 1 *N* $KHCO_3$ solution (50 mL) and dried with $MgSO_4$. After evaporation of the solvent in vacuo, the crude product is purified by chromatography on silica gel (40 g) in dichloromethane–methanol 20:1 → 15:1 to give pure **36**: yield: 341 mg (91%), $[\alpha]_D$ 37.4° (c 0.5, $CH_2Cl_2$); $R_f$ 0.35 ($CHCL_3$–$CH_3OH$ 10:1).

## E. Exchange of Ether-type for Ester-type Protection in the Saccharide Portion: Generation of Stability Against Acids [19,26]

*2-Acetamido-4-O-(3,4,6-tri-O-acetyl-2-deoxy-2-phthalimido-β-D-glucopyranosyl)-3-O-acetyl-6-O-(α-L-fucopyranosyl)-2-deoxy-β-D-glycopyranosyl Azide*

To a solution of 2-acetamido-4-*O*-(3,4,6-tri-*O*-acetyl-2-deoxy-2-phthalimido-β-D-glucopyranosyl)-3-*O*-acetyl-6-*O*-[2,3,4-tri-*O*-(4-methoxybenzyl)-α-L-fucopyranosyl]-2-deoxy-β-D-glucopyranosyl azide [26] **37** (1.1 g, 0.907 mmol) in acetonitrile–water (9:1) is added ceric ammonium nitrate (3 g, 5.47 mmol), and the mixture is stirred until the reaction is complete (about 3 h, TLC monitoring in $CHCl_3$–$CH_3OH$ 10:1). After addition of acetonitrile (30 mL) and concentration in vacuo, the remainder is subjected to chromatography on silica gel (50 g) in dichloromethane–methanol 30:1 → 6:1. The crude product ($R_f$ 0.15 in $CHCl_3$/$CH_3OH$ 10:1), containing inorganic material stemming from the oxidizing reagent is used for further conversion.

*2-Acetamido-4-O-(2-acetamido-3,4,6-tri-O-acetyl-2-deoxy-β-D-glycopyranosyl)-3-O-acetyl-6-O-(2,3,4-tri-O-acetyl-α-L-fucopyranosyl)-2-deoxy-β-D-glucopyranosyl Azide* **38** [26]

The crude product obtained in the preceding experiment is dissolved in ethanol (40 mL). after addition of hydrazine hydrate (100%, 10 mL), the solution is stirred at 80°C for 1 h (removal of the phthaloyl group). Acetone (30 mL) is added, and the mixture is concentrated in vacuo. Codistillation with acetone (30 mL) is repeated twice to remove hydrazine as the azine. The remainder is dried under high vacuum and dissolved in pyridine–acetic anhydride (2:1, 50 mL). After stirring for 18 h, concentration in vacuo, and codistillation with toluene (30 mL), the crude **33** is purified by flash chromatography on silica gel (70 g). Elution with acetone and then with dichloromethane–methanol 30:1 yields pure **33**: overall yield: 581 mg (72%); mp 137°C; $[\alpha]_D$ −72.5° (*c* 0.25, $CH_2Cl_2$); $R_f$ 0.4 ($CHCl_3$–$CH_3OH$ 10:1).

## REFERENCES

1. R. J. Ivatt, ed. *The Biology of Glycoproteins*, Plenum Press, New York, 1984.
2. H. Lis and N. Sharon, Protein glycosylation. Structural and functional aspects, *Eur. J. Biochem.* *218*:1 (1993).
3. J. Martinez, A. A. Pavia, and F. Winternitz, Synthèse d'un *O*-glycopeptide par allongement de la chain peptidique da côté N-terminal d'un glycosylaminoacide, *Carbohydr. Res. 50*:148, (1976) and references cited therein.
4. H. G. Garg and R. W. Jeanloz, The synthesis of 2-acetamido-3,4,5-tri-*O*-acetyl-*N*-[*N*-(benzyloxycarbonyl)-L-aspart-1- and 4-oyl]-2-deoxy-β-D-glucopyranosylamine, *Carbohydr. Res. 23*: 437 (1972), and references cited therein.
5. H. Kunz, Synthesis of glycopeptides. Partial structures of biological recognition components, *Angew. Chem. Int. Ed. Engl. 26*:294 (1987).
6. H. Paulsen, Progress in the chemical synthesis of complex oligosaccharides, *Angew. Chem. Int. Ed. Engl. 21*:155 (1982).
7. R. R. Schmidt, New methods of glycoside and oligosaccharide syntheses—are there alternatives to the Koenigs–Knorr method, *Angew. Chem. Int. Ed. Engl. 25*:212 (1986).

8. H. Kunz, Glycopeptides of biological interest. A challenge for chemical synthesis, *Pure Appl. Chem.* 65:1223 (1993).
9. H. G. Garg, K. von dem Bruch, and H. Kunz, Developments in the synthesis of glycopeptides containing glycosyl L-asparagine, L-serine and L-threonine, *Adv. Carbohydr. Chem. Biochem.* 50:277 (1994).
10. P. Schultheiss-Reimann and H. Kunz, O-Glycopeptide synthesis using 9-fluorenylmethoxycarbonyl (Fmoc)-protected synthetic units, *Angew. Chem. Int. Ed. Engl.* 22:62 (1983).
11. S. Peter, T. Bielefeldt, M. Meldal, K. Bock, and H. Paulsen, Multiple-column solid-phase glycopeptide synthesis, *J. Chem. Soc. Perkin Trans 1*:1163 (1992).
12. H. Kunz and S. Birnbach, Synthesis of O-glycopeptides of the tumor associated $T_N$- and T-antigen type and their binding to bovine serum albumin, *Angew. Chem. Int. Ed. Engl.* 25:360 (1986).
13. B. Lüning, T. Norberg, and J. Tejbrant, Synthesis of mono- and disaccharide amino-acid derivatives for use in solid phase peptide synthesis, *Glycoconjugate J.* 6:5 (1989).
14. L. Otvos, Jr., L. Urge, M. Hollosi, K. Wroblewski, G. Graczy, G. D. Fasman, and J. Thurin, Automated solid-phase synthesis of N-glycoproteins antennae into T cell epitopic peptides, *Tetrahedron Lett.* 31:5889 (1990).
15. H. Kunz and P. Schultheiss-Reimann, unpublished; P. Schultheiss-Reimann, Glycopeptidsynthese mit der 9-Fluorenylmethoxycarbonyl-Schutzgruppe, Dissertation, Universität Mainz, Germany, 1984.
16. H. Paulsen and K. Adermann, Synthese von O-Glycopeptid-Sequenzen des N-Terminus von Interleukin, *Liebigs Ann. Chem.* p. 751 (1989).
17. H. Kunz and B. Dombo, Solid-phase synthesis of peptides and glycopeptides on polymeric support with allylic anchor groups, *Angew. Chem. Int. Ed. Engl.* 27:711 (1988).
18. W. Kosch, J. März, and H. Kunz, Synthesis of glycopeptide derivatives of peptide T on a solid-phase using an allylic linkage, *React. Polym.* 22:181 (1994).
19. H. Kunz and C. Unverzagt, Protecting group-dependent stability of intersaccharide bonds—synthesis of a fucosyl-chitobiose glycopeptide, *Angew. Chem. Int. Ed. Engl.* 27:1697 (1988).
20. V. Bencomo and P. Sinay, Synthesis of M and N active glycopeptides. Part of the N-terminal region of human glycophorin A, *Glycoconjugate J.* 1:5 (1984).
21. S. Lavielle, N. C. Ling, R. Saltmann, and R. Guillemin, Synthesis of a glycotripeptide and glycosomatostatin containing the 3-O-(2-acetamido-2-deoxy-β-D-glucopyranosyl)-L-serine residue, *Carbohydr. Res.* 89:229 (1981).
22. H. Kunz and H. Waldmann, The allyl group as mildly and selectively removable carboxy-protecting group for the synthesis of labile O-glycopeptides, *Angew. Chem. Int. Ed. Engl.* 23:71 (1984).
23. H. Kunz and C. Unverzagt, The allyloxycarbonyl (Aloc) moiety—conversion of an unsuitable into a valuable amino protecting group for peptide synthesis, *Angew. Chem. Int. Ed. Engl.* 23:436 (1984).
24. M. Ciommer and H. Kunz, Synthesis of glycopeptides with partial structure of human glycophorin using the fluorenylmethoxycarbonyl/allyl ester protecting group combination, *Synlett* p. 593 (1991).
25. H. Kunz, H. Waldmann, and J. März, Synthese von N-Glycopeptid-Partialstrukturen der Verknüpfungsregion sowohl der Transmembran-Neuraminidase eines Influenza-Virus als auch des Faktors B des menschlichen Komplementsystems, *Liebigs Ann. Chem.* p. 45 (1989).
26. C. Unverzagt and H. Kunz, Synthesis of glycopeptides and neoglycopeptides containing the fucosylated linkage region of N-glycoproteins, *Bioorg. Med. Chem.* 2:1189 (1994).
27. H. Kunz and J. März, Synthesis of glycopeptides with Lewis[a] antigen side chain and HIV peptide T sequence using the trichloroethoxycarbonal/allyl ester protecting group combination, *Synlett* p. 591 (1992).
28. K. von dem Bruch and H. Kunz, Synthesis of N-glycopeptide clusters with Lewis[x] antigen side chains and their coupling to carrier proteins, *Angew. Chem. Int. Ed. Engl.* 33:101 (1994).

29. P. Braun, H. Waldmann, and H. Kunz, Chemoenzymatic synthesis of glycopeptides carrying the tumor associated $T_N$ antigen structure, *Bioorg. Med. Chem.* 1:197 (1993).
30. S. Hanessian and J. Banoub, Chemistry of the glycosidic linkage. An efficient synthesis of 1,2-*trans*-di-saccharides, *Carbohydr. Res.* 53:C13, (1977).
31. S. J. Danishefsky, K. F. McClure, J. T. Randolph and R. B. Ruggeri, A strategy for the solid-phase synthesis of oligosaccharides, *Science* 260:1307 (1993).
32. L. A. Carpino and G. Y. Han, The 9-fluorenylmethoxycarbonyl amino-protecting group, *J. Org. Chem.* 37:3404 (1972).
33. O. Seitz and H. Kunz, A novel allylic anchor for solid-phase synthesis. Synthesis of protected and unprotected O-glycosylated mucin-type glycopeptides, *Angew. Chem. Int. Ed. Engl.* 34:803 (1995).
34. H. Kunz and G. Braum, unpublished results; G. Braum, Synthese von Glycopeptiden mit photoaktivierbaren Gruppen im Saccharidteil, Dissertation, Universität Mainz, Germany, 1991.
35. M. Dejter-Juszynsky and H. M. Flowers, Koenigs–Knorr reaction. II—synthesis of an α-L-linked disaccharide from tri-O-benzyl-L-fucopyranosyl bromide, *Carbohydr. Res.* 18:219 (1971).
36. R. U. Lemieux and J. I. Haymi, The mechanism of the anomerization of the tetro-O-acetyl-D-glucosyl chlorides, *Can. J. Chem.* 43:2162 (1965).
37. B. Belleau and G. Malek, A new convenient reagent for peptide synthesis, *J. Am. Chem. Soc.* 90:1651 (1968).
38. W. Günther and H. Kunz, Synthesis of a β-mannosyl–chitobiose–asparagine conjugate—a central core region unit of the N-glycoproteins, *Angew. Chem. Int. Ed. Engl.* 29:1050 (1990).
39. J. M. Lacombe, A. A. Pavia, and R. M. Rocheville, Un nouvel agent de glycosylation: l'Anhydride trifluoromethanesulfonique. Synthèse des α et β O-glycosyl-L-serine, -L-threonine et -L-hydroxyproline, *Can. J. Chem.* 59:473 (1981).
40. H. Paulsen, M. Rauwald, and U. Weichert, Glycosydation of oligosaccharide thioglycosides to O-glycoprotein segments, *Liebigs Ann. Chem.* p. 75 (1986).
41. P. Fügedi and P. J. Garegg, A novel promoter for the efficient construction of 1,2-*trans* linkages in glycoside synthesis, using thioglycosides as glycosyl donors, *Carbohydr. Res.* 149:C9 (1986).
42. H. Waldmann and H. Kunz, Allylester als selektiv abspaltbare Carboxylschutzgruppen in der Peptid- und Glycopeptidsynthese, *Liebigs Ann. Chem.* p. 1712 (1983).
43. H. Paulsen and M. Paal, Blocksynthese von O-Glycopeptiden und anderen T-Antigen Strukturen, *Carbohydr. Res.* 135:71 (1984).
44. R. W. Roeske, Preparation of *t*-butyl esters of free amino acids, *J. Org. Chem.* 28:1251 (1963).

# 12

## Oligosaccharide Synthesis with Trichloroacetimidates

**Richard R. Schmidt and Karl-Heinz Jung**
*Universität Konstanz, Konstanz, Germany*

|     |                                                               |     |
| --- | ------------------------------------------------------------- | --- |
| I.  | Introduction                                                  | 283 |
|     | A. Activation through anomeric oxygen-exchange reactions      | 285 |
|     | B. Activation through retention of the anomeric oxygen        | 286 |
| II. | The Trichloroacetimidate Method                               | 289 |
|     | A. Trichloroacetimidate formation (activation step)           | 289 |
|     | B. Glycosylation reactions (glycosylation step)               | 290 |
| III.| Experimental Procedures                                       | 296 |
|     | A. Synthesis of *O*-glycosyl trichloroacetimidates            | 296 |
|     | B. Glycosylation reactions with *O*-glycosyl trichloroacetimidates | 298 |
|     | References and Notes                                          | 308 |

## I. INTRODUCTION

Glycoside synthesis is a very common reaction in nature, thus providing a great variety of compounds, such as various types of oligosaccharides, or glycoconjugates with lipids (glycolipids), with proteins (glycoproteins or proteoglycans), and with many other naturally occurring compounds. The important biological implications of the attachment of sugar moieties and especially of complex oligosaccharide structures to an aglycon are only now becoming more and more obvious, thus creating a growing general interest in the field [1–7]. The great structural variety available to sugars [1a] complicates the synthesis of glycosides and, even more so, of complex oligosaccharide structures. It has only recently become possible to develop methods for the synthesis of such complex compounds [1–4]. The results obtained in this endeavor are summarized in Scheme **1**, which will be shortly discussed; however, the main emphasis in this chapter will be devoted to the *tri-*

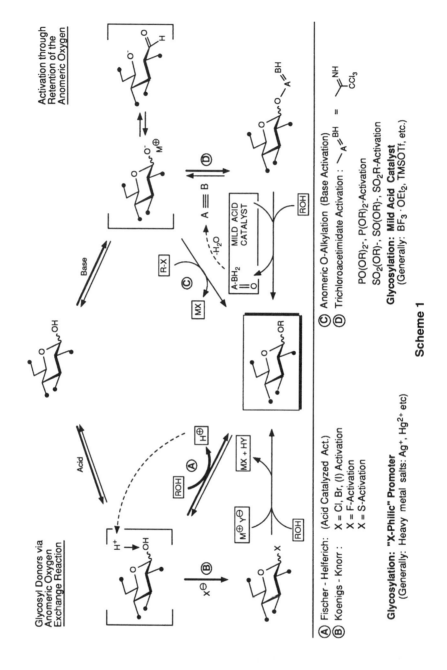

Scheme 1

*chloroacetimidate method*, which has become a very competitive, widely applicable procedure for glycoside bond formation [1,4].

The methods for glycoside bond formation developed thus far usually consist of an activation of the anomeric center (activation step); this activated species is then used to release the glycosyl donor, with the help of a promoter or a catalyst, to generate the glycosidic bond to the acceptor (glycosylation step). In principle, for the activation step, two different approaches are employed: (1) activation through anomeric oxygen-exchange reactions and (2) activation through retention of the anomeric oxygen [1]. Obviously, this basic difference has wide-ranging implications as will be discussed later. The latter method has been essentially developed in our laboratory [1].

## A. Activation Through Anomeric Oxygen-Exchange Reactions

*The Fischer–Helferich Method*

Still the simplest glycosylation procedure available is the *Fischer–Helferich method* [1,8] (see Scheme 1, A); it consists of a convenient direct anomeric oxygen-exchange reaction in a cyclic hemiacetal following mechanistically a typical acid-catalyzed acetal formation reaction. This method has great merits in the synthesis of simple alkyl glycosides for which an excess glycosyl acceptor can be employed, thereby inhibiting self-condensation of unprotected sugar moieties and thus also acting as a solvent for the starting materials and the product(s). However, because of these limitations and because of the reversibility of this method, it has not gained importance in the synthesis of complex oligosaccharides and glycoconjugates. For this, irreversible methods are required, which generally can be gained by preactivation of the anomeric center, thereby generating strong glycosyl donor properties in the presence of mild promoters or catalysts.

*The Koenigs–Knorr Method*

One of these methods that leads to strong glycosyl donor properties in the activated species is exchange of the anomeric hydroxyl group by bromine and chlorine, respectively, in the *activation step* (Koenigs–Knorr method [9], see Scheme 1, B). Thus an α-haloether is generated that can be readily activated in the *glycosylation step* by halophilic promoters, typically heavy-metal salts, thus resulting in irreversible glycosyl transfer to the acceptor. This method is the basis of a very valuable technique for the synthesis of complex oligosaccharides and glycoconjugates, which has been extensively reviewed [1–4]. It has been continually developed and widely applied (e.g., for the synthesis of 1,2-*trans*-glycosides by the silver triflate tetramethylurea method) [10]. In spite of the generality of the method, the requirement of at least an equimolar amount (often up to 4 eq) of metal salt as promoter (frequently incorrectly termed a "catalyst") and problems concerning the disposal of waste material (e.g., mercury salts) could be limiting factors for large-scale preparations. Therefore, alternative methods are of great interest.

*Methods Related to the Koenigs–Knorr Method*

Other *anomeric oxygen-exchange reactions* in the activation step have been recently quite extensively investigated. Thus, closely related to the classic Koenigs–Knorr method is the introduction of fluorine as the leaving group (see Scheme 1, B; X = F) [3,11–13] which, owing to the stability of the C–F bond, leads to much more stable glycosyl donor intermediates. Because of the difference in halophilicity of this element compared with bromine and chlorine, further promoter systems, besides silver salts, were found useful in glycosylation reactions [14]. However, because of the generally lower glycosyl donor properties [1d] and because also at least equimolar amounts of promoter are required, these intermediates generally have no real advantages over the corresponding glycosyl bromides or chlorides.

Anomeric oxygen-exchange reactions by thio groups (see Scheme 1, B; X = SR) have recently attracted considerable attention for the generation of glycosol donors [2,15]. Thioglycosides, thus obtained, offer sufficient temporary protection of the anomeric center, thereby enabling various ensuing chemical modifications of the glycosyl donor without affecting the anomeric center. Additionally, they present several alternative possibilities for the generation of glycosyl donor properties. Besides various thiophilic heavy-metal salts [16], which also exhibit the foregoing disadvantages, iodonium, bromonium, and chloronium ions are also highly thiophilic; however, with counterions, such as bromide and chloride, a subsequent Koenigs–Knorr type reaction is encountered [15,17]. Therefore, a poor nucleophile is required as counterion of the halonium ion, for instance use of $N$-iodo- or $N$-bromosuccinimide, enabling direct reaction with the acceptor as nucleophile. However, owing to the strong activation required for thioglycosides, often low $\alpha,\beta$-selectivities are obtained for nonneighboring group-assisted reactions [16d,18] and, especially, for sugars with lower glycosyl donor properties. Because of the high nucleophilicity of the sulfur atom in thioglycosides, methyl trifluoromethanesulfonate (methyl triflate) was successfully employed for their activation [19]. However, again low $\alpha,\beta$-selectivities, the health hazard of this reagent, the requirement of at least equimolar amounts, and possible formation of methylation products and other nucleophilic centers (for instance, amide groups or the hydroxy group of the acceptor) are major disadvantages of this procedure. Consequently, commercially available dimethyl (methylthio)sulfonium triflate (DMTST), readily obtained from dimethyl disulfide and methyl triflate, was extensively used as a thiophilic reagent because it seems to give better results than methyl triflate [10]. However, the basic drawbacks of the overall method are also associated with this promoter system.

### B. Activation Through Retention of the Anomeric Oxygen

The requirements for glycoside synthesis—high chemical and stereochemical yield, applicability to large-scale preparations, with avoidance of large amounts of waste materials, by having a glycosyl transfer from the activated intermediate through a catalytic process— were not effectively met by any of the methods described in the foregoing for the synthesis of complex oligosaccharides and glycoconjugates. However, the general strategy for glycoside bond formation seems to be correct:

1. The first step should consist of an activation of the anomeric center under formation of a stable glycosyl donor (*activation step*).
2. The second step (*glycosylation step*) should consist of a sterically uniform, high-yielding glycosyl transfer to the acceptor; however, by a truly catalytic process, where diastereocontrol may be derived from the glycosyl donor anomeric configuration (by inversion or retention), by anchimeric assistance, by influence of the solvent, by thermodynamics, or by any other effects.

Because only simple means meeting these requirements will lead to a generally accepted method, we decided to investigate, instead of acid activation, base activation of sugars, thereby simply generating at first an anomeric oxide structure from a pyranose or a furanose (see Scheme 1, C and D) [1].

*Anomeric O-Aklylation Method*

Alkylation of the anomeric oxide, generated by base addition to pyranoses or furanoses should directly and irreversibly lead to glycosides. This process was termed by us "anomeric O-alkylation method" [1] (see Scheme 1, C). In the beginning, this process was considered unlikely to fulfill all of the requirements for glycoside and saccharide synthesis. Even when all remaining functional groups are blocked by protecting groups, the ring-chain tautomerism between the anomeric forms and the open-chain form (Scheme 2)

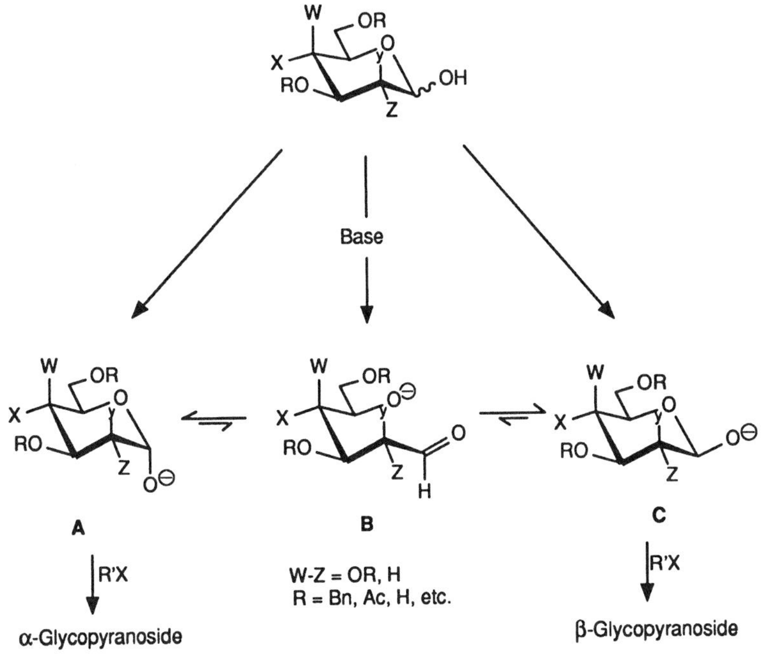

**Scheme 2**

already gives three sites for attack of the alkylating agent. In addition, base-catalyzed elimination in the open-chain form of the sugar could become an important side reaction. Therefore, the yield, the regioselectivity, and the stereoselectivity of the anomeric O-alkylation was not expected to be outstanding.

Surprisingly, no studies employing this simple method for the synthesis of complex glycosides and glycoconjugates had been reported before our work. Only a few scattered examples with simple alkylating agents, for instance, excess of methyl iodide or dimethyl sulfate, have been found [1a]. However, in our hands, direct anomeric O-alkylation of O-benzyl-, O-acyl-, or O-alkylidene-protected sugars in the presence of a base and triflates (R-X = R-OTf) of various primary and secondary alcohols, including sugars as alkylating agents, has become a very convenient method for glycoside bond formation [1,21]. Also O-unprotected sugars and less reactive alkylating agents have recently been successfully employed in this reaction, furnishing directly, and often with very high anomeric stereocontrol, the desired glycosidic products [22]. Potential decomposition reactions, partic-

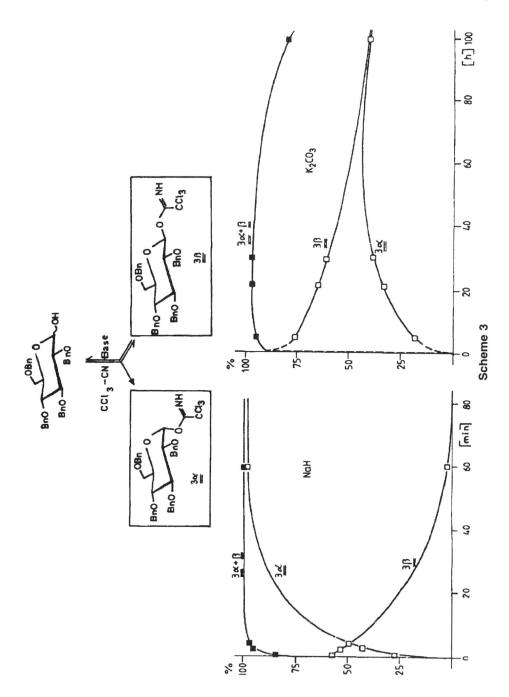

Scheme 3

ularly by the acyclic form (see Scheme 2) or *O*-alkylation reactions at O-5, can usually be avoided. The high diastereocontrol of pyranoses is based on the enhanced nucleophilicity of equatorial oxygen atoms (owing to steric effects and the stereoelectronic *kinetic anomeric effect*) [1,23], thereby favoring equatorial glycoside bond formation; however, thermodynamic reaction control favors the axial glycoside (*thermodynamic anomeric effect*). Chelation control can also become a dominant factor in determining α or β-selection.

*The Trichloroacetimidate Method*

As an alternative to the anomeric *O*-alkylation method, the anomeric oxide can also be used to generate glycosyl donors by addition to appropriate triple bond systems A≡B or cumulenes A = B = C (or by condensation with Z − A = BH systems, where Z represents a leaving group). The most successful methods developed thus far using these types of reactions are trichloroacetimidate (see Scheme 1 D, −A = BH = −C(CCl$_3$)=NH) [1] and phosphate and phosphite formation (A=BH = PO(OR)$_2$, P(OR)$_2$) [1,24]. The analogous glycosyl sulfates, sulfonates, and sulfites have not yet been as successful and, therefore, not as extensively investigated [1,25]. All these methods are particularly tempting because nature has a similar approach for generating glycosyl donors, namely *glycosyl phosphate* formation in the activation step and *(Lewis) acid catalysis* in the glycosylation step. As will be outlined in this chapter for the trichloroacetimidate method, the *activation step* consists of a simple base-catalyzed addition of the anomeric hydroxy group to trichloroacetonitrile and the *glycosylation step* requires only catalytic amounts of a simple (Lewis) acid for the generation of strong glycosyl donor properties. Thus, a very economic and efficient glycosylation procedure is available. Details of this method will be discussed in the following sections of this chapter.

## II. THE TRICHLOROACETIMIDATE METHOD

### A. Trichloroacetimidate Formation (Activation Step)

Electron-deficient nitriles are known to undergo direct and reversible base-catalyzed addition of alcohols to the triple-bond system, thereby providing *O*-alkyl imidates [1,26]. This imidate synthesis has the advantage that the free imidates can be directly isolated as stable adducts, which are less sensitive to hydrolysis than the corresponding salts. Therefore, base-catalyzed transformation of the anomeric oxygen atom into a good-leaving group should be possible, for instance, by addition to trichloroacetonitrile in the presence of base. Because of the experience in anomeric *O*-alkylation, this addition reaction seemed quite likely to occur. Additionally, for achievement of stereocontrolled activation of the anomeric oxygen atom, anomerization at the anomeric center had to be taken into account. Thus, in a reversible activation step, and with the help of kinetic and thermodynamic reaction control (provided essentially by different acidity and by the kinetic and thermodynamic anomeric effect, respectively), frequently both anomers turned out to be accessible at wish. This is exhibited in Scheme 3 for tetra-*O*-benzyl-D-glucose, showing that from an α and β-1-oxide mixture the equatorial (β)-trichloroacetimidate is generated preferentially or even exclusively in a very rapid and reversible addition reaction. However, this product anomerizes in a slow, base-catalyzed reaction (which can be speeded up by stronger bases) through retroreaction, anomerization of the 1-oxide ion, and renewed tricholoracetonitrile addition to form the thermodynamically more stable axial ($\alpha$)-trichloroacetimidate [23] [see Sec. III.A., Eqs. (1) and (2)]. Similar results were obtained for various other O-glycosyl trichloroacetimidates. Thus, with different bases ($K_2CO_3$, $CaCO_3$, NaH, DBU, or other) both O-activated anomers can be isolated, often in pure form and in high yields, as shown for some further typical examples [1] (see under Sec. III.A., Eqs. (3)–(5)]. Both anomers are commonly thermally stable and can be prepared and stored in any quantity. The high yields obtained display that the expected instability of open-chain aldehydic intermediates in basic media (see Schemes 1 and 2) and undifferentiated reactions of other available oxide groups are usually no major problem in the activation process [1].

## B. Glycosylation Reactions (Glycosylation Step)

*General Aspects*

The formation of stable anomeric oxygen-activated intermediates by base catalysis requires a different catalytic system for the generation of glycosyl donor properties in the glycosylation step. Therefore, after base-catalyzed generation of O-glycosol trichloroacetimidates (activation step), mild acid treatment in the presence of acceptors, thus leading to the desired glycosides in an irreversible manner under the reaction conditions, constitutes the required glycosylation step in this method. The water liberated on glycoside bond formation is thereby transferred in the two separate steps to the activating species A≡B = $CCl_3CN$ under formation of stable, nonbasic trichloroacetamide (O = A − $BH_2$; see Scheme 1), providing the driving force for the glycosylation reaction [1]. This is exhibited in Scheme 1 and in Scheme 4 for the formation of N-acetyllactosamine from protected O-galactosyltrichloroacetimidate as donor and 4-O-unprotected N-acetylglucosamine as acceptor [25]. This process very much resembles enzymatic N-acetyllactosamine generation [27]; however, the trichloroacetimidate-based process is obviously more simple, although protection and deprotection steps have to be taken into account. Because of the very low basicity of the liberated trichloroacetamide, the (Lewis) acid required for activation of the basic O-glycosyl trichloroacetimidates, generating an extremely powerful leaving group, is released and is ready for further activation of unreacted glycosyl donors, thereby leading to a truly catalytic process.

The general significance of O-glycosol trichloroacetimidates lies in their ability to act as strong glycosyl donors under relatively mild acid catalysis. This has been overwhelmingly confirmed in various laboratories, and the scope and limitations of this method can be readily derived from these investigations [1]. Representative examples of the application of the trichloroacetimidate method to various important glycoside bond formations are compiled in the procedures in Section III.B., Eqs. (6)–(23).

*$\alpha/\beta$ Selection in the Glycosylation Step*

The general structure of hexoses (and pentoses) and the importance of the hydroxy group next to the anomeric center for anomeric stereocontrol led to differentiation of four structural situations (Scheme 5). The ease of formation of these different anomeric linkages is mainly dependent on the strength of the anomeric effect, which is comparatively stronger in the $\alpha$-*manno*-type than in the $\alpha$-*gluco*-type sugars, and on possible neighboring group

# Oligosaccharide Synthesis with Trichloroacetimidates

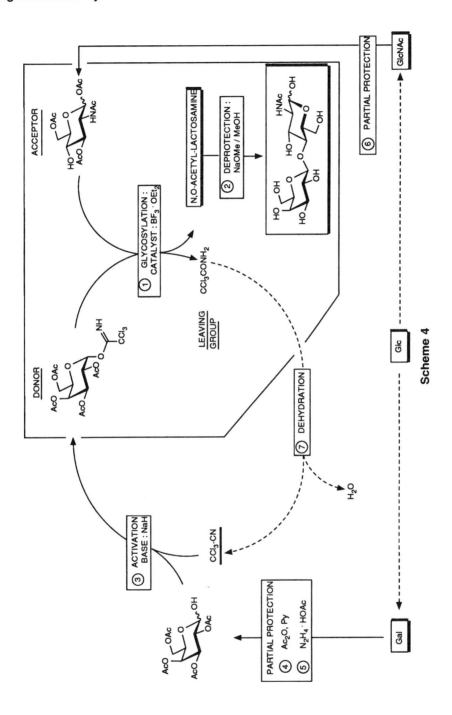

Scheme 4

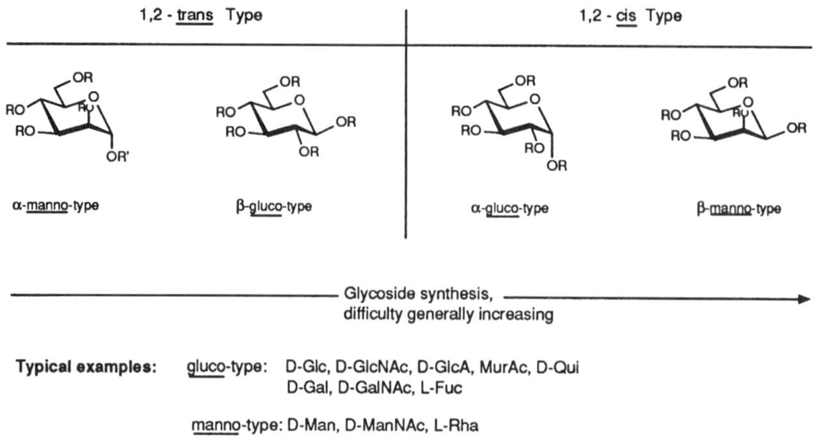

## Scheme 5

**1,2 - trans Type**: α-manno-type, β-gluco-type
**1,2 - cis Type**: α-gluco-type, β-manno-type

Glycoside synthesis, difficulty generally increasing →

Typical examples:
- gluco-type: D-Glc, D-GlcNAc, D-GlcA, MurAc, D-Qui, D-Gal, D-GalNAc, L-Fuc
- manno-type: D-Man, D-ManNAc, L-Rha

participation of protective groups at the C-2 functionality. Thus, the ease of glycoside bond formation decreases from the α-*manno*- by the β-*gluco*-, to the α-*gluco*- and, finally, to the β-*manno*-type (see Scheme 5).

Neighboring group participation of 2-*O*-acyl- (or 2-*N*-acyl)-protecting groups is usually also the dominating effect in anomeric stereocontrol in the trichloroacetimidate method, thus readily furnishing α-*manno*- and β-*gluco*-type (or 1,2-*trans*-)glycosides [1] [see Sec. III.B., Eqs. (7), (10), (12–15), 17–19), (21)]. When nonparticipating protective groups are selected, in nonpolar solvents, and supported by low temperatures (which are generally available owing to the high glycosyl donor properties of *O*-glycosyl trichloroacetimidates) $S_N2$-type reactions can be quite frequently carried out; hence, α-trichloroacetimidates yield β-glycosides and β-trichloroacetimidates α-glycosides (Scheme 6). This has been demonstrated for α-*gluco*- and β-*gluco*-type (1,2-*cis*- and 1,2,-*trans*-)

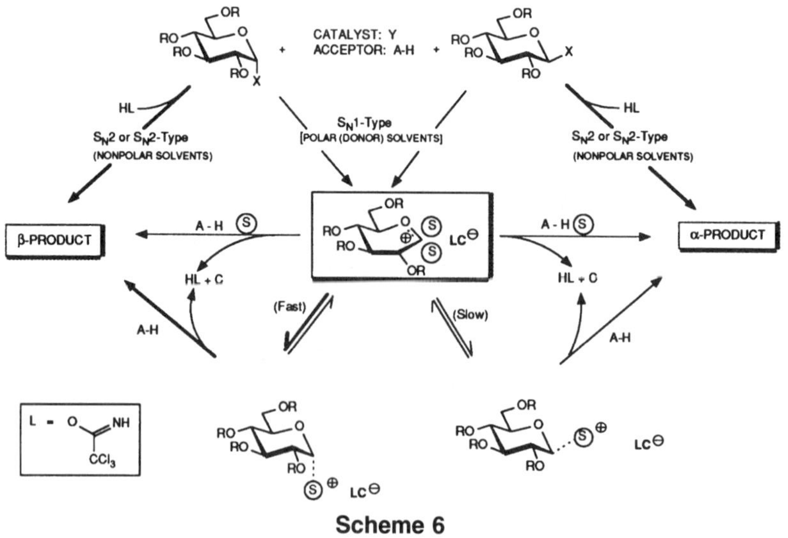

## Scheme 6

glycosides with relatively weak Lewis acid catalysts ($BF_3 \cdot OEt_2$, and such). Strong acidic catalysts (TMSOTf, TfOH), especially at higher temperatures and in more polar solvents, support formation of the thermodynamically more stable product (i.e. α-*manno*- and α-*gluco*-type glycosides [see Sec. III.B, Eqs. (10), (16), (20)]. Thereby the O-glycosyl trichloroacetimidate is transformed into a (possibly tight or solvent separated) half-chair carbenium ion (see Scheme **6**) which because of stereoelectronic reasons is preferentially attacked from the axial (α-)face, thus, for kinetic reasons, leading to the α-product with a chair conformation, which is also thermodynamically favored. Therefore, under $S_N1$-type reaction conditions for kinetic and for thermodynamic reasons the α-*manno*- and the α-*gluco*-type glycosides are preferentially formed [1].

Of particular interest is the influence of solvents under $S_N1$-type conditions, as has been well studied for ethers [1,2,8]. The participation of ethers results, owing to the reverse anomeric effect [28] under $S_N1$-type conditions, in the generation of equatorial oxonium ions (see Scheme **6**, $S = OEt_2$), which again favor the formation of thermodynamically more stable axial α- products [see Sec. III.B, Eqs. (6), (11), (16)].

The dramatic effect of nitriles as participating solvents in glycoside synthesis has only recently been observed by us (*nitrile effect*) [1d,29]; this led to a more complex picture of the influence of solvents (see Scheme **6**). The reaction is controlled by intermediate nitrilium–nitrile conjugate formation. Provided a good leaving group L, such as the trichloroacetimidate group, and low temperatures (up to $-80°C$) are employed (and, for instance, TMSOTf as catalyst), then the glycosyl donor cleaves off its activating group in the presence of nitriles as solvents. Consequently, a highly reactive carbenium ion is formed, which is then attacked by nitrile molecules preferentially from the axial α-face to afford kinetically controlled axial α-nitrilium–nitrile conjugates, which because of their alkylating properties lead to the equatorial β-product. On the other hand, the equatorial β-nitrilium–nitrile conjugate is thermodynamically more stable owing to the reverse anomeric effect (see the influence of ethers), thus favoring the axial α-product. However, this equilibration seems to be much slower for nitriles than for ethers as solvents; therefore, in nitriles the equatorial β-product can become the exclusive product, as found for O-benzyl-protected azidoglucosyl donors [30] and various other cases [24,31] (see examples Sec. III.B, Eq. (8), (9)]. With this procedure some success in the difficult β-*manno*-type glycoside bond formation has also been gained [29c]. For the efficacy of the nitrile effect, obviously the carbenium ion stability and the relative rates of nitrile interception both axially and equatorially play decisive roles in diastereocontrol; however, these factors cannot yet be fully rationalized. Obviously, the α/β ratio of the products is practically independent of the configuration of the glycosyl donor and its leaving group; as long as good leaving groups are employed (as particularly granted for trichloroacetimidates), this leads to the formation of common intermediates that govern product control (as, for instance, clearly shown for sialylation reactions with different leaving groups in nitriles as solvent).

*Reactivity of O-Glycosyl Trichloroacetimidates*

The relative reactivities of different glycosyl donors has been compared several times. A particularly illustrative example exhibiting the high reactivity of trichloroacetimidates is shown in Scheme **7** [32] in which the O-glycosyl trichloroacetimidate affords, even at $-30°C$ and with TMSOTf as catalyst, higher yields of the desired disaccharide—a constituent of the important glycoconjugate shown in Scheme **7**—than the corresponding thioglycosides, which are activated by various thiophilic promoter systems at room temperature.

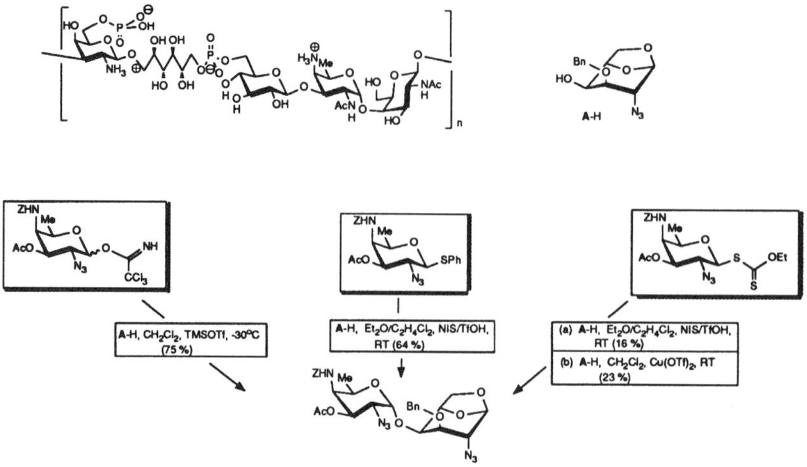

**Scheme 7**

However, the high reactivity of O-glycosyl trichloroacetimidates can also lead to side reactions (mainly with the counterions) or even to decomposition of the donor before reaction with the acceptor, if not compensated for by the protective group pattern. For instance, electron-withdrawing protective groups will stabilize labile glycosyl donors [2a,33]. Besides stabilization of labile glycosyl donors by protective groups (as, for instance, trichloroacetimidates of deoxysugars), also high concentrations or often more effectively application of the *inverse procedure* (i.e., addition of the donor to an acceptor–catalyst solution) can lead to highly improved product yields [1d,34]. Obviously, complex or better cluster formation between acceptor and catalyst seems to lead, with the incoming donor, to intramolecular glycoside bond formation. Thus, in a ternary complex of acceptor molecules, catalyst, and donor, the product is formed, thereby avoiding donor decomposition by the catalyst because the acceptor is not immediately present.

*Glycosylation of Brønsted Acids*

The direct glycosylation of Brønsted acids is a particularly advantageous property of O-glycosyl trichloroacetimidates. Simple Brønsted acids are able to substitute the trichloroacetimidate group at room temperature in high yields [1]. Because of anomerization of possible β-products formed at the beginning of the reaction, owing to the presence of strong acids, generally only α-products are finally isolated. Carboxylic acids as weaker acids usually react, presumably in an eight-membered transition state [1d], under inversion of configuration at the anomeric center, to give 1-O-acyl compounds without addition of any catalyst. The uncatalyzed glycosyl transfer from O-glycosyl trichloroacetimidates to phosphoric acid mono- and diesters opens a simple route to glycosyl phosphates, glycophosphonucleosides, and glycophospholipids [35]; this route was extensively used, because these compounds are of interest in biological glycosyl transfer and as constituents of cell membranes. The anomeric configuration of the product is determined by the acidity of the phosphoric acid derivative applied. Weak acids provide again, in compliance with the proposed eight-membered transition state, the inversion products; stronger acids lead directly owing to anomerization, to the thermodynamically more stable axial α-product [see examples in Sec. III.B, Eq. (22), (23)]. O-Glycosyl phosphonates and phosphinates can similarly be obtained [36].

## Other Nucleophiles as Glycosyl Acceptors

Besides oxygen, other nucleophiles have also been successfully employed as acceptors. For instance, *N*-, *C*-, *S*-, and *P*-glycosyl derivatives have been obtained from corresponding nucleophiles and *O*-glycosyl trichloroacetimidates in the presence of acid catalysts [1]. For instance, aryl-*C*-glycosides have been recently synthesized from phenols in an efficient Fries-type rearrangement reaction [37].

## Scope of O-Glycosyl Trichloroacetimidates

For glycosyl donors requiring high and highest activation (sugar uronates of aldoses, common aldoses, and deoxyaldoses), *O*-glycosyl trichloroacetimidates are highly efficient (Scheme 8) and, owing to simple base and acid catalysis for the activation and glycosyla-

**IMPORTANT SACCHARIDE MOIETIES**

| ALDOPYRANOSYL-URONATE | ALDOPYRANOSYL | DEOXYALDOPYRANOSYL | KETOPYRANOSYL | 3-DEOXY-2-GLYCULOPYRANOSYL-ONATE |
|---|---|---|---|---|
| GlcUA | Glc, Gal, Man, Xyl GlcNAc, GalNAc MurNAc | Fuc, Rha, Qui 2-Deoxyglc | Fru, Sor | Kdo, Neu5Ac |

**PREFERRED APPLICATION OF GLYCOSYL TRICHLOROACETIMIDATES AND PHOSPHITES**

◄─────────────── TRICHLOROACETIMIDATES ─────────── ─ ─ ─ ─ ─ ─ ─ ─

─ ─ ─ ─ ─ ─ ─ ─ ─────────── PHOSPHITES ─────────────►

**Scheme 8**

tion step, respectively, this method is in general also the most economical procedure [1]. Also, partially protected acceptors and donors have been successfully employed in this method [1f]. For sugars requiring relatively low activation to generate glycosyl donor properties (ketoses, 3-deoxy-2-glyculosonates: Neu5Ac, KDO) other methods were more successful [38]. In these cases, the *phosphite moiety* [24,31] (see Scheme 8) provided the same advantageous properties as the trichloroacetimidate group.

## Related Methods

The β-glycosyl imidates prepared from α-glycosyl halides and N-substituted amides (particularly *N*-methylacetamide), using 3 eq of silver oxide have been relatively unstable and unreactive in acid-catalyzed glycosylations; therefore, only a few applications have become known [1a,2a,39]. Reactions with the Vilsmaier–Haack reagent [1a], with 2-fluoro-1-methylpyridinium tosylate [40], and with 2-chloro-3,5-dinitropyridine [40] for anomeric *O*-activation have also been investigated and, also, the isourea group has been introduced by carbodiimide reactions [41]. However, none of these groups gives results in terms of yield, diastereocontrol, performance, or general applicability that are competitive with the tricholoracetimidate method.

## Conclusions

The requirements for efficient glycosylation methods are largely fulfilled by the trichloroacetimidate method. This is clearly indicated by the many examples of successful applications of this method. The results can be summarized as follows.

Features of the *activation step* are (1) convenient base-catalyzed trichloroacetimidate formation; (2) controlled access to α- or β-compounds by choice of the base; (3) thermal stability of α- and β-trichloroacetimidates up to room temperature; if required, silica gel chromatography can be performed.

Features of the *glycosylation step* are (1) catalysis by (Lewis) acids under mild conditions; (2) irreversible reaction; (3) other glycosidic bonds are not affected; (4) usually high chemical yield; (5) stereocontrol of glycoside bond formation is generally good to excellent: (a) participating protective groups give 1,2-*trans*-glycopyranosides, thus yielding β-glycosides of Glc, GlcN, Mur, GlcA, Gal, GalN, Qui, Xyl, and α-glycosides of Man, Rha; (b) with nonparticipating groups, $BF_3 \cdot OEt_2$ as catalyst in nonpolar solvents and at low temperatures favors inversion of anomer configuration, yielding β glycosides of Glc, GlcN, Gal, GalN, Xyl, Mur, GlcA, Qui, whereas TMSOTf as catalyst favors the thermodynamically more stable anomer, yielding α-glycosides of Glc, GlcN, Gal, GalN, Man, Fuc, Mur; (c) the solvent ether favors under $S_N1$-type conditions α-glycoside, and nitriles frequently favor β-glycoside bond formation.

Thus, *O*-glycosyl trichloroacetimidates exhibit in terms of ease of formation, stability, reactivity, and general applicability outstanding glycosyl donor properties; they resemble in various aspects the natural nucleoside diphosphate sugar derivatives. Thus, this method has become a very competitive alternative to the existing techniques.

## III. EXPERIMENTAL PROCEDURES*

### A. Synthesis of O-Glycosyl Trichloroacetimidates

*Synthesis of an* O-*Benzyl-Protected* α-*Trichloroacetimidate with NaH [42]*

(1)

To a solution of compound **1** (12.0 g, 22.2 mmol) [Eq. (1)] in dry $CH_2Cl_2$ (100 mL) were added $Cl_3CCN$ (10 mL, 99 mmol, 4.46 eq) and NaH (50 mg, 2.08 mmol, 0.09 eq) with stirring. After 15 min thin layer chromatography (TLC) indicated α/β ≈ 1:3. For anomerization and completion of the reaction more NaH (700 mg, 29.2 mmol, 1.3 eq) was added. After 2 h, the mixture was filtered (Celite), and the solution was evaporated in vacuo. Short-column chromatography (petroleum ether–$Et_2O$, 3:2) yielded **2** (14.6 g, 96%) as a colorless oil, which crystallized slowly: mp 77°C, $R_f$ 0.71 (petroleum ether–$Et_2O$, 1:1), $[\alpha]_{578}$ +61.5° (c 1, $CHCl_3$).

*Synthesis of an* O-*Acetyl-Protected* β-*Trichloroacetimidate with* $K_2CO_3$ *[23b]*

To a solution of compound **3** (3.0 g, 8.61 mmol) [Eq. (2)] and $Cl_3CCN$ (2.5 mL, 24.8 mmol, 2.9 eq) in dry $CH_2Cl_2$ (20 mL) was added freshly dried and powdered $K_2CO_3$ (2 g) at room temperature with stirring. After 2 h TLC indicated α/β = 1:3–1:2). The mixture was diluted with dry $CH_2Cl_2$ (80 mL), and $K_2CO_3$ was removed by centrifugation. The solution was evaporated and the residue was purified by short column (2 × 2-cm) chromatography

---

*Optical rotations were measured at 20°C.

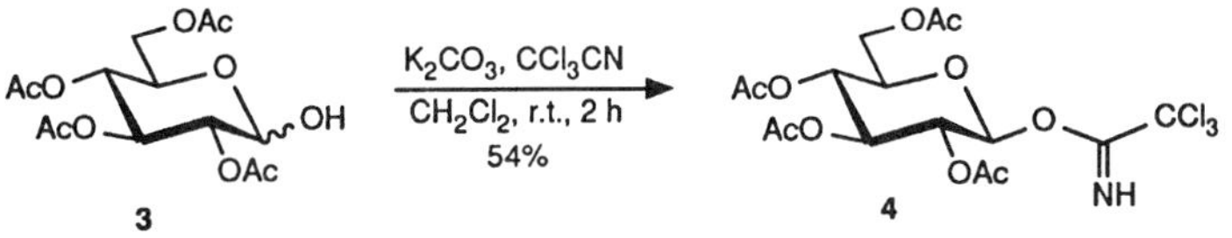

$$\text{3} \xrightarrow[\text{CH}_2\text{Cl}_2, \text{r.t.}, 2\text{ h}]{\text{K}_2\text{CO}_3, \text{CCl}_3\text{CN}} \text{4} \quad (2)$$
$$54\%$$

(CH$_2$Cl$_2$–Et$_2$O, 1:1). Crystallization from Et$_2$O–petroleum ether or CHCl$_3$–Et$_2$O yielded **4** (2.29g, 54%): mp 154°–155°C, $R_f$ 0.65 (ether), $R_f$ 0.80 (petroleum ether–EtOAc, 1:2), [α]$_{578}$ +8.3° (*c* 1, CHCl$_3$).

*Synthesis of α-Tricholoracetimidate of Mannose with DBU [43]*

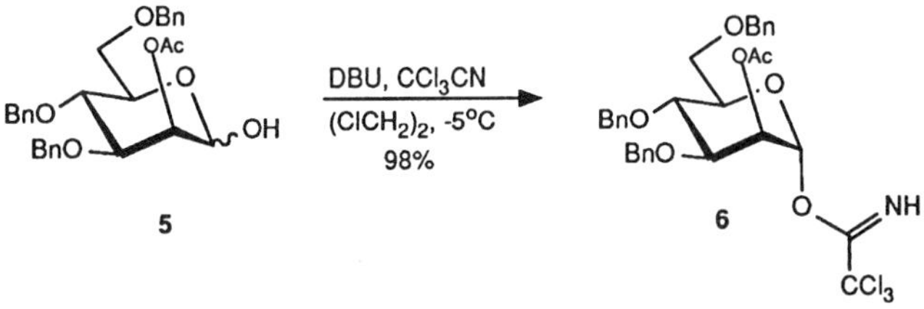

$$\text{5} \xrightarrow[\text{(ClCH}_2)_2, -5°\text{C}]{\text{DBU, CCl}_3\text{CN}} \text{6} \quad (3)$$
$$98\%$$

To a solution of compound **5** (123 mg, 249 μmol) [Eq. (3)] in 1 mL of (ClCH$_2$)$_2$ were added successively Cl$_3$CCN (0.3 mL, 3 mmol, 12 eq) and DBU (10 μL, 0.07 mmol, 0.28 eq) at −5°C, under argon. After stirring for 10 min, the mixture was directly chromatographed on SiO$_2$ (toluene–EtOAc, 6:1) to give **6** (156 mg, 98%): $R_f$ 0.57 (toluene–EtOAc, 3:1), [α]$_D$ +36.3° (*c* 2.2, CHCl$_3$).

*Synthesis of a α-Trichloroacetimidate of Azidogalactose with NaH [44]*

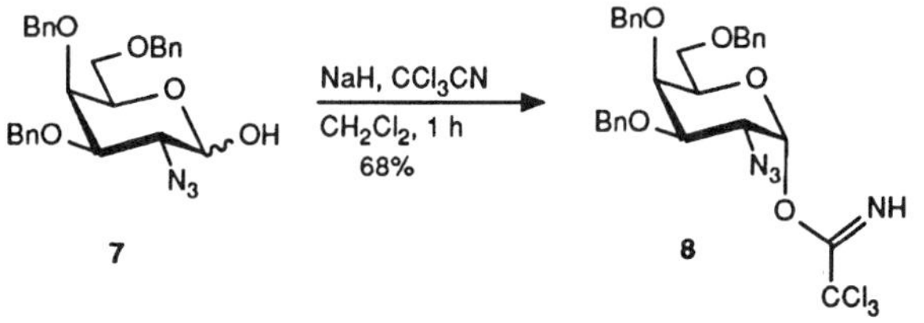

$$\text{7} \xrightarrow[\text{CH}_2\text{Cl}_2, 1\text{ h}]{\text{NaH, CCl}_3\text{CN}} \text{8} \quad (4)$$
$$68\%$$

To a solution of compound **7** (500 mg, 1.05 mmol) [Eq. (4)] and Cl$_3$CCN (1 mL, 9.9 mmol, 9.4 eq) in dry CH$_2$Cl$_2$ (15 mL) was added NaH (30 mg, 1.25 mg, 1.2 eq). After stirring for 1 h the mixture was filtered (Celite), and the solution was evaporated. Chromatography on basic alumina (petroleum ether–Et$_2$O, 2:1) yielded **8** (445 mg, 68%) as a colorless oil: $R_f$ 0.54 (petroleum ether–EtOAc, 2:1), [α]$_{578}$ +89° (*c* 1, CHCl$_3$).

*Synthesis of a Trichloroacetimidate of a Complex Oligosaccharide—the Dimeric Repeating Unit of X-Antigens (Le$^x$) [30b]*

To a solution of compound **9** (5.41 g, 2.67 mmol) [Eq. (5)] in dry CH$_2$Cl$_2$ (150 mL) were added Cl$_3$CCN (20 g, 138 mmol, 52 eq) and DBU (5 drops). After 30 min the mixture was concentrated in vacuo, and the residue was chromatographed (petroleum ether–MeOAc, 1:1, + 1% triethylamine) to give **10** (5.8 g, 100%) in the ratio α/β = 1:2 as an amorphous mass: **10α**: $R_f$ 0.56 (petroleum ether–MeOAc, 1:1), [α]$_D$ −46.0°C (*c* 1, CHCl$_3$); **10β**: $R_f$ 0.62 (petroleum ether–MeOAc, 1:1), [α]$_D$ −62.0° (*c* 1, CHCl$_3$).

(LE$^x$) [REF. 30b]

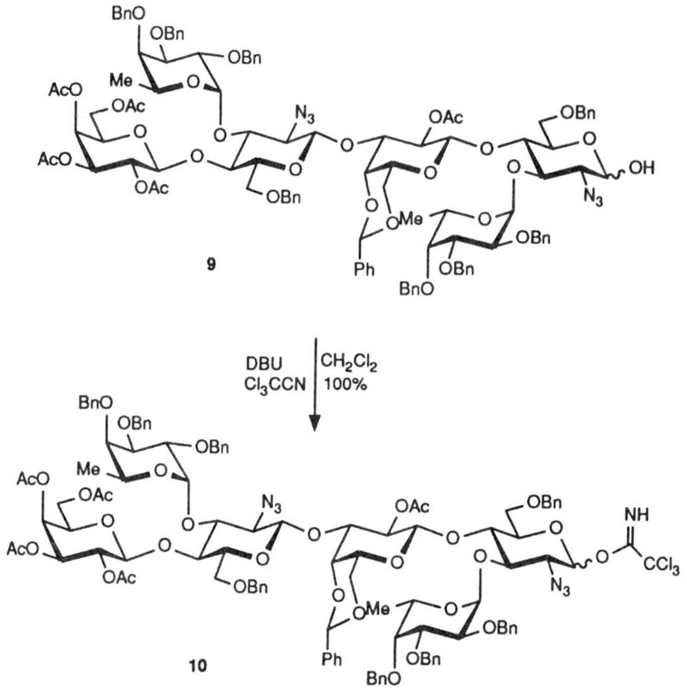

**9**

$$\downarrow \begin{array}{c} \text{DBU} \\ \text{Cl}_3\text{CCN} \end{array} \Big| \begin{array}{c} \text{CH}_2\text{Cl}_2 \\ 100\% \end{array} \qquad (5)$$

**10**

## B. Glycosylation Reactions with O-Glycosyl Trichloroacetimidates

*Synthesis of a Galβ(1-4)Glc Disaccharide: A Suitably Protected Lactose Building Block [45]*

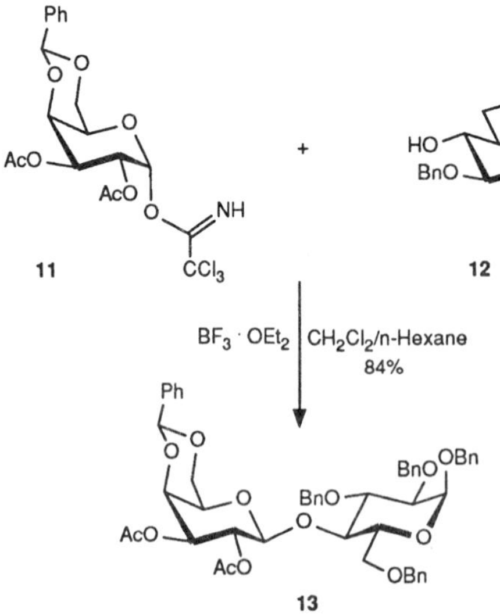

**11** + **12**

$$\downarrow \begin{array}{c} \text{BF}_3 \cdot \text{OEt}_2 \end{array} \Big| \begin{array}{c} \text{CH}_2\text{Cl}_2/n\text{-Hexane} \\ 84\% \end{array} \qquad (6)$$

**13**

To a solution of the acceptor **12** (10.87 g, 20 mmol) [Eq. (6)] and the trichloroacetimidate **11** (15.90 g, 32 mmol, 1.6 eq) in dry *n*-hexane (25 mL) under nitrogen was added at 0°C

BF$_3$·OEt$_2$ (4.0 mL, 32.5 mmol, 1.62 eq). After stirring at room temperature for 30 min, the mixture was poured into ice-cold saturated aqueous NaHCO$_3$ with vigorous stirring. The aqueous layer was extracted with Et$_2$O (2 × 50 mL), and the combined organic layers were washed with water, dried (MgSO$_4$), and evaporated. Chromatography (petroleum ether–EtOAc, 65:35) and crystallization from EtOAc–hexane yielded **13** (14.7 g, 84%): mp 135°-137°C, $R_f$ 0.17 (petroleum ether–EtOAc, 2:1), [α]$_D$ 52.4° (c 1, CHCl$_3$).

*Synthesis of a Galα(1-3)Gal Glycoside: A Tumor-Associated Antigen Structure [46]*

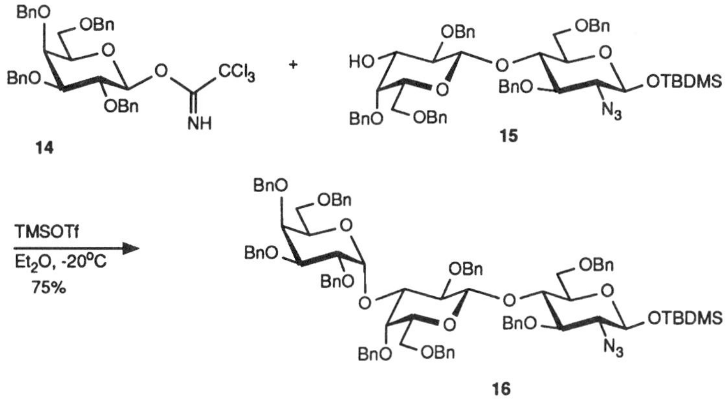

Carefully dried **15** (2.06 g, 2.21 mmol) [Eq. (7)] and trichloroacetimidate **14** (3.03 g, 4.42 mmol, 2 eq) in dry Et$_2$O (44 mL) was treated at −20°C under argon dropwise with 0.033 M TMSOTf in Et$_2$O (1.5 mL, 50 μmol, 0.023 eq). After 5 h, solid NaHCO$_3$ was added, and the mixture was filtered and concentrated. The residue was purified by flash-chromatography (petroleum ether–EtOAc, 8:2) to give **16** (2.40 g, 75%) as a syrup: $R_f$ 0.23 (petroleum ether–EtOAc, 6:4), [α]$_D$ +17.5° (c 1, CHCl$_3$).

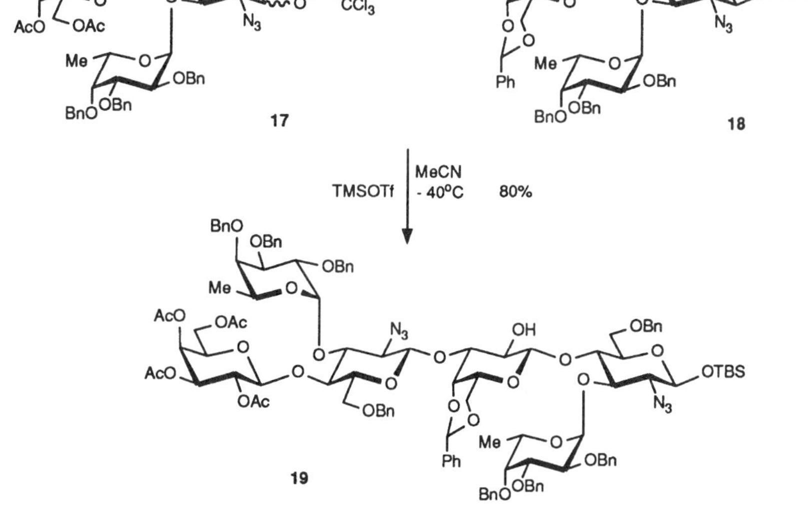

*Synthesis of a Hexasaccharide by Formation of the GlcN$_3$(1–3)Gal Glycoside Bond: Intermediate in the Synthesis of Lewis X-Antigens (Le$^x$) [30b]*

A solution of the trichloroacetimidate **17** (4.75 g, 4.0 mmol) and the acceptor **18** (4.84 g, 4.5 mmol, 1.12 eq) [Eq. (8)] in dry MeCN (60 mL) was treated at −40°C with a 0.05-*M* solution of TMSOTf (0.8 mL, 40 μmol, 0.01 eq); 10 min later solid NaHCO$_3$ (0.5 g) was added. The mixture was filtered and concentrated in vacuo. Short-column chromatography (petroleum ether–EtOAc, 3:2) of the residue gave a colorless foam that was subsequently crystallized from Et$_2$O–petroleum ether to give **19** (6.72 g, 80%): $R_f$ 0.38 (toluene–MeOAc, 4:1), $[\alpha]_D$ −56.5° (*c* 1, CHCl$_3$), mp 195°C.

*Synthesis of a GalN$_3$β(1–4)Gal Glycoside: A Building Block for Glycosphingolipids of the Ganglio-Series [45]*

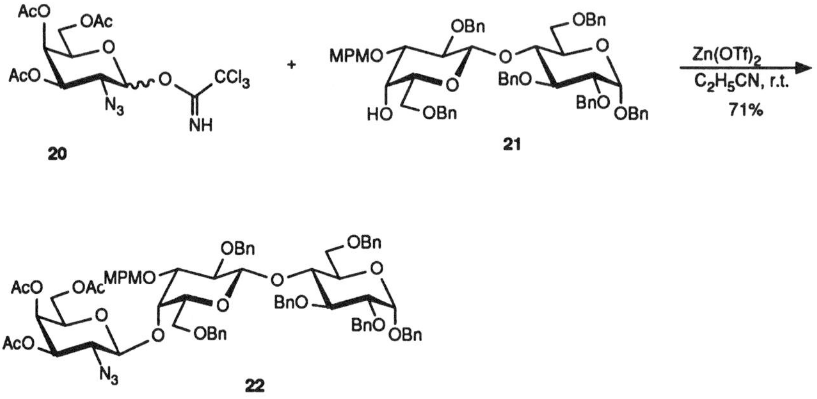

(9)

To a solution of the trichloroacetimidate **20** (595 mg, 1.25 mmol) and the acceptor **21** (2.06, 2.00 mmol, 1.6 eq) [Eq. (9)] in dry propionitrile (6 mL) was added at room temperature Zn(OTf)$_2$ (0.1 *M* in propionitrile, 125 μL, 12.5 μmol, 0.01 eq) with stirring. After 30 min, Et$_2$O was added. The solution was washed with saturated aqueous NaHCO$_3$ and water. The organic solution was dried (Na$_2$SO$_4$) and evaporated. Purification of the residue by flash chromatography (toluene–acetone, 20:1 or toluene–EtOAc, 8:1) yielded a small amount of the α-isomeric trisaccharide (0.30 g, 18%): $R_f$ 0.22 (petroleum ether–EtOAc, 2:1), $[\alpha]_D$ + 83.2° (*c* 1, CHCl$_3$), and the trisaccharide **22** (1.19 g, 71%): $R_f$ 0.16 (petroleum ether–EtOAc, 2:1), $[\alpha]_D$ +14.2° (*c* 0.5, CHCl$_3$).

*Synthesis of a Manα(1–2)Man Glycoside: A Building Block in the Synthesis of GPI-Anchors [47]*

A solution of the acceptor **23** (10.9 g, 12.5 mmol) and the trichloroacetimidate **6** (12.6 g, 16 mmol, 1.3 eq) [Eq. (10)] in dry Et$_2$O (200 mL) under nitrogen was stirred with molecular sieves (4-Å) at room temperature for 10 min. Then TMSOTf (0.4 mL, 2.2 mmol, 0.18 eq) was added, and stirring was continued for 20 min. After neutralization with solid NaHCO$_3$, the mixture was filtered and the solution was evaporated in vacuo. Short-column chromatography (petroleum ether–EtOAc, 5:1) followed by MPLC (toluene–EtOAc, 20:1) yielded **24** (17.2 g, 92%) as a colorless oil: $R_f$ 0.58 (petroleum ether–EtOAc, 8:2), $[\alpha]_D$ +34° (*c* 2, CHCl$_3$).

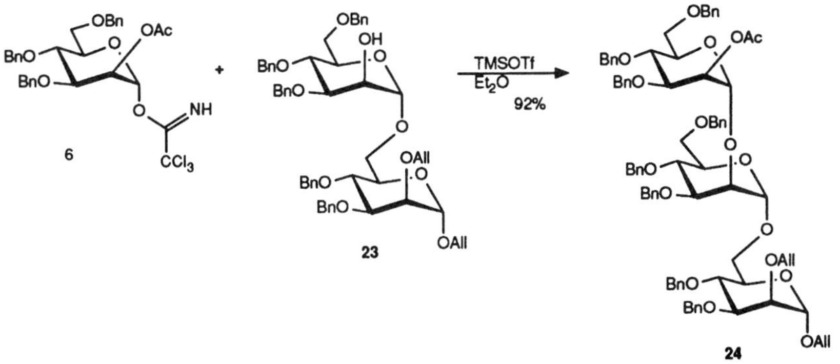

*Synthesis of a Fucα(1–3)Glc Disaccharide: Fucosylation Step in the Synthesis of Lewis X-Antigens (Le$^x$) [30b]*

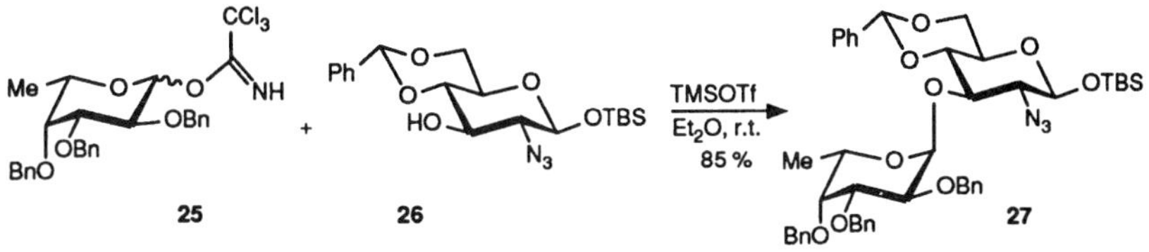

The acceptor **26** (2.93 g, 7.19 mmol) [Eq. (11)] was dissolved in a minimum amount of dry Et$_2$O. First, a 0.1-*M* solution of TMSOTf (1 mL, 0.1 mmol, 0.014 eq) and, thereafter, a concentrated solution of the trichloroacetimidate **25** (6.5 g, 11.2 mmol, 1.5 eq) in dry Et$_2$O were added with stirring at room temperature. After 5 min, the solution was treated with solid NaHCO$_3$ (ca. 0.5 g), filtered, and concentrated in vacuo. The residue was eluted from a column of silica gel (petroleum ether–MeOAc, 7:1) to give **27** as an amorphous mass: $R_f$ 0.54 (petroleum ether–MeOAc, 5:1), [α]$_D$ −76.0° (*c* 1, CHCl$_3$).

*Synthesis of a Quinovose Disaccharide: Building Unit in the Synthesis of Saponins [48]*

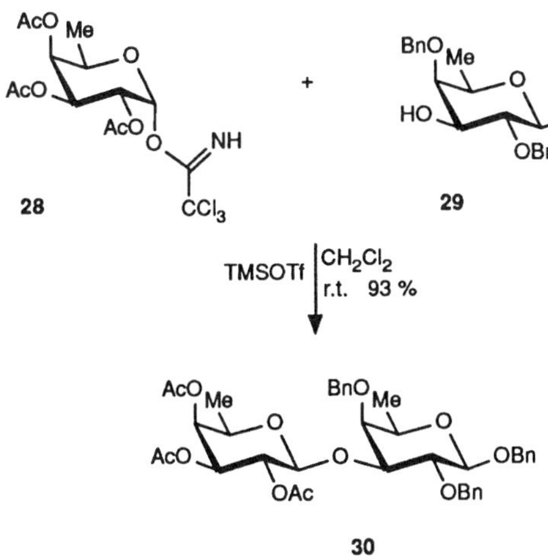

The trichloroacetimidate **28** (1.76, 3.64 mmol) and the acceptor **29** (1.58 g, 3.64 mmol, 1.0 eq) [Eq. (12)] were dissolved in dry $CH_2Cl_2$ (10 mL) and the solution was stirred in the presence of molecular sieves (4 Å, 0.5 g) under argon at room temperature for 10 min. Then TMSOTf (0.01-$M$ solution in $CH_2Cl_2$, 0.7 mL, 0.002 eq) was added dropwise during 20 min. After 40 min, the reaction mixture was treated with saturated aqueous $NaHCO_3$ (5 mL). The aqueous layer was further extracted with $CH_2Cl_2$ (2 × 20 mL). The united organic layers were dried with $MgSO_4$ and concentrated. The residue was separated by flash chromatography (petroleum ether–EtOAc, 3:1) to afford **30** (2.56 g, 93%): $R_f$ 0.24 (petroleum ether–EtOAc, 4:1), $[\alpha]_D$ −13.4° ($c$ 1.8, $CHCl_3$).

*Synthesis of a β-Xylopyranoside: Building Unit in the Synthesis of Saponins [48]*

A solution of the trichloroacetimidate **31** (320 mg, 0.76 mmol) and the acceptor **32** (640 mg, 1.31 mmol, 1.72 eq) in dry $CH_2Cl_2$ (5 mL) [Eq. (13)] was stirred in the presence of molecular sieves (4 Å, 0.3 g) under argon at room temperature for 10 min. TMSOTf (0.01 M in $CH_2Cl_2$, 0.6 ml, 0.006 mmol, 0.0079 eq) was added in three parts during 40 min. The reaction was monitored by TLC every 15 min. After 1 h, the reaction mixture was treated with saturated aqueous $NaHCO_3$ (5 mL), the organic layer separated, and the aqueous layer further extracted with $CH_2Cl_2$ (2 × 15 mL). The combined organic layers were dried with $MgSO_4$ and concentrated. Separation of the residue by flash-chromatography (petroleum ether–EtOAc, 3:1 and then 2:1) gave unreacted **32** (320 mg, 0.655 mmol), compound **33** (290 mg, 51% based on **31**) and the isomeric β(1→4)disaccharide (107 mg, 19%). MPLC on silica gel (toluene–acetone, 5:1) was carried out to afford an analytically pure sample of **33**: $R_f$ 0.28 (petroleum ether–EtOAc, 2:1), $[\alpha]_D$ −35.3° ($c$ 1.3, $CHCl_3$).

*Synthesis of a Heptose Disaccharide: Building Unit for Synthetic Bacterial Lipopolysaccharide Antigens [49]*

A mixture of the trichloroacetimidate **34** (528.6 mg, 0.936 mmol, 2 eq), the acceptor **35** (312.6 mg, 0.473 mmol), and molecular sieves (4 Å, 700 mg) in $CH_2Cl_2$ (10 mL) [Eq. (14)] was stirred for 10 min, with exclusion of moisture. Then TMSOTf (0.04-$M$ solution in $CH_2Cl_2$, 1.2 mL, 0.047 mmol, 0.1 eq) was added at room temperature. After 1 h, (TLC: petroleum ether–EtOAc, 1:1) more TMSOTf solution (1.2 mL, 0.047 mmol, 0.1 eq) was added and stirring was continued for 1 h. Then the mixture was diluted with $CH_2Cl_2$

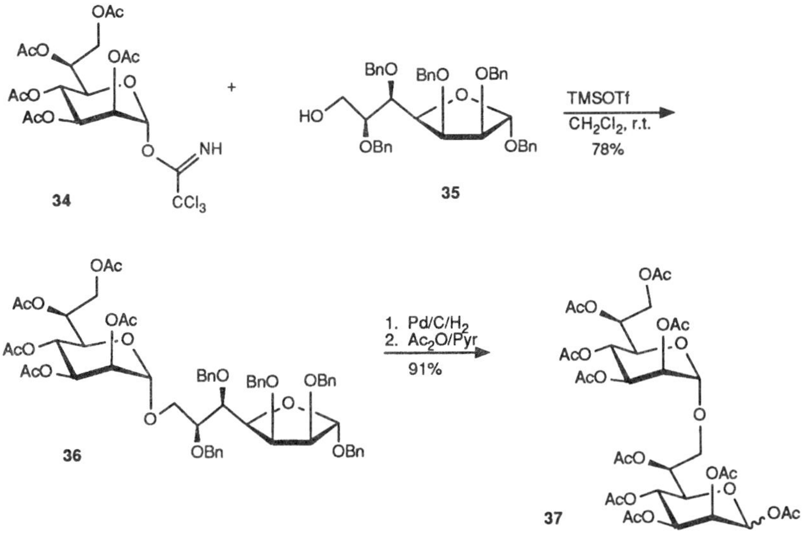

(10 mL), neutralized with Et₃N (20 μL), and filtered. The solution was evaporated in vacuo. MPLC (toluene–EtOAc, 5:1) yielded **36** (392.6 mg, 78%) as a syrup: $[\alpha]_D$ +45.0° (c 1, CHCl₃).

A mixture of **36** (392.6 mg, 0.369 mmol) and Pd/C (10% Pd, 390 mg) in dioxane (3 mL) was shaken under hydrogen (1 atm). After 4 h (TLC: CHCl₃–MeOH–H₂O, 40:10:1) the mixture was filtered, and the solution was evaporated and coevaporated with toluene several times. The residue was dissolved in pyridine (5 mL) and acetic anhydride (1 mL). After 15 h at room temperature (TLC: toluene–acetone, 2:1), toluene (5 mL) was added, and the mixture was evaporated and coevaporated with toluene in vacuo several times. Chromatography on silica gel (petroleum ether–acetone, 3:1) yielded an α/β mixture (276.3 g, 91%, α/β = 32:1) as a syrup. A small quantity of the pure α-anomer **37** was obtained by rechromatography on silica gel (petroleum ether–acetone, 3:1): $[\alpha]_D$ +33.7° (c 1, CHCl₃); the NMR data were in agreement with the reported [50] data.

*Synthesis of a 2-Deoxy-β-D-Glucopyranoside [51]*

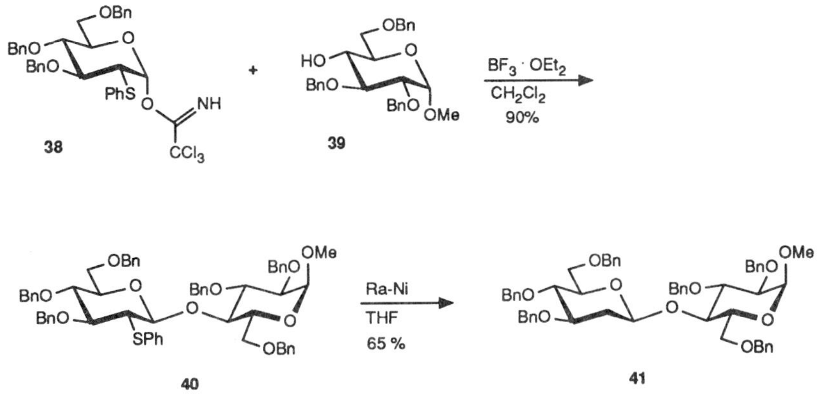

A solution of the trichloroacetimidate **38** (675 mg, 1 mmol) and acceptor **39** (557 mg, 1.2 mmol, 1.2 eq) in dry CH₂Cl₂ (20 mL) [Eq.(15)] is stirred for 30 min at room temperature in the presence of molecular sieves (4-Å, 0.5 g). After cooling to 0°C a solution of $BF_3 \cdot OEt_2$ (0.1 M solution in $Et_2O-CH_2Cl_2$ 1:1, 1 mL, 0.1 eq) is added dropwise within 5 to 10 min. The reaction is monitored by TLC (petroleum ether–EtOAc, 2:1). When the reaction is completed (~ 15 min), $NaHCO_3$ (0.5 g) is added and stirring continued for 10 min. The reaction mixture is filtered, and the solid material is washed with $CH_2Cl_2$ (2 × 10 mL). The solvent is removed in vacuo, and the residue thus obtained is purified by flash chromatography on silica gel using petroleum ether–EtOAc (2:1) as eluent to yield **40** (890 mg, 90%) as an oil: $R_f$ 0.60 (petroleum ether–EtOAc, 2:1), $[\alpha]_D$ +12.5° (*c* 4, EtOAc).

To a solution of compound **40** (495 mg, 0.5 mmol) in dry THF (30 mL) was added Raney-nickel (W II, ~ 4 g) at room temperature. The reaction is monitored by TLC (petroleum ether–EtOAc, 2:1). When the reaction is completed the mixture is filtered, and the solid is washed with THF (3 × 10 mL). The solvent is removed in vacuo, and the colorless residue thus obtained is purified by flash chromatography on silica gel using petroleum ether–EtOAc (2:1) as eluent to yield **41** (292 mg, 65%) as an oil: $R_f$ 0.55 (petroleum ether–EtOAc, 2:1), $[\alpha]_D$ +39.0° (*c* 1, EtOAc).

*Glycosylation of Inositol: Synthesis of the Terminal Carbohydrate Fragment in GPI-Anchors [47]*

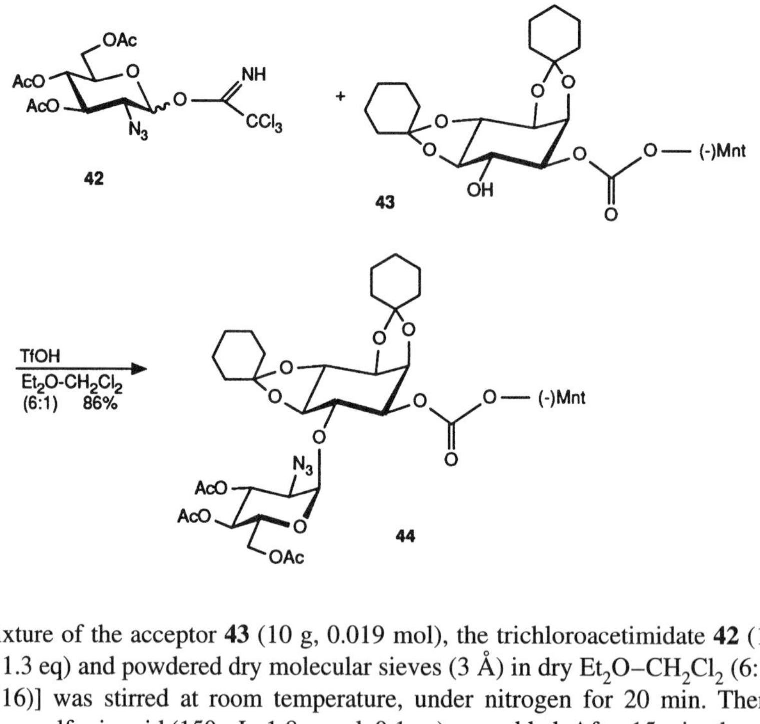

A mixture of the acceptor **43** (10 g, 0.019 mol), the trichloroacetimidate **42** (12 g, 0.025 mol, 1.3 eq) and powdered dry molecular sieves (3 Å) in dry $Et_2O-CH_2Cl_2$ (6:1, 140 mL) [Eq.(16)] was stirred at room temperature, under nitrogen for 20 min. Then trifluoromethanesulfonic acid (150 µL, 1.8 mmol, 0.1 eq) was added. After 15 min, the mixture was neutralized with solid $NaHCO_3$ and evaporated under reduced pressure. The residue was extracted with toluene–petroleum ether (1:1) and filtered. The solution was evaporated in vacuo. Purification by flash chromatography (toluene–EtOAc, 25:1) yielded **44** (13.8 g, 86%) as a colorless foam, which was crystallized from MeOH: mp 157°C, $R_f$ 0.35 (petroleum ether–EtOAc, 8:2), $[\alpha]_D$ +59° (*c* 1, $CHCl_3$).

*Glycosylation of Azidosphingosine: Synthesis of Cerebrosides [52]*

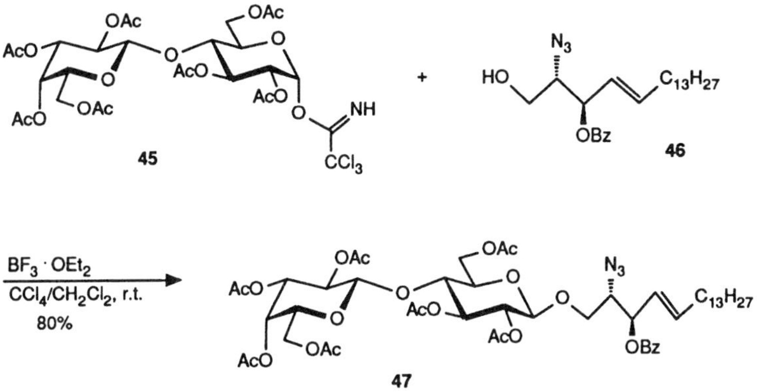

$$\text{45} + \text{46} \xrightarrow[\text{CCl}_4/\text{CH}_2\text{Cl}_2, \text{r.t.}]{\text{BF}_3 \cdot \text{OEt}_2} \text{47} \quad 80\% \tag{17}$$

The trichloroacetimidate **45** (1.0 g, 2.33 mmol) and the azidosphingosine derivative **46** (2.75 g, 3.66 mmol, 1.57 eq [Eq. (17)] were dissolved in dry $\text{CCl}_4$ (50 mL) and the solution stirred in the presence of molecular sieves (4 Å, 0.1g) for 15 min at room temperature: 0.1 $M$ $\text{BF}_3\cdot\text{OEt}_2$ in dry $\text{CH}_2\text{Cl}_2$ (1.1 mL, 0.11 mmol, 0.047 eq) was added within 4 h. The reaction mixture was diluted with petroleum ether (150 mL) and then treated with saturated aqueous $\text{NaHCO}_3$ (10 mL). The organic extract was dried ($\text{MgSO}_4$) and then concentrated. The residue was chromatographed on silica gel (petroleum ether–EtOAc, 3:2) by MPLC to yield **47** (2.10 g, 85%) as a colorless oil: $R_f$ 0.46 (petroleum ether–EtOAc, 3:2), $[\alpha]_D$ −22.3° ($c$ 3.8, $\text{CHCl}_3$).

*An Approach to Polysialogangliosides Containing Neu5Acα(2-8)Neu5Ac: Synthesis of $GD_3$ [53]*

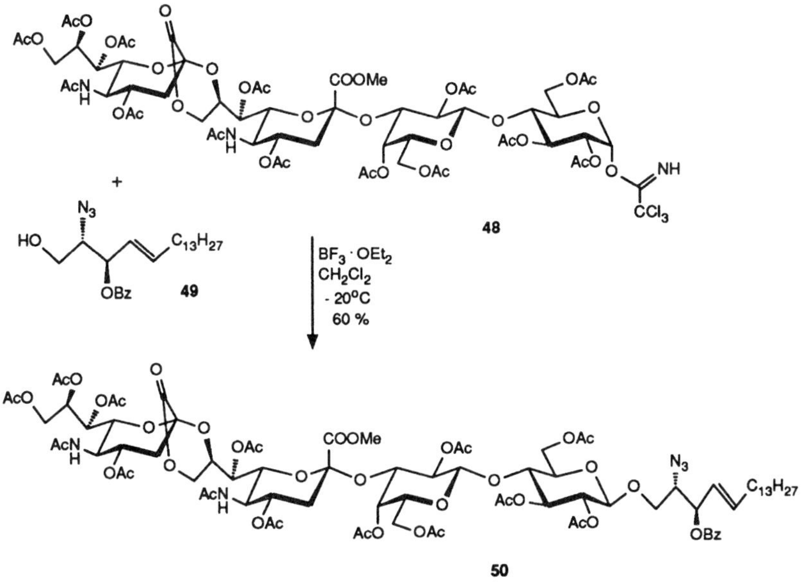

$$\text{48} + \text{49} \xrightarrow[\substack{\text{CH}_2\text{Cl}_2 \\ -20°\text{C} \\ 60\%}]{\text{BF}_3 \cdot \text{OEt}_2} \text{50} \tag{18}$$

To a solution of the trichloroacetimidate **48** (120 mg, 0.076 mmol) and the azidosphingosine derivative **49** (60 mg, 0.15 mmol, 2 eq) in dry $\text{CH}_2\text{Cl}_2$ (1 mL) [Eq. (18)] was added molecular sieves (4 Å, 2 g), and the mixture was stirred for 6 h at room temperature, then cooled to −20°C. BF$_3$·OEt$_2$ (20 μL, 0.16 μmol, 0.002 eq) was added and the mixture was stirred for a further 6 h at −20°C. The course of the reaction was monitored by TLC. The solids were filtered off and washed with CH$_2$Cl$_2$, and the combined filtrate and washings were concentrated. Column chromatography (CH$_2$Cl$_2$–MeOH, 25:1) of the residue on silica gel (20 g) gave **50** (84 mg, 60%) as an amorphous solid: [α]$_D$ −1.6° (c 1.1, CHCl$_3$).

*Glycosylation of Ceramide: Synthesis of GM$_3$ [54]*

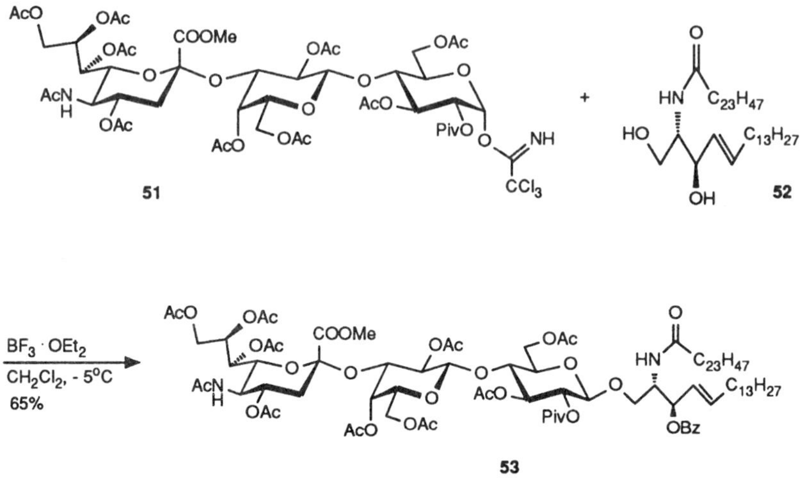

$$\text{(19)}$$

To a stirred mixture of the trichloroacetimidate **51** (22 mg, 18 μmol), the ceramide derivative **52** (20 mg, 26 mmol, 1.4 eq) [Eq. (19)], and powdered molecular sieves (4 Å, 200 mg) in CHCl$_3$ (0.5 mL, freshly purified by passing through an active alumina column) was added BF$_3$·OEt$_2$ (3 μL, 23 μmol, 1.3 eq) at −5°C under argon. The mixture was stirred for 3 h at −5°C, then for 12 h at 20°C, diluted with CHCl$_3$, and filtered through Celite. The filtrate was neutralized with Et$_3$N, and the solvent was evaporated in vacuo. The residue was chromatographed over SiO$_2$ (CHCl$_3$–MeOH, 49:1) to give **53** (21 mg, 65%): $R_f$ (HPTLC) 0.41 (CHCl$_3$–MeOH, 97:3), [α]$_D$ +3.9° (c 1.4, CHCl$_3$).

*Glycosylation of a Threonine Derivative Suitable for Glycopeptide Synthesis [55]*

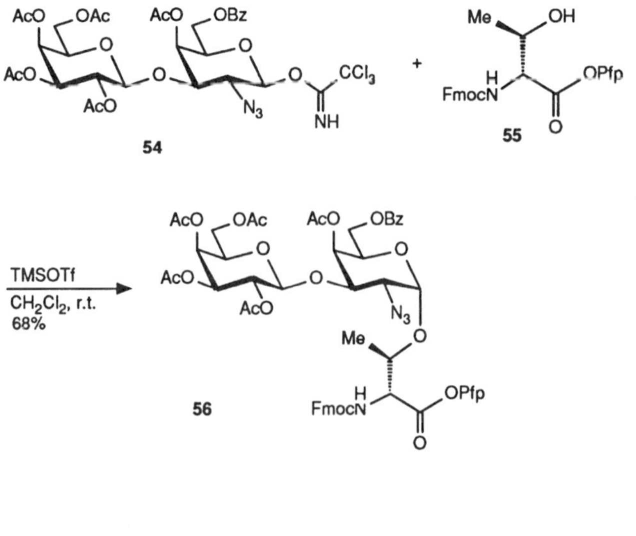

$$\text{(20)}$$

The trichloroacetimidate **54** (0.5 g, 0.6 mmol) and threonine derivative **55** (0.4 g, 0.8 mmol, 1.3 eq) [Eq. (20)] were suspended in $CH_2Cl_2$ (3 mL) at room temperature. A freshly prepared 0.1-*M* solution of TMSOTf in $CH_2Cl_2$ (0.3 mL, 0.03 mmol, 0.05 eq) was added slowly. After 2 h, the mixture was neutralized by the addition of pyridine (3 μL, 0.039 mmol). Solvents were removed under reduced pressure. After chromatography over Florisil (toluene–acetone, 9:1) pure **56** (442 mg, 68%) was obtained: $R_f$ 0.61 (toluene–acetone, 3:1), $[\alpha]_D$ +19.4° (*c* 1, $CHCl_3$).

*Total Synthesis of Amphotericin B with the Trichloroacetimidate of Mycosamine [56]*

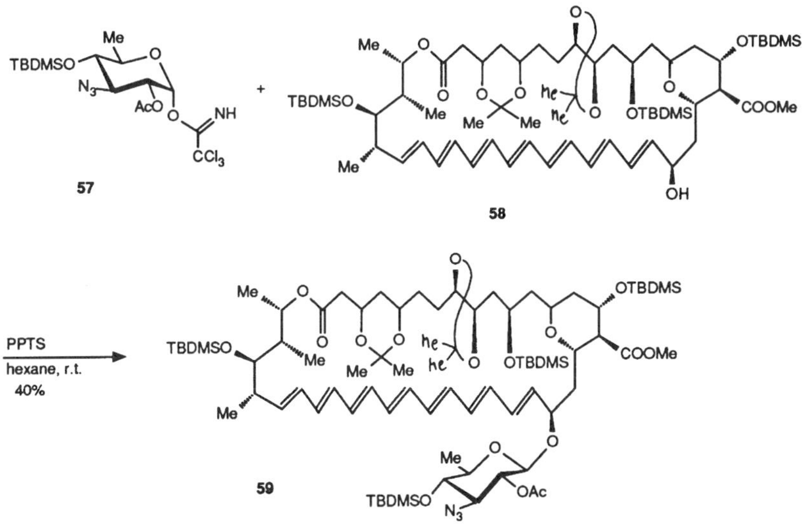

(21)

The amphoteronolide derivative **58** (340 mg, 0.277 mmol) and the trichloroacetimidate **57** (407 mg, 0.831 mmol, 3 eq) [Eq. (21)] were dissolved in dry hexane (118 mL) under an argon atmosphere. To the magnetically stirred solution was added PPTS (20 mg, 0.083 mmol, 0.3 eq) at room temperature, and stirring was continued for 4 h. The mixture was then treated with saturated aqueous $NaHCO_3$ (20 mL) and diluted with $Et_2O$ (100 mL), and the organic phase was separated. The organic solution was washed with brine (20 mL), dried ($MgSO_4$), and concentrated. Flash column chromatography (silica, petroleum ether–$Et_2O$, 95:5 → 50:50) gave, in order of elution, recovered trichloroacetimidate **57** (232 mg, 43%), *ortho*-ester (84 mg, 39%), glycoside **59** (86 mg, 40%), and recovered aglycon derivative **58** (169 mg, 50%). Glycoside **59**: $R_f$ 0.22 (silica, petroleum ether–$Et_2O$, 80:20), $[\alpha]_D$ +152.9° (*c* 0.35, $CHCl_3$).

*Stereospecific Synthesis of β- and α-D-Glucopyranosyl Phosphates [35a]*

To a solution of the α-trichloroacetimidate **2** (685 mg, 1.0 mmol) [Eq. (22)] in dry $CH_2Cl_2$ (20 mL) was added recrystallized dibenzyl phosphate (278 mg, 1.0 mmol, 1 eq) at room temperature. After 1 h the solution was evaporated in vacuo. Column chromatography on silica gel (toluene–acetone, 9:1) yielded **61** (745 mg, 93%): $R_f$ 0.54 (toluene–acetone, 9:1), $[\alpha]_{578}$ +25.7° (*c* 1, $CHCl_3$), mp 44°C; reported [57] mp. 45°–46°C. Synthesis of **63**:

1. From **61**: The β-phosphate **61** (801 mg, 1.0 mmol) was dissolved in dry $CH_2Cl_2$, which was saturated with dry HCl at room temperature. After 4 h TLC (toluene–acetone, 9:1) showed complete anomerization from **61** to **63**. Evaporation in

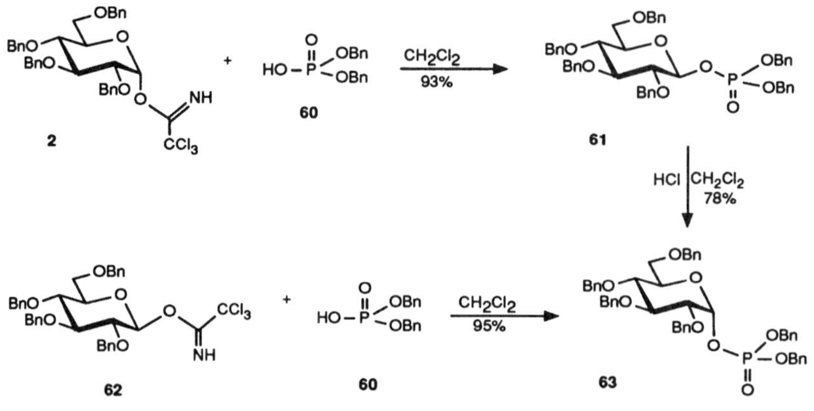

vacuo and chromatography on silica gel (toluene–acetone, 9:1) yielded **63** (625 mg, 78%) and a small amount (5%) of α-halogenose.

1. From **60** and **62**: The reaction of the β-trichloroacetimidate **62** (685 mg, 1.0 mmol) and dibenzyl phosphate (**60**) (278 mg, 1 mmol, 1 eq) was done as described for the synthesis of **61** to yield **63** (761 mg, 95%): $R_f$ 0.53 (CHCl$_3$–Et$_2$O, 20:1), $[\alpha]_{578}$ +57.0° (c 1, CHCl$_3$).

*Synthesis of β-L-Fucopyranosyl Phosphate [58]*

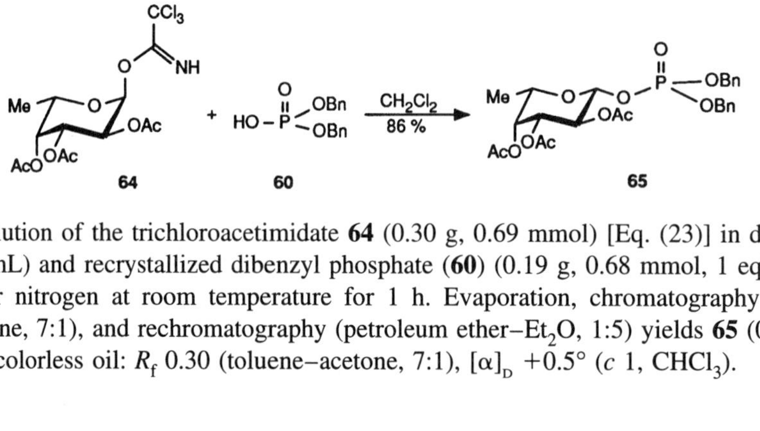

A solution of the trichloroacetimidate **64** (0.30 g, 0.69 mmol) [Eq. (23)] in dry CH$_2$Cl$_2$ (12 mL) and recrystallized dibenzyl phosphate (**60**) (0.19 g, 0.68 mmol, 1 eq) is stirred under nitrogen at room temperature for 1 h. Evaporation, chromatography (toluene–acetone, 7:1), and rechromatography (petroleum ether–Et$_2$O, 1:5) yields **65** (0.36, 86%) as a colorless oil: $R_f$ 0.30 (toluene–acetone, 7:1), $[\alpha]_D$ +0.5° (c 1, CHCl$_3$).

## REFERENCES AND NOTES

1. (a) R. R. Schmidt, Neue Methoden zur Glycoside- und Oligosaccharidsynthese—gibt es Alternativen zur Koenigs-Knorr-Methode? *Angew. Chem.* 98:213 (1986); New methods of glycoside and oligosaccharide syntheses—are there alternatives to the Koenigs–Knorr method? *Angew. Chem. Int. Ed. Engl.* 25:212 (1986). (b) Recent developments in the synthesis of glycoconjugates, *Pure Appl. Chem.* 61:1257 (1989); (c) R. R. Schmidt, in *Comprehensive Organic Synthesis*, Vol. 6, B. M. Trost, I. Fleming, and E. Winterfeldt, eds., Pergamon Press, Oxford, 1991, p. 33. (d) R. R. Schmidt, in *Carbohydrates—Synthetic Methods and Application in Medicinal Chemistry*, A. Hasegawa, H. Ogura, and T. Suami, eds., Kodanasha Scientific, Tokyo, 1992, p. 66. (e) R. R. Schmidt and W. Kinzy, Anomeric-oxygen activation for glycoside synthesis: The trichloroacetimidate method, *Adv. Carbohydr. Chem. Biochem.* 50: 21 (1994).

(f) R. R. Schmidt, in *Modern Methods in Carbohydrate Synthesis*, S. H. Khan and R. A. O'Neill, eds, Harwood Academic, in press.

2. (a) H. Paulsen, Fortschritte bei der selektiven chemischen Synthese komplexer Oligosaccharide, *Angew. Chem. 94*:184 (1982); Progress in the selective chemical synthesis of complex oligosaccharides, *Angew. Chem. Int. Ed. Engl. 21*:155 (1982). (b) Synthesen, Konformationen und Röntgenstrukturanalysen von Saccharidketten der Core-Regionen von Glycoproteinen, *Angew. Chem. 102*:851 (1990); Synthesis, conformation, and x-ray analysis of saccharide chains of glycoprotein core regions, *Angew. Int. Ed. Engl. 29*:823 (1990).

3. (a) H. Kunz, Synthese von Glycopeptiden, Partialstrukturen biologischer Erkennungskomponenten, *Angew. Chem. 99*:297 (1987); New synthetic methods. 67. Synthesis of glycopeptides, partial structures of components of biological recognition systems, *Angew. Chem. Int. Ed. Engl. 26*:294 (1987); (b) Glycopeptides of biological interest: a challenge for chemical synthesis, *Pure Appl. Chem. 65*:1223 (1993).

4. (a) K. Toshima and K. Tatsuta, Recent progress in *O*-glycosylation methods and its application to natural products synthesis, *Chem. Rev. 93*:1503 (1993); (b) J. Banoub, P. Boullanger, and D. Lafont, Synthesis of oligosaccharides of 2-amino-2-deoxy sugars, *Chem. Rev. 92*:1167 (1992).

5. P. Kovac, *Synthetic Oligosaccharides—Indispensable Probes for the Life Sciences*, ACS Symposium Series 560, American Chemical Society, Washington, DC 1994; and papers therein.

6. S. Hakomori, Aberrant glycosylation in tumors and tumor-associated carbohydrate antigens, *Adv. Cancer Res. 52*: 257 (1989).

7. T. A. Springer and I. A. Lasky, Sticky sugars for selectins, *Nature 349*:196 (1991).

8. G. Wulff and H. Röhle, Erkenntnisse und Probleme der *O*-Glycosidsynthese, *Angew. Chem. 86*:173 (1974); Results and problems of *O*-glycoside syntheses, *Angew. Chem. Int. Ed. Engl. 13*: 157 (1974).

9. W. Koenigs and E. Knorr, Über einige Derivate des Traubenzuckers und der Galactose, *Ber. Dtsch. Chem. Ges. 34*:957 (1901).

10. (a) S. Hanessian and J. Banoub, Chemistry of the glycosidic linkage. An efficient synthesis of 1,2-trans-disaccharide, *Carbohydr. Res. 53*: C13 (1977); (b) S. Hanessian and J. Banoub, Preparation of 1,2-trans-glycosides in the presence of silver trifluoromethanesulfonate, in *Methods of Carbohydrate Chemistry*, Vol. 8, R. L. Whistler and J. N. BeMiller, eds., Academic Press, New York, 1980, p. 247.

11. (a) T. Mukaiyama, Y. Murai, and S. Shoda, An efficient method for glucosylation of hydroxy compounds using glucopyranosyl fluoride, *Chem. Lett.* p. 431 (1981). (b) S. Hashimoto, M. Hayashi, and R. Noyori, Glycosylation using glucopyranosyl fluorides and silicon-based catalysts. Solvent dependency of the stereoselection, *Tetrahedron Lett. 25*:1379 (1984).

12. K. C. Nicolaou, A. Chucholowski, R. E. Dolle, and J. L. Randall, Reactions of glycosyl fluorides. Synthesis of *O*-, *S*-, and *N*-glycosides, *J. Chem. Soc. Chem. Commun.* p. 1155 (1984).

13. H. Kunz and W. Sager, Stereoselective glycosylation of alcohols and silyl ethers using glucosyl-fluorides and boron trifluoride etherate, *Helv. Chim. Acta 68*:283 (1985).

14. K. Suzuki, H. Maeta, and T. Matsumoto, An improved procedure for metallocene-promoted glycosylation. Enhanced reactivity by employing 1:2 ratio of $Cp_2HfCl_2$-$AgClO_4$, *Tetrahedron Lett. 30*:4853 (1989), and references therein.

15. (a) P. Fügedi, P. J. Garegg, H. Lönn, and T. Norberg, Thioglycosides and glycosylating agents in oligosaccharide synthesis, *Glycoconjugate J. 4*:97 (1987). (b) T. Norberg and M. Walding, Synthesis of 2-(*p*-trifluoroacetamidophenyl)ethyl-*O*-α-L-fucopyranosyl-(1-3)-*O*-(2-acetamido-2-deoxy-β-D-glucopyranoside, a tetrasaccharide fragment of a tumor-associated glycosphingo-lipid, *Glycoconjugate J. 5*:137 (1988).

16. (a) R. J. Ferrier, R. W. Hay, and N. Vethaviyasar, A potentially versatile synthesis of glycosides, *Carbohydr. Res. 27*:55 (1973). (b) T. Mukaiyama, T. Nakatsuka, and S. Shoda, An efficient glucosylation of alcohol using 1-thioglucoside derivative, *Chem Lett. (5)*:487 (1979). (c) J. van Cleve, Reinvestigation of the preparation of cholesteryl (2,3,4,6-tetra-*O*-benzyl-α-D-glyco-pyranoside, *Carbohydr. Res. 70*:161 (1979). (d) S. Hanessian, C. Bacquet, and N. Lehong,

Chemistry of the glycoside linkage. Exceptionally fast and efficient formation of glycosides by remote activation, *Carbohydr. Res. 80*:C17 (1980). (e) K. Wiesner, T. Y. R. Tsai, and H. Jiu, On cardioactive steroids. XVI. Stereoselective β-glycosylation of digitoxose: the synthesis of digitoxin, *Helv. Chim. Acta 60*:300 (1985). (f) R. B. Woodward (and 48 collaborators), Asymmetric total synthesis of erythromycin. 3. Total synthesis of erythromycin, *J. Am Chem. Soc. 103*:3215 (1981). (g) P. G. M. Wuts and S. S. Bigelow, Total synthesis of oleandrose and the avermecin disaccharide, benzyl α-L-oleandrosyl-α-L-4-acetoxyoleandroside, *J. Org. Chem. 43*:3489 (1983).

17. S. Koto, T. Uchida, and S. Zen, Synthesis of isomaltose, isomaltotetraose, and isomaltooctaose, *Chem. Lett. (11)*:1049 (1972).
18. K. C. Nicolaou, S. P. Seitz, and D. P. Papahatijs, A mild and general method for the synthesis of O-glycosides, *J. Am. Chem. Soc. 105*:2430 (1983).
19. H. Lönn, Synthesis of a tetra- and a nonasaccharide which contain α-L-fucopyranosyl groups and are part of the complex type of carbohydrate moiety of glycoproteins, *Carbohydr. Res. 139*:115 (1985).
20. P. Fügedi and P. J. Garegg, A novel promoter for the efficient construction of 1,2-*trans* linkages in glycoside synthesis, using thioglycosides as glycosyl donors, *Carbohydr. Res. 149*:C9 (1986).
21. R. R. Schmidt and M. Reichrath, Einfache, hochselektive α- und β-Disaccharidsynthesen aus 1-O-metallierten D-Ribofuranosen, *Angew. Chem. 91*:497 (1979). Simple, highly selective α- and β-disaccharide synthesis from 1-O-metalized D-ribofuranoses, *Angew. Chem. Int. Ed. Engl. 18*:466 (1979). W. Klotz and R. R. Schmidt, Anomeric *O*-alkylation of *O*-acetyl-protected sugars, *J. Carbohydr. Chem. 13*:1093 (1994), and references therein.
22. R. R. Schmidt and W. Klotz, Glycoside bond formation via anomeric *O*-alkylation: How many protective groups are required? *Synlett* p. 168 (1991). W. Klotz and R. R. Schmidt, Anomeric *O*-alkylation of *O*-unprotected hexoses and pentoses —convenient synthesis of decyl, benzyl and allyl glycosides, *Liebigs Ann. Chem.*: 683 (1993), and references therein.
23. (a) R. R. Schmidt and J. Michel, Glycosylimidates. Part 11. Direct *O*-glycosyl trichloroacetimidate formation. Nucleophilicity of the anomeric oxygen atom, *Tetrahedron Lett. 25*:821 (1984). (b) R. R. Schmidt, J. Michel, and M. Roos, Direkte Synthese von *O*-α- und *O*-β-Glycosylimidaten, *Liebigs. Ann. Chem.* p. 1343 (1984).
24. T. J. Martin and R. R. Schmidt, Efficient sialylation with phosphites as leaving group, *Tetrahedron Lett. 33*:6123 (1992); T. J. Martin, R. Brescello, A. Toepfer, and R. R. Schmidt, Synthesis of phosphites and phosphates of neuraminic acid and their glycosyl donor properties—convenient synthesis of $GM_3$, *Glycoconjugate J. 10*:16 (1993), and references therein.
25. R. R. Schmidt, University Konstanz, unpublished results.
26. J. U. Nef, Über das zweiwertige Kohlenstoffatom. Die Chemie des Cyans und des Isocyans, *Liebigs Ann. Chem. 287*:265 (1895).
27. C.-H. Wong, S. L. Haynie, and G. M. Whitesides, Enzyme catalyzed synthesis of *N*-acetyllactosamine with in situ regeneration of uridine 5′-diphosphate glucose and uridine 5′-diphosphate galactose, *J. Org. Chem. 47*:5416 (1982).
28. R. U. Lemieux, Effects of unshared pairs of electrons and their solvation on conformational equilibria, *Pure Appl. Chem. 25*:527 (1971).
29. (a) R. R. Schmidt and E. Rücker, Stereoselective glycosidations of uronic acids, *Tetrahedron Lett. 21*:1421 (1980). (b) R. R. Schmidt and J. Michel, *O*-(α-D-Glucopyranosyl)trichloroacetimidate as a glycosyl donor, *J. Carbohydr. Chem. 4*:141 (1985). (c) R. R. Schmidt, M. Behrendt, and A. Toepfer, Nitriles as solvents in glycosylation reactions: Highly selective β-glycoside synthesis, *Synlett* p. 694 (1990). (d) Y. D. Vankar, P. S. Vankar, M. Behrendt, and R. R. Schmidt, Synthesis of β-*O*-glycosides using enol ether and imidate derived leaving groups. Emphasis on the use of nitriles as a solvent, *Tetrahedron 47*:9985 (1991).
30. (a) A. Toepfer and R. R. Schmidt, An efficient synthesis of the Lewis (Le[x]) antigen family, *Tetrahedron Lett. 33*:5161 (1992). (b) A. Toepfer, W. Kinzy, and R. R. Schmidt, Efficient synthesis of the Lewis antigen X (Le[x]) family, *Liebigs Ann. Chem.* p. 449 (1994).

31. T. Müller, R. Schneider, and R. R. Schmidt, Utility of glycosyl phosphites as glycosyl donors—fructofuranosyl and 2-deoxyhexopyranosyl phosphites in glycoside bond formation, *Tetrahedron Lett.* 35:4763 (1994).
32. P. Smid, W. P. A. Joerning, A. M. G. van Duuren, G. J. P. H. Boons, G. A. Van der Marel, and J. H. van Boom, Stereoselective synthesis of a dimer containing an α-linked 2-acetamido-4-amino-2,4,6-trideoxy-D-galactopyranose (SUGp) unit, *J. Carbohydr. Chem.* 11:849 (1992).
33. R. Windmüller and R. R. Schmidt, Efficient Synthesis of *lactoneo* series antigens H, Lewis x (Le$^x$), and Lewis Y (Le$^Y$), *Tetrahedron Lett.* 35:7927 (1994), and references therein.
34. (a) R. Bommer, W. Kinzy, and R. R. Schmidt, Synthesis of the octasaccharide moiety of the dimeric Le$^x$ antigen, *Liebigs Ann. Chem.*, p. 425 (1991). (b) R. R. Schmidt and A. Toepfer, Glycosylation with highly reactive glycosyl donors: Efficiency of the inverse procedure, *Tetrahedron Lett.* 32:3353 (1991).
35. (a) R. R. Schmidt and M. Stumpp, Glycosyl phosphate aus Glycosyl (trichloracetimidaten), *Liebigs Ann. Chem.* p. 680 (1984). (b) M. Hoch, E. Heinz, and R. R. Schmidt, Synthesis of 6-deoxy-6-sulfo-α-D-glucopyranosyl phosphate, *Carbohydr. Res.* 191:21 (1989). (c) E. Heinz, H. Schmitt, M. Hoch, K.-H. Jung, H. Binder, and R. R. Schmidt, Synthesis of different nucleoside 5′-diphospho-sulfoquinonvoses and their use for studies on sulfolipid biosynthesis in chloroplasts, *Eur. J. Biochem.* 184:445 (1989).
36. A. Esswein and R. R. Schmidt, Direct *O*-glycosylation of compounds containing a PO(OH) moiety, *Liebigs Ann. Chem.*, p. 675 (1988).
37. J.-A. Mahling and R. R. Schmidt, Aryl C-glycosides from *O*-glycosyltrichloroacetimidates and phenol derivates with trimethylsilyl trifluormethanesulfonate (TMSOTf) as catalyst, *Synthesis* p. 325 (1993). J.-A. Mahling, K.-H. Jung, and R. R. Schmidt, Synthesis of flavone C-glycosides viexin, isovitexin, and isoembigenin, *Liebigs Ann. Chem.* p. 461 (1995); J.-A. Mahling and R. R. Schmidt, Synthesis of a coumarin C-glucoside, *Liebigs Ann. Chem.* p. 467 (1995).
38. R. R. Schmidt, Chemical synthesis of silylated glycoconjugates, in *Synthetic Oligosaccharides–Indispensible Probes for the Life Sciences*, P. Kovac, ed., American Chemical Society, Washington DC, 1994, p. 276.
39. (a) J.-R. Pougny and P. Sinaÿ, Reaction d'imidates de glucopyranosyle avec l'acetonitrile. Applications synthetiques, *Tetrahedron Lett.* p. 4073 (1976). (b) J.-R. Pougny, J.-C. Jacquinet, M. Nassr, M.-L. Milat, and P., Sinaÿ, A novel synthesis of 1,2-*cis*-disaccharides, *J. Am. Chem. Soc.* 99:6762 (1977). (c) P. Sinaÿ, Recent advances in glycosylation reactions, *Pure Appl. Chem.* 50:1437 (1978).
40. T. Mukaiyama, Y. Hashimoto, Y. Hayashi, and S.-I. Shoda, Stereoselective synthesis of α-ribonucleosides from 1-hydroxy sugars by using 2-fluoropyridinium tosylate, *Chem. Lett*: 557 (1984).
41. R. U. Lemieux, Some implications in carbohydrate chemistry of theories relating to the mechanisms of replacement reactions, *Adv. Carbohydr. Chem. Biochem.* 9:1 (1954). H. Tsutsumi and Y. Ishido, Syntheses of 1-*O*-acylaldose derivatives via the corresponding *O*-glycosylpseudoureas, *Carbohydr. Res.* 111:75 (1982). S. Horvat, L. Varga, and J. Horvat, A new utility of *O*-glycosylpseudoureas for synthesis of glucopeptides and (1→6)-disaccharides, *Synthesis*:209 (1986).
42. R. R. Schmidt and M. Stumpp, Synthese von 1-thioglycosiden, *Liebigs Ann. Chem.* p. 1249 (1983).
43. F. Yamazaki, S. Sato, T. Nukada, Y. Ito, and T. Ogawa, Synthesis of α-D-Manp-(1→3)-[β-D-GlcpNAc-(1→4)]-[α-DManp-(1→6)]-β-D-Manp-(1→4)-β-D-GlcpNAc-(1→4)-[α-L-Fucp-(1-6)]-D-GlcpNAc, a core glycoheptaose of a "bisected" complex-type glycan of glycoproteins, *Carbohydr. Res.* 201:31 (1990).
44. G. Grundler and R. R. Schmidt, Anwendung des Trichloracetimidat-Verfahrens auf 2-Azidoglucose- und -2-Azidogalactose-Derivate, *Liebigs Ann. Chem.* p. 1826 (1984).
45. T. Stauch, Dissertation, Universität Konstanz, 1995.
46. R. Schaubach, J. Hemberger, and W. Kinzy, Tumor-associated antigen synthesis. Synthesis

of the Gal-α-(1→3)-Gal-β-(1→4)-GlcNAc epitope. A specific determinant for metastatic progress? *Liebigs Ann. Chem.*: p. 607 (1991).

47. T. G. Mayer, Bernd Kratzer, and R. R. Schmidt, Synthese eines GPI-Ankers der Hefe (*Saccharomyces cerevisiae*), *Angew. Chem. 106*:2289 (1994). Synthesis of a GPI anchor of yeast (*Saccharomyces cerevisiae*), *Angew. Chem. Int. Ed. Engl. 33*:2177 (1994). Thomas G. Mayer, Dissertation, Universität Konstanz, 1995, submitted.
48. Zi-Hua Jiang and R. R. Schmidt, Synthesis of the hexasaccharide moiety of pectinioside E, *Liebigs Ann. Chem.* p. 975 (1992).
49. H. Paulsen and E. C. Höffgen, Dastellung synthetischer Antigene von LD-Heptose-haltigen Trisacchariden der Core-Region von Lipopolysacchariden durch Copolymerisation mit Acrylamid, *Liebigs Ann. Chem.* p. 543 (1993).
50. H. Paulsen and C. Heitmann, Synthese von Strukturen der inneren Core-Region von Lipopolysacchariden, *Liebigs Ann. Chem.* p. 1061 (1988).
51. R. Preuss and R. R. Schmidt, A convenient synthesis of 2-deoxy-β-D-glucopyranosides, *Synthesis* p. 694 (1988).
52. P. Zimmermann, R. Bommer, T. Bär, and R. R. Schmidt, Azidosphingosine glycosylation in glycosphingolipid synthesis, *J. Carbohydr. Chem. 7*:435 (1988).
53. H. Ishida, Y. Ohta, Y. Tsukada, M. Kiso, and A. Hasegawa, A synthetic approach to polysialogangliosides containing α-sialyl-(2→8) sialic acid: Total synthesis of ganglioside $GD_3$, *Carbohydr. Res. 246*:75 (1993).
54. M. Numata, M. Sugimoto, Y. Ito, and T. Ogawa, An efficient synthesis of ganglioside $GM_3$: Highly stereocontrolled glycosylations by use of auxiliaries, *Carbohydr. Res. 203*: 205 (1990).
55. J. Rademann and R. R. Schmidt, Solid-phase synthesis of a glycosylated hexapeptide of human sialophorin, using the trichloroacetimidate method, *Carbohydr. Res. 269*: 217 (1995).
56. K. C. Nicolaou, R. A. Daines, Y. Ogawa, and T. K. Chakraborty, Total synthesis of amphotericin B. 3. The final stages, *J. Am. Chem. Soc. 110*:4696 (1988).
57. L. V. Volkova, M. G. Luchinskaya, N. M. Karimova, and R. P. Evstigneeva, Synthetic studies in the area of glycosphingolipids. IV Synthesis of phosphatidyl-1-glucose, *Zh. Obstrch. Khim. 42*:1405 (1972) [*Chem. Abstr. 77*:114723 (1972)].
58. R. R. Schmidt, B. Wegmann, and K.-H. Jung, Stereospecific synthesis of α- and β-L-fucopyranosyl phosphates and of GDP-fucose via trichloroacetimidate, *Liebigs Ann. Chem.* p. 121 (1991).

# 13

# Oligosaccharide Synthesis from Glycosyl Fluorides and Sulfides

## K. C. Nicolaou and Hiroaki Ueno
*The Scripps Research Institute and University of California at San Diego, La Jolla, California*

| | | |
|---|---|---|
| I. | Introduction | 314 |
| II. | Methods | 314 |
| | A. The Koenigs–Knorr method | 314 |
| | B. The trichloroacetimidate method | 315 |
| | C. The glycal method | 316 |
| | D. The *n*-pentenyl glycoside method | 316 |
| | E. The glycosyl phoshite method | 316 |
| | F. The enzymatic approach method | 316 |
| | G. Oligosaccharide synthesis from glycosyl fluorides and sulfides and the two-stage activation procedure | 317 |
| | H. Application of two-stage activation procedure to complex molecule synthesis | 321 |
| III. | Experimental Procedures | 329 |
| | A. Synthesis of avermectin $B_{1a}$ | 329 |
| | B. Synthesis of hexasaccharide 14 | 330 |
| | C. Synthesis of trimeric Le[x] | 331 |
| | D. Synthesis of sialyl dimeric Le[x] | 332 |
| | E. Synthesis of sulfated Le[a] tetrasaccharide | 333 |
| | F. Synthesis of globotriaosylceramide (Gb$_3$) | 333 |
| | G. Synthesis of elfamycin | 334 |
| | H. Synthesis of NodRm-IV factors | 335 |
| | References | 336 |

## I. INTRODUCTION

The biological relevance of carbohydrates and carbohydrate-containing compounds is increasingly becoming evident. Compounds of this class include glycoamine and macrolide antibiotics, anthracycline and enediyne anticancer antibiotics, and numerous oligosaccharides implicated in cell–cell recognition, cellular-immune response, cell oncogenic transformation and inflammation, and other cell biology phenomena [1]. As glycobiology moves forward, the need for efficient methods to render oligosaccharides available for further investigations becomes more acute. Despite the long history of the field, oligosaccharide synthesis lacks behind peptide [2] and oligonucleotide [3] synthesis in terms of efficiency and automation. In this chapter, we will briefly review the main synthesis methods available for glucoside bond formation [4] and then proceed to discuss the two-stage activation procedure based on the chemistry of glycosyl fluorides and sulfides.

## II. METHODS

### A. The Koenigs–Knorr Method

This glycosidation methods dates back to 1901 [5]. As in all glycosidation reactions, this method achieves activation of the anomeric center by decomposition of glycosyl halides, normally bromides and chlorides, in the presence of heavy metals (usually silver or mercury). Scheme 1 depicts the general mechanism of this reaction which involves in situ

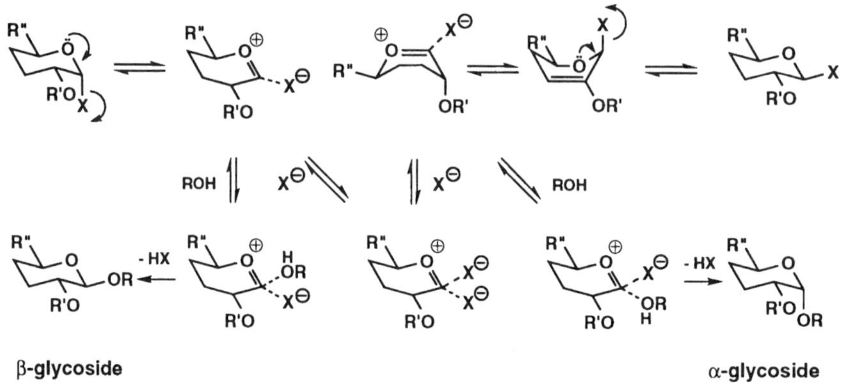

**Scheme 1** Mechanism of the Koenigs–Knorr glycosidation reaction.

generation of the more reactive β-halide, followed by glycosidation with inversion to afford the α-glycoside [6]. Construction of β-glycosides may be accomplished by neighboring group participation of 2-acyl substituents [7] (Scheme 2).

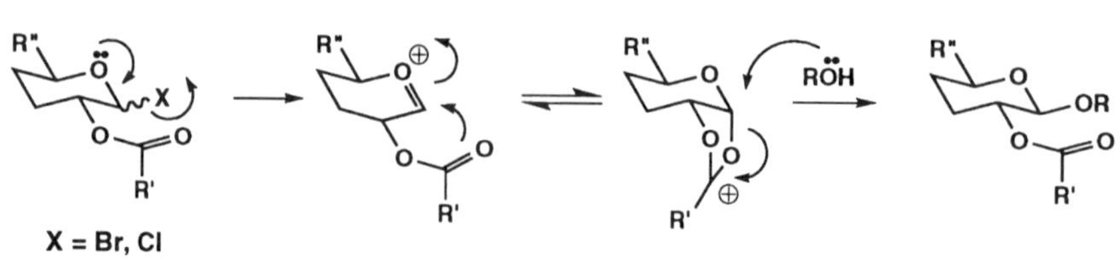

**Scheme 2** Neighboring group participation in the formation of 1,2-*trans*-glycosides (often β-glycosides).

## B. The Trichloroacetimidate Method

Perhaps more than any other, this method has enjoyed considerable success in recent years, owing to its mildness and efficiency. Developed by Schmidt [8], this reaction (Scheme 3)

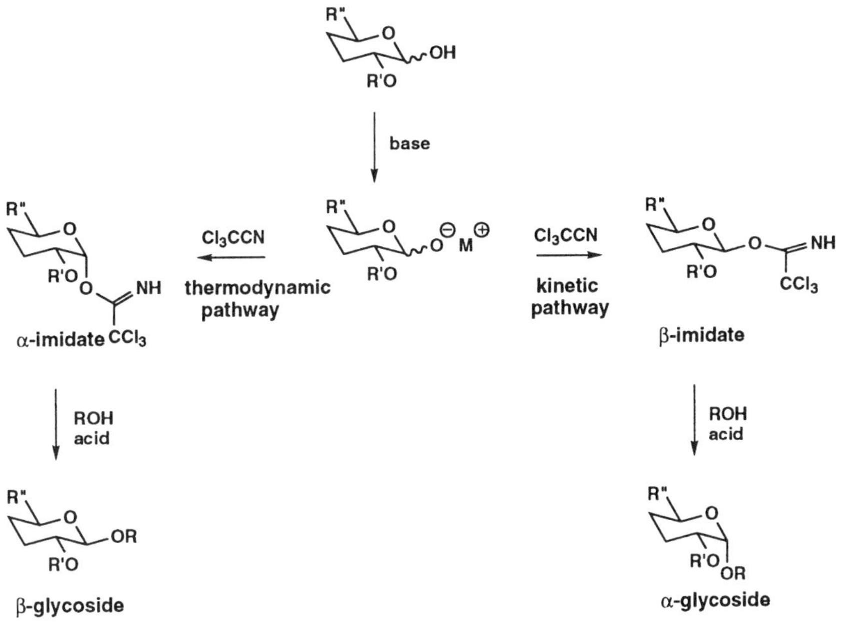

**Scheme 3** The preparation of trichloroacetimidates and their glycosidation reaction [10].

involves the preparation and use of trichloroacetimidates as glycosyl donors under Lewis or protic acid conditions. The glycosidation usually proceeds with inversion of configuration unless a 2-participating group is present, in which case 1,2-*trans* systems are obtained [9]. The requisite trichloroacetimidates can be selectively produced from the corresponding lactols under kinetically (β-anomers) or thermodynamically (α-anomer) controlled conditions [10].

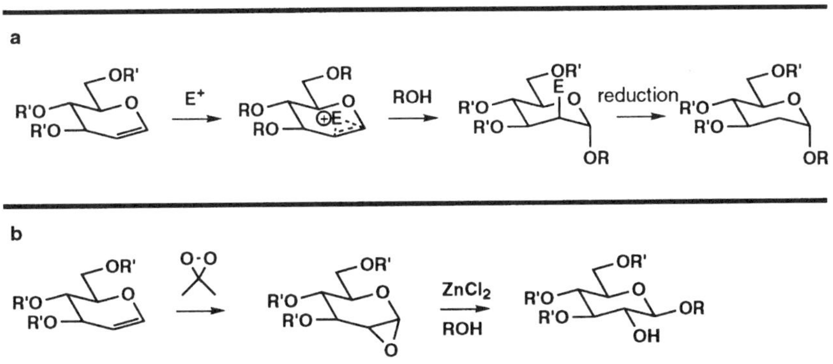

**Scheme 4** Glycal-based glycosidation reactions. (a) $E^+$ = I(collidine)$_2$ClO$_4$ [11], *N*-iodosuccinimide [12], PhSeCl [13], PhSCl [14]. (b) Danishefsky method [15].

## C. The Glycal Method

Scheme 4 presents several variations of the glycal glycosidation method, as pioneered by Lemieux [11], Thiem [12], Sinaÿ [13], Ogawa [14], and Danishefsky [15]. In these methods, an electrophile (E$^+$) is used to attack the electron-rich glycal as a means to activate the anomeric center, which then reacts with a carbohydrate acceptor to afford the glycoside. Removal of the substituent at C-2 then may lead to 2-deoxy glycosides. Of particular note is the Danishefsky method (see Scheme 4), in which a glycal epoxide is initially formed and subsequently activated with a Lewis acid for coupling to glycosyl acceptors. This method has been demonstrated to work on a solid support as have a number of other methods [16].

## D. The *n*-Pentenyl Glycoside Method

Introduced by Fraser-Reid in 1988 [17], this method (Scheme 5) depends on electrophilic addition to the olefin, followed by intramolecular displacement by the ring oxygen and eventual expulsion of the pentenyl chain to form an oxonium species. Trapping with a glycosyl acceptor then leads to the desired glycoside.

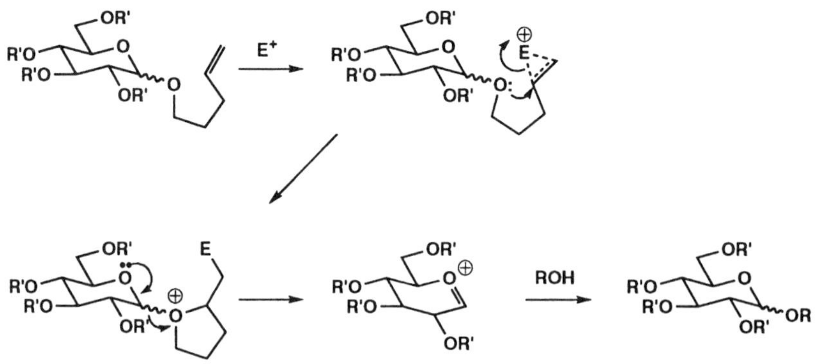

**Scheme 5** The *n*-pentenyl glycoside-based glycosidation method, (E$^+$ = I(collidine)$_2$ClO$_4$) [17].

## E. The Glycosyl Phosphite Method

A relatively new method, this involves glycosyl phosphites as glycosyl donors (Scheme 6). Independently reported by Schmidt [18] and Wong's groups [19], this method has already found important applications in the synthesis of sialyl derivatives [19].

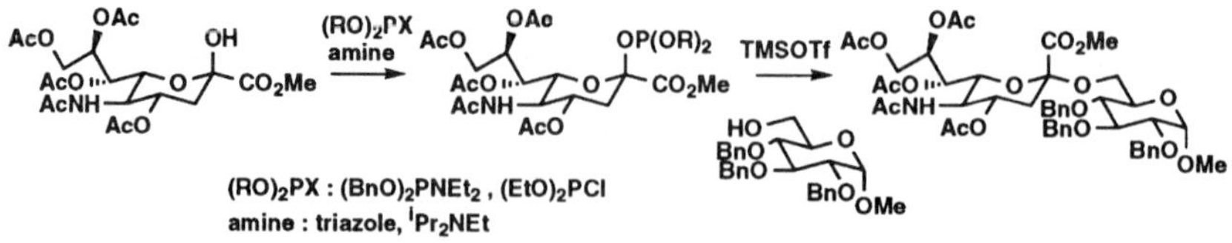

**Scheme 6** Phosphite-based glycosidation method [18,19].

## F. The Enzymatic Approach Method

In recent years enzymatic methods for certain glycosidations have been applied to the construction of a number of oligosaccharides [20,21]. With their proved advantages of

# Chemistry of Glycosyl Fluorides and Sulfides

specificity, stereoselectivity, and no protecting group requirements, these methods appear to have considerable potential in the synthesis of complex oligosaccharides [22].

## G. Oligosaccharide Synthesis from Glycosyl Fluorides and Sulfides and the Two-Stage Activation Procedure

The two-stage activation procedure for the synthesis of complex oligosaccharides developed in these laboratories [23] combines the chemistries of glycosyl fluorides and sulfides. It is, therefore, instructional to discuss first, the use of these two classes of glycosyl donors in oligosaccharide synthesis, before describing the two-stage activation method.

*Glycosyl Fluorides as Glycosyl Donors*

Glycosyl fluorides were first introduced as glycosyl donors by Mukaiyama in 1981 [24]. This group demonstrated the activation and coupling of these carbohydrate intermediates to glycosyl acceptors in the presence of silver perchlorate and tin dichloride. Subsequent reports expanded the repertoire of activators of glycosyl fluorides to include $BF_3 \cdot Et_2O$ [25], TMSOTf [26], $Cp_2HfCl_2$ [27], and $Yb(OTE)_3$ [28]. The mechanism of activation of these glycosyl donors is presumed to be similar to that involved in the classic Koenigs–Knorr process (see Scheme 1), and so is the stereoselectivity of the reaction. Thus, in the presence of a participating group at C-2, β-glycosides are formed [6], whereas in the absence of such moiety, the α-anomers are the predominant products [7]. Several advantages of glycosyl fluorides over their bromide and chloride counterparts should be noted. These include (1) ease of formation by a variety of mild methods (Scheme 7) [23,29,30]; (2) relatively high stability toward silica gel chromatography and storage; and (3) plethora of activating methods and reagents for coupling with glycosyl acceptors [25–28].

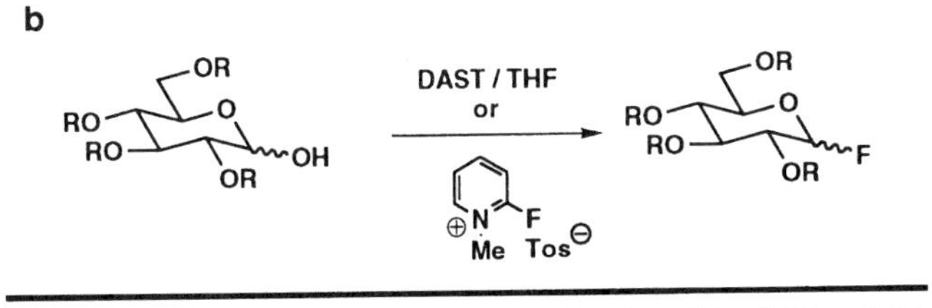

**Scheme 7** Methods for glycosyl fluorides using (a) thioglycosides [23] and (b) lactols [29,30].

*Thioglycosides as Glycosyl Donors*

Thioglycosides were first introduced in glycosidation reactions by Ferrier in 1973 [31], who used the ethylthio group at the anomeric position. Subsequently, Hanessian [32] demonstrated the usefulness of the 2-pyridylthiol. Currently, the phenylthio and the ethylthio carbohydrate derivatives are the most commonly used glycosyl donors from this class of compounds. Thioglycosides are quite versatile and useful intermediates in oligosaccharide synthesis owing to their ease of preparation, stability, and rich chemistry. One of their distinct advantages is that their thio group can serve as a temporary protecting group stable to almost all glycosidation conditions currently in use. Thioglycosides react readily with bromine to afford glycosyl bromides [33], chlorine to furnish glycosyl chlorides [34], and N-bromosuccinimide (NBS) in the presence of diethylaminosulfur triflouride (DAST) or water to give glycosyl fluorides [23] or lactols [35], respectively. Most significantly the thioglycoside group may be introduced early in the synthetic sequence [35,36], and can be maintained throughout the scheme, being quite stable toward most conditions used in oligosaccharide construction.

Common usage of thioglycosides in glycosidation reactions is a relatively new development [37] compared with the Koenigs–Knorr method. The accepted mechanism of the thioglycoside-based glycosidation reaction is similar to that of the Koenigs–Knorr reaction (see Scheme 1). Thus, the thiophilic reagent used as an activator initiates displacement of the sulfur by the lone pair of electrons of the ring oxygen to form the oxonium species, which is then trapped by the glycosyl acceptor in the usual fashion as shown in Scheme **8a**. A variety of thiophiles "X" have been employed for the activation of

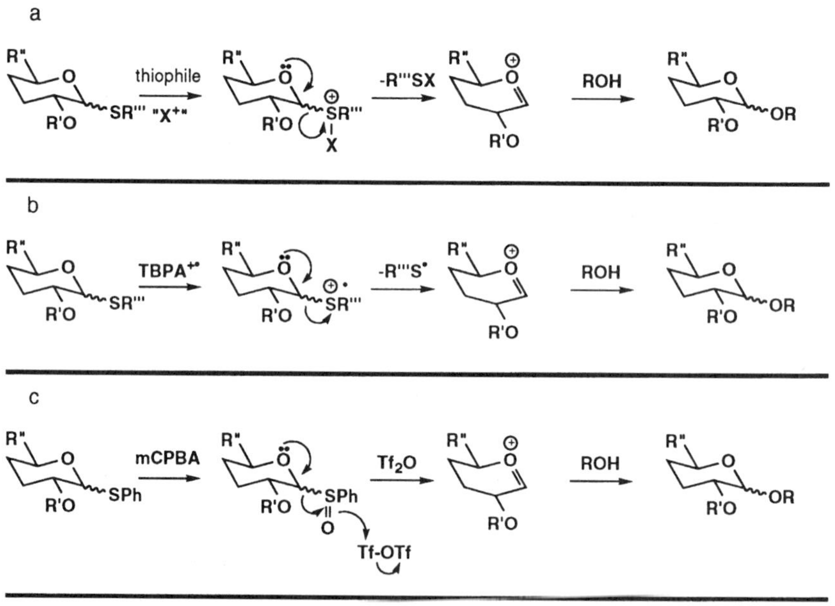

**Scheme 8** Thioglycoside-based glycosidation reactions. (a) thiophile = N-bromosuccinimide (NBS) [35], methyl triflate [38], dimethyl(methylthio)sulfonium triflate (DMTST) [39], or N-iodosuccinimide (NIS)-triflic acid [40]. (b) TBPA$^{+\cdot}$ = tris (4-bromophenyl)ammoniumyl hexachloroantimonate [42]. (c) mCPBA = m-chloroperbenzoic acid [43].

thioglycosides. The initial studies involved heavy-metal salts, such as those of mercury, copper, and lead [37]. These reagents, however, suffer from low reactivity and inconvenience and have not been extensively employed.

Recently, several more effective activators have been introduced, including NBS [35], methyl triflate [38], dimethyl(methylthio)sulfonium triflate (DMTST) [39], N-iodosuccinimide-triflic acid (NIS-TfOH) [40], and trimethylsilyl triflate (TMSOTf) [41]. The

# Chemistry of Glycosyl Fluorides and Sulfides

NIS–TfOH method is thought to involve iodonation at sulfur, followed by replacement with triflic acid to give the highly reactive glycosyl triflate, which then serves as the glycosyl donor.

A rather novel method of triglycoside activation, developed by Sinaÿ [42], involves a single-electron transfer from sulfur to the activating agent tris(4-bromophenyl)ammoniumyl hexachloroantimonate (TBPA$^+$). The generated glycosyl radical cation suffers cleavage to a thiyl radical, leaving behind an oxonium species which then undergoes glycosidation in the expected fashion (see Scheme **8b**). Another new method of glycosidation using thioglycosides is that developed by Kahne [43] in which the sulfur is first oxidized by the corresponding sulfoxide and then activated further by addition of triflic anhydride to yield, in the presence of glycosyl acceptor, glycosides as depicted in Scheme **8c**. An analogous method developed by Ley [44] involves oxidation of the sulfur atom to the sulfone stage before further activation and coupling to hydroxyl components. Both the Kahne and the Ley procedures are related to chemistry developed previously for the synthesis of oxocenes using Lewis acids to activate and displace sulfoxides and sulfones adjacent to an oxygen atom [45].

*The Two-Stage Activation Procedure*

The strategy of combining the chemistry of thioglycosides with that of glycosyl fluorides for the synthesis of oligosaccharides, known as the two-stage activation procedure, was first suggested by us in 1984 [23]. Scheme **9** depicts this strategy, which employs stable

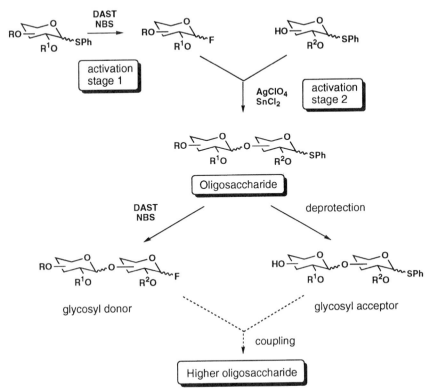

**Scheme 9** The two-stage activation procedure for glycoside bond formation [23].

phenylthioglycosides as key building blocks. In activation stage 1, the phenylthio group is converted to the more reactive glycosyl fluoride by treatment with NBS and DAST. In activation stage 2, the glycosyl fluoride is coupled to a glycosyl acceptor, which may carry a phenylthio group at the anomeric position for further propagation of the oligosaccharide chain.

The disaccharide so obtained may then be deprotected at the desired position to furnish a carbohydrate acceptor, or converted to a glycosyl fluoride which may be used as a new glycosyl donor, with NBS–DAST treatment. Reiteration of the process can produce chain elongation, whereas capping with a suitable group may lead to the desired target oligosaccharide. This strategy inherits all the advantages of the glycosyl fluoride and thioglycoside methods. Thus, the requisite thioglycosides are prepared by a variety of methods under standard glycosidation conditions and conventional functional group manipulations. The glycosyl fluorides, on the other hand, are synthesized under mild conditions from thioglycosides, directly or indirectly, through the corresponding lactols and are normally coupled under mild conditions without damaging other preexisting glycosyl bonds or sensitive functionality. The two-stage activation procedure appears particularly suited for solid-phase oligosaccharide synthesis [16,46] and automation owing to its repetitive nature and mild activation conditions. This innovation remains to be seen, however. Nevertheless, the solution version of this method has been extensively tested and proved itself amply in the synthesis of complex and sensitive glycosides and oligosac-

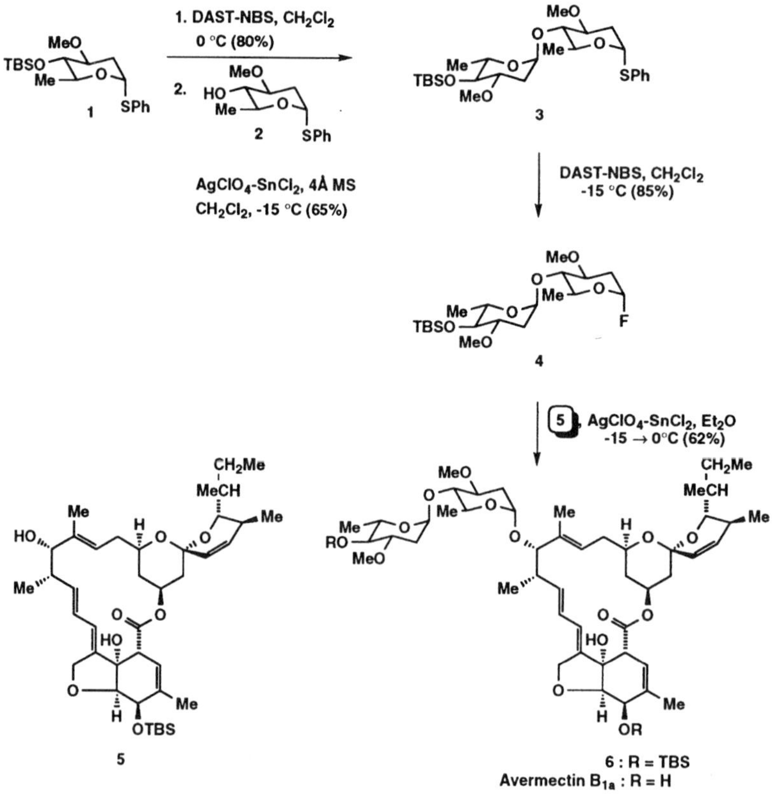

**Scheme 10** Synthesis of avermectin $B_{1a}$ [23].

# Chemistry of Glycosyl Fluorides and Sulfides

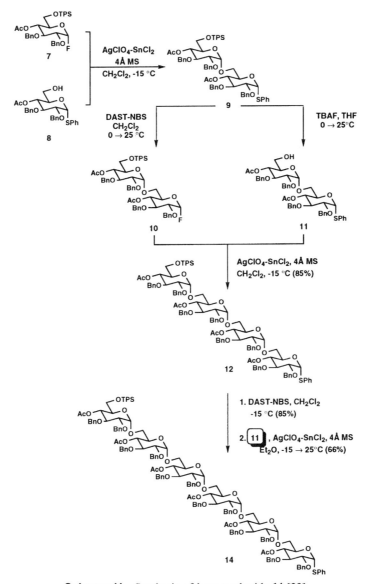

**Scheme 11** Synthesis of hexasaccharide **14** [23].

charides. The following sections demonstrate the scope and generality as well as the experimental techniques of the two-stage activation procedure for complex oligosaccharide synthesis.

## H. Application of Two-Stage Activation Procedure to Complex Molecule Synthesis

Numerous applications of the two-stage activation procedure have already been recorded. In this chapter, we highlight schematically the synthesis of the following complex mole-

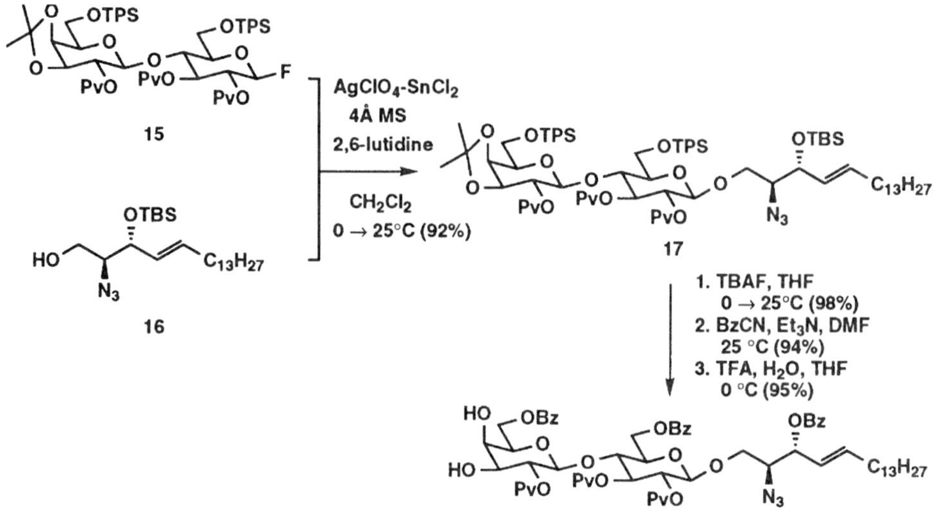

**Scheme 12** Construction of Lactosyl Sphingosine Acceptor **18** (Ref. 47).

cules: avermectin $B_{1a}$ (Scheme **10**) [23], hexasaccharide (**14**, Scheme **11**) [23], trimeric Le$^x$ (**28**, Schemes **12–14**) [47], sialyl dimeric Le$^x$ (**36**, Scheme **15**) [48], sulfated Le$^a$ tetrasaccharide (**43**, Scheme **16**) [49], globotriaosylceramide (Gb$_3$) (**50**, Scheme **17**) [50], elfamycin (**59**, Scheme **18**) [51], and NodRm-IV factor (**68**, Scheme **19**) [52]. We also provide typical experimental procedures for the preparation of glycosyl fluorides and their couplings to glycosyl acceptors.

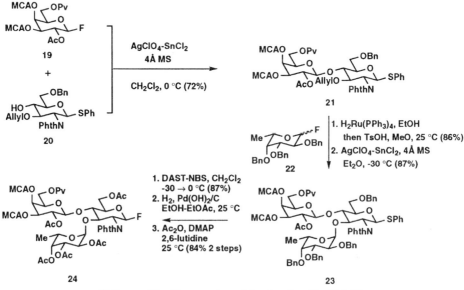

**Scheme 13** Construction of the Le$^x$ fluoride **24** [47].

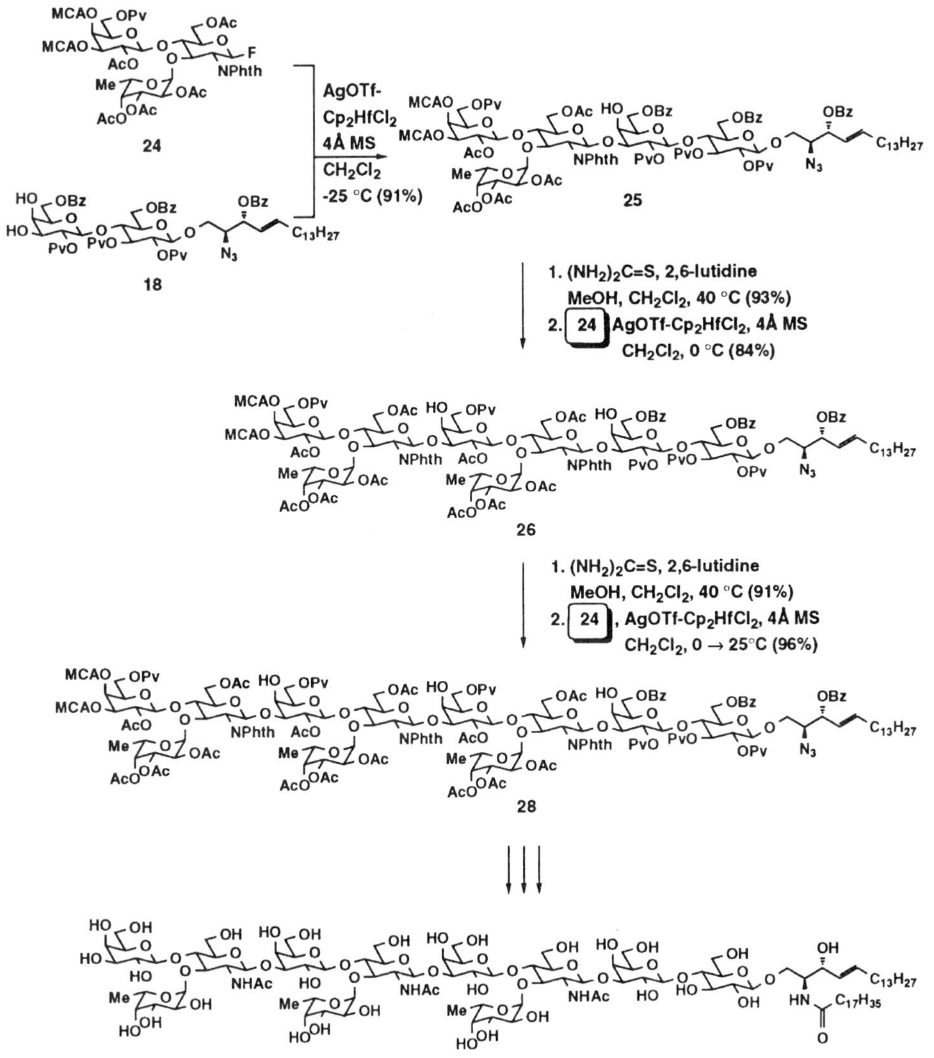

**Scheme 14** Synthesis of trimeric Le^x **29** [48].

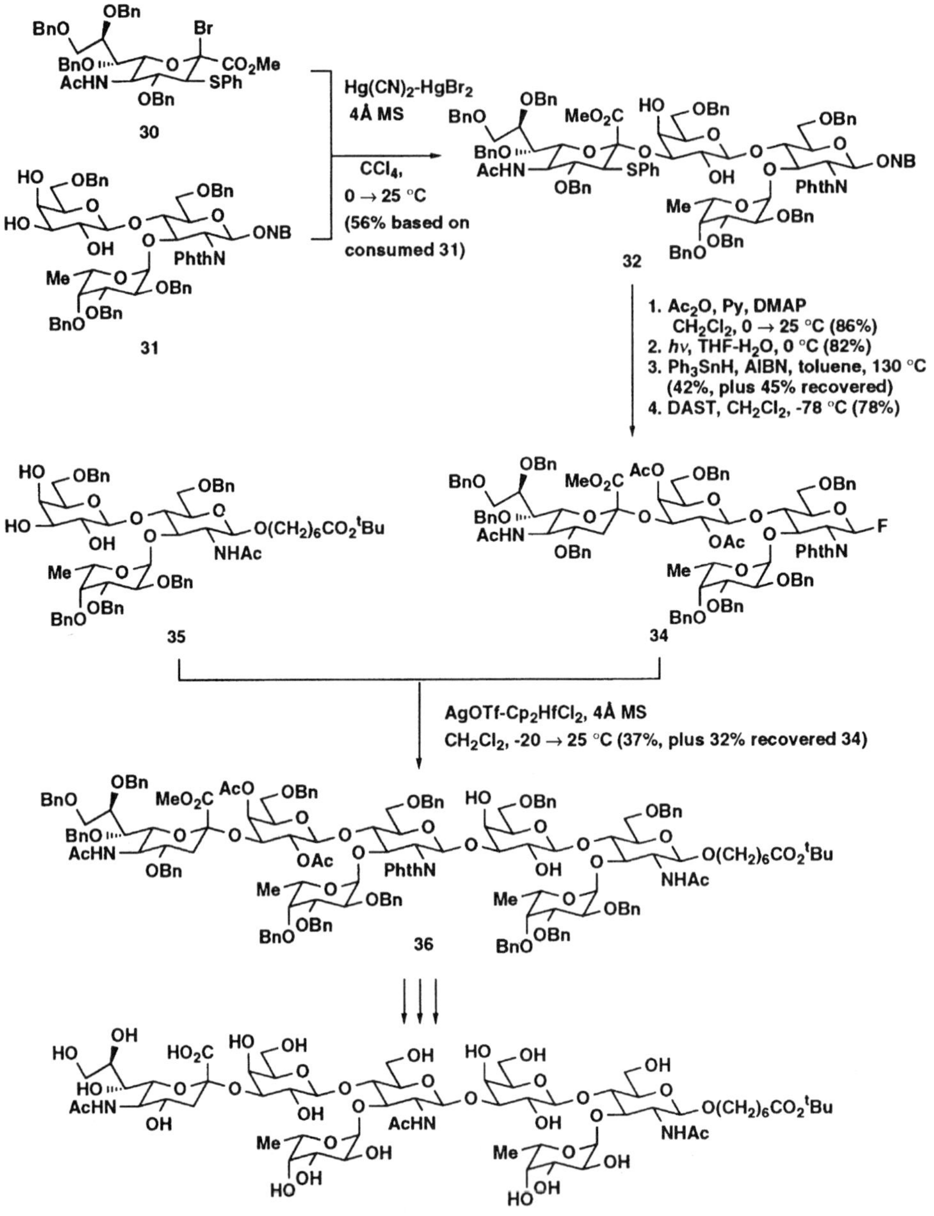

**Scheme 15** Synthesis of sialyl dimeric Le$^x$ **37** [48].

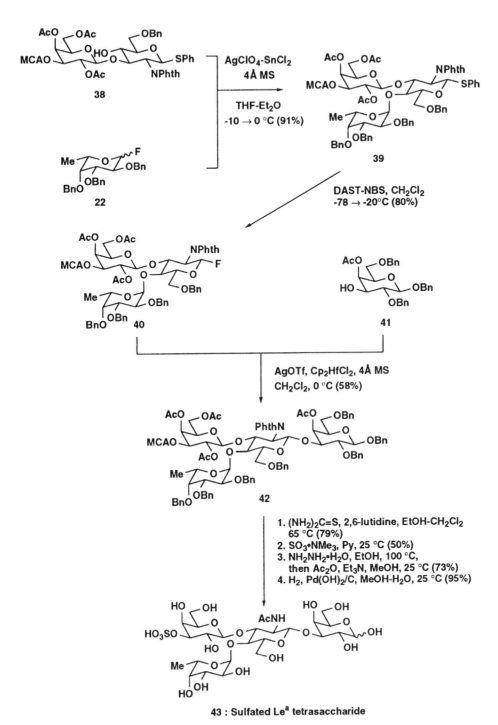

**Scheme 16** Synthesis of sulfated Le$^a$ tetrasaccharide **43** [48].

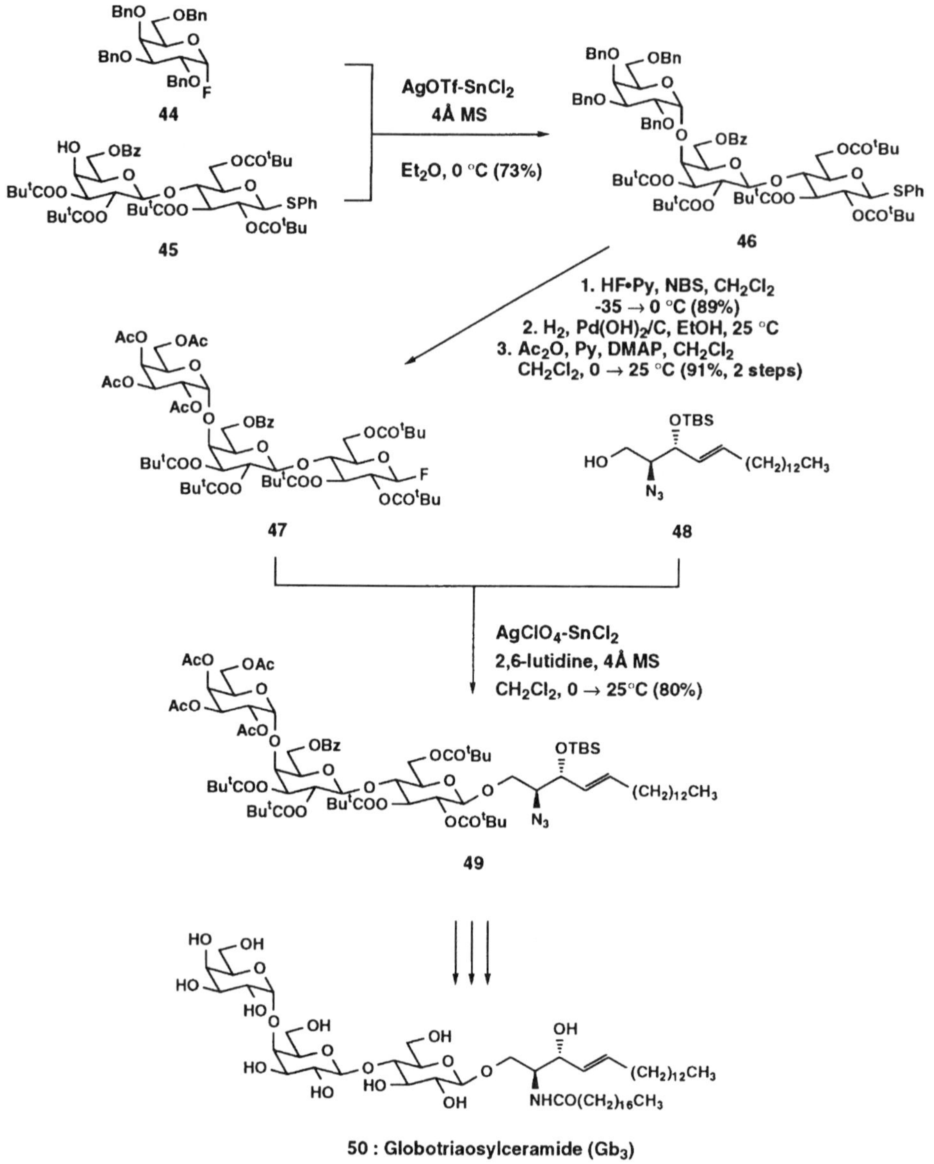

**Scheme 17** Synthesis of globotriaosylceramide (Gb$_3$) **50** [50].

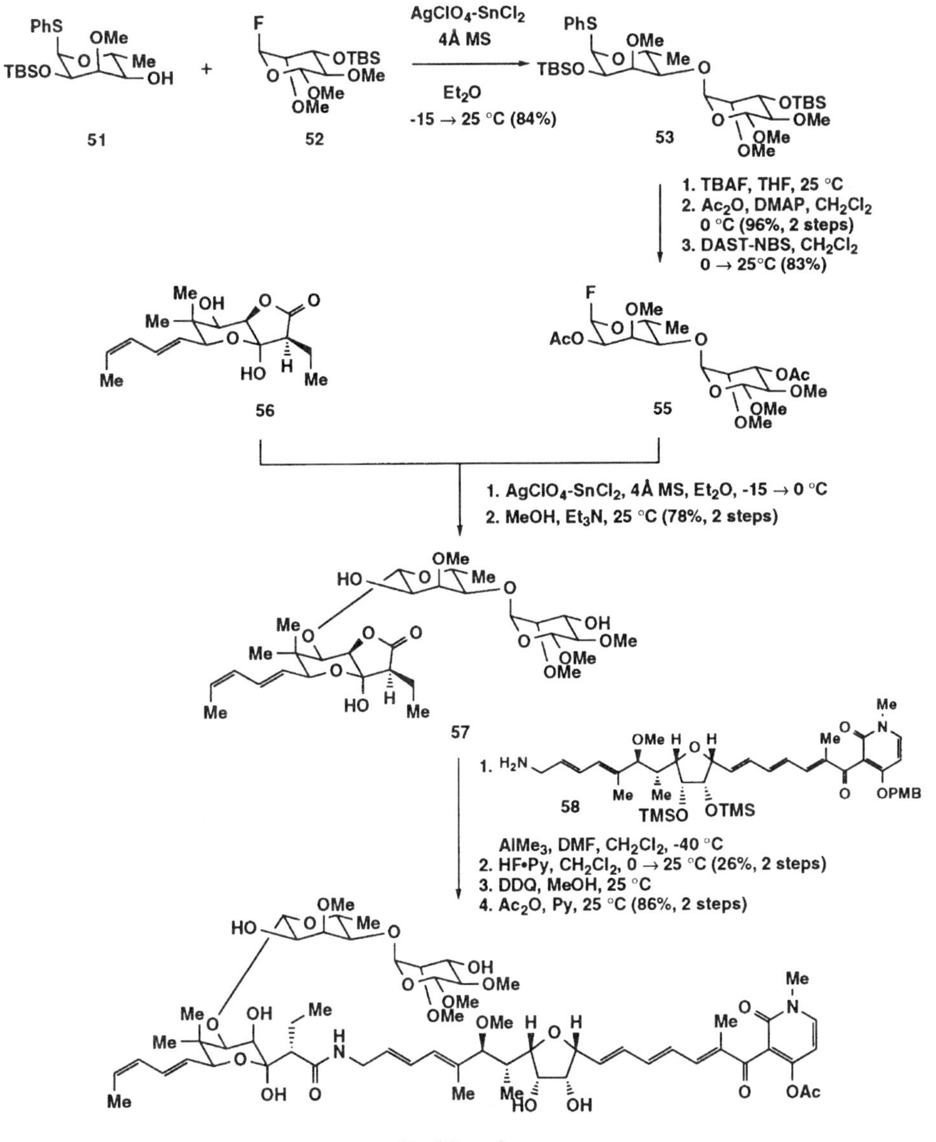

# Chemistry of Glycosyl Fluorides and Sulfides

**Scheme 18** Synthesis of elfamycin **59** [51].

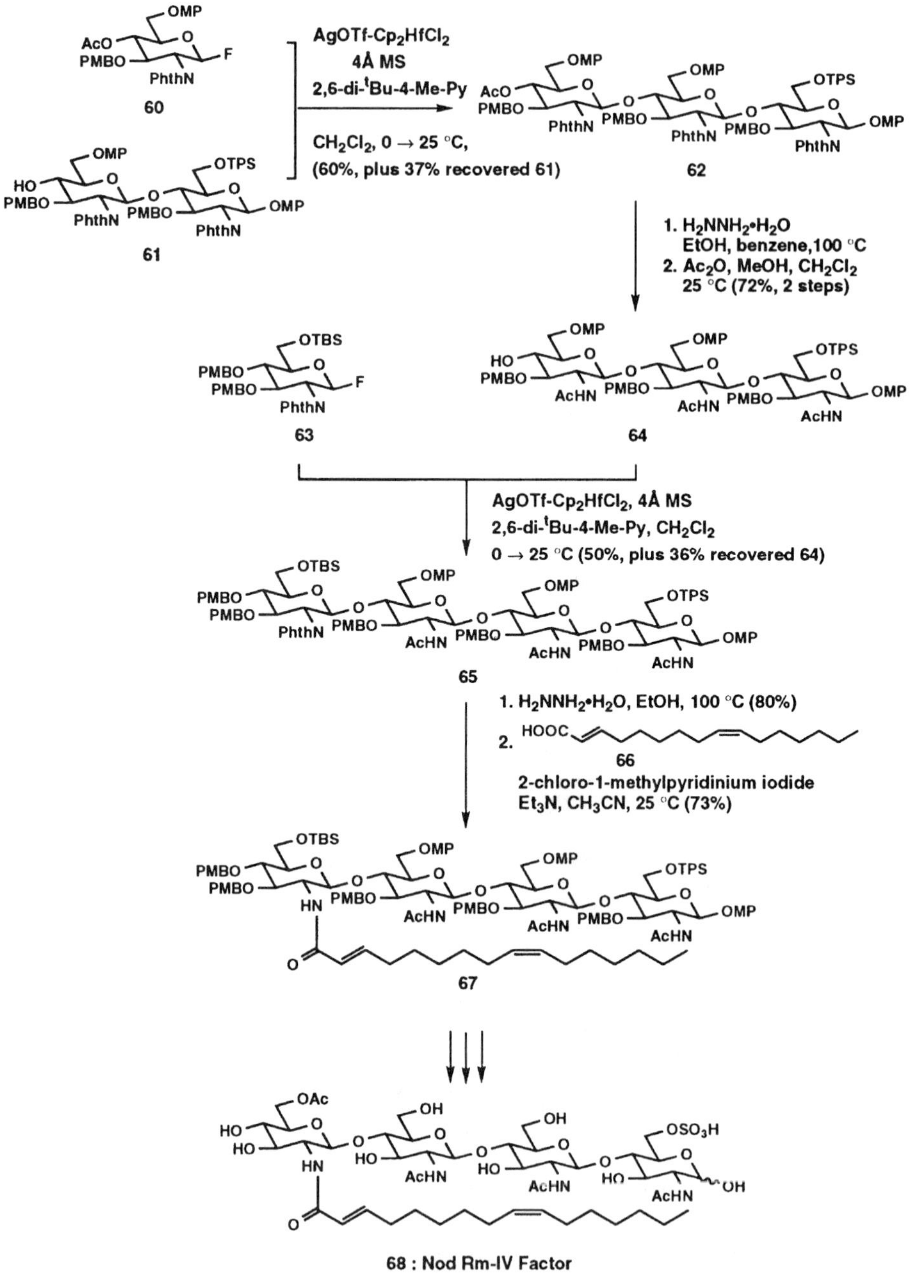

**Scheme 19** Synthesis of NodRm-IV factor **68** [52].

## III. EXPERIMENTAL PROCEDURES*

### A. Synthesis of Avermectin B$_{1a}$ [23]

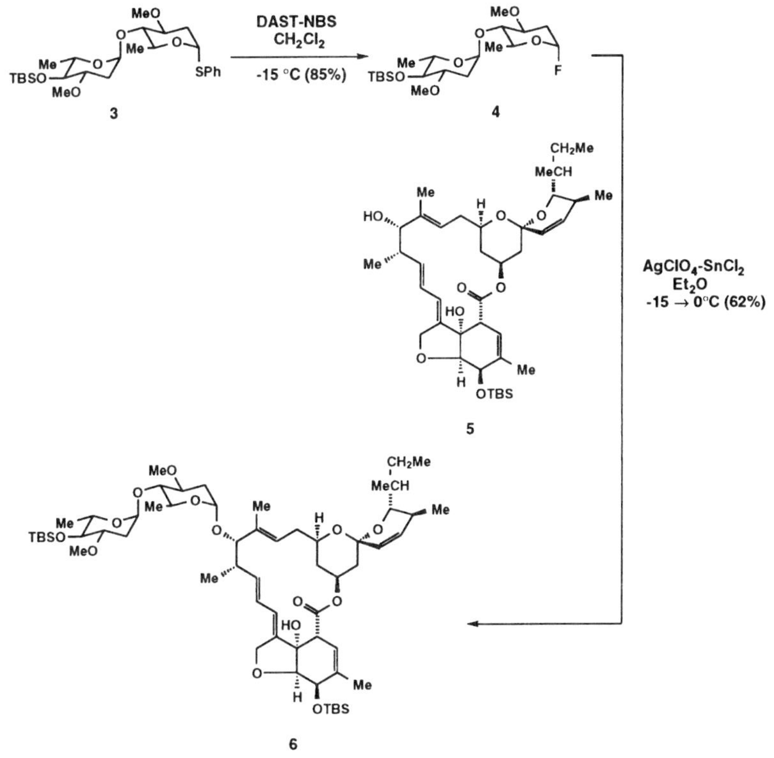

*Preparation of Disaccharide Fluoride* **4**

To a solution of phenylthio disaccharide **3** (56 mg, 0.11 mmol) in CH$_2$Cl$_2$ (2 mL) at $-15°$C was added DAST (21 µL, 0.16 mmol), followed by *N*-bromosuccinimide (25 mg, 0.14 mmol). After stirring at $-15°$C for 15 min, the reaction mixture was poured onto saturated aqueous NaHCO$_3$ solution (5 mL) and extracted with ether (3 × 10 mL). The organic extracts were combined, washed with brine (5 mL), and dried (MgSO$_4$). Evaporation of the solvent followed by flash chromatography (silica, ether–petroleum ether) gave the disaccharide fluoride **4** (40 mg, 85%) with an anomeric ratio of 5:1 (α/β).

*Coupling of Disaccharide Fluoride* **4** *with Avermectin Aglycon* **5**

To a suspension of AgClO$_4$ (12 mg, 0.06 mmol), SnCl$_2$ (11 mg, 0.06 mmol), and crushed 4-Å molecular sieves (100 mg, dried) in dry ether (3.5 mL) under argon at $-15°$C was added alcohol **5** (52 mg, 0.074 mmol) in ether (1 mL). A solution of the disaccharide fluoride **4** (24 mg, 0.057 mmol) in ether (1 mL) was added and stirring was continued at 0°C for 16 h. The reaction mixture was diluted with ether (30 mL) and filtered through Celite. The filtrate was washed with saturated aqueous NaHCO$_3$ solution (5 mL) and brine (5 mL),

---

*Optical rotations were measured at 22°–25°C.

and dried (MgSO$_4$). Evaporation of the solvent and purification by flash chromatography (silica, ether–petroleum ether) afforded avermectin B$_{1a}$ bis(silyl ether) **6** (38 mg, 62%): [α]$_D$ +29.35° (c 0.51, CH$_2$Cl$_2$).

## B. Synthesis of Hexasaccharide 14 [23]

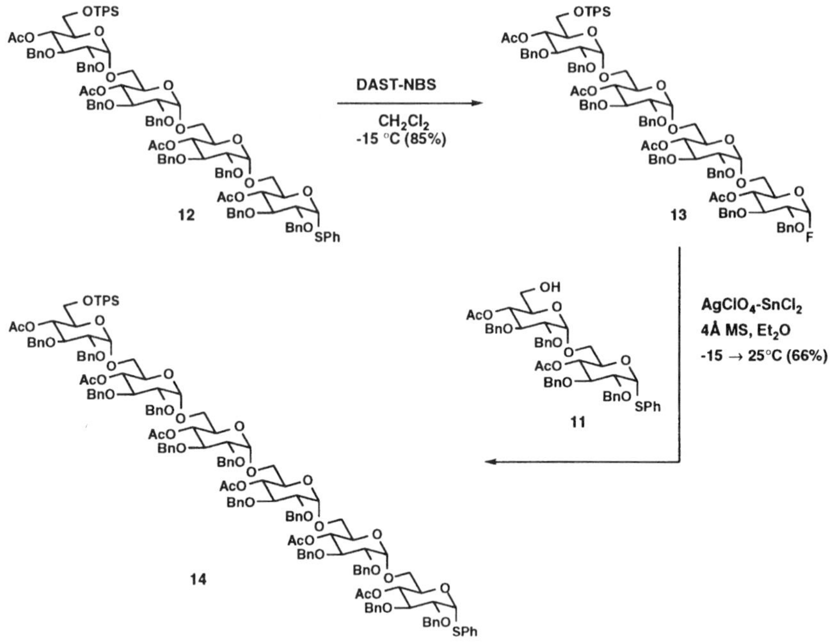

*Preparation of Tetrasaccharide Fluoride* **13**

To a solution of phenylthio tetrasaccharide **12** (570 mg, 0.3 mmol) in CH$_2$Cl$_2$ (3 mL) was added DAST (60 μL, 0.46 mmol) at −15°C. After stirring for 2 min, *N*-bromosuccinimide (70 mg, 0.39 mmol) was added and stirring was continued at −15°C for 25 min. The mixture was diluted with CH$_2$Cl$_2$ (25 mL) and poured onto an ice-cooled, saturated aqueous NaHCO$_3$ solution (3 mL). The organic layer was washed with saturated aqueous NaHCO$_3$ solution (3 mL) and brine (3 mL), dried (MgSO$_4$), concentrated, and purified by flash chromatography (silica, ether–petroleum ether) to afford tetrasaccharide fluoride **13** (463 mg, 85%): [α]$_D$ +63.94° (c 0.011, CH$_2$Cl$_2$).

*Preparation of Hexasaccharide* **14**

To a suspension of AgClO$_4$ (95 mg, 0.46 mmol), SnCl$_2$ (87 mg, 0.46 mmol), and crushed 4-Å molecular sieves (200 mg, dried) in ether (3.5 mL) at −15°C was added alcohol **11** (224 mg, 0.25 mmol) in ether (3.5 mL). After 2 min, tetrasaccharide fluoride **13** (454 mg, 0.25 mmol) in ether (3.5 mL) was added in one portion, and the reaction mixture was allowed to stir at −15°C for 2 h and at room temperature for 10 h. The reaction mixture was diluted with ether (50 mL) and filtered through Celite. The filtrate was washed with saturated aqueous NaHCO$_3$ solution (2 mL) and brine (2 mL), dried (MgSO$_4$), and evaporated to give a yellow syrup. Purification by flash chromatography (silica, ether–petroleum ether) afforded hexasaccharide **14** (442 mg, 66%): [α]$_D$ +66.50° (c 1.9, CH$_2$Cl$_2$).

## C. Synthesis of Trimeric Le$^x$ [47]

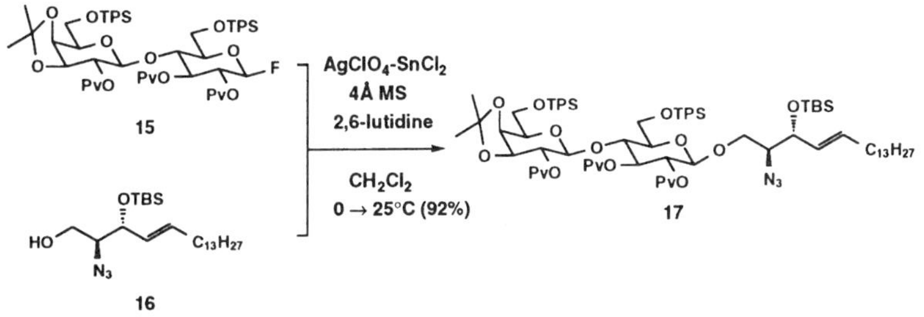

*Coupling of Disaccharide Fluoride* **15** *with Alcohol* **16**

To a suspension of AgClO$_4$ (443 mg, 2.14 mmol), SnCl$_2$ (406 mg, 2.14 mmol), and crushed 4-Å molecular sieves (400 mg) in CH$_2$Cl$_2$ (10 mL) was added a solution of fluoride **15** (1.00 g, 1.07 mmol) and alcohol **16** (376 mg, 0.867 mmol) in CH$_2$Cl$_2$ (5 mL) and 2,6-lutidine (100 μL, 0.856 mmol) at 0°C. The mixture was stirred at below 5°C for 4 h and allowed to warm to room temperature. The reaction mixture was diluted with CH$_2$Cl$_2$ (60 mL) and filtered through Celite. The filtrate was washed with saturated aqueous NaHCO$_3$ solution (2 × 20 mL) and brine (20 mL), dried (MgSO$_4$), and concentrated. Flash chromatography(silica, EtOAc–petroleum ether) provided **17** (6.47 g, 92%): mp 118°–120°C (from CH$_3$OH, water), [α]$_D$ −5.94° (*c* 3.32, CHCl$_3$).

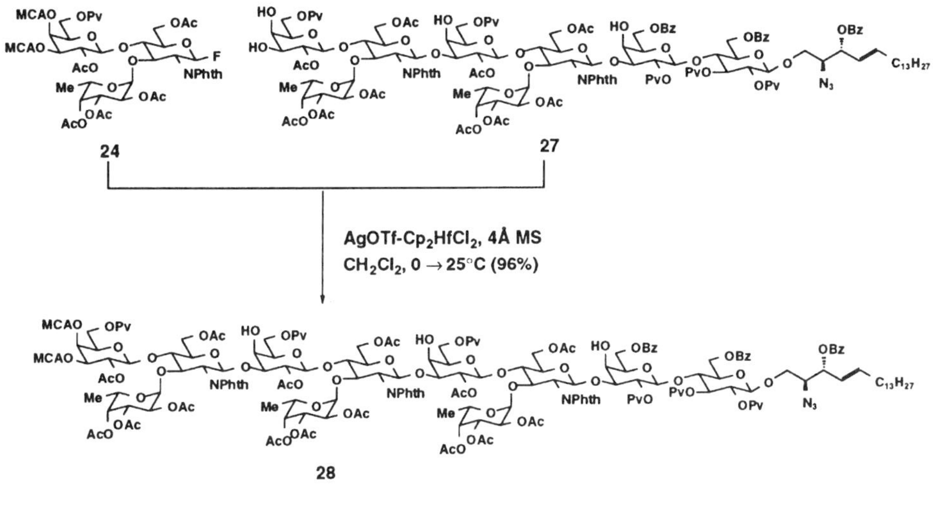

*Coupling of Trisaccharide Fluoride* **24** *with Octasaccharide Tetraol* **27**

To a suspension of AgOTf (295 mg, 1.15 mmol), Cp$_2$HfCl$_2$ (436 mg, 1.15 mmol), and crushed 4-Å molecular sieves (500 mg) in CH$_2$Cl$_2$ (1.5 mL), was added a solution of trisaccharide fluoride **24** (467 mg, 0.459 mmol) and octasaccharide tetraol **27** (230 mg, 0.077 mmol) in CH$_2$Cl$_2$ (1.5 mL) at 0°C. The mixture was stirred at below 5°C for 4 h and allowed to stir at room temperature for 1 h. The reaction mixture was diluted with CH$_2$Cl$_2$ (20 mL) and filtered through Celite. The filtrate was washed with saturated aqueous NaHCO$_3$ solution (2 × 10 mL) and brine (10 mL), and dried (MgSO$_4$), and concentrated. Flash chromatography (silica, acetone–petroleum ether-CH$_2$Cl$_2$) provided **28** (299 mg, 96%): [α]$_D$ −36.5 (*c* 1.02, CHCl$_3$).

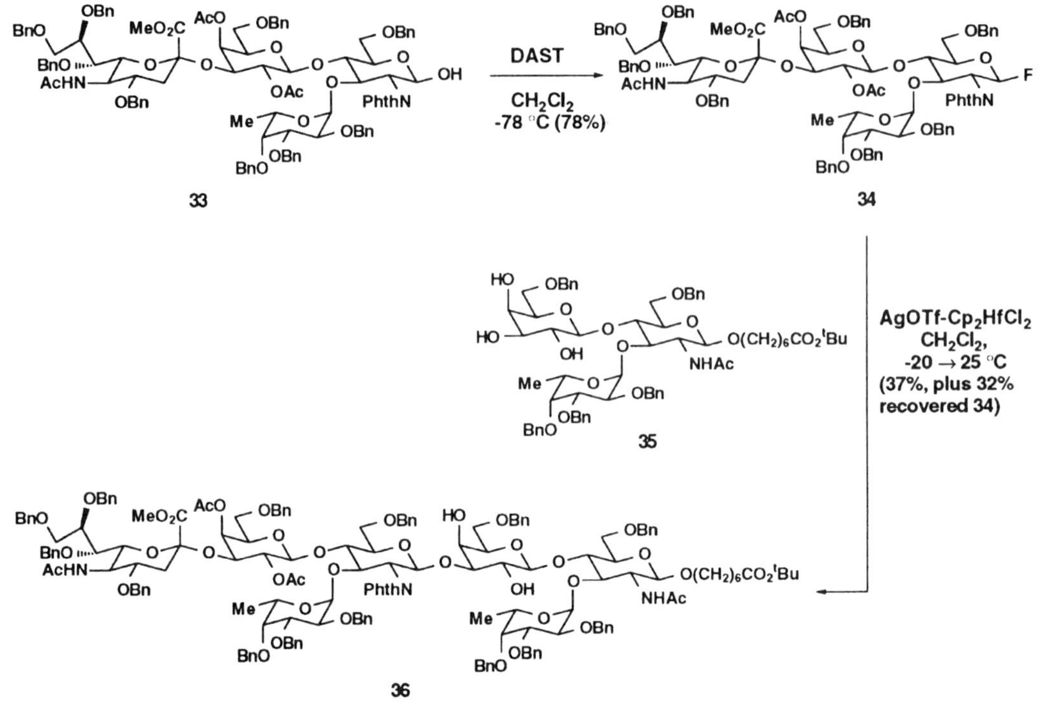

### D. Synthesis of Sialyl Dimeric Le<sup>x</sup> [48]

*Preparation of Tetrasaccharide Fluoride 34*

To a solution of lactol **33** (250 mg, 0.14 mmol) in CH$_2$Cl$_2$ (7 mL) was added DAST (35 μL, 0.27 mmol) at −78°C. After stirring at −78°C for 15 min, the reaction mixture was diluted with CH$_2$Cl$_2$ (100 mL) and washed with saturated aqueous NaHCO$_3$ solution (50 mL) and brine (50 mL). The organic layer was dried (MgSO$_4$), concentrated, and purified by flash column chromatography (silica, EtOAc–petroleum ether) to give fluoride **34** (195 mg, 78%) as a white foam. The fluoride was obtained as a 4:1 (β/α) mixture of anomers and was used for the following step without separation.

*Coupling of Fluoride 34 with Alcohol 35*

To a suspension of AgOTf (50 mg, 0.19 mmol), Cp$_2$HfCl$_2$ (80 mg, 0.21 mmol), and crushed 4-Å molecular sieves (0.3 g) in CH$_2$Cl$_2$ (4 mL) at −20°C, was added a solution of the anomeric mixture of fluoride **34** (ca β/α = 4:1, 79 mg, 0.043 mmol) followed by alcohol **35** (126 mg, 0.11 mmol) in CH$_2$Cl$_2$ (4 mL) at −20°C. The mixture was stirred and allowed to warm from −20°C to 25°C over 18 h. The reaction mixture was diluted with EtOAc (50 mL) and filtered through Celite. The filtrate was washed with saturated aqueous NaHCO$_3$ solution (50 mL) and brine (50 mL). The organic layer was dried (MgSO$_4$), concentrated, and purified by flash column chromatography (silica, EtOAc–petroleum ether) to afford the heptasaccharide **36** (48 mg, 37%) as a white foam: [α]$_D$ −9.4° (*c* 1.0, CHCl$_3$), along with unreacted fluoride **34** (25 mg, 32%).

# Chemistry of Glycosyl Fluorides and Sulfides

## E. Synthesis of Sulfated Le$^a$ Tetrasaccharide [49]

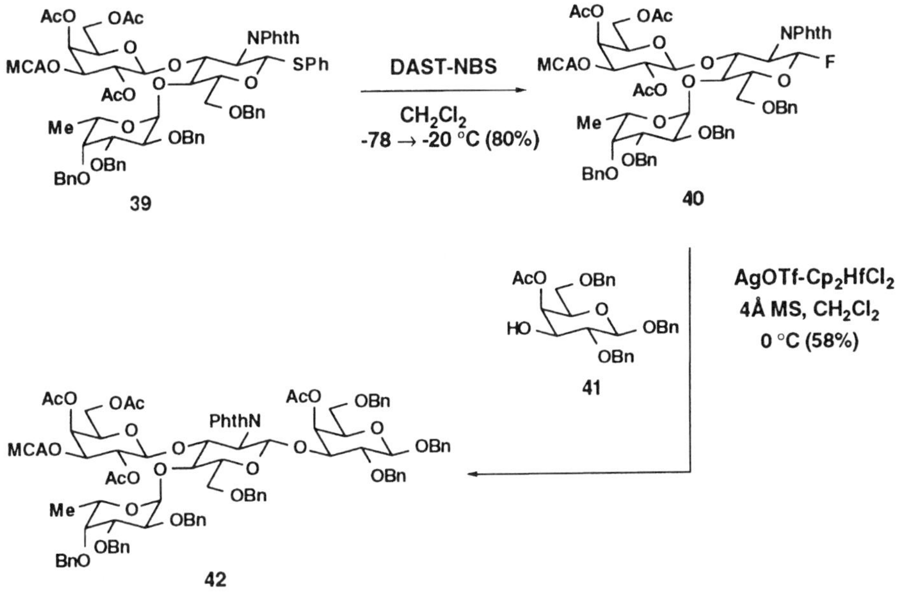

*Preparation of Trisaccharide Fluoride* **40**

To a solution of thioglycoside **39** (250 mg, 0.20 mmol) in CH$_2$Cl$_2$ (3 mL) was added DAST (80 μL, 0.62 mmol) at −78°C. After stirring for 20 min, the mixture was treated with N-bromosuccinimide (45 mg, 0.25 mmol) and allowed to warm to −20°C over 2 h. The reaction mixture was then diluted with EtOAc (25 mL) and washed with saturated aqueous NaHCO$_3$ solution and brine. The organic layer was dried (MgSO$_4$), concentrated, and purified by flash column chromatography (silica, ether–CH$_2$Cl$_2$) to give trisaccharide fluoride **40** (190 mg, 80%): [α]$_D$ −2.0° (c 1.0, CHCl$_3$).

*Coupling of Trisaccharide Fluoride* **40** *with Alcohol* **41**

To a suspension of AgOTf (173 mg, 0.70 mmol), Cp$_2$HfCl$_2$ (260 mg, 0.70 mmol), and crushed 4-Å molecular sieves (1.0 g, dried) in CH$_2$Cl$_2$ (5 mL) was added a solution of fluoride **40** (270 mg, 0.23 mmol) and alcohol **41** (335 mg, 0.70 mmol) in CH$_2$Cl$_2$ (5 mL) at 0°C. The reaction mixture was stirred at 0°C for 1 h and filtered through Celite with CH$_2$Cl$_2$ (25 mL). The filtrate was washed with saturated aqueous NaHCO$_3$ solution (50 mL) and brine (50 mL), dried (MgSO$_4$), concentrated, and purified by flash column chromatography (silica, ether–CH$_2$Cl$_2$) to give tetrasaccharide **42** (220 mg, 58%): [α]$_D$ −7.9° (c 1.0, CHCl$_3$).

## F. Synthesis of Globotriaosylceramide (Gb$_3$) [50]

*Preparation of Trisaccharide Fluoride* **47**

To a solution of thioglycoside **46** (830 mg, 0.56 mmol) in CH$_2$Cl$_2$ (5.6 mL) was added HF·pyridine complex (0.56 mL) followed by N-bromosuccinimide (110 mg, 0.62 mmol) at −35°C. The mixture was allowed to warm to 0°C over 1 h. The reaction mixture was diluted with EtOAc (100 mL) and poured onto saturated aqueous NaHCO$_3$ solution (50 mL). The organic layer was washed with saturated aqueous NaHCO$_3$ solution (25 mL) and brine (25 mL), dried (MgSO$_4$), concentrated, and purified by flash column chromatography (silica, EtOAc–petroleum ether) to afford trisaccharide fluoride (690 mg, 89%) as a white foam: $[\alpha]_D$ +30.2° (c 1.0, CHCl$_3$).

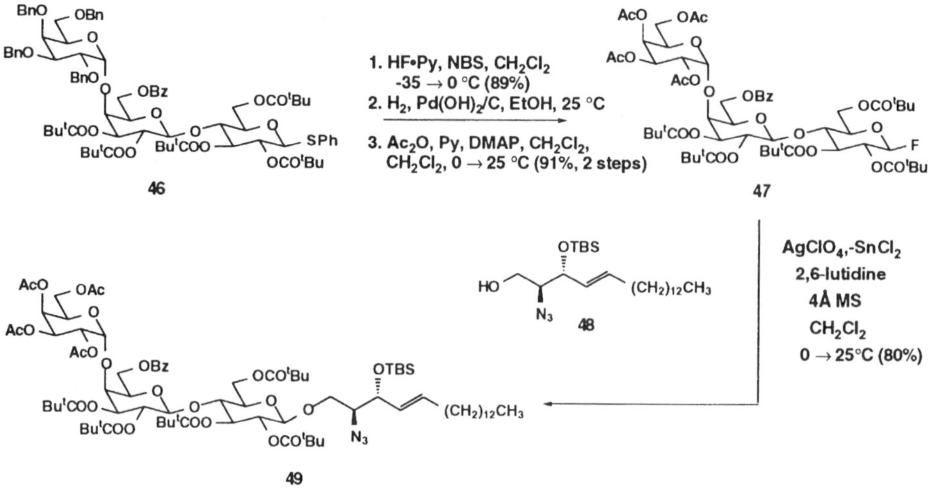

A round-bottomed flask containing a solution of resulting fluoride (690 mg, 0.50 mmol) in EtOH (7.5 mL) and 10% Pd(OH)$_2$ on charcoal (354 mg) was repeatedly evacuated and flushed with hydrogen. The reaction mixture was stirred under hydrogen for 24 h at room temperature. At the end of this period, the reaction mixture was diluted with MeOH and filtered through Celite and evaporated. The residual white solid was azeotropically dried with benzene (2 × 15 mL) and dissolved in pyridine (5 mL). 4-Dimethylamino-pyridine (10 mg) and Ac$_2$O (283 μL) were added, and the mixture was stirred overnight. The reaction mixture was diluted with EtOAc and washed with water and brine. The organic layer was dried (MgSO$_4$), concentrated, and purified by flash chromatography (silica, EtOAc–petroleum ether) to give trisaccharide fluoride **47** (543 mg, 91%, two steps) as a white foam: $[\alpha]_D$ +62.8° (c 1.0, CHCl$_3$).

*Coupling of Trisaccharide Fluoride **47** with Alcohol **48***

To a suspension of AgClO$_4$ (62 mg, 0.30 mmol), SnCl$_2$ (57 mg, 0.30 mmol), and crushed 4-Å molecular sieves (300 mg) in CH$_2$Cl$_2$ (1.0 mL), was added a solution of trisaccharide **47** (179 mg, 0.15 mmol) at 0°C. After stirring for 10 min, a solution of alcohol **48** (660 mg, 1.5 mmol) in CH$_2$Cl$_2$ (2.0 mL) and 2,6-lutidine (17 μL) were added, and stirring was continued at room temperature for 12 h. The reaction mixture was diluted with CH$_2$Cl$_2$ (25 mL) and filtered through Celite. The filtrate was diluted with EtOAc (25 mL) and washed with aqueous NaHCO$_3$ solution (25 mL) and brine (25 mL). The organic layer was dried (MgSO$_4$), concentrated, and purified by flash column chromatography (silica, ether–petroleum ether) to give **49** (194 mg, 80%) as a white foam: $[\alpha]_D$ +27.5° (c 1.6, CHCl$_3$).

## G. Synthesis of Elfamycin [51]

*Preparation of Disaccharide Fluoride **55***

To a solution of thioglycoside **54** (30 mg, 0.057 mmol) in CH$_2$Cl$_2$ (2 mL) was added DAST (0.013 mL, 0.08 mmol) and followed by *N*-bromosuccinimide (14 mg, 0.08 mmol) at 0°C.

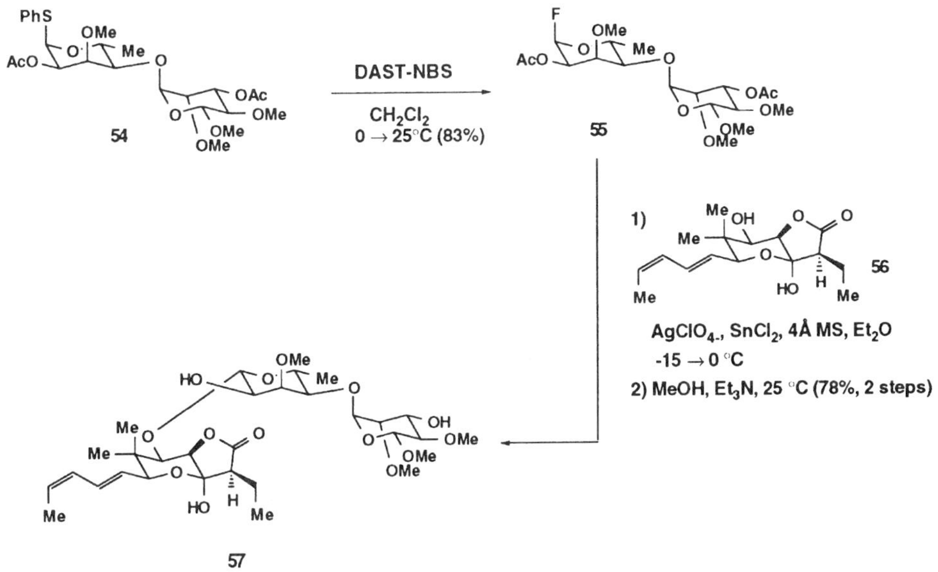

# Chemistry of Glycosyl Fluorides and Sulfides

The mixture was stirred and allowed to warm to room temperature. After stirring 1 h, the reaction mixture was diluted with ether (10 mL) and poured onto saturated aqueous $NaHCO_3$ solution (2 mL). The organic layer was washed with brine, dried ($MgSO_4$), concentrated, and purified by flash column chromatography (silica, ether–petroleum ether) to afford disaccharide fluoride **55** (21 mg, 83%): $[\alpha]_D$ −23.94° (c 1.42, $CHCl_3$).

*Preparation of Trisaccharide* **57**

To a suspension of $AgClO_4$ (12 mg, 0.06 mmol), $SnCl_2$ (11 mg, 0.06 mmol), and crushed 4-Å molecular sieves (50 mg) in ether (1 mL), was added a solution of alcohol **56** (18 mg, 0.06 mmol) in ether (0.5 mL) followed after 2 min by fluoride **55** (22 mg, 0.05 mmol) in ether (1.0 mL) at −15°C. The mixture was stirred and allowed to warm to 0°C over 6 h. The reaction mixture was diluted with ether (10 mL) and filtered through Celite. The filtrate was washed with saturated aqueous $NaHCO_3$ solution (3 × 3 mL), dried ($MgSO_4$), and concentrated to give crude glycoside. This crude glycoside was then azeotropically dried with benzene (3 × 5 mL), dissolved in dry methanol (1 mL), and treated with a catalytic amount of triethylamine. Stirring at room temperature for 48 h, removal of methanol in vacuo, and purification by flash chromatography (silica, methanol–ether) provided pure trisaccharide **57** (24 mg, 78%, two steps): $[\alpha]_D$ −37.4° (c 0.17, $CHCl_3$).

## H. Synthesis of NodRm-IV Factors [52]

*Coupling of Glycosyl Fluoride* **63** *with Trisaccharide* **64**

To a suspension of AgOTf (420 mg, 1.6 mmol), $Cp_2HfCl_2$ (610 mg, 1.6 mmol), and crushed 4-Å molecular sieves (3 g, flame-dried) in $CH_2Cl_2$ (15 mL), was added a solution of fluoride **63** (1.07 g, 1.6 mmol), alcohol **64** (500 mg, 0.32 mmol), and 2,6-di-*tert*-butyl-4-methylpyridine (16 mg, 0.08 mmol) in $CH_2Cl_2$ (15 mL) at 0°C. The mixture was stirred at 0°C for 2 h and then at 25°C for 16 h. The reaction mixture was filtered through Celite and the filtrate was diluted with EtOAc (50 mL) and washed with aqueous $NaHCO_3$ solution (20 mL) and brine (20 mL). The organic layer was dried ($MgSO_4$), concentrated, and purified by flash column chromatography (silica, acetone–dichloromethane) to give tetrasaccharide **65** (354 mg, 50%): $[\alpha]_D$ −14.4° (c 1.0, CHCl$_3$), and starting alcohol **64** (180 mg, 36%).

## REFERENCES

1. N. Sharon and H. Lis, Carbohydrates in cell recognition, *Sci. Am.* 268:82 (1993).
2. S. B. H. Kent, Chemical synthesis of peptides and proteins, *Annu. Rev. Biochem.* 547:957 (1988).
3. M. J. Gait, *Oligonucleotide Synthesis*, M. J. Gait, ed., IRL Press, Oxford, 1984.
4. (a) H. Paulsen, Advances in selective chemical syntheses of complex oligosaccharide, *Angew. Chem. Int. Ed. Engl.* 21:155 (1982); (b) R. R. Schmidt, New methods for the synthesis of glycosides and oligosaccharides—are there alternatives to the Koenigs–Knorr method? *Angew. Chem. Int. Ed. Engl.* 25:212 (1986); (c) K. Toshima and K. Tatsuta, Recent progress in *O*-glycosylation methods and its application to natural product synthesis, *Chem. Rev.* 93:1503 (1993).
5. W. Koenigs and E. Knorr, Uber einige Derivate des Traubenzuckers und der Galactose, *Ber.* 34:957 (1901).
6. R. U. Lemieux and J. Hayami, The mechanism of the anomerization of the tetra-*O*-acetyl-D-glucopyranosyl chlorides, *Can. J. Chem.* 43:2162 (1965).
7. K. Igarashi, The Koenigs–Knorr reaction, *Adv. Carbohydr. Chem. Biochem.* 34:243 (1977).
8. R. R. Schmidt and J. Michel, Facile synthesis of α- and β-*O*-glycosyl imidates; preparation of glycosides and disaccharides, *Angew. Chem. Int. Ed. Engl.* 19:731 (1980).
9. R. H. Amvam-Zollo, Dissertation, Universität Orleans (1983).
10. R. R. Schmidt and J. Michel, Direct *O*-glycosyl trichloroacetimidate formation. Nucleophilicity of the anomeric oxygen atom, *Tetrahedron Lett.* 25:821 (1984).
11. R. U. Lemieux and A. R. Morgan, The synthesis of β-D-glucopyranosyl 2-deoxy-α-D-*arabino*-hexopyranoside, *Can. J. Chem.* 43:2190 (1965).
12. J. Theim, H. Karl, and J. Schwentner, Synthese α-verknüphter 2′-deoxy-2′-iododisaccharide, *Synthesis* p. 696 (1978).
13. G. Janrand, J.-M. Beau, and P. Sinaÿ, Glycosyloxyselenation–deselenation of glycals: A new approach to 2′-deoxy-disaccharide, *J. Chem. Soc. Chem. Commun.* p. 572 (1981).
14. Y. Ito and T. Ogawa, Sulfenate esters as glycosyl acceptors: A novel approach to the synthesis of 2-deoxyglycosides, *Tetrahedron Lett.* 28:2723 (1987).
15. R. L. Halcomb and S. J. Danishefsky, On the direct epoxydation of glycals: Application of a reiterative strategy for the synthesis of β-linked oligosaccharide, *J. Am. Chem. Soc.* 111:6661 (1989).

16. S. J. Danishefsky, K. F. McClure, J. T. Randolf, and R. B. Ruggeri, A strategy for the solid-phase synthesis of oligosaccharides, *Science* 260:1307 (1993) and references cited therein.
17. B. Frasier-Reid, P. Konradsson, D. R. Mootoo, and U. Udodong, Direct elaboration of pent-4-enyl glycosides into disaccharides, *J. Chem. Soc. Chem. Commun.* p. 823 (1988).
18. T. J. Martin and R. R. Schmidt, Efficient sialylation with phosphite as leaving group, *Tetrahedron Lett.* 33:6123 (1992).
19. (a) H. Kondo, Y. Ichikawa, and C.-H. Wong, β-Sialyl phosphite and phosphoramidite: Synthesis and application to the chromoenzymatic synthesis of CMP-sialic acid and sialyl oligosaccharides, *J. Am. Chem. Soc.* 114:8748 (1992); (b) M. M. Sim, H. Kondo, and C.-H. Wong, Synthesis and use of glycosyl phosphites: An effective route to glycosyl phosphates, sugar nucleotides, and glycosides, *J. Am. Chem. Soc.* 115:2260 (1993).
20. (a) C.-H. Wong, Enzymatic catalyst in organic chemistry, *Science* 244:1145 (1989); (b) Y. Ichikawa, G. C. Look, and C.-H. Wong, Enzyme-catalyzed oligosaccharide synthesis, *Anal. Biochem.* 202:215 (1992).
21. Y. Ichikawa, J. L.-C. Liu, G.-J. Shen, and C.-H. Wong, A highly efficient multienzyme system for the one-step synthesis of sialyl trisaccharide: In situ generation of sialic acid and N-acetyllactosamine coupled with regeneration of UDP-glucose, UDP-galactose, and CMP-sialic acid, *J. Am. Chem. Soc.* 113:6300 (1991).
22. Y. Ichikawa, Y.-C. Lin, D. P. Dumas, G.-J. Shen, E. Garcia-Junceda, M. A. Willimas, R. Bayer, C. Ketcham, L. E. Walker, J. C. Paulson, and C.-H. Wong, Chemical–enzymatic synthesis and confrontational analysis of sialyl Lewis x and derivatives, *J. Chem. Soc. Chem.* 114:9283 (1992).
23. K. C. Nicolaou, R. E. Dolle, D. P. Papahatjis, and J. L. Randall, Practical synthesis of oligosaccharides. Partial synthesis of avermectin $B_{1a}$, *J. Am. Chem. Soc.* 106:4189 (1984).
24. T. Mukaiyama, Y. Murai, and S. Shoda, An efficient method for glucosylation of hydroxy compounds using glucopyranosyl fluoride, *Chem. Lett.* p. 431 (1981).
25. (a) H. Kunz, Stereoselective glycosylation of alcohols and silyl ethers using glycosyl fluorides and boron trifluoride etherate, *Helv. Chim. Acta* 68:283 (1985); (b) H. Kunz and H. Waldmann, Directed stereoselective synthesis of α- and β-N-acetyl neuraminic acid–galactose disaccharides using 2-chloro and 2-fluoro derivatives of neuraminic acid allyl ester, *J. Chem. Soc. Chem. Commun.* p. 638 (1985).
26. S. Hashimoto, M. Hayashi, and R. Noyori, Glycosylation using glucopyranosyl fluorides and silicon-based catalysts. Solvent dependency of the stereoselection, *Tetrahedron Lett.* 25:1379 (1984).
27. K. Suzuki, H. Maeta, T. Matsumoto, and G. Tsuchihashi, New glycosidation reaction 2. Preparation of 1-fluoro-D-desosamine derivative and its efficient glycosidation by the use of $Cp_2HfCl_2$-$AgClO_4$ as the activator, *Tetrahedron Lett.* 29:3571 (1988).
28. S. Hosono, W.-S. Kim, H. Sasai, and M. Shibasaki, A new glycosidation procedure utilizing rare earth salts and glycosyl fluorides, with or without the requirement of Lewis acids, *J. Am. Chem. Soc.* (in press).
29. W. Rosenbrook, D. A. Riley, and P. A. Lartey, A new method for the synthesis of glycosyl fluorides, *Tetrahedron Lett.* 26:3 (1985).
30. G. H. Posner and S. R. Haines, A convenient, one-step, high-yield replacement of an anomeric hydroxyl group of a fluorine atom using DAST. Preparation of glycosyl fluorides. *Tetrahedron Lett.* 26:5 (1985).
31. (a) R. J. Ferrier, R. W. Hay, and N. Vethaviyasar, A potentially versatile synthesis of glycosides, *Carbohydr. Res.* 27:55 (1973); (b) R. J. Ferrier and S. R. Haines, A synthesis of 1,2-*trans*-related glycofuranosyl acetates, *Carbohydr. Res.* 127:157 (1984).
32. S. Hanessian, C. Bacquet, and N. Lehong, Chemistry of the glycosidic linkage. Exceptionally fast and efficient formation of glycosides by remote activation, *Carbohydr. Res.* 80:C17 (1980).
33. F. Weygand, H. Ziemann, and H. J. Bestmann, Eine neue Method zur Dorstellung von Acetobrom zuckern, *Chem. Ber.* 91:2524 (1958).
34. M. L. Wolfrom and W. Groebke, Tetra-*O*-acylglycosyl chlorides from 1-thioglycosides and their conversion to penta-*O*-acyl esters, *J. Org. Chem.* 28:2986 (1963).

35. K. C. Nicolaou, S. P. Seitz, and D. P. Papahatjis, A mild and general method for the synthesis of O-glycosides, *J. Am. Chem. Soc. 105*:2430 (1983).
36. S. Hanessian and Y. Guindon, Chemistry of the glycosidic linkage. Direct conversion of glycosides into 1-thioglycosides by use of [alkyl (or aryl)thio] trimethylsilanes, *Carbohydr. Res. 86*:C3 (1980).
37. P. Fügedi, P. J. Garegg, H. Lönn, and T. Norberg, Thioglycosides as glycosylating agents in oligosaccharide synthesis, *Glycoconjugate J. 4*:97 (1987).
38. H. Lönn, Synthesis of a tri- and a hepta-saccharide which contain α-L-fucopyranosyl groups and are part of the complex type of carbohydrate moiety of glycoproteins, *Carbohydr. Res. 139*:105 (1985).
39. P. Fügedi and P. J. Garegg, A novel promoter for the efficient construction of 1,2-*trans*-linkages in glycosides synthesis, using thioglycosides as glycosyl donors, *Carbohydr. Res. 149*:C9 (1986).
40. G. H. Veeneman, S. H. van Leeuwen, and J. H. van Boom, Iodonium ion promoted reactions at the anomeric centre. II. An efficient thioglycoside mediated approach toward the formation of 1,2-*trans*-linked glycosides and glycosidic esters, *Tetrahedron Lett. 31*:1331 (1990).
41. Y. Ito and T. Ogawa, Sulfenate esters as glycosyl acceptors: A novel approach to O-glycosides from thioglycosides and sulfenate esters, *Tetrahedron Lett. 28*:4701 (1987).
42. P. Sinaÿ, Recent advances in glycosylation reactions, *Pure Appl. Chem. 63*:519 (1991).
43. D. Kahne, S. Walker, Y. Cheng, and D. Van Engen, Glycosylation of unreactive substrates, *J. Am. Chem. Soc. 111*:6881 (1989).
44. (a) D. S. Brown, S. V. Ley, and S. Vile, Preparation of cyclic ether acetals from 2-benzenesulfonyl derivatives: A new mild glycosidation procedure, *Tetrahedron Lett. 29*:4873 (1988); (b) D. S. Brown, S. V. Ley, S. Vlle, and M. Thompson, Use of 2-phenylsulfonyl cyclic ethers in the preparation of tetrahydropyran and tetrahydrofuran acetals and in some glycosidation, *Tetrahedron 47*:1329 (1991).
45. K. C. Nicolaou, M. E. Duggan, and C.-K. Hwang, New synthetic technology for the construction of oxocenes, *J. Am. Chem. Soc. 108*:2468 (1986).
46. (a) J. M. Frechet and C. Schuerch, Solid-phase synthesis of oligosaccharides III. Preparation of some derivatives of di- and trisaccharides via a simple alcoholysis reaction, *Carbohydr. Res. 22*:399 (1972); (b) G. H. Veeneman, S. Notermans, R. M. J. Liskamp, G. A. van der Marcel, and J. H. van Boom, Solid-phrase synthesis of a naturally occurring β-(1→5)-linked D-galacto furanosyl heptamer containing the artificial linkage arm L-homoserine, *Tetrahedron Lett. 28*:6695 (1987); (c) S. P. Douglas, D. M. Whitfield, and J. J. Krepinsky, Polymer-supported solution synthesis of oligosaccharides, *J. Am. Chem. Soc. 113*:5095 (1991).
47. K. C. Nicolaou, T. J. Caulfield, H. Kataoka, and N. A. Stylianides, Total synthesis of the tumor-associated Le[x] family of glycosphingolipids, *J. Am. Chem. Soc. 112*:3693 (1990).
48. K. C. Nicolaou, C. W. Hummel, and Y. Iwabuchi, Total synthesis of sialyl dimeric Le[x], *J. Am. Chem. Soc. 114*:3126 (1992).
49. K. C. Nicolaou, N. J. Bockovich, and D. R. Carcanague, Total synthesis of sulfated Le[x] and Le[a]-type oligosaccharide selectin ligands, *J. Am. Chem. Soc. 115*:8843 (1990).
50. K. C. Nicolaou, T. J. Caulfield, and H. Kataoka, Total synthesis of globotriaosylceramide (Gb$_3$) and lysoglobotriaosylceramide, *Carbohydr. Res. 202*:177 (1990).
51. (a) R. E. Dolle and K. C. Nicolaou, Total synthesis of elfamycins: Aurodox and elfotomycin. 1. Strategy and construction of key intermediates, *J. Am. Chem. Soc. 107*:1691 (1985); (b) R. E. Dolle and K. C. Nicolaou, Total synthesis of elfamycins: Aurodox and efrotomycin. 2. Coupling of key intermediates and completion of the synthesis, *J. Am. Chem. Soc. 107*:1695 (1985).
52. K. C. Nicolaou, N. J. Bockovich, D. R. Carcanague, C. W. Hummel, and L. F. Even, Total synthesis of the NodRm-IV factors, the rhizobium nodulation signals, *J. Am. Chem. Soc. 114*:8701 (1992).

# 14

## Oligosaccharide Synthesis by *n*-Pentenyl Glycosides

**Bert Fraser-Reid and Robert Madsen**
*Duke University, Durham, North Carolina*

|      |                                                                                                           |     |
| ---- | --------------------------------------------------------------------------------------------------------- | --- |
| I.   | Introduction                                                                                              | 339 |
| II.  | Methods                                                                                                   | 341 |
| III. | Experimental Procedures                                                                                   | 348 |
|      | A. Preparation of NPGs by Fischer glycosidation                                                           | 348 |
|      | B. Preparation of NPGs by Koenigs–Knorr procedure                                                         | 348 |
|      | C. Preparation of NPGs from glycosyl acetates                                                             | 349 |
|      | D. Preparation of NPGs by orthoester rearrangement                                                        | 349 |
|      | E. NPG coupling with NIS–catalytic Et$_3$SiOTf                                                            | 350 |
|      | F. NPG coupling with NIS–stoichiometric Et$_3$SiOTf                                                       | 350 |
|      | G. NPG coupling with IDCP                                                                                 | 351 |
|      | H. Conversion of NPGs into vicinal dibromides                                                             | 351 |
|      | I. Reductive debromination of vicinal dibromides                                                          | 351 |
|      | J. Conversion of *n*-pentenyl 1,2-orthoesters into glycosyl bromides and coupling by the Koenigs–Knorr procedure | 352 |
|      | K. Coupling of *n*-pentenyl 1,2-orthoesters with NIS/Et$_3$SiOTf                                          | 353 |
|      | L. Preparation of glycoproteins                                                                           | 353 |
|      | References and Notes                                                                                      | 354 |

## I. INTRODUCTION

The past two decades have seen intensive research into the biological roles of oligosaccharides, and it is now clear that the multitude of chiral centers give rise to an array of structural variations that endow oligosaccharides with enormous potential for encoding biological information. Oligosaccharides are now known to be involved in functions ranging from cell–cell recognition to viral and bacterial adhesion [1]. Investigation of their biological roles requires the availability of synthetic oligosaccharides or analogues thereof;

hence, it is not coincidental that there has been a parallel explosive growth in oligosaccharide synthetic methods [2].

The key reaction in oligosaccharide synthesis is the process in which a glycosyl donor and acceptor are coupled together under the agency of an appropriate promoter. In recent years, several novel and very effective glycosyl donors and promoters have been developed [2]. As a result, syntheses of oligosaccharides comprising five units are now considered to be routine, in contrast with 15 years ago when the synthesis of a trisaccharide was still a major task.

The most commonly used glycosyl donors for oligosaccharide assembly are shown in Scheme 1 [2]. They can be divided into two groups: (1) the chlorides, bromides, fluorides,

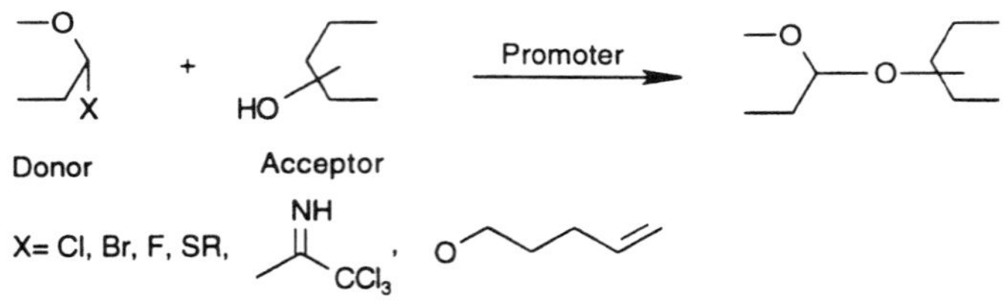

**Scheme 1**

and trichloroacetimidates, which serve only as activators of the anomeric center and must be prepared just before the coupling event; and (2) the alkyl thio and *n*-pentenyloxy groups, which serve dual functions. Thus, the latter are stable to most protecting group manipulations and, therefore, can be installed at an early stage in the synthesis, initially serving as protective groups for the anomeric center and, subsequently, as activators thereof on being presented with an electrophilic promoter [2].

Regioselectivity in glycosidations is controlled by the deployment of protective groups on the glycosyl acceptor such that (usually) only the hydroxyl groups of interest remains unprotected. The α–β stereoselectivity is highly dependent on the protecting group at the C-2 position of the glycosyl donor [2]. Participating groups such as esters and phthalimides, will engender a 1,2-*trans* relation in the product, whereas 1,2-*cis* relation by corollary, can be best obtained with nonparticipating groups, such as ethers and azides at C-2. However, for mannosyl and rhammosyl donors, α-linked products predominate, regardless of the nature of the protective group in the 2-position [2].

On the other hand, the synthesis of β-mannosides continues to be a major obstacle. The most commonly used procedure involves mannosyl bromides as donors and insoluble silver salts as promoters [3]. However, the coupling often gives appreciable amounts of the undesired α-anomer. Recent approaches to β-mannosides are based on participation from the 2-position [4], and the desired β-anomers can thereby be obtained exclusively. However, these procedures often require multiple steps for preparation of the key intermediate(s).

Fundamentally, there are two basic strategies for oligosaccharide assembly: stepwise (linear) and block (convergent) approaches. In the stepwise strategy, one monosaccharide is added at a time to the growing oligosaccharide chain. In the block strategy simpler units, typically di- or trisaccharides, are prefabricated and coupled together. The latter approach is often used for large oligosaccharides, whereas the stepwise approach is the basis for solid-phase synthesis of oligosaccharides. Although an automated solid-phase procedure is the ultimate goal of many oligosaccharide syntheses and interesting achievements have been reported in the area [5], larger oligosaccharides still have to be prepared in solution.

## II. METHODS

*n*-Pentenyl glycosides (NPGs) [6,7] **1**, belong to a special class of glycosyl donors in which the activating groups at the anomeric center can be installed at the outset to protect the anomeric center during synthetic manipulations elsewhere in the molecule, but can be chemospecifically activated on command, to provide a leaving group, thereby generating a glycosyl donor [2] that is ready for coupling to an acceptor (Scheme **2**) [6,7].

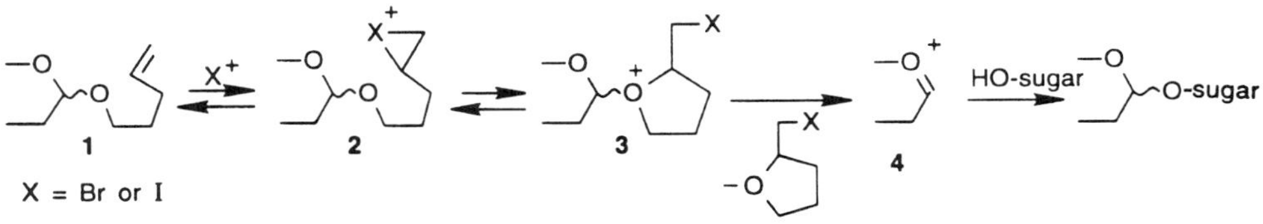

**Scheme 2**

One major advantage of NPGs is that the number of manipulative steps at the anomeric center can be minimized. The NPGs also serve to reduce the number of starting materials that have to be prefabricated, because a given NPG, often with slight modification in protecting group characteristics, can be used for several different sequential couplings. Another stepsaving device emanates from the regiospecific coupling of two NPGs under the armed–disarmed protocol (discussed later). NPGs are particularly advantageous for coupling of large segments of oligosaccharides (e.g., two tetrasaccharides), for with such targets, multiple steps pose special problems for existing protecting groups.

Preparation of NPGs may be carried out by standard procedures for making alkyl glycosides (Scheme **3**) [6]. The most direct method is the Fischer glycosidation (**5** → **1**, Scheme **3**), in which the aldose is treated with *n*-pentenyl alcohol [8] and an acid catalyst

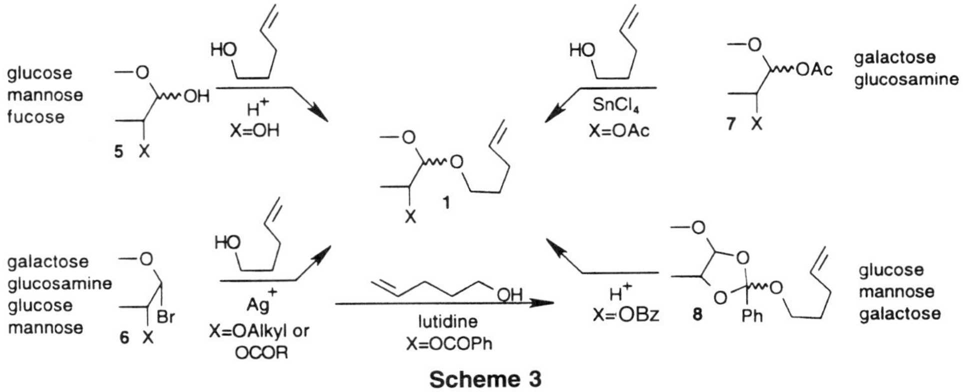

**Scheme 3**

[9]. At the completion of the reaction, the excess *n*-pentenyl alcohol can be recovered by distillation and saved for future glycosidation. As expected, the reaction gives a mixture of anomers. For example, the α/β ratio for glucose is 2:1, whereas mannose gives almost exclusively the α-product [6].

The Fischer glycosidation works very efficiently with glucose, mannose, and fucose, but gives poor yields with galactose and glucosamine. In the latter two instances, the NPGs can be prepared using standard Koenigs–Knorr **6** → **1**, or glycosyl acetate **7** → **1** procedures. In the former, silver triflate [10] is the promoter of choice for perbenzoylated glycosyl bromides [6], whereas silver carbonate is often used with the peracetylated glycosyl bromides [11].

In the absence of a silver salt, perbenzoylated glycosyl bromides can be converted into the corresponding *n*-pentenyl 1,2-orthoesters [6], which can also serve as precursor for NPGs [12] (Scheme **4**, **10** → **1**). The advantage of using these orthoesters is that several

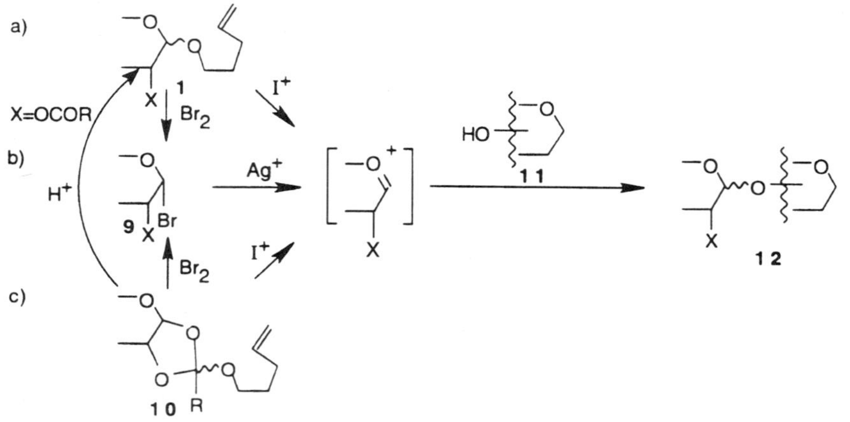

**Scheme 4**

base-promoted protecting group transformations can be carried out before the acid-induced rearrangement to an NPG occurs, which leaves a benzoate at C-2 (see Scheme **4**). Many NPGs are crystalline and, similar to other alkyl glycosides, are stable compounds that can be stored indefinitely at room temperature.

There are two various strategies for coupling of NPGs to acceptors (see Scheme **4**) [7]. Direct coupling under the agency of an halonium ion is the most widely used procedure (see Scheme **4a**). However, in special circumstances conversion into the glycosyl bromide, followed by traditional Koenigs–Knorr coupling (see Scheme **4b**), or use of an *n*-pentenyl 1,2-orthoester (see Scheme **4c**) can be advantageous.

The promoter of choice for direct coupling of NPGs is NIS/Et$_3$SiOTf [13]. The reaction is usually conducted in methylene chloride as solvent at room temperature [14]. Under these conditions, the coupling reaction is very fast, often being completed within the time it takes to sample the mixture by thin-layer chromatography (TLC). The process can be rationalized by an acid-induced heterolysis of NIS, whereby a very potent source of iodonium ion is generated (Scheme **5**). Usually, only a catalytic amount of Et$_3$SiOTf is

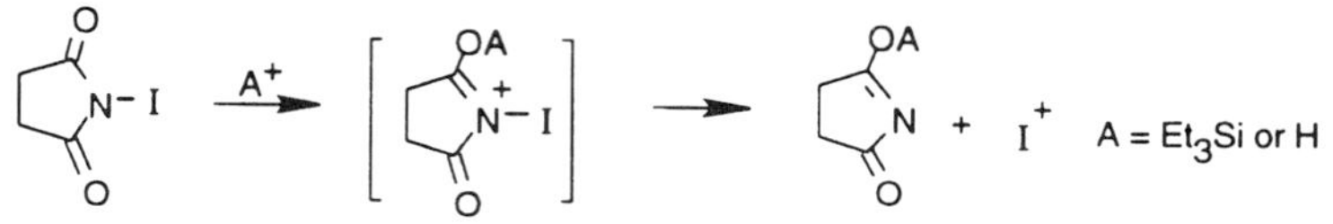

**Scheme 5**

required. Triflic acid can also be used to catalyze the heterolysis, although operationally it is more inconvenient to use than Et$_3$SiOTf [13].

## Synthesis by n-Pentenyl Glycosides

When n-halosuccinimides (NBS, NIS) are used alone without added acid catalyst, the NPG couplings are relatively slow, often requiring hours or days for completion [15]. A promoter of intermediate potency that has been useful in some cases is iodonium dicollidine perchlorate (IDCP) [16]. IDCP is not commercially available, but it is easily prepared as a stable crystalline salt by the procedure of Lemieux and Morgan [17]. It has found success for coupling some reactive (armed) NPGs [16], but is not potent enough for use with unreactive (disarmed) NPGs. For this, NIS/Et$_3$SiOTf must be employed (Scheme 6).

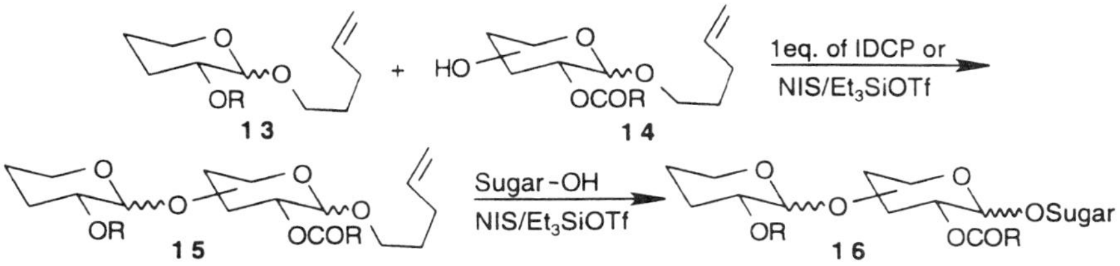

**Scheme 6**

Another problem is that some IDCP-promoted reactions have a tendency to stall, leaving substantial amounts of unreacted starting materials [6]. This is probably due to an inhibiting effect on the collidine liberated during the course of the reaction.

It was the use of NBS for NPG couplings that led to the discovery of the armed–disarmed protocol for oligosaccharide assembly [16]. Thus, it was observed that acyl-protected NPGs reacted more slowly than their alkyl-protected counterparts. These reactivity differences were analyzed, based on the inductive effect of the C-2 protecting group [18]. However, reactivity differences between substrates would not have been observed if NIS/Et$_3$SiOTf had been the promoter because, under these conditions, both NPGs would have reacted by the time the TLC sample had been taken.

However, although the original observation might not have been detectable with NIS/Et$_3$SiOTf, the success of the armed–disarmed strategy is independent of the promoter used (see, e.g., Scheme 10) [6]. Thus, even with NIS/Et$_3$SiOTf, where coupling occurs within minutes, only the cross-coupled products (e.g., **15**) is observed. Because the latter can also serve as glycosyl donors, they can be used directly for the next coupling event to obtain **16** [19].

In cases where the nature of the protecting groups does not allow the armed–disarmed strategy to be applied, two NPGs can still be coupled together by use of an intermediate dibromination step (see Scheme 7). Thus, depending on how the reaction is to

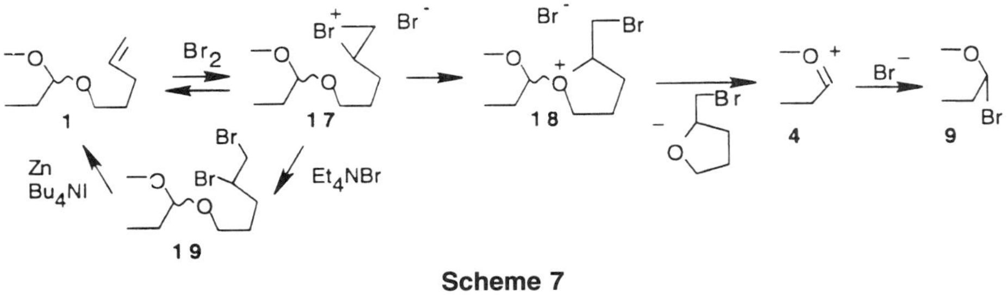

**Scheme 7**

be carried out, one can obtain either a glycosyl bromide **9** [20] or a vicinal dibromide **19** [20] (see Scheme 7). In the absence of other nucleophiles, the cyclic bromonium ion **17** leads to the oxocarbenium ion **4**, which is trapped by Br⁻, the only available nucleophile in the reaction mixture, to give the glycosyl bromide **9**. However, if excess Br⁻ is present (e.g., by the addition of Et$_4$NBr), the intramolecular reaction **17** → **18** can be overwhelmed by the bimolecular process, leading to the vicinal dibromide **19** [21]. This dibromide is a latent [22] NPG in which glycosyl donor activity has been temporarily sidetracked. Restoration of the active NPG is achieved by reductive debromination using zinc [18], samarium iodide [23a], or iodide ion [23b].

The dibromination option is useful as an indirect way of coupling two NPGs together [6] (Scheme 8). Thus, if the donor NPG **20** is titrated with bromine solution, the glycosyl

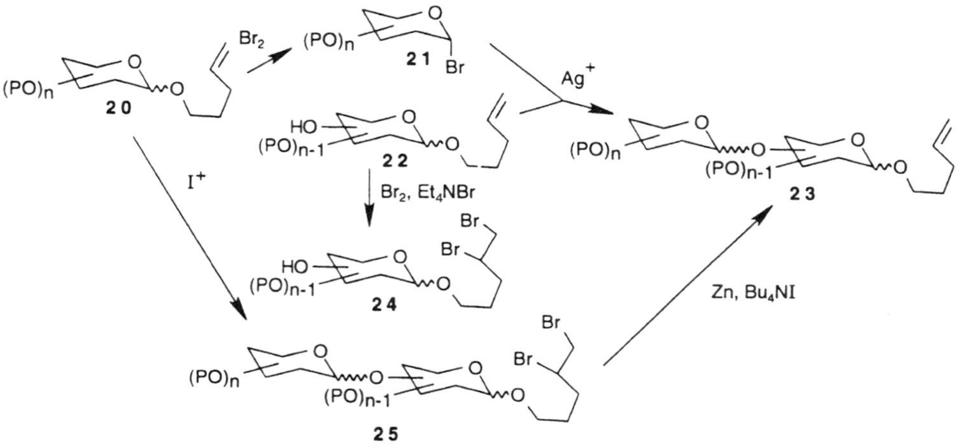

**Scheme 8**

bromide **21** can be obtained. Coupling to the NPG acceptor **22** can be carried out in situ under traditional Koenigs–Knorr conditions to obtain disaccharide **23**. Alternatively, the acceptor NPG **22** can be converted into the vicinal dibromide **24**. A coupling with the donor NPG **20** can then be accomplished under standard conditions to give **25**. Restoration of the double bond then leads to previously described **23**. The overall result from both options is the same as if two NPGs, **20** and **22** had been coupled directly under the armed–disarmed protocol.

The option of sidetracking NPG activity, as shown in Scheme 8 [18], has the effect of reducing the number of de novo precursors that must be prepared and, in turn, this makes for a more rapid synthesis. This is exemplified by assembly of the nonasaccharide portion of the high-mannose glycoprotein [14] (Scheme 9). Thus because of sidetracking, only two starting materials, **26** and **27**, are necessary to obtain the nonasaccharide **34**. The promoter for all couplings was NIS in the presence of a catalytic amount of Et$_3$SiOTf. Direct coupling of **26** and **27** gave disaccharide **28** which, after deprotection and further coupling to **27**, was elaborated into the pentasaccharide **29a**. Reductive debromination was then carried out to restore the pentenyl double bond **29a** → **29b**.

Preparation of the other segment of the nonasaccharide began with **27**. A portion of this material was converted into the dibromo acceptor **30**, which was then coupled to **27**, paving the way for the trisaccharide **32a**. Regeneration of the double bond and coupling to **26**, followed by deprotection, then gave tetrasaccharide **33b**. Finally the crucial coupling between the blocks **29b** and **33b** was carried out to give the target nonasaccharide **34** in 57% yield.

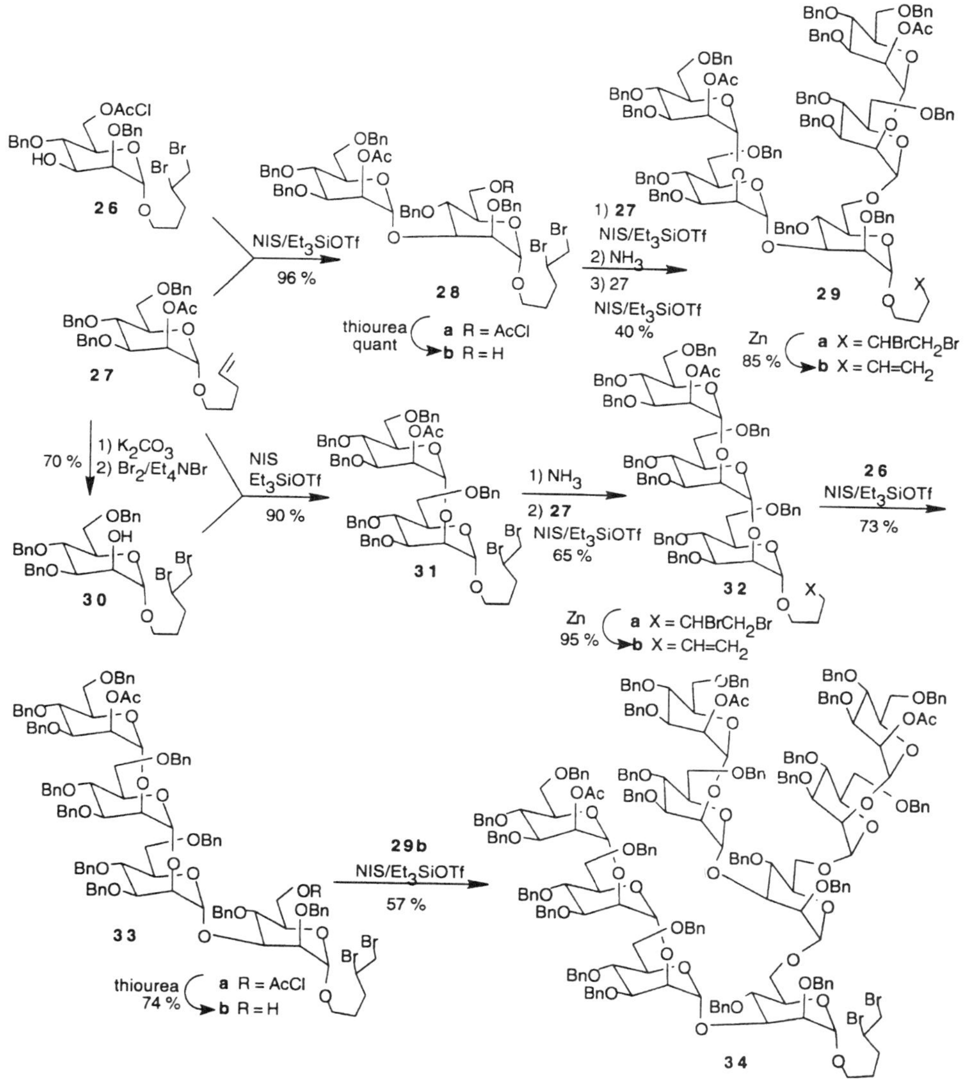

# Synthesis by n-Pentenyl Glycosides

**Scheme 9**

Only 14 steps are necessary beginning with the monosaccharide precursors **26** and **27** in this convergent assembly of nonamannan **34**. The synthetic manipulations are simplified to four ready reactions: dibromination, debromination, deesterification, and coupling [14].

Stepwise assembly of oligosaccharides with NPGs can also be accomplished as exemplified in the synthesis of the blood group substance B tetrasaccharide [24] (Scheme **10**). The four synthons **35**, **36**, **38**, and **40**, are prefabricated from n-pentenyl galactoside, N-acetyl glucosamine, and n-pentenyl fucoside, respectively. Coupling is carried out under the agency of NIS/Et$_3$SiOTf, regardless of whether the donor is armed or disarmed. Owing to neighboring group participation, coupling of NPG **35** and acceptor **36** gives the β-disaccharide **37** as the only product. After deprotection, coupling with NPG **38** affords trisac-

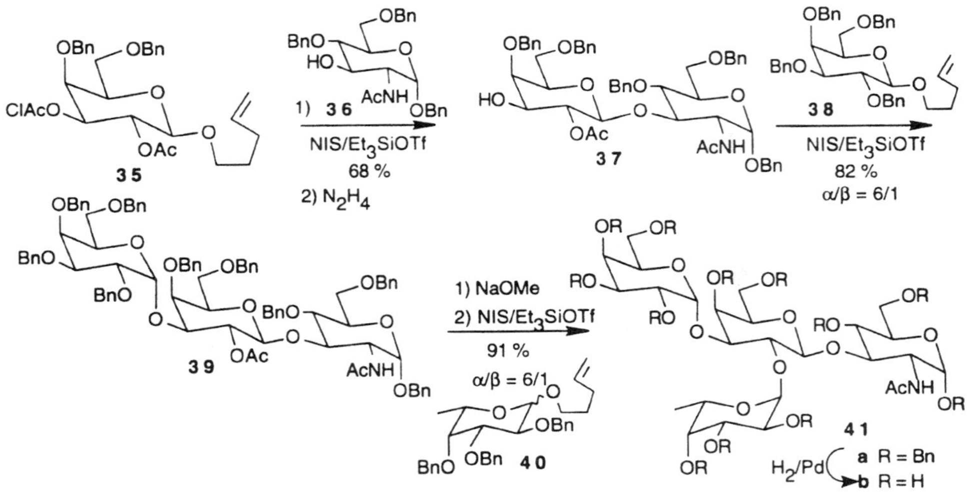

**Scheme 10**

charide **39** as a separable 6:1 mixture of α–β anomers. Subsequent deprotection and coupling with NPG **40** then gives tetrasaccharide **41a**, also as a 6:1 mixture of anomers.

A characteristic of this stepwise approach is that the chemical transformations on the growing oligosaccharide chain are kept to a minimum. Thus, once the protected monosaccharide components have been prepared, the synthetic manipulations are restricted to coupling followed by deprotection, coupling–deprotection ... and so on [24].

A special class of *n*-pentenyl glycosyl donors are *n*-pentenyl 1,2-orthoesters (Scheme **11**)

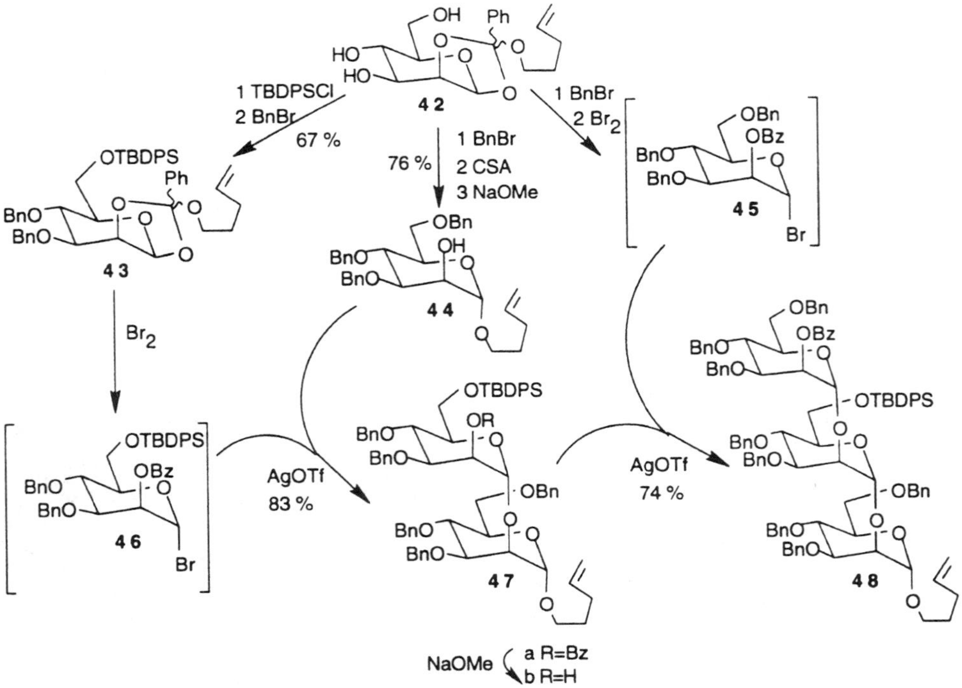

**Scheme 11**

# Synthesis by n-Pentenyl Glycosides

that can undergo some of the same coupling reactions as regular NPGs. Thus, they give glycosides or coupling products under the agency of NIS/Et$_3$SiOTf, and titration with molecular bromine leads to the formation of glycosyl bromides. An example of their utility is the rapid assembly of the trimannan **48** [12] from a single starting material, orthoester **42**.

Benzylation of a portion of **42**, followed by the acid-catalyzed rearrangement and debenzoylation, afforded the acceptor NPG **44**. A second portion of **42** was silylated and benzylated to give the differentially protected orthoester **43**. The armed–disarmed protocol could not be applied for direct coupling of **43** and **44** because the acceptor (i.e., **44**) is not disarmed. As an alternative, orthoester **43** was titrated with bromine to give the glycosyl bromide **46**, and this was used in situ for coupling to **44** under modified [10] Koenigs–Knorr conditions. The product **47a** was then debenzoylated, affording disaccharide acceptor **47b**.

A third portion of orthoester **42** now served as precursor for bromide **45**, which was coupled to **47b** to furnish the target trimannan **48**. This material was used as a building block in the synthesis of the Thy-1 GPI anchor [25].

In the mechanistic outline to Scheme 2, the intermediate oxocarbenium ion was trapped by an oxygen nucleophile to give an O-glycoside. Further experimentation has revealed that N-nucleophiles can also be applied, leading to N-glycosyl products [26]. Thus, if the reactions are carried out in the presence of acetonitrile, a Ritter reaction ensues affording nitrilium ion **49** (Scheme 12). This relatively unstable intermediate can be trapped

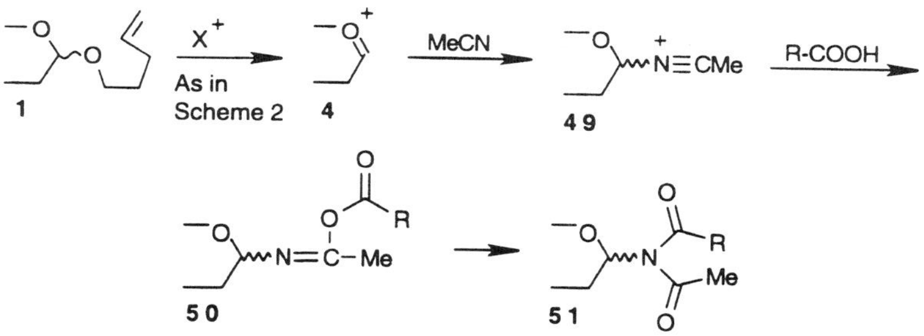

**Scheme 12**

by a carboxylic acid present in the medium, to afford an imidic anhydride **50**, which rearranges spontaneously to give the glycosyl imide **51** [27].

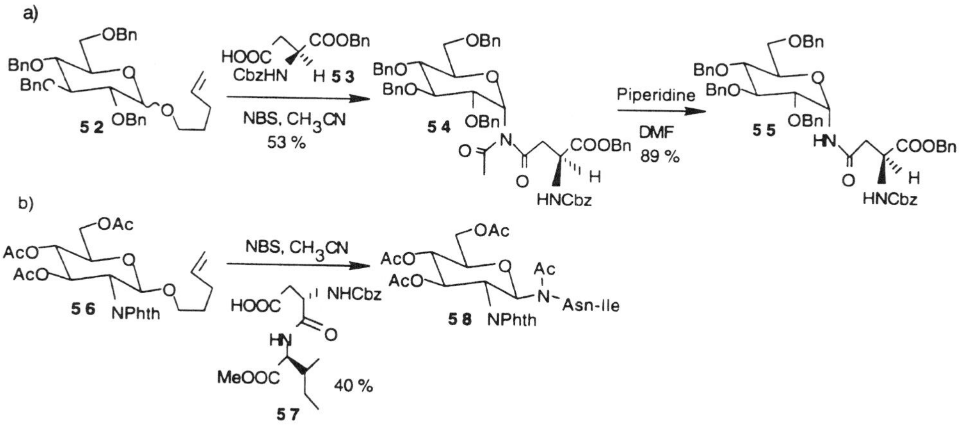

**Scheme 13**

If the carboxylic acid concerned is a protected aspartic acid, an N-linked asparagine results, an observation that suggested that the procedure could be used for the preparation of glycopeptides. Accordingly, coupling of NPG **52** and aspartic acid **53** in acetonitrile solvent yields the *N,N*-diacyl derivative **54**, which can be chemoselectively *N*-deacetylated with piperidine in DMF to give asparagine **55** [19] (Scheme 13a).

Coupling to higher peptides can also be achieved. Thus, reaction of NPG **56** with the aspartoyl isoleucine derivative **57** gave glycopeptide **58** [28]. However, simultaneous deprotection of the *N*-acetyl and *N*-phthaloyl groups in **58** is problematic without cleaving the asparagine linkage. Chemoselective conditions for these deprotections are currently being developed in our laboratory [29].

## III. EXPERIMENTAL PROCEDURES*

### A. Preparation of NPGs by Fischer Glycosidation [6]

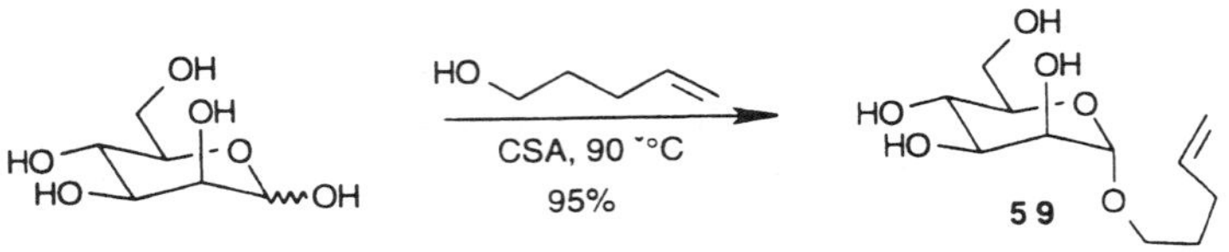

D-Mannose (15 g, 0.083 mol) was added to 4-penten-1-ol (120 mL, 100 g, 1.16 mol), and camphorsulfonic acid (200 mg) was added. The solution was heated overnight in an oil bath at 90–100°C under argon with stirring. The bulk of the pentenol was distilled under vacuum using a dry ice condenser and saved for the future glycosidations. The remaining mixture was basified with $Et_3N$ (ca. 10 drops), poured into water, and extracted with $CH_2Cl_2$. Evaporation of the aqueous phase gave a residue, which was passed through a short column of silica gel using EtOAc as eluant. Evaporation yielded materials (19.6 g, 95%) for which the $^1H$ NMR spectrum showed signals for only one anomer, H1 being a singlet at $\delta = 4.80$ consistent with 4-pentenyl-α-D-mannopyranoside **59**; $R_f = 0.30$ (20% MeOH in EtOAc), $[\alpha]_D + 54.4°$ (*c* 1.0, $CHCl_3$).

### B. Preparation of NPGs by Koenigs–Knorr Procedure [30]

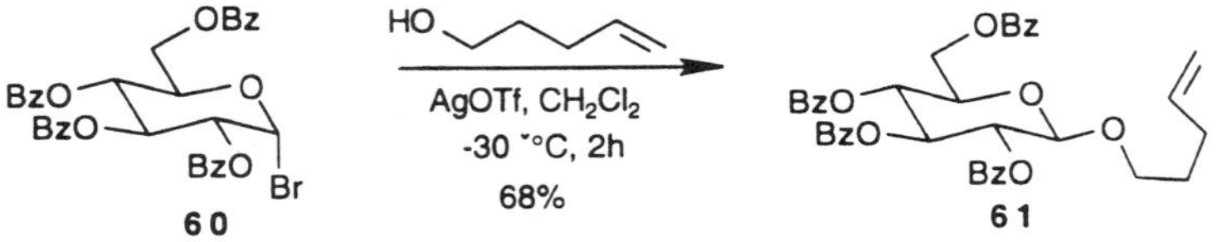

To a mixture of 4-penten-1-ol (0.5 mL, 4.84 mmol), AgOTf [10] (1.2 g, 4.67 mmol) and powdered, activated 4-Å molecular sieves (1 g) in $CH_2Cl_2$ (10 mL) was added at −30°C, through a syringe, a solution of 2,3,4,6-tetra-*O*-benzoyl-α-D-glucopyranosyl bromide **60** [31] (2.5 g, 3.79 mmol) in $CH_2Cl_2$ (4 mL) during 10 min. The reaction mixture was stirred at −30°C for 2 h, quenched with saturated aqueous $NaHCO_3$, and then filtered through Celite. The molecular sieves were washed with $CH_2Cl_2$ (20 mL). The organic phase was washed

---

*Optical rotations were measured at 20°–25°C.

# Synthesis by n-Pentenyl Glycosides

with saturated aqueous $NaHCO_3$ (20 mL), dried and concentrated. The residue was purified by flash chromatography (4:1 petroleum ether–EtOAc) to afford **61** (1.71 g, 68%): $R_f$ 0.55, mp 113°–114°C, $[\alpha]_D$ + 18.7° (c 1.0, $CHCl_3$).

## C. Preparation of NPGs from Glycosyl Acetates [6]

1,3,6-Tri-*O*-acetyl-2,4-di-*O*-benzyl-α-D-mannopyranose **62** (4.11 g, 8.45 mmol), 4-pentenol (1.80 mL, 17.4 mmol) and activated 4-Å molecular sieves in $CH_2Cl_2$ were stirred at room temperature under argon for 30 min and then tin(IV) chloride (1.0 mL, 8.5 mmol) was added. After 6 h, solid $NaHCO_3$ and $Na_2SO_4 \cdot 10H_2O$ were added and the mixture was left stirring overnight. Filtration through Celite, concentration and flash chromatography (80:20 light petroleum–EtOAc) yielded 3.46 g (80%) of **63**: $[\alpha]_D$ + 28.5° (c 1, $CHCl_3$).

## D. Preparation of NPGs by Orthoester Rearrangement [12]

The bromosugar **64** [31] in $CH_2Cl_2$ (200 mL) was treated with 2,6-lutidine (28 mL, 0.284 mol) and 4-penten-1-ol (14.11 mL, 0.137 mol). After 5 days, the reaction mixture was diluted with $CH_2Cl_2$, washed with saturated aqueous $NaHCO_3$ and brine, respectively, dried, concentrated, and flash chromatographed (4:1 petroleum ether–EtOAc) to afford 52 g (86%) of the desired orthoester **65a**. The orthoester **65a** (27.8 g, 0.042 mol) was dissolved in 230 mL of MeOH–$CH_2Cl_2$ (8:1), and treated with NaOMe (1 g, 0.021 mol) for 28 h. After completion (TLC), the solvent was removed in vacuo and the crude material was flash chromatographed (9:1, $CH_2Cl_2$–MeOH) affording 14 g (95%) of the triol **65b** ($R_f$ 0.4). A portion of the triol (7 g, 0.020 mol) in DMF (300 mL) was then treated with NaH (4.8 g of 60% oil dispersion, 0.115 mol) and BnBr (11.82 mL, 0.099 mol), respectively. After 19 h, the reaction mixture was cooled to 0°C quenched with MeOH, diluted with $Et_2O$ (800 mL), washed with $H_2O$ and saturated aqueous $NH_4Cl$, respectively, dried, concentrated, and flash chromatographed (4:1 petroleum either–EtOAc) to afford 11.5 g (94%) of **65c**. Orthoester **65c** (8.94 g, 14.35 mmol) was treated with camphorsulfonic acid (63 mg) in $CH_2Cl_2$ (2 mL) at 50°C for 6 h. $Et_3N$ (0.10 mL) was then added to the reaction mixture and the solution concentrated in vacuo. The crude residue was dissolved in 35 mL of MeOH–$CH_2Cl_2$ (7:1) and treated with NaOMe (600 mg, 11.4 mmol) at room temperature for 2 h. The solution was then cooled to 10°C, neutralized by the addition of 1% HCl in $CH_3OH$, and evaporated in vacuo. Flash chromatography (3:1 petroleum ether–EtOAc) of the residue gave the alcohol **66** (6.01 g, 81%): $R_f$ 0.25, $[\alpha]_D$ +47.6° (c 1.4, $CHCl_3$).

## E. NPG Coupling with NIS–Catalytic Et₃SiOTf [14]

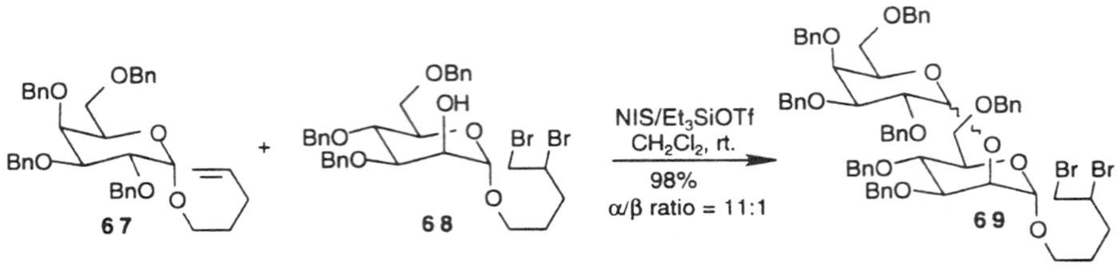

The pentenyl galactoside **67** (791 mg, 1.3 mmol) and the acceptor **68** (679.9 mg, 1.0 mmol) were combined, rotoevaporated twice with dry toluene, and then vacuum dried for 12 h. A solution of this mixture in dry $CH_2Cl_2$ (7 mL) was treated under argon with NIS (351 mg, 1.56 mmol) followed by dropwise addition of $Et_3SiOTf$ (0.059 mL, 0.26 mmol). After 10 min, TLC (4:1 petroleum ether–EtOAc) showed the reaction to be complete, the mixture was diluted with $CH_2Cl_2$ (40 mL) and washed successively with 10% aqueous $Na_2S_2O_3$ (70 mL) and $NaHCO_3$ solution (30 mL). The organic layer was then dried ($Na_2SO_4$), filtered, and evaporated. The residue was flash chromatographed (87% petroleum ether–13% EtOAc → 85% petroleum ether–15% EtOAc) to give **69** as a light yellow oil (1.1815 g, 98% yield based on **68**) α/β ratio, 11:1 (HPLC). For **69α**: $[\alpha]_D$ +26.8 (c 1.0, $CHCl_3$).

## F. NPG Coupling with NIS–Stoichiometric Et₃SiOTf [25]

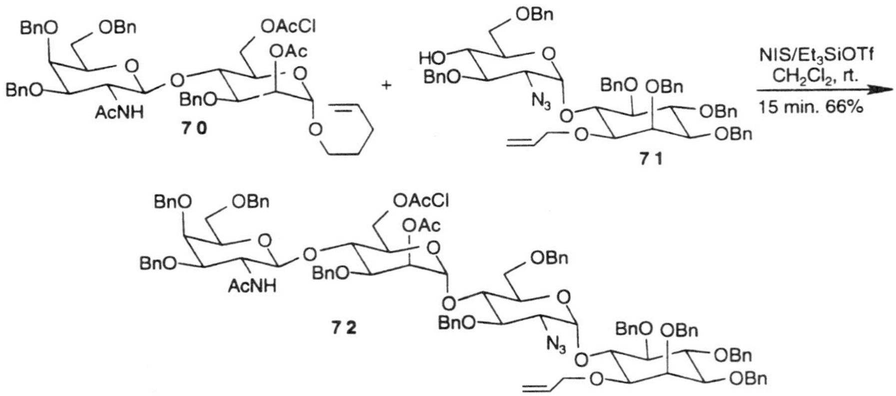

To a solution of **70** (824 mg, 0.886 mmol) and **71** (1.03 g, 1.09 mmol) in $CH_2Cl_2$ (11 mL) were added powdered NIS (240 mg, 1.07 mmol) and $Et_3SiOTf$ (280 μL, 1.24 mmol) and the mixture was stirred at room temperature for 15 min before quenching with $Et_3N$. The solution was diluted with $CH_2Cl_2$ (50 mL) and washed successively with 10% aqueous $Na_2S_2O_3$ (50 mL) and saturated aqueous $NaHCO_3$ (50 mL). The dried crude product was concentrated and flash chromatographed (3:2 petroleum ether–EtOAc) to give 1.05 g (66% based on **70**) of **72** and 465 mg of triethylsilylated **71** [32] which, by treatment with $Bu_4NF$ in THF, was converted into 380 mg of recovered **71**. For **72**: $R_f$ 0.53, $[\alpha]_D$ +55.6° (c 1.1, $CHCl_3$).

## G. NPG Coupling with IDCP [19]

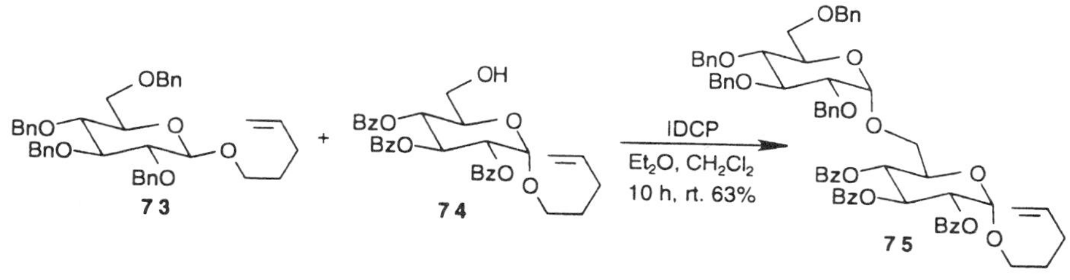

To the sugar alcohol **74** (576 mg, 1.03 mmol), flame-dried 4-Å molecular sieves (2 g) and IDCP (773 mg, 1.65 mmol) was added a solution of the NPG **73** (762 mg, 1.25 mmol) in dry $Et_2O$–$CH_2Cl_2$ (37.5 mL, 4:1). The mixture was stirred under argon in the dark at room temperature. After 10 h, 10% aqueous $Na_2S_2O_3$ (10 mL) was added, the mixture was filtered, and the 4-Å molecular sieves were washed with $CHCl_3$ (100 mL). The combined filtrate and washings were washed with cold (0°C) 1% aqueous HCl solution (100mL), and the aqueous phase was extracted with $CHCl_3$ (3 × 100 mL). The combined organic extracts were washed with cold (0°C) 10% aqueous $NaHCO_3$ (100 mL), dried, and the solvent was removed under reduced pressure. Flash column chromatography (light petroleum ether–EtOAc, 9:1 then 4:1) of the residue gave slightly impure **75** (808 mg, 73%). Further flash column chromatography ($CHCl_3$, 0.75% EtOH) yielded pure **75** (701 mg, 63%) isolated as an oil: $[\alpha]_D$ +69° (c 0.83, $CHCl_3$).

## H. Conversion of NPGs into Vicinal Dibromides [6,14]

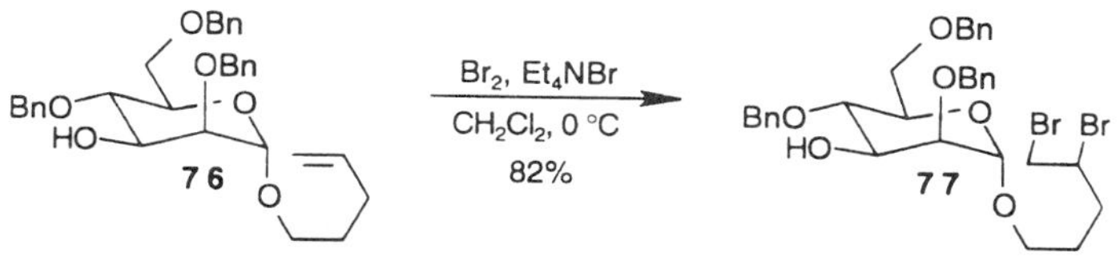

The pentenyl mannoside **76** (3.90 g, 9.11 mmol) and $Et_4NBr$ (1.47 g, 4.55 mmol) were dissolved in $CH_2Cl_2$ (20 mL). The mixture was cooled to 0°C. Bromine (1.46 g, 9.11 mmol) was added dropwise while allowing the solution to decolorize between addition of successive drops. When the brown color persisted, the reaction was quenched with 10% $Na_2S_2O_3$. The organic phase was dried ($Na_2SO_4$) and the solvent was removed by rotary evaporation. The residue was purified by column chromatography on silica gel (15% → 30%, EtOAc–light petroleum) affording **77**, 4.37 g (82%): $R_f$ 0.16 (30%, EtOAc–light petroleum), $[\alpha]_D$ +21.4° (c 1.0, $CHCl_3$).

## I. Reductive Debromination of Vicinal Dibromides [14,23]

*With Zn*

To the dibromide **78** (285.6 mg, 0.180 mmol) in EtOAc (3 mL) and EtOH (20 mL) were added Zn (60 mg, 0.918 mmol) and $Bu_4NI$ (67 mg, 0.181 mmol). The mixture was sonicated [33] for 18 h and then filtered through Celite and evaporated to dryness. The residue was purified by flash chromatography (3:7 EtOAc-petroleum ether) to afford **79** (231.7 mg, 95%): [α]$_D$ +21.2° (*c* 1.1, CHCl$_3$).

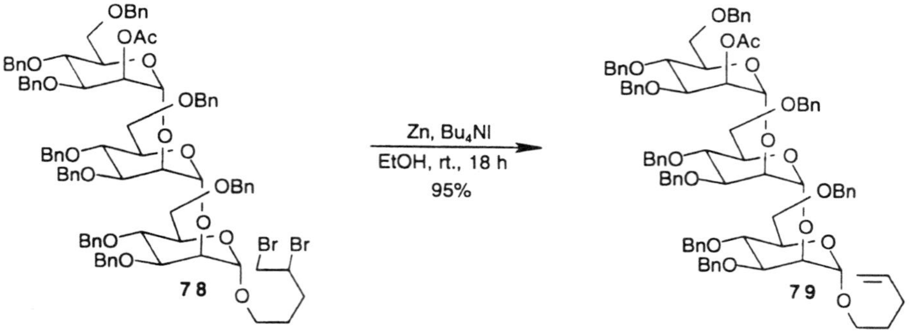

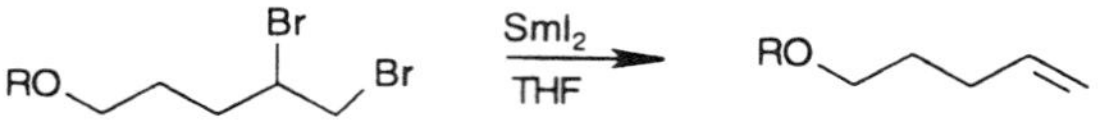

*With SmI$_2$*

The starting material (0.14 mmol) was dissolved in degassed, distilled THF (5 mL) under argon and samarium diiodide (0.1 *M* in THF, 4 mL) was added by syringe. The reaction mixture was stirred for 1 h at room temperature, making sure that the blue color persisted for at least 30 min. If it did not, more SmI$_2$ solution was added to assure complete reaction. The mixture was filtered to remove solids and the solvent was removed by rotary evaporation. The residue was taken up in CH$_2$Cl$_2$, washed with water, and the organic phase was dried over Na$_2$SO$_4$, and the solvent removed by rotary evaporation.

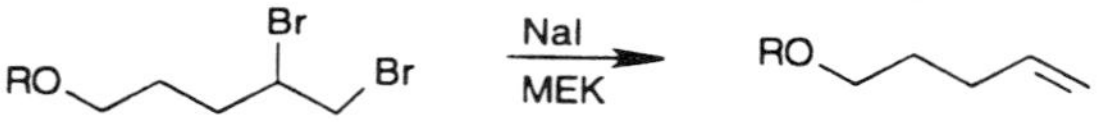

*With NaI*

The starting material (0.081 mmol) was dissolved in methyl ethyl ketone (10 mL), and sodium iodide (20 eq) was added. The reaction mixture was heated to reflux for 4 h, and was then taken up in EtOAc and washed with 10% sodium thiosulfate. The organic phase was dried over Na$_2$SO$_4$, and the solvent was removed by rotary evaporation.

## J. Conversion of *n*-Pentenyl 1,2-Orthoesters into Glycosyl Bromides and Coupling by the Koenigs–Knorr Procedure [12]

To the orthoester **80** (10.4 g, 13.5 mmol, 2.52 Eq.) was added dry CH$_2$Cl$_2$ (20 mL) and, after the solution cooled to 0°C, dropwise addition of 2.5 *M* Br$_2$ in CH$_2$Cl$_2$ was carried out until a faint yellow color persisted. TLC (4:1, petroleum ether–EtOAc) indicated the consumption of the starting material ($R_f$ 0.6) and a new product, presumed to be **81** ($R_f$ 0.7) was formed. The orange solution was then concentrated under reduced pressure, and the resulting syrup was dried for 10 min in vacuo. In the meantime the alcohol **82** (4.66 g, 8.98 mmol, 1 Eq.), dissolved in 10 mL of CH$_2$Cl$_2$, was transferred to a slurry of AgOTf (3.93 g, 15.29 mmol, 1.7 Eq.) in 15 mL of CH$_2$Cl$_2$ containing 2.0 g of powdered, activated 4-Å molecular sieves and stirred for 5 min at −60°C. A solution of **81** in 9 mL of CH$_2$Cl$_2$ was then added dropwise to the slurry under argon, and the temperature was raised to −40°C. After 10 min,

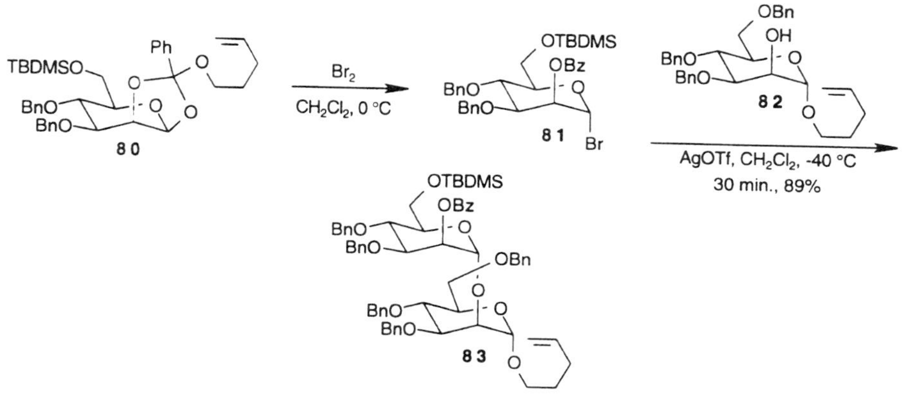

TLC indicated complete consumption of alcohol **82**. The reaction was quenched by vigorous stirring with saturated aqueous NaHCO$_3$, diluted with CH$_2$Cl$_2$, and then filtered through Celite. The organic solution was washed with brine, dried, and concentrated. Flash chromatography (4:1 petroleum ether–EtOAc) afforded 10.81 g (89%) of **83**: $R_f$ 0.63, $[\alpha]_D$ +7.9° (c 1.1, CHCl$_3$).

### K.  Coupling of *n*-Pentenyl 1,2-Orthoesters with NIS/Et$_3$SiOTf [12]

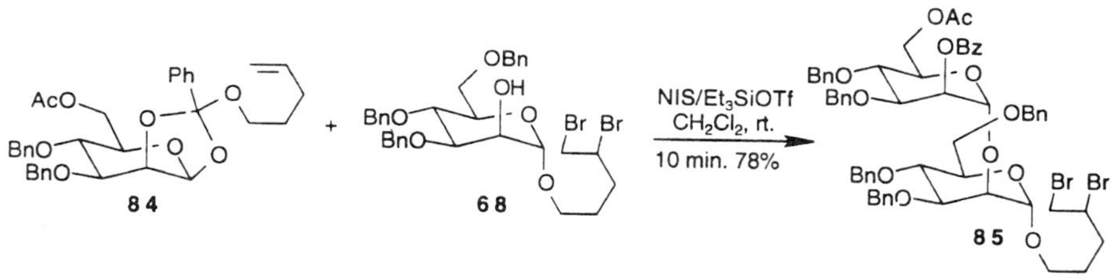

The alcohol **68** (2.70 g, 4.15 mmol) was dissolved in CH$_2$Cl$_2$ (10 mL). NIS (943 mg, 5.39 mmol) and the orthoester **84** (3.10 g, 5.39 mmol) in CH$_2$Cl$_2$ (10 mL) were added, respectively. After 5 min a catalytic amount of Et$_3$SiOTf (120 µL, 0.530 mmol) was added by syringe. The reaction mixture was stirred for 10 min, diluted with CH$_2$Cl$_2$, washed with 10% aqueous Na$_2$S$_2$O$_3$ and saturated aqueous NaHCO$_3$, dried, concentrated, and flash chromatographed (3:1 petroleum ether–EtOAc) to afford 3.61 g (78%) of **85**: $R_f$ 0.53, $[\alpha]_D$ +11.3° (c 1.1, CHCl$_3$).

### L.  Preparation of Glycoproteins [19,28]

To a solution of **86** (1.25 g, 1.65 mmol) in dry acetonitrile (30 mL) were added the aspartic acid derivative **87** (652 mg, 1.83 mmol) and *N*-bromosuccinimide (444 mg, 2.50 mmol). The mixture was stirred in the dark at room temperature under argon. After 1 h, aqueous Na$_2$S$_2$O$_3$ (10 mL) was added, the bulk of organic solvent was evaporated under reduced pressure, and the residue was partitioned between water (100 mL) and chloroform (100 mL). The aqueous layer was extracted with chloroform (3 × 100 mL). The combined organic layers were washed with water (2 × 75 mL) and dried, and the solvent was removed under reduced pressure. Flash column chromatography (light petroleum–ethyl acetate, 4:1) of the residue gave **88** (1.08 g, 61%), isolated as an oil: $[\alpha]_D$ −11° (c 0.54, CHCl$_3$). A solution of **88** (1.06 g, 0.99 mmol) in dry N,N-dimethylformamide (15 mL) and dry piperidine (0.35 mL, 3.54 mmol) was stirred under argon for 3.75 h. The solvent was then evaporated at 0.1 mmHg and the residue partitioned between water (100 mL) and chloroform (100 mL). The aqueous layer was extracted with chloroform (4 × 100 mL). The combined organic solvents were dried and concentrated under reduced pressure. Flash column chromatography (toluene–ethyl acetate 10:1) of the residue gave **89** (908 mg, 89%): mp 128°–129°C (from dichloromethane–light petroleum), $[\alpha]_D$ +29° (c 0.64, CHCl$_3$).

## REFERENCES AND NOTES

1. A. Varki, Biological roles of oligosaccharides; all of the theories are correct, *Glycobiology* 3:97, (1993); H. Lis and N. Sharon, Protein glycosylation-structural and functional aspects, *Eur. J. Biochem.* 218:1, (1993) and references therein.
2. For some recent reviews on oligosaccharide synthesis see: B. Fraser-Reid, R. Madsen, A. S. Campbell, C. S. Roberts, and J. R. Merritt, Chemical synthesis of oligosaccharides, *Carbohydrates*, S. M. Hecht, ed., IRL Press, Oxford, 1996; S. H. Khan and O. Hindsgaul, In *Frontiers in Molecular Biology* (M. Fukuda and O. Hindsgaul, eds.) IRL Press, Oxford, 1994, p. 206; K. Toshima and K. Tatsuta, Recent progress in O-glycosylation methods and its application to natural products synthesis, *Chem. Rev.* 93:1503 (1993); J. Banoub, P. Boullanger, and D. Lafont, Synthesis of oligosaccharides of 2-amino-2-deoxy sugars, *Chem. Rev.* 92:1167 (1992).
3. F. Barresi and O. Hindsgaul, Synthesis of β-D-mannose containing oligosaccharides, *Modern Methods in Carbohydrate Synthesis* (S. H. Khan and R. A. O'Neill, eds.). Harwood Academic Publishers, Amsterdam, 1996, Chap 11.
4. F. Barresi and O. Hindsgaul, The synthesis of β-D-mannopyranosides by intramolecular aglycon delivery; scope and limitations of the existing methodology, *Can. J. Chem.* 72:1447, (1994); Y. Ito and T. Ogawa, A novel approach to the stereoselective synthesis of β-mannosides, *Angew. Chem. Int. Ed. Engl.* 33:1765 (1994); F. W. Lichtenthaler and T. Schneider-Adams, 3,4,6-Tri-O-benzyl-α-D-*arabino*-hexopyranos-2-ulosyl bromide: A versatile glycosyl donor for the efficient generation of β-D-mannopyranosidic linkages, *J. Org. Chem.* 59:6728 (1994); K. K.-C. Liu and S. J. Danishefsky, Route from glycals to mannose β-glycosides, *J. Org. Chem.* 59:1892 (1994); G. Stork and G. Kim, Stereocontrolled synthesis of disaccharides via the temporary silicon connection, *J. Am. Chem. Soc.* 114:1087 (1992); P. J. Garegg, Saccharides of biological importance: Challenges and opportunities for organic synthesis, *Acc. Chem. Res.* 25:575 (1992).
5. J. J. Krepinsky, Advances in polymer-supported solution synthesis of oligosaccharides, *Modern Methods in Carbohydrate Synthesis*, (S. H. Khan and R. A. O'Neill, eds.), Harwood Academic Publishers, Amsterdam, 1996, Chap. 9.

6. B. Fraser-Reid, U. E. Udodong, Z. Wu, H. Ottosson, J. R. Merritt, C. S. Rao. C. Roberts, and R. Madsen, n-Pentenyl glycosides (NPGs) in organic chemistry: A contemporary example of serendipity, *Synlett*, 927 (1992).
7. R. Madsen and B. Fraser-Reid, n-Pentenyl glycosides in oligosaccharide synthesis, *Modern Methods in Carbohydrate Synthesis* (S. H. Khan and R. A. O'Neill, eds.), Harwood Academic Publishers, Amsterdam, 1996, Chap. 7.
8. L. A. Brooks and H. R. Snyder, 4-Penten-1-ol, *Org. Synth. Coll.* 3:698 (1955).
9. P. Konradsson, C. Roberts, and B. Fraser-Reid, Conditions for modified Fischer glycosidation with n-pentenol and other alcohols, *Recl. Trav. Chim. Pays-Bas.* 110:23 (1991).
10. S. Hanessian and J. Banoub, Chemistry of the glycosidic linkage. An efficient synthesis of 1,2-*trans*-disaccharides, *Carbohydr. Res.* 53:C13 (1977); F. Arcamone, S. Penco, S. Redaelli, and S. Hanessian, Synthesis and antitumor activity of 4'-deoxydaunorubicin and 4'-deoxy-adriamycin, *J. Med. Chem.* 19:1424 (1976).
11. G. Legler and E. Bause, Epoxy alkyl oligo-(1→4)β-D-glucosides as active site-directed inhibitors of cellulases, *Carbohydr. Res.* 28:45 (1973).
12. C. Roberts, R. Madsen, and B. Fraser-Reid, Studies related to synthesis of glycophosphatidyl inositol membrane bound protein anchors: Part V. n-Pentenyl orthoesters for mannan components, *J. Am. Chem. Soc.* 117:1546 (1995).
13. P. Konradsson, D. R. Mootoo, R. E. McDevitt, and B. Fraser-Reid, Iodonium ion generated in situ from N-iodosuccinimide and trifluoromethane-sulfonic acid promotes direct linkage of "disarmed" pent-4-enyl glycosides, *J. Chem. Soc. Chem. Commun.* p. 270 (1990).
14. J. R. Merritt, E. Naisang, and B. Fraser-Reid, n-Pentenyl mannoside precursors for synthesis of the nonamannan component of high mannose glycoproteins, *J. Org. Chem.* 59:4443 (1994).
15. B. Fraser-Reid, P. Konradsson, D. R. Mootoo, and U. E. Udodong, Direct elaboration of pent-4-enyl glycosides into disaccharides, *J. Chem. Soc. Chem. Commun.* p. 823 (1988).
16. D. R. Mootoo, P. Konradsson, U. E. Udodong, and B. Fraser-Reid, "Armed" and "disarmed" n-pentenyl glycosides in saccharide couplings leading to oligosaccharides, *J. Am. Chem. Soc.* 110:5583 (1988).
17. R. U. Lemieux and A. R. Morgan, The synthesis of β-D-glucopyranosyl-2-deoxy-α-D-*arabino*-hexopyranoside, *Can. J. Chem.* 43:2190 (1965).
18. B. Fraser-Reid, Z. Wu, U. E. Udodong, and H. Ottosson, Armed/disarmed effects in glycosyl donors: Rationalization and sidetracking, *J. Org. Chem.* 55:6068 (1990).
19. A. J. Ratcliffe, P. Konradsson, and B. Fraser-Reid, Application of n-pentenyl glycosides in the regio- and stereo- controlled synthesis of α-linked N-glycopeptides, *Carbohydr. Res.* 216:323 (1991).
20. P. Konradsson and B. Fraser-Reid, Conversion of 4-pentenyl glycosides into glycosyl bromides, *J. Chem. Soc. Chem. Commun.* p. 1124 (1989).
21. Some strongly disarmed NPGs, e.g., n-pentenyl α-mannosides, also give the dibromides in absence of $Et_4NBr$.
22. R. Roy, F. O. Andersson and M. Letellier, "Active" and "latent" thioglycosyl donors in oligosaccharide synthesis. Application to the synthesis of α-sialosides. *Tetrahedron Lett.* 33:6053 (1992).
23. J. R. Merritt, J. S. Debenham, and B. Fraser-Reid, Methods for debrominating 4,5-dibromopentanyl glycosides. *J. Carbohydr. Chem.* 15:65 (1996).
24. U. E. Udodong, C. S. Rao, and B. Fraser-Reid, n-Pentenyl glycosides in the efficient assembly of the blood group substance B tetrasaccharide, *Tetrahedron* 48:4713 (1992).
25. R. Madsen, U. E. Udodong, C. Roberts, D. R. Mootoo, P. Konradsson, and B. Fraser-Reid, Studies related to synthesis of glycophosphatidyl inositol membrane bound protein anchors: Part VI. Convergent assembly of sub-units, *J. Am. Chem. Soc.* 117:1554 (1995).
26. A. J. Ratcliffe and B. Fraser-Reid, Oxidative hydrolysis of conformationally restrained pent-4-enyl glycosides: Formations of N-acetyl-α-D-glucopyranosylamines, *J. Chem. Soc. Perkin Trans. 1*:1805 (1989).

27. A. J. Ratcliffe and B. Fraser-Reid, Generation of α-D-glucopyranosyl acetonitrilium ions. Concerning the reverse anomeric effect, *J. Chem. Soc. Perkin Trans 1*:747 (1990).
28. A. L. Handlon and B. Fraser-Reid, A convergent strategy for the critical β-linked chitobiosyl-*N*-glycopeptide core, *J. Am Chem. Soc. 115*:3796 (1993).
29. J. S. Debenham, R. Madsen, C. Roberts, and B. Fraser-Reid, Two new orthogonal amine protecting groups that can be cleaved under mild or neutral conditions, *J. Am. Chem. Soc. 117*: 3302 (1995).
30. R. Madsen and B. Fraser-Reid, Acetal transfer via halonium-ion induced reactions of di-pent-4-enyl acetals: Scope and mechanism. *J. Org. Chem. 60*:772 (1995).
31. R. K. Ness, H. G. Fletcher, Jr., and C. S. Hudson, The reaction of 2,3,4,6-tetrabenzoyl-α-D-glucopyranosyl bromide and 2,3,4,6-tetrabenzoyl-α-D-mannopyranosyl bromide with methanol. Certain benzoylated derivations of D-glucose and D-mannose. *J. Am. Chem. Soc. 72*:2200 (1950).
32. Silylation of the glucosyl acceptor by $Et_3SiOTf$ has been observed only in some cases when the donor or the acceptor contained an *N*-acetate.
33. With mono- and disaccharides, reflux can be applied instead of sonication.

# 15

# Chemical Synthesis of Sialyl Glycosides

**Akira Hasegawa and Makoto Kiso**
*Gifu University, Gifu, Japan*

| | | |
|---|---|---|
| I. | Introduction | 358 |
| II. | Regio- and α-Stereoselective Sialyl Glycoside Syntheses Using Thioglycosides of Sialic Acids in Acetonitrile | 359 |
| | A. Monosialyl glycoside synthesis | 359 |
| | B. Oligosialyl glycoside synthesis | 362 |
| | C. Reaction mechanism | 362 |
| III. | Applications to Systematic Synthesis of Gangliosides and Sialyloligosaccharides | 364 |
| | A. Sialyl galactose donors as building blocks | 364 |
| | B. Applications | 364 |
| IV. | Experimental Procedures | 370 |
| | A. Sialylation of 3-*O*-benzoyl-galactose acceptor | 370 |
| | B. Sialylation of 6-*O*-benzoyl-galactose acceptor | 370 |
| | C. Sialylation of penta-*O*-benzyl-lactose acceptor | 371 |
| | D. Sialylation of Lewis X–relevant trisaccharide | 371 |
| | E. Synthesis of sialyl lacto- and neolactotetraose derivatives | 372 |
| | F. Synthesis of ganglioside $GD_2$ oligosaccharide | 372 |
| | G. Synthesis of ganglioside $GQ_{1b}$ oligosaccharide | 373 |
| | H. Synthesis of ganglioside $GM_2$–relevant oligosaccharide | 374 |
| | I. Synthesis of ganglioside $GT_2$–relevant oligosaccharide | 374 |
| | J. Synthesis of ganglioside $GQ_{1b\alpha}$ oligosaccharide | 375 |
| | References | 375 |

## I. INTRODUCTION

Sialic acid-containing glycoconjugates, such as gangliosides and sialoglycoproteins, have been recognized to play important roles in many biological processes [1–5]. Sialyl-oligosaccharide chains of these glycoconjugates are exposed as ligands to the external environment, capable of expressing various biological functions, not only serving as receptors for hormones, viruses and bacterial toxins, but also as mediators in cell differentiation, proliferation, oncogenesis, immunity, and so on. For example, specific recognitions of sialyl-oligosaccharides are involved in the initial step of invasion of a mammalian cell by influenza virus [6] or *Trypanosoma cruzi* [7]. The sialyl Lewis X carbohydrate epitope, found on neutrophils, monocytes, and tumor cells, has been identified [8–10] as the ligand for selectins, a family of cell adhesion receptors that are implicated in the leukocyte traffic or extravasation to sites of inflammation, platelet adhesion, and probably tumor metastasis [11]. Various biological functions of oligo- and polysialyl glycoconjugates, such as ganglioside GQ1b [12] and polysialoglycoproteins [13,14], have also been demonstrated.

The most representative sialic acid, *N*-acetylneuraminic acid (Neu5Ac, **1**; Fig.1) is usually attached to O-3 or O-6 of galactose, or O-6 of *N*-acetylgalactosamine with an α(2→3)- or α(2→6)-linkage, and to O-8 of another Neu5Ac with an α(2→8)-linkage, giving diverse structures and functions. Therefore, the systematic understanding of structure–function relations of sialyl-oligosaccharides at the molecular level necessitates an efficient method for regio- and α-stereoselective glycoside synthesis of sialic acids.

Since 1965, a number of attempts for obtaining sialyl glycosides have been achieved [15,16], mainly by the classic Koenigs-Knorr method using *N*-acetylneuraminyl halides **2** as glycosyl donors. However, the yield and stereoselectivity of the glycosides with sugar

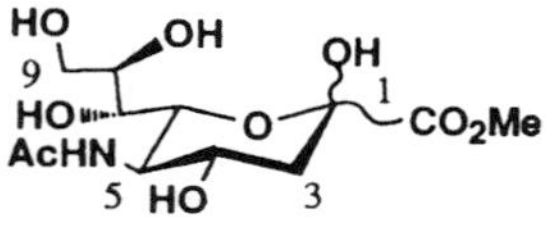


**1** ( Neu5Ac )

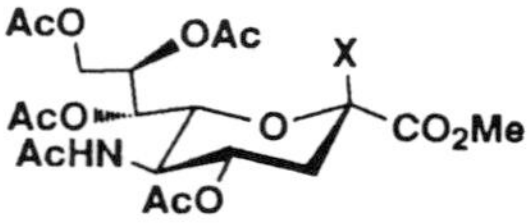


**2**

X = Cl, Br etc

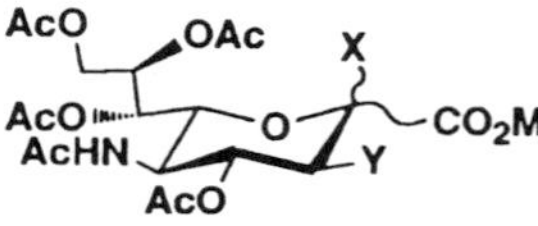


**3**

X = Cl, Br etc
Y = SPh, SePh

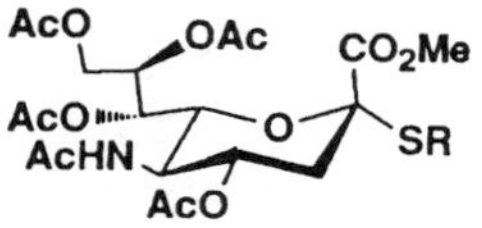


**4** R = Na
**5** R = Me

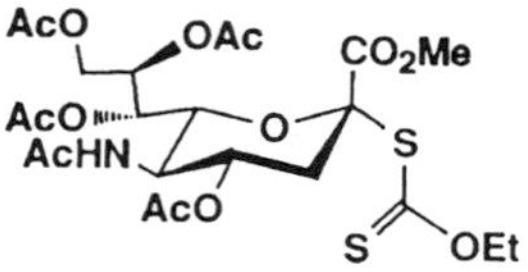


**6**

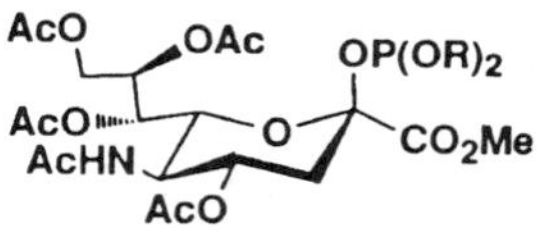


**7**

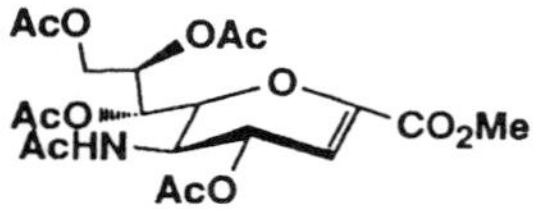


**8**

**Figure 1**

# Synthesis of Sialyl Glycosides

acceptors were generally poor. As critical problems, there are several disadvantages for the stereoselective synthesis of α-sialyl glycosides. First, the anomeric position (C-2) of Neu5Ac is not only sterically hindered, but is also electronically disfavored by the carboxylic acid function. Second, the lack of a substituent at C-3 precludes the suitable neighboring participation leading to α-glycoside. Third, the thermodynamically favored product is β-glycoside. These combined factors disfavor the desired α-glycoside formation, especially with secondary sugar hydroxyls. Particularly annoying is the competitive elimination owing to the deoxy center at C-3, giving the 2,3-dehydro derivative **8** as the major by-product.

The use of 2-halo-3-β-substituted neuraminyl derivatives **3** is an efficient approach for overcoming these difficulties [17,18] and, in fact, many attempts have been reported [15,19]. However, additional multiple steps for preparation of glycosyl donor and removal of the C-3 substituent to arrive at the desired products are required, even though improvements in the yield and stereospecificity of the glycosides are achieved.

We have developed [20,21] a stereoselective synthesis of a variety of S-glycosides of Neu5Ac by use of sodium salt of methyl 5-acetamido-4,7,8,9-tetra-O-acetyl-3,5-dideoxy-2-thio-D-*glycero*-α-D-*galacto*-2-nonulopyranosonate **4** and the β-anomer, from which the methyl α-2-thioglycoside of Neu5Ac **5** was readily prepared [22]. Thioglycosides are stable in many organic operations, and capable of specific activations by suitable thiophilic promoters in very mild conditions, being widely used in O-glycoside syntheses [23].

Our preliminary attempts [22,24] by use of **5** as a glycosyl donor and dimethyl(methylthio)sulfonium triflate (DMTST) [25] as the promoter, with various alcohols and sugar acceptors, indicated that the glycosylation in acetonitrile afforded predominantly α-glycosides. Also, the use of lightly protected sugar acceptors (e.g., **14, 16**, or **18** in Fig. 2), in which several OH groups are unprotected, efficiently gave the desired sialyl-α(2→3)- and sialyl-α(2→6)-sugar derivatives (e.g., **20, 22**, or **24**). This method was successfully extended [26,27] towards systematic ganglioside syntheses as described later. On the other hand, a similar glycosylation of **5** using benzeneselenenyl triflate as a promoter in dichloroethane resulted in the predominant production of β-glycosides [28]. Given these preliminary findings, several new attempts were made employing Neu5Ac glycosyl xanthates **6** [29,30] or glycosyl phosphites **7** [31,32] as alternative glycosyl donors in acetonitrile medium.

In the following section, our recent development of the regio- and α-stereoselective sialyl glycoside syntheses using thioglycosides of sialic acids in acetonitrile is described.

## II. REGIO- AND α-STEREOSELECTIVE SIALYL GLYCOSIDE SYNTHESES USING THIOGLYCOSIDES OF SIALIC ACIDS IN ACETONITRILE

### A. Monosialyl Glycoside Synthesis

Facile and highly regio- and α-stereoselective sialyl glycoside syntheses have been achieved by using, as glycerol donors, the 2-thioglycosides (**11–13**) of Neu5Ac or 3-deoxy-D-*glycero*-D-*galacto*-2-nonulosonic acid (KDN), in which the acetamido group of Neu5Ac is replaced by a hydroxyl; and, as glycerol acceptors, the suitably protected 2-(trimethylsilyl)ethyl (SE) glycosides (**14–19**) of D-galactose or lactose, in the presence of thiophilic promoters, such as DMTST or N-iodosuccinimide-trifluoromethanesulfonic acid (NIS/TfOH) [33,34] in acetonitrile (see Fig. 2).

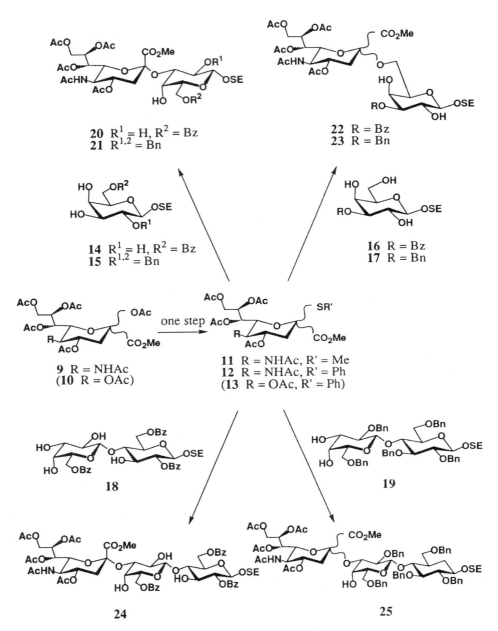

**Figure 2** An efficient regio- and α-stereoselective synthesis of mono-sialyl glycosides. (SE = $-CH_2CH_2SiMe_3$.)

In the initial studies, as briefly described earlier, the methyl α-2-thioglycoside of Neu5Ac **5** [22,24], or the ~1:1 anomeric mixture **11** [26,27], which can be almost quantitatively prepared from **9** in one step, was coupled with lightly protected sugar acceptors such as **14**, **16**, and **18** in the presence of DMTST in acetonitrile to give exclusively the sialyl α(2→3)- or sialyl α(2→6)-D-galactose or lactose derivatives (**20**, **22**, and **24**) in 50–70% yields even in large-scale reactions [26].

Iodonium ion-promoted glycosylation [33,34] is another attractive approach for oligosaccharide synthesis. We have applied this approach to the glycosylations involving 2-thioglycosides (**11** and **12**) of Neu5Ac and reported [27,35,36] the comparative reactivities of DMTST and NIS–TfOH in acetonitrile or dichloromethane. The results are partly summarized in Table 1. Notably, with both DMTST and NIS-TfOH in acetonitrile, the O-3 of galactose in **14**, **15**, and **18** was regioselectively sialylated exclusively in α-configuration, giving **20**, **21**, and **24** in 47–70% yields (see Table 1 entries 1–6), whereas the penta-O-benzyl lactose acceptor **19** afforded an anomeric mixture (α/β 4:1 ~ 6:1) (see Table 1 entries 15–17). The yields with NIS–TfOH are relatively higher than with DMTST. A hindered primary hydroxyl, O-6 of **16**, was also exclusively α-sialylated in acetonitrile, whereas an anomeric mixture was produced with a less-hindered primary hydroxyl, O-6 of **17** (see Table 1, entries 10–12).

Generally, glycosylation with reactive primary hydroxyls (O-6 of glucose, glucosamine, and galactosamine derivatives) have resulted in anomeric mixtures (α/β 4:1 ~ 5:1, data not shown) [37] when compared with those with secondary hydroxyls, which is

**Table 1** DMTST[a] and NIS/TfOH[b] Promoted Sialylation of Primary and Secondary Sugar Hydroxyls Using Methyl or Phenyl 2-Thioglycoside of Neu5Ac

| Entry | Acceptor | Donor | Promoter | Solvent | Product | Isolated yield (%) | |
|---|---|---|---|---|---|---|---|
| | | | | | | α | β |
| 1 | 14 | 11 | DMTST | MeCN | 20 | 52 | 0 |
| 2 | 14 | 11 | NIS | MeCN | 20 | 61 | 0 |
| 3 | 14 | 12 | NIS | MeCN | 20 | 70 | 0 |
| 4 | 15 | 12 | NIS | MeCN | 21 | 70 | 0 |
| 5 | 18 | 11 | DMTST | MeCN | 24 | 47 | 0 |
| 6 | 18 | 12 | NIS | MeCN | 24 | 55 | 0 |
| 7 | 16 | 11 | DMTST | MeCN | 22 | 70 | 0 |
| 8 | 16 | 11 | NIS | MeCN | 22 | 59 | 0 |
| 9 | 16 | 11 | NIS | $CH_2Cl_2$ | 22 | 49 | 25 |
| 10 | 17 | 11 | DMTST | MeCN | 23 | 50 | 15 |
| 11 | 17 | 11 | NIS | MeCN | 23 | 51 | 26 |
| 12 | 17 | 12 | NIS | MeCN | 23 | 50 | 25 |
| 13 | 17 | 11 | NIS | $CH_2Cl_2$ | 23 | 43 | 50 |
| 14[c] | 17 | 11 | NIS | $CH_2Cl_2$ | 23 | 32 | 50 |
| 15 | 19 | 11 | DNTST | MeCN | 25 | 30 | 8 |
| 16 | 19 | 11 | NIS | MeCN | 25 | 59 | 10 |
| 17 | 19 | 12 | NIS | MeCN | 25 | 58 | 11 |

[a]Reactions were performed at −15°C.
[b]Reactions were performed at −35°C.
[c]Reactions were performed at −20°C.

apparently due to the differences in nucleophilicity. Thus, the reactivity of acceptor nucleophiles strongly affect the stereoselectivity.

On the other hand, the reactions in dichloromethane (see Table 1 entries 9, 13, and 14) were of poor stereoselectivity, although the overall yield was high. Furthermore, the formation of β-glycoside increased with rise of reaction temperature (see Table 1, entry 14). The use of a KDN glycosyl donor **13** that is prepared by treatment [38] of **10** with thiophenol and $BF_3$ etherate in dichloromethane, has also given results similar to those described for Neu5Ac [39].

This versatile method, designed for the regio- and α-stereoselective sialyl glycoside synthesis, has been further developed to the oligosialyl glycoside synthesis.

## B. Oligosialyl Glycoside Synthesis

Oligo- and polysialyl glycosides are common in nature, having α(2→8) glycosidic linkage between two sialyl residues. An efficient method for obtaining oligosialyl glycosides is the use of dimeric or trimeric Neu5Ac glycosyl donor, which can be readily prepared from naturally available colominic acid [40].

The lactonized phenyl 2-thioglycosides of dimeric (**26**) [41] and trimeric (**27**) [42] sialic acids, which were prepared from the corresponding oligosialic acids according to the procedure employed for the monosialyl derivatives, were each coupled with the suitably protected sugar acceptors (**14, 16, 17,** and **19**) by use of NIS–TfOH as a promoter in acetonitrile, as just described for the monosialyl glycoside synthesis (Fig. 3).

As summarized in Table 2, the regio- and stereoselectivity of the products appears to be comparable with that obtained for monosialylation (see Table 1). The moderate yields (30–48%) of the desired α-glycosides are quite appreciable, considering the bulkiness of the donor and the steric hindrance on the glycosylation site.

## C. Reaction Mechanism

A possible reaction mechanism is shown in Fig. 4. When the glycosyl donors (**11, 13, 26,** or **27**) are specifically activated by the thiophilic promoter (DMTST or NIS–TfOH) at low temperature, the cyclic oxocarbenium ion (**c**), necessary for almost all known glycosylation reactions, is formed by the initially produced intermediates (**a**) or (**b**), and, subsequently, reacts with acetonitrile to generate the β-acetonitrilium ion (**d**), which then undergoes $S_N 2$ displacement at the anomeric center by sugar-nucleophiles to give predominantly α-glycosides. However, the conformational preference of the positively charged β-acetonitrilium ion (**d**) of Neu5Ac has been claimed to be shifted toward the equatorial conformer (**e**), based on a concept of so-called reverse anomeric effect.

On the other hand, it has been demonstrated that the α-D-glucopyranosyl acetonitrilium ions are stereospecifically generated from the corresponding oxocarbenium ions in dry acetonitrile [43–45]. These results indicate that the acetonitrilium ions are axially oriented by an anomeric effect. Recently, it has been concluded from the conformational analyses of glucopyranosyl-ammonium and glucopyranosyl-imidazolium derivatives that the reverse anomeric effect does not exist [46,47].

Therefore, it seems most plausible that the stereospecific generation of the β-acetonitrilium ion (**d**) of Neu5Ac from the oxocarbenium ion (**c**) holds the key of the α-preponderant formation of sialyl glycosides in acetonitrile. Thus, the reactive nucleophiles, such as OH-6 of **17** have a chance to attack on other intermediates (**a**) ~ (**c**) before the complete

# Synthesis of Sialyl Glycosides

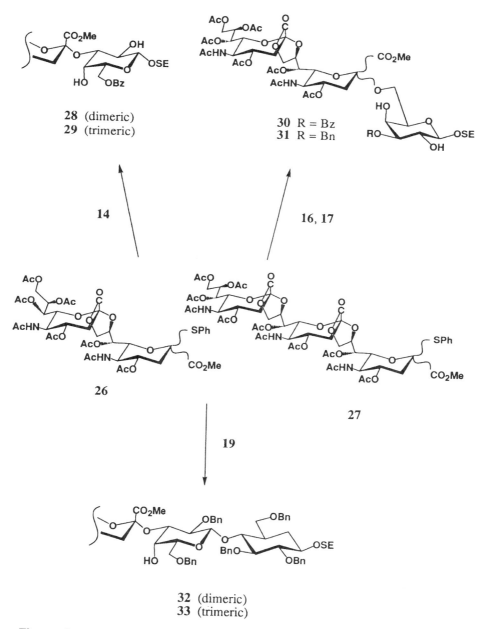

**Figure 3** An efficient regio and α-stereoselective synthesis of oligo-sialyl glycosides.

**Table 2** NIS–TfOH-Promoted Oligo-sialylation[a] of Sugar Acceptors by Phenyl 2-thioglycosides of Dimeric and Trimeric Neu5Ac in Acetonitrile

| Entry | Acceptor | Donor | Product | Isolated yield (%) | |
|---|---|---|---|---|---|
| | | | | α | β |
| 1 | 14 | 26 | 28 | 43 | 0 |
| 2 | 14 | 27 | 29 | 30 | 0 |
| 3 | 19 | 26 | 32 | 31 | 0 |
| 4 | 19 | 27 | 33 | 49 | 0 |
| 5 | 16 | 26 | 30 | 33 | 10 |
| 6 | 17 | 26 | 31 | 48 | 16 |

[a]Reactions were performed at −35°C.

generation of (**d**) to give an anomeric mixture. In the chlorinated solvent, however, nucleophiles react only with (**a**) ~ (**c**) to afford substantial amounts of the β-glycosides nonstereoselectively.

This reaction mechanism could be applicable to the method employing xanthate (**6**) or phosphite (**7**) as the glycosyl donor in acetonitrile medium.

## III. APPLICATIONS TO SYSTEMATIC SYNTHESIS OF GANGLIOSIDES AND SIALYLOLIGOSACCHARIDES

### A. Sialyl Galactose Donors as Building Blocks

The highly efficient method for α-sialyl glycoside synthesis just described has been further extended to the systematic synthesis of gangliosides and their analogues by using the monosialyl- and oligosialyl-galactose donors (**34–36**) as the building blocks. The highly structural diversity of oligosaccharide chains in complex native gangliosides necessitates an efficient procedure for introducing the sialyl-galactose segments into the sugar chains in a tailor-made manner. Because our approach involves the use of 2-(trimethylsilyl)ethyl (SE) group protection [48] at the reducing end, the channel of reaction scheme for preparing a variety of sialyl-oligosaccharides and their conjugates is the selective anomeric transformations into further glycosyl donors.

The selected sialyl galactose derivatives (**20, 23**, and **28**) were readily converted by (1) benzoylation of the remaining hydroxyls, (2) selective transformation of the OSE group at the reducing end into the corresponding β-acetate, and finally (3) introduction of the methylthio (SMe) group (Fig. 5). The use of these glycosyl donors already holding the sialyl-α(2→3)- or sialyl-α(2→6)-linkage excludes any possibility of the β-sialyl glycoside formation as far as the anomerization does not take place.

### B. Applications

The facile and highly versatile approach, combining the selective glycosyl donors, such as **11–13, 26, 27**, and **34–36**, with the suitably protected mono- and oligosaccharides as glycosyl acceptors, has been extensively applied to the synthesis of numerous gangliosides,

# Synthesis of Sialyl Glycosides

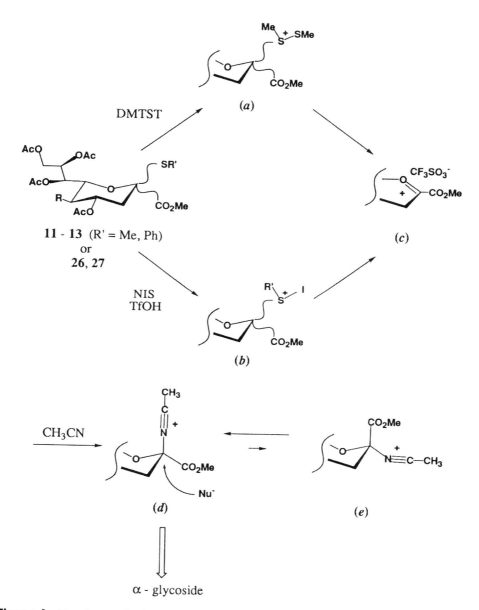

**Figure 4** Reaction mechanism suggested for α-predominant glycoside synthesis of sialic acids in acetonitrile.

sialyl-oligosaccharides, and their derivatives and analogues to elucidate their biological functions [36,49].

Gangliosides $GM_4$ [50], $GM_3$ [35,51], $GD_3$ [41], and their analogues [35] have been synthesized directly from **20**, **22**, **24**, and **32**, respectively, as shown in Fig. 6. KDN-gangliosides $GM_4$ and $GM_3$ were obtained from the KDN-α(2→3)-D-galactose or -lactose

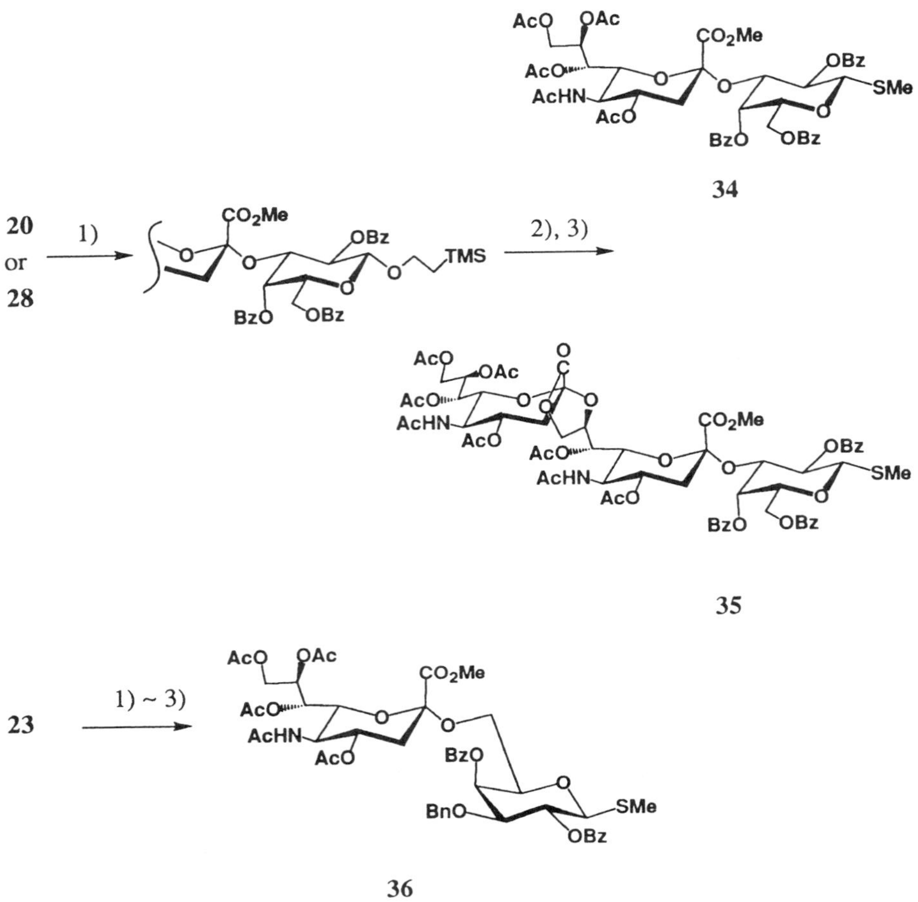

**Figure 5** Mono- and oligo-sialyl galactose donors as building blocks. (1) Bz$_2$O, DMAP, Pyr., (2) Ac$_2$O, BF$_3$·OEt$_2$, (3) TMSSMe, TMSOTf.

derivative that corresponds to **20** or **24** [39]. Further glycosylations of **25α** and **32** with galactosamine, then with galactose, sialyl-α(2→3)-galactose (**34**) or disialyl-α(2→3)-galactose donor (**35**) gave a ganglio series: gangliosides GM$_2$ [52], GM$_{1b}$ [53], GM$_1$ [54], and GD$_{1a}$ [55], GD$_2$ [56], GD$_{1b}$ [57], GT$_1$b [57]. The first total synthesis of ganglioside GQ$_{1b}$, which is one of the most complex gangliosides, has recently been achieved [57,58] by use of two disialyl-α(2→3)-galactose donors (**35**). The α-series gangliosides such as GD$_{1α}$ [59], GM$_{1α}$ [60], and GQ$_{1bα}$ [61], have also been successfully synthesized by a combination of the selective glycosyl donors. The tumor-associated, sialyl lacto- and neolacto-tetraosyl ceramides [62–64], sialyl-Lewis X [65–67], sialyl-Lewis A [68], and their various analogues and derivatives [36,69,70], have been successfully employed for the biological investigations related to influenza virus [6,71], tumor antigens [11], and especially to selectin binding [10, 72–76]. The KDN-lacto- and neolacto-tetraosyl ceramides, and KDN-Lewis X ganglioside were also synthesized [77] by use of KDN-α(2→3)-D-

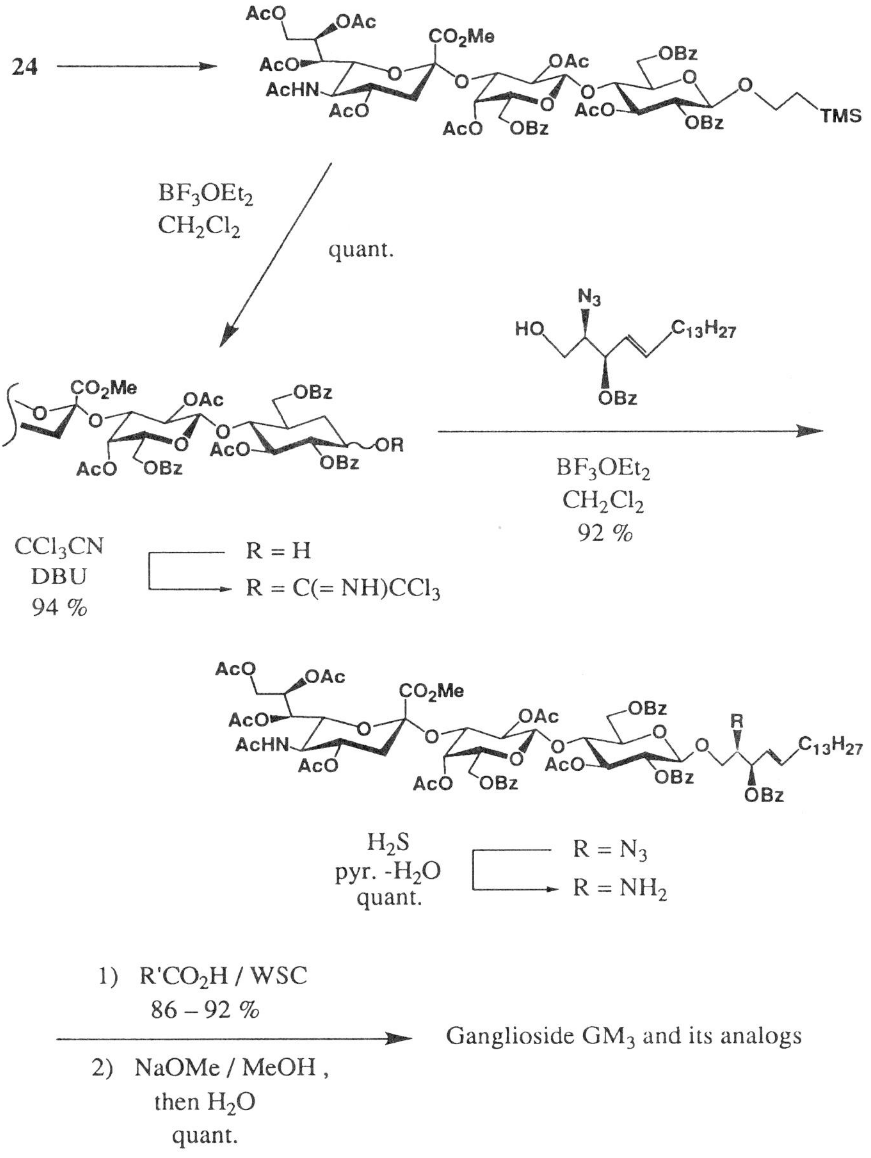

# Synthesis of Sialyl Glycosides

**Figure 6** A synthetic route to ganglioside $GM_3$.

galactose donor corresponding to **34**. Recent reports [e.g. 78–82] have also demonstrated that the use of thioglycosides of sialic acid as glycosyl donors in acetonitrile medium was highly efficient for the synthesis of α-sialyl glycosides.

As the application of our approach, a total synthesis of sialyl dimeric Lewis[x] (sialyl SSEA-1) ganglioside has successfully been achieved [83], as shown in Figures 7 and 8.

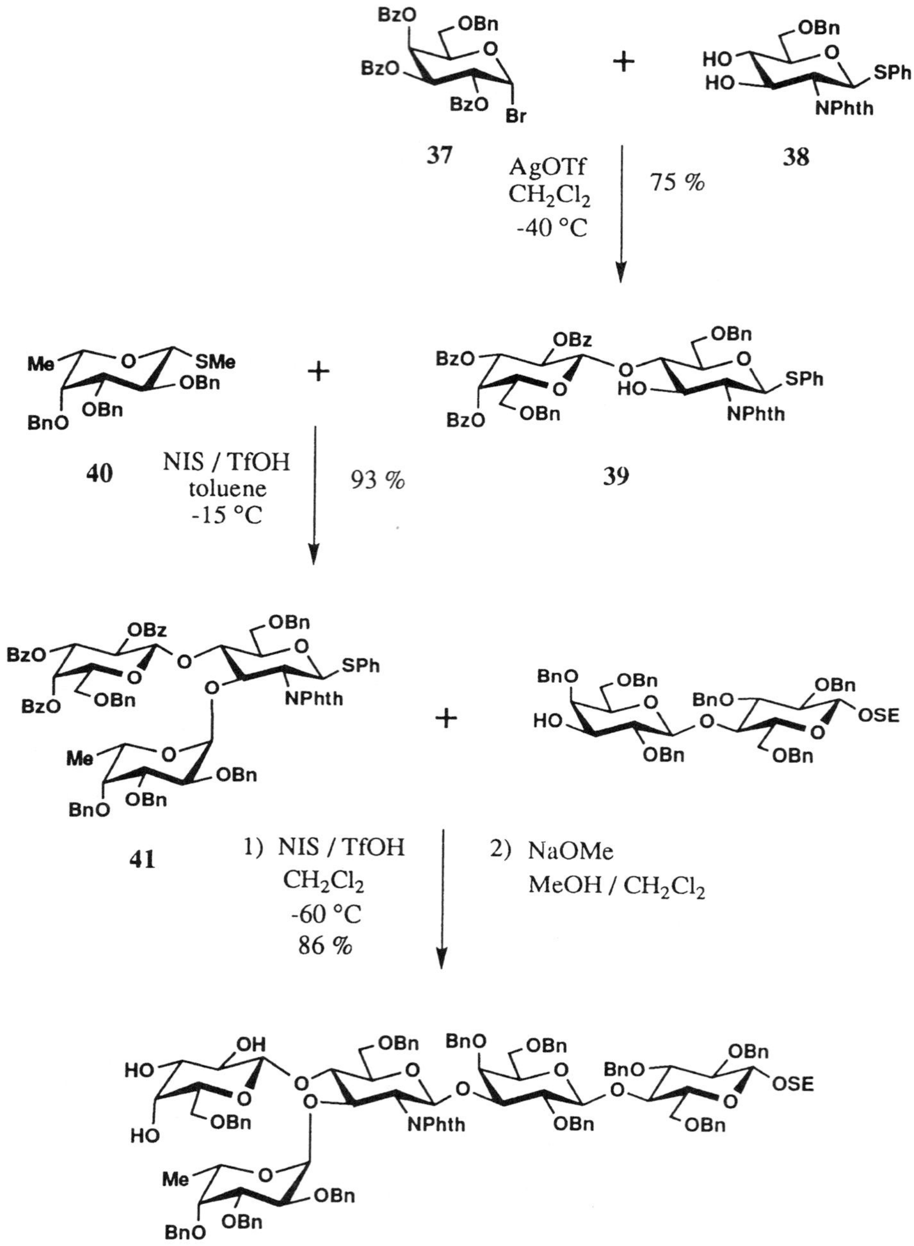

BzO
OBn
O
BzO
BzO
Br
+
OBn
O
HO
HO
SPh
NPhth
37
AgOTf
CH2Cl2
-40 °C
75 %
38
Me
O
SMe
OBn
OBn
BnO
+
OBz
BzO
O
O
BzO
OBn
OBn
O
HO
SPh
NPhth
40
NIS / TfOH
toluene
-15 °C
93 %
39
OBz
OBn
BzO
O
O
O
SPh
BzO
OBn
O
NPhth
Me
O
OBn
BnO
OBn
+
BnO
OBn
O
HO
OBn
BnO
O
OBn
OSE
O
OBn
41
1) NIS / TfOH
CH2Cl2
-60 °C
86 %
2) NaOMe
MeOH / CH2Cl2
OH
HO
O
HO
OBn
OBn
O
O
NPhth
BnO
OBn
O
OBn
BnO
O
OBn
OSE
OBn
Me
O
OBn
BnO
OBn


42

**Figure 7**

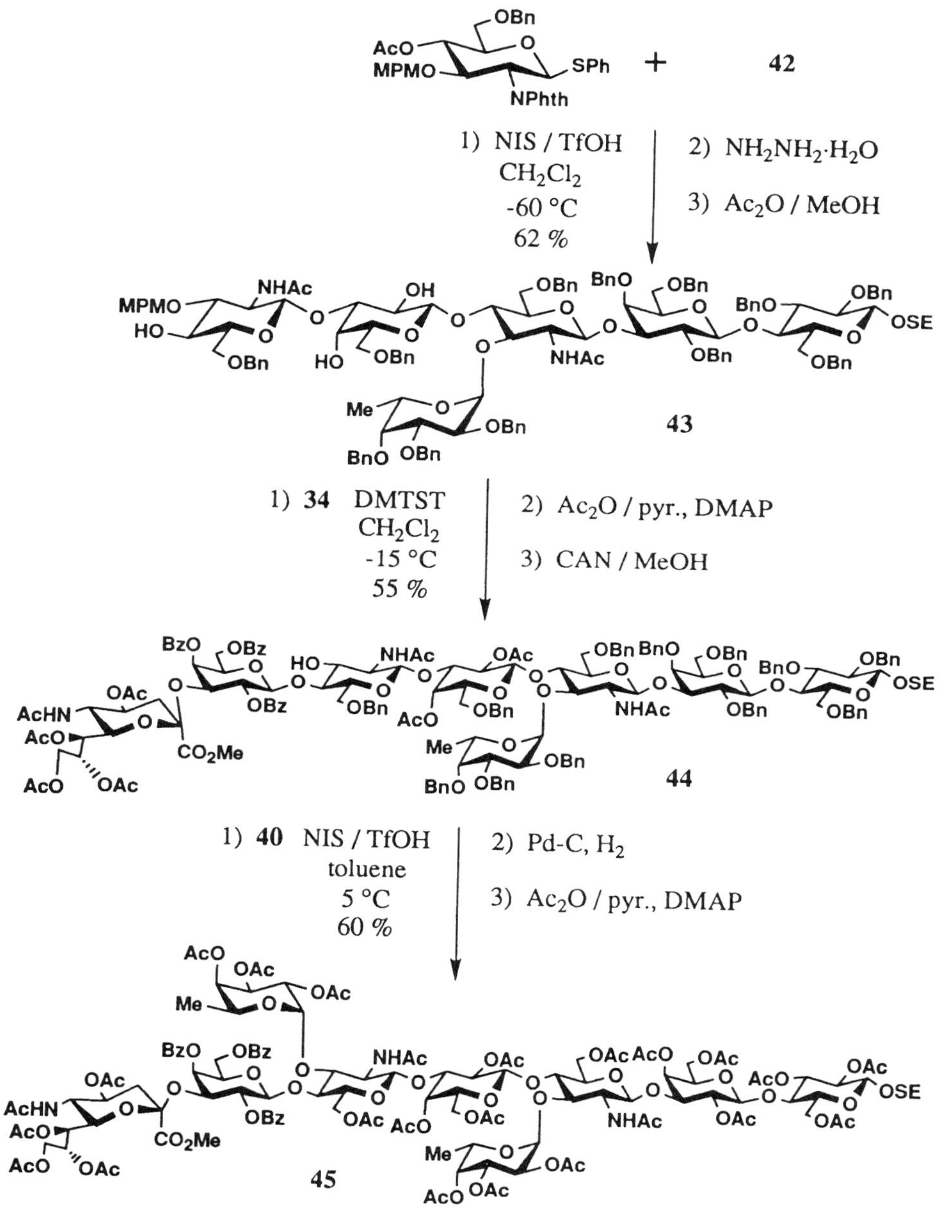

# Synthesis of Sialyl Glycosides

**Figure 8**

Regioselective glycosylation of **38** with 2,3,4-tri-*O*-benzoyl-6-*O*-benzyl-α-D-galactopyranosyl bromide (**37**) and the following 3-*O*-α-L-fucosylation of **39** gave the Lewis$^x$ (Le$^x$) trisaccharide donor **41**, which was then coupled with 2-(trimethylsilyl)ethyl *O*-(2,4,6-tri-*O*-benzyl-β-D-galactopyranosyl)-(1→4)-2,3,6-tri-*O*-benzyl-β-D-glycopyranoside [63]. The resulting pentasaccharide **42** was further glycosylated with the suitably protected glucosamine residue, regio- and β-stereoselectively, to give **43** (see Fig. 8). The use of sialyl-α(2-3)-galactose derivative **34** as the next glycosyl donor afforded octasaccharide **44**, from which sialyl dimeric Le[x] nonasaccharide **45** was prepared by the final fucosylation with **40**. The anomeric transformation of **45** was achieved according to the procedure shown in Fig. 6 to afford the target ganglioside.

The experimental procedures for the selected sialyl-oligosaccharide syntheses are described in the following section.

## IV. EXPERIMENTAL PROCEDURES

### A. Sialylation of 3-O-Benzoyl-galactose Acceptor [24,26]

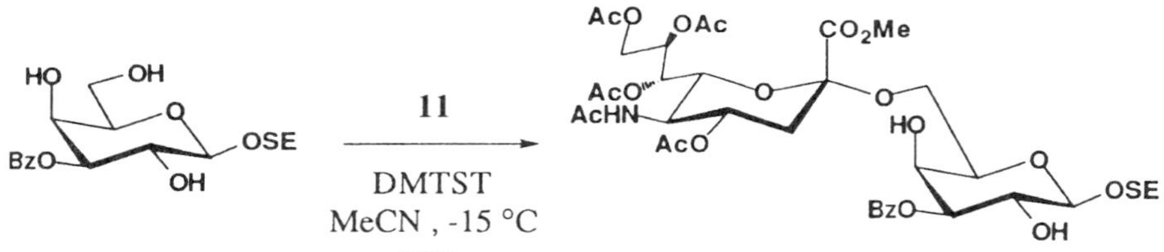

**16** + **11** →(DMTST, MeCN, −15 °C, 70%)→ **22α** (α only)

To a solution of **16** (1.0 g, 2.6 mmol) and **11** (2.7 g, 5.2 mmol) in dry MeCN (20 mL) was added powdered 3-Å molecular sieves (MS-3 Å; 3g), and the mixture was stirred for 6 h at room temperature, then cooled to −30°C. A mixture (6.5 g, 62% DMTST by weight) of DMTST (3.0 Eq. relative to the donor) and MS-3 Å (3 g) was added, and stirring was continued for 17 h at −15°C. Methanol (1 mL) was added to the mixture, and it was neutralized with triethylamine. The solids were filtered off and washed thoroughly with $CH_2Cl_2$, and the combined filtrate and washings was concentrated. The residue was taken up into $CH_2Cl_2$, and successively washed with 1-$M$ aqueous $Na_2CO_3$ and water, dried ($Na_2SO_4$), and concentrated. Column chromatography (EtOAc–hexane, 1:1) of the residue on silica gel afforded amorphous **22α** (1.56 g, 70%): $[\alpha]_D$ −6.4° ($c$ 0.4, $CHCl_3$).

### B. Sialylation of 6-O-Benzoyl-galactose Acceptor [27,36]

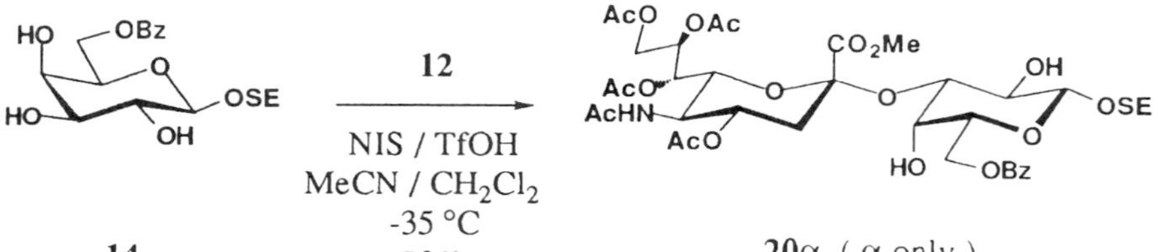

**14** + **12** →(NIS / TfOH, MeCN / $CH_2Cl_2$, −35 °C, 70%)→ **20α** (α only)

To a solution of **12** (10.7 g, 18.3 mmol) and **14** (3.84 g, 10 mmol) in dry MeCN (50 mL) and $CH_2Cl_2$ (5 mL) was added powdered MS-3 Å (20 g), and the mixture was stirred overnight at room temperature, then cooled to −35°C. $N$-Iodosuccinimide (NIS; 8.28 g, 36.8 mmol) and trifluoromethanesulfonic acid (TfOH; 540 mg, 3.6 mmol) were added to the stirred mixture at −35°C under $N_2$ atmosphere, and the stirring was continued for 2.5 h at −35°C. The resulting dark brown suspension was neutralized with triethylamine and filtered. The solids were washed thoroughly with $CH_2Cl_2$, and the combined filtrate and washings was concentrated. The residue was taken up into $CH_2Cl_2$, and successively washed with 1-$M$ aqueous $Na_2S_2O_3$, saturated aqueous $NaHCO_3$ and water, dried ($Na_2SO_4$), and concentrated. Column chromatography (EtOAc–hexane, 1:1) of the residue on silica gel yielded amorphous **20** α (6.0 g, 70%): $[α]_D$ −6.0° (c 2.0, $CHCl_3$).

## C  Sialylation of Penta-O-benzyl-lactose Acceptor [27,52]

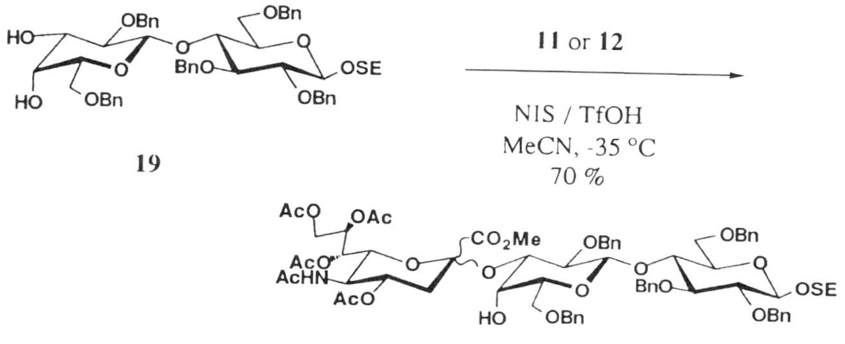

To a solution of **19** (1.0 g, 1.25 mmol) and **11** (1.42 g, 2.49 mmol) in dry MeCN (10 mL) was added powdered MS-3 Å (3.0 g), and the mixture was stirred for 5 h at room temperature, then cooled to −35°C. To the cooled mixture were added, with stirring, NIS (1.12 g, 4.98 mmol) and TfOH (44 μL, 0.50 mmol), and the reaction was continued for 2 h at −35°C. The workup as described for **20α** (see experiment B) and column chromatography (toluene–MeOH, 50:1) on silica gel gave the α-glycoside **25α** (980 mg, 59%), $[α]_D$ +4.2° (c 0.9, $CHCl_3$) and the β-glycoside **25β** (166 mg, 10%): $[α]_D$ +2° (c 1.0, $CHCl_3$), as amorphous masses.

When the sialylation of **19** (1.20 g, 1.34 mmol) was performed with **12** (1.19 g, 2.01 mmol) in dry MeCN (20 mL) in the presence of NIS (695 mg, 3.09 mmol) and TfOH (48 μL, 0.54 mmol) for 6 h at −35°C, as just described, the α-glycoside **25α** (1.10 g, 60%) and the β-glycoside **25β** (199 mg, 11%) were obtained.

## D.  Sialylation of Lewis X–Relevant Trisaccharide [84]

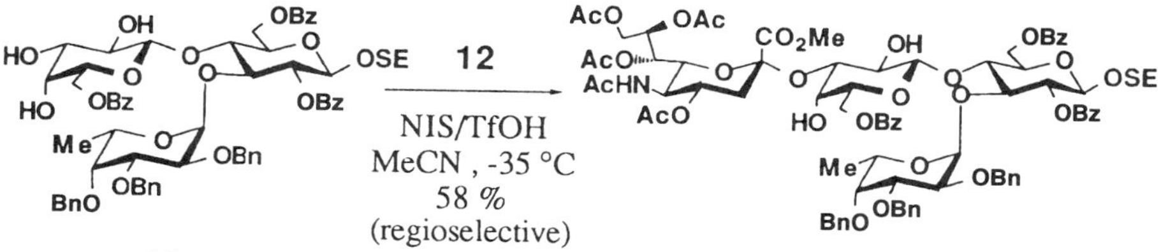

To a solution of **46** (500 mg, 0.43 mmol) and **12** (420 mg, 0.72 mmol) in 5 mL of dry MeCN was added powdered MS-3 Å (3 g), and the mixture was stirred for 5 h at room temperature, then cooled to −35°C. NIS (562 mg, 2.5 mmol) and TfOH (64 μL, 0.73 mmol) were added to the stirred mixture at −35°C, and the stirring was continued overnight at −35°C. The resulting reaction mixture was worked up as described previously. Column chromatography (toluene–MeOH, 50:1) on silica gel gave **47** (410 mg, 58%) as an amorphous mass: $[\alpha]_D$ −9.0° (c 1.8, CHCl$_3$).

### E. Synthesis of Sialyl Lacto- and Neolactotetraose Derivatives [68]

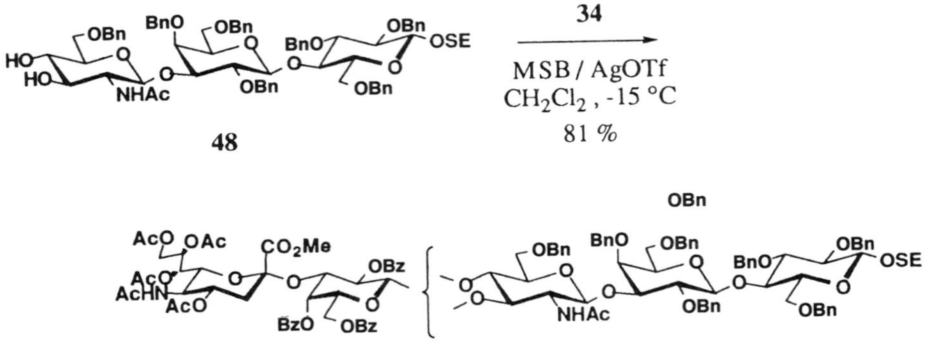

**49** β (1→3), 45%
**50** β (1→4), 36%

To a solution of **48** (1.34 g, 1.05 mmol) and **34** (1.36 g, 1.37 mmol) in dry CH$_2$Cl$_2$ (36 mL) were added powdered MS-4 Å (6.0 g), and the mixture was stirred for 8 h at room temperature. Silver trifluoromethanesulfonate (539 mg, 2.10 mmol) was added to the mixture, which was then cooled to −15°C. Methyl sulfenyl bromide (MSB) solution [85] (2.2 mL, 2.20 mmol) was injected in two equal portions at an interval of 30 min, and the mixture was stirred for 24 h at −15°C. Methanol (1 mL) and triethylamine (0.5 mL) were added to the mixture, and the precipitates were removed by filtration and washed with CH$_2$Cl$_2$. The combination of filtrate and washings was washed with water, dried (Na$_2$SO$_4$), and concentrated. Column chromatography of the residue on silica gel was performed using (a) 6:1 and (b) 4:1 CH$_2$Cl$_2$–acetone as the eluants. Eluant (a) gave the β(1→3)-linked product **49** (1.06 g, 45%), and eluant (b) afforded the β(1→4)-linked product **50** (847 mg, 36%) as an amorphous mass, respectively: **49**, $[\alpha]_D$ +14° (c 1.23, CHCl$_3$); **41** $[\alpha]_D$ +26° (c 1.38, CHCl$_3$).

### F. Synthesis of Ganglioside GD$_2$ Oligosaccharide [56]

To a solution of **51** (1.0 g, 0.82 mmol) and **26** (1.5 g, 1.6 mmol) in dry MeCN (5 mL) was added MS-3 Å (1.1 g), and the mixture was stirred for 5 h at room temperature, then cooled to −25°C. To the cooled mixture were added NIS (750 mg, 3.3 mmol) and TfOH (30 μL, 0.34 mmol), and stirring was continued for 24 h at −25°C. The solids were filtered off, and washed with CH$_2$Cl$_2$. The combination of filtrate and washings was successively washed with 1 M Na$_2$CO$_3$ and 1 M Na$_2$S$_2$O$_3$, dried (Na$_2$SO$_4$), and concentrated. Column chromatography (CH$_2$Cl$_2$–MeOH, 80:1) of the residue on silica gel gave **52** (835 mg, 50%) as an amorphous mass: $[\alpha]_D$ −31° (c 0.7, CHCl$_3$)

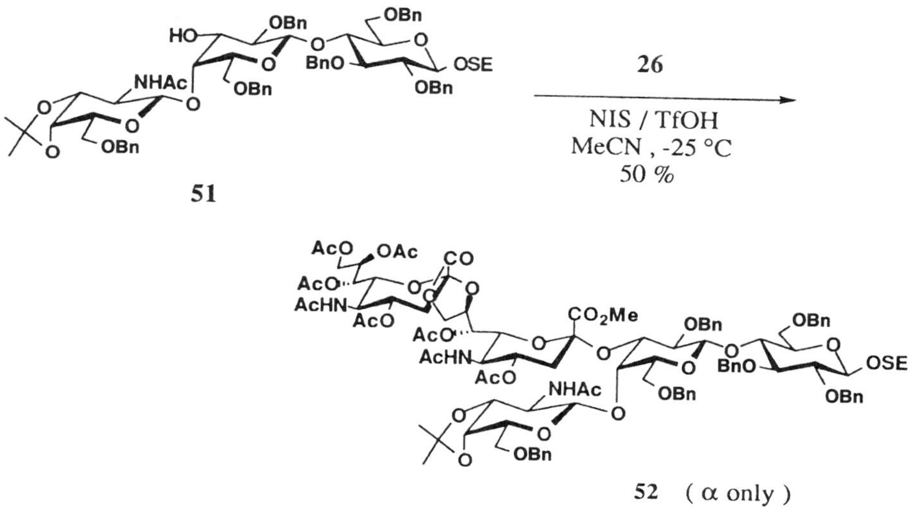

# Synthesis of Sialyl Glycosides

**51** + **26** → **52** (α only)

Reagents: NIS / TfOH, MeCN, −25 °C, 50 %

## G. Synthesis of Ganglioside GQ$_{1b}$ Oligosaccharide [57,58]

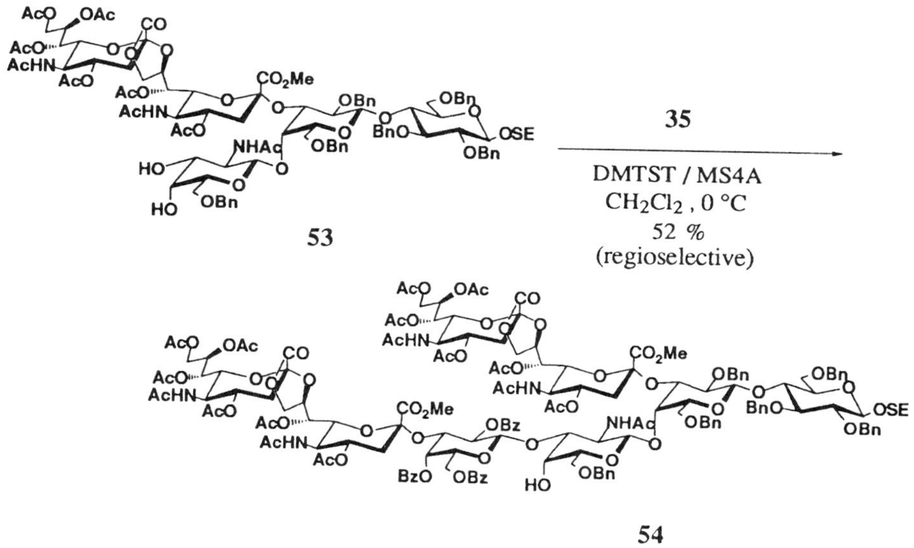

**53** + **35** → **54**

Reagents: DMTST / MS4A, CH$_2$Cl$_2$, 0 °C, 52 % (regioselective)

To a solution of **53** (500 mg, 0.25 mmol) and **35** (500 mg, 0.38 mmol) in CH$_2$Cl$_2$ (9.6 mL) was added MS-4 Å (1.9 g), and the mixture was stirred for 5 h at room temperature, then cooled to 0°C. To the cooled mixture was added DMTST (200 mg, 0.76 mmol), and the stirring was continued for 2 days at 0°C. The solids were filtered off, and washed with 1 M Na$_2$CO$_3$ and water, dried (Na$_2$SO$_4$), and concentrated. Column chromatography (CH$_2$Cl$_2$–MeOH, 20:1) of the residue on silica gel afforded **54** (430 mg, 52%) as an amorphous mass: $[\alpha]_D$ −17° (c 0.9, CHCl$_3$).

## H. Synthesis of Ganglioside GM$_2$-Relevant Oligosaccharide [61]

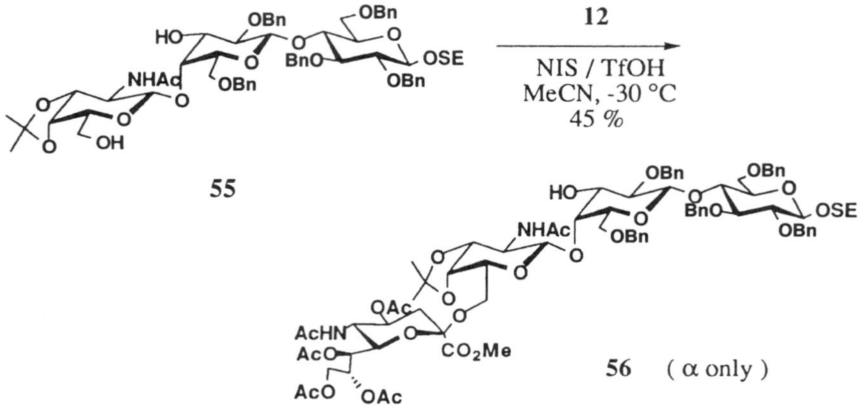

To a solution of **12** (308 mg, 0.53 mmol) and **55** (300 mg, 0.26 mmol) in dry MeCN (20 mL) was added powdered MS-3 Å (1.0 g), and the mixture was stirred for 5 h at room temperature, then cooled to −30°C. NIS (240 mg, 1.1 mmol) and TfOH (5 μL, 0.06 mmol) were added, and stirring was continued for 2 h at −30°C. The solids were removed by filtration and washed with CH$_2$Cl$_2$. The combination of filtrate and washings was successively washed with 1 M Na$_2$CO$_3$ and 1 M Na$_2$S$_2$O$_3$, dried (Na$_2$SO$_4$), and concentrated. Column chromatography (CH$_2$Cl$_2$–MeOH, 50:1) of the residue on silica gel gave **56** (191 mg, 45%), as amorphous mass: [α]$_D$ +14° (c 1.7, CHCl$_3$).

## I. Synthesis of Ganglioside GT$_2$-Relevant Oligosaccharide [61]

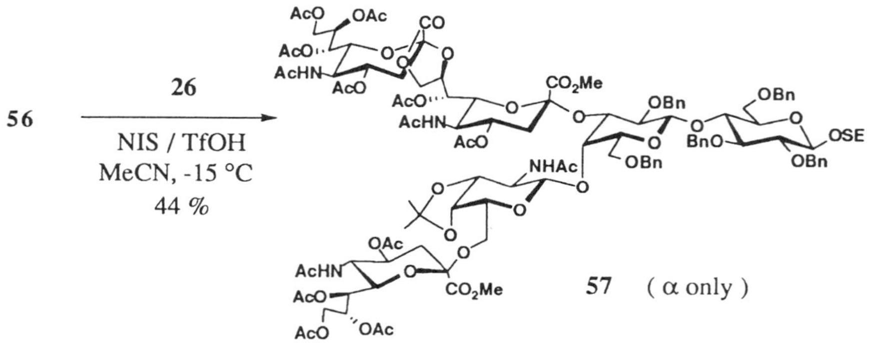

A mixture of **26** (1.75 g, 1.86 mmol), **56** (1.50 g, 0.93 mmol), and powdered MS-3 Å (4 g) was stirred for 5 h at room temperature, then cooled to −15°C. To the mixture were added NIS (963 mg, 4.3 mmol) and TfOH (40 μL, 0.45 mmol), and the stirring was continued for 2 days at −15°C. The workup as just described for **56**, and column chromatography (CH$_2$Cl$_2$–MeOH, 20:1) on silica gel afforded **57** (1.01 g, 44%) as an amorphous mass: [α]$_D$ −14° (c 0.4, CHCl$_3$)

## J. Synthesis of Ganglioside GQ$_{1b\alpha}$ Oligosaccharide [61]

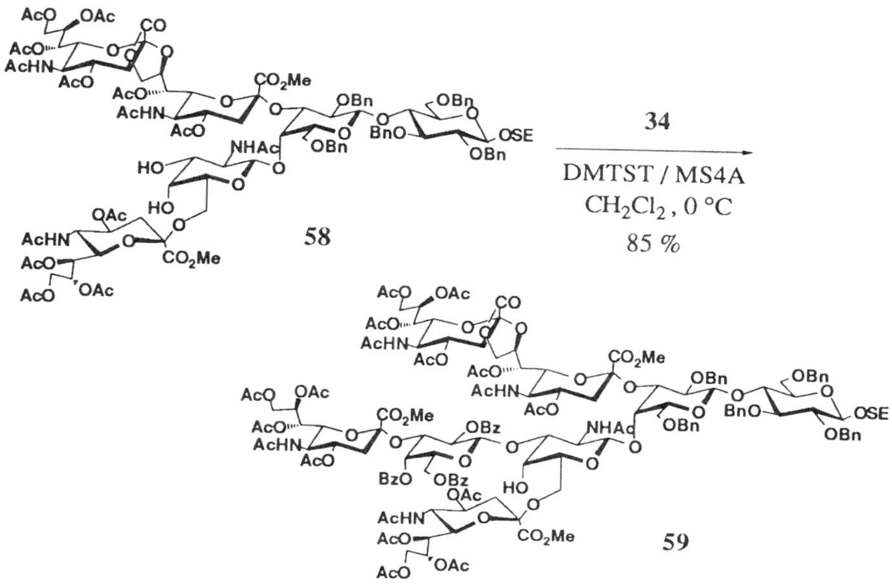

To a solution of **34** (415 mg, 0.42 mmol) and **58** (400 mg, 0.17 mmol) in dry $CH_2Cl_2$ (20 mL) was added MS-4 Å (2.0 g), and the mixture was stirred for 5 h at room temperature, then cooled to 0°C. DMTST (200 mg, 0.76 mmol) was added and the stirring was continued for 2 days at 0°C. The workup as described for **22α** (see experiment A) and column chromatography ($CH_2Cl_2$–MeOH, 20:1) on silica gel gave **59** (475 mg, 85%), as an amorphous mass: $[\alpha]_D$ −7° (c 1.4, $CHCl_3$).

## REFERENCES

1. W. Reutter, E. Köttgen, C. Bauer, and W. Gerok, Biological significance of sialic acid. *Sialic Acids: Chemistry, Metabolism and Function, Cell Biology Monographs*, Vol. 10 (R. Schauer, ed.), Springer-Verlag, New York, 1982, p. 263.
2. H. Wiegandt, Gangliosides. *Glycolipids, New Comprehensive Biochemistry*, Vol. 10 (H. Weigandt, ed.), Elsevier, Amsterdam, 1985, p. 199.
3. A. Singhal and S. Hakomori, Molecular changes in carbohydrate antigens associated with cancer, *BioEssays* 12:223 (1990).
4. K. Furukawa and A. Kobata, Cell surface carbohydrates—their involvement in cell adhesion. *Carbohydrates: Synthetic Methods and Applications in Medicinal Chemistry*. (H. Ogura, A. Hasegawa, and T. Suami, eds.), Kodansha/VCH, Tokyo, 1992, p. 369.
5. N. Sharon and H. Lis, Carbohydrates in cell recognition, *Sci. Am.* 268:74 (1993).
6. G. Xu, T. Suzuki, H. Tahara, M. Kiso, A. Hasegawa, and Y. Suzuki, Specificity of sialyl-sugar chain mediated recognition by the hemagglutinin of human influenza B virus isolates, *J. Biochem. (Tokyo)* 115:202 (1994), and references therein.
7. F. Vandekerckhove, S. Schenkman, L. Pontes de Carvalho, S. Tomlinson, M. Kiso, M. Yoshida, A. Hasegawa, and V. Nussenzweig, Substrate specificity of the *Trypanosoma cruzi* trans-sialidase, *Glycobiology* 2:541 (1992).

8. L. A. Lasky, Selectins: Interpreters of cell-specific carbohydrate information during inflammation, *Science* 258:964 (1992), and references therein.
9. R. Feizi, Oligosaccharides that mediate mammalian cell–cell adhesion, *Curr. Opin. Struct. Biol.* 3:701 (1993), and references therein.
10. B. K. Brandley, M. Kiso, S. Abbas, P. Nikrad, O. Srivastava, C. Foxall, Y. Oda, and A. Hasegawa, Structure–function studies of selectin carbohydrate ligands. Modification to fucose, sialic acid and sulphate as a sialic acid replacement, *Glycobiology* 3:633 (1993).
11. A. Takada, K. Ohmori, T. Yoneda, K. Tsuyuoka, A. Hasegawa, M. Kiso, and R. Kannagi, Contribution of carbohydrate antigens sialyl Lewis A and sialyl Lewis X to adhesion of human cancer cells to vascular endothelium, *Cancer Res.* 53:354 (1993).
12. Y. Nagai and S. Tsuji, Significance of ganglioside-mediated glycosignal transduction in neuronal differentiation and development, *Prog. Brain Res.* 101:119 (1993).
13. C. Zuber, P. M. Lackie, W. A. Cattarall, and J. Roth, Polysialic acid is associated with sodium channels and the neural cell adhesion molecule N-CAN in adult rat brain, *J. Biol. Chem.* 267:9965 (1992).
14. C. Sato, K. Kitajima, I. Tazawa, Y. Inoue, S. Inoue, and F. A. Troy, Structural diversity in the $\alpha$-2$\rightarrow$8-linked polysialic acid chains in salmonid fish egg glycoproteins, *J. Biol. Chem.* 268:23675 (1993).
15. K. Okamoto and T. Goto, Glycosidation of sialic acid, *Tetrahedron* 46:5835 (1990).
16. M. P. DeNinno, The synthesis and glycosidation of N-acetylneuraminic acid, *Synthesis*, p. 583 (1991).
17. Y. Itoh and T. Ogawa, Highly stereoselective glycosylation of N-acetylneuraminic acid aided by a phenylthio substituent as a stereocontrolling auxiliary, *Tetrahedron Lett.* 29:3987 (1988).
18. T. Kondo, H. Abe, and T. Goto, Efficient synthesis of 2$\alpha$-glycoside of N-acetylneuraminic acid via phenylsulfenyl chloride adduct of 2-deoxy-2,3-dehydro-N-acetylneuraminic acid methyl ester tetra-O-acetate, *Chem. Lett.*, p. 1657 (1988).
19. Y. Itoh, S. Nunomura, S. Shibayama, and T. Ogawa, Studies directed toward the synthesis of polysialogangliosides: The regio- and stereocontrolled synthesis of rationally designed fragment of the tetrasialoganglioside $GQ_{1b}$, *J. Org. Chem.* 57:1821 (1992).
20. A. Hasegawa, J. Nakamura, and M. Kiso, Synthesis of alkyl $\alpha$-glycosides of 2-thio-N-acetylneuraminic acid, *J. Carbohydr. Chem.* 5:11 (1986).
21. O. Kanie, J. Nakamura, M. Kiso, and A. Hasegawa, Stereoselective synthesis of 5'-S-(5-acetamido-3,5-dideoxy-D-*glycero*-$\alpha$- and -$\beta$-D-*galacto*-2-nonulopyranosylonic acid)-5'-thiocytidine, *J. Carbohydr. Chem.* 6:105 (1987).
22. O. Kanie, M. Kiso, and A. Hasegawa, Glycosylation using methylthioglycosides of N-acetylneuraminic acid and dimethyl(methylthio)sulfonium triflate, *J. Carbohydr. Chem.* 7:501 (1988).
23. K. Toshima and K. Tatsuta, Recent progress in O-glycosylation methods and its application to natural products synthesis, *Chem. Rev.* 93:1503 (1993).
24. T. Murase, H. Ishida, M. Kiso, and A. Hasegawa, A facile regio- and stereoselective synthesis of $\alpha$-glycosides of N-acetylneuraminic acid, *Carbohydr. Res.* 184:C1 (1988).
25. P. Fügedi and P. J. Garegg, A novel promoter for the efficient construction of 1,2-*trans* linkages in glycoside synthesis, using thioglycosides as glycosyl donors, *Carbohydr. Res.* 149:C9 (1986).
26. A. Hasegawa, H. Ohki, T. Nagahama, H. Ishida, and M. Kiso, A facile, large-scale preparation of the methyl 2-thioglycoside of N-acetylneuraminic acid, and its usefulness for the $\alpha$-stereoselective synthesis of sialoglycosides, *Carbohydr. Res.* 212:277 (1991).
27. A. Hasegawa, T. Nagahama, H. Ohki, K. Hotta, H. Ishida, and M. Kiso, Reactivity of glycosyl promoters in $\alpha$-glycosylation of N-acetylneuraminic acid with the primary and secondary hydroxyl groups in the suitably protected galactose and lactose derivatives, *J. Carbohydr. Chem.* 10:493 (1991).
28. Y. Ito and T. Ogawa, Benzeneselenenyl triflate as a promoter of thioglycosides: A new method for O-glycosylation using thioglycosides, *Tetrahedron Lett.* 29:1061 (1988).

29. A. Marra and P. Sinäy, A novel stereoselective synthesis of N-acetyl-α-neuraminosyl-galactose disaccharide derivatives, using anomeric S-glycosyl xanthates, *Carbohydr. Res. 195*:303 (1990).
30. H. Lönn and K. Stenvall, Exceptionally high yield in glycosylation with sialic acid. Synthesis of a GM$_3$ glycoside, *Tetrahedron Lett. 33*:115 (1992).
31. T. J. Martin and R. R. Schmidt, Efficient sialylation with phosphite as leaving group, *Tetrahedron Lett. 33*:6123 (1992).
32. M. M. Sim, H. Kondo, and C.-H. Wong, Synthesis and use of glycosyl phosphites: An effective route to glycosyl phosphates, sugar nucleotides, and glycosides, *J. Am. Chem. Soc. 115*:2260 (1993).
33. G. H. Veeneman, S. H. van Leeuwen, and J. H. van Boom, Iodonium ion promoter reactions at the anomeric centre. II. An efficient thioglycoside mediated approach toward the formation of 1,2-*trans* linked glycosides and glycosidic esters, *Tetrahedron Lett. 31*:1331 (1990).
34. P. Konradsson, U. E. Udodong, and B. Fraser-Reid, Iodonium promoted reactions of disarmed thioglycosides, *Tetrahedron Lett. 31*:4313 (1990).
35. M. Kiso and A. Hasegawa, Synthesis of ganglioside GM$_3$ and analogs containing modified sialic acids and ceramides, *Methods Enzymol. 242*:171 (1994).
36. A. Hasegawa, Synthesis of sialo-oligosaccharides and their ceramide derivatives as tools for elucidation of biologic functions of gangliosides, *Synthetic Oligosaccharides—Indispensable Probes for the Life Sciences, ACS Symp. Ser.* 560 (P. Kovác, ed.), ACS, Washington DC, 1994, p. 184.
37. A. Hasegawa, M. Ogawa, H. Ishida, and M. Kiso, α-Predominant glycoside synthesis of N-acetylneuraminic acid with the primary hydroxyl group in carbohydrates using dimethyl-(methylthio)sulfonium triflate as a glycosyl promoter, *J. Carbohydr. Chem. 9*:393 (1990).
38. A. Marra and P. Sinäy, Stereoselective synthesis of 2-thioglycosides of N-acetylneuraminic acid, *Carbohydr. Res. 187*:35 (1989).
39. T. Terada, M. Kiso, and A. Hasegawa, Synthesis of KDN-gangliosides GM$_4$ and GM$_3$, *J. Carbohydr. Chem. 12*:425 (1993).
40. R. Roy and R. A. Pon, Efficient synthesis of α(2→8)-linked N-acetyl and N-glycolylneuraminic acid disaccharides from colominic acid, *Glycoconjugate J. 7*:3 (1990).
41. H.-K. Ishida, Y. Ohta, Y. Tsukada, M. Kiso, and A. Hasegawa, A synthetic approach to polysialogangliosides containing α-sialyl-(2→8)-sialic acid: Total synthesis of ganglioside GD$_3$, *Carbohydr. Res. 246*:75 (1993).
42. H.-K. Ishida, H. Ishida, M. Kiso, and A. Hasegawa, α-Stereocontrolled, glycoside synthesis of trimeric sialic acid with galactose and lactose derivatives, *J. Carbohydr. Chem. 13*:655 (1994).
43. A. J. Ratcliffe and B. Fraser-Reid, Generation of α-D-glycopyranosylacetonitrilium ions. Concerning the reverse anomeric effect, *J. Chem. Soc. Perkin Trans. 1*:747 (1990).
44. R. R. Schmidt, M. Behrendt, and A. Toepfer, Nitriles as solvents in glycosylation reactions: Highly selective β-glycoside synthesis, *Synlett* p. 694 (1990).
45. I. Braccini, C. Derouet, J. Esnault, C. Hervé du Penhoat, J.-M. Mallet, V. Michon, and P. Sinäy, Conformational analysis of nitrilium intermediates in glycosylation reactions, *Carbohydr. Res. 246*:23 (1993), and references therein.
46. C. L. Perrin and K. B. Armstrong, Conformational analysis of glucopyranosylammonium ions: Does the reverse anomeric effect exist? *J. Am. Chem. Soc. 115*:6825 (1993).
47. M. A. Fabian, C. L. Perrin, and M. L. Sinnott, Absence of reverse anomeric effect: Conformational analysis of glucosylimidazolium and glucosylimidazole, *J. Am. Chem. Soc. 116*:8398 (1994), and references therein.
48. K. Jansson, S. Ahlforts, T. Frejd, J. Kihlberg, G. Magnusson, J. Dahmén, G. Noori, and K. Stenvall, 2-(Trimethylsilyl)ethyl glycosides. Synthesis, anomeric deblocking, and transformation into 1,2-*trans* 1-*O*-acyl sugars, *J. Org. Chem. 53*:5629 (1988).
49. A. Hasegawa and M. Kiso, Systematic synthesis of gangliosides toward the elucidation and

biomedical application of their biological functions, *Carbohydrates—Synthetic Methods and Applications in Medicinal Chemistry* (H. Ogura, A. Hasegawa, and T. Suami, eds.), Kodansha/VCH, Tokyo, 1992, p. 243.

50. T. Murase, A. Kameyama, K. P. R. Kartha, H. Ishida, M. Kiso, and A. Hasegawa, A facile, regio and stereoselective synthesis of ganglioside $GM_4$ and its position isomer, *J. Carbohydr. Chem.* 8:265 (1989).

51. T. Murase, H. Ishida, M. Kiso, and A. Hasegawa, A facile, regio- and stereoselective synthesis of ganglioside $GM_3$, *Carbohydr. Res. 188*:71 (1989).

52. A. Hasegawa, T. Nagahama, H. Ohki, and M. Kiso, A facile total synthesis of ganglioside $GM_2$, *J. Carbohydr. Chem. 11*:699 (1992).

53. H. Prabhanjan, A. Kameyama, H. Ishida, M. Kiso, and A. Hasegawa, Regio- and stereoselective synthesis of ganglioside $GM_{1b}$ and some positional analogs, *Carbohydr. Res. 220*:127 (1991).

54. A. Hasegawa, T. Nagahama, and M. Kiso, A facile, systematic synthesis of ganglio-series gangliosides: Total synthesis of gangliosides $GM_1$ and $GD_{1a}$, *Carbohydr. Res. 235*:C13 (1992).

55. A. Hasegawa, H.-K. Ishida, T. Nagahama, and M. Kiso, Total synthesis of gangliosides $GM_1$ and $GD_{1a}$, *J. Carbohydr. Chem. 12*:703 (1993).

56. H.-K. Ishida, Y. Ohta, Y. Tsukada, Y. Isogai, H. Ishida, M. Kiso, and A. Hasegawa, A facile total synthesis of ganglioside $GD_2$, *Carbohydr. Res. 252*:283 (1994).

57. H.-K. Ishida, H. Ishida, M. Kiso, and A. Hasegawa, Total synthesis of ganglioside $GQ_{1b}$ and the related polysialogangliosides, *Tetrahedron Asymm. 5*:2493 (1994).

58. H.-K. Ishida, H. Ishida, M. Kiso, and A. Hasegawa, Total synthesis of ganglioside $GQ_{1b}$, *Carbohydr. Res. 260*:C1 (1994).

59. H. Prabhanjan, K. Aoyama, M. Kiso, and A. Hasegawa, Synthesis of disialoganglioside $GD_{1\alpha}$ and its positional isomer, *Carbohydr. Res. 233*:87 (1992).

60. K. Hotta, S. Komba, H. Ishida, M. Kiso, and A. Hasegawa, Synthesis of $\alpha$-series ganglioside $GM_{1\alpha}$, *J. Carbohydr. Chem. 13*:665 (1994).

61. K. Hotta, H. Ishida, M. Kiso, and A. Hasegawa, First total synthesis of a cholinergic neuron-specific ganglioside $GQ_{1b\alpha}$, *J. Carbohydr. Chem. 14*:491 (1995).

62. A. Kameyama, H. Ishida, M. Kiso, and A. Hasegawa, Total synthesis of sialyl lactotetraosyl ceramide, *Carbohydr. Res. 193*:C1 (1989).

63. A. Kameyama, H. Ishida, M. Kiso, and A. Hasegawa, Stereoselective synthesis of sialyl-lactotetraosylceramide and sialylneolactotetraosylceramide, *Carbohydr. Res. 200*:269 (1990).

64. A. Hasegawa, K. Hotta, A. Kameyama, H. Ishida, and M. Kiso, Total synthesis of sialyl-$\alpha(2\rightarrow 6)$-lactotetraosylceramide and sialyl-$\alpha(2\rightarrow 6)$-neolactotetraosylceramide, *J. Carbohydr. Chem. 10*:439 (1991).

65. A. Kameyama, H. Ishida, M. Kiso, and A. Hasegawa, Total synthesis of tumor-associated ganglioside, sialyl Lewis X, *J. Carbohydr. Chem. 10*:549 (1991).

66. A. Kameyama, H. Ishida, M. Kiso, and A. Hasegawa, Synthesis of sialyl-$\alpha(2\rightarrow 6)$-Lewis X, *J. Carbohydr. Chem. 10*:729 (1991).

67. A. Hasegawa, T. Ando, A. Kameyama, and M. Kiso, Stereocontrolled synthesis of sialyl Lewis X epitope and its ceramide derivative, *J. Carbohydr. Chem. 11*:645 (1992).

68. A. Kameyama, H. Ishida, M. Kiso, and A. Hasegawa, Total synthesis of tumor-associated ganglioside, sialyl-Le[a], *J. Carbohydr. Chem. 13*:641 (1994).

69. M. Yoshida, A. Uchimura, M. Kiso, and A. Hasegawa, Synthesis of chemically modified sialic acid-containing sialyl-Le[x] ganglioside analogues recognized by the selectin family, *Glycoconjugate J. 10*:3 (1993).

70. A. Hasegawa, T. Ando, M. Kato, H. Ishida, and M. Kiso, Synthesis of deoxy-L-fucose-containing sialy Lewis X ganglioside analogues, *Carbohydr. Res. 257*:67 (1994).

71. Y. Suzuki, T. Nakao, T. Ito, N. Watanabe, Y. Toda, X. Guiyun, T. Suzuki, T. Kobayashi, Y. Kimura, A. Yamada, K. Sugawara, H. Nishimura, F. Kitame, K. Nakamura, E. Deya, M. Kiso, and A. Hasegawa, Structural determination of gangliosides that bind to influenza A, B, and C

viruses by an improved binding assay: Strain-specific receptor epitopes in sialo-sugar chains, *Virology 189*:121 (1992).

72. M. J. Polley, M. L. Phillips, E. Wayner, E. Nudelman, A. K. Singhal, S. Hakomori, and J. C. Paulson, CD62 and endothelial cell–leukocyte adhesion molecule 1 (ELAM-1) recognize the same carbohydrate ligand, sialyl-Lewis X, *Proc. Natl. Acad. Sci. USA 88*:6224 (1991).

73. D. Tyrrell, P. James, N. Rao, C. Foxall, S. Abbas, F. Dasgupta, M. Nashed, A. Hasegawa, M. Kiso, D. Asa, J. Kidd, and B. K. Brandley, Structural requirements for the carbohydrate ligand of E-selectin, *Proc. Natl. Acad. Sci. USA 88*:10372 (1991).

74. C. Foxall, S. R. Watson, D. Dowbenko, C. Fennie, L. A. Lasky, M. Kiso, A. Hasegawa, D. Asa, and B. K. Brandley, The three members of the selectin receptor family recognize a common carbohydrate epitope, the sialyl Lewis X oligosaccharide, *J. Cell Biol. 117*:895 (1992).

75. M. Larkin, T. J. Ahern, M. S. Stoll, M. Shaffer, D. Sako, J. O'Brien, C.-T. Yuen, A. M. Lawson, R. A. Childs, K. M. Barone, P. R. Langer-Safer, A. Hasegawa, M. Kiso, G. R. Larsen, and T. Feizi, Spectrum of sialylated and nonsialylated fuco-oligosaccharides bound by the endothelial–leukocyte adhesion molecule E-selectin, *J. Biol. Chem. 267*:13661 (1992).

76. P. Kotovuori, E. Tontti, R. Pigott, M. Shepherd, M. Kiso, A. Hasegawa, R. Renkonen, P. Nortamo, D. C. Altieri, and C. G. Gahmberg, The vascular E-selectin binds to the leukocyte integrins CD11/CD18, *Glycobiology 3*:131 (1993).

77. T. Terada, M. Kiso, and A. Hasegawa, Synthesis of KDN-lactotetraosylceramide, KDN-neolactotetraosylceramide, and KDN-Lewis X ganglioside, *Carbohydr. Res. 259*:201 (1994).

78. S. Sabesan, S. Neira, F. Davidson, J. Duus, and K. Bock, Synthesis and enzymatic and NMR-studies of novel sialoside probes: Unprecedented, selective neuraminidase hydrolysis of and inhibition by C-6-(methyl)-Gal sialosides, *J. Am. Chem. Soc. 116*:1616 (1994).

79. S. Hanessian and H. Prabhanjan, Design and synthesis of glycomimetic prototypes-a model sialyl Lewis$^x$ ligand for E-selectin, *Synlett* p. 868 (1994).

80. W. Stahl, U. Sprengard, G. Kretzschmar, and H. Kunz, Synthesis of deoxy sialyl Lewis$^x$ analogues, potential selectin antagonists, *Angew. Chem. Int. Ed. Engl. 33*:2096 (1994).

81. R. K. Jain, R. Vig, R. Rampal, E. V. Chandrasekaran, and K. L. Matta, Total synthesis of 3'-O-sialyl, 6'-O-sulfo Lewis$^x$, Neu5cα2 → 3(6-O-SO$_3$Na)Galβ1 → 4(Fucα1→3)-GlcNAcβ-OMe: A major cappling group of GLYCAM-I, *J. Am. Chem. Soc. 116*:12123 (1994).

82. A. Hasegawa and M. Kiso, Synthesis of sialyl Lewis X ganglioside and analogs, *Methods Enzymol. 242*:158 (1994).

83. A. Kameyama, T. Ehara, Y. Yamada, H. Ishida, M. Kiso, and A. Hasegawa, A total synthesis of sialyl dimeric Le$^x$ ganglioside, *J. Carbohydr. Chem.*, Ottawa Symposium Issue (1995) in press.

84. A. Hasegawa, K. Fushimi, H. Ishida, and M. Kiso, Synthesis of sialyl Lewis X analogs containing azidoalkyl groups at the reducing end, *J. Carbohydr. Chem. 12*:1203 (1993).

85. F. Dasgupta and P. J. Garegg, Alkyl sulfenyl triflate as activator in the thioglycoside-mediated formation of β-glycosidic linkages during oligosaccharide synthesis, *Carbohydr. Res. 177*:c13 (1988). Also see *Carbohydr. Res. 202*:225 (1990).

# 16

## Glycoside Synthesis Based on the Remote Activation Concept: An Overview

**Stephen Hanessian**
*University of Montreal, Montreal, Quebec, Canada*

| | | |
|---|---|---|
| I. | Introduction | 381 |
| II. | The Challenges of the Glycosidic Bond | 382 |
| III. | The Remote Activation Concept | 383 |
| IV. | New Generations of Glycosyl Donors | 386 |
| | References | 387 |

## I. INTRODUCTION

From a strictly chemical point of view, the synthesis of glycosides still presents a formidable challenge to synthetic chemists in spite of major advances in the area [1]. Unlike peptidic bonds, the formation of the glycosidic linkage is subject to various factors that include, among others, electronic, stereoelectronic, conformational, substituent, and reactivity effects generally associated with incipient oxocarbenium ions derived from carbohydrates.

These factors are mostly concerned with the so-called glycosyl donor molecule. Additional parameters to be considered in the formation of $O$-glycosides involve the nature of the alcohol acceptor, the polarity of the solvent, and the type of catalyst or promoter needed to activate the leaving group at the anomeric carbon of the donor molecule.

Although some methods for glycoside synthesis are more popular than others (e.g., so-called imidate method [2]; see also Chap. 11, and metal ion-promoted reactions [3,4]), there is no universal protocol that can be applied to any combinations of donors and acceptors without consideration of their substituent patterns, configurations, or position of the hydroxy group. The analogy of peptides with oligosaccharides is an appropriate one for comparison, because only a few amino acid–coupling steps can be considered as being potentially problematic (e.g., proline). Fortunately, the threat of racemization, which looms over the peptidic bond, is absent in the formation of glycosidic linkages. Finally, although the automated synthesis of polypeptides of impressive molecular weights is a routine operation, thanks to the revolutionary effect of the Merrifield solid-phase techniques [5,6], the analogous technique in oligosaccharide synthesis is still in a highly exploratory stage [7]. Thus, the solid-phase synthesis of a tetrasaccharide with a predetermined anomeric configuration at every glycosidic linkage and attached to a different hydroxy group in each acceptor molecule, would be an achievement of monumental proportions today, and one with a lasting effect, if only it could be generalized.

## II. THE CHALLENGES OF THE GLYCOSIDIC BOND

If we take the synthesis of a simple trisaccharide molecule, shown in Scheme 1, as a target, we can enumerate the choices, challenges and potential problems listed in the following:

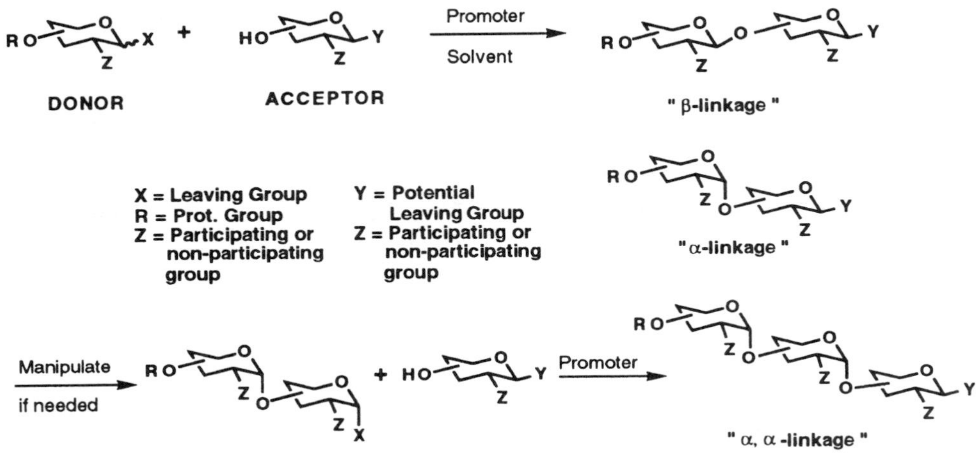

**Scheme 1**

Choices

1. Choice of X and Z in the donor
2. Choice of Y and Z in the acceptor
3. Choice of promoter or catalyst
4. Choice of solvent and temperature
5. Choice of protective groups

Challenges and problems

1. Anomeric selectivity for 1,2-*cis* or 1,2-*trans* linkages
2. Site selectivity and reactivity of acceptor OH groups (e.g., axial, equatorial, primary; D-*gluco*, D-*galacto*, C-3, C-4, or other)
3. Configurational, substituent, steric, and electronic effects in the donor and acceptor (e.g., D-glucopyranosyl and D-galactopyranosyl donors with identical substituents sometimes give different α/β ratios with the *same* acceptor alcohol
4. Stoichiometry relative to donor and acceptor equivalents used
5. Selective activation of anomeric groups (e.g., X, Y exhibit different reactivities as in orthogonal groups; Y can be exchanged to X; Y can be activated after O-protective group manipulation or change of promoter)

6. Iteration and sequential glycosidation in a stepwise manner or by block synthesis (e.g., disaccharide donor and disaccharide acceptor for tetrasaccharide synthesis)
7. Prospects for solid-phase oligosaccharide synthesis with minimum manipulation of protective groups or eliminating such a need altogether
8. Special cases (e.g., sialyl glycosides and oligosaccharides; glycosides of macrolide aglycones)

## III. THE REMOTE ACTIVATION CONCEPT

The venerable and pioneering Fischer glycosidation [8], the first example of a heavy–metal-catalyzed glycosidation reported by Michael [9], as well as recent accomplishments in the field [1] have relied in major part on the activation of a substituent *directly* attached to the anomeric carbon. Thus, protonation of the anomeric hydroxy group in free sugars, as in a Fischer glycosidation, and the treatment of an O-substituted glycosyl halide with heavy-metal salts, or a thioether with a cationic specie, a Lewis acid, or a thiophilic reagent, are examples of direct activation that lead to the dissociation of the anomeric group, with the concomitant formation of an ion-paired oxocarbenium ion, or a dioxolenium ion if neighboring group participation is possible. Subsequent attack by the hydroxy group of the acceptor at the anomeric carbon leads to the intended glycoside.

The well-known effect of C-2 participating (e.g., ester) and nonparticipating (e.g., O-benzyl) groups [1] can be used in predicting the sterochemical outcome of the new glycosidic linkage (e.g., 1,2-*trans* or 1,2-*cis* as major or exclusive products).

Nearly 20 years ago, we considered the prospects of designing anomeric leaving groups that would be adaptable to the synthesis of glycosides without the need for O-protection in the donor [10]. A successful achievement of such a challenging task was expected to have several important consequences in the general area of glycoside synthesis. For example, one could envisage:

1. Glycosidation of simple and complex aglycones without the need for protection of the hydroxy groups in the donor (e.g., macrolide antibiotics)
2. Iterative assembly of saccharide units
3. Possible extension to solid-phase synthesis of glycosides, avoiding protection–deprotection steps
4. Consideration of newer generations of leaving groups from lessons learned

The design of a prototypical leaving group that would activate the anomeric carbon atom in an *unprotected* glycosyl donor presents major challenges and problems, namely:

1. Reactions should be effective with a choice of reagents or catalysts in a relatively short time period.
2. Glycosyl donors should react only with acceptor hydroxy groups and not with themselves, thereby avoiding oligomerization.
3. The anomeric configuration of the newly formed bond should be reasonably well controlled (i.e., major 1,2-*cis* or 1,2-*trans* glycosides).
4. The glycosyl donor should be easily accessible.
5. The glycosylation reaction should be fairly general for a variety of alcohol acceptors.

Interestingly, the original Fischer glycosidation satisfies the first four conditions admirably well using the simplest of glycosyl donors; namely, the free sugar itself.

However, it is not applicable when the alcohol cannot be used as the solvent, as in a cholesteryl α-D-glycopyranoside, for example. In this instance, the alternative glycosylation method with O-benzyl protective groups could be problematic because of the debenzylation step in the presence of the Δ-5 double bond in the aglycone.

The first generation of an O-unprotected glycosyl donor that could be activated based on the remote activation concept was exemplified by 2-pyridylthio β-D-glucopyranoside **1** [10], which could be easily prepared from "acetobromoglucose" [11]. Treatment of **1** with a variety of alcohols in the presence of mercuric nitrate in acetonitrile solution led, *within a few minutes*, to the corresponding O-glycosides, with isolated yields and α/β ratios varying from methanol (70:30, 95%), to 2-propanol (62:38; 77%), and cyclohexanol (51:49, 75%); Scheme **2**. Identical results were obtained when 2-pyrimidinylthio β-D-

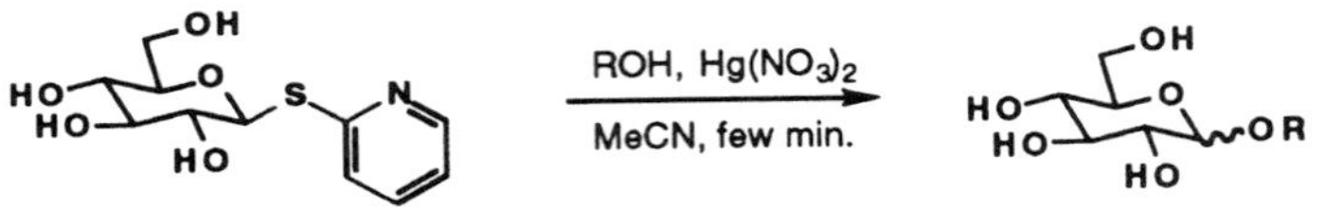

R = Me, Et, Pr, 2-Pr

2,2-diMePr, Cyclohexyl

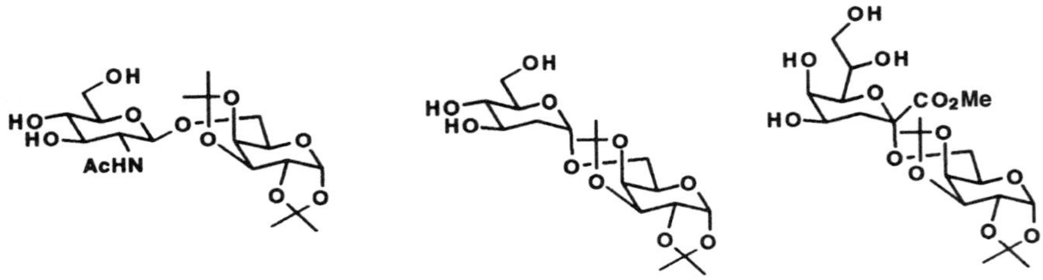

**Scheme 2**

glucopyranoside [12] was used as a glycosyl donor [10]. The synthesis of disaccharides was also possible, but the reaction times were longer (1–3 h). Several disaccharides were thus prepared from the corresponding 2-pyridylthio glycosides according to the foregoing protocol [13].

The conceptual basis for the choice of the leaving group was based on the various modes of activation, shown in Scheme **3**. Thus, with the mercuric salt, the sulfur or nitrogen atoms could be coordinated to generate a reactive 1,2-*trans*-oriented ion pair or loose complex, which could undergo an $S_N2$-like attack by the alcohol. Alternative modes of activation could involve electrophilic species, such as NBS and primary organic halides. Indeed, treatment of the donor with 1-chloropentane and 2-propanol gave, after refluxing for 18 h, the anomeric 2-propyl D-glucopyranosides in 70% yield. With NBS and methanol, it was possible to obtain an 80% yield of methyl α-D-glucopyranoside and the β-anomer (4:1; see Scheme **3**).

Ferrier and co-workers [14] had reported that treatment of phenylthio α-D-glucopyranoside with 2-propanol and mercuric chloride at reflux for 96 h gave a 55% yield of the corresponding 2-propyl α-D-glucopyranoside after acetylation. The exceptionally

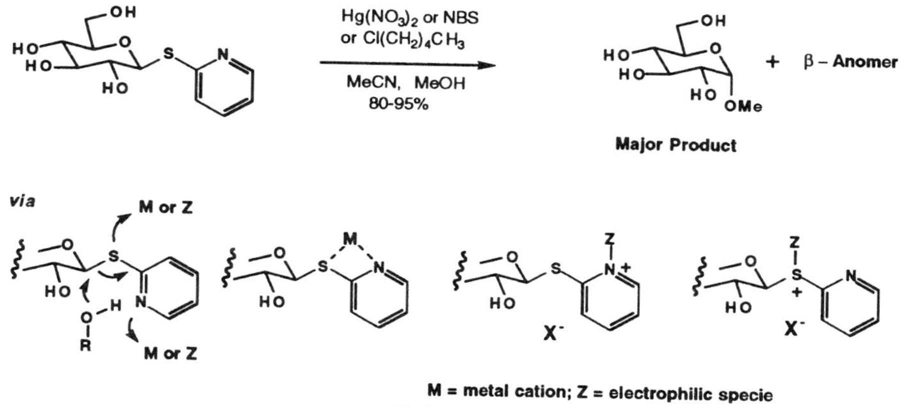

# Glycoside Synthesis Remote Activation Concept

**Scheme 3**

M = metal cation; Z = electrophilic specie

faster glycosidations with 2-pyridylthio and related glycosides [10,13] demonstrates the importance of the pyridyl moiety; hence, our original choice and design of this leaving group [10].

Several anomeric activation methods, developed in recent years for the glycosylation of alcohols and sugars with O-protected donors, also take advantage of electrophilic, thiophilic, or Lewis acidic reagents [1].

In spite of the successful implementation of the concept of remote activation of an anomeric 2-pyridylthio group, there emerged various issues and questions. The absence of self-condensation products could be ascribed to the use of excess acceptor in all cases, and possibly to the coordination of the mercuric salt to the free hydroxy groups of the donor, thereby diminishing their reactivity. However, the necessity to use an excess of acceptor (> 5 eq), the formation of anomeric mixtures of glycosides, and the use of a mercuric salt, limited the generality of this method of glycoside synthesis.

The notion of glycoside synthesis with unprotected glycosyl donors [10] has instigated the exploration of some new approaches using the trichloroacetimidate group [15], electrochemistry [16], electron transfer [17], and Lewis acid catalysis [18]. The scope of applicability of these newer methods remains to be tested.

Several other glycosidation methods can also be considered as proceeding by remote activation. For example, treatment of O-benzyl glycosyl trichloroacetimidates with $BF_3 \cdot Et_2O$ [2; see also Chap. 11], or O-benzyl pentenyl glycosides with NBS [19; see also Chap. 13] leads to activation at an atom not directly attached to the anomeric center.

O-Glycoside synthesis from 2-pyridylthio glycosides is also possible with O-benzyl-protected donors in the presence of methyl iodide [20], silver triflate [21], or mercuric salts [10,13,22]. The versatility of this leaving group has also been demonstrated in the synthesis of C-glycosides [23].

Perhaps the most impressive applications of O-glycosylations with 2-pyridylthio glycosides are in the area of macrolide antibiotics. Thus, Woodward and co-workers [24] successfully completed the total synthesis of erythromycin by adopting the remote activation concept in their choice of 2-pyridylthio and 2-pyrimidinylthio glycosyl donors for the attachment of D-desosamine and L-cladinose units to the aglycone erythronolide A. The 2-pyridylthio donor technology was also successfully used in the glycosylation of avermectin $B_{1a}$ aglycone [25–28]. More recently, the glycosylation of another macrolide aglycone, mycinolide, was accomplished from O-protected glycosyl fluorides in the presence of $Cp_2HfCl_2$-$AgClO_4$ [29].

## IV. NEW GENERATIONS OF GLYCOSYL DONORS

Further studies aimed at the design of glycosyl-leaving groups based on the concept of remote activation in our laboratories led to novel and versatile donor types. Thus, the continued desire to achieve glycoside synthesis with unprotected donors under conditions that lead to oligosaccharide synthesis in solution and in solid phase prompted us to study a series of anomeric heterocyclic-leaving groups. After some exploratory work, we found that the 3-methoxy-2-pyridyloxy (MOP) group was an excellent leaving group in the presence of a catalytic quantity of methyl triflate or triflic acid in solvents such as nitromethane, acetonitrile, or DMF [see Chap. 17] (Scheme 4). Simple alcohols, sugar

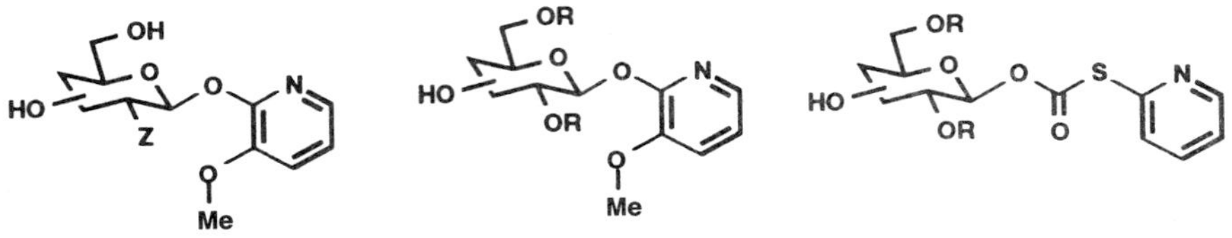

**Scheme 4** MOP (**A** and **B**) and TOPCAT (**C**) glycosyl donors.

alcohols, and related acceptors could be used with unprotected 1,2-*trans*-MOP donors to prepare the corresponding *O*-glycosides with a preponderance of 1,2-*cis*-linkages in good to excellent yields. The reactions could be extended to the syntheses of glycosides on solid supports [30].

The versatility of the MOP-leaving group was further demonstrated in the case of donors with ester- and ether-type O-protective groups [see Chap. 18]. The possibility of modulating the reactivity of the MOP group with donor *and* acceptor molecules that contain different protective groups in conjunction with catalysts, such as $Cu(OTf)_2$, greatly expands the scope of this new method of glycoside and oligosaccharide synthesis.

Another practical development in the remote activation concept for anomeric-leaving groups was achieved with the invention of the 2-thiopyridyl carbonate (TOPCAT) group [see Chap. 19] (see Scheme **4**). Treatment of such donors with a variety of alcohol acceptors in the presence of AgOTf give the corresponding glycosides and oligosaccharides, with excellent selectivity, depending on the O-protective groups used.

The combination of TOPCAT donors with MOP acceptors allows versatility in anomeric reactivity because the TOPCAT group can be activated selectively in the presence of AgOTf [see Chap. 20]. This technology provides oligosaccharides that contain a MOP group at the reducing end; hence, the possibility for further exploiting the remote activation concept by selective "capping" with alcohols of choice.

Details pertaining to the design, preparation, reactivity, and oligosaccharide synthesis technology based on the MOP and TOPCAT leaving groups are discussed in the following four chapters [Chaps. 17–20].

## REFERENCES

1. For recent reviews, see (a) G. Boons, Strategies in oligosaccharide synthesis, *Tetrahedron* 52:1095 (1996); (b) Recent developments in chemical oligosaccharide synthesis, *Contemp. Org. Synth.* 3:173 (1996). (c) K. Toshima and K. Tatsuta, Recent progress in *O*-glycosylation methods and its application to natural product syntheses, *Chem. Rev.* 93:1503 (1993); (d) J. Banoub, P. Boullanger, and D. Lafont, Synthesis of oligosaccharides of 2-amino-2-deoxy sugars, *Chem. Rev.* 92:1167 (1992); (e) P. Kovac, Synthetic oligosaccharides—indispensable probes for the life sciences, ACS Symposium Series 500, American Chemical Society, Washington DC, 1994.
2. R. R. Schmidt and W. Kinzy, Anomeric oxygen activation for glycoside synthesis—the trichloroacetimidate method, *Adv. Carbohydr. Chem. Biochem.* 50:21 (1994), and references cited therein.
3. K. Igarashi, The Koenigs-Knorr reaction, *Adv. Carbohydr. Chem. Biochem.* 34:243 (1977).
4. (a) S. Hanessian and J. Banoub, Preparation of 1,2-*trans*-glycosides in the presence of silver triflate, *Methods Carbohydr. Chem.* 8:247 (1980); (b) S. Hanessian and J. Banoub, Chemistry of the glycosidic linkage. An efficient synthesis of 1,2-*trans*-disaccharides, *Carbohydr. Res.* 53:C13 (1977).
5. J. Jones, *The Chemical Synthesis of Peptides*, Oxford Science Publications, Clarendon Press, Oxord, 1991.
6. R. B. Merrifield, Solid phase peptide synthesis. I. The synthesis of a tetrapeptide, *J. Am. Chem. Soc.* 85:2149 (1963); for recent reviews see: (a) H. Benz, The role of solid-phase fragment condensation (SPFC) in peptide synthesis, *Synthesis* p. 337 (1994); (b) P. Lloyd-Williams, F. Alberico, and E. Giralt, Convergent solid phase peptide synthesis, *Tetrahedron* 49:11065 (1993).
7. For some examples, see: (a) J. J. Krepinsky, in *Modern Methods in Carbohydrate Synthesis*, (S. H. Khan and R. A. O'Neill, eds.), Harwood Academic Publishers, Switzerland, 1995, Chap. 12; S. P. Douglas, D. M. Whitfield, and J. J. Krepinsky, Polymer-supported solution synthesis of oligosaccharides using a novel versatile linker for the synthesis of D-mannopentane a structural unit of D-mannans of pathogenic yeasts, *J. Am. Chem. Soc.* 117:2116 (1995); (b) L. Yan, C. M. Taylor, R. Goodnow, Jr., and D. Kahne, Glycosylation on the Merrifield resin using anomeric sulfoxides, *J. Am. Chem. Soc.* 116: 6953 (1994); (c) J. T. Randolph and S. J. Danishefsky, An iterative strategy for the assembly of complex, branched oligosaccharide domains on a solid support: A concise synthesis of the Lewis[b] domain in bioconjugatable form, *Angew. Chem. Int. Ed. Engl.* 33:1470 (1994); (d) S. J. Danishefsky, K. F. McClure, J. T. Randolph, and R. B. Ruggeri, A strategy for the solid phase synthesis of oligosaccharides, *Science.* 260:11307 (1993); (e) S. P. Douglas, D. M. Whitfield, and J. J. Krepinsky, Polymer-supported solution synthesis of oligosaccharides, *J. Am. Chem. Soc.* 113:5095 (1991); for earlier references see citations in foregoing publications.
8. E. Fischer, Ueber der Glycoside der Alkohole, *Ber.* 26:2400 (1893).
9. A. Michael, On the synthesis of helicin and phenolglucoside, *Am. Chem. J.* 1:305 (1879).
10. S. Hanessian, C. Bacquet, and N. Lehong, Chemistry of the glycosidic linkage. Exceptionally fast and efficient formation of glycosides by remote activation, *Carbohydr. Res.* 80:C17 (1980).
11. G. Wagner and H. Pischel, Glucoside Von Mercaptopyridinen und deren oxydations-Produkte, *Archiv. Pharm.* 296:576 (1963).
12. G. Wagner and F. Suss, Uber Glucoside von 2-Hydroxpyrimidin/Pyrimidon-(2) und 2-Mercaptopyrimidin/Thiopyrimidon-(2), *Pharmazie* 23:8 (1968).
13. S. Hanessian and H. Hori, unpublished results.
14. R. J. Ferrier, R. W. Hay, and N. Vethaviyasar, A potentially versatile synthesis of glycosides, *Carbohydr. Res.* 27:55 (1973).
15. (a) F. Cinget and R. R. Schmidt, Synthesis of unprotected *O*-glycosyl trichloroacetimidates and their reactivity towards some glycosyl acceptors, *Synlett*, p. 168 (1993); (b) R. Haeckel, C. Troll,

H. Fisher, and R. R. Schmidt, Synthesis of unprotected *O*-glycosyl trichloroacetimidates structure assignment and new results, *Synlett*, p. 84 (1994).
16. R. Noyori and I. Kurimoto, Electrochemical glycosylation method, *J. Org. Chem. 51*:4320 (1986).
17. G. Balavoine, A. Gref, J.-C. Fischer, and A. Lubineau, Anodic glycosylation from aryl thioglycosides, *Tetrahedron Lett. 31*:5761 (1990).
18. V. Ferrieres, J.-N. Bertho, and D. Plusquellec, A new synthesis of *O*-glycosides from totally O-unprotected glycosyl donors, *Tetrahedron Lett. 36*:2749 (1995).
19. For a review, see B. Fraser-Reid, U. E. Udodong, Z. Wu, H. Ottosson, J. R. Meritt, C. S. Rao, C. Roberts, and R. Madsen, *n*-Pentyl glycosides in organic chemistry: A contemporary example of serendipity, *Synlett*, p. 927 (1992).
20. G. V. Reddy, V. R. Kulkarni, and H. B. Mereyala, A mild general method for the synthesis of α-linked disaccharides, *Tetrahedron Lett. 30*:4283 (1989).
21. Z. Zhiyuan and G. Magnusson, Synthesis of double-chain bis-sulfone neoglycolipids of the 2″-,3″-,4″-, and 6″-deoxyglobotrioses, *Carbohydr. Res. 262*:79 (1994).
22. J. W. Van Cleve, Reinvestigation of the preparation of cholesteryl 2,3,4,-6-tetra-*O*-benzyl-α-D-glucopyranoside, *Carbohydr. Res. 70*:161 (1979).
23. A. O. Stewart and R. M. Williams, *C*-Glycosidation of pyridyl thioglycosides, *J. Am. Chem. Soc. 107*: 4289 (1985); D. Craig and V. R. Munasinghe, Stereoselective template-directed *C*-glycosidation, *J. Chem. Soc. Commun* p. 901 (1993).
24. R. B. Woodward, et al., Asymmetric total synthesis of erythromycin. 3. Total synthesis of erythromycin, *J. Am. Chem. Soc. 103*:3215 (1981).
25. (a) S. Hanessian, A. Ugolini, D. Dubé, P. J. Hodges, and C. André, Synthesis of (+)-avermectin $B_{1a}$, *J. Am. Chem. Soc. 108*:2776 (1986). (b) S. Hanessian, A. Ugolini, P. J. Hodges, P. Beaulieu, D. Dubé, and C. André, Progress in natural product chemistry by the chiron and related approaches—synthesis of avermectin $B_{1a}$. *Pure Appl. Chem. 59*:299 (1987).
26. J. D. White, G. L. Bolton, A. P. Dantanarayana, C. M. J. Fox, R. N. Hiner, R. W. Jackson, K. Sakuma, and U. S. Warrier, Total synthesis of the antiparasitic agent avermectin $B_{1a}$, *J. Am. Chem. Soc. 117*:1908 (1995).
27. P. G. M. Wuts and S. S. Bigelow, Total synthesis of oleandrose and the avermectin disaccharide, benzyl α-L-oleandrosyl-α-L-4-acetoxyoleandrolide, *J. Org. Chem. 48*:3489 (1983).
28. T. A. Blizzard, G. M. Margiatto, H. Mrozik, and M. H. Fisher, A novel fragmentation reaction of avermectin aglycons, *J. Org. Chem. 58*:3201 (1993).
29. T. Matsumoto, H. Maeta, K. Suzuki, and G. Tsuchihashi, First total synthesis of mycinamicin IV and VII. Successful application of new glycosidation reaction, *Tetrahedron Lett. 29*:3575 (1988).
30. S. Hanessian, B. Lou, and H. Wang, Synthesis of *O*-glycosides on solid phase utilizing 3-methoxy-2-pyridyloxy glycosyl donors, unpublished results; S. Hanessian, Stereocontrolled glycosidation, Canadian patent applied for July 19, 1993.

# 17

# Glycoside and Oligosaccharide Synthesis with Unprotected Glycosyl Donors Based on the Remote Activation Concept

**Boliang Lou**
*Cytel Corporation, San Diego, California*

**Gurijala V. Reddy, Heng Wang, and Stephen Hanessian**
*University of Montreal, Montreal, Quebec, Canada*

| | | |
|---|---|---|
| I. | Introduction | 390 |
| II. | Glycoside and Oligosaccharide Synthesis Using 3-Methoxy-2-pyridyloxy (MOP) *O*-Unprotected Glycosyl Donors | 391 |
| | A. The method: design and concept | 391 |
| | B. Activation of MOP glycosides and mechanism | 391 |
| | C. Preparation of MOP glycosides | 394 |
| | D. Selective activation and iterative oligosaccharide synthesis | 395 |
| | E. Conclusion | 397 |
| III. | Experimental Procedures | 398 |
| | A. Preparation of MOP glycosides | 398 |
| | B. General procedures for the synthesis of 1,2-*cis*-glycosides from unprotected MOP donors | 401 |
| | C. Disaccharide synthesis by selective activation | 403 |
| | D. Trisaccharide synthesis by selective activation | 406 |
| | E. Tetrasaccharide synthesis by selective activation | 408 |
| | References | 410 |

## I. INTRODUCTION

Over the past two decades, there has been a steady and rapidly escalating interest in the important roles that carbohydrates play at the molecular level in many biological processes [1]. Their potential therapeutic value has already been addressed in the context of drug development [2]. Carbohydrate constituents of glycoconjugates on cell surfaces often act as elements for signaling, for transport, and for molecular recognition that are critical in cellular communication. The exciting achievements in the field of glycobiology, and parallel activities in the areas of antibiotic, antitumor, and antiviral therapies, have heralded a new carbohydrate-based bioscience that is deemed to be as important as proteins and nucleic acids. However, compared with peptides and oligonucleotides, which can be readily produced in an automated manner, complex carbohydrate oligomers are much more difficult to synthesize because of their multifunctionality and stereochemical variations. In spite of great strides toward the stereocontrolled synthesis of glycosides, much more needs to be done in this area. Thus, the development of efficient technologies for the construction of simple and complex glycosides remains as a challenging objective for synthetic chemists today.

The two major problems for the chemical synthesis of oligosaccharides [3; see Chap. 12] are stereoselectivity of glycosidic bond formation, and the need to manipulate the hydroxy groups in donor and acceptor molecules using protection and deprotection strategies. As a result, numerous new glycosylation methods, dealing with different leaving groups and promoters, as well as protecting groups, have been reported [4]. Generally, stereocontrolled synthesis of 1,2-*trans* glycosides can be achieved by neighboring group participation from the C-2 position of glycosyl donors. The reaction with glycosyl donors bearing nonparticipating groups gives a mixture of 1,2-*cis* and 1,2-*trans* glycosides in which the anomeric configuration depends on the solvents used and the nature of glycosyl donors and acceptors. The chemical synthesis of oligosaccharides has, so far, relied on the use of *O*-protected glycosyl donors in block or stepwise fashion. Regioselectivity in glycosylation is usually achieved by the differentiation of hydroxy groups by selective protection, leaving a single position available for glycosidic bond formation of the acceptor.

The notion of glycoside synthesis with *unprotected* donors may have enormous appeal and obvious practical value for the following important reasons: (1) The number of steps in glycoside synthesis can be reduced; (2) *O*-unprotected glycosyl donors could be more reactive than the parent structures because of the absence of electronic effects by protective groups, which could deactivate the anomeric-leaving group; (3) the stereochemistry of the glycosidic bond could be different owing to the absence of protective groups; (4) the prospects of developing an iterative or even solid-phase synthesis of glycosides may also be seriously contemplated. Only a few examples of non–Fischer-type glycoside synthesis with unprotected glycosyl donors have been reported in the literature [5]. Ferrier [6] used phenyl 1-thio-β-D-glucopyranoside as a glycosyl donor, which, when reacted with 2-propanol in the presence of mercuric oxide for 96 h, gave the desired 2-propyl α-D-glucopyranoside in 55% yield after acetylation. In 1980, Hanessian and coworkers [7] demonstrated that *O*-unprotected 2-thiopyridyl β-D-glucopyranoside can be activated by $Hg(NO_3)_2$, NBS, or alkyl halides, such as 1-chloropentane in the presence of an excess of a simple alcohol acceptor to afford the desired glycoside with good to modest α-selectivity and in good yield.

# II. GLYCOSIDE AND OLIGOSACCHARIDE SYNTHESIS USING 3-METHOXY-2-PYRIDYLOXY (MOP) O-UNPROTECTED GLYCOSYL DONORS

## A. The Method: Design and Concept

Our previous work using unprotected 2-thiopyridyl glycosides [7] led us to examine several substituted heterocyclic compounds as anomeric-leaving groups. A highly stereoselective 1,2-*cis*-glycosylation using 2-pyridyl β-D-glucopyranoside with 2-propanol in the presence of Lewis acid catalysts led us to focus on this particular system. The glycosylation reactions with several pyridyl β-D-glucopyranosides **1–3**, by activation with stoichiometric MeOTf in the presence of a large excess of 2-propanol proceeded with excellent α-selectivities and gave very good yields, as demonstrated in Scheme 1 [8]. The 3-methoxy-2-pyridyloxy

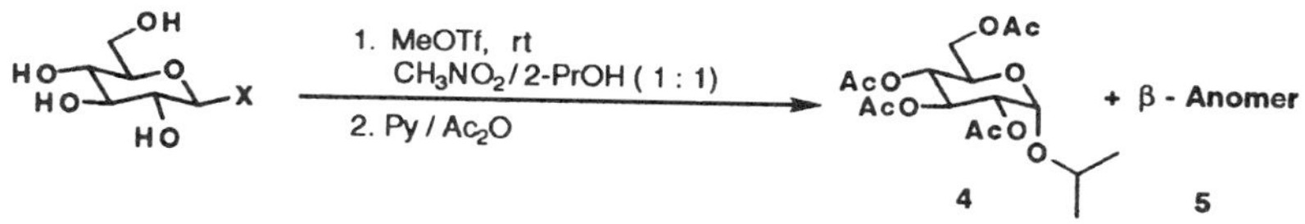

| X | | MeOTf (equiv.) | Time | α : β | Yield(%) |
|---|---|---|---|---|---|
| (2-pyridyloxy) | 1 | 1.1 | 45 min | 9 : 1 | 78 |
| (MOP) | 2 | 1.0 | <5 min | 8 : 1 | 79 |
| (4-methoxy-2-pyridyloxy) | 3 | 1.0 | 2.5 h | 8 : 1 | 88 |

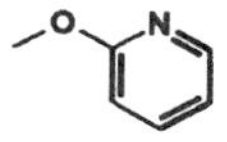

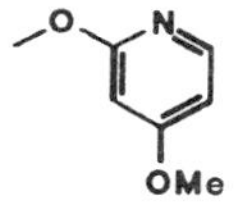

**Scheme 1** A novel anomeric-leaving group: 3-methoxy-2-pyridyloxy (MOP) group.

(MOP) group was the most reactive and efficient leaving group in the glycosylation reaction, compared with the others. The design of the MOP-type group was predicated after finding a promoter capable of activating the heterocycle; hence, the anomeric carbon, leading to the formation of glycosides in the presence of an acceptor alcohol in an $S_N2$-like process.

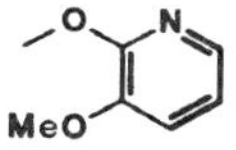

## B. Activation of MOP Glycosides and Mechanism

Only a catalytic amount of MeOTf is required to activate unprotected MOP donors toward simple or complex alcohols in DMF or $CH_3NO_2$ as solvent. The major products are the corresponding 1,2-*cis*-glycosides, which are also formed in good yields, as shown in Scheme 2.

Unprotected MOP glycosyl donors are preparatively stable under neutral or basic conditions, but they can be readily activated by MeOTf in the presence of an excess of

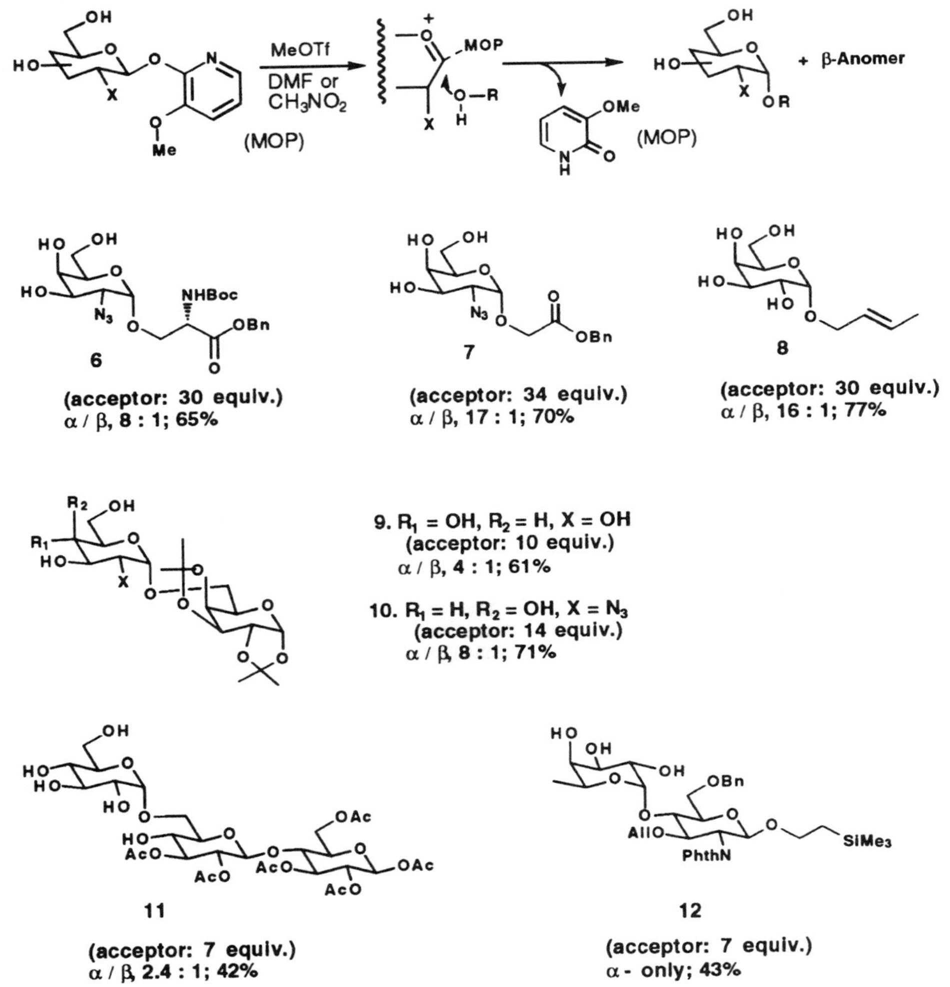

**Scheme 2** Synthesis of 1,2-*cis* glycosides with unprotected MOP-glycosyl donors.

acceptor. Of necessity, the acceptor alcohol should be used in excess (at least 7 eq) to enhance the coupling process and suppress side reactions, such as hydrolysis of unprotected MOP donors, 1,6-anhydro formation, and self-condensation. It is noteworthy that DMF, which is seldom used as a solvent for conventional glycosylation methods [9], is admirably suitable here, increasing the solubility of donors, and providing high α-selectivity.

As expected, unprotected MOP donors containing a 2-acetamido group give 1,2-*trans*-glycosides (Scheme 3), presumably by the intermediacy of an oxazolinium ion **15**.

Interestingly, self-coupling of glycosyl donors was rarely observed under the conditions of the reactions. 1,6-Anhydro-D-glucose and anomerized glycosyl MOP donors were isolated when the MOP glycosyl donor **2** was treated with MeOTf in the absence of alcohol. Under the same condition, 2-acetamido glycosyl donor **13** was converted into the oxazoline **14** as shown in Scheme 3. The actual catalyst for activating the MOP-leaving group in the glycosylation appears to be triflic acid (TfOH), which is generated in situ by reaction of

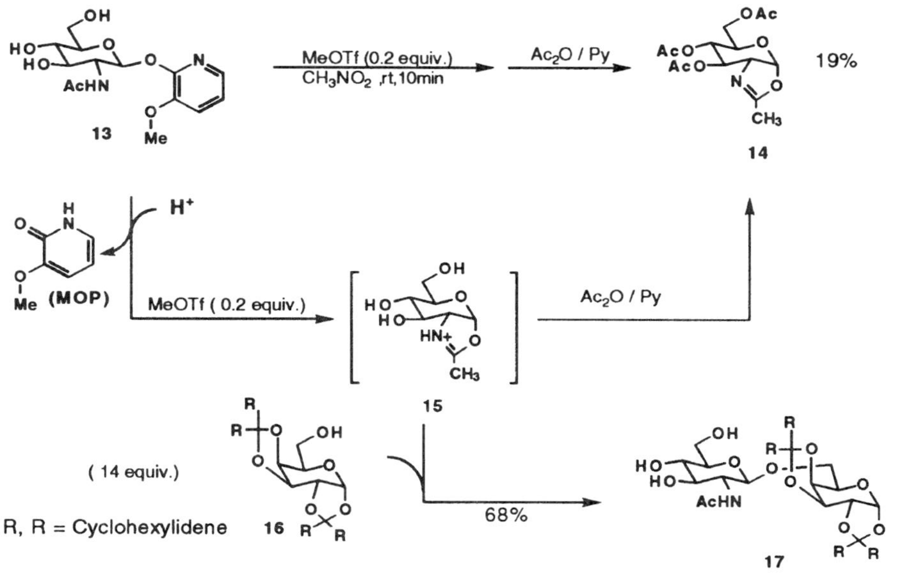

# Synthesis with Unprotected Glycosyl Donors

**Scheme 3** 1,2-*trans*-Glycoside synthesis using an unprotected MOP glycosyl donor containing a 2-acetamido group.

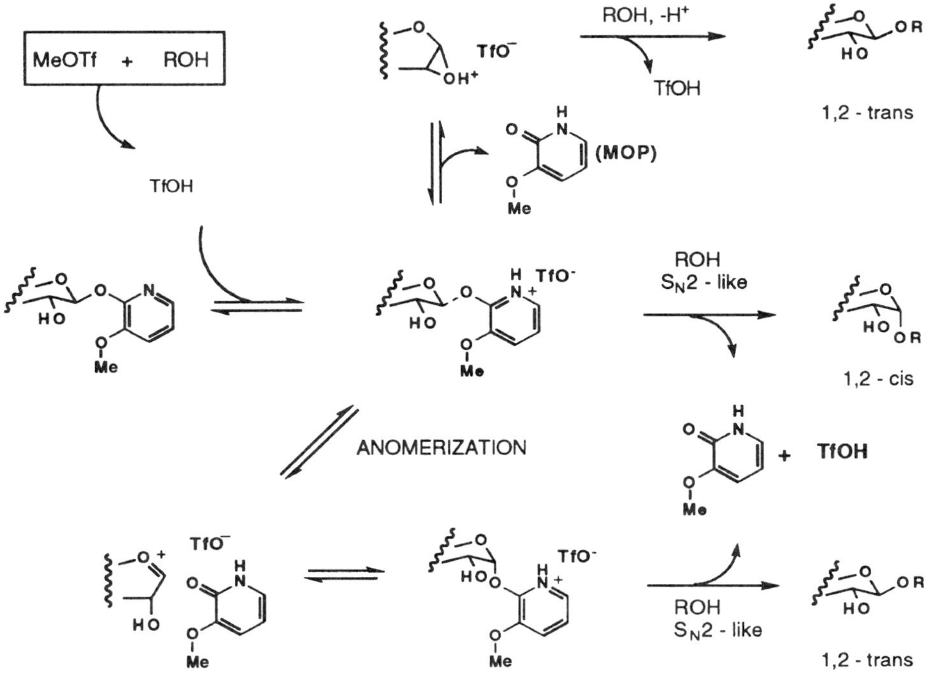

**Scheme 4** Possible mechanistic pathways.

MeOTf with excess alcohol in solution. In fact, glycosylation could be effected equally well using TfOH instead of MeOTf as a catalyst. As predicated in the design of the MOP group, activation most likely proceeds through the reaction of the nitrogen in the pyridine ring with TfOH to generate a tightly bound oxonium ion pair specie in which the β-orientated MOP group shields the β-side. An $S_N2$-like attack from the α-face by the alcohol releases 3-methoxy-2(1*H*)-pyridone and gives the 1,2-*cis*-glycoside with regeneration of the catalyst (Scheme **4**). The released heterocycle can be easily detected by thin-layer chromatography (TLC), which provides a convenient analytic tool to follow the progress of the glycosylation reactions.

As proposed in Scheme **4**, the minor 1,2-*trans*-glycoside could arise by an $S_N2$-type substitution from a small amount of MOP α-glycoside generated through anomerization. To confirm this hypothesis, treatment of MOP α-glycoside with an excess of 2-propanol in the presence of MeOTf, gave 2-propyl β-D-glucopyranoside predominantly (85:15, β/α). 1,2-*trans*-Glycosides could also be produced through the formation of an epoxide intermediate, followed by the ring opening by the nucleophile at the anomeric position. A less tightly bound ion pair, or the intervention of solvent molecules could also explain the formation of the 1,2-*trans*-isomers as the minor products.

## C. Preparation of MOP Glycosides

Unprotected MOP glycosyl donors are easily prepared in multigram scale from the corresponding peracetyl glycosyl halides, followed by de-*O*-acetylation. They are usually crystalline and have excellent shelf life. MOP donors **2**, **13**, **18**, **19**, **20**, and **22**, shown in Scheme **5**, are typical examples.

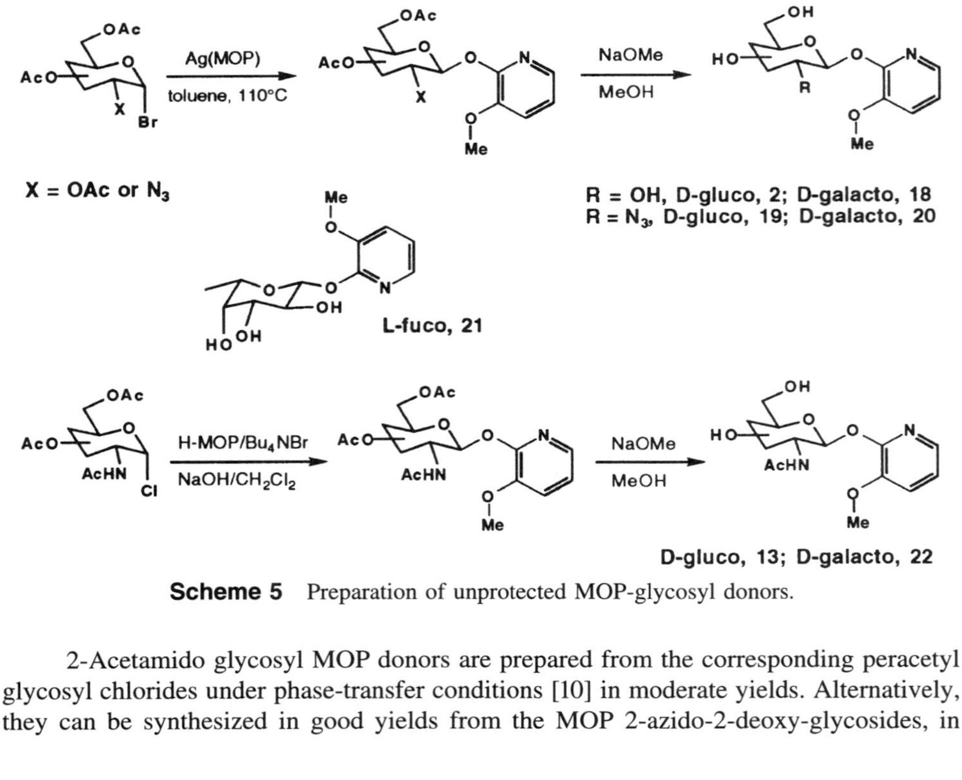

**Scheme 5** Preparation of unprotected MOP-glycosyl donors.

2-Acetamido glycosyl MOP donors are prepared from the corresponding peracetyl glycosyl chlorides under phase-transfer conditions [10] in moderate yields. Alternatively, they can be synthesized in good yields from the MOP 2-azido-2-deoxy-glycosides, in which the azido group is first reduced to an amine by hydrogenation, followed by *N*-acetylation.

The preparation of MOP α-D-galactopyranosyl and MOP α-D-glucopyranosyl donors is achieved by anomerizing the corresponding peracetyl MOP β-D-glycosides in the presence of $HgBr_2$ at high temperature. MOP 2-azido-2-deoxy α-D-galactopyranosyl donor can be prepared in good yield by reaction of the peracetylated β-azidoglycosyl chloride [11] with silver 3-methoxy-2-pyridoxide.

## D. Selective Activation and Iterative Oligosaccharide Synthesis

Unprotected MOP glycosyl donors exhibit much higher reactivities compared with their O-substituted analogues. Even one protective group (TBDMS or benzyl group) significantly deactivates this anomeric-leaving group. For example, completion of the glycosylation with (3-methoxy-2-pyridyl) 6-*O*-TBDMS-β-D-glucopyranoside and with (3-methoxy-2-pyridyl) 2,3,4,6-tetra-*O*-benzyl-β-D-glucopyranoside in the presence of 2-propanol required 20 min and 5 h, respectively, compared with only few minutes using the corresponding unprotected MOP donor. Most important was the finding that *O*-acyl–protected MOP glycosides are inert in the presence of MeOTf or TfOH [12].

It is clear, therefore, that the selective activation of unprotected MOP glycosides as

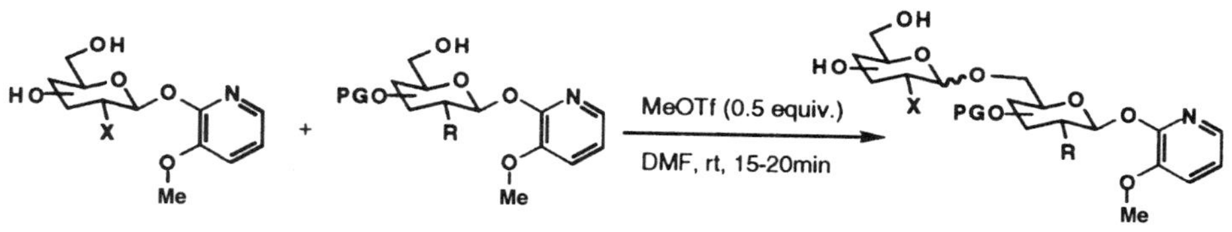


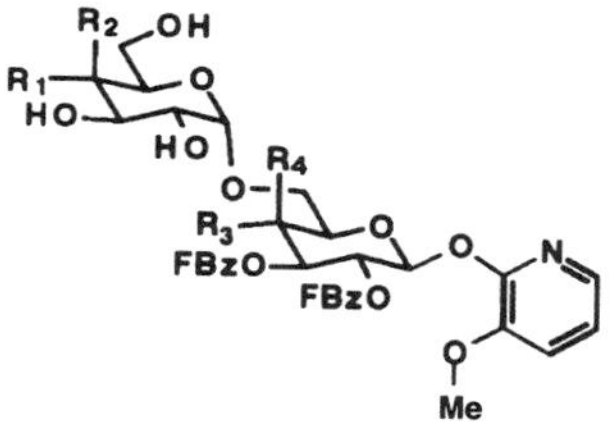

**23.** $R_1$ = OH, $R_2$ = H, $R_3$ = OFBz, $R_4$ = H; α/β, 4 : 1, 68%

**24.** $R_1$ = H, $R_2$ = OH, $R_3$ = OFBz, $R_4$ = H; α/β, 6 : 1, 65%

**25.** $R_1$ = H, $R_2$ = OH, $R_3$ = H, $R_4$ = OFBz; α/β, 6 : 1, 62%

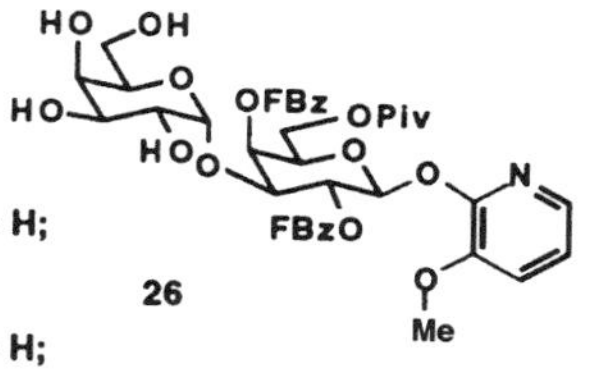

**26**

α-only, 41%

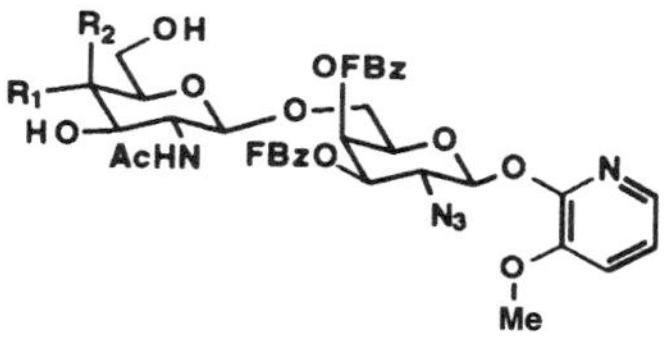

**27.** $R_1$ = OH, $R_2$ = H β-only, 63%

**28.** $R_2$ = OH, $R_1$ = H β-only, 69%

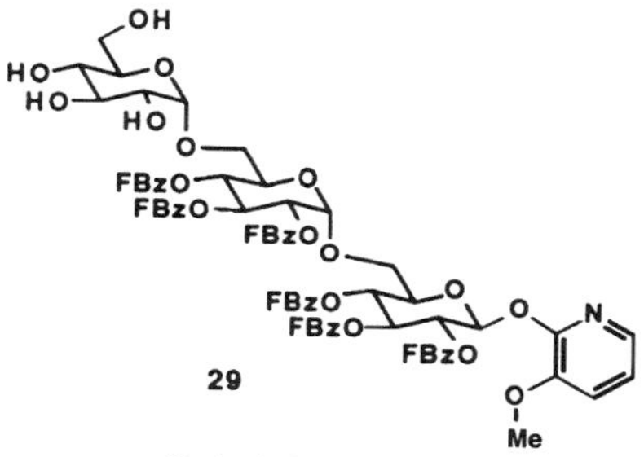

**29**

α/β, 4 : 1, 56%

**Scheme 6** Selective activation; unprotected MOP donors versus protected MOP glycosides.

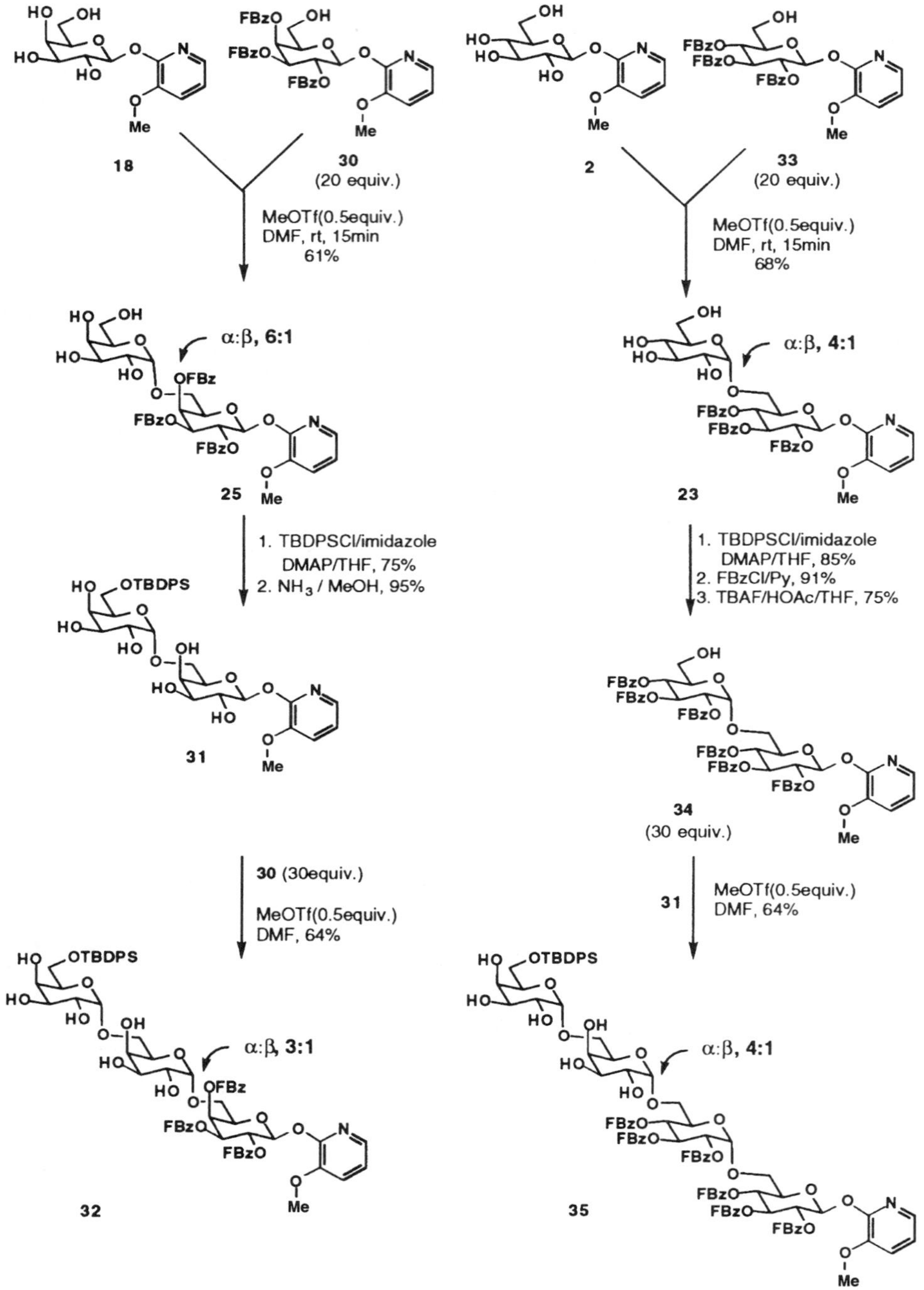

**Scheme 7** Iterative oligosaccharide synthesis.

# Synthesis with Unprotected Glycosyl Donors

glycosyl donors can be achieved in the presence of partially esterified MOP acceptors to give oligosaccharides. Scheme **6** illustrates examples of this strategy using unprotected MOP donors and partially *p*-fluorobenzoylated MOP acceptors in the presence of catalytic amounts of MeOTf. Di- and trisaccharides with 1,3- and 1,6-linkages are produced in reasonable to good yields and with 1,2-*cis*-selectivity. The *p*-fluorobenzoate group was particularly suitable as a protective group to deactivate acceptor MOP glycosides because of better solubility in DMF and ease of deprotection under very mild conditions.

To achieve efficient oligosaccharide synthesis, an excess of acceptor is still required, which can be impractical in certain circumstances. Nevertheless, the products are easily isolated by normal flash chromatography on silica gel column, and a large amount of gel is unnecessary because the excess of acceptors are easily recovered by using less polar eluants. Alternatively, the products can be isolated after *O*-acetylation.

This method has been successfully extended to the iterative synthesis of oligosaccharides. The disaccharides, derived from selective activations of unprotected MOP donors and partially esterified MOP acceptors, can be converted into reactive donors simply by deacylation. The newly generated unprotected MOP disaccharide donor can be subjected to another cycle of glycosylation by repeating the same process. Schemes **7** and **8** illustrate

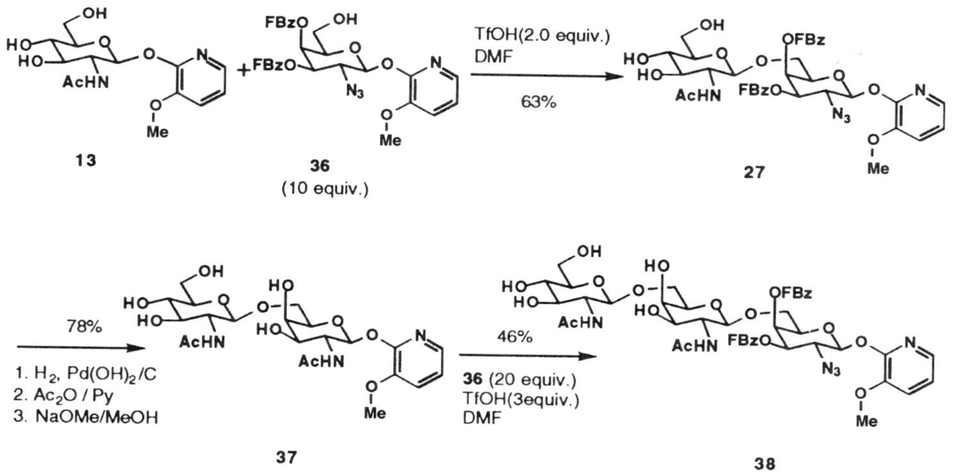

**Scheme 8** Iterative synthesis of 2-deoxy-2-aminosugars.

examples of iterative synthesis of tri- and tetrasaccharides with 1,2-*cis*- and 1,2-*trans*-selectivities.

Linear 1,6-α-linked D-gluco and D-galacto oligosaccharides are conveniently and rapidly assembled by using this MOP selective activation technology. Occasionally, the introduction of a TBDPS group in the MOP disaccharide donor, as in **31**, enhances its solubility. Compounds related to **32** and **35** can be found in the glyceroglycolipids [13]. Biologically relevant 1,2-*trans*-2-acetamido-2-deoxy oligosaccharides such as **38** [3d] are easily accessible using the same strategy by employing 2-acetamido-2-deoxy derivatives.

## E. Conclusion

We have described the first successful examples of oligosaccharide synthesis using unprotected glycosyl donors. Unprotected MOP donors, including MOP 2-azido-2-deoxyglycosides, can be readily activated by MeOTf in the presence of variable excesses of acceptors, and they give 1,2-*cis*-glycosides as major products. MOP 2-acetamido-2-deoxyglycosides give exclusively 1,2-*trans*-products. The technology offers a valuable solution to the heretofore difficult problem of iterative assembly of 1,2-*cis*-linked oligosaccharides and other glycosides, avoiding *O*-benzyl protective groups. The selective MOP donor activation relative to *O*-acyl MOP acceptors, the mildness of conditions, the simplicity of reagent design and the convenience of following the progress of the reaction simply by monitoring the release of the chromophoric heterocycle are definite assets of this novel glycotechnology. Direct applications can be envisaged in solid-phase synthesis [14], combinatorial chemistry [15], protein glycosylation [16], and in related technologies. The design of the MOP glycosyl donor, its chemical reactivity, selectivity, and adaptability to oligosaccharide and general glycoside synthesis should pave the way to new generations of glycosyl-leaving groups, and further innovations in this exciting field.

## III. EXPERIMENTAL PROCEDURES*

### A. Preparation of MOP Glycosides

*Silver 3-Methoxy-2-pyridoxide*

3-Methoxy-2-(1*H*)-pyridone (12 g, 96 mmol) was dissolved in 280 mL of water containing 3.84 g (96 mmol) of sodium hydroxide. A solution of silver nitrate (16.32 g, 96 mmol) in 64 mL of water was added. The mixture was vigorously stirred at room temperature for 10 min. After filtration, the solid was washed with MeOH and ether, dried in vacuo to give 21 g of silver salt in 94% yield.

*3-Methoxy-2-pyridyl β-D-glucopyranoside (2)*

The mixture of 2,3,4,6-tetra-*O*-acetyl-α-D-glucopyranosyl bromide (20 g, 48.7 mmol), and silver 3-methoxy-2-pyridoxide (18 g, 77.6 mmol) in 260 mL of toluene was refluxed for 1 h. The mixture was filtered over Celite, washed with $CH_2Cl_2$, and evaporated. Purification by flash chromatography on silica gel (EtOAc–hexane–$CH_2Cl_2$, 1:1:1), gave 3-methoxy-2-pyridyl 2,3,4,6-tetra-*O*-acetyl-β-D-glucopyranoside (17 g, 77%): mp 102°–103°C, $[\alpha]_D$ +4.5° (*c* 0.8, $CHCl_3$).

A solution of this product (7.80 g) and 0.2 mL of 25% NaOMe–MeOH solution in 35 mL of MeOH–THF (6:1) was stirred until the reaction was completed. The mixture was cautiously neutralized with Amberlite IR-120 ($H^+$) ion-exchange resin. After filtration and concentration, the desired product was obtained in quantitative yield as a white solid. Recrystallization in EtOH gave the title compound as white crystals: mp 168°C, $[\alpha]_D$ −15.8° (*c* 0.6, MeOH).

---

*Optical rotations were measured at 20°–22°C.

# Synthesis with Unprotected Glycosyl Donors

### 3-Methoxy-2-pyridyl β-L-fucopyranoside (21)

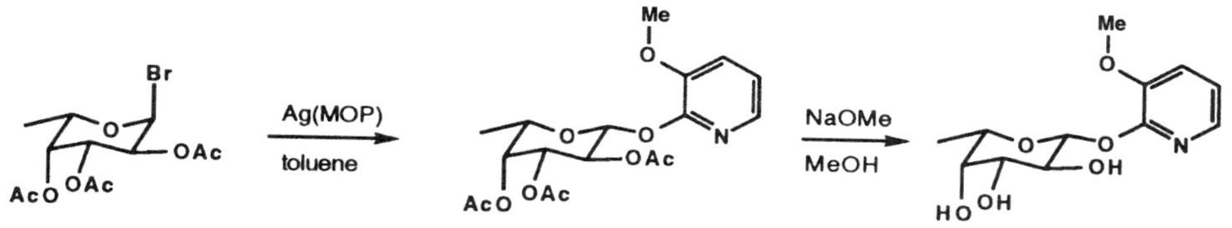

A mixture of 2,3,4-tri-O-acetyl-α-L-fucopyranosyl bromide (1.4 g, 3.91 mmol), and silver 3-methoxy-2-pyridoxide in 75 mL (1.45 g, 6.25 mmol) of toluene was refluxed for 40 min. The mixture was cooled to room temperature, filtered over Celite, and washed with dichloromethane. The residue was washed with dichloromethane, the filtrate was evaporated and the product was purified by flash chromatography on silica gel (EtOAc–hexane, 2:3). First eluted was 3-methoxy-2-pyridyl 2,3,4-tri-O-acetyl-α-L-fucopyranoside (0.15 g, 10%), followed by 3-methoxy-2-pyridyl 2,3,4-tri-O-acetyl-β-L-fucopyranoside (1.02 g, 65%).

A solution of 1.0 g of this product and 0.025 mL of 25% methanolic NaOMe (10 mL) was stirred at room temperature until the reaction was completed. The mixture was cautiously neutralized with Amberlite IR-120 (H$^+$) ion-exchange resin. After filtration and concentration, the material was passed through a small bed of silical gel pretreated with 2% triethylamine in hexane. Elution with chloroform–methanol (8:1) containing 1% of triethylamine gave the pure title compound as a pale yellow solid (0.61 g, 90%): mp 93°C, [α]$_D$ +4.0° (c 0.9, MeOH).

### 3-Methoxy-2-pyridyl 2-azido-2-deoxy-β-D-galactopyranoside (20)

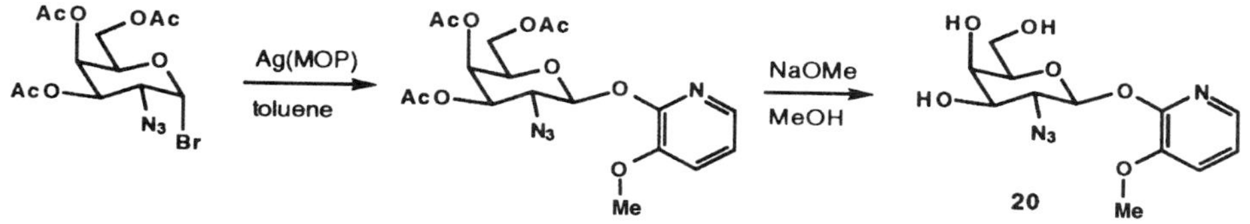

A mixture of 3,4,6-tri-O-acetyl-2-azido-2-deoxy-α-D-galactopyranosyl bromide (1.7 g, 4.3 mmol) (prepared by a known literature procedure [11]) and 2.0 g (8.6 mmol) of silver salt in 30 mL of dry toluene was heated at 100–110°C for 1.5 h. The suspension was cooled to room temperature, filtered over Celite, then washed with CH$_2$Cl$_2$, and concentrated to give the crude product. Purification by flash chromatography on silica gel using EtOAc–hexane (2:1) as the eluant, gave 3-methoxy-2-pyridyl 3,4,6-tri-O-acetyl-2-azido-2-deoxy-β-D-galactopyranoside (1.2 g, 63.5%): mp 53°–55°C, [α]$_D$ −6.8° (c 1.10, CHCl$_3$).

This product (1.2 g) was dissolved in 10 mL of methanol and treated with a catalytic amount of 1 M NaOMe solution in MeOH at room temperature until the reaction was completed. The solution was neutralized with Amberlite IR-120 (H$^+$) ion-exchange resin. Filtration and concentration gave the glycoside **20**, which was recrystallized in EtOH in 95% yield: mp 135°–137°C, [α]$_D$ +7.8° (c 0.7, MeOH).

### 3-Methoxy-2-pyridyl 2-acetamido-2-deoxy-β-D-galactopyranoside (22)

**Method 1.** To a suspension of 1.0 g of N-acetyl-D-galactosamine in 8 mL of pyridine was added 3 mL of acetic anhydride, and the mixture was stirred at room temperature overnight. Dilution with 50 mL of CH$_2$Cl$_2$ and conventional workup gave 1.8 g of crude product.

A portion of this crude product (0.55 g) was dissolved in 3 mL of acetic anhydride, the solution was cooled to 0°C, and saturated with HCl(g). The mixture was stirred at room temperature for 12 h, at which time TLC indicated most of the starting material had been consumed. The mixture was poured into ice-water, extracted with CHCl$_3$, washed with saturated NaHCO$_3$ and brine. The organic layer was dried, filtered, and concentrated to give 440 mg of the crude acetylated glycosyl chloride.

To a solution of 300 mg of 3-methoxy-2(1H)-pyridine and 386 mg of Bu$_4$NBr in 5 mL of 1 N NaOH was added a solution of 440 mg of the crude chloride in 5 mL of CH$_2$Cl$_2$. The mixture was vigorously stirred at room temperature until the chloride completely disappeared (1 h). The water layer was separated, extracted with CH$_2$Cl$_2$; the combined organic layer was processed as usual, and the product was purified by flash chromatography on silica gel to give the pure acetylated glycoside (100 mg, 18%).

**Method 2.** A mixture of 250 mg (0.57 mmol) of 3-methoxy-2-pyridyl 3,4,6-tri-O-acetyl-2-azido-2-deoxy-β-D-galactopyranoside, 50 mg of Pd(OH)$_2$/C, and 2 mL of EtOH was stirred under hydrogen for 5 h, at which time the starting material completely disappeared on TLC. Filtration over Celite, followed by concentration, gave the crude product, which was treated with 2 mL of pyridine and 0.12 mL of Ac$_2$O at 0°C overnight. After standard workup, purification by flash chromatography on silica gel with EtOAc to EtOAc–MeOH (7:1) gave the desired product as a white solid (200 mg, 77%): mp 148°–150°C, [α]$_D$ +31.2° (c 0.33, CHCl$_3$).

This product was dissolved in 3 mL of dry methanol, 14 μL of 25% methanolic NaOMe solution in MeOH was added, and the mixture was stirred at room temperature until no acetate remained. The precipitated solid was filtered, washed with MeOH and ether to give 120 mg of the desired compound **22** (83.3%) as a white solid: mp 179°C, [α]$_D$ +29.3° (c 0.85, H$_2$O–MeOH, 1:1).

*3-Methoxy-2-pyridyl-α-D-glucopyranoside*

A mixture of 60 mg of 3-methoxy-2-pyridyl 2,3,4,6-tetra-O-acetyl-β-D-glucopyranoside and 120 mg of HgBr$_2$ in 3 mL of xylene was refluxed for 5 h. TLC showed most of the starting material was converted into the α-isomer. The solution was cooled to room temperature, filtered over Celite and washed with CH$_2$Cl$_2$. The filtrate was treated with 10% aqueous Na$_2$S$_2$O$_3$, dried over Na$_2$SO$_4$, and concentrated. The residue was purified by flash chromatography on silica gel (EtOAc–hexane, 1:4 to 1:2) to give 29 mg of the α-glycoside

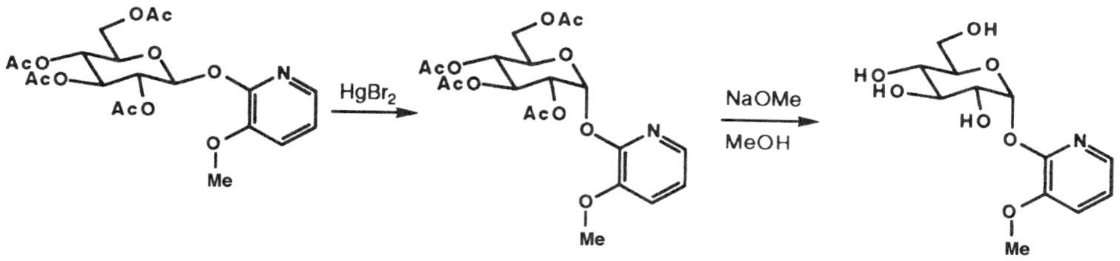

# Synthesis with Unprotected Glycosyl Donors

(48%): $[\alpha]_D$ +117° (c 0.91, CHCl$_3$). Deacetylation under standard conditions (NaOMe–MeOH) gave the corresponding unprotected MOP glycoside: mp 142°–144°C, $[\alpha]_D$ +81.6° (c 0.5, MeOH).

*General Procedure for the Preparation of Partially Acylated MOP Glycosides: 3-Methoxy-2-pyridyl 2,3,4-tri-O-p-fluorobenzoyl-β-D-glucopyranoside (33)*

To a cooled solution of 3-methoxy-2-pyridyl β-D-glucopyranoside (3.2 g, 11.15 mmol) in 60 mL of dry pyridine, was added dropwise 3.1 mL (1.2 Eq.) of TBDMSOTf over 30 min at 0°C. Occasionally, it was necessary to add more TBDMSOTf to complete the reaction. After addition of 4.3 mL (36.40 mmol) of p-fluorobenzoyl chloride, the reaction mixture was kept at 0°C overnight, then poured into cold saturated NaHCO$_3$ and extracted with EtOAc. Purification by flash chromatography on silica gel using EtOAc–hexane (1:4) as the eluant gave 7.95 g of product in 93% yield: mp 93°–94°C, $[\alpha]_D$ +36.8° (c 1.6, CHCl$_3$).

A mixture of 0.95 g (1.24 mmol) of this product, 2.5 mL of 1 M TBAF–THF solution, 0.43 mL (7.5 mmol) of acetic acid, and 7 mL of THF was stirred at 0°C for 1 h. The cooling bath was removed and the mixture was stirred for 1 day. After concentration, the residue was treated with ice-water and EtOAc, and the organic layer was processed as usual. Purification by flash chromatography on silica gel with EtOAc–hexane (1:1 to 2:1) gave 710 mg (88%) of the title product **33** as a white solid: mp 140°–142°C, $[\alpha]_D$ +48.5° (c 0.84, CHCl$_3$).

## B. General Procedures for the Synthesis of 1,2-*cis*-Glycosides from Unprotected MOP Donors

*3-Methoxy-2-pyridyl β-D-glucopyranoside (2) as a Glycosyl Donor*

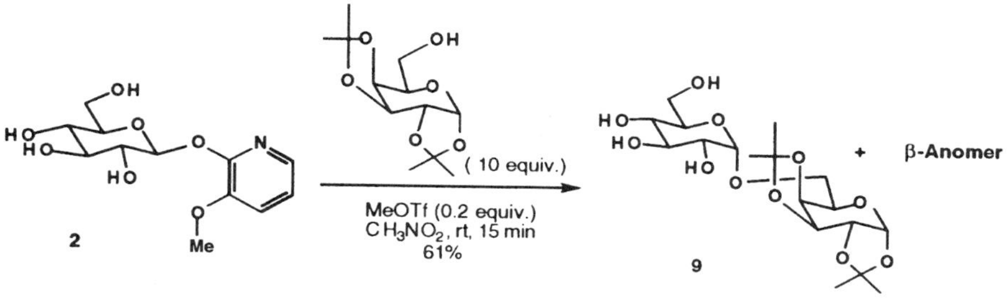

To a mixture of 3-methoxy-2-pyridyl β-D-glucopyranoside **2** (32.8 mg, 0.114 mmol) and 300 mg (1.15 mmol) of 1,2:3,4-di-*O*-isopropylidene-α-D-galactopyranose in 1.5 mL of $CH_3NO_2$ was added 23 μL (0.023 mmol) of 1 $M$ MeOTf solution in $CH_3NO_2$. The mixture was stirred at room temperature for 15 min. After addition of pyridine (1 drop), the solvent was removed in vacuo, and the residue was purified by flash chromatography on silica gel EtOAc–hexane (2:1) then $CHCl_3$–MeOH (10:1) to give 260 mg of recovered acceptor, 23.6 mg of the expected α-isomer and 6 mg of the β-isomer in 61% yield (α/β, 4:1). For the α-anomer: mp 55°–57°C, $[α]_D$ +23.3° (*c* 0.78, $CHCl_3$); peracetate: syrup, $[α]_D$ +44.8° (*c* 2.8, $CHCl_3$); for the β-anomer: syrup, $[α]_D$ −60.3°C (*c* 0.3, $CHCl_3$); peracetate: syrup, $[α]_D$ −42° (*c* 0.82, $CHCl_3$).

*3-Methoxy-2-pyridyl β-L-fucopyranoside* (**21**) *as a Glycosyl Donor*

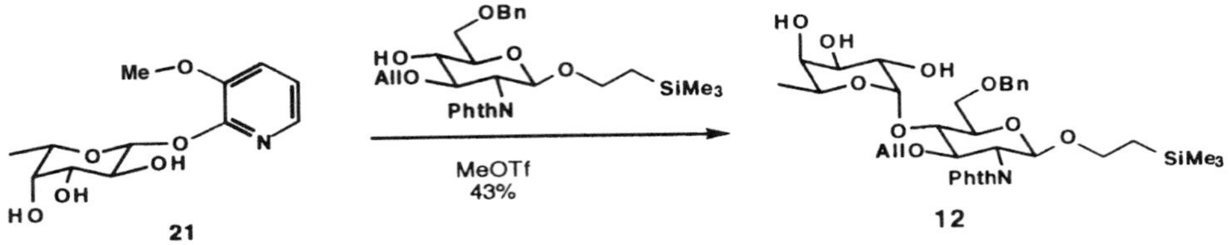

To a solution of 2-(trimethylsilyl)ethyl 3-*O*-allyl-6-*O*-benzyl-2-phthalimido-2-deoxy-β-D-glucopyranoside (250 mg, 0.49 mmol) and 18 μL of 1 $M$ MeOTf solution in 0.5 mL of dry dichloromethane was added dropwise 18.8 mg (0.07 mmol) of 3-methoxy-2-pyridyl β-L-fucopyranoside in 0.5 mL of solvent over 15 min. The mixture was stirred at room temperature for 30 min, and processed as described in the foregoing. The residue was purified by flash chromatography on silica gel column using EtOAc–hexane, (2:1) as eluant to recover the excess of acceptor. Elution of the column with EtOAc–hexane, (1:1) containing 10% methanol gave 20 mg of the title product in 43% yield as a syrup.

*3-Methoxy-2-pyridyl 2-azido-2-deoxy-β-D-galactopyranoside* (**20**) *as a Glycosyl Donor*

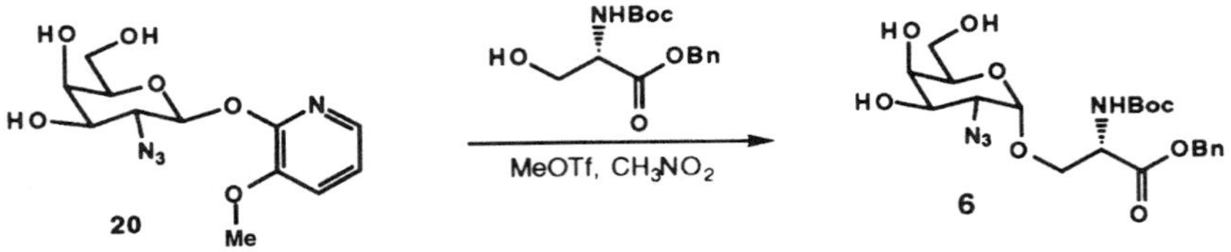

3-Methoxy-2-pyridyl 2-azido-2-deoxy-β-D-galactopyranoside (40 mg, 0.128 mmol) was added to a mixture of 1.13 g (3.84 mmol) of *N-tert*-butyloxycarbonyl L-serine benzyl ester and 0.8 mL of dry $CH_3NO_2$. This was treated with 26 μL of 1 $M$ MeOTf solution in $CH_3NO_2$, and the solution was stirred at room temperature until the glycosyl donor was consumed. After addition of pyridine (1 drop), the solvent was removed in vacuo, and the residue was purified by flash chromatography on silica gel using EtOAc–hexane (1:1) to recover the excess aglycon. Elution with EtOAc–$CHCl_3$–MeOH (2:2:1) gave 35.5 mg of product **6** as a colorless syrup: $[α]_D$ +73.5° (*c* 0.52, MeOH) and 4.5 mg of the β-isomer (α/β, 8:1; 65%), in addition to 6 mg of the carbamate as a by-product.

## 3-Methoxy-2-pyridyl 2-acetamido-2-deoxy-β-D-glucopyranoside (13) as a Glycosyl Donor

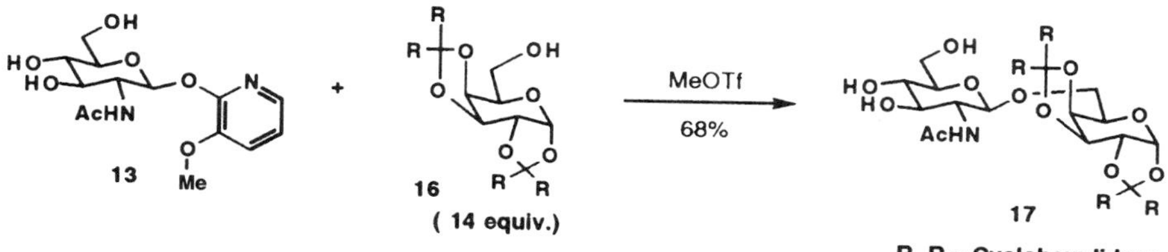

To a suspension of 3-methoxy-2-pyridyl 2-acetamido-2-deoxy-β-D-glucopyranoside (13.4 mg, 0.041 mmol) **13** in 1.3 mL of dry $CH_3NO_2$ were added 208 mg of 1,2:3,4-di-*O*-cyclohexylidene-α-D-galactopyranose and 23 μL of 1 *M* MeOTf solution in $CH_3NO_2$. The mixture was stirred vigorously at room temperature (the solid disappeared immediately) for 1 h. After addition of 1 drop of pyridine, the solvent was removed in vacuo, and the residue was treated with pyridine and acetic anhydride. The usual workup and purification by flash chromatography on silica gel gave 18.5 mg of 1,2:3,4-di-*O*-cyclohexylidene-6-*O*-(3,4,6-tri-*O*-acetyl-2-acetamido-2-deoxy-β-D-glucopyranosyl)-α-D-galactopyranoside **17** in 68% yield: mp 135°–137°C, $[\alpha]_D$ −49.5° (*c* 0.55, $CHCl_3$).

### C. Disaccharide Synthesis by Selective Activation

*Glycosylation of 3-Methoxy-2-pyridyl 2,3,4-tri-O-p-fluorobenzoyl-β-D-galactopyranoside (30) with 3-Methoxy-2-pyridyl β-D-galactopyranoside (18)*

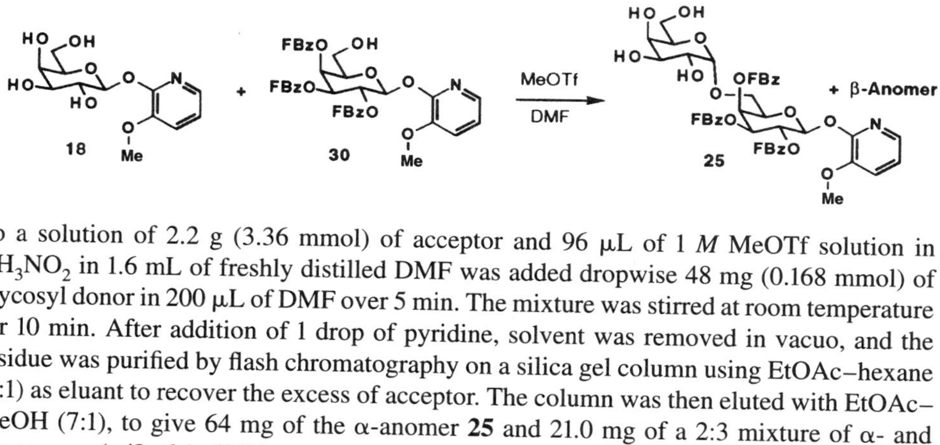

To a solution of 2.2 g (3.36 mmol) of acceptor and 96 μL of 1 *M* MeOTf solution in $CH_3NO_2$ in 1.6 mL of freshly distilled DMF was added dropwise 48 mg (0.168 mmol) of glycosyl donor in 200 μL of DMF over 5 min. The mixture was stirred at room temperature for 10 min. After addition of 1 drop of pyridine, solvent was removed in vacuo, and the residue was purified by flash chromatography on a silica gel column using EtOAc–hexane (2:1) as eluant to recover the excess of acceptor. The column was then eluted with EtOAc–MeOH (7:1), to give 64 mg of the α-anomer **25** and 21.0 mg of a 2:3 mixture of α- and β-anomers (α/β, 6:1, 62%).

For the α-anomer: mp 120–121°C, $[\alpha]_D$ +190.9° (*c* 1.0, $CHCl_3$); peracetate: mp 110–112°C, $[\alpha]_D$ +201.5° (*c* 0.62, $CHCl_3$); for the β-anomer peracetate: mp 116°C, $[\alpha]_D$ +117.4° (*c* 0.82, $CHCl_3$).

*Glycosylation of 3-Methoxy-2-pyridyl 2,3,4-tri-O-p-fluorobenzoyl-β-D-glucopyranoside (33) with 3-Methoxy-2-pyridyl β-D-glucopyranoside (2)*

The same procedure as the foregoing gave the α- and β-products that were separated by flash chromatography to give **23** and its anomer (α/β, 4:1; 68%). For the α-anomer: mp

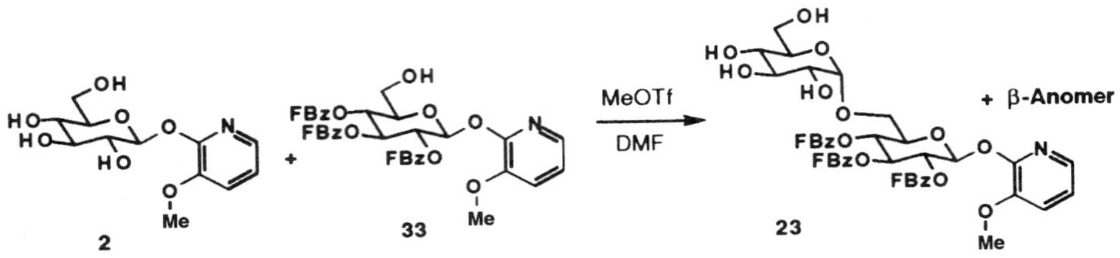

88–90°C, [α]$_D$ +70.3° (c 0.7, CHCl$_3$); for the β-anomer: mp 96–98°C, [α]$_D$ +28.9° (c 0.55, CHCl$_3$).

*Glycosylation of 3-Methoxy-2-pyridyl 2,3,4-tri-O-p-fluorobenzoyl-β-D-glucopyranoside (33) with 3-Methoxy-2-pyridyl β-D-galactopyranoside (18)*

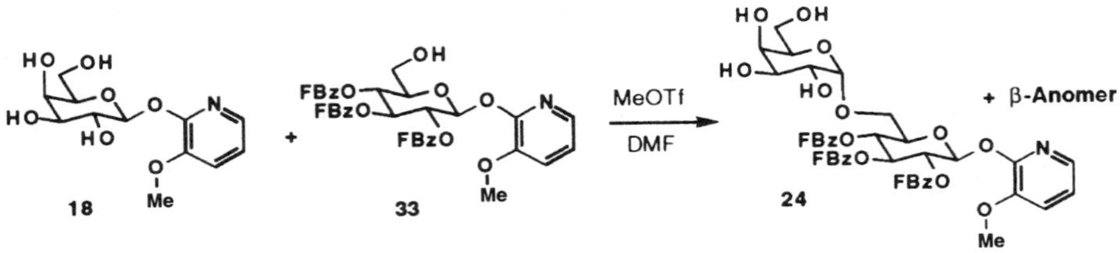

The same procedure gave the α- and β-products (α/β, 6:1) that were separated after acetylation. For the α-anomer peracetate: mp 107–108°C, [α]$_D$ +111.9° (c 0.53, CHCl$_3$); for the β-anomer peracetate: mp 84–85°C, [α]$_D$ +16.8° (c 0.3, CHCl$_3$).

*Glycosylation of 3-Methoxy-2-pyridyl 2,4-di-O-p-fluorobenzoyl-6-O-pivaloyl-β-D-galactopyranoside with 3-Methoxy-2-pyridyl β-D-galactopyranoside (18)*

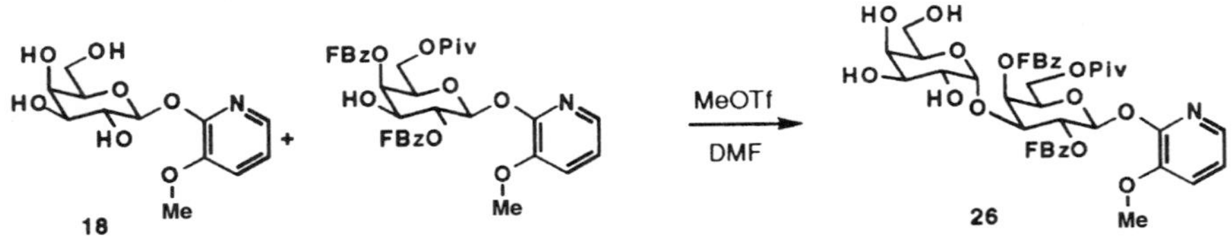

To a mixture of 3-methoxy-2-pyridyl β-D-galactopyranoside (7 mg, 0.025 mmol), 430 mg (0.70 mmol) of 3-methoxy-2-pyridyl 2,4-di-O-p-fluorobenzoyl-6-O-pivaloyl-β-D-galactopyranoside: mp 103–104°C, [α]$_D$ +57.4° (c 0.86, CHCl$_3$), and 0.25 mL of dry DMF was added 12 µL of 1 M MeOTf in CH$_3$NO$_2$. After 20 min, 1 drop of pyridine was added, and the mixture was concentrated in vacuo. The residue was purified by flash chromatography on silica gel column with EtOAc–hexane (2:1) to recover the excess of acceptor. Elution with EtOAc–MeOH (7:1) gave the desired disaccharide **26** (7.7 mg, 41%): [α]$_D$ +111.6° (c 0.4, CHCl$_3$).

*Glycosylation of 3-Methoxy-2-pyridyl-2-azido-2-deoxy-3,4-di-O-p-fluorobenzoyl-β-D-galactopyranoside (36) with 3-Methoxy-2-pyridyl 2-acetamido 2-deoxy-β-D-glucopyranoside (21)*

To a suspension of glycosyl donor **21** (45 mg, 0.138 mmol) and 762 mg (1.38 mmol) of acceptor **38** in 0.75 mL of dry DMF was added 24.3 µL (0.275 mmol) of TfOH at room temperature. The mixture was stirred for 40 min, at which time a homogeneous solution

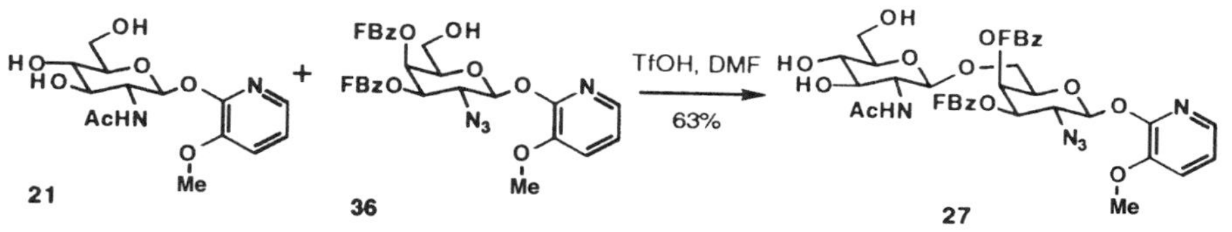

# Synthesis with Unprotected Glycosyl Donors

was observed. After addition of 2 drops of pyridine, the mixture was concentrated in vacuo to provide a residue that was purified by flash chromatography on silica gel column using EtOAc–MeOH (7:1) as the eluant, to give 65.5 mg (63%) of disaccharide **27** as a white solid: mp 126°–127°C, $[\alpha]_D$ +47.7° (c 0.85, CHCl$_3$).

*Glycosylation of 3-Methoxy-2-pyridyl-2-azido-2-deoxy-3,4-di-O-p-fluorobenzoyl-β-D-galactopyranoside (**36**) with 3-Methoxy-2-pyridyl 2-acetamido 3-deoxy-β-D-galactopyranoside (**22**)*

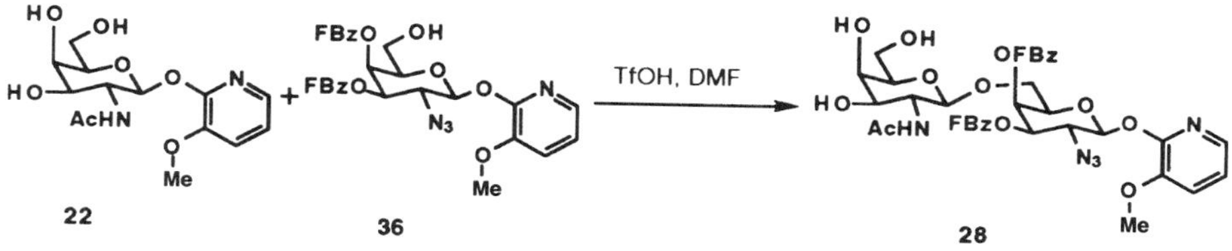

The same procedure described in the foregoing was followed using excess of acceptor **36** (20 Eq.). The disaccharide was isolated by flash chromatography on silica gel column using the same eluant in 69% yield: mp 114°–115°C, $[\alpha]_D$ +43° (c 0.5, CHCl$_3$).

*3-Methoxy-2-pyridyl 3,4-di-O-p-fluorobenzoyl-6-O-(3,4,6-O-triacetyl-2-acetamido-2-deoxy-β-D-glucopyranosyl)-2-acetamido-2-deoxy-β-D-galactopyranoside*

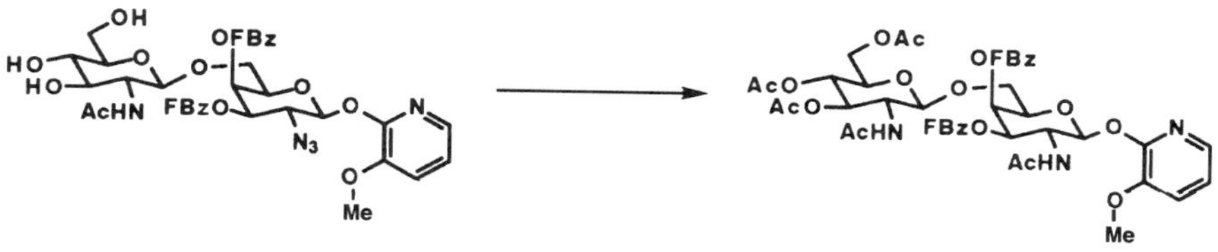

A mixture of 28 mg of azido compound **27**, 8 mg of 20% Pd(OH)$_2$/C in 1.5 mL of EtOAc–MeOH (1:1) was stirred at room temperature under an atmosphere of hydrogen until the starting material completely disappeared by TLC analysis. The catalyst was removed by filtration over Celite, washed with MeOH, and concentrated. The residue was treated with Ac$_2$O-pyridine at room temperature for 4 h, then the mixture was poured into cold aqueous NaHCO$_3$, extracted with EtOAc, and the organic layer was processed as usual. Purification by flash chromatography on silica gel column using EtOAc–MeOH (10:1) as the eluant provided the title product (26 mg, 78%): mp 116°–118°C, $[\alpha]_D$ +106.3° (c 0.2, CHCl$_3$).

*3-Methoxy-2-pyridyl 6-O-(2-acetamido-2-deoxy-β-D-glucopyranosyl)-2-acetamido-2-deoxy-β-D-galactopyranoside (**37**)*

A solution of 3-methoxy-2-pyridyl 3,4-di-O-p-fluorobenzoyl-6-O-(3,4,6-O-triacetyl-2-acetamido-2-deoxy-β-D-glucopyranosyl)-2-acetamido-2-deoxy-β-D-galactopyranoside

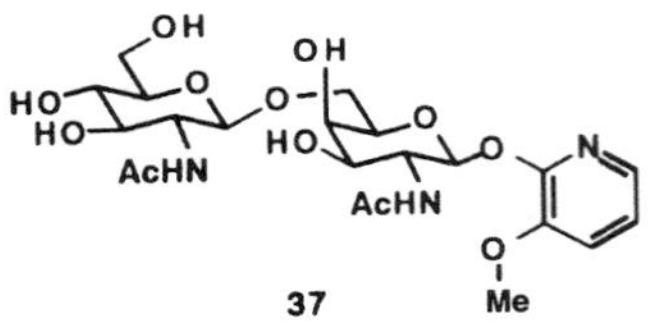

(26 mg) in 1.5 mL of half-saturated NH$_3$–MeOH was stirred at room temperature for 24 h. The mixture was concentrated to give a residue that was washed with diethyl ether to remove methyl *p*-fluorobenzoate. The product **37** (14.5 mg) was obtained in 95% yield: mp 182°C (dec.), [α]$_D$ −10.5° (*c* 0.4, H$_2$O).

## D.  Trisaccharide Synthesis by Selective Activation

*3-Methoxy-2-pyridyl 2,3,4-tri-O-p-fluorobenzoyl-6-O-(6-O-t-butyl-diphenylsilyl-α-D-galactopyranosyl)-β-D-galactopyranoside*

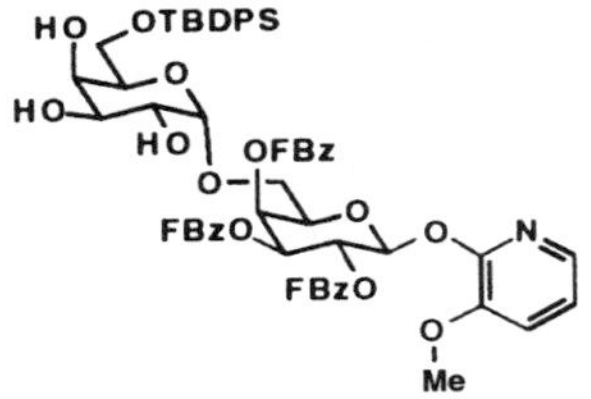

A mixture of 3-methoxy-2-pyridyl 2,3,4-tri-*O*-*p*-fluorobenzoyl-6-*O*-(α-D-galactopyranosyl)-β-D-galactopyranoside **25** (30 mg, 0.0368 mmol), 10.4 μL (0.04 mmol) of TBDPSCl, 3 mg of imidazole, a catalytic amount of DMAP, and 0.3 mL of dry THF was stirred at room temperature for 4.5 h. After concentration, the residue was treated with cold saturated aqueous NaHCO$_3$ and EtOAc. The organic layer was processed as usual, and the product was purified by flash chromatography on silica gel, with MeOH–EtOAc–CHCl$_3$ (1:4:4) as the eluant to give the desired product as a white solid (31 mg, 80% yield): mp 105°–106°C, [α]$_D$ +150.5° (*c* 0.84, CHCl$_3$).

*3-Methoxy-2-pyridyl-6-O-(6-O-t-butyldiphenylsilyl-α-D-galactopyranosyl)-β-D-galactopyranoside* (**31**)

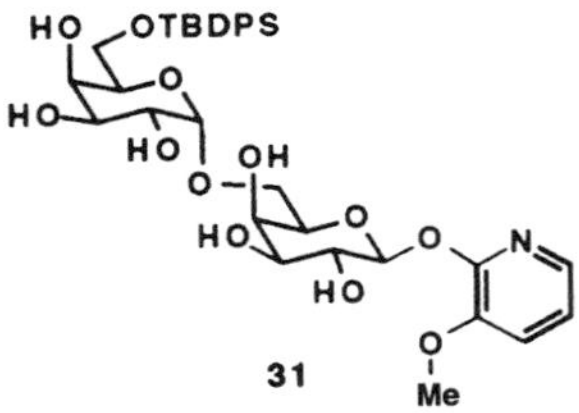

A solution of the preceding compound (30 mg) in 1.5 mL of half-saturated NH$_3$–MeOH was stirred at room temperature overnight. After concentration, the residue was washed with diethyl ether twice to remove methyl *p*-fluorobenzoate. The title product (18 mg) was obtained as a white solid in 95% yield: mp 103°–104°C, [α]$_D$ +34.8° (*c* 1.2, MeOH).

*Glycosylation of 3-methoxy-2-pyridyl 2,3,4-tri-O-p-fluorobenzoyl-β-D-galactopyranoside (30) with 3-Methoxy-2-pyridyl 6-O-(6-O-t-butyldiphenylsilyl-α-D-galactopyranosyl)-β-D-galactopyranoside (31)*

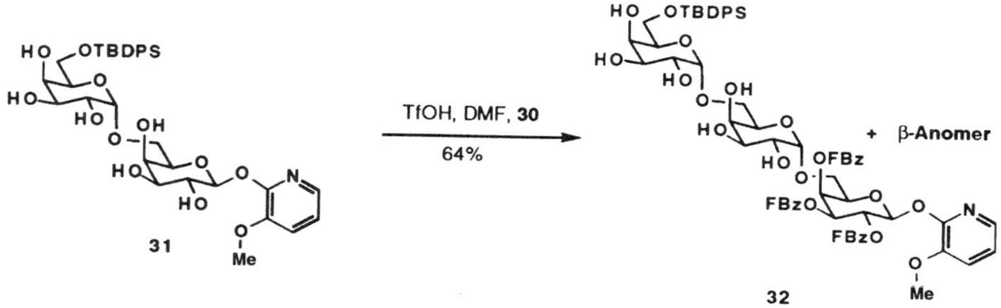

To a solution of glycosyl donor 31 (12 mg, 0.018 mmol) and 342 mg (0.52 mmol) of glycosyl acceptor 30 in 0.3 mL dry DMF was added 9 µL of 1 M MeOTf solution in $CH_3NO_2$ under argon. The mixture was kept at room temperature with stirring for 45 min, at which time TLC showed the complete disappearance of the donor. After addition of 1 drop of pyridine, the solvent was removed in vacuo, and the residue was dissolved in a small amount of $CH_2Cl_2$, and purified by flash chromatography on a silica gel column with EtOAc–hexane (2:1) to recover the excess of acceptor. The column was eluted with EtOAc–$CHCl_3$–MeOH (2:2:1) as the eluant to provide 5 mg of α-linked trisaccharide and a mixture of α- and β-products (8.5 mg), with 1.6:1 ratio (α/β, 3:1; 64% yield). For the α-anomer: mp 117–119°C, $[α]_D$ +140.8° (c 0.5, $CHCl_3$).

*Glycosylation of 3-Methoxy-2-pyridyl 2,3,4-tri-O-p-fluorobenzoyl-6-O-(2,3,4-tri-O-p-fluorobenzoyl-α-D-glucopyranosyl)-β-D-glucopyranoside (34) with 3-Methoxy-2-pyridyl β-D-glucopyranoside (2)*

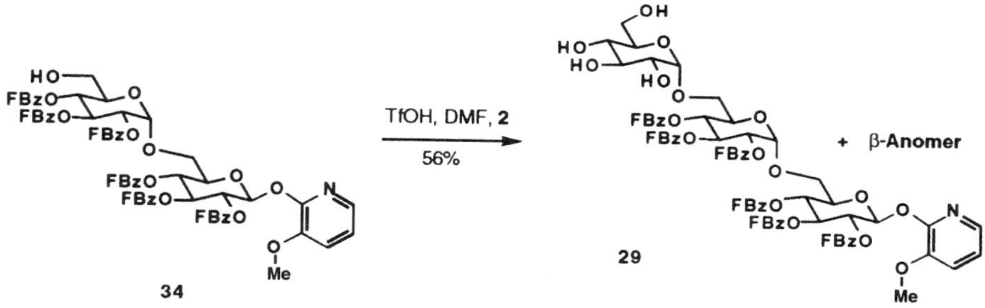

The same procedure described in the foregoing was followed using 30 eq of the acceptor 34, and the desired trisaccharides were isolated by flash chromatography on silica gel column using EtOAc–MeOH (7:1) as the eluant in 56% yield. Pure α- and β-anomers were obtained by chromatography on preparative silica gel thin plate using MeOH–$CHCl_3$–EtOAc (1:2:2) as the eluant. For the α-anomer: mp 129–130°C, $[α]_D$ +119° (c 0.5, $CHCl_3$).

*Glycosylation of 3-Methoxy-2-pyridyl 3,4-di-O-p-fluorobenzoyl-2-azido-2-deoxy-β-D-galactopyranoside (36) with 3-Methoxy-2-pyridyl 6-O-(2-acetamido-2-deoxy-β-D-glucopyranosyl)-2-azido-2-deoxy-β-D-galactopyranoside (37)*

To a mixture of glycosyl donor 37, (8.6 mg, 0.016 mmol) and 270 mg (0.48 mmol) of acceptor 36 in 0.3 mL of dry DMF was added 4.2 µL (0.048 mmol) of TfOH under argon.

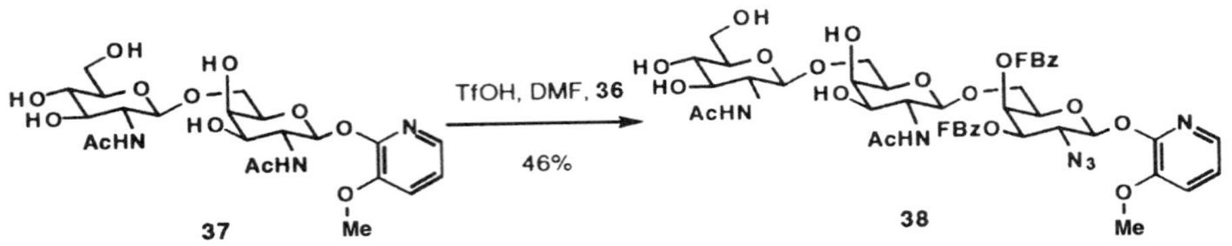

The resulting mixture was vigorously stirred at room temperature for 45 min. After addition of 2 drops of pyridine and concentration, the residue was acetylated. After usual workup, purification by flash chromatography on silica gel using EtOAc–hexane (2:1) gave the unreacted acetylated acceptor. Elution with EtOAc–CHCl$_3$–MeOH (4:4:1) gave a crude product that was further purified on silica gel thin plates (MeOH–EtOAc, 1:10) to give 8.7 mg (46%) of the trisaccharide **38**: [α]$_D$ +67.8° (c 0.2, CHCl$_3$).

### E. Tetrasaccharide Synthesis by Selective Activation

*3-Methoxy-2-pyridyl 2,3,4-tri-O-p-fluorobenzoyl-6-O-(6-O-t-butyldiphenylsilyl-α-D-glucopyranosyl)-β-D-glucopyranoside*

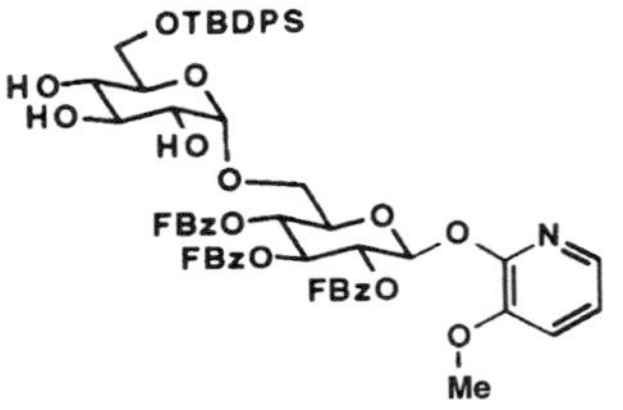

To a mixture of 3-methoxy-2-pyridyl 2,3,4-tri-*O*-*p*-fluorobenzoyl-6-*O*-(α-D-glucopyranosyl)- β-D-glucopyranoside (80 mg, 0.10 mmol), 10 mg (0.15 mmol) of imidazole, 2 mg of DMAP, and 1 mL of dry THF, was added 31 μL (0.12 mmol) of TBDPSCl. The resulting mixture was stirred at room temperature until the starting material completely disappeared. After concentration, the residue was treated with ice-water and EtOAc, and the organic layer was processed as usual. Purification by flash chromatography on silica gel column using MeOH–EtOAc (1:10) as the eluant provided the desired product (88 mg, 85%): mp 105°–106°C, [α]$_D$ +56.9° (c 0.3, CHCl$_3$).

*3-Methoxy-2-pyridyl 2,3,4-tri-O-p-fluorobenzoyl-6-O-(2,3,4-tri-O-p-fluorobenzoyl-6-O-t-butyldiphenylsilyl-α-D-glucopyranosyl)-β-D-glucopyranoside*

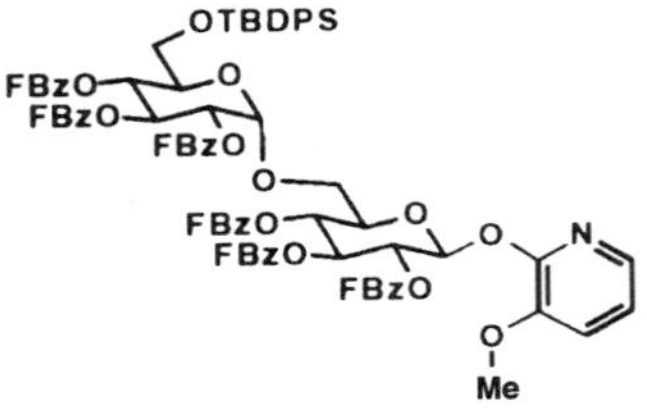

To a cooled solution of the preceding compound (70 mg, 0.066 mmol) in 1 mL of dry pyridine was added 39 μL (0.331 mmol) of *p*-fluorobenzoyl chloride at 0°C. After 30 min, the mixture was stirred at room temperature overnight, then it was poured into cold aqueous NaHCO$_3$, extracted with EtOAc, and the organic layer was processed as usual. Purification by flash chromatography on silica gel column using EtOAc–hexane (1:3) as the eluant gave the title product as a white solid (86 mg, 91%): mp 109°–110°C, [α]$_D$ +85.1° (c 0.5, CHCl$_3$).

*3-Methoxy-2-pyridyl 2,3,4-tri-O-p-fluorobenzoyl-6-O-(2,3,4-tri-O-p-fluorobenzoyl-α-D-glucopyranosyl)-β-D-glucopyranoside (34)*

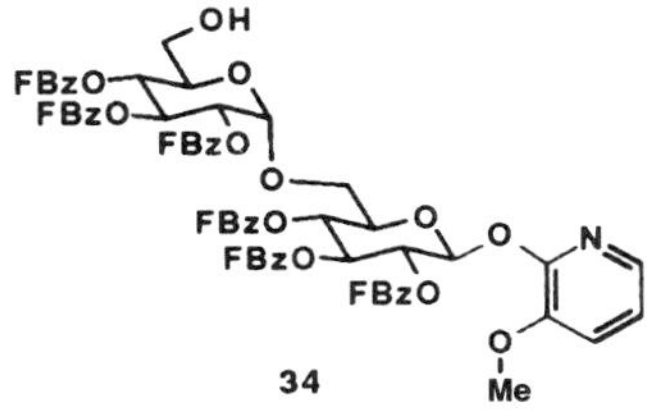

A mixture of the preceding product (84 mg, 0.060 mmol), 0.24 mL of 1 M TBAF in THF solution and 27.4 μL of HOAc was stirred at room temperature until the reaction was completed. After dilution with EtOAc, the mixture was processed as usual, and the residue was purified by flash chromatography on silica gel column using EtOAc–hexane (2:1) to provide the desired product as a white solid (53 mg, 76%): mp 121°–123°C, [α]$_D$ +88° (c 0.61, CHCl$_3$).

*Glycosylation of 3-Methoxy-2-pyridyl 2,3,4-tri-O-p-fluorobenzoyl-6-O-(2,3,4-tri-O-p-fluorobenzoyl-α-D-glucopyranosyl)-β-D-glucopyranoside (34) with 3-Methoxy-2-pyridyl-6-O-(6-O-t-butyldiphenylsilyl-α-D-galactopyranosyl-)β-D-galactopyranoside (31)*

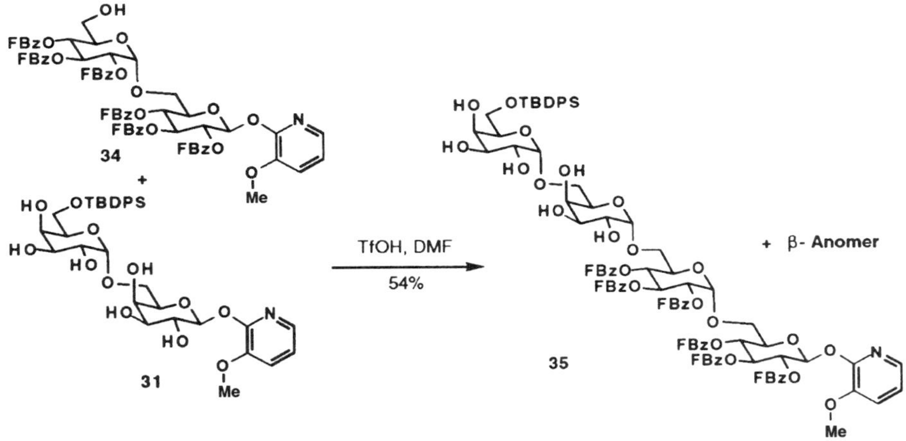

To a mixture of 4 mg (5.82 μmol) of 3-methoxy-2-pyridyl 6-O-(6-O-t-butyldiphenylsilyl-α-D-galactopyranosyl)-β-D-galactopyranoside 31 and 206 mg (0.17 mmol) of 3-methoxy-2-pyridyl 2,3,4-tri-O-p-fluorobenzoyl-6-O-(2,3,4-tri-O-p-fluorobenzoyl-α-D-glucopyrano-syl)-β-D-glucopyranoside 34 in 0.15 mL of dry DMF, was added 3 μL of 1 M MeOTf in CH$_3$NO$_2$. After stirring at room temperature for 50 min, 1 drop of pyridine was added and the mixture was concentrated in vacuo. The residue was purified by flash chromatography on silica gel using EtOAc–hexane (2:1) to recover the excess of acceptor. Elution with MeOH–EtOAc (7:1) gave 5.4 mg of tetrasaccharides in 54% yield (α/β, 4.5:1), [α]$_D$ +105.1° (c 0.6, CHCl$_3$).

# REFERENCES

1. (a) Y. C. Lee and R. T. Lee, Carbohydrate–protein interactions: Basis of glycobiology, *Acc. Chem. Res.* 28:321 (1995); (b) A. Varki, Biological roles of oligosaccharides: All of the theories are correct, *Glycobiology* 3:97 (1993); (c) A. Kobata, Glycobiology: An expanding research area in carbohydrate chemistry, *Acc. Chem. Res.* 26:319 (1993); (d) N. Sharon and H. Lis, Carbohydrates in cell recognition, *Sci. Am.* 268:82 (1993); (e) J. A. Lasky, Selectins: Interpreters of cell-specific carbohydrate information during inflammation, *Science* 258:964 (1992); J. A. Lasky, Selectin–carbohydrate interactions and the initiation of inflammatory response, *Annu. Rev. Biochem.* 64:113 (1995); (f) J. C. Paulson, Selectin/carbohydrate-mediated adhesion of leukocytes, *Adhesion: Its Role in Inflammatory Disease* (H. Harlan and D. Liu, eds.), H. Freeman, New York, 1992, p. 19; (g) S. Hakomori, Aberrant glycosylation in cancer cell membranes as focused on glycolipids: Overview and perspectives, *Cancer Res.* 45:2405 (1985).
2. (a) Z. J. Witczak, Carbohydrates as drugs and potential therapeutics, *Curr. Med. Chem.* 1:392 (1995); (b). J. H. Musser, P. Fugedi, and M. B. Anderson, Carbohydrate-based therapeutics, *Burger's Medicinal Chemistry and Drug Discovery*, Vol. 1, 5th ed., 1995, p. 901; (c) M. Petitou, Drugs based on carbohydrates: Past and future, *Indispensable Probes for the Life Sciences* (P. Kovac, ed.), ACS Symposium Series 560, American Chemical Society, Washington DC, 1994, p. 19; (d) J. H. Musser, Carbohydrates as drug discovery leads, *Annu. Rep. Med. Chem.* 27:301 (1992).
3. For recent reviews on oligosaccharides synthesis: (a) R. R. Schmidt and W. Kinzy, Anomeric-oxygen activation for glycoside synthesis—the trichloroacetimidate method, *Methods Carbohydr. Chem. Biochem.* 50:21 (1994); (b) S. H. Khan and O. Hindsgaul, Chemical synthesis of oligosaccharides, *Molecular Glycobiology* (M. Fukuda and O. Hindsgaul, eds.), IRL Press, Oxford, 1994, p. 206; (c) K. Toshima and K. Tatsuta, Recent progress in O-glycosylation methods and its application to natural product synthesis, *Chem. Rev.* 93:1503 (1993); (d) J. Banoub, P. Boullanger, and D. Lafont, Synthesis of oligosaccharides of 2-amino-2-deoxy-sugars, *Chem. Rev.* 92:1167 (1992); (e) D. Bundle, Synthesis of oligosaccharides related to bacterial O-antigens, *Top. Curr. Chem.* 54:1 (1990); (f) G. Boons, Recent developments in oligosaccharide synthesis, *Contemp. Org. Synth.* 3:173 (1996).
4. For representative glycosylation methods: (a) O. Kanie, Y. Ito, and T. Ogawa, Orthogonal glycosylation strategy in oligosaccharide synthesis, *J. Am. Chem. Soc.* 116:12073 (1994); (b) H. Kondo, S. Aoki, Y. Ichikawa, R. L. Halcomb, H. Ritzen, and C.-H. Wong, Glycosyl phosphites as glycosylation reagents: scope and mechanism, *J. Org. Chem.* 59:864 (1994); (c) A. Marra, J. Esnault, A. Veyrieres, and P. Sinaÿ, Isopropenyl glycosides and congeners as novel classes of glycosyl donors: Theme and variations, *J. Am. Chem. Soc.* 114:6354 (1992); (d) R. Roy, F. O. Andersson, and M. Letellier, "Active" and "latent" thioglycosyl donors in oligosaccharide synthesis. Application to the synthesis of α-sialosides, *Tetrahedron Lett.* 33:6053 (1992); (e) G. Stork and G. Kim, Stereocontrolled synthesis of disaccharides via the temporary silicon connection, *J. Am. Chem. Soc.* 114:1087 (1992); (f) F. Barresi and O. Hindsgaul, Synthesis of β-mannopyranosides by intramolecular aglycon delivery, *J. Am. Chem. Soc.* 113:9376 (1991); (g) G. H. Veeneman, S. H. van Leewen, and J. H. van Boom, Iodonium ion promoted reactions at the anomeric centre. II. An efficient thioglycoside mediated approach toward the formation of 1,2-*trans*-linked glycosides and glycosidic esters, *Tetrahedron Lett.* 31:1331 (1990); (h) D. Kahne, S. Walker, Y. Cheng, and D. V. Engen, Glycosylation of unreactive substrates, *J. Am. Chem. Soc.* 111:6881 (1989); (i) R. L. Halcomb and S. Danishefsky, On the direct epoxidation of glycals: Application of a reiterative strategy for the synthesis of β-linked oligosaccharides, *J. Am. Chem. Soc.* 111:6656 (1989); (j) K. Suzuki, H. Maeta, T. Matsmoto, and G. Tsuchihashi, New glycosidation reaction 2. Preparation of 1-fluoro-D-desosamine derivatives and its efficient glycosidation by the use of $Cp_2HfCl_2$–$AgClO_4$ as the activator, *Tetrahedron Lett.* 29:3571 (1988); (k) D. R. Mootoo, V. Date, and B. Fraser-Reid, *n*-Pentenyl glycosides permit the

chemospecific liberation of the anomeric center, *J. Am. Chem. Soc. 110*:2662 (1988); (l) P. Fugedi and P. J. Garegg, A novel promoter for the efficient construction of 1,2-*trans*-linkages in glycosides synthesis, using thioglycosides as glycosyl donors, *Carbohydr. Res. 149*:C9 (1986); (m) H. Lönn, Synthesis of a tri- and a heptasaccharide which contain α- L-fucopyranosyl groups and are part of the complex type of carbohydrate moiety of glycoproteins, *Carbohydr. Res. 139*:105 (1985); (n) H. Kunz and W. Sager, Stereoselective glycosylation of alcohols and silyl ethers using glycosyl fluorides and boron trifluoride etherate, *Helv. Chim. Acta 68*:283 (1985); (o) S. Hashimoto, M. Hayashi, and R. Noyori, Glycosylation using glucopyranosyl fluorides and silicon-based catalysts. Solvent dependency of the stereoselection, *Tetrahedron Lett. 25*:1379 (1984); (p) K. C. Nicolaou, S. P. Seitz, and D. P. Papahatjis, A mild and general method for the synthesis of *O*-glycosides, *J. Am. Chem. Soc. 105*:2430 (1983); (q) T. Mukaiyama, Y. Murai, and S. Shoda, An efficient method for glucosylation of hydroxy compounds using glucopyranosyl fluoride, *Chem. Lett.* p. 431 (1981); (r) R. R. Schmidt and J. Michel, Facile synthesis of α- and β-*O*-glycosyl imidates: Preparation of glycosides and disaccharides, *Angew. Chem. Int. Ed. Engl. 19*:731 (1980); (s) S. Hanessian and J. Banoub, Chemistry of the glycosidic linkage. An efficient synthesis of 1,2-*trans*-disaccharides, *Carbohydr. Res. 53*:C13 (1977); *ACS Symp. Ser. 39*:36 (1976); (t) R. Lemieux, K. B. Hendricks, R. V. Stick, and K. James, Halide ion catalyzed glycosidation reactions—synthesis of α-linked disaccharides, *J. Am. Chem. Soc. 97*:4056 (1975).
5. For recent examples, see references cited in Chapter 16.
6. R. J. Ferrier, R. W. Hay, and N. Vethaviyasar, A potentially versatile synthesis of glycosides, *Carbohydr. Res. 27*:55 (1973); see also, J. W. VanCleve, *Carbohydr. Res. 70*:161 (1979).
7. S. Hanessian, C. Bacquet, and N. Lehong, Chemistry of the glycosidic linkage. Exceptionally fast and efficient formation of glycosides by remote activation, *Carbohydr. Res. 80*:C17 (1980).
8. For a previous study of glycoside synthesis using *O*-benzylated 2-pyridyloxy glycosyl donors in the presence of Lewis acids, see A. V. Nicolaev and N. K. Kochetkov, Use of (2-pyridyl)-2,3,4,6-*O*-tetra-*O*-benzyl-β-D-glucopyranoside in the synthesis of 1,2-*cis*-bound disaccharides, *Isv. Akad. Nauk SSSR Ser. Khim* p. 2556 (1986).
9. DMF has been used as cosolvent in some glycosylation reactions, see P. V. Nikrad, H. Beierbeck, and R. U. Lemieux, Molecular recognition X. A novel procedure for the detection of the intermolecular hydrogen bonds present in a protein–oligosaccharide complex, *Can. J. Chem. 70*:241 (1992), and references cited therein.
10. R. Roy and F. Tropper, Stereospecific synthesis of aryl β-D-*N*-acetylglucopyranosides by phase transfer catalysis, *Synth. Commun. 20*:2097 (1990).
11. R. U. Lemieux and R. M. Ratcliffe, The azidonitration of tri-*O*-acetyl-D-galactal, *Can. J. Chem. 57*:1244 (1979).
12. It is well known that ester derivatives of glycosyl donors are deactivated owing to an inductive effect: see H. Paulsen, A. Richter, V. Sinnwell, and W. Stenzel, *Carbohydr. Res. 64*:339 (1974). For recent applications to oligosaccharide synthesis: (a) G. H. Veeneman and J. H. van Boom, An efficient thioglycoside-mediated formation of α-glycosidic linkages promoted by iodonium dicollidine perchlorate, *Tetrahedron Lett. 31*:275 (1990); (b) R. W. Friesen and S. Danishefsky, On the controlled oxidative coupling of glycals: A new strategy for the rapid assembly of oligosaccharides, *J. Am. Chem. Soc. 111*:6656 (1989); (c) D. R. Mootoo, P. Konradsson, U. Udodong, and B. Fraser-Reid, "Armed" and "disarmed" *n*-pentenyl glycosides in saccharides couplings leading to oligosaccharides, *J. Am. Chem. Soc. 110*:5583 (1988).
13. (a) T. Ogawa and T. Horisaki, Synthesis of 2-*O*-hexadecanoyl-1-*O*-hexadecyl-[α-Glc-6-SO$_3$Na-(1-6)-α-Glc-(1-6)-α-Glc-(1-3)-*sn*-glycerol: A proposed structure for the glyceroglucolipids of human gastric secretion and of the mucous barrier of rat-stomach antrum, *Carbohydr. Res. 123*:C1 (1983); (b) B. L. Slomiany, A. Slomiany, and G. B. J. Glass, Characterization of two major neutral glyceroglucolipids of the human gastric content, *Biochemistry 16*:3954 (1977); (c) B. L. Slomiany, A. Slomiany, and G. B. J. Glass, Glycolipids of the human gastric content: Structure of the sulfated glyceroglucolipid, *Eur. J. Biochem. 78*:33 (1977).

14. S. Hanessian, B. Lou, and H. Wang, unpublished results.
15. For recent reviews on combinatorial chemistry: (a) N. K. Terrett, M. Gardner, D. W. Gordon, R. J. Kobylecki, and J. Steel, Combinatorial synthesis—the design of compound libraries and their application to drug discovery, *Tetrahedron 51*:8135 (1995), and references cited therein; (b) G. Lowe, Combinatorial chemistry, *Chem. Soc. Rev.* p. 309 (1995); (c) E. R. Felder, The challenge of preparing and testing combinatorial compound libraries in the fast lane, at the front end of drug development, *Chimia 48*:531 (1994); (d) P. C. Andrews, D. M. Leonard, W. C. Cody, and T. K. Awayer, Multiple and combinatorial peptide synthesis: Chemical development and biological application, *Peptide Synthesis and Purification Protocols* (B. M. Dunn and M. W. Pennington, eds.), Humana Press, Clifton NJ, 1994, p. 305; (e) M. A. Gallop, R. W. Barrett, W. J. Dower, S. P. W. Fodor, and E. M. Gordon, Application of combinatorial technologies to drug discovery. 1. Background and peptide combinatorial libraries, *J. Med. Chem. 37*:1233 (1994); (f) E. M. Gordon, R. W. Barrett, W. J. Dower, S. P. W. Fodor, and M. A. Gallop, Application of combinatorial technologies to drug discovery. 2. Combinatorial organic synthesis, library screening strategies, and future directions, *J. Med. Chem. 37*:1385 (1994); (g) W. H. Moos, G. D. Green, and M. R. Pavia, Recent advances in the generation of molecular diversity, *Annu. Rep. Med. Chem. 28*:315 (1993).
16. For selected reviews, see: (a) H. Lis and N. Sharon, Protein glycosylation-structural and functional aspects, *Eur. J. Biochem. 218*:1 (1993); (b) P. Stanley, Glycosylation engineering, *Glycobiology 2*:99 (1992); (c) R. A. Dwek, Glycobiology: Toward understanding the function of sugars, *Chem. Rev. 96*:683 (1996).

# 18

# Oligosaccharide Synthesis by Remote Activation: *O*-Protected 3-Methoxy-2-pyridyloxy (MOP) Glycosyl Donors

**Boliang Lou**
*Cytel Corporation, San Diego, California*

**Hoan Khai Huynh and Stephen Hanessian**
*University of Montreal, Montreal, Quebec, Canada*

|   |   |   |
|---|---|---|
| I. | Introduction | 414 |
| II. | *O*-Protected 3-Methoxy-2-pyridyloxy Glycosyl Donors | 415 |
|   | A. 1,2,-*cis*-disaccharides | 415 |
|   | B. 1,2-*trans*-glycosides | 418 |
| III. | Applications to the Synthesis of T Antigen and Sialyl Le$^x$ | 419 |
| IV. | Experimental Procedures | 422 |
|   | A. General procedure for the preparation of benzylated MOP glycosyl donors | 422 |
|   | B. General procedure for glycosylation with protected MOP glycosyl donors using MeOTf as promoter | 422 |
|   | C. General procedures for glycosylation with protected MOP glycosyl donors using Cu(OTf)$_2$ as promoter | 422 |
|   | D. General procedures for the synthesis of 1,2-*trans*-disaccharides | 423 |
|   | References | 427 |

## I. INTRODUCTION

The synthesis of oligosaccharides entered a new era of practicality in the mid-1970's with the introduction of silver triflate [1] as an activator of anomeric halides, and with the halide ion-catalyzed procedure [2]. The stereocontrolled synthesis of 1,2-*trans*- glycosides can be achieved by activation of the corresponding glycosyl halide bearing an acyl protecting group at the C-2 position in the presence of a heavy-metal salt (Koenigs-Knorr glycosidation) [3]. Neighboring group participation leads to an acyloxonium cation intermediate that reacts with an appropriate acceptor to give the 1,2-*trans*-glycoside directly, or through the intermediacy of an orthoester [4] (Scheme 1). When a nonparticipating benzyl group is present at C-2, the products are usually mixtures of 1,2-*trans*- and *cis*-glycosides. Employing Lemieux's halide ion-catalyzed procedure [2], the thermodynamically more stable 1,2-*cis*-glycosyl bromide is converted to the 1,2-*trans*-anomer in situ by replacement with a bromide ion source. The 1,2-*trans*-bromide being more reactive than the 1,2-*cis*-isomer, is attacked by the nucleophile by an ion pair species, with inversion of anomeric configuration to generate a 1,2-*cis*-glycosidic linkage (see Scheme 1). However, the broad applications of

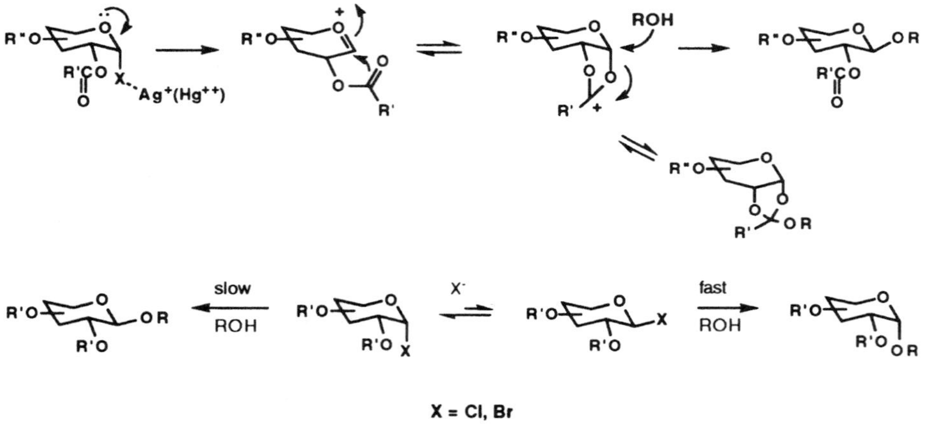

**Scheme 1**  1,2-*trans*-glycoside synthesis by Koenigs–Knorr-type glycosidations, and 1,2-*cis*-glycosides by the halide ion catalyzed method.

these methods to the synthesis of complex oligosaccharides are often hampered owing to the instability of the glycosyl bromides and chlorides, and the conditions involved for their preparation. To this end, significant progress has been made in the exploration of novel and efficient anomeric activation methods over the past 15 years [5; see also Chap. 12].

The trichloroacetimidate method introduced by Schmidt [6] is a method frequently used in glycoside synthesis today. Its major advantages are that the glycosyl trichloroacetimidate donors are easily prepared from the corresponding *O*-substituted reducing sugars, and they can be readily activated by Lewis acids, such as $BF_3$ etherate, to generate oxocarbenium ion intermediates under mild conditions. On the other hand, the relative lability of the leaving group may be a drawback in certain instances (storage, chromatography, and such).

Thioglycosides are also used as glycosyl donors [7]. In contrast with glycosyl bromides and trichloroacetimidates, thioglycosides are more stable under most protection and deprotection conditions. Unlike imidates and bromides, which have to be prepared just before the glycosylation step, anomeric thiomethyl- and thiophenyl-leaving groups may be introduced at earlier stages in the donors. A new method that uses glycosyl sulfoxides as donors has been developed by Kahne [8]. In this method, the anomeric sulfoxide group is activated by triflic anhydride, which generates the oxocarbenium ion intermediate. The method has been suggested to be particularly useful for the glycosylation of unreactive hydroxy groups, and its potential in polymer-supported glycosylation has been shown [9].

Glycosyl fluorides offer a similar advantage as thioglycoside donors in terms of stability. Glycosylations with glycosyl fluorides can be promoted by strong Lewis acids, such as $AgClO_4$–$SnCl_2$ [10], $BF_3$ [11], TMSOTf [12], and $Cp_2HfCl_2$–$AgClO_4$ [13]. Application of the fluoride method to complex oligosaccharides synthesis has been shown by its combination with thioglycoside activation in a two-stage activation procedure developed by Nicolaou and his co-workers [14].

$n$-Pentenyl glycoside donors developed by Fraser-Reid [15], can be activated by electrophilic reagents such as NIS. The reaction may be accelerated in the presence of TfOH or TMSOTf which catalyzes heterolysis of NIS to generate an iodonium ion intermediate. The method has found applications in the synthesis of some complex oligosaccharides. Efforts have been directed toward developing new glycosylation methods based on glycals as donors because the double bond can be readily activated with a wide range of electrophilic reagents [16]. More importantly, the activating reagents usually attack from α-side of the double bond so that the acceptor alcohol approaches from the opposite side to give 1,2-*trans* products in the D-gluco series. For example, Danishefsky's group has capitalized on these notions in the development of methods for the stereoselective epoxidation of the glycal, followed by ring opening at the anomeric position by sugar acceptors in the presence of an appropriate Lewis acid, to form 1,2-*trans* glycosides with high selectivity [17]. The potential of this method in solid-phase oligosaccharide synthesis was recently disclosed by the same group [18]. Glycosyl 2-propenylcarbonates [19], and glycosyl phosphites [20] have been introduced for the stereocontrolled formation of 1,2-*trans*-glycosidic linkages, even in the absence of neighboring participating groups at the C-2 position of the donors.

Despite the great deal of progress made in the chemical synthesis of oligosaccharides, there is still room for innovation and improvement over existing methods. The need for rapid and efficient construction of oligosaccharides has dramatically increased owing to the recent advances in glycobiology and related research fields [21].

## II. *O*-PROTECTED 3-METHOXY-2-PYRIDYLOXY GLYCOSYL DONORS

### A. 1,2-*cis*-Disaccharides

The utility of 3-methoxy-2-pyridyl (MOP) glycosides as glycosyl donors in the synthesis of simple and complex saccharides was discussed in the preceding chapter [22]. One of the major attributes of the method is the mild conditions needed to activate the MOP group, and its adaptability to the sterocontrolled synthesis of 1,2-*cis*-oligosaccharides, without the need to protect the other hydroxy groups in the donors. A prerequisite, however, is the necessity for an excess of acceptor. Therefore, it was important to investigate the extension of the MOP-based glycosylation method in the presence of benzyl-and acyl-type protective groups. The *O*-protected MOP donors are readily prepared from the unprotected MOP glycosides under the normal conditions, or from the corresponding per-*O*-acyl glycosyl halides (Scheme **2**). An additional practical amenity in the use of an MOP-leaving group

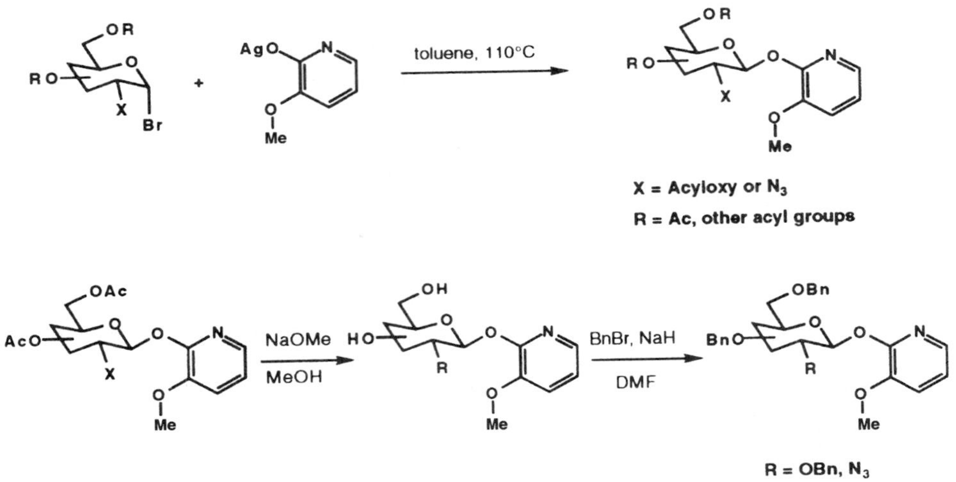

**Scheme 2** Preparation of *O*-protected MOP glycosyl donors.

is the ease of detecting the 3-methoxy 2(*H*)-1-pyridone that is released as the glycosylation reaction progresses.

MeOTf was first chosen as a promoter for the activation of per-*O*-benzyl MOP donors such as **1a**, as it provided the best results in unprotected MOP glycosylations [see Chap. 17]. As shown in Scheme 3, the coupling of tetra-*O*-benzyl MOP D-glucopyranosyl

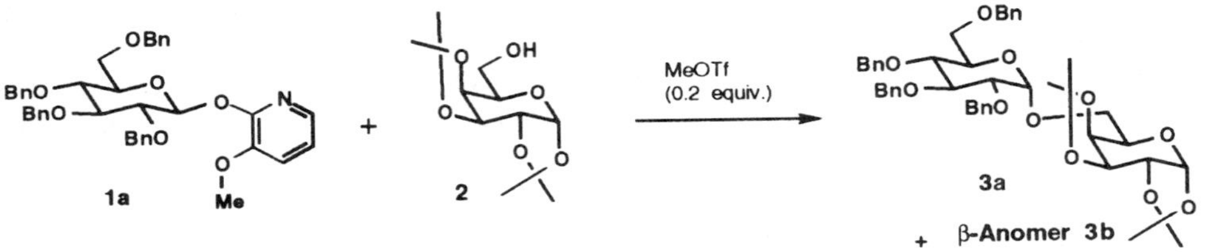

| Acceptor | Solvent | Temp. | time | Yield | 3a : 3b |
|---|---|---|---|---|---|
| 3.4 equiv. | CH₃NO₂ | rt | 12h | 56% | 5 : 4 |
| 1.5 equiv. | Et₂O | rt | 24h | 66% | 5.7 : 1 |

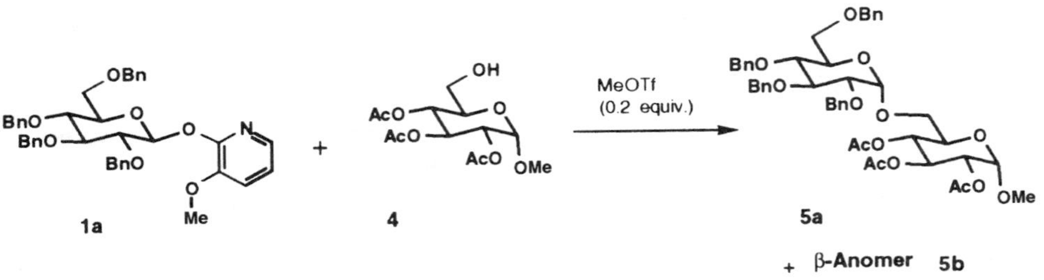

| Acceptor | Solvent | Temp. | time | Yield | 5a : 5b |
|---|---|---|---|---|---|
| 1.5 equiv. | CH₃NO₂/Et₂O | rt | 12h | 58% | 2.3 : 1 |
| 1.5 equiv. | Et₂O | rt | 20h | 64% | 5.1 : 1 |

**Scheme 3** Disaccharide synthesis with *O*-benzyl MOP glycosyl donors using MeOTf as catalyst.

donor **1a** with 1,2:3,4-di-*O*-isopropylidene-α-D-galactopyranose **2** was promoted by a catalytic amount of MeOTf (0.2 eq), to give the disaccharide with α/β ratios of 5:4 and 5.7:1 in $CH_3NO_2$ and $Et_2O$ as solvents, respectively [22]. Similarly, the reaction of **1a** with acceptor **4** in $Et_2O$ afforded predominantly the corresponding α-disaccharide. The use of a large excess of acceptor is unnecessary in these cases. During our studies of the MOP group, we observed that occasionally the donor **1a** was anomerized to the corresponding MOP α-glycoside **1b** and that this anomer was much less reactive under the condition of reaction. Furthermore, MeOTf was not suitable for activation of *O*-acyl-protected MOP glycosides. Faced with these problems, we directed our efforts at finding a catalyst–activator that would be suitable for *O*-benzyl and *O*-acyl protective groups, with little if any anomerization. These requirements were satisfied with copper triflate [$Cu(OTf)_2$]; [23].

Glycosylation with MOP α-glycoside **1b** in the presence of $Cu(OTf)_2$ proceeded at a *faster* rate than that of the corresponding MOP β-glycoside **1a**, to give predominantly the α-disaccharide in good yield (Scheme 4.) No anomerization of either **1a** or **1b** occurred

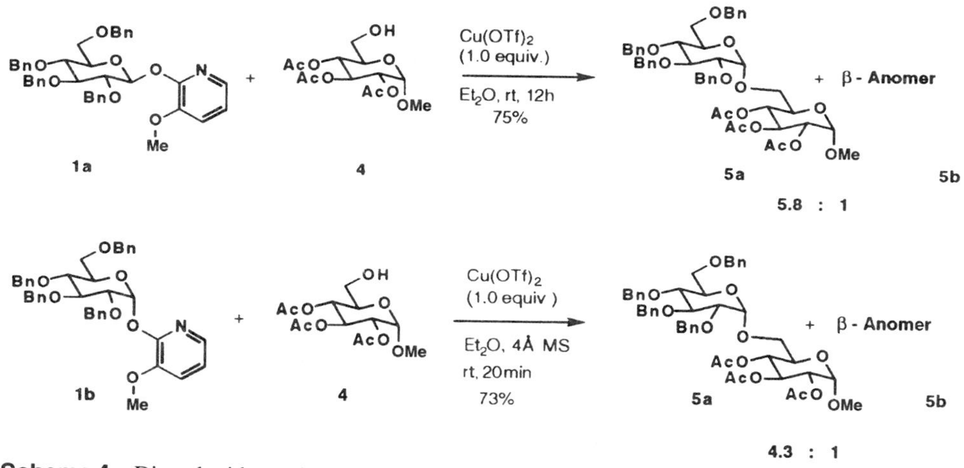

**Scheme 4** Disaccharide synthesis with *O*-benzyl MOP glycosyl donors using $Cu(OTf)_2$ as promoter.

under the reaction conditions. Although a stoichiometric amount of $Cu(OTf)_2$ has to be used for the completion of the reaction, in comparison with the catalytic amounts required for MeOTf, the reactions are cleaner, and lead to products in high yields with good 1,2-*cis*-selectivity in the reactions studied.

The scope of the reaction was further investigated with compounds **6–9** in which a free OH group was present at C-2, C-3, and C-4, respectively. The condensation of donor **1a** with **7** in a mixed solvent ($CH_2Cl_2$–$Et_2O$, 1:4) in the presence of $Cu(OTf)_2$ at room temperature for 15 h gave the corresponding disaccharide in 60% yield (α/β 10:1) (Scheme 5). The same reaction with acceptors **6** and **9** with C-2 and C-4 hydroxy groups, respectively, provided only moderate selectivities. The diminished reactivity C-4 hydroxy groups, particularly in acetylated hexopyranosides is known [24].

High α-selectivities were observed when the D-galactosyl donor **10** was used for the coupling with secondary hydroxy groups in carbohydrate acceptors including **9** (see Scheme 5). With the trichloroacetimidate method [5], the yields of disaccharides derived from **6** and **7** were 85% (10:1, α/β) and 87% (only α), respectively. The condensation of **10**

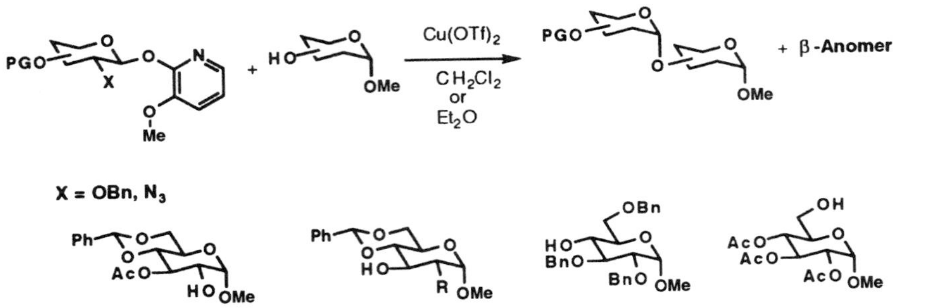


| | 6 | R = OAc, 7; NHCbz, 8 | 9 | 4 |
|---|---|---|---|---|
| MOP β-Glc(1a) | 2.4 :1, 64% | R = OAc, 10 : 1, 60% | 3 : 1, 70% | 5 : 1, 75% |
| MOP β-Gal(10) | α-only, 85% | R = OAc, α-only, 60% | α-only, 75% | 3 : 1, 85% |
| MOP α-($N_3$)Gal(11b) | | R = NHCbz, α-only, 53% | 6 : 1, 63% | 3 : 1, 85% |

**Scheme 5** Synthesis of 1,2-*cis*-glycosyl disaccharides using MOP glycosyl donors.

with primary alcohol **4** under the same conditions ($CH_2Cl_2$, room temperature) gave the desired disaccharide with an α/β ratio of 3:1, and in good yield. Glycosylations with the 2-azido-2-deoxy D-galactosyl donor **11b** were carried out in the same way, and they led to high α-selectivities, especially for secondary hydroxy groups (see Scheme **5**).

## B. 1,2-*trans*-Glycosides

Acetonitrile is frequently used as the solvent of choice to enhance the formation of 1,2-*trans*-glycosidic linkages, by the intermediacy of a kinetically formed α-glycosyl nitrilium ion [25]. Accordingly, coupling of **1a** with **4** in $CH_3CN$ gave a reversed stereoselectivity (α/β, 1:3) compared with the reaction carried out in $Et_2O$ or $CH_2Cl_2$ (Scheme **6**).

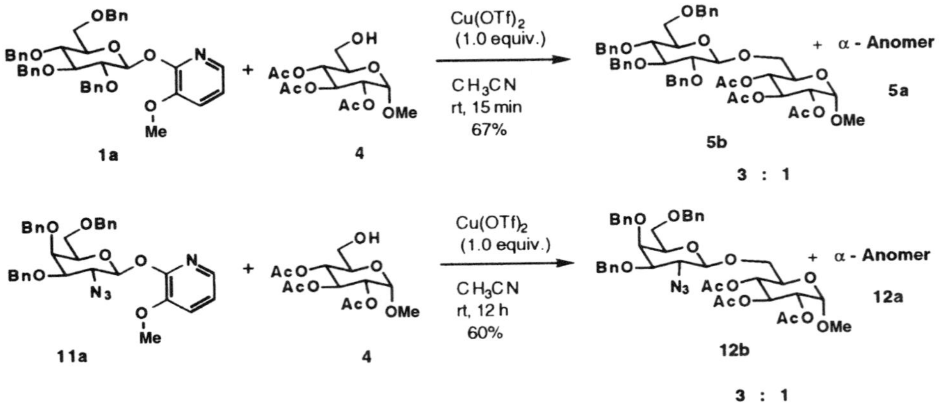


**Scheme 6** 1,2-*trans*-Disaccharide synthesis using $CH_3CN$ as solvent.

Similarly, glycosylation reactions with **11a** under these conditions resulted in the formation of 1,2-*trans*-disaccharides as major products. Thus, the stereochemical outcome of glycosylations using MOP glycosyl donors containing a nonparticipating group at C-2 relies heavily on the solvent used. By simply modulating the solvent system, one can synthesize the 1,2-*cis* or 1,2-*trans*-glycosides from a common MOP glycosyl donor, with good to excellent selectivities as shown in Schemes **4–6**.

Scheme **7** shows examples of 1,2 *trans*-glycosides synthesized from MOP glycosyl donors that have neighboring participating groups. In the presence of $Cu(OTf)_2$, the

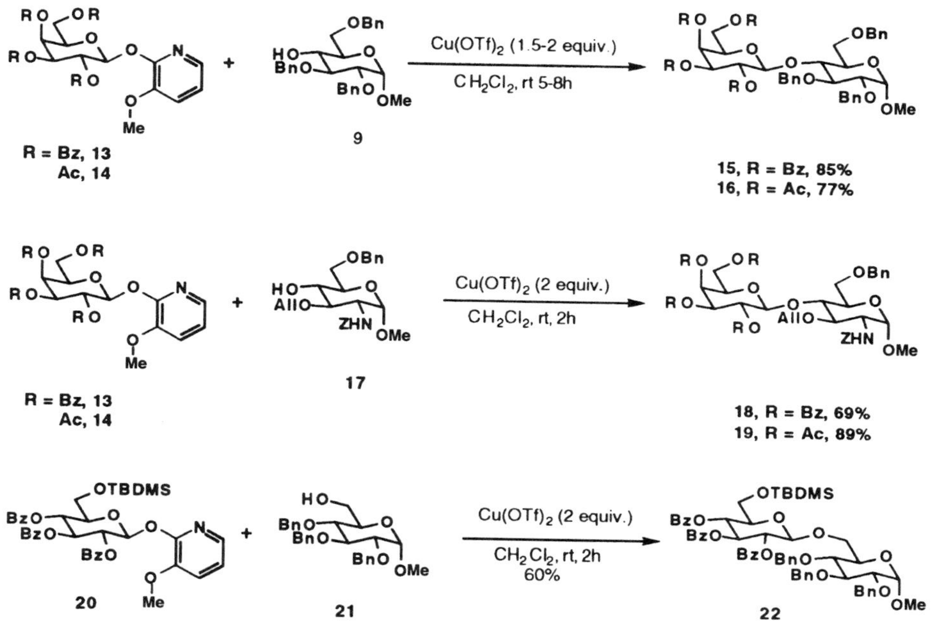

**Scheme 7** 1,2-*trans*-Disaccharide synthesis using MOP-leaving group.

glycosylations of hindered alcohols **9** and **17** with MOP donors **13** and **14**, were carried out at room temperature to afford 1,2-*trans*-disaccharides **15–19** in very good yields. The glycosyl donor **20** in which a bulky protective group is placed at the C-6 gave the expected product **22** in acceptable yield.

## III. APPLICATIONS TO THE SYNTHESIS OF T ANTIGEN AND SIALYL Le[x]

The structure Gal(1-3)GalNAcα1-*O*-Ser or -Thr and GalNAcα1-*O*-Ser or -Thr are characteristic of the glycoproteins of the so-called tumor-associated T and Tn antigens [26]. It is interesting to synthesize the glycopeptides containing immunologically relevant T or Tn structures, because their coupling with carrier proteins could give conjugates for the induction of antibodies against these antigens [27]. Carbohydrate antigens with T-and Tn-active structures are of considerable clinical interest.

The key step in the synthesis of mucin-type glycopeptides is the formulation of an α-glycosidic linkage between a GalNAc residue and amino acids, such as serine or threonine. The nonparticipating azido group is usually placed at the C-2 of the glycosyl donor to enhance α-stereoselectivity. It is subsequently reduced and acetylated to form the acetamido group. Trichloroacetimidate and halides (bromide and chloride) are frequently used as leaving groups for the synthesis of glycopeptide linkages [28; also see Chap. 11].

We have successfully applied the MOP activation method for the synthesis of carbohydrate entities with clustered T and Tn structures [29]. Scheme 8 shows an example

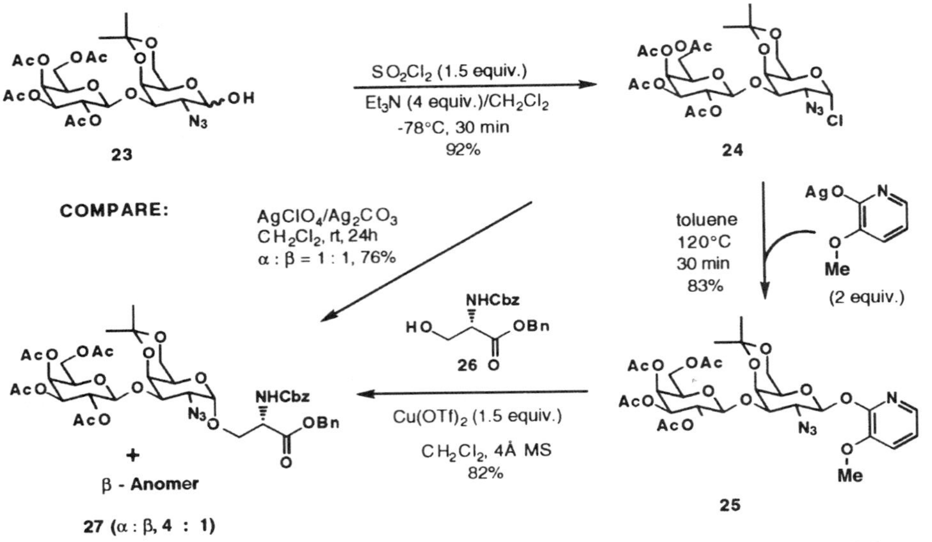

**Scheme 8** Preparation of T-antigen type *O*-serine glycoside using an MOP glycosyl donor.

related to the preparation of disaccharide **27**, which is the core structure of T antigen. The reducing sugar **23** [30] was converted into the glycosyl chloride **24** at −78°C, in presence of sulfuryl chloride and triethylamine, in 92% yield. The reaction of **24** with the serine derivative **26** under the usual Koenigs-Knorr conditions (AgClO$_4$–Ag$_2$CO$_3$), gave the product as an α/β mixture in a ratio of 1:1. In contrast, the treatment of MOP disaccharide donor **25**, prepared from **24** under the standard conditions with **26** in the presence of Cu(OTf)$_2$ at room temperature, afforded the expected 1,2-*cis*-glycoside as a major product and in high yield (α/β, 4:1, 82%).

Le$^x$-antigenic trisaccharide (α-L-Fuc(1-3)-[β-D-Gal-(1-4)-β-D-GlcNAc]) and its sialylated structure (SLe$^x$) are terminal components of a number of glycoconjugates on cell surfaces [31]. SLe$^x$ serves as a ligand for the endothelial leukocyte molecule-1 (E-selectin) [32], which mediates the initial stages of adhesion of leukocytes to activated endothelial cells, and pays a critical role during inflammatory responses [33]. Le$^x$-based carbohydrates have shown promise in therapeutic investigations related to the inflammatory process. As a result, extensive efforts have been directed toward the synthesis of SLe$^x$ and related molecules [34; also see Chap. 15].

Schemes **9** and **10** show some examples in the successful use of the MOP-based

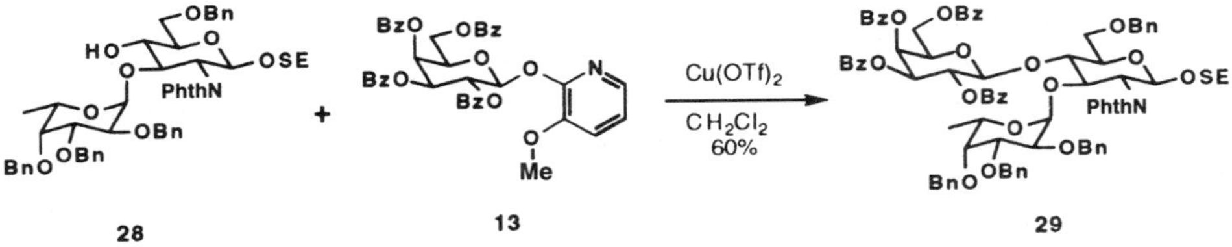

**Scheme 9** Synthesis of a Le$^x$ derivative using an MOP glycosyl donor.

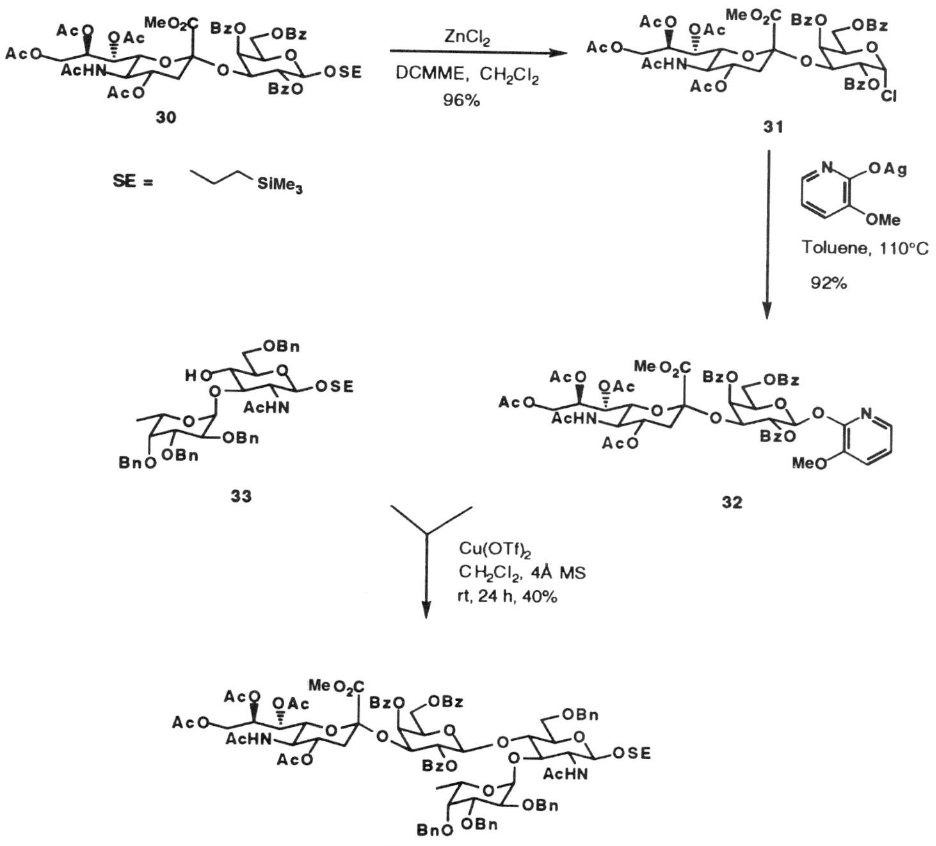

# 3-Methoxy-2-pyridyloxy Glycosyl Donors

**Scheme 10** Synthesis of a sialyl Le$^x$ derivative using an MOP disaccharide donor.

method for the construction of Le$^x$ and SLe$^x$ analogues. The sterically hindered hydroxy group of **28** was successfully glycosylated by employing 3-methoxy-2-pyridyl 2,3,4,6-tetra-*O*-benzoyl β-D-galactopyranoside **13** in the presence of Cu(OTf)$_2$ to give the trisaccharide **29** in 60% yield. This result led to the assembly of SLe$^x$ in a convergent, blockwise manner from building blocks **32** and **33** that were prepared as shown in Scheme **10**. Disaccharide **30** obtained by a known method [35], was converted into the corresponding chloride **31** [36], which was then heated with silver 3-methoxy-2-pyridoxide in toluene to afford the disaccharide donor **32** in 88% overall yield. The coupling of **32** and **33** was achieved in the presence of Cu(OTf)$_2$, to give the tetrasaccharide **34** in 40% isolated yield.

In conclusion, we have described efficient and stereocontrolled syntheses of simple and complex oligosaccharides in good to excellent α/β anomeric selectivities using the MOP-leaving group. Comparative glycosylations of MOP and trichloroacetimidate donors using carbohydrate acceptors show excellent correlations between efficiency and selectivity. Glycosylations with MOP donors offer the possibility of monitoring the progress of the reactions by TLC or other analytic methods based on the easy detection of 3-methoxy 2(*H*)-1-pyridone, which is released on glycosylation.

## IV. EXPERIMENTAL PROCEDURES*

### A. General Procedure for the Preparation of Benzylated MOP Glycosyl Donors

*3-Methoxy-2-pyridyl 2,3,4,6-tetra-O-benzyl-β-D-glucopyranoside (**1a**)*

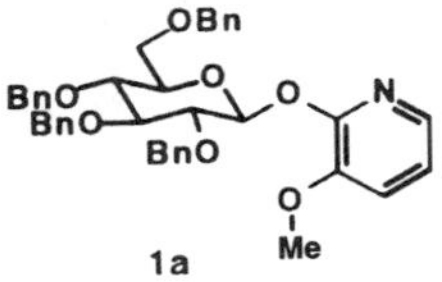

To a cooled solution of 3-methoxy-2-pyridyl β-D-glucopyranoside (800 mg, 2.8 mmol) in 15 mL of DMF were added 600 mg (15 mmol) of 60% NaH and 1.5 mL (12.6 mmol) of benzyl bromide at 0°C with efficient stirring. The temperature was allowed to reach room temperature and the reaction mixture was stirred overnight. MeOH (2 mL) was added to destroy the excess sodium hydride, the mixture was poured into ice-water, extracted with $CH_2Cl_2$ and processed as usual. Concentration and purification by flash chromatography gave 1.6 g of the title product as a syrup in 89% yield: $[\alpha]_D$ +12.4° (c 1.2, $CHCl_3$).

### B. General Procedure for Glycosylation with Protected MOP Glycosyl Donors Using MeOTf as Promoter

*Glycosylation of Methyl 2,3,4-tri-O-acetyl-α-D-glucopyranoside (**4**) with 3-Methoxy-2-pyridyl 2,3,4,6-tetra-O-benzyl-β-D-glucopyranoside (**1a**)*

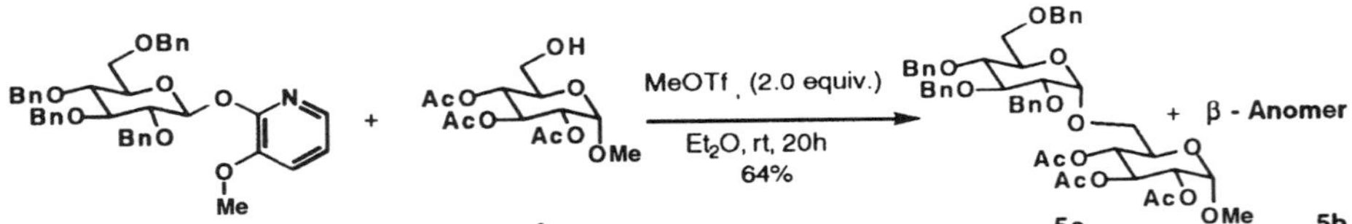

To a stirred solution of the glycosyl donor **1a** (24 mg, 0.037 mmol), 18 mg (0.056 mmol) of methyl 2,3,4-tri-O-acetyl-α-D-glucopyranoside **4** in 1 mL of $Et_2O$ was added 7.4 μL of 1 M MeOTf in $CH_3NO_2$ and the mixture was stirred at room temperature for 20 h. After addition of 1 drop of pyridine, concentration followed by purification by flash chromatography on silica gel (EtOAc–hexane, 1:2) gave 20 mg of a mixture of α- and β-anomers in 64% yield (α/β, 5.1:1).

### C. General Procedures for Glycosylation with Protected MOP glycosyl Donors Using Cu(OTf)₂ as Promoter

*Glycosylation of Methyl 2,3,4-tri-O-acetyl-α-D-glucopyranoside (**4**) with 3-Methoxy-2-pyridyl 2,3,4,6-tetra-O-benzyl-β-D-glucopyranoside (**1a**)*

To a mixture of the glycosyl donor **1a** (26.4 mg, 0.041 mmol), 19.6 mg (0.061 mmol) of methyl 2,3,4-tri-O-acetyl-α-D-glucopyranoside **4**, 1 mL of $Et_2O$, and activated 4-Å molecule sieve (MS) was added 15 mg (0.041 mmol) of $Cu(OTf)_2$. The mixture was stirred for

---

*Optical rotations of were measured at 22°–25°C.

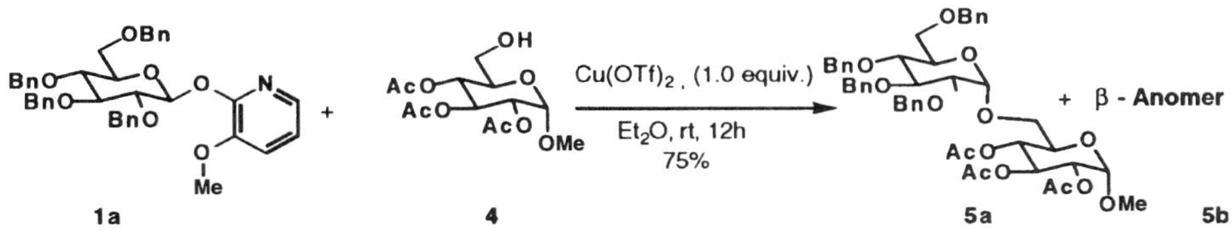

12 h, 2 drops of pyridine were added and the solvent was removed. Purification by flash chromatography on silica gel gave 24 mg of desired products **5a** and **5b** (α/β, 5.8:1) in 75% yield.

*Glycosylation of Methyl 2-O-acetyl-4,6-benzylidene-α-D-glucopyranoside (7) with 3-Methoxy-2-pyridyl 2,3,4,6-tetra-O-benzyl-β-D-galactopyranoside (1a)*

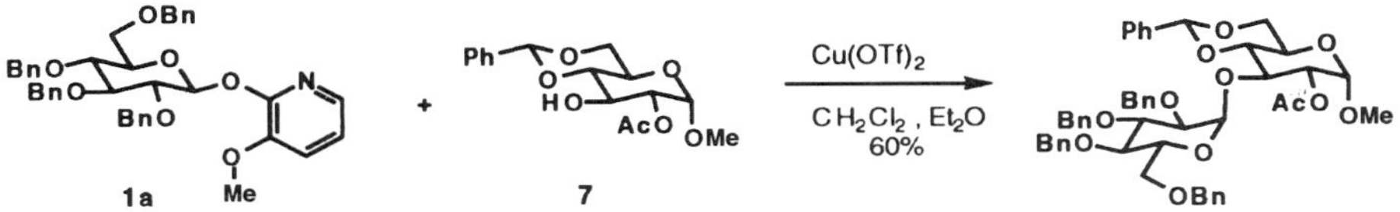

To a mixture of the glycosyl donor **1a** (130.8 mg, 0.20 mmol), 101.2 mg (0.31 mmol) of the acceptor **7**, activated 4-Å MS, and 2 mL of the mixed solvent of $Et_2O/CH_2Cl_2$ (5:1, v/v), was added 75.0 mg (0.21 mmol) of $Cu(OTf)_2$ under argon at room temperature. The mixture was stirred for 15 h, 1 drop of pyridine was added, then the mixture was concentrated. Purification by flash chromatography on silica gel gave 102.5 mg (60%) of the desired disaccharide (α/β, 10:1): for the α-anomer, $[α]_D$ +76.1° (c 1.1, $CHCl_3$).

*Glycosylation of Methyl 3-O-acetyl-4,6-benzylidene-α-D-glucopyranoside (6) with 3-Methoxy-2-pyridyl 2,3,4,6-tetra-O-benzyl-β-D-galactopyranoside (10)*

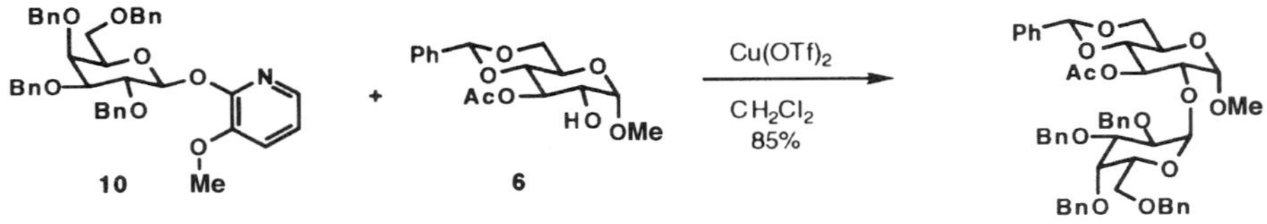

To a mixture of the glycosyl donor **10** (115.6 mg, 0.8 mmol), 38.7 mg (0.12 mmol) of the acceptor **6**, activated 4-Å MS, and 3 mL of dry $CH_2Cl_2$, was added 66.2 mg (0.18 mmol) of $Cu(OTf)_2$ under argon at room temperature. The mixture was stirred for 6 h, then concentrated after 1 drop of pyridine was added. The residue was purified by flash chromatography on silica gel to give 95 mg (85%) of the desired disaccharide as the α-anomer only: 55–57°C, $[α]_D$ +59.9° (c 1.12, $CHCl_3$).

## D. General Procedures for the Synthesis of 1,2-*trans*-Disaccharides

*Glycosylation of Methyl 2,3,6-tri-O-benzyl-α-D-glucopyranoside (9) with 3-Methoxy-2-pyridyl 2,3,4,6-tetra-O-acetyl-β-D-galactopyranoside (14)*

To a mixture of the glycosyl donor **14** (90.6 mg, 0.20 mmol), 61.6 mg (0.13 mmol) of glycosyl acceptor **9**, activated powdered 4-Å MS, and 3 mL of dry $CH_2Cl_2$, was added 108 mg (0.23 mmol) of Cu(OTf)$_2$ under argon. The resulting mixture was stirred at room temperature 5 h, and then concentrated after addition of 1 drop of pyridine. The residue was purified by flash chromatography on silica gel column to provide 81 mg (77%) of the desired disaccharide **16**: mp 54°–56°C, [α]$_D$ +14.4° (*c* 3.5, CHCl$_3$).

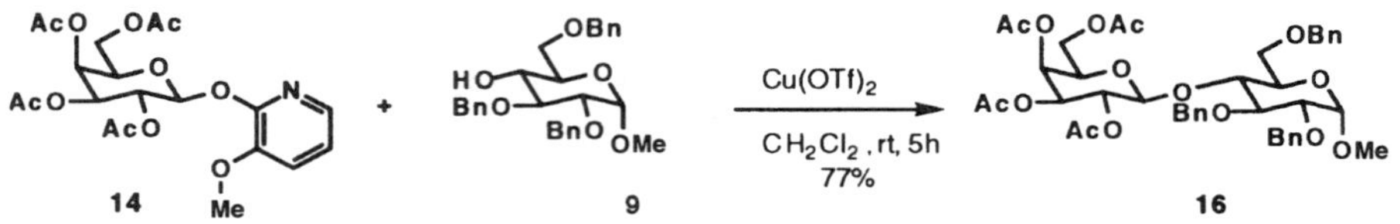

*Glycosylation of Methyl 2,3,4-tri-O-benzyl-α-D-glucopyranoside (21) with 3-Methoxy-2-pyridyl 2,3,4-tri-O-benzoyl-6-O-t-butyldimethylsilyl-β-D-glucopyranoside (20)*

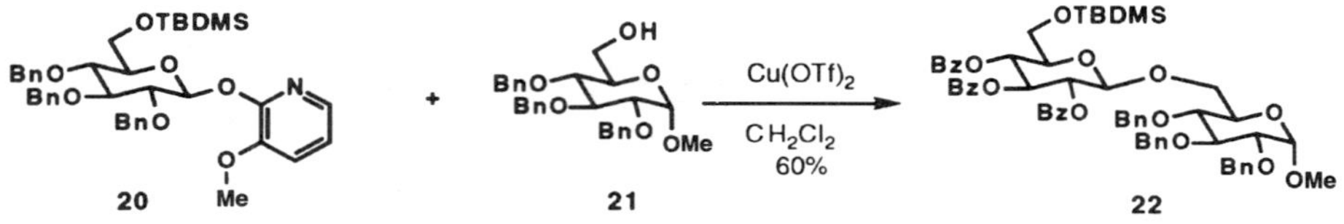

To a mixture of the glycosyl donor **20** (30 mg, 0.042 mmol), 23.4 mg (0.050 mmol) of glycosyl acceptor **21**, 1.5 mL of dry CH$_2$Cl$_2$, and powdered 4-Å MS, was added 30.4 mg (0.084 mmol) of Cu(OTf)$_2$ under argon. The resulting mixture was stirred at room temperature for 2 h, and then concentrated, after addition of 1 drop of pyridine. The residue was purified by flash chromatography on silica gel column to provide 26 mg (60%) of disaccharide **22**: [α]$_D$ +11.3° (*c* 0.71, CHCl$_3$).

*2-Azido-3-O-(2,3,4,6-tetra-O-acetyl-β-D-galactopyranosyl)4,6-O-isopropylidene 2-deoxy-α-D-galactopyranosyl Chloride (24)*

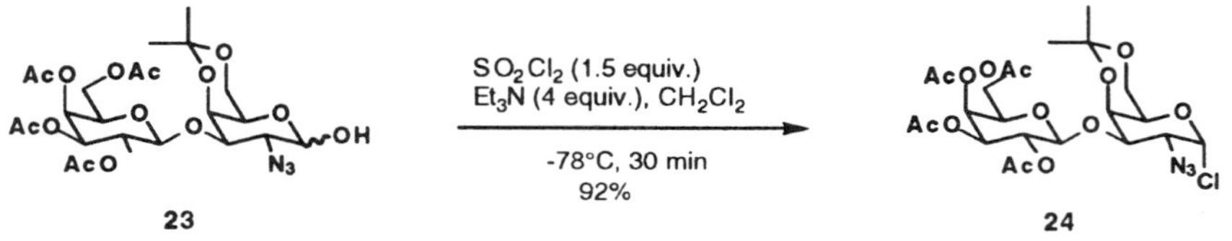

To a solution cooled at −78°C containing 357 mg (0.596 mmol) of 2-azido-3-O-(2,3,4,6-tetra-O-acetyl-β-D-galactopyranosyl)-4,6,O-isopropylidene-2-deoxy-D-galactopyranose **23** and 332 μL (2.38 mmol) of Et$_3$N in 10 mL of CH$_2$Cl$_2$, was added dropwise 89.3 μL (0.894 mmol) of sulfuryl chloride over 10 min. The mixture was stirred at the same temperature for 30 min, 5 mL of saturated NaHCO$_3$ was added, and the organic layer was processed as usual. After drying in vacuo for 5 h, the crude product was purified by flash chromatography on silica gel column (EtOAc–hexane, 1:1 to 2:1) to afford 337 mg of pure α-chloride **24** in 92% yield.

*3-Methoxy-2-pyridyl 2-azido-3-O-(2,3,4,6-tetra-O-acetyl-β-D-galactopyranosyl)-4,6-O-isopropylidene-2-deoxy-β-D-galactopyranoside (25)*

A mixture of 2-azido-3-O-(2,3,4,6-tetra-O-acetyl-β-D-galactopyranosyl)-4,6-O-isopropylidene-2-deoxy-α-D-galactopyranosyl chloride **24** (315 mg, 0.46 mmol), 237 mg (1.02

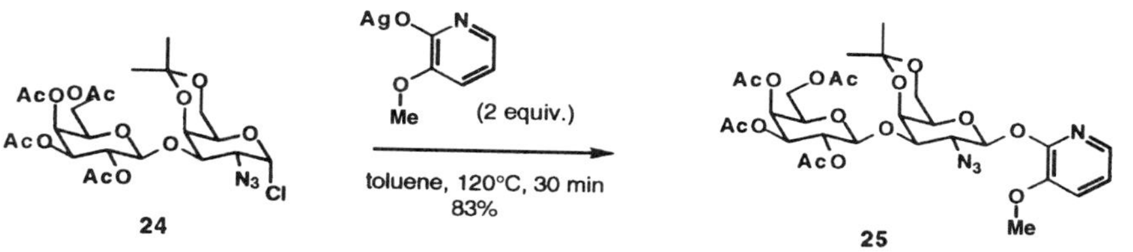

mmol) of silver 3-methoxy-2-pyridoxide, and 10 mL of toluene was heated at 120°C with stirring for 30 min. The mixture was filtered, concentrated, and purified by flash chromatography on silica gel column with EtOAc–hexane (1:1) to give 285 mg (83%) of the title product **25** and 14 mg of α-isomer: mp 195°C, $[\alpha]_D$ +1.8° (c 1.54, $CHCl_3$).

*Glycosylation of* N*-carbobenzyloxy-*L*-serine Benzyl Ester Using the MOP Donor,* **25**

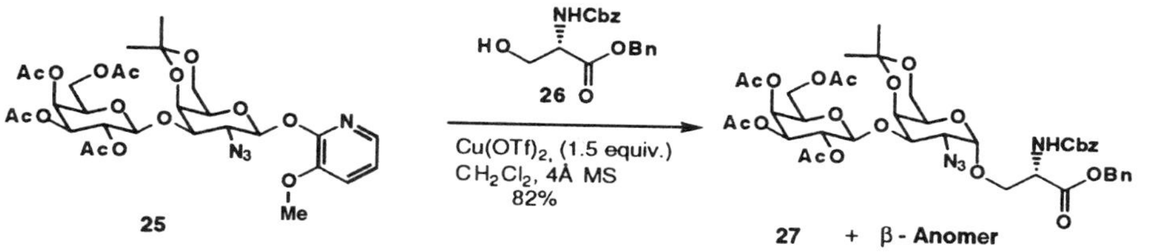

To a mixture of 3-methoxy-2-pyridyl 2-azido-3-*O*-(2,3,4,6-tetra-*O*-acetyl-β-D-galactopyranosyl)-4,6-*O*-isopropylidene-2-deoxy-β-D-galactopyranoside **25** (33 mg, 0.048 mmol), 23 mg (0.073 mmol) of *N*-carbobenzyloxy-L-serine benzyl ester **26**, 1.6 mL of $CH_2Cl_2$, and activated powdered 4 Å MS, was added 25.4 mg (0.070 mmol) of $Cu(OTf)_2$. The mixture was stirred at room temperature until the glycosyl donor had been consumed (2 h). After addition of 1 drop of pyridine, the solution was concentrated to give a residue that was chromatographed using EtOAc–hexane (1:2 to 2:1) to give 28 mg of the α-anomer **27** and 7 mg of the β-anomer (82% yield).

*2-(Trimethylsilyl)ethyl O-(2,3,4,6-tetra-O-benzoyl-β-*D*-galactopyranosyl)-(1→4)[(2,3,4-tri-O-benzyl-α-*L*-fucopyranosyl)-(1→3)]-2-acetamido-6-O-benzyl-2-deoxy-β-*D*-glucopyranoside,* **29**

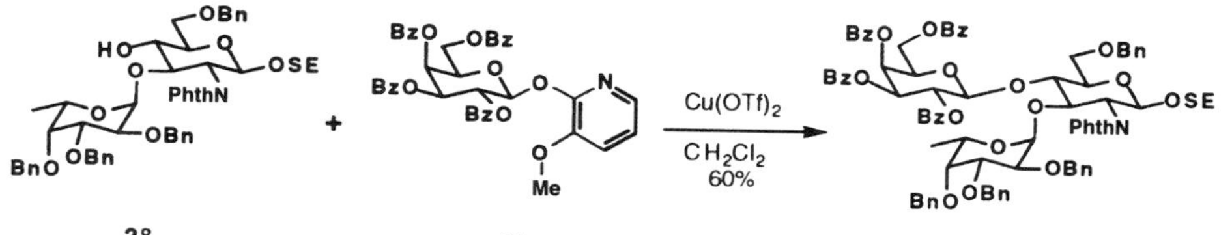

To a solution of the glycosyl donor **13** (29 mg, 0.041 mmol) and the disaccharide donor **28** (16.8 mg; 0.021 mmol) in dichloromethane (1.5 mL) was added activated powdered 4 Å MS (30 mg). The solution was stirred at room temperature overnight under argon, and $Cu(OTf)_2$ (30 mg; 0.0826 mmol) was then added. Stirring was continued overnight (12 h), the reaction was quenched with a few drops of pyridine, the mixture was concentrated, and the residue was purified by flash chromatography on silica gel column with hexane–ethyl acetate–dichloromethane (2:1:0.5) to afford the title trisaccharide **29** (18.0 mg; 60%).

O-*(Methyl 5-acetamido-4,7,8,9-tetra-O-acetyl-3,5-dideoxy-D-*glycero-α-*D-*galacto-*2-nonulopyranosylonate)-(2→3)-2,4,6-tri-O-benzoyl-β-D-galactopyranosyl Chloride* (**31**)

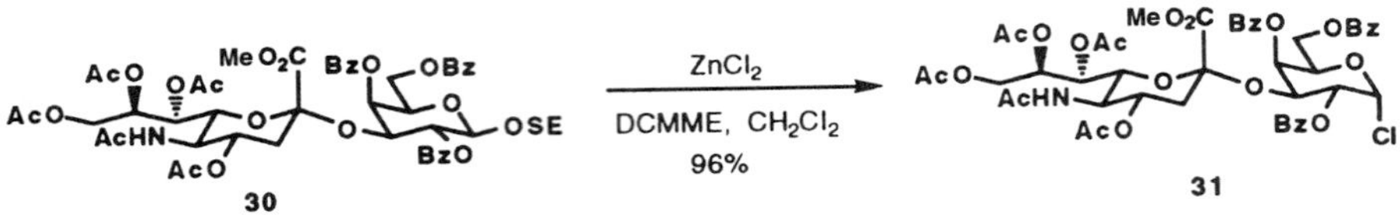

To a solution of 2-(trimethylsilyl)ethyl (methyl 5-acetamido-4,7,8,9-tetra-*O*-acetyl-3,4-dideoxy-D-*glycero*-α-D-*galacto*-2-nonulopyranosylonate)-(2→3)-2,4,6-tri-*O*-benzoyl-β-D-galactopyranoside **30** (210 mg, 0.18 mmol) and zinc chloride (30 mg, 0.220 mmol) in dichloromethane (2 mL) was added α,α-dichloromethyl methyl ether (40 μL, 0.434 mmol) at 0°C. The reaction mixture was stirred at room temperature for 4 h, then diluted with dichloromethane, and washed successively with cold dilute aqueous sodium carbonate solution and water, dried over Na$_2$SO$_4$, and concentrated to give the title product (187 mg, 96%): mp 75°C, [α]$_D$ +42.36° (*c* 1.015, CHCl$_3$).

*3-Methoxy-2-pyridyl* O-*(methyl 5-acetamido-4,7,8,9-tetra-O-acetyl-3,5-dideoxy-D-*glycero-α-*D-*galacto-*2-nonulopyranosylonate)-(2→3)-2,4,6-tri-O-benzoyl-β-D-galactopyranoside* (**32**)

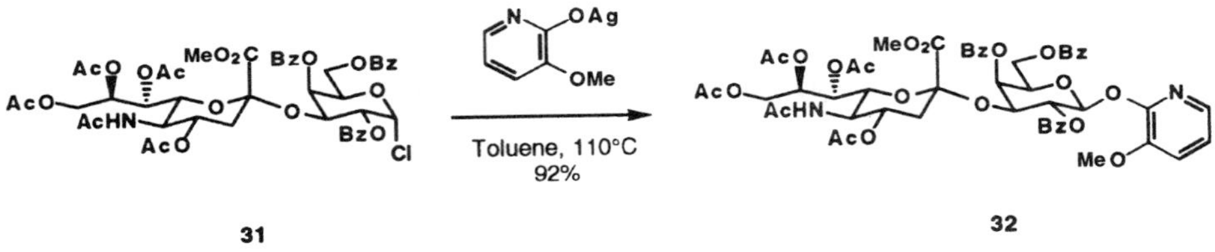

A mixture of the previously obtained glycosyl chloride (186 mg, 0.19 mmol), silver 3-methoxy-2-pyridoxide, and 10 mL of toluene was heated at 110°C for 2 h with stirring. The mixture was filtered, concentrated, and purified by flash chromatography on silica gel column with chloroform–methanol (20:1) to give the desired product **32** (187 mg, 92%): mp 118–120°C.

*2-(Trimethylsilyl)ethyl* O-*(methyl 5-acetamido-4,7,8,9-tetra-O-acetyl-3,5-dideoxy-D-*glycero-α-*D-*galacto-*2-nonulopyranosylonate)-(2→3)-(2,4,6-tri-O-benzoyl-β-D-galactopyranosyl)-(1→4)-O-[2,3,4-tri-O-benzyl-α-L-fucopyranosyl-(1→3)]-2-acetamido-6-O-benzyl-2-deoxy-β-D-glucopyranoside* (**34**)

To a solution of 3-methoxy-2-pyridyl *O*-(methyl 5-acetamido-4,7,8,9-tetra-*O*-acetyl-3,5-dideoxy-D-*glycero*-α-D-*galacto*-2-nonulopyranosylonate)-(2→3)-2,4,6-tri-*O*-benzoyl-β-D-galactopyranoside **32** (64 mg, 0.60 mmol) and 2-(trimethylsilyl)ethyl *O*-(2,3,4-tri-*O*-benzyl-α-L-fucopyranosyl)-(1→3)-*O*-benzyl-2-deoxy-2-acetamido-β-D-glucopyranoside **33** (150 mg, 0.18 mmol) in dichloromethane (5 mL) was added activated powdered 4 Å MS (100 mg). The solution was stirred at room temperature overnight under argon, and Cu(OTf)$_2$ (43 mg, 0.12 mmol) was then added. Stirring was continued overnight (12 h), the reaction was quenched with a few drops of pyridine, the mixture was concentrated, and the

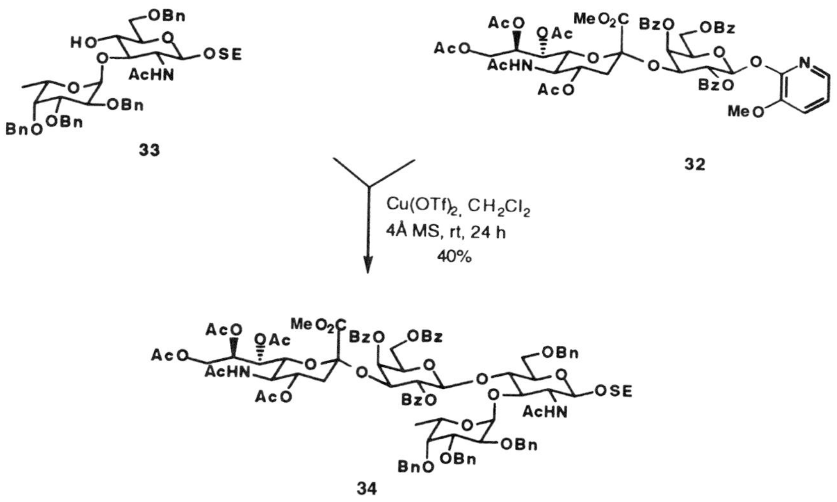

# 3-Methoxy-2-pyridyloxy Glycosyl Donors

residue was purified by flash chromatography on silica gel column with benzene–acetone (3:1) to afford the title tetrasaccharide **34** (37 mg, 40%).

## REFERENCES

1. S. Hanessian and J. Banoub, Chemistry of the glycosidic linkage. An efficient synthesis of 1,2-*trans*-disaccharides, *Carbohydr. Res.* 53:C13 (1977); *ACS Symp. Ser.* 39:36 (1976).
2. R. Lemieux, K. B. Hendricks, R. V. Stick, and K. James, Halide ion catalyzed glycosidation reactions. Synthesis of α-linked disaccharides, *J. Am. Chem. Soc.* 97:4056 (1975).
3. K. Igarashi, The Koenigs-Knorr reaction, *Adv. Carbohydr. Chem. Biochem.* 34:243 (1977); see also Ref. 1.
4. N. K. Kochetkov, A. J. Khorlin, and A. F. Bochkov, A new method of glycosylation, *Tetrahedron* 23:693 (1967).
5. For recent reviews of oligosaccharide synthesis see: (a) R. R. Schmidt and W. Kinzy, Anomeric-oxygen activation for glycoside synthesis—the trichloroacetimidate method, *Adv. Carbohydr. Chem. Biol. Chem.* 50:21 (1994); R. R. Schmidt, New methods for the synthesis of glycosides and oligosaccharides—are there alternatives to the Koenigs-Knoor method? *Angew. Chem. Int. Ed. Engl.* 25:212 (1986); (b) S. H. Khan and O. Hindsgaul, Chemical synthesis of oligosaccharides, *Molecular Glycobiology* (M. Fukuda and O. Hindsgaul, eds.), IRL Press, Oxford, 1994, p. 206; (c) K. Toshima and K. Tatsuta, Recent progress in *O*-glycosylation methods and its application to natural product synthesis, *Chem. Rev.* 93:1503 (1993); (d) J. Banoub, P. Boullanger, and D. Lafont, Synthesis of oligosaccharides of 2-amino-2-deoxy-sugars, *Chem. Rev.* 92:1167 (1992); (e) D. Bundle, Synthesis of oligosaccharides related to bacterial O-antigens, *Top. Curr. Chem.* 54:1 (1990); (f) G. Boons, Recent developments in chemical oligosaccharide synthesis, *Contemp. Org. Synth.* 3:173 (1996).
6. R. R. Schmidt and J. Michel, Facile synthesis of α- and β-*O*-glycosyl imidates: Preparation of glycosides and disaccharides, *Angew. Chem. Int. Ed. Engl.* 19:731 (1980).
7. (a) R. Roy, F. O. Andersson, and M. Letellier, "Active" and "latent" thioglycosyl donors in oligosaccharide synthesis. Application to the synthesis of α-sialosides, *Tetrahedron Lett.* 33:6053 (1992); (b) G. H. Veeneman, S. H. van Leewen, and J. H. van Boom, Iodonium ion promoted reactions at the anomeric centre. II. An efficient thioglycoside mediated approach

toward the formation of 1,2-*trans*-linked glycosides and glycosidic esters, *Tetrahedron Lett.* *31*:1331 (1990); (c) P. Fügedi and P. J. Garegg, A novel promoter for the efficient construction of 1,2-*trans*-linkages in glycosides synthesis, using thioglycosides as glycosyl donors, *Carbohydr. Res. 149*:C9 (1986); (d) H. Lönn, Synthesis of a tri- and a heptasaccharide which contain α-L-fucopyranosyl groups and are part of the complex type of carbohydrate moiety of glycoproteins, *Carbohydr. Res. 139*:105 (1985); (e) K. C. Nicolaou, S. P. Seitz, and D. P. Papahatjis, A mild and general method for the synthesis of *O*-glycosides, *J. Am. Chem. Soc. 105*:2430 (1983).

8. D. Kahne, S. Walker, Y. Cheng, and D. V. Engen, Glycosylation of unreactive substrates, *J. Am. Chem. Soc. 111*:6881 (1989).

9. L. Yan, C. M. Taylor, R. Goodnow, Jr., and D. Kahne, Glycosylation on the Merrifield resin using anomeric sulfoxides, *J. Am. Chem. Soc. 116*:6953 (1994).

10. T. Mukaiyama, Y. Murai, and S. Shoda, An efficient method for glucosylation of hydroxy compounds using glucopyranosyl fluoride, *Chem. Lett. 431* (1981).

11. H. Kunz and W. Sager, Stereoselective glycosylation of alcohols and silyl ethers using glycosyl fluorides and boron trifluoride etherate, *Helv. Chim. Acta 68*:283 (1985).

12. S. Hashimoto, M. Hayashi, and R. Noyori, Glycosylation using glucopyranosyl fluorides and silicon-based catalysts. Solvent dependency of the stereoselection, *Tetrahedron Lett. 25*:1379 (1984).

13. K. Suzuki, H. Maeta, T. Matsumoto, and G. Tsuchihashi, New glycosidation reaction 2. Preparation of 1-fluoro-D-desosamine derivative and its efficient glycosidation by the use of $Cp_2HfCl_2$–$AgClO_4$ as the activator, *Tetrahedron Lett. 29*:3571 (1988).

14. K. C. Nicolaou, R. E. Dolle, D. P. Papahatjis, and J. L. Randall, Practical synthesis of oligosaccharides. Partial synthesis of avermectin $B_{1a}$, *J. Am. Chem. Soc. 106*:4189 (1984).

15. D. R. Mootoo, V. Date, and B. Fraser-Reid, *n*-Pentenyl glycosides permit the chemospecific liberation of the anomeric center, *J. Am. Chem. Soc. 110*:2662 (1988).

16. (a) Y. Ito and T. Ogawa, Sulfenate esters as glycosyl acceptors: A novel approach to the synthesis of 2-deoxyglycosides, *Tetrahedron Lett. 28*:2723 (1987); (b) G. Janorand, J.-M. Beau, and P. Sinaÿ, Glycosyloxyselenation–desalination of glycals: A new approach to 2′-deoxydisaccharides, *J. Chem. Soc. Chem. Commun.* p. 572 (1981); (c) R. U. Lemieux and A. R. Morgan, The synthesis of β-D-glucopyranosyl 2-deoxy-α-D-*arabino*-hexopyranoside, *Can J. Chem. 43*:2190 (1965).

17. R. L. Halcomb and S. Danishefsky, On the direct epoxidation of glycals: Application of a reiterative strategy for the synthesis of β-linked oligosaccharides, *J. Am. Chem. Soc. 111*:6656 (1989).

18. S. J. Danishefsky, K. F. McClure, J. T. Randolph, and R. B. Ruggeri, A strategy for the solid-phase synthesis of oligosaccharides, *Science 260*:1307 (1993).

19. A. Marra, J. Esnault, A. Veyrieres, and P. Sinaÿ, Isopropenyl glycosides and congeners as novel classes of glycosyl donors: Theme and variations, *J. Am. Chem. Soc. 114*:6354 (1992).

20. H. Kondo. S. Aoki, Y. Ichikawa, R. L. Halcomb, H. Ritzen, and C.-H. Wong, Glycosyl phosphites as glycosylation reagents: Scope and mechanism, *J. Org. Chem. 59*:864 (1994).

21. (a) Y. C. Lee, Synthetic oligosaccharides in glycobiology: An overview in synthetic oligosaccharides, *Indispensable Probes for the Life Sciences* (P. Kovac, ed.), ACS Symposium Series 560, American Chemical Society, Washington DC, 1994, p. 2; (b) A. Kobata, Glycobiology: An expanding research area in carbohydrate chemistry, *Acc. Chem. Res. 26*:319 (1993).

22. For a previous study of glycoside synthesis using *O*-benzylated 2-pyridyloxy glycosyl donors in the presence of Lewis acids, see A. V. Nicolaev and N. K. Kochetkov, Use of (2-pyridyl)-2,3,4,6-*O*-tetra-*O*-benzyl-β-D-glucopyranoside in the synthesis of 1,2-*cis*-bound disaccharides, *Isv. Akad. Nauk SSSR. Ser. Khim.*: 2556 (1986).

23. For an example of glycoside synthesis with *O*-benzylated 2-benzothiazolyl-1-thio-D-glucopyranoside in the presence of $Cu(OTf)_2$, see T. Mukaiyama, T. Nakatsuka, and S. Shoda, An efficient glucosylation of alcohol using 1-thioglucoside derivative, *Chem. Lett.* p. 87 (1979).

24. For the reactivity of the C-4 hydroxyl group in pyranose acceptors, see Ref. 5b, p. 167; see also Ref. 1 in this chapter.

25. R. R. Schmidt, M. Behrendt, and A. Toepfer, Nitriles as solvents in glycosylation reactions: Highly selective β-glycoside synthesis, *Synlett* p. 694 (1990)
26. (a) G. F. Springer, T and Tn, general carcinoma autoantigens, *Science* 224:1198 (1984); (b) S. Hakomori, Tumor-associated carbohydrate antigens, *Annu. Rev. Immunol.* 2:103 (1984).
27. G. D. MacLean, M. Reddish, R. R. Koganty, T. Wang, S. Ganchi, M. Smolenski, J. Samuel, J. M. Nabholtz, and B. M. Longenecker, Immunization of breast cancer patients using a synthetic sialyl–Tn glycoconjugate plus Delox adjuvant, *Cancer Immunol. Immunother,* 36:215 (1993).
28. (a) J. Rademann and R. R. Schmidt, Solid-phase synthesis of a glycosylated hexapeptide of human sialophorin, using the trichloroacetimidate method, *Carbohydr. Res.* 269:217 (1995); (b) For recent reviews on chemical synthesis of glycopeptide containing T and Tn antigen structures: H. Kunz, Synthesis of glycopeptides, partial structures of biological recognition components, *Angew. Chem. Int. Ed. Engl.* 26:294 (1987); H. G. Garg, K. von Dem Bruch, and H. Kunz, Developments in the synthesis of glycopeptides containing glycosyl L-asparagine, L-serine, and L-threonine, *Adv. Carbohydr. Chem. Biochem.* 50:277 (1994).
29. S. Hanessian, D. Qiu, H. Prabhanjan, G. V. Reddy, and B. Lou, Synthesis of clustered D-GalNAc(Tn) and D-Galβ(1-3)GalNAc(T) antigenic motifs using a pentaerythritol scaffold, *Can. J. Chem.* (in press).
30. For the preparation of the compound **23**, see Ref. 29.
31. W. M. Watkins, P. O. Skacel, and P. H. Johnson, Human fucosyltransferases involved in the biosynthesis of X (Gal-β-1-4[Fuc-α-1-3]GlcNAc) and sialyl-X (NeuAc-α-2-3Gal-β-1-4[Fuc-α-1-3]GlcNAc) antigenic determinants, *Carbohydrate Antigens* (J. Garegg and A. A. Lindberg, eds.), ACS Symposium Series 519, American Chemical Society, Washington DC 1993, p. 34.
32. (a) For a recent review, see J. B. Lowe, Carbohydrate recognition in cell–cell interaction, *Molecular Glycobiology* (M. Fukuda and O. Hindsgaul, eds.), IRL Press, Oxford, 1994, p. 163; (b) D. V. Erbe, S. R. Watson, L. G. Presta, B. A. Walitzky, C. Foxall, B. K. Brandley, and L. A. Lasky, P- and E-selectin use common sites for carbohydrate ligand recognition and cell adhesion, *J. Cell Biol.* 120:1227 (1993); (c) M. J. Polley, M. L. Phillips, E. Wayner, E. Nudelman, A. K. Singhal, S. Hakomori, and J. C. Paulson, CD62 and endothelial cell leukocyte adhesion molecule 1 (ELAM-1) recognize the same carbohydrate ligand, sialyl Lewis X, *Proc. Natl. Acad. Sci. USA* 88:6224 (1991); (d) G. Walz, A. Aruffo, W. Kolanus, M. Bevilacqua, and B. Seed, Recognition by ELAM-1 of the sialyl-Le[x] determinant on myeloid and tumor cells, *Science* 250:1132 (1990); (e) M. L. Phillips, E. Nudelman, F. C. A. Gaeta, M. Perez, A. K. Singhal, S. Hakomori, and J. C. Paulson, ELAM-1 mediates cell adhesion by recognition of a carbohydrate ligand, sialyl Le[x], *Science* 250:1130 (1990).
33. (a) J. A. Lasky, Selectin-carbohydrate interactions and the initiation of inflammatory response, *Annu. Rev. Biochem.* 64:113 (1995); (b) J. C. Paulson, Selectin carbohydrate-mediated adhesion of leukocytes, *Adhesion: Its Role in Inflammatory Disease* (J. Harlan and D. Liu, eds.), H. Freeman, New York, 1992, p. 19; (c) J. A. Lasky, Selectins: Interpreters of cell specific carbohydrate information during inflammation, *Science* 258:964 (1992).
34. (a) M. Iida, A. Endo, S. Fujita, M. Numata, Y. Matsuzaki, M. Sugimoto, S. Nunomura, and T. Ogawa, Total synthesis of glycononaosyl ceramide with a sialyl dimeric Le[x] sequence, *Carbohydr. Res.* 270:C15 (1995); (b) R. K. Jain, R. Vig, R. Rampal, E. V. Chandrasekaran, and K. L. Matta, Total synthesis of 3'-O-sialyl, 6'-O-sulfo lewis[x], NeuAcα2-3(6-O-SO$_3$Na)Galβ1-4(Fucα1-3)-GlcNAcβ-OMe: A major capping group of GLYCAM-I, *J. Am. Chem. Soc.* 116:12123 (1994); (c) Y. Ichikawa, Y.-C. Lin, D. P. Dumaa, G.-J. Shen, E. Garcia-Juncenda, M. A. Williams, R. Bayer, C. Ketcham, L. E. Walker, J. C. Paulson, and C.-H Wong, Chemical-enzymatic synthesis and conformational analysis of sialyl Lewis[x] and derivatives, *J. Am. Chem. Soc.* 114:9283 (1992); (d) K. C. Nicolaou, C. W. Hummel, Y. Iwabuchi, Total synthesis of sialyl dimeric Le[x], *J. Am. Chem. Soc* 114:3126 (1992); (e) A. Kameyama, H. Ishida, M. Kiso, and A. Hasegawa, Total synthesis of sialyl lewis X, *Carbohydr. Res,* 209:C1 (1991); (f) K. C. Nicolaou, C. W. Hummel, N. J. Bockovich, and C. H. Wong, Stereocontrolled Synthesis of Sialyl Le[x], the oligosaccharide binding ligand to ELAM-1, *J. Chem. Soc. Chem. Commun.* 10:870 (1991); (g) M. M. Palcic, A. P. Venot, R. M. Ratcliffe, and O. Hindsgaul, Enzymic synthesis of

oligosaccharides terminating in the tumor-associated sialyl-Lewis-a-determinant, *Carbohydr. Res. 190*:1 (1989); (h) S. J. Danishefsky, J. Gervay, J. M. Peterson, F. E. McDonald, K. Koseki, D. A. Griffith, T. Oriyama, and S. P. Marsden, Application of glycals to the synthesis of oligosaccharides: Convergent total syntheses of the Lewis X trisaccharide sialyl Lewis X antigenic determinant and higher congeners, *J. Am. Chem. Soc. 117*:1940 (1995).

35. A. Kameyama, H. Ishida, M. Kiso, and A. Hasegawa, Stereoselective synthesis of sialyl–lactotetraosylceramide and sialyl neolactotetraosylcermide, *Carbohydr. Res. 200*:269 (1990).

36. K. P. Ravindranathan and H. J. Jennings, A facile, one-step procedure for the conversion of 2-(trimethylsilyl)ethyl glycosides to their glycosyl chlorides, *Tetrahedron Lett. 31*:2537 (1990).

# 19

# Oligosaccharide Synthesis by Remote Activation: *O*-Protected Glycosyl 2-thiopyridylcarbonate Donors

**Boliang Lou**
Cytel Corporation, San Diego, California

**Hoan Khai Huynh and Stephen Hanessian**
University of Montreal, Montreal, Quebec, Canada

| | | | |
|---|---|---|---|
| I. | Introduction | | 432 |
| | A. | Design of novel anomeric activating groups | 432 |
| | B. | Glycosyl 2-pyridylcarbonates as donors | 433 |
| II. | Methods: Glycosyl 2-thiopyridylcarbonates (TOPCAT) as Glycosyl Donors | | 434 |
| | A. | Preparation of glycosyl donors | 434 |
| | B. | Activation of TOPCAT donors and synthesis of 1,2-*cis*-disaccharides | 434 |
| | C. | 1,2-*trans*-disaccharides | 435 |
| | D. | Application to the synthesis of sialyl Le$^x$ | 436 |
| | E. | Conclusion | 438 |
| III. | Experimental Procedures | | 439 |
| | A. | General procedure for one-pot glycosylation with glycosyl-2-pyridylcarbonates | 439 |
| | B. | General procedure for the preparation of glycosyl 2-thiopyridylcarbonate donors | 440 |
| | C. | General procedures for the synthesis of 1,2-*cis*-disaccharides using TOPCAT glycosyl donors | 440 |
| | D. | General procedure for the synthesis of 1,2-*trans*-glycosides using the TOPCAT-leaving group | 441 |
| | References | | 447 |

## I. INTRODUCTION

Thioglycosides are widely used as a major class of glycosyl donors in oligosaccharide synthesis because they are usually stable under the various conditions used for chemically manipulating of hydroxy groups. Thioglycosides can be selectively activated with a variety of thiophilic reagents [1; see also Chap. 11]. In 1980, we introduced the concept of remote activation of an anomeric group in glycoside synthesis based on $O$-unprotected 2-thiopyridyl glycosyl donors which, when treated by $Hg(NO_3)_2$ in the presence of an excess of alcohol, afforded the desired glycosides in good yield, with good to moderate α-selectivities [2]. The method is applicable in the glycosylation of aglycones derived from complex antibiotics, as in the total syntheses of erythromycin [3] and avermectin $B_{1a}$ [4].

An extension of the remote activation concept for the purpose of designing novel anomeric-leaving groups, has been reported by Kobayashi and his co-workers [5]. Thus, glycosyl 2-pyridylcarboxylates serve as glycosyl donors in the presence of $Cu(OTf)_2$ or $Sn(OTf)_2$ and various sugar acceptors, to give disaccharides in very good yields with high α- or β-stereoselectivities.

The potential usefulness of glycosyl carbonates as glycosyl donors for the glycosylation reaction was first explored by Descotes [6]. Heating glycosyl ethyl carbonates with a large excess of simple alcohols afforded the desired glycosides, but the coupling failed with a sugar acceptor under the same condition. Recently, Sinaÿ [7] reported on the 2-propenyl carbonate-leaving group that was particularly useful for the synthesis of 1,2-*trans*-glycosides when $CH_3CN$ was employed as solvent.

### A. Design of Novel Anomeric Activating Groups

The successful applications of the MOP technology in the synthesis of oligosaccharides in our laboratories [see Chap. 18] led us to explore the design and reactivity of other anomeric derivatives of pyridine by the use of the remote activation concept. We designed novel classes of glycosyl donors containing a carbonate or a thiocarbonate group that bridges a 2-pyridyl group and the anomeric center of the glycosyl moiety (Scheme 1). It was anticipated that such glycosyl donors should be more reactive than the 2-pyridyl glycosides in the presence of a suitable transition metal ion, such as $Cu^{2+}$ or $Ag^+$. The proposed bidentate activation through the chelation of a carbonyl or thiocarbonyl group and nitrogen

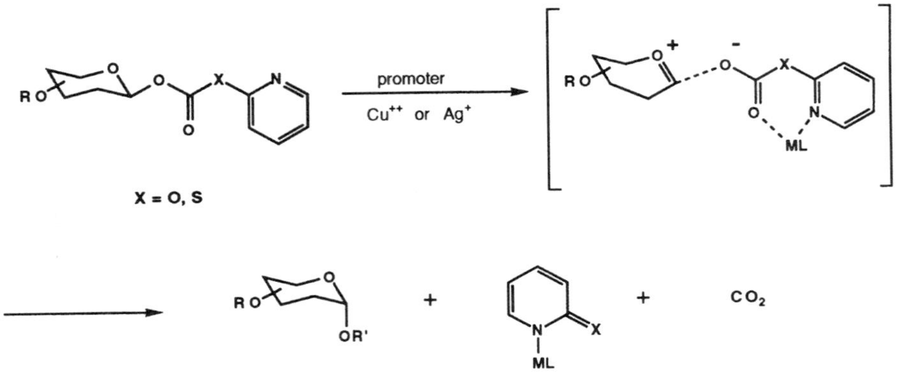

**Scheme 1** The design of glycosyl 2-pyridylcarbonate or glycosyl 2-thiopyridylcarbonate (TOP-CAT) donors based on the remote activation concept.

with the metal cation (remote activation), would lead to an oxocarbenium intermediate with the formation of carbon dioxide and a metal–pyridine complex as shown in Scheme 1. Subsequent nucleophilic attack by an alcohol would form the glycoside bond.

## B. Glycosyl 2-pyridylcarbonates as Donors

Alkyl 2-pyridylcarbonates, which can be readily prepared by the reaction of alcohols with bis(2-pyridyl) carbonate **2**, have been used as intermediates for the synthesis of some biologically interesting carbamates [8]. Following the reported procedure, we prepared bis(2-pyridyl) carbonate **2** in very high yield by the treatment of 2-hydroxy pyridine with triphosgene. A coupling reaction with 2,3,4,6-tetra-O-benzyl-D-glucopyranose **1** proceeded in the presence of Et$_3$N to give the expected glycosyl 2-pyridylcarbonate **3** as a mixture of α and β anomers, but only in 29–37% isolated yields (Scheme 2). The low yield was

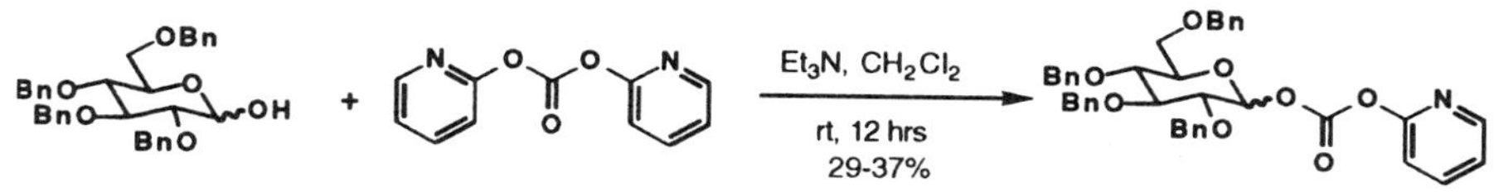

| ROH | Promoter (equiv.) | Solvent | Temp.(°C) | Time | α : β | Yield(%) |
|---|---|---|---|---|---|---|
| 5 | Cu(OTf)$_2$ (3) | Et$_2$O | -20 - 0 | 10min | 3.6 : 1 | 60 |
| 4 | Cu(OTf)$_2$ (2.5), HOTf (0.5) | Et$_2$O | -20 - 0 | 10min | 2.5 : 1 | 63 |
| 4 | Cu(OTf)$_2$ (2.5), HOTf (0.5) | CH$_3$CN | -20 - 0 | 1h | 1 : 6 | 60 |

**Scheme 2** Glycosyl 2-pyridylcarbonate as a novel glycosyl donor.

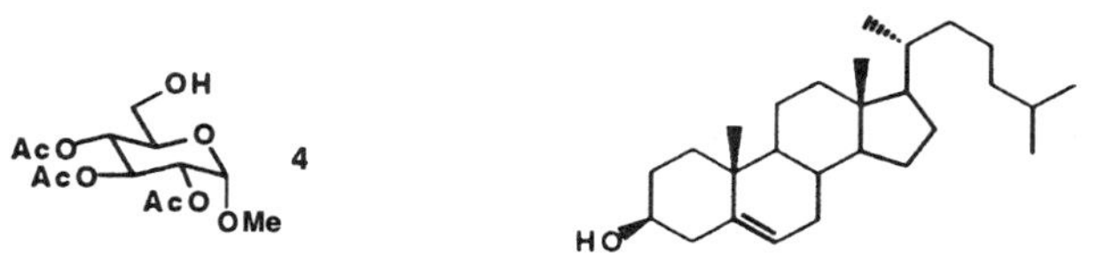

attributed to the instability of the donor **3** during the purification by silica gel chromatography. However, treatment of the glycosyl pyridylcarbonate **3** with cholesterol in the presence of Cu(OTf)$_2$ at room temperature for 10 min, gave the glycoside **7** in 60% yield with an α/β ratio of 3.6:1. The corresponding phenylcarbonate failed to give the glycoside under the same conditions.

By omitting the chromatographic purification step, 2-pyridylcarbonates can be used as glycosyl donors in one-pot glycosylations, resulting in high overall yields and reasonably good selectivities (see Scheme 2). In some cases, (e.g., **4** as an acceptor), it was necessary to add triflic acid (0.5 eq) to the reaction mixture to avoid the formation of an α-glycosyl carbonate as a by-product. The use of $CH_3CN$ as solvent, resulted in the formation of 1,2-*trans*-glycosides as major products through a kinetically formed α-glycosidic nitrilium intermediate [9].

## II. THE METHODS: GLYCOSYL 2-THIOPYRIDYLCARBONATES (TOPCAT) AS GLYCOSYL DONORS

### A. Preparation of Glycosyl Donors

2-Thiopyridyl chloroformate has been reported as an efficient reagent to activate carboxylic acids to give the corresponding 2-thiopyridyl esters [10]. Treatment of 2,3,4,6-tetra-*O*-benzyl-D-glucopyranose with freshly prepared 2-thiopyridyl chloroformate in the presence of $Et_3N$ afforded the corresponding glucosyl 2-thiopyridylcarbonate in very high yield as an anomeric mixture (α/β, 1:2).

Encouraged by the reactivity of bis(2-pyridyl) carbonate **2**, we also explored the use of bis(2-thiopyridyl) carbonate **8** in the formation of anomeric 2-thiopyridyl carbonates. The reagent **8**, prepared in quantitative yield from triphosgene and 2-pyridinethiol, was used to prepare several glycosyl 2-thiopyridylcarbonates (TOPCAT) in high yields (Scheme **3**).

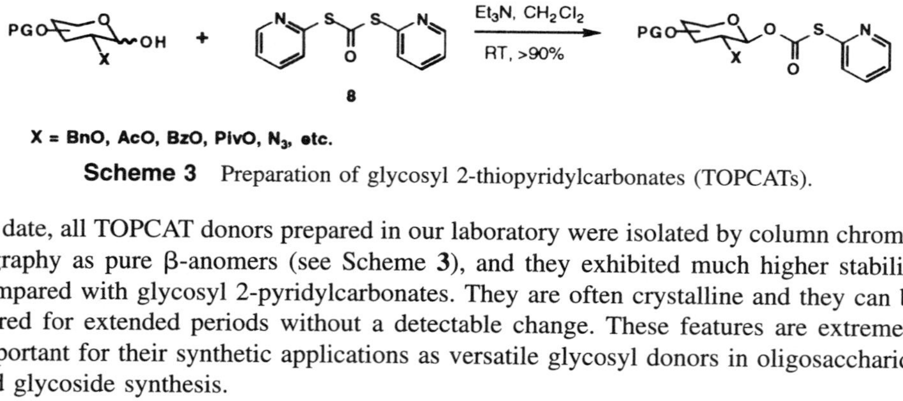

X = BnO, AcO, BzO, PivO, $N_3$, etc.

**Scheme 3** Preparation of glycosyl 2-thiopyridylcarbonates (TOPCATs).

To date, all TOPCAT donors prepared in our laboratory were isolated by column chromatography as pure β-anomers (see Scheme 3), and they exhibited much higher stability compared with glycosyl 2-pyridylcarbonates. They are often crystalline and they can be stored for extended periods without a detectable change. These features are extremely important for their synthetic applications as versatile glycosyl donors in oligosaccharide and glycoside synthesis.

### B. Activation of TOPCAT Donors and Synthesis of 1,2-*cis*-Disaccharides

As originally designed, the activation of the TOPCAT anomeric-leaving group with glycosyl acceptors was promoted by AgOTf under mild conditions. Scheme **4** summarizes the results of stereocontrolled α-glycosylations using TOPCAT donors, **9–11**, with sugar acceptors containing OH groups at C-2, C-3, C-4, and C-6, respectively. The results were similar to our previous finding in the MOP-based glycoside synthesis [see Chap. 18]. Glycosylations with the glycosyl donors **10** and **11**, led to 1,2-*cis*-disaccharides exclusively or in high preponderance. As with the *O*-benzyl-protected MOP donors [see Chap. 18], glucosylations were still α-selective, but with diminished ratios compared with the D-galacto isomer.

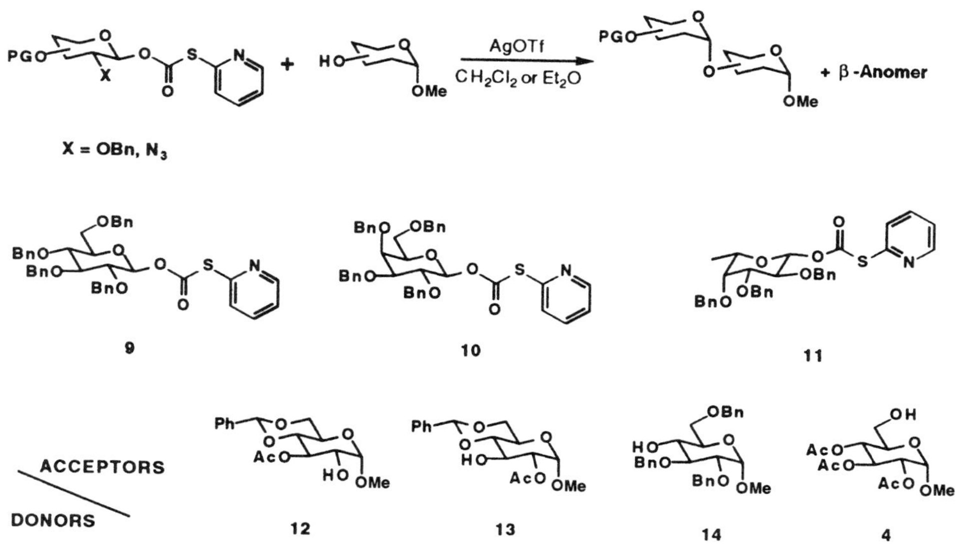


| DONORS | | 12 | 13 | 14 | 4 |
|---|---|---|---|---|---|
| 9 | α / β | 3 : 1, 93% | 9 : 1, 92% | 1.5 : 1, 70% | 3 : 1, 69% |
| 10 | α / β | 14 : 1, 87% | α-only, 90% | | |
| 11 | α / β | α-only, 96% | α-only, 95% | | |

**Scheme 4** Synthesis of 1,2-*cis*-disaccharides using TOPCAT glycosyl donors.

The acid-labile protective groups, such as benzylidene, acetate, esters, and silyl ethers, present in some sugar acceptors, are stable under the reaction conditions. Acetyl migration, which may occur in the strong Lewis acid-mediated glycosylation reactions [11], was not observed in TOPCAT-mediated glycosylations. Furthermore, β-glycosyl TOPCAT derivatives were more reactive toward AgOTf than were the corresponding α-isomers.

The results of this study clearly indicate that the glycosylation proceeds by an oxocarbenium intermediate that is generated by the activation of the 2-thiopyridylcarbonate with silver ions, possibly through a bidentate species (see Scheme **1**). We propose that the complex of silver thiopyridylcarbonate is subsequently decomposed in the reaction mixture into silver thiopyridine salt and $CO_2$. This was supported by the observation of the occasional presence of a trace amount of 2-thiopyridyl glycoside. The high α-selectivities obtained in both galactosylation and fucosylation of hindered alcohol acceptors could be due to a steric effect exerted by the pseudoaxial benzyloxy group at C-4 that hampers the nucleophile approaching from the β-side. However, electronic and conformational effects may be equally if not more important to explain the difference in selectivities that are consistently observed when comparing D-gluco and D-galacto donors.

## C. 1,2-*trans*-Disaccharides

1,2-*trans*-Disaccharides were successfully prepared by employing participating groups at C-2 in the TOPCAT donors. The results shown in Scheme **5** demonstrate that various acyl groups (acetyl, benzoyl, pivaloyl) are suitable to facilitate the formation of 1,2-*trans*-glycosides. The condensation of 2,3,4,6-tetra-*O*-acetyl-β-D-galactopyranosyl TOPCAT **17**

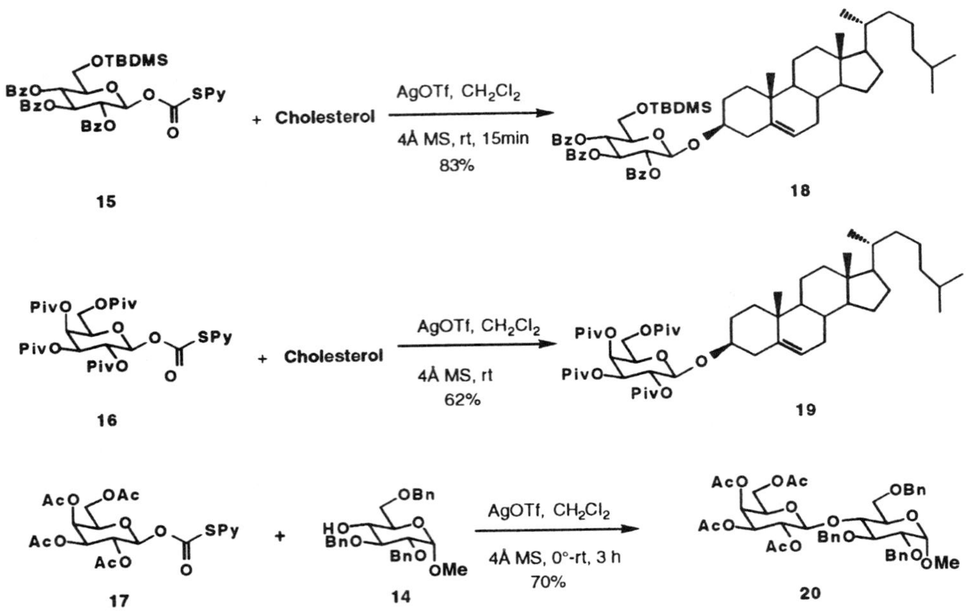

**Scheme 5** 1,2,-*trans*-glycoside synthesis using the TOPCAT-leaving group.

with methyl 2,3,6-tri-*O*-benzyl-α-D-glucopyranoside **14** in $CH_2Cl_2$ in the presence of AgOTf gave the desired β-linked disaccharide **20** as the only isomer in 70% yield. This suggests the possible use of this method for the synthesis of highly functionalized lactose and lactosamine analogues, which are useful building blocks for the construction of complex oligosacchardies.

## D. Application to the Synthesis of Sialyl Le$^x$

In the previous chapter, we briefly discussed the relevance of sialyl Le$^x$ and we showed a synthesis of a protected derivative that was assembled from a MOP glycosyl donor [see Chap. 18]. This complements several other recently reported syntheses. Herein we describe a straightforward synthesis of sialyl Le$^x$ employing the TOPCAT activation method as shown in Scheme **6**.

The key intermediates in our synthetic route are sialyl-Gal TOPCAT donor **22**, Fuc-TOPCAT donor **11** and the GlcNAc acceptor **23**. The disaccharide donor **22** was prepared in high yield by the treatment of the reducing sugar **21** [12] with bis(2-thiopyridyl)carbonate. The subsequent glycosylation of acceptor **23** with **22** was carried out in the presence of AgOTf in $CH_2Cl_2$ at 0°C for 5 h to give the trisaccharide **24** in 73% yield with the expected β-configuration exclusively for the newly formed glycosidic linkage. Rhodium-catalyzed migration of the allylic double-bond, followed by the treatment with $HgCl_2$–HgO led to the trisaccharide precursor **25** in 80% overall yield. The fucosyl unit was introduced in a highly stereoselective manner by employing the β-fucosyl TOPCAT **11** as glycosyl donor, to give the desired fully protected SLe$^x$ tetrasaccharide **26** in over 41% isolated yield. It is of particular interest that tetramethylurea (TMU), which we first used as an acid scavenger in a AgOTf-promoted glycosylation reaction [13], was also effective to increase the yield of fucosylation in this and other cases, especially when sterically hindered acceptors were used.

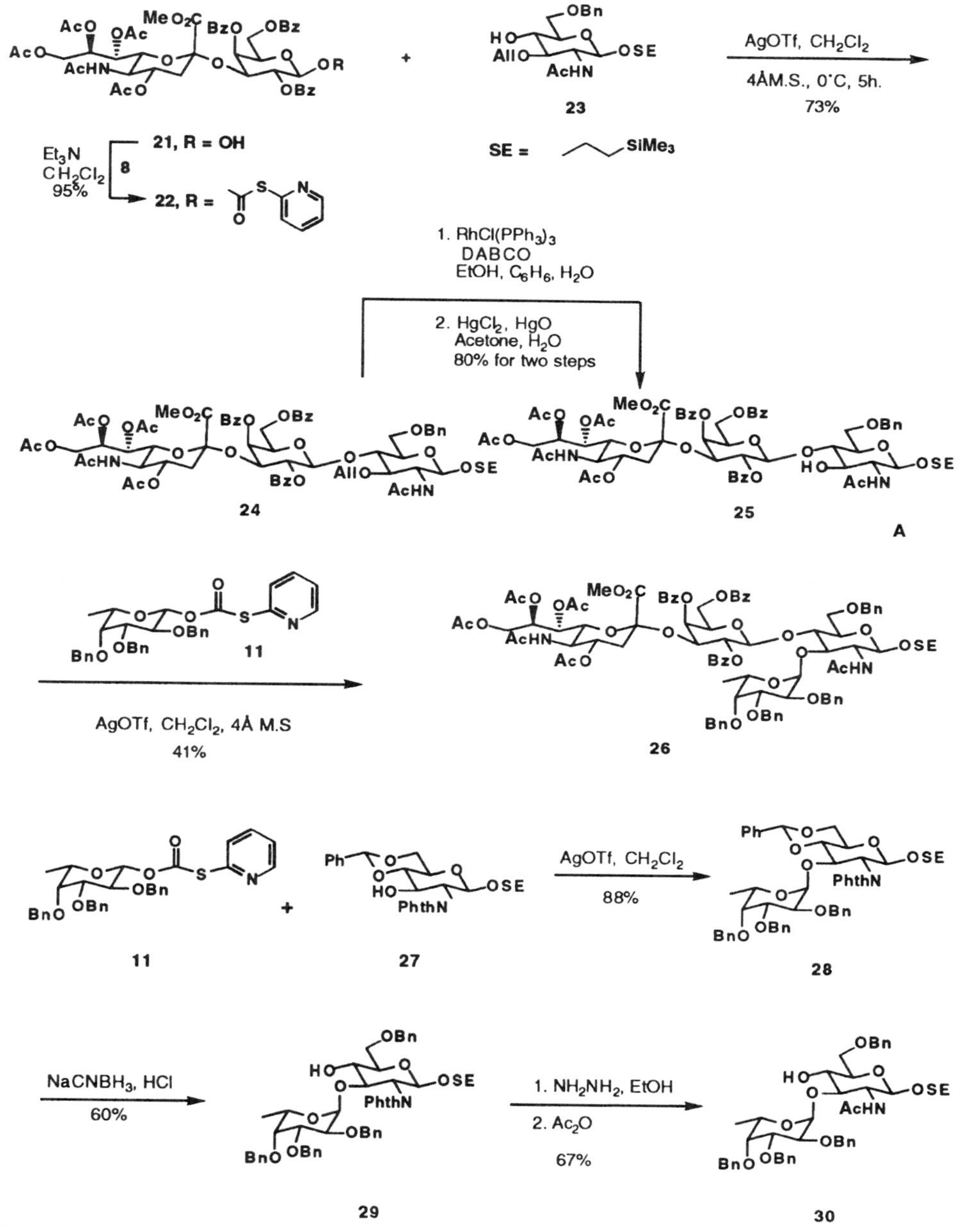

**Scheme 6** Synthesis of a sialyl Le^x derivative and intermediate disaccharide by TOPCAT donors.

The synthesis of a Le$^x$ trisaccharide **34** was also accomplished by using the TOPCAT-based method (Scheme 7). Coupling of the readily available donor **31** with **23** under the

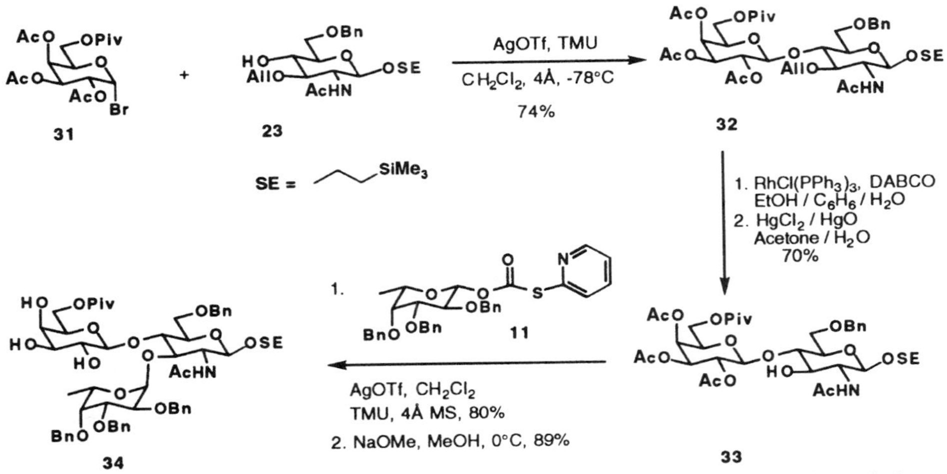

**Scheme 7** Synthesis of Le$^x$ trisaccharide derivative via TOPCAT-mediated glycosylation.

previously developed conditions [13] furnished the disaccharide **32** in 74% yield with no observation of any orthoester. Removal of the allyl group as for **25** generated the disaccharide **33** in 70% yield. Condensation of **33** and fucosyl TOPCAT donor **11** proceeded in the presence of AgOTf and TMU at room temperature overnight, to give the corresponding α-linked trisaccharide in 80% yield. Hydrolysis of the acetate groups led to **34** in high overall yield. This trisaccharide, containing a pivaloyl group at the C-6 of the Gal unit, may be useful in other glycosylations at the C-3 with appropriate glycosyl donors to give analogues of SLe$^x$.

## E. Conclusion

A new method of anomeric activation was discovered based on the remote activation concept [2]. Anomeric 2-thiopyridylcarbonate (TOPCAT) derivatives are easily prepared from the O-protected reducing sugars to give stable, well-defined crystalline glycosyl 2-thiopyridylcarbonates. Activation with AgOTf in $CH_2Cl_2$ generates reactive oxocarbenium ion intermediates that can react with alcohol acceptors to give 1,2-*cis* or 1,2-*trans*-glycosides depending on the nature of the C-2 substituent in the donor. The method is complementary to the pentenyl glycoside [14; see also Chap. 14] and related protocols, such as the trichloroacetimidate method [15]. In the latter type activation, the TOPCAT donors have the advantage of being stable to chromatography and during storage without a detectable change. The TOPCAT activation method is useful for the synthesis of simple and more complex oligosaccharide-type products. It can also be used in conjunction with MOP-acceptors that are relatively stable to AgOTf; hence, the possibility for selective activation of TOPCAT donors, and the option for iterative oligosaccharide synthesis, as discussed in Chapter 20. TOPCAT donors compare favorably with the same trichloroacetimidate counterparts in efficiency and selectivity of glycosylation. Table 1 compares the results of disaccharide syntheses with TOPCAT and imidate donors.

**Table 1** Glycosylation with TOPCAT and Imidate Donors

| ACCEPTOR \ DONOR | TOPCAT (BnO/OBn glucose-O-C(=O)-S-Py) | TOPCAT (BnO/OBn gal-O-C(=O)-S-Py) | IMIDATE (BnO/OBn glucose-O-C(=NH)-CCl₃) | IMIDATE (BnO/OBn gal-O-C(=NH)-CCl₃) |
|---|---|---|---|---|
| 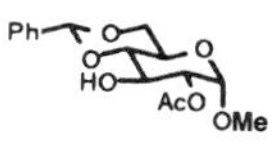 Ph-O-sugar-HO-AcO-OMe — α:β | α | α | α | α |
| Yield (%) | 90 | 95 | 87 | 90 |
| 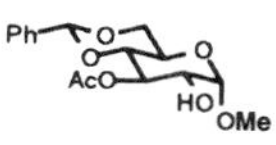 Ph-O-sugar-AcO-HO-OMe — α:β | 14:1 | α | 10:1 | α |
| Yield (%) | 87 | 96 | 85 | 85 |

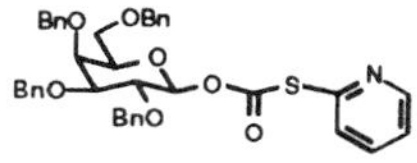

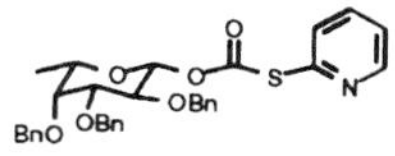

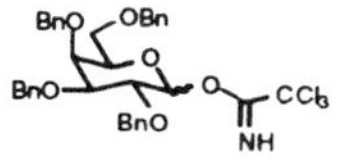

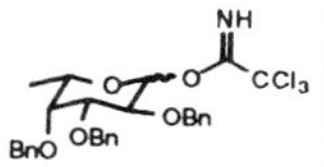

## III. EXPERIMENTAL PROCEDURES*

### A. General Procedure for One-Pot Glycosylation with Glycosyl-2-pyridylcarbonates

*Cholesterol as an Acceptor*

$$\text{1} \xrightarrow[\text{rt, 45 min}]{\text{(PyO)}_2\text{CO (1.2 equiv.)}, \text{ DMAP (0.3 equiv.)}} \text{3} \xrightarrow{\text{ROH}} \text{6 or 7}$$

To a solution of 2,3,4,6-tetra-*O*-benzyl-D-glucopyranose **1** (30 mg, 0.056 mmol) and 15 mg (0.069 mmol) of di(2-pyridyl) carbonate in 1 mL of ether was added 2 mg of DMAP. The mixture was stirred at room temperature until it was homogeneous (45 min to 1h), and then cooled to −20°C. Cu(OTf)$_2$ (60 mg, 0.166 mmol) and cholesterol (26 mg, 0.067 mmol) were added in order. The mixture was allowed to reach room temperature, and the stirring was continued for 10 min. Addition of pyridine (2 drops), concentration, and purification by flash chromatography on silica gel (EtOAc–hexane, 1:2) gave the desired glycosides (30 mg, 60%; α/β, 3.6:1).

*Methyl 2,3,4-tri-O-acetyl-α-D-glucopyranoside as an Acceptor*

To a solution of 2,3,4-tetra-*O*-benzyl-D-glucopyranse **1** (30 mg, 0.056 mmol) and 15 mg (0.069 mmol) of di(2-pyridyl) carbonate in 1 mL of ether was added 2 mg of DMAP. The

---

*Optical rotations were measured at 22–25°C.

mixture was stirred at room temperature until it was homogeneous, then cooled to −20°C; 28 μL of 1 M triflic acid solution in CH$_3$NO$_2$, 50 mg (0.138 mmol) of Cu(OTf)$_2$, and 27 mg (0.084 mmol) of methyl 2,3,4-tri-O-acetyl-α-D-glucopyranoside were added, in order. After the cold bath was removed, the mixture was stirred at room temperature for 10 min. The reaction mixture was quenched with 2 drops of pyridine. Concentration and purification by flash chromatography on silica gel (EtOAc–hexane, 1:2) gave the mixture of α- and β-disaccharides (29 mg, 63% yield; α/β, 2.5:1).

*Preparation of Di(S-2-Pyridyl) thiocarbonate (**8**)*

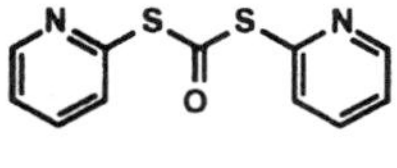

To a solution cooled at 0°C of 8.83 g (80.0 mmol) of 2-mercaptopyridine in 400 mL of dry CH$_2$Cl$_2$, was added 3.94 g (13.3 mmol) of triphosgene. Triethylamine (12 mL, 86 mmol) was added dropwise over 15 min, and the mixture was stirred at this temperature for 30 min, and then at room temperature for 1 h. The mixture was concentrated, treated with cold saturated aqueous NaHCO$_3$, and extracted with 400 mL of EtOAc. After being washed with water and brine, the organice layer was dried over MgSO$_4$, filtered, and concentrated to give the product as a yellow solid that was dried in vacuo overnight (9.58 g, 97%). Pale yellow needle-shaped crystals were obtained by recrystallization from 2-propanol: mp 45°–47°C.

### B. General Procedure for the Preparation of Glycosyl 2-thiopyridylcarbonate Donors

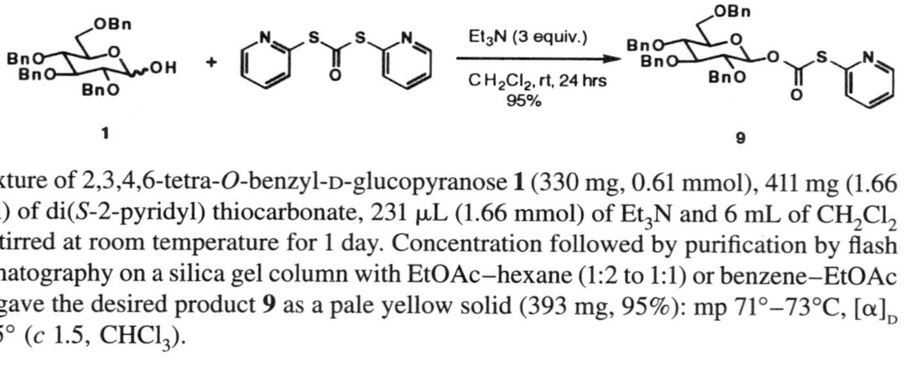

A mixture of 2,3,4,6-tetra-O-benzyl-D-glucopyranose **1** (330 mg, 0.61 mmol), 411 mg (1.66 mmol) of di(S-2-pyridyl) thiocarbonate, 231 μL (1.66 mmol) of Et$_3$N and 6 mL of CH$_2$Cl$_2$ was stirred at room temperature for 1 day. Concentration followed by purification by flash chromatography on a silica gel column with EtOAc–hexane (1:2 to 1:1) or benzene–EtOAc (5:1) gave the desired product **9** as a pale yellow solid (393 mg, 95%): mp 71°–73°C, [α]$_D$ +15.5° (c 1.5, CHCl$_3$).

### C. General Procedures for the Synthesis of 1,2-*cis*-Disaccharides Using TOPCAT Glycosyl Donors

*Glycosylation of Methyl 3-O-acetyl-4,6-benzylidene-α-D-glucopyranoside (**12**) with 2,3,4,6-Tetra-O-benzyl-β-D-glucopyranosyl 2-thiopyridylcarbonate (**9**)*

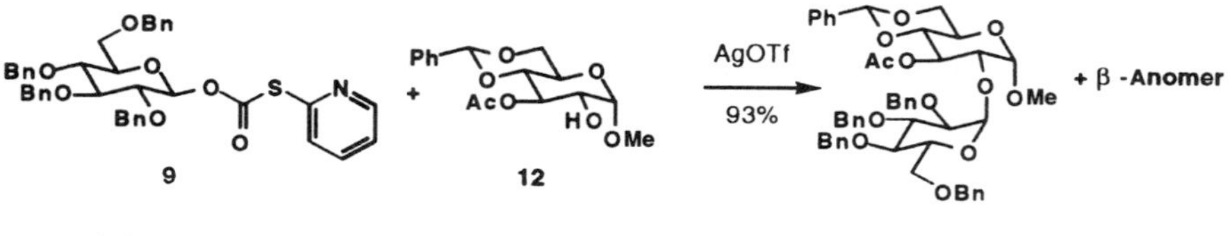

A mixture of the glycosyl donor **9** (126 mg, 0.185 mmol) glycosyl acceptor **12**, (40.3 mg, 0.124 mmol), and activated powdered 4 Å (molecular sieves) (200 mg) in 6 mL of Et$_2$O–

CH$_2$Cl$_2$ (5:1, v/v) was stirred overnight under argon at room temperature and then cooled to 0°C. Silver triflate (143 mg, 0.555 mmol) was added to the reaction mixture, and the stirring was continued for 5 h at 0°C. The suspension was treated with a few drops of pyridine, filtered through Celite and concentrated. Purification by flash chromatography on silica gel column with hexane–EtOAc–CH$_2$Cl$_2$ (4:1:1) gave 98 mg of the desired α- and β-disaccharides (α/β, 3:1, 93%).

*Glycosylation of Methyl 3-O-acetyl-4,6-benzylidene-α-D-glucopyranoside (12) with 2,3,4,6-Tetra-O-benzyl-β-D-galactopyranosyl 2-thiopyridylcarbonate (10)*

A mixture of the glycosyl donor **10**, (126 mg, 0.185 mmol), glycosyl acceptor **12**, (40.3 mg, 0.13 mmol), and activated powdered 4-Å MS (200 mg) in 6 mL of Et$_2$O–CH$_2$Cl$_2$ (5:1 v/v) was stirred overnight under argon at room temperature, and then cooled to 0°C. Silver triflate (142.8 mg, 0.555 mmol) was added to the reaction mixture, the stirring was continued for 5 h at 0°C. The suspension was treated with a few drops of pyridine, filtered through Celite and concentrated. Purification by flash chromatography on silica gel column with hexane–EtOAc–CH$_2$Cl$_2$ (4:1:1) gave 92 mg of the desired disaccharide in 87% yield (α/β, 14:1). For the α-anomer: mp 55°–57°C, [α]$_D$ +59.0° (c 1.12, CHCl$_3$)

*Glycosylation of Methyl 3-O-acetyl-4,6-benzylidene-α-D-glucopyranoside (12) with 2,3,4-Tri-O-benzyl-β-L-fucopyranosyl 2-thiopyridylcarbonate (11)*

A mixture of the glycosyl donor **11** (141 mg, 0.247 mmol) glycosyl acceptor **12** (40 mg, 0.13 mmol), and activated powdered 4-Å MS (200 mg) in 6 mL Et$_2$O–CH$_2$Cl$_2$ (5:1, v/v) was stirred overnight under argon at room temperature, and then cooled to 0°C. Silver triflate (190.4 mg, 0.740 mmol) was added to the reaction mixture, the stirring was continued for 4 h at 0°C. The suspension was treated with a few drops of pyridine, filtered through Celite, and concentrated. Purification by flash chromatography on silica gel column with hexane–EtOAc–CH$_2$Cl$_2$ (4:1:1) gave 90 mg of the desired disaccharide (96%, α-anomer only): mp 152°–154°C, [α]$_D$ −5.6° (c 1.5, CHCl$_3$).

## D. General Procedure for the Synthesis of 1,2-*trans*-Glycosides Using the TOPCAT-Leaving Group

*Glycosylation of Cholesterol with 6-O-t-Butyldimethylsilyl-2,3,4-tri-O-benzoyl-β-D-glucopyranosyl 2-thiopyridylcarbonate*

To a mixture of glycosyl donor **15** (30 mg, 0.040 mmol), 17 mg (0.044 mmol) of cholesterol, 1 mL of dry CH$_2$Cl$_2$, and activated powdered 4-Å MS, was added 21 mg (0.082 mmol) of silver triflate. The resulting suspension was stirred at room temperature until the reaction was completed. One drop of pyridine was added, and the mixture was filtered through Celite and washed with $CH_2Cl_2$. Concentration of the filtrate and purification by flash chromatography on silica gel column with EtOAc–hexane (1:4) gave the desired glycoside **18** (32.5 mg, 83%): mp 188°–190°C, $[\alpha]_D$ +13.2° (c 1.1, $CHCl_3$).

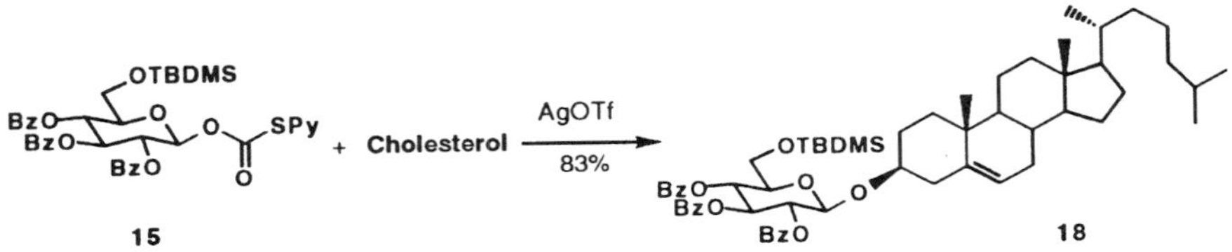

*O-(Methyl 5-acetamido-4,7,8,9-tetra-O-acetyl-3,5-dideoxy-D-glycero-α-D-galacto-2-nonulopyranosylonate)-(2→3)-2,4,6-tri-O-benzoyl-β-D-galactopyranose* (**21**)

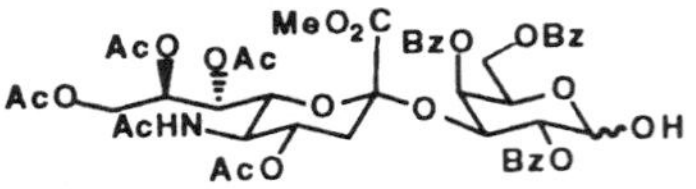

To a solution of 2-(trimethylsilyl)ethyl O-(methyl 5-acetamido-4,7,8,9-tetra-O-acetyl-3,5-dideoxy-D-*glycero*-α-D-*galacto*-2-nonulopyranosylonate)-(2→3)-2,4,6-tri-O-benzoyl-β-D-galactopyranose (295 mg, 0.28 mmol) in 2 mL of dichloromethane was added trifluoroacetic acid (2 mL) at 0°C, and the stirring was continued for 2 h at 0°C. Ethyl acetate (3 mL) and toluene (3 ml) were added and the solvents were evaporated. A second portion of toluene was added and the evaporation was repeated. Purification by flash chromatography on silica gel column with dichloromethane–methanol (20:1) gave the hemiacetal **21** (267 mg, quantitative): mp 85°C.

*O-(Methyl 5-acetamido-4,7,8,9-tetra-O-acetyl-3,5-dideoxy-D-glycero-α-D-galacto-2-nonulopyranosylonate)-(2→3)-2,4,6-tri-O-benzoyl-β-D-galactopyranosyl 2-thiopyridylcarbonate* (**22**)

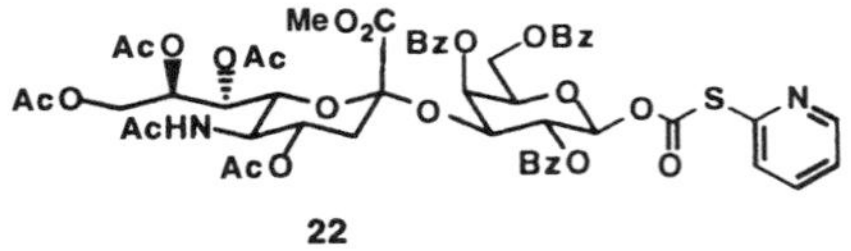

A mixture of hemiacetal disaccharide **21** (248 mg, 0.258 mmol), di(S-2-pyridyl) thiocarbonate (191 mg; 0.77 mmol), triethylamine (110 μL, 0.77 mmol), and 5 mL of dichloromethane was stirred at room temperature for 30 h. Concentration and purification by flash chromatography on silica gel column with 20:1 dichloromethane–methanol gave the desired product **22** (270 mg, 95%): mp 150°C, $[\alpha]_D$ +53.75° (c 0.8, $CHCl_3$).

*2-(Trimethylsilyl)ethyl O-(methyl 5-acetamido-4,7,8,9-tetra-O-acetyl-3,5-dideoxy-D-glycero-α-D-galacto-2-nonulopyranosylonate)-(2→3)-(2,4,6-tri-O-benzoyl-β-D-galactopyranosyl)-(1→4)-2-acetamido-6-O-benzyl-2-deoxy-β-D-glucopyranoside* (**25**)

To a solution of trisaccharide **24** (82.5 mg, 0.060 mmol) in 5 mL of ethanol–benzene–water (5:2:1) was added successively tri(triphenylphosphine)–rhodium(I) chloride (50 mg, 0.054 mmol) and 1,4-diazabicyclo[2,2,2]octane (7 mg, 0.063 mmol). The mixture was stirred for 24 h at 85°C, concentrated, and purified by flash chromatography on silica gel column with 20:1 chloroform–methanol to give the propenyl ether. This was dissolved in acetone–water 9:1 (5 mL), then mercury oxide (20 mg, 0.092 mmol) and a solution of mercury chloride (33 mg, 0.12 mmol) in 2 mL of acetone were added successively to the reaction mixture. Stirring was continued overnight at room temperature, then 10 mL of dichloromethane was added, the reaction mixture was filtered through Celite, and the residue was washed successively with acetone and dichloromethane. The filtrates and washings were combined and concentrated. The residue was dissolved in ether, washed with 10% potassium iodide solution, dried over $Na_2SO_4$, and concentrated. Purification by flash chromatography on silica gel with $CH_2Cl_2$–MeOH (20:1) gave the desired product **25** (64 mg, 80%): mp 92°–94°C, $[\alpha]_D$ +39.6° (c 0.78, $CHCl_3$).

*2-(Trimethylsilyl)ethyl O-(methyl 5-acetamido-4,7,8,9-tetra-O-acetyl-3,5-dideoxy-D-glycero-α-D-galacto-2-nonulopyranosylonate)-(2→3)-(2,4,6-tri-O-benzoyl-β-D-galactopyranosyl)-(1→4)-2-acetamido-3-O-allyl-6-O-benzyl-2-deoxy-β-D-glucopyranoside (24)*

To a solution of glycosyl donor **22** (150 mg, 0.136 mmol) and glycosyl acceptor **23** (170 mg, 0.377 mmol) was added activated powdered 4-Å MS (170 mg). The mixture was stirred overnight under argon at room temperature, cooled to 0°C, and silver triflate (105.2 mg, 0.410 mmol) was added. The reaction mixture was stirred for 5 h at 0°C then at room temperature, the course of the reaction being monitored by TLC. The suspension was treated with a few drops of pyridine, filtered through Celite and concentrated. Purification by flash chromatography on silica gel column with EtOAc–$CHCl_3$–MeOH (10:2:0.2) gave the title trisaccharide **24** (139 mg, 73%): mp 120°–122°C, $[\alpha]_D$ +9.75° (c 0.79, $CHCl_3$).

*2-(Trimethylsilyl)ethyl O-(methyl 5-acetamido-4,7,8,9-tetra-O-acetyl-3,5-dideoxy-D-glycero-α-D-galacto-2-nonulopyranosylonate)-(2→3)-(2,4,6-tri-O-benzoyl-β-D-galactopyranosyl)-(1→4)-O-[2,3,4-tri-O-benzyl-α-L-fucopyranosyl-(1→3)]-2-acetamido-6-O-benzyl-2-deoxy-β-D-glucopyranoside (26)*

To a solution of glycosyl acceptor **25** (20.6 mg, 0.015 mmol) in dichloromethane (2 mL) was added activated powdered 4-Å MS (50 mg). The solution was stirred at room temperature under argon for 1 h, then cooled to 0°C. Silver triflate (160 mg, 0.62 mmol) was added and the stirring was continued for 30 min. A solution of the glycosyl donor **11** (300 mg, 0.53 mmol) in 3 mL of dichloromethane was added dropwise to the reaction mixture at 0°C. After 2 h, a few drops of pyridine were added, and the mixture was filtered through Celite, then concentrated. Purification of the residue by flash chromatography on silica gel with benzene–acetone (3:1) gave the tetrasaccharide **26** (11 mg; 41%).

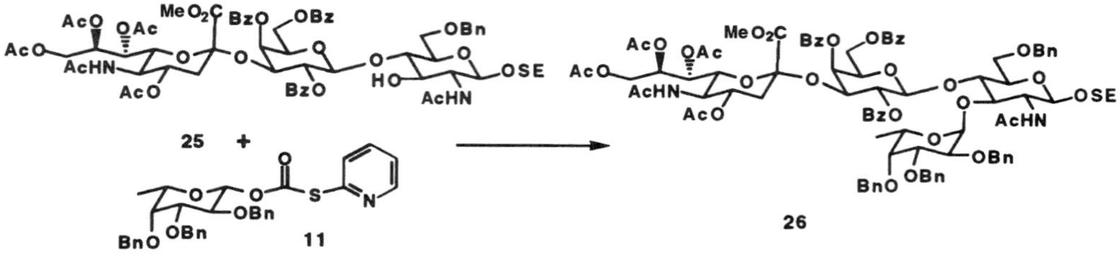

*2-(Trimethylsilyl)ethyl O-(2,3,4-tri-O-benzyl-α-L-fucopyranosyl)-(1→3)-2-phthalimido-4,6-O-benzylidene-2-deoxy-β-D-glucopyranoside (**28**)*

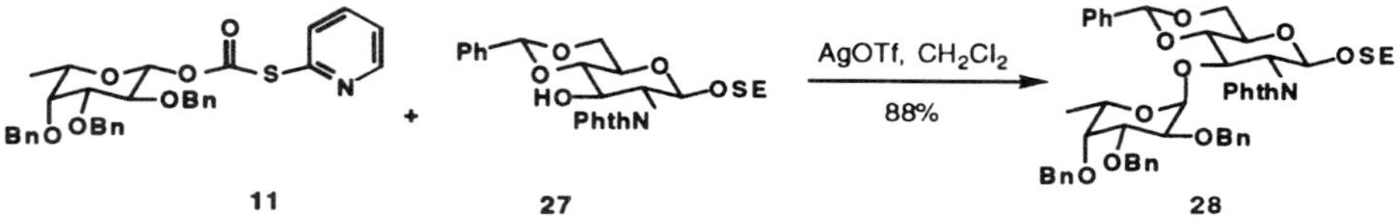

To a solution of glycosyl acceptor **27** (500 mg, 1 mmol) and the glycosyl donor **11** (947 mg, 1.66 mmol) in dichloromethane (10 mL) was added activated powdered 4-Å MS (500 mg). The mixture was stirred overnight under argon at room temperature, cooled to 0°C, then treated with silver triflate (1.55 g, 6.0 mmol). After stirring for 1 h, the mixture was treated with a few drops of pyridine, filtered through Celite, and concentrated. Purification by flash chromatography on silica gel column with hexane–EtOAc–CH$_2$Cl$_2$ (8:2:2) gave the desired product 28 (716 mg; 88%): mp 67°C, [α]$_D$ −30.72° (c 1.1, CHCl$_3$).

*2-(Trimethylsilyl)ethyl O-(2,3,4-tri-O-benzyl-α-L-fucopyranosyl)-(1→3)-O-benzyl-2-deoxy-2-phthalimido-β-D-glucopyranoside (**29**)*

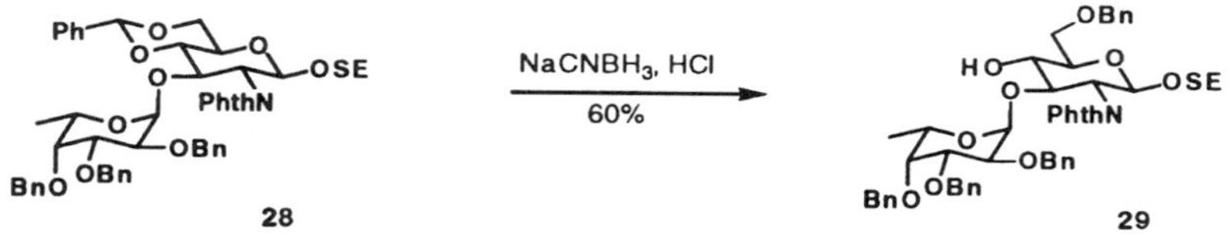

To a solution of the disaccharide **28** (500 mg, 0.55 mmol) and sodium cyanoborohydride (230 mg, 3.70 mmol) in tetrahydrofuran (25 mL) was added powdered 4-Å MS (1 g). The solution was stirred for 20 min at 0°C, then a solution of hydrogen chloride saturated in ether (2 mL) was added dropwise. Stirring was continued for 3 h at 0°C, and the course of the reaction was monitored by TLC. The suspension was filtered through Celite, and the filtrate was processed as usual. Purification by flash chromatography on silica gel with hexane–EtOAc (3:1) afforded the desired product **29** (300 mg, 60%).

*2-(Trimethylsilyl)ethyl O-(2,3,4-tri-O-benzyl-α-L-fucopyranosyl)-(1→3)-O-benzyl-2-deoxy-2-acetamido-β-D-glucopyranoside (**30**)*

A solution of the disaccharide **29** (280 mg, 0.3 mmol) in 3 mL of hydrazine monohydrate and 10 mL of ethanol was heated at 95°C for 2 h. The solution was concentrated, the residue was dissolved in methanol (10 mL), and acetic anhydride (3 mL) was added at 0°C. The solution was stirred for 2 h at room temperature, concentrated, toluene (5 mL) was added, and then evaporated. The solid residue was dissolved in dichloromethane, and the solution was processed as usual. Purification by flash chromatography on silica gel with EtOAc–hexane (2:1) gave the desired product **30** (170 mg, 67%).

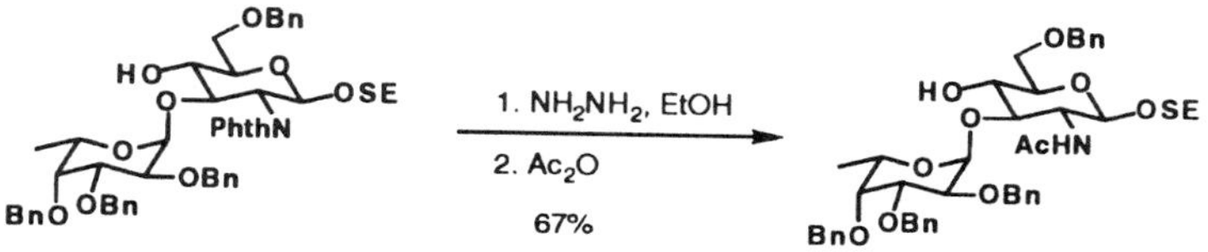

*2-(Trimethylsilyl)ethyl O-(2,3,4-tri-O-acetyl-6-O-pivaloyl-β-D-galactopyranosyl)-(1→4)-2-acetamido-3-O-allyl-6-O-benzyl-2-deoxy-β-D-glucopyranoside (32)*

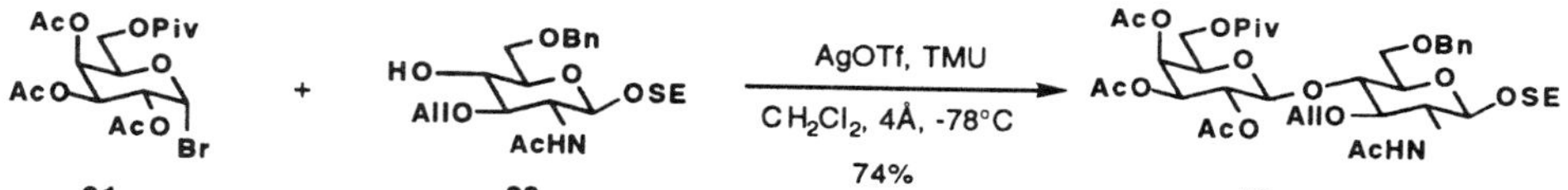

To a solution of the glycosyl acceptor **23** (400 mg, 0.89 mmol) in dry dichloromethane (10 mL) was added activated powdered 4-Å MS (400 mg), and the mixture was stirred for 1 h at room temperature, then cooled to −78°C. Silver triflate (1 g, 3.89 mmol) and tetramethylurea (316 µL, 2.56 mmol) were added successively to the reaction mixture at −78°C, and the stirring was continued for 30 min. A solution of 2,3,4-tri-O-acetyl-6-O-pivaloyl-α-D-galactopyranosyl bromide **31** (1.2 g, 2.65 mmol) in 10 mL of dichloromethane was added to the reaction mixture. After stirring for 3 h at −78°C, the precipitate was filtered off, and washed with dichloromethane. The filtrate and washings were combined, and the solution was processed as usual. Purification by flash chromatography on silica gel with EtOAc-hexane (1:1) gave the title disaccharide **32** (540 mg, 74%): mp 69°C, [α]$_D$ −15.7° (*c* 1.05, CHCl$_3$).

*2-(Trimethylsilyl)ethyl O-(2,3,4-tri-O-acetyl-6-O-pivaloyl-α-D-galactopyranosyl)-(1→4)-2-acetamido-6-O-benzyl-2-deoxy-β-D-glucopyranoside (33)*

To a solution of the preceding compound **32** (394 mg, 0.478 mmol) in 8 mL of ethanol–benzene–water (5:2:1) was added successively tri(triphenylphosphine)–rhodium(I) chloride (88.5 mg, 0.096 mmol) and 1,4-diazabicyclo[2,2,2]octane (26 mg, 0.23 mmol). The mixture was stirred overnight (12 h) at 85°C. Concentration and purification by flash chromatography on a silica gel column with 1:1 hexane–EtOAc gave the propenyl ether.

This was dissolved in 10 mL of acetone–water (10:1), mercury oxide (200 mg, 0.92 mmol), and a solution of mercury chloride (400 mg, 1.47 mmol) in 4 mL of acetone were added successively to the reaction mixture. The stirring was continued for 5 h at room temperature, 10 mL of dichloromethane was added, the mixture was filtered through Celite, and the residue was washed with acetone and dichloromethane. The filtrates and washings were combined and concentrated to give a residue that was dissolved again in ether, then washed with 10% potassium iodide solution, dried over $Na_2SO_4$, and concentrated. Purification by flash chromatography on silica gel with hexane–EtOAc (1:1) gave the desired product **33** (260 mg, 70%): mp 163°C, $[\alpha]_D$ −2.7° (c 1.1, $CHCl_3$).

*2-(Trimethylsilyl)ethyl O-(2,3,4-tri-O-acetyl-6-O-pivaloyl-β-D-galactopyranosyl)-(1→4)-[2,3,4-tri-O-benzyl-α-L-fucopyranosyl)-(1→3)]-2-acetamido-6-O-benzyl-2-deoxy-β-D-glucopyranoside*

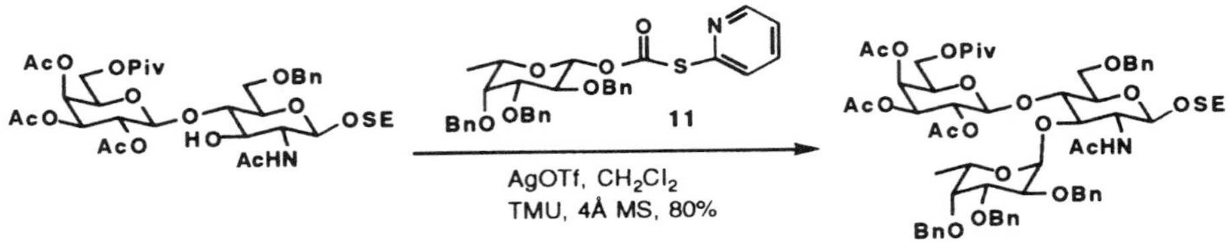

A mixture of the preceding compound **33** (143.2 mg, 0.182 mmol), Fuc-TOPCAT glycosyl donor **11** (313 mg, 0.548 mmol), activated powdered 4-Å MS (400 mg), and tetramethylurea (66 μL, 0.548 mmol) in 10 mL of dichloromethane was stirred overnight under argon at room temperature, and then cooled to 0°C. Silver triflate (423 mg, 1.6 mmol) was added to the reaction mixture, the stirring was continued for 24 h at room temperature. The suspension was treated with a few drops of pyridine, filtered through Celite, and concentrated. Purification by flash chromatography on silica gel column with hexane–EtOAc–$CH_2Cl_2$ (1:1:1) gave the title compound (175 mg, 80%): mp 83°C, $[\alpha]_D$ −32.3° (c 0.77, $CHCl_3$).

*2-(Trimethylsilyl)ethyl O-(6-O-pivaloyl-β-D-galactopyranosyl)-(1→4)-[(2,3,4-tri-O-benzyl-α-L-fucopyranosyl)-(1→3)]-2-acetamido-6-O-benzyl-2-deoxy-β-D-glucopyranoside (**34**)*

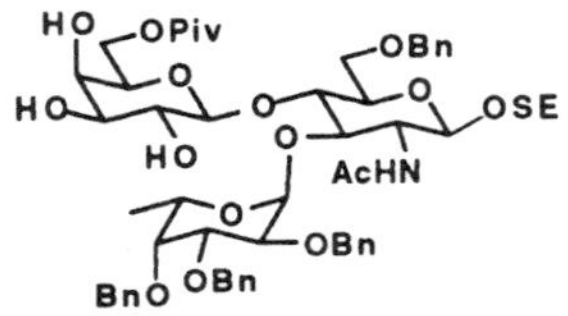

To a solution of the preceding compound (174 mg, 0.145 mmol) in 10 mL of dry methanol was added dropwise 50 μL of 10% sodium methoxide in MeOH at 0°C. The solution was stirred for 3 h at 0°C, then neutralized with Amberlite IR-120 ($H^+$). Filtration and concentration gave **34** (139.3 mg, 89%): mp 97°C, $[\alpha]_D$ −41.5° (c 0.82, $CHCl_3$).

## REFERENCES

1. (a) G. H. Veeneman, S. H. van Leewen, and J. H. van Boom, Iodonium ion promoted reactions at the anomeric centre. II. An efficient thioglycoside mediated approach toward the formation of 1,2-*trans*-linked glycosides and glycosidic esters, *Tetrahedron Lett. 31*:1331 (1990); (b) R. Roy, F. O. Andersson, and M. Letellier, "Active" and "latent" thioglycosyl donors in oligosaccharide synthesis. Application to the synthesis of α-sialosides, *Tetrahedron Lett. 33*:6053 (1992); (c) P. Fügedi and P. J. Garegg, A novel promoter for the efficient construction of 1,2-*trans*-linkages in glycosides synthesis, using thioglycosides as glycosyl donors, *Carbohydr. Res. 149*:C9 (1986); (d) H. Lönn, Synthesis of a tri- and a heptasaccharide which contain α-L-fucopyranosyl groups and are part of the complex type of carbohydrate moiety of glycoproteins, *Carbohydr. Res. 139*:105 (1985); (e) K. C. Nicolaou, S. P. Seitz, and D. P. Papahatjis, A mild and general method for the synthesis of O-glycosides, *J. Am. Chem. Soc. 105*:2430 (1983).
2. S. Hanessian, C. Bacquet, and N. Lehong, Chemistry of the glycosidic linkage. exceptionally fast and efficient formation of glycosides by remote activation, *Cabohydr. Res. 80*:C17 (1980).
3. R. B. Woodward, et al. Asymmetric total synthesis of erythromycin. 3. Total synthesis of erythromycin, *J. Am. Chem. Soc. 103*:3215 (1981).
4. (a) J. D. White, G. L. Bolton, A. P. Dantanarayana, C. M. J. Fox, R. N. Hiner, R. W. Jackson, K. Sakuma, and U. S. Warrier, Total synthesis of the antiparasitic agent avermectin $B_{1a}$, *J. Am. Chem. Soc. 117*:1908 (1995); (b) T. A. Blizzard, G. M. Margiatto, H. Mrozik, and M. H. Fisher, A novel fragmentation reaction of avermectin aglycons, *J. Org. Chem. 58*:3201 (1993); (c) S. Hanessian, A. Ugolini, P. J. Hodges, P. Beaulieu, D. Dubé, and C. André, Progress in natural product chemistry by the Chiron and related approaches—synthesis of avermectin $B_{1a}$, *Pure Appl. Chem. 59*:299 (1987); (d) S. Hanessian, A. Ugolini, D. Dubé, P. J. Hodges, and C. André, Synthesis of (+)avermectin $B_{1a}$, *J. Am. Chem. Soc. 108*:2776 (1986); (e) P. G. M. Wuts and S. S. Bigelow, Total synthesis of oleandrine and the avermectin disaccharide, benzyl α-L-oleandrosyl-α-L-4-acetoxyoleandrolide, *J. Org. Chem. 48*:3489 (1983).
5. K. Koide, M. Ohno, and S. Kobayashi, A new glycosylation reaction based on a "remote activation concept": Glycosyl 2-pyridinecarboxylate as a novel glycosyl donor, *Tetrahedron Lett. 32*:7065 (1991).
6. M. Boursier and G. Descotes, Activation du carbone anomère des sucres par le groupe carbonate et application en synthèse osidique, *C. R. Acad. Sci. Ser 2: 308*:919 (1989).
7. A. Marra, J. Esnault, A. Veyrieres, and P. Sinaÿ, Isopropenyl glycosides and congeners as novel classes of glycosyl donors: Theme and variations, *J. Am. Chem. Soc. 114*:6354 (1992).
8. A. K. Ghosh, T. T. Duong, and S. P. McKee, Di(2-pyridyl) carbonate promoted alkoxycarbonylation of amines: A convenient synthesis of functionalized carbamates, *Tetrahedron Lett. 32*:4251 (1991).
9. R. R. Schmidt, M. Behrendt, and A. Toepfer, Nitriles as solvents in glycosylation reactions: Highly selective β-glycoside synthesis, *Synlett* p. 694 (1990).
10. E. J. Corey and D. A. Clark, A new method for the synthesis of 2-pyridinethiol carboxylic esters, *Tetrahedron Lett.* p. 2875 (1979).
11. T. Ziegler, P. Kovác, and C. P. J. Glaudemans, Transesterification during glycosylation promoted by silver trifluoromethanesulfonate, *Liebigs Ann. Chem.* p. 613 (1990) and references cited therein.
12. Preparation of **21** from the corresponding 2-(trimethylsilyl)ethyl glycoside by a known method, see K. Jansson, S. Ahlfors, T. Frejd, J. Kihlberg, and G. Magnusson, 2-(Trimethylsilyl)ethyl glycosides synthesis, anomeric deblocking, and transformation into 1,2-*trans*-1-*O*-acyl sugars, *J. Org. Chem. 53*:5629 (1988).
13. S. Hanessian and J. Banoub, Chemistry of the glycosidic linkage. An efficient synthesis of 1,2-*trans*-disaccharides, *Carbohydr. Res. 53*:C13 (1977); *Am. Chem. Soc. Symp. Ser. 39*:36 (1976).
14. For a review, see B. Fraser-Reid, U. E. Udodong, Z. Wu, H. Ottosson, J. R. Meritt, C. S. Rao, C.

Roberts, and R. Madsen, *n*-Pentenyl glycosides in organic chemistry: A contemporary example of serendipity, *Synlett* p. 927 (1992).

15. For a review, see R. R. Schmidt and W. Kinzy, Anomeric-oxygen activation for glycoside synthesis—the trichloroacetimidate method, *Advan. Carbohydr. Chem. Biochem. 50*:21 (1994); R. R. Schmidt, New methods for the synthesis of glycosides and oligosaccharides—are there alternatives to the Koenigs-Knorr method? *Angew. Chem. Int. Ed. Engl. 25*:212 (1986); see also Chap. 12.

# 20

## Oligosaccharide Synthesis by Selective Anomeric Activation with MOP- and TOPCAT-Leaving Groups

**Boliang Lou**
*Cytel Corporation, San Diego, California*

**Elisabeth Eckhardt**
*Boehringer Mannheim GmbH, Penzberg, Germany*

**Stephen Hanessian**
*University of Montreal, Montreal, Quebec, Canada*

| | | |
|---|---|---|
| I. | Introduction | 450 |
| | A. Results and discussion | 454 |
| | B. Conclusion | 455 |
| II. | Experimental Procedures | 457 |
| | A. 2,3,4,6-Tetra-*O*-benzyl-β-D-glucopyranosyl 2-thiopyridylcarbonate | 457 |
| | B. 6-*O*-*t*-Butyldimethylsilyl-2,3,4-tri-*O*-benzyl-β-D-glucopyranose | 457 |
| | C. 6-*O*-*t*-Butyldimethylsilyl-2,3,4-tri-*O*-benzyl-β-D-glucopyranosyl 2-thiopyridylcarbonate | 458 |
| | D. 3-Methoxy-2-pyridyl 3,4-di-*O*-acetyl-2-azido-2-deoxy-β-D-galactopyranoside | 458 |
| | E. 3-Methoxy-2-pyridyl 2,3,4-tri-*O*-benzoyl-β-D-glucopyranoside | 458 |
| | F. 3-Methoxy-2-pyridyl 2,3,4-tri-*O*-benzyl-β-D-glucopyranoside | 459 |
| | G. Glycosylation of 3-methoxy-2-pyridyl 2,3,4-tri-*O*-benzoyl-β-D-glucopyranoside with 2,3,4,6-tetra-*O*-benzyl-β-D-glucopyranosyl 2-thiopyridylcarbonate | 459 |
| | H. Glycosylation of 3-methoxy-2-pyridyl 3,4-di-*O*-acetyl-2-azido-2-deoxy-β-D-galactropyranoside with 2,3,4,6-tetra-*O*-benzyl-β-D-glucopyranosyl 2-thiopyridylcarbonate | 460 |

I. Glycosylation of 3-methoxy-2-pyridyl 2,3,4-tri-*O*-benzyl-β-D-glucopyranoside with 2,3,4,6-tetra-*O*-benzyl-β-D-glucopyranosyl 2-thiopyridylcarbonate .......... 460

J. Glycosylation of 3-methoxy-2-pyridyl 2,3,4-tri-*O*-benzoyl-β-D-glucopyranoside with 2,3,4-tri-*O*-benzyl-6-*O*-*t*-butyldimethysilyl-β-D-glucopyranosyl 2-thiopyridylcarbonate .......... 460

K. Glycosylation of 3-methoxy-2-pyridyl 3,4-di-*O*-acetyl-2-azido-2-deoxy-β-D-galactopyranoside with 2,3,4-tri-*O*-benzyl-6-*O*-*t*-butyldimethysilyl-β-D-glucopyranosyl 2-thiopyridylcarbonate .......... 461

L. Glycosylation of 3-methoxy-2-pyridyl 2,3,4-tri-*O*-benzyl-β-D-glucopyranoside with 2,3,4-tri-*O*-benzyl-6-*O*-*t*-butyldimethylsilyl-β-D-glucopyranosyl 2-thiopyridylcarbonate .......... 461

M. 3-Methoxy-2-pyridyl 2,3,4-tri-*O*-benzyl-α-D-glucopyranosyl-(1→6)-2,3,4-tri-*O*-benzoyl-β-D-glucopyranoside .......... 462

N. 3-Methoxy-2-pyridyl 2,3,4-tri-*O*-benzyl-6-*O*-*tert*-butyldimethylsilyl-α-D-glucopyranosyl-(1→6)-2,3,4-tri-*O*-α-D-glucopyranosyl-1(1→6)-2,3,4-tri-*O*-benzoyl-β-D-glucopyranoside .......... 462

O. α-D-Glucopyranosyl azide .......... 462

P. 2,3,4-tri-*O*-benzyl-6-*O*-*t*-butyldimethylsilyl-α-D-glucopyranosyl azide .......... 463

Q. 2,3,4-Tri-*O*-benzyl-α-D-glucopyranosyl azide .......... 463

R. 2,3,4,6-Tetra-*O*-benzyl-α-D-glucopyranosyl-(1→6)-2,3,4-tri-*O*-benzoyl-β-D-glucopyranosyl-(1→6)-2,3,4-tri-*O*-benzyl-α-D-glucopyranosyl azide .......... 463

S. 2,3,4-Tri-*O*-benzyl-t-*O*-*t*-butyldimethylsilyl-α-D-glucopyranosyl-(1→6)-2,3,4-tri-*O*-benzoyl-β-D-glucopyranosyl-(1→6)-2,3,4-tri-*O*-benzyl-α-D-glucopyranosyl azide .......... 464

References .......... 464

## I. INTRODUCTION

Although much effort has been devoted to the development of highly stereocontrolled methods for glycosylation over the past two decades [1; see also Chap. 12], the issue dealing with strategies for the rapid and efficient assembly of oligosaccharides has been a relatively more recent area of interest. The conventional approach to assemble oligosaccharides in a stepwise manner relies on the condensation of an "activated" glycosyl X donor and an "unactivated" glycosyl Y acceptor, in which the anomeric substitutent Y must remain intact under the coupling conditions. The subsequent transformation of Y into X leads to an activated disaccharide donor that can be subjected to further extension of the oligosaccharide as illustrated in Equation (1) (Scheme 1).

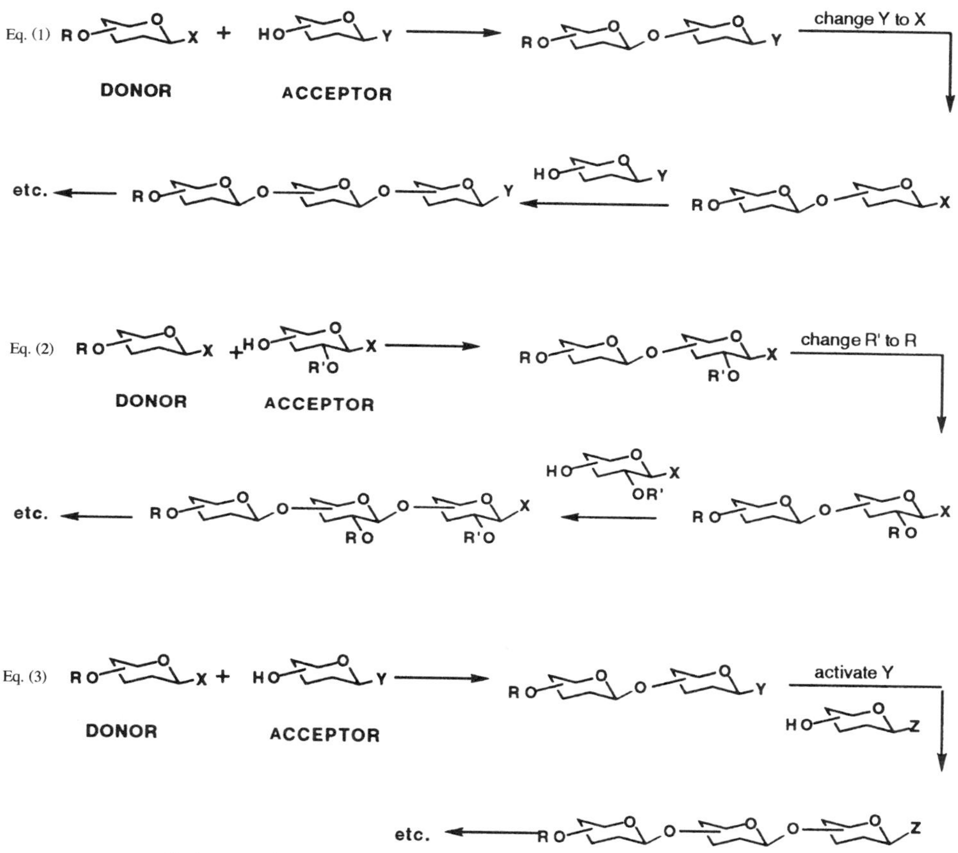

**Scheme 1** Strategies for iterative oligosaccharide synthesis.

An example of this strategy consists of a two-stage activation method as reported by Nicolaou [2]. Thus, a glycosyl fluoride derived from the corresponding phenylthioglycoside by simple treatment with NBS–DAST, is allowed to couple with a phenylthioglycoside to give a disaccharide. Conversion into the disaccharide fluoride as described in the foregoing allows an iterative process to be considered. Similarly, Danishefsky [3] reported an iterative strategy for the stereocontrolled construction of β-linked 1,6-glycal acceptors and repetition of the process. The method has been explored for applications to solid-phase synthesis of 1,6-linked oligosaccharides [4].

In an alternative iterative strategy [see Scheme 1, Eq. (2)], two glycosyl units that have the same unique leaving group at the anomeric positions are able to function as a donor and an acceptor, respectively, by taking advantage of differential reactivities owing to the nature of the protective groups. Usually, ether-protected glycosyl donors, such as pentenyl glycosides [5], thioglycosides [6], and glycals, can be selectively activated and coupled to the corresponding acyl-protected acceptors to form the disaccharide. Subsequently, deacylation and O-alkylation generate activated disaccharide donors. Thus, iterative oligosaccharide synthesis can be achieved by repeating these procedures. Ester protective groups decrease the reactivity of glycosyl donors owing to an inductive effect [7]. It is of particular interest that the O-acylated unactivated glycosyl donors can be made to react under more drastic conditions, producing a 1,2-*trans*-glycosidic linkage. For instance, an O-acyl–protected pentenyl glycoside or thioglycoside can be activated in the presence of NIS–TfOH and engaged in glycoside synthesis, whereas they remain unreactive in the presence of I (collidine)$_2$ClO$_4$ [8].

Recently, we established a novel protocol for the iteration of glycosidic sugar units, based on the selective activation of *unprotected* glycosyl donors relative to O-acyl glycosyl acceptors both bearing the same 3-methoxy-2-pyridyloxy (MOP) as an anomeric substituent [see Chap. 17]. The method avoids O-benzyl protective groups and offers a more direct route to iterative 1,2-*cis*-linked oligosaccharides and other glycosides as the major anomers. Because of the mildness of conditions and the simplicity of reagent design, the method is adaptable to an automated synthesis of glycosides and certain oligosaccharides on a solid-phase medium [9].

The third strategy for the sequential construction of the O-glycosidic bonds involves glycosylation of partners that have different-leaving groups, one of which may be activated preferentially over the other. As illustrated in Eq. (3) of Scheme 1, this allows the glycosylation products to be used as donors for the next coupling reaction, without any manipulation of the anomeric center or protective groups. This strategy offers the most straightforward way to build oligosaccharides in the least number of steps. Examples of this strategy involve an "active–latent" thioglycosyl donor [11], selective activations of selenoglycosides over thioglycosides [11], and arylsulfenyl glycosides over thioglycosides [12], a one-step synthesis applicable in special cases [13], one-pot glycosylation [14], and other methods [15]. More recently, an "orthogonal" glycosylation strategy was described by Ogawa and co-workers [16], which combines the use of phenyl thioglycosides and glycosyl fluorides as both donors and acceptors, resulting in an improvement of the two-stage activation method [2].

Two novel leaving groups for O-glycoside and pyrimidine nucleoside synthesis [17] have been developed in our laboratory. These are the 3-methoxy-2-pyridyloxy (MOP) [see Chap. 18] and the 2-thiopyridylcarbonate (TOPCAT) groups [see Chap. 19], which can be activated in the presence of Cu(OTf)$_2$ and AgOTf, respectively. A practical finding in conjunction with our studies was that TOPCAT glycosyl donors could be selectively activated by using AgOTf in the presence of MOP-glycosyl donors. The merging of the TOPCAT and MOP-based technologies led us to a paradigm for the iterative construction of oligosaccharides (Scheme 2). Thus, a TOPCAT O-protected donor can be activated with AgOTf in the presence of a partially O-protected MOP acceptor to give a 1,2-*cis*- or 1,2-*trans*-linked disaccharide, depending on the nature of the C-2 substituent in the donor. The iteration can be continued using Cu(OTf)$_2$ as a promoter for the activation of the MOP group. Alternatively, the MOP disaccharide resulting from the initial coupling can be partially deprotected and the product used as an acceptor in another AgOTf–TOPCAT donor-mediated glycosylation to provide a MOP trisaccharide and so on. Because MOP glycosyl donors can form O-glycosides in the absence of protective groups [see Chap. 18], each MOP donor can be O-deprotected and coupled with an alcohol acceptor in the presence of MeOTf to give unprotected oligosaccharides with a "capping" acceptor alcohol at the reducing end.

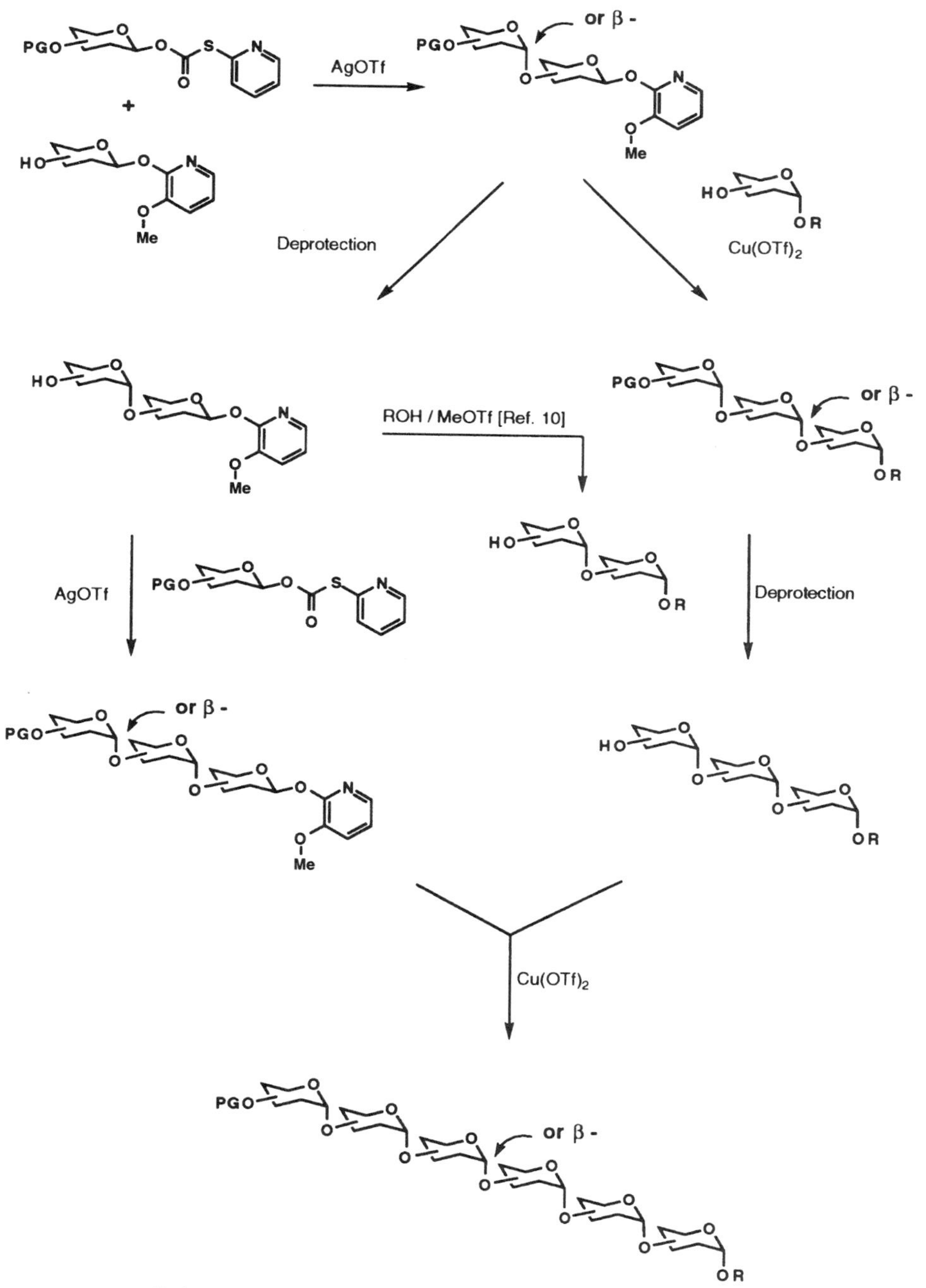

# Activation with MOP- and TOPCAT-Leaving Groups

**Scheme 2** Combination of TOPCAT and MOP activations.

## A. Results and Discussion

2,3,4,6-Tetra-*O*-benzyl-β-D-glucopyranosyl 2-thiopyridylcarbonate **1** was chosen as a representative donor to couple with the *O*-acetylated MOP acceptor **3** bearing a free hydroxy group at the C-6 position. The reaction proceeded smoothly at room temperature in the presence of AgOTf, and in a mixed solvent ($CH_2Cl_2$–$Et_2O$, 1:4), to give the disaccharide **6** in 67% yield and an α/β ratio of 3:1 (Scheme **3**). Glycosylation with the partially *O*-benzoylated MOP acceptor **4** was completed in a similar way to give **7** (α/β, 3:1; 64%). In principle, MOP disaccharides **6** and **7** can be de-*O*-acylated and engaged in glycoside synthesis in the presence of MeOTf as previously described [see Chaps. 16–18].

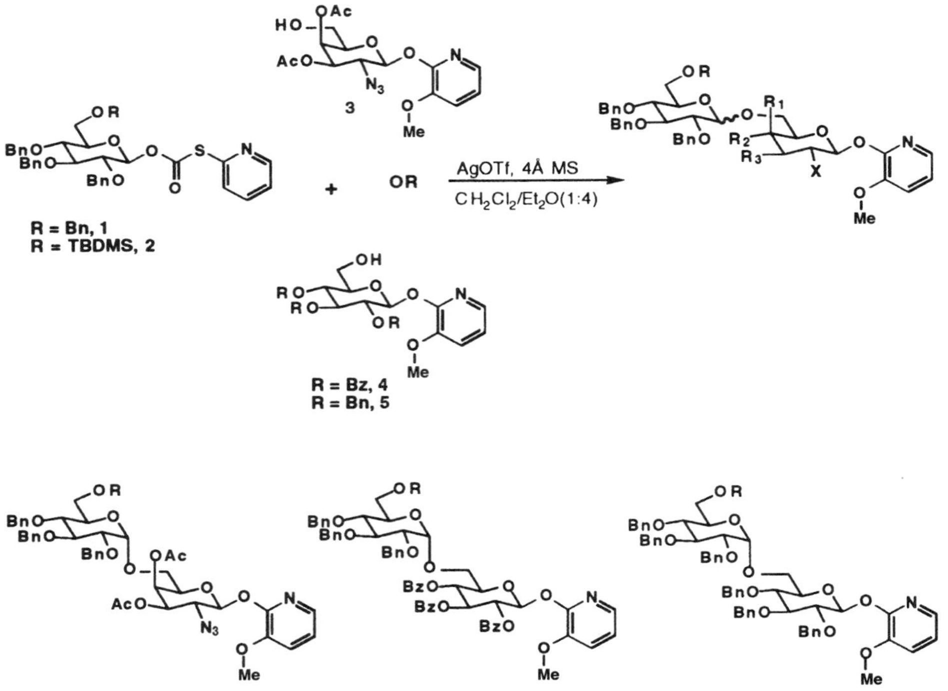


**R = Bn, 6** α:β, **3:1, 67%**
**R = TBDMS, 9** α:β, **4:1, 54%**

**R = Bn, 7** α:β, **3:1, 64%**
**R = TBDMS, 10** α:β, **10:1, 74%**

**R = Bn, 8** α:β, **5:1, 71%**
**R = TBDMS, 11** α:β, **10:1, 77%**

**Scheme 3** Selective activation of TOPCAT glycosyl donors in the presence of MOP acceptors: TOPCAT versus MOP.

Encouraged by the observed selectivity, we next considered the use of *O*-benzylated MOP glycosides as acceptors. Normally, these exhibit higher reactivity than the corresponding *O*-acylated counterparts; hence, the possibility of a lack of selectivity in the attempted glycosylations. MOP 2,3,4-tri-*O*-benzyl-β-D-glucopyranoside **5** behaves as an excellent acceptor when allowed to react with **1** in the presence of AgOTf to give the desired disaccharide **8** in 71% yield, with an α/β ratio of 5:1. Thus, the TOPCAT donor can be activated preferentially relative to both ether- and acyl-protected MOP acceptors, leading to disaccharides capable of being further utilized in the synthesis of 1,2-*cis*- or 1,2-*trans*-oligosaccharides or glycosides in general, by taking advantage of the MOP anomeric extremity.

# Activation with MOP- and TOPCAT-Leaving Groups

We next addressed the question of anomeric stereoselectivity in the formulation of disaccharides and oligosaccharides. The increase of 1,2-*cis*-selectivity, using glycosyl donors bearing a bulky protective group at C-6, has been demonstrated previously [18]. Thus, the influence of a 6-*O*-TBDMS group was studied as exemplified by the reaction of the glycosyl donor **2** with the acceptor **3** in the presence of AgOTf. The expected disaccharide was isolated in 54% yield, with no significant improvement in the selectivity. However, the reaction with acceptors **4** or **5** with donor **2**, led to 1,2-*cis*-disaccharides **10** and **11** with α/β rations of 11:1 or 10:1, respectively, and in good yields. These examples illustrate the difficulties in predicting ratios of α/β-anomers even if the structural variations appear to be minor (compare **9** and **10** and **11**). A combination of electronic, steric, and other subtle effects must contribute to these differences, and this aspect of variable reactivity among carbohydrate derivatives needs further attention.

Scheme **4** illustrates typical examples of TOPCAT–MOP combinations in the assembly of oligosaccharides. The removal of the TBDMS group in **10** under the normal fluoride ion-catalyzed conditions generated the free hydroxy group in **12** which was ready for the use as an acceptor. Glycosylation of **12** with **2** as a donor in the presence of AgOTf gave the expected trisaccharide **13** in 71% yield (α/β, 8:1). As with **10** and **12**, the trisaccharide **13** could be extended at either end, thereby acting as a donor or an acceptor.

*Synthesis of Nephritogenoside Core Trisaccharide*

To demonstrate the usefulness of TOPCAT–MOP combinations in the activation of glycosyl donors, we chose the trisaccharide core structure of nephritogenoside as a target for synthesis. Nephritogenoside, a glycopeptide located in the rat glomerular basement membrane, is active in the induction of glomerulonephritis in homologous animals [19]. The biological activity and the unusual structure of this glycopeptide, in which the reducing end of the trisaccharide is α-*N*-glycosidically linked to an asparagine residue of the peptide, has been the subject of synthetic studies [20]. Application of the TOPCAT–MOP selective activation procedures allowed us to accomplish a very short route to this core structure as shown in Scheme **4**.

Treatment of the unprotected MOP donor **14** with an excess of $TMSN_3$ in the presence TMSOTf gave, after workup, the crystalline α-D-glucopyranosyl azide **15** in excellent yield and anomeric selectivity [21]. This reaction most likely involves the intermediacy of a persilylated oxocarbenium ion that is attacked almost exclusively from the stereoelectronically favored axial trajectory to give the persilylated glycosyl azide, that eventually leads to **15** after workup. Previous syntheses of **15** have used *O*-acetyl protection [22]. Regioselective silylation at the C-6 hydroxy group of **15** and benzylation, followed by desilylation using TBAF afforded the desired 2,3,4-tri-*O*-benzyl-α-D-glucopyranosyl azide **16**. Coupling with MOP disaccharide donors **7** and **10** individually in the presence of $Cu(OTf)_2$ in $CH_2Cl_2$ as solvent gave the trisaccharides **17** and **18**, respectively, in good yields with only a β-configuration for the newly formed glycosidic bond in each case.

## B. Conclusion

We have shown that the selective activation of the TOPCAT donors is possible with AgOTf in the presence of MOP acceptors. In a typical iterative process, one can use *O*-benzyl TOPCAT donors with partially ether or ester *O*-protected MOP acceptors for the synthesis of di-, tri-, or oligosaccharides with predetermined anomeric configurations in the major

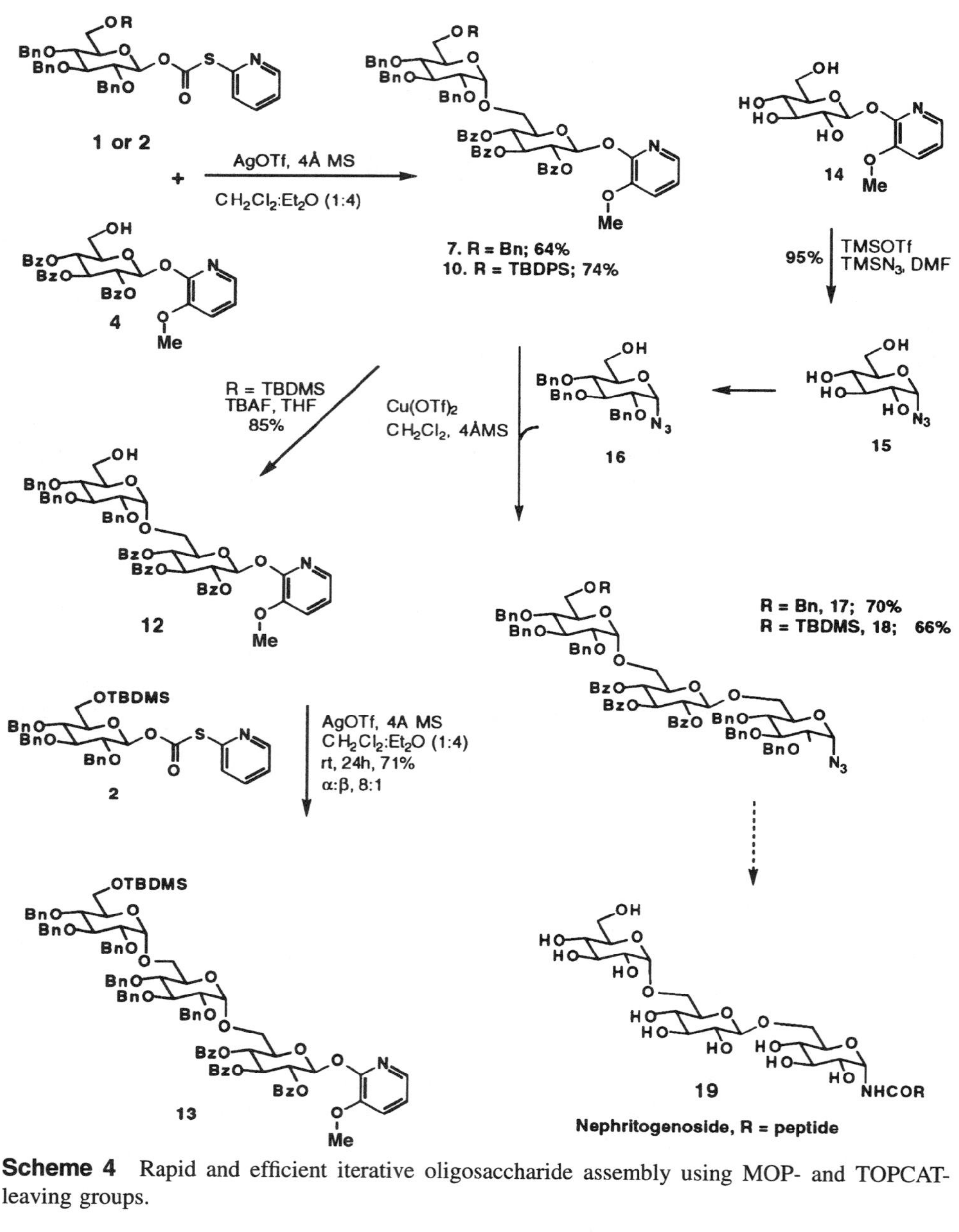

**Scheme 4** Rapid and efficient iterative oligosaccharide assembly using MOP- and TOPCAT-leaving groups.

products. Activation of the MOP group at the reducing end portion of these products with Cu(OTf)$_2$, and reaction with acceptor alcohols allows the assembly of complex glycosides or oligosaccharides. Extension of these glycosylations to linkages other than 1,6- are obviously needed to fully appreciate the general usefulness of combining TOPCAT and MOP anomeric activating groups for the synthesis of oligosaccharides.

## II. EXPERIMENTAL PROCEDURES

### A. 2,3,4,6-Tetra-O-benzyl-β-D-glucopyranosyl 2-thiopyridylcarbonate (1)

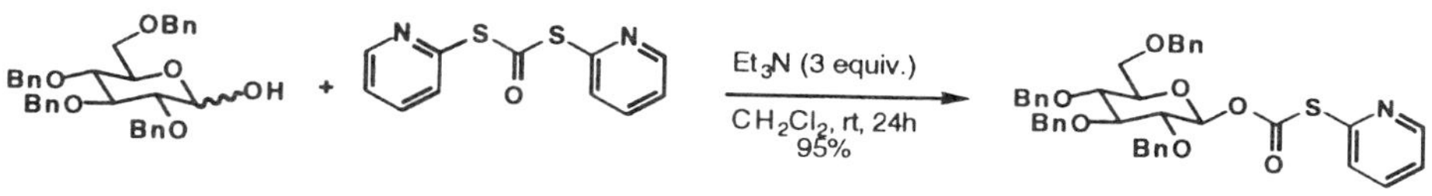

A mixture of 2,3,4,6-tetra-*O*-benzyl-D-glucopyranose (330 mg, 0.61 mmol), 411 mg (1.66 mmol) of di(*S*-2-pyridyl)thiocarbonate, 231 µL (1.66 mmol) of Et$_3$N, and 6 mL of CH$_2$Cl$_2$ was stirred at room temperature for 1 day. Concentration followed by purification by flash chromatography on silica gel column with EtOAc–hexane (1:2 to 1:1) or benzene–EtOAc (5:1) gave the desired product **1** as a pale yellow solid (393 mg 95%): mp 71–73°C, [α]$_D$ +15.5° (*c* 1.5, CHCl$_3$).

### B. 6-O-*t*-Butyldimethylsilyl-2,3,4-tri-O-benzyl-β-D-glucopyranose

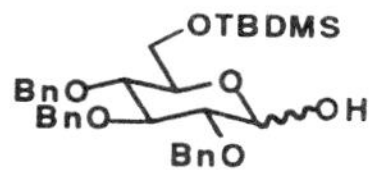

*Method A*

To a solution cooled at 0°C of 2,3,4-tri-*O*-benzyl-D-glucopyranose (98 mg, 0.217 mmol), 51 µL (0.52 mmol) of 2,6-lutidine, and 2 mL of CH$_2$Cl$_2$, was added dropwise 70 µL (0.30 mmol) to TBDMSOTf. The resulting mixture was stirred at this temperature for 15 min, at which time TLC showed the reaction was completed (occasionally an additional portion of TBDMSOTf had to be added). Concentration, followed by purification by flash chromatography with EtOAc–hexane (1:4), gave 120 mg (98%) of product (α/β, 1.5:1), [α]$_D$ +22.7° (*c* 3.4, CHCl$_3$).

*Method B*

To a solution of 2.34 g (3.87 mmol) of allyl 2,3,4-tri-*O*-benzyl-6-*t*-butyldimethylsilyl-β-D-glucopyranoside in 45 mL of EtOH, 20 mL of benzene, and 6 mL of water were added 358 mg (0.38 mmol) of RhCl(PPh$_3$)$_3$ and 384 mg (3.54 mmol) of DABCO. The mixture was refluxed for 5 h at which time thin-layer chromatography (TLC) indicated the starting material had been consumed (EtOAc–hexane, 1:9). After being cooled to room temperature, the mixture was concentrated, treated with ice-water, and extracted with EtOAc. The organic layer was washed with water once. Filtration followed by concentration gave a residue that was purified by flash chromatography on silica gel using EtOAc–hexane (1:15 to 1:8) to afford 1.43 g (60%) of product. This was dissolved in 36 mL of the mixed solvent (acetone–H$_2$O, 10:1), and mercuric oxide (643 mg) then mercuric chloride (643 mg) were added. The resulting suspension was stirred at room temperature until the starting material completely disappeared (TLC). After dilution with EtOAc, the mixture was washed with 5% KI aqueous solution and water. The organic layer was processed as usual, and the residue was chromatographed on silica gel using EtOAc–hexane (1:4 to 1:2) as the eluant to give the title compound (1.05 g, 79%).

## C. 6-O-t-Butyldimethylsilyl-2,3,4,tri-O-benzyl-β-D-glucopyranosyl 2-thiopyridylcarbonate (2)

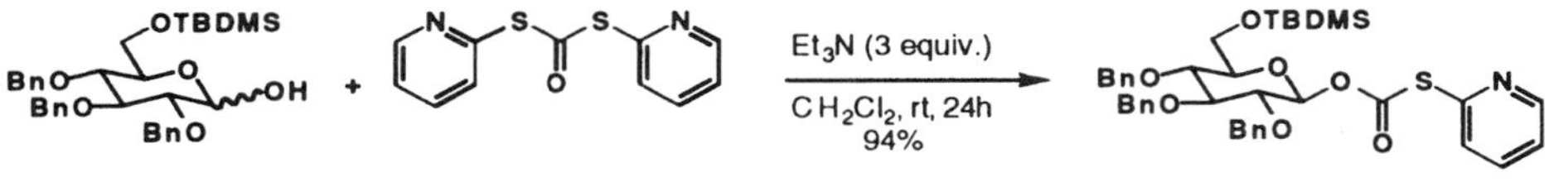

The same procedure as the preparation of **1** gave the title product (94%): mp 90–92°C, $[\alpha]_D$ +3.8° (c 1.8, CHCl$_3$).

## D. 3-Methoxy-2-pyridyl 3,4-di-O-acetyl-2-azido-2-deoxy-β-D-galactopyranoside (3)

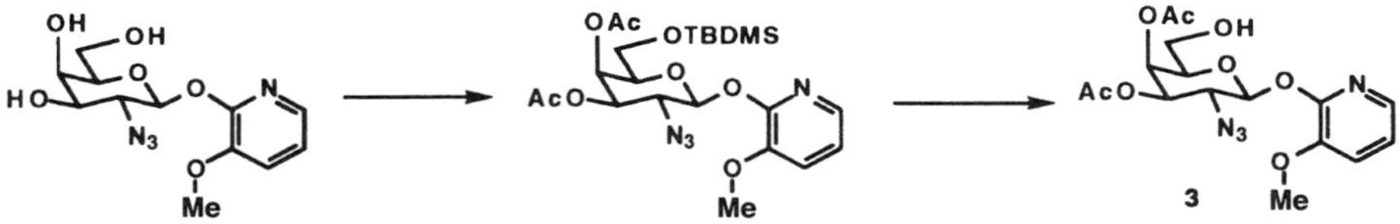

To a cooled solution of 3-methoxy-2-pyridyl 2-azido-2-deoxy-β-D-galactopyranoside (1.5 g, 4.8 mmol), in 20 mL of pyridine, 1.2 mL (5.23 mmol) of t-butyldimethylsilyl trifluoromethanesulfate was added dropwise at 0°C over 30 min. The mixture was stirred for 30 min, Ac$_2$O (2 mL) was then added. The solution was kept at 0°C for 15 h. The reaction mixture was poured into ice-water and extracted with CH$_2$Cl$_2$. The extraction was processed as usual. Purification by flash chromatography on silica gel using 2:1 (v/v) hexane–EtOAc as the irrigant gave 3-methyl-2-pyridyl 3,4-di-O-acetyl-6-O-t-butyldimethylsilyl-2-azido-2-deoxy-β-D-galactopyranoside as a colorless syrup (2.21 g, 90%): $[\alpha]_D$ −0.62° (c 6.2, CHCl$_3$).

To a cooled solution of the preceding compound (342 mg, 0.67 mmol) in 1.5 mL of THF, 115 μL (2.0 mmol) of acetic acid and 0.8 mL of 1.0 M Bu$_4$NF in THF were added at 0°C. The mixture was stirred for 24 h at 0°C, and the solvents were evaporated. The residue was treated with ice-water, and extracted with CH$_2$Cl$_2$. The organic layer was processed as usual. Concentration followed by purification by flash chromatography on silica gel gave the title product (248 mg, 93%) as a white solid: mp 58–60°C, $[\alpha]_D$ +3.8° (c 1.0, CHCl$_3$).

## E. 3-Methoxy-2-pyridyl 2,3,4-tri-O-benzoyl-β-D-glucopyranoside (4)

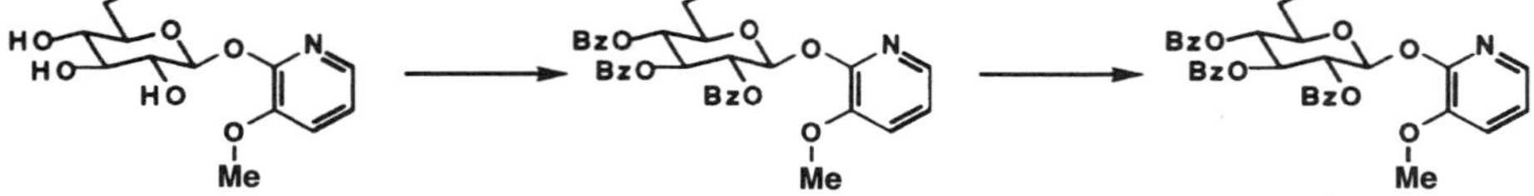

To a mixture cooled at 0°C containing 2.34 g (8.15 mmol) of 3-methoxy-2-pyridyl β-D-glucopyranoside in 50 mL of dry pyridine was added dropwise 2.1 mL (8.97 mmol) of t-butyldimethylsilyl triflate over 15 min. The resulting mixture was stirred at 0°C for another 15 min at which time TLC indicated the silylation was completed. After 3.1 mL (27

## Activation with MOP- and TOPCAT-Leaving Groups

mmol) of benzoyl chloride was added, the mixture was kept at 0°C for 12 h. Pyridine was removed in vacuo, and then the residue was treated with cold saturated NaHCO$_3$ and extracted with CH$_2$Cl$_2$. The organic layer was washed with brine, dried over MgSO$_4$, and concentrated. Purification by flash chromatography on a silica gel column with EtOAc–hexane (1:4) provided 4.6 g (79%) of the product as a white solid: mp 62°–64°C, [α]$_D$ +38.3° (c 1.0, CHCl$_3$).

A solution of this product (3.8 g, 5.3 mmol) and 1.37 mL (24 mmol) of acetic acid in 10 mL of THF was treated with 8 mL (8 mmol) of 1 M TBAF in THF solution at 0°C. The resulting mixture was stirred at this temperature for 30 min, then at room temperature for 24 h. After concentration, the residue was treated with ice-water, extracted with CH$_2$Cl$_2$, and the organic layer was processed as usual. The residue was purified by flash chromatography on silica gel using EtOAc–hexane (1:1 to 3:1) as the eluant to afford the desired product as a white solid (3.0 g, 94%): mp 161°–162°C, [α]$_D$ +49.1° (c 0.85, CHCl$_3$).

### F. 3-Methoxy-2-pyridyl 2,3,4-tri-O-benzyl-β-D-glucopyranoside (5)

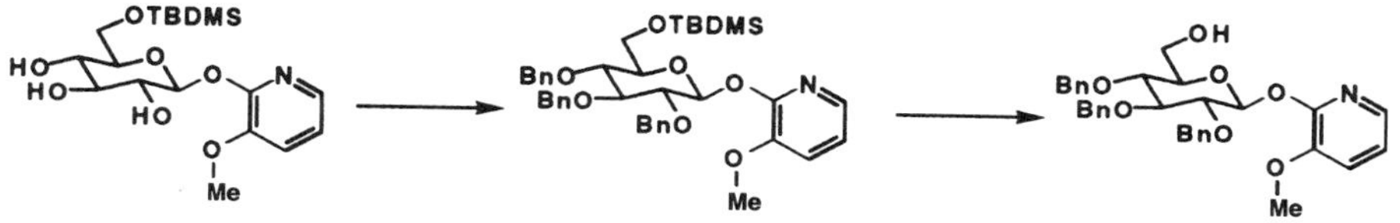

3-Methoxy-2-pyridyl 6-O-t-butyldimethylsilyl-β-D-glucopyranoside (76 mg) was dissolved in 1 mL of DMF, and 34 mg of 60% sodium hydride was added to this solution. After addition of 76.7 μL of benzyl bromide, the mixture was stirred at room temperature for 12 h, then poured into ice-water, extracted with CH$_2$Cl$_2$, and processed as usual. Flash chromatography on silica gel (eluant: EtOAc–hexane 1:4) gave 102 mg of 3-methoxy-2-pyridyl 2,3,4-tri-O-benzyl-6-O-t-butyldimethylsilyl-β-D-glucopyranoside in 80% yield. This product (62 mg) was dissolved in 1 mL of THF and treated with 140 μL of Bu$_4$NF solution in THF at room temperature for 3 h. The solution was poured into ice-water and extracted with CH$_2$Cl$_2$. Concentration and purification by flash chromatography on silica gel gave 44 mg of the title compound (85%): mp 104°–106°C, [α]$_D$ +9.3° (c 0.75, CHCl$_3$).

### G. Glycosylation of 3-Methoxy-2-pyridyl 2,3,4-tri-O-benzoyl-β-D-glucopyranoside (4) with 2,3,4,6-Tetra-O-benzyl-β-D-glucopyranosyl 2-thiopyridylcarbonate (1)

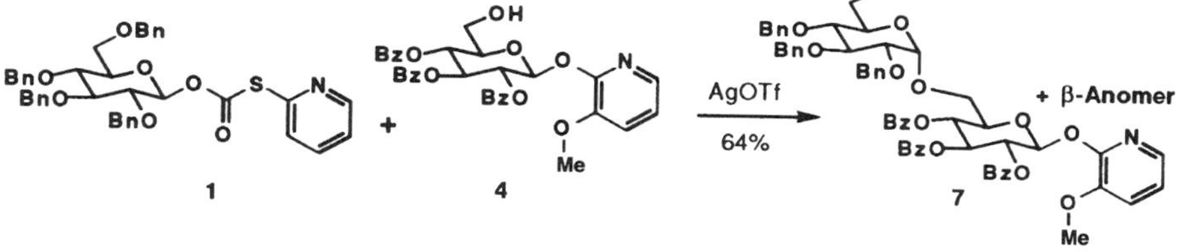

To a mixture of the glycosyl donor **1** (50 mg, 0.074 mmol), the glycosyl acceptor **4** (30 mg, 0.05 mmol), powdered 4-Å MS (100 mg), ether (1.6 mL), and CH$_2$Cl$_2$ (0.4 mL), was added AgOTf (38 mg, 0.148 mmol). The resulting suspension was stirred at room temperature for 6 h, at which time TLC indicated the reaction was completed. After addition of 1 drop of pyridine, the mixture was filtered over Celite. Concentration, followed by purification on silica gel with EtOAc–hexane (1:2), gave 36 mg of product in 64% yield (α/β, 3:1). The α- and β-anomers were separated by a second chromatographic purification with EtOAc–hexane (1:4 to 1:2). For the α-anomer: mp 67°–69°C, $[\alpha]_D$ +56.0° (c 1.35, CHCl$_3$); for the β-anomer: mp 62°–64°C, $[\alpha]_D$ +40.5° (c 0.8, CHCl$_3$).

### H. Glycosylation of 3-Methoxy-2-pyridyl 3,4-di-O-acetyl-2-azido-2-deoxy-β-D-galactopyranoside (3) with 2,3,4,6-tetra-O-benzyl-β-D-glucopyranosyl 2-thiopyridylcarbonate (1)

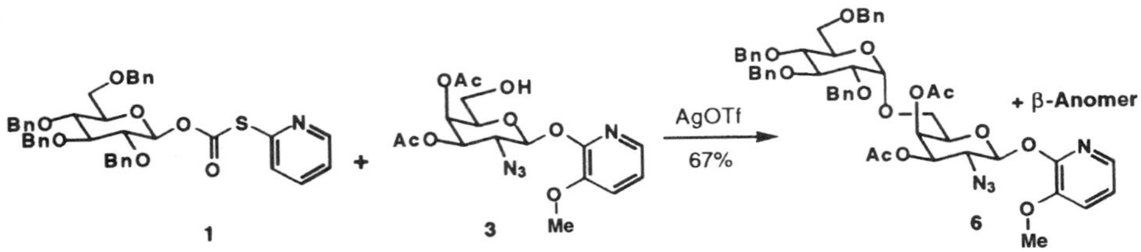

The same procedure described in the foregoing was followed. The α- and β-products could not be separated by flash chromatography using benzene–EtOAc (5:1) as the eluant (α/β, 3.2:1, 67%). A second chromatographic purification using CH$_2$Cl$_2$–EtOAc (10:1) as the eluant afforded the two anomeric products. For the α-anomer: $[\alpha]_D$ +29.1° (c 2.9, CHCl$_3$); for the β-anomer: $[\alpha]_D$ −2.5° (c 1.4, CHCl$_3$).

### I. Glycosylation of 3-Methoxy-2-pyridyl 2,3,4-tri-O-benzyl-β-D-glucopyranoside (5) with 2,3,4,6-Tetra-O-benzyl-β-D-glucopyranosyl 2-thiopyridylcarbonate (1)

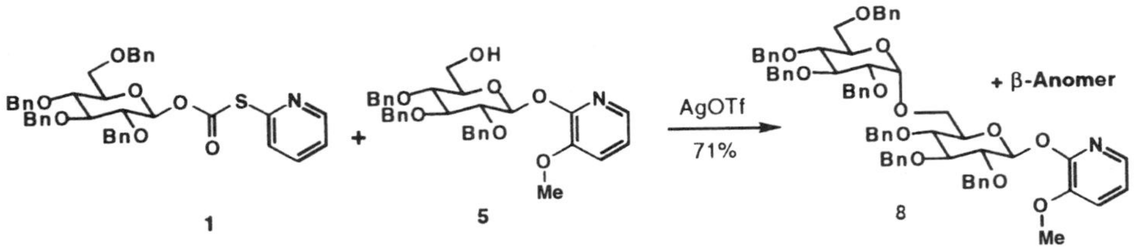

The same procedure as before gave the α- and β-products that could not be separated by flash chromatography on silica gel using EtOAc–hexane (1:2 to 2:3) as eluant (α/β, 5:1, 71%).

### J. Glycosylation of 3-Methoxy-2-pyridyl 2,3,4-tri-O-benzoyl-β-D-glucopyranoside (4) with 2,3,4-tri-O-benzyl-6-O-t-butyldimethylsilyl-β-D-glucopyranosyl 2-thiopyridylcarbonate (2)

To a solution of the glycosyl donor **2** (160 mg, 0.228 mmol), 92 mg (0.154 mmol) of the glycosyl acceptor **4** and 8 mL of the mixed solvent (CH$_2$Cl$_2$–Et$_2$O, 1:4) was added 500 mg of activated powdered 4Å MS. After the mixture was stirred at room temperature for 30 min, 117 mg (0.455 mmol) of AgOTf was added and the resulting suspension was stirred until the donor **2** disappeared by TLC analysis. After addition of a few drops of pyridine, the mixture was filtered through Celite and washed with $CH_2Cl_2$. Concentration, followed by purification by flash chromatography on silica gel (EtOAc–hexane, 1:2), provided the desired disaccharide **10** along with the β-anomer (130 mg, 74%; α/β, 10:1). The α- and β-anomers could not be separated chromatographically (α/β ≈ 10:1, determined by $^1$H NMR).

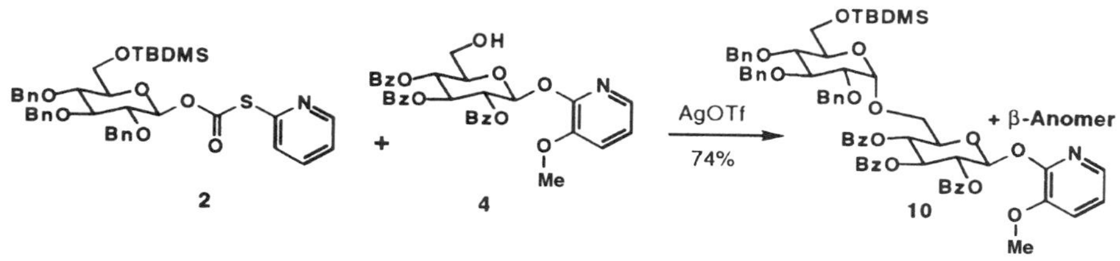

## K. Glycosylation of 3-Methoxy-2-pyridyl 3,4-di-O-acetyl-2-azido-2-deoxy-β-D-galactopyranoside (3) with 2,3,4-tri-O-benzyl-6-O-t-butyldimethylsilyl-β-D-glucopyranosyl 2-thiopyridylcarbonate (2)

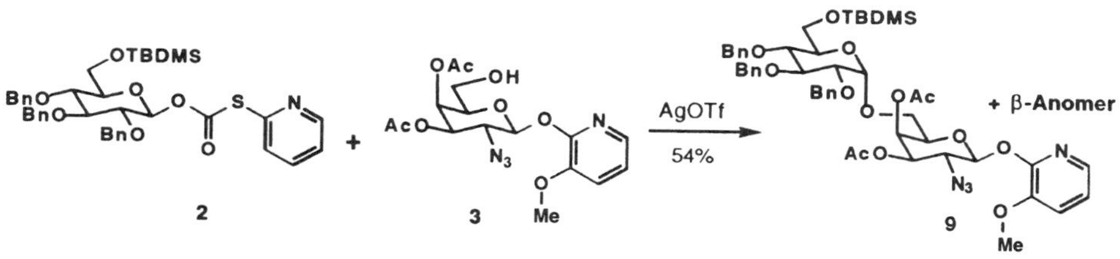

The same procedure described in the foregoing was followed to give the α-linked disaccharide **9** and the β-anomer, which could not be separated by chromatography on silica gel (α/β, 4:1, 54%).

## L. Glycosylation of 3-Methoxy-2-pyridyl 2,3,4-tri-O-benzyl-β-D-glucopyranoside (5) with 2,3,4-tri-O-benzyl-6-O-t-butyldimethylsilyl-β-D-glucopyranosyl 2-thiopyridylcarbonate (2)

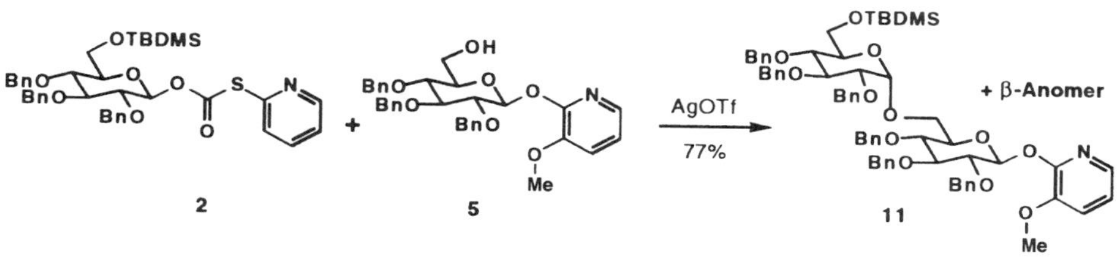

The same procedure as before gave a mixture of the α-linked disaccharide **11** and the β-anomer, which could not be separated by flash chromatography on silica gel using various eluants (α/β, 10:1, 77%).

**M. 3-Methoxy-2-pyridyl 2,3,4-tri-*O*-benzyl-α-D-glucopyranosyl-(1→6)-2,3,4-tri-*O*-benzoyl-β-D-glucopyranoside (12)**

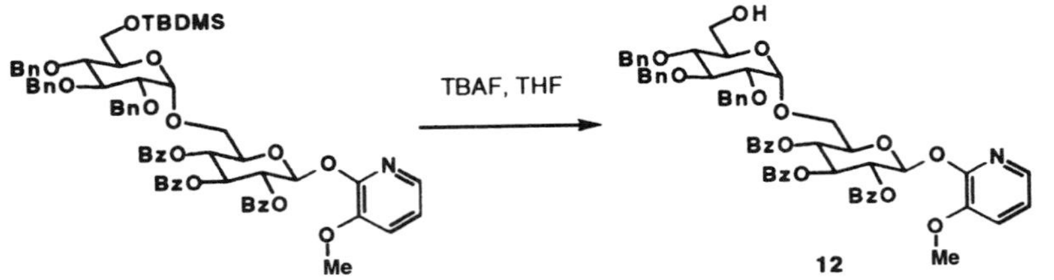

To a cooled solution of 3-methoxy-2-pyridyl 2,3,4-tri-*O*-benzyl-6-*O*-*tert*-butyldimethyl-silyl-α-D-glucopyranosyl-(1→6)-2,3,4-tri-*O*-benzoyl-β-D-glucopyranoside **10** (134 mg, 0.117 mmol); (α/β, 10:1) in 0.6 mL of THF were added 52 μL of acetic acid and 0.46 mL of 1 *M* TBAF in THF. The resulting mixture was stirred at 0°C for 30 min, then at room temperature for 2 days, at which time TLC indicated the reaction was completed (EtOAc–hexane, 1:2). After the solvent was evaporated, EtOAc was added, the solution was washed with water and brine, and the organic layer was processed as usual. The residue was purified by flash chromatography on silica gel using EtOAc–hexane (1:1) as the eluant to afford 98 mg of α-linked product **12** and 10 mg of the β-anomer (α/β, 10:1; 89%).

**N. 3-Methoxy-2-pyridyl 2,3,4-tri-*O*-benzyl-6-*O*-*tert*-butyldimethylsilyl-α-D-glucopyranosyl-(1→6)-2,3,4-tri-*O*-benzyl-α-D-glucopyranosyl-(1→6)-2,3,4-tri-*O*-benzoyl-β-D-glucopyranoside (13)**

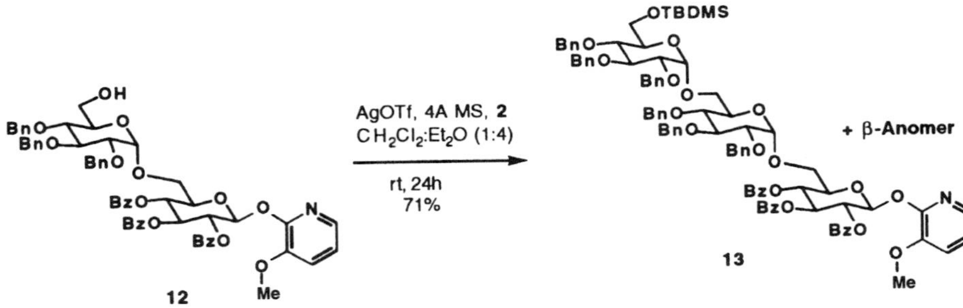

A mixture of glycosyl donor **2** (31 mg, 0.043 mmol), 30 mg (0.029 mmol) of acceptor **12**, 0.4 mL of dry CH$_2$Cl$_2$, 1.6 mL of dry Et$_2$O and 120 mg of powdered 4-Å MS was stirred at room temperature for 2 h. After 22.4 mg (0.0432 mmol) of AgOTf was added, the resulting suspension was stirred at room temperature for 24 h. One drop of pyridine was added, the mixture was filtered over Celite and the pad was washed with CH$_2$Cl$_2$. Concentration of the filtrate, followed by purification by flash chromatography on silica gel using EtOAc–hexane (1:2) as the eluant, gave 32.4 mg of the desired trisaccharide in 71% yield as an inseparable mixture of anomers (α/β ≈ 8:1).

**O. α-D-Glucopyranosyl Azide (15)**

3-Methoxy-2-pyridyl β-D-glucopyranoside (285 mg, 1.0 mmol) was dissolved in DMF (4.4 mL), and TMSN$_3$ (1.33 mL, 10 mmol) and TMSOTf (95 μL, 0.5 mmol) were added.

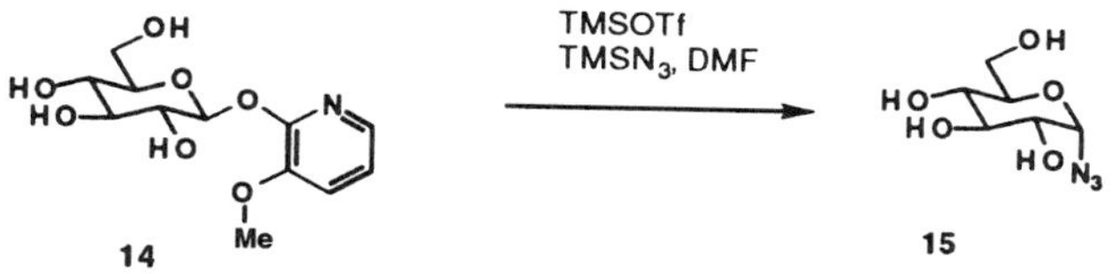

# Activation with MOP- and TOPCAT-Leaving Groups

[Scheme: compound **14** → compound **15** with TMSOTf, TMSN$_3$, DMF]

The resulting mixture was stirred at room temperature for 5 h, water (100 μL) was then added. After stirring for 5 min, the mixture was treated with 1 mL of pyridine. The solvent was evaporated, and the residue was purified by flash chromatography with EtOAc–MeOH (7:1), then with MeOH–EtOAc–CHCl$_3$ (1:4:4) as eluant to give the title compound **15** (196 mg, 95%): mp 175°C, [α]$_D$ +254.7° (c 1.3, MeOH) [22].

## P. 2,3,4-Tri-O-benzyl-6-O-t-butyldimethylsilyl-α-D-glucopyranosyl Azide

[Scheme showing conversion of glucopyranosyl azide to OTBDMS then to tri-O-benzyl OTBDMS glucopyranosyl azide]

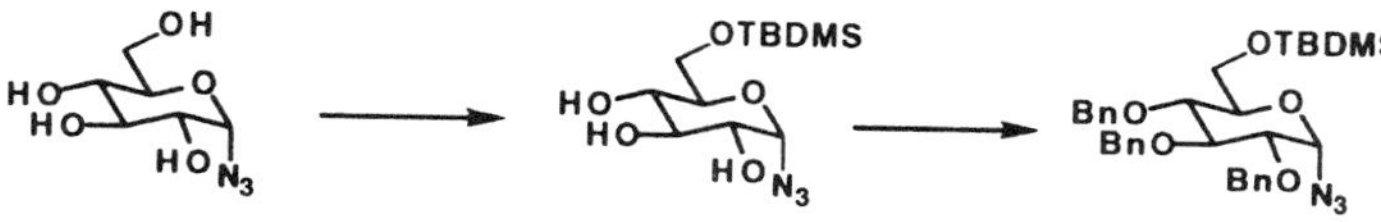

To a solution of 275 mg (1.34 mmol) of the previously-obtained glucosyl azide in 7 mL of DMF were added 201 mg (2.95 mmol) imidazole and 222 mg (1.47 mmol) of TBDMSOTf at 0°C. The mixture was stirred at 0°C for 4 h, EtOAc was then added, and the solution was poured into ice-water. The water layer was extracted with EtOAc and the combined organic layers were processed as usual. Concentration gave the crude product, which was dissolved in 7 mL of DMF; 130 mg (5.38 mmol) of sodium hydride and 540 μL (4.52 mmol) of benzyl bromide were added at 0°C. After stirring for 30 min, EtOAc was added and the organic layer was processed as usual. Purification by flash chromatography with CCl$_4$–acetone (50:1) gave the title compound (499 mg, 63%): [α]$_D$ +72.8° (c 1.6, CHCl$_3$).

## Q. 2,3,4-Tri-O-benzyl-α-D-glucopyranosyl Azide (16)

[Scheme: OTBDMS compound → compound **16**]

To a solution of the preceding product (342 mg, 0.67 mmol) in 4mL of THF was added 1.35 mL of 1 M Bu$_4$NF in THF at 0°C. The mixture was stirred for 24 h, and the solvent was then evaporated. The residue was dissolved in EtOAc, the organic layer washed with saturated NaHCO$_3$ and processed as usual. Chromatographic purification (CCl$_4$–acetone, 15:1) gave the title compound **16** as a crystalline solid (235 mg, 74%): mp, 68°C (EtOH); [α]$_D$ + 96.0° (c 0.9, CHCl$_3$).

## R. 2,3,4,6-Tetra-O-benzyl-α-D-glucopyranosyl-(1→6)-2,3,4-tri-O-benzoyl-β-D-glucopyranosyl-(1→6)-2,3,4-tri-O-benzyl-α-D-glucopyranosyl azide (17)

A solution of 25 mg (0.051 mmol) of the preceding compound **16** in 1 mL of dichloromethane was stirred in the presence of powdered 4-Å molecular sieves and 26 mg (0.073

mmol) of Cu(OTf)$_2$. After 10 min, a solution of 81.7 mg (0.073 mmol) of 3-methoxy-2-pyridyl 2,3,4,6-tetra-*O*-benzyl-α-D-glucopyranosyl-(1→6)-2,3,4-tri-*O*-benzoyl-β-D-glucopyranoside **7** in 2 mL of dichloromethane was added dropwise over 2 h. A few drops of pyridine were added, the solvent was evaporated and the residue was chromatographed with CCl$_4$–acetone (10:1) as eluant to give 52 mg (70%) of the trisaccharide **17**: [α]$_D$ +38.4° (*c* 1.1, CHCl$_3$).

## S. 2,3,4-Tri-*O*-benzyl-6-*O*-*t*-butyldimethylsilyl-α-D-glucopyranosyl-(1→6)-2,3,4-tri-*O*-benzoyl-β-D-glucopyranosyl-(1→6)-2,3,4-tri-*O*-benzyl-α-D-glucopyranosyl Azide (18)

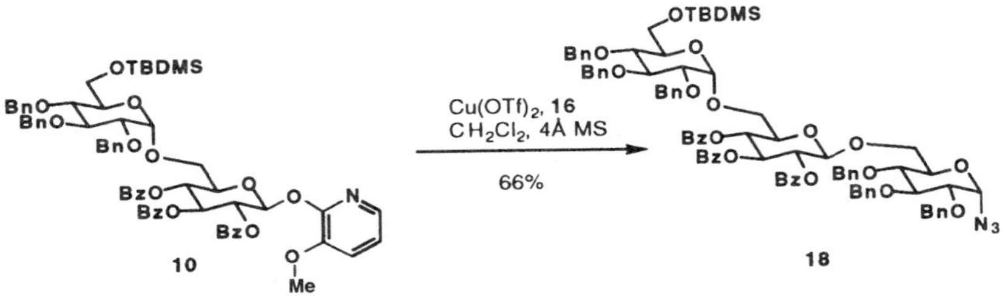

A solution of 2,3,4-tri-*O*-benzyl-α-D-glucopyranosyl azide **16** (21 mg, 0.044 mmol) in 0.6 mL of dichloromethane was stirred in the presence of powdered 4-Å molecular sieves and 24 mg (0.067 mmol) of Cu(OTf)$_2$. After 10 min, a solution of 100 mg (0.067 mmol) of 3-methoxy-2-pyridyl 2,3,4-tri-*O*-benzyl-6-*O*-*t*-butyldimethylsilyl-α-D-glucopyranosyl-(1→6)-2,3,4-tri-*O*-benzoyl-β-D-glucopyranoside **10** in 1.6 mL of dichloromethane was added dropwise over 5 h. A few drops of pyridine were added, the solvent was evaporated, and the residue was chromatographed with CCl$_4$–acetone (10:1) as eluant to give the title compound (44 mg, 66%).

## REFERENCES

1. For selected reviews on oligosaccharides synthesis: (a) R. R. Schmidt and W. Kinzy, Anomeric oxygen activation for glycoside synthesis—the trichloroacetimidate method, *Adv. Carbohydr. Chem. Biochem.* 50:21 (1994), and references cited therein; (b) S. H. Khan and O. Hindsgaul, Chemical synthesis of oligosaccharides, *Molecular Glycobiology* (M. Fukuda and O. Hindsgaul, eds.), IRL Press, Oxford, 1994, p. 206; (c) K. Toshima and K. Tatsuta, Recent progress in *O*-glycosylation methods and its application to natural product synthesis, *Chem. Rev.* 93:1503 (1993); (d) J. Banoub, P. Boullanger, and D. Lafont, Synthesis of oligosaccharides of 2-amino-2-deoxy-sugars, *Chem. Rev.* 92:1167 (1992); (e) D. Bundle, Synthesis of oligosaccharides related to bacterial *O*-antigens, *Top. Cur. Chem.* 54:1 (1990); (f) R. R. Schmidt, New methods for the

synthesis of glycosides and oligosaccharides—are there alternatives to the Koenigs-Knorr method? *Angew. Chem. Int. Ed. Engl. 25*:212 (1986); (g) G. Boons, Recent developments in chemical oligosaccharide synthesis, *Contemp. Org. Synth. 3*:173 (1996).

2. K. C. Nicolaou, R. E. Dolle, D. P. Papahatjis, and J. L. Randall, Practical synthesis of oligosaccharides. Partial synthesis of avermectin $B_{1a}$, *J. Am. Chem. Soc. 106*:4189 (1984).

3. R. L. Halcomb and S. J. Danishefsky, On the direct epoxidation of glycals: Application of a reiterative strategy for the synthesis of β-linked oligosaccharides, *J. Am. Chem. Soc. 111*:6656 (1989).

4. S. J. Danishefsky, K. F. McClure, J. T. Randolph, and R. B. Ruggeri, A strategy for the solid-phase synthesis of oligosaccharides, *Science 260*:1307 (1993).

5. D. R. Mootoo, P. Konradsson, U. Udodong, and B. Fraser-Reid, "Armed" and "disarmed" *n*-pentenyl glycosides in saccharide coupling leading to oligosaccharides, *J. Am. Chem. Soc. 110*:5583 (1988).

6. G. H. Veeneman and J. H. van Boom, An efficient thioglycoside-mediated formation of glycosidic linkages promoted by iodonium dicollidine perchloride, *Tetrahedron Lett. 31*:275 (1990).

7. H. Paulsen, A. Richter, V. Sinnwell, and W. Stenzel, Darstellung selecktiv Blockierter 2-Azido-2-desoxy-D-gluco und -D-Galactopyranosylhalogenide: Reackitivitat und $^{13}C$ NMR-Spektren. *Carboyhdr. Res. 64*:339(1978).

8. P. Konradsson, D. R. Mootoo, R. E. McDevitt, and B. Fraser-Reid, Iodonium ion generated in situ from *N*-iodosuccinimide and trifluoromethanesulphonic acid promotes direct linkage of "disarmed" pent-4-enyl glycosides, *J. Chem. Soc. Chem. Commun.* p. 270 (1990).

9. S. Hanessian, B. Lou, and H. Wang, unpublished results.

10. R. Roy, F. O. Andersson, and M. Letellier, "Active" and "latent" thioglycosyl donors in oligosaccharide synthesis. Application to the synthesis of α-sialosides, *Tetrahedron Lett. 33*:6053 (1992).

11. S. Mehta and B. M. Pinto, Novel glycosidation methodology. The use of phenyl selenoglycosides as glycosyl donors and acceptors in oligosaccharide synthesis, *J. Org. Chem. 58*:3269 (1993).

12. L. A. J. M. Sliedregt, G. A. van der Marel, and J. H. van Boom, Trimethylsilyl triflate mediated chemoselective condensation of arylsulfenyl glycosides, *Tetrahedron Lett. 35*:4015 (1994).

13. S. Raghavan and D. Kahne, A one-step synthesis of the ciclamycin trisaccharide, *J. Am. Chem. Soc. 115*:1580 (1993).

14. H. Yamada, T. Harada, and T. Takahashi, Synthesis of an elicitor-active hexaglucoside analogue by a one-pot, two step glycosidation procedure, *J. Am. Chem. Soc. 116*:7919 (1994).

15. (a) S. Houdier and P. J. A. Vottero, Synthesis of benzylated cycloisomaltotri- and -hexaoside, *Angew. Chem. Int. Ed. Engl. 33*:354 (1994); (b) H. K. Chenault and A. Castro, Glycosyl transfer by isopropenyl glycosides: Trisaccharide synthesis in one-pot by selective coupling of isopropenyl and *n*-pentenyl glycopyranosides, *Tetrahedron Lett. 35*:9145 (1994); (c) R. K. Jain and K. L. Matta, Synthesis of oligosaccharides containing the X-antigenic trisaccharide (α-L-Fuc*p*-(1-3)-[β-D-Gal*p*-(1-4)-β-D-Glc*p*NAc) at their nonreducing ends, *Carbohydr. Res. 226*:91 (1992); (d) M. Mori, Y. Ito, J. Uzawa, and T. Ogawa, Stereoselectivity of cycloglycosylation in *manno*-oligose series depends on carbohydrate chain length: Syntheses of *manno* isomers of β-and γ-cyclodextrins, *Tetrahedron Lett. 31*:3191 (1990); (e) H. Lönn, Synthesis of a tri- and a heptasaccharide which contain α-L-flucopyranosyl groups and are part of the complex type of carbohydrate moiety of glycoprotiens, *Carbohydr. Res. 139*:105 (1985).

16. O. Kanie, Y. Ito, and T. Ogawa, Orthogonal glycosylation strategy in oligosaccharide synthesis, *J. Am. Chem. Soc. 116*:12073 (1994).

17. (a) S. Hanessian, J. J. Conde, and B. Lou, The stereocontrolled synthesis of 1,2-*cis*-furanosyl nucleosides via a novel anomeric activation, *Tetrahedron Lett. 36*:5865 (1995); (b) S. Hanessian, J. J. Conde, and B. Lou, The stereocontrolled synthesis of 1,2-*trans*-hexopyranosyl nucleosides via a novel anomeric activation, *Tetrahedron* (in press).

18. S. Houdier and P. Vottero, Steric effect of a bulky 6-substituent in the $I^+$-promoted glycosylation with pent-4-enyl and thioethyl glycosides, *Carbohydr. Res. 232*:349 (1992).
19. S. Shibata and K. Miura, A third glycopeptide (nephritogenoside) isolated from glomerular basement membrane, *J. Biochem. 89*:1737 (1981).
20. (a) H. Zhang, Y. Wang, and W. Voelter, A new strategy for the synthesis of the nephritogenoside trisaccharide unit using phenylsulfenyl donors, *Tetrahedron Lett. 36*:1243 (1995); (b) T. Teshima, K. Nakajima, M. Takahaski, and T. Shiba, Total synthesis of nephritogenic glycopeptide, nephritogenoside, *Tetrahedron Lett. 33*:363 (1992); (c) M. Sasaki, K. Tachibana, and H. Nakanishi, An efficient and stereocontrolled synthesis of the nephritogenoside core structure, *Tetrahedron Lett. 32*:6873 (1991); (d) A. J. Ratcliffe, R. Konradsson, and B. Fraser-Reid, *n*-Pentenyl glycosides as efficient synthons for promoter-mediated assembly of *N*-α-linked glycoproteins, *J. Am. Chem. Soc. 112*:5665 (1990); (e) T. Takeda, A. Utsuno, N. Okamoto, and Y. Ogihara, Synthesis of the α and β anomer of an *N*-triglycosyl dipeptide, *Carbohydr. Res. 207*:71 (1990).
21. S. Hanessian, E. Eckhardt, and B. Lou, unpublished results; for a review see, Z. Györgydeák, L. Szilágyi, and H. Paulsen, Synthesis, structure and reactions of glycosyl azides, *J. Carbohydr. Res. 12*:139 (1993).
22. T. Takeda, Y. Sugiura, Y. Ogihara, and S. Shibata, The nephritogenic glycopeptide from rat glomerular basement membrane: Synthesis of α-D-glucopyranosylamine derivatives, *Can. J. Chem. 58*:2600 (1980).

# IV
# Enzymatic Synthesis of Sialic Acid, KDO, and Related Deoxyulosonic Acids, and of Oligosaccharides

**THEMES**: *Aldolase-catalyzed Reactions; Glycosyl Transferase-catalyzed Reactions*

As a Ph.D. student working under the direction of the late M. L. Wolfrom at the Ohio State University, I had successfully completed most of my requirements, including a thesis dealing with the reaction of carbonyl sugar derivatives with organometallic reagents. This was the first example of the application of Cram's rules to explain stereochemical control, currently known as a "chelation-controlled" mechanism. I had synthesized a derivative of L-idose, then (and even now) considered a "rare" sugar. A last requirement before graduation was a preceptorial examination given internally by "Doc." My assignment was a take-as-long-as-you-want session to discuss the chemistry of L-gulose, another rare sugar. Soon after graduation, I discovered at Parke-Davis in Ann Arbor, Michigan, that rare sugars were not just those you could not buy, but also those you could not easily make. Hence, my introduction to sialic acid and 3-deoxy-D-*manno*-2-octulosonic acid (KDO), compounds needed in my project, which I could make in minuscule quantities and after much effort. Today, L-idose is still a rare sugar, but sialic acid and KDO can be obtained in gram or kilogram quantities, obviously for a price.

The difference between then and now can be attributed in major part to the advent of enzymes in organic synthesis. Aldolases are nature's catalysts for the "poor man's" chemical aldol reaction. These remarkable enzymes can produce rare and important deoxyulosonic acids such as sialic acid and KDO on a small or large scale.

The natural synthesis of oligosaccharides is performed in a stepwise manner using glycosyl transferases that are specific to the donor sugar. Unlike peptide or protein synthesis, which takes place on a template on a ribosomal unit, oligosaccharides are assembled by specific glycosyl transferases. The enzymatic synthesis of complex oligosaccharies, previously accessible only through multistep chemical processing, has revolutionized this area of glycotechnology. Although the methods are not universal, they work remarkably well in those instances for which the transferases are available and when cofactors can be regenerated in situ. Research in this area is enabling chemists to have access to rare oligosaccharides of therapeutic potential. One pertinent example is the enzymatic synthesis of sialyl Lewis$^x$, a tetrasaccharide expressed on the surface of leukocytes and implicated in some fascinating interactions with proteins of endothelial cells. A new research area has emerged in the synthesis of sialyl Lewis$^x$ analogues and mimetics (glycomimetics), in an effort to develop a therapeutic agent against pain, inflammation, and other disease states.

My prediction is that if carbohydrate-like molecules become contenders as drug candidates, enzyme-based technology for their synthesis on an industrial scale may be the method of choice.

The following two chapters on the use of enzymes in the synthesis of rare sugars, and in the assembly of oligosaccharides should dispel all fears and prejudices of the uncommitted and the skeptics in this area.

*Stephen Hanessian*

# 21

# Enzymatic Synthesis of Carbohydrates

**Claudine Augé and Christine Gautheron-Le Narvor**
*Institut de Chimie Moléculaire d'Orsay, Université Paris-Sud, Orsay, France*

|     |                                                              |     |
|-----|--------------------------------------------------------------|-----|
| I.  | Introduction                                                 | 469 |
| II. | Methods                                                      | 471 |
|     | A. Sialic acid aldolase                                      | 471 |
|     | B. Other aldolases using pyruvate                            | 475 |
| III.| Experimental Procedures                                      | 478 |
|     | A. N-Acetylneuraminic acid                                   | 478 |
|     | B. 9-O-Acetyl-N-acetylneuraminic acid                        | 479 |
|     | C. 3-Deoxy-D-*glycero*-D-*galacto*-nonulosonic acid (KDN)    | 479 |
|     | D. 3-Deoxy-D-*manno*-octulosonic acid (KDO)                  | 480 |
|     | E. 3-9-Dideoxy-L-glycero-L-*galacto*-nonulosonic acid        | 480 |
|     | F. 3-Deoxy-D-*threo*-hexulosonic acid                        | 481 |
|     | G. 3-Deoxy-D-*arabino*-heptulosonic acid                     | 481 |
|     | References                                                   | 482 |

## I. INTRODUCTION

Enzymes are being recognized as efficient catalysts for many of the stereospecific and regioselective reactions necessary for carbohydrate synthesis [1–4]. Thus, natural and cloned glycosyltransferases are now being used in oligosaccharide synthesis, whereas aldolases belonging to the class of lyases have been applied in preparative synthesis of common monosaccharides and analogues. These enzymes, which have mostly catabolic function in vivo, catalyze reversible stereoselective aldol reactions between a nucleophilic donor and an electrophilic acceptor. Excluding deoxyribose aldolase that catalyzes aldol condensation between two aldehydes [5], aldolases can be classified into three major classes, according to the type of donor substrate required: pyruvate-dependent, phosphoenolpyruvate-dependent, and dihydroxyacetone phosphate-dependent lyases. The latter class has been extensively investigated; the most widely used enzyme is fructose diphosphate aldolase (FDPA), which catalyzes the reversible aldol addition of dihydroxyacetone phosphate (DHAP) and D-glyceraldehyde-3-phosphate to form 1,6-diphosphate fructose, thus creating a new C3–C4 bond with D-*threo* stereochemistry (Scheme 1) [6].

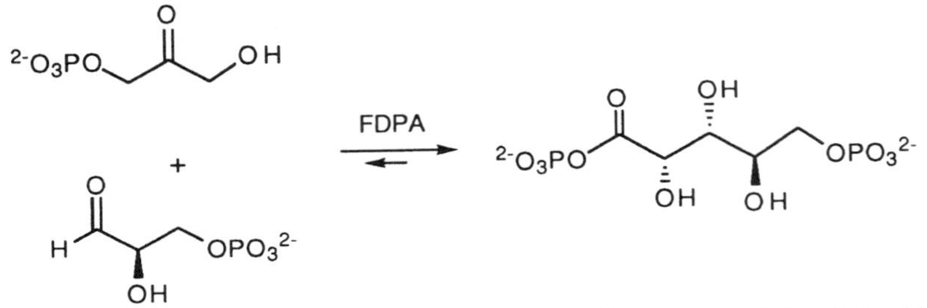

**Scheme 1** Reversible aldol addition reaction catalyzed by fructose diphosphate aldolase.

Besides FDPA, three other aldolases using DHAP as the donor are known; each aldolase generates a new C3–C4 bond with a different stereochemistry: D-*erythro* for fuculose-1-phosphate aldolase, L-*threo* for rhamnulose 1-phosphate aldolase, and D-*erythro* for tagatose 1,6-diphosphate aldolase [7]. These aldolases accept a great variety of electrophilic substrates, which has been widely exploited in synthesis of sugar analogues [8,9].

This chapter deals with the other group of aldolases that catalyzes the reversible aldol reaction of pyruvate as the nucleophilic donor and a sugar as the electrophilic acceptor. Table 1 lists the main aldolases using pyruvate that have been examined for synthetic

**Table 1** Main Aldolases Using Pyruvate[a]

| | Substrates | Products | Ref. |
|---|---|---|---|
| Sialic acid aldolase | 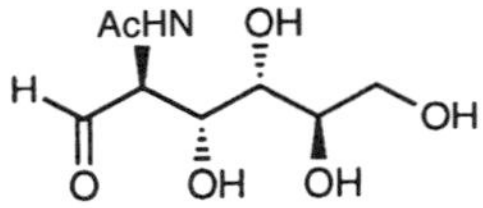 | 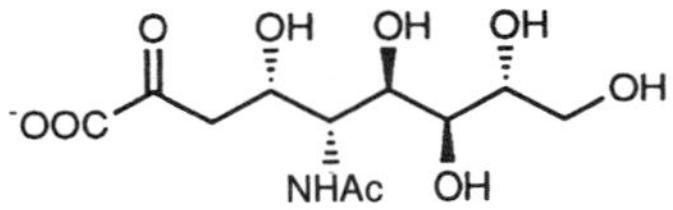 | 10 |
| KDO aldolase | 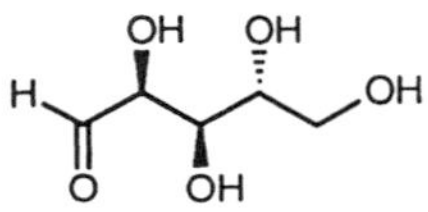 | 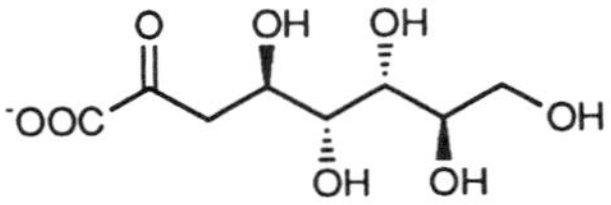 | 11 |
| KDPG aldolase | 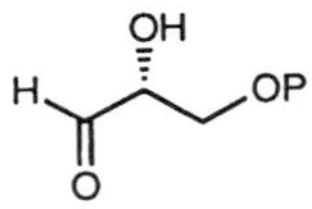 | 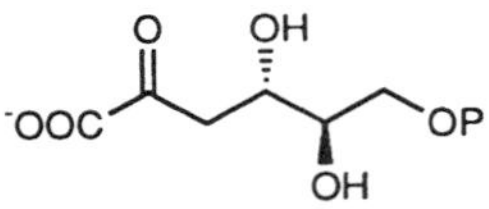 | 12 |
| KDG aldolase | 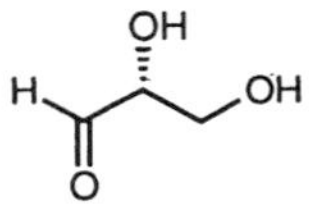 | 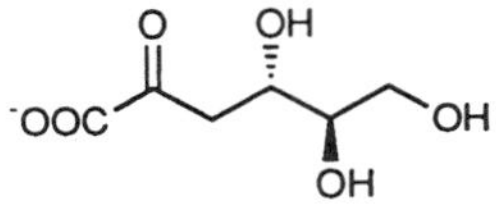 | 13 |
| KDgal aldolase | 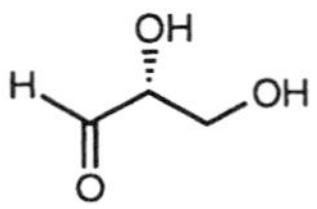 | 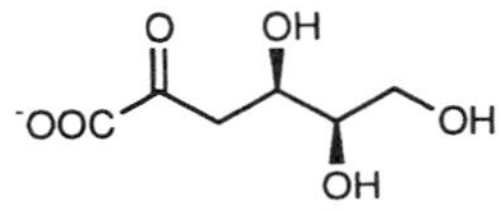 | 14 |

[a]The references refer to the reactions these aldolases catalyze *in vivo*.

purposes. This type of aldolases generates a new chiral center on C-4 and leads to 3-deoxy-2-keto-ulosonic acids.

The most representative example of this family is N-acetylneuraminic acid (Neu5Ac 1; Scheme 2), a nine-carbon sugar, ubiquitous in the animal kingdom and especially

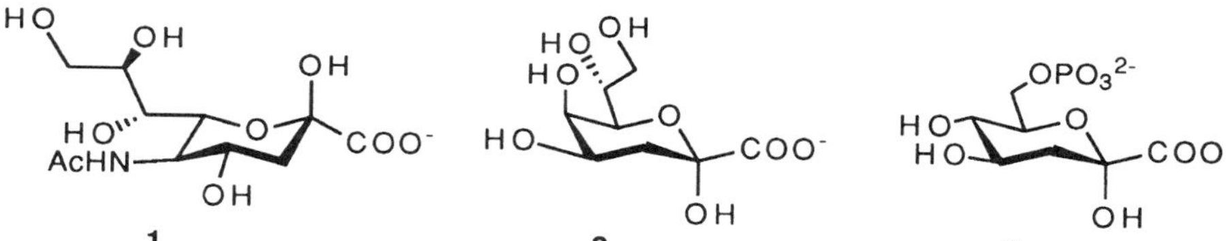

**Scheme 2** The most important 3-deoxy-2-keto-ulosonic acids.

involved in mammals in a great variety of recognition phenomena, including infection, cell adhesion, and metastasis [15]. Natural sialic acids are derived from N-acetylneuraminic acid, the parent molecule by esterification, mostly acetylation of the hydroxyl functions.

In bacterial glycoconjugates, 3-deoxy-D-manno-2-octulosonic acid (KDO; 2) occurs as the essential component in all lipopolysaccharides (LPS) of gram-negative bacteria [16]. Because the incorporation of KDO is a vital step in LPS biosynthesis, the synthesis of KDO analogues has become of great interest for developing a new class of antibiotic against gram-negative bacteria. Moreover the seven-carbon homologue, 2-deoxy-arabino-heptulosonic acid phosphate (DAHP) is the first intermediate in the biosynthetic route of aromatic amino acids by the shikimate pathway in plants and bacteria [17].

Enzymatic synthesis relying on the use of aldolases offers several advantages. As opposed to chemical aldolization, aldolases usually catalyze a stereoselective aldol reaction under mild conditions; there is no need for protection of functional groups and no cofactors are required. Moreover, whereas high specificity is reported for the donor substrate, broad flexibility toward the acceptor is generally observed. Finally, aldolases herein discussed do not use phosphorylated substrates, contrary to phosphoenolpyruvate-dependent aldolases involved in vivo in the biosynthetic pathway, such as KDO synthetase or DAHP synthetase [18,19].

## II. METHODS

### A. Sialic Acid Aldolase

Sialic acid aldolase (SA; EC 4.1.3.3), also named N-acetylneuraminate pyruvate lyase, has been extensively used by our group in its immobilized form, first for the synthesis of large amounts of N-acetylneuraminic acid [20] and then for many natural and unnatural sialic acids [21]. SA catalyzes the reversible aldol reaction of N-acetylmannosamine and pyruvate to give N-acetylneuraminic acid; with an optimum pH for activity of 7.5 and an equilibrium constant of 12.7 $M^{-1}$ in the synthetic direction (Scheme 3) [10].

Therefore, to achieve high conversion of the substrate a tenfold excess of pyruvate is usually needed. The enzymes from *Clostridium perfringens* and *Escherichia coli* are commercially available from Toyobo; the *E. coli* enzyme has been cloned and overexpressed, which has considerably reduced its cost [22,23]. Sodium borohydride inactivates the enzyme in the presence of either sialic acid or pyruvate, indicating that the enzyme belongs to the Schiff-base–forming class 1 aldolase. This aldolase was supposed to be a

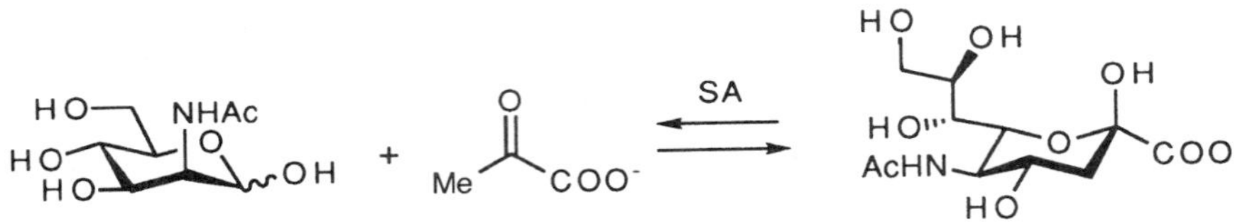

**Scheme 3** Reversible aldol addition reaction catalyzed by sialic acid aldolase.

trimer, similar to the KDPG aldolase of *Pseudomonas putida* and KHG aldolase of *E. coli*, but the recently reported crystallographic structure of the enzyme has shown a tetramer [24]. Lysine 165 has been evidenced in the active site pocket and is presumably involved in formation of the enamine intermediate with pyruvate.

In large-scale synthesis of *N*-acetylneuraminic acid **1**, a mixture of *N*-acetylglucosamine **4** and *N*-acetylmannosamine **5** (for numbering of the compounds see Sec. III) prepared in an inexpensive way by alkaline epimerization of the former sugar, is incubated in the presence of immobilized SA and pyruvate in excess. The immobilization procedure involves the covalent attachment of the enzyme to a support—either agarose activated with cyanogen bromide or epoxide-containing acrylamide beads (Eupergit C)—other groups used cross-linked copolymer of acrylamide and acryloxysuccinimide (PAN) [25]. Besides enzyme stabilization, immobilization allows an easy recovery of the enzyme, which can be reused in several successive batch reactions with very little decrease of enzymatic activity. Alternatively the soluble enzyme can be used enclosed in a dialysis bag [26]. Moreover, the use of a continuous-flow stirred tank, equipped with an ultrafiltration membrane retaining the soluble enzyme (so-called enzyme membrane reactor) has been successfully applied to large-scale processes, in particular in the combined use of sialic acid aldolase and *N*-acylglucosamine 2 epimerase [27].

6-*O*-Acetyl-*N*-acetylmannosamine **6**, prepared from *N*-acetylmannosamine either by chemical acetylation [28] or by transesterification catalyzed by subtilisin [31], led to 9-*O*-acetyl Neu5Ac **7** [29], a receptor of influenza C virus occurring on human erythrocytes. Several other 9-*O*-substituted Neu5Ac derivatives could also be prepared [29–32].

The sialic acid aldolase-catalyzed condensation of D-mannose **8** and pyruvate led, in an excellent yield, to the synthesis of KDN **9** [33], a natural deaminated neuraminic acid first isolated from rainbow trout eggs [34] and then discovered in other species. The discovery that sialic acid aldolase accepts as substrates D-mannose substituted on the 2-position, even by bulky substituents such as phenyl, azido, or bromine, opened the route to novel unnatural sialic acid derivatives [35–39]. Pentoses also are substrates. *N*-Substituted neuraminic acids could be prepared either directly from the corresponding *N*-substituted mannosamine, such as *N*-thioacyl derivatives [40], or after reduction and acylation of 5-azido-KDN [41]. Recently, *N*-carbobenzyloxy-D-mannosamine was converted, in a good yield, into the *N*-carbobenzyloxy-neuraminic acid, further used as a precursor of a derivative of castanospermine [42].

Table 2 lists 37 sialic acid derivatives and analogues that have been synthesized with sialic acid aldolase by our group and others. This is a nice illustration of the great synthetic potential of the enzyme. In all of these examples, sialic acid derivatives or analogues with equatorial hydroxyl on C-4 are formed, corresponding to a new 4$S$ chiral center and attack of the pyruvate from the *si* face of the aldehyde. In summary sialic acid aldolase accepts modifications on the 2, 4, 5, and 6 positions of the substrate without any change of its stereoselectivity.

On the other hand, substrates epimerized on C-3 afford KDO-type products (in the

# Enzymatic Synthesis of Carbohydrates

**Table 2** Natural Sialic Acids and Analogues Synthesized with Sialic Acid Aldolase

| $R_1$ | $R_2$ | $R_3$ | $R_4$ | $R_5$ | $R_6$ | $R_7$ | Ref. |
|---|---|---|---|---|---|---|---|
| NHAc | H | OH | H | $CH_2OAc$ | OH | H | 28,29 |
| NHAc | H | OH | H | $CH_2OCOCHOHCH_3$ | OH | H | 29 |
| NHAc | H | H | H | $CH_2OH$ | OH | H | 44 |
| NHAc | H | $OCH_3$ | H | $CH_2OH$ | OH | H | 29,44 |
| NHAc | H | OH | H | $CH_2OCH_3$ | OH | H | 29 |
| NHAc | H | OH | H | $CH_2F$ | OH | H | 30 |
| NHAc | H | OH | H | $CH_2OH$ | H | H | 43 |
| NHAc | H | OH | H | $CH_2OP(O)Me_2$ | OH | H | 31 |
| NHAc | H | OH | H | $CH_2OCOCH_2NHBoc$ | OH | H | 32 |
| NHAc | H | $N_3$ | H | $CH_2OH$ | H | H | 45 |
| OH | H | OH | H | $CH_2OH$ | OH | H | 33 |
| H | H | OH | H | $CH_2OH$ | OH | H | 35 |
| $N_3$ | H | OH | H | $CH_2OH$ | OH | H | 36 |
| Ph | H | OH | H | $CH_2OH$ | OH | H | 37 |
| Br | H | OH | H | $CH_2OH$ | OH | H | 38 |
| OH | H | OH | H | H | OH | H | 35 |
| H | OH | OH | H | $CH_2OH$ | OH | H | 35 |
| H | OH | OH | H | H | OH | H | 35 |
| OH | H | H | H | $CH_2OH$ | OH | H | 35 |
| $CH_2NHAc$ | H | OH | H | $CH_2OH$ | OH | H | 39 |
| OH | H | H | OH | $CH_2OH$ | OH | H | 46 |
| H | OH | H | OH | $CH_2OH$ | OH | H | 47 |
| H | OH | H | OH | Me | OH | H | 47 |
| H | OH | H | OH | H | OH | H | 46 |
| OH | H | OH | H | $CH_2OH$ | H | H | 43 |
| OH | H | H | OH | $CH_2OH$ | H | OH | 48 |
| OH | H | OH | H | $CH_2OH$ | H | OH | 46 |
| OH | H | H | F | $CH_2F$ | OH | H | 31 |
| H | F | OH | H | $CH_2OH$ | OH | H | 31 |
| H | H | OH | H | H | OH | H | 43 |
| $NHCOCH_2OH$ | H | OH | H | $CH_2OH$ | OH | H | 29 |
| $NHCOCH_2OH$ | H | OH | H | $CH_2OAc$ | OH | H | 29 |
| $NHCOCH_2OAc$ | H | OH | H | $CH_2OH$ | OH | H | 29 |
| NHCbz | H | OH | H | $CH_2OH$ | OH | H | 42 |
| NHC(S)H | H | OH | H | $CH_2OH$ | OH | H | 40 |
| $NHC(S)CH_3$ | H | OH | H | $CH_2OH$ | OH | H | 40 |
| $NHC(S)CH_2CH_3$ | H | OH | H | $CH_2OH$ | OH | H | 40 |

$^5C_2$ conformation, unlike Neu5Ac occurring in the $^2C_5$ conformation), because the conformation of the aldol condensation products is due to the stereochemistry of the substrate on C-3. But in addition to that, the 3-position on the substrate is critical for the enzyme stereoselectivity. Indeed we first observed that condensation of D-arabinose with pyruvate in tenfold excess led to two diastereomers, 4-*epi*-KDO and KDO in the 56:44 ratio (Scheme 4) [35].

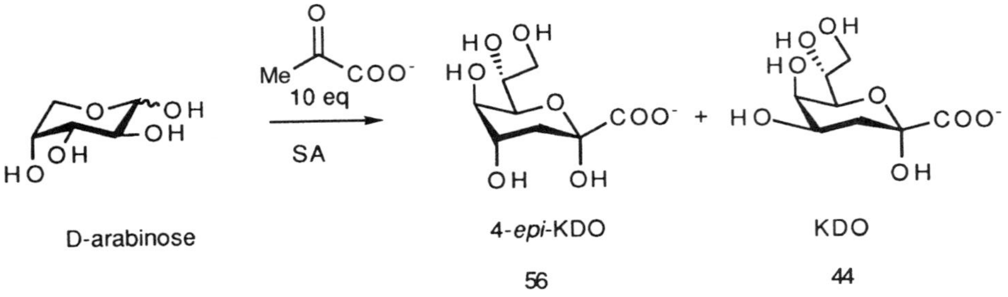

**Scheme 4** Lack of steroselectivity observed in the sialic acid aldolase-catalyzed addition of pyruvate with D-arabinose.

By reversing the ratio of acceptor/donor, namely by using D-arabinose in a 25-fold excess, the percentage of KDO, the product of the inverted enzyme stereoselectivity, could be increased up to 83% [48]. The reason for that is still unclear, but this interesting result was exploited in the preparation of KDO **2**, which was achieved using the enzyme membrane reactor technique [48]. Compound **2** could be separated from its epimer according to the procedure previously described [16].

Moreover a complete inversion of stereoselectivity has been reported with L-mannose and L-rhamnose [49]. Table 3 lists KDO analogues with nine carbons that could be prepared as pure compounds using sialic acid aldolase. They are derived from L-hexoses belonging to L series with the *R* configuration at the 3- and 2-positions. In each case, KDO-type derivatives with equatorial hydroxyl on C-4 are formed, corresponding to a 4*R* chiral center and attack of the pyruvate from the *re* face of the aldehyde. Thus L-KDN

**Table 3** Pure KDO Analogues Synthesized Using Sialic Acid Aldolase

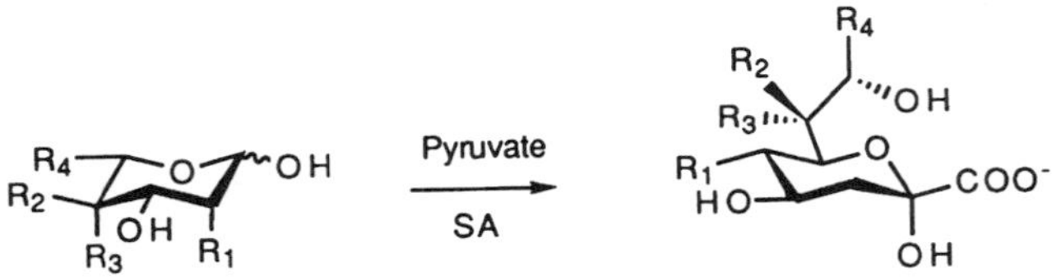

| $R_1$ | $R_2$ | $R_3$ | $R_4$ | Ref. |
|---|---|---|---|---|
| NHAc | OH | H | $CH_2OH$ | 46 |
| OH | OH | H | $CH_2OH$ | 49 |
| OH | OH | H | $CH_3$ | 49 |
| H | OH | H | $CH_2OH$ | 46 |
| H | OH | H | Me | 46 |
| OH | H | OH | $CH_2OH$ | 46 |

# Enzymatic Synthesis of Carbohydrates

**Table 4** KDO Analogues Synthesized as Diastereomeric Mixtures Using Sialic Acid Aldolase

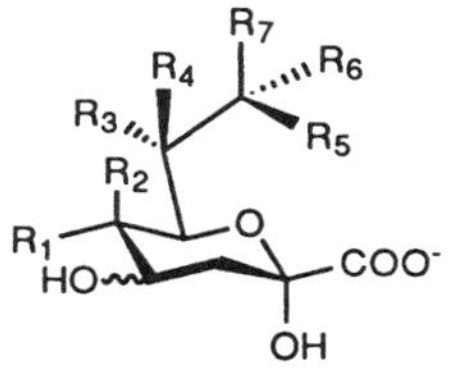

| $R_1$ | $R_2$ | $R_3$ | $R_4$ | $R_5$ | $R_6$ | $R_7$ | Ref. |
|---|---|---|---|---|---|---|---|
| H | OH | OH | H | H | H | H | 35,48 |
| H | OH | OH | H | OH | H | $CH_2OH$ | 48 |
| OH | H | H | OH | OH | H | $CH_2OH$ | 46 |
| OH | H | OH | H | H | H | H | 48 |
| H | OH | H | OH | H | H | H | 48 |
| H | OH | OH | H | H | OH | Me | 47 |
| $N_3$ | H | H | OH | H | OH | $CH_2OH$ | 46 |
| OH | H | H | OH | H | H | H | 47 |

($R_1$ = OH, $R_3$ = H, $R_2$ = OH, $R_4$ = $CH_2OH$ in Table 3), the enantiomeric compound of the one just reported could be easily prepared. Aldol condensation products were obtained as diastereomeric mixtures from L-sugars, such as L-fucose, L-xylose, L-lyxose, and D-sugars epimeric to D-mannose relative to the 3-position, such as D-allose and D-gulose [46–48]. Table 4 lists the corresponding aldol condensation products isolated as diastereomeric mixtures. Also, 3-deoxy-D-mannose by condensation with pyruvate gave a diastereomeric mixture of 6-deoxy-KDN furanose derivatives [43]. All these results confirm that sialic acid aldolase, similar to other aldolases, exhibits broad specificity toward the electrophilic acceptor; on the other hand, only pyruvate was reported acceptable as the donor [10]. But very recently, in contradiction to that, 3-fluoro-Neu5Ac and 3-fluoro-KDN could be prepared by the sialic acid aldolase-catalyzed condensation of 3-fluoropyruvate and N-acetylmannosamine or D-mannose (Scheme 5) [47].

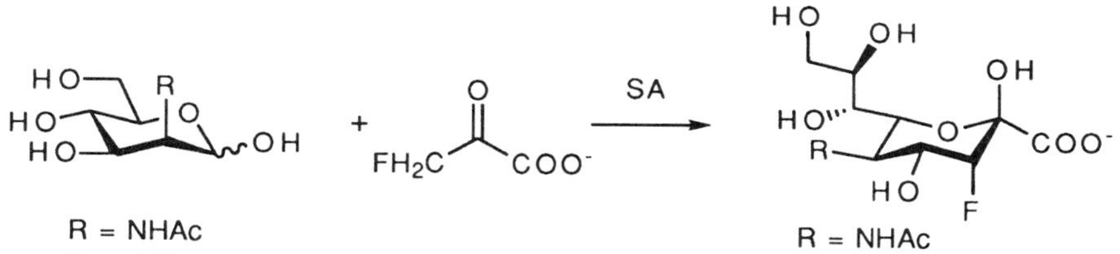

**Scheme 5** Aldol addition of fluoropyruvate with N-acetylmannosamine or mannose catalyzed by sialic acid aldolase.

## B. Other Aldolases Using Pyruvate

### KDO Aldolase

The KDO aldolase (EC 4.1.2.23), which catalyzes the reversible condensation of pyruvate with D-arabinose to form KDO **2** (Scheme 6), was first isolated from, *Enterobacter cloacae*

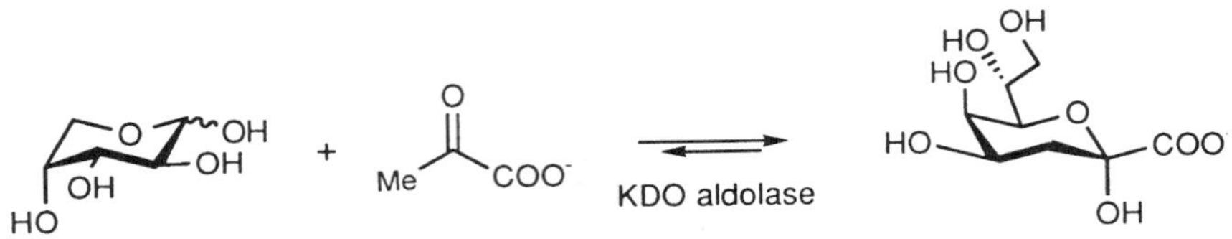

**Scheme 6** Reversible aldol addition reaction catalyzed by KDO aldolase.

(*Aerobacter cloacae*) grown on synthetic KDO [11], but no further use of the enzyme has been reported.

Recently, the enzyme from *Aureobacterium barkerei*, strain KDO 372, has been partially purified and used in the preparative synthesis of KDO. The pyruvate attacks from the *re* face of aldehyde, creating a new *R* chiral center at C-4; thus, the facial selectivity of KDO aldolase is complementary to the one of sialic acid aldolase. Moreover as the enzyme accepts some flexibility toward the acceptor, several other analogues with 6-, 7-, or 9-carbon atoms could be prepared, such as L-KDN and D-DAH [50].

*KDPG Aldolase*

The 2-keto-3-deoxy-6-phosphogluconate aldolase (KDPG aldolase; EC 4.1.2.14) catalyzes the cleavage of the dehydration product of 6-phosphogluconate, (KDPG), into glyceraldehyde-3-phosphate and pyruvate in the Entner–Doudoroff pathway (Scheme 7, R = $PO_3H_2$). This

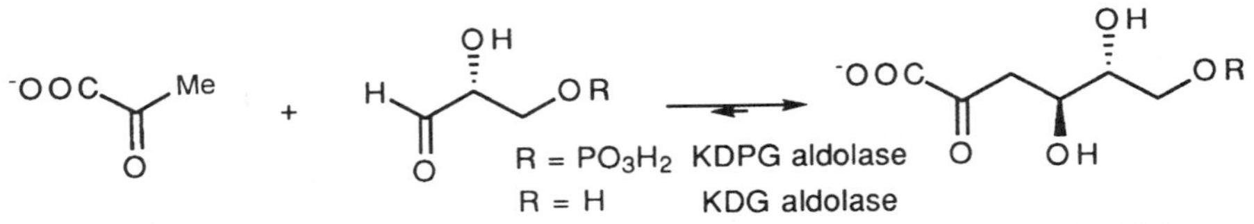

**Scheme 7** Reversible aldol addition reaction catalyzed by KDPG and KDG aldolases.

pathway is widely distributed among prokaryotes, and this aldolase has been purified from several bacteria. The *Pseudomonas fluorescens* enzyme has been investigated for synthetic purposes [51]. KDPG aldolase stereospecifically generates a new chiral center at C-4 with an *S*-configuration, and the equilibrium constant favors the condensation reaction ($K = 10^3$ $M^{-1}$), but substrate specificity studies have shown that nonphosphorylated substrates are accepted at much lower rates than the natural one, which limits the synthetic potential of the enzyme.

*KDG and KDGal Aldolases from Fungi*

In the metabolism of some filamentous fungi, a modified Entner–Doudoroff pathway involving nonphosphorylated intermediates has been reported [13]. Thus, the KDG aldolase induced in *Aspergillus niger*, grown on 2% D-gluconate as the sole source of carbon, catalyzes the reversible aldol reaction between pyruvate and D-glyceraldehyde, leading to KDG (See Scheme 7; R = H). However, the reaction was not completely stereoselective, suggesting either a lack of stereoselectivity of the enzyme or the occurrence of two distinct aldolases with complementary facial stereoselectivity [52]. Another fungus, *Aspergillus terreus*, grown on D-galactonate, has been reported to be able to metabolize D-galactonate, according to Scheme **8**.

This reaction was used in the synthetic direction: condensation of D-glyceraldehyde with pyruvate in the presence of *A. terreus* extracts led to a diastereomerically pure com-

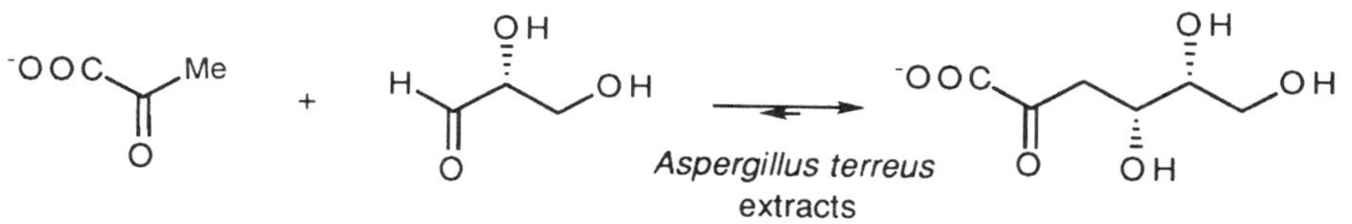

# Enzymatic Synthesis of Carbohydrates

**Scheme 8** Reversible aldol addition reaction catalyzed by *Aspergillus terreus* KDGal aldolase.

pound, identified as 3-deoxy-D-*threo*-2-hexulosonic acid (KDGal) [53]. The KDGal aldolase occurring in *A. terreus* creates a new asymmetric center of *R*-configuration, resulting from the *re* face attack of the aldehyde, thereby exhibiting the same selectivity as KDO aldolase. Several other aldehydes were tested as electrophiles. Table 5 reports the rates of formation of the corresponding keto-deoxy-ulosonic acids relative to D-glyceraldehyde. For comparison, the relative rates for KDPG aldolase are also given. Thus, clearly, the involvement of nonphosphorylated substrates is a major advantage of this fungal aldolase. The high relative rates measured with other hydroxylated aldehydes allowed the preparation of biologically significant compounds, in particular DAH **3** that could be easily obtained from D-erythrose on 0.5-g scale in 60% yield; likewise, its 5-epimer could be prepared from D-threose in 85% yield. The availability in this KDGal aldolase through cloning techniques would certainly allow one to extend its synthetic applications.

As representative examples the following enzymatic syntheses are reported in the experimental section:

1. The large-scale synthesis of *N*-acetylneuraminic acid
2. The synthesis of 9-*O*-acetyl *N*-acetylneuraminic acid
3. The synthesis of KDN
4. The preparation of KDO by using a large excess of D-arabinose
5. The condensation with L-rhamnose, as an example of the complete, inverted stereoselectivity of the sialic acid aldolase
6. The synthesis of 3-deoxy-D-*threo*-hexulosonic acid
7. The synthesis of 3-deoxy-D-*arabino*-heptulosonic acid

**Table 5** Compared Relative Activities of KDGal Aldolase and KDPG Aldolase Toward Various Acceptors

| Acceptor | Relative rate[a] for KDGal aldolase | Relative rate[b] for KDPG aldolase[c] |
|---|---|---|
| D-Glyceraldehyde-3-P |  | 100 |
| D-Glyceraldehyde | 100 | 0.08 |
| Glycolaldehyde | 115 | 0.012 |
| D-Erythrose | 85 | 0.012 |
| D-Threose | 54 |  |
| D,L-Lactaldehyde | 50 | 0.2 |
| D-Arabinose | 0 |  |
| Acetaldehyde | 0 |  |
| Propionaldehyde | 0 |  |

[a]The rates are reported relative to D-glyceraldehyde
[b]The rates are reported relative to D-glyceraldehyde-3-phosphate.
[c]From Ref. 51.

## III. EXPERIMENTAL PROCEDURES*

### A. N-Acetylneuraminic Acid [21]

4 R$_1$ = NHAc, R$_2$ = H
5 R$_1$ = H, R$_2$ = NHAc

SA 67%

2-Acetamido-2-deoxy-D-glucopyranose (**4**; 85 g) was dissolved in water (400 mL); the pH of the solution was adjusted to 11 with 5 $M$ sodium hydroxide and the solution was left for 1 day at room temperature. The mixture was deionized with Dowex 50-X8 (H$^+$) resin and evaporated to dryness under vacuum. The residue was taken up with ethanol (300 mL) and heated on the steam bath with stirring. On cooling **4** crystallized (66 g); the mother liquor was concentrated, and a second crop of **4** (5 g) was obtained. Concentration of the second mother liquor gave a third crop (3.5 g) of **4**. The recovered **4** was re-treated in the same way. Both residual syrups afforded a mixture of **4** and **5** (17.1 g) containing 88% of **5** according to NMR-spectral analysis.

To the foregoing mixture of **4** and **5** (20 mmol) were added immobilized sialic acid aldolase (50 mL of gel, 68 U†), sodium pyruvate (180 mmol), 1,4-dithiothreitol (0.2 mmol), and sodium azide (20 mg) in 0.05 $M$ potassium phosphate buffer, pH 7 (150 mL). The suspension was gently stirred under nitrogen for 4 days at 37°C, the reaction being monitored by thin-layer chromatography (TLC) in 7:3 $n$–propanol–water. The gel was removed by filtration, washed with the buffer, and the filtrate and washing were chromatographed on Dowex 1X8 (HCO$_2^-$) resin, using a (0–2 M) formic acid gradient as the eluant; fractions containing compound **1** were pooled and freeze-dried (3.7 g, 67% yield): [α]$_D$ −31° ($c$ 2.15, H$_2$O); reported [54] [α]$_D$ −33° ($c$ 0.8, H$_2$O).

### Immobilization of Sialic Acid Aldolase on Eupergit 250L

Eupergit 250L (Röhm Chemie, 400 mg) was added to a solution of sialic acid aldolase (8 mg, 64 U) in 1 $M$ potassium phosphate buffer pH 7.4 (3.2 mL) containing 0.04 $M$ sodium pyruvate and 0.02% NaN$_3$; the suspension was stirred for 3 days at room temperature under N$_2$. The gel was washed with 0.1 $M$ potassium phosphate buffer pH 7 (10 mL) and stored at 4°C in this buffer in the presence of 0.04 $M$ pyruvate and 10$^{-3}$ $M$ dithiothreitol. A unit yield of 40% was usually obtained for immobilization.

### Immobilization of Sialic Acid Aldolase on Agarose

Ultrogel A4 (Sepracor; 4 mL) was washed with twice-distilled water (200 mL) and 2 $M$ phosphate buffer pH 11 (200 mL), collected, and suspended in 5 $M$ phosphate buffer pH 12 (4 mL). The suspension was diluted with twice-distillated water (4 mL), cooled to 4°C,

---

*Optical rotations were measured at 22°–25°C.

†One unit is the amount of enzyme that transforms 1 μmol of substrate per minute.

## Enzymatic Synthesis of Carbohydrates

and an aqueous solution of CNBr (100 mg/mL) was added within 2 min. The mixture was stirred for 10 min at 5°–10°C, and the insoluble material was washed with cold water until neutrality, and then with 0.1 $M$ sodium carbonate pH 8 containing 0.5 $M$ sodium chloride and immediately used for coupling with enzyme. SA (10 mg, 130 U) dissolved in 0.05 $M$ potassium phosphate buffer pH 7 (6 mL) was gently stirred overnight at 4°C under $N_2$ with freshly activated Ultrogel (40 mL) in 0.1 $M$ sodium carbonate pH 8 (100 mL), containing 0.5 $M$ sodium chloride and 0.04 $M$ sodium pyruvate to protect the enzyme active site. Then the gel was washed with 1 $M$ NaCl (1 L), twice-distilled water (1 L), 0.05 $M$ potassium phosphate pH 7 (500 mL), and stored at 4°C under $N_2$ in this buffer in the presence of pyruvate (0.04 $M$) and dithiothreitol ($10^{-3}$ $M$). An unit yield of 60% was usually obtained for immobilization.

### B. 9-O-Acetyl-N-acetylneuraminic Acid [29]

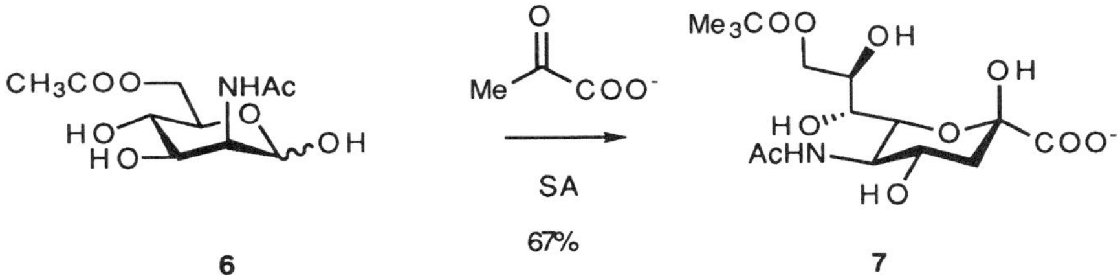

The immobilized enzyme suspension (8 mL, 32 U) was gently stirred with compound **6** (0.576 g, 2.2 mmol), sodium pyruvate (20 mmol), dithiothreitol (0.2 mmol), and $NaN_3$ (4 mg) in 0.05 $M$ potassium phosphate pH 7 at 37°C under nitrogen for 1 day. After filtration of the gel, chromatography of the solution on Dowex-1X8 formate (elution with 0–2 $M$ formic acid gradient) afforded pure **7** (0.52 g, 67%): $[\alpha]_D$ −16° ($c$ 0.83, $H_2O$); reported [55] $[\alpha]_D$ −10° ($c$ 1, $H_2O$).

### C. 3-Deoxy-D-*glycero*-D-*galacto*-nonulosonic Acid (KDN) [33]

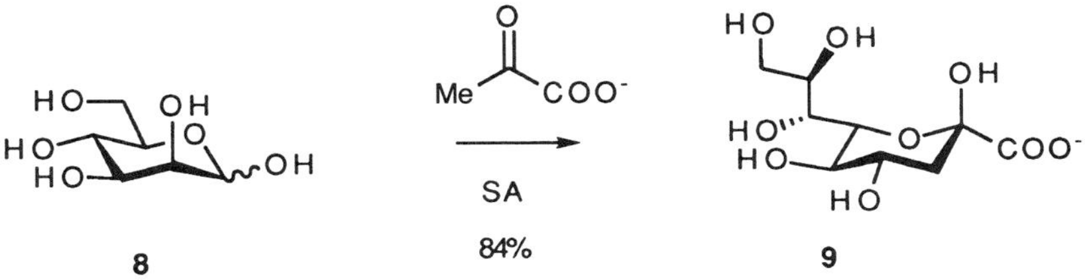

At 0.1-$M$ solution of D-mannose (**8**, 1 mmol) was treated with sodium pyruvate (10 Eq.) in the presence of sialic acid aldolase (15 U) covalently bound to 4% agarose, in 0.05 $M$ potassium phosphate buffer, pH 7.2 (20 mL) containing 0.01 $M$ dithiothreitol and 0.02% sodium azide, at 37°C under nitrogen, with gentle stirring for 1 day. After filtration of the gel, chromatography of the solution on Dowex-1X8 formate (elution with 0–2 $M$ formic acid gradient) afforded pure **9**, which was characterized as its ammonium salt, after treatment with Dowex-50 ($H^+$), neutralization to pH 7 with dilute ammonia, and freeze-drying (84% yield): $\alpha_D$ −41° ($c$ 1, $H_2O$).

## D. 3-Deoxy-D-*manno*-octulosonic Acid [48]

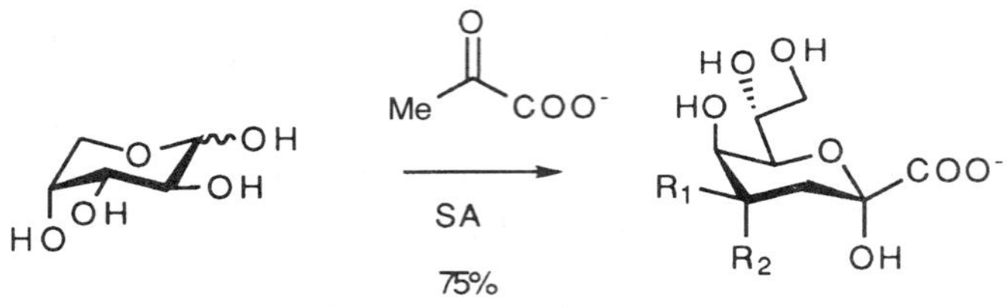

**10** → (Me-CO-COO⁻, SA, 75%) → **2** R₁ = OH, R₂ = H; **11** R₁ = H, R₂ = OH

Synthesis of KDO **2** was performed in the enzyme membrane reactor (EMR). The enzyme, in its soluble form, is retained in the reactor by an ultrafiltration membrane that is permeable only to the substrates and products. The laboratory-scale version of the EMR used here is commercially available (volume 10 mL, membrane diameter 62 mm; Bioengineering, Wald, Switzerland). An Amicon YM 5 membrane was used. The main advantages of this technique are no mass-transfer limitations, no loss of enzyme activity during immobilization steps, and high volumetric activity of enzymes. Substrates are pumped through a sterile filter at a constant flow, corresponding to a constant residence time. The setup was sterilized before use by heating to 121°C for 20 min in an autoclave. By use of phosphate buffer (0.05 $M$, pH 7.5), bovine serum albumin (10 mg) was pumped in the reactor of pretreatment of the membrane, followed by sialic acid aldolase (72 mg, 1800 U). Substrate solution (0.02 $M$ pyruvate, 0.5 $M$ D-arabinose **10**, pH 7.5) was pumped through the reactor with a flow of 5 mL h$^{-1}$, resulting in a residence time of 120 min. Temperature was kept at 25°C. For a total running time of 160 h pyruvate conversion was 75% (ratio **2/11**, 5:1 from HPLC). Both diastereoisomers could be separated by anion-exchange chromatography on Dowex 1X8, (200–400 mesh), hydrogen carbonate form, by using gradient elution with aqueous ammonium hydrogen carbonate (0–0.2 $M$). The first eluted fractions, containing KDO, were lyophilized deionized with Dowex 50 (H$^+$), neutralized to pH 7 with dilute ammonia, and again freeze-dried, affording **2** (1.8 g, 45%): $[\alpha]_D$ +41° ($c$ 1.7, H$_2$O); reported [16] $[\alpha]_D$ +40.3° ($c$ 1.93, H$_2$O); reported [56] $[\alpha]_D$ +42.3° ($c$ 1.7, H$_2$O). A mixture of **2** and **11** (0.28 g, 7%) was obtained in the next fractions, and then pure diastereomer **11** (0.22 g, 6%).

## E. 3-9-Dideoxy-L-*glycero*-L-*galacto*-nonulosonic Acid [49]

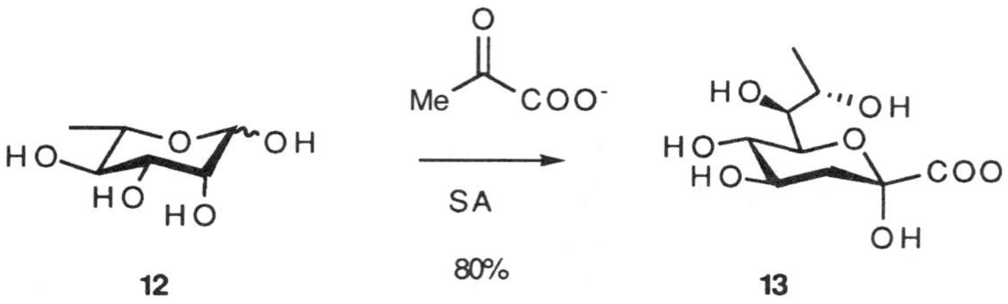

**12** → (Me-CO-COO⁻, SA, 80%) → **13**

A 0.1-$M$ solution of L-rhamnose **12** (1 mmol) in 0.05 $M$ potassium buffer, pH 7.2, containing 0.01 $M$ dithiothreitol, sodium pyruvate (3 Eq.), and 10 U NeuAc aldolase was incubated at 37°C (total volume, 10 mL) for 2 days. The reaction was monitored by TLC (*n*-propanol–water, 7:3 v/v). The product was isolated by anion-exchange chromatography on Dowex 1X8 (100–200 mesh, $HCO_2^-$); (30 × 2 cm) using a gradient of formic acid (0–2 $M$) as eluant. Fractions containing **13** were pooled and freeze-dried: 200 mg (80% yield), $[\alpha]_D$ +60° (*c* 1.2, $H_2O$).

## F. 3-Deoxy-D-*threo*-hexulosonic Acid [53]

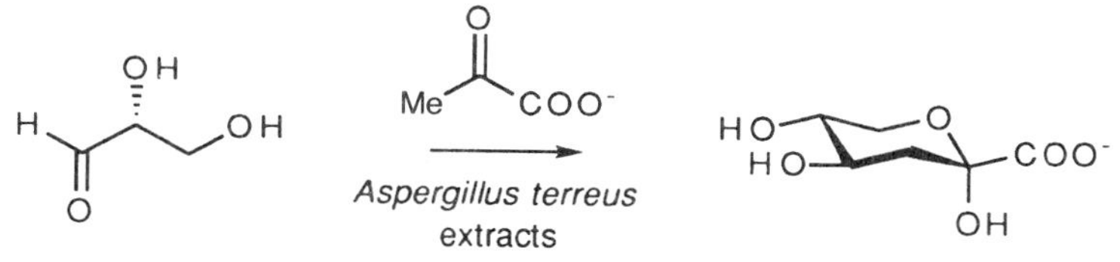

**14**      55%      **15**

D-Glyceraldehyde **14** (90 mg, 1 mmol) was incubated with sodium pyruvate (110 mg, 1 Eq.) and enzymatic extract (0.5 U, 10 mL) in 0.02 $M$ potassium phosphate buffer pH 8 at 27°C for 2 days. The product was isolated by anion-exchange chromatography on AG 1X8 resin ($HCO_3^-$, 100–200 mesh) using a 0- to 0.4-$M$ ammonium bicarbonate linear gradient. Pure fractions containing compound **15** were pooled, freeze-dried, deionized with AG 50X8 ($H^+$), neutralized with dilute ammonia, and again freeze-dried, affording **15** as its ammonium salt (102 mg, 55%): $[\alpha]_D$ +9° (*c* 1, $H_2O$); reported [57] $[\alpha]_D$ +7.9° (*c* 1.65, $H_2O$).

*Aspergillus terreus* (NRRL 265, available from the United States Department of Agriculture, Peoria, IL) was usually grown on 2% D-galactonate as the sole source of carbon, but when the fermentation was conducted on a larger scale, the concentration of D-galactonate could be lowered to 1% without any significant decrease in enzymatic activity; the mycelium was recovered after 3 days, then ground with the French Press in 0.02-$M$ potassium phosphate buffer pH 8. The crude extract was heated at 50°C for 15 min, and the supernatant obtained after centrifugation was used as the enzymatic extract.

## G. 3-Deoxy-D-*arabino*-heptulosonic Acid [53]

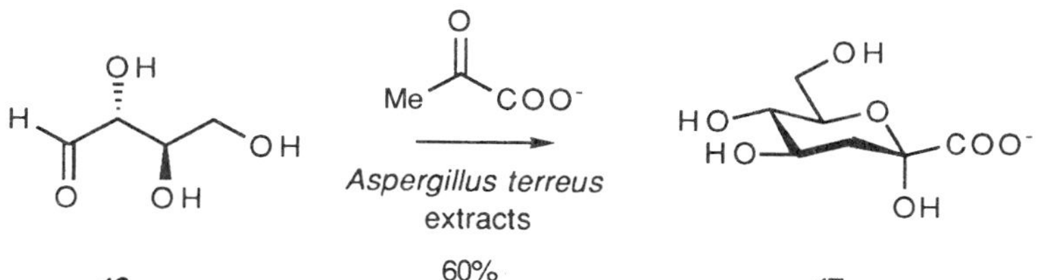

**16**      60%      **17**

D-Erythrose **16** (500 mg 4.2 mmol) was incubated with sodium pyruvate (449 mg, 1 Eq.) and enzymatic extract (5.5 U, 42 mL) in 0.02 $M$ potassium phosphate buffer pH 8.0 at 27°C for 2 days. Anion-exchange chromatography on AG 1X8 resin ($HCO_3^-$, 100–200 mesh) and elution with a 0- to 0.4-$M$ ammonium bicarbonate linear gradient gave pure fractions containing compound **17**; these fractions were pooled, freeze-dried, deionized with AG 50X8 ($H^+$), neutralized with dilute ammonia, and again freeze-dried, affording **17** as its ammonium salt (530 mg, 60%): $[\alpha]_D$ +33.5° (c 2, $H_2O$); reported [58] $[\alpha]_D$ +33° (c 1, $H_2O$).

## REFERENCES

1. E. J. Toone, E. S. Simon, M. D. Bednarski, and G. M. Whitesides, Enzyme-catalyzed synthesis of carbohydrates, *Tetrahedron 45*:5365 (1989).
2. D. G. Drueckhammer, W. J. Hennen, R. L. Pederson, C. F. Barbas, C. M. Gautheron, T. Krach, and C.-H. Wong, Enzyme catalysis in synthetic carbohydrate chemistry, *Synthesis* p. 499 (1991).
3. N. J. Turner, Recent advances in the use of enzyme-catalysed reactions in organic synthesis, *Natural Prod. Rep. 9*:1 (1994).
4. C.-H. Wong and G. M. Whitesides, *Enzymes in Synthetic Organic Chemistry* (J. E. Baldwin and P. D. Magnus, eds.), Tetrahedron Organic Chemistry Series; vol. 12, Pergamon, London, 1994, p. 195.
5. H. J. M. Gijsen and C.-H. Wong, Unprecedented asymmetric aldol reactions with three aldehyde substrates catalyzed by 2-deoxyribose-5-phosphate aldolase, *J. Am. Chem. Soc. 116*:8422 (1994).
6. M. D. Bednarski, E. S. Simon, N. Bischofberger, W.-D. Fessner, M.-J. Kim, W. Lees, T. Saito, H. Waldmann, and G. M. Whitesides, Rabbit muscle aldolase as a catalyst in organic synthesis, *J. Am. Chem. Soc. 111*:627 (1989).
7. W. D. Fessner and O. Eyrisch, One-pot synthesis of tagatose 1,6-bisphosphate by diastereoselective enzymatic aldol addition, *Angew. Chem. Int. Ed. Engl. 31*:56 (1992).
8. W. D. Fessner, G. Sinerius, A. Schneider, M. Dreyer, G. E. Schulz, J. Badia, and J. Arguilar, Diastereoselective enzymatic aldol additions: L-Rhamnulose and L-fuculose 1-phosphate aldolases from *E. coli*, *Angew. Chem. Int. Ed. Engl. 30*:555 (1991).
9. C. H. von der Osten, A. Sinskey, C. F. Barbas, R. L. Pederson, Y.-F. Wang, and C.-H. Wong, Use of a recombinant bacterial fructose-1,6-diphosphate aldolase in aldol reactions: Preparative syntheses of 1-deoxynojirimycin, 1-deoxymannorjirimycin, 1,4-imino-D-arabinitol, and fagomine, *J. Am. Chem. Soc. 111*:3924 (1989).
10. Y. Uchida, Y. Tuskada, and T. Sugimori, Purification and properties of *N*-acetylneuraminate lyase from *Escherichia coli*, *J. Biochem. 96*:507 (1984).
11. M. A. Ghalambor and E. C. Heath, The biosynthesis of cell wall lipopolysaccharide in *Escherichia coli*, *J. Biol. Chem. 241*:3222 (1966).
12. H. P. Meloche, J. M. Ingram, and W. A. Wood, 2-Keto-3-deoxy-6-phosphogluconic aldolase (crystalline), *Methods Enzymol. 9*:520 (1966).
13. A. M. Allam, M. M. Hassan, and T. A. Elzainy, Formation and cleavage of 2-keto-3-deoxygluconate by 2-keto-3-deoxygluconate aldolase of *Aspergillus niger*, *J. Bacteriol. 124*:1128 (1975).
14. A. M. Elshafei and O. M. Abdel-Fatah, Nonphosphorolytic pathway for D-galactonate catabolism in *Aspergillus terreus*, *Enzyme Microb. Technol. 13*:930 (1991).
15. R. Schauer, Chemistry, metabolism and biological functions of sialic acids, *Adv. Carbohydr. Chem. Biochem. 40*:131 (1982).
16. F. M. Unger, The chemistry and biological significance of 3-deoxy-D-*manno*-2-octulosonic acid (KDO), *Adv. Carbohydr. Chem. Biochem. 38*:323 (1981).
17. J. Frost and J. R. Knowles, 3-Deoxy-D-*arabino*-heptulosonic acid 7-phosphate: Chemical synthesis and isolation from *Escherichia coli* auxotrophs, *Biochemistry 23*:4465 (1984).
18. M. D. Bednarski, D. C. Crans, R. Dicosimo, E. S. Simon, P. D. Stein, and G. M. Whitesides, Synthesis of 3-deoxy-D-*manno*-2-octulosonate-8-phosphate (KDO-8-P) from D-arabinose: Generation of D-arabinose-5-phosphate using hexokinase, *Tetrahedron Lett. 29*:427 (1988).

19. L. M. Reimer, D. L. Conley, D. L. Pompliano, and J. W. Frost, Construction of an enzyme-targeted organophosphonate using immobilized enzyme and whole cell synthesis, *J. Am. Chem. Soc. 108*:8010 (1986).
20. C. Augé, S. David, and C. Gautheron, Synthesis with immobilized enzyme of the most important sialic acid, *Tetrahedron Lett. 25*:4663 (1984).
21. S. David, C. Augé, and C. Gautheron, Enzymic methods in preparative carbohydrate chemistry, *Adv. Carbohydr. Chem. Biochem. 49*:176 (1991).
22. K. Aisaka, S. Tamura, Y. Arai, and T. Uwajima, Hyperproduction of N-acetylneuraminate lyase by the gene-cloned strain of *Escherichia coli*, *Biotechnol. Lett. 9*:633 (1987).
23. K. Aisaka, A. Igarashi, K. Yamagushi, and T. Uwajima, Purification, crystallisation and characterization of N-acetylneuraminate lyase from *Escherichia coli*, *Biochem. J. 276*:541 (1991).
24. T. Izard, M. C. Lawrence, R. L. Malby, G. G. Lilley, and P. M. Colman, The three-dimensional structure of N-acetylneuraminate lyase from *Escherichia coli*, *Structure 2*:361 (1994).
25. M.-J. Kim, W. J. Hennen, H. M. Sweers, and C.-H. Wong, Enzymes in carbohydrate synthesis: N-Acetylneuraminic acid aldolase catalyzed reactions and preparation of N-acetyl-2-deoxy-D-neuraminic acid derivatives, *J. Am. Chem. Soc. 110*:6481 (1988).
26. M. D. Bednarski, H. K. Chenault, E. Simon, and G. M. Whitesides, Membrane-enclosed enzymatic catalysis (MEEC): A useful, practical new method for the manipulation of enzymes in organic synthesis, *J. Am. Chem. Soc. 109*:1283 (1987).
27. U. Kragl, D. Gygax, O. Ghisalba, and C. Wandrey, Enzymatic two-step synthesis of N-acetylneuraminic acid in the enzyme membrane reactor, *Angew. Chem. Int. Ed. Engl. 30*:827 (1992).
28. C. Augé, S. David, C. Gautheron, and A. Veyrières, Synthesis with an immobilized enzyme of N-acetyl-9-O-acetylneuraminic acid, a sugar reported as a component of embryonic and tumor antigens, *Tetrahedron Lett. 26*:2439 (1985).
29. C. Augé, S. David, C. Gautheron, B. Cavayé, and A. Malleron, Preparation of six naturally occurring sialic acids with immobilized acylneuraminate pyruvate lyase, *New J. Chem. 12*:733 (1988).
30. H. S. Conradt, A. Bünsch, and R. Brossmer, Preparation of 9-fluoro-9-deoxy-N-[2-$^{14}$C]acetylneuraminic acid, *FEBS Lett. 170*:295 (1984).
31. J. L. C. Liu, G.-J. Shen, Y. Ichikawa, J. F. Rutan, G. Zapata, W. F. Vann, and C.-H. Wong, Overproduction of CMP-sialic acid synthetase for organic synthesis, *J. Am. Chem. Soc. 114*:3901 (1992).
32. W. Fitz and C.-H. Wong, Combined use of subtilisin and N-acetylneuraminic acid aldolase for the synthesis of a fluorescent sialic acid, *J. Org. Chem. 59*:8279 (1994).
33. C. Augé and C. Gautheron, The use of an immobilized aldolase in the first synthesis of a natural deaminated neuraminic acid, *J. Chem. Soc., Chem. Commun.* p. 860 (1987).
34. D. Nadano, M. Iwasaki, S. Endo, K. Kitajima, S. Inoue, and Y. Inoue, A naturally occurring deaminated neuraminic acid, 3-deoxy-D-*glycero*-D-*galacto*-nonulosonic acid (KDN), *J. Biol. Chem. 261*:11550 (1986).
35. C. Augé, B. Bouxom, B. Cavayé, and C. Gautheron, Scope and limitations of the aldol condensation catalyzed by immobilized acylneuraminate pyruvate lyase, *Tetrahedron Lett. 30*:2217 (1989).
36. C. Augé, S. David, and A. Malleron, An inexpensive route to 2-azido-2-deoxy-D-mannose and its conversion into an azido analog of N-acetylneuraminic acid, *Carbohydr. Res. 188*:201 (1989).
37. C. Augé, C. Gautheron, S. David, A. Malleron, B. Cavayé, and B. Bouxom, Sialyl aldolase in organic synthesis: From the trout egg acid, 3-deoxy-D-*glycero*-D-*galacto*-2-nonulosonic acid (KDN), to branched-chain higher ketoses as possible new chirons, *Tetrahedron 46*:201 (1990).
38. S. David, B. Cavayé, and A. Malleron, Some derivatives of 3-deoxy-D-*glycero*-D-*galacto*-non-2-ulosonic acid (KDN), *Carbohydr. Res. 260*:233 (1994).
39. K. Koppert and R. Brossmer, Synthesis of the C-5 homologue of N-acetylneuraminic acid by enzymatic chain elongation of 2-C-acetamidomethyl-2-deoxy-D-mannose, *Tetrahedron Lett. 33*:8031 (1992).

40. R. Iseke and R. Brossmer, Synthesis of 5-*N*- and 9-*N*-thioacylated sialic acids, *Tetrahedron* 50:7445 (1994).
41. A. Schrell and G. M. Whitesides, Synthesis of the α-methyl ketoside of 5-amino neuraminic acid methyl ester and its corresponding 5-myristoyl and 5-cyclopropanoyl derivative, *Liebigs Ann. Chem.* p. 1111 (1990).
42. P. Zhou, H. M. Salled, and J. F. Honek, Facile chemoenzymatic synthesis of 3-(hydroxymethyl)-6-epicastanospermine, *J. Org. Chem.* 58:264 (1993).
43. S. David, A. Malleron, and B. Cavayé, Aldolases in organic synthesis: Acylneuraminate–pyruvate lyase accepts furanoses as substrates, *New J. Chem.* 16:751 (1992).
44. R. L. Halcomb, W. Fitz, and C.-H. Wong, Enzymatic synthesis of 7-deoxy-*N*-acetylneuraminic acid and 7-*O*-methyl-*N*-acetylneuraminic acid, *Tetrahedron Asymm.* 5:2437 (1994).
45. D. C. M. Kong and M. von Itzstein, The first synthesis of a C-7 nitrogen-containing sialic acid analogue, 5-acetamido-7-azido-3,5,7-trideoxy-D-*glycero*-D-*galacto*-2-nonulopyranosonic acid (7-azido-7-deoxy-Neu5Ac), *Tetrahedron Lett.* 36:957 (1995).
46. C.-H. Lin, T. Sugai, R. L. Halcomb, Y. Ichikawa, and C.-H. Wong, Unusual stereoselectivity in sialic acid aldolase-catalyzed aldol condensations: Synthesis of both enantiomers of high-carbon monosaccharides, *J. Am. Chem. Soc.* 114:10138 (1992).
47. C. Augé and N. Le Goff, unpublished results.
48. U. Kragl, A. Gödde, C. Wandrey, N. Lubin, and C. Augé, New synthetic applications of sialic acid aldolase, a useful catalyst for KDO synthesis. Relation between substrate conformation and enzyme stereoselectivity, *J. Chem. Soc. Perkin Trans.* 1:119 (1994).
49. C. Gautheron-Le Narvor, Y. Ichikawa, and C.-H. Wong, A complete change of stereoselectivity in sialic acid aldolase reactions: A novel synthetic route to the KDO type of nine-carbon L sugars, *J. Am. Chem. Soc.* 113:7816 (1991).
50. T. Sugai, G.-J. Shen, Y. Ichikawa, and C.-H. Wong, Synthesis of 3-deoxy-D-*manno*-2-octulosonic acid (KDO) and its analogs based on KDO aldolase-catalyzed reactions, *J. Am. Chem. Soc.* 115:413 (1993).
51. S. T. Allen, G. R. Heintzelman, and E. J. Toone, Pyruvate aldolase as reagents for stereospecific aldol condensation, *J. Org. Chem.* 57:426 (1992).
52. C. Augé and V. Delest, Microbiological aldolisations. Synthesis of 2-keto-3-deoxy-D-gluconate, *Tetrahedron Asymm.* 4:1165 (1993).
53. V. Delest and C. Augé, The use of *Aspergillus terreus* extracts in the preparative synthesis of 2-keto-3-deoxy-ulosonic acids, *Tetrahedron Asymm.* 6:863 (1995).
54. J. W. Cornforth, M. E. Firth, and A. Gottschalk, The synthesis of *N*-acetyneuraminic acid, *Biochem. J.* 68:57 (1958).
55. H. Ogura, K. Furuhata, S. Sato, and K. Anazawa, Synthesis of 9-*O*-acyl and 4-*O*-acetyl sialic acids, *Carbohydr. Res.* 167:77 (1987).
56. C. Hershberger, M. D. Davis, and S. B. Binkley, Chemistry and metabolism of 3-deoxy-D-*manno*-octulosonic acid, *J. Biol. Chem.* 243:1585 (1968).
57. R. Kuhn, D. Weiser, and H. Fischer, Basen-katalysiert Umwandlungen der *N*-Phenyl-D-hexosaminsaürenitrile, *Liebigs Ann. Chem.* 628:207 (1959).
58. R. Ramage, A. M. MacLeod, and G. W. Rose, Dioxolanes as synthetic intermediates. Part 6. Synthesis of 3-deoxy-D-*manno*-2-octulosonic acid (KDO), 3-deoxy-D-*arabino*-2-heptulosonic acid (DAH) and 2-keto-3-deoxy-D-gluconic acid (KDG), *Tetrahedron* 47:5625 (1991).

# 22

# Oligosaccharide Synthesis by Enzymatic Glycosidation

### Wolfgang Fitz and Chi-Huey Wong
*The Scripps Research Institute, La Jolla, California*

| | | |
|---|---|---|
| I. | Introduction | 486 |
| II. | Enzymatic Glycosidation | 486 |
| | A. Biological background | 486 |
| | B. Using glycosyltransferases of the Leloir pathway in synthesis | 488 |
| | C. Using glycosidases in synthesis | 491 |
| III. | Experimental Procedures | 494 |
| | A. Formation of the galactosyl β1,4-linkage using β1,4-galactosyltransferase with regeneration of UDP-Gal | 494 |
| | B. Formation of the galactosyl β1,3-linkage using β-galactosidase | 496 |
| | C. Synthesis of CMP-NeuAc | 496 |
| | D. Formation of the sialyl α2,3-linkage using α2,3-sialyltransferase without regeneration of CMP-NeuAc | 497 |
| | E. Formation of the sialyl α2,3-linkage using α2,3-sialyltransferase with regeneration of CMP-NeuAc | 497 |
| | F. Formation of the sialyl α2,6-linkage using α2,6-sialyltransferase with regeneration of CMP-NeuAc | 499 |
| | G. Formation of the fucosyl α1,3-linkage using α1,3-fucosyltransferase without regeneration of GDP-Fuc | 499 |
| | H. Formation of the mannosyl α1,2-linkage using α1,2-mannosyltransferase with regeneration of GDP-Man | 500 |
| | I. Combined use of β-galactosidase and α2,6-sialyltransferase with regeneration of CMP-NeuAc | 500 |
| | References | 502 |

## I. INTRODUCTION

Complex carbohydrates are components of a broad range of molecular structures in nature. They are frequently found as components of cell surface glycoproteins and glycolipids, playing an important role in cellular communication processes [1] and as points of attachment for antibodies and other proteins. The major classes of cell surface glycolipids include the glycosphingolipids (GSLs) and glycoglycerolipids. Particularly significant are gangliosides [2], or sialic acid-containing glycosphingolipids, which are especially abundant on neural cell surfaces [3]. These compounds play a role in the differentiation of cell types and in the regulation of cell growth. Oligosaccharides and polysaccharides also serve as receptor sites for bacteria [4] and viral particles [5]. The saccharide moieties of glycoproteins are also involved in modulating protein folding and in the sorting and trafficking of proteins to appropriate cellular sites [1]. Carbohydrates often occur in minute quantities and are difficult to isolate in a pure form, characterize, and in particular, synthesize in amounts sufficient for therapeutic study and biological evaluation. The pace of development of carbohydrate-derived therapeuticals, therefore, has been slower than with other classes of biomaterials (i.e., peptides or nucleic acids) that are more easily accessible by automatic solid-phase synthesis. One of the major difficulties in oligosaccharide synthesis is the task of coupling building blocks in a stereoselective manner. Particularly difficult are the synthesis of β-mannosides, for which some elegant solutions have only recently appeared [6,7], and the synthesis of glycosides of sialic acids [8]. A second difficulty encountered in oligosaccharide synthesis originates from the polyfunctionality of these compounds, which necessitates the use of elaborate protective group chemistry if complex carbohydrates are to be synthesized through conventional methods. Although considerable progress has been made recently toward the development of more sophisticated synthetic methods directed at the synthesis of glycoconjugates (see Chap. 9), the chemical synthesis of biologically interesting complex carbohydrates on a scale larger than 1 mmol still appears to be a problem not yet solved. Alternatively, the enzymatic approach avoids the problems encountered in the chemical synthesis and is increasingly considered to be useful for the practical synthesis of certain complex oligosaccharides and glycoconjugates, albeit some drawbacks are also encountered.

## II. ENZYMATIC GLYCOSIDATION

### A. Biological Background

Two classes of enzymes are involved in the biosynthesis of oligosaccharides in nature: the enzymes of the Leloir pathway and those of non-Leloir pathways. The Leloir pathway enzymes are responsible for the synthesis of most glycoproteins and other glycoconjugates in mammalian systems. Glycoproteins can be classified as either N-linked or O-linked, depending on the type of attachment of the carbohydrate moiety to the protein. The N-linked are characterized by a β-glycosidic linkage between a GlcNAc residue and the δ-amide nitrogen of an asparagine. The less common O-linked glycoproteins contain an α-glycosidic linkage between a N-acetylgalactosamine (GalNAc; or xylose) and the hydroxyl group of a serine or threonine. The addition of oligosaccharide chains to glycoproteins occurs cotranslationally for both O-linked and N-linked types in the endoplasmic reticulum and the Golgi apparatus [9]. The N-linked oligosaccharides all contain the same core structure of N-acetylglucosamine (GlcNAc) and mannose residues, the similarity of which stems from their origin. Their biosynthesis starts with the formation of a dolichyl pyrophosphoryl oligosaccharide in the endoplasmic reticulum by the action of GlcNAc-transferases and mannosyltransferases. This structure is further glucosylated, presumably to signal for transfer of the oligosaccharide to the polypeptide. The entire oligosaccharide moiety is then transferred to an asparagine residue of the growing peptide chain by the enzyme oligosaccharyltransferase [9]. The asparagine is typically part of the amino acid sequence Asn-X-Ser(Thr), where X is neither proline or aspartic acid. Before transport into the Golgi apparatus, the glucose residues and some mannose residues are removed by the action of glucosidase I and II and a mannosidase to reveal a core pentasaccharide (peptide-Asn-(GlcNAc)$_2$-(Man)$_3$). This structure is further processed by mannosidases and glycosyltransferases present in the Golgi apparatus to produce either the high-mannose type, the complex type, or the hybrid type oligosaccharides. Monosaccharides are then added sequentially to this core structure to provide the fully elaborated oligosaccharide chain.

The biosynthesis of O-linked oligosaccharides follows a different scheme. In contrast with the dolichyl pyrophosphate-mediated synthesis of N-linked oligosaccharides, the glycosyltransferases necessary for the synthesis of O-linked oligosaccharides are located in the Golgi apparatus [9]. Monosaccharide residues are added sequentially to the growing oligosaccharide chain.

With the exception of erythrocytes, all mammalian cells contain the necessary elements for glycosylation. In certain secretory cells, however, the preponderance of transferases is greater [10]. In contrast with the non-Leloir enzymes, which typically employ glycosyl phosphates as activated donors, the glycosyltransferases of the Leloir pathway in mammalian systems use as glycosyl donors monosaccharides that are activated as glycosyl esters of nucleoside mono- or diphosphates [9]. The Leloir glycosyltransferases use primarily nine nucleoside mono- or diphosphate sugars as monosaccharide donors for the synthesis of most oligosaccharides: UDP-Glc, UDP-GlcNAc, UDP-Gal, UDP-GalNAc, GDP-Man, GDP-Fuc, UDP-Xyl, UDP-GlcUA, and CMP-NeuAc (Fig. 1). Most of these sugar nucleoside phosphates are biosynthesized in vivo from the corresponding monosaccharides. The initial step is a kinase-mediated phosphorylation to produce a glycosyl phosphate. This glycosyl phosphate then reacts with a nucleoside triphosphate (NTP), catalyzed by a nucleoside diphosphosugar pyrophosphorylase, to afford an activated nucleoside diphosphosugar. Other sugar nucleoside phosphates, such as GDP-Fuc and UDP-GlcUA, are biosynthesized by further enzymatic modification of these existing key sugar nucleotide phosphates. Another exception is CMP-NeuAc, which is formed by the direct reaction of NeuAc with CTP.

Many other monosaccharides, such as the anionic or sulfated sugars of heparin and chondroitin sulfate, are also found in mammalian systems, but they usually are a result of modification of a particular sugar after it is incorporated into an oligosaccharide structure [1]. A very diverse array of monosaccharides (e.g., xylose, arabinose, KDO) and oligosaccharides is also present in microorganisms, plants, and invertebrates [1,11]. Interfering with the biosynthesis of the characteristic carbohydrate components of the bacterial cell wall, especially 3-deoxy-D-*manno*-2-octulogonic acid (KDO), lipid A, and heptulose, is an attractive approach to the development of antibacterial agents. The enzymes responsible for the biosynthesis of these compounds, however, have not been extensively exploited for synthesis, although they follow the same principles as do those in mammalian systems.

Both transferases from the Leloir and non-Leloir pathways have been employed for the in vitro synthesis of oligosaccharides and glycoconjugates. Glycosidases have also been exploited for synthesis. The function of glycosidases in vivo is to cleave glycosidic bonds,

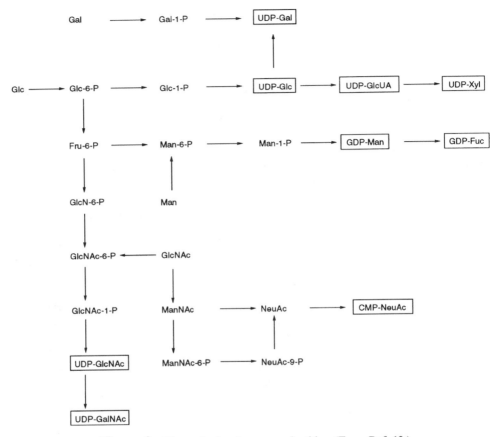

**Figure 1** Biosynthesis of sugar nucleotides. (From Ref. 12.)

but under appropriate conditions they can be useful synthetic catalysts. Each group of enzymes has certain advantages and disadvantages for synthesis. Glycosyltransferases are highly specific in the formation of glycosides; however, the availability of many of the necessary transferases is limited. Glycosidases have the advantage of wider availability and lower cost, but they are not specific or high-yielding in synthetic reactions. In the following, examples of the application of Leloir-type transferases and glycosidases will be described.

### B. Using Glycosyltransferases of the Leloir Pathway in Synthesis

Glycosyltransferases are highly regiospecific and stereospecific in the formation of new glycosidic linkages. They also are usually substrate-specific. Under in vitro conditions, however (i.e., when the substrates are provided in high concentrations), minor chemical modifications are tolerated on both the donor and acceptor components. Several examples of this will be given later. In the past, the preparative use of glycosyltransferases has been limited owing to a lack of availability of these enzymes. Additionally, glycosyltransferases are membrane-bound enzymes and, therefore, are relatively unstable and difficult to handle. However, the recent isolation of many of these enzymes, as well as advances in genetic engineering and recombinant techniques, are rapidly alleviating these drawbacks.

# Synthesis by Enzymatic Glycosidation

Over 20 transferases have now been cloned [12]. Other glycosyltransferases can be obtained from readily available tissue sources in at least milliunit amounts, which is sufficient for milligram-scale synthesis. These are tabulated in a recent review [13].

*Synthesis of Sugar Nucleotides*

Sugar nucleoside phosphates can be prepared either chemically or enzymatically (Table 1, and references cited therein). Most of the chemical methods involve the reaction of an activated nucleoside monophosphate (NMP), usually a phosphoramidate, such as phosphorimidazolidate and phosphoromorpholidate, with a glycosyl phosphate to produce a sugar nucleoside diphosphate. The enzymatic preparation of sugar nucleotides usually involves the coupling of a glycosyl phosphate with a nucleoside triphosphate catalyzed by the appropriate sugar nucleotide pyrophosphorylase.

Several chemical and enzymatic methods are available for the synthesis of glycosyl phosphates. The required nucleoside triphosphates (NTPs) are most conveniently prepared by enzymatic routes. In general, these methods involve the sequential use of two kinases to transform NMPs to NTPs, by the corresponding NDPs (Fig. 2, see a recent review [12] for more details).

The enzymatic preparation of the activated sugar nucleotide may also involve a cofactor regeneration system. An example of this is an economic one-pot procedure, in which $N$-acetylneuraminic acid (NeuAc) is generated in situ from $N$-acetylmannosamine (ManNac) and pyruvate with sialic acid aldolase and then converted irreversibly to CMP-NeuAc ([14], see also Sec. III).

**Table 1** Synthesis of Sugar Nucleotides

| Cofactor | Method |
|---|---|
| UDP-Glc | From Glc-1-P and UTP with UDP-glucose pyrophosphorylase |
| UDP-Gal | From Gal-1-P and UTP with UDP-galactose pyrophosporylase |
| | From UDP-glucose with UDP-glucose epimerase |
| | From Gal-1-P and UDP-Glc with UDP-galactose uridyl transferase |
| | From UMP and galactose using cells of *Torulopsis candida* |
| | By chemical synthesis |
| UDP-GlcNAc | From GlcNAc-1-P and UTP with UDP-GlcNAc pyrophosphorylase |
| | From GlcNH$_2$-1-P and UTP with UDP-GlcNAc pyrophosphorylase, followed by $N$-acetylation |
| UDP-GalNAc | From UDP-Glc and Gal-1-P with UDP-glucose:galactosylphosphate uridyltransferase |
| GDP-Man | From Glc and GMP with Baker's yeast |
| | From Man-1-P and GTP with GDP-Man pyrophosphorylase |
| GDP-Fuc | From GDP-Man with a crude enzyme preparation from *Agrobacterium radiobacter* |
| | From Fuc-1-P and GTP with GDP-Fuc pyrophosphorylase |
| | By chemical synthesis |
| UDP-GlcUA | By oxidation of UDP-Glc with UDP-Glc dehydrogenase from bovine liver or guinea pig liver |
| CMP-NeuAc | From NeuAc and CMP with CMP-NeuAc synthetase |
| | By chemical synthesis |

*Source*: Ref. 12, see references cited therein.

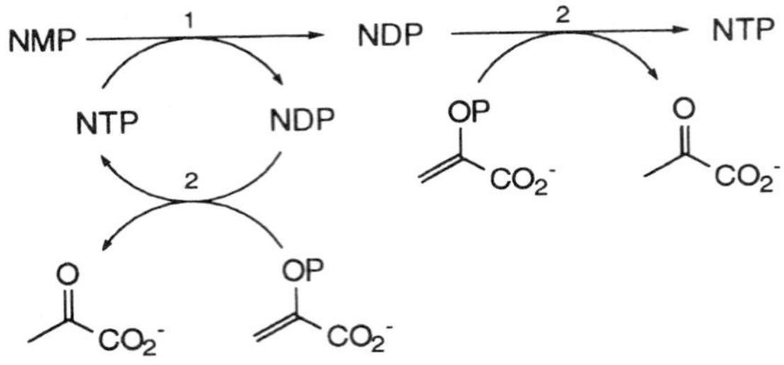

**Figure 2** Synthesis of nucleoside triphosphates: (1) Adenylate kinase (EC 2.7.4.3, N = A,C,U) or guanylate kinase (EC 2.7.4.8, N = G), or nucleoside monophosphate kinase (EC 2.7.4.4, N = U). (2) Pyruvate kinase (EC 2.7.1.40). NMP, nucleoside monophosphate; NDP, nucleoside diphosphate; NTP, nucleoside triphosphate. (From Ref. 12.)

*Substrate Specificity of Glycosyltransferases*

The substrate specificities of glycosyltransferases have been extensively reviewed recently [12]. From the vast amount of data now available, it is clear that these enzymes can be used not only with their natural substrates, but also can be employed for the synthesis of a variety of analogues. β1,4-Galactosyltransferase (EC 2.4.1.22) is readily available, and its substrate specificity has been well studied. The enzyme catalyzes the transfer of galactose from UDP-Gal (**2**) to the 4-position of β-linked GlcNAc residue to produce the Galβ1,4GlcNAc substructure. In the presence of lactalbumin, however, glucose is the preferred acceptor, resulting in the formation of lactose, Galβ1,4Glc. Galactosyltransferase uses as acceptor substrates *N*-acetylglucosamine and glucose and β-glycosides thereof, 2-deoxyglucose, D-xylose, 5-thioglucose, *N*-acetylmuramic acid, and *myo*-inositol [15]. Modifications at the 3- or 6-position of the acceptor GlcNAc are also tolerated [16]. Both α- and β-glycosides of glucose are acceptable, but the presence of lactalbumin is required for galactosyl transfer onto α-glycosides. A particularly interesting example is the β,β-1,1-linked disaccharide **3**, in which the anomeric hydroxyl of 3-acetamido-3-deoxyglucose (**1**) served as the acceptor moiety (Fig. 3) [17].

For the donor substrate, galactosyltransferase also transfers various galactose analogues from their respective UDP-derivatives, thereby providing an enzymatic route to oligosaccharides that terminate in β1,4-linked residues other than galactose [18]. Although the rate of the enzyme-catalyzed transfer of many of these unnatural donor substrates is rather slow, this method is useful for milligram-scale synthesis.

Both α2,6- and α2,3-sialyltransferases have been used for oligosaccharide synthesis [19]. Sialyltransferases generally transfer *N*-acetylneuraminic acid to either the 3- or 6-position of terminal Gal or GalNAc residues. Some sialyltransferases accept CMP-NeuAc analogues that are derivatized at the 9-position of the sialic acid side chain [20]. Analogues of the acceptors Galβ1,4GlcNAc and Galβ1,3GalNAc, in which the acetamido function was replaced by other functional groups, were also accepted by the enzymes [21]. Recently, α2,8-linked homopolymers of sialic acid have been prepared using α2,8-sialyltransferase [22].

Several fucosyltransferases have been isolated and used for in vitro synthesis. The Lewis A α1,4-fucosyltransferase transfers unnatural fucose derivatives in from their GDP esters [23]. The enzyme α1,3-fucosyltransferase has been used to L-fucosylate the 3-position of the GlcNAc of *N*-acetyllactosamine and of sialyl α2,3-*N*-acetyllactosamine [24]. Several acceptor substrates with modifications in the GlcNAc residue could also be

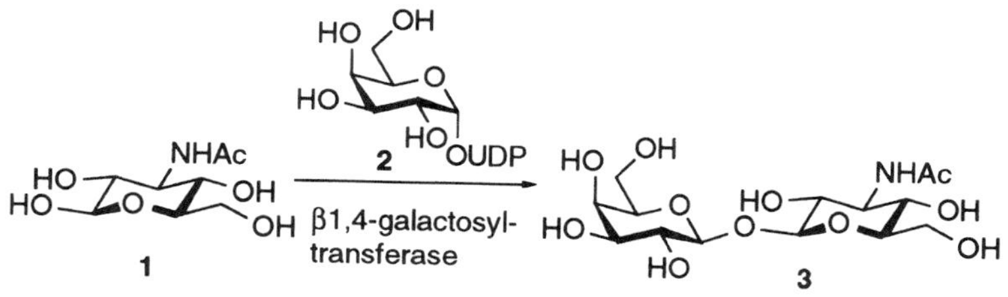

# Synthesis by Enzymatic Glycosidation

**Figure 3** Synthesis of a β,β-1,1-linked disaccharide with galactosyltransferase. (From Ref. 17.)

fucosylated [25]. A related enzyme, α1,3/4-fucosyltransferase, has been used on preparative scale to fucosylate the GlcNAc 3-position of Galβ1,4GlcNAc and the GlcNAc 4-position of Galβ1,3GlcNAc [26]. The corresponding sialylated substrates have also been employed as acceptors.

N-Acetylglucosaminyl transferases I–VI catalyze the addition of GlcNAc residues to the core pentasaccharide of asparagine glycoproteins [27], thereby controlling the branching pattern of N-linked glycoproteins in vivo. Each of the enzymes transfers a β-GlcNAc residue from the donor UDP-GlcNAc to a mannose or other acceptor. GlcNAc-transferases have been used to transfer nonnatural residues onto oligosaccharides [28].

Different mannosyltransferases transfer mannose and 4-deoxymannose from their respective GDP adducts to acceptors. α1,2-Mannosyltransferase was employed to transfer mannose to the 2-position of various derivatized α-mannosides and α-mannosyl peptides to produce the Manα1,2Man structural unit [29,30]. A recent report indicates that mannosyltransferases from pig liver accept GlcNAcβ1,4GlcNAc phytanyl pyrophosphate, an analogue in which the dolichol chain of the natural substrate is replaced by the phytanyl moiety [31].

*In Situ Cofactor Regeneration*

Though analytic- and small-scale synthesis using glycosyltransferases is extremely powerful, the high cost of sugar nucleotides and the product inhibition caused by the released nucleoside mono- or diphosphates present major obstacles to large-scale synthesis. A simple solution to both of these problems is to use a scheme in which the sugar nucleotide is regenerated in situ from the released nucleoside diphosphate. In situ cofactor regeneration offers several advantages. First, a catalytic amount of nucleoside diphosphate and a stoichiometric amount of monosaccharide can be used as starting materials, rather than a stoichiometric quantity of sugar nucleotide, thus tremendously reducing costs. Second, product inhibition by the released NDP is minimized owing to its low concentration in solution. Third, isolation of the product is greatly facilitated.

General regeneration systems are represented in Figure 4 for glycosyltransferases that use UDP-glycosides and CMP-glycosides [12]. The development of these regeneration systems, as well as the more recent development of regeneration schemes for GDP-Man [29], and GDP-Fuc [24] should facilitate the more widespread use of glycosyltransferases for oligosaccharide synthesis. Several examples of this are given in the experimental section.

## C. Using Glycosidases in Synthesis

Glycosidases cleave glycosidic bonds in vivo, but they can be employed as synthetic catalysts under appropriate conditions. They are readily available and inexpensive.

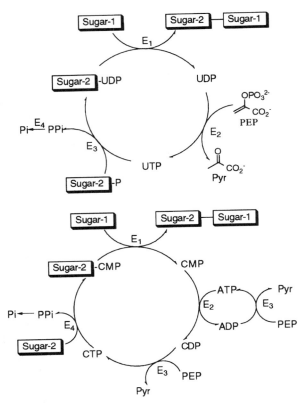

**Figure 4** (a) Regeneration systems for UDP-sugar cofactors. $E_1$, glycosyltransferase; $E_2$, pyruvate kinase; $E_3$, UDP-sugar pyrophosphorylase; $E_4$, pyrophosphatase. (b) Regeneration systems for CMP-sugar cofactors. $E_1$, glycosyltransferase; $E_2$, myokinase; $E_3$, pyruvate kinase; $E_4$, CMP-sugar synthetase.

Glycosidase-mediated synthesis of glycosides can be performed either under thermodynamically or under kinetically controlled conditions. The formation of glycosides from nonactivated precursors is an endergonic process ($\Delta G = +4$ kcal/mol) and the equilibrium approach suffers from poor yields, usually not exceeding 15%, and of the formation of side products. Both features may render the isolation and purification of the desired products difficult. Glycosidase-catalyzed glycoside synthesis under kinetically controlled conditions, on the other hand, relies on the formation of a reactive intermediate from an activated glycosyl donor. The reactive species is then trapped with exogenous nucleophiles to form a new glycosidic bond. Suitable glycosyl donors for this transglycosylation reaction include di- or oligosaccharides, aryl glycosides, and glycosyl fluorides. The reaction must be carefully monitored, and stopped, when the glycosyl donor is consumed, to minimize glycoside hydrolysis. The kinetically controlled approach has been applied to primarily the retaining glycosidases, but using glycosyl fluorides as glycosyl donors, inverting glycosidases have been used to afford products having the configuration at the anomeric position that is opposite that of the donor [32].

## Synthesis by Enzymatic Glycosidation

From the large amount of glycosidase-catalyzed glycosidations reported so far (see a recent review [12] for an extensive tabulation), it is obvious that the reactions are stereoselective processes, in contrast with many chemical methods. The regioselectivity, however, is not necessarily absolute or predictable. In general, the primary hydroxy group of the acceptor reacts preferentially over secondary hydroxy groups. Some control of selectivity could be achieved by the selection of an appropriate donor-acceptor combination [33].

A second possibility to control the regioselectivity involves the use of glycosidases from different species. The galactosyl β1,3-linkage, for example, is an important structural motif of several complex carbohydrates with interesting biological functions, such as the mucin oligosaccharide, sialyl Lewis a and the T-antigen. There is no β1,3-galactosyltransferase available for the synthesis of these molecules, but the β-galactosidase from testes catalyzes the formation of β1,3-linkages [12]. Alternatively, β1,3-linked disaccharides can be prepared using β-galactosidase from *Escherichia coli* using p-$NO_2$-Ph-β-galactopyranoside (**4**) as the donor and glycals (e.g., **5**) as acceptors (Fig. 5). The obtained glycal disaccharide (e.g., **6**) can be converted into *N*-acetylated 2-amino disaccharides (e.g., **7**) through azidonitration followed by reduction and *N*-acetylation. The minor products produced in this preparation were then hydrolyzed by the *E. coli* β-galactosidase, which preferentially hydrolyzes β1,6-linked galactosyl residues. The overall yield of the β1,3-linked disaccharides was about 10–20% [34].

*Bacillus circulans* β-galactosidase prefers 1,4-glycosidic bond formation and has been used in the large-scale synthesis of *N*-acetyllactosamine (**7**). Coupled with a α2,6-sialyltransferase, the disaccharide **7** was converted to the sialyldisaccharide, which is no longer subject to the glycosidase-catalyzed hydrolysis, thus improving the yield (see experimental section) [35].

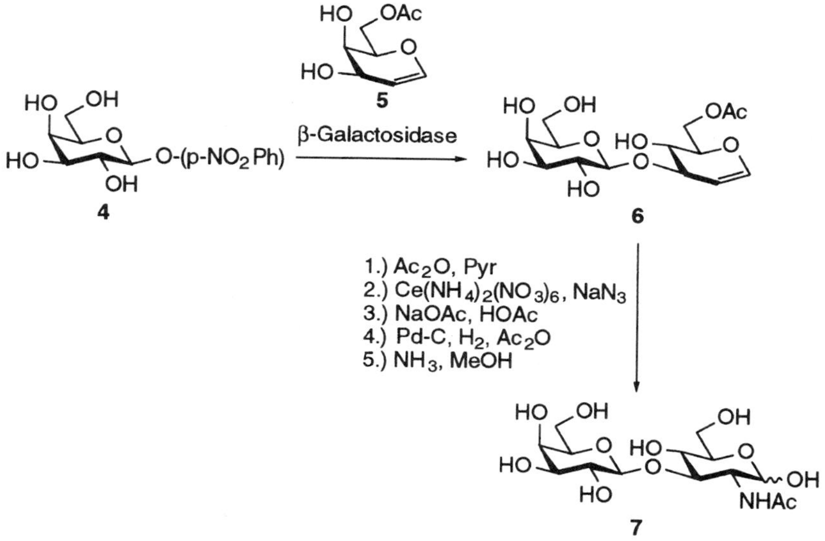

**Figure 5** Synthesis of β1,3-linked disaccharides. (From Ref. 34.)

## III. EXPERIMENTAL PROCEDURES

### A. Formation of the Galactosyl β1,4-Linkage Using β1,4-Galactosyltransferase with Regeneration of UDP-Gal [24]

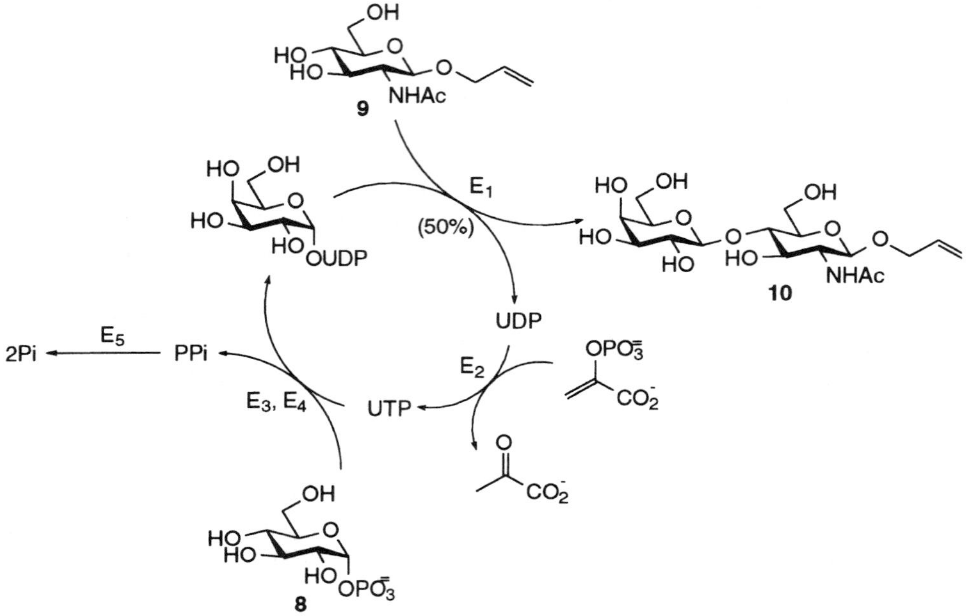

Two multienzyme systems for the synthesis of LacNAc (**7**) and analogues using β1,4-galactosyltransferase (GalT) have been developed with in situ cofactor regeneration. The first starts with Glc-1-P and requires UDP-Glc pyrophosphorylase (EC 2.7.7.9; UDPGP) and UDP-Glc 4-epimerase (EC 5.1.3.2; UDPGE; see foregoing scheme). UDP-galactose is generated from UDP-Glc with UDPGE. The thermodynamics of this reaction, however, favors the formation of UDP-Glc, and Glc-1-P has to be prepared separately.

*LacNAcβOallyl* (**10**)

A mixture of **9** (2.00 g, 7.65 mmol), Glc-1-P (**8**; 2.74 g, 7.65 mmol), PEP potassium salt (1.6 g, 7.65 mmol), NAD$^+$ (193 mg, 0.25 mmol), MnCl$_2$·4H$_2$O (79 mg, 0.4 mmol), MgCl$_2$·6H$_2$O (163 mg, 0.8 mmol), DTT (306 mg, 2 mmol), KCl (1.0 g, 15 mmol), NaN$_3$ (20 mg, 0.31 mmol), and UDP (90 mg, 0.19 mmol) in HEPES buffer (200 mL, 100 m$M$, pH 7.5) was adjusted to pH 7.5 with 10 $N$ and 1 $N$ NaOH, and the enzymes UDPGE (10 U), UDPGP (20 U), pyruvate kinase (100 U), GalT (5 U), and inorganic pyrophosphatase (100 U) were added to the solution. The mixture was gently stirred under an argon atmosphere at room temperature for 5 days. The mixture was concentrated and chromatographed on silica gel (eluting with CHCl$_3$–EtOAc–MeOH; 5:2:2 to 5:2:3) to give a disaccharide, which was further purified with Sephadex G-25 (eluting with water) to give LacNAcβOallyl **10** (1.7 g, 50%). In the second regeneration system Gal is used instead of Glc-1-P as a donor

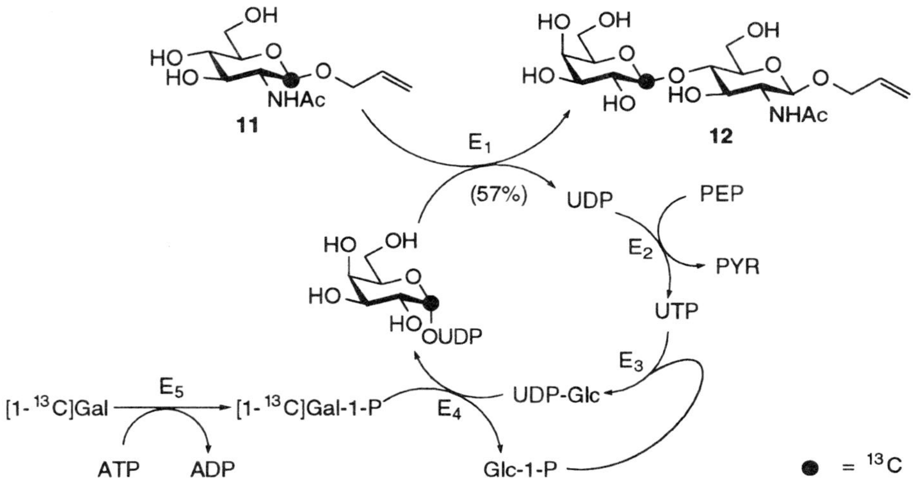

# Synthesis by Enzymatic Glycosidation

$E_1$: β1,4-galactosyltransferase; $E_2$: pyruvate kinase; $E_3$: UDP-Glc pyrophosphorylase
$E_4$: Galactose-1-phosphate uridyltransferase; $E_5$: galactokinase

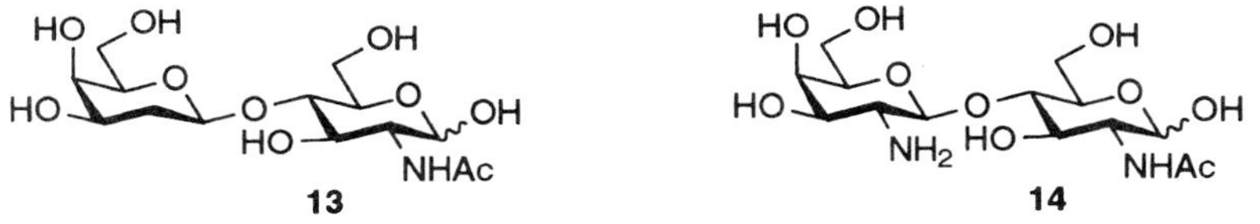

precursor (see foregoing scheme). This procedure requires UDPGP, galactokinase (EC 2.7.1.6; GK) and Gal-1-P uridyltransferase (EC 2.7.7.12; Gal-1-P UT). Galactokinase is specific for galactose, thus allowing the direct production of Gal-1-P, which is converted to UDP-Gal with Gal-1-P UT and UDP-Glc. This system also allows the regeneration of UDP-2-deoxy-D-galactose and UDP-galactosamine, thereby permitting the preparation of analogues such as 2'-deoxy-LacNAc 13 and 2'-amino-2'-deoxy-LacNAc 14.

## 1-[$^{13}$C]Galβ1,4GlcNAcβOallyl (12)

A solution of **11** (1.15 g, 4.4 mmol), [1-$^{13}$C]Gal (800 mg, 4.4 mmol), PEP potassium salt (1.82 g, 8.8 mmol), UDP (90 mg, 0.19 mmol), ATP (100 mg, 0.18 mmol), cysteine (116 mg, 0.96 mmol), DTT (183 mg, 1.2 mmol), $MgCl_2·6H_2O$ (244 mg, 1.2 mmol), $MnCl_2·4H_2O$ (118 mg, 0.6 mmol), KCl (179 mg, 2.4 mmol), and Glc-1-P (77 mg, 0.22 mmol) in HEPES buffer (120 mL, 100 m$M$, pH 7.5) was adjusted with 10 $N$ and 1 $N$ NaOH to pH 7.5, and the enzymes GK (10 U), pyruvate kinase (200 U), inorganic pyrophosphatase (10 U), Gal-1-P UT (10 U), UDPGP (10 U), and GalT (10 U) were added to the solution. The mixture was gently stirred under an argon atmosphere at room temperature for 3 days. The mixture was concentrated in vacuo, and the residue was chromatographed on silica gel (eluting with EtOAc–MeOH 2:1) to give a disaccharide, which was further purified with a column of Sephadex G-25 (eluting with water) to give **12** (1.06 g, 57%).

## B. Formation of the Galactosyl β1,3-Linkage Using β-Galactosidase [34]

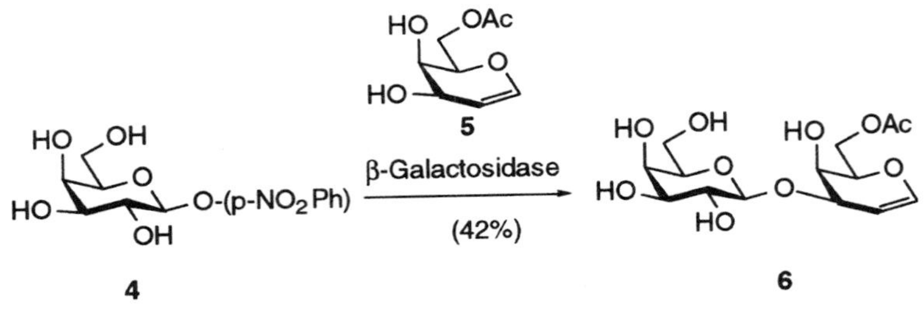

*Compound 6*

A solution of 6-O-acetyl galactal **5** (72 mg, 0.38 mmol) and β-4-nitrophenyl galactopyranoside **4** (80 mg, 0.26 mmol) was prepared in 0.07 $M$ PIPES–0.1 $M$ NaOAc–0.2 $M$ EDTA solution (3 mL) and acetone (100 μL) at 23°C. β-Galactosidase (*E. coli*, 75 U) was added and the reaction mixture was maintained for 26 h. The mixture was concentrated, and the residue purified by silica gel chromatography (eluting with $CHCl_3$–MeOH–EtOAc 5:2:2) to afford **6** (38 mg, 42%) and recovered **5** (39 mg, 54%).

## C. Synthesis of CMP-NeuAc [14]

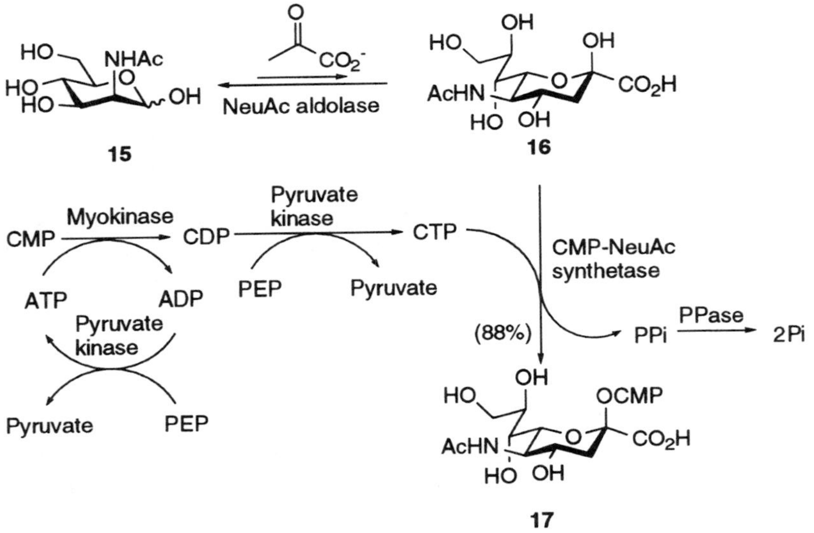

In this one-pot procedure NeuAc **16** is generated from ManNAc **15** and pyruvic acid in situ with sialic acid aldolase and then converted irreversibly to CMP-NeuAc **17**. CMP is converted to CDP with myokinase and ATP. The released ADP is converted to ATP with pyruvate kinase and PEP. CDP is then converted to CTP also with pyruvate kinase and phosphoenolpyruvate (PEP). The formed CTP reacts with NeuAc catalyzed by NeuAc synthetase to give **17**.

## Synthesis by Enzymatic Glycosidation

*CMP-N-Acetylneuraminic Acid* **(17)**

To HEPES buffer (100 mL, 200 m$M$, pH 7.5) were added ManNAc **15** (1.44 g, 6 mmol), PEP sodium salt (1.88 g, 8 mmol), pyruvic acid sodium salt (1.32 g, 12 mmol), CMP (0.64 g, 2 mmol), ATP (11 mg, 0.02 mmol), pyruvate kinase (300 U), myokinase (750 U), inorganic pyrophosphatase (3 U), N-acetylneuraminic acid aldolase (100 U), and CMP-sialic acid synthetase (1.6 U). The reaction mixture was stirred at room temperature for 2 days under argon, until CMP was consumed. The reaction mixture was concentrated by lyophilization and directly applied to a Bio-Gel P-2 column (200–400 mesh, 3 × 90 cm), and eluted with water at a flow rate of 9 mL/h at 4°C. The CMP-NeuAc fractions were pooled, applied to Dowex-1 (formate form), and eluted with an ammonium bicarbonate gradient (0.1–0.5 $M$). The CMP-NeuAc fractions free of the nucleotides were pooled and lyophilized. Excess ammonium bicarbonate was removed by addition of Dowex 50W-X8 (H$^+$ form) to the stirred solution of the residual powder until pH 7.5. The resin was filtered off and the filtrate was lyophilized to yield the ammonium salt of CMP-NeuAc **17** (1.28 g, 88%).

### D. Formation of the Sialyl α2,3-Linkage Using α2,3-Sialyltransferase Without Regeneration of CMP-NeuAc [36]

Galβ1,4GlcNAcβ-O— [structure with HO, Galβ1,4GlcNAcβ-O, OEt, OH] **18**

CMP-NeuAc, α2,3-Sialyl-transferase (86%) →

NeuAcα2,3Galβ1,4GlcNAcβ-O— [structure with HO, NeuAcα2,3Galβ1,4GlcNAcβ-O, OEt, OH] **19**

*Heptasaccharide 19*

To a solution of CMP-NeuAc (116 mg, 0.166 mmol), BSA (5% solution, 0.12 mL), sodium cacodylate (1.8 mL, 1 $M$, pH 6.5), water (5.1 mL), MnCl$_2$ (0.6 mL, 1 $M$), alkaline phosphatase (EC 3.1.3.1; 30 μL, 1 U/μL) and the pentasaccharide **18** (52 mg, 55 μmol) was added α2,3-sialyltransferase (EC 2.4.99.6; 1 U). The reaction mixture was tipped for 4 days, after which time second batches of CMP-NeuAc (116 mg, 0.166 mmol), alkaline phosphatase (30 μL, 1 U/μL), MnCl$_2$ (0.2 mL, 1 $M$), and α2,3-sialyltransferase (1 U) were added and the mixture was tipped for another 5 days. The mixture was filtered and chromatographed (Bio Gel P-2, 0.1 $M$ NH$_4$HCO$_3$) to afford **19** (75 mg, 86%) as a white solid after lyophilization.

### E. Formation of the Sialyl α2,3-Linkage Using α2,3-Sialyltransferase with Regeneration of CMP-NeuAc [24]

The regeneration system for CMP-NeuAc can be employed both for α2,3-sialyltransferase-catalyzed reactions and for reactions mediated by α2,6-sialyltransferase. The system starts with NeuAc, the glycosyl acceptor, PEP, and catalytic amounts of ATP and CMP. CMP is converted to CDP by nucleoside monophosphate kinase (EC 2.7.4.4; NMK) in the presence of ATP, which is regenerated from the by-product ADP, catalyzed by PK in the presence of PEP, then to CTP with PEP by PK. The CTP thus formed reacts with NeuAc, catalyzed by

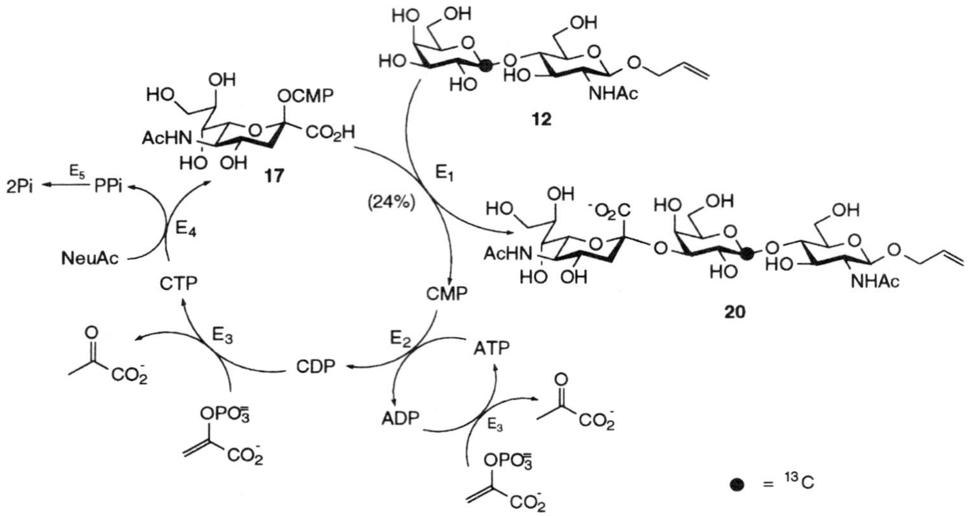

E$_1$: α2,6-sialyltransferase; E$_2$: nucleoside monophosphate kinase or adenylate kinase; E$_3$: pyruvate kinase; E$_4$: CMP-NeuAc synthetase; E$_5$: pyrophosphatase

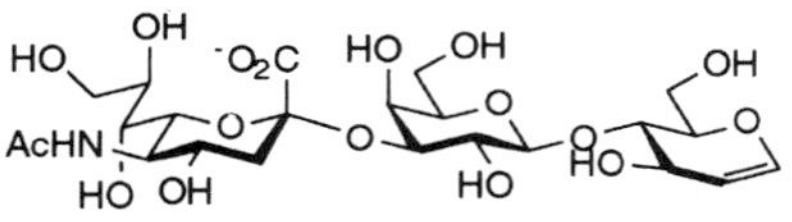

CMP-NeuAc synthetase (EC 2.7.7.43) to produce CMP-NeuAc. The by-product pyrophosphate (PPi) is hydrolyzed to phosphate (Pi) by inorganic pyrophosphatase (PPase). Sialylation is accomplished with α2,3-sialyltransferase (α2,3NeuAcT) or α2,6-sialyltransferase (α2,3NeuAcT), respectively. The released CMP is again converted to CDP, to CTP, and finally to CMP-NeuAc. The UDP-Gal and CMP-NeuAc regeneration schemes have been combined in a one-pot reaction and applied to the synthesis of sialyl Lewis X.

*Sialotrisaccharide* **20**

A solution of 1-[$^{13}$C]Galβ1,4GlcNAcβOallyl **12** (210 mg, 0.50 mmol), N-acetylneuraminic acid (160 mg, 0.52 mmol), PEP sodium salt (120 mg, 0.51 mmol), MgCl$_2$·6H$_2$O (20 mg, 0.1 mmol), MnCl$_2$·4H$_2$O (4.9 mg, 0.025 mmol), KCl (7.5 mg, 0.10 mmol), CMP (16 mg, 0.05 mmol), ATP (2.7 mg, 0.005 mmol), and mercaptoethanol (0.34 μL) in HEPES buffer (3.5 mL, 200 m$M$, pH 7.5) was adjusted with 1 $N$ NaOH to pH 7, and the enzymes NMK (5 U), PK (100 U), PPase (10 U), CMP-NeuAc synthetase (0.4 U), and α2,3NeuAcT (0.1 U) were added to the solution. The mixture was gently stirred under an argon atmosphere at room temperature for 3 days. The mixture was concentrated, and the residue chromatographed on silica gel (eluting with EtOAc-iPrOH-H$_2$O 2:2:1) to give a trisaccharide, which was further purified with Bio Gel P-2 (eluting with water) to give **20** (88 mg, 24%).

By using the same procedure, NeuAcα2,3Galβ1,4Glucal **21** was obtained in 21% yield starting from Galβ1,4Glucal.

## F. Formation of the Sialyl α2,6-Linkage Using α2,6-Sialyltransferase with Regeneration of CMP-NeuAc [37]

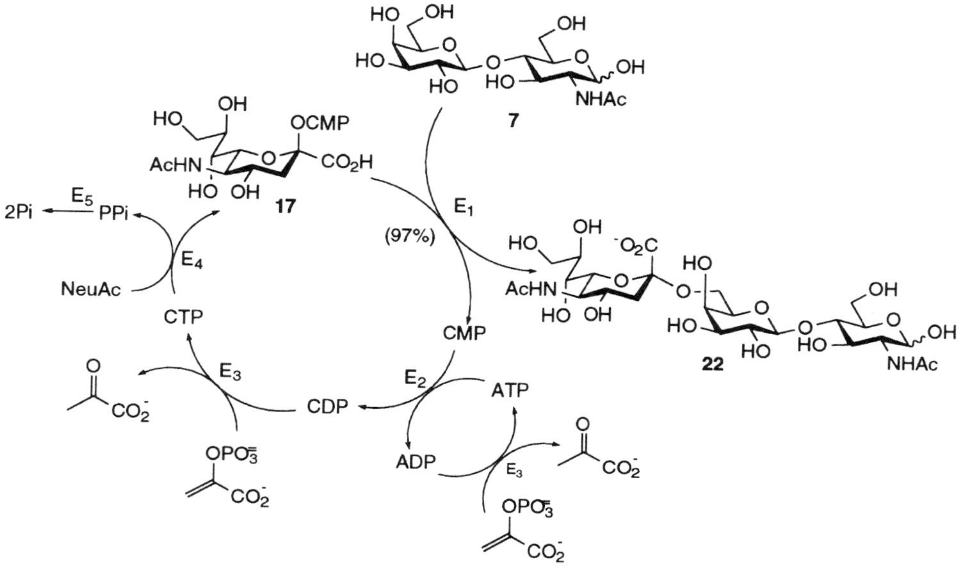

$E_1$: α2,6-sialyltransferase; $E_2$: nucleoside monophosphate kinase or adenylate kinase; $E_3$: pyruvate kinase; $E_4$: CMP-NeuAc synthetase; $E_5$: pyrophosphatase

*NeuAcα2,6Galβ1,4GlcNAc* **22**

To HEPES buffer (150 mL, 0.2 $M$, pH 7.5) were added NeuAc (0.92 g, 3.0 mmol), LacNAc (**7**) (1.1 g, 13 mmol), CMP (30–300 μmol), ATP (3–30 μmol), PEP sodium salt (2.8 g, 6.0 mmol), $MgCl_2 \cdot 6H_2O$ (0.61 g, 3.0 mmol), $MnCl_2 \cdot 4H_2O$ (0.15 g, 0.80 mmol), KCl (0.22 g, 3 mmol), nucleoside monophosphate kinase (450 U), pyruvate kinase (6000 U), inorganic pyrophosphatase (300 U), CMP-NeuAc synthetase (24 U), and α2,6-sialyltransferase (4 U), and the reaction was conducted at room temperature for 2 days under argon. The reaction mixture was concentrated to 20 mL by lyophilization and applied to a Bio Gel P-2 column with water as the mobile phase. The trisaccharide-containing fractions were collected and lyophilized to give pure NeuAcα2,6Galβ1,4GlcNAc **22** (2.0 g, 97%).

## G. Formation of the Fucosyl α1,3-Linkage Using α1,3-Fucosyltransferase Without Regeneration of GDP-Fuc [24]

*Tetrasaccharide* **23**

A solution of α1,3-fucosyltransferase (2 mL, 0.02 U) was added to a solution of the trisaccharide **20** (23 mg, 0.031 mmol) and GDP-Fuc (24 mg, 0.036 mmol) in HEPES buffer (3 mL, 200 m$M$, pH 7.5) containing 5 m$M$ ATP, 20 m$M$ $MnCl_2$, and the mixture was gently stirred under an argon atmosphere at room temperature for 5 days. The mixture was concentrated and the residue chromatographed on silica gel (eluting with EtOAc–iPrOH–$H_2O$ 2:2:1) to give a tetrasaccharide, which was further purified with Bio Gel P-2 (eluting

23

24

25

with water). The eluant was passed through a column of Dowex 50W-X8 ($H^+$ form), eluted with water, neutralized with 1 *N* NaOH, and lyophilized to give **23** (18 mg, 70%).

Compounds **24** and **25** were prepared using the same procedure.

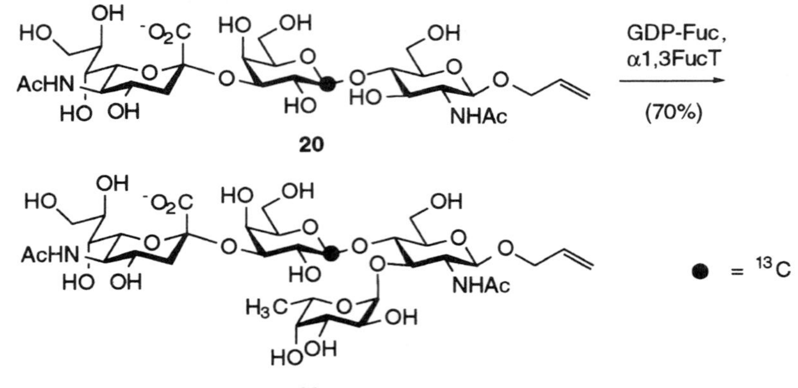

### H. Formation of the Mannosyl α1,2-Linkage Using α1,2-Mannosyltransferase with Regeneration of GDP-Man [29]

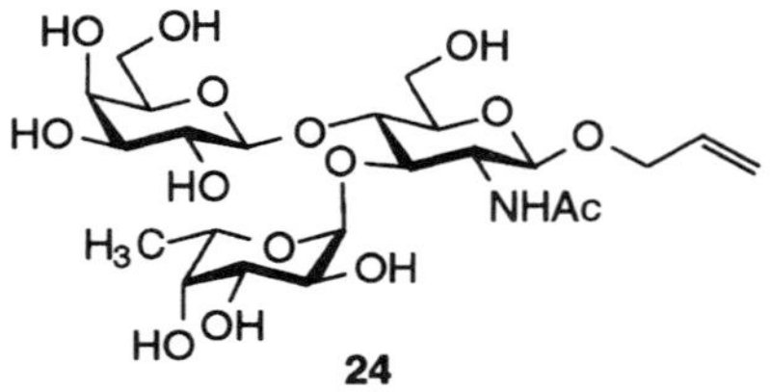

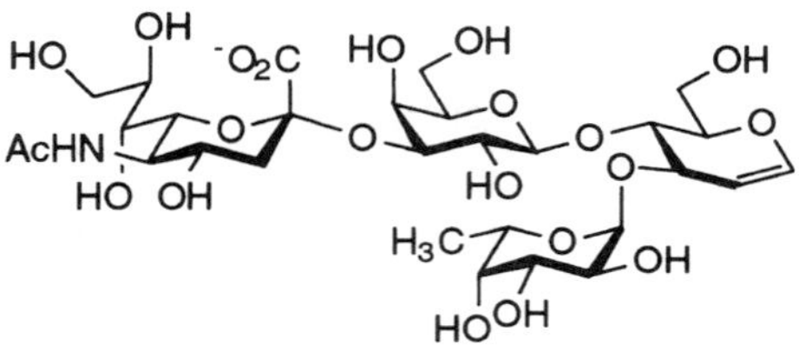

*Cbz-Thr(αManNAcα1,2ManNAc)-Val-OMe* **27**

A reaction mixture (2 mL, 100 m*M* Tris, pH 7.5, 5% acetone, 10 m*M* $MgCl_2$, 10 m*M* $MnCl_2$, 5 m*M* EDTA, 5 m*M* $NaN_3$, 1 m*M* ATP, 0.01 m*M* theophylline, 0.03 m*M* 2,3-dimercapto-propanol, and 0.05 m*M* phenylmethylsulfonyl fluoride) containing Cbz-Thr(α-ManNAc)-Val-OMe **26** (100 mg, 95 m*M*), mannose 1-phosphate (60 mg, 100 m*M*), GDP (9 mg, 10 m*M*), phosphoenolpyruvate (PEP; 47 mg, 100 m*M*), pyruvate kinase (PK; 50 U), dried yeast cells (50 mg), α1,2-mannosyltransferase (0.4 U), and inorganic pyrophosphatase (1 U) was slightly stirred at room temperature for 60 h and then centrifuged. The supernatant was lyophilized and the residue extracted with MeOH. MeOH was removed and the residue purified by silica gel chromatography ($CHCl_3$–MeOH–$H_2O$ 12:6:1) to afford Cbz-Thr(αManNAcα1,2ManNAc)-Val-OMe **27** (31 mg, 41%).

### I. Combined Use of β-Galactosidase and α2,6-Sialyltransferase with Regeneration of CMP-NeuAc [35]

*NeuAcα2,6LacNAc* **22**

To HEPES buffer (1.07 mL, 0.2 *M*, 20 m*M* $MgCl_2$, 5.3 m*M* $MnCl_2$, 20 m*M* KCl, pH 7.5) containing NeuAc (12.3 mg, 20 m*M*), lactose (180 mg, 250 m*M*), GlcNAc (265 mg,

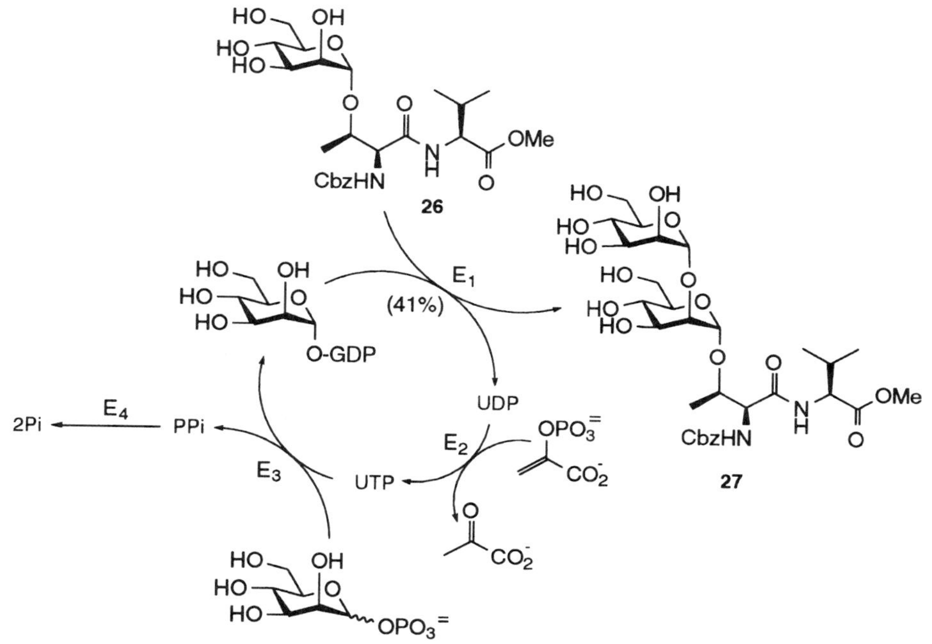

# Synthesis by Enzymatic Glycosidation

E$_1$: α1,2-mannosyltransferase; E$_2$: pyruvate kinase;
E$_3$: GDP-Man pyrophosphorylase; E$_4$: inorganic pyrophosphatase.

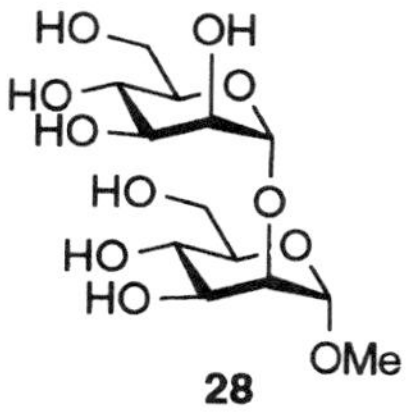

600 mM), phosphoenolpyruvate trisodium salt (30 mg, 40 mM), CMP (2 mM), and ATP (0.2 mM) were added myokinase (EC 2.7.4.3; MK; 6 U), pyruvate kinase (EC 2.7.1.40; PK; 80 U), inorganic pyrophosphatase (EC 3.6.1.1; PPase; 4 U), CMP-NeuAc synthetase (EC 2.7.7.43; 0.32 U), α2,6 sialyltransferase (EC 2.4.99; 0.052 U), and crude β-galactosidase from *Bacillus circulans* (EC 3.2.1.23; 1 mg). The total volume was adjusted to 2 mL. The reaction was conducted for 91 h under argon at room temperature. The reaction mixture was centrifuged and the supernatant was directly applied to a Bio Gel P-2 column (43 × 2 cm) with water as the eluant. The trisaccharide-containing fractions were pooled and lyophilized to give NeuAcα2,6LacNAc **22** (6.8 mg, 26%). No other sialylated product could be detected.

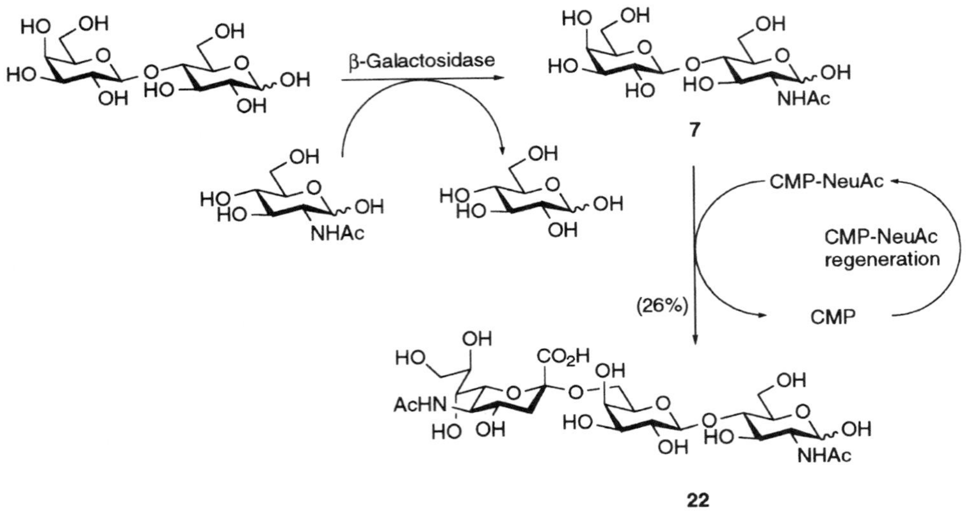

## REFERENCES

1. A. Varki, Biological roles of oligosaccharides: All of the theories are correct, *Glycobiology* 3:97 (1993).
2. R. W. Ledeen, E. L. Hogan, G. Tettamanti, A. J. Yates, and R. K. Yu, eds., *New Trends in Ganglioside Research*, Liviana Press, Padova, 1988.
3. G. van Echten and K. Sandhof, Modulation of ganglioside synthesis in primary cultured neurons, *J. Neurochem.* 52:207 (1989).
4. K.-A. Karlsson, Animal glycosphingolipids as membrane attachment sites for bacteria, *Annu. Rev. Biochem.* 58:309 (1989).
5. T. J. Pritchett, R. Brossmer, U. Rose, and J. C. Paulson, Recognition of monovalent sialosides by influenza virus H3 hemagluttinin, *Virology* 160:502 (1987).
6. F. Barresi and O. Hindsgaul, Synthesis of β-mannopyranosides by intramolecular aglycon delivery, *J. Am. Chem. Soc.* 113:9376 (1991).
7. G. Stork and G. Kim, Stereocontrolled synthesis of disaccharides via the temporary silicon connection, *J. Am. Chem. Soc.* 114:10067 (1992).
8. M. P. DeNinno, The synthesis and glycosidation of *N*-acetylneuraminic acid, *Synthesis* p. 583 (1991).
9. R. Kornfeld and S. Kornfeld, Assembly of asparagine-linked oligosaccharides, *Annu. Rev. Biochem.* 54:631 (1985).
10. J. Roth, Subcellular organization of glycosylation in mammalian cells, *Biochim. Biophys. Acta* 906:405 (1987).
11. L. Anderson and F. M. Unger, eds., *Bacterial Lipopolysaccharides*, ACS Symposium Series 231. American Chemical Society, Washington DC, 1983.
12. C. H. Wong, R. L. Halcomb, Y. Ichikawa, and T. Kajimoto, Enzymes in organic synthesis: Application to the problems of carbohydrate recognition (part 2), *Angew. Chem. Int. Ed. Engl.* 34:1010 (1995).
13. M. M. Palcic, Glycosyltransferases in glycobiology, *Methods Enzymol.* 230:145 (1994).
14. J. L.-C. Liu, G.-J. Shen, J. F. Rutan, G. Zapata, W. F. Vann, and C.-H. Wong, Overproduction of CMP-sialic acid synthetase for organic synthesis, *J. Am. Chem. Soc.* 114:3901 (1992).

15. H. A. Nunez and R. Barker, Enzymatic synthesis and carbon-13 nuclear magnetic resonance conformational studies of disaccharides containing β-D-galactopyranosyl and β-D-[1-$^{13}$C]galactopyranosyl residues, *Biochemistry* 19:489 (1980).
16. C. H. Wong, Y. Ichikawa, T. Krach, C. Gautheron-Le Narvor, D. P. Dumas, and G. C. Look, Probing the acceptor specificity of β-1,4-galactosyltransferase for the development of enzymatic synthesis of novel oligosaccharides, *J. Am. Chem. Soc.* 113:8137 (1991).
17. T. Wiemann, V. Sinwell, and J. Thiem, Xylose: The first ambient acceptor substrate for galactosyltransferase from bovine milk, *J. Org. Chem.* 59:6744 (1994).
18. H. Yuasa, O. Hindsgaul, and M. M. Palcic, Chemical-enzymatic synthesis of 5′-thio-N-acetyllactosamine: The first disaccharide with sulfur in the ring of the nonreducing sugar, *J. Am. Chem. Soc.* 114:5891 (1992).
19. S. Sabesan and J. C. Paulson, Combined chemical and enzymatic synthesis of sialyloligosaccharides and characterization by 500-MHz $^1$H and $^{13}$C NMR spectroscopy, *J. Am. Chem. Soc.* 108:2068 (1986).
20. A. Buensch, J. C. Paulson, and R. Brossmer, Activation and transfer of novel synthetic 9-substituted sialic acids, *Eur. J. Biochem.* 168:595 (1987).
21. Y. Ito, J. J. Gaudino, and J. C. Paulson, Synthesis of bioactive sialosides, *Pure Appl. Chem.* 65:753 (1993).
22. R. D. McCoy, E. R. Vimr, and F. A. Troy, CMP-NeuNAc:poly-α-2,8-sialosyl sialyltransferase and the biosynthesis of polysialosyl units in neural cell adhesion molecules, *J. Biol. Chem.* 260:12695 (1985).
23. U. B. Gokhale, O. Hindsgaul, and M. M. Palcic, Chemical synthesis of GDP-fucose analogs and their utilization by the Lewis α-1,4-fucosyltransferase, *Can. J. Chem.* 68:1063 (1990).
24. Y. Ichikawa, Y.-C. Lin, D. P. Dumas, G.-J. Shen, E. Garcia-Junceda, M. A. Williams, R. Bayer, C. Ketcham, L. E. Walker, J. C. Paulson, and C.-H. Wong, Chemical-enzymatic synthesis and conformational analysis of sialyl Lewis$_x$ and derivatives, *J. Am. Chem. Soc.* 114:9283 (1992).
25. C.-H. Wong, D. P. Dumas, Y. Ichikawa, K. Koseki, S. J. Danishefsky, B. W. Weston, and J. B. Lowe, Specificity, inhibition and synthetic utility of a recombinant human α-1,3-fucosyltransferase, *J. Am. Chem. Soc.* 114:7321 (1992).
26. D. P. Dumas, Y. Ichikawa, C.-H. Wong, J. B. Lowe, and R. P. Nair, Enzymatic synthesis of sialyl Le$^x$ and derivatives based on a recombinant fucosyltransferase, *Bioorg. Med. Chem. Lett.* 1:425 (1991).
27. I. Brackhausen, E. Hull, O. Hindsgaul, H. Schachter, R. N. Shah, S. W. Michnick, and J. P. Carver, Control of glycoprotein synthesis, *J. Biol. Chem.* 264:11211 (1989).
28. G. Strivastava, G. Alton, and O. Hindsgaul, Combined chemical–enzymic synthesis of deoxygenated oligosaccharide analogs: Transfer of deoxygenated D-GlcNAc residues from their UDP-GlcNAc derivatives using N-acetylglucosaminlytransferase I, *Carbohydr. Res.* 207:259 (1990).
29. P. Wang, G.-J. Shen, Y.-F. Wang, Y. Ichikawa, and C.-H. Wong, Enzymes in oligosaccharide synthesis: Active-domain overproduction, specificity study, and synthetic use of an α1,2-mannosyltransferase with regeneration of GDP-Man, *J. Org. Chem.* 58:3985 (1993).
30. G. F. Herrmann, P. Wang, G.-J. Shen, E. Garcia-Junceda, S. H. Khan, K. L. Matta, and C.-H. Wong, Large-scale production of recombinant α-1,2-mannosyltranferase from *E. coli* for the study of acceptor specificity and use of the recombinant whole cells in synthesis, *J. Org. Chem.* 59:6356 (1994).
31. S. L. Flitsch, H. L. Pinches, J. P. Taylor, and N. J. Turner, Chemo–enzymatic synthesis of a lipid-linked core trisaccharide of N-linked glycoproteins, *J. Chem. Soc. Perkin Trans.* 1:2087 (1992).
32. T. Kasumi, Y. Tsumuraya, C. F. Brewer, H. Kersters-Hilderson, M. Claeyssens, and E. S. Hehre, Catalytic versatility of *Bacillus pumilus* β-xylosidase: Glycosyl transfer and hydrolysis promoted with α- and β-D-xylosyl fluoride, *Biochemistry* 26:3010 (1987).
33. D. H. G. Crout, D. A. MacManus, J.-M. Ricca, S. Singh, P. Critchley, and W. T. Gibson, Biotransformations in the peptide and carbohydrate fields, *Pure Appl. Chem.* 64:1079 (1992).

34. G. C. Look, Y. Ichikawa, G. J. Shen, P. W. Cheng, and C.-H. Wong, A combined chemical and enzymatic strategy for the construction of carbohydrate-containing antigen core units, *J. Org. Chem. 58*:4326 (1993).
35. G. F. Herrmann, Y. Ichikawa, C. Wandrey, F. C. A. Gaeta, J. C. Paulson, and C.-H. Wong, A new multi-enzyme system for a one-pot synthesis of sialyl oligosaccharides: Combined use of β-galactosidase and α(2,6)-sialyltransferase coupled with regeneration in situ of CMP-sialic acid, *Tetrahedron Lett. 34*:3091 (1993).
36. S. A. DeFrees, W. Kosch, W. Way, J. C. Paulson, S. Sabesan, R. L. Halcomb, D.-H. Huang, Y. Ichikawa, and C.-H. Wong, Ligand recognition by E-selectin: Synthesis, inhibitory activity, and conformational analysis of bivalent sialyl Lewis x analogs, *J. Am. Chem. Soc. 117*:66 (1995).
37. Y. Ichikawa, G.-J. Shen, and C.-H. Wong, Enzyme-catalyzed synthesis of sialyl oligosaccharide with in situ regeneration of CMP-sialic acid, *J. Am. Chem. Soc. 113*:4698 (1991).

# V

# Synthesis of C-Glycosyl Compounds

**THEMES**: *Free Radical–Mediated Reactions; Lewis Acid, Transition Metal–Mediated Reactions*

The backbone of organic synthesis resides in our ability to make carbon–carbon bonds. The annals of synthesis methods dating back over a century illustrate this feature with innumerable examples. While organic chemists were busy laying down the foundations of modern physical organic chemistry in the 1950s and 1960s, some carbohydrate chemists were beginning to detach themselves from the "*O*-methylation/hydrolysis analysis," era of polysaccharide chemistry. Although much elegant methodology was also developed by carbohydrate chemists during this period, it involved, to a great measure, peripheral chemistry, such as sugar-to-sugar conversions. Thus, all of the exotic aminodeoxy, deoxy, and branched-chain sugars isolated from the hydrolysis of antibiotics and other products of fermentation were masterfully synthesized—mostly from other sugars. Great contributions were made in forming hetero-atom bonds via $S_N2$ displacement reactions and via neighboring group participation for which sugar molecules were a veritable playground. In fact it was up to the late B. R. Baker, and other non-carbohydrate chemists by training, such as Leon Goodman, to translate S. Winstein's teachings to practice in the synthesis of natural and non-natural amino sugars. The first documented example of neighboring group participation of an acetoxy group was that of H. S. Isbell in a paper published in *Journal of the National Bureau of Standards* in 1928. Unfortunately, it was obscured by later publications by Winstein and others in more fashionable journals. It was Hans Paulsen who showed applications of H. Meerwein's acyloxonium ion chemistry to carbohydrate chemistry in the mid-1960s by a "one-step" D-glucose-to-D-idose conversion.

Historically, the formation of carbon–carbon bonds in carbohydrate chemistry was done out of necessity, as in the synthesis of a branched-chain sugar. The synthesis of *C*-glycosyl compounds was a greater challenge because of the need to activate the anomeric carbon atom in a cyclic structure. The basicity of the traditional organometallic reagents, coupled with the propensity for glycal formation via elimination, rendered this approach impractical. Nevertheless, C. D. Hurd and W. A. Bonner were successful in preparing glycosylarenes in the late 1940s and early 1950s. The advent of Lewis acid catalysis and other innovations involving the use of masked anions such as ketene acetals and enol ethers greatly expanded the possibilities in *C*-glycoside synthesis. My first Ph.D. student, André Pernet, now vice-president for research and development in a major Midwestern pharmaceutical laboratory, demonstrated in 1971 that malonate anion and related ketene silyl acetals were excellent nucleophiles for the synthesis of *C*-glycosyl and *C*-ribosyl compounds. While in my laboratory in 1973, Tomoya Ogawa, presently at Riken (Tokyo) and a world-renowned carbohydrate chemist today, showed that ketene acetals and enol ethers were versatile reagents for the synthesis of *C*-glycosyl compounds. Since then, the field has expanded enormously, particularly with the advent of new methodology and the potential importance of *C*-glycosyl compounds in medicinal chemistry.

Because of the delicate balance of functionality in a sugar derivative, it is not surprising that logic and mechanistic insight would join forces to allow for radical-mediated reactions to take place at the anomeric carbon atom. Free-radical reaction conditions are mild and neutral, thus allowing esters or even unprotected hydroxy groups to be used. In addition, stereoelectronic factors play an important role in determining anomeric configuration in free-radical reactions, favoring an "antiperiplanar" approach to the lone pair of the ring oxygen in a preferred conformation.

Viewed (and drawn) in a "terpene-like" perspective, *C*-glycosyl compounds are in fact functionalized tetrahydrofurans and tetrahydropyrans. As such, they can be considered as versatile chirons for the synthesis of a variety of natural products containing cyclic ether motifs, such as in the ionophores.

The next two chapters give the reader an excellent perspective and laboratory-proven methods for the synthesis of *C*-glycosyl compounds in general, and glycosylarenes in particular, utilizing free-radical reactions and Lewis acid–catalyzed oxonium ion chemistry, respectively.

*Stephen Hanessian*

# 23

## C-Glycosyl Compounds from Free Radical Reactions

**Bernd Giese and Heinz-Georg Zeitz**
*University of Basel, Basel, Switzerland*

| | | |
|---|---|---|
| I. | Introduction | 507 |
| | A. Carbohydrate radicals in synthesis | 507 |
| | B. Stereochemistry and conformations of carbohydrate radicals | 509 |
| II. | Intermolecular Methods | 510 |
| | A. The tin hydride method | 510 |
| | B. The fragmentation method | 511 |
| | C. The cobalt method | 513 |
| | D. The vitamin $B_{12a}$-catalyzed method | 514 |
| | E. Methods for the formation of β-C-glycosyl compounds | 515 |
| III. | Intramolecular Methods | 516 |
| IV. | Experimental Procedures | 517 |
| | A. Intermolecular methods | 517 |
| | B. Intramolecular methods | 522 |
| | References | 524 |

## I. INTRODUCTION

### A. Carbohydrate Radicals in Synthesis

The synthesis of *C*-glycosyl compounds, commonly known as *C*-glycosides, in ionic reactions relies on the electrophilicity of the anomeric center and, therefore, involves the attack of an appropriate *C*-nucleophile. An "umpolung" method has been developed, and is described in the previous chapter. But instead of going from a carbocation to a carbanion, one can also consider homolytic or radical reactions to reverse the philicity (Scheme **1**).

Alkoxyalkyl radicals behave like nucleophiles because of the high-lying SOMO which can interact with the LUMO of an electron-poor alkene, such as acrylonitrile,

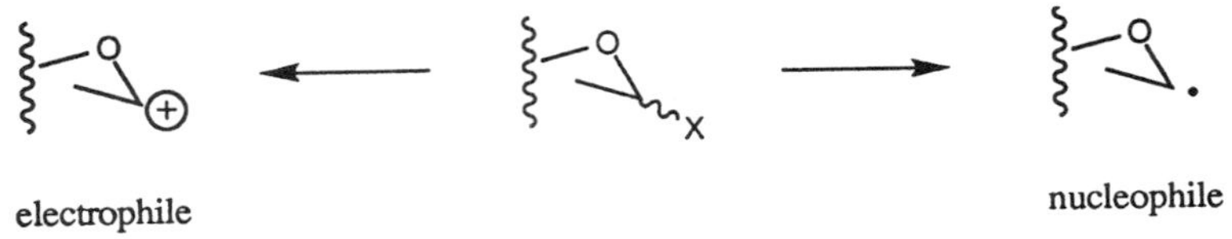

**Scheme 1**

acrolein, fumarodinitrile, maleic anhydride, or such; to form a C–C bond. The presence of electron-withdrawing substituents in the alkene lowers the LUMO energy and increases the addition rate by reducing the SOMO–LUMO difference [1]. Thus, the addition of a glycosyl radical to an electron-poor olefin constitutes an extremely appealing strategy for the construction of C-glycosides or C-disaccharides.

The synthesis of the C-disaccharide **1** of kojibiose from the glucosyl radical **2** and the alkene **3** is a good example for the C–C bond formation by a radical approach (Scheme 2) [2–4].

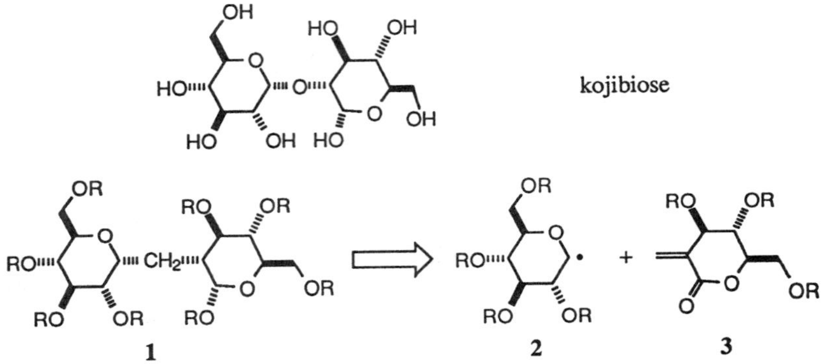

**Scheme 2**

Glycosyl radicals can also be used to mimic the enzymatic aldol reaction between phosphoenolpyruvate and carbohydrates [5]. Whereas respective ionic reactions fail in the absence of enzymes, alkenes **4–6** are suitable synthons for pyruvate in radical C–C bond-forming reactions (Scheme 3).

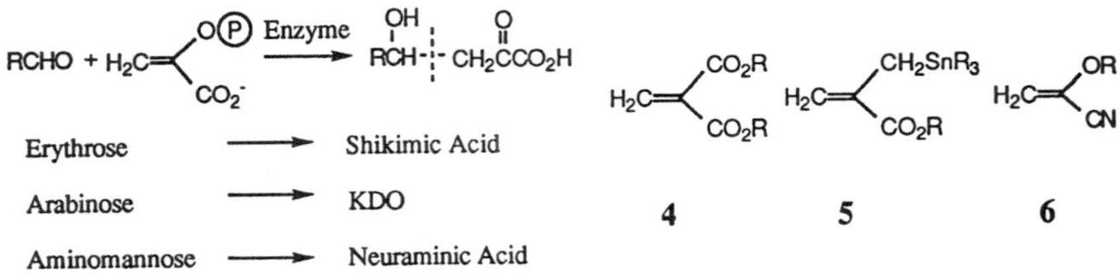

**Scheme 3**

The reactions are complementary to the approach of conjugate addition of an organometallic reagent to an electron deficient alkene. If the required functionality at the anomeric position is introduced before the radical step, one also has access to C-glycosides by trapping glycosyl radicals with hydrogen atom-transfer reagents. The advantage of free radical chemistry includes neutral conditions so that protection of alcohols, amines, and carbonyl groups is often unnecessary. In contrast with ionic intermediates, in most radicals β-oxygen and β-nitrogen substituents are not cleaved or do not rearrange.

## B. Stereochemistry and Conformations of Carbohydrate Radicals

One important aspect of free radical chemistry at the anomeric center is the predictable stereoselectivity of hexopyranosyl radicals. These radicals react with acrylonitrile to give axial-substituted adducts preferentially, whereas cyclohexyl radicals and carbohydrates with the radical center at C-2, C-3, and C-4, and equatorial β-substituents react in the equatorial mode (Scheme 4).

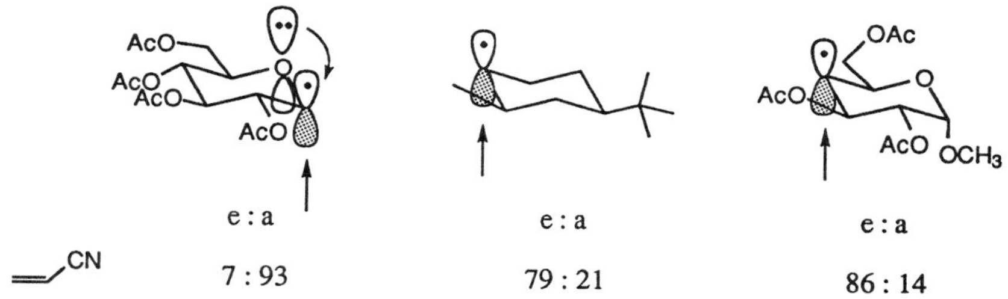

**Scheme 4**

This different behavior can be explained by the conformation of the radicals and by stereoelectronic effects [6]. Electron spin resonance (ESR) investigations have revealed that the D-glucopyranosyl radical **7** does not adopt the $^4C_1$ conformation **7a**, but is distorted into the $B_{2,5}$ shape **7b** (Scheme 5) [7,8]. The equatorial-like attack at the boat conformer

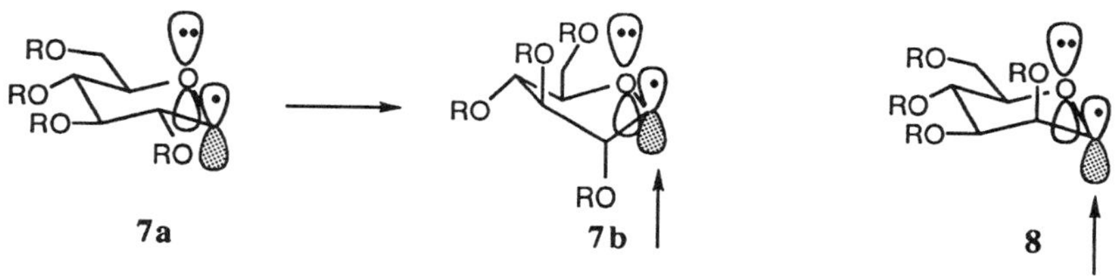

**Scheme 5**

leads to the predominant formation of the observed α-C-glycosides. In addition, during the attack the stabilizing conjugative interaction between the lone pair at the ring oxygen and the SOMO is maintained.

In contrast, the mannosyl radical **8** does not undergo such a conformational change, and the observed α-attack results from the shielding effect of the axial C-2 substituent in the chair conformation and the stereoelectronic effects mentioned earlier. In radicals **7b** and **8** the C–O bonds adjacent to the radical center are coplanar with the singly occupied orbital. This reminds us of the anomeric effect in which an interaction between the nonbonding electron pair of the ring oxygen and the LUMO of the C–O bond stabilizes the conformation.

A pentopyranosyl radical is much more flexible than a hexopyranosyl radical. Because the "alkyl-anchor" at C-5 is absent, the radical is now so flexible that several species of similar energy can coexist. According to ESR spectroscopic data, the arabinopyranosyl radical **9** exists as an equilibrium between the $^4C_1$ **9a** and the $B_{0,3}$ **9b** conformation, which both realize a coplanar arrangement of the C–O bond and the SOMO [9] (Scheme 6). Reactions with alkenes are unselective. However, the arabinofuranosyl radical **10** reacts with high stereoselectivity [9]. This is due to its $^2E$ conformation in which the si-side of the radical center is sterically hindered by the large benzoyl group.

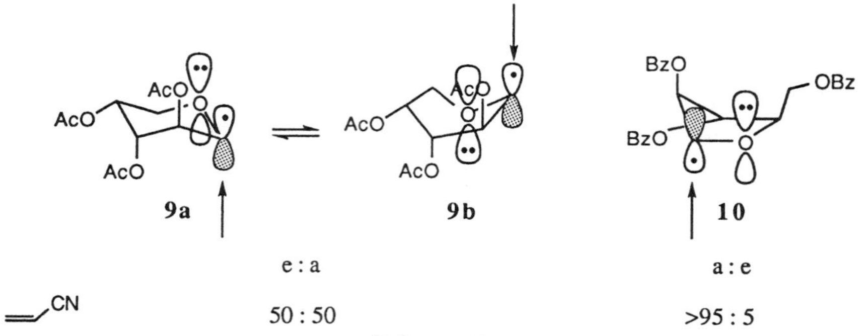

9a ⇌ 9b          10

CH₂=CHCN

e : a
50 : 50

a : e
>95 : 5

**Scheme 6**

## II. INTERMOLECULAR METHODS

### A. The Tin Hydride Method

The tin hydride method is very valuable for forming a C–C bond by addition of a radical to a C–C multiple bond in intermolecular or intramolecular systems [1,10,11]. Since the general principles of the chain reaction are known, this method is easy to apply. Therefore, one can design the reaction conditions carefully, taking into account the rates of the competing reactions and how the reactivity of the radical and the alkene can be influenced by substituents. Actually, formation of *C*-glycosides is the first example of the tin hydride method [12,13].

The reaction process, outlined in Scheme 7, involves the initial formation of the radical **11** by atom abstraction of the trialkyltin radical **15** from the glycosyl precursor **17**.

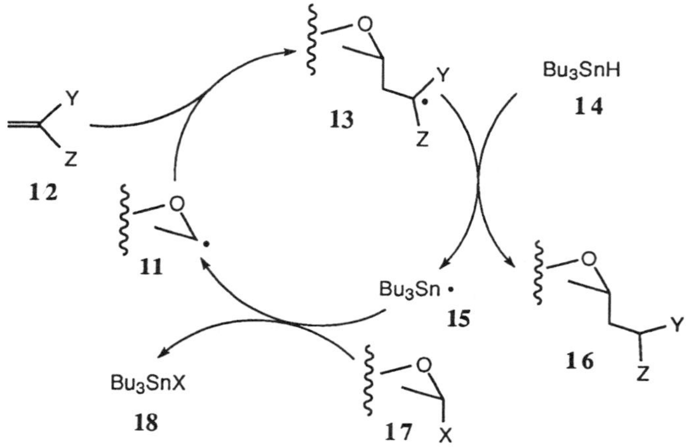

**Scheme 7**

The anomeric radical **11** adds to the olefin **12** to give the intermediate **13**. Interception of this adduct radical **13** by tin hydride **14** yields the saturated product **16** and the organometallic radical **15**. Eventually, the latter reacts with the precursor **17** to produce the chain-carrying radical **11** and the tin compound **18**.

The wide versatility of the tin hydride method in carbohydrate chemistry exists because anomeric radicals can be generated from many functional groups at the anomeric position. However, in practice, only those precursors in which the abstraction of X in **17** by the stannyl radical **15** is sufficiently rapid to compete with hydrostannylation of the alkene, are useful. Thus, the reaction has been applied to bromides, phenylselenides, thiocarbonyl esters, and to tertiary nitro sugars. Table 1 gives some representative examples that were chosen to illustrate both the variety of useful radical-leaving groups and the structural complexity compatible with this method in radical glycosidation reactions.

One important application of the tin hydride method in carbohydrate chemistry is the synthesis of *C*-disaccharides as encountered in Scheme 2.

The tin hydride method suffers from one major disadvantage, namely the efficiency of the reagent as a hydrogen atom donor. For successful synthesis, alkenes have to be reactive enough, otherwise direct reduction of the starting precursor becomes a considerable side reaction. In practice, the yields are increased by slow addition of a solution of tin hydride and a radical initiator into the reaction mixture containing an excess of alkene. However, a delicate balance must be maintained. If a large excess of olefin is used, polymerization can compete. 2,2-Azobisisobutyronitrile is the most commonly employed initiator, with a half-life time for unimolecular scission of 1 h at 80°C.

## B. The Fragmentation Method

A useful sequence for the formation of *C*-glycosides is the fragmentation method [18]. The mechanism, depicted in Scheme **8**, is based on the addition of the carbohydrate radical **11** to

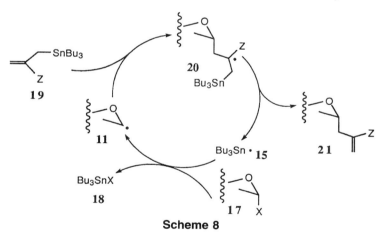

**Scheme 8**

the allyl tin compound **19**. The radical **11** is generated, as in the tin hydride method, by atom abstraction of the stannyl radical **15** from the precursor **17**. The formed adduct radical **20** undergoes a β-bond cleavage, producing the allylated compound **21** and the chain-propagating tin radical **15**.

Because no tin hydride is present, intermediate radicals are only slowly intercepted by hydrogen atom abstraction. Thus, the fragmentation method is a clever alternative that avoids this course, and low concentrations are not required. As a result of this, even relatively unreactive precursors, such as glycosyl chlorides and phenylsulfides, can be used. Therefore, the method is compatible with the same molecular complexity and an extended spectrum of functionality as found in the tin hydride method.

Use of the appropriately substituted glycosyl chloride **22** in a fragmentation reaction

**Table 1** The Tin Hydride Method for the Addition of Glycosyl Radicals to Alkenes

| Substrate | Alkene | Products | Yield (%) | Ref. |
|---|---|---|---|---|
| 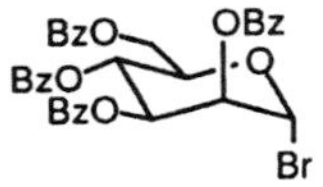 (BzO-protected mannosyl bromide) | 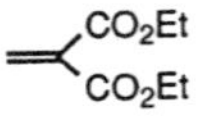 CH$_2$=C(CO$_2$Et)$_2$ | 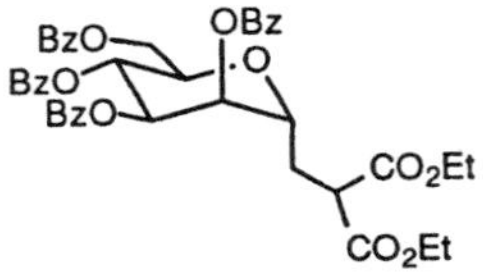 (BzO-protected C-glycoside with CH$_2$CH(CO$_2$Et)$_2$) | 99 | 14 |
| 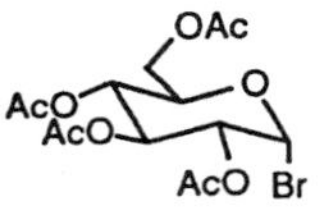 (AcO-protected glycosyl bromide) | 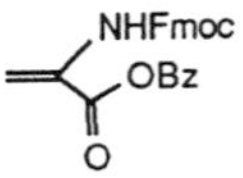 CH$_2$=C(NHFmoc)(C(O)OBz) | 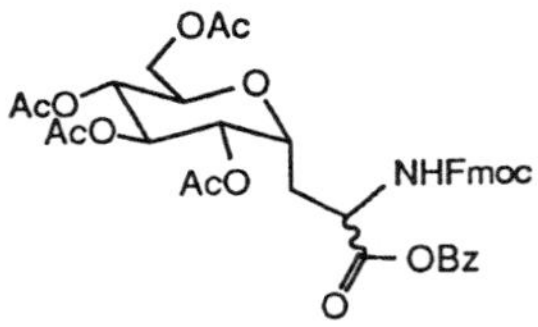 (AcO-protected C-glycoside with CH$_2$C(NHFmoc)(C(O)OBz)) | 75 | 15 |
| 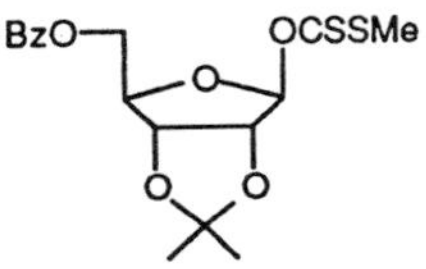 (BzO, OCSSMe furanose) |  MeO$_2$C-CH=CH-CO$_2$Me | 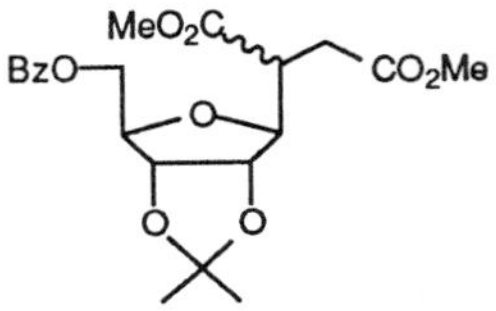 (BzO furanose with CH(CO$_2$Me)CH$_2$CO$_2$Me) | 62 | 16 |
| 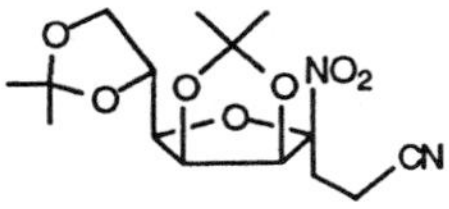 (diacetone sugar with NO$_2$ and CN) | 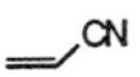 CH$_2$=CHCN | 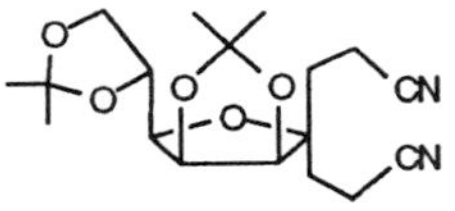 (diacetone sugar with CN and CH$_2$CH$_2$CN) | 54 | 17 |

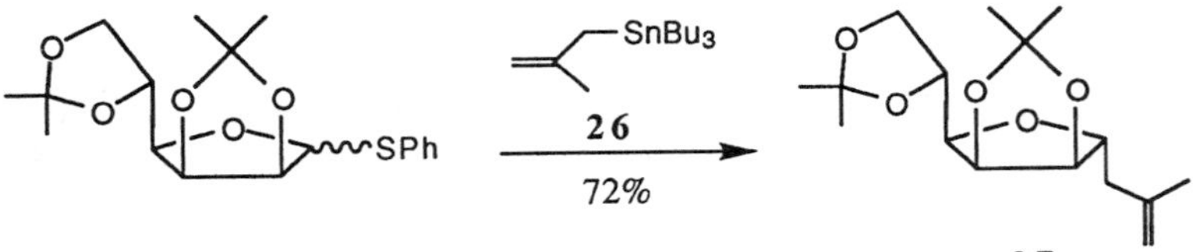

with **23** permits the formation of the protected C-glycosides **24** of N-acetylneuraminic acid [19,20]. The reaction is readily extended to methallylstannanes **26** [21].

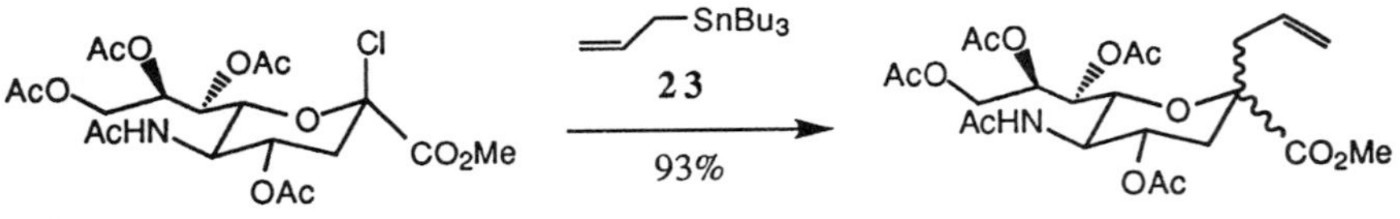

Substitution at the terminal position of the allylstannane, as in crotonyltributyl stannane, however, is not tolerated, because hydrogen abstraction from the allylic position is a competing reaction [21]. An extension of the method involves the coupling of the anomeric radical precursors **28** with the allyltributyltin reagent **29** [14]. In the reagent **29** the double bond is activated toward addition of nucleophilic radicals by the electron-withdrawing t-butoxy carbonyl group. The obtained product **30** has been useful en route to 3-deoxy-D-*manno*-2-octulosonic acid (KDO).

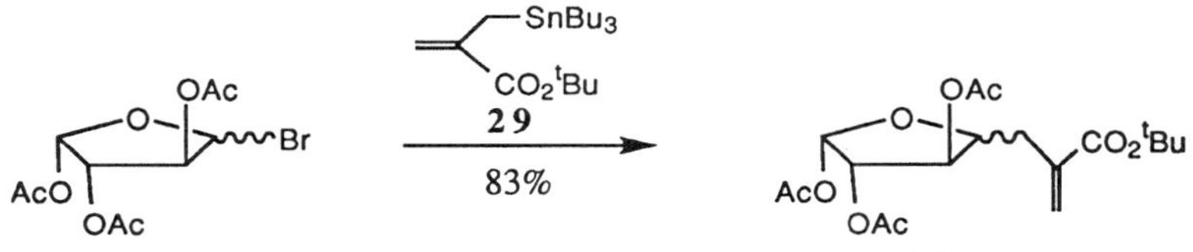

## C. The Cobalt Method

Another possible precursor to conduct free radical reactions is the glycosyl–cobalt(III) dimethylglyoximato complex **33** [22,23]. These organometallic compounds can readily be prepared by the displacement of the halide atom in **17** with the highly nucleophilic cobalt(I) anion **32**. The latter can be generated from the dimeric Co(II) complex **31** under reducing conditions.

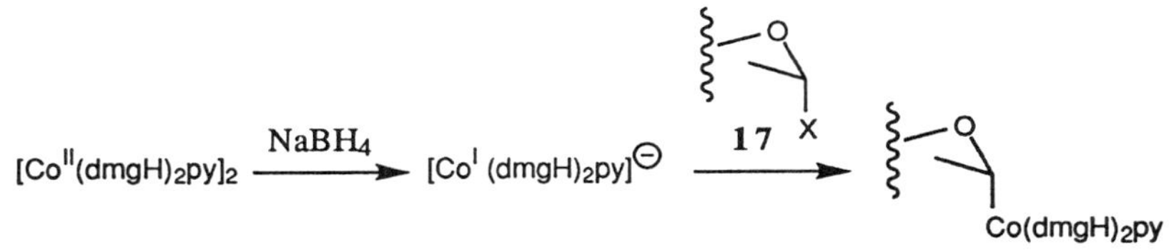

Because such alkylation proceeds by $S_{RN}1$ mechanism, even cobalt complexes derived from unreactive (in an $S_N2$ sense) halides can be formed. The cobalt complexes are air-stable compounds, but are affected by direct daylight. The incorporated Co–C bond is weak and, therefore, photolysis of **33** sets free the anomeric radical **11**. In the presence of olefins **12** this radical adds to the double bond, followed by subsequent combination to give the insertion product **35** (Scheme 9).

Depending on the substituents, either elimination reactions or solvolysis leads to the product **36** or **37**, respectively. The formation of addition product **37** can be explained by heterolytic cleavage under formation of a Co(III) complex and a carbanion, which is then protonated by the solvent. Owing to the carbanion-stabilizing ability of the cyano group this pathway is pronounced in the reaction with acrylonitrile. The product **39** is formed

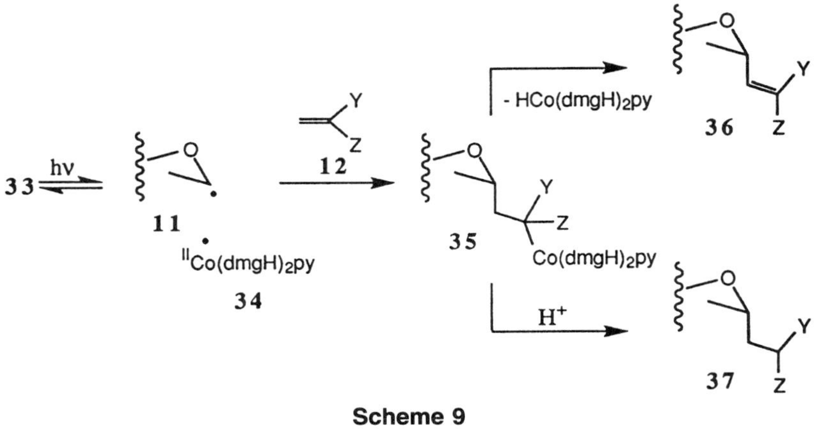

**Scheme 9**

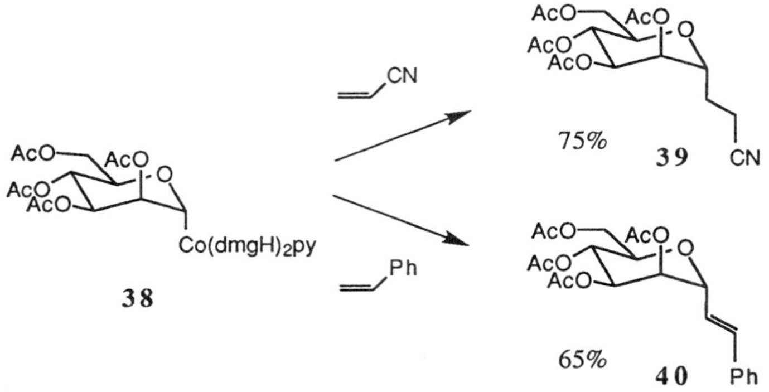

selectively. In contrast, attack to styrene yields the substitution product **40** exclusively [22]. In view of suitable biomimetic synthons of phosphoenolpyruvate, ethoxyacrylonitrile **41** has been used in the reaction with the cobalt complex **38** [22]. The substitution product **42** gives aldol reaction products after hydrolysis.

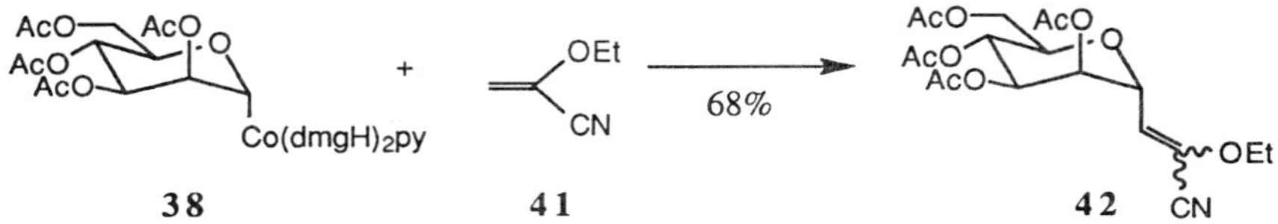

## D. The Vitamin B$_{12a}$-Catalyzed Method

A catalytic variation of the cobalt method is the vitamin B$_{12a}$-mediated reaction to provide C-glycosides [24]. The intermediate in this chain mechanism is the sugar cobalamine **45** formed by reaction of the precursor **43** with vitamin B$_{12a}$ **44** (the brackets in Scheme **10** represent the corrin system) [25].

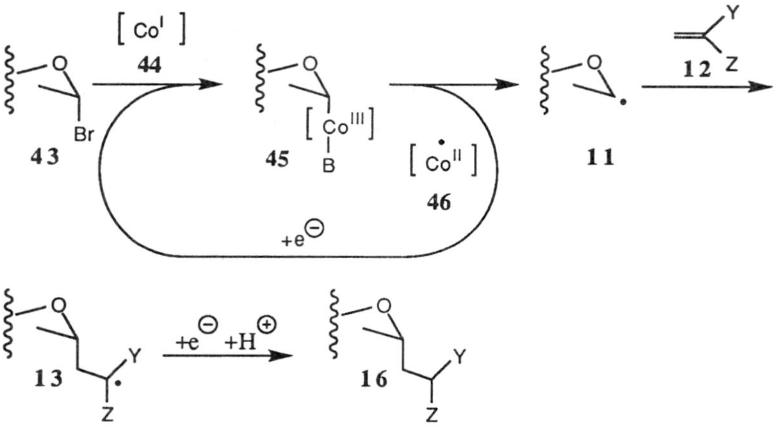

**Scheme 10**

Cleavage of the Co–C in bond in **45** leads to the glycosyl radical **11** which is trapped by the electron-poor alkene **12**. Reduction of the adduct radical **13**, followed by protonation, gives the addition product **16**. The sugar cobalamine **45** is regenerated by reduction (**46** → **44**) and alkylation (**44** → **45**).

Application of this method opens a way to synthesize *C*-glycosides **48** and **49**, similar to those obtained by the tin hydride method.

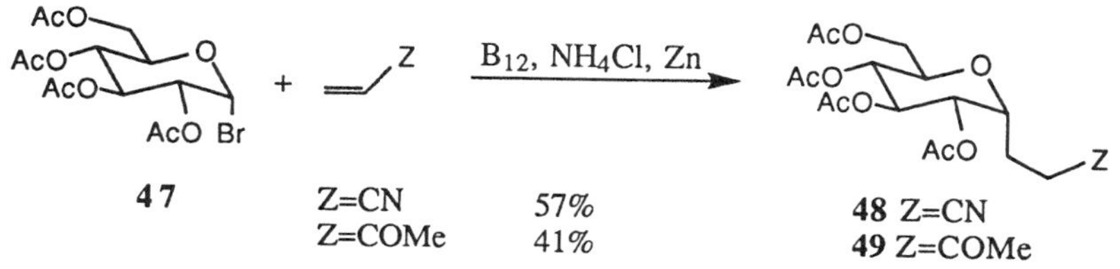

$$\text{47} + \text{CH}_2=\text{CHZ} \xrightarrow{\text{B}_{12}, \text{NH}_4\text{Cl}, \text{Zn}} \text{48/49}$$

|  | Z=CN | 57% | 48 Z=CN |
|---|---|---|---|
|  | Z=COMe | 41% | 49 Z=COMe |

Owing to the reductive reaction conditions, addition products are formed exclusively. However, only sugar bromides such as **47** have been reported as radical precursors. Both cobalt methods are attractive from an environmental point of view because the need of tin hydride has been excluded.

## E. Methods for the Formation of β-*C*-Glycosyl Compounds

The previously described reactions are based on the trapping of anomeric radicals with olefins, yielding *C*-glycosides. As mentioned in the introduction, these compounds can also be obtained by interception of C-1-substituted glycosyl radicals with hydrogen atom-transfer reagents. For example, the reduction of the tertiary nitro sugar **50** with tributyltin hydride **14** in the absence of olefins is such an approach to provide the *C*-glycoside **51** [26].

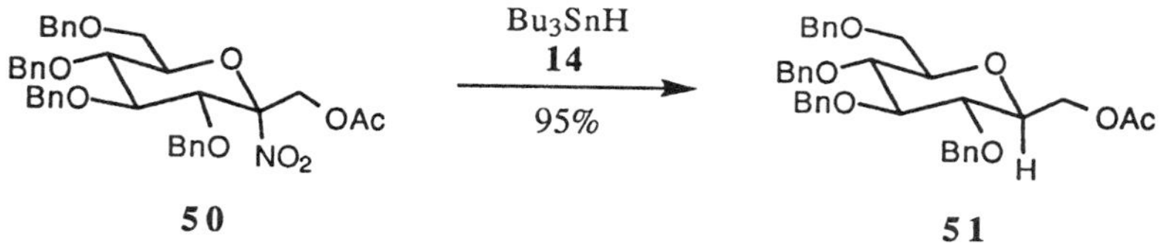

$$\text{50} \xrightarrow[95\%]{\text{Bu}_3\text{SnH} \ \mathbf{14}} \text{51}$$

An alternative method is the deoxygenation of the anomeric carboxylic acid **52**, producing **55** [27]. The functionalized precursor, in particular the *O*-acyl-thiohydroxamate ester **54**, which is formed in situ after exposure of the carboxylic acid **52** to the salt **53**, is decarboxylated by the Barton method.

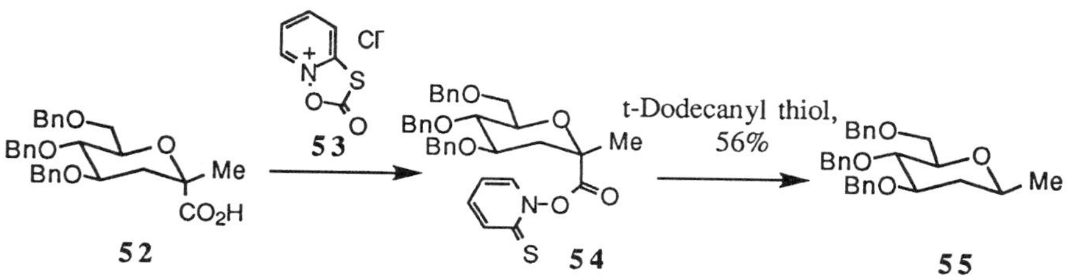

$$\text{52} \xrightarrow{\mathbf{53}} \text{54} \xrightarrow[56\%]{\text{t-Dodecanyl thiol}} \text{55}$$

In both reactions the intermediate glycosyl radicals are trapped from the α-face by hydrogen atom donors. Thus, by taking advantage of the stereoelectronic and conformational effects of hexopyranosyl radicals, these strategies enable the production of β-*C*-glycosides.

## III. INTRAMOLECULAR METHODS

In addition to intermolecular reactions, *C*-glycosides can also be synthesized by intramolecular sequences. A radical cyclization is a very fast reaction, in particular 5-*exo–trig* cyclizations. Thus, intermediate anomeric radicals have only a short time window of reactivity before undergoing the desired cyclization.

The first investigation to realize a *C*-glycosidation through a cyclization reaction has been conducted by using the 3-allyl-3-deoxyglucose derivative **56** [28]. The formation of the bicyclic dideoxysugar **58** can be explained by the formation of the boat conformer **57**. Only in this shape is the radical center close enough for the cyclization onto the double bond.

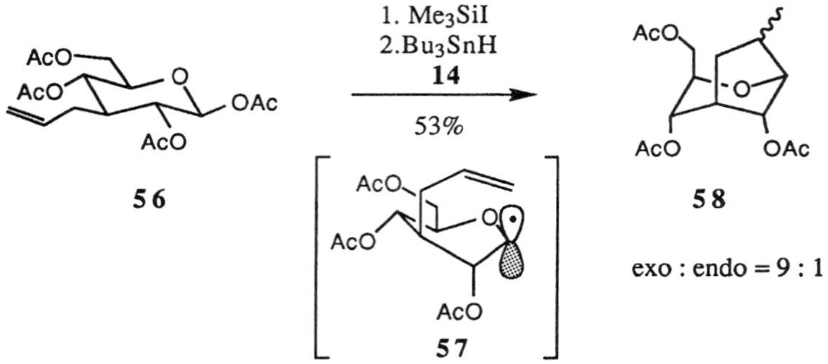

A remarkable extension of this strategy achieves the stereospecific introduction of a C–C bond at the anomeric center by the cyclization of an acceptor group tethered to a suitable hydroxyl group of the sugar moiety through a temporary function. Table 2

**Table 2** Radical Cyclization Reactions Using Temporary Silicon or Acetal Connections

| Precursor | Conditions (overall yield %) | Products | Ref. |
|---|---|---|---|
| 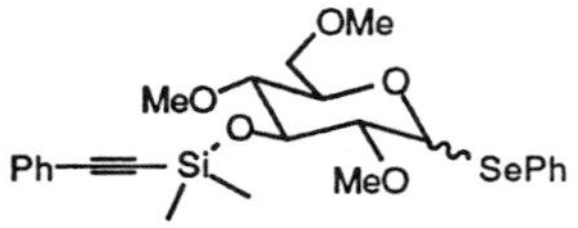  | 1. Bu$_3$SnH AIBN<br>2. TBAF (73%) | 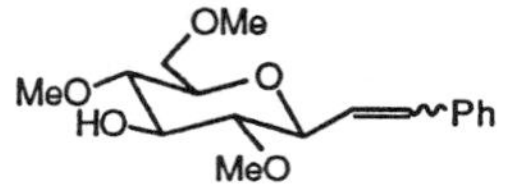  | 29 |
| 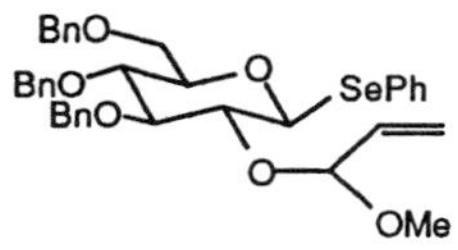  | 1. Bu$_3$SnH AIBN<br>2. *m*-CPBA, BF$_3$, Et$_2$O<br>3. LiAlH$_4$<br>4. *t*-BuPh$_2$SiCl, imidazole<br>5. NaH, MeI (41%) | 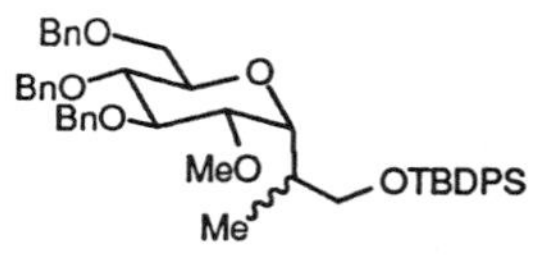  | 30 |
| 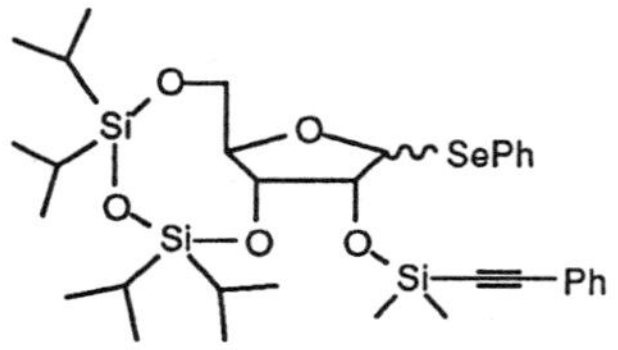  | 1. Bu$_3$SnH AIBN<br>2. TBAF<br>3. Acetylation (95%) | 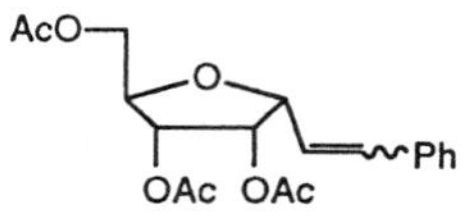  | 29 |

# C-Glycosyl Compounds by Free Radical Reactions

illustrates some examples showing that either silicon or acetal connections have been used in such reaction sequences. Furthermore, the examples demonstrate that owing to the geometric requirements α- or β-C-glycosides can be synthesized.

A similar process is possible using samarium diiodide. With this reagent carbohydrate radicals are generated through a one-electron reduction of, for example, anomeric phenyl sulfones. The so-induced cyclization of the tethered sugar **59** gives compound **60**, which has been transformed to a C-disaccharide [31].

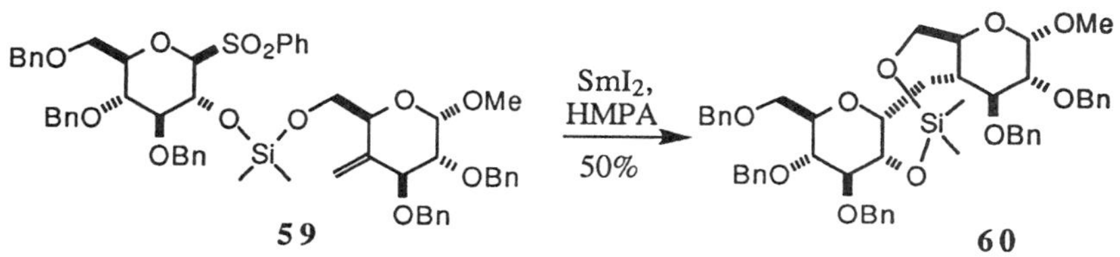

Thus, this method takes advantage of a relatively rapid 9-*endo*–*trig* cyclization, rather than undergoing other possible reactions with samarium iodide [32].

## IV. EXPERIMENTAL PROCEDURES*

### A. Intermolecular Methods

*The Tin Hydride Method*

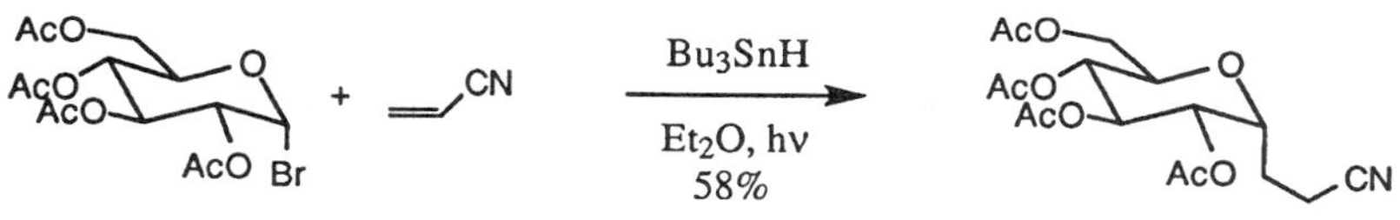

**C-Glycosidation of Pyranosyl Compounds in Large-Scale Production [12,33].**
A boiling solution of the α-D-glucopyranosyl bromide **47** (20.6 g, 50.0 mmol), acrylonitrile (13.5 g, 250 mmol), and tributyltin hydride (16.0 g, 55.0 mmol) in 100 mL of diethyl ether under nitrogen was irradiated with a mercury high-pressure lamp. After 4 h, the mixture was filtered, further acrylonitrile (6.6 g, 120 mmol) and tributyltin hydride (5.8 g, 20.0 mmol) were added to the filtrate, which was further irradiated. When the starting bromide **47** had completely reacted, the cold mixture was filtered, and the combined precipitates gave the product **48** (8.6 g, 45%) after crystallization from diethyl ether as white crystals: mp 121°–123°C, $[\alpha]_D$ +66.2° (c 1.0, CHCl$_3$). The filtrates were evaporated, the residue was dissolved in 50 mL of acetonitrile, and the solution extracted with pentane (3 × 50 mL). The acetonitrile solution was evaporated. Column chromatography on silica gel (hexanes–EtOAc, 1:1) gave the β-isomer (950 mg, 5%) and additional α-glycoside **48** (2.5 g, 13%). The ratio of α/β-glycoside before workup was 93:7 (GLPC).

---

*Optical rotations were measured at 20°–25°C.

**C-Glycosidation of Pyranosyl Compounds using α,α-Disubstituted Olefins [14].**
To a solution of the α-D-mannopyranosyl bromide **61** (3.30 g, 5.0 mmol) in 15 mL of dry

**61** + **62** $\xrightarrow[\text{Toluene, 67°C}]{\text{Bu}_3\text{SnH}\;\text{AIBN}}$ **63**
99%

toluene under argon, were added dropwise a solution of methylene malonic ester **62** (1.03 g, 6.0 mmol) in 8 mL of toluene and, simultaneously, a solution of tributyltin hydride (1.74 g, 6.0 mmol) in 8 mL of toluene containing AIBN (100 mg, 0.61 mmol) over 1 h at 67°C. After completion of the addition the mixture was kept at 67°C for 30 min. The toluene was distilled off under reduced pressure, and the residue was taken up in 40 mL of acetonitrile. The acetonitrile layer was extracted with pentane (5 × 20 mL) and then concentrated. Flash chromatography on silica gel (pentane–diethyl ether, 1:3) afforded the product **63** (3.73 g, 99%), a single anomer, as a colorless oil: $[\alpha]_D$ −56.5° (c 1.0, $CHCl_3$).

**C-Glycosidation of Furanosyl Compounds [9].** To a boiling solution of the α-D-arabinofuranosyl bromide **64** (2.63 g, 5.00 mmol) and acrylonitrile (1.45 g, 5.0 mmol) in

**64** + CH$_2$=CHCN $\xrightarrow[\text{Benzene, }\Delta]{\text{Bu}_3\text{SnH}\;\text{AIBN}}$ **65**
80%

50 mL of dry benzene under argon, was added dropwise a solution of tributyltin hydride (1.75 g, 6.00 mmol) in 10 mL of solvent containing AIBN (160 mg, 1.00 mmol) over 1.5 h. After completion of the addition, the mixture was allowed to stir for 30 min at 80°C. The solvent was distilled off under reduced pressure, and the residue was taken up in 50 mL of acetonitrile. The acetonitrile layer was extracted with pentane (3 × 30 mL) and then concentrated. Flash chromatography on silica gel (pentane–diethyl ether, 1:1) afforded the product **65** (2.0 g, 80%), a single anomer, as a colorless oil.

**C-Glycosidation of Pyranosyl Compounds Forming a C-Disaccharide [2–4].** To a boiling solution of the methylene lactone **66** (610 mg, 1.37 mmol) in 5 mL of 1,2-dimethoxyethane under nitrogen, were added, within 30 min, a solution of tributyltin hydride (873 mg, 3.0 mmol) in 10 mL of 1,2-dimethoxyethane containing AIBN (49 mg, 0.3 mmol) and, simultaneously, a solution of the α-D-glucopyranosyl bromide **47** (1.15 g, 2.80 mmol) in 18 mL of 1,2-dimethoxyethane. After completion of the addition, the mixture was allowed to stir for 15 min at boiling temperature. The solvent was evaporated under reduced

**47** + **66** $\xrightarrow[\text{Dimethoxy-ethane, }\Delta]{\text{Bu}_3\text{SnH, AIBN}}$ **67**
73%

## C-Glycosyl Compounds by Free Radical Reactions

pressure, and the residue was taken up in 20 mL of acetonitrile. The acetonitrile layer was extracted with pentane (4 × 20 mL) and then concentrated. Chromatography on silica gel (pentane–diethyl ether, 1:1) gave the addition product **67**, a single anomer, containing some impurities. Recrystallization from *t*-butylmethylether–hexane afforded the product **67** (770 mg, 73%), as white crystals: mp 109°C, $[\alpha]_D$ +78.9° (*c* 1.0, CHCl$_3$).

**C-Glycosidation of Pyranosyl Compound Forming a C-Disaccharide [34].** To a boiling solution of the α-D-glucopyranosyl bromide **47** (748 mg, 1.82 mmol), methylidene-

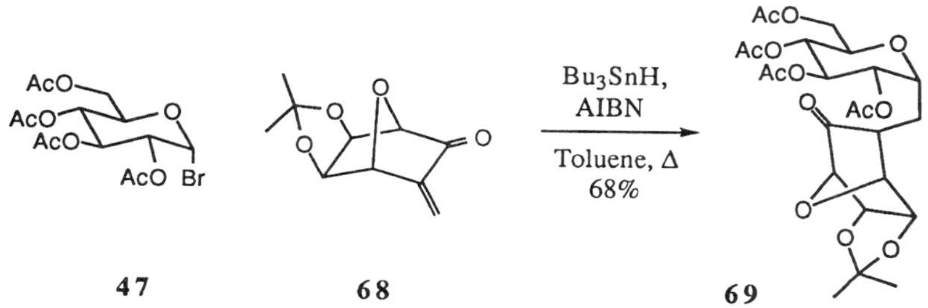

oxabicyclo-heptanone **68** (275 mg, 1.4 mmol), and AIBN (15 mg, 0.09 mmol), in 50 mL of dry toluene, was added over a period of 1 h a solution of tributyltin hydride (611 mg, 2.1 mmol) in 3 mL of toluene containing AIBN (19 mg, 0.12 mmol). After completion of the addition the solvent was distilled off under reduced pressure (0.05 torr) and the residue was filtered through a column of silica gel (EtOAc). After solvent evaporation, the residue was purified by column chromatography on silica gel (light petroleum–EtOAc, 1.5:1) to afford the product **69** 504 mg (68%), two anomers of a ratio of α/β, 5.5:1, as a colorless solid: mp 142°–143°C after crystallization from EtOAc–light petroleum, $[\alpha]_{589}$ +46° (*c* 1.5, CH$_2$Cl$_2$). A first fraction of the chromatography gave 226 mg of a mixture of reduced starting material and 1,3,4,6-tetra-*O*-acetyl-2-deoxy-α-D-*arabino*-hexopyranose.

**C-Glycosidation of Tertiary Nitro Sugars [17].** A solution of the β-D-*manno*-4-*nonulo*-4,7-furanonitrile **70** (345 mg, 1.0 mmol), acrylonitrile (1 g, 20.0 mmol), tributyltin

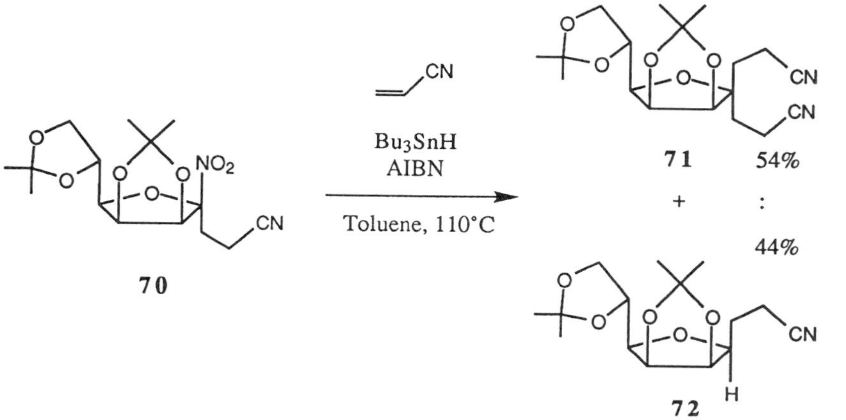

hydride (1.6 g, 5.0 mmol) in 30 mL of dry toluene containing AIBN (100 mg, 0.6 mmol) was refluxed for 1 h. The solvent was distilled off under reduced pressure, and the residue was taken up in 50 mL of acetonitrile. The acetonitrile layer was extracted with pentane (5 × 50 mL) and then concentrated. Chromatography on silica gel (diethyl ether) gave the reduction product **72**, (136 mg, 44%): $R_f$, 0.45, and the desired product **71**, (197 mg, 54%), as an oil, $R_f$, 0.29.

*The Fragmentation Method*

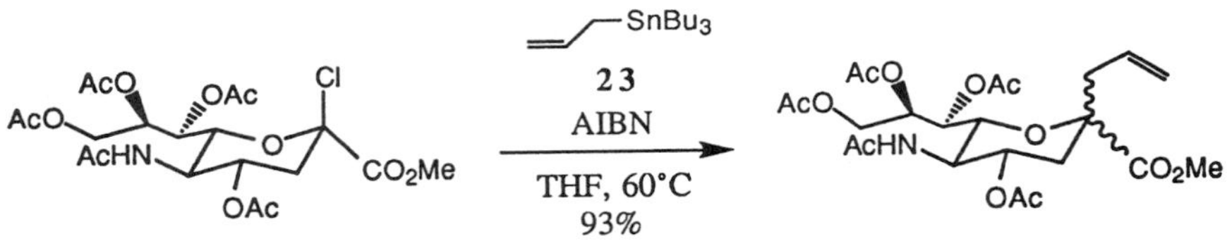

**Allylation of Pyranosyl Compounds with Allylstannane [19,20].** A solution of the methyl-(β-D-galactopyranosyl chloride) onate **22** (953.5 mg, 1.8 mmol) in 30 mL of dry, oxygen-free THF, was heated to 60°C. To this solution under nitrogen, were added the tributylallyl tin **23** (5.2 mL, 14.7 mmol) and AIBN (15 mg, 0.1 mmol). After 20 h, the solvent was evaporated, and the residue was taken up in 5 mL of acetonitrile. The acetonitrile layer was extracted with 20 mL of *n*-hexane and then concentrated. Chromatography on silica gel (toluene–acetone, 4:1) afforded the product **24** (889 mg, 93%), containing two anomers. Nuclear magnetic resonance (NMR) spectroscopy showed a ratio of the two anomers of 1.8:1.

**Allylation of Pyranosyl Compounds Using Activated Allylstannanes [35].** A solution of the 2-deoxy-2-*N*-phthalimido-β-D-mannopyranosyl phenylselenide **73** (580 mg, 1.0 mmol) in 5 mL of dry toluene containing the allytin compound **74** (2.0 g, 5.0 mmol) was treated with argon for 15 min, to remove molecular oxygen, and then heated to 90°C. To this mixture was added 2 mL of lauroyl peroxide (0.05 *M* in toluene, 0.1 mmol, 0.1 Eq.) by syringe pump (1 mL/1 h). After completion of the addition, the solvent was evaporated under reduced pressure, and the residue was dissolved in 10 mL of acetonitrile. The acetonitrile layer was extracted with pentane (5 × 5 mL) and then concentrated. Chromatography on silica gel (pentane–EtOAc, 3:2) afforded the product **75** (398 mg, 75%), a single anomer, as a colorless oil.

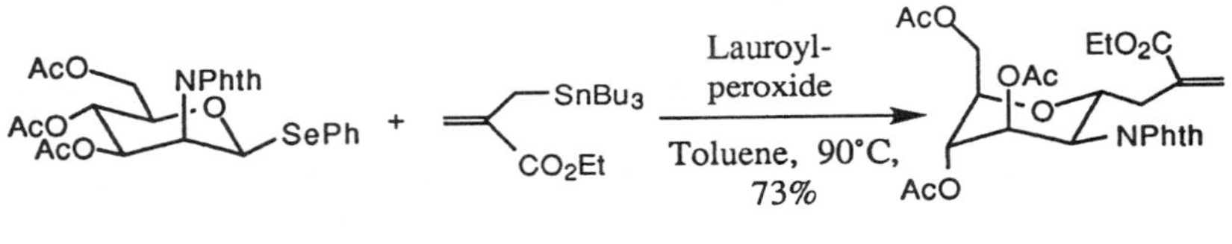

*The Cobalt Method for C-Glycosidation Reactions [22,23]*

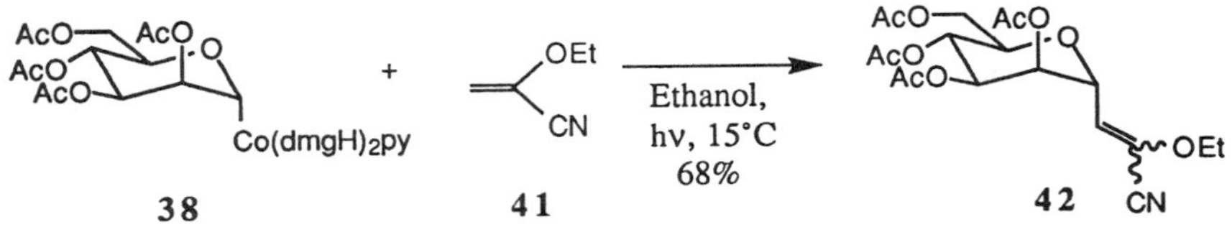

A solution of the α-D-mannopyranosyl cobaloxime **38** (700 mg, 1.0 mmol) in 50 mL of ethanol was treated for 30 min with argon. The α-ethoxy acrylonitrile (1.94 g, 20 mmol)

was added, and the solution was irradiated with a 300-W sun lamp at 15°C. After 46 h, the solvent was distilled off under reduced pressure. Flash chromatography on silica gel (pentane–diethyl ether, 1:3) afforded the *trans*-isomer of **42** (187 mg, 44%), a single anomer, as white crystals: mp 99°C, $[\alpha]_D$ +48.4° (*c* 1.01, $CHCl_3$), $R_f$, 0.30; and the *cis*-isomer of **42** (104 mg, 24%), a single anomer, as a colorless oil: $[\alpha]_D$ +24.4° (*c* 1.04, $CHCl_3$); $R_f$, 0.20.

*Vitamin $B_{12}$-Catalyzed Method for C-Glycosidation Reactions [24]*

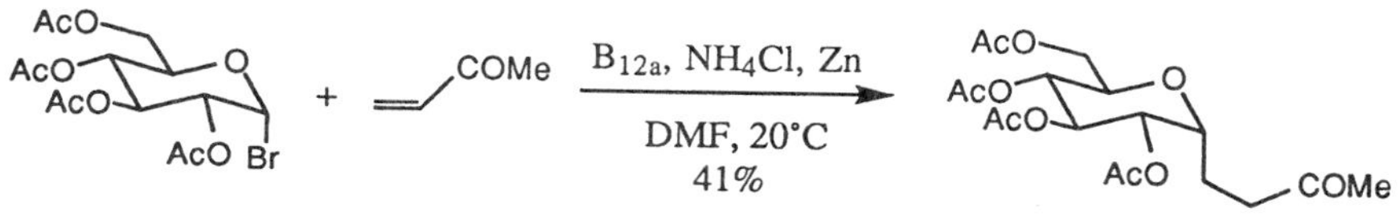

A solution of vitamin $B_{12a}$ (70 mg, 0.05 mmol), ammonium chloride (390 mg, 7.28 mmol), and activated zinc wool (2.30 g, 35 mmol), which was wound around the stirring bar in 20 mL of degased DMF under argon, was stirred at room temperature until the mixture turned dark green. Methylvinyl ketone (1.02 g, 14.6 mmol) was added followed by α-D-glucopyranosyl bromide **47** (1.00 g, 2.43 mmol). During the addition, the green solution turned brown. After completion of the addition, the mixture was allowed to stir at room temperature for 3 h. The solvent was removed under reduced pressure and the residue dissolved in dichloromethane (100 mL), extracted with cold aqueous $NH_4OH$ (2 × 100 mL), and with brine (2 × 50 mL). The organic layer was dried over $Na_2SO_4$, filtered, and concentrated. The crude reaction mixture was taken up in diethyl ether (ca. 3 mL) and treated with pentane until a precipitate formed and then with a few drops of dichloromethane. This suspension was kept at 0°, and the precipitate was recrystallized from dichloromethane–diethyl ether to give the product **49** (403 mg, 41%), a single anomer, as colorless crystals: mp 99°–99.5°C, $[\alpha]_D$ −70.0° (*c* 2.0, $CHCl_3$).

*Formation of β-C-Glycosyl Compounds*

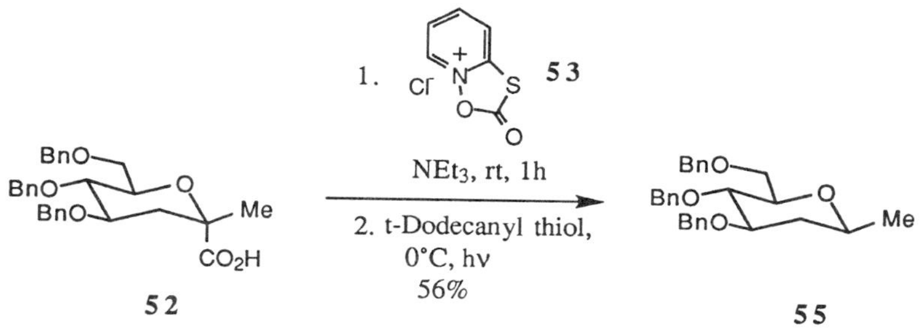

**Deoxygenation of Anomeric Carboxylic Acids Using the Barton Method [27].** The acid **52** (186 mg, 0.39 mmol) was dissolved in 3 mL of methylene dichloride and the solution was stirred in the dark under argon at room temperature with triethylamine (0.07 mL, 0.49 mmol) and the salt **53** (93 mg, 0.49 mmol). After 1 h, *t*-dodecanyl thiol (0.46 ml, 2.0 mmol) was added to the yellow solution, the reaction mixture was cooled to 0°C and photolyzed with a 500-W tungsten lamp for 1 h. Evaporation under reduced pressure and chromatography on silica gel (40–60 light petroleum–diethyl ether, 85:15) gave the product **55** (94 mg, 56%), a single anomer, as an oil: $[\alpha]_D$ +23° (*c* 1.0, $CCl_4$).

**Reduction of Tertiary Nitro Sugars [26].** To a boiling solution of the 2-deoxy-2-nitro-D-*gluco*-heptulopyranose **50** (285 mg, 0.44 mmol) in 15 mL of dry benzene under argon, was added dropwise a solution of tributyltin hydride (640 mg, 2.2 mmol) in 5 mL of benzene containing AIBN (3 mg, 0.14 mmol), 1–2 h. When TLC indicated the disappearance of the nitro compound **50**, the cooled mixture was concentrated. Chromatography on silica gel (hexane–EtOAc, 4:1) afforded the product **51** (250 mg, 95%), a single anomer, as an oil: $[\alpha]_D$ −4.4° (*c* 1.3, CHCl$_3$).

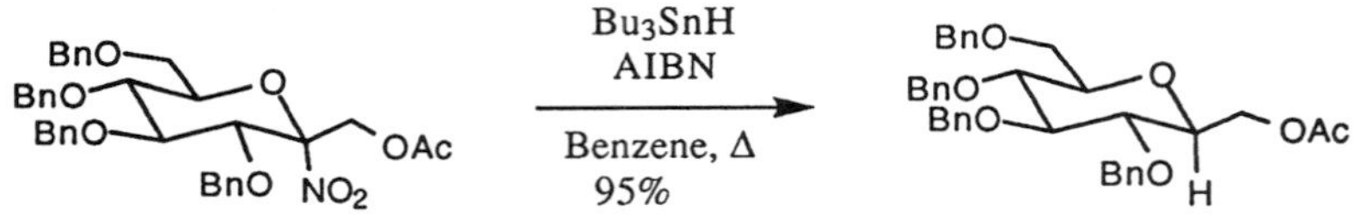

## B.  Intramolecular Methods

*Intramolecular C-Glycosidation with Tin Hydride [28]*

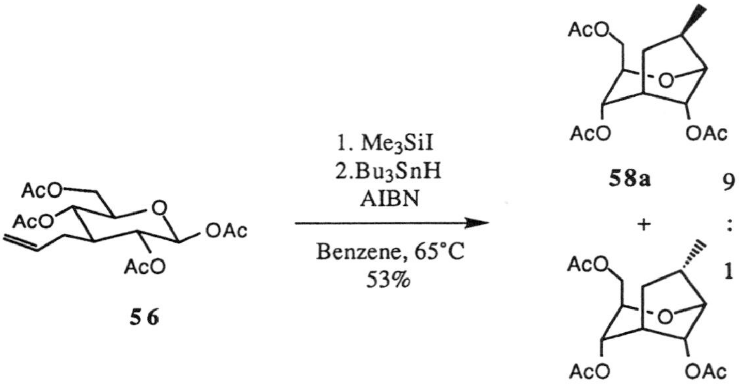

**Method A.** A solution of the 3-allylglucose **56** (440 mg, 1.18 mmol) and trimethylsilyl iodide (260 mg, 1.30 mmol) in 20 mL of dry benzene, under argon, was heated to 65°C. After 40 min, 20 mL benzene was added followed by the dropwise addition of a solution of tributyltin hydride (525 mg, 1.80 mmol) in 5 mL of benzene containing AIBN (35 mg, 60 mmol) over 2 h. After completion of the addition, the mixture was kept at 65°C for 1 h. The solvent was evaporated under reduced pressure, the residue was dissolved in 100 mL of diethyl ether, and the solution treated with 2 g of wet potassium fluoride for 10 h at 20°C. The mixture was filtered over silica gel, the solvent distilled off, and the remaining oil was flash chromatographed twice (first pentane–diethyl ether–dichloromethane, 55:35:10, then hexane–EtOAc, 7:3, $R_f$, 0.13) to give 195 mg (53%) of the cyclized products **58a** and **58b**. NMR spectroscopy showed a ratio of the *exo-/endo-*isomer **58a/58b** of 9:1.

**Method B [30].** The starting material **76** (2.80 g, 4.24 mmol) was dissolved in 14.2 mL of degased benzene. To this solution were added tributyltin hydride (1.235 g, 4.24 mmol) and AIBN (35 mg, 0.21 mmol). The mixture was heated under reflux for 12 h. After this time, some starting material **76** was still present (TLC, hexane–EtOAc, 3:1). Additional

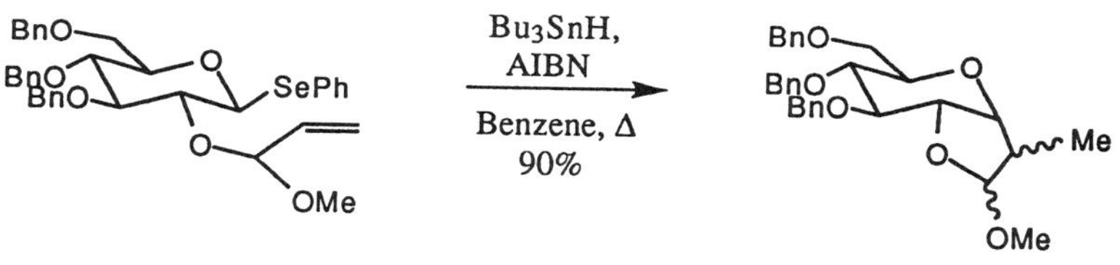

**76** → **77**

tributyltin hydride (370 mg, 1.27 mmol) and AIBN (35 mg, 0.21 mmol) were added. After a total of 17 h of reflux no starting material **76** could be detected by TLC. The solvent was evaporated under reduced pressure, and the residue was taken up in 60 mL of acetonitrile. The acetonitrile layer was extracted with *n*-hexane (5 × 60 mL) and then concentrated. Chromatography on silica gel (using a hexane–EtOAc gradient: 16:1 to 9:1) afforded the product **77** (1.927 g, 90%), four isomers, as an oil.

*Intramolecular C-Glycosidation Forming a C-Disaccharide with Samarium Diiodide [31]*

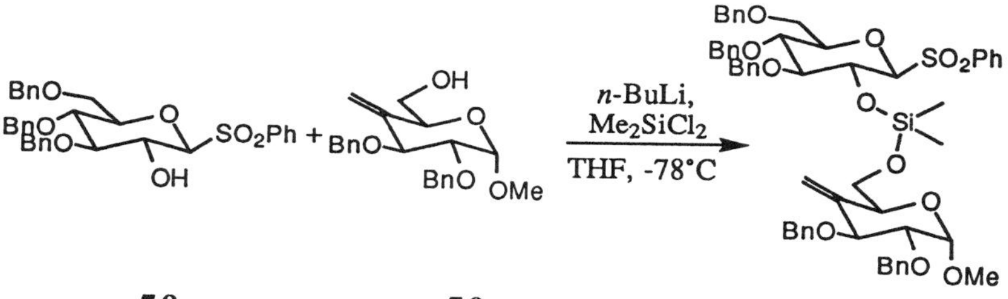

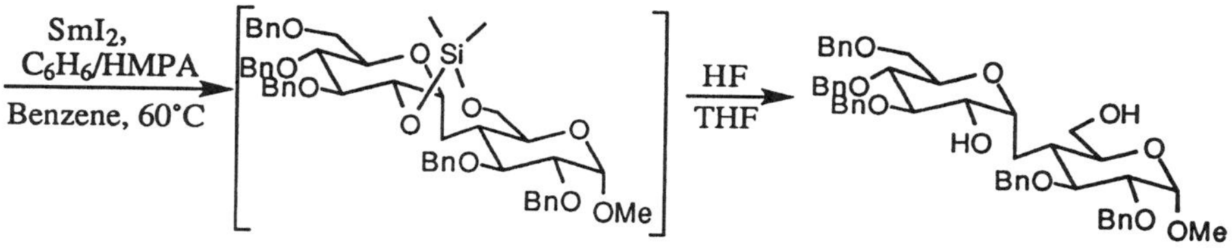

To a solution of the sulfone **78** (500 mg, 0.87 mmol) in 6 mL of dry tetrahydrofuran, under argon in a Schlenk tube, was added at −78°C a solution of *n*-butyllithium in hexane (1.04 mmol). After 1 min, dimethyldichlorosilane (0.53 mL, 4.3 mmol) was added at −78°C and the temperature of the solution was then raised to room temperature over a 2-h period. The solvent was distilled off under reduced pressure and the residue was dried under vacuum for 45 min. A solution of the primary alcohol **79** (89 mg, 1.3 mmol) in 9 mL of dry tetrahydrofuran was added at 0°C to the residue. After completion of the addition, the mixture was allowed to stay for 1 h at room temperature. The solvent was distilled off under reduced pressure and the residue was taken up in 10 mL of diethyl ether. The ether solution was washed with water, dried (MgSO$_4$), and then concentrated to a residue of **59**, a portion of which was directly used for the next two steps.

Into a solution of residue **59** (101 mg, 0.1 mmol) in 20 mL of dry toluene, kept at 60°C, was syringed, during 18 h and under argon, a freshly prepared solution of samarium diiodide in benzene–HMPA (9:1, v/v 6.3 mL, 0.51 mmol) which has been diluted with 3.8 mL of dry benzene. The solvents were distilled off under reduced pressure, and the residue was taken up in 10 mL of diethyl ether. The ether solution was washed with 10% aqueous solution of sodium bisulfite, then water, dried (MgSO$_4$), and concentrated. The crude product was dissolved in 1.5 mL of tetrahydrofuran and treated during 30 min at room temperature with 1.5 mL of a 40% aqueous solution of HF. The solution was neutralized with solid sodium carbonate, and concentrated. Flash chromatography on silica gel (cyclohexane–ethyl acetate, 3:1 to 1:2) afforded the product **80** (40.6 mg, 50%), a single isomer, as an amorphous solid. It was characterized by its diacetate $[\alpha]_D$ +36° (*c* 4.0, CHCl$_3$).

## REFERENCES

1. B. Giese, Formation of carbon–carbon bonds by addition of radicals to alkenes, *Angew. Chem. Int. Ed. Engl. 22*:753 (1983).
2. B. Giese, M. Hoch, C. Lamberth, and R. R. Schmidt, Synthesis of methylene bridged *C*-disaccharides, *Tetrahedron Lett. 29*:1375 (1988).
3. B. Giese, W. Damm, T. Witzel, and H.-G. Zeitz, The influence of substituents at prochiral centers on the stereoselectivity of enolate radicals, *Tetrahedron Lett. 34*:7053 (1993).
4. B. Giese and T. Witzel, Synthesis of *C*-disaccharides by radical C–C bond formation, *Angew. Chem. Int. Ed. Engl. 25*:450 (1986).
5. B. Giese, Stereoselective syntheses with carbohydrate radicals, *Pure Appl. Chem. 60*:1655 (1988).
6. B. Giese, J. Dupuis, K. Gröninger, T. Hasskerl, M. Nix, and T. Witzel, *Substituent Effects in Radical Chemistry* (H. G. Viehe, et al., eds.), D. Riedel Publishing, 1986, p. 283.
7. J. Dupuis, B. Giese, D. Rüegge, H. Fischer, H.-G. Korth, and R. Sustmann, Conformation of glycosyl radicals: Radical stabilization by β-CO bonding, *Angew. Chem. Int. Ed. Engl. 23*:896 (1984).
8. H.-G. Korth, R. Sustmann, J. Dupuis, and B. Giese, Electron spin resonance spectroscopic investigation of carbohydrate radicals. Part 2. Conformation and configuration in pyranos-1-yl radicals, *J. Chem. Soc. Perkin Trans. 2*:1453 (1986).
9. B. Giese and T. Linker unpublished results.
10. B. Giese, *Radicals in Organic Synthesis: Formation of Carbon Carbon Bonds*, Pergamon Press, Oxford, 1986.
11. B. Giese, Syntheses with radicals. Carbon–carbon coupling via organotin and -mercury compounds, *Angew. Chem. Int. Ed. Engl. 24*:553 (1985).
12. B. Giese and J. Dupuis, Diastereoselective synthesis of *C*-glycopyranosides, *Angew. Chem. Int. Ed. Engl. 22*:622 (1983).
13. R. M. Adlington, J. E. Baldwin, A. Basak, and R. P. Kozyrod, Application of radical addition reactions to the synthesis of a *C*-glycoside and a functionalised amino-acid, *J. Chem. Soc. Chem. Commun.* p. 944 (1983).
14. B. Giese, T. Linker, and R. Muhn, Biomimetic chain elongation of carbohydrates via radical carbon–carbon bond formation, *Tetrahedron 45*:935 (1989).
15. H. Kessler, V. Wittmann, M. Köck, and M. Kottenhahn, Synthesis of *C*-glycopeptides via radical addition of glycosyl bromides to dehydroalanine derivatives, *Angew. Chem. Int. Ed. Engl. 31*:902 (1992).

16. Y. Araki, T. Endo, M. Tanji, J. Nagasawa, and Y. Ishido, Additions of a ribofuranosyl radical to olefins. A formal synthesis of showdomycin, *Tetrahedron Lett. 29*:351 (1988).
17. J. Dupuis, B. Giese, J. Hartung, M. Leising, H.-G. Korth, and R. Sustmann, Electron transfer from trialkyltin radicals to nitrosugars: The synthesis of *C*-glycosides with tertiary anomeric carbon atoms, *J. Am. Chem. Soc. 107*:4332 (1985).
18. G. E. Keck and J. B. Yates, Carbon–carbon bond formation via the reaction of trialkylallylstannanes with organic halides, *J. Am. Chem. Soc 104*:5829 (1982).
19. H. Paulsen and P. Matschulat, Synthese von *C*-Glycosiden der *N*-Acetylneuraminsäure und weiteren Derivaten, *Liebigs Ann. Chem.* p. 487 (1991).
20. J. O. Nagy and M. D. Bednarski, The chemical–enzymatic synthesis of a carbon glycoside of *N*-acetyl neuraminic acid, *Tetrahedron Lett. 32*:3953 (1991).
21. G. E. Keck, E. J. Enholm, J. B. Yates, and M. R. Wileg, One electron C–C bond forming reactions via allylstannanes. Scope and limitations, *Tetrahedron 41*:4079 (1985).
22. A. Ghosez, T. Göbel, and B. Giese, Syntheses and reactions of glycosylcobaloxims, *Chem. Ber. 121*:1807 (1988).
23. B. Giese, A. Ghosez, T. Göbel, J. Hartung, O. Hüter, A. Koch, K. Kroder, and R. Springer, Radical reactions with organocobalt complexes, *Free Radicals in Synthesis and Biology* (F. Minisci, ed.), Kluwer Academic Publishers, 1989, p. 97.
24. S. Abrecht and R. Scheffold, Vitamin $B_{12}$-katalysierte Synthese von *C*-Glycosiden, *Chimia 39*:211 (1985).
25. R. Scheffold, Vitamin $B_{12}$, Katalysatoren für die C–C-Bindungsknüpfungen in organisch–chemischen Synthesen, *Chimia 39*:203 (1985).
26. F. Baumberger and A. Vasella, Deoxynitrosugars—stereoelectronic control in the reductive denitration of tertiary nitro ethers. A synthesis of "*C*-glycosides," *Helv. Chim. Acta 66*:2210 (1983).
27. D. Crich and L. B. L. Lim, Diastereoselective free-radical reactions, part 2. Synthesis of 2-deoxy-β-*C*-pyranosides by diastereoselective hydrogen-atom transfer, *J. Chem. Soc. Perkin Trans. 1*:2205 (1991).
28. K. S. Gröninger, K. F. Jäger, and B. Giese, Cyclization reaction with allyl-substituted glucose derivatives, *Liebigs Ann. Chem.* p. 731 (1987).
29. G. Stork, H. S. Suh, and G. Kim, The temporary silicon connection method in the control of regio- and stereochemistry. Application to radical-mediated reactions. The stereospecific synthesis of *C*-glycosides, *J. Am. Chem. Soc. 113*:7054 (1991).
30. A. De Mesmaeker, P. Hoffmann, B. Ernst, P. Hug, and T. Winkler, Stereoselective C–C bond formation in carbohydrates by radical cyclization reaction—IV. Application for the synthesis of α-(*C*)-glucosides, *Tetrahedron Lett. 30*:6311 (1989).
31. A. Chenede, E. Perrin, E. D. Rekai, and P. Sinay, Synthesis of *C*-disaccharides by a samarium diiodide induced intramolecular coupling reaction of tethered sugars, *Synlett* p. 420 (1994).
32. P. de Poully, A. Chenede, J.-M. Mallet, and P. Sinay, $SmI_2$-promoted chemistry at the anomeric center of carbohydrates. Reductive formation and reaction of glycosyl samarium (III) reagents, *Bull. Soc. Chim. Fr. 130*:256 (1993).
33. B. Giese, J. Dupuis, M. Leising, M. Nix, and H. J. Lindner, Synthesis of *C*-pento, -hexo, and -heptulo-pyranosyl compounds via radical C–C bond-formation reaction, *Carbohydr. Res. 171*:329 (1987).
34. R. Mampuga Biwala and P. Vogel, Synthesis of α (1→2)-, α (1→3)-, α (1→4)-, α (1→5)-*C*-linked disaccharides through 2,3,4,6-tetra-*O*-acetylglucopyranosyl radical additions to 3-methylidene-7-oxabicyclo[2.2.1]heptan-2-one derivatives, *J. Org. Chem. 57*:2076 (1992).
35. S. Abel, T. Linker, and B. Giese, Ring opening of *C*-glycosides under mild conditions, *Synlett*, p. 171 (1991).

# 24

# Synthesis of Glycosylarenes

### Keisuke Suzuki and Takashi Matsumoto
*Tokyo Institute of Technology, Tokyo, Japan*

| | | |
|---|---|---:|
| I. | Introduction | 528 |
| II. | Methods | 530 |
| | A. Friedel–Crafts coupling | 530 |
| | B. $O \rightarrow C$-glycoside rearrangement | 531 |
| III. | Experimental Procedures | 535 |
| | A. Friedel–Crafts-type $C$-glycosidation of glycosyl acetates | 535 |
| | B. Friedel–Crafts-type $C$-glycosidation of trichloroacetimidates | 536 |
| | C. $C$-glycosidation of olivosyl fluorides by using $Cp_2HfCl_2$–$AgClO_4$ as the promoter | 536 |
| | D. $C$-glycosidation of glycosyl acetates (digitoxosylation) by using $SnCl_4$ as the promoter | 537 |
| | E. $C$-glycosidation of glycosyl acetates by using $BF_3 \cdot OEt_2$ as the promoter | 537 |
| | F. $C$-glycosylation of 2-iodoresorcinol derivative (14) with fucofuranosyl acetate (15) by using $Ph_3SiCl$–$AgClO_4$ as the promoter | 538 |
| | G. $C$-glycosylation of anthracene derivative (17) with olivosyl fluoride (12) | 539 |
| | H. $C$-glycosylation of angucycline-type skeleton (33) with olivoyl acetate (32) | 539 |
| | I. Synthesis of *para*-aryl-$C$-riboside (23) | 540 |
| | References | 541 |

## I. INTRODUCTION

Carbohydrates directly bound to an aromatic moiety through a C–C bond are designated *glycosylarenes* or *C-aryl glycosides*, which, among more general *C*-glycosides found in nature [1], are currently attracting increasing interest in relation to a rapidly emerging class of bioactive compounds armed with this unique structure [2]. Aquayamycin (**1**), for which the structure was elucidated by Ohno et al. in 1970 [3], and other compounds **2–4** represent this class of novel compounds (Fig. 1).

These compounds exhibit diverse biological activities, such as antibacterial, antitumor, enzyme inhibitory effects, and inhibition of platelet aggregation, and they are currently attracting considerable synthetic interests [2].

In the synthesis of *C*-aryl glycosides, one must consider the following two issues pertaining to selectivity:

1. *The regioselectivity of the aromatic substitution*: As seen in Figure 1, the *C*-aryl glycosidic bonds are often located *ortho* to a phenolic hydroxyl group. However, other patterns of *C*-glycoside substitution are also possible as shown in Figure 2. Synthetically, such regiochemical control is particularly challenging when polycyclic aromatics are concerned.

2. *The stereoselectivity of the anomeric center*: The anomeric effect [4] is not a dominant stereoelectronic factor in *C*-aryl glycosides, and they are better viewed as a tetrahydropyran possessing a large C-1 substituent (i.e., the aryl group). Figure 3 shows an intriguing example that illustrates the extent to which the C-1 aryl group favors the equatorial position. Compound α-**5** adopts a $^1C_4$ conformation even by forcing three substituents at C-3, C-4, and C-5 to take the axial positions. Thus, the thermodynamic stability of the anomers are substantially different, and if there is a path for equilibration,

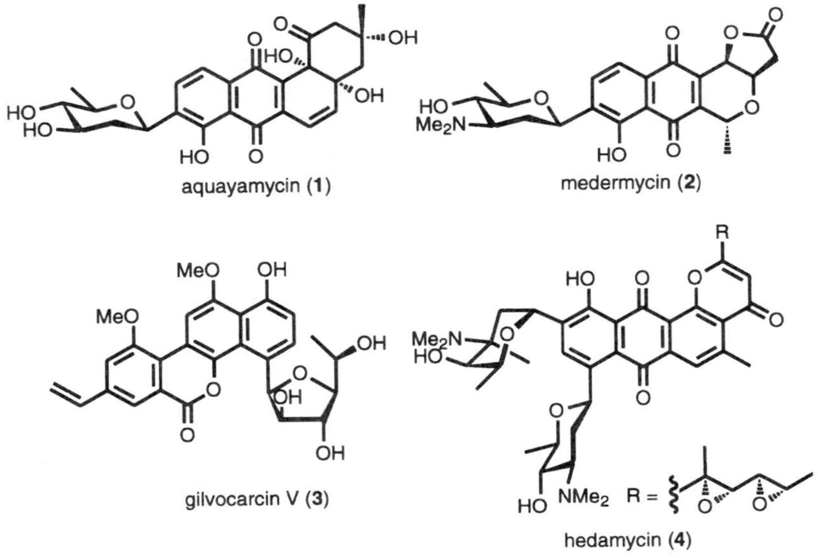

**Figure 1**

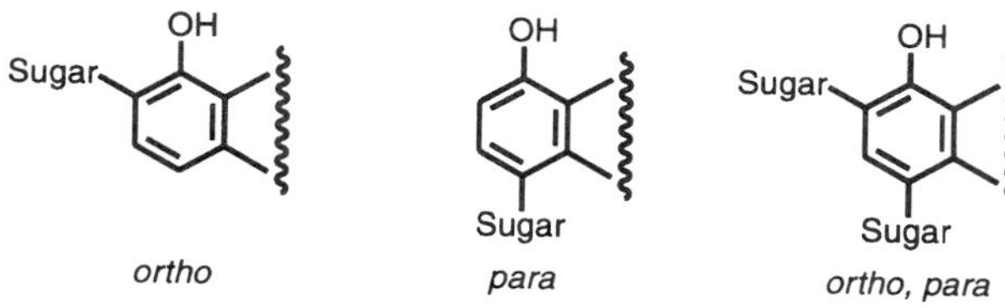

# Synthesis of Glycosylarenes

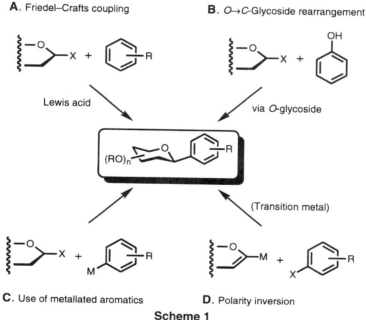

Figure 2

the more stable anomer (β-**5**) would be highly enriched. Such an aspect will be discussed in Section II.B.

Several methods are presently known for the formation of a *C*-aryl glycosidic bond, which fall roughly into four categories (**A**–**D**; Scheme 1).

Scheme 1

Method **A** is the Friedel–Crafts reaction based on the Lewis acidic activation of a glycosyl donor, followed by the capture of the generated oxonium species by an aromatic compound. Method **B** is a closely related approach, but with a substantially broader scope. It differs from **A** simply in that the glycosyl acceptor is a phenol, rather than a fully protected arene, as discussed further in Section II.B. Method **C** relies on the coupling of metallated aromatics with appropriate sugar derivatives, including glycosyl halides and sugar lactones [2]. Method **D** is characterized by the polarity inversion concept, making use of the nucleophilic glycosyl moiety to couple with appropriate aromatics, with or without the aid of a transition metal catalyst [2]. Herein we discuss two approaches, **A** and **B**, with particular attention to the latter.

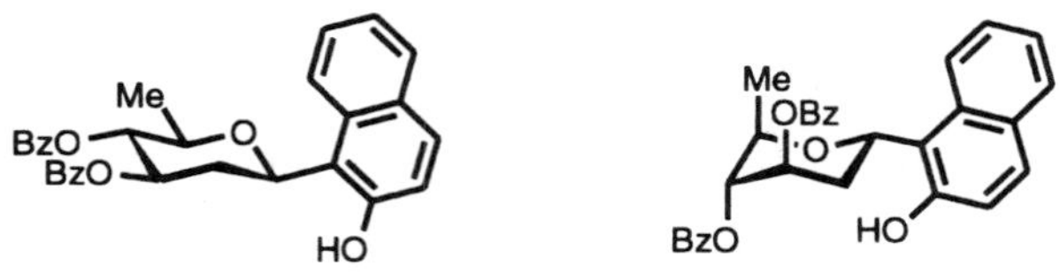

**Figure 3**

## II. METHODS

### A. Friedel–Crafts Coupling

Friedel–Crafts coupling of an aromatic compound with an activated glycosyl donor is a classic, standard method for *C*-aryl glycoside synthesis [1,2]. Early interests were primarily centered at *C*-nucleoside synthesis, so that data were accumulated for the ribofuranosyl series with degradable aromatics to construct the nucleic base (Scheme 2) [5]. Equation (1)

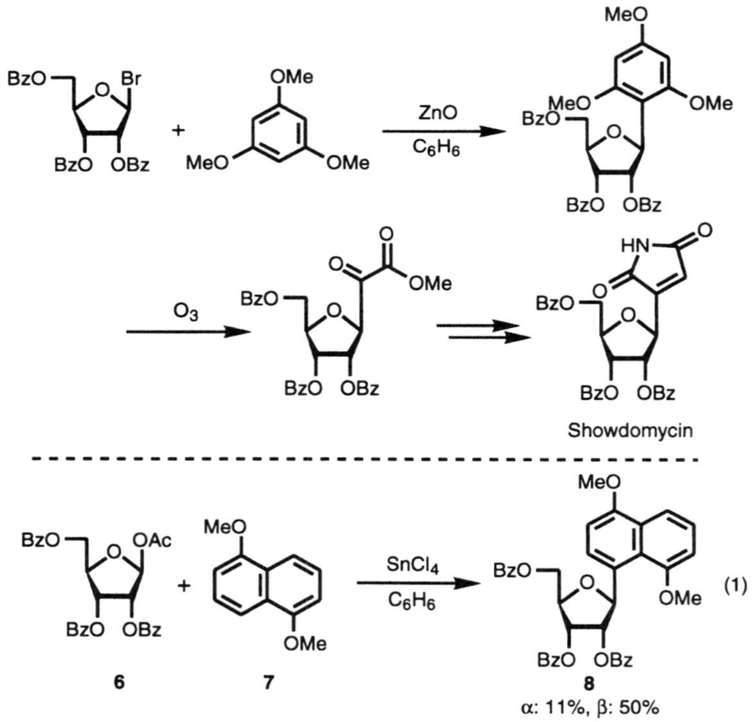

**Scheme 2**

shows a typical more recent example (see also Section III.A). It is notable that the reaction often gives a mixture of the α- and β-anomers, even in the presence of a neighboring group [6].

As a result of recent progress in *O*-glycoside synthesis, a variety of new glycosyl donors and their specific activation methods are now available, some of which have also been applied to *C*-aryl glycoside synthesis. Table 1 shows some examples for the coupling

# Synthesis of Glycosylarenes

**Table 1** Examples of Coupling of Perbenzylated D-Glucosyl Donors with 1,3,5-Trimethoxybenzene

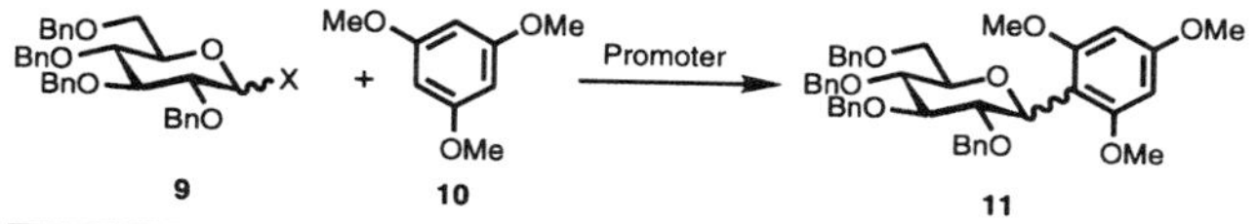

| Donor (X) | Promoter | Yield (%) | α/β | Ref. |
|---|---|---|---|---|
| O(C=NH)CCl$_3$ | BF$_3$·OEt$_2$ | 76 | β | 7 |
| SPy | AgOTf | — | 5/1[a] | 8 |
|  |  | 63 | β |  |
| F | Cp$_2$ZrCl$_2$–AgClO$_4$ | 80 | β | 9 |

[a]Reaction performed in Et$_2$O.

of perbenzylated D-glucosyl donors **9** with 1,3,5-trimethoxybenzene **10**. The reaction works well with the trichloroacetimidate donor activated by BF$_3$·OEt$_2$ (run 1, see also Sec. III.B) [7], with the thiopyridyl donor activated by Ag(I) salt [8], and with the fluoride donor activated by a cationic zirconocene complex [9]. The β-anomer is obtained as a sole product in this example, although the stereochemical reversal is possible by the proper choice of the solvent [8].

As in electrophilic substitution reactions, the reaction occurs preferentially at the position with the highest HOMO coefficient, although the steric factor is also influential [9], as shown in Figure 4.

## B. *O→C*-Glycoside Rearrangement

The second approach uses a *phenol* as the starting material, rather than a fully protected aromatic compound. Such a subtle change of glycosyl acceptor provides the process with great advantages in terms of the reactivity as well as the regio- and stereoselectivity. Because the reaction proceeds through the intermediacy of an *O*-glycoside, we describe this reaction by the trivial name of "*O→C*-glycoside rearrangement" [10–15].

*Reaction Profile*

As shown in Scheme 3, after activation with a Lewis acid, a glycosyl donor rapidly reacts with phenol at low temperature (typically at −78°C) to give an *O*-glycoside (step 1). In contrast to alkanols, in *O*-glycosidation reactions, phenols are very reactive [12], so that less reactive donors (e.g., glycosyl acetates) as well as reactive donors (e.g., glycosyl fluorides) work well [11b]. Simply by raising the temperature, the *O*-glycoside undergoes an in situ conversion to the corresponding *C*-glycoside (step 2).

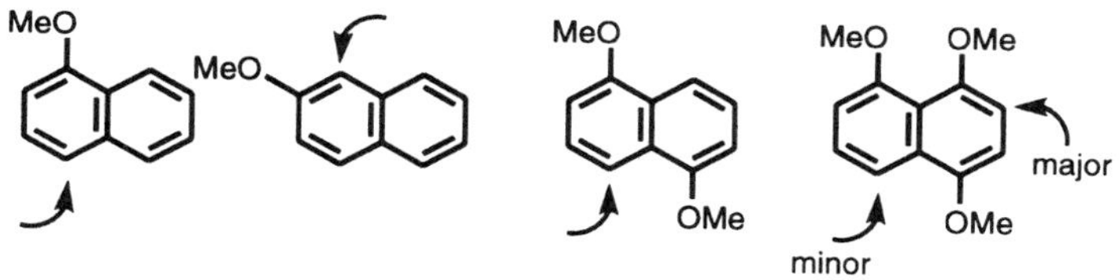

**Figure 4**

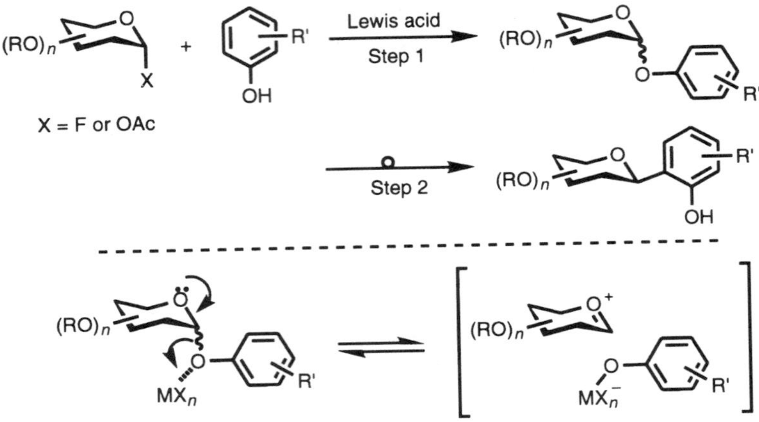

**Scheme 3**

*Mechanism*

Conversion of the *O*-glycoside to the *C*-glycoside (step 2) proceeds through an oxonium–phenolate ion pair, generated by the Lewis acid activation of the *O*-glycoside, which undergoes a Friedel–Crafts coupling (or aromatic aldol reaction) to give the *C*-glycoside irreversibly. Significantly, if the "carbon-trapping" fails, the ion pair can revert to the *O*-glycoside to enable the repeated generation of the key species in step 2.

The particular ease of *O*-glycosidation (see step 1) cannot be overlooked as an important factor in the process, serving to trap the highly reactive oxonium species to form an *O*-glycoside. This is in contrast with the Friedel–Crafts approach **A**, in which the attack of the aromatic generally needs considerable energy, and the oxonium species often undergoes various side reactions.

*Stereoselectivity and the Promoter*

Various types of Lewis acids can be used in this reaction. They can differ in the reactivity and, in particular, the stereoselectivity as illustrated in Scheme 4 (see Sec. III.C) [13]. In the

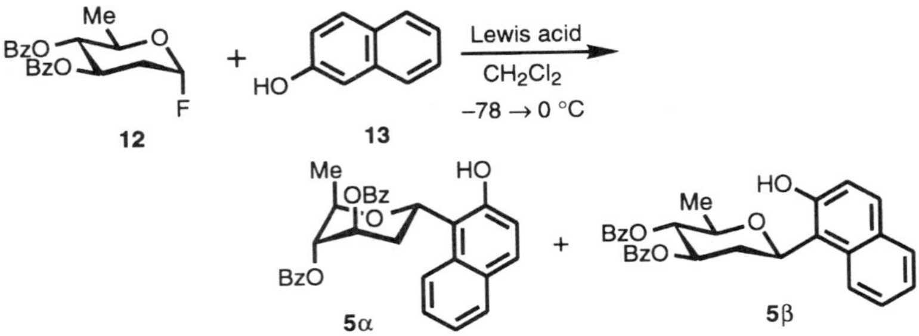

| Lewis acid | Yield of **5** (%) | α/β |
|---|---|---|
| $BF_3 \cdot OEt_2$ | 70 | 3.4/1 |
| $Cp_2HfCl_2$–$AgClO_4$ | 98 | <1/>99 |

**Scheme 4**

# Synthesis of Glycosylarenes

reaction of glycosyl fluoride **12** and β-naphthol, the behavior of two Lewis acids, BF$_3$·OEt$_2$ and Cp$_2$HfCl$_2$–AgClO$_4$, is considerably different. Besides the reactivity, the stereoselectivity is significantly different. BF$_3$·OEt$_2$ leads mainly to the α-anomer, whereas the Hf-promoter gives the β-anomer almost exclusively.

Mechanistically, an additional step accounts for this difference, as shown in Scheme 5. Whatever the α/β ratio is, the kinetically formed *C*-glycoside may undergo an α→β-

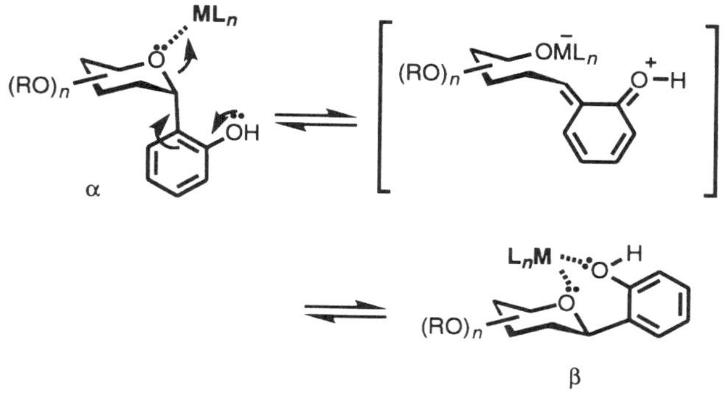

**Scheme 5**

anomerization, through a quinone methide generated by the Lewis acid, and the process of ring opening–reclosure leads to thermodynamic control.

If the preference is not obvious, as in the furanosyl series shown in Scheme 6, an interesting stereochemical divergence is seen [14]. This particular reaction was studied in

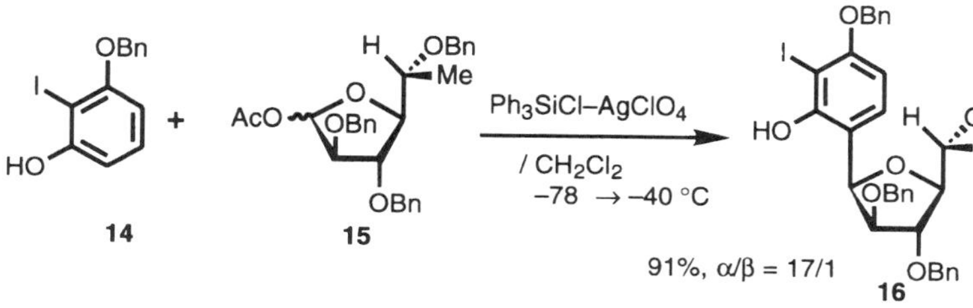

| Lewis acid | Reaction temp. (°C) | Yield (%) | α/β |
|---|---|---|---|
| SnCl$_4$–AgClO$_4$ | −78 to −40 | 60 | 5.1/1 |
| SnCl$_4$–AgClO$_4$ | −78 to −20 | 69 | 1/58 |

**Scheme 6**

detail as one of the key steps in the total synthesis of the gilvocarcins, which revealed the following aspects [14]. The reaction, starting at −78°C, followed by gradual warming, led to *C*-glycoside formation. It turned out that the α/β selectivity heavily depends on the final temperature, and as little as a 20°C difference can dramatically reverse the selectivity. Also notable is the strong dependence of the equilibrium on the Lewis acid in terms of the center metal, the ligand, and the anion (see Sec. III.F.) [14].

*Regioselectivity*

The most useful feature of this reaction is that the *C*-aryl glycoside bond forms regioselectively at the position *ortho* to the phenolic hydroxyl. As shown in Scheme 7, the *ortho*-

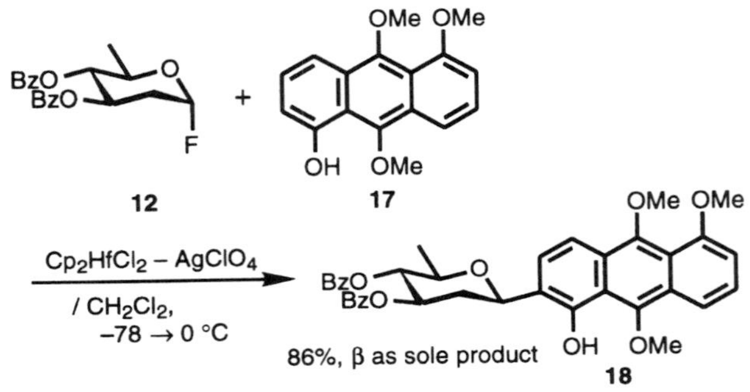

**Scheme 7**

selectivity holds for various phenolic compounds including a polyaromatic, such as anthrol **17** (see Secs. III.G and III.H) [13,15].

Scheme **8** illustrates the case of a *"para"* C-glycoside. In terms of the *ortho*-selectivity, the behavior of a monoprotected resorcinol **20** is not an exception to give

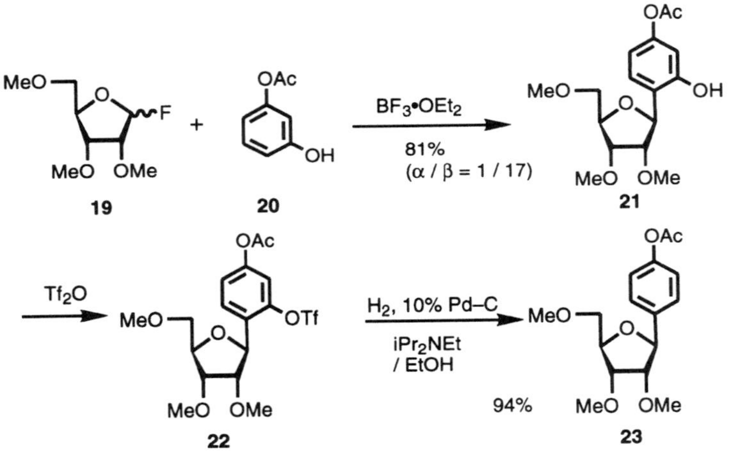

**Scheme 8**

C-glycoside **21** with an additional oxygen functionality at the *para*-position [11c]. Hydrogenolysis of the derived triflate **22** gives the deoxygenated product **23** having a *para*-oxygen function (Sec. III.I) [11c].

The triflate can serve as a precursor to various elaborated aryl C-glycosides, such as **24** and **25** [11c] (Fig. 5).

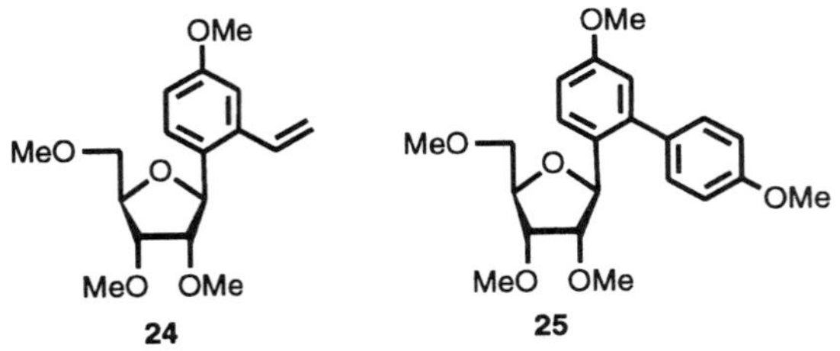

**Figure 5**

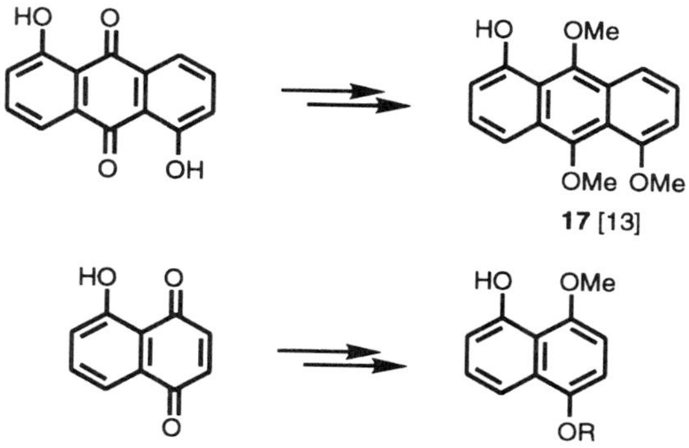

**17** [13]

**26**: R = Me, **27**: R = Ac [16]

**Figure 6**

*Applicability*

The O → C-glycoside rearrangement comprises two Lewis acid-promoted processes; the O-glycoside formation and subsequent conversion to a C-glycoside. Therefore, the aromatic moieties that take part in the reaction should be electron-rich. Otherwise, the reaction stops at the stage of the O-glycoside formation or never proceeds. Hydroxylated quinones in Figure 6 represent such compounds, but the reductive protection technique offers suitable substrates, such as **17**, **26**, and **27**, for the formation of C-aryl glycosides with the present approach [13,15,16].

## III. EXPERIMENTAL PROCEDURES*

### A. Friedel–Crafts-type C-Glycosidation of Glycosyl Acetates [6]

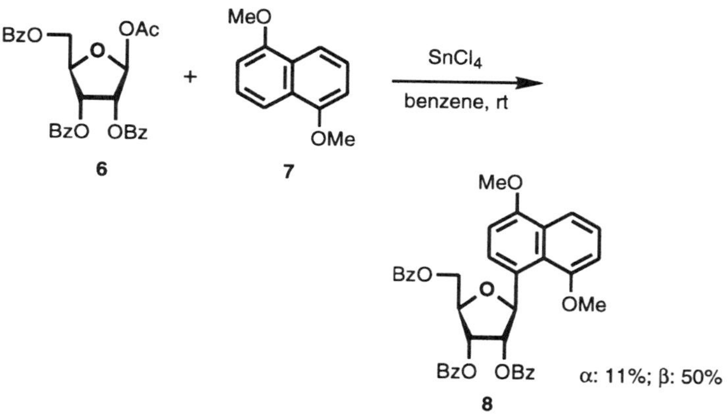

α: 11%; β: 50%

A solution of 1-O-acetyl-2,3,5-tri-O-benzoyl-β-D-ribofuranose **6** (1.00 g, 2 mmol) and 1,5-dimethoxynaphthalene **7** (0.530 g, 2.8 mmol) in a 0.5-$M$ solution of stannic chloride in

---

*Optical rotations were measured at 20°–25°C.

benzene (10 mL), was stirred for 4 h at room temperature. The solution was diluted with EtOAc (15 mL), washed with 10% acetic acid (3 × 10 mL) and water, dried over Na$_2$SO$_4$, and evaporated in vacuo. The residue was purified by column chromatography on silica gel (120 g) with benzene as eluant; and the solvent was evaporated in vacuo to give a solid. Recrystallization from EtOAc–MeOH gave 4-(2,3,5-tri-*O*-benzoyl-β-D-ribofuranosyl)-1,5-dimethoxynaphthalene **8β** as colorless needles (0.620 g, 50%): mp 169°–171°C, [α]$_D$ +14.4° (*c* 0.52, CHCl$_3$). Further elution with benzene gave a solid, which was recrystallized from MeOH to give 4-(2,3,5-tri-*O*-benzoyl-α-D-ribofuranosyl)-1,5-dimethoxynaphthalene **8α** as colorless needles (0.130 g, 11%): mp 129°–130°C, [α]$_D$ −67.8° (*c* 0.5, CHCl$_3$).

## B. Friedel–Crafts-type *C*-Glycosidation of Trichloroacetimidates [7]

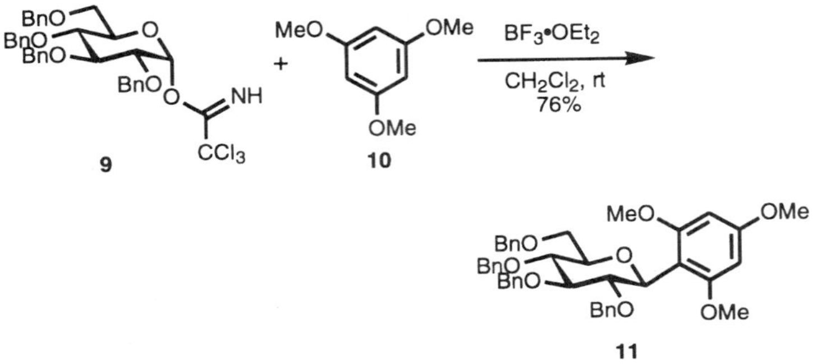

A solution of trichloroacetimidate **9** (2.06 g, 3 mmol) and 1,3,5-trimethoxybenzene **10** (0.50 g, 3 mmol) in dichloromethane (20 mL) was treated with 0.5 *M* boron trifluoride etherate in anhydrous dichloromethane (6 mL, 3 mmol) at room temperature. After 3 h, excess saturated solution of NaHCO$_3$ in water was added. The organic layer was separated, dried (Na$_2$SO$_4$), and evaporated. The oily residue was purified on silica gel by flash chromatography (4:1 v/v, petroleum ether–ethyl acetate) gave 2,4,6-trimethoxy-1-(2,3,4,6-tetra-*O*-benzyl-β-D-glucopyranosyl)benzene **11** (1.55 g, 76%): [α]$_{578}$ + 5.4° (*c* 1.0, chloroform).

## C. *C*-Glycosidation of Olivosyl Fluorides by Using Cp$_2$HfCl$_2$–AgClO$_4$ as the Promoter [13]

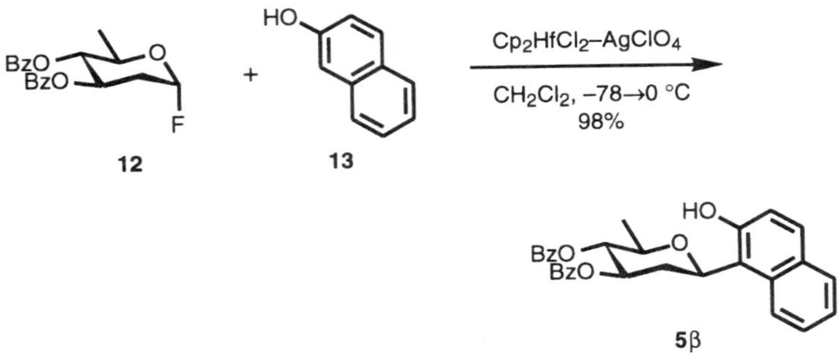

To a stirred mixture of Cp$_2$HfCl$_2$ (56.9 mg, 150 μmol), AgClO$_4$ (31.1 mg, 150 μmol), 2-naphthol **13** (21.6 mg, 150 μmol), and 4-Å powdered molecular sieves (ca. 350 mg) in

# Synthesis of Glycosylarenes

CH$_2$Cl$_2$ (0.5 mL) was added fluoride **12** (17.9 mg, 50.0 μmol) in CH$_2$Cl$_2$ (2 mL) at −78°C. The temperature was gradually raised to 0°C during 2 h, and the reaction was quenched by adding saturated NaHCO$_3$ aqueous solution. After the mixture was filtered through a Celite pad, and the products were extracted with EtOAc (× 2). The combined organic layer was washed with brine and dried over Na$_2$SO$_4$. Removal of the solvents in vacuo and purification by silica gel preparative thin-layer chromatography (TLC) (CCl$_4$–Et$_2$O, 9:1) afforded the C-glycoside **5β** as a colorless oil (23.6 mg, 97.9%): [α]$_D$ − 3.6° (c 1.2, CHCl$_3$).

*Caution*: AgClO$_4$ was obtained from Kojima Chemical Co., and used without further purifcation. Heating or drying should be avoided because AgClO$_4$ is potentially explosive [17].

## D. *C*-Glycosidation of Glycosyl Acetates (Digitoxosylation) by Using SnCl$_4$ as the Promoter [11b]

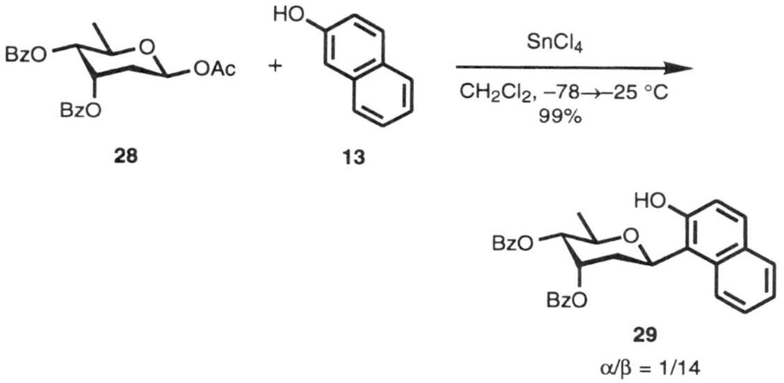

To a stirred mixture of SnCl$_4$ (53 mg, 0.20 mmol), 2-naphthol (18 mg, 0.12 mmol) and 4-Å powdered molecular sieves (ca. 100 mg) in CH$_2$Cl$_2$ (0.5 mL), was added acetate **28** (25.5 mg, 0.0640 mmol) in CH$_2$Cl$_2$ (1.5 mL) at −78°C. The temperature was gradually raised to −25°C during 30 min, and the reaction was quenched by adding saturated NaHCO$_3$ aqueous solution. The mixture was filtered through a Celite pad, and the products were extracted with EtOAc (× 2). The combined organic layer was washed with brine and dried over Na$_2$SO$_4$. Removal of the solvents in vacuo and purification by silica gel preparative TLC (CCl$_4$–Et$_2$O, 85:15) afforded *C*-glycoside **29β** as a colorless oil (28.6 mg, 93%): [α]$_D$ −19.1° (c 1.20, CHCl$_3$); and **29α** as a colorless oil (2.0 mg, 6.5%): [α]$_D$ −80.8° (c 0.76, CHCl$_3$).

## E. *C*-Glycosidation of Glycosyl Acetates by Using BF$_3$·OEt$_2$ as the Promoter [11b]

To a stirred mixture of BF$_3$·OEt$_2$ (27 mg, 0.19 mmol), 2-naphthol (16 mg, 0.11 mmol), and 4-Å powdered molecular sieves (ca. 100 mg) in CH$_2$Cl$_2$ (1.5 mL), was added acetate **30** (22.4 mg, 0.0562 mmol) in CH$_2$Cl$_2$ (1.5 mL) at −78°C. The temperature was gradually raised to room temperature during 45 min, and the reaction was quenched by adding saturated NaHCO$_3$ aqueous solution. The mixture was filtered through a Celite pad, and the products were extracted with EtOAc (× 2). The combined organic layer was washed with brine and dried over Na$_2$SO$_4$. Removal of the solvents in vacuo and purification by silica gel preparative TLC (hexane–Et$_2$O, 55:45) afforded C-glycoside **31β** as a colorless oil (25.5 mg, 94%): [α]$_D$ − 124° (c 1.61, CHCl$_3$).

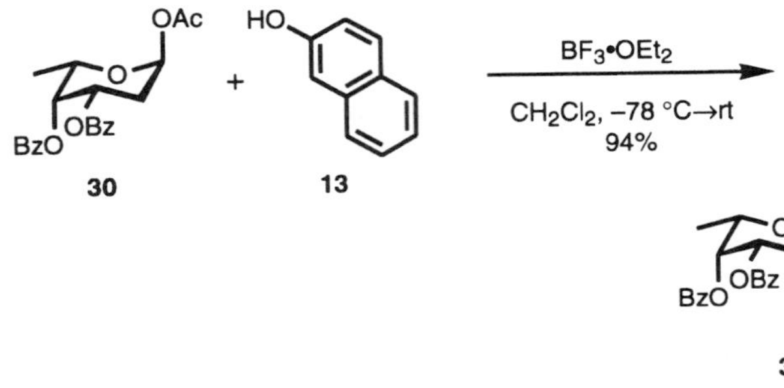

### F. C-Glycosylation of 2-Iodoresorcinol Derivative (14) with Fucofuranosyl Acetate (15) by Using Ph$_3$SiCl–AgClO$_4$ as the Promoter [14]

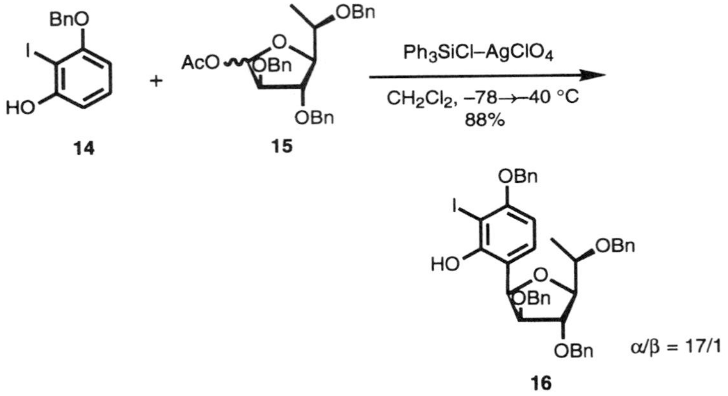

The promoter was prepared in situ by stirring the mixture of Ph$_3$SiCl (310 mg, 1.05 mmol) and AgClO$_4$ (221 mg, 1.07 mmol) in the presence of 4-Å powdered molecular sieves (ca 1.0 g) in CH$_2$Cl$_2$ (20 mL) for 30 min at room temperature. To this suspension at −78°C was added a solution of phenol **14** (257 mg, 0.788 mmol) in CH$_2$Cl$_2$ (6 mL) and glycosyl acetate **15** (250 mg, 0.525 mmol) in CH$_2$Cl$_2$ (6 mL). The reaction mixture was gradually warmed to −40°C during 1 h, and the stirring was continued for 10 min. The reaction was quenched by the addition of saturated aqueous NaHCO$_3$ solution. The mixture was acidified with 2 N HCl, filtered through a Celite pad and extracted with EtOAc. The combined organic extracts were washed successively with saturated aqueous NaHCO$_3$ solution and brine, dried (Na$_2$SO$_4$), and concentrated in vacuo. The residue was purified by silica gel preparative PTLC (CCl$_4$–Et$_2$O, 8:2 and hexane–EtOAc, 8:2) to afford an anomeric mixture of C-glycosides **16α,β** (343 mg, 88.0%, α/β, 17:1). Further purification by PTLC (hexane–EtOAC, 8:2, double development) afforded C-glycosides **16α** as a colorless oil (323 mg, 82.9%): [α]$_D$ −35° (c 3.1, CHCl$_3$); and **16β** as a colorless oil (19.0 mg, 4.9%); [α]$_D$ −4.8° (c 2.8, CHCl$_3$).

## G. *C*-Glycosylation of Anthracene Derivative (17) with Olivosyl Fluoride (12) [13]

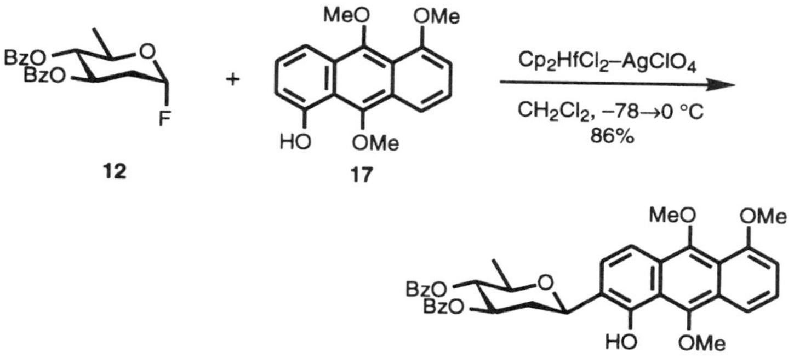

To a stirred mixture of Cp$_2$HfCl$_2$ (76.0 mg, 0.200 μmol), AgClO$_4$ (41.5 mg, 200 μmol), anthrol **17** (57.0 mg, 200 μmol), and 4-Å powdered molecular sieves (ca. 500 mg) in CH$_2$Cl$_2$ (1 mL), was added the fluoride **12** (23.9 mg, 66.8 μmol) in CH$_2$Cl$_2$ (1.5 mL) at −78°C. The temperature was gradually raised to 0°C during 50 min, and the reaction was quenched by adding saturated NaHCO$_3$ aqueous solution. The mixture was filtered through a Celite pad, and the products were extracted with EtOAc (× 2). The combined organic layer was washed with brine, dried over Na$_2$SO$_4$, and filtered. The solvents were removed in vacuo, and the residue was purified by silica gel flash chromatography (hexane–EtOAc, 8:2) to afford the *C*-glycoside **18** as a yellow oil (35.5 mg, 86%): [α]$_D$ +35° (*c* 1.5, CHCl$_3$).

## H. *C*-Glycosylation of Angucycline-type Skeleton (33) with Olivoyl Acetate (32) [15]

To a suspension of Cp$_2$HfCl$_2$ (263 mg, 0.693 mmol), AgClO$_4$ (288 mg, 1.39 mmol), and powdered 4-Å molecular sieves (ca. 600 mg) in CH$_2$Cl$_2$ (5 mL) were added a solution of phenol **33** (162 mg, 0.463 mmol) in CH$_2$Cl$_2$ (8 mL) and D-olivosyl acetate **32** (343 mg, 0.927 mmol) in CH$_2$Cl$_2$ (9 mL) at −78°C. After the mixture was stirred for 15 min, the reaction was stopped by adding pH 7 phosphate buffer. The mixture was acidified by 2 *N* HCl, filtered through a Celite pad, and the products were extracted with EtOAc. The combined extracts were washed with brine and were dried over Na$_2$SO$_4$. Concentration in vacuo, followed by purification by silica gel flash chromatography (hexane–EtOAc, 4:1 and then CCl$_4$–EtOAc, 86:14) gave C-glycoside **34** as a foam, (210 mg, 69%): mp 87°–88°C, [α]$_D$ +40.1° (c 1.17, CHCl$_3$).

## I. Synthesis of *para*-aryl-*C*-Riboside 23 [11c]

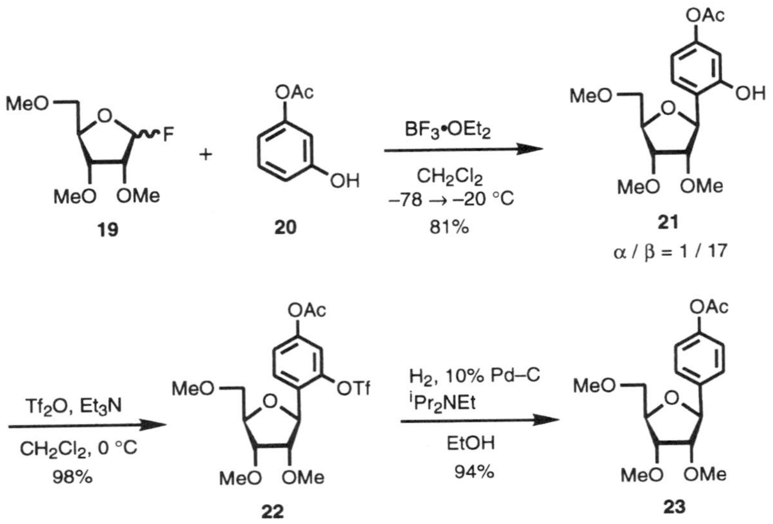

*Reaction of Resorcinol Monoacetate **20** and Glycosyl Fluoride **19***

To a mixture of resorcinol monoacetate **20** (142 mg, 0.932 mmol), BF$_3$·OEt$_2$ (132 mg, 0.932 mmol), and 4-Å powdered molecular sieves (ca. 200 mg) in CH$_2$Cl$_2$ (3 mL), was added a solution of glycosyl fluoride **19** (60.3 mg, 0.311 mmol) in CH$_2$Cl$_2$ (2 mL) at −78°C. The reaction was gradually warmed to −10°C during 30 min and quenched with 1 N HCl. The mixture was filtered through a Celite pad and extracted with EtOAc. The combined organic layer was washed with brine and then dried over Na$_2$SO$_4$. After removal of the solvent in vacuo, the residue was purified by silica gel PTLC (CH$_2$Cl$_2$–Et$_2$O, 8:2) to afford C-glycosides **21β** (77.8 mg, 76.8%): [α]$_D$ −24° (c 1.7, CHCl$_3$); **21α** (4.5 mg 4.4%): [α]$_D$ +13° (c 1.1, CHCl$_3$); and **21αβ** (10.9 mg, 10.8%).

*Trifluoromethanesulfonylation of Phenol **21β***

To a mixture of phenol **21β** (70.4 mg, 0.216 mmol) and Et$_3$N (28.4 mg, 0.281 mmol) in CH$_2$Cl$_2$ (3 mL) was added Tf$_2$O (91.3 mg, 0.324 mmol) at 0°C. The reaction was terminated immediately by adding saturated aqueous NaHCO$_3$, and the mixture was extracted with CH$_2$Cl$_2$. The combined organic layer was washed with brine and then dried over Na$_2$SO$_4$. After removal of the solvent in vacuo, the residue was purified by silica gel PTLC (hexane–EtOAc, 5:5) to afford triflate **22** (96.0 mg, 97.1%).

*Hydrogenolysis of Triflate **22***

A suspension of triflate **22** (92.9 mg, 0.203 mmol), catalytic amount of 10% Pd/C, and i-Pr$_2$NEt (52.4 mg, 0.405 mmol) in EtOH (5 mL), was stirred under H$_2$ (1 atm) at room temperature for 40 min. After changing the atmosphere to argon, the mixture was filtered through a Celite pad, and the solvent was removed in vacuo. The residue was diluted with $CH_2Cl_2$, washed with water, and then dried over $Na_2SO_4$. After removal of the solvent in vacuo, the residue was purified by silica gel PTLC (hexane–EtOAc 5:5) to afford acetate **23** as a colorless oil (58.9 mg, 93.6%): $[\alpha]_D$ −14° ($c$ 2.6, $CHCl_3$).

## REFERENCES

1. Reviews: (a) M. H. D. Postema, Recent developments in the synthesis of *C*-glycosides, *Tetrahedron* 48:8545 (1992); (b) S. Hanessian and A. G. Pernet, Synthesis of naturally occurring *C*-nucleosides, their analogs, and functionalized *C*-glycosyl precursors, *Adv. Chem. Biochem.* 33:111 (1976).
2. (a) U. Hacksell and G. D. Daves, Jr., The chemistry and biochemistry of *C*-nucleosides and *C*-arylglycosides, *Prog. Med. Chem* 22:1(1985); (b) K. Suzuki and T. Matsumoto, Total synthesis of aryl *C*-glycoside antibiotics, *Recent Progress in the Chemical Synthesis of Antibiotics and Related Microbial Products*, Vol 2. (G. Lukacs ed.), Springer-Verlag, Berlin, 1993, p. 352.
3. M. Sezaki, S. Kondo, K. Maeda, H. Umezawa, and M. Ohno, The structure of aquayamycin, *Tetrahedron* 26:5171 (1970).
4. (a) W. A. Szarek and D. Horton, eds., *Anomeric Effect: Origin and Consequences*, ACS Symposium Series 87, American Chemical Society; Washington DC, 1979; (b) A. J. Kirby, *The Anomeric Effect and Related Stereoelectronic Effects at Oxygen*, Springer-Verlag, Berlin, 1983; (c) E. Juaristi and G. Cuevas, Recent studies of the anomeric effect. Tetrahedron report no. 315, *Tetrahedron* 48:5019 (1992); (d) G. R. J. Thatcher, ed., *The Anomeric Effect and Associated Stereoelectronic Effects*, ACS Symposium Series 539, American Chemical Society, Washington DC, 1993.
5. (a) L. Kalvoda, J. Farkas, and F. Sorm, Synthesis of showdomycin, *Tetrahedron Lett.* 26:2297 (1970); (b) L. Kalvoda, Acid-catalyzed *C*-ribofuranosylation of benzene derivatives; some novel conversions of *C*-ribofuranosyl derivatives, *Collection Czechoslov. Chem. Commun.* 38:1679 (1973).
6. N. Hamamichi and T. Miyasaka, Synthesis of methyl- and methoxy-substituted β-D-ribofuranosylnaphthalene derivatives by Lewis acid catalyzed ribofuranosylation, *J. Org. Chem.* 56:3731 (1991).
7. R. R. Schmidt and G. Effenberger, *C*-Glucosylarenes from *O*-α-D-glucosyl trichloroacetimidates. Structure of bergenin derivatives, *Carbohydr. Res.* 171:59 (1987).
8. A. O. Stewart and R. M. Williams, *C*-Glycosidation of pyridyl thioglycosides, *J. Am. Chem. Soc.* 107:4289 (1985).
9. T. Matsumoto, M. Katsuki, and K. Suzuki, $Cp_2ZrCl_2$–$AgClO_4$: Efficient promoter in the Friedel–Crafts approach to *C*-aryl glycosides, *Tetrahedron Lett.* 30:833 (1989).
10. Reviews: (a) K. Suzuki and T. Matsumoto, $Cp_2MCl_2$–AgX (M = Zr, Hf): A reagent for glycosidation, *J. Synth. Chem. Soc. Jpn.* 48:1026 (1990); (b) K. Suzuki and T. Matsumoto, New methods in glycoside synthesis by using early transition metal-derived reagent, *J. Synth. Chem. Soc. Jpn.* 51:718 (1993); (c) K. Suzuki, Novel Lewis acid catalysis in organic synthesis, *Pure Appl. Chem.* 66:1557 (1994).
11. T. Matsumoto, M. Katsuki, and K. Suzuki, New approach to *C*-aryl glycosides starting from phenol and glycosyl fluoride. Lewis acid-catalyzed rearrangement of *O*-glycoside to *C*-glycoside, *Tetrahedron Lett.* 29:6935 (1988); (b) T. Matsumoto, T. Hosoya, and K. Suzuki, Improvement in $O \rightarrow C$-glycoside rearrangement approach to *C*-glycosides: Use of 1-*O*-acetyl sugar as stable but efficient glycosyl donor, *Tetrahedron Lett.* 31:4629 (1990); (c) T. Matsumoto, T. Hosoya, and K. Suzuki, $O \rightarrow C$-Glycoside rearrangement of resorcinol derivatives. Versatile intermediates in the synthesis of aryl *C*-glycosides, *Synlett*, p. 709 (1991).

12. T. Matsumoto, M. Katsuki, and K. Suzuki, Rapid $O$-glycosidation of phenols with glycosyl fluoride by using the combinational activator, $Cp_2HfCl_2$ and $AgClO_4$, *Chem. Lett.*, p. 437 (1989).
13. (a) T. Matsumoto, M. Katsuki, H. Jona, and K. Suzuki, Synthetic study toward vineomycins. Synthesis of $C$-aryl glycoside sector via $Cp_2HfCl_2$–$AgClO_4$-promoted tactics to $C$-aryl glycoside synthesis, *Tetrahedron Lett. 30*:6185 (1989); (b) T. Matsumoto, M. Katsuki, H. Jona, and K. Suzuki, Convergent total synthesis of vineomycinone $B_2$ methyl ester and its C(12)-epimer, *J. Am. Chem. Soc. 113*:6982 (1991).
14. (a) T. Matsumoto, T. Hosoya, and K. Suzuki, Total synthesis and absolute stereochemical assignment of (+)-gilvocarcin M, *J. Am. Chem. Soc. 114*:3568 (1992); (b) T. Hosoya, E. Takashiro, T. Matsumoto, and K. Suzuki, Total syntheses of the gilvocarcins, *J. Am. Chem. Soc. 116*:1004 (1994).
15. (a) T. Matsumoto, T. Sohma, S. Kurata, H. Yamaguchi, and K. Suzuki, Benzyne–furan cycloaddition approach to the angucyclines: Total synthesis of antibiotic C104, *Synlett*, p. 263 (1995); (b) T. Matsumoto, T. Sohma, H. Yamaguchi, S. Kurata, and K. Suzuki, Total synthesis of antibiotic C104: Benzyne–furan cycloaddition approach to the angucyclines, *Tetrahedron 51*:7347 (1995).
16. T. Hosoya, Y. Ohashi, T. Matsumoto, and K. Suzuki, On the stereochemistry of aryl $C$-glycosides: Unusual behavior of bis-TBDPS-protected aryl $C$-olivosides, *Tetrahedron Lett. 37*:663 (1996).
17. S. R. Brinkley, Jr., The instability of silver perchlorate, *J. Am. Chem. Soc. 62*:3524 (1940).

# VI
# Carbocycles from Carbohydrates

**THEMES**: *Functionalized Cyclopentanes, Functionalized Cyclohexanes*

It has been said that inositol is a sugar with carbon in the ring. Indeed it is interesting to reflect on the effect of such a simple atom-for-atom replacement on the chemical and enzymatic reactivities of a hexose, when compared with a rather inert cyclitol.

Nature has given us a plethora of compounds that contain one or more carbocyclic ring systems. Its generosity continues with the discovery of therapeutically important molecules, such as paclitaxel (Taxol) and the ginkolides. Monocyclic carbocycles can also be appreciated in the structures of the prostaglandins, the inositol phosphates, and aminoglycoside antibiotics, to mention only three representatives of biogenetically diverse molecules. From a viewpoint of synthetic strategy, the construction of carbocyclic molecules has followed a traditional approach, with occasional diversions. Thus cyclohexane-based structures are easily derived from appropriately substituted dienes and dienophiles in a Diels–Alder reaction. The construction of cyclopentane and cyclohexane motifs might, for example, involve a two-step Michael addition–cyclization sequence, or the venerable Robinson annulation reaction. A more "natural" approach to the construction of such carbocycles could take advantage of chiral templates, such as terpenes (e.g., carvone or functionalized cyclopentanes arising from the fragmentation of camphor and its derivatives).

One may also ask if the six-carbon framework of D-glucose could be used as a chiral template for the construction of a functionalized cyclohexane. It would simply involve the equivalent of an intramolecular condensation of an anionic C-6 atom with the existing masked aldehyde at C-1. Indeed, intramolecular aldol reactions of 6-nitro sugars to the corresponding nitrocyclohexanols was successfully carried out in the early 1950s by none other than H. O. L. Fischer, the prodigal son himself, who was then at Berkeley. Today, there are various mild and efficient methods for the synthesis of five- and six-membered carbocycles from sugars precursors. It is also interesting to comment on the biosynthesis of some of the cyclohexane-type products that could arise from the corresponding aromatic precursors by enzymatic reactions. It is more interesting to question the biogenetic source of aromatic compounds in nature. This will bring us right back to D-glucose, recognized by nature as an excellent source of six contiguous carbon atoms ready for the sacrificial abdication of five stereogenic centers, but for a good cause.

The propensity for β-elimination of an anionic sites at C-6 of a hexose derivative requires the judicious choice of reaction conditions, protective groups, and aldehyde activation modes for intramolecular carbocyclization. In addition, one must appreciate that a favorable trajectory to approach the anionic site through an energetically allowed conformer must be ascertained, to maximize carbocyclization, rather than the occurrence of undesirable side reactions. This can be avoided by exploiting free-radical conditions and electron-transfer chemistry that are mild enough to allow intramolecular reactions to occur between donor and acceptor sites. Sugar molecules can be easily groomed to include the functionality necessary for such intramolecular ring closures. They have been nicely exploited by carbohydrate and, particularly, by noncarbohydrate chemists in the synthesis of carbocycles.

The next two chapters demonstrate the power of anionic, free-radical and electron-transfer processes in the systematic construction of highly functionalized five- and six-membered carbocycles.

*Stephen Hanessian*

# 25

# Functionalized Carbocylic Derivatives from Carbohydrates: Free Radical and Organometallic Methods

## T. V. RajanBabu
*The Ohio State University, Columbus, Ohio*

| | | |
|---|---|---|
| I. | Introduction | 546 |
| II. | Cyclopentanes | 546 |
| | A. The hex-5-enyl radical cyclization | 546 |
| III. | Cyclohexanes | 554 |
| | A. The hept-6-enyl radical cyclization | 554 |
| IV. | Functionalized Carbocylic Compounds by Organometallic Methods | 555 |
| V. | Experimental Procedures | 558 |
| | A. Carbocycles from unsaturated halo sugars by hex-5-enyl radical cyclization | 558 |
| | B. Highly functionalized cyclopentanes from hexopyranose derivatives | 559 |
| | C. A hept-5-enyl radical cyclization route to functionalized cyclohexanes | 562 |
| | D. Trimethyltin radical–initiated cyclization of 1,6-dienes | 562 |
| | E. Ti(III)-mediated epoxy–olefin cyclization route to carbocyclic compounds from carbohydrates | 563 |
| | F. Ketyl–olefin cyclization mediated by samarium(II) iodide | 563 |
| | G. Pauson–Khand reactions | 563 |
| | H. Zirconocene-mediated cyclization of sugar eneynes | 564 |
| | I. Intramolecular cyclization of an allyl zirconium derivative | 565 |
| | References and Notes | 565 |

## I. INTRODUCTION

From the days of Emil Fisher, carbohydrates have played an important role in the development of organic chemistry [1]. Considering such a long historical relation and the remarkable progress made in the functional group manipulations of carbohydrates, studies aimed the usefulness of these compounds as a source of carbocyclic compounds are of recent origin. Most of the developments have appeared in the past 15 years or so, and several excellent reviews on the subject are available [2]. Carbohydrates, being the most ubiquitous source of chirality in nature, are ideal starting materials for many enantiomerically pure natural products [3], especially those that are highly oxygenated. Several biologically important compounds, such as antiviral carbocyclic nucleosides, macrocyclic antibiotics, aminocyclitol antibiotics, glycosidase inhibitors, inositols, and C-glycosides, are represented among this class of compounds [2b]. Historically, as with many other areas of organic chemistry, the first reported methods for the carbohydrate to carbocyclic conversions depended on carbanionic intermediates; the reader is referred to the excellent review by Ferrier [2a] and the references cited therein, for a detailed discussion on this subject. Typical among these methods are intramolecular alkylation and intramolecular condensations of aldehydes with enolates, phosphonate, and nitro-stabilized anions. Cycloaddition reactions, including intramolecular 1,3-dipolar additions and [4 + 2]-cyclo additions have also been used.

The explosive growth in free radical and organometallic chemistry has prompted an intense interest in these methods for the conversion of carbohydrates to carbocyclic compounds. These methods are generally complementary to the traditional approaches that rely on highly polar intermediates, discussed earlier because, under the reaction conditions, different functional group compatibilities often exist. For example, although polar groups, such as the carbonyl group, play a central role in most ionic and even in many pericyclic carbon–carbon bond-forming processes (e.g., in the activation of $\pi$-systems), in free radical and organometallic methods, unactivated olefins and acetylenes can act as reaction partners. Unlike carbanionic reaction conditions, under conditions of free radical generation, a $\beta$-leaving group and a relatively acidic hydrogen such as −OH or −NHC(O)R are tolerated. Often reactions can be carried out with no hydroxyl-protecting groups, or with protecting groups that are incompatible with carbanionic intermediates. Thanks to the ancillary ligands that are often bound to the metal mediating the transformation, such processes often exhibit remarkable stereochemical control in the formation of new bonds.

## II. CYCLOPENTANES

### A. The Hex-5-enyl Radical Cyclization

Of all the radical reactions, the *exo*-1,5-cyclization of a hex-5-enyl radical to cyclopentylmethyl radical and its subsequent trapping by various reagents have attracted the most attention from synthetic chemists (Scheme 1) [4–7]. Starting materials that are most often used for the "tin method" (initiation of the chain by trialkyl tin radical) are halides, sulfides, selenides, or thionocarbonates. The generation and cyclization of the radical proceeds under exceptionally mild neutral conditions, and these conditions are compatible with a wide variety of common functional groups. A prototypical example of an application in carbohydrate chemistry is shown in Scheme 2 [8]. Readily available 2,3-di-*O*-isopropylideneribonolactone **1** was converted into the bromoacrylate **2** in three steps. Radical

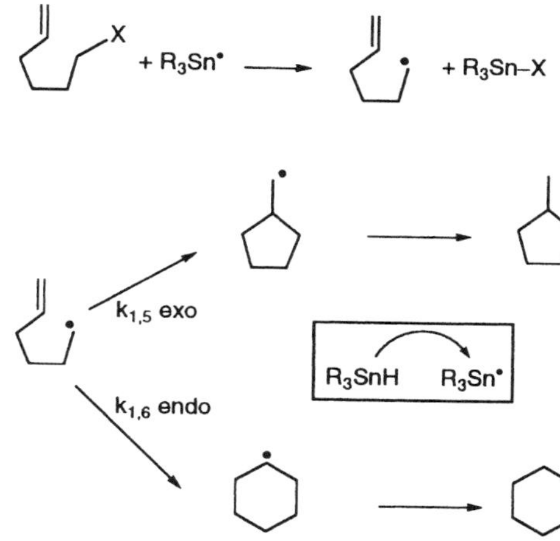

# Methods for Forming Carbocylic Derivatives

**Scheme 1** Hex-5-enyl Radical Cyclization at 60°C, $k_{1,5} = 10^5–10^6$ s$^{-1}$; $k_{1,5}/k_{1,6} = 50$.

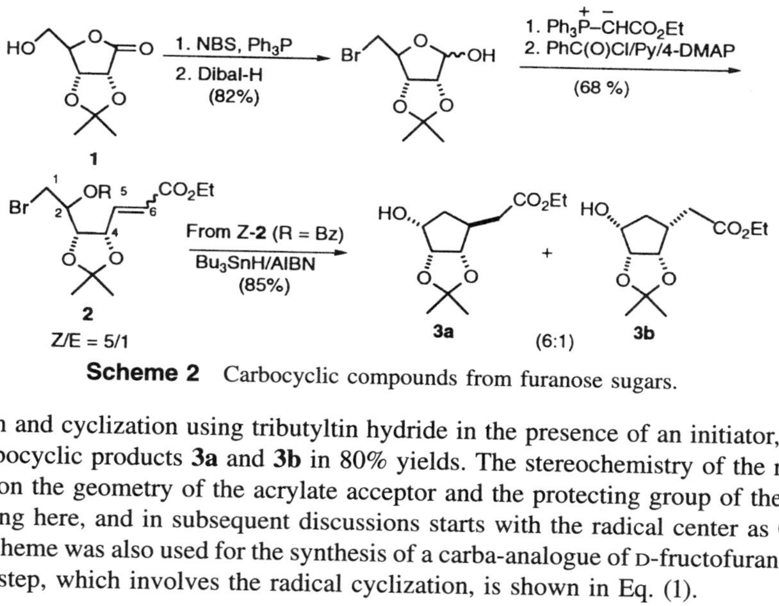

**Scheme 2** Carbocyclic compounds from furanose sugars.

formation and cyclization using tributyltin hydride in the presence of an initiator, AIBN, gave carbocyclic products **3a** and **3b** in 80% yields. The stereochemistry of the reaction depends on the geometry of the acrylate acceptor and the protecting group of the at C-2 (numbering here, and in subsequent discussions starts with the radical center as C-1). A related scheme was also used for the synthesis of a carba-analogue of D-fructofuranose [9]. The key step, which involves the radical cyclization, is shown in Eq. (1).

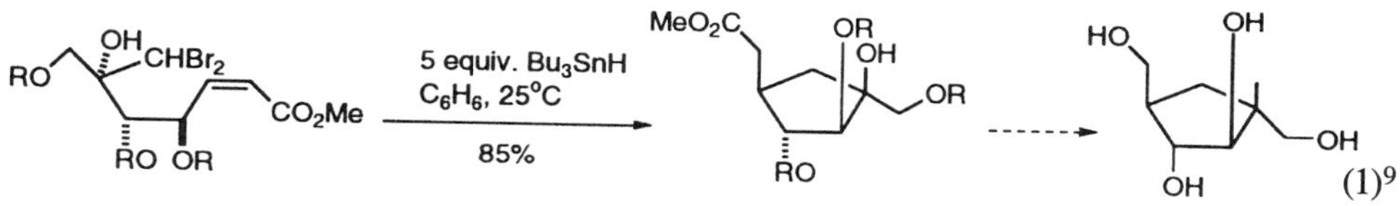

$$(1)^9$$

Carba-D-fructofuranose

A versatile protocol for the generation and cyclization of secondary radicals from hexopyranose sugars is shown in Scheme 3 [10]. The Wittig reaction of reducing sugars with two eq of an alkylidene phosphorane readily provide hex-5-ene-1-ols, which were converted into hex-5-enyl radicals by the 1-*H*-imidazole-1-carbothioate. The cyclization reaction is carried out in refluxing benzene or toluene with tributyltin hydride and AIBN, according to

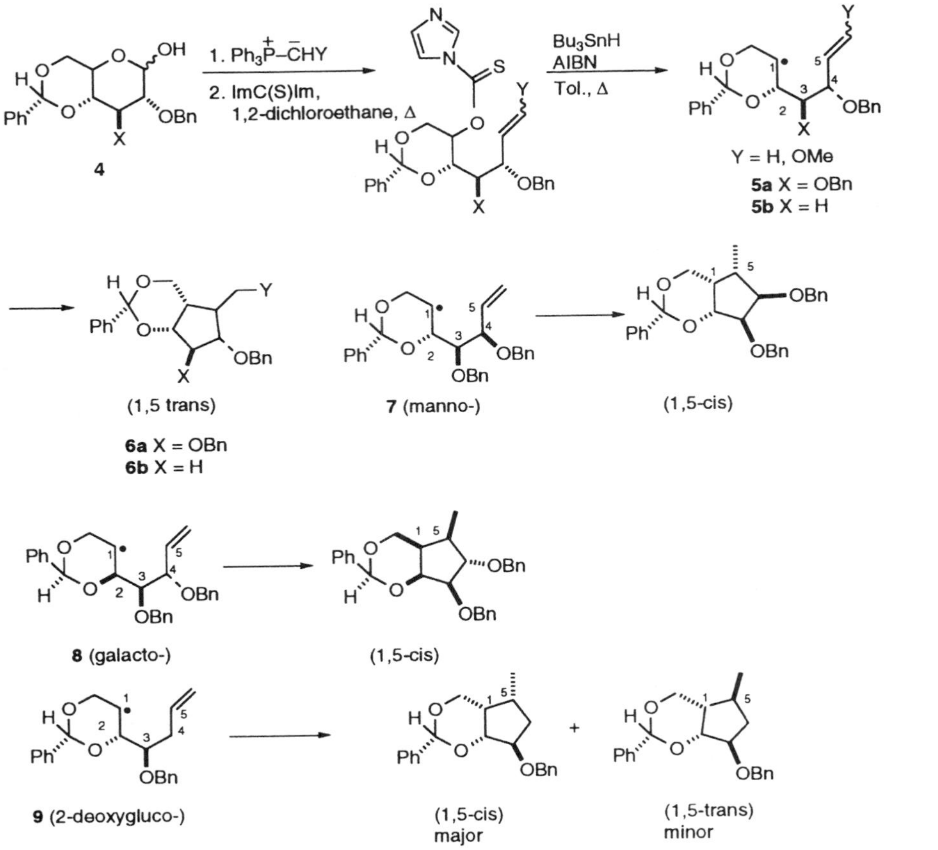

**Scheme 3** 1,2-Dialkylcyclopentanes from hexopyranose sugars.[12]

the Barton protocol [11]. In addition to the obvious synthetic usefulness in the construction of densely functionalized cyclopentanes, the generation of a secondary radical in this fashion allowed an examination of several stereochemical aspects of the hex-5-enyl radical cyclization. Hex-5-enitols, with varying configuration at every carbon atom, became now readily available, and the effect of these configurations on the stereochemical course of the reaction (e.g., 1,2 or 1,5-stereochemistry) could be explored. Thus, in the example shown in Scheme 3, the radicals **5a** and **5b** from a 4,6-benzylideneglucose (**4**) underwent a stereospecific cyclization to give *exclusively* the 1,5-*trans*-products **6a** and **6b**, respectively. The stereochemistry of the double bond (when Y = OMe) had no effect on this outcome. The mannose- (**7**) and galactose- (**8**) derived radicals gave almost exclusively the 1,5-*cis*-products. The $C_4$ deoxy system (**9**) gave a mixture of both 1,5-*cis*-and *trans*-products, with the former predominating. Thus, the stereochemistry of the newly formed carbon–carbon bond is controlled by the configuration of the C-4 center of the hexenyl tether [12]. This unprecedented sterochemical control can be rationalized (Scheme **4**) by a cyclic transition state, for which the conformation, "chair-like" or "boat-like" is determined by steric and stereoelectronic effects of the allylic substituents [7]. For example, in the gluco system, a favorable conformation of the C-3 to C-6 segment (4-*H*- in the same plane as the double bond, see Figure 10) of the hex-5-enyl chain which avoids 1,3- strain may be responsible

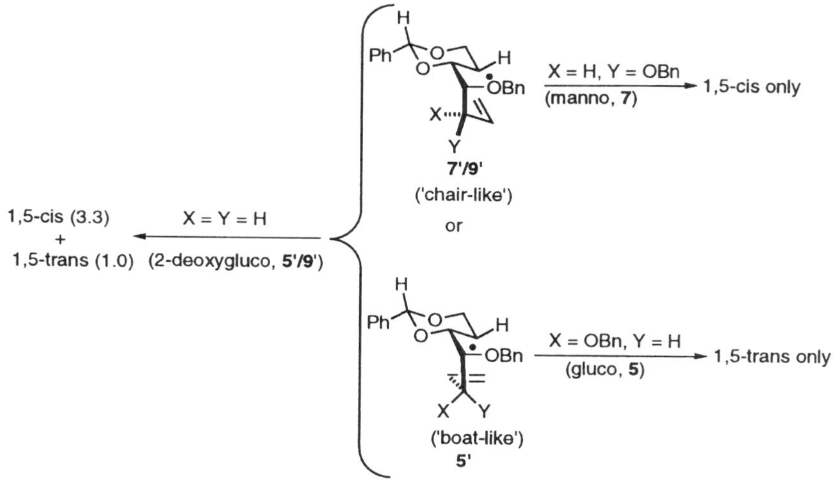

# Methods for Forming Carbocylic Derivatives

**Scheme 4** Transition-state models for sugar hex-5-enyl radical cyclizations.[7]

for the seemingly high-energy boat-like transition state **5'**. No such allylic strain exists in the chair-like transition state corresponding to the "manno" radical **7'**, and a 1,5-*cis*-product results. With no substituent at C-4 (i.e., with no control element present), a mixture of 1,5-*cis*- and *trans*-products are formed, and the anticipated *cis*-product from a chair-like transition state **9'**, predominates. Acyclic radicals, in which the 4,6-*O*-benzylidene group in the gluco system is replaced with di-*O*-benzyl-protecting groups (Figure 11), give a mixture of products in which, as expected [13], the 1,5-*cis*-products predominate [12].

This stereochemical control in hex-5-enyl radical cyclizations can be used for the synthesis of highly functionalized cyclopentanes with vicinal *trans*- or *cis*-dialkyl-substituents. The synthesis of a versatile prostaglandin intermediate, Corey lactone **12**, from the intermediate **6a** (Y = OMe) has been described [14].

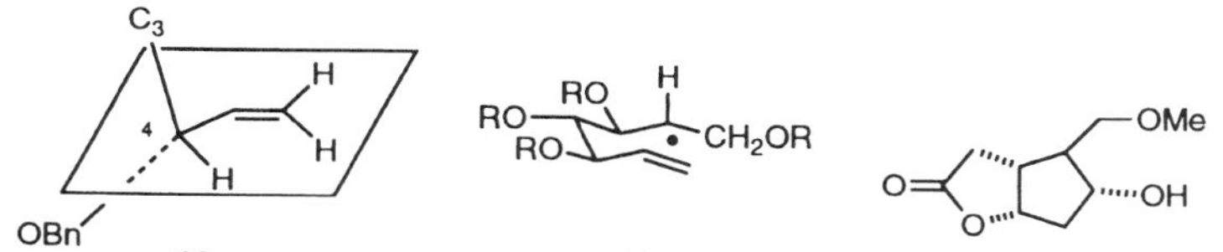

A useful modification of the Barton deoxygenation of secondary alcohols involves the use of *O*-phenylthionocarbonates developed by Robins et al. [15]. Application of this method for the generation and cyclization of a hex-5-ynyl radical is shown is Scheme 5. The precursors are readily prepared from D-ribose by a Grignard addition, followed by selective alcohol derivatizations. The major *exo*-isomer has been converted into carba-α-D-ribofuranose [16].

The phenylthionocarbonate procedure was also used for the cyclization of a 5-oxime-ether radical (Scheme 6) [17]. The stereochemical outcome of this reaction is almost identical with that observed for a closely related 6-methoxyhex-5-enyl radical cyclization [12,14]. A related glucosamine-derived radical cyclization has been employed for the synthesis of allosamizoline **13** [18]. Other examples in this area include the cyclization of

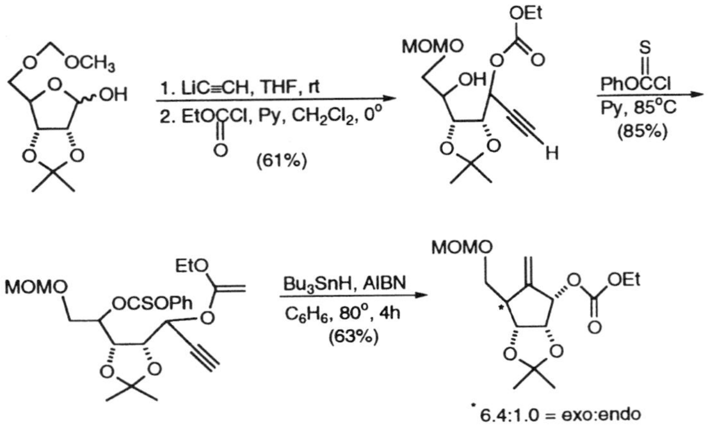

**Scheme 5** Cyclization of a hex-5-ynyl radical.[16]

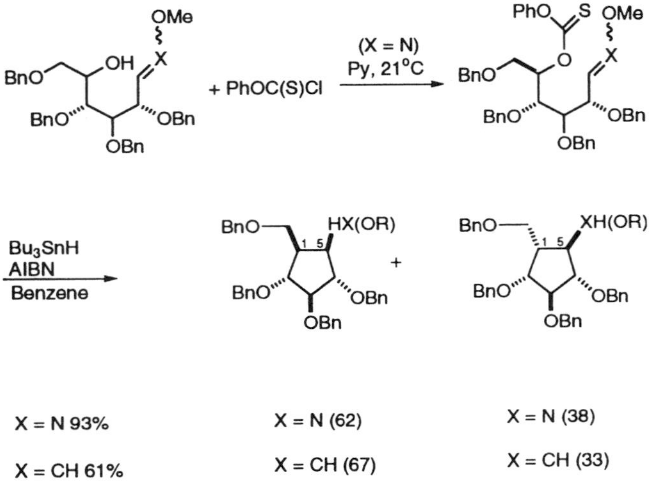

**Scheme 6** Cyclization oxime and vinylether radicals[12,17]

an oximeether structurally related to the bromoacrylate **2** [19] and a pyranoside annulation [20].

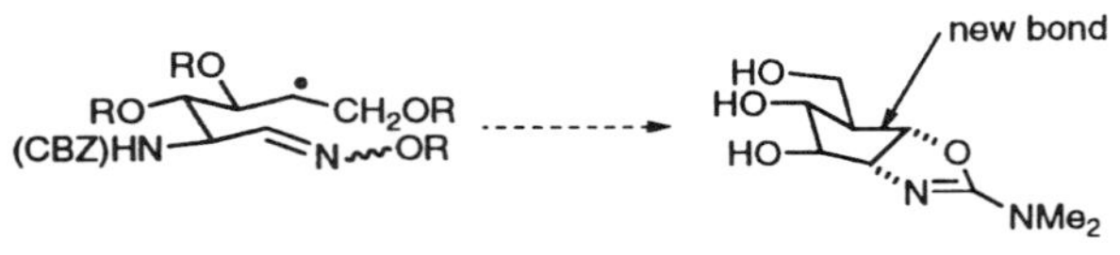

**13** Allosamizoline[18]

Fraser-Reid and co-workers have deveoped an ingeneous strategy using $C_6$-chain-extended sugars, in which several reactive latent functional groups, ready for further elaboration, are still preserved in the cyclization product [21]. Thus, D-glucal–derived iodoacrylate **14** (Scheme 7) undergoes cyclization in the presence of tributyltin hydride and AIBN in refluxing benzene to give two products in a ratio of 1.8:1 in an overall yield of 91%. Side-chain manipulation also allowed these workers to prepare iodoacrylates, such as **15**), which undergo an *exo*-hept-6-enyl radical cyclization (see later discussion) with surprising

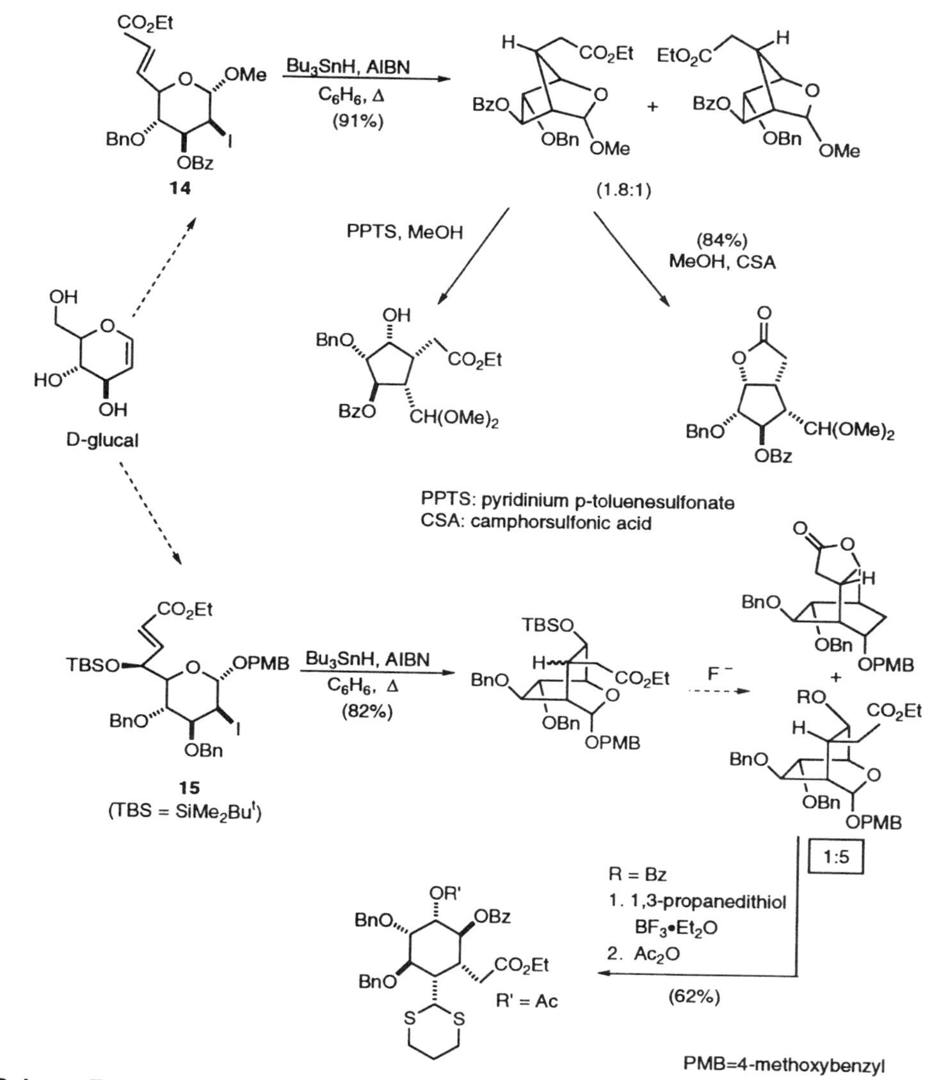

# Methods for Forming Carbocylic Derivatives

**Scheme 7** Formation of bicyclic compounds via hex-5-enyl and hept-6-enyl radical cyclizations.[21]

efficiency. Related reactions using benzenethiol adducts of an unsaturated lactone(s) **16** are also known (Scheme **8**) [22]. Three other examples of radical trapping on the side chain are shown in Eqs. 2 [23], 3 [24], and 4 [20]. The example in Eq. (3) verifies the previously obtained electron spin resonance (ESR) evidence for the remarkable stabilization of α-oxyradicals of the type **17** (conformation shown) by a β-acetoxy group [25].

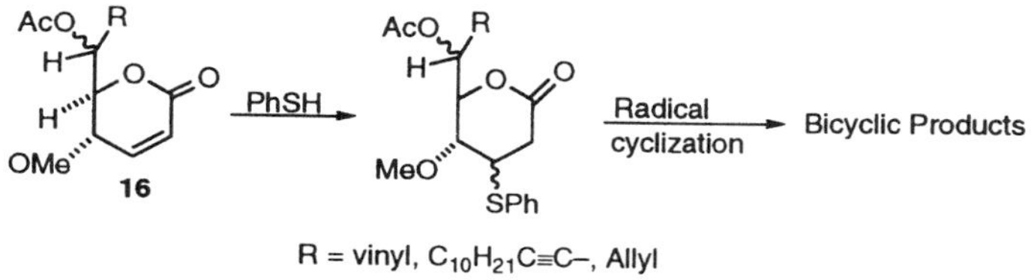

**Scheme 8** Radicals from unsaturated sugar lactones.[22]

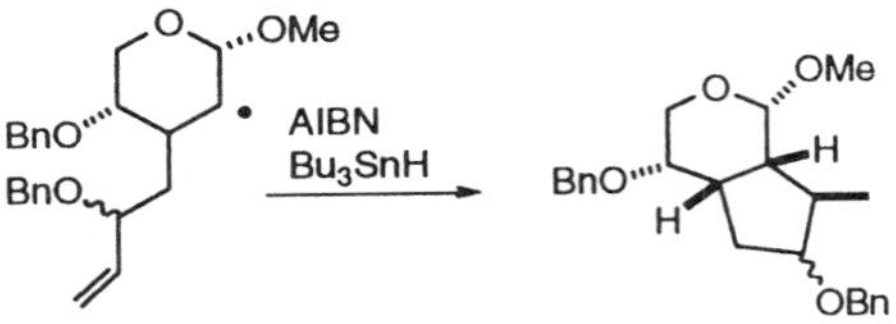


(2)[23]

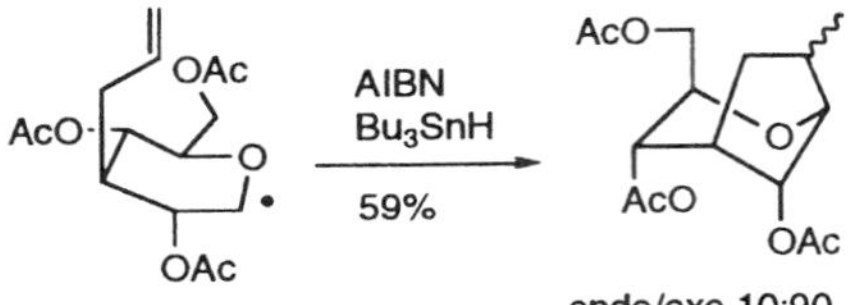


**17**

endo/exo 10:90

(3)[24]

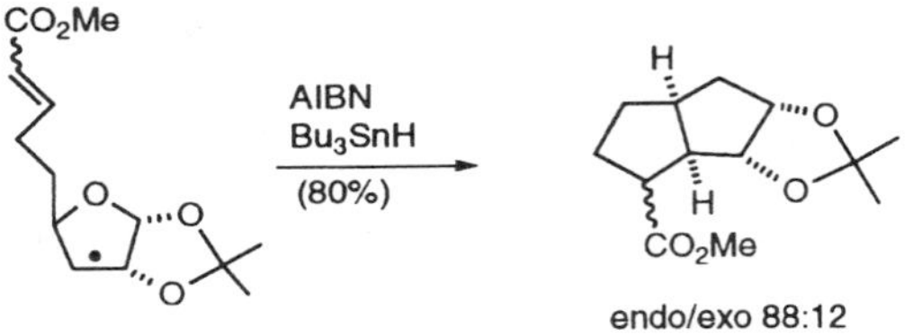


endo/exo 88:12

(4)[20]

Addition of radicals to an appropriately placed aldehyde group has been investigated by Fraser-Reid and co-workers Eq. (5); [26,27]. Thus, the substrate **18** undergoes radical

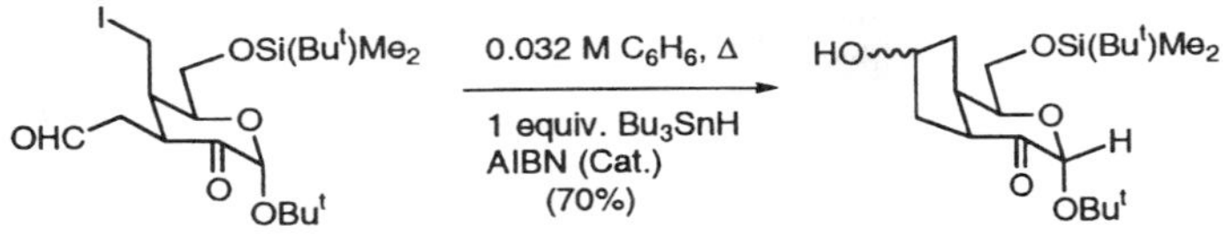


**18** **19** (5:1)

(5)[27]

cyclization to **19** in 70% yield. The course of this reaction, which is almost at the boundary of what is thermodynamically feasible, is affected by subtle stereochemical and structural features of the substrate [27,28].

The same group also studied the addition of trialkyltin radical to an acetylene and cyclization of the resulting vinyl radical in the context of serial radical cyclizations [Eq. (6);

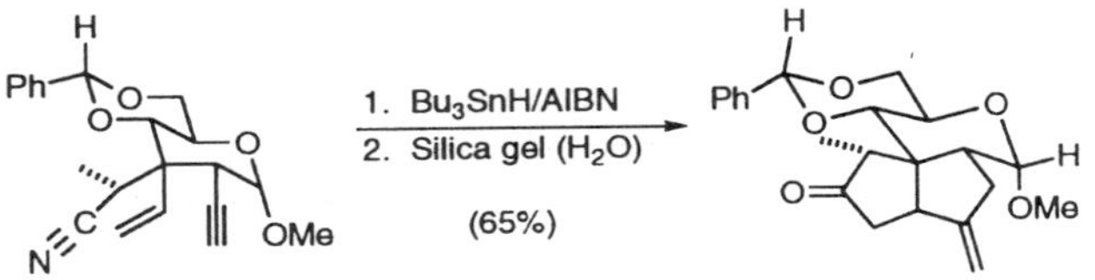


(6)[29]

29]. Tandem addition of trimethyltin followed by cyclization in a 1,6-heptadiene system [Eq. (7)] proceeds with surprising efficiency [30]. Oxidative destannylation of the primary product gives a synthetically useful dimethyl acetal. An acetylene-terminated tandem addition is shown in Eq. (8) [31].

Carbohydrate substrates have often been used to probe the stereochemical features

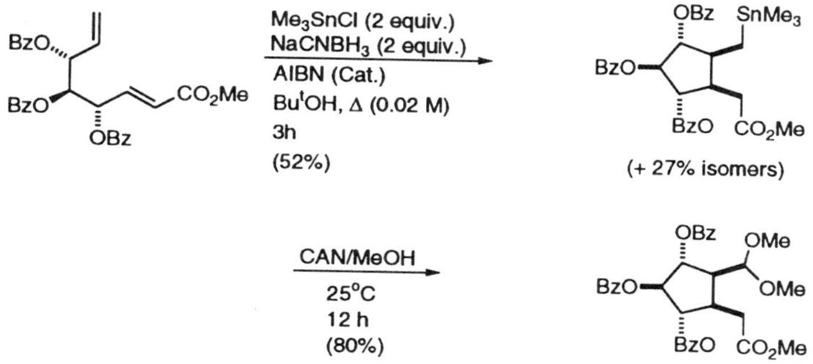

# Methods for Forming Carbocylic Derivatives

$$(7)^{30}$$

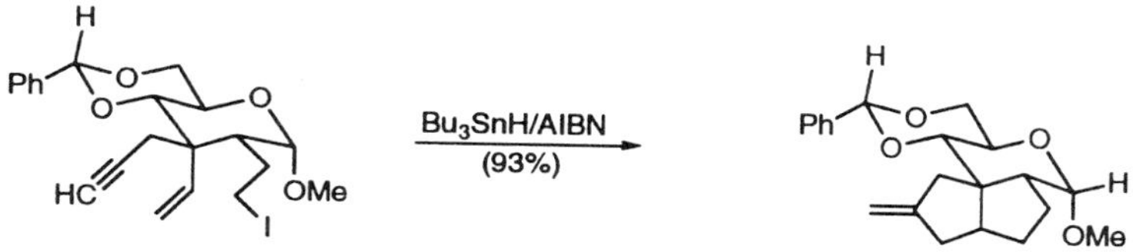

$$(8)^{29}$$

and compatibilities of new reagents. A notable illustration is the application of the Cp$_2$TiCl-mediated epoxy–olefin cyclization [31,32] shown in Scheme **9**. A selective homolytic

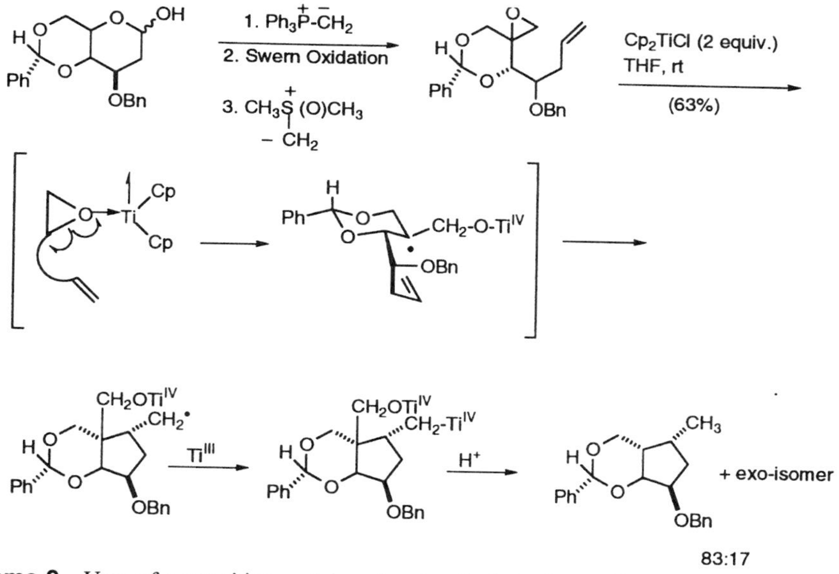

**Scheme 9** Uses of a transition-metal centered radical for the cyclization of an epoxyolefin.[32]

cleavage takes place to the tertiary radical, which undergoes facile cyclization to a highly functionalized cyclopentylmethyl radical. This radical is further reduced by a second equivalent of the Ti reagent. Such reductive protocol giving an organometallic intermediate, is a radical departure from the traditional termination sequence that, in most instances, results in an unactivated carbon residue by H-atom abstraction. The stereochemical outcome (1,5-*cis*, is the major product in this instance) is another confirmation of the models developed using structurally related secondary radicals (see Scheme **4**) [7]. As expected the open-chain radicals gave a mixture of products.

Samarium diiodide has been used by Enholm and co-workers for the generation and cyclization of ketyl radicals from aldehydic substrates [Eq. (9) and (10); 33]. As noted, the

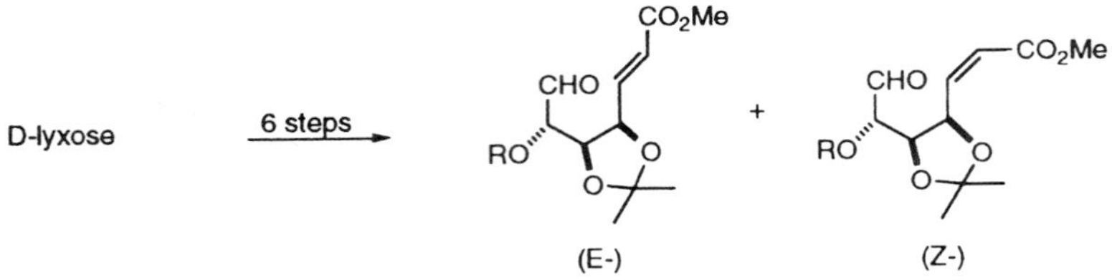

$$(9)^{33}$$

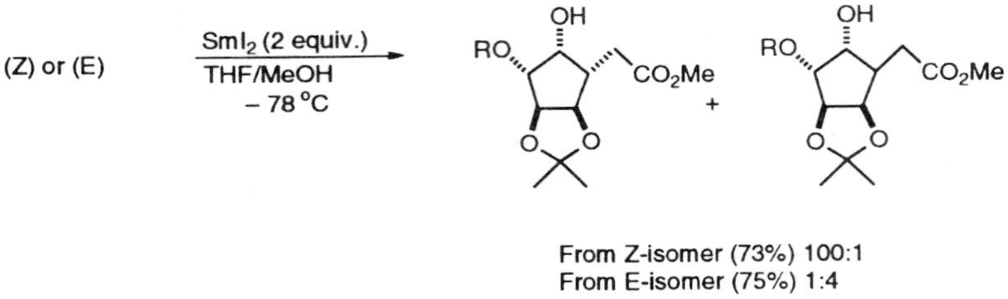

$$(10)^{33}$$

From Z-isomer (73%) 100:1
From E-isomer (75%) 1:4

reaction proceeds with remarkable stereoselectivity when Z-acrylates are used. This protocol has been used for the synthesis of the highly oxygenated C-ring (**20**) of the fungal metabolite anguidine [34].

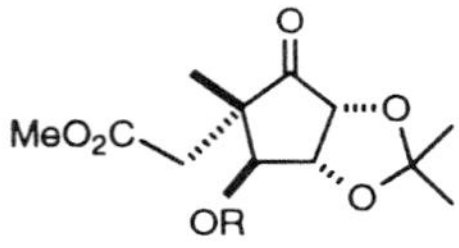

**20**[34]

## III. CYCLOHEXANES

### A. The Hept-6-enyl Radical Cyclization

Typically a hept-6-enyl radical will cyclize much slower than the corresponding hex-5-enyl radical, and often significant proportion of the primary radical is trapped before cyclization, and the olefinic product results. However, with the appropriate substituents or an activated acceptor, hepenyl radical cyclization can be used to prepare 6-membered carbocycles from carbohydrate precursors. For example, Redlich et al. [35] reported that 1,2-dideoxyhept-1-enitol derivatives can be cyclized to carbahexose derivatives [Eq. (11)]. Even though structural requirements (protecting groups, configuration of atoms in the carbon chain, and

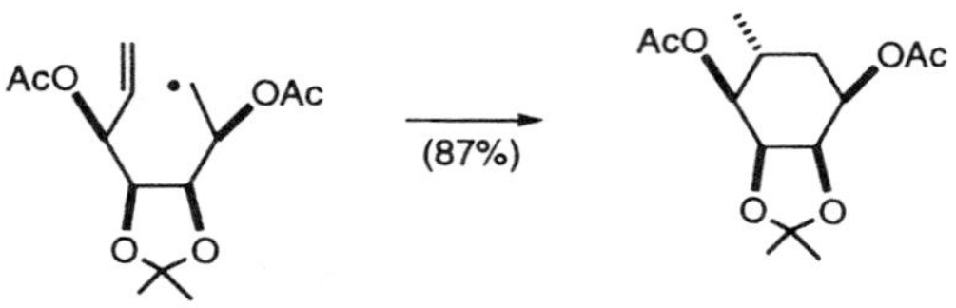

$$(11)^{35}$$

nature of the radical acceptor) [35,36] appear to be stringent for efficient cyclization to take place, as illustrated in Eq. (12) [37], a reactive radical acceptor, such as an acrylate, will

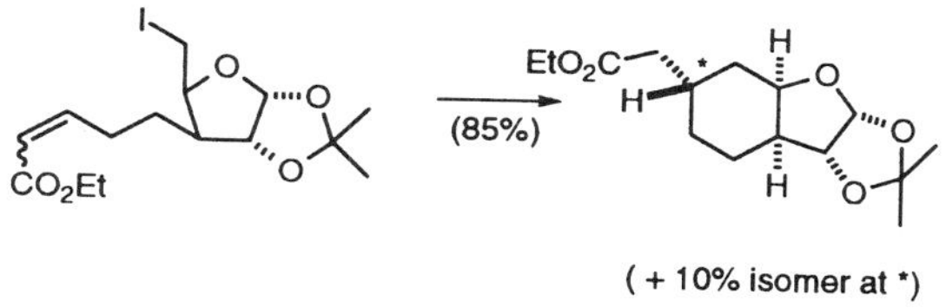

# Methods for Forming Carbocylic Derivatives

(12)[37]

(+ 10% isomer at *)

facilitate the intramolecular radical addition vis-à-vis H-atom abstraction by the initially formed radical. Two other examples of this type of heptenyl radical cyclization were discussed in Schemes **7** and **8**. Enol ethers and oximes can also act as acceptors in heptenyl radical cyclizations [36].

The surprising efficiency with which an aldehyde group acts as a radical acceptor [see Eq. (5)] was indeed first realized [27] in the context of a heptenyl analogue [Eq. (13)]. Note that the minor product arises as a result of two consecutive hex-5-enyl cyclizations [26,27].

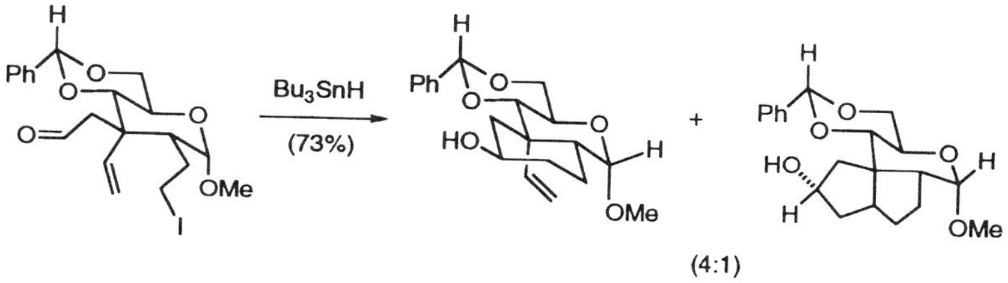

(13)[26]

(4:1)

Cyclization of an acyl radical generated from a carbohydrate-derived hept-6-enoic acid selenoester (to a cyclohexanone) has been studied even though the full potential of this reaction is yet to be established [38].

An electron-transfer method using low-valent McMurray-type titanium reagent has been reported for the synthesis of an inositol derivative **21** [Eq. (14); 39].

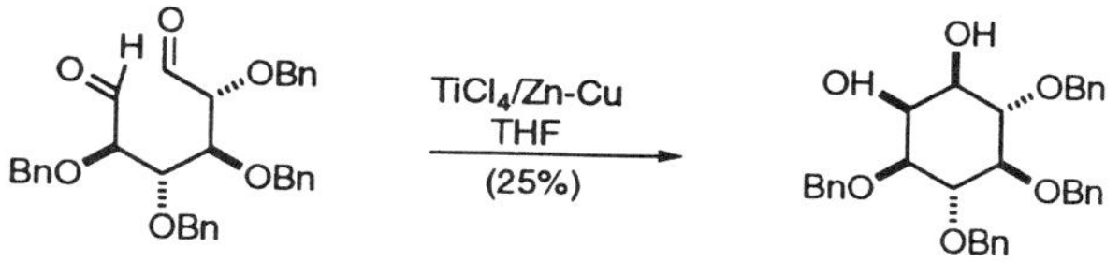

(14)[39]

**21**

## IV. FUNCTIONALIZED CARBOCYLIC COMPOUNDS BY ORGANOMETALLIC METHODS

Organometallic methods, with the possible exception of those involving the stoichiometric generation of enolates and other stabilized carbanionic species [40], have seldom been used in carbohydrate chemistry for the synthesis of cyclohexane and cyclopentane derivatives. The present discussion will not cover these areas. The earliest of the examples using a catalytic transition metal appears in the work of Trost and Runge [41], who reported the Pd-catalyzed transformation of the mannose-derived intermediate **22** to the functionalized cyclopentane **23** in 98% yield (Scheme **10**). Under a different set of conditions, the same substrate gives a cycloheptenone **24**. Other related reactions are the catalytic versions of the Ferrier protocol for the conversion of methylene sugars to cyclohexanones (see Chap. 26) [40,42,43].

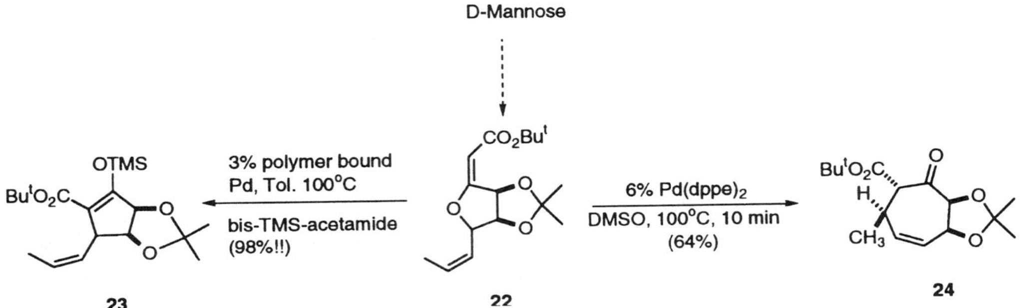

**Scheme 10** Allyl palladium intermediates for the synthesis of carbocyles.[41]

An application of the Pauson–Khand reaction for the synthesis of a carbaprostacyclin analogue (Scheme 11) [44] illustrates the power of organometallic methods for the activation of olefins and acetylenes.

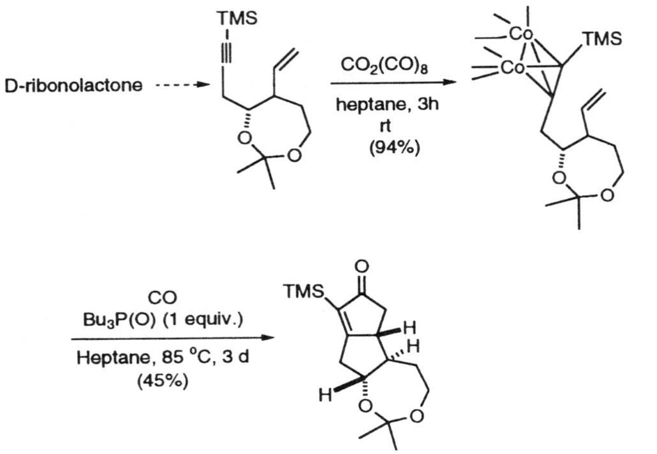

**Scheme 11** An application of the Pauson–Khand reaction.[44a]

Chemistry of low-valent titanium and zirconium has produced a number of powerful methods for the transformation of carbohydrates to carbocyclic compounds. The Ti(III)-mediated generation of a radical from epoxides and its subsequent cyclization [32] was discussed earlier under free radical methods (see Scheme 9). As shown in Scheme 12,

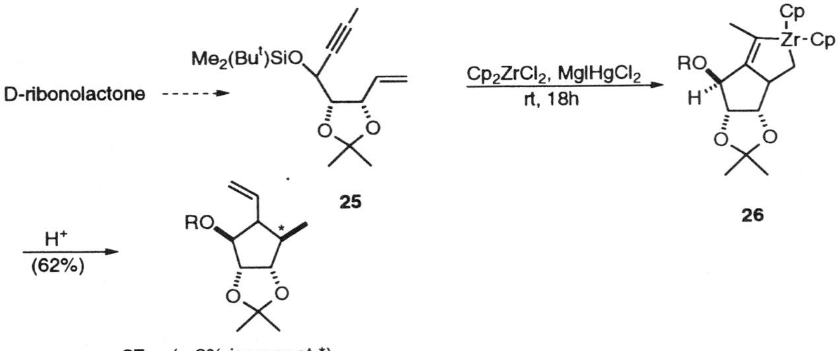

**Scheme 12** Low-valent zirconium-mediated cyclization of eneynes.[45]

### Methods for Forming Carbocyclic Derivatives

eneyne **25**, readily prepared from D-ribonolactone, undergoes stereospecific cyclization in the presence of in situ generated bis-(cyclopentadienyl)zirconium to give a metallacycle **26**, which can be cleaved with electrophilic reagents [45]. Protonation yields the highly functionalized allylic alcohol **27**. Intermediates that are similar to **27** are useful for the stereoselective introduction of exocyclic side chains using the allylic alcohol functionality. Eneynes and appropriately substituted dienes undergo cycloisomerization in the presence of Pd(0) catalysts as illustrated in Equations (15) and (16) [46]. The starting materials for

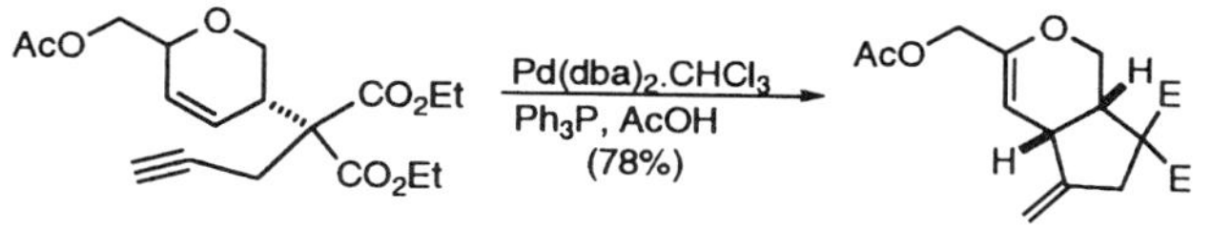

$$(15)^{46}$$

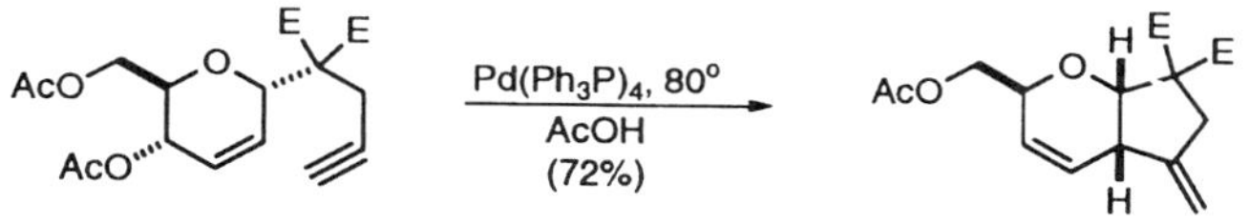

$$(16)^{46}$$

some of these transformations are made by a Pd-catalyzed alkylation of sugar-derived allylic carbonates [47].

A remarkable $Cp_2Zr$-initiated ring contraction of vinyl furanosides and pyranosides was recently reported by Taguchi et al. [48]. Thus, the readily available 5-vinylpyranoside **28** undergoes (presumably through a reductive cleavage of the allylic C–O bond) a highly stereoselective ring contraction to a single *cis*-2-vinylcyclopentanol **29**. A related reaction

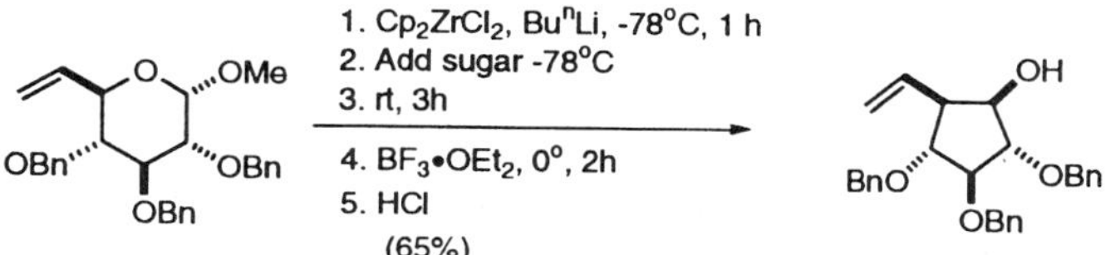

$$(17)^{48}$$

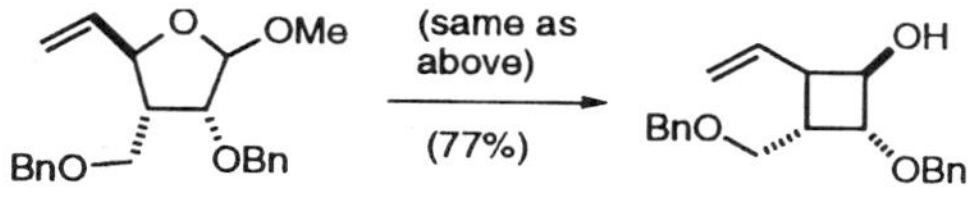

$$(18)^{48}$$

was also observed for the furanonide **30**, which gives a vinylcyclobutanol with equally good stereochemical control. Based on nuclear magnetic resonance (NMR) experiments and protonation studies, the involvement of a chair-like transition state **32**, in which the steric interactions of the cyclopentadienyl ligand with the ring substituents are minimized, has been proposed as a rationale for this control.

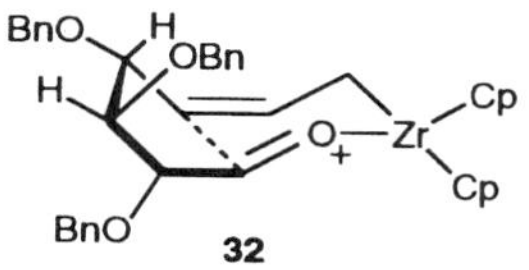

Rhodium-catalyzed hydroacylation of appropriately substituted olefinic aldehydes gives cyclopentanone and cyclohexanone, respectively (Scheme 13) [49].

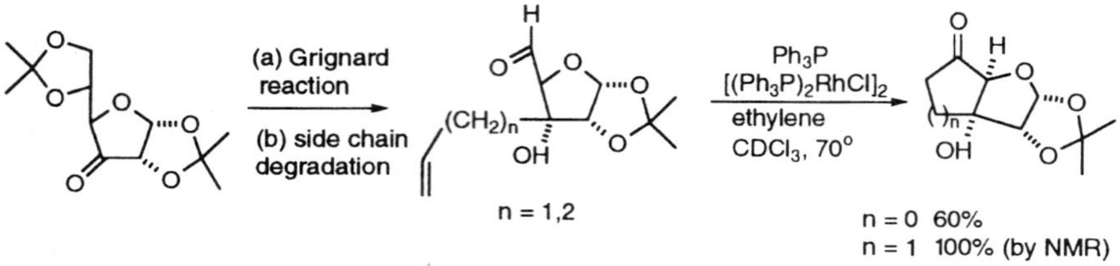

**Scheme 13** Rh-catalyzed intramolecular hydroacylation route to carbocycles.[49]

## V. EXPERIMENTAL PROCEDURES*

### A. Carbocycles from Unsaturated Halo Sugars by Hex-5-enyl Radical Cyclization [8]

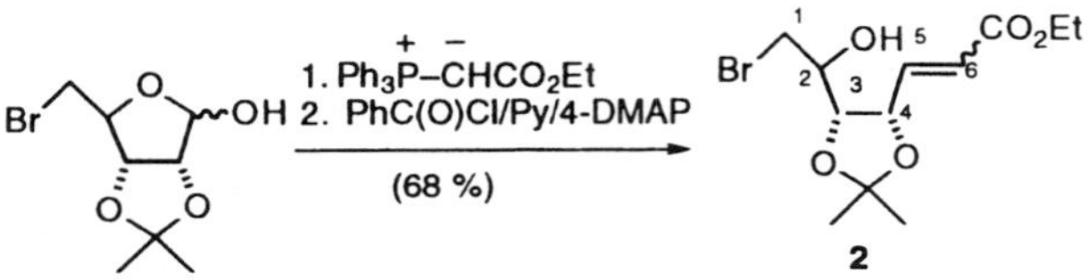

To a stirred solution of 2.77 g (10.9 mmol) of the lactol in 45 mL of dry $CH_2Cl_2$ at 25°C, under nitrogen, was added 4.55 g (13.1 mmol) of carbethoxymethylenetriphenylphosphorane and 26.6 mg (0.22 mmol) of benzoic acid. After being stirred at room temperature for 26 h, the reaction mixture was poured into ether (200 mL), washed with saturated $NaHCO_3$ (2 × 50 mL), and dried over $MgSO_4$. Solvent was removed in vacuo to afford a yellow oil. Flash column chromatography (50 × 158 mm; 29% ethyl acetate–hexanes), to remove triphenylphosphine oxide, followed by MPLC (65 g of $SiO_2$ and 60 g of $SiO_2$ in series; 20% ethyl acetate–hexanes) afforded 2.66 g of pure Z isomer (76%) and 0.33 g of pure E isomer (9%). Z isomer: $[\alpha]_D$ +87.38° (c 1.26, $CHCl_3$); E isomer: $[\alpha]_D$ +29.10° (c 1.26, $CHCl_3$).

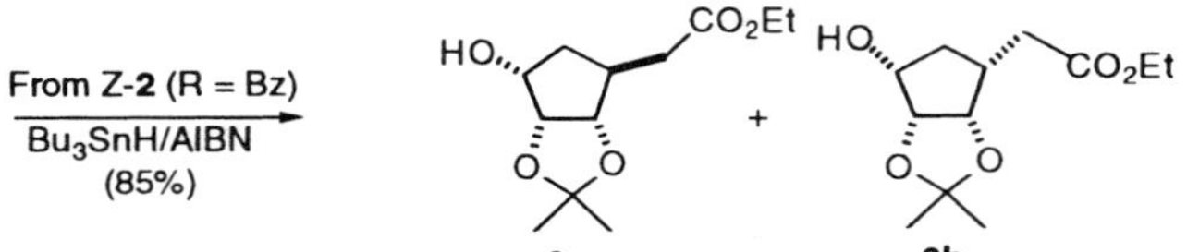

To a stirred solution of 3.1 g (9.5 mmol) of the Z-alcohol, in 50 mL of dry pyridine, was added 1.63 mL (14.06 mmol) of benzoyl chloride and 23 mg (0.19 mmol) of 4-dimethylaminopyridine. The resulting mixture was stirred at room temperature for 21 h and then diluted into ether (200 mL). The ethereal solution was washed with 1 M $H_2SO_4$ (2 × 80 mL) and saturated $NaHCO_3$ (2 × 80 mL), and dried over $MgSO_4$. Removal of solvent followed by flash chromatography afforded 3.65 g (90%) of the benzoyl-protected alcohol: $[\alpha]_D$ +67.48° (c 0.65, $CHCl_3$).

To a solution of 1.7 g (4.0 mmol) of the benzoate, in 130 mL of dry benzene, was added 1.2 mL (4.4 mmol) of tri-n-butyltin hydride and 32.8 mg (0.2 mmol) of 2,2′-azobis(2-

---

*Optical rotations were measured at 25°C.

methylpropionitrile) (AIBN). The reaction mixture was heated to reflux for 14 h. The benzene was removed in vacuo; the residue was diluted into acetonitrile under reduced pressure, followed by flash column chromatography (50 × 158 mm; 20% ethyl acetate–hexanes) yielded 1.20 g (85%) of the desired cyclic benzoates (in 10:1 ratio of *exo/endo* isomers: $[\alpha]_D$ +30.7° (*c* 0.468, $CHCl_3$).

## B. Highly Functionalized Cyclopentanes from Hexopyranose Derivatives

*Fully Functionalized 1,2-trans-dialkylcyclopentanes [12]:*
*[(2R)-(2α,4aβ,5β,6α,7β,7aβ)]-Hexahydro-5-methyl-2-phenyl-6-7-bis-(phenylmethoxy)cyclopenta-1,3-dioxin*
**6a**, Y = H)

**General Procedure for the Wittig Reaction.** A three-necked flask, fitted with a dropping funnel, thermocouple lead, and serum stopper, was thoroughly flame dried and was charged with a suspension of recrystallized, powdered, and dried phosphonium salt in anhydrous tetrahydrofuran (THF) (0.5 *M*). The mixture was cooled to −20°C and, from the funnel, 1.96 eq of 1.6 *M n*-BuLi in hexane was added. After all the butyl lithium was added, the dropping funnel was washed down with more THF. The mixture was stirred at −20°C to room temperature until all the solid disappeared (~1 h). A solution of the pyranose (1.0 eq, 0.5 *M* in THF) was added to the reaction mixture at −20°C from the dropping funnel, and the mixture was stirred for 16 h while the temperature gradually came to room temperature. A dry condenser was connected to the flask and the reaction mixture was heated to 50°C for 15 min. It was subsequently cooled to room temperature, and excess of reagent-grade acetone was added. After stirring for 5 min, ether (120 mL/mmol of sugar) was added and the precipitated solid was filtered off with the aid of Celite. The Celite pad was washed with excess ether, and the combined ether solutions were washed with saturated sodium bicarbonate, sodium chloride, and water. It was dried and concentrated. The products were collected by flash chromatography on silica gel, using ethyl acetate–hexane as the solvent. With the methyl vinyl ethers the *Z*- and *E*-isomers can be separated by careful chromatography before further reactions.

**General Procedure for Radical Generation and Cyclization** A flame-dried single-necked flask, with a reflux condenser, was charged with a 0.2- to 0.3-*M* solution of the enitol in distilled, dry 1,2-dichloroethane. To this solution was added 2 eq of thiocarbonyl bisimidazole (99% + pure Fluka), and the mixture was refluxed under nitrogen until all starting material disappeared, as judged by thin-layer chromatography (TLC). In certain instances when the reaction was incomplete after two h, an additional 1 eq of thiocarbonylbisimidazole was added, and the reaction was heated further. The product was extracted into methylene chloride after adding excess water to destroy the thiocarbonylbisimidazole. The combined $CH_2Cl_2$ layer was washed with ice-cold 1 *N* HCl, saturated NaHCO$_3$, and brine. The product was purified by flash chromatography on silica gel using ethyl acetate–hexane solvent system. The yields of the product in general are about 75–85%.

The foregoing product was transferred into a single-necked flask and was further dried azeotropically with toluene. It was dissolved in freshly distilled toluene to make a 0.1- to 0.2-$M$ solution, and 10–20 mg of azo-bis-isobutyronitrile (AIBN) per millimole of starting material and 0.5 eq of tributyltin hydride were added. The mixture was brought to reflux and a solution of 1.5 eq more of tributyltin hydride and AIBN (10–20 mg) dissolved in toluene were added from a syringe in about 2 h. After all, the hydride had been added the reaction was further refluxed for 1 h and subsequently cooled. Excess ether was added, and the organic layer was washed with 1 $N$ HCl, saturated NaHCO$_3$ and KF. The dried organic extract was concentrated and the products were isolated by chromatography on silica gel.

The starting enitol was prepared by the Wittig reaction in 82% yield from 2,3-bis-$O$-(phenylmethyl)-4,6-$O$-(phenylmethylene)-D-glucopyranose (**4**) [50]. The corresponding 1-$H$-imidazole-1-carbothioate, prepared in 68% yield, was subjected to the deoxygenation reaction to obtain a single compound (**6a**, Y = H) in 57% yield: mp 76–78°C, [α]$_D$ −10.4 ± 0.8° (*c* 1, CHCl$_3$).

*Fully Functionalized 1,2-**cis**-dialkylcyclopentanes [12]: ([2R-(2α,4aβ,5α,6β,7β,7aβ)]-hexahydro-5-methyl-2-phenyl-6,7-bis(phenylmethoxy)-cyclopenta-1,3-dioxin)*

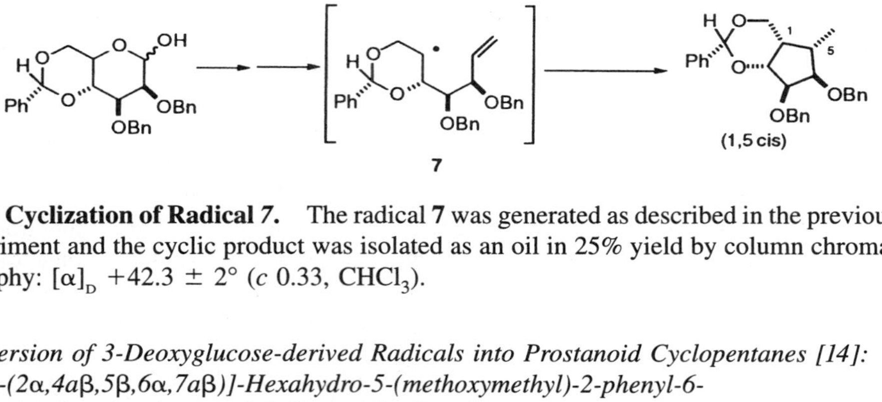

**Cyclization of Radical 7.** The radical **7** was generated as described in the previous experiment and the cyclic product was isolated as an oil in 25% yield by column chromatography: [α]$_D$ +42.3 ± 2° (*c* 0.33, CHCl$_3$).

*Conversion of 3-Deoxyglucose-derived Radicals into Prostanoid Cyclopentanes [14]: [(2R)-(2α,4aβ,5β,6α,7aβ)]-Hexahydro-5-(methoxymethyl)-2-phenyl-6-(phenylmethoxy)cyclopenta-1,3-dioxin (**6b**, Y = OMe)*

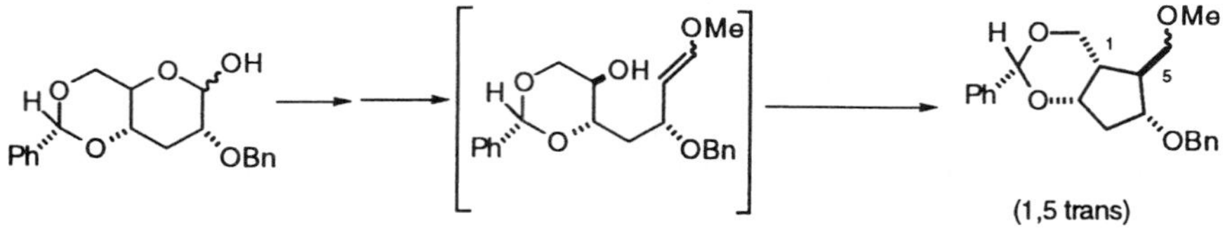

A three-necked flask, fitted with a dropping funnel, thermocouple lead, and serum stopper, was thoroughly flame-dried and was charged with 2.66 g (7.75 mmol) of methoxymethyltriphenylphosphonium chloride (recrystallized from ethyl acetate–chloroform and dried at 100°C/1 mm) and 40 mL of anhydrous THF. The mixture was cooled to −20°C, and from a dropping funnel 4.75 mL of 1.6$M$ $n$-butyl lithium in hexane was added. After all the butyl lithium was added, the dropping funnel was washed down with 5 mL of THF. The mixture was stirred at $-20°C$ to room temperature until all the solid disappeared (~ 1 h). The benzylidene sugar (1.09 g, 3.09 mmol) dissolved in 8 mL of THF was added to the reaction mixture from the dropping funnel at $-20°C$ and the reaction was warmed to room temperature and was further stirred overnight (16 h). The flask was attached to a dry condenser, and the mixture was maintianed at 50°C for 15 min. The mixture was cooled to room temperature and 20 mL of reagent-grade acetone was added. After the mixture was stirred for 5 minutes, 500 mL of ether was added, and the precipated solid was filtered off with the aid of Celite. The Celite pad was washed with 100 mL of ether. The combined ether portion was washed with 80 mL each of saturated sodium bicarbonate, sodium chloride, and water, dried and concentratedl The product **5b** was collected as a mixture of Z- and E-enolethers (0.987 g, 84 %) by chromatography on silica gel, using 40 to 50% ethyl acetate–hexane as the solvent. It was dried azeotropically by using toluene. The last traces of the solvent were removed on a high-vacuum pump to give a mixture of the desired products.

A mixture of 3.72 g (20.8 mmol) of thiocarbonyl-bis-imidazole and 6.44 g (17.4 mmol) of the enol ethers in 60 mL of 1,2-dichloroethane was refluxed for 3 h under nitrogen. An additional 9.3 g of thiocarbonyl-bis-imidazole was added, and refluxing was continued for one hour longer. A check of TLC (30% ethyl acetate–hexane, silica) indicated complete consumption of the starting material. Fifty milliliters of water and 500 mL of dichloromethane were added and the mixture was shaken thoroughly for 2 min in a separatory funnel. The organic layer was quickly washed with 100 mL of water, dried ($MgSO_4$), and concentrated. Filtration through a silica pad using 1:1 ethyl acetate–hexane followed by evaporation of the solvents yielded 6.13 g (74%) of the expected product, which was used for the subsequent reaction.

A solution of 6.13 g (12.8 mmol) of the 1-$H$-imidazole-1-carbothioate, 5.15 mL (19.1 mmol) of tri-$n$-butyltin-hydride and 0.12 g AIBN in 120 mL of dry toluene was refluxed for 1 h. An additional 0.5 eq of $Bu_3SnH$ and 60 mg of AIBN were added and the refluxing was continued for 1 h longer. The reaction mixture was added to 400 mL of ether, and it was washed with 80 mL each of saturated KF, 1 $N$ HCl and saturated $NaHCO_3$. The organic layer was washed with three 50-mL portions of saturated potassim fluoride and dried over anhydrous $MgSO_4$. Concentration and chromatography of the crude mixture yielded 3.59 g (58% from the enitol) of the desired product (**6b**, Y = OMe): $[\alpha]_D$ $-23.8 \pm 0.8°$ (c 1.0, $CHCl_3$).

*Functionalized Cyclopentanes from Bridged Pyranosides [21]*

A 5-mM solution of the Z-iodide in dry toluene was degased with argon and heated to reflux. Tri-$n$-butyltin hydride (1.3–1.5 eq) and (10 mol%) AIBN in toluene were added by syringe pump over 2–4 h. The solvent was evaporated under reduced pressure and the product was isolated by flash chromatography. Cyclization of 85 mg (0.706 mmol) of substrate gave 58 mg (97%) of the bicyclic product, which was carried on to the next step. A solution of the bicyclic compound (33 mg, 0.115 mmol) and camphor sulfonic acid (12 mg) in methanol (3 mL) was stirred at 40°C under argon until all of the starting material and the intermediate acetonide-dimethylacetal were consumed. The addition of triethyl amine and evaporation of the solvents gave the crude diol, which was dissolved in ethyl acetate and treated with dimethylaminopyridine (catalytic amount) and acetic anhydride (excess) at room temperature. An ice-cold solution of sodium bicarbonate was added, and the aqueous phase was extracted with ethyl acetate. Drying and evaporation of the organic solvent, followed by filtration through silica gel afforded the lactone (32 mg, 88%) as a colorless oil: $[\alpha]_D$ −32.0° (c 1.5, $CHCl_3$).

## C. A Hept-5-enyl Radical Cyclization Route to Functionalized Cyclohexanes [36]

To 244 mg (0.64 mmol) of the starting bromide (E/Z ratio 1:7) in dry toluene (0.015 M) under reflux was added 2.4 eq of tributyltin hydride and AIBN (catalytic amount) in 3 h through a syringe pump. The reaction mixture was cooled and the solvent was evaporated. The residue was dissolved in ether and 10% aqueous KF solution was added, and the mixture was stirred for 18 h. The organic phase was separated, dried, and evaporated. After flash chromatography (hexane–ethyl acetate, 90:10) of the residue gave the 134 mg (80%) of the product: mp 75–77°C $[\alpha]_D$ −61° (c 1.2, $CHCl_3$). Minor amounts of the noncyclized reduction product and the isomeric compound with an $\alpha$-$CH_2CO_2Me$ were also isolated.

## D. Trimethyltin Radical–Initiated Cyclization of 1,6-Dienes [30]

To a solution of the diene [51] (1 mmol) in anhydrous t-butanol (20 mL, 0.05 M) under argon, are added trimethyltin chloride (400 mg, 2 mmol), sodium cyanoborohydride (190 mg, 3 mmol), and AIBN (15 mg, 10%). The solution is refluxed for 1–20 h (until the diene disappears as monitored by TLC using $KMnO_4$ spraying agent). The solution is cooled to room temperature, quenched with 5% ammonia solution, stirred, and concentrated under reduced pressure. The residue is dissolved in ether, washed three times with brine, dried ($MgSO_4$), and processed as usual. Flash chromatography afforded the trimethylstannyl derivative in 52% yield: $[\alpha]_D$ −20° (c 0.6, $CHCl_3$).

To a solution of the trimethylstannyl compounds (1 mmol) in methanol (20 mL, 0.05 M) is added ceric ammonium nitrate (10 mmol, commercial grade) at 25°C, and the solution is stirred until the aldehyde has been converted to dimethyl acetal (10 h). The reaction mixture is poured into ether, washed three times with water, dried ($MgSO_4$), and concentrated under reduced pressure. The residue is then purified by flash chromatography to afford the dimethyl acetal derivative in 79% yield: $[\alpha]_D$ $-7°$ (c 1.2, CHCl$_3$)

### E. Ti(III)-Mediated Epoxy–Olefin Cyclization Route to Carbocyclic Compounds from Carbohydrates (see Scheme 9) [31]

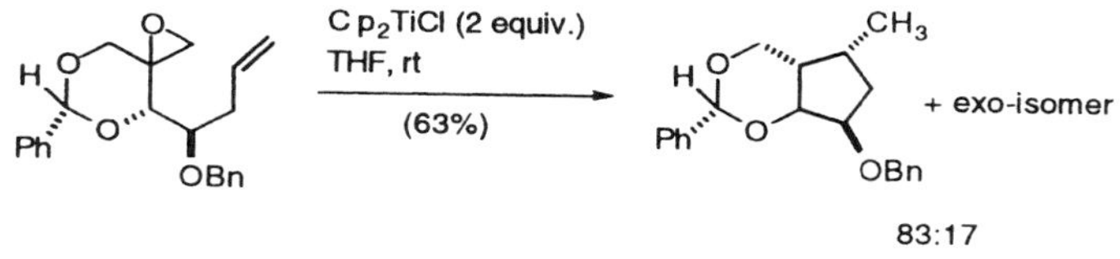

A solution of Cp$_2$TiCl [52] (0.43 g, 2 mmol) in THF was added dropwise to a solution of the epoxide (1.0 mmol) in 20 mL of THF at room temperature. A solution of 1 N HCl in ether (4 mL) was added, and the mixture was stirred for 10 min. The precipitated solid was removed and the filtrate was added to excess (5 mL) of aqueous saturated sodium dihydrogenphosphate. The organic layer was separated and the aqueous layer was extracted with ether. The product was isolated as a mixture of exo- and endo-Me isomers in 70% yield. The structures were confirmed by $^{13}$C-NMR, $^1$H-NMR, chemical shift correlation mapping, nOe measurements, and attached proton test (APT) experiments. No $[\alpha]_D$ was recorded because the product was isolated as a mixture.

### F. Ketyl–Olefin Cyclization Mediated by Samarium(II) Iodide [33]

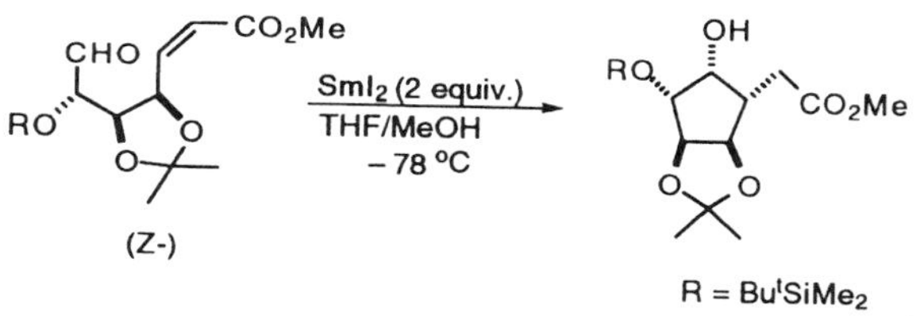

The aldehyde precursor, with the cis-alkene derived from lyxose (42.0 mg, 0.12 mmol) was drawn into a syringe with a solution of 1 mL of THF–methanol (3:1 v/v) and added dropwise to a cooled solution ($-78°C$) of SmI$_2$ (3.6 mL, 0.1 M in THF), over 5 min. The solution was kept at $-78°C$ for 1 h. When the reaction was complete, as indicated by TLC analysis, it was quenched with aqueous saturated sodium bicarbonate solution (1 mL) and extracted with ether. Chromatography using 50:50 ether–hexane gave 31.0 mg (73%) of pure syn-cyclic alcohol.

### G. Pauson–Khand Reactions

*A Carbacyclin Intermediate by Pauson–Khand Reaction [44a]*

To the Co complex (1.28 g, 2.32 mmol) in heptane (23 mL, purged with carbon monoxide for 3 h before use) was added tri-n-butylphosphine oxide (506 mg, 2.32 mmol). The solution was sealed in a screw-cap resealable tub under an atmosphere of CO and heated to 85°C (over glyme heated at reflux) for 71 h. After cooling, the solution was applied directly to a bed of Fluorisil and eluted with ethyl acetate–petroleum ether 95:5 to 50:50) giving the tricyclic enone 304 mg, 45%) as a colorless oil: $[\alpha]_D$ 22 $+116°$ (c 2.47, CHCl$_3$).

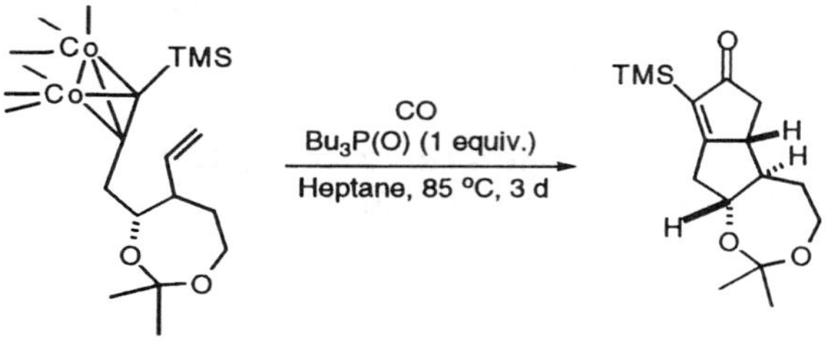

*Pauson–Khand Reaction of a Sugar-Derived Eneyne [44b]*

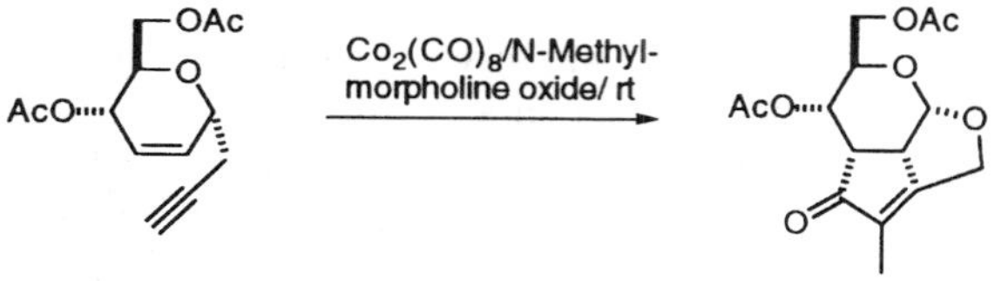

To a solution of the precursor in $CH_2Cl_2$, dicobaltoctacarbonyl (1.1 eq) was added in one portion at room temperature. The mixture was stirred for about 3 h and then, anhydrous NMO (6.3 eq) was slowly added and stirred for 5 h at room temperature. Part of the solvent was removed, the suspension was adsorbed in silica gel, and submitted to flash chromatography. Elution with hexane–ethyl acetate mixtures gave pure product in 66% yield: $[\alpha]_D$ $-17°$ ($c$ 2.3, $CHCl_3$).

## H. Zirconocene-Mediated Cyclization of Sugar Eneynes [45]

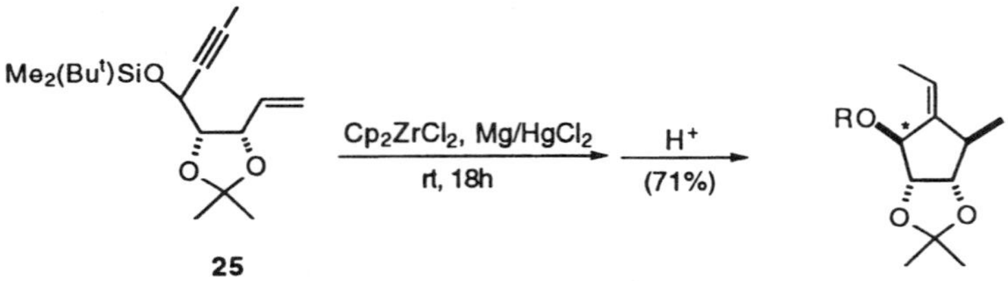

5-*O*-*t*-Butyldimethylsilyl-D-ribonolactone was converted into the eneyne by the following sequence: (1) Dibal reduction to the lactol, (2) Wittig reaction with $Ph_3P^+CH_2^-$, (3) removal of the Si-protecting group by treatment with $Bu_4N^+F^-$, (4) periodate cleavage of the diol, (5) propynyl lithium addition, (6) protection of the secondary alcohol as the TBDMS ether.

A mixture of Mg turnings (0.32 g, 13 mmol) and $HgCl_2$ (0.36 g, 1.3 mmol) in THF (15 mL) was stirred for 15 min. A solution of bis(cyclopentadienyl) zirconium dichloride (0.71 g, 2.43 mmol) and the sugar eneyne in THF was added dropwise [53]. After the mixture was stirred overnight, unreacted Mg was filtered off under nitrogen and the mixture was quenched with 10% $H_2SO_4$ (30 mL). The mixture was extracted with ether ($2 \times 25$ mL), washed with sodium bicarbonate (25 mL), and dried ($MgSO_4$). The solvent was removed under reduced pressure. Flash chromatography (95:5 hexane–ethyl acetate) afforded the product (114 mg, 71%) as a mixture (92:8) of two isomers as determined by gas chromatography.

## I. Intramolecular Cyclization of an Allyl Zirconium Derivative [48]

Reagents and conditions:
1. $Cp_2ZrCl_2$, $Bu^nLi$, -78°C, 1 h
2. sugar -78°C
3. rt, 3h
4. $BF_3 \cdot OEt_2$, 0°, 2h
5. HCl
(65%)

Compound 28 → Compound 29

A zirconocene–butene complex ("$Cp_2Zr$") was generated in situ [53] by adding $n$-BuLi (1.56 $M$ in hexane, 5.2 mmol) to a solution of $Cp_2ZrCl_2$ (760 mg, 2.6 mmol) in toluene (8 mL) at $-78°C$ under an argon atmosphere, and the mixture was stirred for 1 h at the same temperature. To the $Cp_2Zr$ solution was added a solution of the methyl glycoside (1 g, 2.17 mmol) in toluene (5 mL) at $-78°C$, the mixture was gradually heated to room temperature, and was stirred for 3 h. To the cooled reaction mixture was added a solution of $BF_3 \cdot OEt_2$ (0.53 mL, 4.43 mmol) in toluene (3 mL), and the mixture was stirred at room temperature for 3 h. After 1 $N$ HCl was added to the reaction mixture, the mixture was extracted with methylene chloride. The combined organic layer was washed with saturated aqueous NaCl and dried over $MgSO_4$. After the filtrate was concentrated in vacuo, the crude product was purified by silica gel column chromatography (hexane–ethyl acetate, 5:1) to give the product (605 mg, 1.41 mmol) in 65% yield: mp 35–36°C, $[\alpha]_D$ $-13.41°$ ($c$ 1.13, $CHCl_3$).

## REFERENCES AND NOTES

1. R. U. Lemieux and U. Spohr, How Emil Fisher was led to the lock and key concept for enzyme specificity, *Adv. Carbohydr. Chem Biochem.* 50:1(1994).
2. For reviews see: (a) R. J. Ferrier and S. Middleton, The conversion of carbohydrates into functionalized cyclopentanes and cyclohexanes, *Chem. Rev.* 93:2779 (1993); (b) K. J. Hale, Monosaccharides: Use in asymmetric synthesis of natural products, *Rodd's Chemistry of Carbon Compounds*. Second Supplement to the 2nd ed. (M. Sainsbury, ed.), Elsevier, Amsterdam, 1993, Vol. 1E/F/G, pp. 315–435; (c) T. D. Inch, Formation of convenient chiral intermediates from carbohydrates and their use in synthesis, *Tetrahedron* 40:3161 (1984); (d) S. Hanessian, *Total Synthesis of Natural Products: The Chiron Approach* Pergamon, New York, 1983.
3. (a) For a list of readily available chiral starting materials, including carbohydrates, see J. W. Scott, Readily available chiral carbon fragments and their use in synthesis, *Asymmetric Synthesis* (J. D. Morrison, ed.), Academic Press, New York, 1984, pp. 1–54. (b) An immensely valuable compendium of carbohydrate intermediates may be found in P. M. Collins, *Carbohydrates*, Chapman & Hall, New York, 1987.
4. A. L. J. Beckwith, Regioselectivity and stereoselectivity in radical reactions, *Tetrahedron* 37:3073 (1981).
5. D. J. Hart, Free radical carbon–carbon bond formation in organic synthesis, *Science* 223:883 (1984).
6. D. P. Curran, The design and applications of free radical chain reactions in organic synthesis I and II, *Synthesis* pp. 417, 489 (1988).
7. T. V. RajanBabu, Stereochemistry of intramolecular free-radical cyclization reactions, *Acc. Chem. Res.* 24:139 (1991).
8. C. S. Wilcox and L. M. Thomasco, New syntheses of carbocycles from carbohydrates. Cyclization of radicals derived from unsaturated halo sugars, *J. Org. Chem.* 50:546 (1985); see also S. Hanessian, D. S. Dhanoa, and P. L. Beaulieu, Synthesis of ω-substituted α,β-unsaturated esters via radical-induced cyclizations, *Can. J. Chem.* 65:1859 (1987).

9. C. S. Wilcox and J. J. Gaudino, New approaches to enzyme regulators. Synthesis and enzymological activity of carbocyclic analogues of D-fructofuranose and D-fructofuranose 6-phosphate, *J. Am. Chem. Soc. 108*:3102 (1986).
10. T. V. RajanBabu, From carbohydrates to optically active carbocycles, I: Stereochemical control in sugar hex-5-enyl radical cyclization, *J. Am. Chem. Soc. 109*:609 (1987).
11. D. H. R. Barton and S. W. McCombie, A new method for deoxygenation of secondary alcohols, *J. Chem. Soc. Perkin Trans. 1*:1574 (1975).
12. T. V. RajanBabu, T. Fukunaga, and G. S. Reddy, Stereochemical control in hex-5-enyl radical cyclizations: From carbohydrates to carbocycles, 3, *J. Am. Chem. Soc. 111*:1759 (1989).
13. Well-known "Beckwith" transition states in which a maximum number of groups are in an equatorial orientation of a "chair-like" transition state provides a satisfactory rationalization for the formation of the products. A. L. J. Beckwith, C. J. Easton, T. Lawrence, and A. K. Serelis, *Aust. J. Chem. 36*:545 (1983); see also D. C. Spellmeyer and K. N. Houk, A force field model for intramolecular radical additions, *J. Org. Chem. 52*:959 (1987); T. V. RajanBabu and T. Fukunaga, Stereochemical control in hex-5-enyl radical cyclizations: Axial vs. equatorial 2-(but-3-enyl)cyclohexyl radicals, *J. Am. Chem. Soc. 111*:296 (1989).
14. T. V. RajanBabu, From carbohydrates to carbocycles. 2. A free radical route to Corey lactone and other prostanoid intermediates, *J. Org. Chem. 53*:4522 (1988).
15. M. J. Robins, J. S. Wilson, and F. Hansske, Nucleic acid related compounds. 42. A general procedure for the efficient deoxygenation of secondary alcohols. Regiospecific and stereoselective conversion of ribonucleosides to 2'-deoxynucleosides, *J. Am. Chem. Soc. 105*:4059 (1983).
16. J. J. Gaudino and C. S. Wilcox, A concise approach to enantiomerically pure carbocyclic robose analogues. Synthesis of (4S, 5R, 6R, 7R)-7-(hydroxymethyl)spiro[2.4]heptane-4,5,6-triol-7-O-(dihydrogen phosphate), *J. Am. Chem. Soc. 112*:4374 (1990).
17. P. A. Bartlett, K. L. McLaren, and P. C. Ting, Radical cyclization of oxime ethers, *J. Am. Chem. Soc. 110*:1633 (1988).
18. N. S. Simpkins, S. Stokes, and A. J. Whittle, An enantiospecific synthesis of allosamizoline, *J. Chem. Soc. Perkin Trans 1*:2471 (1992).
19. J. Marco-Contelles, L. Martínez, and A. M. Grau, Carbocycles from carbohydrates. A free radical route to (1R, 2R, 3S, 4R)-4-amino-1,2,3-cyclopentanetriol derivatives, *Tetrahedron Asymm. 2*:961 (1991).
20. J. Marco-Contelles, P. Ruiz, B. Sánchez, and M. L. Jimeno, A new synthetic route to chiral, multiply functionalized cyclopentane rings, *Tetrahedron Lett. 33*:5261 (1992).
21. R. A. Alonso, G. D. Vite, R. E. McDevitt, and B. Fraser-Reid, Radical cyclization routes to bridged pyranosides as precursors of densely functionalized cycloalkanes, *J. Org. Chem. 57*:573 (1992).
22. J. C. López, A. M. Gómez, and S. Valverde, A novel entry to cyclohexanes and cyclopentanes from carbohydrates via inversion of radical reactivity in hex-2-enono-δ-lactones, *J. Chem. Soc., Chem. Commun.* 613 (1992).
23. H. Hashimoto, K. Furuichi, and T. Miwa, Cyclopentane-annelated pyranosides: A new approach to chiral iridoid synthesis, *J. Chem. Soc. Chem. Commun.* 1002 (1987).
24. K. S. Gröninger, K. F. Jäger, and B. Giese, Cyclization reactions with of allyl-substituted glucose derivatives, *Liebigs Ann. Chem.* 731 (1987).
25. H.-G. Korth, R. Sustmann, J. Dupuis, and B. Giese, EPR spectroscopic investigations of carbohydrate radicals II, *J. Chem. Soc. Perkin Trans 2*:1453 (1986).
26. R. Tsang and B. Fraser-Reid, Serial radical cyclization via a vinyl group immobilized by a pyranoside. A route to bis-annulated pyranosides, *J. Am. Chem. Soc. 108*:2116 (1986).
27. R. Tsang and B. Fraser-Reid, Surprisingly efficient intramolecular addition of carbon radicals to carbonyl groups: A ready route to cycloalkanols, *J. Am Chem. Soc. 108*:8102 (1986).
28. R. Walton and B. Fraser-Reid, Studies on intramolecular competitive addition of carbon radicals to aldehydo and alkenyl groups, *J. Am. Chem. Soc. 113*:5791 (1991).
29. H. Pak, J. K. Dickson, and B. Fraser-Reid, Serial radical cyclization of branched carbohydrates.

## Methods for Forming Carbocylic Derivatives

2. Claisen rearrangements routes to multiply substituted pyranoside diquinanes, *J. Org. Chem.* 54:5357 (1989).

30. S. Hanessian and R. Léger, Expedient assembly of carbocyclic, heterocyclic, and polycyclic compounds by Me₃Sn radical mediated carbocyclizations of dienes and trienes: A novel oxidative cleavage of the C–Sn bond, *J. Am. Chem. Soc.* 114:3115 (1992).

31. W. A. Nugent and T. V. RajanBabu, Transition metal-centered radicals in organic synthesis—Ti(III)-induced cyclization of epoxyolefins, *J. Am. Chem. Soc.* 110:8561 (1988).

32. T. V. RajanBabu and W. A. Nugent, Selective generation of free radicals from epoxides using a transition-metal radical. A powerful new tool for organic synthesis, *J. Am. Chem. Soc.* 116:986 (1994).

33. E. J. Enholm and A. Trivellas, Samarium(II) iodide mediated transformation of carbohydrates to carbocycles, *J. Am. Chem. Soc.* 111:6463 (1989).

34. E. J. Enholm, H. Satici, and A. Trivellas, Samarium(II)-iodide mediated carbocycles from carbohydrates: Application to synthesis of the C ring of anguidine, *J. Org. Chem.* 54:5841 (1989).

35. H. Redlich, W. Sudau, A. K. Szardenings, and R. Vollerthun, Radical cyclization of hept-1-enitols, *Carbohydr. Res.* 226:57 (1992).

36. J. Marco-Contelles, C. Pozuelo, M. L. Jimeno, L. Martínez, and A. M. Grau, 6-*exo* Free radical cyclization of acyclic carbohydrate intermediates: A new synthetic route to enantiomerically pure polyhydroxylated cyclohexane derivatives, *J. Org. Chem.* 57:2625 (1992).

37. B.-W. A. Yeung, J. L. Marco-Contelles, and B. Fraser-Reid, A radical cyclization route to Collum's key intermediate for (+)-phyllantocin: Annulated furanoses via radical cyclizations, *J. Chem. Soc. Chem. Commun.* 1160 (1989).

38. D. Batty and D. Crich, Acyl radical cyclizations in synthesis. Part 4. Tandem processes: The 7-*endo*/5-*exo* serial cyclizations approach to enantiomerically pure bicyclo[5.3.0]decan-2-ones, *J. Chem. Soc. Perkin Trans 1*:3193 (1992).

39. Y. Watanabe, M. Mitani, and S. Ozaki, Synthesis of optically active inositol derivatives starting from D-glucurono-6,3-lactone, *Chem. Lett.* 123 (1987).

40. For a discussion of these methods see Ref. 2a. Also see Chapter 26 in this volume.

41. B. M. Trost and T. A. Runge, Pd-catalyzed 1,3-oxygen-to-carbon alkyl shifts: A cyclopentanone synthesis, *J. Am. Chem. Soc.* 103:7559 (1981).

42. A. S. Machado, A. Olesker, and G. Lukacs, Synthesis of two tetrasubstituted cyclohexanones from 6-deoxyhex-5-enopyranoside derivatives, *Carbohydr. Res.* 135:231 (1985).

43. S. Adam, Palladium(II) promoted carbocyclization of aminodeoxyhex-5-enopyranosides, *Tetrahedron Lett.* 29:6589 (1988).

44. (a) P. Magnus and D. P. Becker, Stereospecific dicobalt octacarbonyl mediated eneyne cyclization for the enantiospecific synthesis of a 6a-carbocycline analog, *J. Am. Chem. Soc.* 109:7495 (1987); (b) J. Marco-Contelles, The Pauson–Khand reaction on carbohydrate templates. I. Synthesis of bis-heteroannulated pyranosides, *Tetrahedon Lett.* 35:5059 (1994).

45. T. V. RajanBabu, W. A. Nugent, T. F. Taber, and P. J. Fagan, Stereoselective cyclization of eneynes mediated by metallocene reagents, *J. Am. Chem. Soc.* 110:7128 (1988).

46. G. J. Engelbrecht and C. W. Holzapfel, Stereoselective palladium catalyzed cyclization on carbohydrate templates—a route to chiral cyclopentanes and some heterocyclic analogs, *Tetrahedron Lett.* 32:2161 (1991).

47. See also T. V. RajanBabu, Pd(0)-catalyzed C-glycosylation: A facile alkylation of trifluoroacetylglucal, *J. Org. Chem.* 50:3642 (1985).

48. H. Ito, Y. Motoki, T. Taguchi, and Y. Hanzawa, Zirconium mediated, highly diastereoselective ring contraction of carbohydrate derivatives: Synthesis of highly functionalized, enantiomerically pure carbocycles, *J. Am. Chem. Soc.* 115:8835 (1993).

49. K. P. Gable and G. A. Benz, Rhodium (I)-catalyzed cyclizations of 3-*C*-alkenyl pentodialdose derivatives, *Tetrahedron Lett.* 32:3473 (1991).

50. A. Liptak, J. Imre, J. Harangi, and P. Nanasi, Hydrogenolysis of the isomeric 1,2:4,6-di-*O*-

benzylidene-α-D-glucopyranose derivatives with the LiAlH$_4$-AlCl$_3$ reagent, *Carbohydr. Res.* *116*:217 (1983).
51. B. Bernet and A. Vasella, Carbocyclische verbindungen aus monosacchariden. I. umsetzungen in der glucosereihe, *Helv. Chim. Acta 62*:1990 (1979).
52. L. Manzer, Dicyclopentadienyltitaniummonochloride, *Inorg. Synth. 21*:84 (1982).
53. Alternately, Cp$_2$Zr can also be prepared by the reaction of Cp$_2$ZrCl$_2$ with two equivalents of *n*-BuLi at −78°C: E.-I Negishi, F. K. Cederbaum and T. Takahashi, Reaction of zirconocene dichloride with alkyl lithiums or alkyl Grignard reagents as a convenient method for generating a "zirconocene" equivalent and its use in zirconium promoted cyclization of alkenes, alkynes dienes, eneynes and diynes, *Tetrahedron Lett. 27*:2829 (1986).

# 26

# The Conversion of Carbohydrates to Cyclohexane Derivatives

### Robert J. Ferrier
Victoria University of Wellington, Wellington, New Zealand

|      |                                                                                                                          |     |
| ---- | ------------------------------------------------------------------------------------------------------------------------ | --- |
| I.   | Introduction                                                                                                             | 570 |
| II.  | Methods                                                                                                                  | 571 |
|      | A. Cyclizations proceeding by nucleophilic additions to carbonyl groups                                                  | 571 |
|      | B. Cyclizations proceeding by nucleophilic attack at noncarbonyl centers                                                 | 580 |
|      | C. Free radical cyclizations                                                                                             | 582 |
|      | D. Cycloaddition reactions                                                                                               | 582 |
| III. | Experimental Procedures                                                                                                  | 585 |
|      | A. Cyclization of a 5,6-dideoxy-6-nitrohexose derivative                                                                 | 585 |
|      | B. Cyclization of a 6-dimethoxyphosphorylhexose derivative                                                               | 586 |
|      | C. Intramolecular aldol cyclization                                                                                      | 586 |
|      | D. Conversion of a 6-deoxyhex-5-enopyranose derivative to a cyclohexanone derivative                                     | 587 |
|      | E. Conversion of a 6-deoxyhex-5-enopyranose derivative to a cyclohex-1-en-3-one derivative                               | 587 |
|      | F. Direct conversion of a 1,6-anhydrohexose derivative to a cyclohexane derivative                                       | 588 |
|      | G. Conversion of a 6,7-dideoxyhept-6-enose derivative to a 7-aza-8-oxabicyclo[4.3.0]nonane derivative                    | 588 |
|      | H. Conversion of a hex-2-enopyranosid-4-ulose to a 4-oxa-2-oxobicyclo[4.4.0]dec-8-ene derivative                         | 589 |
|      | I. Conversion of a 3,5,6-trideoxyhex-3,5-dienofuranose derivative to a cyclohexene derivative bearing three carbon- and one oxygen-bonded substituents | 589 |
|      | References                                                                                                               | 590 |

## I. INTRODUCTION

The plethora of functionalized chiral cyclohexane derivatives found in natural products has served as a potent stimulus to the development of methods for their synthesis from readily available carbohydrates. Although the important inositols and inosamines and their derivatives are obvious targets, there are a host of other natural materials for which cyclohexane ring systems can be fashioned by this approach, including various alkaloids; antibiotics; terpenes; anthracyclinones; families of compounds, such as the milbemycins and avermectins; and such specific compounds as cryptosporin, compactin, tacrolimus (FK 506), and phyllanthocin. Apart from these, many carbocyclic analogues of noncarbocyclic compounds are sought for specific biological purposes—notably carbasugars (analogues having the ring oxygen atoms replaced by methylene groups), which mimic natural sugars but, being devoid of hemiacetal or acetal character, do not take part in many typical carbohydrate reactions and are thus of interest, for example, as specific enzyme inhibitors.

The conversion of 6-deoxy-6-nitro-D-glucose and -L-idose by base-catalyzed cyclizations to deoxynitroinositols was effected in 1948 as the first carbohydrate to carbocyclic compound transformation [1]. With the extensive range of functionalized cyclohexane derivatives that occur naturally, and beckon as synthetic challenges, it now seems surprising that it was not until 30 years later that much further headway was made with synthesizing them from carbohydrates. Since the 1970s, very appreciable interest has been taken in the biological importance of many of the foregoing compounds, and concomitant progress has been made with the development of suitable methods for their production from sugars.

Review coverage of the chemistry involved in the synthesis of enantiomerically pure natural products from carbohydrates is now extensive [2]. It is relevant to note that whereas, in 1983, the numbers of carbocyclic target compounds were relatively limited, as outlined in an authoritative monograph written by Hanessian [3], a comprehensive review of the methods available for making cyclopentane and cyclohexane derivatives from carbohydrates published 10 years later [4] cited 338 references, more than 80% of which were dated 1980 or later. The attention afforded the synthesis of carbocyclic products also features prominently in a 1993 review of the use of sugars in the preparation of enantiomerically pure natural products [2].

For the biosynthesis of inositols and of the benzene rings of aromatic α-amino acids nature employs nucleophilic centers at C-6 of a D-glucose–derived hexos-5-ulose phosphate [5] and at C-7 of a 3,7-dideoxyhept-2,6-diulosonic acid [6], respectively, ions **1** and **2**

being the derived nucleophilic species involved in the critical ring-forming steps. These (enol–endo)-*exo–trig* processes [7] serve as models for many of the carbohydrate to cyclohexane synthetic methods that have been developed, the first-used nitrosugar method [1] (see foregoing), however, involving the stabilized carbanions **3** and thereby giving rise to an (enol–exo)-*exo–trig*-like reaction.

For the preparation of the anions **3** the base-catalyzed addition of nitromethane to the

# Formation of Cyclohexane Derivatives from Sugars

aldehyde **4** was used followed by acid-catalyzed cleavage of the isopropylidene acetal ring. The two epimeric nitro products, of which **3** are the carbanions, are formed, and during the carbocyclization reaction, a further two chiral centers are produced; therefore, eight isomeric deoxynitroinositols can be obtained. Although appreciable stereoselectivity ensures that there are favored isomers, it was this type of problem, added to the fact that carbonyl compounds, such as those from which compounds **1–3** are derived, are relatively unstable by being subject to β-elimination reactions, that held up progress until more selective routes involving the use of mild reagents and conditions were developed.

Many newer methods for generating cyclohexane derivatives from carbohydrates still depend on the intramolecular attack of nucleophilic carbon species at electrophilic centers, and the range of options is now extensive. Thus, the nucleophiles may be carbanions stabilized by carbonyl, phosphonate, nitro, or dithio groups, and they may bond to carbonyl carbon atoms, or to those that carry appropriate leaving groups or are contained in epoxide rings, or as β-centers of α,β-unsaturated carbonyl systems. Otherwise, the nucleophilic activity at the γ-centers of allylsilanes or α-positions of vinyl silanes may be used to react with electrophilic carbon atoms.

Free radical reactions are used less frequently to effect six- rather than five-membered ring closures, because carbohydrate-derived hept-6-enyl radicals are less available than are hex-5-enyl analogues, these being the species required for normal *exo* ring closure processes. Such reactions can, however, be used with good efficiency to produce cyclohexane ring systems bearing a carbon substituent (see Chap. 25).

Further reactions that are highly suited to the synthesis of cyclohexane derivatives, such as cycloaddition processes, 1,3-dipolar additions, and Diels–Alder cyclizations, have been used extensively. In the latter set, carbohydrate-based dienes or dienophiles have been employed and, in addition, intramolecular processes have provided highly suitable means of synthesizing complex polycyclic systems.

## II. METHODS

### A. Cyclizations Proceeding by Nucleophilic Additions to Carbonyl Groups

*Nitro Group Activation of the Nucleophile*

The original method of nucleophile activation remains useful—particularly, when specific nitro compounds can be employed. Although, in such cases, two new asymmetric centers are still generated during the ring-forming step, base-catalyzed epimerization often ensures that the nitro groups adopt the equatorial orientation in the major ring-closed products. That further selectivity can be obtained is nicely illustrated by the ring closure of the aldehyde **5**, made from D-mannose following nitromethane addition, refunctionalization at C-2 and C-3, and sodium periodate oxidation of the terminal diol of 1,2-dideoxy-4,5-*O*-isopropylidene-3-*O*-mesyl-1-nitro-D-*manno*-heptitol, which occurs on treatment with sodium methoxide in methanol [8]. The products are the nitroalcohols **6** and **7** (Scheme **1**), the former with the substituents at the new asymmetric centers, both equatorial, being obtainable in low yield by direct crystallization. However, when crystallization is effected in the presence of sodium methoxide in ethanol, compounds **6** and **7** interconvert by reversal of the ring closure step, the former crystallizes preferentially and can thus be obtained in 70% yield. It serves as a source of 2-deoxystreptamine **8**, which is a component of many aminoglycoside antibiotics. An advantage of this route to 2-deoxystreptamine, which is achiral, is that the chiral precursors that are used give access to chiral derivatives.

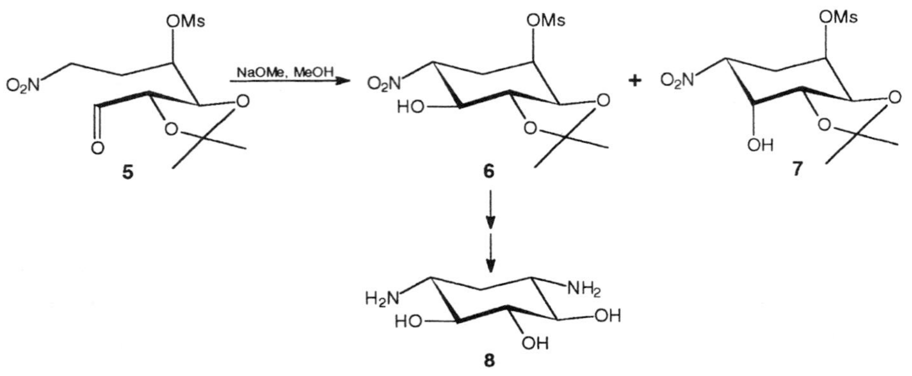

## Scheme 1

The foregoing ring-closure reaction requires that O-protecting groups be stable to basic conditions; however, the sulfonate ester group survives both the cyclization and the subsequent further treatment with base.

### Phosphonate Group Activation of the Nucleophile

Treatment of the triflate **9** with the sodio derivative of *t*-butyl dimethoxphosphorylacetate in dimethylformamide (DMF), at room temperature for 20 h in the presence of a crown ether, gives the epimeric phosphonates **10** in 81% yield [9]. Cleavage of the glycosidic bond by hydrogenolysis then affords the hemiacetals **11** which, with sodium hydride in tetrahydrofuran (THF), give the C-6 phosphonium ylid, which reacts with C-1 in the aldehydo form to afford the alkene **12** in 73% (from the glycoside **10**; Scheme 2). The cyclization

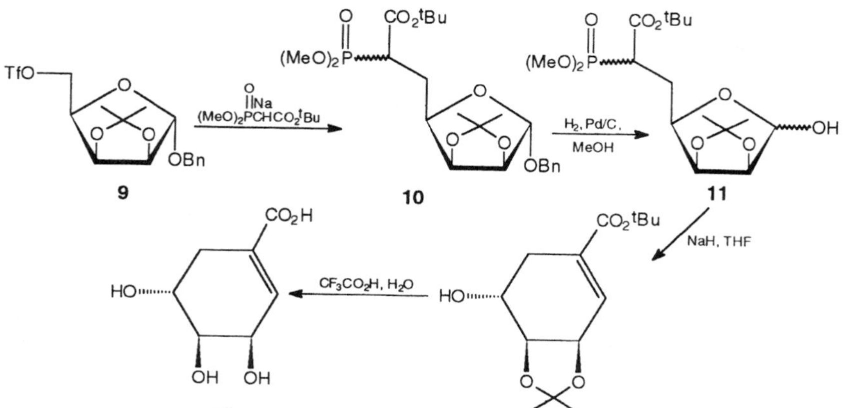

## Scheme 2

step is an example of the Wadsworth-Emmons reaction [10] (a modification of the Wittig reaction) applied intramolecularly. With aqueous trifluoroacetic acid, the ester and acetal groups are cleaved, and (−)-shikimic acid **13** is obtained in a yield that represents a 39% efficient conversion from D-mannose.

For the preparation of the phosphonates, the triflate group was required in the starting material **9**; neither the corresponding tosylate nor deoxyiodo compound underwent the required displacement reaction. For the cyclization step it is necessary that the protecting

## Formation of Cyclohexane Derivatives from Sugars

groups be stable to base; notably, however, the carboxylate ester group withstood the conditions used.

A related approach directed toward cyclohexane derivatives affords products without carbon substituents on the rings. Ozonolysis of the 6-deoxyhex-5-enopyranoside derivative **14**, followed by silylation, gives the unusual "pseudolactone" **15** in high yield, and this reacts with lithio dimethyl methylphosphonate to yield the cyclohexenone **18** (Scheme 3),

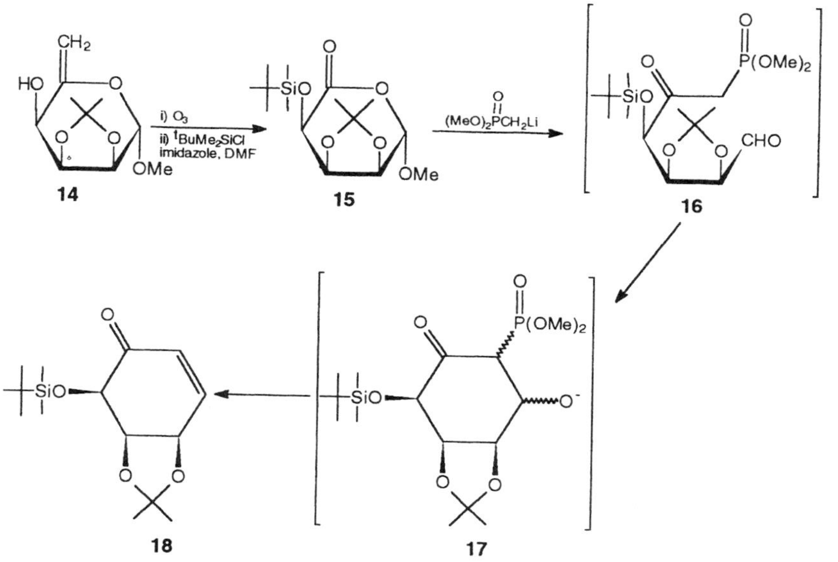

**Scheme 3**

the nucleophilic phosphonate initially effecting displacement of the ring oxygen atom from C-5 and generating a C-1 hemiacetal from which the intermediate aldehydo β-ketophosphonate **16** is formed. Attack of the α-phosphonate active center at C-1 then gives the intermediate anion **17** which, following protonation, undergoes β-elimination of dimethyl phosphate to afford the enone **18** in 62% yield [11]. This is an example of a reaction that proceeds by generation of a nucleophile and simultaneous release of an electrophilic center. Other examples are given later in this section.

This reaction is potentially more complex than indicated by the conversion **15 → 18** because, with the 4-epimeric pseudolactone **19**, the only product formed is the vinyl phosphonate **21** (80%; Scheme 4). This has been rationalized in terms of the relative

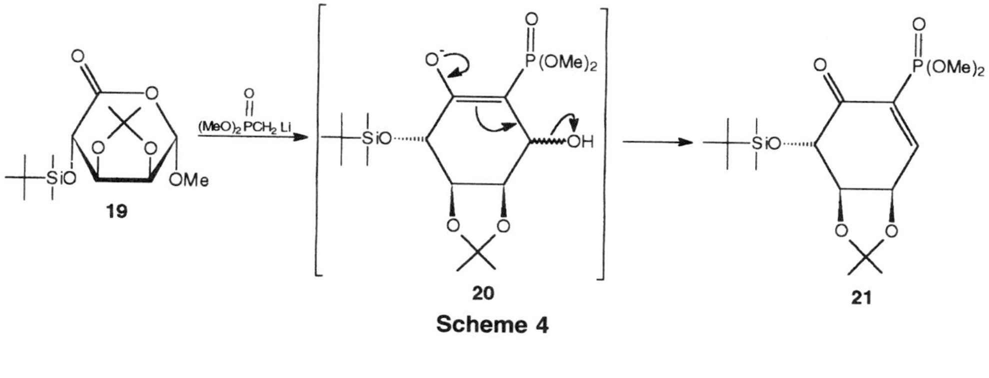

**Scheme 4**

significances of the different elimination paths followed by the interchangeable anions of the types represented by **17** and **20**, the former conceivably proceeding by way of a Wittig-like transition state to give the alkene **18**, and the latter by loss of hydroxide ion [12]. Conformational factors presumably determine the choice of elimination routes followed in the two cases.

*Carbonyl Group Activation of the Nucleophile*

Treatment with base of the hexos-5-ulose **23**, prepared from the acetal **22**, gives mainly the inosose **24** with all the hydroxyl groups equatorial (Scheme 5) [13], and similarly the

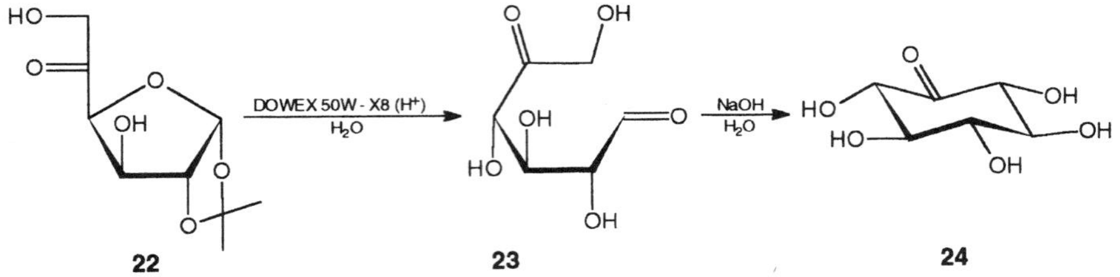

**Scheme 5**

6-phosphate ester of **23** has been converted to corresponding cyclitol phosphates [14] in a reaction that mimics the cyclization of the biochemical intermediate **1**. Such base-catalyzed chemical cyclizations are, however, not favored because of poor selectivities during the ring closures and also because the initial products are subject to epimerization under the conditions in which they are formed. Reactions of this type are best restricted to carbonyl-containing derivatives, which are limited in the number of isomeric products they can give. This, in turn, limits the range of the applicability of this approach to cyclohexanes.

Efficient Dieckmann-like cyclization occurs in the doubly branched-chain hepturonic acid derivative **25**, which gives the tricyclic product **26** in 91% yield when treated with potassium *t*-butoxide in benzene (Scheme 6) [15]. This finding reveals that, of the two

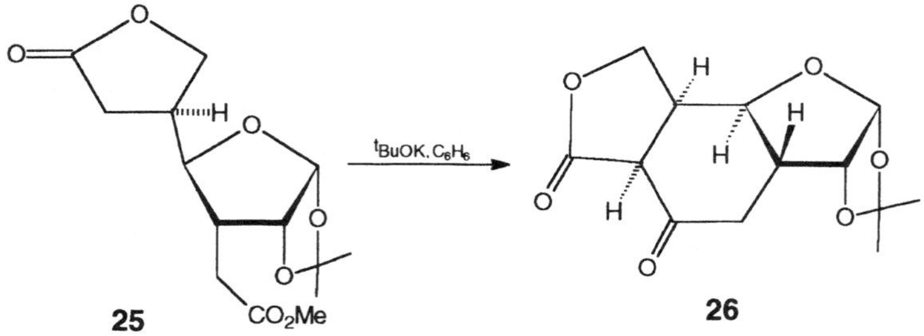

**Scheme 6**

possible carbanions, that generated at the carbon atom α-related to the lactone carbonyl group dominates the reaction in causing displacement of methoxide from the ester carbonyl center.

A related reaction of more general applicability occurs with the octos-3,7-diulose derivative **27**, which was made from the C-3 epimer of dialdose acetal **4** by treatment with (2-oxopropylidene)triphenylphosphorane in a Wittig chain extension reaction, followed by hydrogenation of the resulting enone and oxidation of the alcohol groups at C-3 and C-7. Cyclilzation to give the crystalline tertiary alcohol **28** in 81% yield is promoted by 1,8-diazabicyclo[5.4.0]undec-7-ene (DBU) in refluxing benzene (Scheme 7) [16].

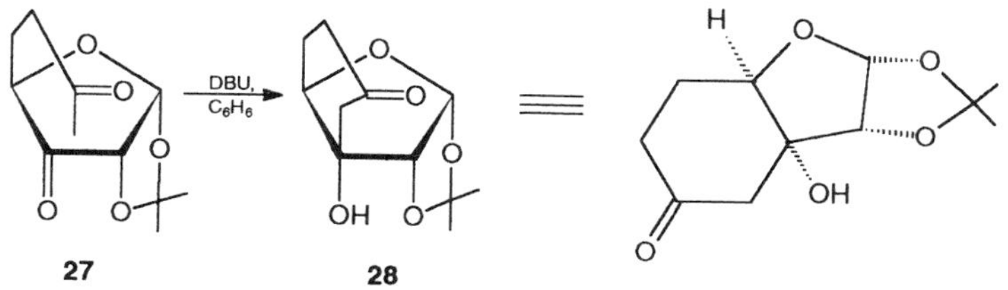

# Formation of Cyclohexane Derivatives from Sugars

**Scheme 7**

An approach that affords cyclohexyl rings, with single carboxylate substituents, and is suitable for the preparation of some carbahexopyranoses (pseudosugars) depends on the initial condensation of O-substituted pentoses in the aldehydo forms with dimethyl malonate to give alkenes [17]. Thus, 2,3,4-tri-O-benzyl-5-O-t-butyldiphenylsilyl-D-ribose **29** (Scheme 8) can be converted to the alkene **30** in 85% yield by application of the

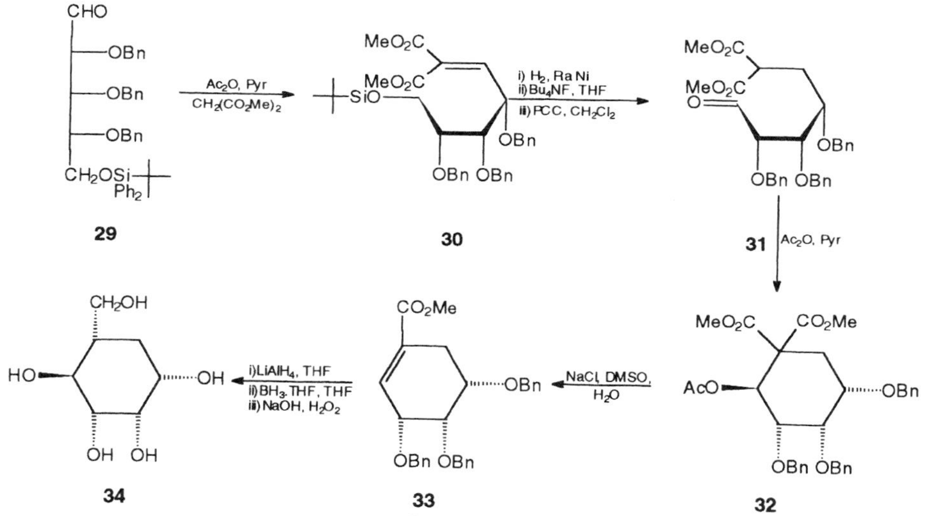

**Scheme 8**

Knoevenagel reaction. Hydrogenation of the double bond, desilylation and oxidation of the released primary alcohol group to the aldehydic function with PCC in dichloromethane in the presence of molecular sieves, gives the branched chain L-*ribo*-heptouronic acid derivative **31**. Treatment with acetic anhydride and pyridine results in cyclization, and acetylation of the resulting alcohol affords the acetate **32** in 69% yield.

By heating **32** in aqueous dimethyl sulfoxide (DMSO), containing sodium chloride, up to 170°C, loss of a methoxycarbonyl group and then β-elimination of acetic acid occur to give alkene **33** in 70% yield. From this, carba-β-L-mannopyranose **34** has been produced by conversion of the ester group to the hydroxymethyl function, hydroboration, and de-O-protection. Thus, the hydroboration step proceeded by *cis*-addition *anti*- to the allylic benzyloxy group.

## Alkenylsilanes in Aldol-like Ring Closures

A reaction that appears to have potential for carbohydrate to substituted cyclohexane conversions, and has not been applied to carbohydrate-derived starting materials, involves the intramolecular bonding of the electrophilic carbon atoms of aldehydes and the nucleophilic centers available from appropriate allylsilane groups [18]. The reactions can be initiated by the attack of the electrophiles on the π-bonds of the alkenyl groups to give ionic intermediates that are desilylated in the presence of nucleophiles, such as fluoride ion, that attack the silicon to give alkene products. Otherwise the reactions may be effected by use of Lewis acids, such as boron trifluoride or tin(IV) chloride, in which case the aldehydo groups will be activated.

In Scheme **9**, such a ring closure applied to the optically pure enal **35** synthesized from noncarbohydrate precursors, is illustrated, the yield of the cyclohexanols **36** being

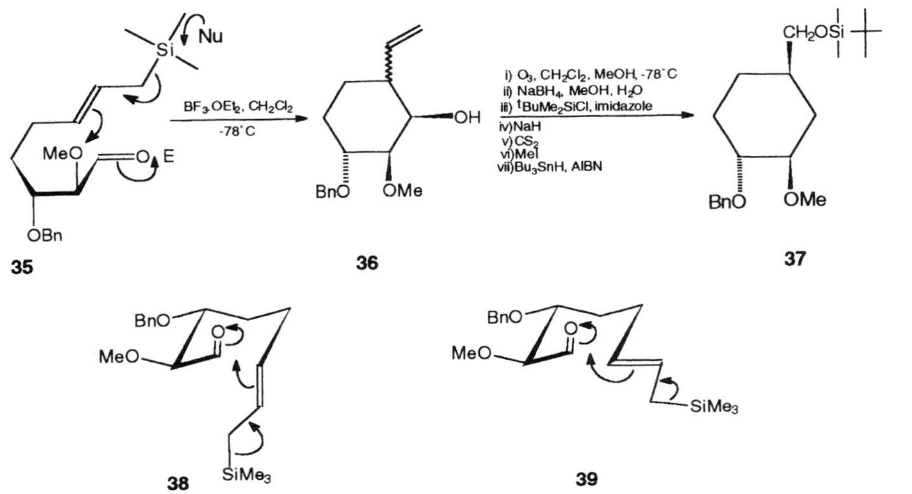

**Scheme 9**

82% and the stereoselectivity being very high during the generation of the chiral alcohol center. This suggests that the transition states leading to the epimers formed (**36**) involves the approach of the nucleophilic carbon atom and the aldehydic group with the latter in the orientation from which axial hydroxyl groups develop, which is in accord with the Felkin model of nucleophilic approach to prochiral aldehyde functions [19]. There is almost no selectivity exhibited in the generation of the tertiary chiral center, suggesting that ring closure involves the silylallyl group in both orientations **38** and **39**.

Reductive ozonolysis of the double bond of the appropriate epimer of **36**, followed by selective silylation of the diol produced, and radical deoxygenation of the secondary alcohol function, lead to **37**, which is a derivative of the cyclohexyl unit of the immunosuppressive agent tacrolimus [20].

Vinylsilanes, similarly to the allyl analogues, also undergo electrophilic attack, allowing them to serve as nucleophilic species [18]. 1,1-Dibromo-6-*O*-*t*-butyldimethylsilyl-1,2-dideoxy-3,4,5-tri-*O*-methyl-L-*arabino*-hex-1-enitol, made from the appropriate *aldehydo*-L-arabinose ether, has been converted to the *aldehydo*-vinylsilane **40** by treatment with *n*-butyllithium and TMEDA in THF at −78°C followed by trimethylsilyl chloride to give the trimethylsilylethyne derivative. Partial hydrogenation of the triple bond gives the C-silylated *cis*-alkene, which on O-desilylation at C-6 with aqueous acetic acid, and oxidation of the resulting alcohol with DMSO–oxalyl chloride affords the hex-5-enose derivative **40** [21].

When cyclization is brought about by use of boron trifluoride etherate in dichloro-

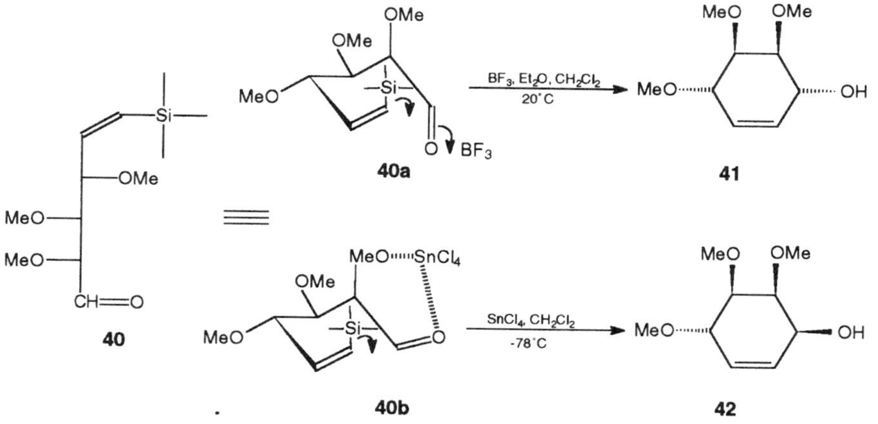

# Formation of Cyclohexane Derivatives from Sugars

**Scheme 10**

methane at room temperature the product derived from **40** (Scheme **10**) is the alcohol **41** formed in 86% yield with more than 30:1 stereoselectivity following coordination of the catalyst to the carbonyl oxygen atom in the *anti* orientation relative to the adjacent methoxy group (**40a**). On the other hand, reaction with tin(IV) tetrachloride in the same solvent at −78°C results in the epimer **42**, in 68% yield, with the same high stereoselectivity. Here, the chelation shown in **40b** controls the stereochemistry of the ring closure reaction.

*Simultaneous Activation of the Nucleophile and Release of the Electrophile During Reaction*

One cyclization procedure that depends on this approach has been described in an earlier section. A further method that affords "double activation" of a methylene group at C-6 of aldose derivatives, and simultaneous release of an aldehydic group, leads to C-6- to C-1-bonding under extremely mild conditions and is compatible with the presence of most O-protecting groups (Scheme **11**). For this reason, and because the cyclization step is normally very efficient and stereoselective, this strategy has been used extensively. It

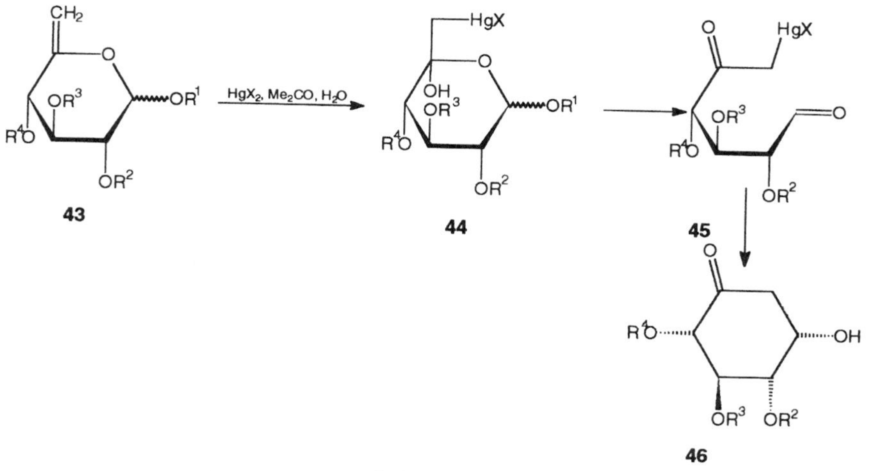

**Scheme 11**

depends on the treatment of 6-deoxy-hex-5-enopyranosyl compounds **43** with mercury(II) salts in aqueous organic media. Normal hydroxymercuration of the enol ether groups gives the hemiacetals **44** ($R^1$ = alkyl or acyl) which equilibrate with the acyclic hexos-5-ulose *aldehydo*-hemiacetals. These are extremely unstable and give the corresponding *aldehydo*-hydrates, which exist largely in the cyclic form **44** ($R^1$ = H) and can be isolated as the dicarbonyl organomercury compounds **45** by dehydration. However, very conveniently, the doubly activated C-6 nucleophiles of **45**, under the conditions of their formation, attack C-1 in aldol fashion to give the hydroxycyclohexanones **46**, usually in high yield and with good stereoselectivity at the new asymmetric centers (see Scheme **11**) [4,22].

As an example, the tetrabenzoate **43** ($R^1 = R^2 = R^3 = R^4$ = Bz, β-anomer) is converted to triester **46** ($R^2 = R^3 = R^4$ = Bz; 93% crystalline, but unpurified) by treatment with mercury(II) acetate in refluxing aqueous acetone for 5 h [23].

That the mercurial species released during the ring-closure step can, at least with some mercury(II) salts, recycle and act catalytically is shown by the finding that mercury(II) trifluoroacetate, used in 0.1 $M$ proportions (and even lower), in aqueous acetone at room temperature catalyzes the reaction of alkene **43** ($R^1$ = Me, $R^2 = R^3$ = Bn, $R^4$ = Ac; α-isomer), and gives the cyclized products in 96% yield as a mixture of the alcohol **46** ($R^2 = R^3$ = Bn, $R^4$ = Ac) and its epimer at the alcohol center, in the ratio 8:1 [24].

The relatively mild conditions of the reaction are consistent with the use of a wide range of ether- and ester-protecting groups—even tosyl groups at O-3 of the original alkenes, which engender products with excellent leaving groups in the β-relation to the carbonyl functions. Such products can be isolated in good yield under normal conditions or, if more severe conditions are employed, the products of β-elimination can be obtained directly. Thus the ketone **46** ($R^2 = R^4$ = Bz, $R^3$ = Ts) and the enone **47** can be obtained

directly in 72 and 27% (only product isolated) from the alkene **43** ($R^1$ = Me, $R^2 = R^4$ = Bz, $R^3$ = Ts; α-isomer) by variation in the conditions used [25]. Otherwise, the alternative type of conjugated enones can be obtained from the first hydroxyketone products by the induction of β-elimination by sulfonylation or acetylation of the hydroxyl groups. Compound **46** ($R^2 = R^3$ = Bn, $R^4$ = Ac) gives the enone **48** in this way [24].

A notable feature of the direct mercury(II) salt-induced cyclization (compounds **43** → **46**) is the pronounced stereoselectivity exhibited in most of the cases reported, the major products having the hydroxyl groups at the new asymmetric centers *trans*-related to the ring substituent in the β-position to it (i.e., at what was C-3 of the starting alkene). Although this could follow as a consequence of coordination between the mercury atom in an intermediate **45** and an electronegative substituent at C-3 (as in **49**), I favor the likelihood that the steric control follows from the coordination illustrated in the chair-like transition state **50** [4]. However, there is no evidence for such coordination, and the high selectivities observed may simply be a consequence of favored mercury enolate attack at the aldehydic centers in the manner that gives rise of axial alcohols (see **38**, **39**, and **40a**).

## Formation of Cyclohexane Derivatives from Sugars

Inversion of the configuration at C-3 to give compounds with axial substituents at this position would have a major destabilizing effect on **50**, and this would favor the conforma-

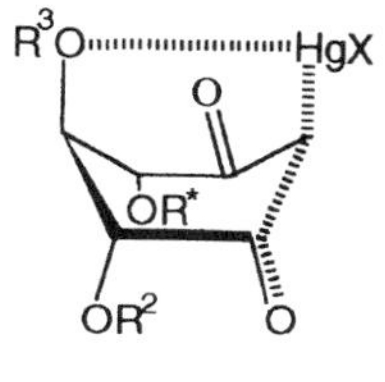

**49** 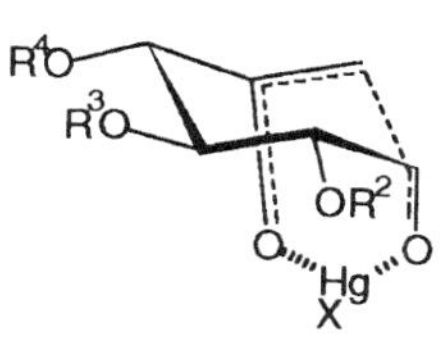    **50** 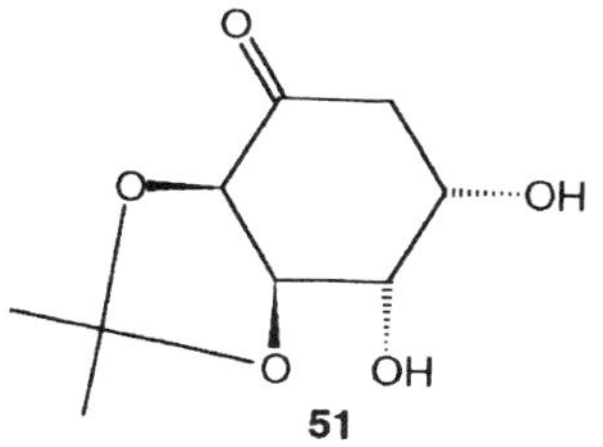    **51**

tionally inverted transition states and the generation of products with β-hydroxyl groups. This accords with findings because, for example, starting alkenes with inverted configurations at C-3 (relative to **43**) result in hydroxycyclohexanones with inverted configuration at the alcohol centers, whereas alkenes with inversions at C-2 or C-4 do not [4].

This carbocylization method has been used successfully with a range of hex-5-enopyranose derivatives, and good yields of hydroxyketones of various configurations have been recorded [4]. It is less satisfactory with the 6-deoxy-5-ene derived from 1,2:3,4-di-*O*-isopropylidene-α-D-galactose, giving only 40% of compound **51** [26], but good yields of the cyclohexanones have been obtained from substituted methyl galactopyranosides [25]. The reaction does not afford cyclopentanones when applied to 5-deoxy-4-enofuranose derivatives [27].

Most often the method gives 2-deoxyinosose derivatives, as is determined by the use of 6-deoxyhex-5-enoses, and these have been used as starting materials for a range of natural products [4]. Applied to the (*Z*)-enol acetate **52** (Scheme 12), obtainable from the

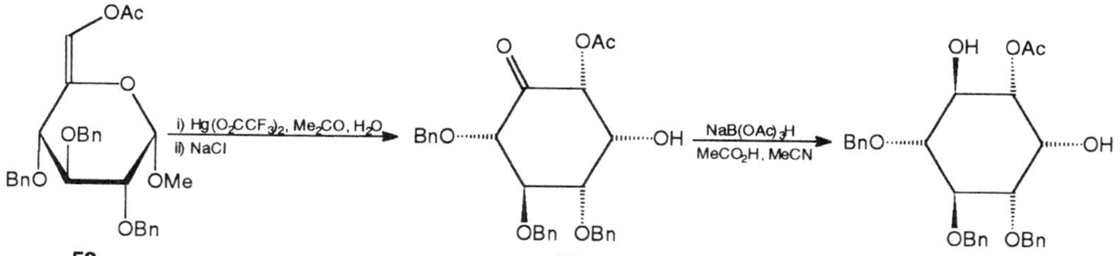

**52**    **53**    **54**

**Scheme 12**

corresponding 6-aldehyde by treatment at 80°C with acetic anhydride in acetonitrile that contains potassium carbonate, the reaction [mercury(II) trifluoroacetate, in aqueous acetone followed by sodium chloride] affords the inosose derivative **53** (59% isolated) and its epimer at the ester center in the ratio 5.7:1 [28]. The new hydroxyl group in **53** is *trans* to the central benzyloxy group as expected (see foregoing), and *cis* relative to the ester group. This is consistent with the intermediacy of the species akin to **50**, the orientation of the adlehydic group determining that the hydroxyl group will be directed "downward" in the structure shown, and the Z-configuration of alkene **52** requires that the ester group also be on the same face of the product. Reduction of **53** with sodium triacetoxyborohydride in acetic acid–acetonitrile gives only the *myo*-inositol derivative **54**.

A very different reaction is mechanistically rather similar to the foregoing process. The unexpected intramolecular rearrangement was encountered when Klemer and Kohla treated 1,6-anhydro-3,4-*O*-isopropylidene-β-D-galactopyranose **55** with *n*-butyllithium in THF, the *C*-butyl-substituted inositol derivative **59** being obtained in 58% yield [29]. This can be accounted for by abstraction of H-5 of the starting material by the strong base to give the carbanion **56**. Rearrangement with ring opening affords the enolate **57**, as shown in Scheme **13**, and in the next step the reaction resembles those involving mercury-containing

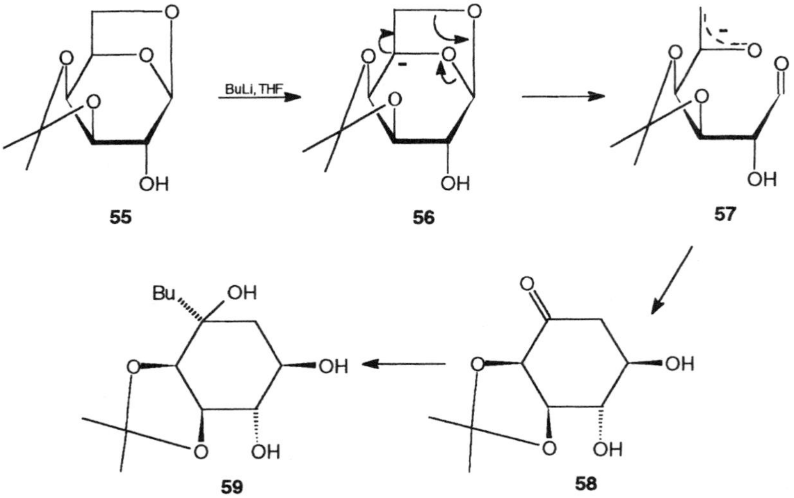

**Scheme 13**

intermediates discussed already in this section with C-6 attacking C-1 and giving a triol reported to be **59** following butyllithium addition to the carbonyl group of **58**. Conformational analysis of the ring closure **57** → **58** step suggests that the generated hydroxyl group in the product would be *trans* to the acetal ring, and the ¹H nuclear magnetic resonance (NMR) spectrum seems to leave this possibility open.

The reaction can be applied with other 1,6-anhydrohexopyranose derivatives [30], but clearly it is restricted by the strongly basic conditions used and by the specific natures of the substrates.

## B. Cyclizations Proceeding by Nucleophilic Attack at Noncarbonyl Centers

*Attack at Epoxide Ring Carbon Atoms*

Aldose dithioacetals, which are devoid of leaving groups, notably oxygen-bonded functions at C-2, on treatment with strong bases, afford stable C-1 carbanions that allow the formation of cyclohexane derivatives when suitable leaving groups are present at C-6.

The dithiane derivative **60** (Scheme **14**) is such a compound, it being made from 2,3:5,6-di-*O*-isopropylidene-D-mannose by treatment with 2-lithio-1,3-dithiane to give a heptose dithioacetal that was refunctionalized at C-2–C-3 by way of the C-1 anion and then converted to the 6,7-epoxide following selective acid-catalyzed cleavage of the 6,7-acetal ring. Treated with *n*-butyllithium it gives, in 70% yield, the cyclized **61**, which is efficiently convertible into validatol **62**, a component of validamycin A, by desulfurization with Raney nickel and de-*O*-protection by use of boron tribromide in dichloromethane [31].

The high stereo- and regioselectivity with which this ring closure can be expected to occur suggests that the procedure holds promise as a means of preparing functionalized

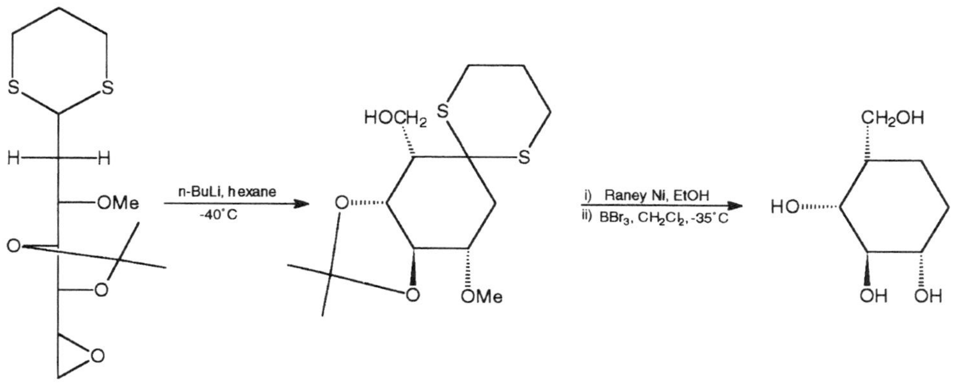

# Formation of Cyclohexane Derivatives from Sugars

**Scheme 14**

cyclohexanes, but shorter routes to the starting materials will be required before it will have wide applicability. Clearly, base-sensitive protecting groups are incompatible with this approach.

Epoxide rings may also be opened by intramolecular nucleophiles derived from allylsilane groups (see Scheme **9**) to permit aldol-like closure onto aldehyde functions. Compound **63** (Scheme **15**), derived by a multistep route from L-arabinose, when activated

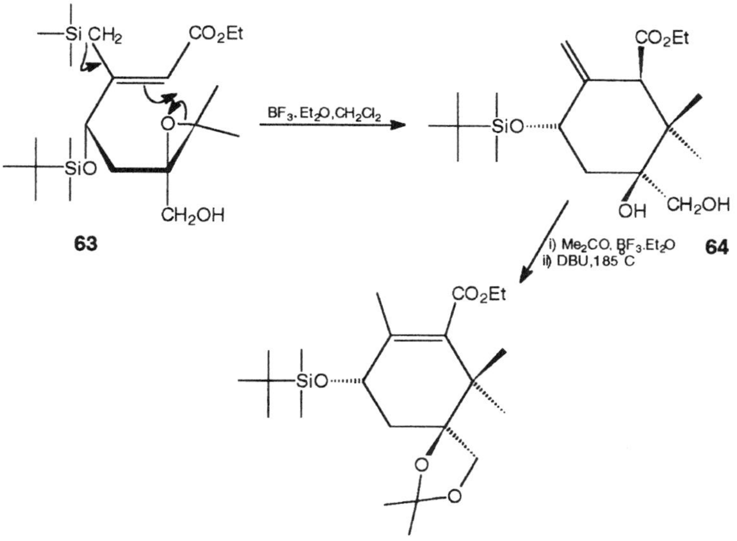

**Scheme 15**

by boron tirfluoride etherate, rapidly gives compound **64** in 80% yield. Acetalation of the diol and allylic rearrangement by high-temperature treatment in basic conditions then gives the cyclohexane derivative **65**, which has the structural features of the A-ring of paclitaxel (Taxol) [32].

*Attack at Alkene Centers*

Tsang and Fraser-Reid effected an unusual ring-forming reaction during synthetic studies aimed at the carbon skeleton of the trichothecene group of terpenes (Scheme 16) [33].

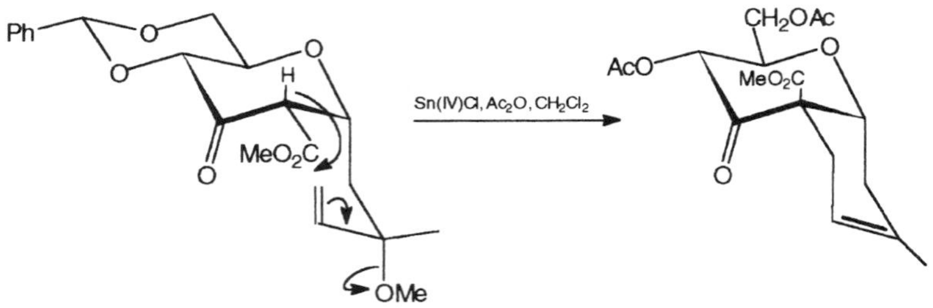

**Scheme 16**

Efforts to cause the carbon nucleophile available at C-2 (carbohydrate numbering) of the osulose derivative **66** to displace the methoxy group with allylic rearrangement and with consequent formation of a tricyclic product by use of Pd(0) catalysts [34] were unsuccessful, but the intended reaction proceeds "smoothly" when tin(IV) chloride is used together with acetic anhydride in dichloromethane. Clearly, the Lewis acid activates the allylic ether group, and the C-2 nucleophile effects its displacement. Concurrently, acetolysis of the benzylidene ring occurs and the product isolated is the *cis*-decalin analogue **67** [33].

## C. Free Radical Cyclizations

Although radical cyclization is more useful for converting carbohydrate derivatives to cyclopentanes than cyclohexanes, radicals at C-1 or C-6 of acyclic aldohexose compounds can be used satisfactorily to generate intramolecular C-1–C-6 bonds when suitable trapping groups are present at C-6 and C-1, respectively. Otherwise, radical centers and radical traps may be located in branches attached to the cyclic carbohydrate backbones, and fused-ring systems containing cyclohexane rings formed by their use. These processes are described in Chapter 25.

## D. Cycloaddition Reactions

Cycloaddition reactions can provide excellent means of producing carbocyclic compounds, and there are many good examples of the application of this approach to the synthesis of functionalized cyclohexanes from carbohydrates [4].

Intramolecular 1,3-dipolar additions of nitrones and nitrile oxides to carbohydrate alkene groups have met with success. Thus, treatment of the unsaturated heptose ether **68** (Scheme 17), which can be made following 1,3-dithianyl anion addition to C-1 of 2,3,4-tri-*O*-benzyl-5,6-dideoxy-D-*xylo*-hex-5-enose, with *N*-methylhydroxylamine in refluxing methanol, affords the nitrone **69** that cyclizes to give the bicyclic isoxazolidine **70** (60% isolated) together with the epimer at the new asymmetric center carrying the methylene carbon atom (16% isolated) [35].

In related fashion, the oximes of the *aldehydo* compound **68**, following reaction with sodium hypochlorite, form the corresponding nitrile oxide **71** and, hence, afford the isoxazolines **72** in 86% yield as a 64:36 mixture (see Scheme 17) [35]. Here, the main

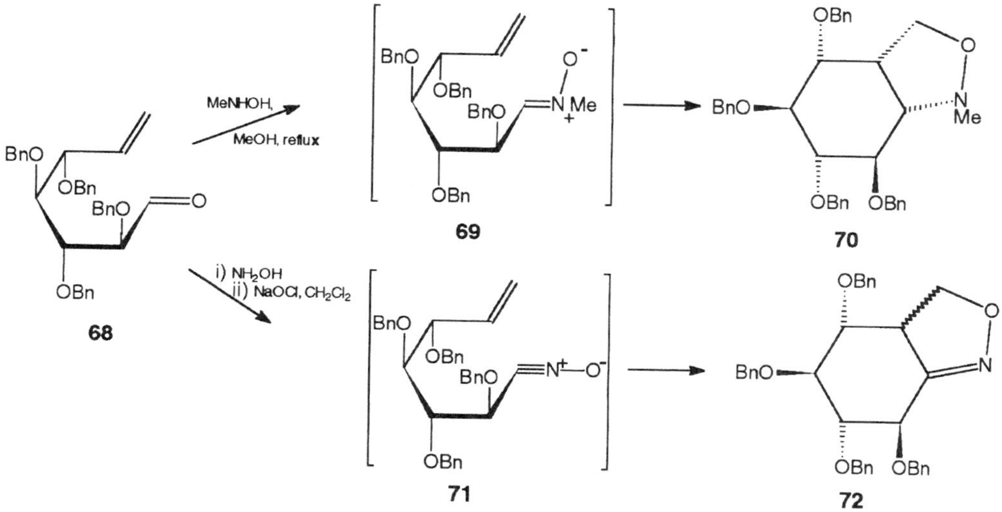

# Formation of Cyclohexane Derivatives from Sugars

**Scheme 17**

epimer has the methylene group *trans*-related to the adjacent benzyloxy group which is the main influence in directing the stereochemistry of the addition process.

Diels–Alder reactions can be applied to the synthesis of cyclohexane rings by use of carbohydrate-derived dienophiles or dienes, and intramolecular processes allow the controlled elaboration of bicyclic systems.

In the simplest case (Scheme **18**), the enone **73** gives the adduct **74** in 81% yield on treatment with 1,3-butadiene at the remarkably low temperature of −40°C in dichloro-

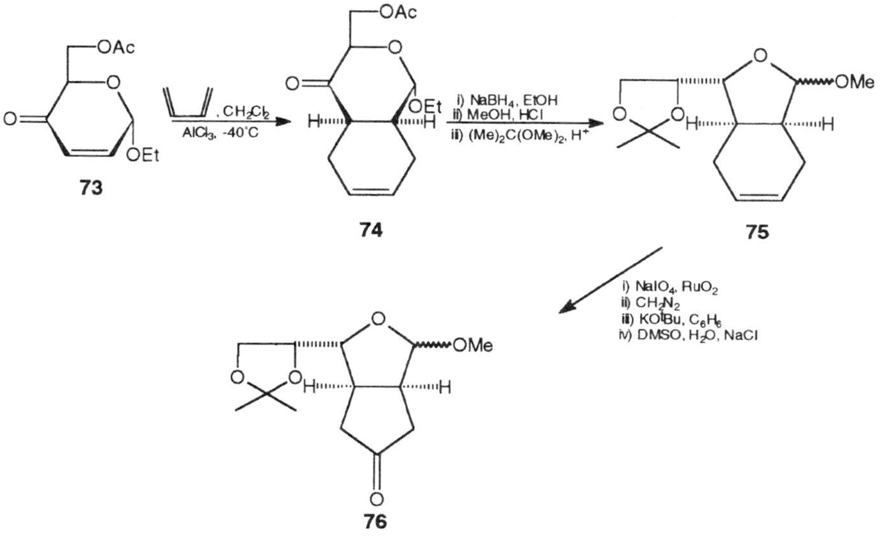

**Scheme 18**

methane containing aluminum chloride as catalyst. The product clearly offers considerable scope for the synthesis of a range of functionalized cyclohexanes, but it also affords access to cyclopentanes, for on reduction of the ketonic group of **74** with sodium borohydride and methanolysis of the product, ring contraction occurs and acetonation of the furanosides formed gives compound **75**. When the alkene group is hydroxylated and the resulting diol cleaved with sodium periodate and catalytic amounts of ruthenium dioxide, dicarboxylic acids are produced that, on methylation and Dieckmann cyclization of the esters, followed by heating in moist DMSO containing sodium chloride to cause demethoxycarbonylation, give the cyclopentanone **76** in good yield [36].

The use of butadienes with oxygen-bonded substituents with the same type of dienophile allows access to benzannelated pyranosides; for example the phenol **77** can be

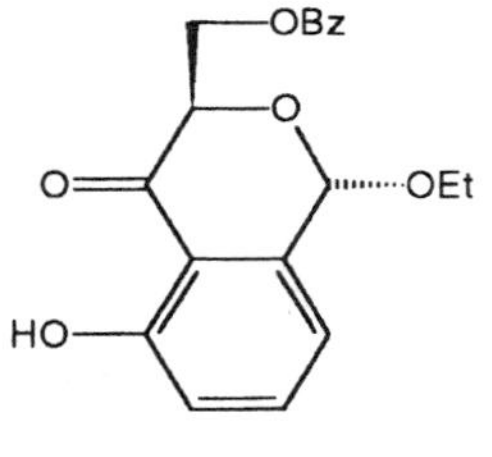

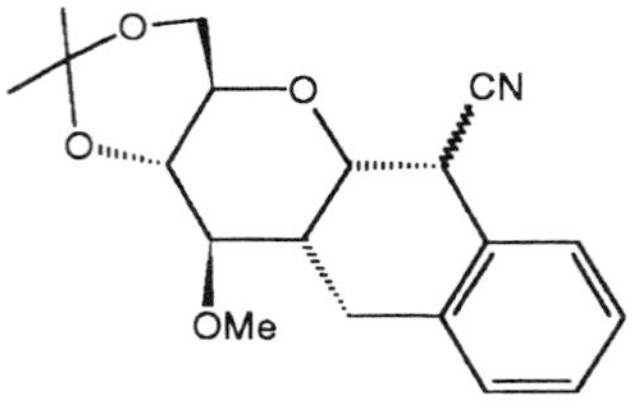

produced in 67% yield by cycloaddition of the 6-*O*-benzoyl analogue of **73** to 1-trimethylsilyloxy-1,3-butadiene, followed by treatment with DDQ in refluxing benzene [37].

For such cycloaddition reactions to occur successfully, it is not mandatory that the alkene groups be electron-deficient, compounds **78**, for example, being obtainable in 65% yield by addition of the corresponding glycal derivative to the reactive intermediate cyano-*o*-xylylene [38]. Compounds related to the aglycons of the antibiotic aureolic acids are obtainable from this product following ring opening of the carbohydrate portion of the product, that process being accompanied by an elimination reaction β to the nitrile group.

Compounds **79** [39] and **80** [40] are examples of products obtained from acyclic carbohydrate alkenes that exemplify types of enantiomerically pure cyclohexane deriva-

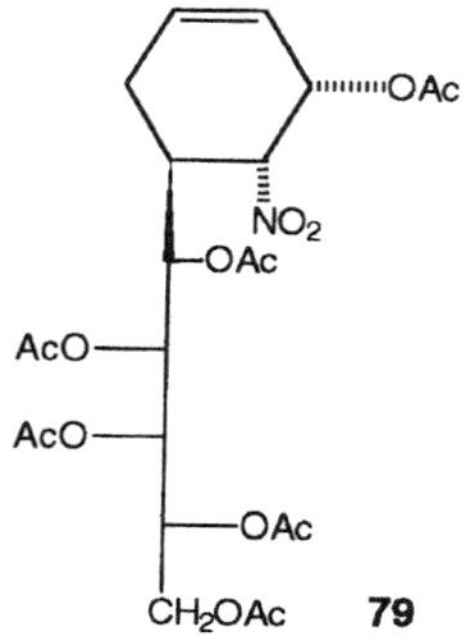

tives available by use of the Diels–Alder approach. The dienophiles used are 1-acetoxy-1,3-butadiene and cyclopentadiene and the dienophiles derivatives of a 1-nitrohept-1-enitol and a hept-2-enonic acid, respectively.

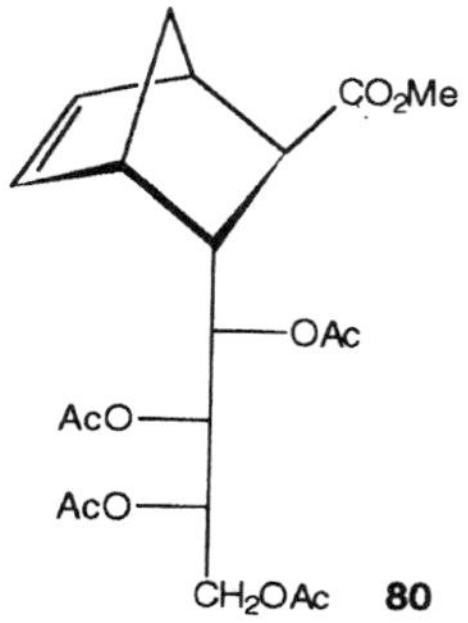

Monosaccharides do not give great scope for producing conjugated dienes within their carbon skeletons; usually when such dienes have been prepared from carbohydrate derivatives, one or two of the double bonds have been in structural features appended to the sugar chains. The conversion of the *C*-vinylglycal derivative **81** exclusively to the tetracyclic **82** by thermal addition to maleic anhydride is an example of the latter approach [41], the methoxyl group of the diene controlling the geometry of the reaction transition state.

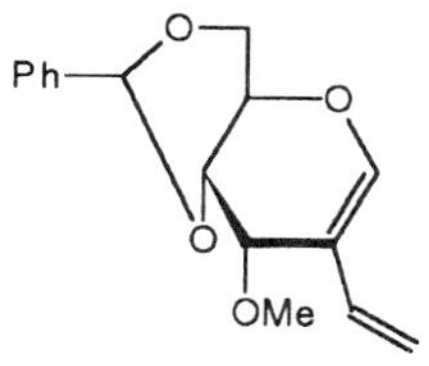

Compound **83** is a diene that comprises four carbon atoms of a hexose chain, and it takes part in the Diels–Alder reaction with maleic anhydride to give 86% of the product **84**, the stereochemistry of the addition being directed by the acetal ring [42].

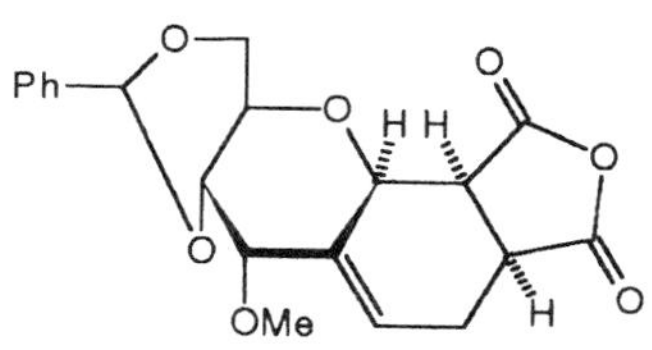

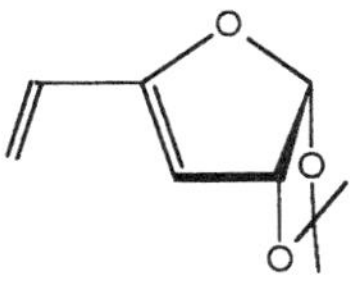

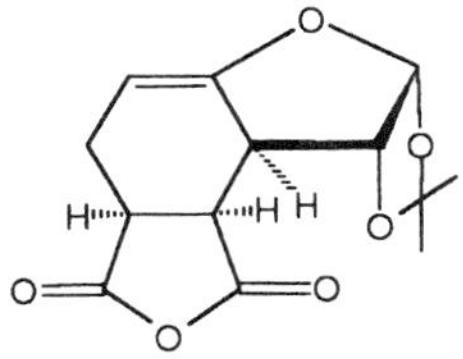

The intramolecular Diels–Alder reaction is exemplified by the thermal rearrangement of the triene **85**, which has been made from 6-*O*-benzoyl-2,3:4,5-di-*O*-isopropylidene-D-glucose by Wittig extension to C-1 using the ylid derived from allyltriphenylphosphonium bromide, followed by debenzoylation, oxidation of the resulting primary alcohol, and Wittig reaction using methyltriphenylphosphonium bromide and *n*-butyllithium. Heating of the triene at 160°C for 16 h in toluene resulted exclusively in the *cis*-fused octahydronaphthalene **86** in 90% yield [43]. An extensive range of reactions of this type have been reviewed [44].

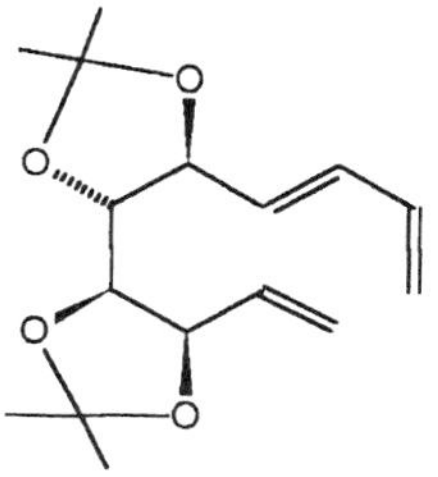

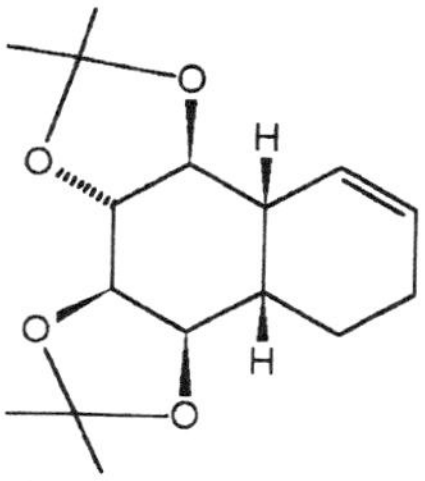

## III. EXPERIMENTAL PROCEDURES*

### A. Cyclization of a 5,6-Dideoxy-6-nitrohexose Derivative [8]

The aldehyde **5** (2.33 g, 7.5 mmol) in dry methanol (100 mL) containing NaOMe (1 g) was kept at 25°C for 30 min. The solution was neutralized by stirring with a carboxylic acid-based cation-exchange resin (e.g., Amberlite IRC-50, H$^+$), the mixture was filtered, and the

---

*Optical rotations were measured at 22°–25°C.

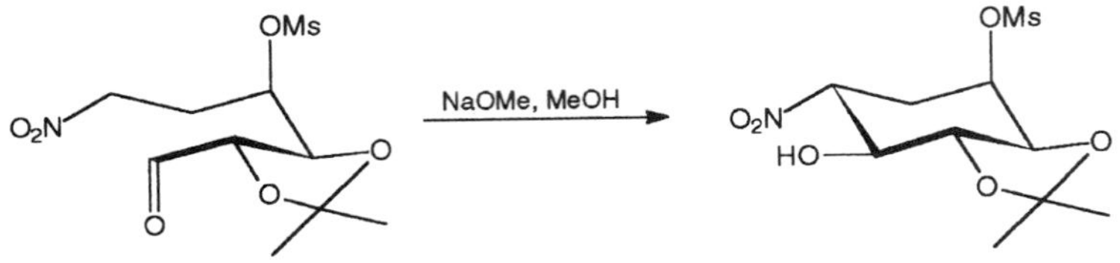

resin was exhaustively washed with methanol. The solvent was removed from the combined methanolic solutions to give compounds **6** and **7** as a foam (2.19 g, 94%), which was dissolved in ethanol (2.5 mL, 99%) and seeded with compound **6**, a sample of which had been obtained following chromatographic separation of a small portion of the mixture of **6** and **7**. After 2 h, ethanol (15 mL) was added, the supernatant was decanted, and the first crop of **6** (209 mg) was collected and washed with ethanol. The combined supernatant and washings were evaporated with addition of $CH_2Cl_2$ to give a syrupy residue that was dissolved in ethanol (2.5 mL) containing NaOMe (10 mg). Seeding with compound **6** and storage at room temperature overnight gave a second crop (1.147 g) of **6**, which was isolated as before. Three repetitions of these procedures, using progressively smaller volumes of solvent, gave additional crops of **6** (191, 125, and 43 mg; total 1.715 g, 78.3%). Recrystallization from ethanol and reprocessing all mother liquors with sodium methoxide gave pure alcohol **6** (1.54 g, 70%): mp 147°–148°C, (dec.), $[\alpha]_D$ −20.7° (c 1.4, $CHCl_3$).

### B. Cyclization of a 6-Dimethoxyphosphorylhexose Derivative [9]

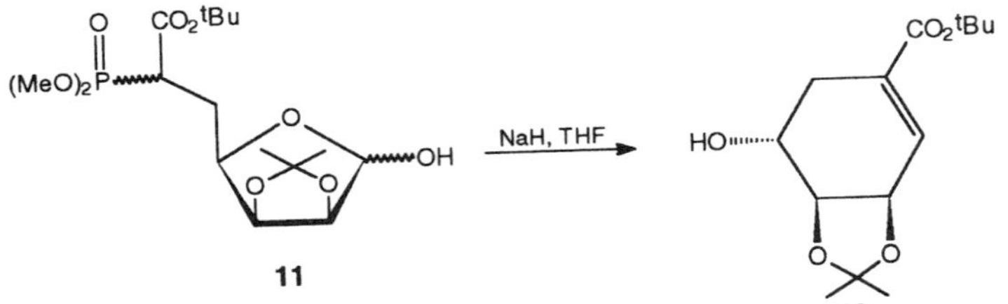

Sodium hydride (50%, 85 mg, 1.77 mmol) was washed with dry ether (× 3) under nitrogen and suspended in dry THF (8 mL). To this suspension a solution of the hemiacetals **11** (0.52 g, 1.31 mmol) in dry THF (6 mL) was added during 5 min. The reaction was exothermic and a white gelatinous precipitate formed. After 45 min, the mixture was cooled to 0°C, quenched with cold aqueous potassium dihydrogen orthophosphate (1 M, 40 mL) and extracted with chloroform (3 × 20 mL). The combined extracts were dried ($Na_2SO_4$), filtered through a pad of silica, and the filtrate was concentrated. Flash chromatography (Merck Kieselgel 60, 230–400 mesh; diethyl ether–hexane, 2:1, v/v) afforded compound **12** (260 mg, 73%) as a syrup that crystallized at 5°C: mp 44°–46°C, $[\alpha]_D$ −88.3° (c 0.7, $CHCl_3$).

### C. Intramolecular Aldol Cyclization [16]

A solution of compound **27** (4.70 g, 20.6 mmol) in benzene (70 mL) containing 1,8-diazabicyclo[5.4.0]undec-7-ene (0.31 mL, 2.06 mmol) was heated under reflux for 3 h, and

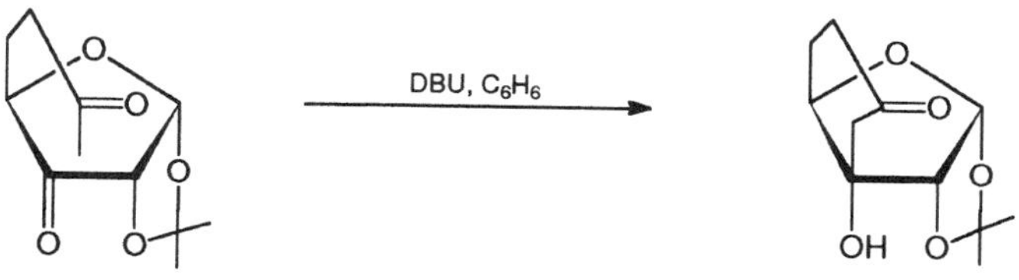

# Formation of Cyclohexane Derivatives from Sugars

**27** → **28**  (DBU, C$_6$H$_6$)

the mixture was concentrated. The residue was partitioned between CH$_2$Cl$_2$ (70 mL) and water (70 mL), the aqueous phase was extracted with CH$_2$Cl$_2$ (2 × 70 mL), the combined extracts were dried (Na$_2$SO$_4$), and the solvent was removed. The residue was chromatographed on a column of silica gel 60 (50 g, ethanol–toluene, 1:40) to give compound **28** (3.81 g, 81%): mp 132°–133°C, [α]$_D$ +50.3° (c 0.88, CHCl$_3$).

## D. Conversion of a 6-Deoxyhex-5-enopyranose Derivative to a Cyclohexanone Derivative [23]

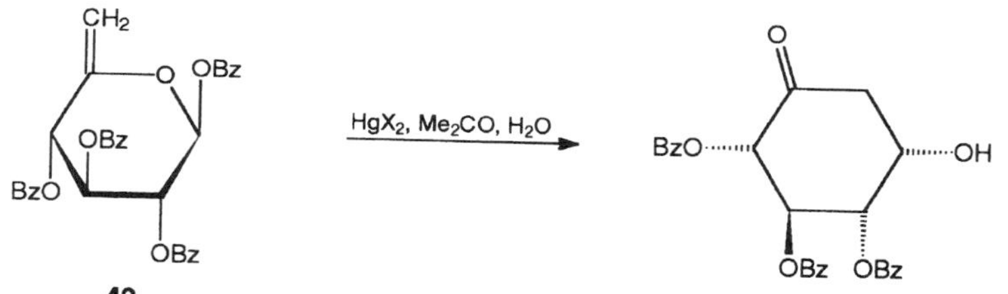

**43** → **46**  (HgX$_2$, Me$_2$CO, H$_2$O)

The alkene **43** (R$^1$ = R$^2$ = R$^3$ = R$^4$ = Bz; β-anomer; 1.0 g, 1.73 mmol) and mercury(II) acetate (1.0 g, 3.14 mmol) were heated in refluxing aqueous acetone (100 mL, 2:5) containing acetic acid (1 mL) for 5 h. The acetone was removed, the residual aqueous mixture was extracted with chloroform, and the extracts were dried (Na$_2$SO$_4$) and reduced to a small volume. Addition of light petroleum caused the hydroxyketone **46** (R$^2$ = R$^3$ = R$^4$ = Bz) to crystallize (0.77 g, 93%). When recrystallized from chloroform–light petroleum, it had mp 184°–187°C, [α]$_D$ −3.5° (c 1, CHCl$_3$).

## E. Conversion of a 6-Deoxyhex-5-enopyranose Derivative to a Cyclohex-1-en-3-one Derivative [24]

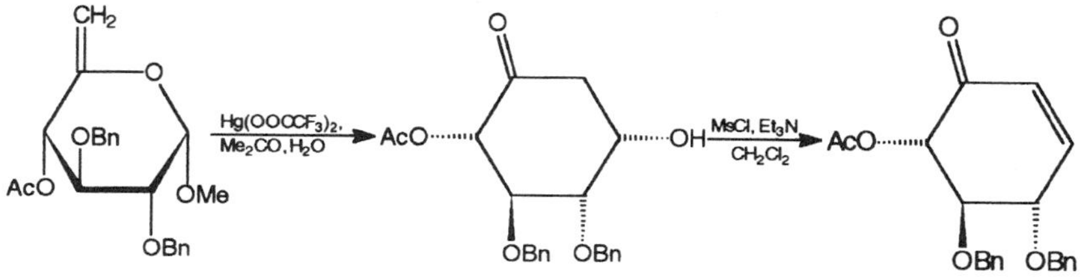

**43** → **46** → **48**  (Hg(OOCCF$_3$)$_2$, Me$_2$CO, H$_2$O; then MsCl, Et$_3$N, CH$_2$Cl$_2$)

To a stirred mixture of alkene **43** (R$^1$ = Me, R$^2$ = R$^3$ = Bn, R$^4$ = Ac; α-isomer; 0.398 g, 1.0 mmol) in acetone, water (10 mL, 2:1) at room temperature, mercury(II) trifluoroacetate (43 mg, 0.1 mmol) was added and stirring was continued for 6 h. The acetone was partly removed by evaporation, the aqueous residue was extracted with ethyl acetate, the extracts were dried ($Na_2SO_4$), and the solvent was removed. Chromatography of the residue on a column of silica gel (ethyl acetate–toluene) gave the hydroxyketones **46** ($R^2 = R^3 = Bn$, $R^4 = Ac$) (0.370 g) in a mixture, 96% (8:1) with its epimer at the hydroxylated center: $[\alpha]_D$ $-24°$ (c 2.2, $CHCl_3$).

To a solution of the unchromatographed epimers (derived from 0.500 g, 1.25 mmol of the alkene) in $CH_2Cl_2$ (15 mL) at 0°C, methanesulfonyl chloride (0.39 mL, 5.0 mmol) and $Et_3N$ (1.34 mL, 9.62 mmol) were added and the mixture was stirred at room temperature for 1.5 h. Further $CH_2Cl_2$ was added and the solution was extracted with sulfuric acid (0.5 M), washed with water, and dried ($Na_2SO_4$). Removal of the solvent gave an oil that was purified on a column of silica gel (18 g; EtOAc–toluene, 1:10) to give the enone **48** (0.397 g, 87%): $[\alpha]_D$ $+86°$ (c 1.3, $CHCl_3$) [45].

## F. Direct Conversion of a 1,6-Anhydrohexose Derivative to a Cyclohexane Derivative [29]

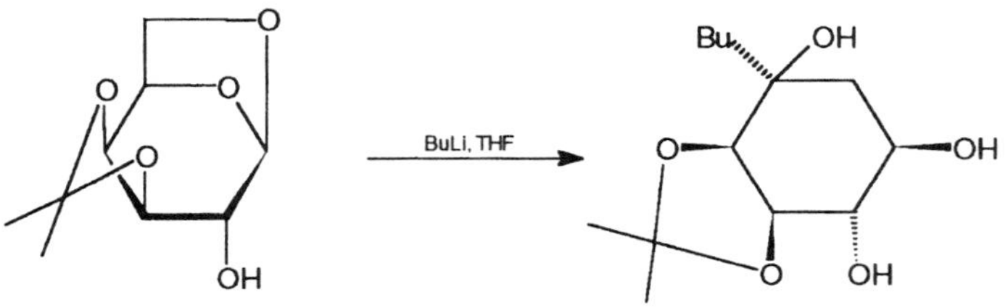

The anhydro compound **55** (2.0 g, 9.9 mmol) was treated in THF (70 mL) at $-10°C$ for 8 h with n-butyllithium (45 mL, 1.6 M, 72 mmol in hexane), which had been added dropwise with stirring under nitrogen. The stirring was continued throughout the reaction. Ammonium chloride solution (10% aqueous) was added, and the aqueous phase was extracted with $CH_2Cl_2$ (5 × 20 mL). The organic extracts were combined, dried ($MgSO_4$), and removal of the solvents gave a brown-colored product (2.20 g, 85%) that was purified by column chromatography using Kieselgel eluted with ethyl acetate–light petroleum (5:1) to give product **59** (1.48 g, 58%): mp 128°C, $[\alpha]_D$ $-11.05$ (c 1.0, $CHCl_3$).

## G. Conversion of a 6,7-Dideoxyhept-6-enose Derivative to a 7-Aza-8-oxabicyclo[4.3.0]nonane Derivative [35]

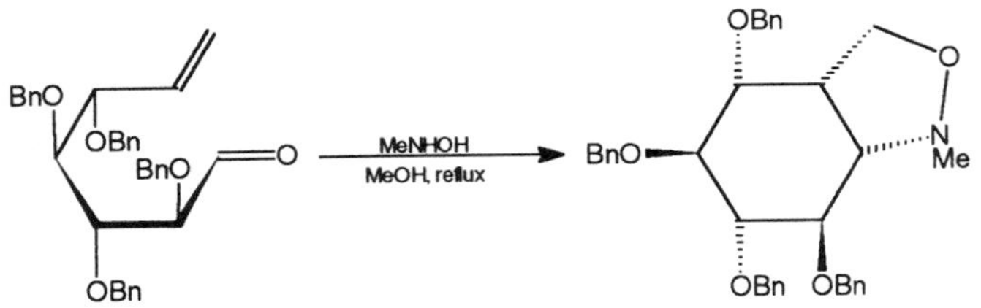

A solution of $CH_3NHOH$ (0.203 g, 4.33 mmol) in MeOH (10 mL) was added with stirring to the enal **68** (1.904 g, 3.55 mmol) in MeOH (40 mL), and the mixture was heated at reflux under nitrogen for 4 h, then stirred at 25°C for 2.5 days. The solution was partly concentrated, the residue was diluted with water, and was extracted twice with EtOAc–cyclohexane. The combined extracts were washed with water–brine and were dried (MgSO$_4$). Concentration in vacuo and separation by flash chromatography (cyclohexane–EtOAc, 3:1) gave compound **70** (1.21 g, 60%) and the epimer (0.321 g, 16%). When recrystallized from ether–pentane, the major product had mp 58.5°–61°C, [α]$_D$ −13.3° (c 1.1, CHCl$_3$); the minor, from the same solvent, had mp 85°–87.5°C.

## H. Conversion of a Hex-2-enopyranosid-4-ulose to a 4-Oxa-2-oxo-bicyclo[4.4.0]dec-8-ene Derivative [36]

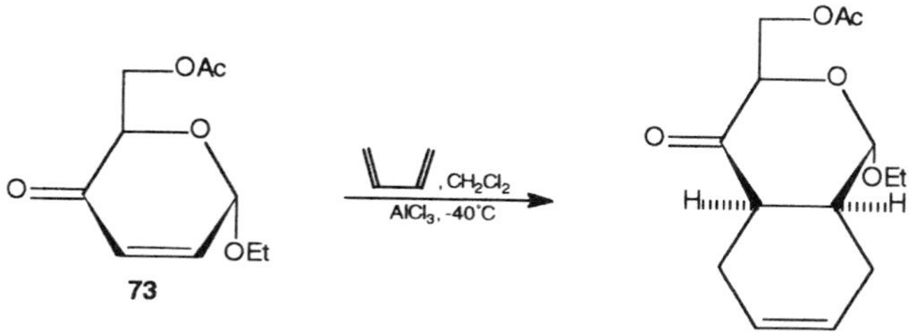

To the enone **73** (200.3 mg, 0.935 mmol) in CH$_2$Cl$_2$ (40 mL) at −78°C under argon, 1,3-butadiene (10 mL) was added followed, with stirring, by aluminum chloride (800 mg, 5.99 mmol). The temperature was allowed to rise to −40°C and, after 2.5 h the mixture was poured into saturated aqueous sodium hydrogen carbonate (100 mL) cooled to 0°C. The products were extracted with CH$_2$Cl$_2$ (2 × 100 mL), and the mixed organic solutions were washed with water (2 × 100 mL), dried (Na$_2$SO$_4$), and the solvent was removed. The residue was fractionated by column chromatography (silica gel; ethyl acetate–petroleum ether, 1:4) and gave the adduct **74** (203.1 mg, 81%): [α]$_D$ +144° (c 6.26, CHCl$_3$).

## I. Conversion of a 3,5,6-Trideoxyhex-3,5-Dienofuranose Derivative to a Cyclohexene Derivative Bearing Three Carbon- and One Oxygen-Bonded Substituents [42]

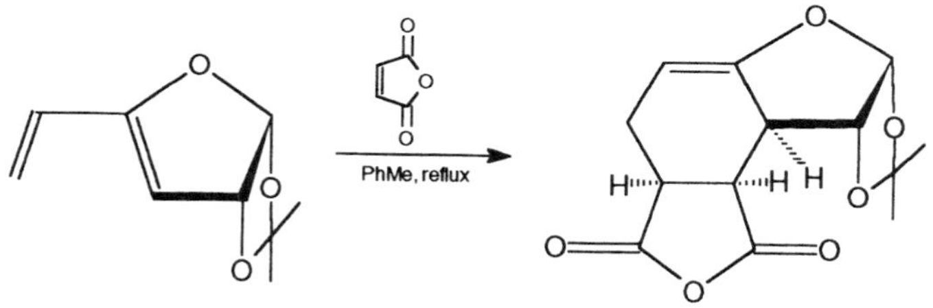

Compound **83** (2.5 g, 14.9 mmol) and maleic anhydride (1.5 g, 15.3 mmol) were heated in refluxing toluene (50 mL) for 10 h. The solvent was removed and the residue was resolved by column chromatography (silica gel; diethyl ether) to give the tricyclic **84** (3.4 g, 86%): mp 170°C (dec.), [α]$_D$ +21.34 (c 1.32, CHCl$_3$).

## REFERENCES

1. J. M. Grosheintz and H. O. L. Fischer, Cyclization of 6-nitrodesoxyaldohexoses to nitrodesoxy-inositols, *J. Am. Chem. Soc. 70*:1476 (1948).
2. K. J. Hale, Monosaccharides: Use in the asymmetric synthesis of natural products, *Rodd's Chemistry of Carbon Compounds*, 2nd ed., Vol. 1 EFG Second Suppl. (M. Sainsbury, ed.), Elsevier Science Publishers, Amsterdam, 1993, p. 315, and references therein.
3. S. Hanessian, *Total Synthesis of Natural Products: The "Chiron" Approach*, Pergamon Press, Oxford, 1983.
4. R. J. Ferrier and S. Middleton, The conversion of carbohydrate derivatives into functionalized cyclohexanes and cyclopentanes. *Chem. Rev. 93*:2779 (1993).
5. J. E. G. Barnett, A. Rasheed, and D. L. Corina, Partial reactions of D-glucose 6-phosphate-1-L-*myo*-inositol 1-phosphate cyclase, *Biochem. J. 131*:21 (1973).
6. P. A. Bartlett and K. Satake, Does dehydroquinate synthase synthesize dehydroquinate? *J. Am. Chem. Soc. 110*:1628 (1988).
7. J. E. Baldwin and M. J. Lusch, Rules for ring closure: Application to intramolecular aldol condensations in polyketonic substrates, *Tetrahedron 38*:2939 (1982).
8. H. H. Baer, I. Arai, B. Radatus, J. Rodwell, and N. Chinh, A chiral approach to 2-deoxystreptamine, *Can. J. Chem. 65*:1443 (1987).
9. G. W. J. Fleet, J. K. M. Shing, and S. M. Warr, Enantiospecific synthesis of shikimic acid from D-mannose, *J. Chem. Soc. Perkin Trans. 1*:905 (1984).
10. W. S. Wadsworth, Synthetic applications of phosphoryl-stabilized anions, *Org. React. 25*:73 (1977).
11. S. Mirza, L.-P. Molleyres, and A. Vasella, Synthesis of a glyoxalase I inhibitor from *Streptomyces griseosporeus*, *Helv. Chim. Acta 68*:988 (1985).
12. M. Aloui, B. Lygo, and H. Trabsa, Observations on the modified Fujimoto-Belleau approach to mannose-derived carbocyclic sugars, *Synlett* p. 115 (1994).
13. D. E. Kiely and H. G. Fletcher, Jr., Cyclization of D-*xylo*-hexos-5-ulose, a chemical synthesis of *scyllo*- and *myo*-inositols from D-glucose, *J. Org. Chem. 34*:1386 (1969).
14. D. E. Kiely and W. R. Sherman, A chemical model for the cyclization step in the biosynthesis of L-*myo*-inositol 1-phosphate, *J. Am. Chem. Soc. 97*:6810 (1975).
15. M. Georges, T. F. Tan, and B. Fraser-Reid, Stereocontrol at "off-template" sites in 1,2-*O*-isopropylidene glycofuranoses, *J. Chem. Soc. Chem. Commun.* p. 1122 (1984).
16. K.-I. Tadano, M. Miyazaki, S. Ogawa, and T. Suami, Construction of an enantiomerically pure *cis*-fused 7-oxabicyclo[4.3.0]nonan-3-one skeleton, *J. Org. Chem. 53*:1574 (1988).
17. K.-I. Tadano, H. Maeda, M. Hoshino, Y. Iimura, and T. Suami, A novel transformation of four aldoses to some optically pure pseudohexopyranoses and a pseudopentofuranose, carbocyclic analogues of hexopyranoses and pentofuranose, *J. Org. Chem. 52*:1946 (1987).
18. I. Fleming, J. Dunoguès, and R. Smithers, The electrophilic substitution of allylsilanes and vinylsilanes, *Org. React. 37*:57 (1989).
19. M. T. Reetz, Chelation or non-chelation control in addition reactions of chiral α- and β-alkoxy carbonyl compounds, *Angew. Chem. Int. Ed. Engl. 23*:556 (1984).
20. M. E. Maier and B. Schoffling, Synthesis of the cyclohexyl fragment of FK-506 by intramolecular ene-reaction, *Tetrahedron Lett. 31*:3007 (1990).
21. M. C. McIntosh and S. M. Weinreb, A strategy for synthesis of conduritols and related cyclitols via stereodivergent vinylsilane-aldehyde cyclizations, *J. Org. Chem. 56*:5010 (1991).
22. R. J. Ferrier, A carbocyclic ring closure of a hex-5-enopyranoside derivative, *J. Chem. Soc. Perkin Trans. 1*:1455 (1979).
23. R. Blattner, R. J. Ferrier, and S. R. Haines, Observations on the conversion of 6-deoxyhex-5-enopyranosyl compounds into 2-deoxyinosose derivatives, *J. Chem. Soc. Perkin Trans. 1*:2413 (1985).

24. N. Chida, M. Ohtsuka, K. Ogura, and S. Ogawa, Synthesis of optically active cyclohexenones from carbohydrates by catalytic Ferrier rearrangement, *Bull. Chem. Soc. Jpn. 64*:2118 (1991).
25. A. S. Machado, A. Olesker, and G. Lukacs, Syntheses of two enantiomeric tetrasubstituted cyclohenanones from 6-deoxyhex-5-enopyranoside derivatives, *Carbohydr. Res. 135*:231 (1985).
26. R. J. Ferrier and S. R. Haines, A route to functionalized cyclopentanes from 6-deoxyhex-5-enopyranosyl derivatives, *Carbohydr. Res. 130*:135 (1984).
27. R. J. Ferrier and S. R. Haines, Alkenes from 4-bromohexofuranose esters; reactions of 5-deoxyald-4-enofuranose derivatives in the presence of mercury(II) ions, *J. Chem. Soc. Perkin Trans. 1*:1689 (1984).
28. S. L. Bender and R. J. Budhu, Biomimetic synthesis of enantiomerically pure D-*myo*-inositol derivatives, *J. Am. Chem. Soc. 113*:9883 (1991).
29. A. Klemer and M. Kohla, One-step stereoselective syntheses of C-branched α-deoxycyclitols from 1,6-anhydrohexopyranoses, *Liebigs Ann. Chem.* p. 1662 (1984).
30. A. Klemer and M. Kohla, Further syntheses of C-branched α-deoxycyclitols starting from 1,6-anhydrohexopyranoses, *Liebigs Ann. Chem.* p. 967 (1986).
31. K. Krohn, G. Börner, and S. Gringard, Synthesis of validatol and 4-*epi*-validatol from mannose and glucose, *J. Org. Chem. 59*:6069 (1994).
32. L. Pettersson, T. Frejd, and G. Magnusson, An enantiospecific synthesis of a taxol A-ring building unit, *Tetrahedron Lett. 28*:2753 (1987).
33. R. Tsang and B. Fraser-Reid, A route to optically active trichothecene skeleton by bisannulation of a pyranose derivative, *J. Org. Chem. 50*:4659 (1985).
34. B. M. Trost, Transition metal templates for selectivity in organic synthesis, *Pure Appl. Chem. 53*:2357 (1981).
35. N. P. Peet, E. W. Huber, and R. A. Farr, Diastereoselectivity in the intramolecular nitrone, oxime, and nitrile oxide cycloaddition reactions, *Tetrahedron 47*:7537 (1991).
36. J. L. Primeau, R. C. Anderson, and B. Fraser-Reid, Routes to cyclohexano and cyclopentano pyranosides from carbohydrate α-enones, *J. Am. Chem. Soc. 105*:5874 (1983).
37. P. J. Card, Synthesis of benzannelated pyranosides, *J. Org. Chem. 47*:2169 (1982).
38. R. W. Franck and T. V. John, Aureolic acid antibiotics: Synthesis of a model aglycon, *J. Org. Chem. 48*:3269 (1983).
39. J. A. Serrano, L. E. Cáceres, and E. Román, Asymmetric Diels–Alder reactions of a chiral sugar nitroalkene, *J. Chem. Soc. Perkin Trans. 1*:941 (1992).
40. D. Horton and T. Usui, Transformations of unsaturated acyclic sugars into enantiomerically pure norbornene derivatives, *Carbohydr. Res. 216*:33 (1991).
41. C. Burnouf, J. C. Lopez, F. G. Calvo-Flores, M. D. Laborde, A. Olesker, and G. Lukacs, π-Facial selectivity in Diels–Alder reactions of 2-*C*-vinylglycals. Stereocontrolled route to annulted *C*-glycopyranosides, *J. Chem. Soc. Chem. Commun.* p. 823 (1990).
42. K.-M. Sun, R. M. Giuliano, and B. Fraser-Reid, Diacetone glucose derived dienes in Diels–Alder reactions. Products and transformations, *J. Org. Chem. 50*:4774 (1985).
43. P. Herczegh, M. Zséby, L. Szilágyi, Z. Dinya, and R. Bognár, High diastereoselection in the intramolecular Diels–Alder reaction of sugar-based 1,7(*E/Z*),9-decatrienes, *Tetrahedron 45*: 5995 (1989).
44. P. Herczegh, M. Zséby, L. Szilágyi, I. Bajza, A. Kovacs, G. Batta, and R. Bognár, Inter- and intramolecular Diels–Alder reactions of sugar derivatives, *ACS Symp. Ser. 494*:112 (1992).
45. N. Chida, M. Ohtsuka, K. Nakazawa, and S. Ogawa, Total synthesis of antibiotic hygromycin A, *J. Org. Chem. 56*:2976 (1991).

# VII
## Total Synthesis of Sugars from Nonsugars

**THEMES**: *Aminodeoxy Sugars, Deoxy and Related Sugars*

Carbohydrate chemists have been regaled over the years in having an abundance of naturally occurring sugars to work with. They have also appreciated the richness of their field in terms of carbon atom and stereochemical diversity. With the ready availability of three to seven carbon sugars, either directly or through degradation, there has seldom been a need to synthesize a sugar de novo. It has been easier and, at times, more convenient to prepare a given "rare" sugar by chemical transformations effected on another "less rare" sugar derivative. Although this may be the predictable course to follow when a full complement of hydroxy groups is required in the target sugar, it may not be the same for a deoxy, a branched-chain, or an aminodeoxy sugar.

The notion of a total synthesis of a sugar or sugar-like molecule is intellectually challenging for several reasons; namely, (1) achieving the highest levels of stereochemical control; (2) developing or adapting new methods of assembly; (3) possible industrial relevance; and (4) exploration and development of new asymmetric processes, especially catalytic ones.

Therefore, it is not surprising that some of the earlier contributions to the total synthesis of racemic sugar-like molecules (e.g., from acetylene) were from noncarbohydrate chemists. The advent of stereochemical control in asymmetric reactions brought players outside the pool of carbohydrate chemists into the arena. In fact, it has become fashionable for synthetic organic chemists to devise new ways to construct sugar molecules under the guise of asymmetric synthesis. If a Diels–Alder reaction is a method of choice for the de novo construction of a cyclohexane derivative, why not consider the analogous hetero (oxygen) Diels–Alder reaction for the synthesis of a hexose-like motif? This strategy was, in fact, originally studied by A. Zamojski, in the late 1960s, extended by S. David and co-workers at Orsay (France) in the mid-1970s, and more recently popularized by S. Danishefsky, who demonstrated applications to the synthesis of some rare higher-carbon sugars in racemic and enantio-pure form. The assembly of a hexopyranose motif by a hetero-Diels–Alder reaction could be a practical strategy. For the method to be competitive with the more traditional routes, however, a high degree of stereochemical control must be assured at the newly formed stereogenic centers. Great strides continue to be made toward this goal.

For the total synthesis of aminodeoxy sugars from nonsugar precursors, it is logical to turn to amino acids as primary sources of chiral templates. The challenge is to find innovative and stereocontrolled methods of chain extension reactions that lead to higher-carbon aminodeoxy sugars.

It is of historical interest to remind the reader that the absolute configuration of D-glyceraldehyde, D-lactic acid, and D-alanine, representing the stereochemical cornerstones of three major classes of naturally occurring compounds, were intercorrelated, starting with the pioneering work of Emil Fischer nearly a century ago and further advanced by K. Freudenberg, M. L. Wolfrom, and C. K. Ingold and their co-workers.

Aspects of the total synthesis of aminodeoxy, deoxy, and related compounds are covered in the next two chapters.

*Stephen Hanessian*

# 27
# Total Synthesis of Amino Sugars

**Janusz Jurczak**
Warsaw University and Institute of Organic Chemistry, Polish Academy of Sciences, Warsaw, Poland

| | | |
|---|---|---|
| I. | Introduction | 595 |
| II. | Methods | 596 |
| | A. $C_1$-elongation | 596 |
| | B. $C_2$-elongation | 596 |
| | C. $C_3$-elongation | 598 |
| | D. $C_4$-elongation | 599 |
| III. | Experimental Procedures | 601 |
| | A. Synthesis of L-daunosamine | 601 |
| | B. Synthesis of α-methyl mannofuranoside and β-methyl allofuranoside | 603 |
| | C. Synthesis of 6-epi-D-purpurosamine B | 604 |
| | D. Synthesis of anhydrogalantinic acid | 606 |
| | E. Synthesis of the diastereoisomer of lincosamine | 608 |
| | F. Synthesis of methyl α-D-lincosaminide | 609 |
| | References | 611 |

## I. INTRODUCTION

Amino sugars constitute the glycosidic subunits of biologically active compounds of significant pharmacological importance [1]. Although "ordinary" sugars are most readily obtained from the members of the same class of compounds, amino sugars can also be synthesized from nonsugar precursors, among which α-amino acids are the starting materials of choice [2].

The main purpose of this chapter is to present the basic transformations leading to chirons useful in the total synthesis of amino sugars. Although the interest in the application of α-amino acids in the total synthesis of natural products has a long tradition, only the past 10–15 years brought impressive expansion in this field [3–5]. The subject of total synthesis of amino sugars is covered in several general reviews [6–8]. Some important particulars were also reviewed: namely, the thiazole route to carbohydrates [9,10], addition of organometallics to α-amino aldehydes [11], reagent tuning in reactions of the carbonyl group [12], amino acid derivatives as chiral auxiliaries [13], and the synthesis of antibiotic amino sugars from α-amino aldehydes [14–16].

Stereoselective elongation of the carbon skeleton is the central point of the synthesis of amino sugars from α-amino acids and their derivatives (Scheme **1**).

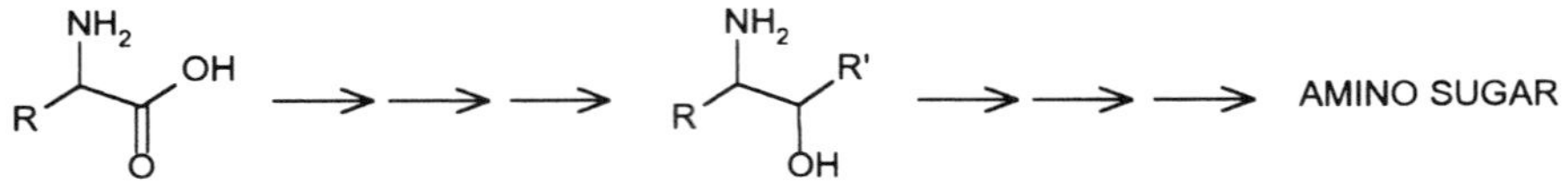

**Scheme 1**

Literature data on the carbon skeleton elongation may be classified according to the number of carbon atoms added. Such a classification, although far from being ideal, is useful from the synthetic viewpoint. Scheme **2** illustrates the usefulness of α-amino acids (in particular α-amino-β-hydroxy acids; e.g., serine, threonine, and their homologues) in the synthesis of amino sugars.

This classification leaves out of account the other applications to α-amino acids (e.g., as chiral auxiliaries). Besides, the approach shown in Scheme **2** excludes higher carbon chain elongation (from $C_5$ up to $C_{16}$), which can be effectively realized using organometallic additions and Wittig-type reactions [e.g., see Ref 17].

Keeping the foregoing limitations in mind, we will discuss the recently published methods for achieving carbon chain elongation. Throughout this chapter, the *syn–anti* convention, as proposed Masamune et al. [18,19] will be followed. The use of the *threo–erythro* descriptors is restricted to sugars to avoid any ambiguity.

## II. METHODS

### A. $C_1$-Elongation

Usually the products of $C_1$-elongation are intermediates, rather than the target amino sugars. The elongation can be repeated iteratively [20]. Cyanohydrin formation belongs to the most typical $C_1$-elongation processes. Addition of trimethylsilyl cyanide to α-amino aldehydes of type **1** in the presence of Lewis acid yielded a mixture of diastereoisomers **2** and **3** [21] (Scheme **3**).

Boron trifluoride, zinc bromide, and tin tetrachloride led to non–chelation-controlled *anti*-adducts **3**, whereas the use of titanium tetrachloride and magnesium bromide resulted in chelation-controlled *syn*-adducts **2**. Both adducts can be used as starting compounds for the iterative synthesis of amino sugars.

Single-carbon elongation can also be achieved with the use of organometallic reagents (MeMgI, $Me_2Zn/TiCl_4$) [22], sulfonium ylides [23], or by Wittig olefination, followed by asymmetric epoxidation [24].

### B. $C_2$-Elongation

Addition of vinylmagnesium bromide **5** to N-monoprotected α-amino aldehydes **4** usually yielded a mixture of diastereoisomers, with a preference for the chelation-controlled

# Synthesis of Amino Sugars

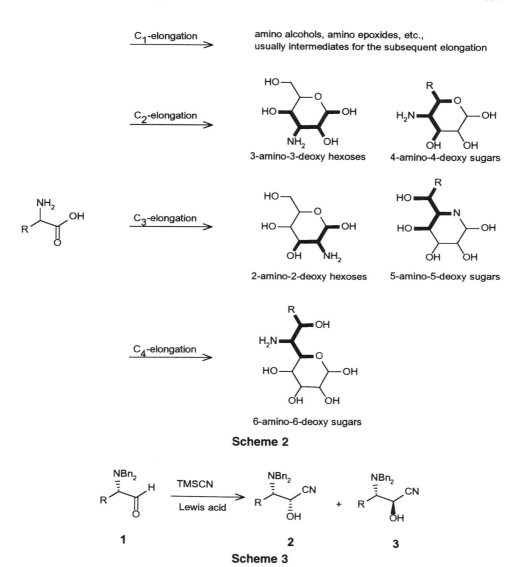

Scheme 2

Scheme 3

*syn*-product, but the diastereoselectivity was low [25]. Higher *syn*-diastereoselectivity was observed in the presence of a stoichiometric amount of zinc chloride [26] (Scheme 4). A similar method was used in the synthesis of L-daunosamine (see Sec. III. A).

Another reagent, recently used for $C_2$-elongation of α-amino aldehydes, is iso-

Scheme 4

cyanoacetic ethyl ester **8**. When reacted with various *N,N*-dibenzyl-protected aldehydes **1** in the presence of sodium cyanide, ester **8** produced diastereoisomeric oxazines **9** and **10**. Subsequent ring-opening reaction with triethylamine afforded a pair of α,γ-diamino-β-hydroxy esters **11** and **12**, with *syn*-configuration of the newly formed chiral centers (98:2 d.e.) and predominating *anti*-configuration relative to the original chiral carbon [27] (Scheme **5**).

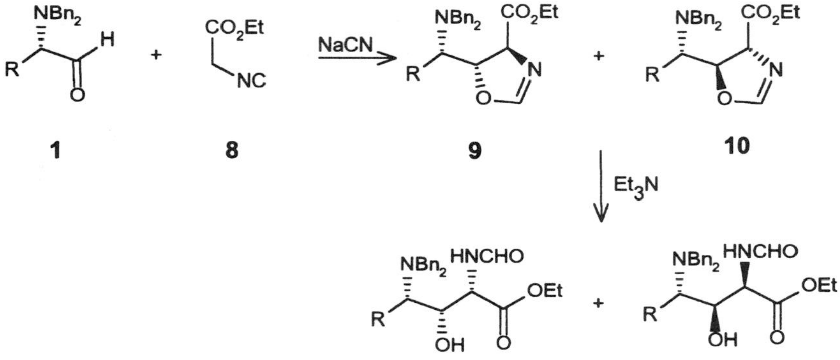


1 8 9 10

11 15 : 85 12

**Scheme 5**

Wittig and Wittig-like reactions are frequently used for $C_2$-elongation of α-amino aldehydes. For example, (carbethoxymethyl)triphenylphosphorane **14** was recently used in preparation of the intermediate **15** for enantioselective synthesis of (+)-allokainic acid [28] (Scheme **6**). A similar method was used in the synthesis of the amino hexose, *N*-acetyl-L-tolyposamine [29].

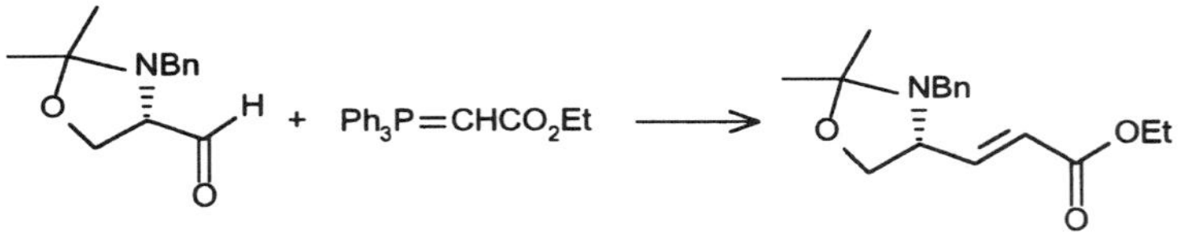


13 14 15

**Scheme 6**

## C. $C_3$-Elongation

Recently, 2-thiazolylcarbonylmethylenetriphenylphosphorane **17** was introduced by Dondoni [9,10,30] as a convenient reagent for $C_3$-elongation of α-hydroxy and α-amino aldehydes. Such a transformation is outlined in Scheme **7**. The product **18** could be further

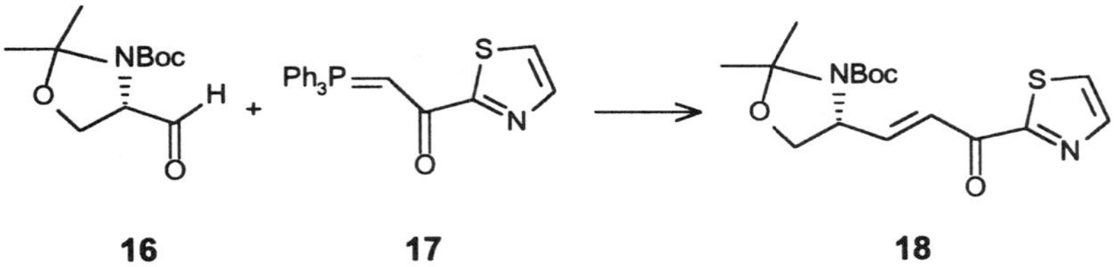


16 17 18

**Scheme 7**

functionalized under stereocontrolled conditions, to afford 3-deoxynojirimycin and 3-deoxymannojirimycin [31,32]. A similar method was used in the synthesis of a series of 4-amino-4-*C*-methyl-L-hexoses [33,34].

Vara Prasad and Rich [35] examined the diastereoselectivity of addition of allylic organometallics **19** to N,N-diprotected α-amino aldehydes **4** (Scheme 8).

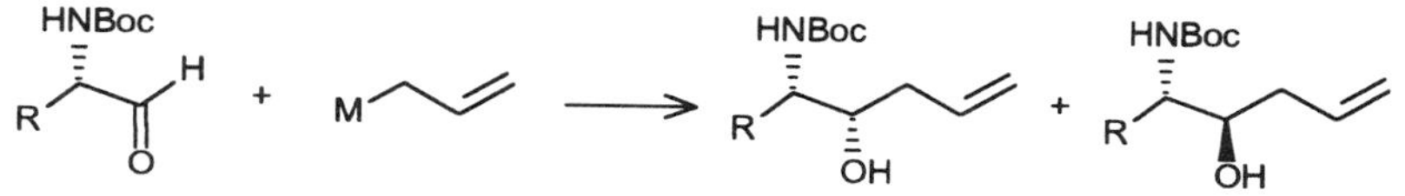

**Scheme 8**

N-Monoprotected aldehydes reacted according to the non–chelation-controlled model. The best diastereocontrol (*syn/anti*, 20.6:1) was achieved with allyltrimethylsilane in the presence of tin tetrachloride. A similar method was used in the synthesis of the aminoglycosidic fragment of calicheamycin $\gamma_1'$ [36].

Three-carbon elongation of amino acid derivatives can also be achieved by addition of lithioacetylenes to an α-amino aldehyde. The addition proceeds according to the chelation and nonchelation models. Diprotected serinal **16** reacts with **22**, giving rise to *anti*-diastereoisomer **24** as a predominating product [37,38] (Scheme 9).

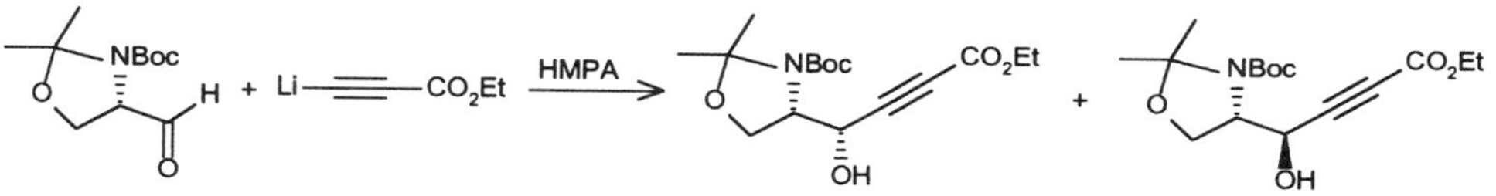

**Scheme 9**

This method was used in the synthesis of thymine polyoxin C [38]. A similar method was also used in the synthesis of 5-aminohexofuranoses (see Sec. III.B).

## D.  $C_4$-Elongation

In this section we refer to types of addition to the carbonyl group, which by their very nature lead to $C_4$-elongation. Examples are found in the addition of furan derivatives and [4+2] cycloaddition. We have recently described [39] the stereoselective addition of 2-furyllithium **26** to N,N-diprotected alaninals. For example, the reaction of **26** with alaninal **25** led to a mixture of diastereoisomers *syn*-**27** and *anti*-**28**, with predominance of the latter isomer (Scheme 10). A similar method was used in the synthesis of methyl α-D-lincosaminide (see Sec. III.F).

2-(Trimethylsilyloxy)furan **29** was used by Casiraghi and co-workers [40,41] for the

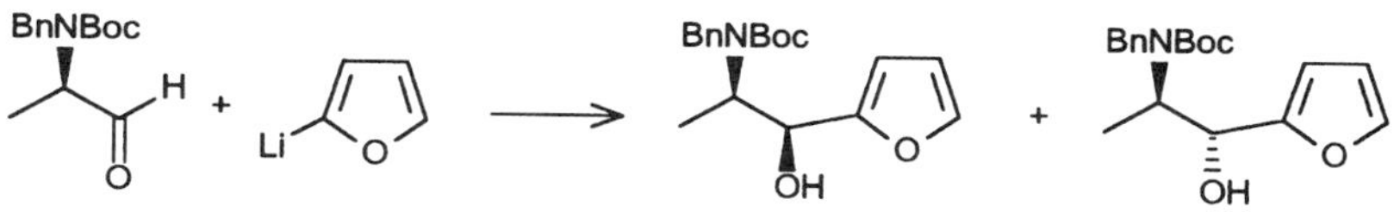

**Scheme 10**

16 29 30 94 : 6 31

**Scheme 11**

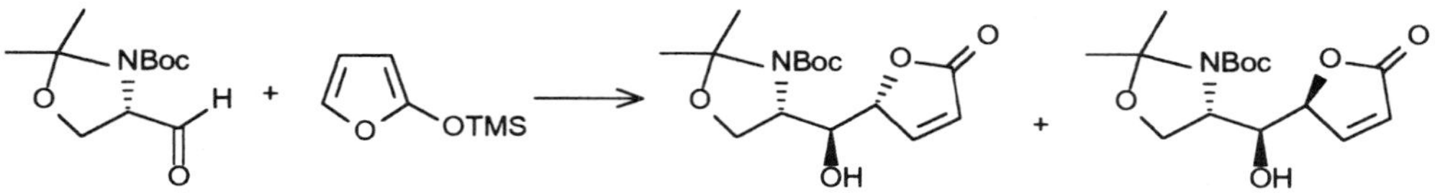

preparation of butenolides **30** and **31** from serinal **16** (Scheme **11**). The reaction proceeds with high stereoselectivity, and two chiral centers are generated simultaneously. The hydroxy group generated in **30** and **31** becomes exclusively *anti* to the inducing center of the aldehyde **16**. The C–O of the butenolide unit in the major product **30** assumes mainly the *anti*-configuration to the hydroxy group (i.e., the compound **30** has an *anti–anti* configuration).

The hetero-Diels–Alder reaction of activated butadienes with carbonyl compounds is a convenient method for the preparation of precursors of sugars. Up to three chiral centers are created simultaneously. The high-pressure [4 + 2]cycloaddition of 1-methoxybuta-1,3-diene **32** to N-mono- and N,N-diprotected alaninals was investigated [42–45]. The Eu(fod)$_3$-mediated reaction of **32** with alaninal **25** gave a mixture of four diastereoisomers, which was then subjected to acidic isomerization, leading to the thermodynamically more stable pair of adducts *syn*-**33** and *anti*-**34**, with predominance of the latter isomer (Scheme **12**). The N-monoprotected alaninals reacted with a moderate *syn*-diastereoselectivity. This method was used in the synthesis of purpurosamines (see Sec. III.C).

25 32 33 6 : 94 34

**Scheme 12**

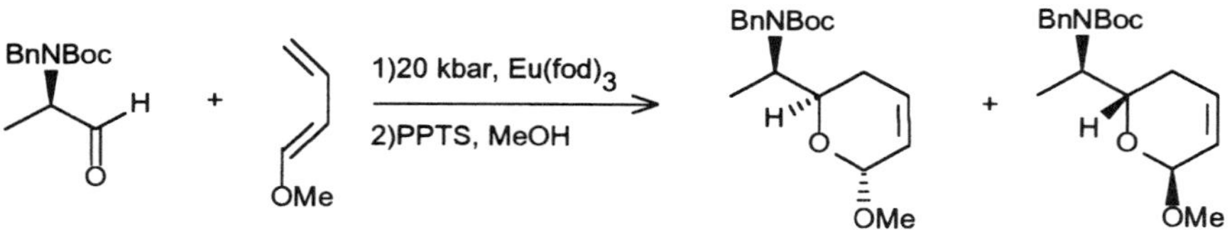

The cyclocondensation of Danishefsky's diene **35** with alaninals of type **25**, contrary to diene **32**, requires neither elevated pressure nor high temperature. The zinc bromide-mediated reaction of **35** with alaninal **25** was followed by acidic workup, resulting in removal of both the trimethylsilyl protection and the ethoxy group [45,46] (Scheme **13**).

25 35 36 7 : 93 37

**Scheme 13**

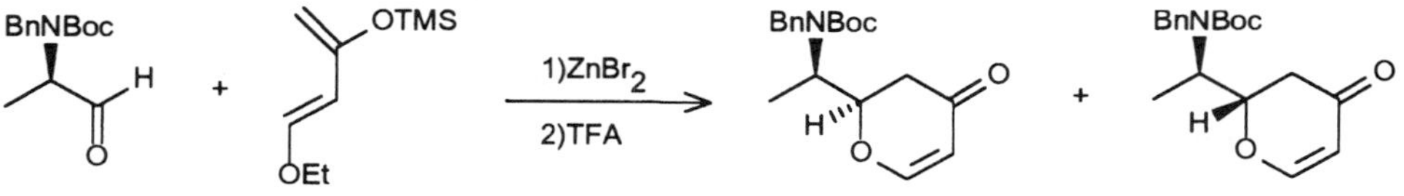

The stereochemical results were similar to those observed for the reaction shown in Scheme **12**. Similar methods were used in the syntheses of anhydrogalantinic acid and a diastereoisomer of lincosamine (see Sec. III. D and E).

As can be seen from the presented set of methods, α-amino acids are versatile

# Synthesis of Amino Sugars

chirons, widely recognized, and easily accessible from natural sources. We feel that the near future will bring many more examples of synthetic applications of this type of chirons.

## III. EXPERIMENTAL PROCEDURES*

### A. Synthesis of L-Daunosamine [47] (Scheme 14)

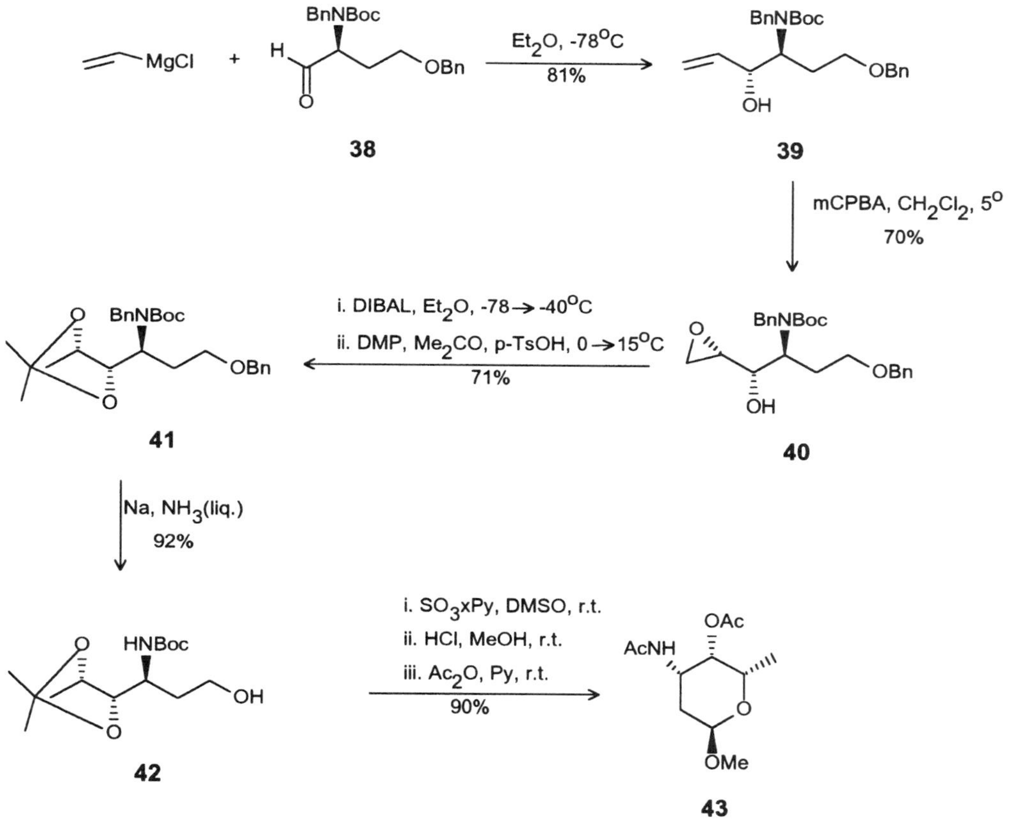

**Scheme 14**

*Compound 39*

L-Homoserinal **38** (800 mg, 2.09 mmol), prepared from L-aspartic acid in a nine-step sequence, was dissolved in 20 mL of dry $Et_2O$. The solution was cooled under argon to $-78°C$, and vinyl magnesium chloride in tetrahydrofuran (THF; 15%, 1.5 mL, 2.5 mmol) was added dropwise. The reaction mixture was stirred at the same temperature for 1 h. Water (125 mL) was then added, and the organic phase was separated. The aqueous solution was extracted with $Et_2O$ (2 × 50 mL), and combined organic layers were washed with water (2 × 30 mL), brine (30 mL), and dried ($MgSO_4$). Evaporation of solvents and flash

---

*Optical rotations were measured at 25°–27°C.

chromatography (hexanes–EtOAc, 85:15) afforded the product **39** (695 mg, 81%) as a colorless oil: $[\alpha]_D$ − 8.2° (*c* 1, CHCl$_3$).

*Compound 40*

Olefin **39** (414 mg, 1.01 mmol) was dissolved in dry CH$_2$Cl$_2$ (10 mL), and mCPBA (85%, 408 mg, 2 mmol) was added. The reaction flask was allowed to stay at 5°C for 4 days, then the reaction mixture was transferred into a separatory funnel, diluted with Et$_2$O (100 mL), and washed with saturated NaHCO$_3$ (3 × 20 mL) and brine (20 mL). The organic layer was dried (MgSO$_4$), concentrated in vacuo, and the residue was purified by flash chromatography (hexanes–EtOAc, 85:15) to give the product **40** (300 mg, 70%) as a colorless oil: $[\alpha]_D$ − 12.3° (*c* 0.7, CHCl$_3$).

*Compound 41*

The epoxide **40** (210 mg, 0.49 mmol) was dissolved in dry Et$_2$O (10 mL). The solution was cooled under argon to −78°C and DIBAL (1.5 *M* toluene solution, 1 mL, 1.5 mmol) was added dropwise. Reduction was carried out at −40°C for 1.5 h. An excess of hydride was decomposed with MeOH and saturated sodium–potassium tartrate (20 mL) was added. After 1.5 h of vigorous stirring, the organic layer was separated, and the aqueous layer was extracted with Et$_2$O (2 × 20 mL). Combined organic layers were washed with water (3 × 20 mL) and brine (20 mL), dried (MgSO$_4$), and concentrated in vacuo. The residue containing crude 2,3-diol was dissolved in acetone (5 mL), and 2,2-dimethoxypropane (DMP; 1.25 mL, 1 mmol) with catalytic amount of *p*-toluenesulfonic acid was added. The reaction mixture was maintained at room temperature for 45 min, diluted with Et$_2$O (50 mL), and washed with saturated NaHCO$_3$ (3 × 20 mL) and brine (20 mL), dried (MgSO$_4$), and concentrated in vacuo. Flash chromatography (hexanes–EtOAc, 85:15) of the residue afforded the product **41** (166 mg, 71%) as a colorless oil: $[\alpha]_D$ − 32.2° (*c* 0.8, CHCl$_3$).

*Compound 42*

Compound **41** (85 mg, 0.18 mmol) was dissolved in dry THF (0.5 mL). The solution was cooled under argon to −40°C, and the reaction flask was protected with a condenser containing dry ice. Liquid ammonia was added slowly (ca. 2 mL), and then a few small pieces of sodium metal to the stable violet color of the reaction mixture. After 15 min of stirring, the reaction was quenched with saturated NH$_4$Cl (1 mL). The cooling bath and condenser were removed, and the mixture was allowed to warm up to room temperature without any extra heating. The residue was diluted with Et$_2$O (20 mL), washed with water (3 × 5 mL), brine (5 mL), dried (MgSO$_4$), and concentrated in vacuo. Flash chromatography (hexanes–EtOAc, 6:4) of the residue afforded product **42** (48 mg, 92%) as a colorless oil: $[\alpha]_D$ − 18.8° (*c* 1, CHCl$_3$).

*Compound 43*

Alcohol **42** (55 mg, 0.19 mmol) was dissolved in DMSO (1.5 mL), and Et$_3$N (130 μL, 0.95 mmol) and SO$_3$–pyridine (120 mg, 0.76 mmol) in DMSO (0.5 mL) were added. Oxidation was carried out at room temperature for 30 min. The reaction mixture was transferred into a separatory funnel, diluted with Et$_2$O (20 mL), washed with saturated NaHCO$_3$ (3 × 10 mL), brine (10 mL), and dried (MgSO$_4$). Solvents were evaporated in vacuo, the residue was dissolved in 0.1 *N* methanolic HCl (6 mL), and the mixture was maintained at room temperature for 12 h. Solvents were carefully evaporated in vacuo, and the oily residue was treated with pyridine (0.5 mL) and Ac$_2$O (100 μL, 1 mmol). The reaction was carried out at room temperature for 12 h. The mixture was then diluted with Et$_2$O (20 mL) and pyridine was washed out with 1 N HCl (2 × 10 mL). The organic layer was neutralized with saturated NaHCO$_3$ (2 × 10 mL), washed with brine (10 mL), and dried (MgSO$_4$). Evaporation and flash chromatography (hexanes–EtOAc–MeOH, 6:4:1) of the residue afforded product **43** (42 mg, 90%): mp 186°–187°C, [α]$_D$ − 207° (c 1.3, CHCl$_3$); reported [48] mp 188°–189°C, [α]$_D$ − 202° (c 1, CHCl$_3$).

### B. Synthesis of α-Methyl Mannofuranoside and β-Methyl Allofuranoside [38] (Scheme 15)

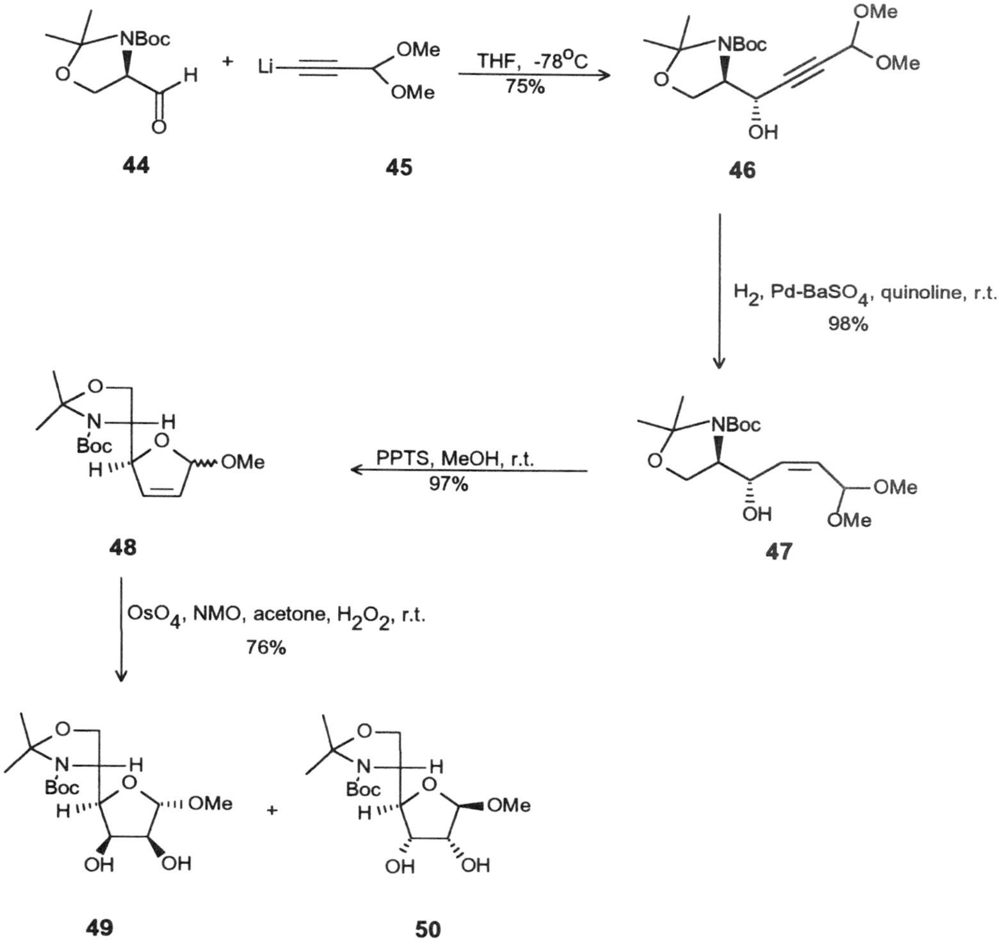

**Scheme 15**

### Compound 46

To a stirred solution of propiolaldehyde dimethyl acetal **45** (13.0 g, 0.13 mol) in dry THF (225 mL) at −78°C was added slowly n-BuLi (2.3 M in hexanes, 47 mL, 0.11 mol) under nitrogen. The suspension was stirred at −78°C for 1 h, and then to this solution was added slowly a solution of aldehyde **44** (13.5 g, 0.0591 mol) in dry THF (75 mL). After stirring at −78°C for 2 h, the resulting solution was slowly poured into ice-cold 1 M NaH$_2$PO$_4$ (2 L, pH 7) with swirling. The mixture was extracted with Et$_2$O (3 × 1 L), washed with brine (1 L), dried (MgSO$_4$), and concentrated in vacuo. Flash chromatography (hexanes–EtOAc, 12:1) of the residue afforded the *erythro*-product **46** (14.5 g, 75%) as an oil: [α]$_D$ + 52° (c 1.2, CHCl$_3$), followed by the *threo*-product (2.38 g, 12%).

*Compound 47*

To a solution of compound **46** (8.38 g, 25.5 mmol) in dry benzene (180 mL), were added synthetic quinoline (1.5 g, 8.9 mmol) and reduced 5% Pd–BaSO$_4$ (2.68 g, 134 mg of Pd, 1.26 mmol). The black suspension was stirred under hydrogen atmosphere for 2 h at room temperature. Then the catalyst was filtered off, washing with benzene, and the filtrate (350 mL) was concentrated in vacuo. Flash chromatography (hexanes–EtOAc, 4:1) of the residue afforded the pure product **47** (8.229 g, 98%) as a colorless oil: [α]$_D$ + 34.2° (c 0.86, CHCl$_3$).

*Compound 48*

To a solution of the olefin **47** (4.2 g, 12.7 mmol) in MeOH (180 mL) was added solid PPTS (326 mg, 1.33 mmol). The solution was stirred at room temperature for 2.5 h. The reaction mixture was poured into a half-saturated NaHCO$_3$ solution (500 mL), extracted with CH$_2$Cl$_2$ (500 mL), and washed with brine (500 mL). Each aqueous layer was reextracted with CH$_2$Cl$_2$ (2 × 500 mL), and all organic layers were combined, dried (MgSO$_4$), and concentrated in vacuo. Flash chromatography (hexanes–EtOAc, 4:1) afforded a 1:1 mixture of anomers **48** (3.7 g, 97%) as a colorless oil: [α]$_D$ − 12.2° (c 0.96, CHCl$_3$).

*Compounds 49 and 50*

To a solution of a mixture of anomers **48** (3.65 g, 12.2 mmol) in acetone (80 mL) was added a stock solution of OsO$_4$ + NMO in water (90 mL, 0.008 M in OsO$_4$, 1.2 M in NMO). The solution was stirred at room temperature for 12 h. The reaction mixture was poured into saturated Na$_2$SO$_3$ solution (500 mL) and extracted with EtOAc (500 mL). The organic layer was washed with a pH-4 buffer solution (500 mL) and brine (500 mL). Each aqueous layer was reextracted with EtOAc (2 × 500 mL), and all organic layers were combined, dried (MgSO$_4$), and concentrated in vacuo. Flash chromatography (hexanes–EtOAc, 1:1) of the residue afforded α-methyl mannofuranoside **49** (1.8 g, 44%) as a colorless oil: [α]$_D$ + 97.8° (c 1.13, CHCl$_3$); and β-methyl allofuranoside **50** (1.3 g, 32%) as a white solid: mp 120°–121°C, [α]$_D$ − 55° (c 1.47, CHCl$_3$).

### C. Synthesis of 6-Epi-D-Purpurosamine B [49] (Scheme 16)

*Compound 52*

A solution of 1- methoxybuta-1,3-diene **32** (1.66 g, 20 mmol), aldehyde **51** (1.73 g, 10 mmol), and Eu(fod)$_3$ (0.104 g, 0.1 mmol) in Et$_2$O (6 mL) was charged into a Teflon ampoule, which was then placed in a high-pressure vessel filled with pentane as a transmission medium. The pressure was slowly elevated to 20 kbar at 50°C. After stabilization of pressure, the reaction mixture was kept under these conditions for 20 h. After cooling and decompression, the solvent was evaporated and the residue was filtered through a short

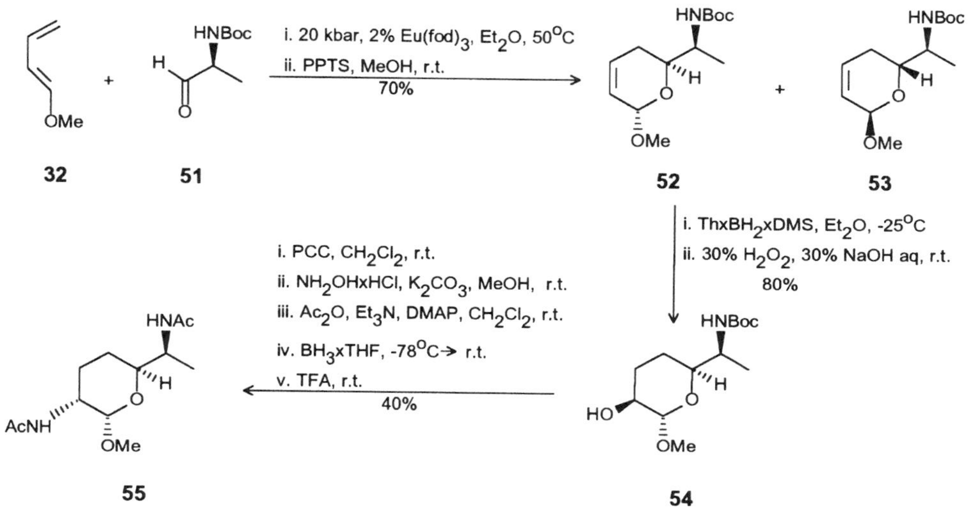

# Synthesis of Amino Sugars

## Scheme 16

silica gel pad (hexanes–EtOAc, 8:2). The filtrate was evaporated to dryness and the residue was dissolved in MeOH (20 mL); to this solution was added PPTS (0.25 g, 1 mmol). The *cis–trans* isomerization was carried out at room temperature for 20 h, then solid NaHCO$_3$ (92 mg, 1.1 mmol) was added, and the mixture was stirred for 1 h. The solvent was evaporated and the residue was treated with Et$_2$O (10 mL). The precipitated inorganic salts were filtered off and the crude mixture was purified by flash chromatography (hexanes–acetone, 95:5 → 9:1) to afford the analytically pure product **52** (1.25 g, 47%) as an oil: $[\alpha]_D$ − 28.5° (*c* 3, CHCl$_3$); and its diastereoisomer **53** (0.62 g, 23%) as an oil: $[\alpha]_D$ − 46.2° (*c* 1.5, CHCl$_3$).

## Compound 54

To a solution of 2,3-dimethyl-2-butene (420 mg, 5 mmol) in Et$_2$O (20 mL), BH$_3$·Me$_2$S (0.35 mL, 3.5 mmol) was added at −5°C. The mixture was stirred at 0°C, then it was cooled to −25°C, and the adduct **2** (514 mg, 2 mmol), dissolved in Et$_2$O (1.5 mL), was added. The reaction mixture was kept at −25°C for 3 h, and the excess borane was decomposed with MeOH (10 mL), followed by addition of a mixture of 30% H$_2$O$_2$ and 30% aqueous NaOH (2 mL, 1:1 v/v). The temperature was raised to 20°C, and stirring was continued for additional 1 h. The postreaction mixture was extracted with EtOAc (3 × 20 mL), the combined extracts were washed with water, dried (MgSO$_4$), and evaporated. Flash chromatography (hexanes–acetone, 8:2 → 7:3) afforded the product **54** (440 mg, 80%) as an oil: $[\alpha]_D$ + 42.3° (*c* 1, CHCl$_3$).

## Compound 55

To a solution of alcohol **54** (275 mg, 1 mmol) in CH$_2$Cl$_2$ (5 mL) were added pyridinium chlorochromate (PCC; 647 mg, 3 mmol) and freshly dried 4-Å molecular sieves (1 g). The heterogeneous mixture was stirred at room temperature for 1.5 h; then Et$_2$O (30 mL) was added, and the postreaction mixture was filtered through a short silica gel pad. The filtrate was evaporated, and the residue was dissolved in dry MeOH (5 mL). Solid hydroxylamine hydrochloride (139 mg, 2 mmol) and solid $K_2CO_3$ (276 mg, 2 mmol) were added in one portion, and the whole mixture was stirred at room temperature for 16 h. Then MeOH was evaporated and the residue was dissolved in $CH_2Cl_2$ (5 mL). To this solution were added $Et_3N$ (202 mg, 2 mmol), $Ac_2O$ (153 mg, 1.5 mmol), and a crystal of DMAP, and the reaction mixture was stirred at room temperature for 15 min. After evaporation of solvents, the oily residue was dissolved in THF (5 mL) and the solution was cooled to $-50°C$. The $BH_3$·THF complex (2 mL of 1M solution) was added, the reaction mixture was stirred at $-50°C$ for 8 h, and then at room temperature for additional 16 h. An excess of borane was decomposed with MeOH (2 mL), solvents were evaporated, and the residue was dissolved in trifluoroacetic acid (TFA; 1 mL). The solution was stirred at room temperature for 1 h, then TFA was evaporated, and to the residue, dissolved in $CH_2Cl_2$ (2 mL), were added $Et_3N$ (151 mg, 1.5 mmol) and $Ac_2O$ (112 mg, 1.1 mmol). The reaction mixture was stirred at room temperature for 1 h, and after evaportion of solvents, a crude product was subjected to flash chromatography ($CHCl_3$–MeOH, 99:1) to afford, after recrystallization from acetone, the product **55** (102 mg, 40%): mp 213°–214°C, $[\alpha]_D + 64°$ (c 0.3, MeOH); reported [50] mp 212°–213°C, $[\alpha]_D + 62.9°$ (c 1, MeOH).

### D. Synthesis of Anhydrogalantinic Acid [51] (Scheme 17)

*Compound 57*

A mixture of aldehyde **56** (1.99 g, 4.07 mmol), diene **35** (2 mL, ca. 8 mmol), and a catalytic amount of $ZnBr_2$ in anhydrous THF (10 mL) was stirred at room temperature overnight. The reaction mixture was diluted with $Et_2O$ (80 mL), washed with saturated $NaHCO_3$ (10 mL) and brine (10 mL), dried ($MgSO_4$), and concentrated in vacuo. The resultant oil was dissolved in $CH_2Cl_2$ and treated with TFA (1 mL) for 5 min. The reaction mixture was then partitioned in a separatory funnel between $Et_2O$ (150 mL) and saturated $NaHCO_3$ (100 mL). The organic layer was washed with brine (50 mL), dried ($MgSO_4$), and concentrated in vacuo. Crystallization of the residue from a mixture of *n*-hexane and $Et_2O$ afforded the product **57** as white crystals (1.89 g, 70%): mp 137°–139°C, $[\alpha]_D - 14.9°$ (c 1, $CHCl_3$).

*Compound 58*

To a solution of pyrone **57** (654 mg, 1.24 mmol) and $CeCl_3$·$7H_2O$ (693 mg, 1.86 mmol) in MeOH (12.5 mL) at $-78°C$, under argon, was added $NaBH_4$ (70.7 mg, 1.86 mmol) in absolute EtOH (2.5 mL). After stirring at $-78°C$ for 1 h, the reaction mixture was allowed to warm to 0°C, whereupon it was diluted with $Et_2O$ (50 mL) and quenched with a pH 7 buffer (25 mL). The reaction mixture was transferred into a separatory funnel, and the aqueous layer was extracted with $Et_2O$ (3 × 25 mL). The organic layers were combined, dried ($MgSO_4$), and concentrated in vacuo to afford the crude alcohol (656 mg, 100%). This only product was dissolved in $CH_2Cl_2$ (12.5 mL), and $Et_3N$ (505 mg, 5 mmol), followed by $Ac_2O$ (204 mg, 2 mmol), and a catalytic amount of DMAP were added. After 5 min, solvents were removed in vacuo. Flash chromatography (hexanes–EtOAc, 8:2), followed by crystallization from a mixture of *n*-hexane and $Et_2O$, afforded the product **58** (602 mg, 85%): mp 131°–132°C, $[\alpha]_D - 14.4°$ (c 1, $CHCl_3$).

*Compound 59*

Acetate **58** (571 mg, 1 mmol) was dissolved in dioxane (50 mL), and 5 m$M$ $H_2SO_4$ (200 mL) and $HgSO_4$ (592 mg, 2 mmol) were added. The reaction mixture was vigorously stirred for

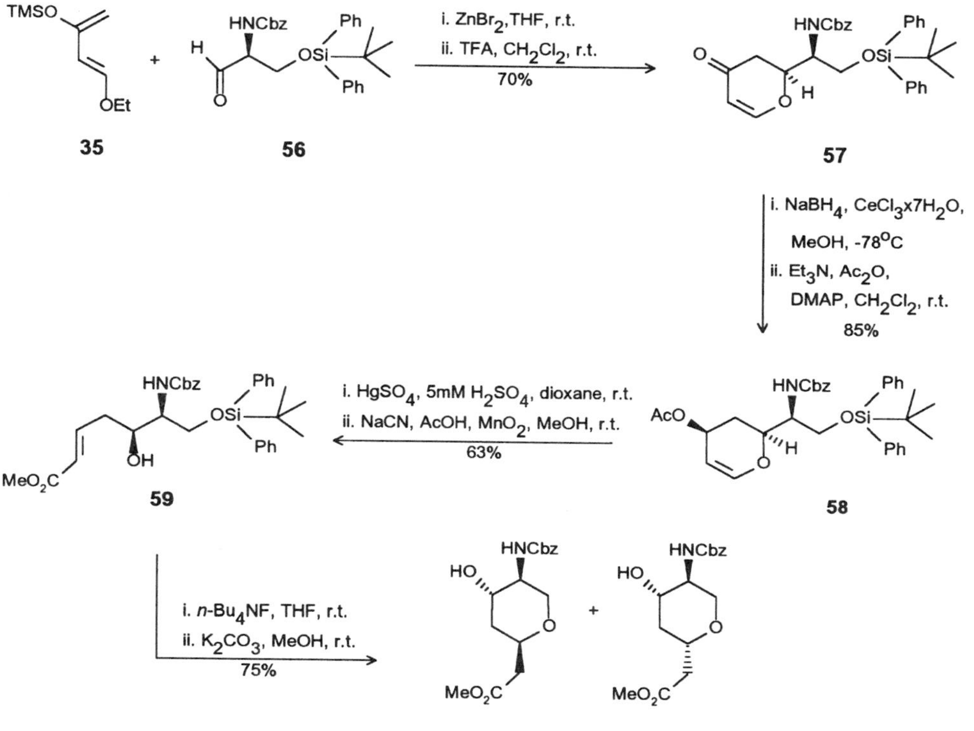

# Synthesis of Amino Sugars

**Scheme 17**

48 h. Then NaHCO$_3$ (252 mg, 3 mmol) was added, and after 10 min the product was extracted with Et$_2$O (3 × 200 mL). Combined extracts were washed with brine (100 mL), dried (MgSO$_4$), and concentrated in vacuo, to afford the crude unstable aldehyde (475 mg, 90%), which was immediately used in the next transformation.

The aldehyde (475 mg, 0.9 mmol) was dissolved in MeOH (100 mL), and NaCN (220.5 mg, 4.5 mmol) was added. Oxidation was carried out over 2 days. Then the reaction mixture was filtered through Celite, and the filtrate was concentrated in vacuo. The oily residue was partitioned between water (50 mL) and Et$_2$O (150 mL). The organic layer was washed with brine (50 mL), dried (MgSO$_4$), and concentrated in vacuo. Flash chromatography (hexanes–EtOAc, 7:3) afforded the product **59** (393 mg, 70%) as an oil: $[\alpha]_D$ + 2.5° (c 1, CHCl$_3$).

## Compounds 60 and 61

Ester **59** (143 mg, 0.255 mmol) was dissolved in THF (4 mL), and then n-Bu$_4$NF (0.5 mL of 1 M solution in THF, 0.5 mmol) was added. After 15 min of stirring, the solvent was evaporated and the residue was dissolved in MeOH (25 mL). Then anhydrous K$_2$CO$_3$ (7 mg, 0.05 mmol) was added and, after 15 h of stirring at room temperature, the solvent was evaporated in vacuo. Flash chromatography (hexanes–EtOAc, 7:3) of the residue afforded the product **60** (31 mg, 38%): mp 126°–127°C, $[\alpha]_D$ +2.6° (c 1.1, CHCl$_3$), followed by its epimer **61** (30 mg, 37%): mp 123°–124°C, $[\alpha]_D$ +12.4° (c 2.5 CHCl$_3$).

## E. Synthesis of the Diastereoisomer of Lincosamine [52] (Scheme 18)

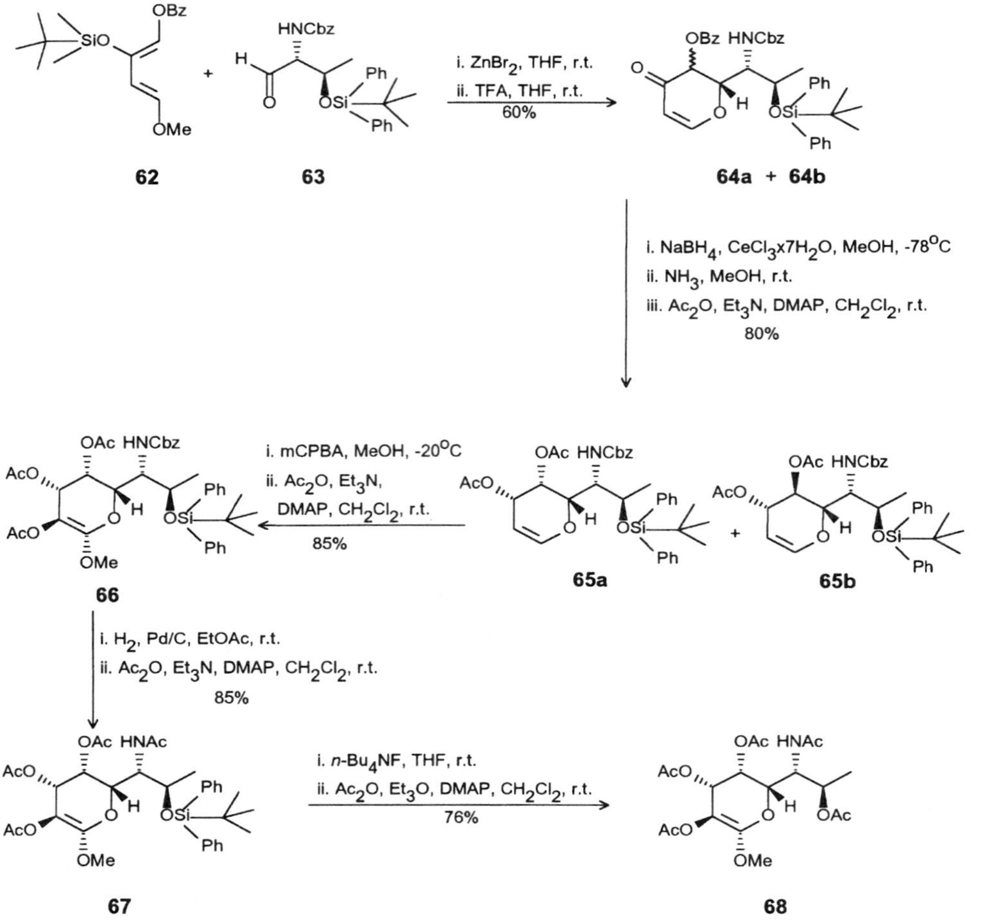

**Scheme 18**

*A Mixture of Diastereoisomers 64a and 64b*

Aldehyde **63** (950 mg, 2 mmol) and anhydrous ZnBr$_2$ (450 mg, 2 mmol) in dry THF (10 mL) were stirred at room temperature for 1 h. To the homogeneous solution was added diene **62** (1.3 g, 4 mmol), and the reaction mixture was stirred at room temperature for 48 h. TFA (1 mL) was then added, and after an additional 5-min stirring, the solution was partitioned between Et$_2$O (30 mL) and a saturated NaHCO$_3$ solution (30 mL). The organic layer was dried (MgSO$_4$), and concentrated in vacuo. Flash chromatography (hexanes–EtOAc, 6:4) of the residue afforded an unseparable mixture (2:1) of pyrones **64a** and **64b** (800 mg, 60%) as an oil.

*Compound 65a*

To a solution of pyrones **64a** and **64b** (2:1, 660 mg, 1 mmol) and CeCl$_3$·7H$_2$O (558.8 mg, 1.5 mmol) in MeOH (10 mL) was added at −78°C under argon, a solution of NaBH$_4$ (56.7 mg,

## Synthesis of Amino Sugars

1.5 mmol) in absolute EtOH (2 mL). After stirring at −78°C for 1 h, the reaction mixture was allowed to warm up to 0°C, whereupon it was diluted with Et$_2$O (40 mL) and quenched with a pH 7 buffer solution (20 mL). The postreaction mixture was transferred into a separatory funnel and the aqueous layer was extracted with Et$_2$O (3 × 20 mL). The organic layers were combined, dried (MgSO$_4$), and evaporated in vacuo. The oil residue was dissolved in MeOH saturated with ammonia, and the reaction mixture was left overnight at room temperature. Then the solvent was evaporated, the residue was dissolved in CH$_2$Cl$_2$ (10 mL), and Et$_3$N (202 mg, 2 mmol) was added, followed by Ac$_2$O (153 mg, 1.5 mmol) and a catalytic amount of DMAP. After 30 min, the solvent was evaporated in vacuo. Flash chromatography (hexanes–EtOAc, 85:15 → 8:2) of the residue afforded the product **65a** (340 mg, 53%) as an oil: [α]$_D$ + 1.5° (c 4, CHCl$_3$); and its diastereoisomer **65b** (170 mg, 27%), [α]$_D$ + 4.2° (c 2, CHCl$_3$).

*Compound 66*

To a cold solution (−20°C) of diacetate **65a** (130 mg, 0.2 mmol) in MeOH (1 mL), solid mCPBA (85%, 120 mg, 0.6 mmol) was added. After 7 days at room temperature, MeOH was evaporated, and the residue was acetylated under standard conditions (see procedure for compound **65a**). After completion of the reaction, the solvent was evaporated in vacuo. Flash chromatography (hexanes–EtOAc, 7:3 → 6:4) of the residue afforded the product **66** (124 mg, 85%): mp 51°–53°C, [α]$_D$ + 11.3° (c 3, CHCl$_3$).

*Compound 67*

Compound **66** (74 mg, 0.1 mmol) dissolved in MeOH (5 mL) was reduced with hydrogen in the presence of 10% palladium–charcoal (15 mg) for 1 h; the catalyst was then filtered off and the solvent was evaporated. The residue was acetylated under standard conditions (see procedure for compound **65a**) to afford the product **67** (55 mg, 85%): mp 197°–199°C, [α]$_D$ + 22.5° (c 2.5, CHCl$_3$).

*Compound 68*

Compound **67** (45 mg, 0.07 mmol) was treated with n-Bu$_4$NF (100 mg, 0.3 mmol) in THF (0.5 mL). After 15 h, the solvent was evaporated, and the oily residue was acetylated under standard conditions (see procedure for compound **65a**). Flash chromatography (hexanes–EtOAc, 1:9) afforded the product **68** (23 mg, 76%): mp 184.5°C, [α]$_D$ + 54.8° (c 1.1, CHCl$_3$).

### F. Synthesis of Methyl α-D-Lincosaminide [53,54] (Scheme 19)

*Compound 70*

To a cold (−30°C) solution of furan (1.8 g, 27.5 mmol) in Et$_2$O (10 mL) was added, under argon, the solution of n-BuLi in Et$_2$O (1.27 M, 4.8 mL, 6 mmol). The reaction mixture was allowed to reach room temperature, stirred for 1 h, and cooled to −78°C. The solution of aldehyde **69** (1.17 g, 4.1 mmol) in THF–n-hexane (20 mL, 1:1 v/v) was added dropwise and the stirring was continued for 2 h. The reaction mixture was washed with saturated aqueous NH$_4$Cl (10 mL), water (10 mL), brine (10 mL), and dried (MgSO$_4$), then concentrated in vacuo. Flash chromatography (hexanes–Et$_2$O–CH$_2$Cl$_2$, 10:2:1) of the oily residue afforded

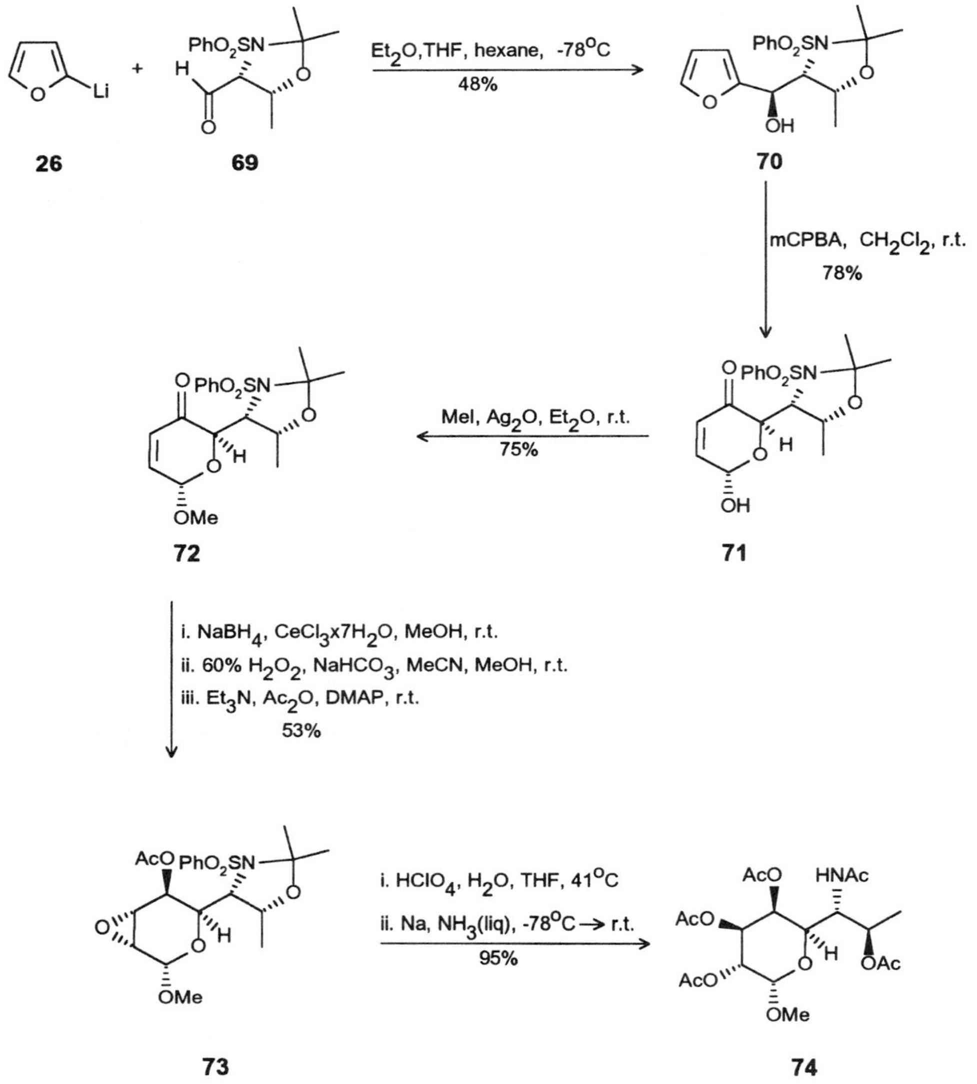

**Scheme 19**

the pure product **70** (0.63 g, 48%): mp 81.5°–82.5°C (from *n*-hexane–Et$_2$O), [α]$_D$ + 40.5° (*c* 0.5, CHCl$_3$), followed by its *syn*-diastereoisomer.

*Compound 71*

To a solution of compound **70** (3.98 g, 11.3 mmol) in CH$_2$Cl$_2$ (70 mL) were added, at room temperature, solid NaHCO$_3$ (5.0 g) and mCPBA (85%, 5.0 g). After allowing the mixture to stir for 24 h, two portions of mCPBA (2 × 0.5 g) were added and stirring was continued until the disappearance of the substrate (TLC). The reaction mixture was diluted with CH$_2$Cl$_2$ (450 mL), washed successively with water (50 mL), saturated NaHSO$_3$ (50 mL), NaHCO$_3$ (50 mL), water (50 mL), brine (50 mL), and dried (MgSO$_4$), then concentrated in vacuo. Crystallization of the residue from *n*-hexane–EtOAc, followed by flash chromatography of the mother liquors, afforded the product **71** (3.26 g, 78%): mp 163.5°–165°C, $[\alpha]_D$ + 41.7° (c 0.2, CHCl$_3$).

*Compound 72*

To a solution of ulose **71** (2.58 g, 7 mmol) in dry Et$_2$O (150 mL) were added iodomethane (18 mL, 289 mmol) and Ag$_2$O (5.0 g, 21.6 mmol). After stirring at room temperature for 24 h, the mixture was filtered through a Celite pad, and concentrated in vacuo. Crystallization of the residue from *n*-hexane–Et$_2$O afforded the product **72** (2.0 g, 75%): mp 148°–151°C, $[\alpha]_D$ + 91.8° (c 1, CHCl$_3$).

*Compound 73*

To a mixture of CeCl$_3$·7H$_2$O (745.2 mg, 2 mmol) in MeOH (5 mL) and uloside **72** (762.4 mg, 2 mmol) at 15°C, was added solid NaBH$_4$ (80 mg, 2 mmol) in small portions. The mixture was left for 30 min, diluted with water (5 mL) and extracted with Et$_2$O (3 × 50 mL). The combined organic layers were washed with water (10 mL), saturated NH$_4$Cl (10 mL), and brine (10 mL), dried (MgSO$_4$), and concentrated in vacuo. Flash chromatography (hexanes–EtOAc, 7:3) afforded the appropriate alcohol (495 mg, 65%) as a colorless thick oil.

A solution of the resulting alcohol (245 mg, 0.64 mmol) in MeOH (25 mL) was stirred with a mixture of acetonitrile (1 mL), 60% H$_2$O$_2$ (1 mL), and NaHCO$_3$ (150 mg) overnight at room temperature. Then, acetonitrile (0.2 mL) and 60% H$_2$O$_2$ (0.2 mL) were added, and the reaction mixture was stirred for an additional 20 h, poured into water (10 mL), and extracted with Et$_2$O (3 × 30 mL). The combined organic layers were washed with water (5 mL), aqueous 1 *M* Na$_2$S$_2$O$_3$ (5 mL), water (5 mL), brine (5 mL), and dried (MgSO$_4$). Evaporation of solvents gave the crude anhydro compound, which was acetylated under standard conditions (Ac$_2$O–pyridine–DMAP) to afford the product **73** (250 mg, 82%): mp 149°–150°C (from Et$_2$O), $[\alpha]_D$ + 26.1° (c 0.5, CHCl$_3$).

*Compound 74*

To a stirred solution of acetate **73** (53.8 mg, 0.122 mmol) in THF (0.5 mL) was added 1 *M* perchloric acid (0.2 mL), and stirring was continued at 41°C for 24 h. The reaction mixture was then brought to pH 7 with saturated NaHCO$_3$, and EtOAc (50 mL) was added. The solution was dried (MgSO$_4$), filtered, and evaporated to give the crude alcohol (45 mg, 98%) as colorless gum.

To a stirred liquid ammonia (ca. 2 mL) at −78°C was added by syringe a solution of the alcohol (50.1 mg, 0.133 mmol) in THF (0.5 mL) and two small lumps of sodium. The blue solution was stirred for 5 min, the excess of sodium was neutralized with NH$_4$Cl (until the colorless solution was obtained), and ammonia was evaporated. The acetylation carried out under standard conditions (Ac$_2$O–pyridine–DMAP) afforded product **74** (57.4 mg, 97%): mp 193°–194.5°C, $[\alpha]_D$ + 151.2° (c 0.5, CHCl$_3$); reported [55], mp 194°–196°C, $[\alpha]_D$ + 148.2° (c 0.5, CHCl$_3$).

# REFERENCES

1. S. Umezawa, Structures and syntheses of aminoglycoside antibiotics, *Adv. Carbohydr. Chem. Biochem. 30*:111 (1974).

2. S. Hanessian, *Total Synthesis of Natural Products. The "Chiron" Approach*, Pergamon Press, Oxford, 1983.
3. G. M. Coppola and H. F. Schuster, *Asymmetric Synthesis: Construction of Chiral Molecules Using Amino Acids*, John Wiley & Sons, New York, 1987.
4. J. Jurczak and A. Gołębiowski, Optically active N-protected α-amino aldehydes in organic synthesis, *Chem. Rev. 89*:149 (1989).
5. M. T. Reetz, New approaches to the use of amino acids as chiral building blocks in organic synthesis, *Angew. Chem. Int. Ed. Engl. 30*:1531 (1991).
6. J. Jurczak and A. Gołębiowski, From α-amino acids to amino sugars, *Studies in Natural Products Chemistry*, Vol. 4., Part C, (Atta-ur-Rahman, ed.), Elsevier, Amsterdam, 1989, p. 111.
7. A. Gołębiowski and J. Jurczak, α-Amino-β-hydroxy acids in the total synthesis of amino sugars, *Synlett* p. 241 (1993).
8. K. Kiciak, U. Jacobsson, A. Gołębiowski, and J. Jurczak, α-Amino acids in the total synthesis of amino sugars, *Pol. J. Chem. 68*:199 (1994).
9. A. Dondoni, Carbohydrate synthesis via thiazoles, *Modern Synthetic Methods* (R. Scheffold, ed.), Verlag Helvetica Chimica Acta, Basel, 1992, p. 377.
10. A. Dondoni and D. Perrone, 2-Thiazolyl α-amino ketones: A new class of reactive intermediates for the stereocontrolled synthesis of unusual amino acids, *Synthesis* p. 1162 (1993).
11. M. T. Reetz, Asymmetric C–C bond formation using organometallic chemistry, *Pure Appl. Chem. 60*:1607 (1988).
12. M. T. Reetz, Metal, ligand and protective group tuning as a means to control selectivity, *Pure Appl. Chem. 64*:351 (1992).
13. H. Waldmann and M. Braun, Amino acid esters as chiral auxiliaries in asymmetric cycloadditions, *Gazz. Chim. Ital. 121*:277 (1991).
14. I. F. Pelyves, C. Monneret, and P. Herczegh, *Synthetic Aspects of Aminodeoxy Sugars of Antibiotics*, Springer Verlag, Heidelberg, 1988.
15. A. Gołębiowski and J. Jurczak, Total synthesis of lincomycin and related chemistry, *Recent Progress in the Chemical Synthesis of Antibiotics*, (G. Lukacs and M. Ohno, eds.), Springer Verlag, Heidelberg, 1990, p. 365.
16. J. Jurczak and A. Gołębiowski, The synthesis of antibiotic amino sugars from α-amino aldehydes, *Antibiotics and Antiviral Compounds, Chemical Synthesis and Modification*, (K. Krohn, H. Kirst, and H. Maas, eds.), VCH, Weinheim, 1993, p. 343.
17. A. Dondoni, G. Fantin, M. Fogagnolo, and P. Pedrini, Stereochemistry associated with the addition of 2-(trimethylsilyl)thiazole to differentially protected α-amino aldehydes. Applications toward the synthesis of amino sugars and sphinogosines, *J. Org. Chem. 55*:1439 (1990).
18. S. Masamune, S. A. Ali, D. L. Snitman, and D. S. Garvey, Highly stereoselective aldol condensation using an enantioselective chiral enolate, *Angew. Chem. Int. Ed. Engl. 19*:557 (1980).
19. S. Masamune, T. Kaiho, and D. S. Garvey, Aldol methodology: Synthesis of versatile intermediates. 3-Hydroxy-2-vinylcarbonyl compounds, *J. Am. Chem. Soc. 104*:5521 (1982).
20. A. Dondoni, G. Fantin, M. Fogagnolo, A. Medici, and P. Pedrini, Iterative, stereoselective homologation of chiral polyalkoxy aldehydes employing 2-(trimethylsilyl)thiazole as a formyl anion equivalent. The thiazole route to higher carbohydrates, *J. Org. Chem. 54*:693 (1989).
21. M. T. Reetz, M. W. Drewes, K. Harms, and W. Reif, Stereoselective cyanohydrin-forming reactions of chiral α-amino aldehydes, *Tetrahedron Lett. 29*:3295 (1988).
22. M. T. Reetz, M. W. Drewes, and A. Schmitz, Stereoselective synthesis of β-amino alcohols from optically active α-amino acids, *Angew. Chem. Int. Ed. Engl. 26*:1141 (1987).
23. M. T. Reetz and J. Binder, Protective group tuning in the stereoselectivite conversion of α-amino aldehydes into aminoalkyl epoxides, *Tetrahedron Lett. 30*:5425 (1989).
24. J. R. Luly, J. F. Dellaria, J. J. Plattner, J. L. Soderquist, and N. Yi, A synthesis of protected aminoalkyl epoxides from α-amino acids, *J. Org. Chem. 52*:1487 (1987).
25. G. J. Hanson and T. Lindberg, Synthesis of new dipeptide analogues containing novel ketovinyl

and hydroxyethylidene isosteres via Grignard addition to chiral α-amino aldehydes, *J. Org. Chem. 50*:5399 (1985).

26. W. J. Thompson, T. J. Tucker, J. E. Schwering, and J. L. Barnes, A stereocontrolled synthesis of *trans*-allylic amines, *Tetrahedron Lett. 31*:6819 (1990).
27. M. T. Reetz, T. Winsch, and K. Harms, Stereoselective synthesis of α, γ-diamino-β-hydroxy amino acid esters: A new class of amino acids, *Tetrahedron Asymm. 1*:371 (1990).
28. A. Barco, S. Benetti, A. Casolari, G. P. Pollini, and G. Spalluto, Enantioselective synthesis of (+)- and (−)-α-allokainic acid, *Tetrahedron Lett. 31*:4917 (1990).
29. G. Guanti, L. Banfi, E. Narisano, and R. Riva, Stereoselective synthesis of *N*-acetyl-L-tolyposamine from (*S*) ethyl β-hydroxybutyrate, *Tetrahedron Lett. 33*:2221 (1992).
30. A. Dondoni and P. Merino, Chemistry of the enolates of 2-acetylthiazole: Aldol reactions with chiral aldehydes to give 3-deoxy aldos-2-uloses and 3-deoxy-2-ulosonic acids. A short total synthesis of 3-deoxy-D-manno-2-octulosonic acid (KDO), *J. Org. Chem. 56*:5294 (1991).
31. A. Dondoni, G. Fantin, M. Fogagnolo, and P. Merino, A divergent route to nojirimycin analogues from L-serinal and 2-acetylthiazole, *J. Chem. Soc. Chem. Commun.* p. 854 (1990).
32. A. Dondoni, G. Fantin, and M. Fogagnolo, 2-Acetylthiazole as a three-carbon homologating reagent of aldehydes. Application toward the synthesis of amino hexoses from L-serinal, *Tetrahedron Lett. 30*:6063 (1989).
33. G. Fronza, C. Fuganti, and G. Pedrocchi-Fantoni, Synthesis of the four configurational isomers of a protected form of *N*-trifluoroacetyl-4-*C*-methyl-2,4,6-trideoxy-4-amino-L-hexose from L-threonin, *J. Carbohydr. Chem. 8*:85 (1989).
34. G. Fronza, C. Fuganti, A. Mele, and G. Pedrocchi-Fantoni, On the use of (4*S*,5*S*)-2-phenyl-4,5-dimethyl-4-formyl-4,5-dihydrooxazole in the synthesis of N-protected 2,3,4,6-tetradeoxy-4-*C*-methyl-4-amino-L-hexose derivatives, *J. Carbohydr. Chem. 10*:197 (1991).
35. J. V. N. Vara Prasad and D. H. Rich, Addition of allylic metals to α-amino aldehydes. Application to the synthesis of statine, ketomethylene and hydroxyethylene dipeptide isosteres. *Tetrahedron Lett. 31*:1803 (1990).
36. K. C. Nicolaou, R. D. Groneberg, N. A. Stylianides, and T. Miyazaki, Synthesis of the CD and E ring systems of the calicheamicin, γ$_1$ oligosaccharide, *J. Chem. Soc. Chem. Commun.* p. 1275 (1990).
37. P. Garner and J. M. Park, Glycosyl α-amino acids via stereocontrolled buildup of a penaldic acid equivalent. An asymmetric synthesis of thymine polyoxin C, *Tetrahedron Lett. 30*:5065 (1989).
38. P. Garner and J. M. Park, Glycosyl α-amino acids via stereocontrolled buildup of a penaldic acid equivalent. A novel synthetic approach to the nucleosidic component of the polyoxins and related substances, *J. Org. Chem. 55*:3772 (1990).
39. J. Raczko, A. Gołębiowski, J. W. Krajewski, P. Gluziński, and J. Jurczak, Stereoselectivite addition of furyllithium to variously N,N-diprotected D-alaninals, *Tetrahedron Lett. 32*:3797 (1990).
40. G. Casiraghi, L. Colombo, G. Rassu, and P. Spanu, Synthesis of enantiomerically pure 2,3-dideoxy-hept-2-enono-1,4-lactone derivatives via diastereoselective addition of 2-(trimethylsiloxy)furan to D-glyceraldehyde and D-serinal-based three-carbon synthons, *Tetrahedron Lett. 30*:5325 (1989).
41. G. Casiraghi, L. Colombo, G. Rassu, and P. Spanu, Total synthesis of 6-deoxy-6-aminoheptopyranuronic acid derivatives, *J. Org. Chem. 56*:6523 (1991).
42. J. Jurczak, A. Gołębiowski, and J. Raczko, High-pressure [4+2]cycloaddition of 1-methoxybuta-1,3-diene to α-amino aldehydes. Influence of N-protecting groups on asymmetric induction, *Tetrahedron Lett. 29*:5975 (1988).
43. A. Gołębiowski, U. Jacobsson, J. Raczko, and J. Jurczak, Total synthesis of purpurosamine B from D-alanine, *J. Org. Chem. 54*:3759 (1989).
44. A. Gołębiowski and J. Jurczak, High-pressure [4+2]cycloaddition of 1-methoxy-1,3-butadiene to N,O-protected D-threoninals and D-*allo*-threoninals, *Tetrahedron 47*:1037 (1991).
45. A. Gołębiowski, J. Raczko, U. Jacobsson, and J. Jurczak, Influence of N-protecting groups

on the stereochemical course of [4+2]cycloaddition of activated dienes to α-amino aldehydes, *Tetrahedron 47*:1053 (1991).
46. J. Jurczak, A. Gołębiowski, and J. Raczko, Influence of the N-protecting group on the stereochemical course of [4+2]cycloaddition of 1-ethoxy-3-[(trimethylsilyl)oxy] buta-1,3-diene to α-amino aldehydes, *J. Org. Chem. 54*:2495 (1989).
47. J. Jurczak, J. Kozak, and A. Gołębiowski, The total synthesis of L-daunosamine, *Tetrahedron 48*:4231 (1992).
48. F. Arcamone, G. Cassinelli, G. Franceschi, R. Mondelli, P. Orezzi, and S. Penco, Struttura e stereochimica della daunomicina, *Gazz. Chim. Ital. 100*:949 (1970).
49. A. Gołębiowski, U. Jacobsson, and J. Jurczak, High-pressure approach to the total synthesis of 6-*epi*-D-purpurosamine B, *Tetrahedron 43*:3063 (1987).
50. T. Suami, Y. Honda, T. Kato, M. Masu, and K. Matsuzawa, Synthesis of methyl 2,6-di-*N*-acetyl-2,3,4,6,7-pentadeoxy-L-*lyxo*-heptopyranoside derivative of 6-*epi*-D-purpurosamine B, *Bull. Chem. Soc. Jpn. 53*:1372 (1980).
51. A. Gołębiowski, J. Kozak, and J. Jurczak, Synthesis of destomic acid and anhydrogalantinic acid from L-serinal, *J. Org. Chem. 56*:7344 (1991).
52. A. Gołębiowski and J. Jurczak, The cyclocondensation reaction of 1-benzoyloxy-2-*tert*-butyldimethylsilyloxy-4-methoxy-1,3-butadiene with N,O-protected D-threoninals and D-*allo*-threoninals, *Tetrahedron 47*:1045 (1991).
53. B. Szechner, O. Achmatowicz, Z. Gałdecki, and A. Fruziński, Synthesis and absolute configuration of four diastereoisomeric 1-(2-furyl)-2-aminobutane-1,3-diols, *Tetrahedron 50*:7611 (1994).
54. B. Szechner and O. Achmatowicz, A totally synthetic route to enantiomerically pure D- and L-6-amino-6,8-dideoxyoctoses: Stereocontrolled synthesis of methyl α-D-lincosaminide, *Pol. J. Chem. 68*:1149 (1994).
55. S. Knapp and P. J. Kukkola, Stereocontrolled lincomycin synthesis, *J. Org. Chem. 55*:1632 (1990).

# 28

## Total Synthesis of Sugars

**Aleksander Zamojski**
*Institute of Organic Chemistry, Polish Academy of Sciences, Warsaw, Poland*

| | | |
|---|---|---|
| I. | Introduction | 615 |
| II. | Methods | 617 |
| III. | Experimental Procedures | 622 |
| | A. Butyl 2-methoxy-5,6-dihydro-2*H*-pyran-6-carboxylate | 622 |
| | B. *trans* 6-Hydroxymethyl-2methoxy-5,6-dihydro-2*H*-pyran | 622 |
| | C. 1,2:5,6-Di-*O*-isopropylidene-3-*O*-(2,3,4-trideoxy-α-L-*glycero*-hex-2-enopyranosyl)-α-D-glucofuranose | 625 |
| | D. 2,3-Dihydro-4*H*-pyran-4-one | 626 |
| | E. 1-Menthyl 2,3,6-tri-*O*-acetyl-4-deoxy-β-L-xylo-hexopyranoside (1-menthyl 2,3,6-tri-*O*-acetyl-4-deoxy-β-L-glucopyranoside) | 627 |
| | F. Ethyl 3,4-di-*O*-benzyl-2-deoxy-6,7-*O*-isopropylidene-α-L-galactoheptopyranoside | 628 |
| | G. 1$R$,4$R$-7-oxabicyclo[2.2.1]hept-5-en-2-one | 630 |
| | H. Methyl 4-*O*-acetyl-6,7-*O*-isopropylidene-D-*glycero*-α-D-*manno*-heptopyranoside | 631 |
| | I. 2-*O*-α-D-rhamnopyranosyl-L-rhamnose | 632 |
| | References and Notes | 634 |

## I. INTRODUCTION

The synthesis of sugars from noncarbohydrate substrates (total synthesis or de novo synthesis) has attracted the attention of organic chemists for more than 130 years [1]. However, only the last 20 years have witnessed a rapid development of methods leading to structures of desired constitution and stereochemistry. Initially, sugars have been obtained as pure diastereoisomers in racemic form and, more recently, as pure enantiomers of desired configuration. This development is certainly correlated with the discovery of highly chemo- and stereoselective methods of modern organic synthesis. The subject of total synthesis of sugars is covered in several surveys. Reviews [2–4] describe the earlier literature and [5–7] give account of the more recent development of the subject.

Total synthesis of carbohydrates (in enantiomeric forms) can be roughly divided into two large areas: syntheses starting from aliphatic precursors and those employing cyclic substrates.

In the first group of syntheses, three carbon substrates (e.g., D- and L-glyceraldehyde; D- and L-lactaldehyde); and tetracarbon, (e.g., L-threose) are often elongated by addition reaction to the carbonyl group—Wittig or aldol reactions (Scheme 1)—and the products obtained are subsequently converted into required structures.

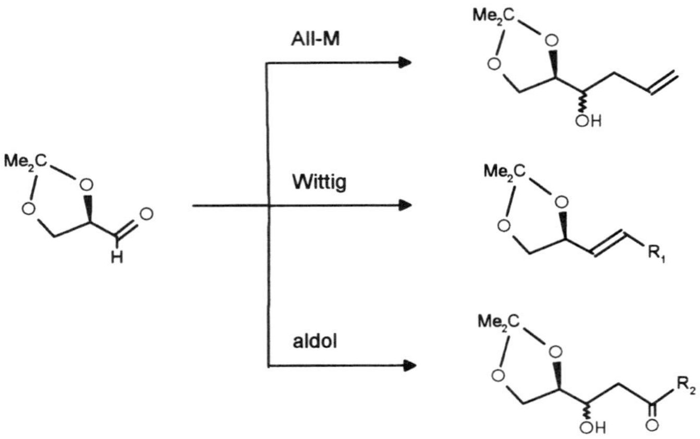

$R_1$ = CHO, $CO_2R$     $R_2$ = OMe, t-Bu, $CMe_2OTMS$

**Scheme 1**

This approach can be illustrated by the synthesis of 2-deoxy-D-ribose **4** elaborated by Harada and Mukaiyama [8] (Scheme **2**).

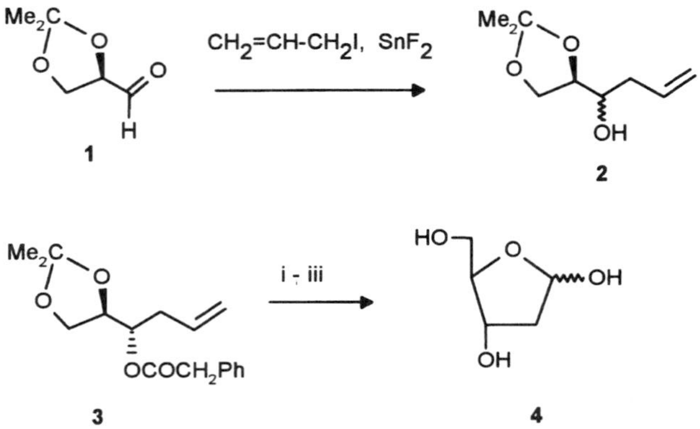

i. $NH_4OH$, ii. AcOH, iii. $O_3$; $Me_2S$

**Scheme 2**

The reaction of 2,3-O-isopropylidene-D-glyceraldehyde **1** with allyltin difluoroiodide yielded two stereoisomeric products **2**, which—after conversion to phenoxyacetyl esters—

were separated and the major (*anti*) stereoisomer **3** was transformed into the target compound **4**.

Sharpless asymmetric epoxidation of open-chain allylic alcohols, an important discovery of the previous decade, was extensively used in carbohydrate synthesis [9–12]. Presently, many aliphatic substrates can be elaborated into sugars [5,6; see also Chap. 9].

An attractive entry to the carbohydrate synthesis is provided by the cycloaddition reaction. Hetero-Diels–Alder reaction, either between an oxa-diene (α,β-unsaturated aldehyde) and an nucleophilic dienophile, or between activated diene and carbonyl compound (usually an aldehyde), leads to dihydropyrans, which can be subsequently functionalized to sugars in the desired manner (Scheme 3).

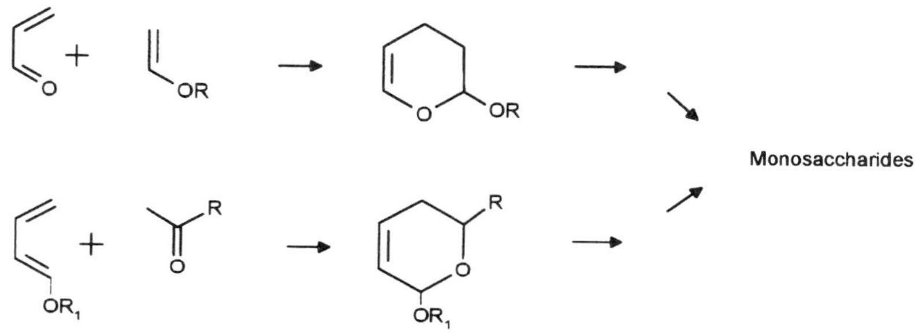

**Scheme 3**

Two approaches, based on furan, have found wide application in carbohydrate synthesis. Cycloaddition reactions of furan with 2-substituted acrylonitrile or acrolein lead to oxabicycloheptanes which, in turn, can be transformed to monosaccharides. On the other hand, furfuryl alcohols can be converted—either by the Clauson–Kaas reaction or by mild oxidation—into 5,6-dihydro-4-pyrones, suitable for easy functionalization to sugars.

Two features of the total synthesis of sugars based on noncarbohydrate substrates should be emphasized: (1) dihydropyrans or oxabicycloheptanes can be transformed into a variety of regio- and stereoisomeric saccharides, and (2) from these substrates sugars belonging to unnatural configurational series can be readily obtained.

These subjects are more extensively covered in Section II.

## II. METHODS

Cycloaddition between dienes having electron-donating substituents [1-alkoxy-1,3-butadiene **1**, 1-(3-glycosyloxy)-, and 1,4-diacetoxy-1,3-butadienes] and carbonyl compounds with electron-withdrawing substituents (alkyl glyoxylates, dialkyl ketomalonates) occurs readily under normal conditions, leading to derivatives of 5,6-dihydro-2*H*-pyran **2a** as mixtures of stereoisomers in good to excellent yields [13–15]. High pressure also enables α-alkoxyaldehydes (e.g., 2,3-*O*-isopropylidene-D-glyceraldehyde) to enter into cycloaddition [16,17]. This leads to optically active cycloadducts.

The basic cycloadduct, butyl 2-methoxy-5,6-dihydro-2*H*-pyran-6-carboxylate **2a**, can be reduced to 6-hydroxymethyl-2-methoxy-5,6-dihydro-2*H*-pyran **2b**. Both 5,6-dihydro-2*H*-pyrans **2a** and **2b** can be converted in several ways to sugars. *cis*-Hydroxylation or epoxidation, followed by basic (or acidic) opening of epoxides with various nucleophilic reagents leads to various 4-deoxyhexoses **3b** [18]. Dihydropyran **2b** can be transformed in a few steps into stereoisomeric alkyl 3,4-dideoxy-hex-3-enopyranosides **4**, thereby opening an access (by *cis*-hydroxylation or epoxidation) to practically all hexoses, normal or substituted [19] (Scheme **4**). Condensation of 1,4-diacetoxy-1,3-butadiene with butyl gly-

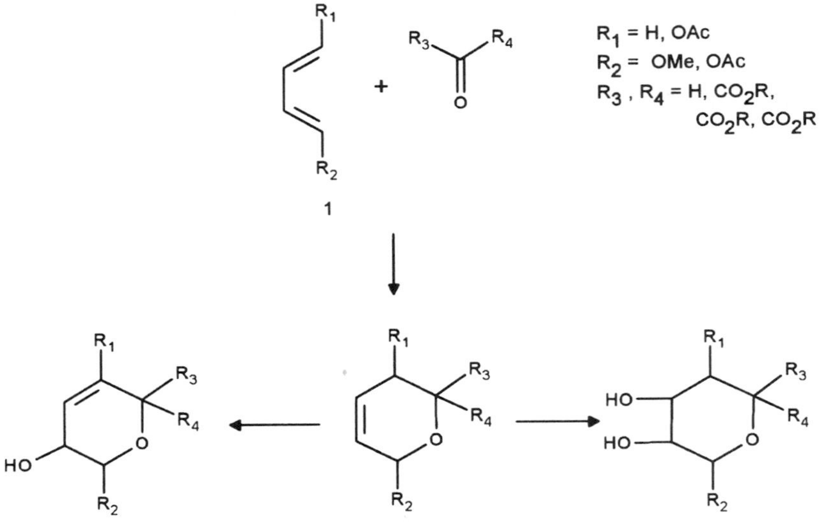

**Scheme 4**

a  $R_1 = R_3 = H$, $R_2 = OMe$, $R_4 = CO_2Bu$
b  $R_1 = R_3 = H$, $R_2 = OMe$, $R_4 = CH_2OH$

oxylate or diethyl ketomalonate directly yields mixtures of stereoisomeric 2,3-unsaturated hexuronic acid esters **2a**, $R_1 = R_2 = $ OAc, $R_3$, $R_4 = $ H, $CO_2Bu$, or $R_3 = R_4 = CO_2Et$), which can be used for glycosidation of, for example, methyl 2,3-*O*-isopropylidene-β-D-ribofuranoside [20]. If the 1,3-butadienyl system is built on an oxygen atom of a monosaccharide, then cycloaddition with an alkyl glyoxylate opens a way—by a similar functionalization of the dihydropyran ring—to disaccharides [21–23].

A general remark should be made here. In the synthesis of sugars from 5,6-dihydro-2*H*-pyrans, often two stereoisomers are produced in each step: *cis*- and *trans*-adducts from the cycloaddition, two products of *cis*-hydroxylation of the double bond, two epoxides from epoxidation of the double bond, two regioisomers from opening of the epoxides with nucleophiles. Usually the products formed can be readily separated by column chromatography and their identification can be made on the basis of 1H nuclear magnetic resonance (NMR) spectra. There are also cases for which single products are easily obtained, e.g., mixtures of *cis*- and *trans*-adducts can be equilibrated under acidic conditions to practically single *trans*-stereoisomers (see Sec. III).

Except for the disaccharide synthesis, in which introduction of natural sugars produced optically active diastereoisomers at the dihydropyran ring stage, most transformations leading to sugars was done with racemic substrates. Alcohol **3** was resolved into enantiomers by crystallization of its 6-*O*-camphanyl ester and subsequent hydrolysis. Recent discoveries of enantioselective catalysts and reagents permit one to obtain basic

# Total Synthesis of Sugars

compounds of this series in the desired enantiomeric form, thereby paving the way to optically active sugars [24,25].

In Section III, the syntheses of 2-methoxy-5,6-dihydro-2*H*-pyrans **2a** and **2b** are described in racemic and in both enantiomeric forms, and by asymmetric synthesis. Also, the synthesis of 1,2:5,6-di-*O*-isopropylidene-3-*O*-(2,3,4-trideoxy-α-L-*glycero*-hex-2-enopyranosyl)-α-D-glucofuranose, a precursor of several disaccharides, is presented.

New possibilities in hetero-Diels–Alder condensation have been opened by the introduction of highly active 1-methoxy-3-trimethylsilyloxy-, 4-benzoyloxy-1-methoxy-3-trimethylsilyloxy-, and 2-acetoxy-1-alkoxy-3-trimethylsilyloxy-1,3-butadienes ("Danishefsky dienes," **5**). These compounds readily react under atmospheric pressure, in the presence of Lewis acids, with normal aldehydes (e.g., acetaldehyde, benzaldehyde, furfural) to furnish 2,3-disubstituted or 2,3,5-trisubstituted derivatives of 2,3-dihydro-4*H*-pyran-4-one **7** capable of readily functionalizing to sugars (Scheme 5) [26]. This approach

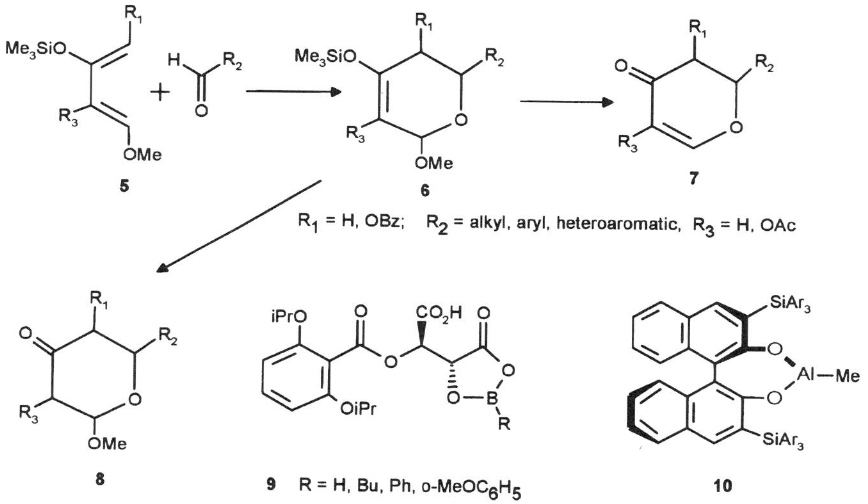

**Scheme 5**

was exploited in numerous syntheses of normal and higher sugars; including syntheses of KDO, hikozamine, and tunicamine [27]. Chiral (acyloxy)boran complexes (**9**) have been used as enantioselective catalysts in the condensation between Danishefsky dienes and aldehydes to yield—in the best examples—2,3-dihydro-4*H*-pyran-4-ones of greater than 92% enantiomeric excess (e.e.) [28]. Similar results can be obtained with chiral organoaluminum catalysts (**10**) based on 3,3′-bis(triarylsilyl)-binaphthol [29].

Soluble lanthanide complexes such as Eu(fod)$_3$ and Yb(fod)$_3$ catalyze the cycloaddition reaction leaving the primary Diels–Alder products (**6**). These compounds can be converted on treatment with triethylamine in methanol to ulosides (**8**) (see Scheme 5) [30].

In Section III, Danishefsky's methodology is illustrated by the synthesis of the basic compound 2,3-dihydro-4*H*-pyran-4-one, and 1-menthyl 2,3,6-tri-*O*-acetyl-4-deoxy-L-glucopyranoside. In this last example, high enantiomeric purity was achieved owing to the matched interaction between the catalyst [(+)-Eu(hfc)$_3$] and the diene containing a chiral (1-menthyl) moiety.

Hetero-Diels–Alder reaction with inverted electron demand between α,β-unsaturated carbonyl compounds (1-oxa-1,3-butadienes **11**; Scheme 6) and enol ethers provides an access to 6-alkoxy-3,4-dihydro-2*H*-pyrans **12** [31,32]. These heterocycles are also useful

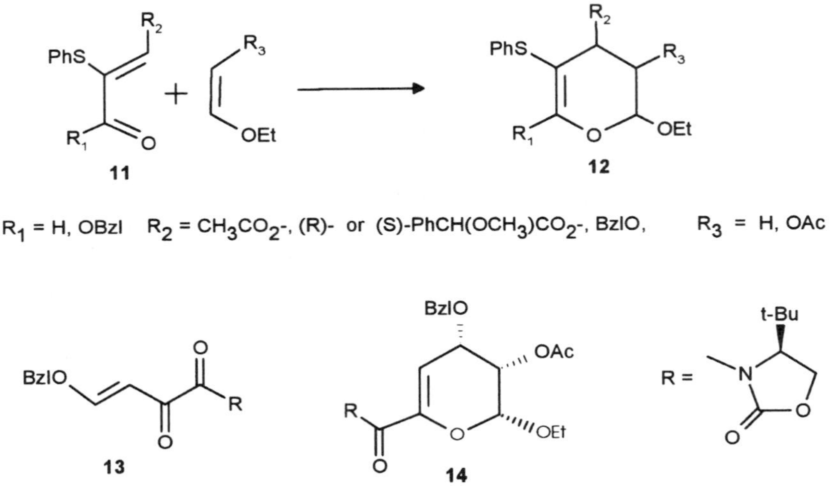

**Scheme 6**

$R_1$ = H, OBzl   $R_2$ = $CH_3CO_2$-, (R)- or (S)-PhCH($OCH_3$)$CO_2$-, BzlO,   $R_3$ = H, OAc

substrates for carbohydrate synthesis (see Scheme **6**). The cycloaddition reaction suffers, in general, from low *endo–exo* selectivity. Asymmetric induction can be achieved by placing optically active moieties in $R_1$ or $R_2$ of the oxadiene [33,34]. However, the degree of enantiopurity achieved is usually not very high.

In Section III, the synthesis of ethyl 3,4-di-*O*-benzyl-2-deoxy-6,7-*O*-isopropylidene-α-L-*galacto*-heptose demonstrates the preparative details of the method.

Recently, substantial progress in stereochemistry of the cycloaddition reaction has been reported [34]. Cycloaddition between optically active oxadiene **13** and 1-acetoxy-2-ethoxy-ethylene, promoted by dimethylaluminum chloride, leads to dihydropyran **14**, with a very high *endo–exo* stereoselectivity (54:1) and in an almost quantitative yield (see Scheme **6**). When trimethylsilyl triflate was used as the promoter in this reaction, the reverse *endo–exo* selectivity (1:5) has been noted. The dihydropyrans obtained served as substrates for the synthesis of β-D- and β-L-mannopyranosides [34].

Furan, which can be considered an 1,4-oxygenated 1,3-butadiene, usually does not react readily with normal dienophiles. However, with 2-acetoxyacrylonitrile (**15**; R = OAc) and in the presence of zinc iodide, furan enters into cycloaddition to form derivatives of 7-oxabicyclo[2.2.1]hept-5-ene **16** in very good yield [35](Scheme 7).

If 2-camphanyloxyacrylonitrile (**15**; R = $C_8H_{11}O_2$COO) is taken for cycloaddition, diastereoisomeric cycloadducts can be separated, and the basic system, 7-oxabicyclo-[2.2.1]hept-5-en-2-one **17**, can be obtained in optically pure form [36]. Another way of obtaining enantiomeric ketones is based on crystallization of a brucine complex obtained from the corresponding cyanohydrines (see Sec. III). Ketone **17** can be converted [e.g., by *cis*-hydroxylation (→**18**), protection of the diol system, and Baeyer-Villiger oxidation] to lactone **19**, the opening of which leads to furanuronic acid **20**. A new development in this field is based in cycloaddition between furan and 2-chloro- or 2-bromoacrolein in the presence of 5 mol% chiral oxazaborolidine **21** as catalyst [37].

The exo/endo (99:1) adducts are of 90–92% optical purity. The enantiomeric bicyclic ketones can be transformed into carbohydrates in a few, stereochemically well-defined steps. In this way, derivatives of D- and L-ribose [38], L-allose, L-talose, and such [39] have been obtained.

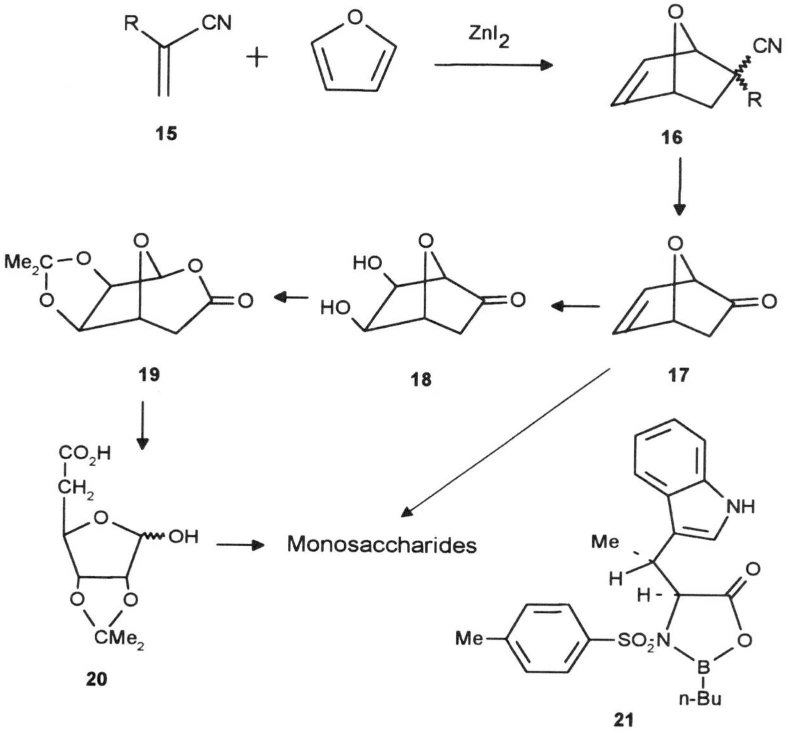

# Total Synthesis of Sugars

**Scheme 7**

Another general approach to sugars starts from (2-furyl)methanols (furfuryl alcohols, **22**). Bromination at low temperature in methanol (Clauson–Kaas reaction) and subsequent mild acidic hydrolysis of 2,5-dimethoxy-2,5-dihydrofurans **23** obtained, leads to 6-hydroxy-2,3-dihydro-6H-pyran-3-ones **24** (Scheme **8**) [40]. The same dihydropyranones can be

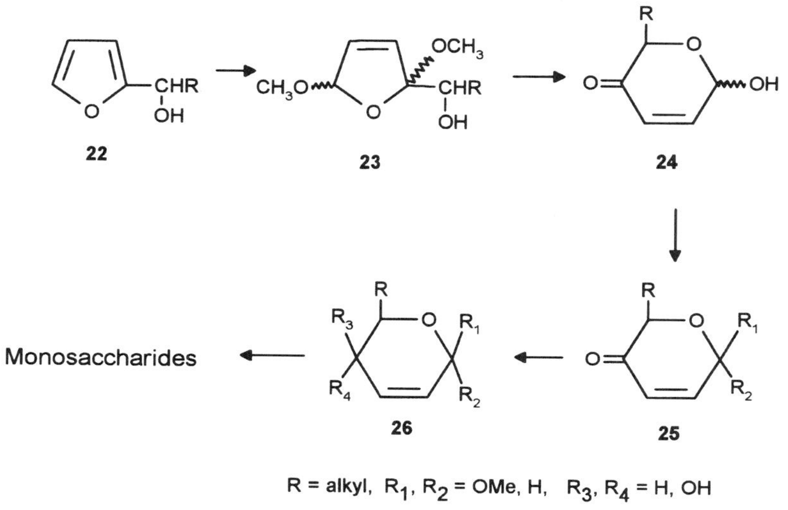

R = alkyl, $R_1$, $R_2$ = OMe, H, $R_3$, $R_4$ = H, OH

**Scheme 8**

obtained by mild oxidation of furfuryl alcohols [41]. These products are convenient substrates in a large series of transformations (→ **25** → **26** → *cis*-hydroxylation or epoxidation, opening of epoxides) leading to a variety of sugars (see Scheme **8**) [42].

The synthetic potential of the method is exemplified by the synthesis of methyl D-*glycero*-D-*manno*-heptopyranoside and 2-*O*-(α-D-rhamnopyranosyl)-α,β-L-rhamnose (see Sec. III).

## III. EXPERIMENTAL PROCEDURES*

### A. Butyl 2-Methoxy-5,6-dihydro-2H-pyran-6-carboxylate (2; R = Bu) [13] (Scheme 9)

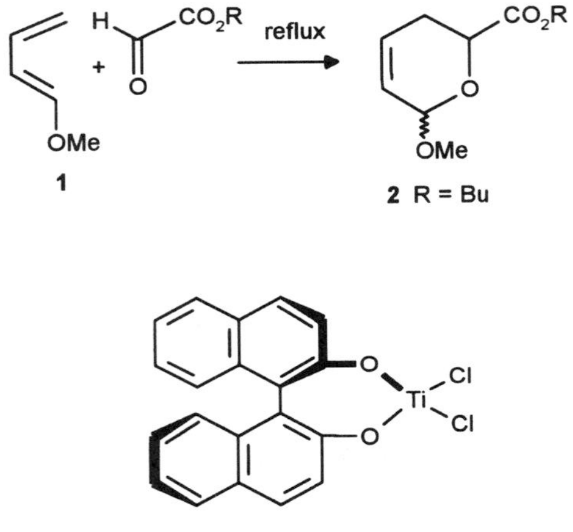

**Scheme 9**

To a mixture of 1-methoxy-1,3-butadiene (**1**; 55.3 g, 0.66 mol) [43] and *n*-butyl glyoxylate (86.0 g, 0.66mol) [44], 0.1 g hydroquinone was added and the solution was refluxed at 100°C for 6 h under argon. Distillation at 88°–90°C/0.5 torr gave **2** (R = Bu) (91.6 g, 65%) as a pale-yellow liquid with a faint odor. By gas chromatography this product is a mixture of *cis* and *trans* (ca. 7:3) isomers. Treatment of the isomer mixture in dichloromethane solution with *p*-toluenesulfonic acid or zinc chloride for 2 h leads to an almost pure *trans*-isomer (ca. 95%).

Methyl 2-methoxy-5,6-dihydro-2H-pyran-6-carboxylate (**2**; R = Me) can be obtained in optically active form by condensation of diene **1** with methyl glyoxylate in the presence of dichlorotitanium (*R*)-binaphthol catalyst (**3**) [45].

### B. *trans* 6-Hydroxymethyl-2-methoxy-5,6-dihydro-2H-pyran (2) [46] (Scheme 10)

To a stirred suspension of lithium aluminum hydride (3.8 g, 0.1 mol) in ether (25mL), a solution of **1** (95% *trans*; 10 g, 0.046 mol) in ether (50 ml) was slowly added maintaining

---
*Optical rotations were measured at 22°–25°C.

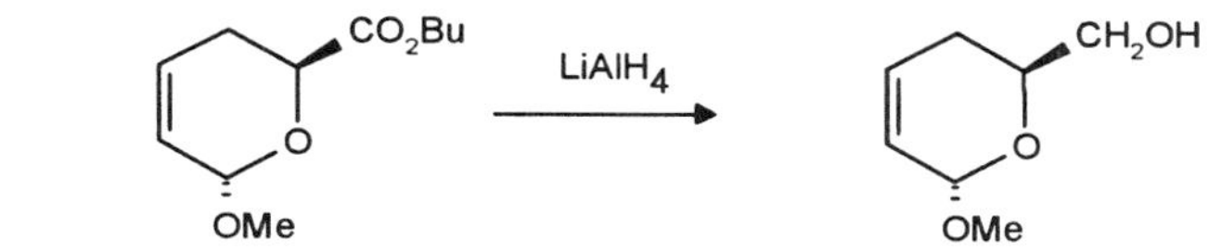

# Total Synthesis of Sugars

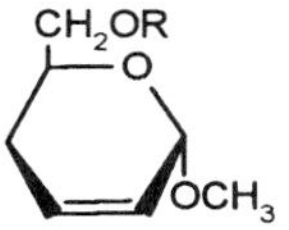

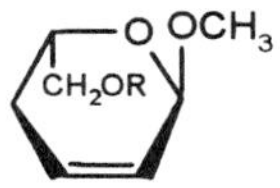

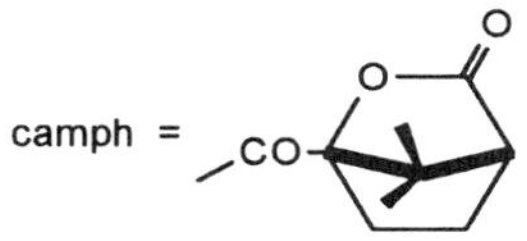

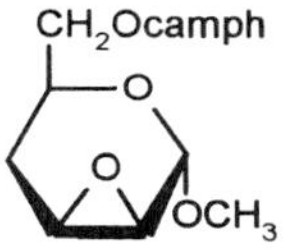

**Scheme 10**

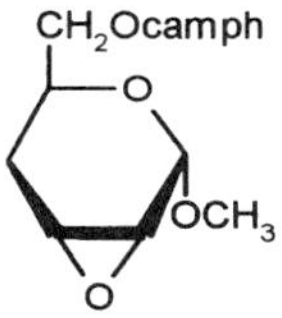

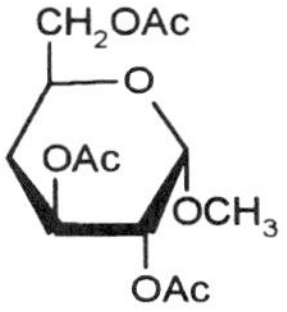

gentle boiling. After 2.5 h (TLC) the excess of the hydride was decomposed with aqueous saturated ammonium chloride solution, the precipitate was filtered off and washed with ether. The combined ether solutions were dried over anhydrous potassium carbonate and concentrated to dryness. The residue was distilled at 69°–71°C/1 torr to afford **2** (4.8 g, 71%) as a colorless liquid.

*Resolution of* trans-6-hydroxymethyl-2-methoxy-5,6-dihydro-2H-pyran (**2**) [47]

To a solution of **3** (8.7 g, 0.06 mol) in anhydrous pyridine (150 mL) 1S-camphanic chloride (18.7 g, 0.086 mol) [48] was added and the mixture was left at room temperature. After 48 h the solution was poured into ice-water and extracted with $CHCl_3$. Combined chloroform extracts were washed with water, dried ($MgSO_4$), and concentrated to dryness. Crude ester was dissolved in 40 mL ether and cooled to a −15°C. The deposited crystals (first fraction, 8.96 g) were crystallized four times from the same solvent, 4.5 g of pure **3**; mp 102°–103.5°C, $[\alpha]_{578}$ − 29.8° (c 2.02, $C_6H_6$).

The mother liquor was concentrated to dryness after separation of the first fraction, and the residue was crystallized from a mixture of ether and *n*-hexane (7:3) at -15°C. The crystals obtained were combined with the residue obtained after crystallization of **3**. Two consecutive crystallizations from the same solvent gave the second pure diastereoisomeric ester **4**: 1.65 g, mp 72°–75°C, $[\alpha]_{578}$ + 26.2° (c 1.8, $C_6H_6$).

A suspension of ester **3** (1 g) in 60 mL of 0.5 *N* KOH solution in water–ethanol (1:3) was refluxed for 1 h. The solvents were evaporated, and the residue was extracted with ethyl acetate, the extract was washed with water, dried ($MgSO_4$), and concentrated to dryness.

The remaining oil was distilled at 65°–67°C/0.4 torr to give 0.332 g (75%) of **5**; $[\alpha]_{578}$ − 80.0° (c 2.0, $C_6H_6$).

Analogous hydrolysis of **4** (0.8 g) afforded **6**, 0.225 g (77%), distilled at 64°–66°C/0.4 torr, $[\alpha]_{578}$ + 79.2° (c 1.97, $C_6H_6$).

2S,6S Configuration of the levorotatory alcohol **5** was determined by the following sequence of reactions [47]:

6-O-Camphanyl ester **3** was epoxidized with m-chloroperoxybenzoic acid and the two stereoisomeric epoxides [of the lyxo **7** and ribo **8** configurations, (1:1.1,90%)] were separated by column chromatography with light petroleum–ethyl acetate (65:35). The ribo-epoxide **8** was hydrolyzed with 1 N $HClO_4$ in dioxane–water solution to methyl 4-deoxy-D-xylo-hexopyranoside, identified as the 2,3,6-tri-O-acetyl derivative **9**: mp 74°C, $[\alpha]_{589}$ + 139.1° (c 0.56, $CHCl_3$); reported [49]: mp 75°–76°C, $[\alpha]_{589}$ + 138° ($CHCl_3$).

Optically active alcohol **6** (94% e.e.) can be obtained by $LiAlH_4$ reduction of trans-methyl 2R,6R-2-methoxy-5,6-dihydro-2H-pyran-6-carboxylate (see Scheme 9, 2 R = Me) [25].

### 2S,6S-6-Benzyloxymethyl-2-methoxy-5,6-dihydro-2H-pyran **4** by Asymmetric Synthesis [24,50] (Scheme **11**)

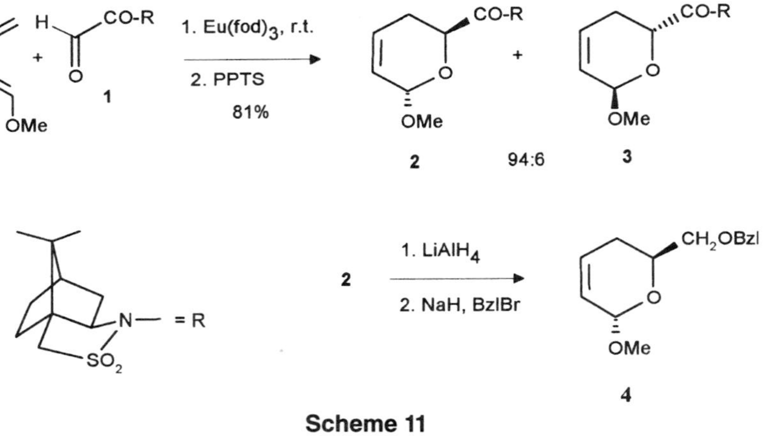

**Scheme 11**

To a solution of 2R-N-glyoxylylbornane-10,2-sultam **1** (100 mg, 0.37 mmol) [51] in dry dichloromethane (5 mL), Eu(fod)$_3$ (8 mg) and trans-1-methoxy-1,3-butadiene (0.2 mL, 2 mmol) were added, and the mixture was stirred at room temperature. After 1 h, the solution was concentrated under reduced pressure to dryness, the residue was dissolved in methanol (10 mL) and isomerized with a few milligrams of pyridine p-toluenesulfonate (PPTS). After stirring overnight, the solution was again concentrated to dryness, and the residue was purified by chromatography on a silica gel column with hexane–ethyl acetate 7:3. Eluted first was **2** (99.7 mg, 75.8%): mp 215°–217°C, $[\alpha]_D$ − 113.4° (c 1.2, $CHCl_3$). Eluted second was the enantiomer of **2**, cycloadduct **3** (6.4 mg, 4.9%): mp 208°–210°C, $[\alpha]_D$ − 100.9° (c 2.23, $CHCl_3$).

Alternatively, instead of chromatographic separation, the main produce **2** could be isolated by crystallization of the crude reaction mixture from dichloromethane. One crystallization afforded the optically pure product.

To a cooled (0°C) solution of **2** [178 mg, 0.5 mmol (from two runs)] in dry THF (10 mL), lithium aluminum hydride (20 mg, 0.5 mmol) was added. The cooling bath was removed and the mixture was allowed to attain room temperature. After 30 min the solution was poured into 20 mL of saturated solution of sodium potassium tartrate, and the two-phase mixture was stirred at room temperature. After 1.5 h the mixture was extracted with ether (3 × 20 mL) and the combined organic extracts were washed with water (20 mL) and brine, dried ($MgSO_4$), and concentrated under reduced pressure at low temperature (cold-water bath) to avoid losses of the volatile 6-hydroxymethyl-2-methoxy-5,6-dihydro-2*H*-pyran.

Sodium hydride (25 mg, 0.52 mmol) was added to a cooled (0°C) solution of the crude reduction product in dry THF (10 mL) and the reaction mixture was stirred for 15 min, whereupon benzyl bromide (0.065 mL, 0.55 mmol) was added. The cooling bath was removed, and the mixture was stirred at room temperature. After 2 h the mixture was diluted with ether (50 mL), washed with water (3 × 20 mL), dried ($MgSO_4$), and concentrated to dryness. The residue was purified by chromatography on a silica gel column with hexane–ethyl acetate 4:1 to yield **4** (58.5 mg, 50%) as a pale yellow oil: $[\alpha]_D$ − 15.6° (*c* 0.6, $CHCl_3$).

### C. 1,2:5,6-Di-*O*-isopropylidene-3-*O*-(2,3,4-trideoxy-α-L-*glycero*-hex-2-enopyranosyl)-α-D-glucofuranose (9) [21] (Scheme 12)

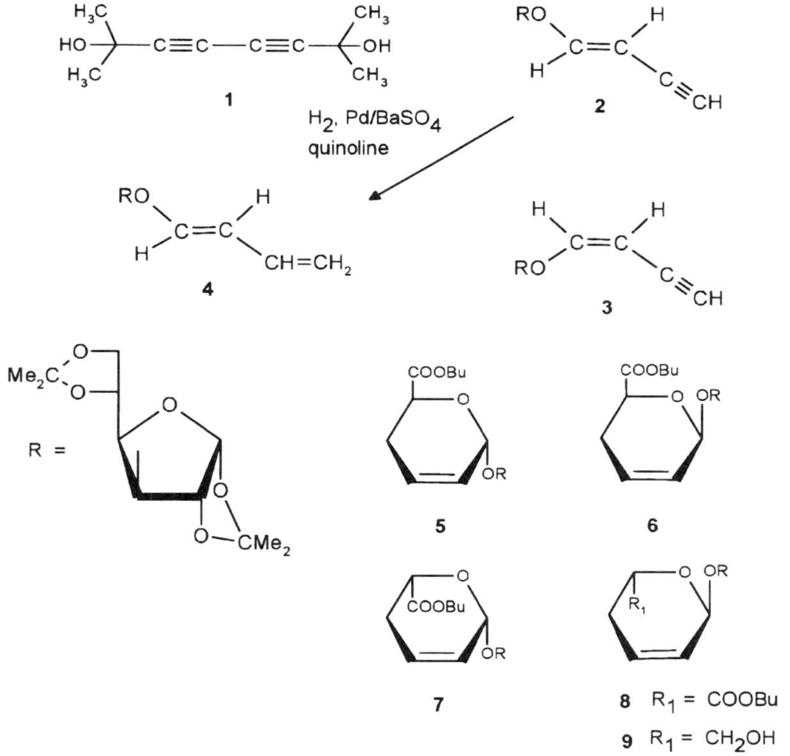

**Scheme 12**

*3-O-(1,3-Butadienyl)-1,2:5,6-di-O-isopropylidene-α-D-glucofuranose* **4**

A solution of the diacetylenic glycol **1** (4.98 g, 30 mmol) [52] in THF (40 mL) was added dropwise to a solution of 1,2:5,6-di-*O*-isopropylidene-α-D-glucofuranose (3.9 g, 15 mmol) and potassium hydroxide (0.1 g) in THF (30 mL) maintained at 80°C under nitrogen. After 16 h at 80°C the mixture was cooled and filtered, and the filtrate concentrated to dryness. An aqueous solution of the residue was extracted with chloroform, and the organic extract was washed ($H_2O$), dried, and applied to a silica gel column. Elution with benzene–ethyl acetate 7:1, gave the mixed enynyl ethers **2** and **3** (4.46 g, 96%).

This mixture was rechromatographed on a silica gel column with ether–light petroleum 1:3 as eluant. Eluted first was **3**, an oil (3.2 g), followed by **2** (3.8 g): mp 83°–85°C (from light petroleum).

A solution of **2** (2 g) in light petroleum (50 mL) was hydrogenated over 5% palladium–barium sulfate catalyst (0.15 g) in the presence of quinoline (0.15 g) for 30 min. The mixture was filtered through Celite and concentrated to an oil **4** (100%): bp 115°C/0.01 torr [53].

*Cycloaddition of Butyl Glyoxylate to* trans-*Dienyl Ether* **4**

A mixture of **4** (1.9 g, 6.1 mmol) and butyl glyoxylate (1 mL) was stored at ambient temperature for 15 days (or 3 days at 60°C). Chromatography and a silica gel column with ether–light petroleum 1:1 gave a mixture (2.49 g, 92.5%) of four stereoisomers **5** (α-D), **7** (β-L), **6** (β-D), and **8** (α-L) in order of increasing polarity.

This mixture was rechromatographed with the same eluant to give a mixture of **5** and **7** (73%): bp 165°C/0.1 torr. Continued elution gave **6** and **8** in the ration 2:1 as nearly pure fractions.

*Isomerization of* **5–8** *in Acidic Medium*

The mixture **5–8** (1 g) was either (1) dissolved in benzene (10 mL) containing toluene-*p*-sulfonic acid (5 mg), or (2) dissolved in anhydrous ether (20 mL) containing boron trifluoride–ether complex (2 drops). In each case, the mixture was stirred for 2 h at ambient temperature, then washed with aqueous sodium hydrogen carbonate and water, dried, and concentrated. The residue was chromatographed on a silica gel column with ether–light petroleum 1:1 to give initially **5** (0.25 g, 25%): $[\alpha]_D$ − 48.3° (*c* 0.77, $CHCl_3$). Continued elution gave α-L-*glycero*-isomer **8** (0.5 g, 55%): $[\alpha]_D$ + 11.6° (*c* 0.78, $CHCl_3$).

*Reduction of* **8**

To a stirred suspension of lithium aluminum hydride (350 mg) in ether (20 mL), a solution of **8** (913 mg) in ether (20 mL) was added dropwise at room temperature. After 2 h ice-water was added and the ethereal layer was separated, dried, and concentrated to an oil, which crystallized, **9** (680 mg, 89%): mp 148°–149°C (from ether).

Disaccharide precursor **9** was converted into 3-(α-L-altropyranosyl)-1,2:5,6-di-*O*-isopropylidene-α-D-glucofuranose by functionalization of the dihydropyran ring [22].

### D. 2,3-Dihydro-4*H*-pyran-4-one (2) [26] (Scheme 13)

To a solution of 1-methoxy-3-trimethylsilyloxy-1,3-butadiene (**1**; $R_1 = R_2 = H$) (2.17 g, 0.12 mol) [54] in THF (50 mL) were added paraformaldehyde (4 g) and zinc chloride (1.71 g,

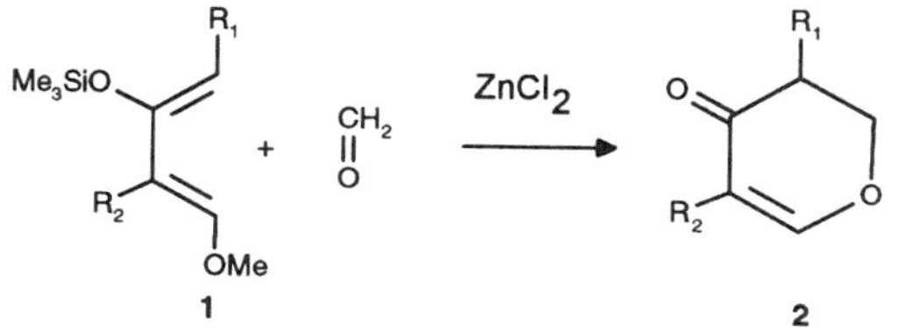

# Total Synthesis of Sugars

Scheme 13

$R_1 = R_2 = H$
$R_1 = H, R_2 = OAc$

0.13 mol). The suspension was refluxed for 3 h, then ice-cold saturated aqueous $NaHCO_3$ was added, and the resulting mixture was extracted with ether. The combined ether extracts were dried ($K_2CO_3$) and concentrated under diminished pressure, leaving a dark yellow oil. Chromatography of the residue with hexane–ethyl acetate (7:3) yielded (**2**; $R_1 = R_2 = H$), 680 mg (55%), volatile, unstable yellow oil.

Similarly prepared were: 3-acetoxy-2,3-dihydro-4*H*-pyran-4-one (**2**; $R_1 = H$, $R_2 = OAc$) [from 1-methoxy-2-acetoxy-3-trimethylsilyloxy-1,3-butadiene (**1**; $R_1 = H, R_2 = OAc$) and paraformaldehyde in 67% yield] and 5-acetoxy-3-benzoyloxy-2,3-dihydro-4*H*-pyran-4-one (from 1-benzoyloxy-2-*t*-butyldimethylsilyloxy-3-acetoxy-4-methoxy-1,3-butadiene and paraformaldehyde in 75% yield).

### E. 1-Menthyl 2,3,6-tri-*O*-acetyl-4-deoxy-β-L-xylo-hexopyranoside (1-Menthyl 2,3,6-tri-*O*-acetyl-4-deoxy-β-L-glucopyranoside) (5) [30] (Scheme 14)

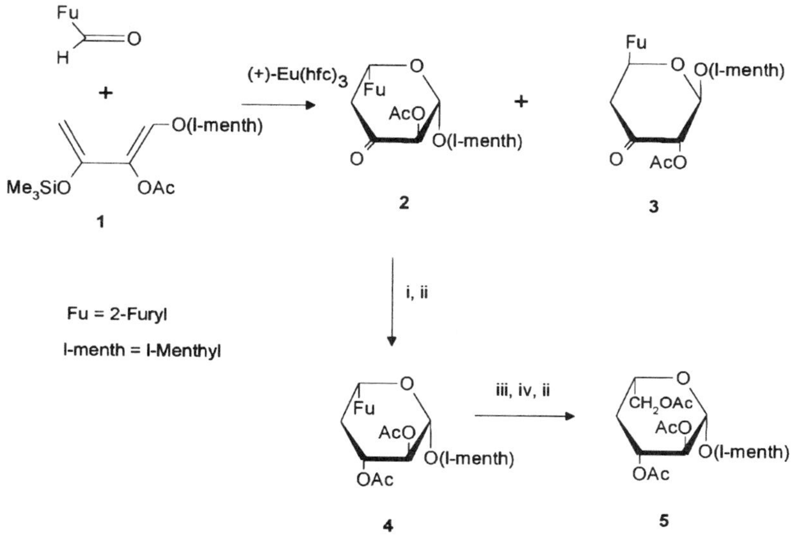

Fu = 2-Furyl
l-menth = l-Menthyl

i. K-Selectride.  ii. $Ac_2O$, $Et_3N$, DMAP.  iii. $O_3$,  iv. $BH_3$-THF.

**Scheme 14**

To a solution of 2-acetoxy-1-(1-menthyloxy)-3-(trimethylsilyloxy)-1,3-butadiene **1**; (530 mg, 1.5 mmol) [55] and 2-furylaldehyde (130 mg, 1.35 mmol) in chloroform (5 mL), (+)-

Eu(hfc)$_3$ (240 mg, 0.2 mmol) [56] catalyst was added and the mixture was left at room temperature. After 24 h, triethylamine (4 mL) and methanol (3 mL) were added. The reaction mixture was stirred for 2 h and was then concentrated under lowered pressure to dryness. The solution of the residue in ethyl acetate was filtered through a short silica gel layer and the eluate was concentrated to dryness to yield a 87:13 mixture of **2** and **3** (503 mg, 96%). Chromatography of this mixture with hexane–ethyl acetate (3:1) gave **2** which was crystallized from hexane, 380 mg (75%): mp 126°–127°C, [α]$_D$ + 65.2° (*c* 1.3, CHCl$_3$).

To a solution of **2** (340 mg, 0.9 mmol) in THF (30 mL) cooled to −78°C K-Selectride (1.34 mL of a 1-*M* solution in THF) was slowly added (1 h) under nitrogen atmosphere. The reaction mixture was stirred for an additional 1 h, quenched with an aqueous saturated solution of NaHCO$_3$ (20 mL) and allowed to attain room temperature. The product was extracted with ethyl acetate (4 × 20 mL), the combined extracts were dried (MgSO$_4$) and concentrated to dryness. The residue was dissolved in dichloromethane (30 mL), then triethylamine (500 μL, 3.6 mmol), acetic anhydride (250 μL, 2.6 mmol), and a crystal of *N,N*-dimethylaminopyridine (DMAP) were added, and the mixture was left under nitrogen at room temperature. After 24 h the solution was concentrated to dryness and the residue was chromatographed on a silica gel column with hexane–ethyl acetate (3:1) to give 2*S*,3*S*,4*R*,6*R*-3,4-diacetoxy-2-(1-menthyloxy)-6-(2-furyl)-tetrahydropyran **4** (290 mg, 79%): mp 127°–129°C, [α]$_D$ − 5.3° (*c* 0.8, CHCl$_3$).

A solution of **4** (110 mg, 0.26 mmol) in a mixture of dichloromethane (25 mL) and methanol (8 mL), cooled to −78°C, was ozonized until the solution turned blue (about 4 min). Excess ozone was removed by a stream of nitrogen, the solution was allowed to attain room temperature and was concentrated to dryness. The crude product was dissolved in THF (10 mL) and a borane–THF complex (1.04 mL of a 1-*M* solution) was added at room temperature under nitrogen. After 24 h the reaction mixture was treated with diluted hydrochloric acid (5 mL) and the solution was extracted with ethyl acetate (4 × 15 mL). The combined extracts were dried (MgSO$_4$) and concentrated to dryness. The residue was dissolved in dichloromethane (20 mL) and acetylated with acetic anhydride (62 μL, 0.66 mmol), triethylamine (200 μL, 1.50 mmol), and a crystal of DMAP, under nitrogen atmosphere. After 12 h the reaction mixture was concentrated under diminished pressure, and the crude product was purified on a silica gel column with hexane–ethyl acetate (3:1) to give **5** (73 mg, 65%) mp 103°–105°C (after crystallization from hexane), [α]$_D$ − 26.6° (*c* 0.7, CHCl$_3$).

## F. Ethyl 3,4-di-*O*-benzyl-2-deoxy-6,7-*O*-isopropylidene-α-L-galactoheptopyranoside (13) [57] (Scheme 15)

A mixture of 1-deoxy-3,4-*O*-isopropylidene-1-phenylthio-L-erythrulose (**1**; 6.0 g, 24 mmol) [57] and *N,N*-dimethylformamide dimethyl acetal (4.25 g, 36 mmol) was stirred for 2 h at 70°C. The liberated methanol and the excess of the reagent were evaporated under reduced pressure and the dark oily residue was purified on a silica gel column with ethyl acetate as eluent to yield **2** (6.21 g, 85%) as a yellow solid: mp 93°C (from diisopropyl ether).

To a suspension of **2** (6.0 g, 19 mmol) in boiling water (100 mL) solid barium hydroxide octahydrate (3.0 g) was slowly added while stirring. The mixture was additionally heated for 5 min and then rapidly cooled with ice to room temperature. The solution was extracted with ether, saturated with sodium chloride, then acidified to pH 6 and

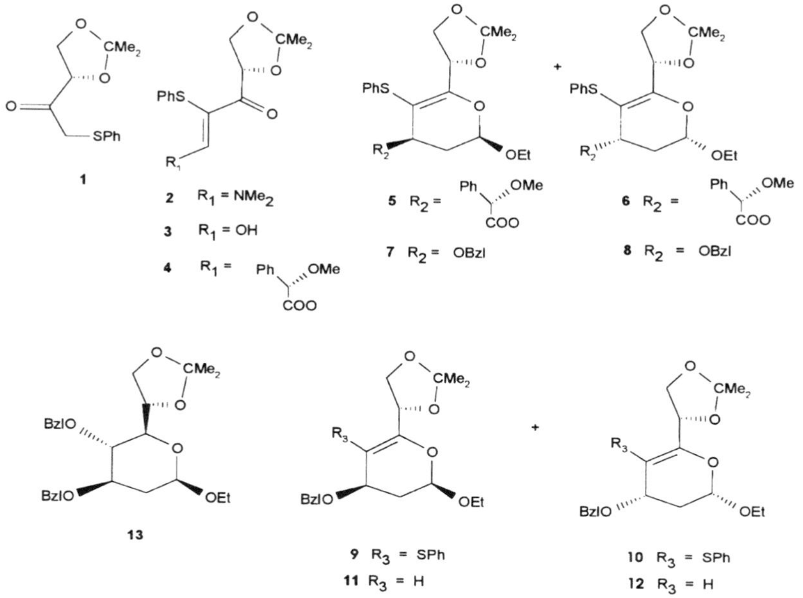

## Scheme 15

extracted with dichloromethane. The CH$_2$Cl$_2$ extract was dried (MgSO$_4$) and concentrated to dryness to yield **3** (4.4 g, 80%), an unstable yellow semisolid.

S-(O-Methyl)mandelic acid chloride was prepared (14-h reflux in 5 mL of benzene) from the S-(O-methyl)mandelic acid (0.33 g, 2.0 mmol) and oxalyl chloride (0.34 mL, 4.0 mmol). The chloride was added to a solution of **3** (0.56 g, 2.0 mmol) and dry pyridine (0.33 mL, 4.0 mmol) in dry toluene (20 mL) at −10°C. The mixture was then allowed to attain 0°C. After 14 h, toluene was added, and the mixture was washed with water (20 mL). Toluene solution was dried (MgSO$_4$) and concentrated to dryness to yield **4** (0.84, 98%), which was used in the next step without purification.

A solution of **4** (0.84 g, 1.9 mmol) in ethyl vinyl ether (10 mL) was heated in a sealed tube at 75°C for 6 h. The excess of ethyl vinyl ether was distilled off, and the residue was chromatographed on a silica gel column with light petroleum–ethyl acetate (4:1) to yield a mixture of **5** and **6** (0.31 g, 30%).

To a suspension of sodium hydride (0.48 g, 20 mmol) in N,N-dimethylformamide (DMF; 20 mL) at −5°C under nitrogen water (0.18 mL, 10 mmol) in 5 mL of DMF was added, followed by a larger portion of **5** and **6** (3.6 g, 9.1 mmol). After 2 h, a solution of benzyl bromide (2.38 mL, 20 mmol) in dry DMF (20 mL) was added and the mixture was left at room temperature. After 24 h, ethanol (2 mL) was added and the mixture was poured into aqueous saturated ammonium chloride solution (100 mL). The product was extracted with ethyl acetate, the extracts were dried (MgSO$_4$) and concentrated to dryness. The residue was purified by chromatography on silica gel column using toluene–ethyl acetate (9:1) for elution to yield **7** and **8** (3.37 g, 83%).

To a well-stirred, cooled suspension of an excess of Raney nickel (W2) in dry THF, the mixture **7** and **8** (3.2 g, 7.2 mmol) was added and the reaction was left for 2 h. Raney nickel was filtered, and the filtrate was concentrated to yield **9** and **10** (2.20 g, 86%).

To a solution of **9** and **10** (2.0 g, 6 mmol) in dry THF (30 mL), cooled to −10°C, under nitrogen atmosphere, borane–THF complex (8 mL of a 1-$M$ solution) was added. The temperature was raised to 0°C. After 20 h, a solution of sodium hydroxide (1.2 g in 10 mL of water and 10 mL of ethanol) was added followed by 20 mL of hydrogen peroxide (35%). The mixture was allowed to attain room temperature and, after 1 h, activated charcoal (0.2 g) was added to destroy the excessive hydrogen peroxide. The mixture was filtered, saturated with sodium chloride, and extracted with ethyl acetate. The extracts were dried (MgSO$_4$) and concentrated to dryness. Chromatography on a silica gel column with light petroleum–ethyl acetate (4:1) yielded **11** (1.54 g, 73%): [α]$_D$ − 16.7° ($c$ 1, CHCl$_3$), followed by **12** (0.31 g, 15%).

Benzylation (NaH, BzlBr in DMF) of **11** yielded **13** (82%): mp 95°–96°C, [α]$_D$ − 4.9° ($c$ 1, CHCl$_3$).

### G. 1$R$,4$R$-7-Oxabicyclo[2.2.1]hept-5-en-2-one (4) [58,59] (Scheme 16)

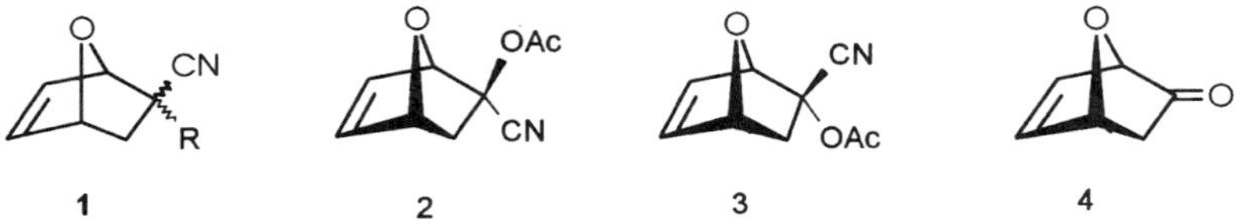

**Scheme 16**

A mixture of racemic 2-*exo* and 2-*endo*-cyano-7-oxabicyclo[2.2.1]hept-5-en-2-yl acetates (**1**; 8.95 g, 50 mmol), obtained by cycloaddition between α-acetoxyacrylonitrile and furan [55,56], was dissolved in methanol (100 mL) under nitrogen. Sodium methoxide (0.25 mL of a 5.4-$M$ solution in methanol) was added and the solution was left at room temperature for 4 h. Under vigorous stirring, brucine (19.75 g, 50 mmol) was added. After 5 min a precipitate began to form. After 12 h the precipitate was collected, washed with ether (100 mL) and dried under vacuum, 16.95 g (38.2 mmole, 63.7%; sample A). The filtrate and ether washing were combined and concentrated to dryness to yield a yellowish solid (sample B). The crude cyanohydrine–brucine complex (sample A; 16.82 g, 37.9 mmol) was dissolved in a minimum amount of hot methanol (250 mL) from which crystallization took place on cooling to 20°C, giving 9.73 g (18.3 mmol) of a solid. This material was again recrystallized from hot methanol (160 mL) to give, after filtration, washing with ether, and drying, 5.56 g (10.44 mmol, 21%) of pure complex (sample C). To a solution of sample C (1.33 g, 2.5 mmol) in chloroform (75 mL), stirred at 20°C under nitrogen, was added acetic anhydride (0.765 g, 7.5 mmol) and pyridine (0.59 g, 7.5 mmol). The solution was stirred for 24 h in the dark and then washed with 1 $N$ hydrochloric acid (20 mL, three times) and 5% aqueous NaHCO$_3$ (10 L, two times). After drying (MgSO$_4$) the solution was concentrated under vacuum and the residue was purified on a silica gel column by flash chromatography using ethyl acetate–ether–pentane (1:1:5) for elution to yield 0.425 g (95%) of **2** and **3** (97:3). Recrystallization of this product from ether–petroleum ether (1:1) gave pure **2**, 0.32 g (71.5%): mp 57.5°–58°C, [α]$_{589}$ + 57.9°, [α]$_{578}$ + 60.9°, [α]$_{546}$ + 69.8° ($c$ 1.69, CHCl$_3$).

To a solution of **2** (200 mg, 1.12 mmol) in dry methanol (1 mL), stirred at 20°C under nitrogen, sodium methoxide (10 μL of a 5.4-$M$ solution in methanol) was added. After 2 h the mixture was treated with 40% aqueous formaldehyde (0.35 mL) and stirred for an additional 1 h. The mixture was diluted with water, saturated sodium chloride solution was added, and the product was extracted with CH$_2$Cl$_2$. The combined extracts were washed with saturated aqueous NaCl, dried (MgSO$_4$), and the solvent was carefully evaporated.

The residue was bulb-to-bulb distilled (90°C/15 torr) to give pure **4** (113 mg, 93%), $[\alpha]_{589}$ + 959°, $[\alpha]_{578}$ + 1008°, $[\alpha]_{546}$ + 1207° (c 0.12, $CHCl_3$).

## H. Methyl 4-*O*-Acetyl-6,7-*O*-isopropylidene-D-*glycero*-α-D-*manno*-heptopyranoside (10) [60] (Scheme 17)

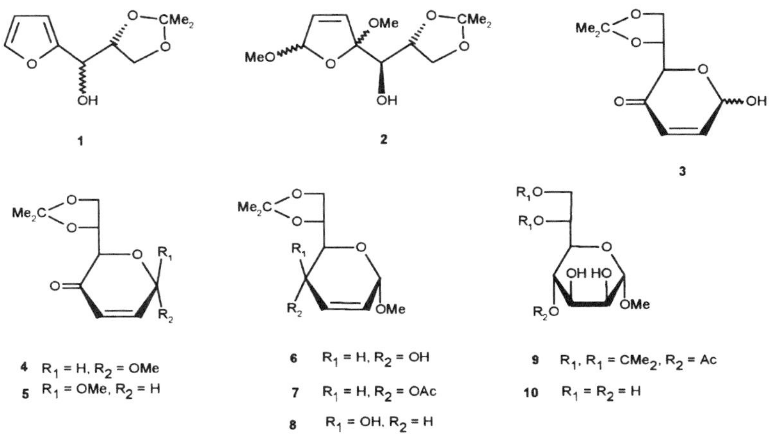

4 $R_1$ = H, $R_2$ = OMe
5 $R_1$ = OMe, $R_2$ = H
6 $R_1$ = H, $R_2$ = OH
7 $R_1$ = H, $R_2$ = OAc
8 $R_1$ = OH, $R_2$ = H
9 $R_1$, $R_1$ = $CMe_2$, $R_2$ = Ac
10 $R_1$ = $R_2$ = H

**Scheme 17**

To a solution of 2,3-*O*-isopropylidene-D-glyceraldehyde (12.6 g, 97 mmol) in furan (20 mL), chloroacetic acid (5.7 g, 60 mmol) dissolved in furan (20 mL) was added and the mixture was refluxed for 8 h. After 12 h at room temperature saturated aqueous $NaHCO_3$ solution (50 mL) was added and the product was extracted with ether (3 × 100 mL). Combined ether extracts were dried ($MgSO_4$), concentrated to dryness, and the residue was chromatographed on a silica gel column with light petroleum–ether–methanol (6:4:0.5) to yield 1*R*,2*R*- and 1*S*,2*R*-1-*C*-(2-furyl)-2,3-*O*-isopropylidene-glycerols **1** (95:5, 7.0 g, 36.5%).

To a solution of **1** (7.0 g, 35 mmol) in absolute methanol (18 mL) and absolute ethyl ether (18 mL), cooled to −78°C (dry ice–acetone), a solution of bromine (3 mL) in methanol (18 mL) was slowly added, and the reaction mixture was left at this temperature for 30 min. Ammonia was bubbled to reach pH 9, the mixture was allowed to attain room temperature and concentrated to dryness. The residue was dissolved in benzene (20 mL) and the suspension was filtered through a layer of aluminum oxide, and the filtrate was concentrated under reduced pressure to yield **2** (8.5 g, 60%), distilled at 105°C/0.05 torr, $[\alpha]_D$ + 14° (*c* 1.3, $CHCl_3$).

A solution of **2** (246 mg, 0.9 mmol) in 2.5 mL of a mixture of acetone, water, and sulfuric acid (96:4:0.25) was left at room temperature for 40 min, whereupon the solution was neutralized with solid lead carbonate (ca. 300 mg). The suspension was filtered, the solid was washed with acetone, and to the filtrate water (10 mL) was added. Acetone was evaporated, and the residue was extracted with chloroform. The $CHCl_3$ extract was dried ($MgSO_4$) and concentrated to dryness to yield **3** (167 mg, 83%).

A larger portion of **3** (466 mg, 2.2 mmol) was dissolved in methyl iodide (20 mL), and freshly prepared silver oxide (2.5 g) was added. The mixture was vigorously stirred in the dark at room temperature. After 24 h, the solid was filtered off, washed with chloroform, and the filtrate and washings were concentrated to dryness to yield a mixture of **4** and **5** (368 mg). This mixture was separated by chromatography on a silica gel column with light petroleum–ether–methanol (7:3:0.5) to give first the α-anomer **4** (126 mg, 25%): $[\alpha]_D$ + 34° (*c* 0.9, CHCl$_3$), followed by β-anomer **5** (180 mg, 36%): $[\alpha]_D$ − 10° (*c* 2.0, CHCl$_3$).

To a solution of the α-anomer **4** (0.76 g, 3.3 mmol) in THF (5 mL), cooled to 0°C, a solution of sodium borohydride (60 mg) in water (1.5 mL) was added. After 40 min the product was extracted with ethyl acetate (3 × 10 mL), and the same volume of benzene was added to the combined extracts. The solution was filtered, and the filtrate was concentrated to dryness. The residue was chromatographed on a silica gel column with light petroleum–ether–methanol (6:4:0.5). The first eluted product was the *ribo*-stereoisomer **6** (383 mg, 50%); 4-*O*-acetyl derivative **7** (Ac$_2$O, C$_5$H$_5$N, DMAP): $[\alpha]_D$ + 107° (*c* 1.1, MeOH).

The second eluted was the *arabino*-stereoisomer **8** (75 mg, 10%): mp 106°–107° (from ethanol), $[\alpha]_D$ − 81° (*c* 1.3, MeOH).

To a solution of **7** (458 mg, 1.7 mmol) and *N*-methyl morpholine *N*-oxide (244 mg) in *t*-butanol (1 mL), a few crystals of osmium tetraoxide dissolved in THF (1.5 mL) were added. After 3 days 40% aqueous solution of NaHSO$_3$ (3 mL) was added, the mixture was stirred for additional 30 min, and extracted with ethyl acetate (3 × 20 mL). The extracts were dried (MgSO$_4$) and concentrated to dryness. The residue was purified by chromatography on a silica gel column with light petroleum–ether–methanol (4:5:0.5) to yield **9** (371 mg, 72%): mp 89.5° (from a mixture of hexane and ethyl acetate), $[\alpha]_D$ + 52° (*c* 1.0, CHCl$_3$).

Deacetylation (MeONa in methanol) followed by hydrolysis with Dowex 50W (H$^+$) resin furnished methyl D-*glycero*-α-D-*manno*-heptopyranoside **10**: $[\alpha]_D$ + 78° (*c* 1.7, MeOH), in 82% yield.

### I. 2-*O*-α-D-Rhamnopyranosyl-L-Rhamnose (14) [61] (Scheme 18)

*1-O-Acetyl-2,3,6-trideoxy-α,β-D-glycero-hex-2-enopyranos-4-ulose* **4** *[40,62]*

1*R*-(2-Furyl)ethanol **1**, $[\alpha]_D$ + 15° (*c* 1.4, CHCl$_3$), was obtained from the racemic alcohol by resolution of its ω-camphanyl ester [61].

To a stirred solution of 1*R*-(2-furyl)ethanol **1** (1.4 g, 12.5 mmol) in absolute methanol (10 mL) and absolute ether (15 mL), cooled to −40°C, a solution of bromine (2.14 g) in absolute methanol (10 mL) was added dropwise, maintaining the temperature at −35°– −40°C. Stirring was continued for 30 min, whereupon dry ammonia was bubbled to reach pH 8.0. After additional 15 min of stirring, ether (100 mL) was added, and the precipitated ammonium bromide was filtered off. The filtrate was concentrated to dryness, the residue was dissolved in a small volume of benzene, and filtered through a short layer of alumina. The filtrate was again concentrated and the oily residue was distilled at 72–75°C/1.2 torr to yield **2** (1.94 g, 89%).

A solution of **2** (1.74 g, 10 mmol) in 1% aqueous H$_2$SO$_4$ (20 mL) was left at room temperature for 70 min, then brought with solid NaHCO$_3$ to pH 5–6. Water was evaporated (⩽ 30°C) under reduced pressure and the residue was extracted with ether, dried (MgSO$_4$), and concentrated to dryness, to give **3** (1.12 g, 100%): mp 62°–65°C (from ether).

This product (415 mg, 3.2 mmol) was acetylated with acetic anhydride (0.5 mL) and pyridine (0.5 mL). After 2 h, the solvents were evaporated to leave crude product, which was purified on a silica gel column with light petroleum–ether (10:1) to give a mixture of α- and β-anomers of 1-*O*-acetyl derivative **4** (436 mg, 79.2%): $[\alpha]_D$ + 90° (*c* 1.95, CHCl$_3$).

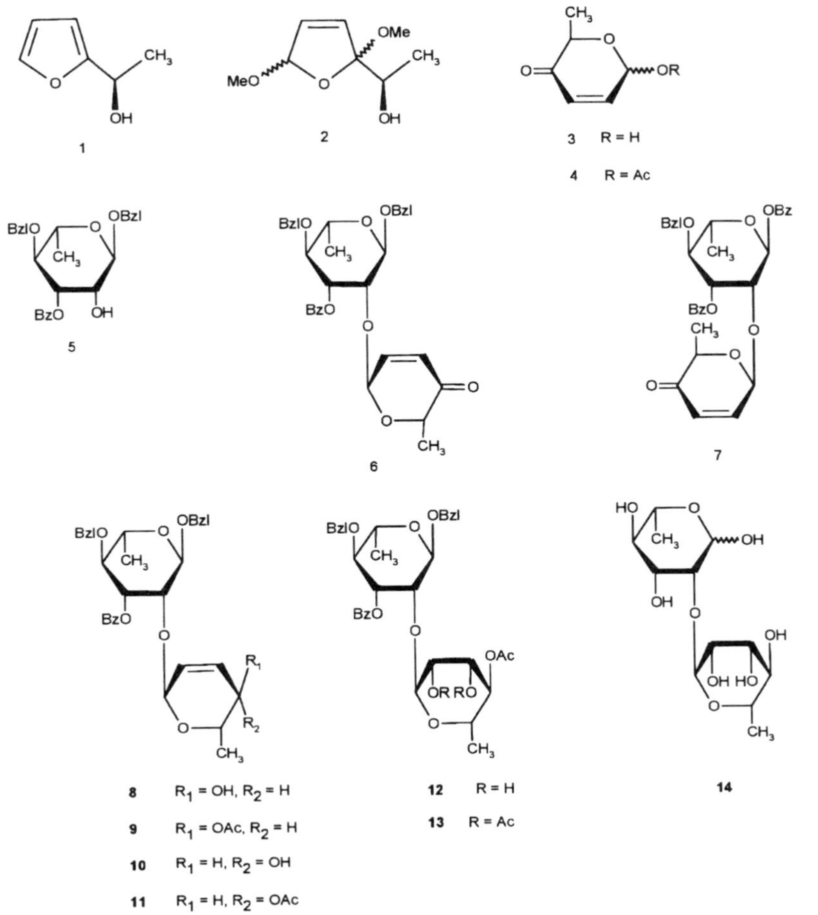

# Total Synthesis of Sugars

**Scheme 18**

8  $R_1 = OH, R_2 = H$
9  $R_1 = OAc, R_2 = H$
10 $R_1 = H, R_2 = OH$
11 $R_1 = H, R_2 = OAc$

12 $R = H$
13 $R = Ac$

*Benzyl 3-O-benzoyl-4-O-benzyl-2-O-(2,3,6-trideoxy-α-D- and β-D-hex-2-enopyranosyl-4-ulose)-α-L-rhamnopyranoside (6 and 7)*

To a solution of **5** (879 mg, 1.96 mmol) and **4** (357 mg, 2.1 mmol) in 1,2-dichloroethane (6 mL) was added stannic chloride (0.14 mL of a 1-*M* solution in 1,2-dichloroethane). After 3 h at room temperature, the mixture was diluted with $CH_2Cl_2$ (15 mL) and quickly washed with aqueous 5% $NaHCO_3$ and twice with water, dried, and concentrated. The oily residue (1.04 g) was subjected to flash chromatography with light petroleum–ethyl acetate (4:1).

Eluted first was **6** (803 mg, 73%), syrup: $[\alpha]_D - 13°$ (*c* 0.8, $CHCl_3$); and eluted second was **7** (109 mg, 9.9%), syrup: $[\alpha]_D\ 0°$ (*c* 1.2, $CHCl_3$).

*Benzyl 4-O-benzoyl-3-O-benzyl-2-O-(2,3,4-tri-O-acetyl-α-D-rhamnopyranosyl)-α-L-rhamnopyranoside* **13**

To a solution of sodium borohydride (2.0 g) in water (5 mL) and THF (15 mL) was added, dropwise, a solution of **6** (1.1 g) in THF (10 mL). The mixture was stirred for 1.5 h, poured into cold water, and extracted with ether, and the extract was dried and concentrated. The residue was chromatographed with light petroleum–ethyl acetate (7:2) to yield first α-*erythro* compound **8** (639 mg, 58%), acetylation of which afforded **9**: $[\alpha]_D + 51°$ (c 1.9, $CHCl_3$). Eluted second was α-*threo* compound **10** (138 mg, 12.5%), acetylation of which gave **11**, as a syrup: $[\alpha]_D - 41°$ (c 1.7, $CHCl_3$).

A solution of osmium tetraoxide (300 mg) in pyridine (1.2 mL) was added to a solution of **9** (551 g) in pyridine (3 mL) and the mixture was stirred at room temperature for 3 days. A solution of $NaHSO_3$ (600 mg) in pyridine (6 mL) and water (9 mL) was then added and the stirring was continued for 2 days. The mixture was diluted with water (100 mL) and extracted with several portions of $CH_2Cl_2$. The combined extracts were washed with water, dried ($MgSO_4$), and concentrated. The resulting thick syrup was chromatographed on a silica gel column with light petroleum–ethyl acetate (1:1) to yield diol **12** (454 mg, 78%), which was acetylated to give **13** as a thick syrup: $[\alpha]_D + 51°$ (c 1.05, $CHCl_3$).

A solution of **13** (300 mg) in ethanol (10 mL) was hydrogenated at 1 atm in the presence of 10% Pd–C. The catalyst was then filtered off and the solution was concentrated to dryness. To a solution of the residue in methanol (20 mL) was added sodium (120 mg), and the mixture was left for 12 h, neutralized with Amberlite IR-120 ($H^+$) resin, and concentrated to dryness. The residue was introduced onto a silica gel column and washed first with benzene to remove methyl benzoate, and then eluted with methanol to give **14** (115 mg, 86%): $[\alpha]_D + 49°$ (c 0.6, $C_2H_5OH$). The hexaacetate of **14** had $[\alpha]_D + 30°$ (c 0.3, $CHCl_3$).

## REFERENCES AND NOTES

1. A. Butlerov, *Liebigs Annalen der Chemie* 120:295 (1861).
2. J. K. N. Jones and W. A. Szarek, *Total Synthesis of Natural Products, Vol. 1*. (J. ApSimon, ed.), Wiley-Interscience, New York, 1973, p. 1.
3. A. Zamojski, A. Banaszek, and G. Grynkiewicz, *Adv. Carbohydr. Chem. Biochem.* 40:1 (1982).
4. A. Zamojski and G. Grynkiewicz, Total Synthesis of Natural Products, Vol. 6. (J. ApSimon, ed.), J. Wiley, New York, 1984, p. 141.
5. G. J. McGarvey, M. Kimura, T. Oh, and J. M. Williams, *J. Carbohydr. Chem.* 3:125 (1984).
6. D. J. Ager and M. B. East, *Tetrahedron* 49:5683 (1993).
7. W. R. Roush, *Trends in Synthetic Carbohydrate Chemistry* (D. Horton, L. D. Hawkins, and G. J. McGarvey, eds.), American Chemical Society, Washington DC, *ACS Symp. Ser.* 386:242 (1989).
8. T. Harada and T. Mukaiyama, *Chem. Lett.* p. 1005 (1981).
9. T. Katsuki, A. W. M. Lee, P. Ma, V. S. Martin, S. Masamune, K. B. Sharpless, D. Tuddenham, and F. J. Walker, *J. Org. Chem.* 47:1373 (1982).
10. A. W. M. Lee, V. S. Martin, S. Masamune, K. B. Sharpless, and F. J. Walker, *J. Am. Chem. Soc.* 104:3515 (1982).
11. P. Ma, V. S. Martin, S. Masumune, K. B. Sharpless, and S. M. Viti, *J. Org. Chem.* 47:1378 (1982).
12. S. Y Ko, A. W. M. Lee, S. Masamune, L. A. Reed III, K. B. Sharpless, and F. J. Walker, *Tetrahedron* 46:245 (1990).
13. A. Konował, J. Jurczak, and A. Zamojski, *Rocz. Chem.* 42:2045 (1968).
14. S. David, J. Eustache, and A. Lubineau, *J. Chem. Soc., Perkin Trans.* 1:2274 (1974).
15. R. R. Schmidt and R. Angerbauer, *Angew. Chem.* 89:822 (1977).

# Total Synthesis of Sugars

16. J. Jurczak, T. Bauer, S. Filipek, M. Tkacz, and K. Zygo, *J. Chem. Soc. Chem. Commun.* p. 540 (1983).
17. J. Jurczak, T. Bauer, and S. Jarosz, *Tetrahedron* 42:6477 (1986).
18. A. Konował and A. Zamojski, *Rocz. Chem.* 45:859 (1971).
19. A. Banaszek, *Bull. Acad. Pol. Sci. Ser. Sci. Chim.* 20:925 (1972); 22:79 (1974).
20. R. Angerbauer and R. R. Schmidt, *Carbohydr. Res.* 89:193 (1981).
21. S. David, J. Eustache, and A. Lubineau, *J. Chem. Soc. Perkin Trans.* 1:2274 (1974).
22. S. David, A. Lubineau, and J.-M. Vatèle, *J. Chem. Soc. Perkin Trans.* 1:1831 (1976).
23. S. David and J. Eustache, *J. Chem. Soc. Perkin Trans.* 1:2230 (1979).
24. T. Bauer, C. Chapuis, J. Kozak, and J. Jurczak, *Helv. Chim. Acta,* 72:482 (1989).
25. M. Terada, K. Mikami, and T. Nakai, *Tetrahedron Lett.* 32:935 (1991).
26. S. Danishefsky and R. R. Webb II, *J. Org. Chem.* 49:1955 (1984).
27. S. J. Danishefsky and M. P. DeNinno, *Angew. Chem. Int. Ed. Engl.* 26:15 (1987) and the literature cited therein.
28. Q. Gao, K. Ishihara, T. Maruyama, M. Mouri, and H. Yamamoto, *Tetrahedron* 50:979 (1994).
29. K. Maruoka, T. Itoh, T. Shirasaka, and H. Yamamoto, *J. Am. Chem. Soc.* 110:310 (1988).
30. M. Bednarski and S. Danishefsky, *J. Am. Chem. Soc.* 108:7060 (1986).
31. S. Apparao and R. R. Schmidt, *Synthesis.* pp. 896, 900 (1987).
32. R. R. Schmidt and M. Maier, *Liebigs Ann. Chem.* pp. 2261 (1985).
33. R. R. Schmidt, *Acc. Chem. Res.* 19:250 (1986) and the literature cited therein.
34. L.-F. Tietze, A. Montenbruck, and C. Schneider, *Synlett.* p. 509 (1994).
35. P. Vogel, D. Fattori, F. Gasparini, and C. Le Drian, *Synlett.* p. 173 (1990) and the literature cited therein.
36. E. Vieira and P. Vogel, *Helv. Chim. Acta* 66:1865 (1983).
37. E. J. Corey and T.-P. Loh, *Tetrahedron Lett.* 34:3979 (1993).
38. J. Wagner, E. Vieira, and P. Vogel, *Helv. Chim. Acta* 71:624 (1988).
39. Y. Auberson and P. Vogel, *Helv. Chim. Acta* 72:278 (1989).
40. O. Achmatowicz, Jr., P. Bukowski, B. Szechner, Z. Zwierzchowska, and A. Zamojski, *Tetrahedron* 27:1973 (1971).
41. G. Piancatelli, A. Scettri, and M. D'Auria, *Tetrahedron Lett.* 25:2199 (1977).
42. O. Achmatowicz, Jr. and R. Bielski, *Carbohydr. Res.* 55:165 (1977) and the literature cited therein.
43. This diene, prepared according to A. E. Montagna and D. H. Hirsch [US Patent 2,902,722 (1959)], is practically a pure *trans*-isomer, whereas the commercially available reagent is a *cis–trans*-isomer mixture.
44. *Org. Synth. Coll.* Vol. *4*:124. Freshly distilled butyl glyoxylate reacts with the diene with exothermic effect.
45. Ref. 25: Condensation (− 55°C, 1h) affords a *cis–trans* isomer mixture of **2** (R = Me) (87:13, 72%). *cis*-Adduct (96% e.e. determined by LIS-NMR analysis using (+)-Eu(DPPM)$_3$ as a chiral reagent) has 2*S*,6*R* configuration.
46. J. Jurczak, A. Konował, and A. Zamojski, *Rocz. Chem.* 44:1587 (1970).
47. A. Konował, J. Jurczak, and A. Zamojski, *Tetrahedron* 32:2957 (1976).
48. H. Gerlach, *Helv. Chim. Acta* 51:1587 (1968). The reagent is commercially available.
49. D. D. Gero and R. D. Guthrie, *J. Chem. Soc.* C:1761 (1967).
50. T. Bauer, C. Chapuis, A. Jezewski, J. Kozak, and J. Jurczak, *Tetrahedron Asymm.* 7:1391 (1996).
51. Obtained by ozonolysis of *N*-crotonoyl- or *N*-fumaroyl-(2*R*)-bornane-10,2-sultam [50]. Preparation of (2*R*)-bornane-10,2-sultam: W. Oppolzer, C. Chapuis, and G. Bernardinelli, *Helv. Chim. Acta* 67:1397 (1984).
52. 2,7-Dimethyl-3,5-octadiyne-2,7-diol (**1**, Scheme **4**) is obtained by CuCl$_2$-promoted oxidative coupling of 2-methyl-3-butyn-2-ol: H. A. Stansbury, Jr. and W. R. Proops, *J. Org. Chem.* 27:320 (1962). Preparation of 2-methyl-3-butyn-2-ol: *Org. Synth.* 3:320 (1955).
53. Alternatively, 3-*O*-butadienyl ethers of 1,2:5,6-di-*O*-isopropylidene-α-D-glucofuranose can be

prepared by a Wittig reaction between the ylide generated from 3-*O*-(triphenylmethyl)-phosphonium salt of 1,2:5,6-di-*O*-isopropylidene-α-D-glucofuranose and acrolein [21].

54. *Org. Synth. 61*:147 (1983); 1-methoxy-3-trimethylsilyloxy-1,3-butadiene is commercially available.
55. S. Danishefsky, M. Bednarski, T. Izawa, and C. Maring, J. Org. Chem. *49*:2290 (1984).
56. (+)-Eu(hfc)$_3$: Tris[3-(heptafluoropropylhydroxymethylene)-(+)-camphorato]europium(III). The reagent is commercially available.
57. L. DeGaudenzi, S. Apparao, and R. R. Schmidt, *Tetrahedron 46*:277 (1990).
58. K. A. Black and P. Vogel, *Helv. Chim. Acta 67*:1612 (1984).
59. A. Warm and P. Vogel, Helv. Chim. Acta *70*:690 (1987). An improved synthesis of (1*S*,4*S*)-7-oxabicyclo[2.2.1]hept-5-en-2-one starting from a new chiral auxiliary, methyl (1*S*,5*R*,7*S*)-3-ethyl-2-oxo-3-aza-6,8-dioxabicyclo[3.2.1]octane-7-*exo*-carboxylate (obtained from di-*O*-acetyl-(*S*,*S*)-tartaric anhydride and *N*-ethylaminoethanol ethyl acetal), was reported: J.-L. Reymond and P. Vogel, *Tetrahedron Asymm. 1*:729 (1990).
60. Dziewiszek, M. Chmielewski, and A. Zamojski, *Carbohydr. Res. 104*:Cl (1982); K. Dziewiszek, Doctoral Dissertation, Institute of Organic Chemistry, Polish Academy of Sciences, Warsaw, 1987.
61. A. Jaworska and A. Zamojski, *Carbohydr. Res. 126*:191, 205 (1984).
62. O. Achmatowicz, Jr. and B. Szechner, *Rocz. Chem. 49*:1715 (1975).

# Index

Acetals, 3
    2-acetamido-2-deoxy-D-glucose, 23
    acidic conditions, 7
    acyclic sugars, 15
    amide acetals, 6
    amino sugars, 23
    Baldwin's rules, 11
    basic conditions, 10
    benzylidene, 18, 20, 23
    catalyst, 7
    copper(II)sulfate, 7
    D-arabinopyranose, 17
    D-galactopyranose, 21
    D-glucofuranose, 18
    D-lyxofuranose, 17
    D-mannopyranose, 22
    D-xylose, 15
    deoxysugars, 24
    $\alpha,\alpha$-dihalotoluenes, 10
    2,2-dimethoxypropane, 8
    $\alpha,\alpha$-dimethoxytoluene, 8
    1,3-dioxanes, 4
    1,3-dioxolanes, 4
    dispiroacetals, 27
    formation, 11
    furanoses, 18
    hexoses, 18
    hydrolysis, 11
    isopropylidene, 15, 16, 17, 18, 19, 20, 21, 22, 23, 24
    kinetic control, 12
    L-fucopyranose, 24
    L-rhamnopyranose, 24
    Lewis acid, 7
    maltose, 26
    2-methoxypropene, 9

[Acetals]
    methyl $\alpha$-D-mannopyranoside, 28
    methyl $\alpha$-L-fucopyranoside, 27
    methyl D-galactopyranoside, 20
    methyl D-glucopyranoside, 19, 27
    methyl D-ribofuranoside, 16
    N-bromosuccinimide, 11
    neutral conditions, 7
    $O$-(dimethylaminoalkylidene), 6
    $O$-alkylboron, 6
    $O$-benzylidene, 5
    $O$-cyanoalkylidene, 6
    $O$-cycloalkylidenes, 5
    $O$-ethylidene, 5
    $O$-isopropylidene, 5
    $O$-methylene, 5
    $O$-silylene, 6
    oligosaccharides, 25
    *ortho*-esters, 11
    oxocarbenium ion, 11
    pentoses, 16
    preparation, 6
    procedures, 15
    pyranoses, 17
    selective hydrolysis, 14
    solvent, 7
    sucrose, 25
    thermodynamic control, 12
    transacetalation, 8
    $\alpha,\alpha$-trehalose, 26
    vicinal diols, 27
    with enol ethers, 8
Aldolases, 469
    classification, 469, 470
(+)-Allokainic acid, 598
Allosamizoline, 549

Amino sugars, 595
　allyltrimethylsilane, 599
　asymmetric epoxidation, 596
　$C_1$-elongation, 596
　$C_3$-elongation, 598, 603
　$C_4$-elongation, 599, 607, 608
　cyanohydrin formation, 596
　L-daunosamine, 597, 601
　*N*-acetyL-L-tolyposamine, 598
　organometallic reagents, 596
　total synthesis, 595
　trimethylsilyl cyanide, 596
　2-(trimethylsilyoxy)furan, 599, 610
　Wittig olefination, 596
Aminoglycoside antibiotics, 134
Amphotericin B, 307
Anguidine, 554
Anthracyclinones, 570
Aquayamycin, 528
Avermectin $A_{1a}$, 225
Avermectin $B_{1a}$, 263, 320, 329, 385, 432

Barton deoxygenation, 549
Barton-McCombie deoxygenation, 154
Bidentate activation, 432, 435
Boromycin, 134
Branched-chain sugars, 207
　aldolization-crotonization, 232
　alkyllithium reagents, 212, 241
　allyl branched sugars, 224
　from allylic acetates, 225
　from anomeric radicals, 224
　from carbohydrate enolates, 229
　from carbonyl sugars, 211
　Claisen rearrangement, 228
　conjugate addition, 217, 243
　cyclobutanes, 237
　cyclohexanes, 238
　cyclopentanes, 237
　cyclopropanes, 237
　deoxygenation, of tertiary alcohols, 209
　Diels-Alder reaction, 233, 239, 252
　dithiane, 214, 216, 219, 242, 244
　doubly branched-chain sugars, 235
　enolate type anions, 214, 242, 248

[Branched-chain sugars]
　from enones, 217
　from enopyranosides, 225
　from expoxides, 216
　Eschenmoser rearrangement, 228, 236, 251
　exocyclic olefins, 231
　FriedeL-Crafts reaction, 234
　from nitroolefins, 218
　from olefinic compounds, 225
　Grignard reagents, 211, 212, 235, 240, 241
　Horner-Wadsworth-Emmons reactions, 231, 250
　IrelanD-Claisen rearrangement, 227
　Knoevanagel reaction, 232
　nitrone-nitrile oxide cycloaddition, 227
　organo-cuprates, 217, 218, 225
　palladium-catalyzed reaction, 226, 247
　Pauson-Khand reaction, 237, 238, 258
　Peterson olefination, 231, 232, 249, 250
　radical reactions, 219, 244, 245, 246
　rearrangements, 227, 247
　samarium iodide-mediated coupling, 230
　$S_N2'$ substitution, 225
　StilL-Wittig rearrangement, 227
　Tebbe reaction, 231
　types, 208
　Vilsmeier-Haack reaction, 234
　with functionalized carbanions, 216
　with organometallic reagents, 216, 243
　Wittig reaction, 231, 249

*C*-aryl glycosides, 528
*C*-glycosides, 528
*C*-glycoside synthesis, 505
　from enol ethers, 505
　by free-radical methods, 507
　from ketene acetals, 505
　from malonate anion, 505
　from 2-pyridylthio glycosides, 385

# Index

C-Glycosyl compounds, 507
  acrylonitrile, 513
  alkene reactivity, 511
  alkoxyalkyl radicals, 507
  allyl tin radical, 511
  allylstannane, 511, 512, 520
  anomeric effect, 509
  anomeric phenyl sulfones, 517
  anomeric radical, 509, 510, 512, 516
  2,2-azobisisobutyronitrile, 511
  Barton decarboxylation, 515, 521
  β-bond cleavage, 511
  α-C-glycosides, 509
  β-C-glycosyl compounds, 515
  C-disaccharides, 508, 511, 517, 518, 519, 523
  cobalt method, 513
  conformations, 509
  α,α-disubstituted olefins, 518
  electron spin resonance (ESR), 509
  9-endo-trig cyclizations, 517
  5-exo-trig cyclizations, 516
  fragmentation method, 511, 512
  glycosyl radicals, 508
  glycosyL-cobalt(III) complex, 513, 520
  hexopyranosyl radical, 509
  hydrogen atom-transfer, 508
  hydrostannylation, 511
  intramolecular methods, 516, 522
  large-scale production, 517
  N,O-protection, 508
  N-acetylneuraminic acid, 512
  one-electron reduction, 517
  pentopyranosyl radical, 509
  radical initiator, 511
  reduction, 511
  samarium diiodide, 517, 523
  SOMO-LUMO interactions, 507, 508
  stereochemistry, 509
  stereoelectronic effets, 509
  sugar cobalamine, 514, 521
  tertiary nitro sugar, 515, 519, 522
  tetheredD acceptors, 516
  tin hydride method, 510, 517, 518, 519, 522, 523
  trialkyltin radical, 510

[C-Glycosyl compounds, 507]
  tributyltin hydride, 510, 517, 518, 519
  umpolung method, 507
  unimolecular scission, 511
  vitamin $B_{12a}$-catalyzed method, 514, 521
Calicheamycin, 263
Carbasugars, 570
Carbocycles, 237, 238, 543, 545, 569
  cyclohexanes, 543
  cyclopentanes, 543
  free radical methods, 545
  organometallic methods, 545
Carbon extension, thiazole-based, 173
  α-aminoaldehydes, 192
  aminohomologation, 182
  C-formylation, 187
  C-glycoside, thiazolyl, 186
  dialdoses, 189
  Lewis acids, effect of, 184
  2-lithiothiazole, 190
  N-benzyl nitrones, 190
  polyalkyoxyaldehydes, 189
  stability, 187
  syn/anti selectivity, 181
  thiazole hydrolysis, 179
  2-(trimethylsilyl)thiazole, 178
Cell adhesion, 358
Ceric ammonium nitrate, 272
Chiron, 506
Chiral templates, 594
Chlorodeoxy sugars, 99, 106, 111
  chloronium ion, 115
  chlorosulfate groups, 112
  chlorosulfnoyloxy group, 112
  2,3-cyclic sulfate, 111
  cyclic sulfates, 111
  dechlorosulfation, 112, 113, 116
  furanoid derivatives, 115
  glycosyl chlorides, 114
  hydrogenation, 115
  paratose, 115, 121
  preparation, 111
  Raney nickel catalyst, 115
  S–Cl bond scission, 112
  steric and polar factors, 112

[Chlorodeoxy sugars]
  sulfuryl chloride, 111
  transition state, 112
L-Cladinose, 385
Claisen rearrangement, 228, 236, 251
Clauson-Kaas reaction, 621
Clindamycin, 107, 108
Compactin, 570
Cryptosporin, 570
Corey lactone, 549
Cyclic, 55
Cyclohexane derivatives, 569
  aldol cyclization, 575, 586
  alkenylsilanes, 575
  C-1 carbanions, 580
  cycloaddition processes, 582
  cycloaddition reactions, 582
  2-deoxystreptamine, 571
  Dieckmann cyclization, 574
  Diels-Alder cyclizations, 571, 582, 584, 585, 589
  1,3-dipolar additions, 582, 583, 588
  dithiane derivative, 580
  double activation, 577
  epoxide opening, 580
  free radical cyclizations, 582
  hydroxycyclohexanones, 579
  hydroxymercuration, 577, 587
  intramolecular cyclization, 571, 572, 573
  Knoevanagel reaction, 575
  *myo*-inositol, 579
  nitrile oxides, 582
  nitroalcohols, 571
  nitroalkane cyclizations, 571, 585, 586
  nitrones, 582
  phosphonate cyclization, 572, 586
  validatol, 580
  vinylsilanes, 576

Danishefsky diene, 619
L-Daunosamine, 597
  total synthesis, 601
Deoxy sugars, 151
Deoxygenation, 151

Deoxygenation, free radical, 151
  azoiso-butyronitrile (AIBN), catalyst, 157
  chain carriers, 155
  dialkyl phosphites method, 156, 158, 160, 161
  diphenylsilane method, 155, 163
  4-fluorophenyoxythiocarbonate, 156
  hindered hydroxyl, 152
  hydrogen atom sources, 155
  hypophosphorous acid method, 158, 162
  industrial scale, 157
  pentafluorophenoxythiocarbonate, 156
  primary alcohols, 154
  radical chain, 153
  relative rates, 156
  secondary alcohols, 152
  $^{119}$Sn NMR, 154
  thiocarbonyl derivative, 153
  thiocarbonylimidazolides, 153
  thioxobenzoates, 153
  tributyltin hydride, 153, 157, 159, 162, 164, 165, 166, 167, 168
  triphenylsilane, 155
  triphenyltin hydride, 154
  tris(trimethylsilyl)silane, 155, 163
  xanthathes, 153
2-Deoxystreptamine, 571
D-Desosamine, 385
Diels-Alder reaction, 233, 239, 252, 571, 582, 584, 585, 589
Diethylaminosulfur trifluoride (DAST reagent), 108, 129, 318
1,3-Dipolar cycloaddition, 210
1,3-Dithiane, 176, 214, 216, 219, 242, 244
Dithioacetals, 35
  *aldehydo* sugars, 37
  aldoses, 40
  alkyl, 40
  as intermediates, 37
  chain extension, 38
  chain-descent, 37
  crystallization, 40
  demercaptalation, 37

# Index

[Dithioacetals]
   deprotection, 37
   dialdoses, 42
   disulfones, 38
   elimination, 38
   Fischer, 37
   formation, 40
   glycuronic acids, 42
   isolation, 40
   ketene dithioacetal, 38
   ketoses, 42
   McDonalD-Fischer degradation, 38
   mechanism, 40
   monobromination, 38
   oxidation, 38
   procedures, 43
   Raney nickel, 38
   reduction, 38
   solubility, 39
   stench, 43
   synthetic applications, 37
   thioglycoside products, 40
   umpolung, 37
   Wolfrom-Karabinos reaction, 38
Drug development, 390

Elfamycin, 327, 334
Entner-Doudoroff pathway, 476
Enzymatic glycosidation, 485
   *Baccilus circulans* enzyme, 493
   background, 486
   CMP-NeuAc, 496
   cofactor generation, 491
   disaccharide, 496
   fucosylation, 499, 500
   fucosyltransferas, 490
   galatosyltransferases, 490
   glycosidases in synthesis, 491, 492, 493
   glycosyltransferases, 489, 490
   Leloir pathway, 486
   mannosylation, 500, 501
   mannosyltransferases, 491
   $N$-acetylglucosaminyl transferases, 491
   N-linked oligosaccharides, 486
   non-Leloir pathway, 486

[Enzymatic glycosidation]
   O-linked oligosaccharides, 487
   sialylation, 497, 498, 499, 501, 502
   sialyltransferases, 490
   substrate specificity, 490
Enzymatic synthesis, 467, 469
   *Aureobacterium barkerei* enzyme, 476
   *Clostridium perfringens* enzyme, 471
   diastereomeric sugars, 474
   Entner-Doudoroff pathway, 476
   *Escherichia coli* enzyme, 471
   KDG aldolase, 470, 476
   KDgal aldolase, 470
   KDN, 472
   KDO aldolase, 470, 475
   KDO analogues, 474
   KDO, 471
   KDPG aldolase, 470, 476
   KHG aldolase, 472
   *Pseudomonas fluorescens* enzyme, 476
   *Pseudomonas putida* enzyme, 472
   pyruvate aldolases, 470
   sialic acid aldolase, 470, 471
   sialic acid analogs, 473
   stereoselectivity, 474
   substrates, 472
   subtilisin, 472
Erythromycin, 385, 432
Erythronolide A, 385
   glycosidation, 385
Eschenmoser rearrangement, 228, 236, 251

Fischer glycosidation, 383
FK 506 (Tacrolimus ), 570
9-Fluorenylmethoxycarbonyl (Fmoc) group, 266, 267, 268, 269, 273, 274, 276
Formyl anion equivalents, 178
FriedeL-Crafts reaction, 234, 529, 530, 535, 536
Fries-type rearrangement, 295
Functionalized carbocycles, 545
   acyl radical, 555
   allosamizoline, 549

Barton deoxygenation, 548, 549
[Functionalized carbocycles]
   bicyclic compounds, 551
   carba-D-fructofuranose, 547
   Corey lactone, 549
   $Cp_2TiCl$ catalyst, 553, 563
   cyclic transition state, 548, 549
   cyclohexanes, 554
   cyclopentanes, 546
   1,2-dialkylcyclopentanes, 548
   electron spin resonance, 552
   eneynes, 556
   epoxy-olefin cyclization, 553
   Ferrier protocol, 555
   free radical chemistry, 546
   hept-6-enyl radical cyclization, 554
   1,6-heptadiene system, 552, 562
   hex-5-enyl radical cyclization, 546, 558
   low-valent titanium, 556
   low-valent zirconium, 556, 564, 565
   McMurray titanium reagent, 555
   organometallic chemistry, 546
   organometallic methods, 555
   oxidative destannylation, 552
   $\alpha$-oxyradicals, 552
   Pauson-Khand reaction, 556, 564
   PD-catalyzed alkylation, 557
   radical trapping, 551
   rhodium-catalyzed hydroacylation, 557, 558
   ring contraction, 557
   samarium diiodide, 554, 564
   tandem cyclization, 552
   tin method, 546
   tributyltin hydride, 547, 558, 559, 561, 562
   Wittig reaction, 547, 559
Furanoses, 185

Gangliosides, 358, 362, 366, 367, 372, 373, 374, 375
   synthesis, 367, 372, 373, 374, 375
Gilvocarcin, 528
Globotriasolylceramide ($Gb_3$), 326, 333
$\alpha$-D-Glucopyranosyl azide, 462
Glycobiology, 263

Glycomimetics, 468
Glycoproteins, biosynthesis, 486, 487
Glycosidases, 487, 491, 492
Glycoside synthesis, 381
   anomeric selectivity, 382
   challenges, 382
   via electrochemistry, 385
   via electron transfer, 385
   inductive effects, 452
   miscellaneous methods, 432
   new glycosyl donors, 386
   orthogonal, 452
   overview, 381
   2-pyridylthio glycosyl donors, 385
   remote activation concept, 381
   solid phase, 382, 386
   thioglycoside activation, 384
   unprotected donors, 389
Glycosyl carbonates, 415, 432, 433
Glycosylarenes, 527
   applicability, 535
   aromatic substitution, 528
   $BF_3 \cdot OEt_2$ promoter, 533
   conformation, 528
   $Cp_2HfCl_2$–$AgClO_2$ promoter, 533, 536, 539
   electron-rich aromatics, 535
   fluoride donor, 531
   FriedeL-Crafts reaction, 529, 530, 535, 536
   Lewis acidic activation, 529
   mechanism, 532
   $Ph_3SiCL$-$AgClO_4$ promoter, 538
   polarity inversion, 529
   quinone methide, 533
   reaction profile, 531
   rearrangement, 531
   regioselectivity, 533
   $\alpha/\beta$ selectivity, 533
   $SnCl_4$ promoter, 537
   stereoelectronic factor, 528
   stereoselectivity promoter, 532
   thermodynamic stability, 528
   thiopyridyl donor, 531
   trichloroacetimidate donor, 531
   1,3,5-trimethoxybenzene, 531
Glycosyltransferases, 469, 487, 488

Golgi apparatus, 486
Halide ion-catalyzed, 414
Halodeoxy sugars, 105, 127
    (chloromethylene)dimethyliminium
        chloride, 109
    aldehydo sugars, 37
    alkoxyphosphonium salt, 108
    anhydro-ring formation, 109
    catalytic hydrogenation, 110
    direct replacement, 107
    elimination, 109
    general methods, 107
    nucleophilic attack, 109
    $O$-benzylidene sugars, 107
    oxiranes, 107
    Raney nickel catalyst, 109
    reactivity, 109
    reduction, 109
    reductive dehalogenation, 109
    stereoelectronic factors, 109
    tributyltin hydride, 110
    triphenylphosphine dichloride, 107
    triphenylphosphine–carbon tetra-
        chloride, 107, 108
Hedamycin, 528
Henry reaction, 176
Hikizimycin, 181
Hikozamine, 619
High-mannose glycoprotein, 344
Horner-Wadsworth-Emmons reaction,
    231, 250, 572
Hydantocidin, 187

Influenza virus, 366
Imidazol-1-sulfonate esters
    acetate, 143
    3,6-anhydro derivatives, 134, 135, 144
    azide, 132, 133, 135, 141, 142
    benzoate, 132, 133, 134, 142, 143,
        144
    C-2 position, 132
    C-3 and C-4 positions, 133
    chloride, 131
    chlorination, 131
    chromatographic purifiction, 135
    compatibility, 135
    crystallinity, 135

[Imidazol-1-sulfonate esters]
    dichloro aminoglycoside, 131
    β-elimination, 131
    fluoride, 133
    hydrolytic stability, 135
    imidazylate, Imz, 130, 136
    intramolecular cyclization, 135
    iodide, 132, 133, 139, 140
    β-lactams, 135
    $N,N'$-sulfuryL-diimidazole, 130
    nitrate, 143
    nonbonded 1,3-diaxial interaction,
        133
    nucleofugal properties, 135
    preparation, 130, 136
    reactions, 131
    reactivity, 130
    remote activation, 130
    ring-contraction, 135
    shelf-life, 135
    unsaturated sugars, 134
Ireland-Claisen rearrangement, 227
Iterative oligosaccharide synthesis, 396,
    451

KDO, 619
Keck reaction, 86
Kiliani cyanohydrin synthesis, 86
Kiliani-Fischer cyanohydrin synthesis,
    174
Knoevanagel reaction, 232, 574
Koenigs-Knorr method, 285, 314, 358,
    414

Leloir pathway, 486
Lincomycin, 107, 108
Lipopolysaccharides (LPS), 471
D-Lividosamine, 134

Masamune-Sharpless homologation, 177
McMurray reaction, 555
Medermycin, 528
3-Methoxy-2-pyridyloxy (MOP)
    glycosyl donors, 391
    activation, 391
    disaccharide, trisaccharide, 454, 460,
        461, 462

[3-Methoxy-2-pyridyloxy (MOP)
    glycosyl donors]
  1,2-cis-glycosides, 391
  glycosides, oligosaccharides, 389,
    413
  mechanism, 391
  preparation, 394, 398, 399, 400, 401
  O-protected, 413, 422
  reaction progress, 394
  O-unprotected, 391
Michael glycosidation, 383
Micinolide, 386
Mitsunobu reaction, 128
Milbemycins, 570
Multienzyme systems, 494

N-acetylneuraminic acid, 358
  dehydro, 358
  glycosides, 358
  oligosaccharides, 357, 413
Nef reaction, 176
Nephritogenoside, 455
NodRM-IV factor, 328, 335
D-Nojirimycin, 185
Nucleophilic displacement reactions, 87,
    105, 127

O- and N-glycopeptides, 265, 486
  allyl ester, 267, 270, 276
  allyl transfer, 267
  allyloxycarbonyl group (Aloc), 267,
    271, 277
  β-elimination, 266
  benzyl ester, 267, 274
  biotinyl conjugate, 270
  chemoselectivity, 268
  compatibility, 268
  enzymatic reactions, 267, 268
  ethyl 2-ethoxy-1,2-dihydroquinoline-
    1-carboxylate (EEDQ), 271
  9-fluorenylmethoxycarbonyl (Fmoc)
    group, 266, 267, 268, 269, 273,
    274, 276
  from glycals, 268
  from glycosyl azides, 268
  from glycosyl trichoroacetimidates,
    268

[O- and N-glycopeptides]
  $N^4$-glycosyl asparagine, 267
  O-glycosyl serine, 266
  O-glycosyl threonine, 266
  O→N acetyl transfer, 270
  orthogonal deprotection, 273
  protecting groups, 268
  silver triflate, promoter, 268, 269
  soliD-phase synthesis, 266
  tert-butyl ester, 267, 274
  tert-butyloxycarbonyl (Boc), 267
  from thioglycosides, 268
  $T_n$ antigen, 270
O-benzylidene acetals, 53
  benzyl ethers, from, 57
  cleavage, 55
  diphenylmethylene acetals, 61
  $LiAlH_4$–$AlCl_3$, 57, 59, 61, 62, 63
  $Me_3NBH_3$–$AlCl_3$, 59, 60, 64
  N-bromosuccinimide, 56
  $NaCNBH_3$/$CF_3COOH$, 61, 65
  $NaCNBH_3$/HCl, 57, 63
  $NaCNBH_3$/HCl/THF, 58, 60
  $NaCNBH_3$/$Me_3SiCl$, 61, 65
  p-methoxybenzylidene acetals, 61
  2'-propenylidene acetals, 61
  reductive cleavage, 57
  regioselective openings, 56
O-glycosyl serine, 266
O-glycosyl threonine, 266
O-protecting groups, 54
  protecting groups, 54
Oligosaccharide and glycoside synthesis,
    283, 313
  anomeric activation, 286, 287, 289
  enzymatic methods, 485
  Fischer-Helferich method, 285
  glycal method, 316
  glycosyl fluoride method, 313, 386
  halide-ion catalyzed, 414
  imidate method, 283
  Koenings-Knorr method, 285, 314,
    414
  pentenyl glycoside method, 339, 316
  phosphite method, 295, 319
  remote activation method, 381, 389,
    413, 431, 449

# Index

[Oligosaccharide and glycoside synthesis]
   silver triflate TMU Method, 414, 436, 438
   sulfoxide method, 319, 415
   thioglycoside method, 286, 313, 359
   unprotected donors, 389

Oligosaccharides from $n$-pentenyl glycosides, 339
   armed-disarmed protocol, 341
   basic strategies, 340
   blood group substance B tetrasaccharide, 345
   bromine promoter, 344, 351
   1,2-$cis$, 340
   disaccharides, 350, 351
   mechanism, 341
   methods, 341
   preparation, 341, 348, 349
   $N$-glycosyl products, 347
   $n$-halosuccinimide promoter, 343, 354
   $n$-pentenyl glycosides (NPGs), 341
   NIS/Et$_3$SiOTf promoter, 343, 350, 353
   nonasaccharide assembly, 344, 345
   1,2-orthoesters, 342, 352, 353
   promoters, 342
   strategies, 342
   tetrasaccharides, 350
   1,2-$trans$, 340

Oligosaccharide synthesis by selective anomeric activation, 449
   AgOTf, promoter, 452
   1,2-$cis$-disaccharides, 455, 456, 459, 460, 461, 462
   Cu(OTf)$_2$, promoter, 452
   disaccharide glycosyl MOP donors, 454, 460, 461
   oligosaccharide assembly, 455, 456, 462, 464
   TOPCAT-MOP combinations, 455

Oligosaccharide synthesis from glycosyl fluorides and sulfides, 313
   advantages, 317
   diethylaminosulfur triflouride (DAST), 318

[Oligosaccharide synthesis from glycosyl fluorides and sulfides]
   elfamycin, 327, 334
   globotriasolylceramide (Gb$_3$), 326, 333
   glycosyl fluoride donors, 317
   glycosyl sulfone donors, 319
   glycosyl sulfoxide donors, 319
   hexasaccharide, 321, 330
   mechanism, 317
   $N$-bromosuccinimide (NBS), 318
   NodRM-IV factor, 328, 335
   octasaccharide, 323
   pentasaccharide, 323
   sialyl dimeric Le$^x$, 324, 332
   single-electron transfer, 319
   stability, 317
   sulfated Le$^x$ tetrasaccharide, 325, 333
   thioglycoside donors, 317
   thiophiles, 318
   trimeric Le$^x$, 323, 331
   trisaccharide, 322
   two-stage activation, 317, 318, 319
   undecasaccharide, 323

Oligosaccharide synthesis with O-protected glycosyl 2-thiopyridyl-carbonate donors, 431
   anomeric activation, 432, 435
   1,2-$cis$-disaccharides, 434, 440, 441, 444
   crystallinity, 434
   Cu(OTf)$_2$ promoter, 433
   design, 432
   glycosyl 2-pyridylcarbonates, 433, 439
   glycosyl 2-thiopyridylcarbonates, 434, 444
   Le$^x$ trisaccharide, 438, 446
   one-pot glycosylations, 433, 439
   reactivity, 435
   selective activation, 438
   $\alpha$-selectivity, 434, 440, 441, 444
   sialyl Le$^x$, 436, 444, 445
   silver triflate, promoter, 434, 435, 436, 440, 441, 445
   stability, 434
   TOPCAT donors, 434, 444

[Oligosaccharide synthesis with O-protected glycosyl 2-thiopyridyl-carbonate donors]
    TOPCAT and imidate donors, 438, 439
    1,2-*trans*-disaccharides, 435, 436, 441, 442, 443
Oligosaccharide synthesis with O-protected (MOP) donors, 413
    1,2-*cis*-disaccharides, 415, 416, 417, 418
    $CH_3CN$, solvent, 418
    copper triflate, promoter, 417, 423, 424, 425, 427
    $Le^x$ trisaccharide, 420
    MeOTf, promoter, 416, 422
    reaction progress, 421
    scope, 417
    sialyl $Le^x$, 419, 426
    solvents, 417, 418
    T-antigen, 420
    1,2-*trans*-disaccharides, 418, 419
Oligosaccharide synthesis with unprotected MOP donors, 389
    catalysts, 391
    1,2-*cis* selectivity, 391
    disaccharide, 392, 395, 397, 401, 402, 403, 404, 405
    glycosyl MOP donors, 391
    iterative synthesis, 396
    mechanism, 393
    reaction progress, 394
    selective activation, 395
    tetrasaccharide, 393, 396, 409
    trisaccharides, 396, 408
Oligosaccharide synthesis with trichloroacetimidates, 283, 289, 315
    activation, 289
    amphoteronolide, 307
    aryl-*C*-glycosides, 295
    Brønsted acids, 294
    carbenium ion, 293
    decomposition, 294
    disaccharides, 298, 301, 302, 303, 304
    formation 289, 296, 297, 298
    Fries-type rearrangement, 295

[Oligosaccharide synthesis with trichloroacetimidates]
    general aspects, 290
    glycopeptide, 306
    glycosyl phosphates, 294, 307, 308
    hexasaccharide, 299
    inverse procedure, 294
    kinetic, thermodynamic effects, 293
    Lewis acid catalysis, 289
    neighboring group participation, 292
    nitrile effect, 293
    1-*O*-acyl compounds, 294
    *O*-glycosyl phosphinates, 294
    *O*-glycosyl phosphonates, 294
    reactivity, 293
    scope, 295
    $\alpha/\beta$ selection, 290
    side-reactions, 293
    $S_N1$-type reaction, 293
    $S_N2$-type reactions, 292
    solvent effect, 293
    stability, 290
    trisaccharides, 299, 300, 301
*Ortho*-ester, 55

Pauson-Khand reaction, 237, 238, 258, 556, 564
Peterson olefination, 231, 232, 249, 250
Phyllanthocin, 570
Protein folding, 486
Pseudomonic acid, 226
Remote activation concept, 381, 383, 389, 413
Rydon reagent, 108

Seldomycin factor 2, 131
Selectins, 358, 366, 420
Shikimate pathway, 471
(−)-Shikimic acid, 572
Showdomycin, 530
Sialyl dimeric $Le^x$, 324
Sialyl glycosides, 357
    β-acetonitrilium ion, 362
    anomeric effect, 362
    benzeneselenyl triflate, 359
    biology, 366, 419, 421, 436, 445

# Index

[Sialyl glycosides]
   donor activation, DMTST, 359, 361, 362, 370, 373, 375
   enzymatic, 467
   gangliosides, 358, 362, 366, 367, 372, 373, 374, 375
   glycosyl phosphites, 359
   glycosyl xanthates, 359
   iodonium ion, 361
   KDN, 359, 362, 366
   Koenigs-Knorr method, 358
   lactones, 363, 366
   mechanism, 362, 365
   methylsulfenyl bromide M5β, 372
   monosialyl, 359
   $N$-iodosuccinimide-trifluromethane-sulfonic acid, 359, 364, 370, 371, 374
   nonstereoselectively, 364
   oligosialyl, 362
   primary hydroxyls, 361
   secondary hydroxyls, 361
   sialyl donors, 358, 359
   sialyl Le$^x$, 358, 367, 371
   sialoglycoproteins, 358
   sialyloligosaccharides, 364
   silver triflate, 372
   α–stereoselective, 359, 360, 361, 367
   stereoselectivity, 362
   synthesis, 371
   thioglycoside method, 359
   thiophilic promoters, 359
Sialyl glycosides, 357
Sialyl Lewis X, 357
   biology, 366, 419, 421, 436, 445
   carbohydrate epitope, 358
   enzymatic synthesis, 467
   chemical synthesis, 371
Silver triflate, 353, 348, 414, 436, 438
$S_N2$ reactions, 85
Solid-phase synthesis, 266, 272
   glycopeptides, 266, 272
   glycosides, 382, 386
Sowden homologation, 176
(+)-Spectinomycin, 75
Stannyl ethers, 69
   coordination, 74

[Stannyl ethers]
   electrophilic substitution, 82
   nucleosides, 74
   $O$-allylation, 74
   $O$-benzylation, 74, 82
   $O$-tosylation, 74
   oxidation, 74
Stannylene acetals, 69
   $O$-allylation, 77
   $O$-benzoylation, 75
   $O$-benzylation, 76
   brominolysis, 79, 80, 81
   constitution, 70
   dibutyltin oxide, 70
   formation, 70
   oxidation, 79, 80, 81
   $O$-silylation, 78
   $O$-sulfation, 77
   $O$-toluenesulfonylation, 76
   trimeric, 75
Still-Wittig rearrangement, 227
Streptomycin, 263
Sugar, 1
   derivatives, 1
   nitromethane adducts, 36
   osazones, 36
   oximes, 36
Sugar synthesis, 593, 615
   cyanohydrin adducts, 36
   cycloaddition reactions, 617
   1,4-diacetoxy-1,3-butadiene, 618
   dichlorotitanium ($R$)-binaphthol catalyst, 622
   dimethylaluminium chloride, 620
   disaccharide synthesis, 618, 625
   from nonsugars, 593
   furfuryl alcohols, 621, 633
   hetero-Diels-Alder reaction, 617, 619, 622, 624
   history, 615
   hydrazones, 36
   lanthanide complexes, 619, 624
   Lewis acids, 619
   total synthesis, 615
   trimethylsilyl triflate, 620
   zinc iodide, 620
Sugar nucleotides, 488

[Sugar nucleotides]
   biosynthesis, 488, 489
Sugarophobia, 1
Sulfated Le$^x$ tetrasaccharide, 325, 333

Tebbe reaction, 231
Tetrabutylammonium azide, 133
Tetramethylurea (TMU), 436
2-Thiopyridylcarbonate (TOPCAT)
   glycosyl donors, 431
   preparation, 434, 440
   design, 432
   reactivity, 435
   selective activation, 438, 449
Thomsen-Friedenreich antigen, 266
T antigen, 419
   synthesis, 419
$T_n$ antigen, 270
Triflates, 87, 128
   anion nucleophilicity, 96
   azidodeoxy sugars, from, 93
   crown ether, 96
   cryptand, 96
   deoxyhalogeno sugars, from, 90
   displacement by azide ion, 101
   displacement by bromide, chloride,
      and iodide ions, 92, 98, 99
   displacement by fluoride ion, 95, 100
   elimination reactions, 94, 95
   nucleoside, 99

[Triflates]
   procedures, 97
   rates of solvolysis, 89
   reactions, 93
   reactivity, 88
   rearrangement reactions, 92, 96
   ring-contraction, 96
   $S_N2$ reactions, 93
   synthesis, 88, 89, 97
   triflic anhydride, 89
   triflyl chloride, 89, 97
   tris(dimethylamino)-sulfur (trimethyl-
      silyl)difluoride (TASF), 96
Trimeric Le$^x$, 323, 331
Triphenylphosphine, 107
   halogenation, 107, 108, 111, 129
Tumor antigens, 366
Tunicamine, 619
Tunicamycin, 181

Umpolung, 86, 507

Validamine, 134
Validamycin A, 580
Vilsmeier-Haack reaction, 129, 234

Wittig reaction, 231, 239, 547, 559
Wolfrom-Karabinos reaction, 38

Zimmerman-Traxler model, 230